Whey

Whey

Composition, Processing, Applications, Health Benefits and Waste Utilization

Edited by

Bhavbhuti M. Mehta
Dairy Chemistry Department, SMC College of Dairy Science, Kamdhenu University (KHETIWADI), Gujarat, India

Mamdouh El-Bakry
Gevekom GmbH, Spain

ELSEVIER

ACADEMIC PRESS
An imprint of Elsevier

Academic Press is an imprint of Elsevier
125 London Wall, London EC2Y 5AS, United Kingdom
50 Hampshire Street, 5th Floor, Cambridge, MA 02139, United States

For accessibility purposes, images in electronic versions of this book are accompanied by alt text descriptions provided by Elsevier. For more information, see https://www.elsevier.com/about/accessibility.

Books and Journals published by Elsevier comply with applicable product safety requirements. For any product safety concerns or queries, please contact our authorised representative, Elsevier B.V., at productsafety@elsevier.com.

Publisher's note: Elsevier takes a neutral position with respect to territorial disputes or jurisdictional claims in its published content, including in maps and institutional affiliations.

Notices

Knowledge and best practice in this field are constantly changing. As new research and experience broaden our understanding, changes in research methods, professional practices, or medical treatment may become necessary.

Practitioners and researchers must always rely on their own experience and knowledge in evaluating and using any information, methods, compounds, or experiments described herein. In using such information or methods they should be mindful of their own safety and the safety of others, including parties for whom they have a professional responsibility.

To the fullest extent of the law, neither the Publisher nor the authors, contributors, or editors, assume any liability for any injury and/or damage to persons or property as a matter of products liability, negligence or otherwise, or from any use or operation of any methods, products, instructions, or ideas contained in the material herein.

ISBN: 978-0-443-27648-4

For information on all Academic Press publications visit our website at https://www.elsevier.com/books-and-journals

Publisher: Mara Conner
Acquisitions Editor: Kelsey Connors
Editorial Project Manager: Ritika Dagar
Production Project Manager: Rajeswari
Cover Designer: Mark Rogers

Typeset by TNQ Tech

Contents

Part 2
Food applications

9. Cheeses and ice cream

Tomáš Gryger, Eva Samková, Itthanan Suttikhana, Simona Janoušek Honesová, Eva Dadáková, Bhavbhuti M. Mehta and Richardos Nikolaos Salek

10. Whey use in bakery and confectionery products

Alejandro Escobedo, Pavel Prieto-Vázquez del Mercado and Ivan Luzardo-Ocampo

11. Infant formulas

Cristine Couto de Almeida, Vania M. Flosi Paschoalin and Carlos Adam Conte Junior

12. Beverages

Anita Rejdlová, Eva Lorencová and Richardos Nikolaos Salek

13. Whey protein and muscle synthesis: Implications for healthy aging and sports performance

Isabel Gigli

14. Edible films and coatings

Essam Hebishy, Suresh Gurusamy and Ayten A. Tas

Part 3
Functional properties and health effects

15. Functional properties of whey proteins in food formulation

Qihui Wu, Usman Amin and Haotian Zheng

16. Whey protein foams: Preparation, types, and rheology

Alberto Tecante and Ricky Frank López-Santiago

17. Digestion of whey and its products

Juan Camilo Charris-Barreto, María Ximena Quintanilla-Carvajal and Bhavbhuti M. Mehta

18. Nutritional applications of whey

Jeferson Silva Cunha, Flaviana Coelho Pacheco, Viviane Lopes Pereira, Isabela Soares Magalhães, Danielly Aparecida de Souza, Irene Andressa, Fabio Ribeiro dos Santos and Bruno Ricardo de Castro Leite Júnior

19. Whey: A source of bioactive peptides

Carla Paulo Vieira, Italo Rennan Sousa Vieira, Abel Alberto Massingue Júnior, Katherine Gutiérrez-Alzate, Joselene Conceição Nunes Nascimento, Arlen Carvalho de Oliveira Almeida, Hannay Crystynah Almeida de Souza, Anamaria M.P. Dos Santos, Thiago da Silveira Álvares and Carlos Adam Conte Junior

20. Lactose Intolerance

Simona Dominici, Nico Donati, Mirko Tarocchi, Filippo Fassio and Maria Sole Facioni

Part 4
Utilization of whey as a dairy waste

21. Whey as a dairy waste: Treatments in effluent plant and disposal methods in environment

Mohan Li, Yumeng Zhang, Xiqing Yue and Junhua Shao

26. Novel whey-based products

Lara Barnum, Siyu Liu and Alice Marciniak

27. Sensory aspects of whey and whey products

Michele Rosset, Mariana Buranelo Egea and Tatiana Colombo Pimentel

Contributors

Omar Amin Alhaj, Department of Nutrition, Faculty of Pharmacy and Medical Sciences, University of Petra, Amman, Jordan

Cristine Couto de Almeida, Federal University of Rio de Janeiro (UFRJ), Department of Biochemistry, Chemistry Institute, Graduate Studies in Food Science, Rio de Janeiro, Brazil

Hannay Crystynah Almeida de Souza, Center for Food Analysis (NAL), Technological Development Support Laboratory (LADETEC), Federal University of Rio de Janeiro (UFRJ), Cidade Universitária, Rio de Janeiro, Brazil; Laboratory of Advanced Analysis in Biochemistry and Molecular Biology (LAABBM), Department of Biochemistry, Federal University of Rio de Janeiro (UFRJ), Cidade Universitária, Rio de Janeiro, Brazil

Usman Amin, Food Rheology Laboratory, Department of Food, Bioprocessing and Nutrition Sciences, North Carolina State University, Raleigh, NC, United States; Southeast Dairy Foods Research Center, North Carolina State University, Raleigh, NC, United States

Irene Andressa, Department of Food Technology, Federal University of Viçosa, University Campus, Viçosa, Brazil

Danielly Aparecida de Souza, Department of Food Technology, Federal University of Viçosa, University Campus, Viçosa, Brazil

Lara Barnum, Department of Food Science, University of Guelph, Guelph, ON, Canada

Thomas Bintsis, Laboratory of Safety and Quality of Milk and Dairy Products, Faculty of Veterinary Medicine, Aristotle University of Thessaloniki, Thessaloniki, Greece

Arlen Carvalho de Oliveira Almeida, Center for Food Analysis (NAL), Technological Development Support Laboratory (LADETEC), Federal University of Rio de Janeiro (UFRJ), Cidade Universitária, Rio de Janeiro, Brazil; Laboratory of Advanced Analysis in Biochemistry and Molecular Biology (LAABBM), Department of Biochemistry, Federal University of Rio de Janeiro (UFRJ), Cidade Universitária, Rio de Janeiro, Brazil; Laboratorio de Inspeção e Tecnologia de Leite e Derivados (LAITLACTEOS), Federal University of Bahia (UFBA), Salvador, Brazil

Christopher Alexis Cedillo-Jiménez, Engineering Faculty, Autonomous University of Queretaro, Querétaro, Mexico

Jayani Chandrapala, School of Science, STEM College, RMIT University, Bundoora, VIC, Australia

Juan Camilo Charris-Barreto, Universidad de La Sabana, Facultad de Ingeniería, Bogotá, Colombia

Carlos Adam Conte Junior, Center for Food Analysis (NAL), Technological Development Support Laboratory (LADETEC), Federal University of Rio de Janeiro (UFRJ), Cidade Universitária, Rio de Janeiro, Brazil; Laboratory of Advanced Analysis in Biochemistry and Molecular Biology (LAABBM), Department of Biochemistry, Federal University of Rio de Janeiro (UFRJ), Cidade Universitária, Rio de Janeiro, Brazil; Federal University of Rio de Janeiro (UFRJ), Department of Biochemistry, Chemistry Institute, Graduate Studies in Food Science, Rio de Janeiro, Brazil

Jeferson Silva Cunha, Department of Food Technology, Federal University of Viçosa, University Campus, Viçosa, Brazil

Eva Dadáková, Department of Applied Chemistry, Faculty of Agriculture and Technology, University of South Bohemia in České Budějovice, České Budějovice, Czech Republic

Thiago da Silveira Álvares, Multidisciplinary Center UFRJ-Macaé, Federal University of Rio de Janeiro, Rio de Janeiro, Brazil

Pavel Prieto-Vázquez del Mercado, Centro Universitario de Ciencias de la Salud (CUCS), Universidad de Guadalajara (UdeG), Guadalajara, Mexico

Simona Dominici, ELLEFREE S.r.l., Polo Tecnologico Lucchese, Lucca, Italy

Nico Donati, Gastroenterology Unit, Department of Experimental and Clinical Biomedical Sciences, "Mario Serio", University of Florence, Florence, Italy

Anamaria M.P. Dos Santos, Center for Food Analysis (NAL), Technological Development Support Laboratory (LADETEC), Federal University of Rio de Janeiro (UFRJ), Cidade Universitária, Rio de Janeiro, Brazil; Laboratory of Advanced Analysis in Biochemistry and Molecular Biology (LAABBM), Department of Biochemistry, Federal University of Rio de Janeiro (UFRJ), Cidade Universitária, Rio de Janeiro, Brazil

Mariana Buranelo Egea, Goiano Federal Institute, Rio Verde, Brazil

Wael F. Elkot, Dairy Science and Technology Department, Faculty of Agriculture & Natural Resources, Aswan University, Aswan, Egypt

Nour Amin Elsahoryi, Department of Nutrition, Faculty of Pharmacy and Medical Sciences, University of Petra, Amman, Jordan

Alejandro Escobedo, Tecnologico de Monterrey, Institute for Obesity Research, School of Engineering & Sciences, Monterrey, Mexico

Maria Sole Facioni, AILI-Associazione Italiana Latto-Intolleranti Aps, Lucca, Italy; Department of Clinical Medicine and Surgery, University of Naples "Federico II," Naples, Italy

Filippo Fassio, SOSD Allergology and Clinical Immunology, San Giovanni di Dio Hospital, Azienda USL Toscana Centro, Florence, Italy

Ana Angélica Feregrino-Pérez, Engineering Faculty, Autonomous University of Queretaro, Campus Juriquilla, Querétaro, Mexico

Juan Fernando García-Trejo, Engineering Faculty, Autonomous University of Queretaro, Campus Juriquilla, Querétaro, Mexico

Isabel Gigli, Agronomy College, National University of La Pampa (UNLPam), La Pampa, Argentina

Cláudia Andréia Gräff, Food Biotechnology Laboratory, University of Vale do Taquari - Univates, Lajeado, Brazil; Graduate Program in Biotechnology, University of Vale do Taquari - Univates, Lajeado, Brazil

Tomáš Gryger, Department of Food Technology, Faculty of Technology, Tomas Bata University in Zlín, Zlín, Czech Republic

Divya Sandamali Gunasekara, SPC Global, Shepparton, VIC, Australia

Suresh Gurusamy, Centre of Excellence in Agri-food Technologies, National Centre for Food Manufacturing, College of Sciences, University of Lincoln, Spalding, United Kingdom

Katherine Gutiérrez-Alzate, Program in Food Science (PGAli), Faculty of Pharmacy, Federal University of Bahia (UFBA), Salvador, Brazil; Laboratorio de Inspeção e Tecnologia de Leite e Derivados (LAITLACTEOS), Federal University of Bahia (UFBA), Salvador, Brazil

Claudia Gutiérrez-Antonio, Engineering Faculty, Autonomous University of Queretaro, Campus Juriquilla, Querétaro, Mexico; Engineering Faculty, Amealco Campus, Autonomous University of Queretaro, Querétaro, Mexico

Carlos Eduardo Guzmán-Martínez, Engineering Faculty, Autonomous University of Queretaro, Querétaro, Mexico

H.S.T. Hapuarachchi, Faculty of Science, University of Colombo, Colombo, Sri Lanka

Essam Hebishy, Centre d'Innovació, Recerca i Transferència en Tecnologia dels Aliments (CIRTTA), TECNIO, Departament de Ciència Animal i del Aliments, Facultat de Veterinària, Universitat Autònoma de Barcelona, Bellaterra, Spain

Simona Janoušek Honesová, Department of Food Biotechnologies and Agricultural Products' Quality, Faculty of Agriculture and Technology, University of South Bohemia in České Budějovice, České Budějovice, Czech Republic

Daniel Neutzling Lehn, Food Biotechnology Laboratory, University of Vale do Taquari - Univates, Lajeado, Brazil; Graduate Program in Biotechnology, University of Vale do Taquari - Univates, Lajeado, Brazil

Bruno Ricardo de Castro Leite Júnior, Department of Food Technology, Federal University of Viçosa, University Campus, Viçosa, Brazil

Mohan Li, College of Food Science, Shenyang Agricultural University, Shenyang, China; State Key Laboratory of Dairy Biotechnology, Shanghai Engineering Research Center of Dairy Biotechnology, Dairy Research Institute, Bright Dairy & Food Co. Ltd., Shanghai, China; School of Life Sciences & Biotechnology, Shanghai Jiao Tong University, Shanghai, China

Siyu Liu, Department of Food Science, University of Guelph, Guelph, ON, Canada

Ricky Frank López-Santiago, Departamento de Alimentos y Biotecnología, Facultad de Química, Universidad Nacional Autónoma de México, Cd. Universitaria, Mexico City, Mexico; Institute of Physical Chemistry, RWTH Aachen University, Aachen, Germany

Eva Lorencová, Department of Food Technology, Faculty of Technology, Tomas Bata University in Zlín, Zlín, Czech Republic

Ivan Luzardo-Ocampo, Tecnologico de Monterrey, Institute for Obesity Research, School of Engineering & Sciences, Monterrey, Mexico

Isabela Soares Magalhães, Department of Food Technology, Federal University of Viçosa, University Campus, Viçosa, Brazil

Alice Marciniak, Department of Food Science, University of Guelph, Guelph, ON, Canada

Sergio Iván Martínez-Guido, Engineering Faculty, Amealco Campus, Autonomous University of Queretaro, Querétaro, Mexico

Abel Alberto Massingue Júnior, Department of Fish Processing and Quality Control Engineering, Higher Institute Polytechnic of Quissico, Division of Fisheries and Aquaculture, Inhambane, Mozambique

Bhavbhuti M. Mehta, Department of Dairy Chemistry, SMC College of Dairy Science, Kamdhenu University, Anand, Gujarat, India

Hari Meletharayil, Dairy Management Inc., Rosemont, IL, United States

Golfo Moatsou, Laboratory of Dairy Research, Department of Food Science and Human Nutrition, Agricultural University of Athens, Athens, Greece

Ekaterini Moschopoulou, Laboratory of Dairy Research, Department of Food Science and Human Nutrition, Agricultural University of Athens, Athens, Greece

Joselene Conceição Nunes Nascimento, Program in Food Science (PGAli), Faculty of Pharmacy, Federal University of Bahia (UFBA), Salvador, Brazil

Flaviana Coelho Pacheco, Department of Food Technology, Federal University of Viçosa, University Campus, Viçosa, Brazil

Marcy Heli Paiva Rodrigues, Food Biotechnology Laboratory, University of Vale do Taquari - Univates, Lajeado, Brazil

Photis Papademas, Department of Agricultural Sciences, Biotechnology and Food Science, Cyprus University of Technology, Limassol, Cyprus

Vania M. Flosi Paschoalin, Federal University of Rio de Janeiro (UFRJ), Department of Biochemistry, Chemistry Institute, Graduate Studies in Food Science, Rio de Janeiro, Brazil

Viviane Lopes Pereira, Department of Food Technology, Federal University of Viçosa, University Campus, Viçosa, Brazil

Tatiana Colombo Pimentel, Paraná Federal Institute, Paranavaí, Brazil

María Ximena Quintanilla-Carvajal, Universidad de La Sabana, Facultad de Ingeniería, Bogotá, Colombia

Valeria Caltzontzin Rabell, Engineering Faculty, Amealco Campus, Autonomous University of Queretaro, Querétaro, Mexico

Alejandra Ramírez-Hernández, Engineering Faculty, Autonomous University of Queretaro, Campus Juriquilla, Querétaro, Mexico

Anita Rejdlová, Department of Food Technology, Faculty of Technology, Tomas Bata University in Zlín, Zlín, Czech Republic

Fabio Ribeiro dos Santos, Department of Food Technology, Federal University of Viçosa, University Campus, Viçosa, Brazil

Araceli Guadalupe Romero-Izquierdo, Engineering Faculty, Amealco Campus, Autonomous University of Queretaro, Querétaro, Mexico

Michele Rosset, Paraná Federal Institute, Colombo, Brazil

Richardos Nikolaos Salek, Department of Food Technology, Faculty of Technology, Tomas Bata University in Zlín, Zlín, Czech Republic

Eva Samková, Department of Food Biotechnologies and Agricultural Products' Quality, Faculty of Agriculture and Technology, University of South Bohemia in České Budějovice, České Budějovice, Czech Republic

Junhua Shao, College of Food Science, Shenyang Agricultural University, Shenyang, China

Mayumi Silva, School of Science, STEM College, RMIT University, Bundoora, VIC, Australia

Itthanan Suttikhana, Department of Applied Chemistry, Faculty of Agriculture and Technology, University of South Bohemia in České Budějovice, České Budějovice, Czech Republic

Mirko Tarocchi, Gastroenterology Unit, Department of Experimental and Clinical Biomedical Sciences, "Mario Serio", University of Florence, Florence, Italy

Ayten A. Tas, Centre of Excellence in Agri-food Technologies, National Centre for Food Manufacturing, College of Sciences, University of Lincoln, Spalding, United Kingdom; Department of Health Professions, Faculty of Health and Education, Manchester Metropolitan University, Manchester, United Kingdom

Alberto Tecante, Departamento de Alimentos y Biotecnología, Facultad de Química, Universidad Nacional Autónoma de México, Cd. Universitaria, Mexico City, Mexico

Zhoutai Teng, Food Rheology Laboratory, Department of Food, Bioprocessing and Nutrition Sciences, North Carolina State University, Raleigh, NC, United States;

Southeast Dairy Foods Research Center, Department of Food, Bioprocessing and Nutrition Sciences, North Carolina State University, Raleigh, NC, United States

Todor Vasiljevic, Victoria University, Melbourne, VIC, Australia

Juan Manuel Vera-Morales, Engineering Faculty, Amealco Campus, Autonomous University of Queretaro, Querétaro, Mexico

Carla Paulo Vieira, Center for Food Analysis (NAL), Technological Development Support Laboratory (LADETEC), Federal University of Rio de Janeiro (UFRJ), Cidade Universitária, Rio de Janeiro, Brazil; Laboratory of Advanced Analysis in Biochemistry and Molecular Biology (LAABBM), Department of Biochemistry, Federal University of Rio de Janeiro (UFRJ), Cidade Universitária, Rio de Janeiro, Brazil

Italo Rennan Sousa Vieira, Center for Food Analysis (NAL), Technological Development Support Laboratory (LADETEC), Federal University of Rio de Janeiro (UFRJ), Cidade Universitária, Rio de Janeiro, Brazil; Laboratory of Advanced Analysis in Biochemistry and Molecular Biology (LAABBM), Department of Biochemistry, Federal University of Rio de Janeiro (UFRJ), Cidade Universitária, Rio de Janeiro, Brazil

Claucia Fernanda Volken de Souza, Food Biotechnology Laboratory, University of Vale do Taquari - Univates, Lajeado, Brazil; Graduate Program in Biotechnology, University of Vale do Taquari - Univates, Lajeado, Brazil

Dilema Wijegunawardhana, School of Science, STEM College, RMIT University, Bundoora, VIC, Australia

Qihui Wu, Food Rheology Laboratory, Department of Food, Bioprocessing and Nutrition Sciences, North Carolina State University, Raleigh, NC, United States; Southeast Dairy Foods Research Center, North Carolina State University, Raleigh, NC, United States

Xiqing Yue, College of Food Science, Shenyang Agricultural University, Shenyang, China

David Gael Zamorano-Maldonado, Engineering Faculty, Autonomous University of Queretaro, Querétaro, Mexico

Yumeng Zhang, College of Food Science, Shenyang Agricultural University, Shenyang, China

Haotian Zheng, Food Rheology Laboratory, Department of Food, Bioprocessing and Nutrition Sciences, North Carolina State University, Raleigh, NC, United States; Southeast Dairy Foods Research Center, Department of Food, Bioprocessing and Nutrition Sciences, North Carolina State University, Raleigh, NC, United States; Southeast Dairy Foods Research Center, North Carolina State University, Raleigh, NC, United States

Part 1

Whey: Composition, its components, and processing

Chapter 1

Introduction: Overview of milk composition

Wael F. Elkot
Dairy Science and Technology Department, Faculty of Agriculture & Natural Resources, Aswan University, Aswan, Egypt

1. Introduction

Milk is a nutrient-rich liquid secreted by the mammary glands of female mammals, serving as the primary source of nourishment for their offspring immediately after birth. In addition to its biological role, the milk of domesticated animals constitutes a significant component of the human diet, consumed either in its fresh fluid form or processed into a wide range of dairy products, including butter, cheese, and yoghurt (Ismail et al., 2024).

Chemically, milk is a heterogeneous mixture that can be characterized as a complex biochemical substance. It consists of emulsified fat globules, the primary milk protein casein, and various minerals in a colloidal state, while lactose, soluble whey proteins, and additional minerals exist in a true solution (Mehta, 2015). This intricate composition provides a balanced array of macronutrients—proteins, lipids, and carbohydrates—as well as essential micronutrients and bioactive compounds that contribute to both neonatal development and human health.

The microbiological definition of milk refers to it as a nutrient-rich biological fluid that serves as an excellent medium for microbial growth. It is a complex secretion from the mammary glands of mammals, primarily composed of water, proteins, fats, lactose, vitamins, and minerals. From a microbiological perspective, milk is considered a highly perishable food due to its susceptibility to contamination by various microorganisms that include: Beneficial microorganisms (e.g., lactic acid bacteria used in fermentation), spoilage microorganisms (e.g., psychrotrophic bacteria that degrade milk quality) as well as pathogenic microorganisms (*Salmonella* spp., *Listeria monocytogenes*, *Escherichia coli*, and *Mycobacterium bovis*), which pose health risks if consumed raw or improperly processed. Proper handling, pasteurization, and refrigeration are essential to control microbial contamination and extend milk shelf life.

The study of milk composition has gained increasing significance due to its implications for both nutrition and dairy processing. Advances in analytical techniques and molecular biology have enhanced our understanding of milk's biochemical properties, functional characteristics, and its role in the development of value-added dairy products. Furthermore, the challenges encountered in dairy manufacturing, including variations in milk composition due to genetic, dietary, and environmental factors, underscore the necessity of continuous research in this field to optimize both production efficiency and product quality (Linehan et al., 2024).

Milk has long been regarded as one of nature's most complete and nutritionally balanced foods. While it's primary function is to provide essential nourishment to neonates of mammalian species, its role in human nutrition extends far beyond infancy. It serves as a crucial dietary component for individuals across all life stages, supporting growth and development in children, fulfilling the nutritional needs of pregnant and lactating women, and contributing to overall health and well-being in adults. Beyond its fundamental nutritional value, milk is a rich source of biologically active compounds, including casein and whey proteins, which play key roles in physiological and biochemical processes. These components exert significant effects on human metabolism, influencing muscle maintenance, immune function, and overall health. Additionally, certain dairy products are specifically recommended in various dietary regimens to address individual nutritional needs. For instance, infants require a diet primarily focused on growth and cognitive development, whereas pregnant and lactating women need increased intake of essential nutrients to support fetal development and milk production (Jonathan et al., 2021). Moreover, food consumption is not solely driven by physiological requirements; it must also satisfy

Whey. https://doi.org/10.1016/B978-0-443-27648-4.00004-9

taste preferences and promote dietary adherence. Milk and its derivatives hold a unique position in this regard, offering both sensory appeal and substantial health benefits. Rich in essential macronutrients, micronutrients, and bioactive compounds, dairy products contribute to bone strength, metabolic regulation, and immune support throughout life.

In conclusion, milk continues to be a fundamental component of a well-balanced diet, offering a rich array of essential nutrients that support growth, development, and sustained health throughout life. From early childhood to adulthood, regular consumption of milk contributes to strong bones, optimal metabolic function, and overall physiological well-being. Its diverse nutritional profile and bioactive properties underscore its irreplaceable role in human nutrition, making it a vital dietary staple for maintaining long-term health and vitality (Foroutan et al., 2019).

2. Milk composition

2.1 Definition of milk

Milk is a nutrient-rich, whitish liquid produced by the mammary secretory cells of female mammals through a physiological process known as lactation, one of the defining characteristics of the mammalian class. The milk synthesized by these glands is stored within the udder, with the initial secretion following parturition referred to as colostrum. This early milk is particularly significant due to its high concentration of immunoglobulins and bioactive compounds, which provide essential protection and nourishment to neonates during the critical early stages of life. The composition of milk is of paramount importance, necessitating proper storage and transportation under optimal conditions to maintain its quality and nutritional integrity. Its primary components include water, fat, lactose, casein, whey proteins, and minerals (often referred to as ash), though their concentrations vary among different species. However, within a given species, the compositional range remains relatively stable (Imperiale et al., 2023).

From a physiological perspective, milk represents the secretion of a normally functioning mammary gland, produced for a specific duration following birth to provide essential sustenance for the early development of offspring. From a physical chemistry standpoint, milk is an opaque, multiphase fluid consisting of a true solution, a colloidal suspension, and an emulsion. The true solution contains lactose, vitamins, organic acids, enzymes, and inorganic salts, while the colloidal phase comprises casein, calcium phosphate, and globular proteins. Additionally, milk fat exists in the form of an oil-in-water emulsion, with fat globules ranging in size from 0.1–22 μm in diameter. These structural and compositional attributes contribute to milk's unique physicochemical properties, making it a vital dietary component for both neonates and humans across all stages of life (Mehta, 2015).

As both a standalone dietary staple and a versatile food ingredient, milk offers an exceptional nutritional profile, making it a fundamental component of the human diet. Recognized by nutrition experts as one of the most nutritionally complete foods, milk provides substantial amounts of essential macronutrients—including high-quality proteins, fats, and carbohydrates—as well as a wide array of vital micronutrients such as minerals and vitamins. In response to evolving dietary preferences and health-conscious trends, low-fat and nonfat milk variants have gained increasing popularity, particularly in fat-reduced and fat-free food formulations. These alternatives offer the nutritional benefits of milk while catering to individuals seeking to manage their fat intake. Whether consumed in its natural form or incorporated into a variety of food products, milk remains an indispensable source of nourishment, supporting overall health and well-being across all age groups (Rivero and Anrique, 2015).

Globally, cow milk holds the highest commercial significance compared to milk from any other mammalian species. In the United States, the term "milk" is legally defined as referring exclusively to cow milk, whereas milk from other species must be specifically labeled to indicate its source, such as "goat milk" or "sheep milk." From a regulatory perspective, milk is defined as the whole, clean lacteal secretion obtained from one or more healthy, well-nourished cows, excluding milk collected within 15 days before calving and 3–5 days postpartum. Colostrum, the nutrient-rich secretion produced immediately after birth, is explicitly excluded from the legal definition of milk due to its distinct composition. The U.S. Public Health Service establishes strict standards for Grade A milk, defining it as "the lacteal secretion practically free from colostrum, obtained by the complete milking of one or more healthy cows, containing no less than 8.25% milk solids-not-fat (MSNF) and at least 3.25% milkfat." These stringent guidelines ensure the safety, quality, and consistency of commercially available milk, reinforcing its role as a key dietary component in human nutrition (Huppertz, 2016).

3. Constituents of milk

Milk serves as a vital resource for the production of a diverse range of dairy products, with its composition varying based on species, breed, diet, and environmental factors. Four primary components dominate in quantitative terms: water, fat, protein, and lactose. In addition to these macronutrients, milk contains minor but essential components, including minerals, enzymes, vitamins, and dissolved gases, all of which contribute to its nutritional and functional properties. As consumer preferences shift toward more innovative and high-quality dairy products, the dairy industry must effectively harness the full potential of this raw material, which, despite its seemingly simple appearance, possesses a highly complex composition. Cow milk, the most commercially significant type, generally contains lower levels of lactose, fat, and protein compared to milk from some other mammalian species. However, its mineral content remains relatively consistent, making it a valuable and well-balanced nutritional source for human consumption (Mourad et al., 2014).

3.1 Major constituents

The composition of milk varies significantly among animal species, breeds within the same species, and even among individual dairy farms. These variations are influenced by multiple factors, including the stage of lactation and the dietary regimen of the animal (Fig. 1.1). While pooled market milk exhibits more uniformity, milk from individual cows tends to show a wider range of fluctuations in its major constituents. Several factors contribute to these compositional differences, including breed, milking intervals, stages of milking, variations among different udder quarters, lactation period, seasonal changes, diet, nutritional status, environmental temperature, overall health, age, estrus cycle, gestation, and physical activity. Although these variations generally average out in commercially processed milk, they follow a distinct seasonal pattern that can significantly impact the properties of dairy products. Notably, milk received in July and August tends to have approximately 10% lower fat and protein content compared to milk obtained in October and November, when these levels peak. Such seasonal fluctuations directly influence the quality and consistency of dairy products, including yoghurt, cheese, and whey protein. For instance, to maintain uniform viscosity in yogurt production throughout the year, dairy processors often adjust formulations by fortifying milk with additional nonfat solids. These adaptations ensure consistent product quality despite natural variations in raw milk composition (Mourad et al., 2014).

3.1.1 Water

Water is the primary component of milk, accounting for approximately 87% in cow milk, 83%–84% in buffalo milk, 85%–87% in goat milk, 86%–89% in camel milk, and 79%–82% in sheep milk. The variation in water content among different types of milk influences their overall composition, including fat, protein, lactose, and mineral concentrations. Higher water content, as seen in camel and goat milk, results in a lower fat and protein concentration, making these types of milk more easily digestible. In contrast, buffalo and sheep milk have lower water content and higher fat and protein levels, contributing to their richer texture and nutritional density. Water plays a crucial role in maintaining milk's physical

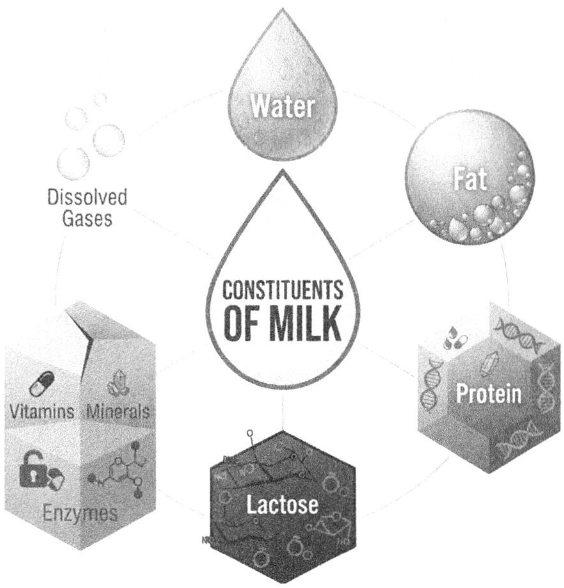

FIGURE 1.1 Milk composition.

properties, acting as a solvent for lactose, minerals, and vitamins, ensuring the homogeneity of milk, and facilitating its digestibility. Additionally, water content influences milk's viscosity, taste, and suitability for processing into dairy products such as cheese and yogurt, where lower water content leads to higher yield and firmer texture (Mourad et al., 2014).

3.1.2 Milk proteins

Milk proteins are primarily categorized into two main groups: caseins and whey proteins. Caseins constitute approximately 80% of the total protein content in milk and play a crucial role in the formation of dairy products such as cheese and yoghurt (Fig. 1.2). Casein is defined as a phosphoprotein that precipitates from raw milk at a pH of 4.6 at 20°C. Structurally, casein consists of amino acid chains linked by peptide bonds (-CO-NH_2-), where the amino and carboxyl groups of adjacent amino acids react to form long-chain polymeric structures. The remaining 20% of milk proteins are classified as whey proteins, which include lactalbumins and lactoglobulins. These proteins remain in the liquid fraction after casein has been precipitated and are highly valued for their nutritional and functional properties, including their role in muscle synthesis and immune support.

3.1.2.1 Casein micelles

Caseins and caseinates play a crucial role not only in biological and nutritional functions but also due to their structural characteristics, charge, and physical properties. One of the key properties of caseins is their insolubility when milk undergoes acidification, leading to a reduction in pH to approximately 4.6. In contrast, whey proteins remain in solution under these conditions. Caseins are naturally present in milk in the form of colloidal particles known as micelles, which are composed of casein complexes and calcium phosphate. These micelles are spherical in shape, with diameters ranging from 50–200 nm, and they contain thousands of protein molecules (Mehta, 2015).

A typical casein micelle contains thousands of casein molecules, most of which form thermodynamically stable complexes with nanoclusters of amorphous calcium phosphate. Like many other unfolded proteins, caseins have an actual or potential tendency to assemble into toxic amyloid fibrils, particularly at the high concentrations found in milk. Fibrils do not form in milk because an alternative aggregation pathway is followed that results in formation of the casein micelle. As a result of forming micelles, nutritious milk can be secreted and stored without causing either pathological calcification or amyloidosis of the mother's mammary tissue. The ability to sequester nanoclusters of amorphous calcium phosphate in a stable complex is not unique to caseins. It has been demonstrated using a number of noncasein secreted phosphoproteins and may be of general physiological importance in preventing calcification of other biofluids and soft tissues. Thus, competent noncasein phosphoproteins have similar patterns of phosphorylation and the same type of flexible, unfolded conformation as caseins. The ability to suppress amyloid fibril formation by forming an alternative amorphous aggregate is also not unique to caseins and underlies the action of molecular chaperones such as the small heat-shock proteins. The open structure of the protein matrix of casein micelles is fragile and easily perturbed by changes in its environment. Perturbations can cause the polypeptide chains to segregate into regions of greater and lesser density. As a result, the reliable determination of the native structure of casein micelles continues to be extremely challenging. The biological functions of caseins, such as their chaperone activity, are determined by their composition and flexible conformation and by how the casein polypeptide chains interact with each other. These same properties determine how caseins behave in the manufacture of many dairy products and how they can be used as functional ingredients in other foods (Holt et al., 2013).

Caseins are classified into four primary fractions: αs1, αs2, β, and κ. Additionally, a γ-fraction is generated from the enzymatic degradation of β-casein by endogenous proteolytic enzymes present in milk. Each of these casein fractions exhibits distinct chemical characteristics. For instance, αs1-casein and αs2-casein contain 8 and 10–13 phosphoserine residues, respectively, while β-casein possesses five phosphoserine residues. Notably, all casein fractions, with the exception of κ-casein, are precipitated in the presence of calcium. κ-Casein, which contains a single phosphoserine unit, is

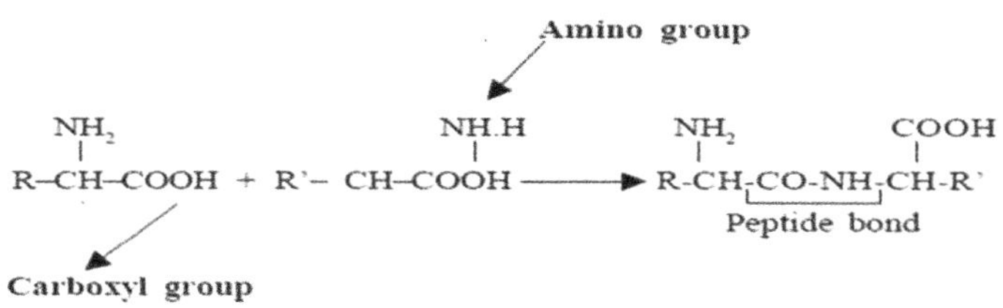

FIGURE 1.2 Structure of milk protein.

resistant to calcium-induced precipitation. Moreover, κ-casein is the only fraction that contains a carbohydrate moiety. Calcium and phosphate play a significant role in casein interactions, with αs1, αs2, and β-caseins complexing with calcium (30 mM) and phosphate (20 mM). In contrast, κ-casein functions as a stabilizing agent for micelles by binding to their surface. This stabilizing function is particularly relevant in cheese production, where κ-casein undergoes enzymatic cleavage by rennin. The hydrolysis of κ-casein results in the destabilization of the micellar structure, ultimately leading to curd formation, which is a fundamental step in the cheese making process (Goff and Hill, 1993).

The casein micelle is a complex structure composed of various casein proteins, salts, and embedded enzymes. The exact proportions of these components vary depending on factors influencing milk composition. Casein micelles play a crucial role in the stability and functionality of milk proteins, particularly in dairy processing.

Casein micelles aggregate through four primary mechanisms: enzymatic action, acidification, heat treatment, and aging. Proteolytic enzymes like rennet cleave κ-casein, destabilizing micelles and triggering coagulation. Acidification to around pH 4.7 leads to casein precipitation and curd formation. Although caseins are generally heat-stable, extended exposure to high temperatures can induce structural changes that compromise micelle integrity. Additionally, aging causes gradual structural alterations within micelles, promoting their aggregation and the formation of gels.

The total casein fraction comprises multiple molecular forms, including αs-casein (50%), β-casein (15%), and γ-casein (3%). These molecules, with an approximate molecular weight of 25,000 Da in their monomeric state, can be separated using electrophoresis and ultracentrifugation. At physiological pH (around 7) and body temperature (37°C), αs-casein and β-casein precipitate in the presence of calcium ions (Ca^{2+}) during coagulation. However, other proteins remain dissolved in the whey fraction alongside lactose and salts. Casein molecules naturally associate into micelles, which play a central role in milk coagulation. During curdling—induced by lactic acid bacteria—κ-casein undergoes degradation, leading to the formation of para-casein. In the presence of Ca^{2+}, micellar structures coagulate and polymerize, forming a gel or curd that releases whey through syneresis. This coagulation process is fundamental to traditional cheese-making, where the rate and extent of curd formation depend on the milk casein content, calcium levels, and acidity. Beyond cheese production, the physicochemical properties of casein micelles influence a wide range of technological applications in the dairy industry, including yogurt production, protein fortification, and the formulation of functional dairy products (El-Hakaoui, 2008).

3.1.2.2 Whey proteins

Whey proteins constitute approximately 20% of the total protein content in bovine milk and remain soluble when the pH is adjusted to 4.6. These proteins are commonly referred to as whey proteins, serum proteins, or non-casein nitrogen. Additionally, whey contains phosphopeptides (PPs) derived from caseins, which, despite being present in whey, are classified as casein derivatives.

Whey proteins can be effectively separated from milk using several established methods employed for casein preparation. These methods include: Isolation of proteins that remain soluble at pH 4.6. Extraction of proteins that remain soluble in saturated sodium chloride (NaCl) solutions. Separation of proteins that remain soluble after the enzymatic coagulation of caseins by rennet. Fractionation of whey proteins from casein micelles using gel filtration or microfiltration techniques. Isolation of proteins that do not sediment during ultracentrifugation, regardless of the presence or absence of added calcium ions (Ca^{2+}). These characteristics make whey proteins highly functional and valuable in various food and industrial applications.

Whey proteins are present in the serum phase of milk in a dissolved state, along with various minor proteins and enzymes. These include lactoperoxidase (LP), lactotransferrin, lysozyme, glycoproteins, serum transferrin, and degradation products derived from caseins. These components contribute to the functional and bioactive properties of whey, playing essential roles in milk's natural defense mechanisms and its nutritional and technological applications (Zhu and Guo, 2020).

Whey proteins such as β—lactoglobulin (β—Lg) and α—lactalbumin (α—La) are cheese byproducts. Whey proteins represent about 20% of milk proteins. Whey proteins are comprised of 50% β—Lg; 12% α—La; 10% immunoglobulins; 5% serum albumin, and 0.23% proteose peptones, Lactoferrin (LF), and LP. β—Lactoglobulin is a major protein in bovine milk, representing ~50% of whey proteins and 12% of the total protein. It was among the first proteins to be crystallized and, since it could be crystallized readily in large amounts, was for long considered to be homogeneous and a typical globular protein. β—Lactoglobulin is the principal whey protein in the milk of the cow, buffalo, sheep and goat, although there are slight interspecies differences (Gobbetti et al., 2002).

The α—La is the second abundant component in whey protein. A number of β—Lg—and α—La—derived peptides have significant angiotensin—converting enzyme (ACE)—inhibitory activity and antimicrobial activity. Other bioactivities of β—Lg—and α—La—derived peptides include antiviral, anticarcinogenic, and hypocholesterolemic. Bovine serum albumin (BSA) has been given little attention in respect to its role in the functional properties of whey protein concentrates and makes up only about 5% of the protein in whey protein concentrates. BSA is one of the proteins with good essential amino acid profiles. Free fatty acids, lipids, and flavor compounds can be easily absorbed to BSA. LF is a glycoprotein (80 kDa) that is a member of the transferring family in milk. LF consists of a single polypeptide chain comprising 673 amino acid residues that form two homologous domains. LF has been shown to have a number of other physiological and biological functions. LF plays an important role in iron uptake in the intestine and is considered to be an important host defense molecule (Wakabayashi et al., 2006).

LF plays an important role in innate defense against antimicrobial activities, antiviral activities, antioxidant activities, immunomodulation, and cell growth modulation. The antimicrobial peptide derived from hydrolyzed LF also exhibits various bioactivities. LF makes an important contribution to the host defense system by eliminating pathogens, viruses, and fungi. The biological activities of LF include iron transport, antimicrobial activity, antifungal activity, antiviral activity, anticancer activity, immunomodulating effects, and anti inflammatory activity (Legrand et al., 2005). It is believed that immunoglobulins represent about 10%–15% of whey proteins. There are five classes of antibodies; IgA, IgD, IgE, IgG, and IgM. Immunoglobulins (IGs) are a family of globular proteins with a number of bioactivities, including antimicrobial and protective activities. IGs form a globular structure in water. IGs have several bioactive functions, such as protection of the gut mucosa against pathogenic microorganism and passive immunity to the ruminant neonate in colostrum. Among IGs, immunoglobulin G (IgG) antibodies show multifunctional activities, including complement activation, opsonization, and agglutination in bacteria. In addition, IgG helps reduce infection by binding to specific sites on the surfaces of most infectious agents or products.

Whey proteins can be used as simple protein supplements as well as novel healthy ingredients in the food industry. Whey proteins provide beneficial health—promoting functions. These include promotion of health conditions and prevention of diseases. It is clear that bioactive components in whey have potential benefits in terms of nutritional, functional, and bioactive food and pharmaceutical products.

3.1.2.3 Comparison of proteins from different milk species

Caseins and whey proteins can be differentiated based on their distinct physical and chemical properties. Caseins and whey proteins, the two primary protein types in milk, differ significantly in their physical and chemical properties. Physically, caseins exist in milk as large, amorphous micellar aggregates that scatter light, making milk appear white and opaque. These micelles are heat-stable and remain intact even under high temperatures. However, caseins are insoluble at their isoelectric point (around pH 4.6), causing them to precipitate and form curds, which is essential in cheese production. In contrast, whey proteins are smaller, globular proteins that remain dissolved in the liquid portion of milk (whey) after caseins precipitate. They are highly soluble across a wide pH range and are heat-sensitive, undergoing denaturation and aggregation at high temperatures (above 70°C).

Chemically, caseins are characterized by their high proline content, which disrupts typical protein folding patterns, and by the presence of phosphate groups, especially in αs1-and αs2-caseins, contributing to their ability to bind calcium and form micelles. They lack disulfide bonds and are particularly susceptible to enzymatic cleavage by chymosin, a key step in rennet-induced coagulation during cheese-making. Whey proteins, on the other hand, contain disulfide bonds and are rich in sulfur-containing amino acids like cysteine, which play a significant role in their functional properties, especially in food processing. Each whey protein, such as β-lactoglobulin and α-lactalbumin, has a unique isoelectric point and contributes distinct nutritional and functional qualities. These differences in structure and behavior underlie their diverse applications in both food and nutritional industries. One of the key structural characteristics of caseins is their inherently disordered molecular conformation, which lacks stabilizing disulfide bonds. This structural disorder contributes to the unique solubility behavior of caseins, making them prone to precipitation under acidic conditions or in the presence of di- and polyvalent ions from various salts. Unlike globular proteins, casein molecules naturally exist in an open and extended conformation, rendering them relatively heat-stable. Consequently, exposure to moderate heat has little to no effect on their structure. However, when subjected to intense heat treatment, caseins undergo significant chemical modifications, including the loss of phosphate groups and the formation of brown pigments due to their interaction with milk sugars, a reaction known as the Maillard reaction. This heat-induced alteration can impact the functional and nutritional properties of caseins, influencing their applications in food processing and dairy technology.

Whey proteins possess a more structured and organized molecular conformation compared to caseins, characterized by the presence of stabilizing disulfide linkages. As a result, they exhibit greater resistance to precipitation under acidic conditions or in the presence of polyvalent ions, maintaining their solubility in such environments. Similar to other globular proteins, whey proteins are susceptible to heat denaturation, which can lead to structural unfolding and subsequent gel formation. Notably, β-lactoglobulin, one of the major whey proteins, interacts with κ-casein when milk is subjected to intense heat treatment, forming protein complexes that can alter the functional properties of dairy products. This heat-induced association plays a crucial role in modifying the texture, stability, and nutritional attributes of processed milk and dairy formulations (Huppertz, 2016).

3.1.3 Milk fat

Milkfat exists as microscopic globules dispersed in an oil-in-water emulsion, with globule sizes ranging from 0.1–22 μm in diameter. This dispersion plays a crucial role in the stability, texture, and sensory properties of dairy products. The primary component of milkfat is triacylglycerol, which constitutes approximately 97%–98% of its total lipid content. In addition, milkfat contains 0.2%–1% phospholipids, 0.2%–0.4% sterols, and trace amounts of free fatty acids, along with fat-soluble vitamins A, D, E, and K, which contribute to its nutritional value. The fatty acid composition of milkfat is a key determinant of its functional properties. It is composed of: 65% saturated fatty acids (SFAs), which contribute to its solid structure at refrigeration temperatures. 32% monounsaturated fatty acids (MUFAs), which enhance fluidity and improve melting properties. 3% polyunsaturated fatty acids (PUFAs), which play a role in human health and oxidative stability. Additionally, milkfat contains fatty acids of varying chain lengths: Short-chain fatty acids (C4–C8): 7%. Medium-chain fatty acids (C10–C14): 15%–20%. Long-chain fatty acids (C16 or higher): 73%–78%.

The cholesterol content of whole milk (3.3% fat) is 14 mg per 100 mL, whereas skim milk contains 2 mg per 100 mL, reflecting the significant reduction in fat content after skimming. The unique composition of milkfat influences its functional properties, such as texture, mouthfeel, emulsification, and flavor release in dairy products. The presence of short- and medium-chain fatty acids contributes to its digestibility and metabolic benefits, while phospholipids and sterols aid in membrane stability and emulsification. These characteristics make milkfat an essential component in a wide range of dairy-based formulations, including butter, cheese, cream, and processed dairy products.

3.1.3.1 Simple lipids

Simple lipids are composed mainly of triglycerides (98% fat), in small quantities, and of sterides and cerebrosides. Glycerides (neutral lipids) consist of triglycerides (98%), diglycerides (A1.5 0.2%) and monoglycerides (traces). Gravimetrically, glyceride fatty acids account for almost90% of the fat. If over 400 compounds have been identified in bovine milk, only 15 of them are present in substantial amounts (>1% of total lipids). The origin of milk fatty acids is two-fold: fatty acids with carbon chain lengths from 4 to 12 carbon atoms are synthesized by the mammary gland from blood precursors, fatty acids with carbon chain lengths 18 (and more) carbon atoms are directly collected in the blood plasma (they are derived from the diet, the fat reserves or synthesis in tissues other than the mammary gland) and fatty acids and 16–14 carbon atoms is derived from de novo synthesis by the udder or a levy in the bloodstream (Jaakamo et al., 2019).

3.1.3.2 Complex lipids in milk: Structure, composition, and functional roles

Complex lipids in milk are distinguished by their association with phosphorus and/or nitrogen, playing critical roles in the structural integrity of fat globules and emulsification. Among these, phospholipids are the most significant, accounting for approximately 1% of total milk fat (0.3–0.5 g/L). Despite their low concentration, phospholipids are essential for stabilizing milk fat globules and maintaining the balance between the fat and aqueous phases of milk (Jaakamo et al., 2019).

Phospholipids exhibit both hydrophilic (water-attracting) and lipophilic (fat-attracting) properties, enabling them to act as emulsifiers. This dual nature allows them to form a bridge between fat and water, ensuring the stability of milk and dairy products. Their distribution in dairy matrices is as follows: Cream and butter: About 60% of total phospholipids. Skim milk and buttermilk: Approximately 40% of total phospholipids.

Lecithin (Phosphatidylcholine), a key emulsifier that contributes to membrane integrity and fat metabolism. Cephalin (Phosphatidylethanolamine), plays a role in cellular function and structural stability. Sphingomyelin, a major component of milk fat globule membranes, crucial for brain and nerve development. Approximately 85% of the fatty acids in phospholipids are long-chain fatty acids, making them essential for structural and metabolic functions. Additionally, minor complex lipids such as gangliosides, glycolipids, and glycosphingolipids contribute to the nutritional and bioactive properties of milk (Jaakamo et al., 2019).

Milk fat composition varies significantly depending on species, breed, and diet. Fat in milk is presented as microscopic globules suspended in water, with triglycerides and fatty acids forming the primary constituents, especially in goat milk. These globules, which vary in size, are synthesized within the mammary secretory epithelium. The synthesis process occurs in the endoplasmic reticulum, where proteins and triglycerides are produced. Triglycerides then accumulate in the cytoplasm as small lipid droplets, which range from 1–5 μm in diameter before being secreted into milk (Jaakamo et al., 2019). Understanding the role of complex lipids in milk is essential for improving dairy product formulations, enhancing nutritional benefits, and optimizing milk-based emulsions.

3.1.4 Milk sugar

Lactose, commonly known as milk sugar, is the primary carbohydrate found in milk. It is a disaccharide composed of one molecule of D-galactose, linked via its semi-acetyl function, and one molecule of D-glucose, connected through its hydroxyl group at the fourth carbon position. This linkage forms a β-D-galactoside 1,4 bond, which can be hydrolyzed by the enzyme β-galactosidase. Structurally, lactose is classified as 4-D-glucopyranosyl-β-D-galactopyranose. Despite being a sugar, lactose has a relatively less sweetness compared to other carbohydrates. Its concentration in milk typically ranges from 4.5–5.2 g per 100 g, showing slight variations. Unlike milk fat content, which can be influenced by dietary factors, the lactose concentration remains relatively stable across different feeding regimes and dairy breeds. Lactose serves as a crucial substrate in the fermentation of milk by lactic acid bacteria, a process essential for producing dairy products such as yogurt and cheese. The extent of lactic acid production in fermented milk depends on multiple factors, including the specific bacterial strain used (as some strains exhibit higher enzymatic activity than others), the fermentation conditions, and the availability of lactose as a substrate. Additionally, the buffering capacity of milk plays a significant role in regulating the fermentation process, affecting the final composition and texture of the product.

3.2 Minor and trace constituents

The typical composition of major milk minerals is shown in Table 1.1.

3.2.1 Minerals

Mineral composition in milk is closely linked to its physicochemical equilibria, which play a crucial role in dairy processing, nutritional quality, and product shelf life. The mineral content of milk includes essential ions such as chloride, phosphate (PO_4^{2-}), and citrates of potassium (K^+), sodium (Na^+), calcium (Ca^{2+}), and magnesium (Mg^{2+}). Although minerals account for less than 1% of milk's total composition, they significantly influence key properties such as heat stability, alcohol-induced coagulation, and structural changes in various dairy products (Zhu and Guo, 2020).

For instance, minerals contribute to the age-thickening phenomenon observed in sweetened condensed milk and the feathering effect in coffee cream. They also affect rennin-induced coagulation in cheese production and play a role in the clustering of fat globules during homogenization. Among these minerals, calcium concentration is particularly important, as it directly impacts the firmness and structural integrity of curd during cheese-making.

The total mineral content of milk is often estimated through ash analysis, which involves incinerating a specific weight of milk to leave behind a white residue. However, the ash content does not provide a precise measure of the total mineral

TABLE 1.1 Major mineral composition of milk.

Milk salt constituents	Concentration (mg/100 mL whole milk)	Concentration range (mg/100 mL)
Calcium	123	110–130
Magnesium	12	9–14
Phosphorus (total)	95	90–100
Phosphorus (in organic)	75	70–80
Sodium	58	35–90
Potassium	141	110–170
Chloride	99	90–100
Sulfate	10	–
Carbonate (as Co_2)	20	–
Citric acid	160	–

concentration due to the decomposition and volatilization of certain components during the heating process. The ash fraction includes carbonates derived from organic compounds, sulfates from proteins, and phosphates, primarily originating from casein, which typically contains about 1.62% phosphate. Additionally, a significant portion of chloride (approximately 45%–50%) and all citric acid are lost during incineration. On average, milk ash content is approximately 0.70%, which corresponds to an estimated mineral concentration of about 0.90%.

Milk contains all the essential minerals required for human nutrition, existing in various forms depending on their chemical properties. Sodium (Na^+), potassium (K^+), and chloride (Cl^-) are present in milk as free ions in true solution. Due to their ionic nature, these minerals can easily pass through membranes during ultrafiltration and electrodialysis processes, allowing for their separation from milk and whey.

Other minerals, such as calcium (Ca^{2+}), magnesium (Mg^{2+}), inorganic phosphorus (PO_4^{3-}), and citrate, exist in both colloidal and diffusible states, with their distribution influenced by the pH of milk. Approximately 20%–30% of calcium and magnesium in their diffusible state are found as free ions, while the remaining portion is bound as salts of citrate and phosphate. As the pH decreases, colloidal minerals progressively transition into their ionic forms. At a pH of 4.4, most of these minerals become diffusible, which has practical applications in modifying the mineral content and ratios of dairy protein concentrates by controlling acidity in milk or whey.

In addition to these macrominerals, milk also contains trace elements, which are present at the parts-per-million (ppm) level. These trace elements contribute to the overall nutritional value of milk and can be quantitatively analyzed, as demonstrated in Table 1.2.

3.2.2 Vitamins

3.2.2.1 Water-soluble vitamins

Milk is a valuable source of several water-soluble vitamins, which play essential roles in maintaining energy metabolism, supporting immune function, and promoting the health of skin, nerves, and blood cells (Table 1.3). Since these vitamins are not stored in significant amounts in the body and are readily excreted, their consistent intake through food sources like milk is important. Cow milk naturally contains measurable amounts of several B-vitamins and vitamin C, although the concentration of some (like vitamin C and B12) is relatively low and may not meet daily requirements without other dietary sources.

TABLE 1.2 Trace minerals of milk.

Constituent	Concentration (μg/100 mL of milk)
Iron	30–60
Zinc	200–600
Copper 200–600	200–600
Manganese	2–5
Iodine	2–6
Fluoride	3–22
Selenium	0.5–6.7
Cobalt	0.05–0.13
Chromium	0.8–1.3
Molybdenum	1.8–12
Nickel	0–5
Silicon	75–700
Vanadium	0–31
Tin	4–50
Arsenic	2–6

TABLE 1.3 Typical concentrations of water-soluble vitamins in cow milk.

Vitamin	Name	Concentration (mL)
B1	Thiamin	0.04–0.05 mg/100
B2	Riboflavin	0.15–0.18 mg/100
B3	Niacin	0.09–0.11 mg/100
B5	Pantothenic acid	0.35–40 mg/100
B6	Pyridoxine	0.04–0.06 mg/100
B7	Biotin	0.003–0.005 mg/100
B9	Folate/Folic acid	0.005–0.01 mg/100
B12	Cobalamin	0.3–0.5 μg/100
C	Ascorbic acid	1.0–2.0 mg/100

3.2.2.2 Fat-soluble vitamins

Fat-soluble vitamins are a group of essential nutrients that dissolve in fat and are stored in the body's fatty tissues and liver. Unlike water-soluble vitamins, they do not need to be consumed daily, as the body can draw on its reserves when needed. Milk contains several important fat-soluble vitamins, each playing a unique role in maintaining health. Vitamin A supports vision, immune function, and cellular growth. Vitamin D enhances calcium absorption, contributing to strong bones and immune defense. Vitamin E functions primarily as an antioxidant, safeguarding cells from oxidative damage. Meanwhile, Vitamin K is vital for proper blood clotting and bone metabolism. Table 1.4 below summarizes the typical concentrations of these vitamins in cow milk.

3.2.3 Enzymes

Enzymes are highly specific globular proteins synthesized by living cells, playing a crucial role in facilitating and accelerating biochemical reactions by lowering the activation energy required for reactants to transition into products. They serve as biological catalysts, ensuring the efficiency and regulation of metabolic pathways essential for cellular function and homeostasis. Each enzyme is characterized by a unique isoelectric point (pI), which represents the pH at which the enzyme carries no net electrical charge. This intrinsic property is crucial for enzyme solubility, interaction with substrates, and overall stability within various cellular environments (Linehan et al., 2024). The structural integrity and catalytic functionality of enzymes are highly sensitive to external factors that may induce denaturation, thereby altering their activity. Key denaturing factors include fluctuations in pH, which can disrupt ionic interactions and hydrogen bonding, leading to changes in enzyme conformation and substrate binding affinity. Similarly, extreme temperatures can cause thermal denaturation, where excessive heat disrupts non-covalent interactions such as hydrogen bonds and hydrophobic interactions, ultimately leading to protein unfolding and loss of enzymatic activity. Conversely, extremely low temperatures may reduce molecular motion, thereby slowing enzymatic reactions without necessarily causing denaturation.

TABLE 1.4 Typical concentrations of fat-soluble vitamins in cow milk.

Vitamin	Common forms	Function	Concentration (mL)
A	Retinol, Beta-carotene	Vision, immune function, cell growth	~28–41 μg/100
D	D2 (Ergocalciferol), D3 (Cholecalciferol)	Calcium absorption, bone health, immunity	~0.03–0.09 μg/100
E	Alpha-Tocopherol	Antioxidant, protects cell membranes	~0.1–0.2 mg/100
K	Phylloquinone, Menaquinone	Blood clotting, bone metabolism	~0.3–0.5 μg/100

Alterations in ionic strength can also have profound effects on enzyme stability by modifying electrostatic interactions, potentially leading to aggregation or dissociation of the enzyme structure. Additionally, exposure to organic solvents can interfere with the enzyme's hydrophobic core, disrupting the native three-dimensional conformation required for its catalytic function. These denaturing factors not only affect enzyme activity but also impact the efficiency of metabolic processes, potentially leading to physiological imbalances or impaired cellular function.

Enzymes are sensitive to environmental factors such as heat, light, and pH, and their activity can be either enhanced by co-enzymes or inhibited by anti-enzymes. In milk, enzymes exist in four phases—water-soluble, casein-bound, lipid-bound, and within microsomal particles—and these forms can shift due to agitation, homogenization, or heat. Key enzymes in milk include peroxidase, which releases oxygen from hydrogen peroxide and improves milk's shelf life but is destroyed above 80°C; amylase, which breaks down starch and is inactivated at 60–65°C; lipase, which causes fat hydrolysis and rancidity, and is destroyed at 63°C for 20 min; catalase, which decomposes hydrogen peroxide and increases with contamination, being inactivated above 90°C; phosphatase, a marker for proper pasteurization as it hydrolyzes phosphoric acid esters and is destroyed during the process; proteases, which break down proteins and are denatured at 80°C; reductases, which reduce organic compounds like methylene blue; and lactases, particularly beta-glycosidase, which convert lactose into glucose and galactose, ultimately forming lactic acid. Understanding these enzymes and their behavior under varying conditions is essential in fields like biotechnology, pharmaceuticals, and diagnostics, where maintaining enzyme stability and activity is key to optimizing processes and ensuring functional outcomes.

4. Conclusion

Milk intended for human consumption must be carefully collected from a healthy, well-nourished female animal, ensuring that it is free from colostrum and that the animal is not subjected to excessive strain. The production of milk is not a constant process, as it is influenced by various factors. While breed and species are the primary determinants of milk yield, individual characteristics such as the animal's health status, nutritional intake, and age also play a crucial role in production variations. Among all dairy sources, cow milk is the most extensively produced and processed on a global scale, serving as the primary raw material for the dairy industry. However, milk from other mammals, including goats, sheep, buffalo, and camels, holds substantial economic significance, particularly in semi-arid regions. This is especially evident in the Mediterranean basin, where these alternative milk sources contribute significantly to local economies and food security.

References

El-Hakaoui, A., 2008. Les principaux systèmes biochimiques alimentaires lait et produits dérivés. Chapitre IV. Universite HassanII—Mohammadia, FSTM, pp. 67–69.

Foroutan, A., Guo, A.C., Vazquez-Fresno, R., Lipfert, M., Zhang, L., Zheng, J., Badran, H., Budinski, Z., Mandal, R., Ametaj, B.N., Wishart, D.S., 2019. Chemical composition of commercial cow's milk. J. Agric. Food Chem. 67 (17), 4897–4914. https://doi.org/10.1021/acs.jafc.9b00204.

Gobbetti, M., Stepaniak, L., De Angelis, M., Corsetti, A., Di Cagno, R., 2002. Latent bioactive peptides in milk proteins: proteolytic activation and significance in dairy processing. Crit. Rev. Food Sci. Nutr. 42 (3), 223–239. https://doi.org/10.1080/10408690290825538.

Goff, H.D., Hill, A.R., 1993. Chemistry and physics. In: Hui, Y.H. (Ed.), Dairy Science and Technology Handbook, 1. VCH Publishers, New York.

Holt, C., Carver, J.A., Ecroyd, H., Thorn, D.C., 2013. Invited review: caseins and the casein micelle: their biological functions, structures, and behavior in foods. J. Dairy Sci. 96 (10), 6127–6146. https://doi.org/10.3168/jds.2013-6831.

Huppertz, T., 2016. Heat stability of milk. In: McSweeney, P.L.H., O'Mahony, J.A. (Eds.), Advanced Dairy Chemistry: Volume 1B: Proteins: Applied Aspects. Springer, New York, pp. 179–196.

Imperiale, S., Morozova, K., Ferrentino, G., Scampicchio, M., 2023. Analysis of milk with liquid chromatography–mass spectrometry: a review. Eur. Food Res. Technol. 249 (4), 861–902. https://doi.org/10.1007/s00217-022-04197-3.

Ismail, H.A., Elgaml, N.B., Taher, E.S., Osman, S.A., El-Sherbiny, M., Ghamry, H.I., Al-Farga, A., Elkot, W.F., 2024. Production and improving the quality of Domiati cheese and increasing its shelf life by using milk exposure magnetic fields. Int. Dairy J. 158. https://doi.org/10.1016/j.idairyj.2024.106033.

Jaakamo, M.J., Luukkonen, T.J., Kairenius, P.K., Bayat, A.R., Ahvenjärvi, S.A., Tupasela, T.M., Vilkki, J.H., Shingfield, K.J., Leskinen, H.M., 2019. The effect of dietary forage to concentrate ratio and forage type on milk fatty acid composition and milk fat globule size of lactating cows. J. Dairy Sci. 102 (10), 8825–8838. https://doi.org/10.3168/jds.2018-15833.

Jonathan, B.M., Tom, F.O., Alan, L.K., Noel, A.M., 2021. Compositional and functional properties of milk and dairy products derived from cows fed pasture or concentrate-based diets. Compr. Rev. Food Sci. Food Saf. 20, 1–32.

Legrand, D., Elass, E., Carpentier, M., Mazurier, J., 2005. Lactoferrin: a modulator of immune and inflammatory responses. Cell. Mol. Life Sci. 62 (22), 2549–2559. https://doi.org/10.1007/s00018-005-5370-2.

Linehan, K., Patangia, D.V., Ross, R.P., Stanton, C., 2024. Production, composition and nutritional properties of organic milk: a critical review. Foods 13 (4). https://doi.org/10.3390/foods13040550.

Mehta, B.M., 2015. Chemical Composition of Milk and Milk Products Handbook of Food Chemistry. Springer, Berlin/Heidelberg, pp. 511–553. https://doi.org/10.1007/978-3-642-36605-5_31.

Mourad, G., Bettache, G., Samir, M., 2014. Composition and nutritional value of raw milk A review. Issues Biol. Sci. Pharmaceut. Res. 2 (10), 115–122.

Rivero, M.J., Anrique, R., 2015. Milk fat depression syndrome and the particular case of grazing cows: a review. Acta Agric. Scand. Sect. A Anim. Sci. 65 (1), 42–54. https://doi.org/10.1080/09064702.2015.1052545.

Wakabayashi, H., Yamauchi, K., Takase, M., 2006. Lactoferrin research, technology and applications. Int. Dairy J. 16 (11), 1241–1251. https://doi.org/10.1016/j.idairyj.2006.06.013.

Zhu, Z., Guo, W., 2020. Recent developments on rapid detection of main constituents in milk: a review. Crit. Rev. Food Sci. Nutr. 1–13.

Chapter 2

Whey production status and types

Thomas Bintsis[1] **and Photis Papademas**[2]
[1]*Laboratory of Safety and Quality of Milk and Dairy Products, Faculty of Veterinary Medicine, Aristotle University of Thessaloniki, Thessaloniki, Greece;* [2]*Department of Agricultural Sciences, Biotechnology and Food Science, Cyprus University of Technology, Limassol, Cyprus*

1. Introduction

Whey (or cheese whey) is the yellowish to greenish liquid expelling from the coagulation of milk during cheesemaking. Cheesemaking is the process that transforms milk into a solid food, that is cheese, through dehydration. Whey has been used for thousands of years as a beverage (Guo and Wang, 2019) and whey cheeses were produced during the last centuries (Gobbetti et al., 2017). With the increasing amount of whey produced from industrial cheesemaking and yogurt manufacture, a great number of valorization techniques have been developed.

Approximately 80%–90% of the milk volume used in cheesemaking is expelled as whey (Buchanan et al., 2023). Considering this approximation, and the fact that world cheese production was 22.34 million tons in 2024 (Statista, 2025), the whey produced yearly is estimated at 180 million tons. Based on a 1%–2% annual increase, this amount is projected to 203–241 million tons in 2030 (Buchanan et al., 2023). In addition, the demand for Greek-style yoghurt, has increased in recent years, and the volume of acid whey generated has increased by 1%–2% annually (Uduwerella et al., 2018, Rocha-Mendoza et al., 2021).

The special characteristics of the main types of whey, namely, (1) sweet whey, (2) acid whey, and (3) second whey are overviewed.

2. Historical aspects of cheesemaking and whey production

The first cheesemaking attempt dates back to the Neolithic era, more than 9000 years ago, perhaps after the livestock domestication and dairying, in the upper Euphrates and Tigris River valleys of Southwest Asia, where sheep and goats were first domesticated (Kindstedt 2018). After several millenia, the Romans officially recognized two classes of cheese for tax purposes: *caseus mollis*, or soft cheese, and *caseus aridus*, or dry cheese. Over time, the cheesemakers realized that heating whey could lead to the coagulation of the serum proteins and thus, a special family of cheeses; whey cheeses were created. Interestingly, the making of Ricotta, which is the first whey cheese ever described in De Re *Rustica* (50 AD), seems to have dominated cheese production on the Italian peninsula during the 2nd millennium BC, as inferred from the abundant occurrence of ceramic devices referred to as "milk boilers" in the archeological records (Gobbetti et al., 2017; Kindstedt, 2018).

During the manufacture of every kind of cheese, whey is produced, and our ancestors had developed different ways to use it as at that time food was precious and nothing could be wasted (Guo and Wang, 2019). As early as the 5th century BC, the Greek physician Hippocrates prescribed whey to his patients for boosting the immune system, gastrointestinal disorders and skin conditions (Bintsis and Papademas, 2023; Smithers, 2015). The habit of drinking whey had been passed down to early 20th century. The introduction of pioneering methods for separation of lactose, spray drying and the development of membrane filtration that enabled the production of many types of whey ingredients by economically efficient processes (Johnson, 2017). In the 20th century, the rapid growth of cheese and yogurt production has driven the explosive expansion of whey production. The whey was produced in such amounts that could not be used as a beverage and, moreover, beverages faced spoilage problems with short shelf-lives.

More than 1500 different types of cheese are produced worldwide and thus, different types of whey are produced (Bintsis and Papademas, 2018). The majority of cheeses are made from milk coagulation with rennet (Fig. 2.1) and the

Whey. https://doi.org/10.1016/B978-0-443-27648-4.00019-0

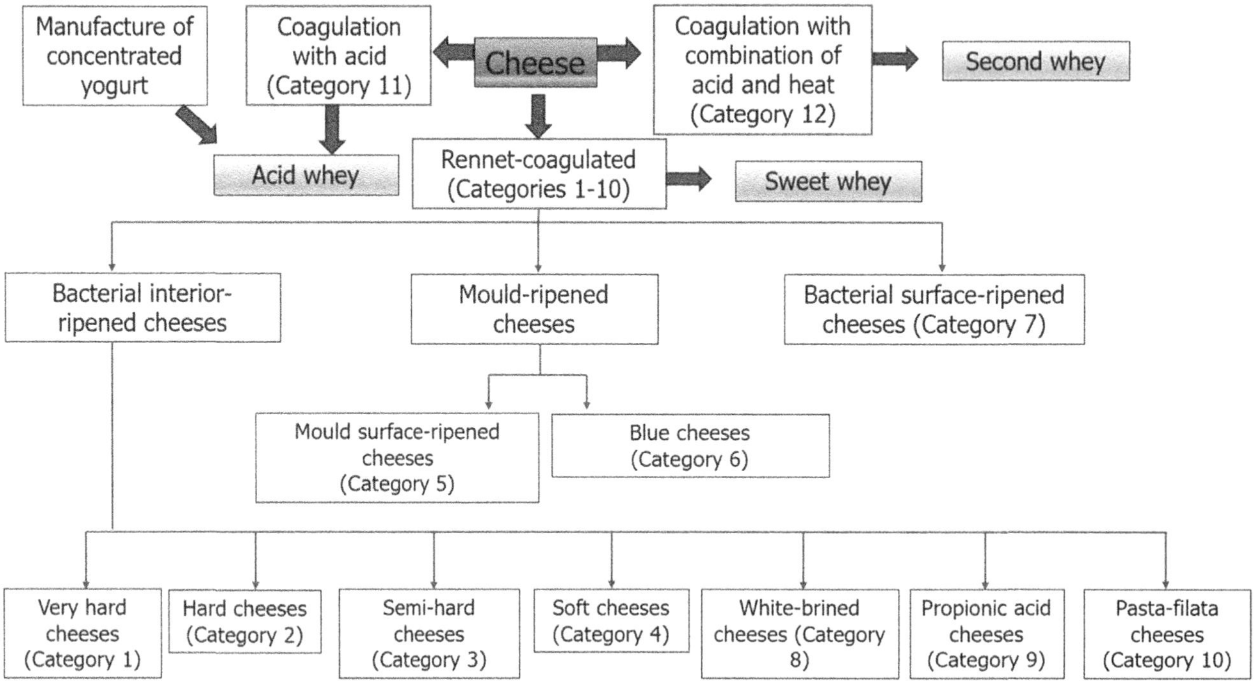

FIGURE 2.1 The classification of cheese categories and the different types of whey produced. *After: Bintsis and Papademas (2018).*

whey produced is called sweet whey (or whey). Whey from the manufacture of acid-coagulated cheeses is termed acid whey; acid whey is also produced from the manufacture of concentrated yogurt. Second whey is a third type of whey and it is the liquid left from the manufacture of cheese after the coagulation with a combination of acid and heat (i.e., whey cheeses) (Fig. 2.1).

The large amounts of whey produced from the industrial cheese manufacture were regularly either fed to animals or considered a waste and directly discarded into receiving water without pretreatment, stored in tanks or discharged in sewage (Prazeres et al., 2012). Whey discarding causes soil and water pollution due to its high organic load and biological oxygen demand (BOD) values are estimated to be 40–60 g/L and chemical oxygen demand (COD), 50–80 g/L), resulting from the presence of lactose and proteins (González Siso, 1996; Papademas and Kotsaki, 2019; Bandara et al., 2023).

During the recent 20 years, and after the expansion in research for the nutritional value of whey proteins, great efforts have been put on whey valorization. In addition, waste regulations have restricted the disposal, resulting increased effluent treatment costs stimulating the development of the whey product marketplace. The increased environmental issues have forced the dairy industry to orientate on efficient waste treatments for the whey (Buchanan et al., 2023); and subsequent development of innovative processes for the valorization of whey (Arshad et al., 2022; Barba, 2021; Buchanan et al., 2023; Lappa et al., 2019; Panghal et al., 2018; Papademas and Kotsaki, 2019; Pires et al., 2021a; Risner et al., 2018; Rocha-Mendoza et al., 2021; Tsermoula et al., 2021). Furthermore, utilization of waste whey as animal feed is now seldom advised as high lactose content or possible acidification of the waste whey, particularly where farms are not in a close vicinity to processing plants may raise an alarm for animal welfare (Buchanan et al., 2023).

3. Characteristics of whey

Whey is the main by-product of cheese-making; this by-product has gained great attention during the last decades (Bintsis and Papademas, 2023; Smithers, 2015). The manufacture of cheese is based on the coagulation of milk and for most cheese types, coagulum is formed as a network of caseins that entraps fat globules and other milk constituents. Milk is the main raw material and it is a stable colloidal dispersion of proteins and emulsified fat in a mineral and lactose-rich aqueous serum (Walstra et al., 2006). Milk proteins are present as casein micelles, that is, large protein assemblies and serum (or whey) proteins, that is, small globular proteins. Casein micelles consist of a hydrophobic core of α_{s1}-, α_{s2}- and β-caseins covered by an amphiphilic κ-casein "hairy" layer, held together by hydrophobic interactions and colloidal calcium phosphate (Walstra et al., 2006; Horne and Lucey, 2017). The casein micelles are negatively charged and repulsive forces hold the micelles in suspension until removed by some external influence (Horne and Lucey, 2017). On the other hand, serum proteins have different amino acid profiles from that of caseins due to a greater amount of sulfur-containing amino groups; serum proteins mostly exist as monomers or dimers that are folded into compact globular three-dimensional arrangements (Walstra et al., 2006).

During the manufacture of cheese, the casein fraction is coagulated either by acidification (acid-coagulated cheeses) or by specific proteolysis (rennet-coagulated cheeses); some minor cheese varieties are produced from the coagulation of serum proteins by heating (Bintsis and Papademas, 2018; Fox and Guinee, 2013). In all cases, whey (or cheese whey) is produced. The composition of the whey is mainly determined by the dairy animal species milk used for cheesemaking while cheesemaking manufacturing steeps are equally important (i.e., curd cutting, scalding, pressing etc.). Whey retains approximately 55% of the milk solids and 20% of the total milk protein (Ryan and Walsh, 2016; Pires et al., 2021a).

The pH values and chemical composition (g/100 g) of whey from different milk and cheese sources are shown in Table 2.1. In general, whey contains total solids (5%–10.8%), of which lactose is the main component (2.6%–6.1%), proteins (0.04%–1.8%), minerals (0.4%–1.9%) and fat (0.01%–1.5%). Acid whey has a low pH (4.3–4.8), whereas, higher values have been reported for sweet whey (5.3–6.3) and second whey (5.5–6.3) (Table 2.1). Moreover, acid whey contains less lactose and more total solids, ash and minerals; the coagulation by acidification causes the demineralization of casein and more soluble calcium and phosphorus are transferred in the acid whey (Giroux et al., 2018).

The serum fraction of the bovine milk protein is typically ~0.7 g/100 g (i.e., ~20% of the total protein); the main milk serum proteins (or whey proteins) are β-lactoglobulin (54%), α-lactalbumin (21%), immunoglobulins (14%) and bovine

TABLE 2.1 pH values and chemical composition (g/100 g) of whey from different milk and cheese sources.

Type of whey	pH	Minerals	Fat	Lactose	Protein	Total solids
		Acid whey				
Bovine acid whey	NA	NA	0.6–0.8	3.8–4.6	0.8–1.0	6.3–7.0
Bovine acid whey (from Cottage)	4.5–4.6	NA	0.03–0.04	4.7–4.9	0.75	5.0–6.0
Bovine acid whey from (from Cream cheese)	4.3–4.8	NA	0.01–0.32	3.0–4.6	0.2–0.5	5.2–5.9
Salty whey						
Bovine salty whey (from hard cheese)	5.1–5.5	NA	0.6–0.8	2.6–3.0	0.3–0.85	8.6–22.4
Second whey						
Bovine second whey	6.1	1.0–1.1	0.1–0.2	3.7–5.0	0.1–0.5	6.1
Buffalo second whey	NA	NA	NA	3.3	0.08	NA
Ovine second whey	5.5–6.3	1.7–1.9	0.2–0.4	4.5–5.7	0.8–1.2	6.9–8.3
Whey (or sweet whey)						
Bovine whey	5.6–6.3	0.5–1.0	0.1–1.0	4.2–6.0	0.5–1.1	6.0–7.0
Bovine whey (from Cheddar cheese)	6.3	NA	0.1	4.8	0.8	
Bovine whey (from Feta-type cheese)	6.3	NA	0.3	4.7	0.8	6.3
Bovine whey (from Graviera cheese)	6.3	NA	0.6	4.9	0.9	6.9
Bovine whey (from hard cheese)	5.4–6.3	0.4–1.0	0.1–0.4	3.1–5.0	0.6–1.2	5.7–8.5
Buffalo whey	NA	NA	0.5	4.9	0.7	6.5
Caprine whey	6.34	0.57	0.5–0.8	4.2–6.7	1.2–0.6	7.1–10.8
Ovine whey	5.3–5.9	1.0–1.8	1.2–1.5	4.3–6.1	1.6–1.8	7.6–10.5
Ovine whey (from Feta cheese)	6.3	NA	0.4	5.3	1.6	7.9
Ovine whey (from Graviera cheese)	6.3	NA	1.3	5.3	1.5	8.7

NA, not available.
After: Alfano et al. (2022), Anand et al. (2013), Bandara et al. (2023), Bintsis and Papademas (2023), Blaschek et al. (2007), Fancello et al. (2024), Jaeggi et al. (2005), Kalatzopoulos (1993), Kandarakis (1986), Macedo et al. (1999, 2015), Monari et al. (2019), Morr and Ha (1993), Pasta et al. (2024), Pires et al. (2021a), Prazeres et al. (2012), Secchi et al. (2012), Yadav et al. (2015), Zhao et al. (2022).

serum albumin (6%) and they exist as globular soluble proteins (Bintsis and Papademas, 2018; Ryan et al., 2016; Deeth and Bansal, 2019). The serum fraction of caprine milk ranges from 0.37–0.7 g/100 g of which 39.2%–72.1% represent β-lactoglobulin and 17.8%–33.3% α-lactalbumin (Amigo and Fontecha, 2011). Ovine serum proteins account for 17%–22% of the total protein. As with other milk types, α-lactalbumin and β-lactoglobulin are the major serum proteins (Ramos and Juarez, 2011). Ovine whey contains more β-lactoglobulin, about the same proportion of α-lactalbumin, and lower proportions of serum albumin and immunoglobulins compared to bovine whey (Hejtmánková et al., 2012). Caprine whey contains twice as much α-lactalbumin and lower levels of β-lactoglobulin, serum albumin, and immunoglobulins compared to bovine whey (Moatsou et al., 2005; Pintado et al., 1999). Similar variations exist in the distribution of serum proteins across these milk sources. Camel whey is characterized by a lower β-lactoglobulin and a higher α-lactalbumin content, comparing with bovine whey. The bioactive potential of cheese whey peptides from different animal origins (i.e., bovine, goat, sheep, buffalo, and camel) has been recently reviewed by Rodrigues et al. (2024). The average β-lactoglobulin/α-lactalbumin ratios were reported to be 2.1, 1.6, and 4.1 for bovine, caprine and ovine whey, respectively (Giroux et al., 2018).

Serum proteins are the main proteins in every type of whey, and these proteins are denatured by heat and acidity, a function i.e. technologically exploited by the production of whey cheeses (e.g., Ricotta), acid-coagulated cheeses (e.g., Quark) and yogurt manufacture. Serum proteins have a higher nutritive value than casein and are considered as one of the richest proteins available; they contain essential amino acids, for example, branched-chain amino acids (isoleucine, leucine, and valine) and sulfur amino acids (cysteine and methionine), and antioxidant properties have been reported (Bintsis and Papademas, 2023). Apart from the serum proteins, whey contains casein macro peptides, lipids, lactose, minerals (e.g., calcium, phosphate, magnesium, sodium and zinc), vitamins (e.g., B2, B6, B12 and C) and folic acid (Deeth and Bansal, 2019; Bandara et al., 2023).

Different types of whey have different characteristics, depending on the specific cheesemaking process from which are produced and a great variation in the chemical composition of whey can be observed. These differences can be attributed to the type of cheese manufactured and thus the type of whey produced; moreover, differences within the same type can be attributed to the kind of milk from which the main cheese is made (Bintsis and Papademas, 2023). Due to the composition of ovine milk, ovine whey consists of significantly higher levels of total solids, lipids, and proteins than bovine whey.

4. Whey production types

There are many differences in the quantity and quality of whey produced during cheesemaking (and strained yogurt-making) mainly arising from different milk types used and from differences in the manufacturing process. Three diverse coagulation types are used in cheesemaking and thus, three basic cheese families can be distinguished: (a) rennet-coagulated, (b) acid-coagulated, and (c) acid/heat-coagulated cheeses (Fig. 2.1).

In the first case, a typical cheese production includes rennet addition to cleave the κ-casein, which is adhered at the surface of the casein micelles, thus destabilizing the micelle structure and initiating the coagulation of milk (Kindstedt, 2013). Most of the cheeses produced worldwide are rennet-coagulated cheeses. Sweet whey is released from the curd of rennet-coagulated cheeses.

In the second case, and also in the manufacture of yogurt, the coagulum is formed by destabilized casein micelles as a result of acidification and pH decrease close to pH 4.6 i.e., the isoelectric point of a-s1 casein. The acidification is either caused by the adventitious lactic acid bacteria, the use of starter cultures or by adding acid (e.g., lactic acid or acetic acid). Acid whey is released from the curd of acid-coagulated cheeses. Similarly, acid whey is released from the straining of concentrated yogurts.

The third case has a different source of coagulation which is due to the denaturation of serum proteins, and subsequent aggregation, which is caused by heating at 85–90°C. Therefore, second whey is the serum left after whey proteins are denatured and aggregated by heating at 85–90°C.

4.1 Sweet whey—Whey from rennet-coagulated cheese

Approximately 75% of total cheese production is produced by rennet coagulation (Fox and Guinee, 2013). The rennet coagulation process involves rennet; that is, the complex of enzymes produced in ruminant stomachs. The functional enzyme is chymosin, an aspartic proteinase that can cleave κ-casein at the specific bond of Phe105-Met106. Chymosin will shave off the hairy layer, so that the subsequently exposed micelles begin to aggregate; once a sufficient number of their κ-casein molecules have been hydrolyzed (Kindstedt, 2013; Bintsis and Papademas, 2018). This cleavage produces a glycomacropeptide, which is the water-soluble fragment and expelled in whey (sweet whey). The remaining N-terminal

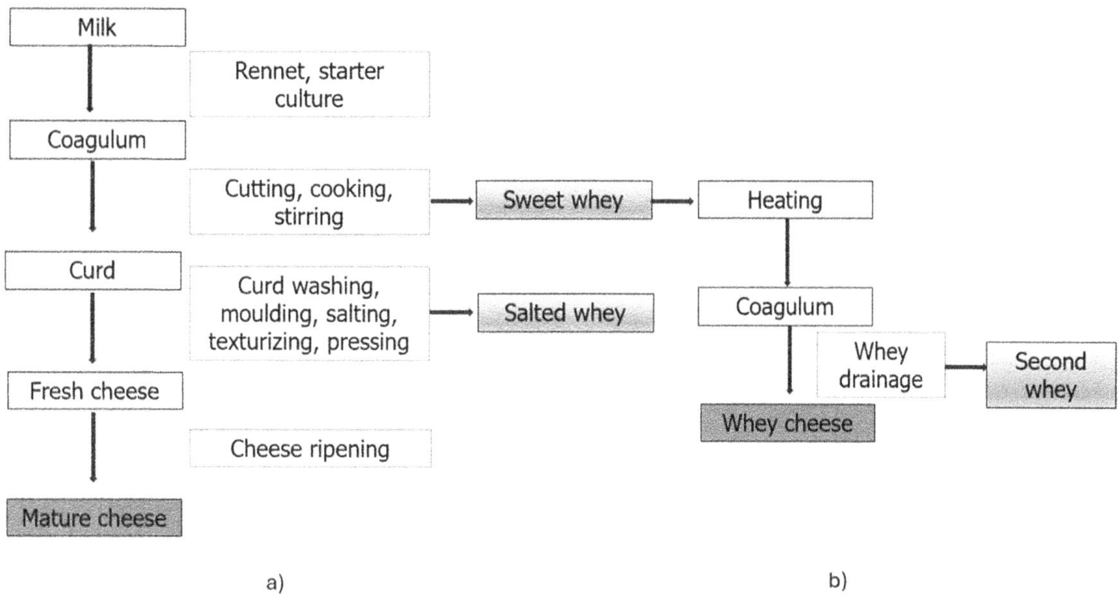

FIGURE 2.2 General flow diagram for the manufacture of (A) rennet curd and (B) whey cheese.

region of the κ-casein, termed the para-κ-casein remains bound in the casein network. While the glycomacropeptide solubilizes in the aqueous medium, para-κ-casein remains attached to the casein micelle that subsequently aggregates and forms a coagulum encompassing fat globules. Aggregation leads to bigger and bigger clusters until eventually the system acquires the solid-like nature of the gel. Fat globules are trapped or emulsified by the casein curd while the serum phase can be strained out. Rennet coagulation is a two-stage process with an enzymatic and a non-enzymatic phase (Horne and Lucey, 2017; Fox and Guinee, 2013).

The coagulum is then cut, cooked (scalded) and subjected to cheese-specific processing steps (depending on the type of cheese i.e. produced) until desired curd properties (e.g., pH, texture and moisture content) are achieved and sufficient whey is expulsed (Fox et al., 2017). Every single step has an impact on the release of whey and thus, on the final moisture content.

The final moisture content of most cheeses is determined primarily by the rate and the duration of the processes causing whey expulsion. After curd making, cutting and drainage, one or more of the following procedures are usually applied: cooking (scalding), molding, washing, salting, texturizing or stretching, and pressing (Fig. 2.2). During several of these process steps, the curd may undergo syneresis and lose considerable amounts of whey. The amount of whey expelled in each single step is determined by the environmental conditions (e.g., temperature). For example, pressing at a higher initial temperature leads to extended syneresis, and more whey expelled. Moreover, the factors are interrelated; for instance, temperature affects the rate of acidification. The curd size after cutting and stirring time also influence syneresis. The scalding (cooking) temperature and the acidity of the curd have a significant influence on the amount of whey expelled.

Salting contributes to the syneresis of the curd, and this is also dependent on temperature (Fagan et al., 2017). Whey removed during this stage due to osmosis is referred to as salty whey and has a pH 5.1–5.5. The characteristics of salty whey are determined by the specific cheese-making process steps; whey accounts for ~2%–5% of whey generated during dry salted cheese-making (Blaschek et al., 2007). The differences in the salt content between sweet and salty whey affect whey protein solubility and their release into the aqueous stream during syneresis. A comparison between the serum protein quantities in salty and sweet whey has shown that salty whey has higher amounts of lactoferrin and lower amounts of α-lactalbumin, β-lactoglobulin, serum albumin and immunoglobulin compared with sweet whey (Blaschek et al., 2007).

4.2 Acid whey—Whey from acid-coagulated cheese and strained yogurt

Acid-coagulated cheese varieties include Cream cheese, Cottage cheese, and Quarg. Concentrated yogurts (e.g., Labneh, Greek-style, Stragisto, Ymer) are closely related to fresh acid-coagulated cheeses. In both cases, the manufacture differs from fermented milks in the removal of a significant amount whey after coagulation (Fig. 2.3) (Bintsis and Papademas, 2022). Cultures of lactic acid bacteria are used as starter cultures for yogurt and most fresh acid cheeses. A common factor

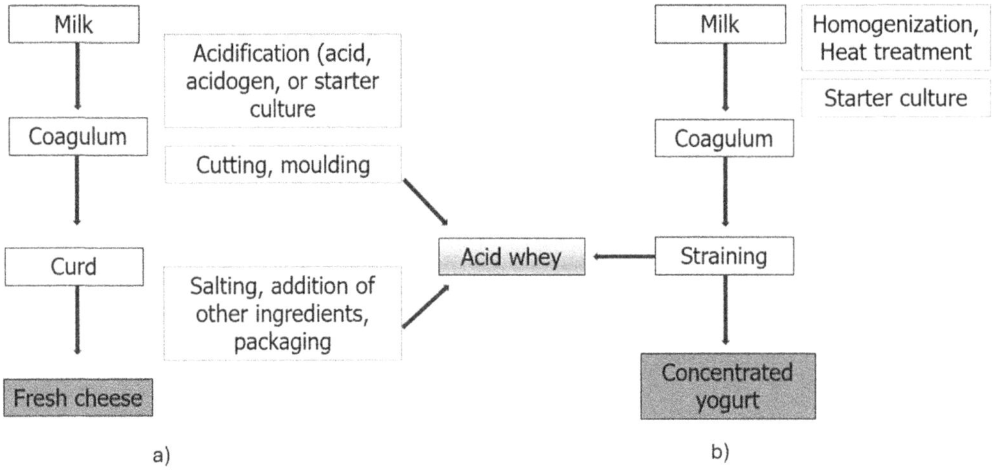

FIGURE 2.3 General flow diagram for the manufacture of (A) acid curd cheese and (B) concentrated yogurt.

in all of these acid cheese products is that the initial step involves the formation of an acid-induced gel, which is then further processed. The acidification is carried out using acids (e.g., lactic acid or acetic acid), acidogens (e.g., glucono-δ-lactone) or starter cultures, that produce lactic acid from the fermentation of lactose. At pH close to 5.0, a decrease in the net negative charge on the casein micelles, thereby reducing electrostatic repulsion and this causes the micelle hair layer of the κ-casein to shrink (Lucey, 2017). On the other hand, colloidal calcium phosphate which binds casein molecules is solubilized into the serum phase (Le Graët and Gaucheron, 1999). The net result is a decrease in both electrostatic repulsion and steric stabilization, two factors that are primarily responsible for micelle stability. Acidification eventually leads to the formation of chains and clusters that are linked together to form a three-dimensional network (Lucey, 2002).

Acid whey, due to the early acidification of the curd of acid-coagulated cheese, has a low pH in comparison to sweet whey (Table 2.1). The lower pH achieved during acid-induced coagulation leads to the dissolution of colloidal calcium phosphate in casein micelles, resulting in a higher calcium concentration in acid whey compared to sweet whey (Nishanthi et al., 2017). Moreover, acid whey does not contain a glycomacropeptide (see rennet-coagulated cheeses) and has high ash content (calcium released from micelle into serum phase), high lactic acid and usually low lactose content, as some of the lactose is converted into lactic acid during acidification, when acidification is carried out by lactic acid bacteria, compared to sweet whey. In general, acid-coagulated curds undergo much less syneresis than the rennet-coagulated and thus, acid-coagulated cheeses usually have a very high moisture content (e.g., higher than 60%).

Acid whey from strained yogurt shows some differences in the composition as the manufacturing process is different from that of acid-coagulated cheese (Nishanthi et al., 2016; Chandrapala et al., 2015). For concentrated yogurt manufacture, coagulation takes place due to the aggregation of serum denatured proteins. The denaturation occurs at high temperatures, <75°C, which takes place before the addition of starter to initiate fermentation. The severe heat treatment of milk results in an association between serum proteins and casein, and this significantly impacts on the properties of the micelles. Due to heating, some of the serum protein (especially β-lactoglobulin) interacts with κ-casein via disulfide-thiol interaction (Lucey, 2002) and become part of the milk curd. Thus, a proportion of serum proteins is entrapped in the casein network, and acid whey has a lower protein content than sweet whey.

4.3 Second whey—Whey from acid/heat-coagulated cheeses

Second whey is the whey released from the production of whey cheese after the coagulation of serum proteins (Fig. 2.3). It is called Scotta in Italy and Sorelho in Portugal (Pires et al., 2021a). Serum proteins are heat sensitive and begin to denature and unfold at temperatures above about 79°C, thus exposing their nonpolar hydrophobic regions; the sensitivity increases when the pH of milk is decreased. Heat-denatured serum proteins lose their capacity to interact with water molecules due to their exposed hydrophobic regions, and this causes them to aggregate with each other and with casein micelles (Kindstedt, 2013).

For the manufacture of traditional whey cheese, the whey from the manufacture of hard or other cheeses is mixed with milk and/or cream and heated at 88–90°C for 40–45 min; heating causes aggregation of the whey proteins and formation of the curd (Papademas et al., 2018). Salt is added in some cases, and the curd is molded, drained and cooled. They may be packed and consumed immediately or left for drying and used for grating (Papademas et al., 2018).

Other whey cheeses are manufactured with similar processes, however, differences in manufacturing parameters may produce cheeses with different characteristics. More specifically, the temperature and the duration of heating and the addition or not of milk, the origin of the whey, the kind of milk are the main factors accounting for thee variability. The Italian Ricotta is the most famous and most widely produced and consumed whey cheese, worldwide (Mangione et al., 2023). For the production of Ricotta, sweet whey is used, mixed with 5%–10% milk, salt is added at 0.1% and citric acid at 0.11 kg/L while heating is carried out to 80–85°C (Pintado et al., 2001). There different varieties of Ricotta exist; from bovine, ovine and/or buffalo milk. Ricotta forte is a special variety, which is ripened for at least 6 months (Faccia et al., 2018). Moreover, Ricotta di Bufala Campana PDO has been registered as a cheese with Protected Designation of Origin (PDO) (Montone et al., 2017). Urda is a well-known whey cheese in Serbia and Romania as well as in other countries such as Israel; Also, Scuta is produced in Serbia; during the heating process, milk and/or cream can be added to the whey, before or after coagulation (Paskaš et al., 2019). In Croatia, Bračka skuta and Skuta Dalmatian are the most popular whey cheeses, while the latter is produced from acidified whey by the addition of up to 10% vinegar or citric acid (Bintsis and Papademas, 2023). Anthotyros, Myzithra, Manouri PDO, Xynomyzithra Kritis PDO, and Urda are produced in Greece; Manouri and Xynomyzithra have been registered as PDO cheeses. In Cyprus, Anari is consumed either fresh or dried. Turkish Lor cheese is a well-known whey cheese produced in Turkey; the raw cheese is strained in a hemp or other straining cloth and press is applied (Kamber, 2007). Arishi is a whey cheese produced in Lebanon by heating the whey at 85°C for 5–10 min (Hassan et al., 2022). Requeijão, produced in Portugal and Requesón, produced in Spain and Mexico are also well-known whey cheeses (Pintado et al., 1996; Villarruel-López et al., 2016). In the North, Gjetost, Mysost and Brunost are produced in Norway; Gjetost is a special whey cheese with a brownish color as the caramelized lactose in concentrated caprine whey is combined with whey proteins and fat (Skeie, 2014). Mesost and Messmör are produced in Sweden, with the whey being boiled to evaporate water (Papademas et al., 2018). Moreover, novel whey cheeses focused on functional properties have developed during the last 10 years (Bintsis and Papademas, 2023).

Second whey has a high concentration in lactose (3.3%–5.7%) and high amounts of minerals and especially calcium (Table 2.1). It is considered as a high pollutant effluent with BOD value of 50 g/L and COD of 80 g/L (Maragkoudakis et al., 2016). The high lactose in second whey is maybe a limiting factor for use as animal feed since it has been shown to cause digestive disorders (Pires et al., 2021a). On the other hand, Sutera et al. (2023) indicated that a low-cost supplemented diet using a dairy by-product as co-feed, such as liquid whey, might be potentially employed in swine production in order to improve animal health.

The low protein content is another characteristic of second whey, however, besides the low concentration, these proteins are a valuable source of bioactive peptides. Recently, Fancello et al. (2024) systematically reviewed the valorization strategies of second whey for the production of food, beverages, microbial starters, single cell proteins, single cell oils, and bioactive metabolites, as well as in the formulation of animal feed, infant formula ingredients and the production of polymers for biodegradable bioplastics, bio-ethanol, bio-butanol and bio-gas/biomethane and hydrogen.

5. Whey products

Historically, whey was used as a refreshing beverage in ancient Greece (Bintsis and Papademas, 2023). Later, the Romans discovered the manufacture of whey cheeses; Ricotta was the first whey cheese and its manufacture was described by Columella, in the first century AD in his book De Re *Rustica* (Mangione et al., 2023). The demand for whey products has been increasing during the last 20 years and value-added products such as whey protein powders, novel whey cheeses, processed cheeses, beverages, infant formulas, yogurts, ice cream, ingredients for pharmaceuticals and food supplements and packaging materials (i.e., edible films) are widely produced from all types of whey.

5.1 Sweet whey products

The main volume of sweet whey is used for the manufacture of traditional whey cheeses (Bintsis and Papademas, 2023) and the different types of whey cheese have been described in Section 4.2. However, a great number of applications have been suggested for alternative utilization of whey proteins and other whey components.

5.1.1 Ingredients

The characteristic of all types of whey is the high-water content (Table 2.1), thus drying or other kind of concentration process is necessary for valorization (de Castro et al., 2017, de wit 1998, Kumar et al., 2018; Pescuma et al., 2015). After drying, whey can be directly used in the food industry or can be transformed into several by-products such as lactose, whey permeate, whey protein concentrate (WPC34-80), whey protein isolates (WPI), whey hydrolyzates, and even pure whey proteins such as β-lactoglobulin, α-lactalbumin, and immunoglobulins; ultrafiltration or diafiltration are common techniques for the separation of whey proteins (Siso, 1996; Smithers, 2008).

Moreover, whey permeate is produced from the production of WPC and WPI as well as purified individual whey proteins (Argenta and Scheer, 2019; Kaur et al., 2020; Pescuma et al., 2015). The nutritional, technological and functional properties of whey proteins (Zhao et al., 2022) enabled extensive research on food and non-food applications. The high concentration of lactose (Table 2.1), present in whey permeate, of all types of whey, as the major carbohydrate component of whey, can act as a carbon source for growth and product formation in numerous biotechnological processes. Fermentation products include animal feed, single cell protein, organic acids, enzymes, carotenoids, bio-preservatives, prebiotics, exopolysaccharides, ethanol, butanol and edible films (Ahmed et al., 2013; Akaraonye et al., 2010; Athanasiadis et al., 2004; Bandara et al., 2023; Blais et al., 2022; Besediuk et al., 2024; Buchanan et al., 2023; Caltzontzin-Rabell et al., 2024; Chaudhary et al., 2023; Dinika et al., 2020; Dopazo et al., 2022; Guimarães et al., 2010; Hassan and Nelson, 2012; Karadag et al., 2014; Koukoumaki et al., 2024; Koutinas et al., 2009; Moreira et al., 2015; Pescuma et al., 2015, Tita et al., 2024, Yadav et al., 2015).

While biogas and other bio-fuels are produced from whey by anaerobic digestion (Casallas-Ojeda et al., 2021; Ramos-Suárez et al., 2024), lactose fermenting yeasts, that is, *Kluyveromyces lactis*, *Kluyveromyces marxianus*, and *Candida pseudotropicalis* (synonym of *Candida kefir*) have been employed for the production of bio-ethanol from sweet whey (Guimarães et al., 2010; Moreira et al., 2015, Zoppellari and Bardi, 2013). The use of whey as an alternative culture medium for the growth and encapsulation of lactic acid bacteria, that is *Leuconostoc mesenteroires* (Okuda et al., 2017), *Lacticaseibacillus paracasei* (Pinilla et al., 2025; Rama et al., 2019) and probiotics, such as *Limosilactobacillus fermentum*, *Lactobacillus delbruekii*, *Lacticaseibacillus rhamnosus* and *Enterococcus faecium* has been suggested (Bauer Estrada et al., 2022). D'Alessandro et al. (2025) investigated the addition of the prebiotic lactobionic acid and *L. rhamnosus* GG and the production of symbiotic fermented milk; sweet whey was used as a substrate to produce the lactobionic acid from fermentation by *Pseudomonas taetrolens* DSM 21104. Recently, O'Donoghue and Murphy (2023) reviewed the non-dairy food applications of whey permeates.

5.1.2 Dairy products

Whey and whey protein products, enriched with bioactive compounds, can be utilized for the functional and nutritional enhancement of foods and the development of novel functional dairy foods with health beneficial effects (Chourasia et al., 2022). The development of ultrafiltration techniques was a breakthrough in industrial cheesemaking, achieving the retention or the reintegration of whey proteins to increase the cheese yield; additionally, different cheese varieties are produced with novel flavor, texture and functional properties (Masotti et al., 2017). On the other hand, the use of milk with higher than pasteurization heat treatment is another way to include denaturated whey proteins; however, it caused prolonged rennet coagulation time and texture defects (Hougaard et al., 2010). The addition or inclusion of whey proteins in cheesemaking process can increase the cheese yield (El-Aidie and Khalifa, 2024; Neelima et al., 2013). The use of high heat-treated milk to increase the cheese yield is more suitable in the manufacture of lactic and fresh cheeses (e.g., Cottage) where the coagulation by acidification dominates over the action of rennet action (Masotti et al., 2017).

The addition of whey concentrates in the production of traditional whey cheeses, such as Ricotta, can increase α-lactalbumin protein yield and recovery rate (Salvatore et al., 2014). Attempts to incorporate whey protein concentrates in fresh cheeses such as Myzithra (Kaminarides, 2015), Queso fresco (Henriques et al., 2012) showed good results in terms of yield and acceptance, although excessive levels may cause detrimental changes in the texture and flavor and the process needs to be optimized. Dimitrellou et al. (2017) evaluated the feasibility of producing a novel whey cheese with extended shelf life using probiotic lactobacilli immobilized on casein. Pires et al. (2021b) evaluated the application of kefir culture and probiotics, that is, *Bifidobacterium* spp., which has previously added and fermented the cream added to the whey for the manufacture of Requeijão cheese. In another study, the addition of *L. delbrueckii* ssp. *bulgaricus* NCFB 2772 was investigated to manufacture a functional acid whey cheese (Carrero-Puentes et al., 2022). Nzekoue et al. (2021) suggested Ricotta cheese as a cheese matrix for vitamin D fortification. An added-value whey cheese through addition of pine nuts was developed by Semeniuc et al. (2015); different concentrations of pine nuts were added to whey cheese, and the whey cheese sample with 4% pine nuts was the most appreciated by the sensory panel. Probiotic bacteria (i.e., *Bifidobacterium animalis* and *Lactobacillus casei*) were incorporated in whey cheese (Madureira et al., 2011, 2013). Moreover, Madureira et al. (2015) combined the application of probiotic culture with several additives (i.e., salt, garlic, aromatic herbs and xanthan); whey cheese has proven to be a good vehicle to deliver probiotic bacteria into the gastrointestinal tract.

Another application for whey produced from cheesemaking is to use it as an ingredient in processed cheese and cheese analogs (Ayed et al., 2023; Masotti et al., 2018). Adding whey products to cheese analogs, that is WPI and WPC80 can control textural and rheological properties and meltability and reduce disodium phosphate content by partially eliminating emulsifying salts (Sołowiej et al., 2014). A whey cheese analog was made from WPC65 and vegetable fat with 15% olive oil, and the product was enriched with thyme leaves, film-packed, and then subjected to heat treatment; the product was characterized as a new whey cheese easy to produce, with improved health enhancing properties, acceptable textural characteristics, a rich aroma profile, and extended shelf life (Zoidou et al., 2016). The effects of adding different levels of whey cream to replace some of the sweet cream present in cream mix on the rheology and texture of cream cheese was investigated; although the use of higher levels of whey cream resulted in defective gels and cheeses, the use of up to 25% of the sweet cream with whey cream resulted in high quality and texture cream cheeses (Brighenti et al., 2021).

Recently, Tomczyńska-Mleko et al. (2025) studied an experimental high-protein hard cheese, with approximately 39% (w/w) protein, produced by mixing a soft cheese with WPI and heating at different temperatures.

Due to nutritional and functional properties, whey proteins have gained an increased interest in fortifying yogurt with these proteins (Wherry et al., 2019). According to Mahomud et al. (2017), addition of whey proteins to yogurt's milk base, at the range of 0.6%–4% w/w, resulted in syneresis reduction. However, due to an undesirable whey flavor and sometimes grainy texture in high protein levels, its content yogurt mix is limited to 1%–2% (Lee and Lucey, 2010). The addition of WPC to the milk base of yogurt sounds to be a more common practice than the WPI addition (Penna et al., 1997). The high heat treatment of the whey during the production of WPC induct negative effect on the water-holding capacity of the yogurt, resulting in the more particulate and heterogeneous texture of the final product (Arab et al., 2023; Karam et al., 2013). The incorporation of WPI, as a fat replacer, increased the water holding capacity and improved the quality of non-fat yogurt (Hashim et al., 2021). Sakkas et al. (2019) studied a number of yoghurt-type gels from skim sheep milk base enriched with WPC. Recently, Galarreta-Morales et al. (2025) studied the sensory characteristics of a whey-based yogurt made from whey with the addition of mango with cushuro jam. Minj and Anand (2020) observed that whey and WPCs may contain bioactive components that help in the growth of probiotic bacterial cultures in both yogurt and the human digestive tract. The researchers observed that the addition of *Bifidobacterium bifidum* to standard yogurt cultures resulted in higher viable *B. bifidum* counts in samples containing sweet whey or whey proteins.

Marnotes et al. (2021) manufactured frozen yoghurts using goat and sheep liquid whey concentrates produced by ultrafiltration and inulin. Whey proteins have been applied in the manufacture of ice cream; Yilsay et al. (2006) reported that ice mixtures made with a 50% addition of WPC or whey-fat formulations, containing vegetable fats, improved the quality of the final product. Roy et al. (2022) developed an ice cream with a creamy, smooth texture, and rich flavor using a mixture of WPC and skim milk powder. Pires et al. (2023) recommended various methods of incorporating WPCs into ice cream blends, offering manufacturers a range of options to select from. Ice cream made from sheep milk was manufactured with addition of whey protein concentrates (Moschopoulou et al., 2021).

Moreover, another option for whey utilization in dairy products is the production of butter using sweet whey; whey cream is separated from the aqueous fraction of whey and processed to obtain whey cream butter (Jinjarak et al., 2006; Nadeem et al., 2015). Costa et al. (2022) reviewed the technology of production of butter from whey cream and the physico-chemical, sensory, and nutritional characteristics of the product.

5.1.3 Beverages

A great number of fermented or non-fermented beverages made from sweet whey has been developed, and many of them are commercially produced in recent years, with special nutritional and functional potential (Bandara et al., 2023; Barone et al., 2022; Buchanan et al., 2023; Cunha et al., 2022; El-Aidie et al., 2020; Panghal et al., 2018; Papademas and Kotsaki, 2020; Pescuma et al., 2015; Turkmen et al., 2019; Yamahata et al., 2020).

Such products are mainly fermented beverages (Marcus et al., 2021; Saglam et al., 2019; Zotta et al., 2020) and probiotic beverages (Kadyan et al., 2021; Rosa et al., 2023; Saglam et al., 2019; Skryplonek et al., 2019, Silva e Alves et al., 2018). Athanasiadis et al. (2004) and Magalhaes et al. (2011) studied whey beverages fermented with kefir grains. Jitpakdee et al. (2022) fermented whey with a microbial community to create a raw material for a functional whey drink with increased low molecular weight peptides. Sabokbar and Khodaiyan (2015) developed a kefir-based beverage using whey and pomegranate, exhibiting high antioxidant activity due to peptides from WPs and polyphenols.

Goveas et al. (2024) studied the development and health benefits of a functional beverage from whey fermented with a bioactive EPS producing *Lactiplantibacillus plantarum* SVP2 strain. Guimarães et al. (2018) developed a prebiotic whey beverage exploring the influence of degree of inulin polymerization. Hernandez-Mendoza et al. (2007) developed a whey-based probiotic product with *Lactobacillus reuteri* and *Bifidobacterium bifidum*. Fermented whey beverage was produced by *Pediococcus pentosaceus* ENM104 and *L. plantarum* SPS109 (Jitpakdee et al., 2022). Moreover, Sohrabi et al. (2016) studied fermented and unfermented whey beverages fortified with vitamin E.

The production of whey beverages has some limitations concerning the flavor that may not be attractive to all consumers; furthermore, the sensitivity of whey components to heat and acidity can affect the taste and texture of the product. The incorporation of fruit and vegetable concentrates, together with sweeteners, stabilizers and emulsifiers has been suggested (Ayed et al., 2023; Tanwar et al., 2022). The addition of fruit and vegetable concentrates, such as mango (Ahmed et al., 2013; Tanwar et al., 2022; Yadav et al., 2016), pineapple (Baba et al., 2016; Islam et al., 2021; Kumar et al., 2017), strawberries (Janiaski et al., 2016), papaya (Panghal et al., 2017), grape (Amaral et al., 2018), pomegranate (Sabokbar and Khodaiyan, 2015), Aloe vera (Sasi, 2015) and carrot (Naik et al., 2023) has been reported. Tita et al. (2024) developed soft beverages from whey and deproteinized whey with different additions of fruits and spices.

The production of alcoholic beverages has been extensively studied as a potential application of whey, resulting in the creation of whey beer, wine, and sparkling wine, usually beverages with low alcohol content (Jeličić et al., 2008; Ryan and Walsh, 2016). They are mostly manufactured using whey permeate, which has a low protein content, along with the addition of yeast strains such as *Kluyveromyces fragilis* or *Saccharomyces lactis* (Krolczyk et al., 2016). The production of an alcoholic beverage from the fermentation of whey by *Kl. marxianus* ATCC 10022 was suggested (Dragone et al., 2009). A recent study demonstrated the production of whey distillate by fermenting whey permeate, from the ultrafiltration of sweet whey from Swiss cheesemaking, by *Kl. marxianus* (Gantamur et al., 2025); the authors concluded that fermented whey permeate has the potential to produce high quality whey-based spirits with enhanced ethanol yields and refined sensory attributes, providing an environmentally and economically sustainable solution.

5.2 Acid and second whey products

Acid whey as mentioned before is characterized by a low protein content (~0.8%) but high lactose (~4.5%) and mineral levels, acid whey disposal is challenging due to its BOD and environmental impact (Smithers, 2015). To address sustainability concerns, various valorization strategies have been developed. Biotechnological conversions utilize acid whey as a substrate for producing lactic acid, bioethanol, biogas, and single-cell proteins through fermentation with microorganisms such as *Lactobacillus* and *Kluyveromyces* species (Panesar et al., 2010; Aghbashlo et al., 2021). Food applications include incorporating acid whey into bakery products, beverages, and dairy-based drinks to enhance nutritional content while reducing waste. Furthermore, acid whey minerals (e.g., calcium, magnesium, and phosphorus) can be recovered for nutraceuticals or used as agricultural fertilizers. Industrial applications include bioplastics (PLA) derived from whey-based lactic acid (Wang et al., 2021). These innovative approaches contribute to a circular bioeconomy, enhancing both environmental sustainability and economic value within the dairy sector.

Bendelja Ljoljić et al. (2023) studied a method of second whey valorization by producing a second whey permeate-derived alcoholic beverage; first they manufactured an Istrian semi-hard cheese made from cow milk, then from the whey they produced a whey cheese (Istrian albumin cheese), and then, a whey distillate by fermenting the second whey with *Kl. marxianus*. Kotsaki et al. (2025) in an effort to valorize side streams from the dairy industry used second whey (deproteinated whey) from Halloumi cheese production to manufacture a novel whey fermented beverage with probiotics and added juice concentrates to improve the sensory characteristics. A static in-vitro digestion system was used to mimic the gastrointestinal human digestion and study the beverage's bioactivities before and after digestion.

6. Conclusion

The main types of whey; that is, the liquid remaining from the coagulation of milk during cheesemaking, are sweet whey, acid whey and second whey. Each type is produced from different cheesemaking technologies and thus, differences in their composition and other characteristics can be observed. The presence of serum (whey) proteins, milk fat and lactose in all types of whey make this by-product a valuable raw material for the dairy industry.

The valorization of all types of whey has become a global trend in the dairy industry during the last 20 years. By reusing whey, the dairy industry reduces its environmental footprint and produces value-added products. Novel technologies, in parallel with the traditional whey processing, and the principles of circular economy, are implemented, to transform all types of whey into a great diversity of dairy products or ingredients valuable to the food industry.

References

Aghbashlo, M., Hosseinzadeh-Bandbafha, H., Tabatabaei, M., et al., 2021. Dairy waste valorization through biorefinery technologies: state-of-the-art review. J. Clean. Prod. 283, 124523.

Ahmed, S.G., Wafaa, H.E., Gamal, F.M., Ahmed, F.S., 2013. Utilization whey in production of functional healthy beverage "Whey-mango Beverages". Am. J. Food Technol. 8, 133–148.

Akaraonye, E., Keshavarz, T., Roy, I., 2010. Production of polyhydroxyalkanoates: the future green materials of choice. J. Chem. Technol. Biotechnol. 85, 732–743.

Alfano, A., D'ambrosio, S., Cimini, D., et al., 2022. No waste from waste: membrane-based fractionation of second cheese whey for potential nutraceutical and cosmeceutical applications, and as renewable substrate for fermentation processes development. Fermentation 8, 514.

Amaral, G., Silva, K., Costa, A., Alvarenga, V., Cavalcanti, R., Esmerino, E.A., Guimares, J.T., Freitas, M.Q., Sant'Ana, A.S., Cunha, R.L., et al., 2018. Whey-grape juice drink processed by supercritical carbon dioxide technology: physical properties and sensory acceptance. LWT 92, 80–86.

Amigo, L., Fontecha, J., 2011. Goat milk. In: Fuquay, J.W., Fox, P.F., McSweeney, P.L.H. (Eds.), Encyclopedia of Dairy Sciences, second ed. Academic Press, London, pp. 484–493.

Anand, S., Khanal, S.N., Marella, C., 2013. Whey and whey products. In: Park, Y.W., Haenlein, G.F.W. (Eds.), Milk and Dairy Products in Human Nutrition: Production, Composition and Health. Wiley-Blackwell, Chichester, pp. 477–497.

Arab, M., Yousefi, M., Khanniri, E., Azari, M., Ghasemzadeh-Mohammadi, V., Mollakhalili-Meybodi, N., 2023. A comprehensive review on yogurt syneresis: effect of processing conditions and added additives. J. Food Sci. Technol. 60 (6), 1656–1665.

Argenta, A.B., Scheer, A.D.P., 2019. Membrane separation processes applied to whey: a review. Food Rev. Int. 36, 1–30.

Arshad, U.-E.-T., Hassan, A., Ahmad, T., Naeem, M., Chaudhary, M.T., et al., 2022. A recent glance on the valorisation of cheese whey for industrial prerogative: high-value-added products development and integrated reutilising strategies. Int. J. Food Sci. Technol. 58, 2001–2013.

Athanasiadis, I., Paraskevopoulou, A., Blekas, G., Kiosseoglou, V., 2004. Development of a novel whey beverage by fermentation with kefir granules. Effect of various treatments. Biotechnol. Prog. 20, 1091–1095.

Ayed, L., M'hir, S., Asses, N., 2023. Sustainable whey processing techniques: innovations in derivative and beverage production. Food Biosci. 53, 102642.

Baba, W.N., Din, S., Punoo, H.A., Wani, T.A., Ahmad, M., Masoodi, F.A., 2016. Comparison of cheese and paneer whey for production of a functional pineapple beverage: nutraceutical properties and shelf life. J. Food Sci. Technol. 53, 2558–2568.

Bandara, T.A., Munasinghe-Arachchige, S.P., Gamlath, C.J., 2023. Fermented whey beverages: a review of process fundamentals, recent developments and nutritional potential. Int. J. Dairy Technol. 76 (4), 737–757.

Barba, F.J., 2021. An integrated approach for the valorization of cheese whey. Foods 10, 564.

Barone, G., O'Regan, J., Kelly, A.L., O'Mahony, J.A., 2022. Interactions between whey proteins and calcium salts and implications for the formulation of dairy protein-based nutritional beverage products: a review. Compr. Rev. Food Sci. Food Saf. 21, 1254–1274.

Bauer Estrada, K., Caldas Abril, M., Bonilla Bravo, V., Ruiz, M., Quintanilla Carvajal, M.X., 2022. Whey as food-grade culture medium on an industrial scale that protects probiotics during in vitro digestion. Front. Food Sci. Technol. 2, 894761.

Bendelja Ljoljić, D., Kalit, S., Kazalac, J., Dolenćić Špehar, I., Mihaljević Žulj, M., Maslov Bandić, L., Tudor Kalit, M., 2023. The potential of using Istrian albumin cheese whey in the production of whey distillate. Fermentation 9, 192.

Besediuk, V., Yatskov, M., Korchyk, N., Kucherova, A., Maletskyi, Z., 2024. Whey—from waste to a valuable resource. J. Agric. Food Res. 18, 101280.

Bintsis, T., Papademas, P., 2018. An overview of the cheesemaking process. In: Papademas, P., Bintsis, T. (Eds.), Global Cheesemaking Technology—Cheese Quality and Characteristics. John Wiley & Sons, Ltd., Chichester, pp. 120–156.

Bintsis, T., Papademas, P., 2022. The evolution of fermented milks, from artisanal to industrial products: a critical review. Fermentation 8, 679.

Bintsis, T., Papademas, P., 2023. Sustainable approaches in whey cheese production: a review. Dairy 4 (2), 249–270.

Blais, H.N., Schroën, K., Tobin, J.T., 2022. A review of multistage membrane filtration approaches for enhanced efficiency during concentration and fractionation of milk and whey. Int. J. Dairy Technol. 75, 749–760.

Blaschek, K.M., Wendorff, W.L., Rankin, S.A., 2007. Survey of salty and sweet whey composition from various cheese plants in Wisconsin. J. Dairy Sci. 90, 2029–2034.

Brighenti, M., Govindasamy-Lucey, S., Jaeggi, J.J., Johnson, M.E., Lucey, J.A., 2021. Effect of substituting whey cream for sweet cream on the textural and rheological properties of cream cheese. J. Dairy Sci. 104 (10), 10500–10512.

Buchanan, D., Martindale, W., Romeih, E., Hebishy, E., 2023. Recent advances in whey processing and valorisation: technological and environmental perspectives. Int. J. Dairy Technol. 76, 291–312.

Caltzontzin-Rabell, V., Feregrino-Pérez, A.A., Gutiérrez-Antonio, C., 2024. Bio-upcycling of cheese whey: transforming waste into raw materials for biofuels and animal feed. Heliyon 10, e32700.

Carrero-Puentes, S., Fuenmayor, C., Jiménez-Pérez, C., Guzmán-Rodríguez, F., Gómez-Ruiz, L., Rodríguez-Serrano, G., Alatorre-Santamaría, S., García-Garibay, M., Cruz-Guerrero, A., 2022. Development and characterization of an exopolysaccharide-functionalized acid whey cheese (requesón) using *Lactobacillus delbrueckii* ssp. *bulgaricus*. J. Food Process. Preserv. 46, e16095.

Casallas-Ojeda, M., Torres-Guevara, L.E., Caicedo-Concha, D.M., Gómez, M.F., 2021. Opportunities for waste to energy in the milk production industry: perspectives for the circular economy. Sustainability 13, 12892.

Chandrapala, J., Duke, M.C., Gray, S.R., Zisu, B., Weeks, M., et al., 2015. Properties of acid whey as a function of pH and temperature. J. Dairy Sci. 98, 4352–4363.

Chaudhary, V., Kaila, P., Verma, D., Singh, T.P., Kothakota, A., Prasath, V.A., Jeevarathinam, G., Kumar, M., Ramniwas, S., Rustagi, S., Pandiselvam, R., 2023. Valorization of dairy wastes into wonder products by the novel use of microbial cell factories. Trends Food Sci. Technol. 142, 104221.

Chourasia, R., Phukon, L.C., Abedin, M.M., Padhi, S., Singh, S.P., Rai, A.K., 2022. Whey valorization by microbial and enzymatic bioprocesses for the production of nutraceuticals and value-added products. Bioresour. Technol. Rep. 19, 101144.

Costa, M.A., Kuhn, D., Rama, G.R., Lehn, D.N., Souza, C.F.V., 2022. Whey butter: a promising perspective for the dairy industry. Braz. J. Food Technol. 25, e2021088.

Cunha, D.S., Coelho, M.C., Ribeiro, S.C., Silva, C.C.G., 2022. Application of *Enterococcus malodoratus* SJC25 for the manufacture of whey-based beverage naturally enriched with GABA. Foods 11 (3), 447.

D'Alessandro, M., Gottardi, D., Franceschini, S., Braschi, G., Siroli, L., Nissen, L., Romano, R., Putignani, L., Gianotti, A., Raddadi, N., Lanciotti, R., Vernocchi, P., Patrignani, F., 2025. Impact of cheese whey enriched in lactobionic acid on the characteristics of fermented milks prepared with or without the probiotic *Lacticaseibacillus rhamnosus* GG. Int. Dairy J.

de Castro, R.J.S., Domingues, M.A.F., Ohara, A., Okuro, P.K., dos Santos, J.G., Brexo, R.P., Sato, H.H., 2017. Whey protein as a key component in food systems: physicochemical properties, production technologies and applications. Food Struct. 14, 17–29.

de Wit, J.N., 1998. Nutritional and functional characteristics of whey proteins in food products. J. Dairy Sci. 81 (3), 597–608.

Deeth, H., Bansal, N., 2019. Whey proteins: an overview. In: Deeth, H., Bansal, N. (Eds.), Whey Proteins: From Milk to Medicine. Academic Press, London, pp. 1–50.

Dimitrellou, D., Kandylis, P., Kourkoutas, Y., Kanellaki, M., 2017. Novel probiotic whey cheese with immobilized lactobacilli on casein. LWT Food Sci. Technol. 86, 627–634.

Dinika, I., Verma, D.K., Balia, R., Utama, G.L., Patel, A.R., 2020. Potential of cheese whey bioactive proteins and peptides in the development of antimicrobial edible film composite: a review of recent trends. Trends Food Sci. Technol. 103, 57–67.

Dopazo, V., Luz, C., Calpe, J., Vila-Donat, P., Rodriguez, L., Meca, G., 2022. Antifungal properties of whey fermented by lactic acid bacteria in films for the preservation of cheese slices. Int. J. Dairy Technol. 75, 619–629.

Dragone, G., Mussatto, S.I., Oliveira, J.M., Teixeira, J.A., 2009. Characterisation of volatile compounds in an alcoholic beverage produced by whey fermentation. Food Chem. 112 (4), 929–935.

El-Aidie, S.A.M., Khalifa, G.S.A., 2024. Innovative applications of whey protein for sustainable dairy industry: environmental and technological perspectives—a comprehensive review. Compr. Rev. Food Sci. Food Saf. 23, e13319.

El-Aidie, S.A.M., Omar, H.H., Ewis, A.M., El-nimer, A.M.M., 2020. Production of nutraceutical probiotic whey—based beverage fortified. J. Food Dairy Sci. 11 (6), 171–177.

Faccia, M., Trani, A., Natrella, G., Gambacorta, G., 2018. Chemical-sensory and volatile compound characterization of ricotta forte, a traditional fermented whey cheese. J. Dairy Sci. 101, 5751–5757.

Fagan, C.C., O'Callaghan, D.J., Mateo, M.J., Dejmek, P., 2017. The syneresis of rennet-coagulated curd. In: McSweeney, P.L.H., Fox, P.F., Cotter, P. D., Everett, D.W. (Eds.), Cheese: Chemistry, Physics and Microbiology, fourth ed. Academic Press, London, pp. 145–177.

Fancello, F., Zara, G., Hatami, F., Scano, E.A., Mannazzu, I., 2024. Unlocking the potential of second cheese whey: a comprehensive review on valorisation strategies. Rev. Environ. Sci. Biotechnol. 23, 411–441.

Fox, P.F., Guinee, T.P., 2013. Cheese science and technology. In: Park, Y.W., Haenlein, G.F.W. (Eds.), Milk and Dairy Products in Human Nutrition: Production, Composition and Health. John Wiley & Sons, Ltd., Chichester, pp. 357–389.

Fox, P.F., Guinee, T.P., Cogan, T.M., McSweeney, P.L.H., 2017. Fundamentals of Cheese Science, 2nd Ed. Springer, New York, pp. 11–26.

Galarreta-Morales, C., Ludena-Urquizo, F., Castillo-Arispe, M., Carrasco-Macedo, J., Ocampo-Ocampo, K., Farfán-Saico, G., Gutierrez-Estrada, S., Liñan-Pérez, J., 2025. Sensory development of whey-based yogurt with Cushuro (*Nostoc sphaericum*), through novel consumer-based sensory evaluation methods. Int. Dairy J. 166, 106254.

Gantamur, M.-A., Sukhbaatar, N., Bilawal, A., Gankhuyag, J., Gao, C., Enkhtuya, E., Jiang, Z., Jiang, J., 2025. Elucidating fermentation characteristics, key volatile compounds, and safety evaluations of whey permeate-derived alcoholic beverages. Food Biosci. 68, 106448.

Giroux, H.J., Veillette, N., Britten, M., 2018. Use of denatured whey protein in the production of artisanal cheeses from cow, goat and sheep milk. Small Rumin. Res. 161, 34–42.

Gobbetti, M., Neviani, E., Fox, P., 2017. The origins of cheesemaking. In: The Cheeses of Italy: Science and Technology. Springer International Publishing AG, Switzerland, pp. 1–12.

González Siso, M.I., 1996. The biotechnological utilization of cheese whey: a review. Bioresour. Technol. 57, 1–11.

Goveas, L.C., Darshini, S.M., Kausthubba, P., Padyana, S., Anvitha, Suvarna, A.R., Lewis, S.P., Vidya, S.M., 2024. Probiotic Lactobacillus plantarum SVP2 fermented bioactive EPS-rich milk whey functional beverage. J. Food Sci. Technol. 62 (8), 1547–1556.

Guimarães, J.T., Silva, E.K., Costa, A.L.R., Cunha, R.L., Freitas, M.Q., Meireles, M.A.A., Cruz, A.G., 2018. Manufacturing a prebiotic whey beverage exploring the influence of degree of inulin polymerization. Food Hydrocoll. 77, 787–795.

Guimarães, P.M.R., Teixeira, J.A., Domingues, L., 2010. Fermentation of lactose to bio-ethanol by yeasts as part of integrated solutions for the valorisation of cheese whey. Biotechnol. Adv. 28, 375–384.

Guo, M., Wang, G., 2019. History of whey production and whey protein manufacturing. In: Guo, M. (Ed.), Whey Protein Production, Chemistry, Functionality, and Applications. John Wiley & Sons, Hoboken, pp. 1–12.

Hashim, M.A., Nadtochii, L.A., Muradova, M.B., Proskura, A.V., Alsaleem, K.A., Hammam, A.R.A, 2021. Non-fat yogurt fortified with whey protein isolate: physicochemical, rheological, and microstructural properties. Foods 10, 1762.

Hassan, A.N., Nelson, B.K., 2012. Invited review: anaerobic fermentation of dairy food wastewater. J. Dairy Sci. 95, 6188–6203.

Hassan, H.F., Tabarani, P., Abiad, M.G., 2022. Microbiological, chemical, and sensory characteristics of Arishi cheese. J. Food Process. Preserv. 46, e16383.

Hejtmánková, A., Pivec, V., Trnková, E., Gragounová, H., 2012. Differences in the composition of total and whey proteins in goat and Ewe milk and their changes throughout the lactation period. Czech J. Anim. Sci. 57 (7), 323–331.

Henriques, M.H.F., Gomes, D.M.G.S., Pereira, C.J.D., Gil, M.H.M., 2012. Effects of liquid whey protein concentrate on functional and sensorial properties of set yoghurts and fresh cheese. Food Bioprocess Technol. 6, 952–963.

Hernandez-Mendoza, A., Robles, V.J., Angulo, J.O., De La Cruz, J., Garcia, H.S., 2007. Preparation of a whey-based probiotic product with *Lactobacillus reuteri* and *Bifidobacterium bifidum*. Food Technol. Biotechnol. 45, 27–31.

Horne, D., Lucey, J.A., 2017. Rennet-induced coagulation of milk. In: McSweeney, P.L.H., Fox, P.F., Cotter, P.D., Everett, D.W. (Eds.), Cheese: Chemistry, Physics and Microbiology, fourth ed. Academic Press, London, pp. 115–143.

Hougaard, A.B., Ardo, Y., Ipsen, R.H., 2010. Cheese made from instant infusion pasteurized milk: rennet coagulation, cheese composition, texture and ripening. Int. Dairy J. 20, 449–458.

Islam, M., Tabassum, S., Harun-ur-Rashid, M., Vegarud, G., Alam, M., Islam, M., 2021. Development of probiotic beverage using whey and pineapple (*Ananas comosus*) juice: sensory and physico-chemical properties and probiotic survivability during in-vitro gastrointestinal digestion. J. Agric. Food Res. 4, 100144.

Jaeggi, J.J., Wendorff, W.L., Romero, J., Berger, Y.M., Johnson, M.E., 2005. Impact of seasonal changes in ovine milk on composition and yield of a hard-pressed cheese. J. Dairy Sci. 88, 1358–1363.

Janiaski, D.R., Pimentel, T.C., Cruz, A.G., Prudencio, S.H., 2016. Strawberry-flavored yogurts and whey beverages: what is the sensory profile of the ideal product? J. Dairy Sci. 99 (7), 5273–5283.

Jeličić, I., Božanić, R., Tratnik, L., 2008. Whey-based beverages-a new generation of dairy products. Mljekarstvo 58, 257–274.

Jinjarak, S., Olabi, A., Jiménez-Florez, R., Sodini, I., Walker, J.H., 2006. Sensory evaluation of whey and sweet cream buttermilk. J. Dairy Sci. 89, 2441–2450.

Jitpakdee, J., Kantachote, D., Kanzaki, H., Nitoda, T., 2022. Potential of lactic acid bacteria to produce functional fermented whey beverage with putative health promoting attributes. LWT 160, 113269.

Johnson, M.E., 2017. A 100-year review: cheese production and quality. J. Dairy Sci. 100 (12), 9952–9965.

Kadyan, S., Rashmi, H.M., Pradhan, D., Kumari, A., Chaudhari, A., Kr, G., 2021. Effect of lactic acid bacteria and yeast fermentation on antimicrobial, antioxidative and metabolomic profile of naturally carbonated probiotic whey drink. LWT 142, 111059.

Kalatzopoulos, G.C., 1993. Cheeses from ewes' and goats' milk. In: Fox, P.F. (Ed.), Cheese: Chemistry, Physics and Microbiology. Springer, Boston, MA, pp. 507–553.

Kamber, U., 2007. The traditional cheeses of Turkey: cheeses common to all regions. Food Rev. Int. 24, 1–38.

Kaminarides, S., 2015. A modified form of Myzithra cheese produced by substituting the fresh cheese whey by dried whey protein concentrate and ovine milk and cream. Small Rumin. Res. 131, 118–122.

Kandarakis, J.G., 1986. Traditional whey cheeses. Bull. Int. Dairy Fed. 202, 118–123.

Karadag, D., Köroğlu, O.E., Ozkaya, B., Cakmakci, M., Heaven, S., 2014. A review on fermentative hydrogen production from dairy industry wastewater. J. Chem. Technol. Biotechnol. 89, 1627–1636.

Karam, M.C., Gaiani, C., Hosri, C., Burgain, J., Scher, J., 2013. Effect of dairy powders fortification on yogurt textural and sensorial properties: a review. J. Dairy Res. 80, 400–409.

Kaur, N., Sharma, P., Jaimni, S., Kehinde, B.A., Kaur, S., 2020. Recent developments in purification techniques and industrial applications for whey valorization: a review. Chem. Eng. Commun. 207 (1), 123–138.

Kindstedt, P.S., 2018. The history of cheese. In: Papademas, P., Bintsis, T. (Eds.), Global Cheesemaking Technology—Cheese Quality and Characteristics. John Wiley & Sons, Ltd., Chichester, pp. 3–19.

Kindstedt, P.S., 2013. The basics of cheesemaknig. Microbiol. Spectr. 1 (1). CM-0002-2012.

Kotsaki, P., Aspri, M., Papademas, P., 2025. Novel whey fermented beverage enriched with a mixture of juice concentrates: evaluation of antimicrobial, antioxidant, and angiotensin I converting enzyme inhibitory (ACE) activities before and after simulated gastrointestinal digestion. Microorganisms 13 (7), 1490.

Koukoumaki, D.I., Gkatzionis, K., Panagiotou, M., Sarris, D., 2024. Consumer perception of novel edible films produced by single cell protein derived from cheese whey bio-valorization. J. Sensory Stud. 39, e70004.

Koutinas, A.A., Papapostolou, H., Dimitrellou, D., Kopsahelis, N., Katechaki, E., Bekatorou, A., Bosnea, L.A., 2009. Whey valorisation: a complete and novel technology development for dairy industry starter culture production. Bioresour. Technol. 100 (15), 3734–3739.

Krolczyk, J.B., Dawidziuk, T., Janiszewska-Turak, E., Solowiej, B., 2016. Use of whey and whey preparations in the food industry—a review. Pol. J. Food Nutr. Sci. 66 (3), 157–165.

Kumar, K., Singh, J., Chandra, S., 2017. Formulation of whey based pineapple herbal beverages and its storage conditions. Chem. Sci. Rev. Lett. 6, 198–203.

Kumar, R., Chauhan, S.K., Shinde, G., Subramanian, V., Nadanasabapathi, S., 2018. Whey proteins: a potential ingredient for food industry—a review. Asian J. Dairy Food Res. 37, 283–290.

Lappa, I.K., Papadaki, A., Kachrimanidou, V., Terpou, A., Koulougliotis, D., et al., 2019. Cheese whey processing: integrated biorefinery concepts and emerging food applications. Foods 8, 347.

Le Graët, Y., Gaucheron, F., 1999. pH-induced solubilization of minerals from casein micelles: influence of casein concentration and ionic strength. J. Dairy Res. 66 (2), 215–224.

Lee, W.J., Lucey, J.A., 2010. Formation and physical properties of yogurt. Asian-Australas. J. Anim. Sci. 23 (9), 1127–1136.

Lucey, J., 2002. Formation and physical properties of milk protein gels. J. Dairy Sci. 85 (2), 281–294.

Lucey, J.A., 2017. Formation, structural properties, and rheology of acid-coagulated milk gels. In: McSweeney, P.L.H., Fox, P.F., Cotter, P.D., Everett, D.W. (Eds.), Cheese: Chemistry, Physics and Microbiology, fourth ed. Academic Press, London, pp. 179–197.

Macedo, A., Duarte, E., Fragoso, R., 2015. Assessment of the performance of three ultrafiltration membranes for fractionation of ovine second cheese whey. Int. Dairy J. 48, 31–37.

Macedo, R.F., Freitas, R.J.S., Pandey, A., Soccol, C.R., 1999. Production and shelf-life studies of low cost beverage with soymilk, buffalo cheese whey and cow milk fermented by mixed cultures of *Lactobacillus casei* ssp. *shirota* and *Bifidobacterium adolescentis*. J. Basic Microbiol. 39, 243–251.

Madureira, A.R., Pintado, M.E., Gomes, A.M.P., Malcata, F.X., 2011. Incorporation of probiotic bacteria in whey cheese: decreasing the risk of microbial contamination. J. Food Protect. 74, 1194–1199.

Madureira, A.R., Soares, J.C., Amorim, M., Tavares, T., Gomes, A.M., Pintado, M.M., Malcata, F.X., 2013. Bioactivity of probiotic whey cheese: characterization of the content of peptides and organic acids. J. Sci. Food Agric. 93, 1458–1465.

Madureira, A.R., Soares, J.C., Pintado, M.E., Gomes, A.M.P., Freitas, A.C., Malcata, X.F., 2015. Effect of the incorporation of salted additives on probiotic whey cheeses. Food Biosci. 10, 8–17.

Magalhaes, K.T., Dias, D.R., de Melo Pereira, G.V., Oliveira, J.M., Domingues, L., Teixeira, J.A., de Almeida e Silva, J.B., Schwan, R.F., 2011. Chemical composition and sensory analysis of cheese whey-based beverages using kefir grains as starter culture. Int. J. Food Sci. Technol. 46, 871–878.

Mahomud, M.S., Katsuno, N., Nishizu, T., 2017. Formation of soluble protein complexes and yoghurt properties influenced by the addition of whey protein concentrate. Innov. Food Sci. Emerg. Technol. 44, 173–180.

Mangione, G., Caccamo, M., Natalello, A., Licitra, G., 2023. Graduate student literature review: history, technologies of production, and characteristics of ricotta cheese. J. Dairy Sci. 106, 3807–3826.

Maragkoudakis, P., Vendramin, V., Bovo, B., Treu, L., Corich, V., Giacomini, A., 2016. Potential use of scotta, the by-product of the ricotta cheese manufacturing process, for the production of fermented drinks. J. Dairy Res. 83, 104.

Marcus, J.F., Demarsh, T.A., Alcaine, S.D., 2021. Upcycling of whey permeate through yeast- and mold-driven fermentations under anoxic and oxic conditions. Fermentation 7, 16.

Marnotes, N.G., Pires, A.F., Díaz, O., Cobos, A., Pereira, C.D., 2021. Sheep's and goat's frozen yoghurts produced with ultrafiltrated whey concentrates. Appl. Sci. 11, 6568.

Masotti, F., Cattaneo, S., Stuknyte, M., De Noni, I., 2017. Technological tools to include whey proteins in cheese: current status and perspectives. Trends Food Sci. Technol. 64, 102–114.

Masotti, F., Cattaneo, S., Stuknyte, M., De Noni, I., 2018. Status and developments in analogue cheese formulations and functionalities. Trends Food Sci. Technol. 74, 158–169.

Minj, S., Anand, S., 2020. Whey proteins and its derivatives: bioactivity, functionality, and current applications. Dairy 1, 233–258.

Moatsou, G., Hatzinaki, A., Samolada, M., Anifantakis, E., 2005. Major whey proteins in ovine and caprine acid wheys from indigenous Greek breeds. Int. Dairy J. 15, 123–131.

Monari, S., Ferri, M., Russo, C., Prandi, B., Tedeschi, T., Bellucci, P., et al., 2019. Enzymatic production of bioactive peptides from scotta, an exhausted by-product of ricotta cheese processing. PLoS One 14 (12), e0226834.

Montone, A.M., De Felice, A., Brunetti, R., Mollica, D., Capo, S., et al., 2017. Evaluation of some microbiological and chemical parameters of Campania buffalo ricotta cheese. Ital. J. Food Saf. 6 (3), 6319.

Moreira, N.L., dos Santos, L.F., Soccol, C.R., Suguimoto, H.H., 2015. Dynamics of ethanol production from deproteinized whey by *Kluyveromyces marxianus*: an analysis about buffering capacity, thermal and nitrogen tolerance. Braz. Arch. Biol. Technol. 58, 454–461.

Morr, C.V., Ha, E.Y.W., 1993. Whey protein concentrates and isolates: processing and functional properties. Crit. Rev. Food Sci. Nutr. 33 (6), 431–476.

Moschopoulou, E., Dernikos, D., Zoidou, E., 2021. Ovine ice cream made with addition of whey protein concentrates of ovine-caprine origin. Int. Dairy J. 122, 105146.

Nadeem, M., Mahud, A., Imran, M., Khalique, A., 2015. Enhancement of the oxidative stability of whey butter through almonds (*Prunus dulcis*) peel extract. J. Food Process. Preserv. 39, 591–598.

Naik, B., Kohli, D., Walter, N., Gupta, A., Mishra, S., Khan, J.M., Saris, P.E.J., Irfan, M., Rustagi, S., Kumar, V., 2023. Whey-carrot based functional beverage: development and storage study. J. King Saud Univ. Sci. 35, 102775.

Neelima, Sharma, R., Rajput, Y.S., Mann, B., 2013. Chemical and functional properties of glycomacropeptide (GMP) and its role in the detection of cheese whey adulteration in milk: a review. Dairy Sci. Technol. 93, 21–43.

Nishanthi, M., Chandrapala, J., Vasiljevic, T., 2017. Compositional and structural properties of whey proteins of sweet, acid and salty whey concentrates and their respective spray dried powders. Int. Dairy J. 74, 49–56.

Nishanthi, M., Vasiljevic, T., Chandrapala, J., 2016. Properties of whey proteins obtained from different whey streams. Int. Dairy J. 66, 76–83.

Nzekoue, F.K., Alesi, A., Vittori, S., Sagratini, G., Caprioli, G., 2021. Development of functional whey cheese enriched in vitamin D3: nutritional composition, fortification, analysis, and stability study during cheese processing and storage. Int. J. Food Sci. Nutr. 72, 746–756.

O'Donoghue, L.T., Murphy, E.G., 2023. Nondairy food applications of whey and milk permeates: direct and indirect uses. Compr. Rev. Food Sci. Food Saf. 22, 2652–2677.

Okuda, N.K., Tulini, F.L., Winkelstroter, L.K., De Martins, E.C.P., 2017. Partial characterisation of a bacteriocin produced by *Leuconostoc mesenteroides* A11 and evaluation of bacteriocin production using whey as culture medium. Int. J. Dairy Technol. 71, 240–244.
Panghal, A., Kumar, V., Dhull, S.B., Gat, Y., Chhikara, N., 2017. Utilization of dairy industry waste-whey in formulation of papaya RTS beverage. Curr. Res. Nutr. Food Sci. J. 5 (2), 168–174.
Panesar, P.S., Kennedy, J.F., Gandhi, D.N., Bunko, K., 2010. Bioutilisation of whey for lactic acid production. Food Chem. 105 (1), 1–14.
Panghal, A., Patidar, R., Jaglan, S., Chhikara, N., Khatkar, S., Gat, Y., Sindhu, N., 2018. Whey valorization: current options and future scenario—a critical review. Nutr. Food Sci. 48, 520–535.
Papademas, P., Kotsaki, P., 2019. Technological utilization of whey towards sustainable exploitation. J. Adv. Dairy Res. 7, 231.
Papademas, P., Kotsaki, P., 2020. Technological utilization of whey towards sustainable exploitation. J. Adv. Dev. Res. 7 (4), 1–10.
Papademas, P., Bintsis, T., Alichanidis, E., Ardö, Y., 2018. Whey cheeses (Heat coagulated). In: Papademas, P., Bintsis, T. (Eds.), Global Cheesemaking Technology—Cheese Quality and Characteristics. John Wiley & Sons, Ltd., Chichester, pp. 446–452.
Paskaš, S., Miocinović, J., Savić, M., Ješić, G., Rašeta, M., Becskei, Z., 2019. Comparison of the chemical composition of whey cheeses: Urda and Ricotta. Maced. Vet. Rev. 42 (2), 151–161.
Pasta, C., Caccamo, M., Petriglieri, R., Difalco, A., Farina, G., et al., 2024. Sicilian whey: utilization of ricotta whey in the production of value-added artisanal beers. Fermentation 10, 19.
Penna, A.L.B., Baruffaldi, R., Oliveira, M.N., 1997. Optimization of yogurt production using demineralized whey. J. Food Sci. 62 (4), 846–850.
Pescuma, M., de Valdez, G.F., Mozzi, F., 2015. Whey-derived valuable products obtained by microbial fermentation. Appl. Microbiol. Biotechnol. 99, 6183–6196.
Pintado, M.E., Lopes da Silva, J.A., Malcata, F.X., 1996. Characterization of Requeijão and technological optimization of its manufacturing process. J. Food Eng. 30, 363–376.
Pinilla, C.M.B., Galland, F., Pacheco, M.T.B., et al., 2025. Unraveling the real potential of liquid whey as media culture and microencapsulation material for lactic acid bacteria. Innov. Food Sci. Emerg. Technol. 100, 103885.
Pintado, M.E., da Silva, J.L., Malcata, F.X., 1999. Comparative characterization of whey protein concentrates from ovine, caprine and bovine breeds. LWT Food Sci. Technol. 32, 231–237.
Pintado, M.E., Macedo, A.C., Malcata, F.X., 2001. Review: technology, chemistry and microbiology of whey cheeses. Food Sci. Technol. Int. 7, 105–116.
Pires, A.F., Gomes, D., Noronha, J., Díaz, O., Cobos, A., Pereira, C., 2023. Evaluation of the characteristics of sheep's and goat's ice cream, produced with UF concentrated second cheese whey and different starter cultures. Foods 11 (24), 4091.
Pires, A.F., Marnotes, N.G., Bella, A., Viegas, J., Gomes, D.M., Henriques, M.H.F., Pereira, C.J.D., 2021a. Use of ultrafiltrated cow's whey for the production of whey cheese with Kefir or probiotics. J. Sci. Food Agric. 101, 555–563.
Pires, A.F., Marnotes, N.G., Rubio, O.D., Garcia, A.C., Pereira, C.D., 2021b. Dairy by-products: a review on the valorization of whey and second cheese whey. Foods 10, 1067.
Prazeres, A.R., Carvalho, F., Rivas, J., 2012. Cheese whey management: a review. J. Environ. Manag. 110, 48–68.
Rama, G.R., Kuhn, D., Beux, S., Maciel, M.J., de Souza, C.F.V., 2019. Potential applications of dairy whey for the production of lactic acid bacteria cultures. Int. Dairy J. 98, 25–37.
Ramos, M., Juarez, M., 2011. Sheep milk. In: Fuquay, J.W., Fox, P.F., McSweeney, P.L.H. (Eds.), Encyclopedia of Dairy Sciences, second ed. Academic Press, London, pp. 494–502.
Ramos-Suárez, J.L., Alvarez-Méndez, S.J., Ritter, A., Mata González, J., Camacho Pérez, A., 2024. A comprehensive evaluation of cheese whey to produce biogas in the Canary Islands. Biomass Bioenergy 180, 107008.
Risner, D., Shayevitz, A., Haapala, K., Meunier-Goddik, L., Hughes, P., 2018. Fermentation and distillation of cheese whey: carbon dioxide equivalent emissions and water use in the production of whey spirits and white whiskey. J. Dairy Sci. 101, 2963–2973.
Rocha-Mendoza, D., Kosmerl, E., Krentz, A., Zhang, L., Badiger, S., et al., 2021. Invited review: acid whey trends and health benefits. J. Dairy Sci. 104, 1262–1275.
Rodrigues, M.H.P., Graff, C.A., Yerovi, D.S.T., Schmitz, C., de Lima, J.C., Timmers, L.F.S., Lehn, D.N., de Souza, C.F.V., 2024. The bioactive potential of cheese whey peptides from different animal origins (bovine, goat, sheep, buffalo, and camel): a systematic review and meta-analysis. Food Res. Int. 196, 115053.
Rosa, L.S., Santos, M.L., Abreu, J.P., Rocha, R.S., Esmerino, E.A., Freitas, M.Q., Mársico, E.T., Campelo, P.H., Pimentel, T.C., Cristina Silva, M., et al., 2023. Probiotic fermented whey-milk beverages: effect of different probiotic strains on the physicochemical characteristics, biological activity, and bioactive peptides. Food Res. Int. 164, 112396.
Roy, S., Hussain, S.A., Prasad, W.G., Khetra, Y., 2022. Quality attributes of high protein ice cream prepared by incorporation of whey protein isolate. Appl. Food Res, 2 (1), 100029.
Ryan, M.P., Walsh, G., 2016. The biotechnological potential of whey. Rev. Environ. Sci. Biotechnol. 15, 479–498.
Sabokbar, N., Khodaiyan, F., 2015. Characterization of pomegranate juice and whey based novel beverage fermented by kefir grains. J. Food Sci. Technol. 52 (6), 3711–3718.
Saglam, H., Sarioglu, T., Karahan, A.G., Oner, Z., 2019. An investigation for the development of whey-based probiotic beverages. Rom. Biotechnol. Lett. 24, 1097–1106.
Sakkas, L., Tzevdou, M., Zoidou, E., Gkotzia, E., Karvounis, A., Samara, A., et al., 2019. Yoghurt-type gels from skim sheep milk base enriched with whey protein concentrate hydrolysates and processed by heating or high hydrostatic pressure. Foods 8, 342.

Salvatore, E., Pes, M., Falchi, G., Pagnozzi, D., Furesi, S., Fiori, M., Roggio, T., Addis, M.F., Pirisi, A., 2014. Effect of whey concen- tration on protein recovery in fresh ovine ricotta cheese. J. Dairy Sci. 97 (8), 4686–4694.

Sasi, K., 2015. Development, quality evaluation and shelf life studies of probiotic beverages using whey and *Aloe vera* juice. J. Food Process. Technol. 6.

Secchi, N., Giunta, D., Pretti, L., Garcia, M.R., Roggio, T., et al., 2012. Bioconversion of ovine scotta into lactic acid with pure and mixed cultures of lactic acid bacteria. J. Indus. Microbiol. Biotechnol. 39, 175–181.

Semeniuc, C.A., Zapartan, L., Stan, L., Pop, C.R., Bors, M.D., Rotar, A.M., 2015. Physicochemical and sensory properties of whey cheese with pine nuts. Bulletin UASVM Food Sci. Technol. 72 (2), 177–181.

Silva e Alves, A.T., Spadoti, L.M., Zacarchenco, P.B., Trento, F.K., 2018. Probiotic functional carbonated whey beverages: development and quality evaluation. Beverages 4, 49.

Siso, G.M.I., 1996. The biotechnological utilization of cheese whey: a review. Bioresour. Technol. 57, 1–11.

Skeie, S.B., 2014. Quality aspects of goat milk for cheese production in Norway: a review. Small Rumin. Res. 122, 10–17.

Skryplonek, K., Dmytrow, I., Mituniewicz-Malek, A., 2019. Probiotic fermented beverages based on acid whey. J. Dairy Sci. 102 (9), 7773–7780.

Smithers, G.W., 2008. Whey and whey proteins-from "gutter-to-gold". Int. Dairy J. 18 (7), 695–704.

Smithers, G.W., 2015. Whey-ing up the options-yesterday, today and tomorrow. Int. Dairy J. 48, 2–14.

Sohrabi, Z., Eftekhari, M.H., Eskandari, M.H., Rezaeianzadeh, A., Sagheb, M.M., 2016. Development and characterization of fermented and unfermented whey beverages fortified with vitamin E. J. Agric. Sci. Technol. 18, 1511–1512.

Sołowiej, B., Cheung, I.W.Y., Li-Chan, E.C.Y., 2014. Texture, rheology and meltability of processed cheese analogues prepared using rennet or acid casein with or without added whey proteins. Int. Dairy J. 37 (2), 87–94.

Statista, 2025. Major cheese producing countries in 2024. Available from: https://www.statista.com/statistics/195809/cheese-production-in-selected-countries-2009/. Accessed 10-03-2025.

Sutera, A.M., Arfuso, F., Tardiolo, G., Riggio, V., Fazio, F., Aiese Cigliano, R., Paytuví, A., Piccione, G., Zumbo, A., 2023. Effect of a co-feed liquid whey-integrated diet on crossbred pigs' fecal microbiota. Animals 13 (11), 1750.

Tanwar, T., Wagh, R.M., Malav, O., Kour, S., Kumar, P., 2022. Preparation of functional beverage from whey-based mango juice. Pharm. Innov. 11, 4710–4716.

Tita, M.A., Moga, V.-M., Constantinescu, M.A., Bătusaru, C.M., Tita, O., 2024. Harnessing the potential of whey in the creation of innovative food products: contributions to the circular economy. Recycling 9, 79.

Tomczyńska-Mleko, M., Solowiej, B.G., Terpilowski, K., Wesolowska-Trojanowska, M., Mleko, S., 2025. Novel high-protein dairy product based on fresh white cheese and whey protein isolate. J. Dairy Sci. 108 (1), 272–281.

Tsermoula, P., Khakimov, B., Nielsen, J.H., Engelsen, S.B., 2021. Whey—the waste-stream that became more valuable than the food product. Trends Food Sci. Technol. 118, 230–241.

Turkmen, N., Akal, C., Özer, B., 2019. Probiotic dairy-based beverages: a review. J. Funct.Foods 53, 62–75.

Uduwerella, G., Chandrapala, J., Vasiljevic, T., 2018. Preconcentration of yoghurt base by ultrafiltration for reduction in acid whey generation during Greek yoghurt manufacturing. Int. J. Dairy Technol. 71 (1), 71–80.

Villarruel-López, A., Castro-Rosas, J., Gómez-Aldapa, C.A., Nuño, K., Torres-Vitela, M.R., et al., 2016. Indicator microorganisms, *Salmonella, Listeria monocytogenes, Staphylococcal* enterotoxin, and physicochemical parameters in requeson cheese. Afr. J. Food Sci. 10 (9), 178–184.

Walstra, P., Wouters, J.T.M., Geurts, T.J., 2006. Colloidal particles of milk. In: Dairy Science and Technology, Second ed. Taylor & Francis Group, Boca Raton, pp. 109–158.

Wang, Q., Wu, J., Zhao, X., 2021. Valorization of dairy waste for sustainable bioplastics production. Bioresour. Technol. 319, 124166.

Wherry, B., Barbano, D.M., Drake, M.A., 2019. Use of acid whey protein concentrate as an ingredient in nonfat cup set-style yogurt. J. Dairy Sci. 102 (10), 8768–8784.

Yadav, D.N., Vishwakarma, R.K., Borad, S., Bansal, S., Jaiswal, A.K., Sharma, M., 2016. Development of protein fortified mango based ready-to-serve beverage. J. Food Sci. Technol. 53 (10), 3844–3852.

Yadav, J.S.S., Yan, S., Pilli, S., Kumar, L., Tyagi, R.D., Surampalli, R.Y., 2015. Cheese whey: a potential resource to transform into bioprotein, functional/nutritional proteins and bioactive peptides. Biotechnol. Adv. 33 (6), 756–774.

Yamahata, N., Toyotake, Y., Kunieda, S., Wakayama, M., 2020. Application of multiple sensory evaluations to produce fermented beverages made from sole whey using *Kluyveromyces marxianus*. Int. J. Food Sci. Technol. 55, 1698–1704.

Yilsay, T.Ö., Yilmaz, L., Bayizit, A.A., 2006. The effect of using a whey protein fat replacer on textural and sensory characteristics of low-fat vanilla ice cream. Eur. Food Res. Technol. 222 (1–2), 171–175.

Zhao, C., Chen, N., Ashaolu, T.J., 2022. Whey proteins and peptides in health-promoting functions—a review. Int. Dairy J. 126, 105269.

Zoidou, E., Andreadaki, I., Massouras, T., Kaminarides, S., 2016. A new whey cheese analogue made from whey protein concentrate and vegetable fat with 15% olive oil. J. Nutr. Med. Diet Care 2, 17.

Zoppellari, F., Bardi, L., 2013. Production of bioethanol from effluents of the dairy industry by *Kluyveromyces marxianus*. New Biotechnol. 30, 607–613.

Zotta, T., Solieri, L., Iacumin, L., Picozzi, C., Gullo, M., 2020. Valorization of cheese whey using microbial fermentations. Appl. Microbiol. Biotechnol. 104, 2749–2764.

Chapter 3

Whey chemistry

Siyu Liu, Lara Barnum and Alice Marciniak
Department of Food Science, University of Guelph, Guelph, ON, Canada e-mail address: amarcini@uoguelph.ca

1. Introduction

Dating back over 3000 years BCE, when Bedouins, nomadic Arab tribes, traversed the desert with milk stored in sacks made of dried animal stomachs, they occasionally observed some yellow liquid on the top. The residual enzymes present in the animal's stomach coagulated the milk proteins, and such observation helped people produce cheese products with enzymatic coagulation. Nowadays, cheese manufacturers have been making a variety of cheeses using different coagulating and aging methods (de Wit, 2001a).

After the casein proteins coagulate under the effect of added mineral/organic acid or chymosin, the translucent liquid leftover is called whey, which is mainly composed of water and the majority of the water-soluble constituents in the native milk (Ghaly and Kamal, 2004; Ferchichi et al., 2005). Whey is a yellowish liquid byproduct of cheese/casein production, and the riboflavin (Vitamin B2) in the milk gives the yellow color (de Wit, 2001a). Based on the milk quality and milk type, the whey may also have a bluish color. Whey could be obtained from any type of milk. Bovine milk is the most common type in the West, and in other parts of the world, other popular options include milk from sheep, goats, and camels (Smithers, 2008). Based on the method applied for casein coagulation, there are two main types of whey: sweet and acid whey. These two different types of whey possess different pH, mineral, and protein compositions (Panesar et al., 2007).

As an outstanding source of lactose, functional proteins, peptides, minerals, and vitamins, many researchers and industries have focused on exploiting whey in various fields (Smithers, 2008). So far, whey has been applied in the pharmaceutical industry, food, cosmetics, cleaning detergents, and personal hygiene products (Manso and López-Fandiño, 2004; Paterson, 2009; de Boer, 2014). Among all the water-soluble nutrients in whey, bioactive proteins such as β-lactoglobulin, α-lactalbumin, and lactoferrin have attracted massive attention due to their versatile functionalities as health promoters for digestion, metabolism, organ development, and the immune system. Additionally, whey protein-derived peptides also enhance the nervous, immune, and cardiovascular systems. The development of processing and extraction techniques allows the food industry to produce whey products with various functionalities catering to consumers' demands (Editorials, 2011).

Despite being an excellent source of nutrients, the massive production of whey still poses a great disposal issue for the dairy industry. Approximately 9 kg of whey will be produced to produce each kg of cheese; the annual cheese production in the world is above 160 million tons, and the production rate is expected to grow in the future (Kosikowski, 1979; Smithers, 2008; Guimarães et al., 2010). The biochemical oxygen demand (BOD) and chemical oxygen demand (COD) of whey are significantly high, varying from 40,000 to 60,000 mg/L and 50,000 to 80,000 mg/L, respectively (Chatzipaschali and Stamatis, 2012). The culprit for the notably high BOD and COD is the major component of whey, lactose (45–50 g/L, composes 70%–72% of the total solids) (Jelen, 2002; Patel and Murthy, 2011). Whey disposal can be extremely damaging to aquatic life (due to the depletion of oxygen) and crop yield (due to its effects on soil structure) (Marwaha and Kennedy, 1988; González Siso, 1996; Ryan and Walsh, 2016). Many developed countries enforce strict laws regarding whey disposal; however, the massive whey production and high pollution potential call for considerable investment of funds as whey takes 100–175 times higher waste load than domestic wastewater (Mockaitis et al., 2006; Smithers, 2008). To safeguard human well-being and the ecosystem from the impacts of whey disposal, and at the same time compensate the dairy industries for the whey disposal costs, more value-added food products and bioproducts are expected to be made from whey (Ryan and Walsh, 2016).

Whey. https://doi.org/10.1016/B978-0-443-27648-4.00005-0

2. Whey types

Followed by casein coagulation by proteolytic enzymes or pH adjustments, the casein curd is separated, and the remaining yellowish liquid is denoted as whey. There are two types of whey: sweet whey (pH is around 5.6) and acid whey (pH < 5). The processing methods used to coagulate the cheese curds determine the type and composition of the whey (Panesar et al., 2007).

2.1 Sweet whey

In the production of soft, semisoft, and hard cheeses (such as Edam and Cheddar cheese) that are made using rennet to coagulate the casein proteins, the major whey byproduct is called sweet whey (Venetsaneas et al., 2009). During the manufacture of cheese and sweet whey, rennet (a mixture of different casein-coagulating enzymes) is added to the standardized and pasteurized milk. The rennet will hydrolyze κ-casein to decrease the steric repulsion between the casein micelles. The negatively charged hydrophilic tail (caseinomacropeptide [CMP], f106–169) of κ-casein will be cleaved by rennet, and when the hydrophobic interactions between the casein micelles take over the steric stabilization maintained by the remaining CMP, casein micelles start to coagulate at pH 6.5 and transform into a gel (De Kruif, 1999; Creamer, 2002; Panesar et al., 2007). After the casein curds are removed, the remaining liquid is the sweet whey. In sweet whey, only a negligible amount of lactose has been converted into lactic acid (Anand et al., 2013) (Fig. 3.1).

2.2 Acid whey

Acid whey is produced when lactic acid bacteria or acid and heat (organic and inorganic acids) are applied to precipitate, rather than coagulate, the caseins to eventually produce industrial casein and fresh acid cheese types such as cottage and ricotta cheese (Jelen, 2002). At neutral pH, the native casein micelles are suspended in the serum through colloidal calcium phosphate (CCP) embedded inside the casein micelles and the "hairy layer" consisting of CMP located at the surface of casein micelles, providing a net negative charge (De Kruif and Holt, 2003). As the pH of milk drops with the addition of the acidic casein coagulant, CCP dissolves into the serum, and the electrostatic repulsion between casein micelles is reduced (the hairy layer shrinks as the inter- and intrachain interactions of CMP are too weak to keep them straight) (De Kruif, 1999). In this case, the casein submicelles coagulate under the hydrophobic effects and form the casein curd (Fig. 3.2).

As shown in Table 3.1, the major differences in the composition between sweet and acid whey include protein, lactose, and mineral content (Jelen, 2002; Kosseva et al., 2009; Tunick, 2009). The acid whey contains less lactose, as some lactose is converted into lactic acid during the fermentation by lactic acid bacteria (Tunick, 2009). In the acid-induced casein precipitation, CCP is solubilized and remains in the acid whey, which gives rise to higher mineral content in the acid whey. Since rennet cleaves the CMP, a significant CMP content can be found in the sweet whey, contributing to the slightly higher protein content. On the other hand, CMP is absent in the acid cheese, as they are still located at the

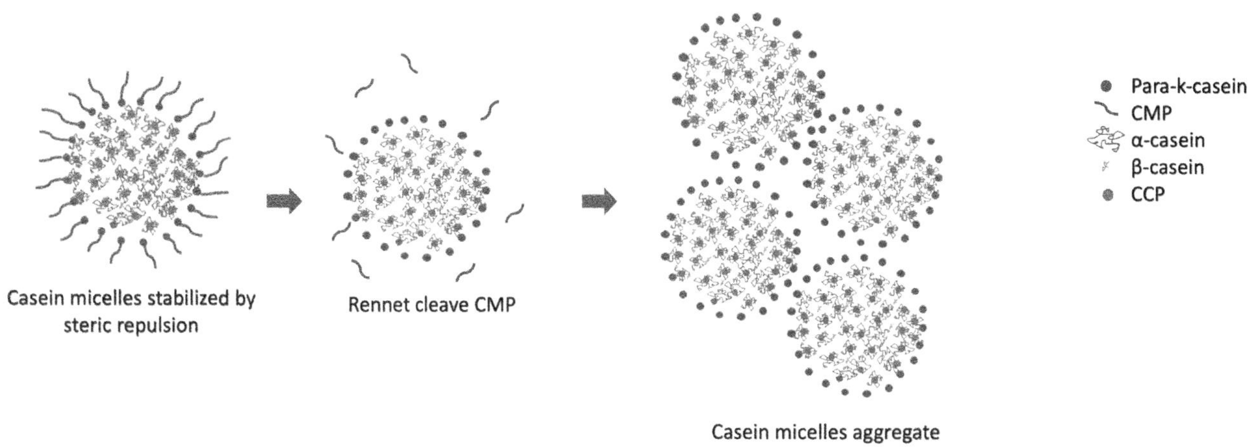

FIGURE 3.1 Mechanism of rennet-induced casein coagulation. *Adapted from Wang and Guo (2019).*

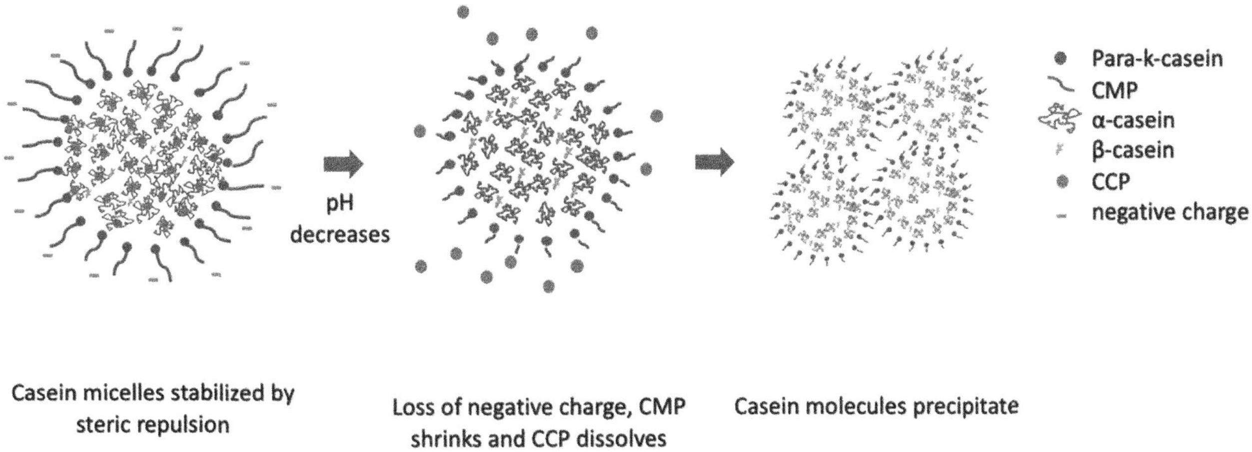

FIGURE 3.2 Mechanisms of acid-induced casein precipitation. *Adapted from Wang and Guo (2019).*

TABLE 3.1 Differences in the whey compositions.

	Sweet whey	Acid whey
pH	~5.6	<5
Total solids (g/L of whey)	63–70	63–70
Protein (g/L of whey)	6–10	6–8
Lactose (g/L of whey)	46–52	44–46
Mineral (g/L of whey)	2.5–4.7	4.3–7.2
Calcium (g/L of whey)	0.4–0.6	1.2–1.6
Phosphate (g/L of whey)	1–3	2–4.5

Adapted from Jelen (2002), Kosseva et al. (2009), and Tunick (2009).

surface of casein micelles (Panesar et al., 2007). Because of its acidic flavor along with low protein and high mineral composition, the application of acid whey is limited (Mawson, 1994; Venetsaneas et al., 2009).

2.3 Typical composition of whey

The resulting volume of whey from cheesemaking is 85%–95% of the total starting volume of milk used. It also contains nearly half of the nutrients from milk, which includes approximately 20% of the total protein present (Farizoglu et al., 2004; Ryan and Walsh, 2016). The major constituents of whey consist of water, which accounts for 93% of its volume, lactose (45–50 g/L which constitutes approximately 70% of the total solids), whey protein (6–8 g/L which makes up 8%–10% of the total solids), minerals (4–5 g/L which constitute around 15% of the total solids), and vitamins (such as B group vitamins) (Kosikowski, 1979; Marwaha and Kennedy, 1988). 0.2%–0.5% milk fat might also be retained in the whey. Whey also includes growth factors. Many parameters should impact the whey composition, such as the origin of cheese milk (stage of lactation, feed, and animal breed, etc.) and pretreatment of the cheese milk (impacts of various milk pasteurization methods) (de Wit, 2001b).

3. Whey proteins

Whey proteins are water-soluble globular proteins present in milk serum, and the major whey proteins are α-lactalbumin, β-lactoglobulin, bovine serum albumin (BSA), and immunoglobulin. The major secondary structures in whey proteins are β-sheets and α-helices. Both secondary and tertiary structures are present in the whey proteins; they are tightly packed, keeping the insoluble amino acids at the interior and most of the polar/charged residues at the surface of the compact protein structure. Such an arrangement explains the good water solubility of whey proteins in milk serum (de Wit, 2001b). Since whey proteins do not precipitate with pH adjustment to 4.6 or get impacted by the addition of chymosin/proteinases, whey proteins can be separated after casein coagulation during the production of industrial casein or cheese (Graham and Phillips, 1979; Dickinson et al., 1993).

Whey proteins are considered an attractive protein option since they contain substantial amounts of essential amino acids and sulfur-containing amino acids (cysteine and methionine), and the amino acid composition is nearly equivalent to the biological optimum (Smithers, 2008; Ismail and Gu, 2010). Many studies proved that the ingestion of whey proteins is beneficial in protein synthesis, muscular degeneration, physical performance, weight management, and body tissue growth for humans (Exl, 2001; Ha and Zemel, 2003; Tipton et al., 2004). Additionally, whey proteins are great sources of branched-chain amino acids (BCAAs), valine, and leucine, accounting for around 26% of the amino acid profile. Compared to other amino acids, higher solubility in the digestive system and faster absorption rate in the digestive tract make BCAA high-value amino acids (Shimomura et al., 2004).

3.1 β-lactoglobulin

β-lactoglobulin is a globular protein that makes up 50% of the whey proteins, originating from the mammary gland in most but not all mammals (Bhattacharjee and Das, 2000; Madureira et al., 2007). β-lactoglobulin is absent in humans, camels, and rodent species, which explains why β-lactoglobulin is considered a major whey protein allergen to infants. Scientists found seven genetic variants, and variants A and B are the most common ones, where the two differences are observed at two positions (positions 64 and 118) of the amino acid sequences (McKenzie, 1971). These variants exhibit differences in biochemical properties, such as heat liability, isoelectric points, and heat-induced aggregation properties (Schokker et al., 2000).

β-lactoglobulin is a small protein (radius of 2 nm) composed of 162 amino acid residues; its molecular weight is 18.20–18.40 kDa (Yadav et al., 2015). Its isoelectric point is 5.1–5.2; β-lactoglobulin and BSA (isoelectric point at pH 5.3) both stay soluble at their isoelectric points because there are more surface hydrophilic amino acid residues compared to nonpolar groups on the surface (Bolder et al., 2007). The quaternary structure of β-lactoglobulin varies based on the pH value of the medium; it exists as a monomer of 18 kDa in an acid or basic environment ($pH < 3$ or $pH > 7.5$), it is a noncovalent dimer of 36 kDa at pH 5–7 (mainly through hydrophobic effects), and between pH 3.5 and 5.2 it presents as an octamer with a molecular weight of 140 kDa (Walstra et al., 2006; Uhrínová et al., 2000; Apenten et al., 2002; Alves and Tavares, 2019).

β-lactoglobulin is sensitive to heat; it denatures above 70°C and might aggregate depending on other parameters such as protein concentration and pH (de Wit, 2001b). Each β-lactoglobulin contains five cysteines; four are involved in two disulfide bonds, and the remaining Cys121 is located at the protein's interior and covered by an α-helix in the native state, preventing it from forming covalent bonds with other molecules through SH/S–S exchange or thiol oxidation (Considine et al., 2007; Chihi et al., 2016; Marciniak et al., 2018). When β-lactoglobulin is exposed to denaturing conditions, such as high-pressure and heat treatment, which unfold the β-lactoglobulin, exposing the hydrophobic amino acid residues, the free thiol group will interact with other molecules (Marciniak et al., 2018). As Fig. 3.3 suggests, β-lactoglobulin consists of sheets predominantly. The β-lactoglobulin monomer is formed with two β-sheets and one α-helix (Godovac-Zimmermann, 1988; Liu et al., 2006).

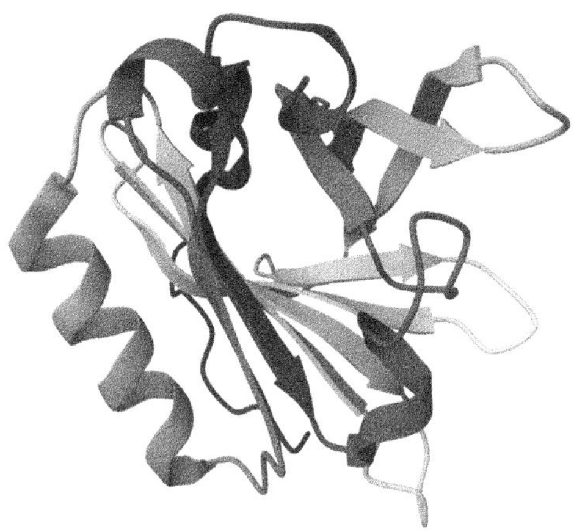

FIGURE 3.3 Ribbon diagram of bovine β-lactoglobulin in monomeric form. *From RCSB PDB (2010) (PDB ID: 3NPO).*

β-lactoglobulin belongs to the lipocalin superfamily, which could function as a transporter of various hydrophobic ligands such as retinoids and fatty acids (Sawyer et al., 2002). β-lactoglobulin functions as a reliable retinol carrier from the cow to the calves, as it remains stable when exposed to proteolytic enzymes and the acid environment in the stomach (de Wit, 1998). The ligand-binding property is achieved through a large central hydrophobic pocket, an invagination referred to as a calyx (derived from the Greek word for "cup"), which is composed of a nine-stranded β-barrel (Godovac-Zimmermann, 1988; Qin et al., 1999). Among the antiparallel β-sheet structures, the flexible loop EF will open at pH > 7, exposing approximately 10% of the overall surface area to the bulk solvent, thus facilitating hydrophobic ligands to bind to the side chains of A and H strands. In a neutral or acidic environment, loop EF closes off the calyx to isolate this area from the aqueous solvent (Fogolari et al., 1998; Qin et al., 1999; Adams et al., 2006).

β-lactoglobulin is an outstanding source of essential, branched-chain, and sulfur-containing amino acids (methionine and cysteine), comprising 48%, 25%, and 5.7% of the total weight of β-lactoglobulin, respectively (Etzel, 2004). Cysteine, which makes up 2.8% by weight of total β-lactoglobulin, has a vital function in protecting against intestinal tumors by prompting glutathione synthesis (McIntosh et al., 1995). Since β-lactoglobulin contains a comparable essential amino acid profile of egg white, it also displays high solubility within a large pH range, especially at acidic pH; therefore, it has been added as an active agent to protein-fortified beverages such as sports drinks and fruit juice (Chatterton et al., 2006). β-lactoglobulin also gives rise to peptides with different bioactivities. Hydrolysis of β-lactoglobulin by chymotrypsin releases a peptide called β-lactotensin, His-Ile-Arg-Leu (f146–149), which helps to lower blood cholesterol levels and also exhibits an antistress effect to cope with painful stimuli and memory loss (Yamauchi et al., 2003; Yamauchi et al., 2006; Ohinata et al., 2007). Lactostatin, Ile-Ile-Ala-Glu-Lys (f71–75) and Ala-Leu-Pro-Met-His (f142–146) are released by trypsin. And it shows hypocholesterolemic effect by either reducing cholesterol absorption in the intestinal system or inducing the gene transcription of a cholesterol-regulating enzyme called 7α-hydroxylase (Nagaoka, 2006). Lactokinins, Tyr-Leu (f102–103) and Ala-Leu-Pro-Met-His-Ile-Arg (f142–148), show a high potential for curing/preventing high blood pressure (Meisel, 2005). β-lactorphin (f102–105) can assist in intestinal absorption (Sipola et al., 2002).

β-lactoglobulin shows excellent water-binding and heat-set gelation characteristics; thus, it has been applied in the food industry to enhance water binding and texture (Editorials, 2011). Additionally, β-lactoglobulin heat-set gel could be either elastic or inelastic, translucent or opaque, since its characteristics can be manipulated with the chemical gelling conditions, such as ionic strength and pH of the solvent (Dufour et al., 1998). Due to the flexibility of β-lactoglobulin heat-set gel, β-lactoglobulin has been enriched with formulated foods such as reformed fish products (Pihlanto, 2011). β-lactoglobulin also demonstrates superior foamability and foam stability even when sugar is present, so β-lactoglobulin can be applied as an affordable alternative to egg white in food products such as meringues (Foegeding et al., 2006).

3.2 α-lactalbumin

α-lactalbumin (14 KDa) is a small calcium-binding protein consisting of 123 amino acid residues; four disulfide bonds (Cys6–Cys120, Cys61–Cys77, Cys73–Cys91, and Cys28–Cys111) contribute to the compact globular structure (Brew and Grobler, 1992). Unlike β-lactalbumin, α-lactalbumin does not contain any free thiol group, which explains why it remains stable under most denaturing conditions (Walstra et al., 2006; Ustunol, 2014; Marciniak et al., 2018). It constitutes the second major protein in whey, making up 20% of overall whey proteins, and 1.2–1.5 g/L of α-lactalbumin is present in the whey (Jackson et al., 2004). Its isoelectric point is at pH 4.6. α-lactalbumin is a valuable source of essential amino acids; essential amino acids make up 63.2% of the overall amino acid content, while casein, the primary protein in bovine milk, which composes approximately 80% of the overall milk protein, only contains 51.4% of essential amino acids (Heine et al., 1991). BCAA such as leucine (10.4%), isoleucine (6.4%), and valine (4.2%) constitute a large portion of α-lactalbumin (Etzel, 2004).

The concentration of α-lactalbumin in bovine milk declines toward the end of the lactation period (Regester and Smithers, 1991). α-lactalbumin is also in charge of normal milk secretion for mammals; it functions as a major osmotic component by moving water into the secretory vesicles in the mammary gland (Chatterton et al., 2006). In an animal study, mice without the gene encoding murine α-lactalbumin secreted milk with no lactose, and most importantly, this milk was too viscous, prohibiting normal milk removal by the developing pups (Stacey et al., 1995).

α-lactalbumin is also the primary whey protein in human milk, composing 30% of the overall protein in human breast milk, and it is a crucial protein for infant nutrition. Due to the high homology (72% similarity) between bovine and human α-lactalbumin, it is of great interest to enrich infant formula with bovine α-lactalbumin (Jackson et al., 2004). The bovine α-lactalbumin addition to baby formula not only provides a protein composition that highly resembles its counterpart in human milk but also offers a series of antibacterial peptides (Kelleher et al., 2003). During digestion, bovine α-lactalbumin releases antibacterial peptides. For example, during the hydrolysis of α-lactoglobulin by trypsin, Glu-Gln-Leu-Thr-Lys (f1–5) and Gly-Tyr-Gly-Gly-Val Ser-Leu-Pro-Glu-Trp-Val-Cys-Thr-Thr-Phe (f17–31) disulfide-bonded to Ala-Leu-Cys-Ser-Glu-Lys (f109–114) are released. Research proved that compared to human breast milk, peptides derived from bovine α-lactalbumin are effective in targeting enteropathogenic *E. coli* O127 and diarrheal incidents (Brück et al., 2003).

There are two major α-lactalbumin genetic variants, variants A and B. The predominant α-lactalbumin in Western cattle is variant B; it is considered as the reference α-lactalbumin (Farrell et al., 2004). Zebu and Droughtmaster can produce both variants (Fox et al., 2015). α-lactalbumin variant A is found in limited amounts of milk secreted by some European Bos taurus breeds. The only difference between variants A and B is that there is a Glutamic acid at position 10 for variant A, while the amino acid at position 10 for variant B is arginine instead (Rapeanu and Nicoleta, 2010). Studies have noted the occurrence of two rare variants, C and D, in Bali cattle (Fox et al., 2015).

The tertiary structure of the native α-lactalbumin constitutes a small lobe and a large α-helical lobe separated by a cleft. The large α-helical lobe encompasses four α helices and two short helices. The small domain comprises an anti-parallel β-sheet (consisting of three polypeptide strands), a few irregular structures, and a small helix (Chrysina et al., 2000). The binding site can be found at the junction of the two lobes (Stuart et al., 1986). One mole of native α-lactalbumin contains 1 mole of calcium ions; the binding of calcium ions contributes significantly to maintaining α-lactalbumin's globular tertiary structure in its native state (Chrysina et al., 2000). The calcium-bounded (holo form) α-lactalbumin is quite thermo-stable and starts to denature around 60°C, while the calcium-free (apo form) α-lactalbumin denatures at approximately 30 °C (Croguennec et al., 2017). Fig. 3.2 shows two ribbon models for the holo and the apo form of α-lactalbumins. The removal of calcium ion occurs when the pH of the solvent is adjusted to the isoelectric point of α-lactalbumin (pH 4.6), causing α-lactalbumin to partially unfold into a form called the molten globule state (Chatterton et al., 2006). When the apo form of α-lactalbumin is exposed at elevated temperatures, it also transforms into the molten globule state (Permyakov et al., 1981). The molten globule state represents a transitional stage between the native and fully denatured states; it remains the native secondary protein structure but with no fixed tertiary structure (Chrysina et al., 2000; Farrell et al., 2004). At the molten globule state, mild heat treatment could result in the irreversible denaturation of α-lactalbumin (Walstra et al., 2006). As a calcium-binding protein, α-lactalbumin also displays a pronounced binding capacity for other metal ions such as divalent and trivalent ions (Heine et al., 1991) (Fig. 3.4).

Administration of α-lactalbumin is beneficial in different ways. The application of α-lactalbumin in dietary regimes has also been associated with improved glycemic control and a lower likelihood of developing type II diabetes (Blouet et al., 2006). The high content of BCAA in α-lactalbumin enhances protein metabolism, and it also powers the muscle, thus enhancing performance (Blomstrand et al., 2006). Additionally, it has been reported that the evening administration of α-lactalbumin can reduce sleep disturbances and improve brain-sustained attention the following morning (Markus et al., 2005). High-calcium meals enriched with α-lactalbumin can also enhance weight loss and control (Hursel et al., 2010).

3.3 Bovine serum albumin

Serum albumin is a major blood serum protein found in the body tissue. It regulates blood osmotic pressure and is also in charge of ligand metabolism, transportation, and distribution (Carter and Ho, 1994). Through the passive leakage from the bloodstream to milk, serum albumin in the milk is denoted as BSA (Madureira et al., 2007). BSA in milk along with serum albumin in the blood exhibit identical immunological and physical characteristics (Coulson and Stevens, 1950; Polis et al., 1950).

BSA is a single-chain globular whey protein, and as one of the major whey proteins constituting 5%–10% of overall whey proteins, its concentration in whey is approximately 0.4 g/L (Liu et al., 2004). BSA is a single polypeptide composed of 583 amino acid residues with a molecular weight of 66 kDa (Hirayama et al., 1990; Carter and Ho, 1994). Its isoelectric point is at pH 4.7–4.9, and its denaturation temperature is around 64°C (Guo and Wang, 2019). The primary structure of BSA exhibits a significant similarity of 76% with human serum albumin; both proteins consist of 35 cysteine residues, which include 17 disulfide bonds along with a free thiol group (Sadowska-Bartosz et al., 2015). All the cysteine residues in BSA are located close to each other, thus giving rise to a protein structure composed of nine loops (Carter and Ho, 1994) (Fig. 3.5).

BSA is made of a heart-shaped three-domain structure stabilized by the intramolecular disulfide bonds, and each of these three homologous domains (domain I, II, and III) consists of two subdomains; thus, there are six subdomains: IA, IB, IIA, IIB, IIIA, and IIIB. The secondary structure contains predominantly α-helix (Carter and Ho, 1994). BSA contains 2 tryptophan moieties, Trp-134 and Trp-213, which reside in the first and second domains, respectively (Kandagal et al., 2006). Trp-134 can be found on the surface hydrophilic region, while Trp-213 resides within a hydrophobic binding pocket (Fonseca et al., 2017). Besides, there are two major binding sizes that have been identified, Sudlow's site I (located in subdomain IIA) and Sudlow's site II (located in subdomain IIIA) (Poór et al., 2014).

Serum albumins carry many physiological functions in the circulatory system; they maintain blood pH and regulate osmotic blood pressure (Klajnert and Bryszewska, 2002). However, their primary functions are transporting thyroid hormones, drugs, and free fatty acids in the serum. For species in which β-lactoglobulin is absent in breast milk, such as humans, BSA's function of binding hydrophobic molecules could be considered as a substitute for β-lactoglobulin. BSA can be applied as an important amino acid source, a cheap and biodegradable drug carrier to enhance the bioavailability of certain drugs (Fonseca et al., 2017). BSA-derived peptides serorphin (Tyr-Gly-Phe-Gln-Asn-Ala) and (f399–404) and albutensin A (Ala-Leu-Lys-Ala-Trp-Ser-Val-Ala-Arg) (f208–216) also demonstrate some biological functions (Meisel, 2005).

3.4 Lactoperoxidase

Lactoperoxidase is a heme-binding protein with a molecular weight of 78 kDa; it contains four/five carbohydrates which constitute 10% of its molecular weight, and a long polypeptide chain with 612 amino acid residues (Kussendrager and Van Hooijdonk, 2000). There are 10 variants with variations in the polypeptide and carbohydrate chains (Deeth and Bansal, 2019). As a member of the peroxidase enzyme family, lactoperoxidase is present in various organisms within the realms of plants and animals. It is present in pretty much all the milk that exists. Followed by xanthine oxidase, lactoperoxidase is the enzyme found in the second highest quantity in bovine milk. Considering xanthine oxidase is part of the milk fat globule membrane, lactoperoxidase is the predominant soluble enzyme in bovine milk, taking up 0.5% of the overall whey proteins (Buys, 2011). Each lactoperoxidase contains a ferriprotoporphyrin IX, and this heme group is linked with Glu375 and Asp275 with covalent bonds (Buys, 2011). Lactoperoxidase's isoelectric point at 9.6 has shown relatively high heat stability at pH 6.7 and stays active during the high-temperature short-time pasteurization (72°C for 15 s) (Griffiths, 1986; Kussendrager and Van Hooijdonk, 2000).

Lactoperoxidases are native enzymes with antimicrobial properties; about 30 mg/L of lactoperoxidase (varies based on the breed, feed, stage of lactation, etc.) can be found in bovine milk. Lactoperoxidase catalyzes the oxidation of thiocyanate (SCN^-) with hydrogen peroxide and produces primarily hypothiocyanate ($OSCN^-$), which can halt or kill many microorganisms, including bacteria (Gram-positive and Gram-negative), viruses, and fungi (Seifu et al., 2005). Its function is critical to protect the mammary gland during lactation and newborns' digestive system against pathogenic microbes (Björck et al., 1975; Seifu et al., 2005). Besides, lactoperoxidase can extend the raw milk shelf-life as it exhibits bactericidal or bacteriostatic properties against both pathogenic and spoilage bacteria (FAO/WHO, 2006).

Lactoperoxidase is of growing interest to scientific research and industry due to its antibacterial properties. It has been incorporated into a series of products such as bleaching agents, cosmetics, and mouthwashes (de Boer, 2014). Both lactoperoxidase and lactoferrin are fundamental proteins characterized by elevated isoelectric points, approximately 9; therefore, they can be extracted from whey using ion exchange chromatography, and this isolation method yields a combination of lactoperoxidase and lactoferrin, which are challenging to isolate (Dionysius et al., 1991).

3.5 Immunoglobulin

Immunoglobulins, a group of large glycoproteins produced by B lymphocytes, function as specialized antibodies in response to unique antigens (Walstra et al., 2006). Colostrum is an excellent reservoir of immunoglobulins (40–200 mg/mL) since their protective bioactivities provide passive immunological protection against toxins, viruses, and pathogens for newborn calves; the concentration of immunoglobulins in whey is only around 0.7 mg/mL (Korhonen et al., 2000; Mehra et al., 2006). Blood immunoglobulins are absent in newborn calves; thus, they need to obtain a significant amount of immunoglobulins from the colostrum to gain passive immunity (Deeth and Bansal, 2019). Immunoglobulins can achieve their protective bioactivities in the following ways: they can prevent bacteria from adhering to epithelial cells by adsorbing to certain sites on the bacteria; they can block bacterial metabolism, thus halting or decreasing the production of toxins; and they can block receptors in the host cells, therefore stopping virus replication (Korhonen and Marnila, 2009).

Immunoglobulins have the highest molecular weight (150–1000 kDa) among all the whey proteins, constituting 8%–10% of the overall whey protein (Lyerly et al., 1991). The immunoglobulin family includes several classes, namely IgM, IgA, IgG, IgE, and IgD. IgM, IgA, and IgG are the main types found in bovine milk and serum, and IgG is the primary isotype that constitutes 80% of all the immunoglobulins in milk/whey (Korhonen et al., 2000; Hurley and Theil, 2011). IgG contains two subclasses, IgG 1 and IgG 2, and IgG 1 is the preliminary type (Chen et al., 2000). Even though immunoglobulins are called whey proteins present in whey and milk serum (the majority of immunoglobulins can be found in whey), some IgA, IgM, and IgG2 are associated with casein proteins and fat globules.

In terms of structure, all immunoglobulins are glycoproteins that are present as either monomers or polymers, and the carbohydrate molecules (the main ones include hexoses and hexosamines) are linked at the C-terminal of the heavy chains (Deeth and Bansal, 2019). All the monomeric IgGs share a similar molecular structure (molecular weight of approximately 160 kDa) consisting of two light chains and two heavy chains. As Fig. 3.6 illustrates, each heavy chain and light chain is linked through a disulfide bond, which leads to a Y-shape. The gray section represents the constant region, and the blue (dark gray in print version) ones represent the variable regions (Mix et al., 2006). The monomeric IgG contains two antigen-binding regions, each composed of the N-terminal of a light chain's variable region and the N-terminal of a heavy chain (Deeth and Bansal, 2019). Both monomeric IgA and IgM exhibit a comparable structure to the monomeric IgG, differing only by the presence of a C-terminal octapeptide on the heavy chains (Marnila and Korhonen, 2011). IgA can be found in monomeric, dimeric, or tetrameric structures, where the dimer (380 kDa) is composed of two IgA monomers connected with a secretory component and a J-chain (15 kDa). IgM (~900 kDa) is a pentamer made up of five IgG-like molecules; these monomeric units are stabilized in a circular arrangement by disulfide bonds and a J-chain, as shown in Fig. 3.4B (Korhonen et al., 2000).

Even though colostrum can offer a significant amount of immunoglobulins, due to its limited availability, whey can be considered a reliable source for immunoglobulins. It has been reported that immunoglobulin administration safeguards humans from the infections of rotavirus and *Escherichia coli* (Fukumoto et al., 1994).

3.6 Caseinomacropeptide

Casein, as the major class of bovine milk proteins, is a phosphoprotein consisting of four main types: αs1-, αs2-, β-, and κ-casein. These caseins coassemble in milk and are presented as colloidal particles, that is, casein micelles, with diameters from 50 to 600 nm (Vasbinder et al., 2003). κ-casein protein is located at the surface of casein micelles; it is composed of 169 amino acids. Amino acids 1–105 constitute a water-insoluble part called para-κ-casein, and the remaining amino acid residues 106–169 (high ratios of polar amino acid residues) make up the hydrophilic C-terminal of κ-casein, called κ-casein glycomacropeptide. More specifically, it is an N-acetylated form of glycomacropeptide, also known as CMP. CMP has a small molecular weight of 8.0 kDa, and galactose is a major constituent of CMP, constituting around 50%–60% of overall CMP (Thomä-Worringer et al., 2006).

CMP is a negatively charged hydrophilic part of κ-casein that reaches out into the surrounding aqueous phase; the 5–10 nm thick layer of CMPs on the surface of the casein micelles is known as the "hairy" layer (De Kruif and Zhulina, 1996). CMP is stable in an acidic environment and remains negatively charged until pH 3 (Brody, 2000). The flexible "hairy" layer provides steric stabilization between casein micelles, which keeps the casein micelles suspended in the milk and avoids casein protein aggregation/precipitation. The hair layer is not quite dense; it is loose enough to allow β-casein to exit and re-enter the casein micelles during temperature changes (Creamer et al., 1977). The bond between κ-casein Phe105–Met106 is readily cleavable by proteolytic enzymes, such as chymosin, a common protease enzyme used for cheese production. Interestingly, the amino acid profile could be altered slightly differently when other proteinases are

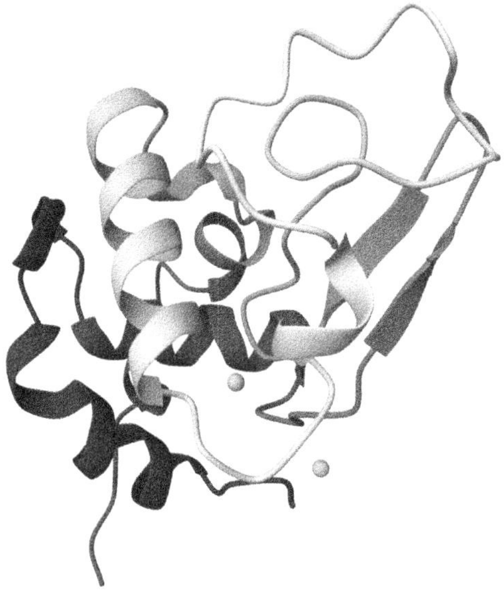

FIGURE 3.4 Ribbon models of α-lactalbumin binded with calcium ion (green [very light gray in print version] dot). *From RCSB PDB (1999) (PDB ID: 1A4V).*

applied. For example, when aspartic proteinases are used, the bond between κ-casein Ser104–Phe105 will be cleaved instead, which produces a CMP with an additional amino acid (Drøhse and Foltmann, 1989).

During certain cheese production, when chymosin is applied, it will decrease steric stabilization between the casein micelles by liberating CMP from the κ-casein (Sandra et al., 2007; Kethireddipalli et al., 2010). Followed by the chymosin-induced hydrolysis, the steric stabilization gets weakened and becomes too weak to keep casein micelles separate from each other. When the majority of κ-casein ($\geq$85%) is hydrolyzed, casein proteins will aggregate and form the curd (Tuinier and De Kruif, 2002). The obtained whey, which in this case is called sweet whey (due to the addition of chymosin), contains a large amount of CMP, which consists of 15% of the overall protein content in the sweet whey (Lim et al., 2007; Neelima et al., 2013).

When the cheese is produced with the addition of acids or through fermentation, CMP does not get cleaved from the κ-casein; hence, it is absent in the acid whey. Instead, the negative charge of CMP decreases as the pH of the milk/cream drops. As the steric repulsion between the casein micelles gets weakened, caseins tend to diffuse to each other; when hydrophobic effects take over, casein protein gel starts to form (De Kruif, 1999).

There is 1.2–1.5 g/L of CMP in sweet whey, and it accounts for 15%–20% of the overall protein content in sweet whey and has been commercialized as a health-promoting bioactive peptide in healthcare and food processing industries (Brody, 2000; Thomä-Worringer et al., 2006). CMP contains a unique amino acid profile; it not only contains a significant amount of BCAAs and essential amino acids but also lacks aromatic amino acids (tryptophan, phenylalanine, and tyrosine) (Pihlanto, 2011). CMP is comprised of essential amino acids (47%) and BCAAs (22.5%) (Etzel, 2004). As one of the rare phenylalanine-free natural proteins, CMP is safe for individuals with phenylketonuria to consume (Pihlanto, 2011). The recovery of CMP from sweet whey can be achieved through numerous methods such as ion-exchange chromatography, solvent partition, and ultrafiltration (Bonnaillie and Tomasula, 2009). The CMP available on the market is usually a high-purity product with a light color, homogeneous powder size, and a bland flavor profile (U.S. Dairy Export Council, 2004).

CMP is associated with other health-promoting benefits such as boosting immune system response, preventing enterotoxigenic *Escherichia coli* infection, and controlling liver diseases (Brody, 2000; Manso and López-Fandiño, 2004; Tidona et al., 2009). CMP could regulate the dental microbial composition and prevent dental cavity formation by prohibiting cariogenic bacteria from adhering to the oral cavity (Vacca Smith and Bowen, 2000). The anticarcinogenic properties make CMP a suitable ingredient for dental hygiene products (Manso and López-Fandiño, 2004). Besides, the administration of CMP has been linked with increased cholecystokinin release (a hormone that regulates gastric emptying), thus assisting in weight loss (Yvon et al., 1994; Burton-Freeman, 2008). Even though CMP-enriched products have emerged in the market lately, targeting appetite control and weight loss, more studies are required to prove the clinical efficacy of CMP consumption (Keogh et al., 2010).

3.7 Proteose peptone

Proteose peptones are nonheat-sensitive proteins/peptides that stay soluble in milk after milk has been heat-treated (95°C for 20 min) in an acid environment (pH 4.7) (Farkye and Shah, 2014). Plasminogen is the inactivated enzyme that is associated with casein micelles. Once its enzymatic activity is activated, that is, with heat treatment, it will turn into plasmin. Plasmin is responsible for β-casein degradation and produces proteose peptones through proteolysis of β-caseins (Walstra et al., 2006). Proteose peptones compose 1.1% of the overall milk protein (0.8–1.2 g/L) and 10% of the overall whey protein (Claeys et al., 2014; Innocente et al., 1998).

There are four major proteose peptones: proteose peptone component 3 (PP3), component 5 (PP5, f1–105 of β-CN), component 8 fast (PP8f, f1–28 of β-CN), and component 8 slow (PP8s, f29–105 of β-CN). PP3 (also known as lactophorin) is the primary proteose peptone which constitutes 25% of the total proteose peptone (Girardet and Linden, 1996; Krissansen, 2013). PP3 is a glycoprotein, and it is associated with the fat globule membrane. At neutral pH, PP3 is bound to the fat globule and thus is absent in sweet whey. For the acid-included cheese, PP3 could be found in the acid whey (Walstra et al., 2006). PPs could be isolated from other proteose peptone components with hydrophobic interaction chromatography and affinity chromatography. Due to the surface activity and small molecular weight (28 kDa), PP3 has been commercialized as an emulsifier in ice cream (Innocente et al., 2002).

3.8 Lactoferrin

Lactoferrin is a monomeric glycoprotein with a molecular weight of 80 kDa, it is a basic protein with an isoelectric point around 8.7. As a member of the transferrin family, which is characterized by binding and transferring ferric iron, around 20–35 mg/L of lactoferrin is present in the whey (Krissansen, 2007; Adlerova et al., 2008). Fig. 3.7 shows the structure of the bovine lactoferrin. The lactoferrin consists of two identical globular units, and each unit can bind 1.4 mg Fe^{3+}/g of protein. The homologous iron-binding site can be found at the interdomain cleft of each lobe (Wang and Guo, 2019; Sharma and Shah,

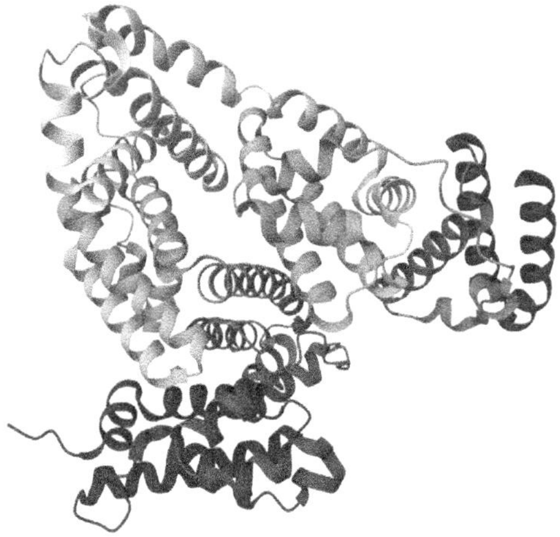

FIGURE 3.5 Structure of bovine serum albumin. *From RCSB PDB (2012) (PDB ID: 4F5S).*

2010). Fe^{3+}ions will be released in an acidic environment when large-scale conformation occurs at the iron-binding sites (Sharma and Shah, 2010). Mucosal epithelial cells produce lactoferrin, and lactoferrin is also synthesized in other mammalian species such as humans. Human lactoferrin shares a similar structure and amino acid composition with bovine-driven lactoferrin; however, the concentration of human milk lactoferrin is significantly higher than that of its counterpart in bovine milk (1–3 g/L lactoferrin in human milk versus 100–400 mg/L lactoferrin in bovine milk) (Sharma and Shah, 2010; Manzoni et al., 2011). Therefore, some whey protein baby formula has been enriched with bovine lactoferrin to imitate the antimicrobial effects present in human milk (Wang and Guo, 2019).

Because of its therapeutic properties, lactoferrin is one of the most important biomedical proteins in cheese whey. Lactoferrin is associated with multiple biological activities, including antimicrobial, antiinflammatory, and antiviral activities against cervical cancer, and it also prompts iron absorption in the intestines (Chen et al., 2007; Ben Ounis et al., 2008).

Lactoferrin offers a robust natural defense mechanism against a broad spectrum of fungal and bacterial pathogens. So far, lactoferrin has shown bacteriostatic and bactericidal effects on both Gram-positive and Gram-negative bacteria. It also exhibits fungicidal properties targeting Candida species (Wakabayashi et al., 2006). Its natural defense against bacteria and fungi is dependent on the iron-sequestering ability. Lactoferrin is able to withhold iron and then prevent the growth of iron-requiring microbes (Lönnerdal and Iyer, 1995). Oral ingestion of lactoferrin boosts gut immunity, decreases tumor growth, and aids in curing tinea pedis. Therefore, bovine lactoferrin has been commercialized as a nutraceutical and added to various food products such as health supplements and infant formula (Ushida et al., 1999; Pihlanto, 2011).

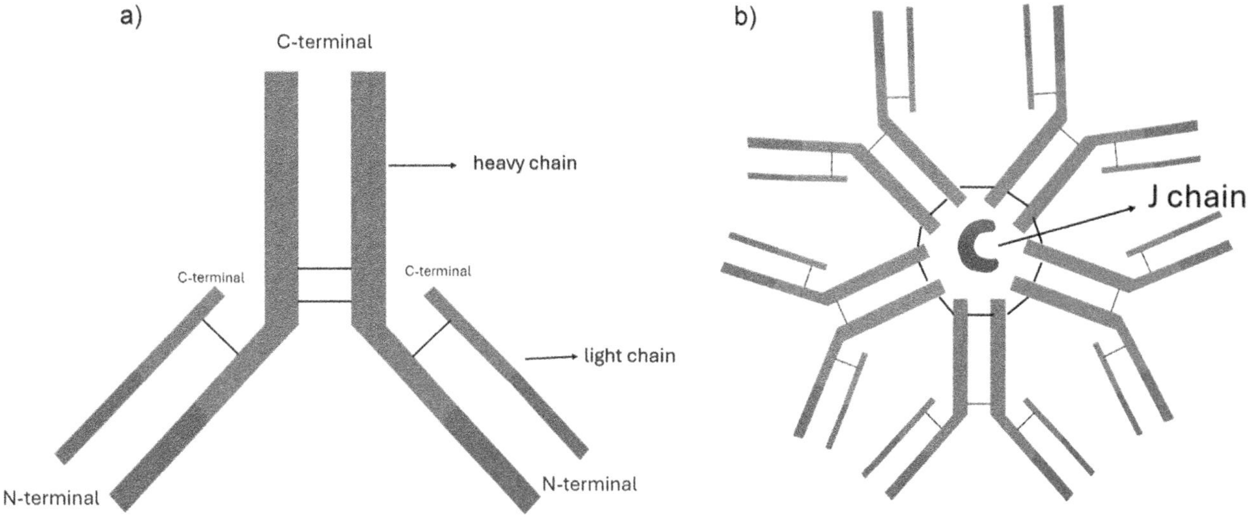

FIGURE 3.6 Schematic structure of (A) monomeric IgG and (B) pentameric IgM, where dashed lines represent disulfide bonds. *Adapted from Walstra et al. (2006).*

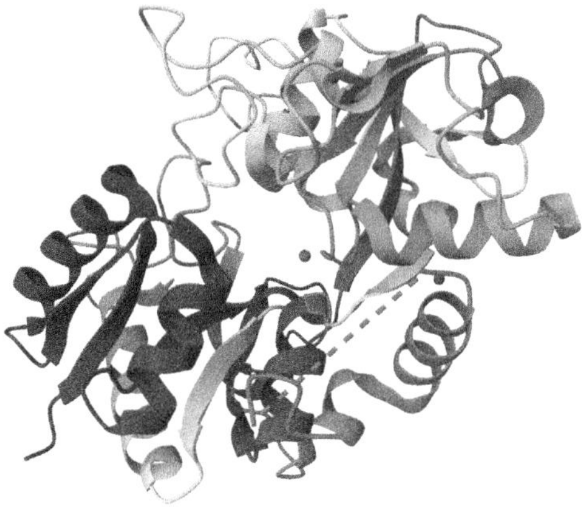

FIGURE 3.7 Ribbon models of lactoferrin binded with iron ion (orange [gray in print version] dot). *From RCSB PDB (2010) (PDB ID: 2AYS).*

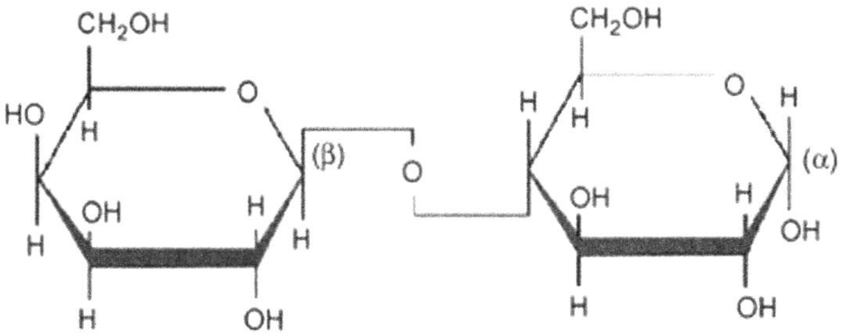

FIGURE 3.8 The structure of α-lactose (4-*O*-β-D-galactopyranosyl-D-glucose) (Schaafsma, 2008).

Sweet whey is an excellent reservoir of lactoferrin (30–100 mg/L), and high purity of lactoferrin powder (containing <5% water) can be recovered from sweet whey. Lactoferrin recovery is usually carried out with cation exchange membranes (Wakabayashi et al., 2006; U.S. Dairy Export Council, 2004) (Figs. 3.6, 3.7, 3.8, and 3.9).

3.9 Growth factor

Growth factors are polypeptides or proteins that are synthesized from the mammary gland tissues; they specify the cellular proliferation and/or cellular differentiation by binding certain receptors to intracellular secondary messengers (Michaelidou and Steijns, 2006). Growth factors also contribute to inhibitory activity on specific cells, depending on cell type (e.g., TGF-β is able to prevent the growth of lymphocytes) (Pouliot and Gauthier, 2006). They also contribute to the development and growth of young calves; therefore, the highest concentration of growth factors is found in colostrum (Gauthier et al., 2006).

As a great source of growth factors, whey contains many growth factors, including epidermal growth factor (EGF), betacellulin (BTC), fibroblast growth factors (FGF1 and FGF2), insulin-like growth factors (IGF-I and IGF-II), transforming growth factors (TGF-β1 and TGF-β2), and platelet-derived growth factor (PDGF). BTC and EGF are associated with wound healing, bone resorption, cell proliferation (embryonic cells, epithelial cells, and epidermal cells), and regulation of hormone production. IGF family includes single-chain polypeptides with a similar structure to insulin; they trigger cellular proliferation and differentiation. Administration of IGFs is expected to improve nitrogen balance, lower cholesterol, and enhance renal functions (Clemmons et al., 1992; Hirschberg et al., 1993; Gauthier et al., 2006). There are five isoforms identified in the TGF-β family; they share similar sequence homologies and biological activities. TGF-β family holds significant importance in embryonic development, lymphocyte inhibition, immune system regulation, connective tissue proliferation, and tissue repair (Massague, 1990). TGF-β2 is the primary isoform in bovine milk, yet its physiological function still remains unclear (Cox and Bürk, 1991). PDGF supports the growth of certain cells (i.e., glial and mesangial) and is associated with embryogenesis, especially for the development of lungs, kidneys, and the central nervous system. Basic FGF serves a vital purpose in all organ systems; it assists in cell differentiation and proliferation in organ systems (Chen et al., 2004).

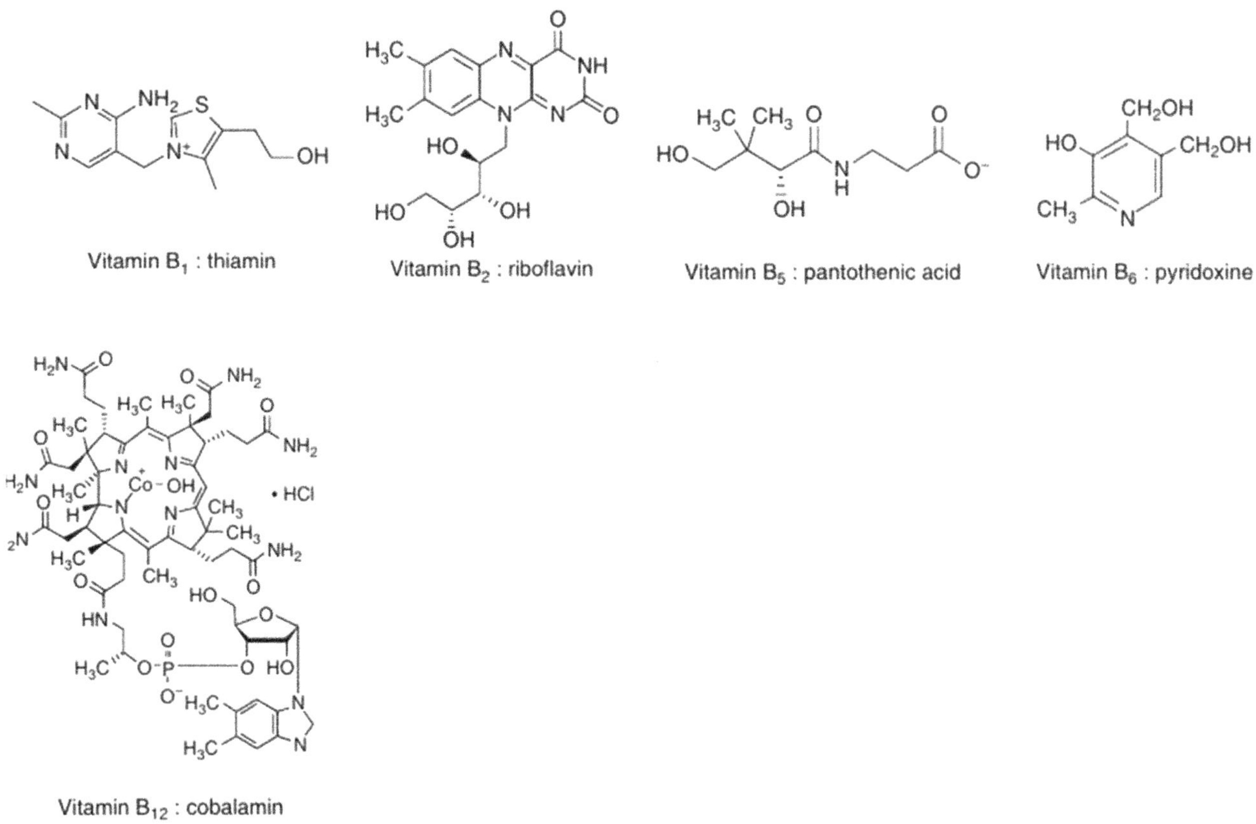

FIGURE 3.9 Vitamin B complex in whey (Graulet et al., 2013).

Although only a trace amount of growth factors exists in bovine milk and whey, they still attract a lot of commercial interest due to the fact that growth factors stay active after extraction (Kelly, 2003). Milk-derived growth factors can be concentrated according to their molecular weight/binding affinity/charge; they have been commercialized and sold as supplements to treat gastrointestinal diseases and skin disorders (Gauthier et al., 2006).

4. Lactose

Lactose (4-*O*-β-D-galactopyranosyl-D-glucose) is a disaccharide with a molecular weight of 342 kDa (Fig. 3.8); it constitutes most of the carbohydrates in the whey (U.S. Dairy Export Council, 2004). Bovine milk contains around 4.6 g lactose/100 mL of milk, and during the cheese-making procedure, almost all the lactose is retained in the whey (Schaafsma, 2008). In acid whey, the presence of lactic acid (metabolic product of lactose) enhances the solubility as well as absorption of calcium by lowering the pH of the intestinal system (Pérez et al., 2008). Intestinal β-galactosidase can hydrolyze lactose into two types of constituent monosaccharides, galactose and glucose. Lactose shows therapeutic effects for intestinal disorders by prompting the acid-forming lactobacilli in the intestine. However, most Asians, Africans, and a minority of Western Europeans might suffer from lactose intolerance, which means these people fail to digest lactose, and the unabsorbed lactose feeds the colon micro-organisms. As these microbes proliferate, gas and lactic acid are produced to cause discomfort such as diarrhea and abdominal cramps (de Wit, 2001b).

The application of membrane filtration technology enables lactose separation from the whey. Nowadays, many crystalline forms of lactose (such as nonhygroscopic α-lactose monohydrate) have been commercialized (Yang and Silva, 1995). Due to its low sweetness (84% less than sucrose), good water binding capacities, and color enhancement, lactose has been applied in the bakery, sausages, and confectionery (Vesa et al., 2000; Ryan and Walsh, 2016). Meanwhile, the food industry also exploits isolated lactose as the substrate to produce a gut health probiotic named galacto-oligosaccharides through the catalysis of galactosyl transferase, which has been added to a series of foods such as infant formulas, ice cream, and yogurt (Schaafsma, 2008).

5. Vitamins

Vitamins are organic molecules, and most milk vitamins are produced by rumen bacteria (Gaucheron, 2013). Vitamins are key organic nutrients with vital biological activity for the body's growth and functionality. They also involve protein catalysis and enzymatic activity boost and assist in metabolism. Vitamins have been denoted as micronutrients, which suggests that even if only a trace amount of vitamin intake is required, they are significantly important for the body's functioning. Since human endogenous metabolism cannot produce them, any deficiency could give rise to severe health problems (Król et al., 2020).

Vitamins are often classified based on their solubility. The majority of vitamins in milk are fat-soluble (vitamins A, D, E, and K), and B-complex and vitamin C belong to the water-soluble group (Graulet et al., 2013). Vitamin A, the primary fat-soluble vitamin in milk, along with other fat-soluble vitamins, including vitamins D, E, and K, is associated with milk fat and, therefore, gets trapped in the cheese curd and is absent in the whey (Fig. 3.9) (Walzem et al., 2002). Vitamins (vitamin B1, B2, B5, B6, and B12) that remain in the whey are water-soluble; these water-soluble vitamins are originally present in the milk serum and then get retained in the whey after the cheese curd formation. B vitamins primarily function in cellular physiological and catabolic metabolism (Gaucheron, 2005). A few studies have linked slight deficiencies in B vitamins with cancer and vascular diseases (Brachet et al., 2004).

6. Minerals

The mineral composition in milk varies based on the feed, health status, lactation stage, and so on (Amalfitano et al., 2024). Minerals present either in the form of ions dissolved in the serum phase or associated with other milk constituents, that is, fat globule membrane and milk proteins (Cashman, 2021). The mineral content in whey ranges from approximately 5–8 g/L depending on the coagulants applied for cheese-making (Anand et al., 2013). When organic or mineral acids are added to coagulate the cheese curds, the retained acid whey contains higher mineral content due to the dissociation of phosphorus and calcium from the casein micelles when the pH decreases to 4.6 (de Wit, 2001a). For the mineral content in whey, the major cations are potassium, calcium, sodium, and magnesium, while the main anions include phosphate, chloride, and citrate (Gaucheron, 2005). Other minerals such as manganese, zinc, and iron, which are identified as essential microminerals (essential intake less than 50 mg per day), are also found in the whey (Amalfitano et al., 2024; Paterson, 2009). Phosphate and calcium enhance teeth and bone growth. The low sodium/potassium ratio (around 0.3) in whey helps to prevent hypertension (de Wit, 2001a).

The presence of minerals in whey could create challenges for its potential use as a food ingredient (i.e., for infant formulas, 90%–95% mineral reduction is required to simulate human milk) and for further processing (i.e., spray drying for whey powder production). Therefore, whey demineralization is a quite common procedure that can be carried out with ion exchange, nanofiltration, and electrodialysis (Gernigon et al., 2011; Chen et al., 2018). The permeate of nanofiltration could be filtered and dried to recover the water-soluble minerals in whey; thus, minerals can be sold as whey salt powder or supplemented into drinks and foods (Harju, 2001; U.S. Dairy Export Council, 2004).

7. Lipids

During the cheese coagulation, the majority of lipids are embedded in the cheese curd; therefore, there is only a trace amount of lipids present in whey (Walzem et al., 2002). Triacylglycerols constitute most lipids in whey. Qualitatively, the concentrations of lipids in whey are triacylglycerols > phospholipids (primary phospholipids are sphingomyelin, phosphatidylcholine, and phosphatidyl) > ethanolamine > diacylglycerols > free fatty acids > cholesterol esters > cholesterol > monoacylglycerol. Compared to milk, the free fatty acids present in whey contain more butyric acid and less caprylic, oleic, and caproic acids (Walzem et al., 2002).

8. Conclusion

Over 160 million tons of whey are produced with cheese production each year, and the disposal of whey into the environment poses threats to both wildlife animals and the environment. As a valuable side product, whey comprises the majority of water-soluble constituents from milk, including whey proteins, peptides, lactose, minerals, vitamins, and minor proteins. Nowadays, emerging isolation techniques allow us to recover nutritional constituents from whey and commercialize them to make value-added products. However, the functions of certain whey proteins remain unclear, and with further research, whey is expected to be further valorized in various products, and the value of whey will be emphasized.

References

Adams, J.J., et al., 2006. Structure of bovine β-lactoglobulin (variant A) at very low ionic strength. J. Struct. Biol. 154 (3), 246–254. https://doi.org/10.1016/J.JSB.2005.12.010.

Adlerova, L., Bartoskova, A., Faldyna, M., 2008. Lactoferrin: a review. Vet. Med. 53 (9), 457–468. https://doi.org/10.17221/1978-VETMED.

Alves, A.C., Tavares, G.M., 2019. Mixing animal and plant proteins: is this a way to improve protein techno-functionalities? Food Hydrocoll. 97, 105171. https://doi.org/10.1016/J.FOODHYD.2019.06.016.

Amalfitano, N., et al., 2024. Detailed mineral profile of milk, whey, and cheese from cows, buffaloes, goats, ewes, and dromedary camels, and efficiency of recovery of minerals in their cheese. J. Dairy Sci. 107 (11), 8887–8907. https://doi.org/10.3168/JDS.2023-24624.

Anand, S., Khanal, S.N., Marella, C., 2013. Whey and whey products. Milk Dairy Prod. Hum. Nutr. Prod. Compos. Health 477–497. https://doi.org/10.1002/9781118534168.CH22.

Apenten, R.K.O., Khokhar, S., Galani, D., 2002. Stability parameters for β-lactoglobulin thermal dissociation and unfolding in phosphate buffer at pH 7.0. Food Hydrocoll. 16 (2), 95–103. https://doi.org/10.1016/S0268-005X(01)00067-4.

Attia, H., et al., 2000. Characterization of the dromedary milk casein micelle and study of its changes during acidification. Lait 80 (5), 503–515. https://doi.org/10.1051/LAIT:2000141.

Ben Ounis, W., et al., 2008. Separation of minor protein components from whey protein isolates by heparin affinity chromatography. Int. Dairy J. 18 (10–11), 1043–1050. https://doi.org/10.1016/J.IDAIRYJ.2008.04.004.

Bhattacharjee, C., Das, K.P., 2000. Thermal unfolding and refolding of β-lactoglobulin. Eur. J. Biochem. 267 (13), 3957–3964. https://doi.org/10.1046/J.1432-1327.2000.01409.X.

Björck, L., et al., 1975. Antibacterial activity of the lactoperoxidase system in milk against pseudomonads and other gram-negative bacteria. Appl. Microbiol. 30 (2), 199–204. https://doi.org/10.1128/AM.30.2.199-204.1975.

Blomstrand, E., et al., 2006. Branched-chain amino acids activate key enzymes in protein synthesis after physical exercise. J. Nutr. 136 (1 Suppl. 1). https://doi.org/10.1093/JN/136.1.269S.

Blouet, C., et al., 2006. Meal cysteine improves postprandial glucose control in rats fed a high-sucrose meal. J. Nutr. Biochem. 18 (8), 519–524. https://doi.org/10.1016/J.JNUTBIO.2006.10.002.

de Boer, R., 2014. Information sheets. In: From Milk By-Products to Milk Ingredients, pp. 208–261. https://doi.org/10.1002/9781118598634.OTH.

Bolder, S.G., et al., 2007. Heat-induced whey protein isolate fibrils: conversion, hydrolysis, and disulphide bond formation. Int. Dairy J. 17 (7), 846–853. https://doi.org/10.1016/J.IDAIRYJ.2006.10.002.

Bonnaillie, L.M., Tomasula, P.M., 2009. Whey protein fractionation. In: Whey Processing, Functionality and Health Benefits, pp. 15–38. https://doi.org/10.1002/9780813803845.CH2.

Brachet, P., et al., 2004. Age-associated B vitamin deficiency as a determinant of chronic diseases. Nutr. Res. Rev. 17 (1), 55–68. https://doi.org/10.1079/NRR200478.

Brew, K., Grobler, J.A., 1992. α-Lactalbumin. In: Fox, P.F. (Ed.), Advanced Dairy Chemistry-1: Proteins, second ed. Elsevier Applied Science Publishers, pp. 191–229.

Brody, E.P., 2000. Biological activities of bovine glycomacropeptide. Br. J. Nutr. 84 (Suppl. 1). https://doi.org/10.1017/S0007114500002233.

Brück, W.M., et al., 2003. rRNA probes used to quantify the effects of glycomacropeptide and alpha-lactalbumin supplementation on the predominant groups of intestinal bacteria of infant rhesus monkeys challenged with enteropathogenic *Escherichia coli*. J. Pediatr. Gastroenterol. Nutr. 37 (3), 273–280. https://doi.org/10.1097/00005176-200309000-00014.

Burton-Freeman, B.M., 2008. Glycomacropeptide (GMP) is not critical to whey-induced satiety but may have a unique role in energy intake regulation through cholecystokinin (CCK). Physiol. Behav. 93 (1–2), 379–387. https://doi.org/10.1016/J.PHYSBEH.2007.09.010.

Buys, E.M., 2011. Enzymes Indigenous to Milk | Lactoperoxidase. In: Encyclopedia of Dairy Sciences: Second Edition, pp. 319–323. https://doi.org/10.1016/B978-0-12-374407-4.00160-6.

Carter, D.C., Ho, J.X., 1994. Structure of serum albumin. Adv. Protein Chem. 45, 153–203. https://doi.org/10.1016/S0065-3233(08)60640-3.

Cashman, K.D., 2021. Milk trace elements: nutritional significance. In: Encyclopedia of Dairy Sciences: Third edition, vol. 5, pp. 953–960. https://doi.org/10.1016/B978-0-12-818766-1.10171-0.

Chatterton, D.E.W., et al., 2006. Bioactivity of β-lactoglobulin and α-lactalbumin—technological implications for processing. Int. Dairy J. 16 (11), 1229–1240. https://doi.org/10.1016/J.IDAIRYJ.2006.06.001.

Chatzipaschali, A.A., Stamatis, A.G., 2012. Biotechnological utilization with a focus on anaerobic treatment of cheese whey: current status and prospects. Energies 5 (9), 3492–3525. https://doi.org/10.3390/EN5093492.

Chen, C.C., Tu, Y.Y., Chang, H.M., 2000. Thermal stability of bovine milk immunoglobulin G (IgG) and the effect of added thermal protectants on the stability. J. Food Sci. 65 (2), 188–193. https://doi.org/10.1111/J.1365-2621.2000.TB15977.X.

Chen, C.-H., et al., 2004. Fibroblast growth factor 2: from laboratory evidence to clinical application. Curr. Vasc. Pharmacol. 2 (1), 33–43. https://doi.org/10.2174/1570161043476500.

Chen, L., et al., 2007. Isolation of lactoferrin from acid whey by magnetic affinity separation. Separ. Purif. Technol. 56 (2), 168–174. https://doi.org/10.1016/J.SEPPUR.2007.01.019.

Chen, G.Q., et al., 2018. A review of salty waste stream management in the Australian dairy industry. J. Environ. Manag. 224, 406–413. https://doi.org/10.1016/J.JENVMAN.2018.07.056.

Chihi, M.L., et al., 2016. Heat-induced soluble protein aggregates from mixed pea globulins and β-lactoglobulin. J. Agric. Food Chem. 64 (13), 2780–2791. https://doi.org/10.1021/ACS.JAFC.6B00087.

Chrysina, E.D., Brew, K., Acharya, K.R., 2000. Crystal structures of apo- and holo-bovine alpha-lactalbumin at 2. 2-A resolution reveal an effect of calcium on inter-lobe interactions. J. Biol. Chem. 275 (47), 37021–37029. https://doi.org/10.1074/JBC.M004752200.

Claeys, W.L., Verraes, C., Cardoen, S., De Block, J., Huyghebaert, A., Raes, K., Dewettinck, K., Herman, L., 2014. Consumption of raw or heated milk from different species: an evaluation of the nutritional and potential health benefits. Food Control 42, 188–201. https://doi.org/10.1016/j.foodcont.2014.01.045.

Clemmons, D.R., Smith-Banks, A., Underwood, L.E., 1992. Reversal of diet-induced catabolism by infusion of recombinant insulin-like growth factor-I in humans. J. Clin. Endocrinol. Metab. 75 (1), 234–238. https://doi.org/10.1210/JCEM.75.1.1619015.

Considine, T., et al., 2007. Interactions of milk proteins during heat and high hydrostatic pressure treatments—a review. Innov. Food Sci. Emerg. Technol. 8 (1), 1–23. https://doi.org/10.1016/J.IFSET.2006.08.003.

Coulson, E.J., Stevens, H., 1950. The serological relationship of bovine whey albumin to serum albumin. J. Biol. Chem. 187 (1), 355–363. https://doi.org/10.1016/S0021-9258(19)50960-7.

Cox, D.A., Bürk, R.R., 1991. Isolation and characterization of milk growth factor, a transforming-growth-factor-β2-related polypeptide, from bovine milk. Eur. J. Biochem. 197 (2), 353–358. https://doi.org/10.1111/J.1432-1033.1991.TB15918.X.

Creamer, L.K., 2002. MILK PROTEINS | Casein Nomenclature, Structure and Association Properties. In: Encyclopedia of Dairy Sciences, pp. 1895–1902. https://doi.org/10.1016/B0-12-227235-8/00318-7.

Creamer, L.K., Berry, G.P., Mills, O.E., 1977. A study of the dissociation of β-casein from the bovine casein micelle at low temperature. New Zealand J. Dairy Sci. Technol. 12, 58–66.

Croguennec, T., Tavares, G.M., Bouhallab, S., 2017. Heteroprotein complex coacervation: a generic process. Adv. Colloid Interface Sci. 239, 115–126. https://doi.org/10.1016/J.CIS.2016.06.009.

Deeth, H., Bansal, N., 2019. Whey proteins: an overview. Whey Proteins: From Milk to Medicine 1–50. https://doi.org/10.1016/B978-0-12-812124-5.00001-1.

Dickinson, E., et al., 1993. A neutron reflectivity study of the adsorption of β-casein at fluid interfaces. Langmuir 9 (1), 242–248. https://doi.org/10.1021/LA00025A046/ASSET/LA00025A046.FP.PNG_V03.

Dionysius, D., Herse, J., Grieve, P., 1991. Extraction of lactoperoxidase and lactoferrin from whey using batch ion exchange techniques. Aust. J. Dairy Technol. 1991;46,72–76.

Drøhse, H.B., Foltmann, B., 1989. Specificity of milk-clotting enzymes towards bovine κ-casein. Biochim. Biophys. Acta Protein Struct. Mol. Enzymol. 995 (3), 221–224. https://doi.org/10.1016/0167-4838(89)90039-3.

Dufour, E., Robert, P., Renard, D., Llamas, G., 1998. Investigation of β-lactoglobulin gelation in water/ethanol solutions. Int. Dairy J. 8 (2), 87–93. https://doi.org/10.1016/S0958-6946(98)00024-7.

Editorials, G., 2011. Whey proteins and peptides. Nutrafoods 10 (2), 29–42. https://doi.org/10.1007/BF03223386.

Etzel, M.R., 2004. Manufacture and use of dairy protein fractions. J. Nutr. 134 (4). https://doi.org/10.1093/JN/134.4.996S.

Exl, B.M., 2001. A review of recent developments in the use of moderately hydrolyzed whey formulae in infant nutrition. Nutr. Res. 21 (1–2), 355–379. https://doi.org/10.1016/S0271-5317(00)00259-1.

FAO/WHO. 2006. Benefits and Potential Risks of the Lactoperoxidase System of Raw Milk Preservation. Available at: https://www.fao.org/4/a0729e/a0729e00.htm#:~:text=The%20activation%20of%20the%20lactoperoxidase%20in%20the%20presence,for%207%E2%80%938%20hours%20under%20ambient%20temperatures%20of%20. (Accessed: 30 January 2025).

Farizoglu, B., et al., 2004. Cheese whey treatment performance of an aerobic jet loop membrane bioreactor. Process Biochem. 39 (12), 2283–2291. https://doi.org/10.1016/J.PROCBIO.2003.11.028.

Farkye, N.Y., Shah, N., 2014. Milk proteins. Wiley Blackwell 6 (9781119944492), 427–458. Available at:

Farrell, H.M., et al., 2004. Nomenclature of the proteins of cows' milk-sixth revision. J. Dairy Sci. 87 (6), 1641–1674. https://doi.org/10.3168/JDS.S0022-0302(04)73319-6.

Fennema, O.R., Damodaran, S., Parkin, K.L., 2017. Amino acids, peptides, and proteins. In: Srinivasan, D., Kirk, L. (Eds.), Fennema's Food Chemistry, fifth ed. CRC Press, Taylor & Francis Group, Boca Raton, pp. 235–356.

Ferchichi, M., et al., 2005. Influence of initial pH on hydrogen production from cheese whey. J. Biotechnol. 120 (4), 402–409. https://doi.org/10.1016/J.JBIOTEC.2005.05.017.

Foegeding, E.A., Luck, P.J., Davis, J.P., 2006. Factors determining the physical properties of protein foams. Food Hydrocoll. 20 (2–3), 284–292. https://doi.org/10.1016/J.FOODHYD.2005.03.014.

Fogolari, F., et al., 1998. Monomeric bovine β-lactoglobulin adopts a β-barrel fold at pH 2. FEBS Lett. 436 (2), 149–154. https://doi.org/10.1016/S0014-5793(98)00936-3.

Fonseca, D.P., Khalil, N.M., Mainardes, R.M., 2017. Bovine serum albumin-based nanoparticles containing resveratrol: characterization and antioxidant activity. J. Drug Deliv. Sci. Technol. 39, 147–155. https://doi.org/10.1016/J.JDDST.2017.03.017.

Fox, P.F., et al., 2015. Milk proteins. Dairy Chem. Biochem. 145–239. https://doi.org/10.1007/978-3-319-14892-2_4.

Fukumoto, L.R., et al., 1994. Isolation of immunoglobulins from cheese whey using ultrafiltration and immobilized metal affinity chromatography. Food Res. Int. 27 (4), 335–348. https://doi.org/10.1016/0963-9969(94)90189-9.

Gaucheron, F., 2005. The minerals of milk. Reprod. Nutr. Dev. 45 (4), 473–483. https://doi.org/10.1051/RND:2005030.

Gaucheron, F., 2013. Milk minerals, trace elements, and macroelements. Milk Dairy Prod. Hum. Nutr. Prod. Compos. Health 172–199. https://doi.org/10.1002/9781118534168.CH9.

Gauthier, S.F., Pouliot, Y., Maubois, J.L., 2006. Growth factors from bovine milk and colostrum: composition, extraction and biological activities. Lait 86 (2), 99–125. https://doi.org/10.1051/LAIT:2005048.

Gernigon, G., et al., 2011. Whey Processing: Demineralization. Encyclopedia of Dairy Sciences: Second Edition, pp. 738–743. https://doi.org/10.1016/B978-0-12-374407-4.00496-9.

Ghaly, A.E., Kamal, M.A., 2004. Submerged yeast fermentation of acid cheese whey for protein production and pollution potential reduction. Water Res. 38 (3), 631–644. https://doi.org/10.1016/J.WATRES.2003.10.019.

Girardet, J.M., Linden, G., 1996. PP3 component of bovine milk: a phosphorylated whey glycoprotein. J. Dairy Res. 63 (2), 333–350. https://doi.org/10.1017/S0022029900031848.

Godovac-Zimmermann, J., 1988. The structural motif of ß-lactoglobulin and retinol-binding protein: a basic framework for binding and transport of small hydrophobic molecules? Trends Biochem. Sci. 13 (2), 64–66. https://doi.org/10.1016/0968-0004(88)90031-X.

González Siso, M.I., 1996. The biotechnological utilization of cheese whey: a review. Bioresour. Technol. 57 (1), 1–11. https://doi.org/10.1016/0960-8524(96)00036-3.

Graham, D.E., Phillips, M.C., 1979. Proteins at liquid interfaces: III. Molecular structures of adsorbed films. J. Colloid Interface Sci. 70 (3), 427–439. https://doi.org/10.1016/0021-9797(79)90050-X.

Graulet, B., et al., 2013. Vitamins in milks. Milk Dairy Prod. Hum. Nutr. Prod. Compos. Health 200–219. https://doi.org/10.1002/9781118534168.CH10.

Griffiths, M.W., 1986. Use of milk enzymes as indices of heat treatment. J. Food Protect. 49 (9), 696–705. https://doi.org/10.4315/0362-028X-49.9.696.

Guimarães, P.M.R., Teixeira, J.A., Domingues, L., 2010. Fermentation of lactose to bio-ethanol by yeasts as part of integrated solutions for the valorisation of cheese whey. Biotechnol. Adv. 28 (3), 375–384. https://doi.org/10.1016/J.BIOTECHADV.2010.02.002.

Guo, M., Wang, G., 2019. Chemistry of whey proteins. In: Guo, M. (Ed.), Whey Protein Production, Chemistry, Functionality, and Applications, first ed. Wiley, Hoboken, pp. 39–57.

Ha, E., Zemel, M.B., 2003. Functional properties of whey, whey components, and essential amino acids: mechanisms underlying health benefits for active people (Review). J. Nutr. Biochem. 14 (5), 251–258. https://doi.org/10.1016/S0955-2863(03)00030-5.

Harju, M., 2001. Milk sugars and minerals as ingredients. Int. J. Dairy Technol. 54 (2), 61–63. https://doi.org/10.1046/J.1471-0307.2001.00016.X.

Heine, W.E., Klein, P.D., Reeds, P.J., 1991. The importance of alpha-lactalbumin in infant nutrition. J. Nutr. 121 (3), 277–283. https://doi.org/10.1093/JN/121.3.277.

Hirayama, K., et al., 1990. Rapid confirmation and revision of the primary structure of bovine serum albumin by ESIMS and Frit-FAB LC/MS. Biochem. Biophys. Res. Commun. 173 (2), 639–646. https://doi.org/10.1016/S0006-291X(05)80083-X.

Hirschberg, R., et al., 1993. Effects of insulin-like growth factor I on renal function in normal men. Kidney Int. 43 (2), 387–397. https://doi.org/10.1038/KI.1993.57.

Hurley, W.L., Theil, P.K., 2011. Perspectives on immunoglobulins in colostrum and milk. Nutrients 3 (4), 442–474. https://doi.org/10.3390/NU3040442.

Hursel, R., Van Der Zee, L., Westerterp-Plantenga, M.S., 2010. Effects of a breakfast yoghurt, with additional total whey protein or caseinomacropeptide-depleted alpha-lactalbumin-enriched whey protein, on diet-induced thermogenesis and appetite suppression. Br. J. Nutr. 103 (5), 775–780. https://doi.org/10.1017/S0007114509992352.

Innocente, N., et al., 1998. Dynamic surface properties of the proteose-peptone fraction of bovine milk. J. Dairy Sci. 81 (7), 1833–1839. https://doi.org/10.3168/JDS.S0022-0302(98)75753-4.

Innocente, N., Comparin, D., Corradini, C., 2002. Proteose-peptone whey fraction as emulsifier in ice-cream preparation. Int. Dairy J. 12 (1), 69–74. https://doi.org/10.1016/S0958-6946(01)00166-2.

Ismail, B.P., Gu, Z., 2010. Whey Protein Hydrolysates: Current Knowledge and Challenges.

Jackson, J.G., et al., 2004. A multinational study of α-lactalbumin concentrations in human milk. J. Nutr. Biochem. 15 (9), 517–521. https://doi.org/10.1016/j.jnutbio.2003.10.009.

Jelen, P., 2002. Whey Processing | Utilization and Products. Encyclopedia of Dairy Sciences, pp. 2739–2745. https://doi.org/10.1016/B0-12-227235-8/00511-3.

Kandagal, P.B., Ashoka, S., Seetharamappa, J., Shaikh, S.M.T., Jadegoud, Y., Ijare, O.B., 2006. Study of the interaction of an anticancer drug with human and bovine serum albumin: Spectroscopic approach. J. Pharmaceut. Biomed. Anal. 41 (2), 393–399. https://doi.org/10.1016/j.jpba.2005.11.037.

Kelleher, S.L., et al., 2003. Glycomacropeptide and α-lactalbumin supplementation of infant formula affects growth and nutritional status in infant rhesus monkeys. Am. J. Clin. Nutr. 77 (5), 1261–1268. https://doi.org/10.1093/AJCN/77.5.1261.

Kelly, G., 2003. Bovine colostrums: a review of clinical uses. Altern Med Rev. 2003 Nov 8 (4), 378-94. Erratum in: Altern Med Rev. 2004 Mar;9(1):69. PMID: 14653766.

Keogh, J.B., et al., 2010. Effect of glycomacropeptide fractions on cholecystokinin and food intake. Br. J. Nutr. 104 (2), 286–290. https://doi.org/10.1017/S0007114510000280.

Kethireddipalli, P., Hill, A.R., Dalgleish, D.G., 2010. Protein interactions in heat-treated milk and effect on rennet coagulation. Int. Dairy J. 20 (12), 838–843. https://doi.org/10.1016/J.IDAIRYJ.2010.06.003.

Klajnert, B., Bryszewska, M., 2002. Fluorescence studies on PAMAM dendrimers interactions with bovine serum albumin. Bioelectrochemistry 55 (1–2), 33–35. https://doi.org/10.1016/S1567-5394(01)00170-0.

Korhonen, H.J., Marnila, P., 2009. Bovine milk immunoglobulins against microbial human diseases. In: Dairy-Derived Ingredients: Food and Nutraceutical Uses, pp. 269–289. https://doi.org/10.1533/9781845697198.2.269.

Korhonen, H., Marnila, P., Gill, H.S., 2000. Milk immunoglobulins and complement factors. Br. J. Nutr. 84 (S1), 75–80.

Kosikowski, F.V., 1979. Whey utilization and whey products. J. Dairy Sci. 62 (7), 1149–1160. https://doi.org/10.3168/JDS.S0022-0302(79)83389-5.

Kosseva, M.R., et al., 2009. Use of immobilised biocatalysts in the processing of cheese whey. Int. J. Biol. Macromol. 45 (5), 437–447. https://doi.org/10.1016/J.IJBIOMAC.2009.09.005.

Krissansen, G.W., 2007. Emerging health properties of whey proteins and their clinical implications. J. Am. Coll. Nutr. 26 (6), 713S–723S. https://doi.org/10.1080/07315724.2007.10719652.

Krissansen, G.W., 2013. Emerging Health Properties of Whey Proteins and Their Clinical Implications. https://doi.org/10.1080/07315724.2007.10719652.

Król, J., et al., 2020. The effect of selected factors on the content of fat-soluble vitamins and macro-elements in raw milk from Holstein-Friesian and Simmental cows and acid curd cheese (Tvarog). Animals 10, 1800. https://doi.org/10.3390/ANI10101800.

de Kruif, C.G., 1999. Casein micelle interactions. Int. Dairy J. 9 (3–6), 183–188. https://doi.org/10.1016/S0958-6946(99)00058-8.

de Kruif, C.G., Holt, C., 2003. Casein micelle structure, functions and interactions. In: Advanced Dairy Chemistry—1 Proteins, pp. 233–276. https://doi.org/10.1007/978-1-4419-8602-3_5.

de Kruif, C.G., Zhulina, E.B., 1996. κ-casein as a polyelectrolyte brush on the surface of casein micelles. Colloids Surf. A Physicochem. Eng. Asp. 117 (1–2), 151–159. https://doi.org/10.1016/0927-7757(96)03696-5.

Kussendrager, K.D., Van Hooijdonk, A.C.M., 2000. Lactoperoxidase: physico-chemical properties, occurrence, mechanism of action and applications. Br. J. Nutr. 84 (Suppl. 1). https://doi.org/10.1017/S0007114500002208.

Lim, K., et al., 2007. Acceptable low-phenylalanine foods and beverages can be made with glycomacropeptide from cheese whey for individuals with PKU. Mol. Genet. Metabol. 92 (1–2), 176–178. https://doi.org/10.1016/J.YMGME.2007.06.004.

Liu, J., et al., 2004. Spectrofluorimetric study of the binding of daphnetin to bovine serum albumin. J. Pharmaceut. Biomed. Anal. 35 (3), 671–677. https://doi.org/10.1016/J.JPBA.2004.02.010.

Liu, X., et al., 2006. Conformational changes of β-lactoglobulin induced by anionic phospholipid. Biophys. Chem. 121 (3), 218–223. https://doi.org/10.1016/J.BPC.2005.12.015.

Lönnerdal, B., Iyer, S., 1995. Lactoferrin: molecular structure and biological function. Annu. Rev. Nutr. 15, 93–110. https://doi.org/10.1146/ANNUREV.NU.15.070195.000521.

Lyerly, D.M., et al., 1991. Passive immunization of hamsters against disease caused by *Clostridium difficile* by use of bovine immunoglobulin G concentrate. Infect. Immun. 59 (6), 2215–2218. https://doi.org/10.1128/IAI.59.6.2215-2218.1991.

Madureira, A.R., et al., 2007. Bovine whey proteins—overview on their main biological properties. Food Res. Int. 40 (10), 1197. https://doi.org/10.1016/J.FOODRES.2007.07.005.

Manso, M.A., López-Fandiño, R., 2004. κ-Casein macropeptides from cheese whey: physicochemical, biological, nutritional, and technological features for possible uses. Food Rev. Int. 20 (4), 329–355. https://doi.org/10.1081/FRI-200033456.

Manzoni, P., Mostert, M., Stronati, M., 2011. Lactoferrin for prevention of neonatal infections. Curr. Opin. Infect. Dis. 24 (3), 177–182. https://doi.org/10.1097/QCO.0B013E32834592E6.

Marciniak, A., et al., 2018. Studying a chaperone-like effect of beta-casein on pressure-induced aggregation of beta-lactoglobulin in the presence of alpha-lactalbumin. Food Hydrocoll. 84, 9–15. https://doi.org/10.1016/J.FOODHYD.2018.05.038.

Markus, C.R., et al., 2005. Evening intake of alpha-lactalbumin increases plasma tryptophan availability and improves morning alertness and brain measures of attention. Am. J. Clin. Nutr. 81 (5), 1026–1033. https://doi.org/10.1093/AJCN/81.5.1026.

Marnila, P., Korhonen, H., 2011. Milk Proteins | Immunoglobulins. In: Encyclopedia of Dairy Sciences: Second Edition, pp. 807–815. https://doi.org/10.1016/B978 0 12 374407 4.00436-2.

Marwaha, S.S., Kennedy, J.F., 1988. Whey—pollution problem and potential utilization. Int. J. Food Sci. Technol. 23 (4), 323–336. https://doi.org/10.1111/J.1365-2621.1988.TB00586.X.

Massague, J., 1990. The transforming growth factor-β family. Annu. Rev. Cell Biol. 6, 597–641. https://doi.org/10.1146/ANNUREV.CB.06.110190.003121/CITE/REFWORKS.

Mawson, A.J., 1994. Bioconversions for whey utilization and waste abatement. Bioresour. Technol. 47 (3), 195–203. https://doi.org/10.1016/0960-8524(94)90180-5.

McIntosh, G.H., et al., 1995. Dairy proteins protect against dimethylhydrazine-induced intestinal cancers in rats. J. Nutr. 125 (4), 809–816. https://doi.org/10.1093/JN/125.4.809.

McKenzie, H.A., 1971. β-Lactoglobulins. In: Milk Proteins: Chemistry and Molecular Biology, vol. II. Academic Press, New York, pp. 257–330.

Mehra, R., Marnila, P., Korhonen, H., 2006. Milk immunoglobulins for health promotion. Int. Dairy J. 16 (11), 1262–1271. https://doi.org/10.1016/J.IDAIRYJ.2006.06.003.

Meisel, H., 2005. Biochemical properties of peptides encrypted in bovine milk proteins. Curr. Med. Chem. 12 (16), 1905–1919. https://doi.org/10.2174/0929867054546618.

Michaelidou, A., Steijns, J., 2006. Nutritional and technological aspects of minor bioactive components in milk and whey: growth factors, vitamins and nucleotides. Int. Dairy J. 16 (11), 1421–1426. https://doi.org/10.1016/J.IDAIRYJ.2006.06.018.

Mix, E., Goertsches, R., Zett, U.K., 2006. Immunoglobulins-basic considerations. J. Neurol. 253 (Suppl. 5). https://doi.org/10.1007/S00415-006-5002-2.

Mockaitis, G., et al., 2006. Anaerobic whey treatment by a stirred sequencing batch reactor (ASBR): effects of organic loading and supplemented alkalinity. J. Environ. Manag. 79 (2), 198–206. https://doi.org/10.1016/J.JENVMAN.2005.07.001.

Moore, S.A., et al., 1997. Three-dimensional structure of diferric bovine lactoferrin at 2.8 Å resolution. J. Mol. Biol. 274 (2), 222–236. https://doi.org/10.1006/JMBI.1997.1386.

Nagaoka, S., 2006. Cholesterol lowering proteins and peptide. In: Mine, Y., Shahidi, F. (Eds.), Nutraceutical Proteins and Peptides in Health and Disease. Taylor Francis Group, pp. 41–68.

Neelima, et al., 2013. Chemical and functional properties of glycomacropeptide (GMP) and its role in the detection of cheese whey adulteration in milk: a review. Dairy Sci. Technol. 93 (1), 21–43. https://doi.org/10.1007/S13594-012-0095-0.

Ohinata, K., et al., 2007. β-Lactotensin, a neurotensin agonist peptide derived from bovine β-lactoglobulin, enhances memory consolidation in mice. Peptides 28 (7), 1470–1474. https://doi.org/10.1016/J.PEPTIDES.2007.06.002.

Panesar, P.S., et al., 2007. Bioutilisation of whey for lactic acid production. Food Chem. 105 (1), 1–14. https://doi.org/10.1016/J.FOODCHEM.2007.03.035.

Patel, S.R., Murthy, Z.V.P., 2011. Waste valorization: recovery of lactose from partially deproteinated whey by using acetone as anti-solvent. Dairy Sci. Technol. 91 (1), 53–63. https://doi.org/10.1051/DST/2010036.

Paterson, A.H.J., 2009. Production and uses of lactose. Adv. Dairy Chem. 3, 105–120. https://doi.org/10.1007/978-0-387-84865-5_4/FIGURES/9.

Pérez, A.V., et al., 2008. Minireview on regulation of intestinal calcium absorption. Emphasis on molecular mechanisms of transcellular pathway. Digestion 77 (1), 22–34. https://doi.org/10.1159/000116623.

Permikov, E.A., 2005. [Alpha]-lactalbumin. Nova Science Publishers, New York.

Permyakov, E.A., Berliner, L.J., 2000. α-Lactalbumin: structure and function. FEBS Lett. 473 (3), 269–274. https://doi.org/10.1016/S0014-5793(00)01546-5.

Permyakov, E.A., et al., 1981. Calcium binding to alpha-lactalbumin: structural rearrangement and association constant evaluation by means of intrinsic protein fluorescence changes. Biochem. Biophys. Res. Commun. 100 (1), 191–197. https://doi.org/10.1016/S0006-291X(81)80081-2.

Pihlanto, A., 2011. Whey proteins and peptides. Nutrafoods 10 (2–3), 29–42. https://doi.org/10.1007/BF03223386.

Polis, B.D., Shmukler, H.W., Custer, J.H., 1950. Isolation of a crystalline albumin from milk. J. Biol. Chem. 187 (1), 349–354. https://doi.org/10.1016/S0021-9258(19)50959-0.

Poór, M., et al., 2014. Quantitation of species differences in albumin–ligand interactions for bovine, human and rat serum albumins using fluorescence spectroscopy: a test case with some Sudlow's site I ligands. J. Lumin. 145, 767–773. https://doi.org/10.1016/J.JLUMIN.2013.08.059.

Pouliot, Y., Gauthier, S.F., 2006. Milk growth factors as health products: some technological aspects. Int. Dairy J. 16 (11), 1415–1420. https://doi.org/10.1016/J.IDAIRYJ.2006.06.006.

Qin, B.Y., et al., 1999. Functional implications of structural differences between variants A and B of bovine beta-lactoglobulin. Protein Sci. Publ. Protein Soc. 8 (1), 75–83. https://doi.org/10.1110/PS.8.1.75.

Rapeanu, G., Nicoleta, S., 2010. An overview of bovine α-lactalbumin structure and functionality. Available at: https://www.academia.edu/18943184/An_overview_of_bovine_%CE%B1_lactalbumin_structure_and_functionality. (Accessed: 29 January 2025).

RCSB Protein Data Bank, 1999. Structure summary for PDB ID: 1A4V. Retrieved August 20, 2025, from https://www.rcsb.org/structure/1A4V.

RCSB Protein Data Bank, 2010a. Structure summary for PDB ID: 2AYS. Retrieved August 20, 2025, from https://www.rcsb.org/structure/2AYS.

RCSB Protein Data Bank, 2010b. Structure summary for PDB ID: 3NPO. Retrieved August 20, 2025, from https://www.rcsb.org/structure/3NPO.

RCSB Protein Data Bank, 2012. Structure summary for PDB ID: 4F5S. Retrieved August 20, 2025, from https://www.rcsb.org/structure/4F5S.

Regester, G.O., Smithers, G.W., 1991. Seasonal changes in the β-lactoglobulin, α-lactalbumin, glycomacropeptide, and casein content of whey protein concentrate. J. Dairy Sci. 74 (3), 796–802. https://doi.org/10.3168/JDS.S0022-0302(91)78227-1.

Ryan, M.P., Walsh, G., 2016. The biotechnological potential of whey. Rev. Environ. Sci. Biotechnol. 15 (3), 479–498. https://doi.org/10.1007/S11157-016-9402-1.

Sadowska-Bartosz, I., et al., 2015. Glycation of bovine serum albumin by ascorbate in vitro: possible contribution of the ascorbyl radical? Redox Biol. 6, 93–99. https://doi.org/10.1016/J.REDOX.2015.06.017.

Sandra, S., Alexander, M., Dalgleish, D.G., 2007. The rennet coagulation mechanism of skim milk as observed by transmission diffusing wave spectroscopy. J. Colloid Interface Sci. 308 (2), 364–373. https://doi.org/10.1016/J.JCIS.2007.01.021.

Sawyer, L., et al., 2002. Milk protein structure—what can it tell the dairy industry? Int. Dairy J. 12 (4), 299–310.

Schaafsma, G., 2008. Lactose and lactose derivatives as bioactive ingredients in human nutrition. Int. Dairy J. 18 (5), 458–465. https://doi.org/10.1016/J.IDAIRYJ.2007.11.013.

Schokker, E.P., Singh, H., Creamer, L.K., 2000. Heat-induced aggregation of β-lactoglobulin A and B with α-lactalbumin. Int. Dairy J. 10 (12), 843–853. https://doi.org/10.1016/S0958-6946(01)00022-X.

Seifu, E., Buys, E.M., Donkin, E.F., 2005. Significance of the lactoperoxidase system in the dairy industry and its potential applications: a review. Trends Food Sci. Technol. 16 (4), 137–154. https://doi.org/10.1016/J.TIFS.2004.11.002.

Sharma, R., Shah, N., 2010. Health benefits of whey proteins. Nutrafoods 9 (4), 39–45. https://doi.org/10.1007/BF03223347.

Shimomura, Y., et al., 2004. Exercise promotes BCAA catabolism: effects of BCAA supplementation on skeletal muscle during exercise. J. Nutr. 134 (Suppl. 6). https://doi.org/10.1093/JN/134.6.1583S.

Sipola, M., et al., 2002. α-lactorphin and β-lactorphin improve arterial function in spontaneously hypertensive rats. Life Sci. 71 (11), 1245–1253. https://doi.org/10.1016/S0024-3205(02)01793-9.

Smithers, G.W., 2008. Whey and whey proteins—from "gutter-to-gold". Int. Dairy J. 18 (7), 695–704. https://doi.org/10.1016/J.IDAIRYJ.2008.03.008.

Stacey, A., et al., 1995. Lactation is disrupted by alpha-lactalbumin deficiency and can be restored by human alpha-lactalbumin gene replacement in mice. Proc. Natl. Acad. Sci. 92 (7), 2835–2839. https://doi.org/10.1073/PNAS.92.7.2835.

Stuart, D.I., et al., 1986. Alpha-lactalbumin possesses a novel calcium binding loop. Nature 324 (6092), 84–87. https://doi.org/10.1038/324084A0.

Thomä-Worringer, C., Sørensen, J., López-Fandiño, R., 2006. Health effects and technological features of caseinomacropeptide. Int. Dairy J. 16 (11), 1324–1333. https://doi.org/10.1016/J.IDAIRYJ.2006.06.012.

Tidona, F., et al., 2009. Bioactive peptides in dairy products. Ital. J. Anim. Sci. 8 (3), 315–340. https://doi.org/10.4081/IJAS.2009.315.

Tipton, K.D., et al., 2004. Ingestion of casein and whey proteins result in muscle anabolism after resistance exercise. Med. Sci. Sports Exerc. 36 (12), 2073–2081. https://doi.org/10.1249/01.MSS.0000147582.99810.C5.

Tuinier, R., De Kruif, C.G., 2002. Stability of casein micelles in milk. J. Chem. Phys. 117 (3), 1290–1295. https://doi.org/10.1063/1.1484379.

Tunick, M.H., 2009. Whey protein production and utilization: a brief history. In: Whey Processing, Functionality and Health Benefits, pp. 1–13. https://doi.org/10.1002/9780813803845.CH1.

Uhrínová, S., et al., 2000. Structural changes accompanying pH-induced dissociation of the β-lactoglobulin dimer†,‡. Biochemistry 39 (13), 3565–3574. https://doi.org/10.1021/BI992629O.

U.S. Dairy Export Council., 2004. Reference Manual for U.S. Whey and Lactose Products. Available at: https://www.thinkusadairy.org/resources-and-insights/resources-and-insights/product-resources/reference-manual-for-us-whey-and-lactose-product (Accessed: 3 January 2025).

USDEC, 2012. Reference manual for US whey and lactose products. Available at: http://www.usdec.org. Accessed January 2, 2025.

Ushida, Y., et al., 1999. Possible chemopreventive effects of bovine lactoferrin on esophagus and lung carcinogenesis in the rat. Jpn. J. Cancer Res. 90 (3), 262–267. https://doi.org/10.1111/J.1349-7006.1999.TB00742.X.

Vacca Smith, A.M., Bowen, W.H., 2000. The effects of milk and kappa-casein on salivary pellicle formed on hydroxyapatite discs in situ. Caries Res. 34 (1), 88–93. https://doi.org/10.1159/000016558.

Vasbinder, A.J., de Kruif, C.G., 2003. Casein–whey protein interactions in heated milk: the influence of pH. Int. Dairy J. 13 (8), 669–677. https://doi.org/10.1016/S0958-6946(03)00120-1.

Venetsaneas, N., et al., 2009. Using cheese whey for hydrogen and methane generation in a two-stage continuous process with alternative pH controlling approaches. Bioresour. Technol. 100 (15), 3713–3717. https://doi.org/10.1016/J.BIORTECH.2009.01.025.

Vesa, T.H., Korpela, R., Marteau, P., 2000. Lactose intolerance. J. Am. Coll. Nutr. 19 (2 Suppl. 1), 165S–175S. https://doi.org/10.1080/07315724.2000.10718086.

Wakabayashi, H., Yamauchi, K., Takase, M., 2006. Lactoferrin research, technology and applications. Int. Dairy J. 16 (11), 1241–1251. https://doi.org/10.1016/J.IDAIRYJ.2006.06.013.

Walstra, P., M Wouters, J.T., Geurts, T.J., 2006. Milk components. In: Dairy Science and Technology, second ed. Taylor & Francis Group, pp. 17–108.

Walzem, R.L., Dillard, C.J., German, J.B., 2002. Whey components: millennia of evolution create functionalities for mammalian nutrition: what we know and what we may be overlooking. Crit. Rev. Food Sci. Nutr. 42 (4), 353–375. https://doi.org/10.1080/10408690290825574.

Wang, G., Guo, M., 2019. Manufacturing technologies of whey protein products. In: Guo, M. (Ed.), Whey Protein Production, Chemistry, Functionality, and Applications, first ed. Wiley, Hoboken, NJ, pp. 13–32.

de Wit, 1998. Nutritional and functional characteristics of whey proteins in food products. J. Dairy Sci. 81, 597–602.

de Wit, J.N., 2001a. From milk to whey. In: Lecturer's Handbook on Whey and Whey Products, first ed. European Whey Products Association, Belgium, pp. 8–15.

de Wit, J.N., 2001b. Nutrients in whey. In: Lecturer's Handbook on Whey and Whey Products, first ed. European Whey Products Association, Belgium, pp. 16–20.

Ustunol, Z., 2014. Physical, chemical, and processing induced changes in proteins. In: Ustunol, Z. (Ed.), Applied Food Protein Chemistry. https://doi.org/10.1002/9781118860588.ch4.

Walstra, P., Wouters, J.T., Geurts, T.J., 2006. 'Milk Components', Dairy Science and Technology, Second Edition. Taylor & Francis Group, Florida, pp. 17–108.

Yadav, J.S.S., et al., 2015. Cheese whey: a potential resource to transform into bioprotein, functional/nutritional proteins and bioactive peptides. Biotechnol. Adv. 33 (6), 756–774. https://doi.org/10.1016/J.BIOTECHADV.2015.07.002.

Yamauchi, R., Ohinata, K., Yoshikawa, M., 2003. β-lactotensin and neurotensin rapidly reduce serum cholesterol via NT2 receptor. Peptides 24 (12), 1955–1961. https://doi.org/10.1016/J.PEPTIDES.2003.10.003.

Yamauchi, R., et al., 2006. Effect of β-lactotensin on acute stress and fear memory. Peptides 27 (12), 3176–3182. https://doi.org/10.1016/J.PEPTIDES.2006.08.009.

Yang, S.T., Silva, E.M., 1995. Novel products and new technologies for use of a familiar carbohydrate, milk lactose. J. Dairy Sci. 78 (11), 2541–2562. https://doi.org/10.3168/JDS.S0022-0302(95)76884-9.

Yvon, M., et al., 1994. Effects of caseinomacropeptide (CMP) on digestion regulation. Reprod. Nutr. Dev. 34 (6), 527–537. https://doi.org/10.1051/RND:19940602.

Chapter 4

Whey of different species and nutritional aspects

Omar Amin Alhaj and Nour Amin Elsahoryi
Department of Nutrition, Faculty of Pharmacy and Medical Sciences, University of Petra, Amman, Jordan

1. Introduction

Whey is the liquid by-product produced during the cheese-making process, when milk curdles and separates into curds and whey (Patel, 2015). This liquid, rich in proteins, lactose, and minerals, has gained considerable attention for its nutritional value and functional properties (Deeth and Bansal, 2019). In 2019, the estimated global market value of whey protein was 8 billion USD, with a projected reach of 18 billion USD by 2027, this is catalyzed by a wave of health consciousness brought on by the COVID-19 pandemic (Kelkar, 2023). Scientific research has increasingly focused on whey proteins and their bioactive peptides as therapeutic nutraceuticals, as these compounds demonstrate significant antiviral and antiinflammatory properties while offering the advantages of extensive availability and well-documented biosafety profiles for viral disease prevention and treatment (Gallo et al., 2022). Furthermore, this protein has the highest protein digestibility and absorption rate available on the market; the protein digestibility corrected amino acid score (PDCAAS) is 1.0, and its biological value (BV) is 104 (Kelkar, 2023). Specifically, whey proteins are an excellent supply of amino acids that are necessary for immune system maintenance, tissue repair, and muscle growth (Alhaj and Kanekanian, 2014; Deeth and Bansal, 2019; Khan and Selamoglu, 2020; Rashmi et al., 2022). Whey is important for human nutrition because of its important composition, particularly for athletes, the elderly, and people with specific medical issues (Alhaj et al., 2024b; Kelkar, 2023). There are a few different forms of whey protein, including powders (Isolates, concentrates, and specialty products), liquids (ready-to-drink nutritional solutions, and shakes), and functional ingredients incorporated in cereals, snacks, and desserts, often to pump up protein content (Kelkar, 2023).

While whey produced from cow's milk accounts for most of the world's total supply, they can also be produced from other mammalian milk from sheep, goats, buffalo, and camels (Al-Haj and Al Kanhal, 2010; Deeth and Bansal, 2019; Guha et al., 2021). It varies considerably, depending on species, but often includes water, lactose, minerals and vitamins (Guha et al., 2021). In fact, the most common and extensive whey protein products are hydrolyzed whey protein (HWP), whey protein isolate (WPI), and whey protein concentrate (WPC) (Jovanović et al., 2005; Kelkar, 2023). Whey proteins are a good source of complete protein as they contain all essential amino acids, such as branched-chain amino acids (BCAAs), including leucine, isoleucine, and valine, which play an important role in energy production and muscle repair during exercise (Boland, 2011; Kelkar, 2023). Whey is a crucial nutrition source for humans, mainly through whey proteins, which comprise about 20% of total milk proteins (Jovanović et al., 2005), and due to its high digestibility, the protein digestibility corrected amino acid score (PDCAAS) is 1.0, and its biological value (BV) is 104 (Kelkar, 2023), whey is considered by several research to be an effective supplement to foster muscle protein synthesis after sport conferring rapid absorption (Lampová et al., 2024; Schoenfeld and Aragon, 2018). In another researches, Lopes Gomes et al. (2017) and Luhovyy et al. (2007) have shown that whey helps with weight management, this is through increasing fat burning, decreasing hunger, and making consumers feel fuller. Whey is reported to have nutritional and functional properties (Deeth and Bansal, 2019) as it contains different concentrations of important proteins, including α-lactalbumin, β-lactoglobulin, serum albumin, immunoglobulins, and Lactoferrin from different animal species as shown in Fig. 4.1; in addition to having other important components each contributing unique nutritional and physiological benefits (Jovanović et al., 2005). Among these milk proteins, α-lactalbumin and β-lactoglobulin stand out as particularly significant biological

Whey. https://doi.org/10.1016/B978-0-443-27648-4.00023-2

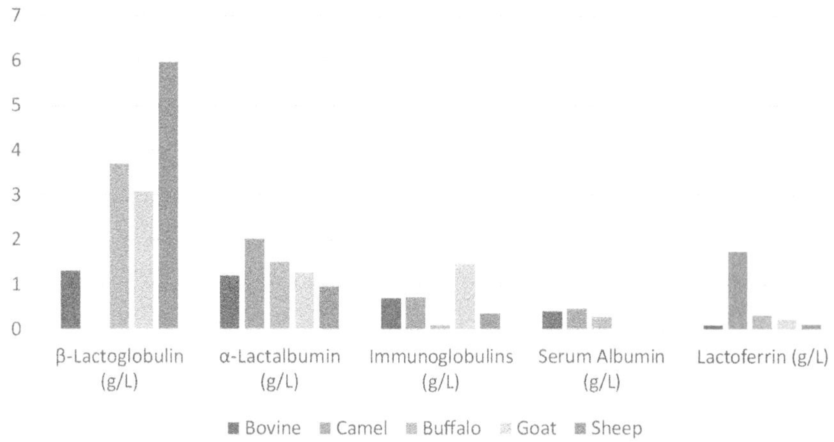

FIGURE 4.1 Comparative analysis of whey proteins in bovine, camel, buffalo, goat, and sheep milk. *Results adopted from Pillai et al. (2024); Omar et al. (2016); Konuspayeva et al. (2007); Sindhu and Arora (2011); Arora and Khetra (2017); Ruprichová et al. (2014); Navarro et al. (2018); Campanella et al. (2009).*

components, renowned for their exceptional digestibility and capacity to enhance muscle function and recovery (Lampová et al., 2024). Research has demonstrated that the immunoglobulins found in whey enhance immune system function by stimulating antibody production and increasing existing immunoglobulin levels (Ha et al., 2021). The nutritional profile of whey extends beyond proteins to include vital minerals such as calcium, magnesium, and phosphorus, which play crucial roles in bone development and various metabolic processes (Amalfitano et al., 2024), these minerals and whey proteins can also change some functional properties and stability of milk (Alhaj et al., 2011, 2022b; Hailu et al., 2016; Elkot, 2019). While cow's milk whey remains the most extensively studied and commonly utilized variety (Alhaj et al., 2007; Brunner, 1981), researchers are increasingly examining whey derived from other species, including goats, sheep, buffalo, and camels, with each type displaying distinct protein compositions and functional properties (Deeth and Bansal, 2019; Arain et al., 2023). Notably, goat's milk whey protein has emerged as a promising alternative for individuals with lactose intolerance or milk protein allergies, owing to its superior digestibility and reduced allergenicity compared with cow's milk whey (Alkaisy et al., 2023). Additionally, buffalo milk whey distinguishes itself through higher protein and fat content relative to cow's milk whey, potentially offering enhanced nutritional benefits (Khetra et al., 2022). Exploring individual animal-specific whey can also reveal unique characteristics of each meat that may suit specific dietary and health needs. This heterogeneity of protein concentration within whey products allows greater customization of whey produced goods to different consumer preferences and or health concerns, thus warranting the examination of this space to be a more niche area of ongoing study.

2. Composition of whey from different species

Dairy products are highly consumed worldwide and tend to be one of the essential products in some diet such as Mediterranean diets (Sikalidis et al., 2021). Globally, the majority of milk production (81%) comes from cows followed by buffalo (15%), but there is also small production from minor species, including goat (2%), sheep (1%), and camel (0.4%) (FAO, 2024). Whereas the remaining percentage of milk production comes from less dominant dairy animals including equines, yaks, and reindeer (FAO, 2024). An increased interest in alternative species of dairy livestock, including sheep, goats, and camels, has come to the attention of researchers in recent years due to their potential health claims, high digestibility, and nutritious value compared with cow's milk (Guha et al., 2021). Differences in the functional properties of these forms of whey protein affect their use in food systems (Alshuniaber et al., 2021). Milk composition including whey from different mammalians were reported to be influenced by several factors including breed, age, diet, stage of lactation, parity, farming practices, physical environment, and season (Alhaj et al., 2022a). All these factors together are responsible for producing a wide variety of milk products. Furthermore, milk composition varies between species depending on the neonate requirement (McGrath et al., 2016). Moreover, vary in terms of its composition depending on genetic and environmental factors (Guha et al., 2021). In general, ruminant milk contains higher total solids (fat, protein, and ash) (Medhammar et al., 2012) compared with milk of a nonruminant nature (mare, donkey, and humans) that contains higher levels of lactose (Martini et al., 2014; Guo et al., 2007). Moreover, the ratio of casein to whey protein is different from one species to another; for instance, horse milk has a ratio close to 50:50, whereas human milk has a different ratio of 40:60, while cow's, goat, sheep, and camel's milk is approximately 80:20.

2.1 The importance and nutritional values of bovine whey

Bovine whey is historically viewed as a by-product, but advances in dairy science and technology research have shown that bovine whey can be a functional food ingredient (Buchanan et al., 2023; Alhaj and Kanekanian, 2014). Bovine whey was reported by several researchers to be an appealing raw material common in food products, dietary additives, and industry applications, due to its nutritional composition and various potential functional properties (Lampová et al., 2024; Mehra et al., 2021; Schoenfeld and Aragon, 2018). Whey is from a bovine source and complex and contains many components, some of which play a key role in its nutritional and functional properties. Bovine liquid whey obtained from cheese making consists of 94.3% water and 5.7% total solids, making water the main component of whey; total solids in whey consist of lactose (4.3%), whey proteins (0.8%), minerals (0.5%), and fat (0.1%) (González-Weller et al., 2023). Moreover, cow's milk whey has lower cholesterol (25.6 mg/100 g) than camel whey (34.5 mg/100g), which makes it a more attractive health food (Konuspayeva et al., 2008; Gorban and Izzeldin, 1999).

2.1.1 Bovine whey proteins

Whey proteins are classified as complete proteins because it has all nine essential amino acids, especially branched-chain amino acids (BCAAs) leucine, isoleucine, and valine (Kelkar, 2023; Boland, 2011). Based on their abundance, whey protein can be fractionated into further components; two of these proteins are also the most abundant ones, namely β-lactoglobulin (1.3 g/L), and α-lactalbumin (1.2 g/L) (Pillai et al., 2024) as shown in Fig. 4.1. The other proteins, such as immunoglobulins (0.7 g/L), bovine serum albumin (0.4 g/L), and lactoferrin (0.1 g/L) as well as a number of other components, including glycomacropeptide (1.2 g/L), protease-peptones (0.2–0.4 g/L), and bovine lactoperoxidase (0.03 g/L) (Pillai et al., 2024), each of which has diverse health benefits covering areas of immune support, antimicrobial, and antioxidant functions (Alhaj and Kanekanian, 2014; Pillai et al., 2024).

2.1.2 Lactose content in bovine whey

Lactose is the main sugar found in whey and represents roughly 4.3% from cheese-making of the composition of bovine whey (González-Weller et al., 2023). The full-term neonate is able to digest up to 60–70 g of lactose/day (Heine et al., 2017). Young infants, however, do not fully absorb all lactose present in the breast milk that they ingest on a daily basis, a condition termed physiological lactose malabsorption. Nonabsorbed lactose is fermented in the colon producing short-chain fatty acids and gases, including hydrogen, carbon dioxide and methane (Heine et al., 2017). Diabetes mellitus lactose intolerance is a common carbohydrate source, as it is a milk disaccharide (Malik and Panuganti, 2023), moreover, several approaches were adopted to overcome this challenge using different mammalian milks (Alhaj et al., 2020; Liu et al., 2021; Almasri et al., 2024). Nevertheless, in such dairy types of processed whey products, such as whey protein isolates, and derived from dairy, the lactose content is often reduced to an insignificant level, which makes it probiotic-safe for people with lactose sensitivity (Li et al., 2023; Alhaj et al., 2024c). Both lactose-free milk and milk that have undergone lactose hydrolysis were reported to help decrease the risk of nutritional lack while providing essential nutrients for lactose-intolerant and nonlactose sensitive people as alternatives to whole milk (Sharp et al., 2021). In addition, low-lactose and lactose-free dairy products are the most economical calcium sources for lactose-intolerant people in comparison with other calcium-rich foods including selected vegetables and mineral water (Taeger and Thiele, 2021).

2.1.3 Minerals content in bovine whey

Whey protein provides several essential minerals, including calcium, magnesium, potassium, sodium, and phosphorus (Gordon, 2014). Whey mineral profiles are affected by the species of dairy animals, but to a lesser degree than milk profiles, since the primary nutrient levels of whey among species do not vary greatly (Al-Haj and Al Kanhal, 2010; Alhaj et al., 2022a; Amalfitano et al., 2024; Gordon, 2014). Although the concentrations of macro-minerals are mostly species-dependent as shown in Fig. 4.2 (Al-Haj and Al Kanhal, 2010). The comparative analysis in Fig. 4.2 and 4.3 showed that mineral contents in bovine whey are comparable with goat whey, but have the lowest mineral contents among other

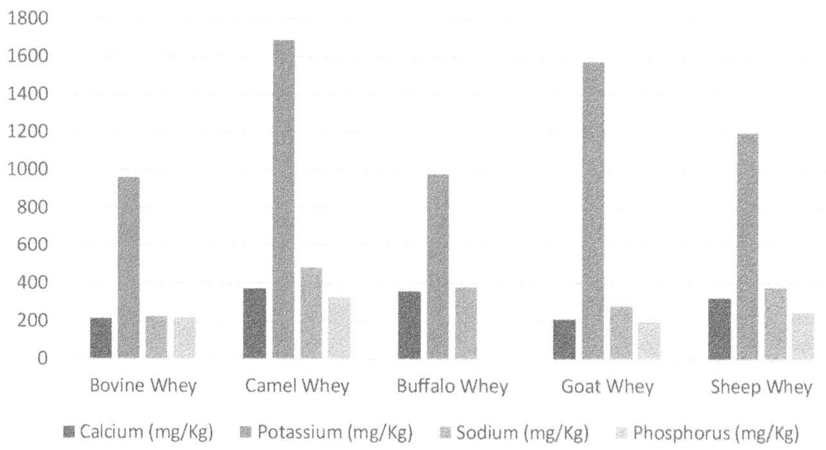

FIGURE 4.2 Comparative analysis of calcium, potassium, sodium, and phosphorus in whey from bovine, camel, buffalo, goat, and sheep milk. *Results adopted from Amalfitano et al. (2024); Bornaz et al. (2009).*

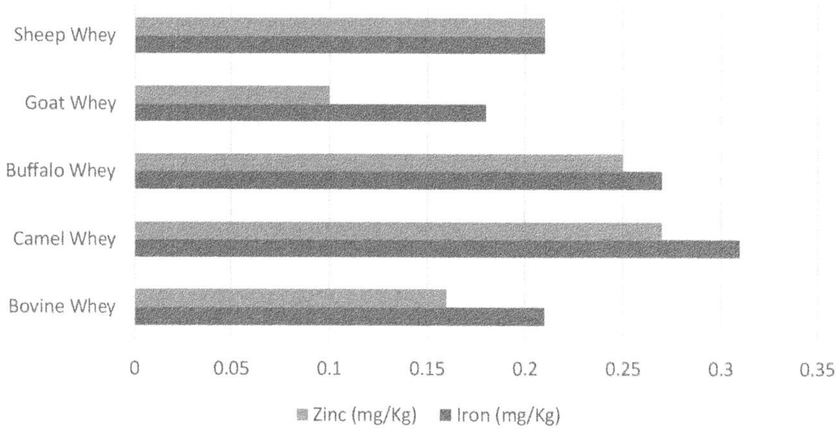

FIGURE 4.3 Comparative analysis of zinc, and iron in whey from bovine, camel, buffalo, goat, and sheep milk. *Results adopted from Amalfitano et al. (2024).*

animal species, including camel, buffalo, and sheep, except for phosphorus. Between whey from ruminants and that from dromedary camels, camel whey features higher sodium (+52%), potassium (+43%), and calcium (+34%) levels, and lower magnesium (−45%), while phosphorus and sulfur levels remain the same (Amalfitano et al., 2024). Moreover, whey from large ruminants (e.g., Bovine subfamily) contains 21% less phosphorus, 17% less sulfur, and 30% less potassium than from small ruminant (e.g., Caprine subfamily) (Amalfitano et al., 2024). Sheep and buffalo whey's macro mineral contents were found to be higher than goats and cow's whey, except for potassium, in the same study. It is where these minerals are indispensable in various of physiological functions for for example, bone health, nerve function and electrolyte balance (Al-Haj and Al Kanhal, 2010). Whey is high in minerals including calcium, which is needed to help prevent conditions such as osteoporosis and other ailments that affect the bones (Kerstetter et al., 2015; Gordon, 2014).

2.1.4 Vitamins content in bovine whey

Whey is also a source of various water-soluble and fat-soluble vitamins. For example, your B vitamins (B1, B2, B6, and B12), all contribute to the energy metabolic process and the formation of red blood cells (Gordon, 2014). Additionally, whey contains small amounts of fat-soluble vitamins, amongst others, A and D, which regulate immune function and bone health (Hewison, 2012).

2.1.5 Bovine whey utilization in human nutrition and industry applications

Bovine whey is employed in human nutrition and industrial applications because of its nutritional composition and functional properties (Madureira et al., 2007). The most widely exploited use of bovine whey is in human nutrition, either in the form of whey protein concentrate (WPC), whey protein isolate (WPI), or the diverse forms of whey protein hydrolyzate (Castro et al., 2019). These protein-rich products are frequently used for sports nutrition, weight management, and by people looking to boost their daily protein intake (Madureira et al., 2007). Whey protein is very helpful in building muscles, for the postworkout recovery phase, and for the preservation of lean body mass. It is also found in some meal replacement drinks, protein bars, and shakes (Castro et al., 2019). The immunoglobulins in whey have been found to boost the immune system, along with lactoferrin, which has possible antiviral activity, so whey has potential appeal as a health supplement due to the presence of bioactive compounds (Ostertag and Hinrichs, 2023).

Apart from protein isolates, whey finds use not only in the food market. Much of the dairy products we consume, including yogurt, ice cream, and a variety of beverages depend on whey as a key ingredient, adding texture, taste, and nutritional content (Beucler et al., 2005). Whey protein has emulsification, stabilization, and foaming functions, and can be used in processed foods, including bakery items, soups, sauces, and dressings (Lajnaf et al., 2018). Whey permeate, a lactose and minerals-rich syrup, is also used as a flavor enhancer in processed foods (Beucler et al., 2005).

Industrial utilization of whey goes beyond food manufacturing. Whey is used in the cosmetic and pharmaceutical fields for its moisturizing and skin-restoring effects (Augustyniak et al., 2023). Whey proteins have also been incorporated into marine-derived packaging composites due to their superior structural properties (Jiang et al., 2024; Chen et al., 2019). Other uses such as the development of functional foods and bioprocesses that can utilize the valuable bioactive ingredients of whey are also studied (Yiğit et al., 2023).

2.2 Camel whey

2.2.1 The importance and nutritional values of camel whey

Camel milk is described as the white treasure in the desert because of its important composition and potential health properties (Alhaj et al., 2024a; Al-Haj and Al Kanhal, 2010; Alhaj, 2020). Cheese production from camel milk generates whey as a by-product, which has attracted considerable interest in recent years owing to its distinctive compositional profile and health benefits (Rocha and Guerra, 2020). The liquid, filled with proteins, vitamins and minerals, is known to be a nutritious option (Konuspayeva, 2020). A global shift to use of alternative dairy products due to lactose intolerance, dietary beliefs or preference for healthier foods has opened up a new opportunity for camel whey (Elsahoryi and Al-Sayyed, 2020; Alhaj et al., 2022a). Camel whey also has an important cultural role, especially in those areas with little water, where camels become the main milk source (Alhaj et al., 2024b, 2024c; Wernery, 2006; El-Agamy, 2006). It has been used for its health-promoting properties for centuries (Alhaj, 2020). Camel whey is rich in bioactive compounds, nutrients and is essential for health. With its proprietary blend formula, it is highly attractive to individuals looking for functional foods that help improve immune function, repair muscle faster, and serve as a potent source of protein (Hassanein, 2023).

Camel whey is the food and nutraceutical fitting alternative for worldwide market growth for health-focused foods (Khalesi et al., 2017). The response to demand for nutritious ingredients can also be met with the development of new products containing camel whey, as quality products become commodities. Therefore, a comprehensive examination of camel whey's composition and nutritional properties is essential to its effective incorporation into numerous culinary applications.

2.2.2 Composition of camel whey

Camel whey contains a bundle of dietary benefits and is a versatile food ingredient; however, consumers need to be aware of its composition to fully appreciate its various applications. The camel whey composition is mainly dependent on camel milk composition. In the most recent comprehensive meta-analysis of camel milk composition, which included 79 studies (Alhaj et al., 2022a). The analysis of this study found that camel milk had significant differences in macronutrient composition, including protein at 3.17% ($p = 0.001$); fat at 3.47% ($p = 0.001$); lactose at 4.28% ($p = 0.001$); and ash at 0.78% ($p = 0.001$). An earlier review, which included 40 studies, reported similar values for protein (3.1%), fat (3.5%), and ash (0.79%), but higher values for lactose (4.4%) (Al-Haj and Al Kanhal, 2010). Another study reported a higher content of macronutrients, including protein, which was 3.35%, fat was 3.82%, and lactose was 4.46%, while ash content remained constant at 0.79% (Konuspayeva et al., 2009). Such variations in camel milk and whey composition are due to the number of studies considered, sample sizes, analytic methods applied, livestock management systems, feeding conditions, water availability, camel breeds, geographical regions, and seasonality (Alhaj et al., 2022a). This variation in camel milk composition shall eventually reflect varying whey composition.

2.2.2.1 Camel whey proteins

Whey proteins are the second protein in camel milk, with about 20%–25%, which consists of 0.63%–0.80% of total protein (Khaskheli et al., 2005; Mehaia et al., 1995). Camel milk has low or deficient β-lactoglobulin content, while α-lactalbumin is the predominant component; this is because camel's whey proteins contain qualitatively different whey proteins versus human and cow milk. In contrast, the β-lactoglobulin (50%) is the predominant component in cow milk whey, while α-lactalbumin (25%) is the second main component (Al-Haj and Al Kanhal, 2010). Camel whey proteins contain α-lactalbumin (2.01 ± 0.02 mg/mL), lactoferrin (1.74 ± 0.06 mg/mL), serum albumin (0.46 ± 0.01 mg/mL) and Immunoglobulins (0.718 ± 0.330 mg/mL) (Omar et al., 2016; Konuspayeva et al., 2007) as shown in Fig. 4.1.

α-lactalbumin from camel milk, on the other hand, was closely associated with α-lactalbumin from goat, human and cow milk (Beg et al., 1985) and corresponded to similar molecular weight (14.43 kDa) (Lajnaf et al., 2017). Variations of the amino acids sequence in camel milk α-lactalbumin compared with other species, such as goat or cow, are also unique (Al-Haj and Al Kanhal, 2010). Camel whey has an important biological value and therefore is an effective means of supplying people who wish to supplement their protein intake (Alhaj et al., 2024b). Moreover, it contains important bioactive peptides that provide potential health benefits including antimicrobial properties, ACE-inhibitory, antioxidant activities, and other functional properties (Alshuniaber et al., 2022; Alhaj, 2020; Vincenzetti et al., 2022; Abdelazez et al., 2024).

2.2.2.2 Fat in camel whey

While camel whey is considered as an important source of proteins, the fat content should not be neglected. Dromedary camel milk (Majaheim and Najdi breeds) has also been reported to have a lower content of carotene and short chain fatty acids than bovine milk (Bakry et al., 2021; Konuspayeva et al., 2008). The whiter color of camel fat (Abu-Lehia, 1989) may be due to a lower carotene concentration in camel milk fat. Furthermore, dromedary camel milk fat was rich in long chain fatty acids, whereas bovine milk fat contained only traces concentration of them (Xiao et al., 2022). Similarly, camel milk has higher mean concentrations of unsaturated fatty acids (43%) (Haddadin et al., 2008). Human milk fat had higher content of unsaturated fatty acids than fat from cows and camels (Shamsia, 2009). Camel milk fat contains a smaller proportion of saturated acids (67.7%) than bovine milk fat (69.9%) (Konuspayeva et al., 2008; Bakry et al., 2021). Polyunsaturated fats, in particular, provide potential cardiovascular health and other physiological functions (Kapoor et al., 2021). On the other hand, camel whey (34.5 mg/100g) was reported to have a higher cholesterol content (25.6 mg/100 g) than cow's milk whey (Gorban and Izzeldin, 1999; Konuspayeva et al., 2008).

2.2.2.3 Minerals content in camel whey

Camel whey has also received attention as a possible source of nutrition and has received several claims as a therapeutic supplement due to its mineral composition. The body also requires macro elements such as calcium, magnesium, potassium, and phosphorus in high amounts, which participate in bone metabolism, muscle contraction and blood pressure regulation (Al-Haj and Al Kanhal, 2010), and are found in high amounts in camel whey (Amalfitano et al., 2024). Camel whey has characteristic macro element composition with several of the elements present at relatively high concentrations compared with other dairy species. The analysis of Bornaz et al. (2009) data revealed that camel whey calcium (43 mg/100 g), and phosphorus (33 mg/100 g) were markedly influenced by animal species, region and dietary factors (Alhaj et al., 2022a). These concentrations were also found (Bornaz et al., 2009) to exceed those in bovine whey calcium (26 mg/100 g), and phosphorus (22 mg/100 g), sheep whey calcium (33 mg/100 g), and phosphorus (25 mg/100 g), goat whey calcium (22 mg/100 g), and phosphorus (20 mg/100 g) as shown in Fig. 4.2, especially for calcium concentration that camel whey has approximately 25% or more. The discrepancies of calcium, and phosphorus is a result of diet or feed of animals, breed of the animal or the analysis methods, which may have affected the accuracy of reported values (Alhaj et al., 2022a; Al-Haj and Al Kanhal, 2010). Camel milk is rich in chloride since camels feed on such plants as Atriplex and Acacia, which are relative to salts (Khaskheli et al., 2005). The whey mineral content depends on the species, though not as much as in milk, because the differences in nutrient levels in whey between different species of mammals are not as substantial (Alhaj et al., 2022a). Like in the case of milk, the effects of species are mostly expressed in macro-mineral concentrations. Compared with ruminant milk whey (Fig. 4.2), dromedary camel whey contains significantly higher Na (+52%), K (+43%), and Ca (+34%), and less Mg (−45%) but similar P and S contents (Amalfitano et al., 2024). Furthermore, microminerals of dromedary camel whey contained higher concentrations of Fe, Zn, B, and particularly Sr than ruminant species but lower in Cu and Si. Like camel milk, those two minerals, Li can be identified in whey above the quantification limit while Mo is in concentration below the limit of quantification (Amalfitano et al., 2024). The comparative analysis in Fig. 4.2 and 4.3 showed that mineral contents in camel whey are the highest among other animal species, including bovine, buffalo, goat, and sheep, particularly for calcium, potassium, sodium, iron, and zinc.

2.2.3 Camel whey: An emerging force in nutraceutical development

The nutraceutical industry is evolving and camel whey is a promising ingredient with huge market potential (Faye and Konuspayeva, 2024). More recently, this versatile component has been increasingly used in comprehensive health supplements and can be a beneficial addition to a manufacturer's product portfolio. One particularly useful function of camel whey lies in the development of protein-rich supplements and meal replacement formulations meant to answer the growing consumer demand for nutritionally complete and convenient dietary options (Badr et al., 2017). Moreover, the high content of minerals in camel whey can make it a promising potential for the management of COIVD-19 (Altooq et al., 2022). Moreover, based on the aforementioned investigation, increased research is finding further health benefits of fermented dairy derivatives, thereby expanding the arena of its use in the nutraceutical sector.

2.2.4 New products from camel whey

Industry innovators are advancing the development of specialized products that leverage camel whey's nutritional properties (Seifu, 2023). A prominent category in this space is wellness-focused beverages that incorporate camel whey, targeting health-conscious consumers (Al-Thaibani et al., 2024). They include smoothies, ready-to-drink protein shakes, and fortified juices, all of which are formulated to provide the health benefits of camel whey in a convenient form (Elsahoryi and Al-Sayyed, 2020; Al-Thaibani et al., 2024). Camel whey is also being recognized by the cosmetics industry owing to its humectant and antiinflammatory properties, making it a supplement to the skin care products (Mehta and Agrawal, 2020). As awareness of camel whey continues to grow, so too is expected a wave of new, creative products, each designed to meet the needs and preferences of different consumers (Seifu, 2023).

Camel whey is emerging as a multifunctional component in food, as well as nutraceutical and product development, thereby demonstrating its potential in enhancing nutrition while providing health benefits.

2.2.5 The promise of camel whey as a nutritional supplement

With increasing demand for alternative protein sources and functional foods, camel whey should pave the way for an exciting new ingredient in the future of nutrition (Al-Thaibani et al., 2024). Its unique and specific compositional characteristics allow it to be an alternative to classical dairy, justifying its entry as one of the most leading healthy foods.

Future research on benefits of camel whey, especially milk and so on for immune support and muscle recovery, will fuel consumer interest and increase its market potential (Faye and Konuspayeva, 2024). These health benefits are likely to increase the demand for camel whey, and encourage manufacturers to introduce novel products that utilize its nutritional properties (scientific validation of these benefits is anticipated) (Alhaj, 2020; Alhaj et al., 2024a).

Camel whey is also compatible with sustainable agricultural practices. Utilizing camel whey helps reduce food waste while increasing the economic viability of camel milk production, which is crucial, as camel herding is integral to many livelihoods in arid regions (Seifu, 2023). This potential makes camel whey attractive to producers and consumers alike, particularly in light of sustainable food systems (Faye and Konuspayeva, 2024). In conclusion, camel whey has great potential for the future of nutrition. The growing awareness regarding the unique aspects of camel milk and the associated health advantages associated with its use adds up to the camel whey's potential to be known by consumers. Camel whey, a by-product of camel milk, has the potential to not only meet changing health and wellness trends for an increasingly nutrition-driven world population but also innovative uses, and its workability with respect to sustainability.

2.3 The importance and nutritional values of buffalo whey

Buffalo milk is a rich source of nutrition, but following soft and hard cheese production, the liquid fraction of buffalo milk whey, which remains is considered the dairy by-product with nutritional and functional properties. More recent studies (Abd El-Salam and El-Shibiny, 2011) have also examined the compositional and nutritional properties of buffalo whey, highlighting increasing awareness of its unique nature and applications from food manufacturing to pharmaceutical processing.

2.3.1 Buffalo whey protein

Sindhu and Arora (2011), González-Weller et al. (2023), and Pandya et al. (2006) accorded buffalo whey with a higher protein concentration of 0.9% than bovine whey of 0.8%. The whey proteins in buffalo are shown in Fig. 4.1, the content of β-lactoglobulin (β-LG), α-lactalbumin (α-LA), serum albumin (0.29 g/L), immunoglobulins (0.1 g/L milk), and lactoferrin (0.32 g/L) in the form of the protein composition of buffalo whey has been reported by Sindhu and Arora (2011), Arora and Khetra (2017). The quantity of β-LG was significantly high in buffalo whey (3.4–4 g/L) compared with bovine concentrations of 3.1–3.8 g/L as indicated by Sindhu and Arora (2011) (Ranjit Aich et al., 2014). In the same study, the α-LA concentration of buffalo whey to be 1.5 g/L, which explains why buffalo whey has a better protein profile than bovine whey (0.7–1.0 g/L).

2.3.2 Fat and lactose in buffalo whey

Amalfitano et al. (2024), buffalo whey shows more fat content (1.3%) than bovine whey (1%), goat whey (0.8%), and sheep whey (1.05%), however less than camel whey (2%). Buffalo whey has higher lactose (4.5%) than bovine whey (3.1%), goat whey (3.3%), and camel whey (3.7%), but less than sheep whey (4.9%) (Amalfitano et al., 2024). Due to this, buffalo whey contains comparatively more fructose than bovine whey and hence may be used as a substrate for fermentation processes (Sindhu and Arora, 2011).

2.3.3 Minerals content in buffalo whey

Ash concentration is greater for buffalo whey (0.42%) than for bovine whey (0.32%) but less than for goat whey (0.44%), sheep whey (0.49%) and camel whey (0.58%) (Amalfitano et al., 2024). Amalfitano et al. (2024) found that the macrominerals, which are essential in buffalo whey were distinct from other species. The sodium content in buffalo whey is higher (380 mg/kg) than that reported for cow whey (225 mg/kg), and goat whey (281 mg/kg), but lower than sheep whey (381 mg/kg), and camel whey (483 mg/kg) as shown in Fig. 4.2. Calcium content in buffalo whey is higher (358 mg/kg) than that reported for cow whey (216 mg/kg), goat whey (213 mg/kg), and sheep whey (323 mg/kg), but lower in camel whey (372 mg/kg) as shown in Fig. 4.2. Moreover, potassium concentration was reported to be higher in buffalo whey (978 mg/kg) than cow whey (962 mg/kg), but lower than goat whey (1574 mg/kg), sheep whey (1198 mg/kg), and camel whey (1689 mg/kg) as shown in Fig. 4.2. Furthermore, the essential microminerals of different species were detailed by Amalfitano et al. (2024). The iron content in buffalo whey is higher than (0.27 mg/kg) than cow whey (0.21 mg/kg), goat whey (0.18 mg/kg), and sheep whey (0.21 mg/kg), but lower in camel whey (0.31 mg/kg) as shown in Fig. 4.3. The zinc content in buffalo whey is higher than (0.25 mg/kg) than cow whey (0.16 mg/kg), goat whey (0.10 mg/kg), and sheep whey (0.21 mg/kg), but lower in camel whey (0.27 mg/kg) as shown in Fig. 4.3. Camel's whey contains a higher mineral composition (Fig. 4.2 and 4.3) than other dairy animals such as cow, goat, and sheep with high retention rate. Such high mineral concentrations could partially account for the poor rennet-coagulating ability of cheese made from this milk as compared with one made from ruminant and/or nonruminant species when traditional bovine rennet is employed in the cheese-making practice (Bittante et al., 2022).

2.4 The importance and nutritional values of sheep/goat whey

Research indicates that sheep/goat whey protein has established itself as an important functional food ingredient, offering substantial nutritional and health advantages (Hernández-Ledesma et al., 2011). Studies have demonstrated that this protein source contains an exceptional profile of essential amino acids that exceeds the nutritional quality of both egg and soy proteins (Kerasioti et al., 2018). The protein concentration is particularly noteworthy, with analysis showing 80 g of protein per 100 g of product (Kerasioti et al., 2018). Within this protein profile, the composition includes β-lactoglobulin as the main component at 47g, followed by α-lactalbumin at 14 g, glycomacropeptide at 13 g, and serum albumin at 3 g (Kerasioti et al., 2018; Hernández-Ledesma et al., 2011).

The nutritional profile extends beyond protein content, featuring substantial mineral concentrations per 100 g: 415 mg of calcium, 397 mg of potassium, 319 mg of phosphorus, 157 mg of sodium, and 73 mg of magnesium (Kerasioti et al., 2018). Additionally, the product contains 10 g of carbohydrates and 4 g of fats, contributing to its comprehensive nutritional profile (Kerasioti et al., 2018). Of particular significance is the presence of sulfur-containing amino acids, specifically cysteine and methionine, as reported by Kerasioti et al. (2018). These amino acids serve a vital function in supporting antioxidant defense mechanisms (Egbujor et al., 2024), functioning either as direct reducing agents or as precursors in the body's production of glutathione (GSH), a crucial antioxidant compound (Wu et al., 2004). This antioxidant capacity positions sheep/goat whey protein as a potentially valuable resource in addressing oxidative stress-related conditions.

2.4.1 Sheep/goat whey proteins

Sheep milk was found to be composed of 76%–83% total casein and 17%–24% total whey proteins (Moatsou et al., 2005). Protein distribution in this protein shows whey protein has a high content that provides value to sheep whey as a dairy by-product. The casein-to-whey protein ratio in sheep milk follows a similar pattern to other ruminant species such as cows, goats, and camels, which typically maintain an approximate 80:20 ratios. This is notably different from nonruminant species such as horses, which have a 50:50 ratios, and humans have a ratio of 40:60 (Martini et al., 2014; Guo et al., 2007). The β-lactoglobulin and α-lactalbumin in sheep are 5.97 g/L, and 0.95 g/L, respectively, while their content in goat are 3.07 g/L, and 1.27 g/L, respectively (Ruprichová et al., 2014) as shown in Fig. 4.1. Moreover, the concentrations of Immunoglobulins and Lactoferrin in sheep are 0.35 g/L, and 0.12 g/L, respectively (Navarro et al., 2018); whereas their content in goat are 1.46 g/L, and 0.22 g/L, respectively (Campanella et al., 2009) as shown in Fig. 4.1.

2.4.2 Lactose and fat content in sheep/goat whey

Amalfitano et al. (2024) reported that highest lactose content (4.9%) contains sheep whey among all these species that were mentioned earlier, higher than that of buffalo, bovine, goat and camel whey (4.5, 3.1, 3.3 and 3.7 respectively). On the other hand, fat content, sheep whey contains 0.8% but less than goat whey (1.05%), but greater than buffalo whey (0.8%) and camel whey (2.0%) (Amalfitano et al., 2024). Sheep whey constitutes a valuable source of energy and nutritional components for dairy processing and product development.

2.4.3 Minerals content in sheep/goat whey

Sheep whey contains distinctive mineral concentrations compared with other species. The ash content in sheep whey (0.49%) is higher than that found in bovine whey (0.32%) and goat whey (0.44%), though lower than camel whey (0.58%) (Amalfitano et al., 2024). Regarding specific minerals, sheep whey contains 381 mg/kg of sodium, comparable with buffalo whey (380 mg/kg) but higher than cow whey (225 mg/kg) and goat whey (281 mg/kg) as shown in Fig. 4.2. Moreover, the potassium content in sheep whey (1198 mg/kg) is notably higher than cow whey (962 mg/kg), though lower than goat whey (1574 mg/kg) and camel whey (1689 mg/kg) (Amalfitano et al., 2024) as shown in Fig. 4.2.

3. Conclusion and future prospective

Whey composition across different animal species has been explored with much valuable implications to scientific research and commercial application. Cow's milk whey remains the leading worldwide producer, but the unique nutritional and functional properties of whey from other species represent compelling opportunities. For example, camel whey is rich in minerals and buffalo whey has higher protein, which offers a basis for some specialty nutritional and functional food uses.

Species-specific whey products have been in growing demand by the market landscape at large, a fact that indicates a growing demand for products that can address specialized dietary needs and health objectives, moreover, having potential development of species-specific functional foods. The diversification in sources of whey provides manufacturers the basis for developing differentiated products and targeted nutritional solutions. The industry is going beyond the traditional applications and is putting innovative approaches in both product development and therapeutic use. An expanded understanding of these species-specific characteristics represents a strong foundation for optimizing these dairy by-products for use to meet a wide spectrum of nutritional and health needs.

Future investigation is indicated in several key areas. Deeper analysis of bioactive compounds from alternative whey sources, and development of specialized processing methods should be additionally prioritized as research priorities. Finally, there is a need for clinical research to validate the health benefits of different whey sources and to assess the economic and environmental sustainability of other sources of whey production systems.

References

Abd El-Salam, M.H., El-Shibiny, S., 2011. A comprehensive review on the composition and properties of buffalo milk. Dairy Sci. Technol. 91, 663–699.

Abdelazez, A., Abd-Elmotaal, H., Abady, G., 2024. Exploring the potential of camel milk as a functional food: physicochemical characteristics, bioactive components, innovative therapeutic applications, and development opportunities analysis. Food Mater. Res. 4.

Abu-Lehia, I.H., 1989. Physical and chemical characteristics of camel milkfat and its fractions. Food Chem. 34, 261–271.

Al-Haj, O., Al Kanhal, H.A., 2010. Compositional, technological and nutritional aspects of dromedary camel milk. Int. Dairy J. 20, 811–821.

Alhaj, O., Kanekanian, A., 2014. Milk-derived bioactive components from fermentation. Milk Dairy Prod. Funct. Foods 237–288.

Alhaj, O.A., 2020. Exploring potential therapeutic properties of camel milk. In: Handbook of Research on Health and Environmental Benefits of Camel Products. IGI Global.

Alhaj, O.A., Abodoleh, G., Abu Jadayil, S., Irshaid, N., Mehta, B.M., Faye, B., Jahrami, H.A., 2024a. A systematic review and meta-analysis of the antihypertensive effects of camel milk hydrolysates. Int. J. Dairy Technol. 77, 1029–1043.

Alhaj, O.A., Ahmad, L., Alenezi, A.F., Abodoleh, G., Ghazzawi, H., Trabelsi, K., Bragazzi, N.L., Mehta, B.M., Faye, B., Jahrami, H.A., 2024b. Does total caloric count of camel milk differ by species, country, season and year of publication: a systematic review, meta-analysis and meta-regression. Int. J. Dairy Technol. 77, 15–34.

Alhaj, O.A., Altooq, N.J., Alenezi, A.F., Janahi, A.I., Janahi, M.I., Humood, A.M., Alrasheed, M.M., Bragazzi, N.L., Jahrami, H.A., Faye, B., 2022a. Camel milk composition by breed, season, publication year, and country: a global systematic review, meta-analysis, and meta-regression. Compr. Rev. Food Sci. Food Saf. 21, 2520–2559.

Alhaj, O.A., Faye, B., Prasad, R., 2020. Handbook of Research on Health and Environmental Benefits of Camel Products.

Alhaj, O.A., Jrad, Z., Oussaief, O., Jahrami, H.A., Ahmad, L., Alshuniaber, M.A., Mehta, B.M., 2024c. The characterization of Lactobacillus strains in camel and bovine milk during fermentation: a comparison study. Heliyon 10.

Alhaj, O.A., Kanekanian, A.D., Peters, A.C., 2007. Investigation on whey proteins profile of commercially available milk-based probiotics health drinks using fast protein liquid chromatography (FPLC). Br. Food J. 109, 469–480.

Alhaj, O.A., Lajnaf, R., Jrad, Z., Alshuniaber, M.A., Jahrami, H.A., Serag El-Din, M.F., 2022b. Comparison of ethanol stability and chemical composition of camel milk from five samples. Animals 12, 615.

Alhaj, O.A., Metwalli, A.A., Ismail, E.A., 2011. Heat stability of camel milk proteins after sterilisation process. J. Camel Pract. Res. 18, 277–282.

Alkaisy, Q.H., Al-Saadi, J.S., Al-Rikabi, A.K.J., Altemimi, A.B., Hesarinejad, M.A., Abedelmaksoud, T.G., 2023. Exploring the health benefits and functional properties of goat milk proteins. Food Sci. Nutr. 11, 5641–5656.

Almasri, R.S., Bedir, A.S., Ranneh, Y.K., El-Tarabily, K.A., Al Raish, S.M., 2024. Benefits of camel milk over cow and goat milk for infant and adult health in fighting chronic diseases: a review. Nutrients 16, 3848.

Alshuniaber, M., Alhaj, O., Abdallah, Q., Jahrami, H., 2021. Effects of camel milk hydrolysate on blood pressure and biochemical parameters in fructose-induced hypertensive rats. Nutr. Food Sci. 52, 292–307.

Alshuniaber, M.A., Alshammari, G.M., Eleawa, S.M., Yagoub, A.E.A., Al-Khalifah, A.S., Alhussain, M.H., Al-Harbi, L.N., Yahya, M.A., 2022. Camel milk protein hydrosylate alleviates hepatic steatosis and hypertension in high fructose-fed rats. Pharm. Biol. 60, 1137–1147.

Al-Thaibani, A., Mostafa, H., Alshamsi, O., Moin, A., Bansal, N., Mudgil, P., Maqsood, S., 2024. Spray drying and ultrasonication processing of camel whey protein concentrate: characterization and impact on bioactive properties. J. Dairy Sci. 107 (11), 8824–8836. https://doi.org/10.3168/jds.2024-24900.

Altooq, N., Humood, A., Alajaimi, A., Alenezi, A.F., Janahi, M., Alhaj, O., Jahrami, H., 2022. The role of micronutrients in the management of COIVD-19 and optimizing vaccine efficacy. Hum. Nutr. Metabol. 27, 200141.

Amalfitano, N., Patel, N., Haddi, M.-L., Benabid, H., Pazzola, M., Vacca, G.M., Tagliapietra, F., Schiavon, S., Bittante, G., 2024. Detailed mineral profile of milk, whey, and cheese from cows, buffaloes, goats, ewes, and dromedary camels, and efficiency of recovery of minerals in their cheese. J. Dairy Sci. 107, 8887–8907.

Arain, M.A., Khaskheli, G.B., Shah, A.H., Marghazani, I.B., Barham, G.S., Shah, Q.A., Khand, F.M., Buzdar, J.A., Soomro, F., Fazlani, S.A., 2023. Nutritional significance and promising therapeutic/medicinal application of camel milk as a functional food in human and animals: a comprehensive review. Anim. Biotechnol. 34, 1988–2005.

Arora, S., Khetra, Y., 2017. Buffalo Milk Cheese. *Cheese*. Elsevier.

Augustyniak, A., Gottardi, D., Giordani, B., Gaffey, J., Mc Mahon, H., 2023. Dairy bioactives and functional ingredients with skin health benefits. J. Funct.Foods 104, 105528.

Badr, G., Ramadan, N.K., Sayed, L.H., Badr, B.M., Omar, H.M., Selamoglu, Z., 2017. Why whey? Camel whey protein as a new dietary approach to the management of free radicals and for the treatment of different health disorders. Iran. J. Basic Med. Sci. 20, 338.

Bakry, I.A., Yang, L., Farag, M.A., Korma, S.A., Khalifa, I., Cacciotti, I., Ziedan, N.I., Jin, J., Jin, Q., Wei, W., 2021. A comprehensive review of the composition, nutritional value, and functional properties of camel milk fat. Foods 10, 2158.

Beg, O.U., Von Bahr-Lindström, H., Zaidi, Z.H., Jörnvall, H., 1985. The primary structure of α-lactalbumin from camel milk. Eur. J. Biochem. 147, 233–239.

Beucler, J., Drake, M., Foegeding, E.A., 2005. Design of a beverage from whey permeate. J. Food Sci. 70, S277–S285.

Bittante, G., Amalfitano, N., Bergamaschi, M., Patel, N., Haddi, M.-L., Benabid, H., Pazzola, M., Vacca, G.M., Tagliapietra, F., Schiavon, S., 2022. Composition and aptitude for cheese-making of milk from cows, buffaloes, goats, sheep, dromedary camels, and donkeys. J. Dairy Sci. 105, 2132–2152.

Boland, M., 2011. Whey proteins. In: Handbook of Food Proteins. Elsevier.
Bornaz, S., Sahli, A., Attalah, A., Attia, H., 2009. Physicochemical characteristics and renneting properties of camels' milk: a comparison with goats', ewes' and cows' milks. Int. J. Dairy Technol. 62, 505–513.
Brunner, J.R., 1981. Cow milk proteins: twenty-five years of progress. J. Dairy Sci. 64, 1038–1054.
Buchanan, D., Martindale, W., Romeih, E., Hebishy, E., 2023. Recent advances in whey processing and valorisation: technological and environmental perspectives. Int. J. Dairy Technol. 76, 291–312.
Campanella, L., Martini, E., Pintore, M., Tomassetti, M., 2009. Determination of lactoferrin and immunoglobulin G in animal milks by new immunosensors. Sensors 9, 2202–2221.
Castro, L.H., S. De Araújo, F.H., M. Olimpio, M.Y., B. De B. Primo, R., T. Pereira, T., F. Lopes, L.A., BS De M. Trindade, E., Fernandes, R., A. Oesterreich, S., 2019. Comparative meta-analysis of the effect of concentrated, hydrolyzed, and isolated whey protein supplementation on body composition of physical activity practitioners. Nutrients 11, 2047.
Chen, H., Wang, J., Cheng, Y., Wang, C., Liu, H., Bian, H., Pan, Y., Sun, J., Han, W., 2019. Application of protein-based films and coatings for food packaging: a review. Polymers 11, 2039.
Deeth, H., Bansal, N., 2019. Whey Proteins: An Overview. Whey Proteins, pp. 1–50.
Egbujor, M.C., Olaniyan, O.T., Emeruwa, C.N., Saha, S., Saso, L., Tucci, P., 2024. An insight into role of amino acids as antioxidants via NRF2 activation. Amino Acids 56, 23.
El-Agamy, E.S.I., 2006. Camel milk. In: Handbook of Milk of Non-bovine Mammals, pp. 297–344.
Elkot, W.F., 2019. Characterization and healthier properties of whey proteins of camel milk: a review. J. Agroaliment. Proc. Technol. 25, 79–88.
Elsahoryi, N.A., Al-Sayyed, H.F., 2020. Manufacture of dairy and non-dairy camel milk products. In: Handbook of Research on Health and Environmental Benefits of Camel Products. IGI Global.
FAO, 2024. Gateway to Dairy Production and Products [Online]. Available: https://www.fao.org/dairy-production-products/production/dairy-animals/en. (Accessed 29 November 2024).
Faye, B., Konuspayeva, G., 2024. Camel Milk Composition and Future Market Potential. CABI Reviews.
Gallo, V., Giansanti, F., Arienzo, A., Antonini, G., 2022. Antiviral properties of whey proteins and their activity against SARS-CoV-2 infection. J. Funct.Foods 89, 104932.
González-Weller, D., Paz-Montelongo, S., Bethencourt-Barbuzano, E., Niebla-Canelo, D., Alejandro-Vega, S., Gutiérrez, Á.J., Hardisson, A., Carrascosa, C., Rubio, C., 2023. Proteins and minerals in whey protein supplements. Foods 12, 2238.
Gorban, A.M., Izzeldin, O.M., 1999. Study on cholesteryl ester fatty acids in camel and cow milk lipid. Int. J. Food Sci. Technol. 34, 229–234.
Gordon, I., 2014. Minerals and vitamins in milk and dairy products. In: Milk and Dairy Products as Functional Foods, pp. 289–313.
Guha, S., Sharma, H., Deshwal, G.K., Rao, P.S., 2021. A comprehensive review on bioactive peptides derived from milk and milk products of minor dairy species. Food Prod. Process. Nutr. 3, 1–21.
Guo, H., Pang, K., Zhang, X., Zhao, L., Chen, S., Dong, M., Ren, F., 2007. Composition, physiochemical properties, nitrogen fraction distribution, and amino acid profile of donkey milk. J. Dairy Sci. 90, 1635–1643.
Ha, D.J., Kim, J., Kim, S., Go, G.W., Whang, K.Y., 2021. Dietary whey protein supplementation increases immunoglobulin G production by affecting helper T cell populations after antigen exposure. Foods 10.
Haddadin, M.S., Gammoh, S.I., Robinson, R.K., 2008. Seasonal variations in the chemical composition of camel milk in Jordan. J. Dairy Res. 75, 8–12.
Hailu, Y., Hansen, E.B., Seifu, E., Eshetu, M., Ipsen, R., Kappeler, S., 2016. Functional and technological properties of camel milk proteins: a review. J. Dairy Res. 83, 422–429.
Hassanein, Y.R., 2023. Health Outcomes of Camel Milk on the Common Viral, Digestive, and Immunity Disorders. University of Bridgeport.
Heine, R.G., Alrefaee, F., Bachina, P., De Leon, J.C., Geng, L., Gong, S., Madrazo, J.A., Ngamphaiboon, J., Ong, C., Rogacion, J.M., 2017. Lactose intolerance and gastrointestinal cow's milk allergy in infants and children–common misconceptions revisited. World Allergy Organ. J. 10, 1–8.
Hernández-Ledesma, B., Ramos, M., Gómez-Ruiz, J.Á., 2011. Bioactive components of ovine and caprine cheese whey. Small Rumin. Res. 101, 196–204.
Hewison, M., 2012. Vitamin D and immune function: an overview. Proc. Nutr. Soc. 71, 50–61.
Jiang, L., Zhang, Z., Qiu, C., Wen, J., 2024. A review of whey protein-based bioactive delivery systems: design, fabrication, and application. Foods 13, 2453.
Jovanović, S., Barać, M., Maćej, O., 2005. Whey proteins-properties and possibility of application. Mljekarstvo 55, 215–233.
Kapoor, B., Kapoor, D., Gautam, S., Singh, R., Bhardwaj, S., 2021. Dietary polyunsaturated fatty acids (PUFAs): uses and potential health benefits. Curr. Nutr. Rep. 10, 232–242.
Kelkar, N., 2023. Clinical applications of whey protein. In: Nutrition Science, Marketing Nutrition, Health Claims, and Public Policy. Elsevier.
Kerasioti, E., Stagos, D., Tsatsakis, A.M., Spandidos, D.A., Taitzoglou, I., Kouretas, D., 2018. Effects of sheep/goat whey protein dietary supplementation on the redox status of rats. Mol. Med. Rep. 17, 5774–5781.
Kerstetter, J.E., Bihuniak, J.D., Brindisi, J., Sullivan, R.R., Mangano, K.M., Larocque, S., Kotler, B.M., Simpson, C.A., Cusano, A.M., Gaffney-Stomberg, E., Kleppinger, A., Reynolds, J., Dziura, J., Kenny, A.M., Insogna, K.L., 2015. The effect of a whey protein supplement on bone mass in older caucasian adults. J. Clin. Endocrinol. Metab. 100, 2214–2222.
Khalesi, M., Salami, M., Moslehishad, M., Winterburn, J., Moosavi-Movahedi, A.A., 2017. Biomolecular content of camel milk: a traditional superfood towards future healthcare industry. Trends Food Sci. Technol. 62, 49–58.
Khan, U.M., Selamoglu, Z., 2020. Nutritional and medical perspectives of whey protein: a historical overview. J. Pharmaceut. care. 7 (4), 112–117.

Khaskheli, M., Arain, M., Chaudhry, S., Soomro, A., Qureshi, T., 2005. Physico-chemical quality of camel milk. J. Agric. Soc. Sci. 2, 164–166.

Khetra, Y., Meena, G., Arora, S., 2022. Buffalo milk and its products: composition, nutrition and benefits. In: Biotechnological Applications in Buffalo Research. Springer.

Konuspayeva, G., Faye, B., Loiseau, G., 2009. The composition of camel milk: a meta-analysis of the literature data. J. Food Compos. Anal. 22, 95–101.

Konuspayeva, G., Faye, B., Loiseau, G., Levieux, D., 2007. Lactoferrin and immunoglobulin contents in camel's milk (*Camelus bactrianus*, *Camelus dromedarius*, and hybrids) from Kazakhstan. J. Dairy Sci. 90, 38–46.

Konuspayeva, G., Lemarie, E., Faye, B., Loiseau, G., Montet, D., 2008. Fatty acid and cholesterol composition of camel's (*Camelus bactrianus*, *Camelus dromedarius* and hybrids) milk in Kazakhstan. Dairy Sci. Technol. 88, 327–340.

Konuspayeva, G.S., 2020. Manufacture and challenges of camel milk cheese. In: Handbook of Research on Health and Environmental Benefits of Camel Products. IGI Global.

Lajnaf, R., Picart-Palmade, L., Attia, H., Marchesseau, S., Ayadi, M., 2017. The effect of pH and heat treatments on the foaming properties of purified α-lactalbumin from camel milk. Colloids Surf. B Biointerfaces 156, 55–61.

Lajnaf, R., Picart-Palmade, L., Cases, E., Attia, H., Marchesseau, S., Ayadi, M., 2018. The foaming properties of camel and bovine whey: the impact of pH and heat treatment. Food Chem. 240, 295–303.

Lampová, B., Doskočil, I., Šmíd, P., Kouřimská, L., 2024. Comparison of cricket protein powder and whey protein digestibility. Molecules 29, 3598.

Li, A., Zheng, J., Han, X., Yang, S., Cheng, S., Zhao, J., Zhou, W., Lu, Y., 2023. Advances in low-lactose/lactose-free dairy products and their production. Foods 12.

Liu, W., Zhou, Y., Sun, H., Li, R., Qin, Y., Yu, L., Chen, Y., Li, Y., Tan, Y., Zhao, R., Zhang, W., Jiang, S., Xu, Y., 2021. Goat milk improves glucose homeostasis via enhancement of hepatic and skeletal muscle AMP-activated protein kinase activation and modulation of gut microbiota in streptozocin-induced diabetic rats. Mol. Nutr. Food Res. 65, e2000888.

Lopes Gomes, D., Moehlecke, M., Lopes da Silva, F.B., Dutra, E.S., D'agord Schaan, B., Baiocchi De Carvalho, K.M., 2017. Whey protein supplementation enhances body fat and weight loss in women long after bariatric surgery: a randomized controlled trial. Obes. Surg. 27, 424–431.

Luhovyy, B.L., Akhavan, T., Anderson, G.H., 2007. Whey proteins in the regulation of food intake and satiety. J. Am. Coll. Nutr. 26, 704S–712S.

Madureira, A.R., Pereira, C.I., Gomes, A.M., Pintado, M.E., Malcata, F.X., 2007. Bovine whey proteins–Overview on their main biological properties. Food Res. Int. 40, 1197–1211.

Malik, T.F., Panuganti, K.K., 2023. Lactose Intolerance. *StatPearls*. StatPearls Publishing.

Martini, M., Altomonte, I., Salari, F., Caroli, A.M., 2014. Monitoring nutritional quality of Amiata donkey milk: effects of lactation and productive season. J. Dairy Sci. 97, 6819–6822.

Mcgrath, B.A., Fox, P.F., Mcsweeney, P.L., Kelly, A.L., 2016. Composition and properties of bovine colostrum: a review. Dairy Sci. Technol. 96, 133–158.

Medhammar, E., Wijesinha-Bettoni, R., Stadlmayr, B., Nilsson, E., Charrondiere, U.R., Burlingame, B., 2012. Composition of milk from minor dairy animals and buffalo breeds: a biodiversity perspective. J. Sci. Food Agric. 92, 445–474.

Mehaia, M.A., Hablas, M.A., Abdel-Rahman, K.M., El-Mougy, S.A., 1995. Milk composition of majaheim, wadah and hamra camels in Saudi Arabia. Food Chem. 52, 115–122.

Mehra, R., Kumar, H., Kumar, N., Ranvir, S., Jana, A., Buttar, H.S., Telessy, I.G., Awuchi, C.G., Okpala, C.O.R., Korzeniowska, M., 2021. Whey proteins processing and emergent derivatives: an insight perspective from constituents, bioactivities, functionalities to therapeutic applications. J. Funct. Foods 87, 104760.

Mehta, R.D., Agrawal, R., 2020. Camel milk disguised cosmeceutical. In: Handbook of Research on Health and Environmental Benefits of Camel Products. IGI Global.

Moatsou, G., Hatzinaki, A., Samolada, M., Anifantakis, E., 2005. Major whey proteins in ovine and caprine acid wheys from indigenous Greek breeds. Int. Dairy J. 15, 123–131.

Navarro, F., Galan-Malo, P., Pérez, M.D., Abecia, J.-A., Mata, L., Calvo, M., Sánchez, L., 2018. Lactoferrin and IgG levels in ovine milk throughout lactation: correlation with milk quality parameters. Small Rumin. Res. 168, 12–18.

Omar, A., Harbourne, N., Oruna-Concha, M.J., 2016. Quantification of major camel milk proteins by capillary electrophoresis. Int. Dairy J. 58, 31–35.

Ostertag, F., Hinrichs, J., 2023. Enrichment of lactoferrin and immunoglobulin G from acid whey by cross-flow filtration. Foods 12, 2163.

Pandya, A.J., Mohamed, M., Khan, H., 2006. Buffalo milk utilization for dairy products. In: Handbook of Milk of Non-bovine Mammals, pp. 215–256.

Patel, S., 2015. Emerging trends in nutraceutical applications of whey protein and its derivatives. J. Food Sci. Technol. 52, 6847–6858.

Pillai, A.T., Morya, S., Kasankala, L.M., 2024. Emerging trends in bioavailability and pharma-nutraceutical potential of whey bioactives. J. Nutr. Metab. 2024, 8455666.

Ranjit Aich, R.A., Subhasis Batabyal, S.B., Joardar, S.N., 2014. Simple purification method for beta-lactoglobulin from buffalo milk.

Rashmi, K., Gupta, M., Tarapdar, S., Duggal, M., Tyagi, S., Kaur, H., Vyas, P., Gupta, A., 2022. Influence of dietary molecules on human health: whey proteins. Res. J. Pharm. Technol. 15, 1910–1914.

Rocha, J.M., Guerra, A., 2020. On the valorization of lactose and its derivatives from cheese whey as a dairy industry by-product: an overview. Eur. Food Res. Technol. 246, 2161–2174.

Ruprichová, L., Králová, M., Borkovcová, I., Vorlová, L., Bedáňová, I., 2014. Determination of whey proteins in different types of milk. Acta Vet. 83, 67–72.

Schoenfeld, B.J., Aragon, A.A., 2018. How much protein can the body use in a single meal for muscle-building? Implications for daily protein distribution. J. Int. Soc. Sports Nutr. 15, 10.

Seifu, E., 2023. Camel milk products: innovations, limitations and opportunities. Food Prod. Process. Nutr. 5, 15.

Shamsia, S., 2009. Nutritional and therapeutic properties of camel and human milks. Int. J. Genet. Mol. Biol. 1, 52–58.

Sharp, E., D'cunha, N.M., Ranadheera, C.S., Vasiljevic, T., Panagiotakos, D.B., Naumovski, N., 2021. Effects of lactose-free and low-lactose dairy on symptoms of gastrointestinal health: a systematic review. Int. Dairy J. 114, 104936.

Sikalidis, A.K., Kelleher, A.H., Kristo, A.S., 2021. Mediterranean diet. Encyclopedia 1, 371–387.

Sindhu, J.S., Arora, S., 2011. Buffalo milk. In: Fuquay, J.W., P. F. F, Mcsweeney, P.L.H. (Eds.), Encyclopedia of Dairy Sciences, 2 ed. Academic press, Oxford, UK.

Taeger, M., Thiele, S., 2021. Additional costs of lactose-reduced diets: lactose-free dairy product substitutes are a cost-effective alternative for people with lactose intolerance. Public Health Nutr. 24, 4043–4053.

Vincenzetti, S., Cammertoni, N., Rapaccetti, R., Santini, G., Klimanova, Y., Zhang, J.-J., Polidori, P., 2022. Nutraceutical and functional properties of camelids' milk. Beverages 8, 12.

Wernery, U., 2006. Camel milk, the white gold of the desert. J. Camel Pract. Res. 13, 15.

Wu, G., Lupton, J.R., Turner, N.D., Fang, Y.-Z., Yang, S., 2004. Glutathione metabolism and its implications for health. J. Nutr. 134, 489–492.

Xiao, Y., Yi, L., Ming, L., He, J., Ji, R., 2022. Changes in milk components, amino acids, and fatty acids of Bactrian camels in different lactation periods. Int. Dairy J. 131, 105363.

Yiğit, A., Bielska, P., Cais-Sokolińska, D., Samur, G., 2023. Whey proteins as a functional food: health effects, functional properties, and applications in food. J. Am. Nutr. Assoc. 42, 758–768.

Chapter 5

Manufacturing processes of whey products

Haotian Zheng[1, 2], Zhoutai Teng[1, 2] and Hari Meletharayil[3]

[1]Food Rheology Laboratory, Department of Food, Bioprocessing and Nutrition Sciences, North Carolina State University, Raleigh, NC, United States; [2]Southeast Dairy Foods Research Center, Department of Food, Bioprocessing and Nutrition Sciences, North Carolina State University, Raleigh, NC, United States; [3]Dairy Management Inc., Rosemont, IL, United States

1. Introduction

Fresh bovine milk is a colloidal oil-in-water (O/W) emulsion system, in which the liquid fraction contains milk proteins, lactose, and minerals; the oil droplets also referred as "milk fat globules" are covered by a sophisticated trilayer system, namely the "milk fat globule membrane" (MFGM) (Zheng et al., 2013a, 2013b, 2014a, 2014b, 2025a; Zheng, 2014). The protein fraction in milk may be processed yielding different types of protein ingredients with different functionalities and they are, in general, available in the market in powder form (An and Zheng, 2024, 2025; Patel et al., 2022a); and the fat fraction may be used for manufacturing butter products (Patel et al., 2022b).

The major type of protein in bovine milk is caseins (Zheng, 2024). The liquid residue after coagulating caseins or removing caseins through microfiltration is the whey fraction. Whey products cover a broad range of consumer products and industrial ingredients derived from whey and milk. The majority whey products may be categorized in the industrial ingredient group, and the consumer products may include whey cheese and beverages composed of whey components. Whey protein ingredients can be used as "building blocks" constructing different basic food structures such as protein dispersions, foams, emulsions, and gels; the functionalities of whey proteins determine their interactions with other food materials (proteins, lipids, and carbohydrates), therefore, impact on the physical stability and textural properties of the referred structures (Zheng, 2018, 2019; Kumar et al., 2018, 2021, 2023; Rivero et al., 2022; Zheng, 2022; Amin et al., 2024, 2025a; Shi and Zheng, 2025; Zheng and Meletharayil, 2025). The referred basic food structures are essential to many consumer food products.

The *Code of Federal Regulations* (CFR) of the United States (Title 21, Chapter I, subchapter B, Part 184, Subpart B, § 184.1979) clearly defined three whey products (1) "whey"; (2) "concentrated whey"; (3) "dry whey". According to the referred content in CFR, "*whey is the liquid substance obtained by separating the coagulum from milk, cream, or skim milk in cheesemaking.*"; "*concentrated whey*" is in liquid form and its total solid content is enriched by partial removal of water from whey meaning there is no change to the relative proportions of the solid constituents in the original whey ("21 CFR 184.1979 – Whey.," 2025). Therefore concentrated whey is different from whey protein concentrates (WPC) and whey protein isolates (WPI), which are enriched with whey proteins. "Dry or dried whey" is the dried form of whey by removing water. Besides these basic whey ingredient products, various ingredients that are derived from whey are available in the ingredient market. The present book chapter uses the list of whey ingredients organized by American Dairy Products Institute (ADPI) as a clue of introducing whey ingredient products and their manufacturing methods (Whey Ingredients Overview, 2025). Whey proteins may be concentrated from two sources: "whey" and skim milk. Whey proteins that are concentrated or isolated from skim milk using microfiltration (MF) coupled with diafiltration (DF) are termed as "milk whey proteins" and these whey protein ingredients are referred as milk whey protein concentrate (mWPC) and milk whey protein isolate (mWPI); these names are used for distinguishing from whey proteins and their ingredients originated from whey (Carter et al., 2021; Milk Whey Protein, 2025). It is recommended to use the distinctive

Whey. https://doi.org/10.1016/B978-0-443-27648-4.00024-4

names between WPC/WPI and mWPC/mWPI when describing manufacturing methods or ingredients used for consumer foods formulation.

The ingredient products derived from (cheese) whey may be summarized in a block chart (Fig. 5.1). To make ingredients and consumer products derived from cheese whey, cheese curd fines and milk fat residues presented in cheese whey are removed by clarification processes. Casein fines such as cheese curd grains present in whey need to be removed first before whey cream separation (Tetra Pak, 2025a). Coarse filters or continuous centrifugal separators for separating solid particles from a liquid may be used for fines removal (Fig. 5.2). The collected casein fines may be further processed and used in processed cheese manufacturing or in cheese milk standardization (Tetra Pak, 2025a). The clarified whey stream is further processed through a typical centrifuge-based fat separator; the resultant products are whey cream stream and liquid nonfat whey. Typically, whey cream contains ~25–30% of fat, this is relatively lower than conventional whipping cream (with ~40% fat content). The whey cream can also be used for cheese milk standardization after appropricate sterilization procedure for inactivation of residual enzymes and microorganisms derived from the cheese curd formation step. Alternatively, whey cream can be used for making other cream products. Valorization of whey cream is of research interest related to the dairy processing industry. A previous work investigated the impact of substituting whey cream for sweet cream on the texture of cream cheese; the work reported that high substitution proportion level (e.g., 75%) resulted in softer texture of cream cheese (Brighenti et al., 2021). The authors suggested that whey cream can be used for manufacturing more spreadable cream cheeses. Moreover, whey cream may be used for manufacturing butter products; a previous work suggested that butter made from whey cream had distinctive sensory attributes compared with butter made from sweet cream featuring more nutty flavor and cardboard odor (Jinjarak et al., 2006). After removing casein fines and separating whey cream, the liquid whey stream can be used for processing whey products including condensed whey, whey protein related ingredients, and byproducts derived from whey protein isolation.

A few previous publications have categorized whey products and elaborated their characteristics (Fox et al., 2017; Kilara, 2015); the manufacturing processes for whey protein-based ingredients were also discussed in other book chapters (Kelly, 2019; Wang and Guo, 2019). Lactose-enriched permeates are byproducts derived from processing whey protein ingredients; the whey permeates can be dried, yielding commercial ingredients with relatively lower value as compared with whey protein ingredients. Moreover, whey permeate derivative materials, for example, lactobionates, have been manufactured as value-added ingredients to increase the economic value of cheese whey permeates (Amin et al., 2025b). However, the present chapter aims to provide more updated and comprehensive knowledge on manufacturing processes for different whey products with a focus on protein-based ingredients. Moreover, the scope of the present chapter focuses on components and products directly and indirectly derived from bovine milks. Other milk sources and whey proteins synthesized through fermentation technology are out of the scope of this chapter.

2. Whey ingredients obtained from concentrating total solids

2.1 Dry whey

Dry whey products are divided into "dry sweet whey" (DSW) and "dry acid whey" (DAW) to indicate the origins of whey derived from rennet-coagulum and acid-coagulum of milk or milk products. DSW and DAW are slightly different regarding the Codex requirements as shown in Table 5.1; the limits are adapted from ADPI's "Dry Whey Standard" (American Dairy Products Institute, 2023a).

Dry wheys are obtained by removing water from liquid whey. Although the theoretical processing principle is simple, the practical processing requires specialized knowledge for manufacturing products with high quality attributes. Regarding a typical manufacturing process, briefly, liquid whey with ~6% dry matter is first concentrated to 18.0%–28.5% dry matter by reverse osmosis (RO) or a combined process involving RO and nanofiltration (NF). Subsequently, the concentrated whey is processed by evaporation and spray drying techniques; for example, evaporation can further increase the dry matter in the preconcentrated liquid whey from ~20% to 45%–65%. After evaporation, the concentrated whey is cooled and is gone through a crystallization step to form lactose crystals (Tetra Pak, 2025a). The precrystallization may be done by seeding the evaporation processed whey concentrate (containing 50%–60% dry matter) with lactose crystals to induce bulk crystallization (Fox et al., 2017). The precrystallization step before drying is critical because it is for converting unstable amorphous state lactose to a non or less hygroscopic form of lactose, which prevents the powder from caking over storage time.

Dry whey powder may contain ~70% lactose, which may be present in a thermodynamically unstable amorphous state and absorb moisture from the surroundings resulting in a decrease in glass-transition temperature lower than the storage temperature. Such hygroscopic nature will result in sticky powder or powder caking after a certain period of storage.

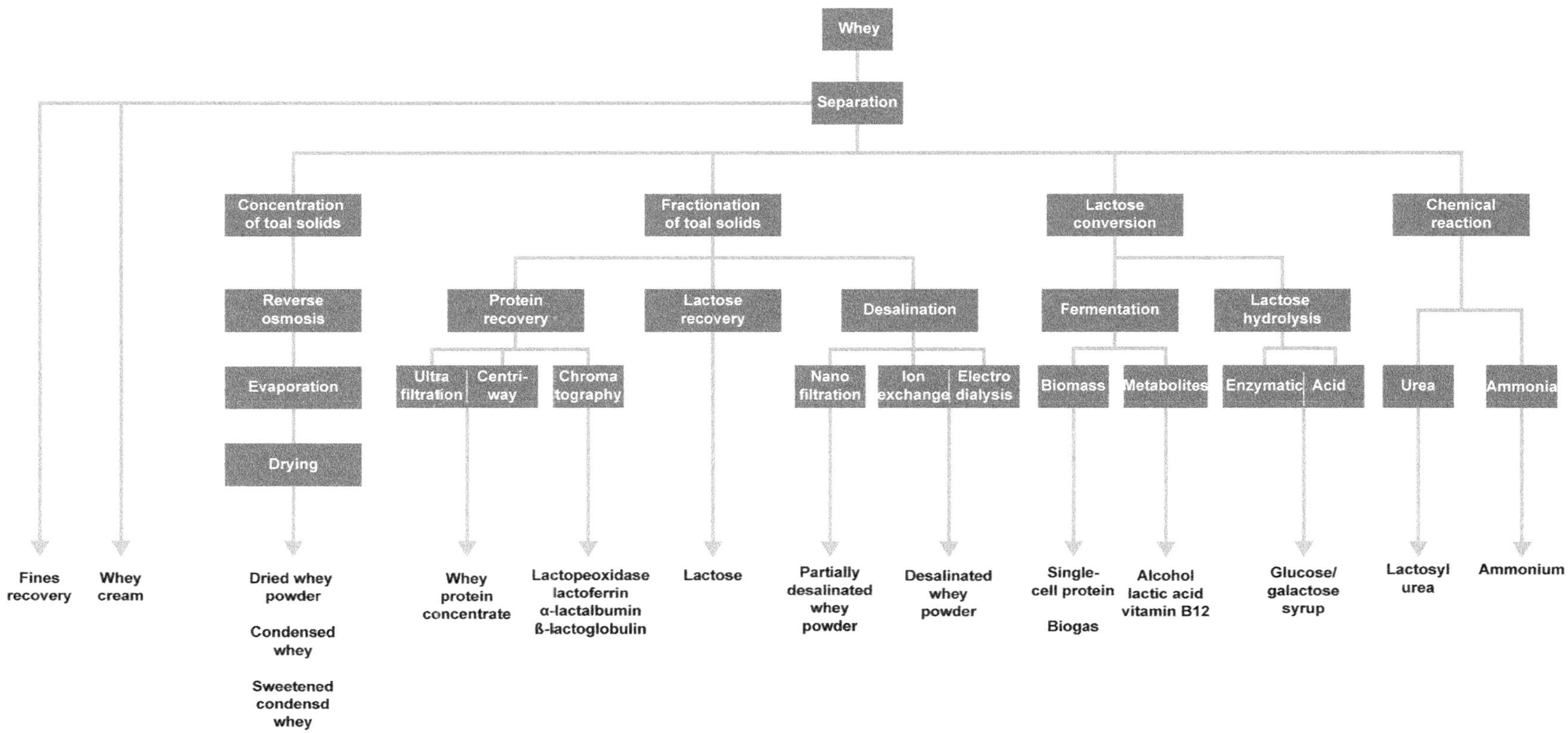

FIGURE 5.1 Schematic representation of whey processing alternatives. *The figure was redrawn, adapted, and modified based on Tetra Pak (2025a).*

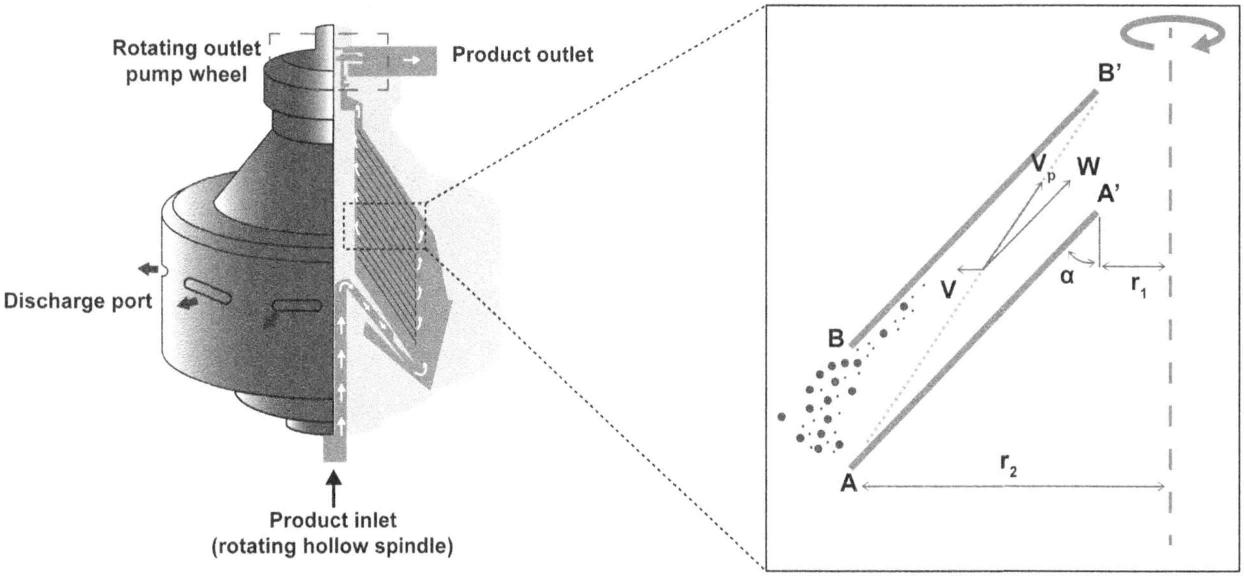

FIGURE 5.2 Schematic illustration of centrifugal clarifier equipment (left side) with an enlarged section showing the movement direction of solid particles within a separation channel (right side); the figures were redrawn, adapted, and modified based on Tetra Pak (2025b). r_1 and r_2 denote the radii of the inner and the outer edges of the separation channel, respectively. The symbol "W" indicates the velocity of a particle moving toward the inner edge along the separation channel. The symbol "V" represents the sedimentation velocity of a solid particle moving toward the periphery, and v_p is the net velocity by summing these two motions. Sedimentable particles are separated from the fluid and are deposited on the wall near the outer edge of separator discs.

TABLE 5.1 Codex requirements for dry whey products (dry sweet whey and dry acid whey), showing limits for key compositional parameters (the information is based on American Dairy Products Institute, 2023a).

		Dry sweet whey	Dry acid whey
Parameter	**Units of measure**	**Limits**	**Limits**
Protein	%	11 minimum	7 minimum
Moisture	%	5.0 maximum	4.5 maximum
Lactose	%	65 minimum	n/a
Ash	%	8.5 maximum	15.0 maximum
Titratable acidity	%	0.16 maximum	0.35 minimum
	pH	6.0 minimum	5.1 maximum

Therefore lactose crystallization is a critical step within the overall procedure of dry whey production for reducing or eliminating the hygroscopic property (Fox et al., 2017; Ibach and Kind, 2007).

Besides the referred precrystallization, postcrystallization may be carried out on dry whey powders containing a high proportion of amorphous materials. A previous work suggested a crystallization step after the spray drying process to convert amorphous powder materials (e.g., lactose) into a crystallized form (Ibach and Kind, 2007); the authors reported that treating dry whey with hot air with temperature of 100°C and relative air humidity of 80% took 5 min for crystallization. Additionally, the referred work utilized a reduced form of the Johnson-Mehl-Avrami (JMA) theory for computing crystallization rate constants according to the following equation

$$ln[-ln(1-\theta)] = n \cdot ln(t - t_{Ind}) + n \cdot ln(k) \tag{5.1}$$

where θ is the fraction of crystallized lactose, n is the reaction order, t is the crystallization time, t_{Ind} is the induction time, k is the crystallization rate constant. θ can be quantified by fitting experimental data in the following equation

$$\theta = \frac{w_{max} - w}{w_{max} - w_{final}} \tag{5.2}$$

where w_{max} and w_{final} are the maximum moisture content at the induction time (t_{Ind}) and the moisture content at the end time of crystallization.

2.2 Reduced lactose whey and reduced minerals whey

Reduced lactose whey (RLW) is processed by partially removing lactose from whey. Typical RLW contains ~52%–58% lactose (not exceeding 60%), ~18%–24% protein, and ~1%–4% fat (American Dairy Products Institute, 2023b). Lactose reduction may be obtained by crystallizing lactose and separating the lactose crystals by centrifugation or filtration (Fox et al., 2017).

Reduced mineral whey (RMW) is processed by partially removing the minerals from liquid whey. Typically, an RMW contains no more than 7% ash, ~11%–15% protein, 70%–82% lactose, 0.5%–1.8% fat, and 3%–4% moisture (American Dairy Products Institute, 2023c). In the original liquid whey, the salt content is about 8%–12% (dry matter basis), for specific food applications, for example, infant formula, liquid whey may be processed for 20%–95% demineralization by one of the following techniques: nanofiltration, electrodialysis, or ion exchange (Tetra Pak, 2025a). These techniques are also used for demineralization for whey permeates derived from whey ultrafiltration.

3. Protein concentrates and isolates

3.1 Whey protein concentrate and whey protein isolate

Due to the increased demand for protein, about 50% of processed whey solids are associated with whey protein concentrates/isolates and their byproducts (Tetra Pak, 2025a). Whey contains about 13% (dry matter basis) protein, which may be concentrated/isolated by tangential membrane filtration and chromatography techniques for removing nonprotein solids; the protein-enriched retentate phase can be spray dried for obtaining protein powder ingredients. Whey protein concentrates (WPC) include a number of different WPC products differing in protein content ranging from 34%–82%. In general, a WPC contains no less than 25% protein. Typically, WPC 34 contains 34%–36% (dry matter basis) of protein with a minimum threshold of $\geq$33.5%; WPC 80 contains 80%–82% of protein with a minimum threshold of $\geq$79.5% (American Dairy Products Institute, 2023d). Whey protein isolates (WPI) are protein powders obtained by removing nonprotein solids and water from whey for reaching to no less than 90% (dry matter basis) protein content. Typically, WPI contains 90%–92% (dry matter basis) of protein with a minimum threshold of $\geq$89.5% (American Dairy Products Institute, 2023e).

To obtain different WPC with different protein contents, different volume concentration factors (VCF) are needed; also for obtaining higher protein contents, for example, 75%–85%, UF + diafiltration (DF) is needed (Kelly, 2019). For instance, for manufacturing WPC 34, clarified liquid whey needs to be concentrated to a VCF of 5.88 by ultrafiltration (UF); this is equivalent to concentrating 100 kg whey (with 0.55% protein and 4.80% of lactose) to 16.7 kg of protein-rich UF retentate resulting in a yield of 83.3 kg of protein-depleted UF permeate; the retentate phase contains 3.25% protein, 5.34% lactose and the permeate phase contains 0.01% protein and 4.69% of lactose (Tetra Pak, 2025a). Comprehensive information on the WPC 34 production process, composition, and related mass balance & protein concentration is provided in Fig. 5.3. For processing WPC 60, a VCF within 13–20 is required; for processing WPC 75–85 (75%–85% of protein), VCF 30–35 followed by DF is needed (Kelly, 2019). UF may result in an increase of total solids of whey to 10%–17%, after that DF may be required for achieving higher protein content for example, >50% on a dry matter basis; the water quantity to be added for DF steps may be 40%–95% of the initial UF permeate volume and the required number of DF steps is determined by the target protein content level, the higher the designated protein content, the more DF steps are needed (Bansal and Bhandari, 2016). Three variables such as VCF, DF water volume, and total DF step number determine the effectiveness of DF for protein concentration (Baldasso et al., 2011). The referred work showed that a relatively smaller volume of DF water in each DF cycle resulted in a more efficient enrichment of protein and reduction of lactose. The WPC with higher protein content, for example, WPC 80, still contains 4%–8% fat (American Dairy Products Institute, 2023d). Fig. 5.4 presents the process flow and DF water addition for WPC 80, along with protein composition and the corresponding protein enrichment and mass reduction across processing stages. Further purification of protein from the retentate of WPC 80 by reducing its lipid content to $\leq$1.0% will result in a new ingredient category, namely, whey protein isolates (WPI). Kelly (2019) elaborated the technology evolution regarding further concentration of proteins

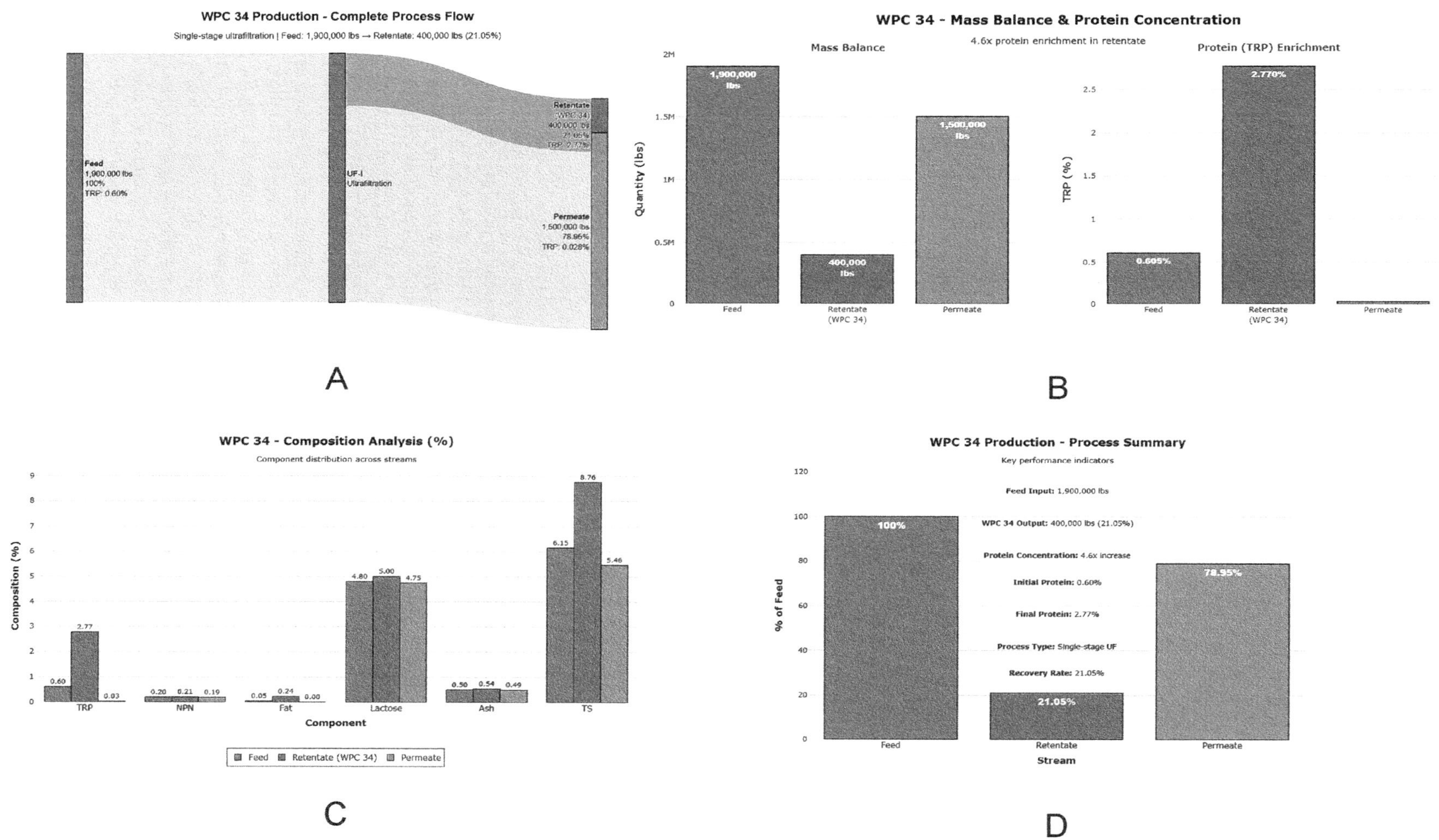

FIGURE 5.3 Overview of whey protein concentrate (WPC) 34 production and process outcomes. All calculations are based on an initial whey feed mass of 1,900,000 lb. (A) Complete process flow for WPC 34 production through single-stage ultrafiltration (UF). (B) Mass balance and true protein (TRP) enrichment across feed, retentate (WPC 34), and permeate streams. (C) Composition analysis of major whey components in feed, retentate (WPC 34), and permeate. NPN is nonprotein nitrogen, TS represents total solids. (D) Process summary highlighting protein enrichment, mass reduction, and stream distribution during WPC 34 production.

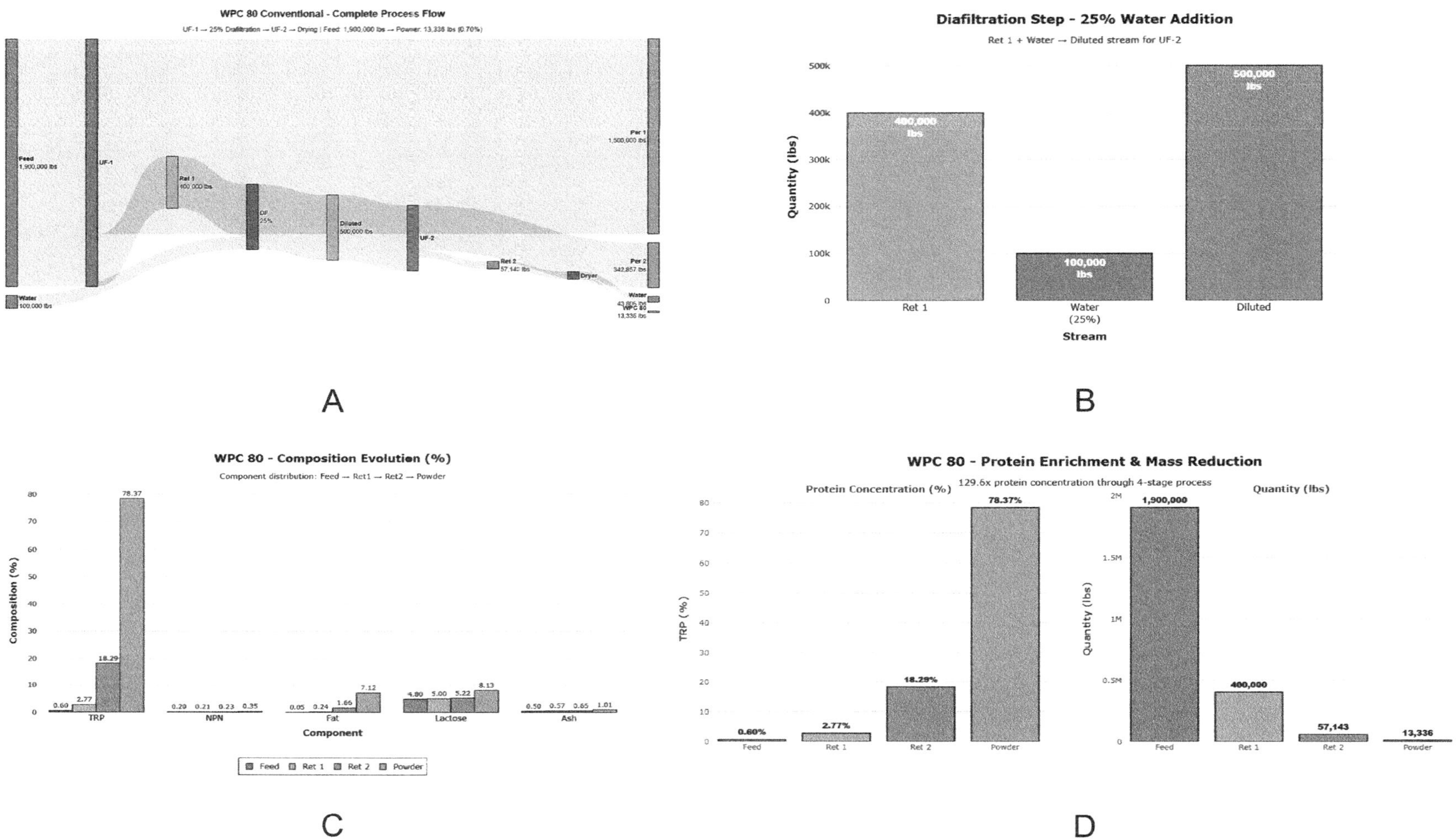

FIGURE 5.4 Overview of whey protein concentrate (WPC) 80 production and process outcomes. All calculations are based on an initial whey feed mass of 1,900,000 lb. (A) Complete process flow for WPC 80 production through ultrafiltration followed by diafiltration (DF). Ret represents to retentate, and Per represents permeate. (B) Diafiltration step showing 25% water addition relative to the retentate mass and the resulting diluted stream. (C) Ccmposition evolution of major whey components in feed, intermediate retentates, and final powder during WPC 80 production. TRP and NPN represent true protein and nonprotein nitrogen, respectively. (D) Protein enrichment and mass reduction across processing stages, illustrating increases in protein concentration and corresponding reductions in stream mass.

from MPC 80 and removal of fat residuals from whey for manufacturing protein isolates with higher protein contents (>90%). A typical WPI contains 90%–92% protein on a dry matter basis; such high protein content may be achieved by removing nonprotein solids through membrane filtration and/or ion exchange techniques (American Dairy Products Institute, 2023e). Industrial WPC/WPI processing plant conducts cross-flow filtration in connected multiple stages; the setup is referred to as "feed and bleed" configuration (Yee et al., 2007). Such configuration could contain up to 12 stages for conducting continuous UF/DF processes. Fig. 5.5 depicts the process flow used for WPI and provides stage-specific information on protein composition together with protein enrichment.

WPC and WPI processing technologies have been elaborated previously (Bansal and Bhandari, 2016; Kelly, 2019). A typical processing flow chart for producing WPC and WPI is shown in Fig. 5.6 (Bansal and Bhandari, 2016). A series of pretreatment processes on crude liquid whey or milk whey are needed in order to further concentrate the whey proteins into WPC and WPI. A centrifugation or an MF-based clarification may be performed for removing casein fines, whey cream, and (cheese) culture cells if there were any. The MF membrane with a pore size >0.1 μm is often used and it is able to separate fat droplets and casein fines from the liquid whey. These may help reduce fouling rate and improve flux during UF and DF processes for processing WPC and WPI. The calcium salts present in acid whey and sweet cheese whey, for example, calcium phosphates, may precipitate and cause membrane fouling issues during the protein concentration process. Musale and Kulkarni (1998) reported that performing a clarification procedure involving increasing whey pH to 7.5 plus removing precipitates significantly improved UF performance for whey protein concentration. It was found that an MF pretreatment for sweet whey using a 0.5 μm ceramic membrane may significantly increase flux and reduce fouling effect during UF (Barukčić et al., 2015).

The clarified liquid whey is then submitted to UF and DF processes for concentrating protein to a designated level corresponding to a specific WPC powdered ingredient for example, WPC 80. The critical technical parameters associated with UF/DF processing for whey protein concentration is discussed by Kelly (2019): in general, the molecular weight cutoff (MWCO) of UF membrane used for isolating proteins from whey is often at 10,000 Da (10 kDa). In UF and DF steps, the apparent permeate flux (J, $L/m^2/h$) and the net transmembrane pressure (TMP, ΔP) are two critical parameters determining the performance of UF and DF. TMP is expressed by the following equation

$$\Delta P = \frac{P1 + P2}{2} - P3 \tag{5.3}$$

where P1, P2, and P3 are feed inlet, feed outlet, and permeate outlet pressures, respectively. These pressures may be measured and monitored during the filtration process.

The permeate flux is expressed according to the D'Arcy's Law (Kelly, 2019; Wang et al., 2024)

$$J = \frac{\Delta P}{\mu R_t} \tag{5.4}$$

where μ is the permeate viscosity. R_t is the total (net) resistance during UF/DF processing and it is expressed as $R_t = R_m + R_g + R_p + R_{cp}$, in which R_m is the intrinsic membrane hydraulic resistance, R_g is the resistance due to membrane gel layer formation due to the deposit material on the membrane surface, R_p is the resistance due to membrane pore blocking by retentate components, and R_{cp} is the resistance caused by "concentration polarization". R_m is the intrinsic resistance determined by the UF membrane characteristics such as the material used, structural features of the membrane construction, thickness, etc. R_{cp} is induced by the concentration gradient of the concentrated constituents close the membrane surface due to the constituents accumulation on the membrane surface; this resistance is different than but is related to R_g. During the UF/DF process, the process induce net resistance, R_{pro}, may be expressed as $R_{pro} = R_g + R_p + R_{cp}$. After UF/DF processing, a cleaning-in-place (CIP) process will be conducted. The CIP may wash off part of the fouling materials (if not all) wrestling in a residual resistance, R_{res}. Therefore, the total combined (net) membrane resistance after CIP, R_t', can be expressed as $R_t = R_m + R_{res}$. A real time flux at three critical filtration stages, for example, before (J_{before}), during (J_{pro}), and after (J_{after}) UF/DF process may be estimated by the following equation:

$$J = \frac{V}{A\Delta t} \tag{5.5}$$

where V (mL) is the permeate volume collected at the specific time interval (Δt), A (m^2) is the membrane effective area. This equation can be used for measuring the referred three critical fluxes (J_{before}, J_{pro}, and J_{after}). ΔP and μ can be measured using pressure transducers mounted on the UF processing equipment and a viscometer respectively. These

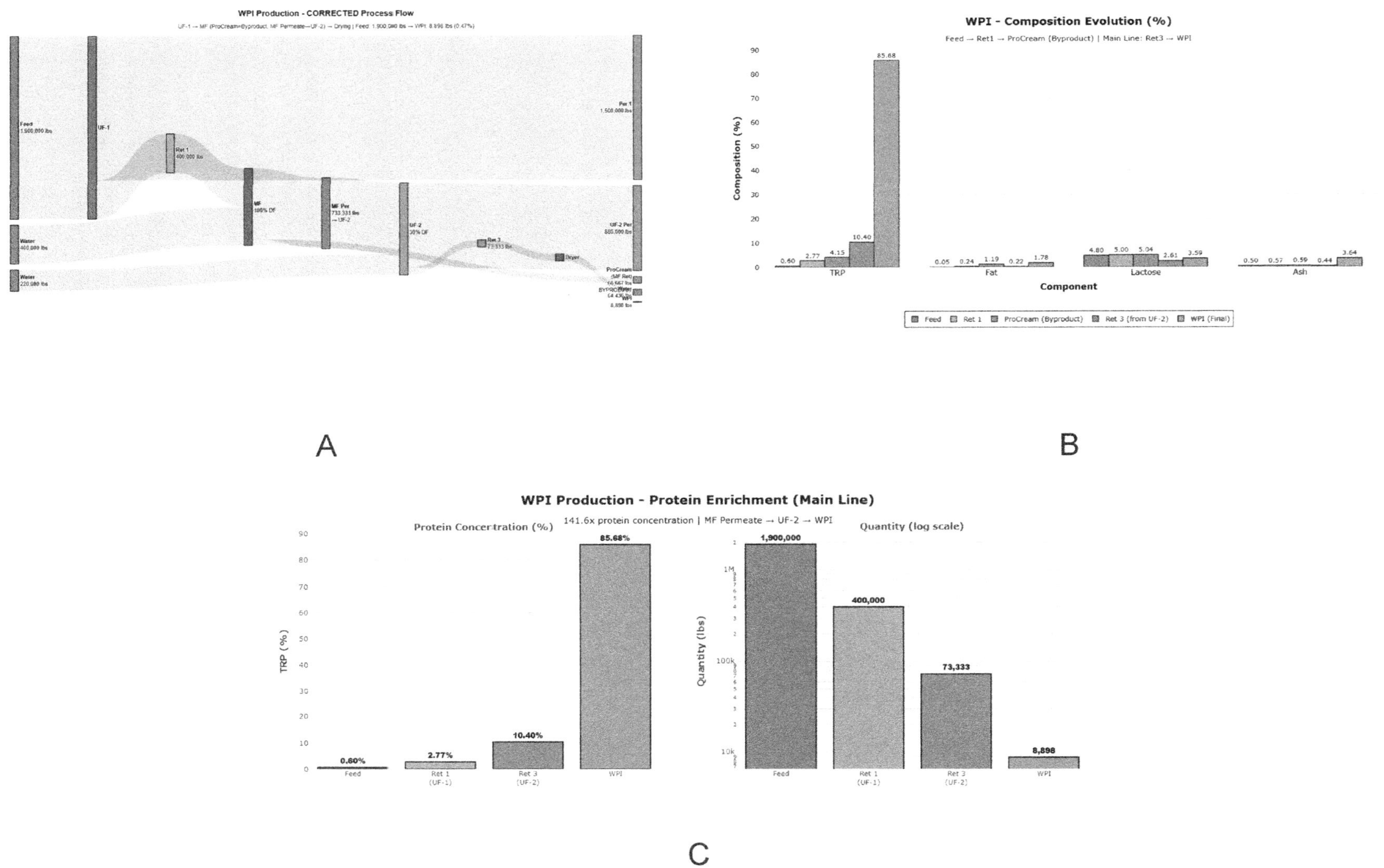

FIGURE 5.5 Overview of whey protein isolate (WPI) production and process outcomes. All calculations are based on an initial whey feed mass of 1,900,000 lb. (A) Complete process flow for WPI production using sequential ultrafiltration, microfiltration and diafiltration steps. Ret represents to retentate, and Per represents permeate. (B) Composition evolution of major whey components across feed, intermediate retentates, and final WPI product. TRP represents true protein. (C) Protein enrichment and mass reduction along the main processing line, illustrating increases in protein concentration and corresponding decreases in stream mass.

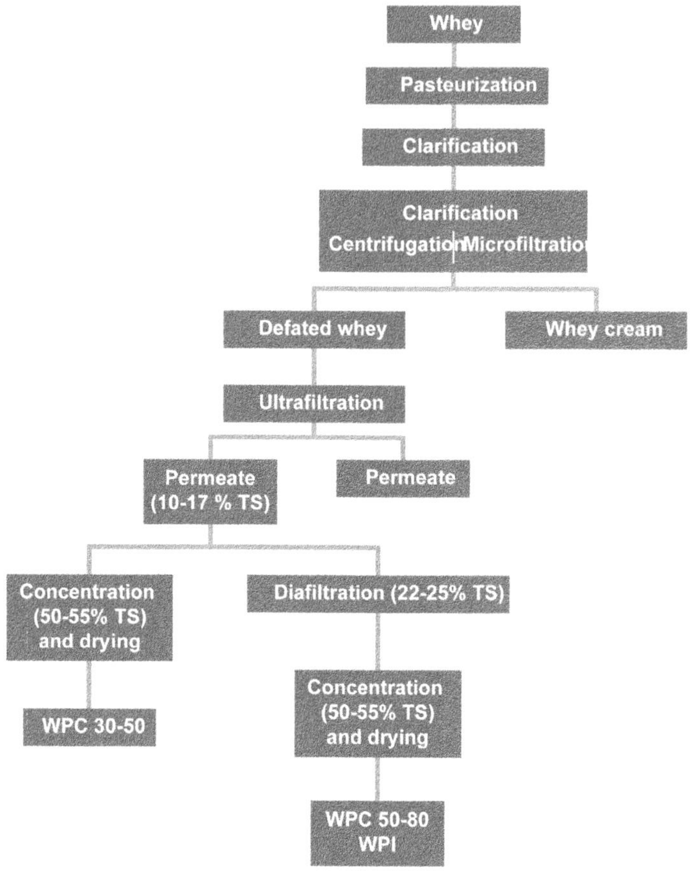

FIGURE 5.6 Process flowchart for the production of whey protein concentrates (WPC) and whey protein isolate (WPI). *The figure was redrawn, adapted, and modified based on Bansal and Bhandari (2016).*

measured and estimated parameters can be fitted into the following equations for estimating the three critical resistances (R_m, R_{pro}, and R_{res})

$$J_{before} = \frac{\Delta P}{\mu R_m} \tag{5.6}$$

$$J_{pro} = \frac{\Delta P}{\mu R_t} = \frac{\Delta P}{\mu (R_m + R_{pro})} \tag{5.7}$$

$$J_{after} = \frac{\Delta P}{\mu R_t'} = \frac{\Delta P}{\mu (R_m + R_{res})} \tag{5.8}$$

For characterizing and comparing the efficiency of UF/DF processing in milk protein concentration, average J and processing induced resistance (e.g., R_{pro} and R_{res}) may be measured and estimated according to the referred methods and equations. These parameters, J and R can be plotted as a function of volume reduction level or UF/DF processing time (Marella et al., 2015; Tarapata et al., 2022; Wang et al., 2024).

The membrane fouling mechanism and approaches that can effectively reduce fouling rate still need to be further investigated. The membrane fouling mechanism in whey protein concentration processes is discussed thoroughly in a review article (Wen-qiong et al., 2019). Besides chemistry and physicochemical fouling mechanisms, biofouling as induced by *Streptococcus thermophilus* and the intrinsic microflora could cause about 29% reduction of normalized flux compared with a system without biofouling (Villeneuve et al., 2021). Moreover Damar et al. (2020) investigated three commercial UF membranes regarding their impacts on permeate flux and fouling behaviors. The authors found that membrane nature such as pore morphology, active layer porosity, roughness, surface free energy, surface charge, and hydrophilicity all showed an impact on a parameter related to protein concentration effectiveness for example, fouling resistance, solute rejection, permeability, etc. In a more recent work, Wang et al. (2024) reported that cation content balance among Ca^{2+}, Na^{+}, K^{+}, Mg^{2+}, and Zn^{2+} had a significant impact on protein adhesion on membrane; the authors particularly stated that the higher Na^{+} ratio would promote membrane fouling effect. Approaches based on different working mechanisms have been investigated for improving antifouling properties and the overall flux during the protein concentration processes. Some instance approaches are (1) using negatively charged UF membranes (Arunkumar and Etzel, 2015); (2) performing cation regulation during UF process (Li et al., 2024); (3) applying pressure surges to the membrane surface (Aslam et al., 2022); (4) blending Mg-Al nanolayered double hydroxide with a polyvinylidene fluoride (PVDF)-based UF membrane (Arefi-Oskoui et al., 2016).

3.2 Native whey protein and milk whey protein

In general, there are two major approaches to obtaining whey protein-rich streams from whole milk or skim milk for manufacturing protein concentrate and isolate ingredients particularly rich in whey proteins; the schematic of pathways are illustrated in Fig. 5.7. The primary approach of obtaining liquid whey is based on coagulating caseins through cheesemaking, and rennet/acid induced casein micelles coagulation. The secondary approach is separating micellar caseins and whey proteins in the skim milk into two liquid streams using microfiltration, in which whey proteins are enriched in the permeate stream. The whey protein fraction as obtained by MF of skim milk meets ADPI's definition of "milk whey protein" (mWP) (American Dairy Products Institute, 2023f). According to the ADPI's standard, milk whey protein concentrates and isolates (mWPC and mWPI) may be obtained by MF and/or chromatography treatments on milk or skim milk; enzymatic coagulation and acid precipitation of caseins cannot be involved in processing mWPC and mWPI (American Dairy Products Institute, 2023f). The three mWP-based protein powder ingredients specified in ADPI standard are mWPC 34, mWPC 80, and mWPI; their minimum protein contents are 33.5%, 79.5%, and 89.5%. Note mWPC 34 protein content is as-is basis, whereas the other higher protein content ingredients are dry matter basis.

Whey proteins obtained from skim milk through MF are also referred to as "milk microfiltrate" and "native whey" in the literature (Muuronen et al., 2021). For the purposes of quality preservation and avoiding protein denaturation, native whey was manufactured at low temperatures (<20°C) and unpasteurized skim milk was used as the starting milk source (Muuronen et al., 2021). Here the referred work is used as an example for explaining the manufacturing procedure for native whey concentrate (NWC). First, MF process was conducted for unpasteurized skimmed milk at low temperatures (10–12°C) with an average TMP (ΔP) of 0.6 bar. The MF membrane used was a PVDF spiral wound membrane with a pore size of 0.1 μm (and 9.9 m^2 membrane area). A 30 min MF circulation step was conducted for each manufacturing trial for normalizing permeate quality between batches/trials. The MF permeate, namely, liquid "native whey" was obtained by reaching to a VCF of 3.3. Subsequently, native whey was further concentrated to native whey concentrate (NWC) using reverse osmosis (RO) equipment with a polymeric spiral wound RO element; the NWC was obtained by reaching to a Brix of 28° for the permeate at a temperature of 10°C; this is equivalent to a VCF of ~4.4. The salt rejection

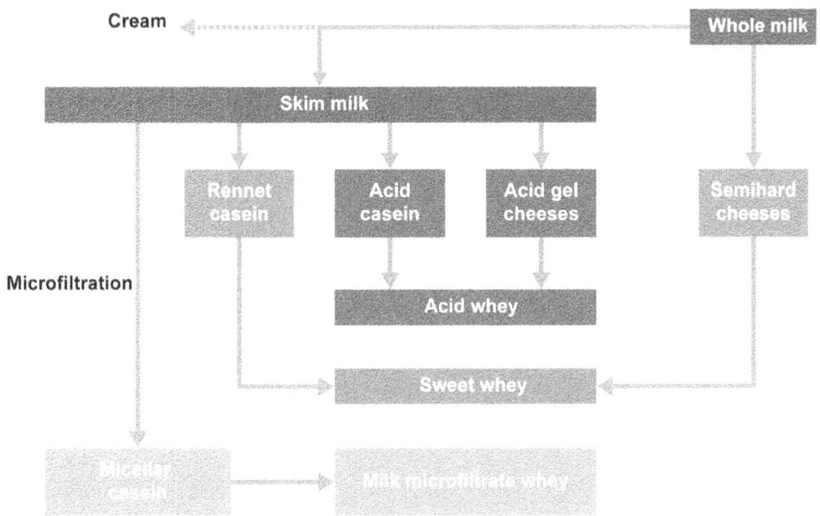

FIGURE 5.7 Classification of whey types: acid, sweet, and milk microfiltrate whey. This is based on major dairy processing pathways. *The figure was redrawn, adapted, and modified based on Kelly (2019).*

rate was 99.4% and the TMP (ΔP) was 50 bar. Muuronen et al. (2021) found that standard "high-temperature short-time" (HTST) pasteurization and spray drying caused negligible denaturation of the major whey proteins in the MF whey protein stream.

3.3 Whey protein phospholipid concentrate (WPPC)

In WPI processing, an MF process is conducted on liquid (cheese/sweet) whey and/or liquid WPC as obtained from an UF + DF process. The schematic MF retentate (MFR) processing pathways as originated from cheese whey is shown in Fig. 5.8. Further concentrating proteins of MFR using UF + DF process to enrich protein and fat (dry basis) contents to ⩾50.0% (w/w) and ⩾12.0% (w/w), respectively yield WPPC (Carter et al., 2021). Two types of whey protein phospholipid concentrate (WPPC) are available in the market: they are WPPC and infant formula grade WPPC (IF-WPPC) according to American Dairy Products Institute (2023g, 2023h). The MFR is enriched in milk fat globule membrane (MFGM) materials including milk polar lipids (e.g., glycerophospholipids and phosphosphingolipids) (Mestawet et al., 2024). The MFGM is a trilayer system, in which glycerophospholipids, phosphosphingolipid, and cholesterol construct its backbone; it also contains larger molecular weight proteins (Zheng, 2014; Zheng et al., 2013b, 2014a, 2014b, 2025b). WPPC contains high level MFGM proteins representing 23% of the total protein and high levels of MFGM phospholipids making up 20% of the total fatty acid pool of milk fat (Ozturk et al., 2025). The polar lipids derived from milk possess characteristic functionalities being beneficial for human health and for constructing delivery systems for example, phytosomes (Huang et al., 2020a, 2020b, 2020c; Ozturk et al., 2025; Zheng et al., 2013a).

Levin et al. (2016) characterized the composition of WPPC ingredients sourced from four different suppliers; it was found that commercial WPPC powder contained both native and aggregated whey proteins, the latter were not dissociated

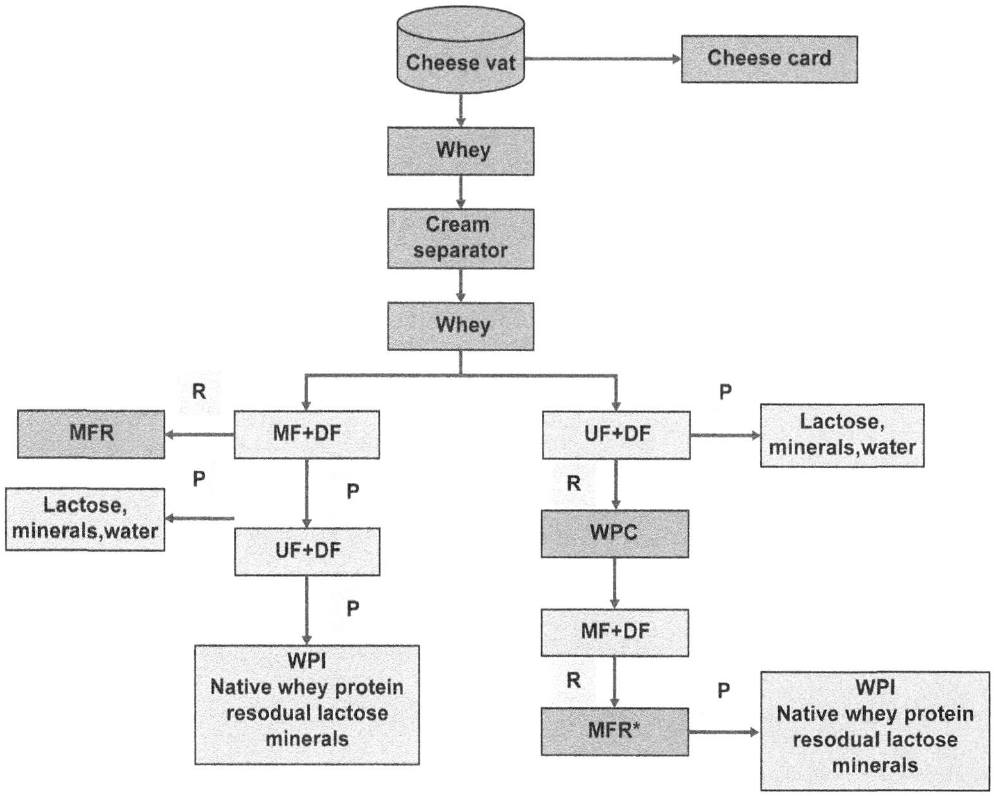

FIGURE 5.8 Process flowchart of the microfiltration retentate (MFR). UF refers to ultrafiltration, DF to diafiltration, and MF to microfiltration. P and R indicate permeate and retentate, respectively. WPC and WPI represent whey protein concentrates and whey protein isolates. *The figure was redrawn, adapted, and modified based on Mestawet et al. (2024).*

under reducing conditions as induced by β-mercaptoethanol suggesting other covalent bonds rather than S-S bridging mediated the protein aggregates. Moreover, the referred work reported that commercial WPPC powders contained ~0.6%–0.7% of free phospholipids (PL) (Levin et al., 2016). It is important to note that the referred PL contents were only based on a sum of phosphatidylcholine (PC) and phosphatidylethanolamine (PE). Sprick et al. (2019) investigated a two-step supercritical fluid extraction (SFE) process for extracting polar lipids from WPPC; the authors reported that the averaged polar lipid content may be reached ~26 g/100 g fat, and sphingomyelin, PC, and PE made up the total polar lipid content.

Carter et al. (2021) documented a practical procedure of MFR manufacturing process from liquid sweet whey. This process is explained here as a reference example. The authors used a polymeric polyvinylidene fluoride (PVDF) MF membrane with a 0.3 μm pore size and 20.5 m^2 surface area. Retentate inlet and outlet pressures were monitored using pressure gauges and these pressures were used for estimating TMP by assuming the permeate outlet pressure was 0 kPa. The MFR process was conducted as a "three-stage" process, in which the first stage was an initial MF process with a mass concentration factor (CF) of 3X; the second and the third stage are DF processes. The deionized (DI) water mass used in each DF stage was the same as the removed amount of permeate in stage 1 (2 mass parts out of 3 mass parts). To start up the process, the permeate exit valve was kept as closed at the beginning of MF process to allow the sweet whey to fill the recirculation loop. Subsequently, the permeate exit valve was gradually opened, this would allow balancing the inlet and outlet pressures for a target pressure difference of <100 kPa (or 1 bar). The MF and DF processes were conducted at 50°C. The obtained MFR can be further concentrated for increasing protein content through UF + DF processes yielding WPPC.

4. Fractionated protein ingredients from whey

Fractionation of individual proteins in whey may yield value added protein ingredients for specialized applications. Whey protein-based ingredients that are enriched in α-lactalbumin (α-lac), β-lactoglobulin (β-lg), and lactoferrin (LF) may be manufactured in large scale and have been available in the commercial ingredient market. Research and development efforts in manufacturing technology are still needed for improving the fractionation efficiency and the purity of enriched protein components. Recently literature review articles elaborated available technologies, technical challenges, and recent developments regarding fractionating targeted whey protein components (Chen et al., 2023; Khalid et al., 2025). In general, the design and the selection of fractionation methods for enriching/isolating a specific protein depends on protein intrinsic properties and their responses to environmental conditions. The intrinsic properties include molecular size, surface charge, isoelectric point, solubility and changeable environmental conditions may be pH, ionic strength, added processing aids, and temperature (Bonnaillie and Tomasula, 2008; Chen et al., 2023). The contents, sizes, and isoelectric points (IEP) are shown in Table 5.2; the table information is obtained from Khalid et al. (2025). The methods used for whey protein fractionation include but are not limited to (1) membrane filtration; (2) electrodialysis coupled with UF; (3) chromatography techniques (Bonnaillie and Tomasula, 2008; Chen et al., 2023). Some example manufacturing methods for processing fractionated whey protein ingredients are explained below.

TABLE 5.2 Content, molecular weight, and physicochemical properties of individual proteins found in whey. The table is redrawn and adapted from Khalid et al. (2025).

Protein	Content (% w/w)	Molecular weight (kDa)	Physical size (nm)	Isoelectric point (*pI*)
β-lactoglobulin (β-lg)	40–50	18.4	3.5	5.35–5.49
α-lactalbumin (α-lac)	12–15	14	1.8	4.5–4.8
Glycomacropeptide (GMP)	12	6.8		4.3–4.6
Immunoglobulins (Ig)	8.0	150–1000	6	5.5–8.3
Bovine serum albumin (BSA)	5.0	66	4	4.7–5.1
Lactoferrin (LF)	1.0	76.5		7.0–9.0
Lactoperoxidase (LPO)	0.5	78		9.2–9.9
Proteose-peptone (PP)	0.19	4–22		3.3–3.7

4.1 Alpha-lactalbumin (α-lac) and beta-lactoglobulin (β-lg) enriched ingredients

Essentially, α-lac and β-lg-enriched whey protein ingredients are WPI and WPC products (American Dairy Products Institute, 2023i). α-lac-based WPI and WPC contain no less than 89.5% and 33.5% total protein (dry matter basis), respectively, in which no less than 28% of the total protein is α-lac. β-lg-WPI and -WPC have the same required limits for total protein content, but the minimum % of β-lg in total protein was 70% and 60% for WPI and WPC, respectively (American Dairy Products Institute, 2023i, 2023j). Therefore, in other words, α-lac and β-lg WPI are WPI ingredients enriched with α-lac and β-lg, respectively. These ingredients are with moderate purities of the enriched components, α-lac and β-lg, respectively.

α-lac and β-lg possess different heat stability in acidic whey. This principle may be utilized for separating the two major whey proteins. Toro-Sierra et al., 2013b) developed a method for separating and isolating α-lac and β-lg from reconstituted WPI dispersions (15% WPI content), respectively. The referred method was able to manufacture enriched protein fractions with significantly higher purities of α-lac and β-lg as required in the referred ADPI standards; the obtained purities of α-lac and β-lg were 91.3% and 97.2%, respectively (Toro-Sierra et al., 2013b). This method contains three major stages and they are explained below:

Stage 1, thermal precipitation of α-lac in a WPI dispersion. The 15% WPI dispersion was adjusted to pH 3.4 using a mixture of trisodium citrate and citric acid. The acidic protein dispersion was heat-treated at 50 ± 1 °C for 120 min by a plate heat exchanger (PHE). The addition of 60 g L^{-1} citrate resulted in the lowest level of native α-lac after heat treatment. A 24 h aging period was used to allow further α-lac protein aggregation.

Stage 2, separating β-lg from precipitated α-lac using membrane separation methods and purification of β-lg. The β-lg removal and concentration processes using MF + UF in diafiltration modes are shown in a schematic figure (Fig. 5.9) (Toro-Sierra et al., 2013b). Ceramic MF membrane with 100 nm cutoff size was used for the separation of native β-lg in a diafiltration mode. The authors showed that a relatively lower TMP of 20 kPa and a wall shear stress of 105 Pa resulted in a higher permeation as compared with a TMP of 50 kPa. The authors showed that seven steps of diafiltration resulted in a removal of the majority of the native β-lg; the MF + DF process resulted in 95% removal of β-lg. The native β-lg was collected in the UF tank and the α-lac was kept in the MF tank (Fig. 5.9). Subsequently, the UF plant was separated from the MF plant; the β-lg-enriched fraction was used for conducting UF-diafiltration (with 10 kDa cutoff membrane) against softened water. This was done for washing out the citrate ions and minerals, therefore, purified β-lg fraction is obtained.

Stage 3, solubilization of α-lac precipitates and purification of α-lac. The α-lac precipitates left in the MF tank were solubilized at higher pH. The pH of MF retentate rich in α-lac was adjusted from 3.4 to 8.0 using 4 M NaOH and $CaCl_2{\cdot}2H_2O$. Subsequently, soluble α-lac was firstly concentrated by UF with 25 kDa cutoff and purified through three-step DF against 10 kDa membrane.

Geng et al. (2015) reported an alternative method for α-lac purification from a commercial ingredient, α-lac-enriched WPC, with a trade name of "Lacprodan® Alpha-20" manufactured by Arla Foods Ingredients, Denmark. The referred method comprises (1) acidic precipitation of denatured whey proteins and caseins; (2) UF + DF prepurification of α-lac

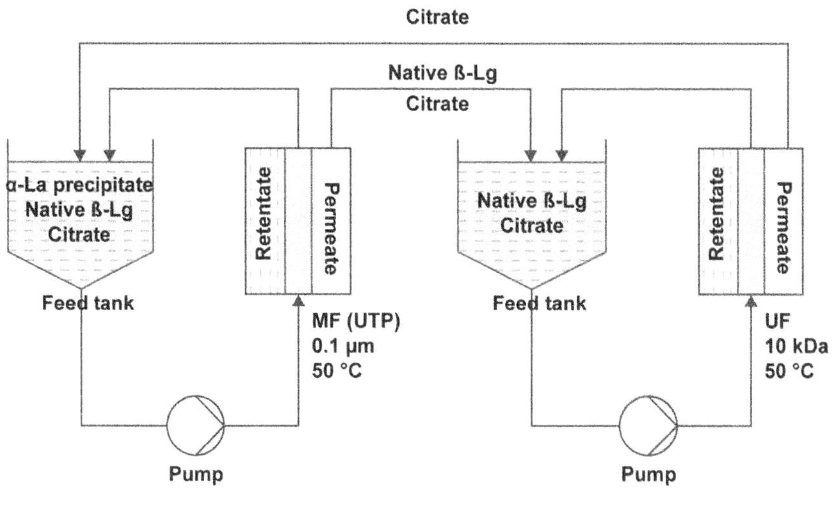

FIGURE 5.9 Schematic illustration of diafiltration mode of microfiltration and ultrafiltration for separating native protein from aggregated protein. *UTP*, uniform trans-membrane pressure. *The figure was redrawn, adapted, and modified based on Toro-Sierra et al. (2013b).*

from WPC; (3) selective binding of β-lg using anion exchange chromatography (AIEX) and collecting α-lac fraction in the 3 times of sodium acetate eluates. The process resulted in 100% and 85% in α-lac purity and recovery, respectively. A 5 L resin was packed in the AIEX column; 1 M NaCl was used for regeneration; and 0.2 M sodium acetate buffer (pH 7) was used for equilibration.

4.2 Lactoferrin

Bovine lactoferrin (LF) may be isolated from fresh bovine milk and/or from fresh whey. The isolated lactoferrin fraction is available in the ingredients market as protein ingredients. According to the ADPI standard, commercial bovine LF ingredients in powder form contain no less than 96% protein (dry basis) and no less than 90% LF (protein basis) (American Dairy Products Institute, 2023k). In general, it was concluded that about 110 g LF may be isolated from 1000 L cheese whey (Vigolo et al., 2024). The referred work found that LF was more abundant in Ricotta cheese whey than in raw milk and cheese whey in terms of protein fraction % of total whey protein. However, regarding mass/volume (w/v) content, raw milk and cheese whey contained 0.16 and 0.11 mg/mL LF, respectively; whereas Ricotta cheese whey contained 0.03 mg/mL LF. Research activities have been conducted on the topic of isolating/separating LF for decades. For instance, LF can be isolated or separated from whey or raw milk by gel filtration techniques (Al-Mashikhi and Nakai, 1987); chelating chromatography (Al-Mashikhi et al., 1988); cation exchange chromatography (Yoshida and Ye-Xiuyun, 1991; Chiu and Etzel, 1997; Hahn et al., 1998; Ye et al., 2000); magnetic affinity separation (Chen et al., 2007); electroseparation (Ndiaye et al., 2010; Wang et al., 2020); multimodal chromatography (Hirsch et al., 2020). Since ion exchange (IEX) chromatography is considered as the mostly used technique for purifying LF from cheese whey, this technique is elaborated below as an example manufacturing process for concentrating LF.

According to the work conducted by Ye et al. (2000), a rennet whey (350 mL) was passed through an IEX column (1.5 × 18 cm) packed with a strong cation exchange resin, sulphopropyl-Toyopearl (SP-TP). Both lactoperoxidase (LPOase) and LF were adsorbed by the SP-TP resin. The process contained a washing phase, which was 200 mL 0.05 M phosphate buffer (pH 6.5). LPOase and LF were eluted with the same phosphate buffer containing a gradient of NaCl (e.g., 0–0.55 M). The authors of the referred work suggested that 60 mL regenerated SP-TP resin is needed for adsorbing LF in 1 L fresh rennet whey.

The LF contents in bovine WPC were found at 0.30% and 0.99% (powder mass basis) for cheese WPC and acid WPC, respectively (Palmano and Elgar, 2002). In general, membrane filtration can be applied as a prepurification step before the LF purification by IEX chromatography. UF + DF processes coupled with different drying techniques (spray drying vs. freeze drying) with different processing parameters (e.g., TMP) may yield feta cheese WPC containing ~2% LF (powder mass basis) (Tsakali et al., 2015). Maciel et al. (2020) reported a manufacturing procedure of concentrating LF from sweet whey. The method is composed of (1) a clarification step by MF; (2) a prepurification step by UF + DF; (3) a purification step by cationic exchange expanded bed chromatography. The method was able to concentrate LF from 0.12 mg/mL as in the original sweet whey to 17.4 mg/mL in the purified fraction, which is corresponding to ~93% purity and ~87% recovery rate.

Previously, other research efforts have also focused on the prepurification step, specifically utilizing membrane filtration. Electrodialysis coupled with a filtration membrane (EDFM) had been practiced for concentrating LF. Ndiaye et al. (2010) conducted an EDFM process for concentrating LF from lactoferrin-enriched-whey solution. Briefly, the overall electrodialysis system was composed of a cationic membrane, an anionic membrane, and a (polyethersulfone) UF membrane with an MWCO at 500 kDa; these membranes constructed three closed loops: (1) electrode loop filled with 250 mL 20 g/L NaCl solution, (2) permeate loop filled with 250 mL 6 g/L KCl solution, and (3) feed loop filled with 200 mL lactoferrin-enriched whey solution at pH 3. The schematic configuration of EDFM is shown in Fig. 5.10, which is adapted based on (Ndiaye et al., 2010). The authors reported that the maximum migration rate of LF into the permeate phase was ~15%, however, the migration rates of other whey proteins such as β-lg and α-lac were high, at ~19% and ~18%, respectively. This resulted in a low LF fraction percentage of the total protein in the permeate (Ndiaye et al., 2010). Therefore, the referred EDFM process was not able to directly obtain an LF fraction meeting the ADPI standard of LF ingredients. Secondary purification must be applied for further concentrating of LF in order to meet the ADPI standard for commercial LF ingredients (American Dairy Products Institute, 2023k). Wang et al. (2020) manufactured an asymmetric polyvinyl alcohol (PVA) membrane in-house for conducting a benchtop scale EDFM process (e.g., 40 mL feed volume). The PVA membrane showed strong rejection for LF indicating a potential for scaling up investigations in the future. In summary, membrane filtration can be used as a pretreatment step before chromatography for yielding high concentration LF products.

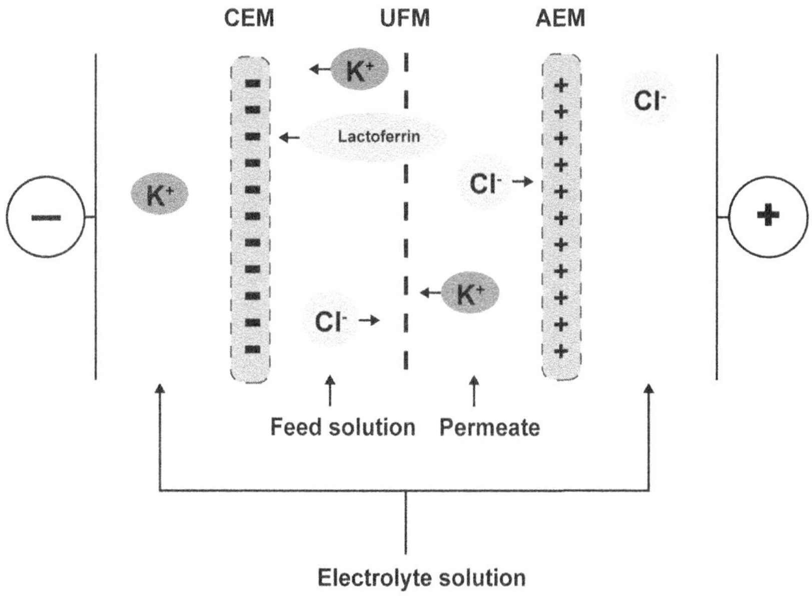

FIGURE 5.10 Schematic illustration of lactoferrin separation in an electrodialysis cell with ultrafiltration membrane. AEM, UFM and CEM represent anion-exchange membrane, ultrafiltration membrane, and cation-exchange membrane, respectively. *The figure was redrawn, adapted, and modified based on Ndiaye et al. (2010).*

4.3 Glycomacropeptide (GMP)

GMP is also known as caseinomacropeptide, which is obtained from chymosin hydrolysis of κ-casein during curd formation step in cheesemaking process. Chymosin hydrolyzes κ-casein at Phe105–Met106 yielding para-κ-casein and GMP; the former is an amino acid (AA) chain (AA 1–105) derived from the N-terminal of κ-casein, the latter is a polypeptide chain (AA 106–169) derived from the C-terminal of κ-casein (Farrell et al., 2004; Zheng, 2024). Para-κ-caseins are associated with casein micelles and are incorporated in cheese curd; whereas, GMP is released into the whey phase. Numerous research efforts have reported the biological activities and potential of health benefits for GMP; moreover, GMP was projected to have growing market demand for commercial applications (Chavda et al., 2025). For instance, GMP is peptide free of phenylalanine, therefore, it may be used in medical foods prepared for phenylketonuria (PKU) patients (Ebrahimi et al., 2024). These motivate activities in isolating GMP from cheese whey.

GMP may be separated from other whey proteins by different approaches (Chavda et al., 2025). Ebrahimi et al. (2024) provided comprehensive reviews regarding different methods for GMP isolation. The referred review articles categorized the isolation methods into three groups: (1) methods based on heat treatment and solvent precipitation; (2) aqueous two-phase system; (3) chromatography-based methods.

The methods in the referred "group 1" can be combined with a method in the "group 3" for obtaining a fraction enriched in GMP. For instance, Nakano et al. (2018) performed two precipitation steps for removing contaminant (non-GMP) materials from sweet whey protein products, WPI solution (1g WPI in 20 mL water). The solution was heated in boiling water for 10 min to form "precipitate I" composed of whey protein aggregates. The heat-treated dispersion was ultracentrifuged at 27,000 × g for 20 min and the supernatant was collected and adjusted to pH 4.6 yielding "precipitate II" after centrifugation. The supernatant derived from "precipitate II" enriched with GMP was adjusted to pH 3.0. This supernatant was used to further isolate GMP by using diethylaminoethyl (DEAE)-Sephacel anion exchange chromatography. GMP was eluted by a linear gradient of NaCl solution (0–1 M). The two precipitation steps removed a significant amount of non-GMP materials up to ~68% of dry mass of the original WPI solution. >70% GMP bound in the IEC column; IEC purification resulted in a yield of ~7.5% of the dry mass of the original WPI sample. The GMP concentrate samples as manufactured in the referred work contained no detectable phenylalanine suggesting a high purity of GMP in the IEC-processed sample. Nakano and Betti (2020) reported that different resin and chitin may offer different GMP binding capacity ranging from ~8 to 52 mg GMP/100 mg dry weight of anion exchanger components; DEAE-Sephacel showed the highest binding capacity.

4.4 Whey protein hydrolysates (WPH)

Enzymatic hydrolysis of WPC resulted in an increase in protein digestibility from 25% to 69%–70% (Sinha et al., 2007). WPH exhibits antioxidant activity in model beverages (Athira et al., 2015), provides nutritional benefits as a functional ingredient in infant formulas (Li and Zhu, 2024), reduces allergenicity (Guadix et al., 2006), increases angiotensin-converting enzyme (ACE) inhibitory activity (O'Loughlin et al., 2014).

WPH may be processed using batch style reactors and continuous membrane reactors (Bansal and Bhandari, 2016). The processes conducted using batch style reactors are commonly reported in literature regarding lab scale research activities on WPH manufacturing and functionality. Guadix et al. (2006) reported a WPH processing method using a continuous stirred tank membrane reactor (CSTMR). The system contained a polyethersulfone plate and frame ultrafiltration module with an MWCO of 3 kDa. The key features of CSTMR include separating hydrolyzate peptides as a product from the substrate and the enzyme; enabling the reuse of enzymes; and controlling the molecular size of peptide products produced through choosing an MWCO of the incorporated membrane as part of the CSTMR system (Guadix et al., 2006).

It has been reported that hydrolyzing heat-treated WPC with 1% pepsin then with 1% trypsin was more effective for producing low-antigenic hydrolysates than native WPC (Kim et al., 2007). This phenomenon may likely be related to the conformational state, which is resistant to proteolysis; heat treatment may induce unfolding and alter the conformation of whey proteins (O'Loughlin et al., 2012). Schmidt and Markwijk (1993) investigated the impact of heat treatment on enzymatic hydrolysis capacities for α-lac and β-lg, respectively. The authors found that α-lac may be hydrolyzed by pepsin but not by papain. The authors also found that heating α-lac at 82°C for 8 min did not significantly affect the hydrolysis capacity for either enzyme; moreover, β-lg was resistant to pepsin but it was hydrolyzed by papain, heating β-lg at the same temperature allowed the protein to be hydrolyzed by pepsin. It was found that pH was a determining factor for the total hydrolysis rate of α-lac commercial ingredients; the hydrolysis was catalyzed by pepsin (Vreeke et al., 2023). Due to the positive effects of heat treatment and predenaturation of whey proteins on their susceptibility to hydrolysis, O'Loughlin et al. (2012) investigated the enzymatic hydrolysis rate of aggregated WPI as induced by heat treatment, the target degree of hydrolysis (DH) was 5%. The authors suggested that whey protein aggregate size and aggregate porosity were determining factors to the rate of enzymatic hydrolysis; a higher heat treatment intensity (80°C for 10 min) resulted in a faster hydrolysis rate. The hydrolysis method performed by the referred work is explained below as an example of a hydrolysis process of WPI:

A 300 mL WPI solution at pH 8 containing 100 g/L protein was used as the substrate solution. A commercial digestive-enzyme complex Corolase PP (E.C. 3.4.21.4.) was used as the proteolysis enzyme. The enzyme:substrate ratio used was 1:100 (w/w, protein equivalent basis). The hydrolysis was performed at 50°C and at pH 8, pH was maintained using the pH-stat method and 2 M NaOH was used as titrant. The hydrolysis reaction was terminated by heat treatment (85°C for 20 min). The DH estimation equation was used for determining the needed volume of base titrant, the equation is shown below according to Butré et al. (2014):

$$DH\% = 100 \times h/h_{\text{tot}} = 100 \times V_b \times N_b \times 1/\alpha \times 1/M_p \times 1/h_{\text{tot}} \tag{5.9}$$

where DH% is the percentage of peptide bonds cleaved per gram of protein (h) to the total peptide bonds per gram of protein (h_{tot}). h_{tot} used for estimating DH of WPI hydrolysis was 8.5 (Butré et al., 2014). V_b is the volume of base titrant in mL, N_b is the normality of the base titrant. M_p is the mass of the protein in the solution in g. α is the average degree of dissociation of the α-amino groups or α-carboxylic groups in a protein and it may be calculated using the equation below (Butré et al., 2014; Deng et al., 2018)

$$\alpha = \frac{1}{1 + 10^{\left(pk_{a,NH_2} - pH\right)}} - \frac{1}{1 + 10^{\left(pH - pk_{a,\text{COOH}}\right)}} \tag{5.10}$$

where pH is the pH maintained during the hydrolysis, pk_{a,NH_2} and $pk_{a,\text{COOH}}$ are the average dissociation constants of α-amino and α-carboxylic groups, respectively. $pk_{a,\text{COOH}}$ is 3.1 and pk_{a,NH_2} can be calculated according to the following equation

$$pk_{a,NH_2} = 7.8 + \frac{(298 - T)}{(298 \times T)} \times 2400 \tag{5.11}$$

where T is the temperature (in Kelvin) maintained during hydrolysis.

The DH of WPH is a critical factor determining bioactive properties, sensory attributes, and techno-functionalities. Eberhardt et al. (2019) performed enzymatic hydrolysis on WPC using a commercial enzyme, Alcalase® (EC 3.4.21.62, serine endopeptidase), the work investigated the impact of DH on WPH functionalities. The authors found that increasing

DH from 8% to 17% increased solubility, antioxidant activity and ACE inhibition activities. However, WPH with a higher DH may possess undesirable bitterness. By evaluating flavor profiles for 22 WPH manufactured from eight manufacturers, Leksrisompong et al. (2010) found positive correlations between DH and two ingredient characteristics, which are (1) bitter taste intensity and (2) concentration of low molecular weight peptides, respectively. The DH of the mentioned commercial WPH samples ranged from ~1% to 47% as characterized by the TNBS method. Singh and Dalgleish (1998) reported that WPH with 10% or 20% DH exhibited maximum emulsifying capacity. The authors also found that WPH-stabilized emulsions were not stable after an ultra-high temperature heat treatment (122°C for 15 min); WPH with relatively lower DH may be able to result in a relatively stable emulsion after a lower intensity heat treatment (90°C for 30 min) (Singh and Dalgleish, 1998).

4.5 Aggregated whey protein ingredients

Microparticulated whey proteins (MWP) in concentrated powder form are available as commercial ingredients. The ingredient resulted in superior heat stability of emulsions in which it was used as an emulsifying agent (Çakır-Fuller, 2015). It may also be considered as a fat replacer or modulating textural properties for cultured foods and frozen desserts (Ipsen, 2017). The manufacturing processes are consolidated and explained in literature (Bansal and Bhandari, 2016; Kelly, 2019).

In general, assembling denatured or unfolded whey proteins into colloidal particles through thermal and/or antisolvent treatments may result in new or improved functional properties (Amin et al., 2025a; Amin and Zheng, 2022; Nicolai, 2016; Shi and Zheng, 2025; Zheng, 2024). Upon heating, whey proteins may be unfolded, subsequently, the exposed reactive sites or AA residues may mediate protein-protein interactions through noncovalent bonds (e.g., hydrophobic attraction and hydrogen bonding) and covalent bonds (e.g., thiol-disulfide interexchange) (Bansal and Bhandari, 2016). Undesirable or uncontrolled protein aggregation may result in destabilized colloidal systems or insoluble particles, which may cause phase separation issues in processed consumer foods. Therefore, the colloidal stability of thermal treatment induced whey protein aggregate particles in model foods is an important concern. For a whey protein solution at pH 7, if a whey protein content, e.g., β-lg, is higher than a critical level (e.g., > 60 g/L) the thermal aggregation may result in four different systems depending on the NaCl concentration (e.g., 0.01–1 M): (1) a stable colloidal sol, (2) a homogenous gel, (3) a phase-separated gel, and (4) a precipitate system (Ako et al., 2009; Ryan et al., 2013).

Singer and Dunn (1990) showed a schematic flow chart of processing microparticulated proteins, for example, a commercial ingredient, namely "Simplesse". According to the flow chart and the referred article, it is known that not only milk proteins but egg white proteins were used for making the protein microparticulates; moreover, other nonprotein materials such as lecithin and pectin can be incorporated. The protein microparticulation concept, namely "Simplesse", was first invented in 1979 at the Canadian Beer company, John Labatt Ltd.; the "Simplesse" trademark was assigned in 1988 and currently, it has been evolved as a whey protein-based ingredient, namely microparticulated WPC as offered by CP Kelco (Kelly, 2019). Microparticulation concept is based on simultaneous treatments combining thermal denaturation/coagulation and mechanical shear. It is common that MWP contains protein aggregates with diameters larger than 1 μm. A previous study, for example, indicated that a substantial volume fraction of protein particles in a dispersion made from a commercial MWP ingredient exceeded 1 μm in diameter (Çakır-Fuller, 2015). Whey protein aggregates with submicron sizes were manufactured from conventional WPI; these particles exhibited superior functionalities compared with the parental WPI in terms of improving physical stabilities for emulsion gel, foam, and heat-treated protein dispersion (Amin et al., 2025a; Shi and Zheng, 2025).

Manufacturing conditions and the characteristics of the original whey protein dispersion determine the structural and functional properties of aggregated whey protein particles. Toro-Sierra et al. (2013a) suggested that whey protein aggregate particle size was slightly increased upon spray drying, however, it was independent of the spray drying temperatures (e.g., inlet: 180–210°C; outlet: 75–105°C). Manufacturing functional whey protein aggregates from a protein dispersion with high protein content is often favored because it can result in a higher yield per processing batch. With this interest, Wolz and Kulozik (2015) studied the thermal denaturation kinetics of whey proteins in concentrated protein systems with a protein content up to 40%. The authors reported that the denaturation rate of β-lg was dependent on temperature of protein content. The same research group conducted a follow-up work, in which the authors investigated the impacts of protein content and shear rate (during thermal aggregation) on aggregate size (Wolz et al., 2016). Interestingly, the authors found that increasing the protein content from 5% to 30% resulted in formation of smaller and stable aggregates; this phenomenon was attributed to the higher viscosity of the high protein content system (Wolz et al., 2016). Detailed processing information of commercial MWP ingredients is often proprietary (Bansal and Bhandari, 2016). Further research is needed to investigate the key factors that influence the yield and functional properties of whey protein aggregates.

5. Remarks and conclusion

This chapter focuses on manufacturing processes of different whey protein ingredients. The chapter provides a general landscape of different whey processing pathways for the sake of concentration and fractionation of proteins in whey. Standards of common commercial protein ingredients are introduced and example manufacturing processes of featured ingredients are explained based on published literature information. Manufacturing treatments such as heat, shear, and added chemicals may alter protein structures, therefore, the protein functionalities and digestibilities may be modified. Such processing induced protein structural changes may be detrimental to ingredient functionality or such changes may be controlled and be utilized for improving protein functionality toward special food applications. Although the manufacturing processes for many whey protein ingredients are relatively mature and have been well studied, continuous and systematic research efforts are encouraged for the following aspects: (1) understanding impacts of processing variables on protein structural and functionality changes; (2) improving the concentration/fractionation efficiency of (bioactive) proteins from whey (e.g., bulk proteins, lactoferrin, and GMP); (3) modifying whey proteins toward specialized and value added WPC/WPI ingredients for constructing dedicated food structures, for example, protein dispersion, O/W emulsion, foam, gel, etc.

References

21 CFR 184.1979—Whey. 2025. Code Fed. Regul. U. S. https://www.ecfr.gov/current/title-21/part-184/section-184.1979. Accessed July 16, 2025.

Ako, K., Nicolai, T., Durand, D., Brotons, G., 2009. Micro-phase separation explains the abrupt structural change of denatured globular protein gels on varying the ionic strength or the pH. Soft Matter 5, 4033–4041. https://doi.org/10.1039/B906860K.

Al-Mashikhi, S.A., Nakai, S., 1987. Isolation of bovine immunoglobulins and lactoferrin from whey proteins by gel filtration techniques. J. Dairy Sci. 70, 2486–2492. https://doi.org/10.3168/jds.S0022-0302(87)80315-6.

Al-Mashikhi, S.A., Li-Chan, E., Nakai, S., 1988. Separation of immunoglobulins and lactoferrin from cheese whey by chelating chromatography. J. Dairy Sci. 71, 1747–1755. https://doi.org/10.3168/jds.S0022-0302(88)79741-6.

American Dairy Products Institute, 2023a. Dry Whey Standard.

American Dairy Products Institute, 2023b. Reduced Lactose Whey Standard.

American Dairy Products Institute, 2023c. Reduced Minerals Whey Standard.

American Dairy Products Institute, 2023d. Whey Protein Concentrate Standard.

American Dairy Products Institute, 2023e. Whey Protein Isolate Standard.

American Dairy Products Institute, 2023f. Milk Whey Protein Standard.

American Dairy Products Institute, 2023g. Whey Protein Phospholipid Concentrate Standard.

American Dairy Products Institute, 2023h. Infant Formula Grade Whey Protein Phospholipid Concentrate Standard.

American Dairy Products Institute, 2023i. Alpha-lactalbumin Standard.

American Dairy Products Institute, 2023j. Beta-lactoglobulin Standard.

American Dairy Products Institute, 2023k. Lactoferrin Standard.

Amin, U., Zheng, H., 2022. Soft matter strategy for creating novel food texturizer: cold set whey protein gels constructed by fractal protein assemblies. In: 2022 American Dairy Science Association® Annual Meeting. American Dairy Science Association via Journal of Dairy Science, Kansas City, MO.

Amin, U., Lin, Y., Zuo, X., Zheng, H., 2024. Soft matter approach for creating novel protein hydrogels using fractal whey protein assemblies as building blocks. Food Hydrocoll. 151, 109828. https://doi.org/10.1016/J.FOODHYD.2024.109828.

Amin, U., Yeung, C.K., Zheng, H., 2025a. Construction of cold-set Mickering emulsion gel using whey protein assembly particles as oil-water interfacial stabilizer and gelling agent: phase stability, nonlinear rheology, and tribology. J. Dairy Sci. https://doi.org/10.3168/JDS.2024-26001.

Amin, U., Yeung, C.K., Zheng, H., 2025b. Enzymatic synthesis of calcium lactobionate from cheese whey permeate as a value-added ingredient. JDS Commun. https://doi.org/10.3168/jdsc.2025-0834.

An, H., Zheng, H., 2024. Investigation of correlations between powder functionalities and powder surface properties for milk protein ingredients. J. Dairy Sci. https://doi.org/10.3168/JDS.2024-25289.

An, H., Zheng, H., 2025. Investigation of the relation between interfacial properties of milk protein ingredients and their emulsifying properties. J. Dairy Sci. https://doi.org/10.3168/JDS.2025-26514.

Arefi-Oskoui, S., Vatanpour, V., Khataee, A., 2016. Development of a novel high-flux PVDF-based ultrafiltration membrane by embedding Mg-Al nanolayered double hydroxide. J. Ind. Eng. Chem. 41, 23–32. https://doi.org/10.1016/j.jiec.2016.06.023.

Arunkumar, A., Etzel, M.R., 2015. Negatively charged tangential flow ultrafiltration membranes for whey protein concentration. J. Membr. Sci. 475, 340–348. https://doi.org/10.1016/j.memsci.2014.10.049.

Aslam, M., Wicaksana, F., Farid, M., Wong, A., Krantz, W.B., 2022. Mitigation of membrane fouling by whey protein via water hammer. J. Membr. Sci. 642, 119967. https://doi.org/10.1016/j.memsci.2021.119967.

Athira, S., Mann, B., Saini, P., Sharma, R., Kumar, R., Singh, A.K., 2015. Production and characterisation of whey protein hydrolysate having antioxidant activity from cheese whey. J. Sci. Food Agric. 95, 2908–2915. https://doi.org/10.1002/jsfa.7032.

Baldasso, C., Barros, T.C., Tessaro, I.C., 2011. Concentration and purification of whey proteins by ultrafiltration. Desalination 278, 381–386. https://doi.org/10.1016/j.desal.2011.05.055.

Bansal, N., Bhandari, B., 2016. Functional milk proteins: production and utilization—whey-based ingredients. In: McSweeney, H.P.L., O'Mahony, A.J. (Eds.), Advanced Dairy Chemistry: Volume 1B: Proteins: Applied Aspects. Springer, New York, pp. 67–98. https://doi.org/10.1007/978-1-4939-2800-2_3.

Barukčić, I., Božanić, R., Kulozik, U., 2015. Influence of process temperature and microfiltration pre-treatment on flux and fouling intensity during cross-flow ultrafiltration of sweet whey using ceramic membranes. Int. Dairy J. 51, 1–7. https://doi.org/10.1016/j.idairyj.2015.07.002.

Bonnaillie, L.M., Tomasula, P.M., 2008. Whey protein fractionation. In: Whey Processing, Functionality and Health Benefits. John Wiley & Sons, Ltd, pp. 15–38. https://doi.org/10.1002/9780813803845.ch2.

Brighenti, M., Govindasamy-Lucey, S., Jaeggi, J.J., Johnson, M.E., Lucey, J.A., 2021. Effect of substituting whey cream for sweet cream on the textural and rheological properties of cream cheese. J. Dairy Sci. 104, 10500–10512. https://doi.org/10.3168/jds.2021-20338.

Butré, C.I., Wierenga, P.A., Gruppen, H., 2014. Influence of water availability on the enzymatic hydrolysis of proteins. Process Biochem. 49, 1903–1912. https://doi.org/10.1016/j.procbio.2014.08.009.

Çakır-Fuller, E., 2015. Enhanced heat stability of high protein emulsion systems provided by microparticulated whey proteins. Food Hydrocoll. 47, 41–50. https://doi.org/10.1016/j.foodhyd.2015.01.003.

Carter, B., DiMarzo, L., Pranata, J., Barbano, D.M., Drake, M., 2021. Efficiency of removal of whey protein from sweet whey using polymeric microfiltration membranes. J. Dairy Sci. 104, 8630–8643. https://doi.org/10.3168/jds.2020-18771.

Carter, B.G., Cheng, N., Kapoor, R., Meletharayil, G.H., Drake, M.A., 2021. *Invited review:* microfiltration-derived casein and whey proteins from milk. J. Dairy Sci. 104, 2465–2479. https://doi.org/10.3168/jds.2020-18811.

Chavda, V., Panchal, M., Ganvit, P., Pasaya, R., Chaudhari, A., Teli, D., 2025. Health potential of glycomacropeptide: a review highlighting isolation, characterization methods, chemistry and biological activities. Int. J. Pept. Res. Ther. 31, 72. https://doi.org/10.1007/s10989-025-10733-y.

Chen, L., Guo, C., Guan, Y., Liu, H., 2007. Isolation of lactoferrin from acid whey by magnetic affinity separation. Sep. Purif. Technol. 56, 168–174. https://doi.org/10.1016/j.seppur.2007.01.019.

Chen, G.Q., Qu, Y., Gras, S.L., Kentish, S.E., 2023. Separation technologies for whey protein fractionation. Food Eng. Rev. 15, 438–465. https://doi.org/10.1007/s12393-022-09330-2.

Chiu, C.K., Etzel, M.R., 1997. Fractionation of lactoperoxidase and lactoferrin from bovine whey using a cation exchange membrane. J. Food Sci. 62, 996–1000. https://doi.org/10.1111/j.1365-2621.1997.tb15023.x.

Damar, I., Cinar, K., Gulec, H.A., 2020. Concentration of whey proteins by ultrafiltration: comparative evaluation of process effectiveness based on physicochemical properties of membranes. Int. Dairy J. 111, 104823. https://doi.org/10.1016/j.idairyj.2020.104823.

Deng, Y., Gruppen, H., Wierenga, P.A., 2018. Comparison of protein hydrolysis catalyzed by bovine, porcine, and human trypsins. J. Agric. Food Chem. 66, 4219–4232. https://doi.org/10.1021/acs.jafc.8b00679.

Eberhardt, A., López, E.C., Ceruti, R.J., Marino, F., Mammarella, E.J., Manzo, R.M., Sihufe, G.A., 2019. Influence of the degree of hydrolysis on the bioactive properties of whey protein hydrolysates using Alcalase®. Int. J. Dairy Technol. 72, 573–584. https://doi.org/10.1111/1471-0307.12606.

Ebrahimi, A., Andishmand, H., Huo, C., Amjadi, S., Khezri, S., Hamishehkar, H., Mahmoudzadeh, M., Kim, K.H., 2024. Glycomacropeptide: a comprehensive understanding of its major biological characteristics and purification methodologies. Compr. Rev. Food Sci. Food Saf. 23, e13370. https://doi.org/10.1111/1541-4337.13370.

Farrell, H.M., Jimenez-Flores, R., Bleck, G.T., Brown, E.M., Butler, J.E., Creamer, L.K., Hicks, C.L., Hollar, C.M., Ng-Kwai-Hang, K.F., Swaisgood, H.E., 2004. Nomenclature of the proteins of cows' milk—sixth revision. J. Dairy Sci. 87, 1641–1674. https://doi.org/10.3168/jds.S0022-0302(04)73319-6.

Fox, P.F., Guinee, T.P., Cogan, T.M., McSweeney, P.L.H., 2017. Whey and whey products. In: Fundamentals of Cheese Science. Springer, Boston, MA, pp. 755–769. https://doi.org/10.1007/978-1-4899-7681-9_22.

Geng, X.L., Tolkach, A., Otte, J., Ipsen, R., 2015. Pilot-scale purification of α-lactalbumin from enriched whey protein concentrate by anion-exchange chromatography and ultrafiltration. Dairy Sci. Technol. 95, 353–368. https://doi.org/10.1007/s13594-015-0215-8.

Guadix, A., Camacho, F., Guadix, E.M., 2006. Production of whey protein hydrolysates with reduced allergenicity in a stable membrane reactor. J. Food Eng. 72, 398–405. https://doi.org/10.1016/j.jfoodeng.2004.12.022.

Hahn, R., Schulz, P.M., Schaupp, C., Jungbauer, A., 1998. Bovine whey fractionation based on cation-exchange chromatography1. J. Chromatogr. A 795, 277–287. https://doi.org/10.1016/S0021-9673(97)01030-3.

Hirsch, D.B., Martínez Álvarez, L.M., Urtasun, N., Baieli, M.F., Lázaro- Martínez, J.M., Glisoni, R.J., Miranda, M.V., Cascone, O., Wolman, F.J., 2020. Lactoferrin purification and whey protein isolate recovery from cheese whey using chitosan mini-spheres. Int. Dairy J. 109, 104764. https://doi.org/10.1016/j.idairyj.2020.104764.

Huang, Z., Brennan, C., Zhao, H., Guan, W., Mohan, M.S., Stipkovits, L., Zheng, H., Liu, J., Kulasiri, D., 2020a. Milk phospholipid antioxidant activity and digestibility: kinetics of fatty acids and choline release. J. Funct.Foods 68, 103865. https://doi.org/10.1016/j.jff.2020.103865.

Huang, Z., Brennan, C.S., Zhao, H., Liu, J., Guan, W., Mohan, M.S., Stipkovits, L., Zheng, H., Kulasiri, D., 2020b. Fabrication and assessment of milk phospholipid-complexed antioxidant phytosomes with vitamin C and E: a comparison with liposomes. Food Chem. 324, 126837. https://doi.org/10.1016/j.foodchem.2020.126837.

Huang, Z., Zheng, H., Brennan, C.S., Mohan, M.S., Stipkovits, L., Li, L., Kulasiri, D., 2020c. Production of milk phospholipid-enriched dairy ingredients. Foods 9, 263.

Ibach, A., Kind, M., 2007. Crystallization kinetics of amorphous lactose, whey-permeate and whey powders. Carbohydr. Res. 342, 1357–1365. https://doi.org/10.1016/j.carres.2007.03.002.

Ipsen, R., 2017. Microparticulated whey proteins for improving dairy product texture. Int. Dairy J. 67, 73–79. https://doi.org/10.1016/j.idairyj.2016.08.009.

Jinjarak, S., Olabi, A., Jiménez-Flores, R., Walker, J.H., 2006. Sensory, functional, and analytical comparisons of whey butter with other Butters1. J. Dairy Sci. 89, 2428–2440. https://doi.org/10.3168/jds.S0022-0302(06)72316-5.

Kelly, P., 2019. Chapter 3—manufacture of whey protein products: concentrates, isolate, whey protein fractions and microparticulated. In: Deeth, H.C., Bansal, N. (Eds.), Whey Proteins. Academic Press, pp. 97–122. https://doi.org/10.1016/B978-0-12-812124-5.00003-5.

Khalid, M., Ghasem, N., Yang, X., Schroen, K., Nazir, A., 2025. Progress in membrane technology for the enrichment of individual whey proteins. Ind. Eng. Chem. Res. 64, 19305–19321. https://doi.org/10.1021/acs.iecr.5c00643.

Kilara, A., 2015. Whey and whey products. In: Dairy Processing and Quality Assurance. John Wiley & Sons, Ltd, pp. 349–366. https://doi.org/10.1002/9781118810279.ch15.

Kim, S.B., Ki, K.S., Khan, M.A., Lee, W.S., Lee, H.J., Ahn, B.S., Kim, H.S., 2007. Peptic and tryptic hydrolysis of native and heated whey protein to reduce its antigenicity. J. Dairy Sci. 90, 4043–4050. https://doi.org/10.3168/jds.2007-0169.

Kumar, L., Brennan, M., Zheng, H., Brennan, C., 2018. The effects of dairy ingredients on the pasting, textural, rheological, freeze-thaw properties and swelling behaviour of oat starch. Food Chem. 245, 518–524. https://doi.org/10.1016/J.FOODCHEM.2017.10.125.

Kumar, L., Brennan, M., Brennan, C., Zheng, H., 2021. Influence of whey protein isolate on pasting, thermal, and structural characteristics of oat starch. J. Dairy Sci. https://doi.org/10.3168/jds.2021-20711.

Kumar, L., Brennan, M., Zheng, H., Kumar, G., Brennan, C., 2023. Protein fortification in oat flour gel using various dairy protein ingredients: an evaluation of textural and pasting properties. Int. J. Dairy Technol. https://doi.org/10.1111/1471-0307.13018.

Leksrisompong, P.P., Miracle, R.E., Drake, M., 2010. Characterization of flavor of whey protein hydrolysates. J. Agric. Food Chem. 58, 6318–6327. https://doi.org/10.1021/jf100009u.

Levin, M.A., Burrington, K.J., Hartel, R.W., 2016. Composition and functionality of whey protein phospholipid concentrate and delactosed permeate. J. Dairy Sci. 99, 6937–6947. https://doi.org/10.3168/jds.2016-10974.

Li, J., Zhu, F., 2024. Whey protein hydrolysates and infant formulas: effects on physicochemical and biological properties. Compr. Rev. Food Sci. Food Saf. 23, e13337. https://doi.org/10.1111/1541-4337.13337.

Li, J., Wang, W., Zhou, J., Tang, C.-C., 2024. Ionic regulation whey protein composition on membrane fouling during ultrafiltration process analyzed by HPLC. Int. J. Biol. Macromol. 254, 127751. https://doi.org/10.1016/j.ijbiomac.2023.127751.

Maciel, K.S., Santos, L.S., Bonomo, R.C.F., Verissimo, L.A.A., Minim, V.P.R., Minim, L.A., 2020. Purification of lactoferrin from sweet whey using ultrafiltration followed by expanded bed chromatography. Sep. Purif. Technol. 251, 117324. https://doi.org/10.1016/j.seppur.2020.117324.

Marella, C., Salunke, P., Biswas, A.C., Kommineni, A., Metzger, L.E., 2015. Manufacture of modified milk protein concentrate utilizing injection of carbon dioxide. J. Dairy Sci. 98, 3577–3589. https://doi.org/10.3168/jds.2014-8946.

Mestawet, A.T., France, T.C., Mulcahy, P.G.J., O'Mahony, J.A., 2024. Microfiltration retentate co-product from whey protein isolate production—composition, processing, applications and potential for value addition. Trends Food Sci. Technol. 153, 104739. https://doi.org/10.1016/j.tifs.2024.104739.

Milk Whey Protein, 2025. ADPI. https://adpi.org/ingredient-resources/milk-whey-protein-whey/. Accessed July 17, 2025.

Musale, D.A., Kulkarni, S.S., 1998. Effect of whey composition on ultrafiltration performance. J. Agric. Food Chem. 46, 4717–4722. https://doi.org/10.1021/jf9801567.

Muuronen, K., Partanen, R., Heidebrecht, H.-J., Kulozik, U., 2021. Effects of conventional processing methods on whey proteins in production of native whey powder. Int. Dairy J. 116, 104959. https://doi.org/10.1016/j.idairyj.2020.104959.

Nakano, T., Betti, M., 2020. Isolation of κ-casein glycomacropeptide from bovine whey fraction using food grade anion exchange resin and chitin as an adsorbent. J. Dairy Res. 87, 127–133. https://doi.org/10.1017/S0022029919000918.

Nakano, T., Ozimek, L., Betti, M., 2018. Separation of bovine κ-casein glycomacropeptide from sweet whey protein products with undetectable level of phenylalanine by protein precipitation followed by anion exchange chromatography. J. Dairy Res. 85, 110–113. https://doi.org/10.1017/S0022029917000838.

Ndiaye, N., Pouliot, Y., Saucier, L., Beaulieu, L., Bazinet, L., 2010. Electroseparation of bovine lactoferrin from model and whey solutions. Sep. Purif. Technol. 74, 93–99. https://doi.org/10.1016/j.seppur.2010.05.011.

Nicolai, T., 2016. Formation and functionality of self-assembled whey protein microgels. Colloids Surf. B Biointerfaces 137, 32–38. https://doi.org/10.1016/j.colsurfb.2015.05.055.

Ozturk, G., Nicholson, C.F., Hartel, R.W., 2025. Toward the future of next-generation dairy foods—a processing perspective and economic analysis*. JDS Commun. 6, 714–719. https://doi.org/10.3168/jdsc.2024-0727.

O'Loughlin, I.B., Murray, B.A., Kelly, P.M., FitzGerald, R.J., Brodkorb, A., 2012. Enzymatic hydrolysis of heat-induced aggregates of whey protein isolate. J. Agric. Food Chem. 60, 4895–4904. https://doi.org/10.1021/jf205213n.

O'Loughlin, I.B., Murray, B.A., Brodkorb, A., FitzGerald, R.J., Kelly, P.M., 2014. Production of whey protein isolate hydrolysate fractions with enriched ACE-inhibitory activity. Int. Dairy J. 38, 101–103. https://doi.org/10.1016/j.idairyj.2013.12.008, 8th NIZO Dairy Conference.

Palmano, K.P., Elgar, D.F., 2002. Detection and quantitation of lactoferrin in bovine whey samples by reversed-phase high-performance liquid chromatography on polystyrene–divinylbenzene. J. Chromatogr. A 947, 307–311. https://doi.org/10.1016/S0021-9673(01)01563-1.

Patel, A.S., Zheng, H., Yeung, C.K., 2022a. The product and its manufacture. In: Encyclopedia of Dairy Sciences. Academic Press, pp. 744–754.

Patel, A.S., Yeung, C.K., Brennan, C., Zheng, H., 2022b. Additives in dairy foods: ingredients and formulation. In: McSweeney, P.L.H., McNamara, J.P. (Eds.), Encyclopedia of Dairy Sciences. Academic Press, Oxford, pp. 663–673. https://doi.org/10.1016/B978-0-12-818766-1.00161-6.

Rivero, W.C., Wang, Q., Salvi, D., Zheng, H., 2022. Plasma-activated Water (PAW) as a Cleaning-In-Place (CIP) Solution for Proteinaceous Fouling Removal. 2022 IFT FIRST Annu. Event Expo.

Ryan, K.N., Zhong, Q., Foegeding, E.A., 2013. Use of whey protein soluble aggregates for thermal stability—a hypothesis paper. J. Food Sci. 78, R1105–R1115. https://doi.org/10.1111/1750-3841.12207.

Schmidt, D., Markwijk, B. van, 1993. Enzymatic hydrolysis of whey proteins. Influence of heat treatment of α-lactalbumin and β-lactoglobulin of their proteolysis of pepsin and papain. Neth. Milk Dairy J. 47, 15–22.

Shi, H., Zheng, H., 2025. Mesoscale fractal whey protein particles derived from microscale linear-shaped protein assemblies (Part 2): foaming properties and heat stability. J. Dairy Sci. https://doi.org/10.3168/jds.2025-27190.

Singer, N.S., Dunn, J.M., 1990. Protein microparticulation: the principle and the process. J. Am. Coll. Nutr. 9, 388–397. https://doi.org/10.1080/07315724.1990.10720397.

Singh, A.M., Dalgleish, D.G., 1998. The emulsifying properties of hydrolyzates of whey proteins. J. Dairy Sci. 81, 918–924. https://doi.org/10.3168/jds.S0022-0302(98)75651-6.

Sinha, R., Radha, C., Prakash, J., Kaul, P., 2007. Whey protein hydrolysate: functional properties, nutritional quality and utilization in beverage formulation. Food Chem. 101, 1484–1491. https://doi.org/10.1016/j.foodchem.2006.04.021.

Sprick, B., Linghu, Z., Amamcharla, J.K., Metzger, L.E., Smith, J.S., 2019. Selective extraction of phospholipids from whey protein phospholipid concentrate using supercritical carbon dioxide and ethanol as a co-solvent. J. Dairy Sci. 102, 10855–10866. https://doi.org/10.3168/jds.2019-16419.

Tarapata, J., Dybowska, B.E., Zulewska, J., 2022. Evaluation of fouling during ultrafiltration process of acid and sweet whey. J. Food Eng. 328, 111059. https://doi.org/10.1016/j.jfoodeng.2022.111059.

Tetra Pak, 2025a. 18 milk and whey fractionation. In: Dairy Processing Handbook (Online Version).

Tetra Pak, 2025b. 7.2 Centrifugal separators and milk standardization. In: Dairy Processing Handbook (Online Book). Tetra Pak.

Toro-Sierra, J., Schumann, J., Kulozik, U., 2013a. Impact of spray-drying conditions on the particle size of microparticulated whey protein fractions. Dairy Sci. Technol. 93, 487–503. https://doi.org/10.1007/s13594-013-0124-7.

Toro-Sierra, J., Tolkach, A., Kulozik, U., 2013b. Fractionation of α-lactalbumin and β-lactoglobulin from whey protein isolate using selective thermal aggregation, an optimized membrane separation procedure and resolubilization techniques at pilot plant scale. Food Bioprocess Technol. 6, 1032–1043. https://doi.org/10.1007/s11947-011-0732-2.

Tsakali, E., Petrotos, K., D'Alessandro, A.G., Mantas, C., Tripolitsiotis, I., Goulas, P., Chatzilazarou, A., van Impe, J.F., 2015. Exploring the effect of ultrafiltration/diafiltration processing conditions on the lactoferrin and immunoglobulin G content of feta whey protein concentrates. J. Food Process. Eng. 38, 363–373. https://doi.org/10.1111/jfpe.12167.

Vigolo, V., Niero, G., Ballancin, E., De Marchi, M., 2024. Variation of lactoferrin and whey protein fractions along different stages of milk technological processing at dairy plant level. Int. Dairy J. 152, 105895. https://doi.org/10.1016/j.idairyj.2024.105895.

Villeneuve, W., Bérubé, A., Chamberland, J., Pouliot, Y., Labrie, S., Doyen, A., 2021. Contribution of biofouling to permeation flux decline and membrane resistance changes during whey ultrafiltration. Int. Dairy J. 117, 105010. https://doi.org/10.1016/j.idairyj.2021.105010.

Vreeke, G.J.C., Vincken, J.-P., Wierenga, P.A., 2023. Quantitative peptide release kinetics to describe the effect of pH on pepsin preference. Process Biochem. 134, 351–362. https://doi.org/10.1016/j.procbio.2023.10.021.

Wang, G., Guo, M., 2019. Manufacturing technologies of whey protein products. In: Whey Protein Production, Chemistry, Functionality, and Applications. John Wiley & Sons, Ltd, pp. 13–37. https://doi.org/10.1002/9781119256052.ch2.

Wang, Q., Chen, G.Q., Kentish, S.E., 2020. Isolation of lactoferrin and immunoglobulins from dairy whey by an electrodialysis with filtration membrane process. Sep. Purif. Technol. 233, 115987. https://doi.org/10.1016/j.seppur.2019.115987.

Wang, W., Zhou, J., Li, J., Cong-Cong, T., 2024. The mechanism of whey protein on membrane surface fouling during ultrafiltration process. Food Biophys. 19, 143–159. https://doi.org/10.1007/s11483-023-09811-9.

Wen-qiong, W., Yun-chao, W., Xiao-feng, Z., Rui-xia, G., Mao-lin, L., 2019. Whey protein membrane processing methods and membrane fouling mechanism analysis. Food Chem. 289, 468–481. https://doi.org/10.1016/j.foodchem.2019.03.086.

Whey Ingredients Overview, 2025. ADPI. https://adpi.org/ingredient-resources/whey-overview/. Accessed July 17, 2025.

Wolz, M., Kulozik, U., 2015. Thermal denaturation kinetics of whey proteins at high protein concentrations. Int. Dairy J. 49, 95–101. https://doi.org/10.1016/J.IDAIRYJ.2015.05.008.

Wolz, M., Mersch, E., Kulozik, U., 2016. Thermal aggregation of whey proteins under shear stress. Food Hydrocoll. 56, 396–404. https://doi.org/10.1016/J.FOODHYD.2015.12.036.

Ye, X., Yoshida, S., Ng, T.B., 2000. Isolation of lactoperoxidase, lactoferrin, α-lactalbumin, β-lactoglobulin B and β-lactoglobulin A from bovine rennet whey using ion exchange chromatography. Int. J. Biochem. Cell Biol. 32, 1143–1150. https://doi.org/10.1016/S1357-2725(00)00063-7.

Yee, K.W.K., Wiley, D.E., Bao, J., 2007. Whey protein concentrate production by continuous ultrafiltration: operability under constant operating conditions. J. Membr. Sci. 290, 125–137. https://doi.org/10.1016/j.memsci.2006.12.026.

Yoshida, S., Ye-Xiuyun, 1991. Isolation of lactoperoxidase and lactoferrins from bovine milk acid whey by carboxymethyl cation exchange chromatography. J. Dairy Sci. 74, 1439–1444. https://doi.org/10.3168/jds.S0022-0302(91)78301-X.

Zheng, H., 2014. Structural Aspects of the Bovine Milk Fat Globule Membrane. University of Otago.

Zheng, H., 2018. Polymers for structure design of dairy foods. In: Gutiérrez, T.J. (Ed.), Polymers for Food Applications. Springer International Publishing, Cham, pp. 509–528. https://doi.org/10.1007/978-3-319-94625-2_19.

Zheng, H., 2019. Introduction: measuring rheological properties of foods. In: Joyner, H.S. (Ed.), Rheology of Semisolid Foods. Springer International Publishing, Cham, pp. 3–30. https://doi.org/10.1007/978-3-030-27134-3_1.

Zheng, H., 2022. Texture. In: Encyclopedia of Dairy Sciences. Academic Press, pp. 616–625.

Zheng, H., 2024. Chemistry and properties of the caseins. In: Casein Structural Properties, Uses, Health Benefits and Nutraceutical Applications, pp. 23–36. https://doi.org/10.1016/B978-0-443-15836-0.00022-6.

Zheng, H., Meletharayil, H., 2025. Constructing aqueous foams from milk components: structure and interfaces. Curr. Opin. Food Sci. 63, 101290. https://doi.org/10.1016/J.COFS.2025.101290.

Zheng, H., Gordon, K.C., Everett, D.W., 2013a. Innovative application of confocal Raman microscopy to investigate the interaction between trans-2-hexenal and bovine milk fat globules. Int. Dairy J. 32, 68–70. https://doi.org/10.1016/j.idairyj.2013.04.011.

Zheng, H., Jiménez-Flores, R., Everett, D.W., 2013b. Bovine milk fat globule membrane proteins are affected by centrifugal washing processes. J. Agric. Food Chem. 61, 8403–8411. https://doi.org/10.1021/jf402591f.

Zheng, H., Jiménez-Flores, R., Everett, D.W., 2014a. Lateral lipid organization of the bovine milk fat globule membrane is revealed by washing processes. J. Dairy Sci. 97, 5964–5974. https://doi.org/10.3168/jds.2014-7951.

Zheng, H., Jiménez-Flores, R., Gragson, D., Everett, D.W., 2014b. Phospholipid architecture of the bovine milk fat globule membrane using giant unilamellar vesicles as a model. J. Agric. Food Chem. 62, 3236–3243. https://doi.org/10.1021/jf500093p.

Zheng, H., Jiménez-Flores, R., Everett, D.W., 2025a. The impact of sphingomyelin and cholesterol on ordered lipid domain formation in the bovine milk fat globule membrane using artificial giant unilamellar vesicles as a model. JDS Commun. 6, 485–489. https://doi.org/10.3168/jdsc.2024-0719.

Zheng, H., Jiménez-Flores, R., Everett, D.W., 2025b. The impact of sphingomyelin and cholesterol on ordered lipid domain formation in the bovine milk fat globule membrane using artificial giant unilamellar vesicles as a model. JDS Commun. 6, 485–489. https://doi.org/10.3168/jdsc.2024-0719.

Chapter 6

Properties and microstructure of whey protein products

Todor Vasiljevic
Victoria University, Melbourne, VIC, Australia

1. Introduction

Whey protein (WP) is a term widely used and understood to refer to the proteins not associated with the casein micelle or other milk particles. However, the proteins found in whey may differ from those in the soluble phase of milk, usually termed milk serum, as their composition depends on the process applied during the separation of the casein. For example, sweet whey, typically produced during the manufacturing of various cheeses, may include proteins associated with starter cultures as well as glycomacropeptide (GMP), a peptide released during the action of rennet on κ-casein. Therefore, the commercially available whey protein concentrates (WPC) and/or isolates may differ in their composition and, consequently, functionality. For decades, whey was regarded merely as a by-product of cheese production and often considered waste. However, due to its high production volume and the nutritional and functional benefits of WPs, the industry has shifted its focus to extracting WPs from liquid whey. In 2014, global whey output was approximately 240 million metric tons, with production rates expected to increase by 3.5% annually (Lagrange et al., 2015). Past research indicates that WPs are among the most nutritionally valuable proteins available (Sousa et al., 2012). As a result, the demand for nutritional markets, such as sports, clinical, and infant nutrition, has prompted unprecedented levels of investment in dairy production.

Whey is increasingly recognized as a highly valuable nutrient source due to its rich components, such as high-gelling β-lactoglobulin (β-LG), human milk equivalent protein α-lactalbumin (α-LA), lactoferrin, immunoglobulins, and its role as a precursor to the probiotic galactoligosaccharides. Scientific and technological advancements have greatly improved whey's significance in the food industry (Bogahawaththa and Vasiljevic, 2022). WP powders, available in varying purity levels and concentrations of major WPs, are versatile ingredients appreciated for their wide range of functionalities, along with their outstanding nutritional and physiological benefits. They are widely utilized in numerous food applications, including protein supplements, baked goods, dairy products, and meat items.

WPs are globular proteins that remain soluble in milk serum even after caseins precipitate at a pH of 4.6. The main WPs are β-lactoglobulin (β-LG), α-lactalbumin (α-LA), bovine serum albumin (BSA), and immunoglobulins (Ig), which are classified according to their concentrations in milk. The minor WPs include lactoferrin (Lf), proteose peptone 3 (PP3), osteopontin, lactoperoxidase (LPO), lysozyme, and cathepsin D, among others (Deeth and Bansal, 2019). β-LG and α-LA represent roughly 50% and 20% of the total WPs in whey and milk serum, respectively (Bogahawaththa and Vasiljevic, 2022).

This chapter aims to describe the origins of WPs and how their processing history affects their structural properties. Different WP sources need distinct processing methods for optimal use. Although the proteins may be identical, environmental factors such as mineral content, pH, and the type of acid in the whey solution influence protein conformation, leading to different functionalities.

2. Sources of whey proteins (WPs)—Native, sweet, salty and acid whey

The manufacturing process for WP products differs based on their source. While certain WPs, especially valuable minor ones like lactoferrin, can be directly extracted from milk, whey is typically the principal source of these proteins. The two

Whey. https://doi.org/10.1016/B978-0-443-27648-4.00003-7

TABLE 6.1 Summary of the findings from previous studies on the composition of sweet whey, acid whey and salt whey.

Component	Sweet whey	Acid whey	Salty whey	References
Total protein (%)	0.4–0.6[2] 0.8[3]	0.8[1] 0.6–0.9[2]	0.6[3]	[1]Lievore et al. (2013), [2]Alsaed et al. (2013), [3]Blaschek et al. (2007)
Lactose (%)	2.7–4.9	4.2–4.9		Alsaed et al. (2013), Josephson et al. (1975), Lievore et al. (2013)
Titratable acidity (%), (As lactic acid)	<0.2	>0.4		Chegini and Taheri (2013)
pH	6.0–6.5	3.5–4.3	5.2–5.4	Alsaed et al. (2013), Kapoor and Metzger (2004)
Ash (%)	0.6–1.9[1]	0.6–1.0[2]	0.4–0.6[1]	[1]Alsaed et al. (2013), [2]Lievore et al. (2013)
Salt (%)	0.1–0.4		2.6–18.9	Blaschek et al. (2007)
Ca^{2+} (g/L) Cl^- (g/L)	0.5–0.7 0.7–1.3	0.9		Jimenez et al. (2012), Josephson et al. (1975)

primary types are "sweet whey" and "acid whey," which are primarily formed from the acid and rennet coagulation of caseins, respectively. Moreover, the use of high salt concentrations during the production of specific cheeses results in what is referred to as "salty whey" (Nishanthi et al., 2017a). Table 6.1 summarizes the characteristics of these three whey types.

The whey produced from hard and semihard cheese manufacturing, along with rennet casein, is referred to as "sweet whey." This type of whey is predominantly utilized in the production of WP products. As shown in Table 6.1, sweet whey has a well-balanced composition, lacking significant salinity or acidity. The total protein content varies between 0.6% and 0.8%, featuring β-LG and α-LA concentrations that closely resemble those found in milk. The compositional traits of sweet whey are similar to those of milk, enabling it to maintain its protein qualities with minimal denaturation and aggregation. Additionally, sweet whey's pH is analogous to that of milk, resulting in a reduced impact from acidity or alkalinity on protein properties. This allows for effective processing of sweet whey and the successful production of shelf-stable whey powders. Sweet whey powder is widely used as an energy supplement and food ingredient due to its excellent functional properties in baked goods, infant formulas, confections, and desserts.

Salty whey is a byproduct of salted hard cheese production. During the manufacturing of these cheese varieties, an additional salting step occurs after the coagulation of caseins. Consequently, the whey produced after this salting step has high levels of sodium salts. This high salinity leads to underutilization of salty whey due to the elevated processing costs. In contrast to sweet whey, salty whey contains about 15–20 times more sodium salts (Table 6.1). It also has a relatively high fat content (Blaschek et al., 2007), primarily due to the ongoing agitation of the cheddar curd after salting, which increases the fat's solubility in a high ionic strength medium. Salty whey is abundant in lactoferrin but has lower amounts of β-LG and α-LA compared to sweet whey (Blaschek et al., 2007). The higher lactoferrin levels stem from the reduced pH during cheese production (Dupont et al., 2006). While the solubility of β-LG and α-LA remains unaffected by the high salt concentration in salty whey, lactoferrin solubility is impacted by the increased salt levels (Blaschek et al., 2007). Despite its unique composition, salty whey is utilized in various products. For example, nanofiltration can remove 90% of the salt from salty whey, recovering 80% of the nonsalt whey solids, which are then added back into the sweet whey stream (Gregory, 1987). In another instance, permeate separated and concentrated through nanofiltration has been used as a salt addition for different cheese varieties (Sanderson et al., 1996). Moreover, salty whey has been incorporated into processed cheese and related products without significantly affecting quality (Kapoor and Metzger, 2004).

The acid precipitation of casein during the production of cottage cheese, ricotta cheese, and mainly Greek yoghurt results in acid whey as a by-product. Over the past decade, consumer demand for acid-coagulated products has tripled, and this trend is expected to continue for several more years. As a result, the industry faces the challenge of managing an annual output of 1.6 billion liters of acid whey as waste. Acid whey has similar protein and lactose levels when compared to sweet whey, but it contains higher amounts of calcium and phosphorus salts (Table 6.1) (Chandrapala et al., 2015). However, the acid whey produced during Greek yoghurt manufacturing has a lower protein content due to the aggregation caused by intense heat treatment during pasteurization (Nishanthi et al., 2017a). Additionally, acid whey has a relatively high concentration of lactic acid (LA), primarily resulting from the fermentation of lactose in milk by lactic acid bacteria. The high solubility of LA leads to its presence in the acid whey stream when straining the casein coagulum. The composition of acid whey can vary depending on its source; for example, acid whey from soft cheese production contains less calcium and phosphate compared to acid whey from Greek yoghurt. Yet, the problematic component, lactic acid, is found in roughly equivalent amounts in the acid whey from both processes. The protein and nonprotein components of acid whey hold potential for various applications in food systems, contingent on their purity. Extracted proteins from acid whey can be mixed with other milk proteins for use in infant formulas (Rudloff and Lönnerdal, 1992). Lactic acid is commonly employed as a food preservative (Datta et al., 1995), while isolated lactose can also be utilized in infant formulas (Chandra et al., 1989). Deacidified and demineralized acid whey is currently incorporated in bakery products (Kosikowski, 1979).

3. Manufacturing of whey products

A variety of WP products, including concentrates, isolates, hydrolyzates, and demineralized powders, are available on the market (see Fig. 6.1). The global WP market is expected to grow from USD 9.78 billion in 2024 to USD 10.65 billion in 2025, projecting a total value of around USD 22.95 billion by 2034. In North America, the WP market is predicted to reach USD 3.72 billion in 2024 (Precedence Research, 2024). These whey products fulfill different roles in diverse food systems, identified as the most effective and efficient ways to consume WP. Figure 6.1 displays the various WP products derived from sweet whey, which are frequently utilized in commercial beverage production, including energy drinks, as well as in baked goods, confections, desserts, and infant formula.

4. Whey protein concentrate and isolate

WPC and whey protein isolate (WPI) powders are widely used in food manufacturing. WPC protein content varies from 35% to 80%, influenced by the protein concentration achieved through membrane processing. In the dairy sector, ultrafiltration (UF) membranes with a molecular weight cutoff (MWCO) of 10 kDa are primarily used for WP separation and concentration, although some processors are now employing tighter 5 kDa membranes as technology progresses in multistage plants.

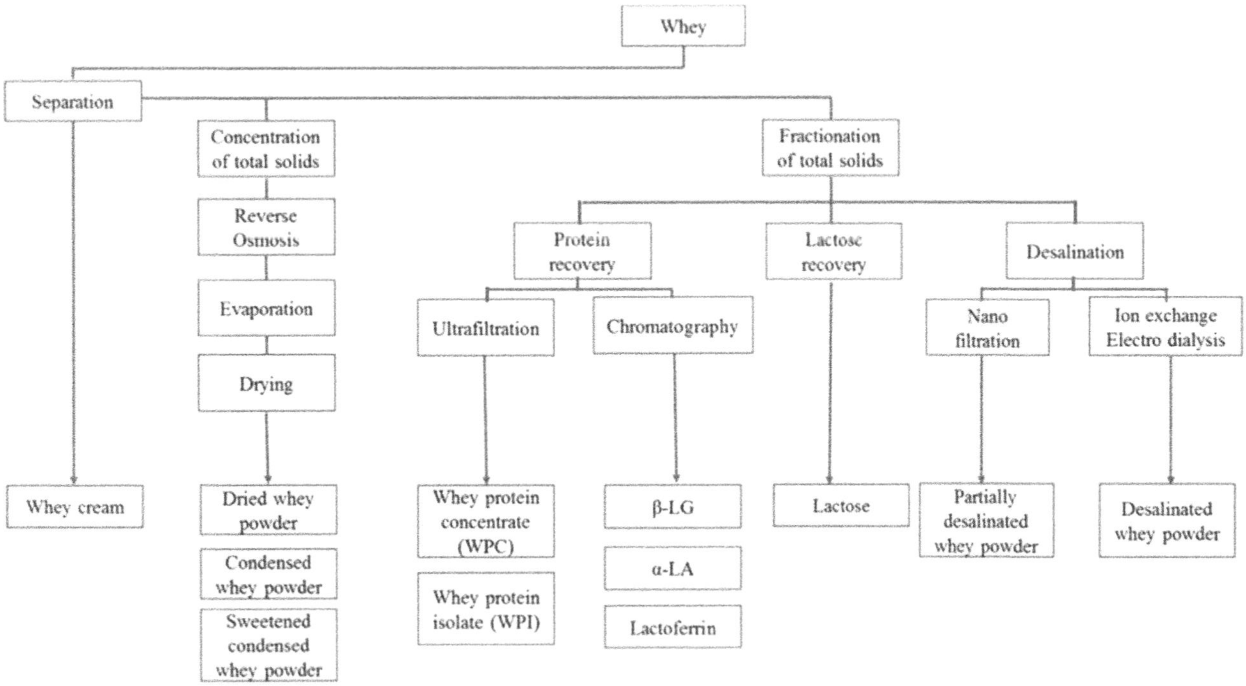

FIGURE 6.1 Production of different whey products.

The protein enrichment level in WPC depends on the volume concentration factor (VCF), which varies from 35% for WPC35% to 80% for WPC80. For WPC35, a VCF between 4.5 and 7.0 is needed, whereas WPC60 requires a VCF of 13–20. To enhance protein concentration, it is essential to pair ultrafiltration (UF) with diafiltration. This combination allows for a VCF of 30–35, resulting in WPC with 75%–85% protein content. The proteins captured and concentrated by ultrafiltration membranes undergo further processing and are spray-dried to produce WPC-based ingredients.

WPC is a product formed by removing the nonprotein elements of whey, resulting in a final dry product that contains at least 25% protein (ADPI, 2016). The proximate compositions of WPC at 35% protein (WPC35), 80% protein (WPC80), and WPI are outlined in Table 6.2. To achieve a protein concentration of 90% or higher on a dry basis, reducing the fat content, which typically ranges from 4% to 8% in WPC80, is essential. Currently, there is a distinct market for WPCs containing between 35% and 80% protein, showcasing the varied applications and market potential. On the lower end, WPC35 can be economically matched with commodity-produced skim milk powder (SMP) as a viable substitute. Formulators who seek specific ingredient functionalities or wish to minimize nonprotein elements in their formulations typically prefer higher-protein WPCs.

The key stages in producing WPC and WPI involve separation, pasteurization, concentration, and spray drying (see Fig. 6.1). To achieve high-quality WPC and WPI, it is crucial to effectively separate cheese fines and residual fat from the liquid whey, which is accomplished using centrifugal clarifiers and separators. The pasteurization process is primarily aimed at reducing or eliminating the microbial load in the whey, which may contain harmful pathogens. During concentration, other nonprotein components are removed to enhance the protein ratio, leading to variations in WPC and WPI depending on the concentration level. Lastly, the main method for drying WPC and WPI commercially is spray drying, which aims to lower the product's moisture content to 4% or less (Fig. 6.2).

TABLE 6.2 Composition difference of WPC and WPI (Abd El-Salam et al., 2009).

Component	WPC35 (%)	WPC80 (%)	WPI
Moisture	3.0–4.5	3.5–4.5	4.5
Ash	2.5–4.5	3.0–4.0	0.3–4.5
Protein	34–36	80–82	90–92
Lactose	48–52	4.0–8.0	0.2–2.0
Fat	3.0–4.5	4.0–8.0	0.2–1.5

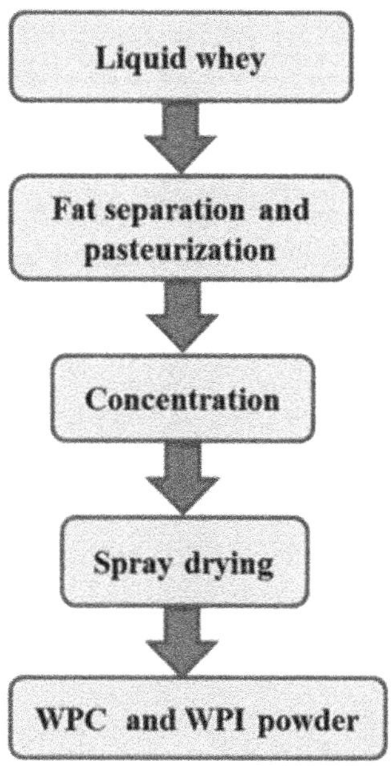

FIGURE 6.2 Primary steps involved in the manufacturing of WPC and WPI.

5. WP properties in different whey streams

Each whey stream possesses a distinct composition (Alsaed et al., 2013; Josephson et al., 1975) that influences its subsequent processing and the quality of the final product. Consequently, these chemical differences play a crucial role in determining the selection of processes and the characteristics of the end products. As shown in Table 6.1, sweet whey has the highest total protein and WP levels, while acid whey typically has lower protein content. These variations stem from the different manufacturing methods used to produce these whey streams. Sweet whey is produced during the manufacturing of rennet-type hard cheeses, where rennet/chymosin is used to trigger coagulation. This chymosin enzyme cleaves the Phe_{105}-Met_{106} bond in κ-casein, producing para-κ-casein and glycomacropeptide (GMP). Most GMP remains in the whey, while para-κ-casein stays bound to the casein micelles and becomes part of the cheese. The total protein content of acid whey ranges from 0.23% to 0.52% (w/w) (Nishanthi et al., 2017a; Chandrapala et al., 2015) and is influenced by its processing history. Acid whey, derived from Greek yoghurt production typically has a lower protein concentration than that produced from soft cheese. In Greek yogurt production, milk is enhanced with milk solids and subjected to high-temperature treatment, typically at 90°C for 5 min (Bong and Moraru, 2014), before being inoculated with the starter culture. In contrast, soft cheese production involves a gentle warming step at around 30°C. This heating causes WPs to denature above 80°C, exposing active thiol groups (Chandrapala et al., 2010) that engage in thiol-disulphide exchange reactions with caseins, particularly κ-casein and αs2-casein (Vasbinder and de Kruif, 2003). The aggregated WPs are then incorporated into the casein coagulum, which reduces the WP content in the acid whey stream. Salty whey has a significantly higher sodium (Na) concentration compared to sweet, acid, and native whey streams (Nishanthi et al., 2017a). This increase is attributed to the addition of sodium chloride (NaCl) during the salting of various cheese types (Grummer et al., 2013; Lu and McMahon, 2015). NaCl enhances the concentration gradient and establishes an osmotic pressure outside the casein coagulum, which aids in the expulsion of whey, a process known as syneresis. During syneresis, the NaCl that settles on the cheese coagulum dissolves into the whey, leading to an increased concentration of Na in the waste stream.

WPs isolated from sweet whey through membrane filtration have become valuable ingredients for gelling, emulsifying, and foaming applications (de Faria et al., 2013). However, extracting WPs from acid and salty whey remains challenging due to filtration membrane fouling. The varying solubility of minerals, along with protein/protein interactions, leads to changes in particle size within the feed streams, which subsequently causes membrane fouling (Koh et al., 2014). The largest particle size is typically found in acid whey (~263 nm), followed by salty whey (~215 nm), while sweet (~180 nm) and native whey (~190 nm) have smaller sizes (Nishanthi et al., 2017a). The ionically charged environment of salty whey decreases intramolecular electrostatic repulsion, encouraging protein unfolding. This condition reduces intermolecular electrostatic repulsion, allowing WPs to arrange more closely, which enhances intermolecular hydrophobic, van der Waals, and thiol interactions that facilitate protein aggregation (Jeyarajah and Allen, 1994). Furthermore, the heat treatment applied during Greek yoghurt production contributes to the formation of protein aggregates in acid whey. At temperatures above 65°C, the free sulfhydryl group of β-LG can initiate sulfhydryl-disulphide interchange reactions, thereby promoting protein/protein aggregation (Rahaman et al., 2015).

The physical functionality of WPs largely depends on their surface properties, which are influenced by the presence of hydrophobic segments, free thiol groups, and related surface charges. Even minor changes can lead to different surface interactions, possibly resulting in protein unfolding and subsequent aggregation. In this regard, three primary types of interactions need evaluation: hydrophobic, disulphide/thiol, and electrostatic interactions, to identify suitable membranes for processing acid and salty whey streams. Surface hydrophobicity (S_0) serves as a key predictor of protein functionality. Variations in S_0 are primarily caused by conformational changes in proteins' secondary structures due to environmental influences during processing (Laiho et al., 2015). Consequently, the S_0 values of proteins in acid whey are significantly lower than those from sweet, salty, or native whey streams (Nishanthi et al., 2017a). Additionally, heat treatment during Greek yogurt production instigates conformational changes due to intermolecular aggregation, driven by hydrophobic interactions that trap hydrophobic amino acid (AA) residues in the aggregate structures (Rahaman et al., 2015). Moreover, the higher levels of lactic acid in acid whey, when compared to sweet, salty, and native whey, impede hydrophobic interactions and promote lactic acid-induced protein aggregation. This is predominantly driven by electrostatic, covalent, and van der Waals forces (Dissanayake et al., 2013a, 2013b), which leads to a more stable, dense spherical structure while limiting the hydrophobic residues available on the protein surface (Rahaman et al., 2015; de la Fuente et al., 2002). Conversely, salty whey exhibits the highest S_0 values among the different whey streams (Nishanthi et al., 2017a). High concentrations of NaCl alter the electric double layer surrounding protein molecules, leading to intermolecular unfolding and revealing previously hidden hydrophobic areas. At the same time, previously obscured negatively charged AA residues, such as Asp_{129}, Asp_{130}, Glu_{131}, Glu_{134}, and Asp_{137}, enhance the protein surface's negative charge, resulting in a significantly higher negative surface potential in salty whey. While surface charge and charge density may fluctuate at elevated NaCl concentrations, the surface remains negatively charged (Israelachvili, 2011). Covalent interactions through thiol/disulphide exchange and thiol oxidation reactions are crucial in determining the extent of protein aggregation across various whey streams. The ability of β-LG to initiate thiol covalent linkages stems from its free thiol group (Cyc_{121}). In addition, this molecule contains two disulphide bonds (Cys_{66}-Cys_{160}) and (Cys_{106}-Cys_{119}) (Sava et al., 2005) that facilitate thiol/disulphide-mediated reactions. Conversely, α-LA has four disulphide bonds but lacks free thiol groups. Consequently, denatured α-LA can only incorporate into aggregate structures through thiol/disulphide exchange reactions with denatured proteins that have free sulfhydryl groups. For this reason, WPs in sweet and native whey tend to have a higher total thiol content than those found in acid and salty whey. However, WPs in acid and salty streams are more prone to engage in thiol/disulphide exchange reactions (Nishanthi et al., 2017a). The sodium-rich environment in salty whey somewhat neutralizes intramolecular electrostatic repulsions, facilitating protein unfolding and exposing formerly hidden free thiol groups on the surface (Jeyarajah and Allen, 1994). This exposure enhances the thiol/disulphide exchange reaction by reducing intermolecular distances. Furthermore, the heat applied during Greek yogurt manufacturing is a primary factor contributing to lower levels of reactive thiol groups in acid whey. Nevertheless, in both acid and salty whey streams, thiol/thiol oxidation reactions are more prevalent than in native whey, underscoring that high salt content in salty whey and severe heat treatment conditions during acid whey production promote thiol oxidation (Jeyarajah and Allen, 1994; Rahaman et al., 2015).

Preserving the native structures of WPs is crucial when processing whey streams. Steps like isolation, concentration, and drying during downstream processing can alter the native WP structure (Koh et al., 2014), impacting their specific functionalities. Thus, maintaining the secondary structure of the WPs in liquid whey waste streams is essential for achieving the desired functional performance. The variations in the main physical properties of WPs across different whey streams are reflected in their conformational characteristics. Analyzing the FTIR interferograms in the Amide I region (1600–1700 cm^{-1}), which relates to C=O stretching vibrations, shows that native and sweet whey exhibit two key peaks at 1637 cm^{-1} and 1654 cm^{-1}, indicating the presence of intermolecular β-sheets and α-helical structures (Fig. 6.3; Nishanthi et al., 2017a). These peaks primarily correspond to the native structure of WPs. In contrast to sweet whey, salty whey displays no peaks linked to the native secondary structures, suggesting a loss of native organization. Instead, a significant peak emerged at 1683 cm^{-1}, pointing to intramolecular antiparallel β-sheet aggregation. Consequently, the diminishing of major secondary structural segments (β-sheet, α-helix) coincided with the intramolecular β-sheet aggregation, which typically indicates the presence of protein aggregates in salty whey. Acid whey typically exhibits less intense peaks at 1648 cm^{-1} (random coils), 1655 cm^{-1} (α-helix), and 1670 cm^{-1} (β-turns), indicating some disruption of the native structure. Additionally, acid whey is also characterized by a peak at 1685 cm^{-1}, indicating aggregation that is likely a result of low pH (Nishanthi et al., 2017a).

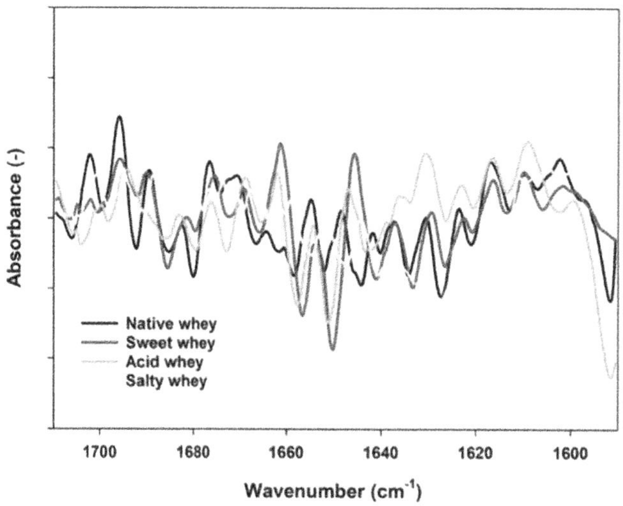

FIGURE 6.3 The second derivative of FTIR interferograms in the amide I region (1600–1700 cm^{-1}) for whey proteins across four whey streams. *Based on the data from Nishanthi et al. (2017a).*

6. Properties of WPs during the concentration step and after spray drying

WP streams are typically processed through membrane techniques, allowing for the selective permeation of smaller solutes such as minerals, acids, and lactose, while retaining larger components like proteins. Environmental changes affect the proteins during both the concentration and drying stages of WP powder production. For instance, increased levels of lactic acid and calcium in the acid whey concentrate (AWC), along with the intense heat treatment used in Greek yogurt pasteurization, contribute to a heightened surface hydrophobicity and a more positive surface charge compared to other whey streams (Nishanthi et al., 2017b). The low acid dissociation constant of lactic acid causes the negatively charged carboxyl groups on the surface of WPs to become protonated, leading to a decrease in their overall negative charge. As the negative surface charge density diminishes, electrostatic interactions are reduced, allowing van der Waals and covalent interactions to dominate (Dissanayake et al., 2013a, 2013b). The dissociation of β-LG dimers into monomers exposes the free thiol group (Cys_{121}) on the exterior of the β-barrel, which then triggers aggregation (Zhu and Damodaran, 1994). Additionally, the disulphide bond (Cys_{66}-Cys_{160}) on the external loop of β-LG becomes exposed, facilitating intermolecular thiol-disulphide exchange reactions that further promote aggregation. This aggregation of WPs in AWC is evidenced by a significantly larger particle size compared to other whey streams (Nishanthi et al., 2017b). High concentrations of sodium in sweet whey concentrate contribute to its lower surface hydrophobicity when compared to other concentrated WP streams. Water molecules interact favorably with salts to alleviate osmotic imbalance, leading to protein/protein attraction via hydrophobic interactions (Hussain et al., 2011), which causes a reduction in surface hydrophobicity. Interestingly, despite increased hydrophobic interactions in sweet whey concentrate, no significant change in particle size is observed (Nishanthi et al., 2017b). In solutions with high ionic strength, a critical size for stable nuclei decreases, resulting in multiple nucleations and numerous small aggregates (Uversky and Lyubchenko, 2013).

During the spray drying process, the high inlet temperature and the atomized feed distribution in the drying chamber lead to rapid moisture evaporation from droplets. This process simultaneously prompts whey components to move toward the surface, replacing the aqueous areas (Kim et al., 2002). The migration rate of whey components varies depending on their surface activity, interaction states, and drying temperatures, resulting in a compositional gradient between the core and the crust that ultimately influences functional performance (Kim et al., 2009a, 2009b). Regardless of the whey source, the particles usually have protein-dominant particle surfaces, indicating that the diffusion of WPs to the surface is substantially greater than that of lactose and minerals (Kim et al., 2003). Salty whey has the least surface protein coverage, suggesting a limited migration of proteins due to the preferential migration of fat to the surface (Faldt et al., 1993). The increased fat content on the surface of the powder obtained from the salty WPC contributes to greater hydrophobicity in salty whey powder particles (Nishanthi et al., 2017b). Denatured WPs exhibit lower surface activity and, therefore, migrate poorly to the particle surface (Kim et al., 2009a, 2009b). The higher presence of denatured WPs in salty whey, characterized by a low surface hydrophobicity and reduced free and total thiol contents compared to other whey concentrates, may also contribute to decreased protein coverage on the particle surface. Compared to other whey powders, the particle surface of salty whey powder contained a higher concentration of minerals, primarily sodium, characterized by a positive surface charge (Nishanthi et al., 2017b).

The SEM images of the powder particles obtained by spray drying of the various WPCs (Fig. 6.4) reveal that the surface of native whey powder appears wrinkled and dented, characterized by various deep and shallow folds, as commonly seen in protein-rich particles (Stevenson and Chen, 1996). Among the four powders compared, sweet whey powder exhibited the largest spherical particles. The formation of such large, smooth particles is facilitated by a high total solid content in the feed and elevated inlet temperatures during the spray drying process (Kim et al., 2009a, 2009b). In contrast, acid whey powder displays a cluster-like particle agglomeration due to its relatively high moisture content, typical of a lactic acid-rich environment. The dense, compact structure of particles in salty whey powder, reminiscent of cauliflower, is attributed to the highly hydrophobic nature of the particle surfaces, which enhances hydrophobic interactions with adjacent powder particles (Nijdam and Langrish, 2006). Additionally, cluster-type protein aggregation may contribute to the compact organization of particles in salty whey powder (Zhang et al., 2016).

In both native whey concentrate and its corresponding powder, the levels of β-sheet and its aggregated forms are comparable, suggesting that spray drying has no substantial impact on β-sheet unfolding or aggregation (Fig. 6.5). Conversely, spray drying completely unraveled the α-helices found in native whey concentrate, reshaping them into random coil structures. On the other hand, the spray drying of sweet whey concentrate results in protein aggregation due to both β-sheet crosslinking and α-helical unfolding. The more pronounced protein aggregation in sweet whey powder is also reflected in larger particle sizes. For AWC, spray drying unfolds α-helices, leading to the formation of random coils. Moreover, the β-sheets in this concentrate engage in intermolecular crosslinking during the spray drying process. At an acidic pH, the destabilization of the β-LG dimer reveals the free Cys_{121} residue on β-strand H, which triggers

FIGURE 6.4 Scanning electron micrographs of representative whey protein powders.

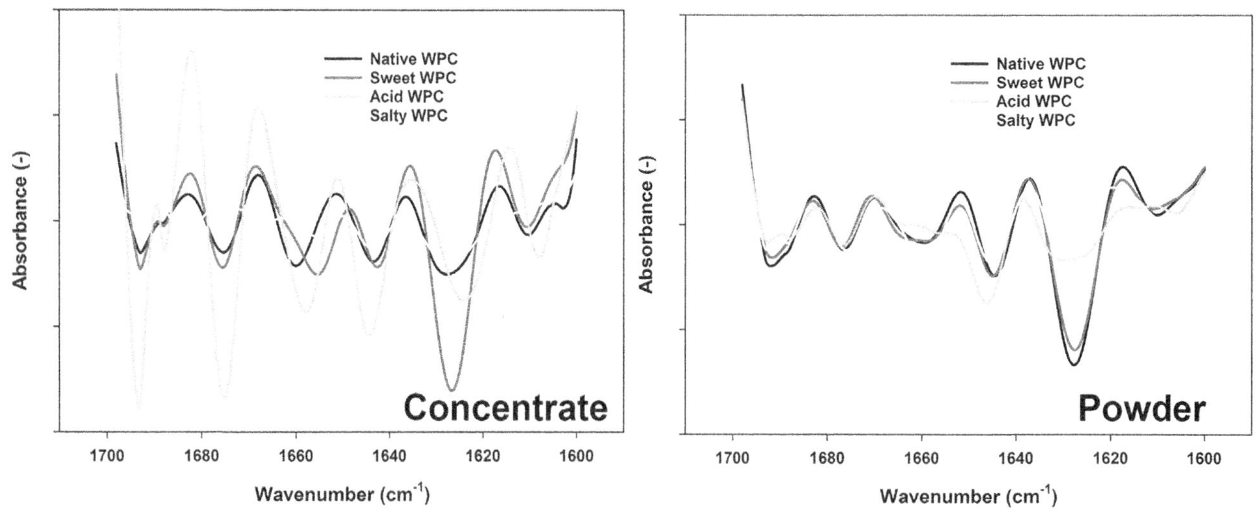

FIGURE 6.5 Second derivative of the FTIR interferograms of whey protein concentrates and powders in the amide I region (1600–1700 cm^{-1}). *Based on the data from Nishanthi et al. (2017b).*

intermolecular β-sheet aggregation (Kontopidis et al., 2004) through covalent bonding. Following the exposure of the free Cys_{121} residue and the subsequent intermolecular β-sheet aggregation, an α-helix located just above the β-strand H becomes destabilized (Kontopidis et al., 2004). Unlike the other three whey concentrates, salty whey concentrate showed a significant absence of secondary structures. In a highly positive ionic environment, the electrostatic status of the dimer interface is disrupted, exposing free Cys_{121} and two disulphide bonds, which can participate in thiol/disulphide-based reactions (Fig. 6.5). The spray drying of salty whey concentrate leads to the emergence of native secondary structures that are absent in the concentrate prior to drying. This may also be facilitated by the spray drying process, which eliminates nearly 96% of water, prompting salt crystallization and thereby reducing its osmotic effect on WPs. The lack of dissolved ionic salt forms allows WPs to reach a negative repulsion state, reversing aggregation, resulting in both native monomers and intermediate structures.

Even with minimal processing during the manufacture of native WPC, disulphide bonds are formed between k-casein and BSA, leading to the creation of protein aggregates (Nishanthi et al., 2017b). Spray drying, in turn, increases the number of these aggregates, incorporating β-LG alongside BSA and caseins. At or above 70°C, thiol-disulphide interchange reactions occur between β-LG and k-casein. Denatured β-LG may also bond with BSA, creating disulphide links that promote polymerization. The FTIR spectrum for native whey powder shows a disruption of α-helices and intermolecular β-sheet aggregation (Fig. 6.5), which may relate to the formation of these aggregates at the expense of native proteins. In sweet whey concentrate and its corresponding powder, protein polymerization through disulphide interactions mainly occurs between κ-casein and β-LG. Spray drying of sweet whey concentrate results in numerous protein aggregates linked by nondisulphide covalent bonds, indicating the formation of early Maillard reaction products likely due to the high inlet temperature used during spray drying (Bernard et al., 2011). As sweet whey concentrate may have a high lactose content, spray drying may lead to the lactosylation of lysine residues in β-LG and α-LA. Unlike native and sweet whey concentrate, AWC has an equal proportion of protein aggregates formed through both covalent-disulphide and nondisulphide bonds. Disulphide bonds mainly arise between β-LG and α-LA. At the same time, some of these aggregates may be formed through the self-aggregation of β-LG, as indicated by the intramolecular β-sheet crosslinking observed in the FTIR (Fig. 6.5). Protein aggregation through covalent-non-disulphide bonds may result from the lactosylation of proteins, producing early Maillard reaction products due to the intense pasteurization during Greek yoghurt manufacturing. In contrast to AWC, nearly all protein aggregates in acid whey powder are generated through covalent-non-disulphide interactions. The spray drying of acid whey powder further lactosylates the proteins into Maillard products. FTIR data for this powder indicates substantial disruption of α-helices, potentially reflecting the formation of these unreducible aggregates. Compared to the other three whey concentrates, most protein aggregates in salty whey concentrate are formed via intramolecular covalent links among lactoferrin molecules (Nishanthi et al., 2017b). The transformation of salty whey concentrate into powder diminishes the presence of disulphide-linked protein aggregates. The moisture loss during spray drying decreases the osmotic influence of salt on WPs, enabling them to revert to their native structure, thus reducing the amount of WPs involved in aggregation.

7. Changes in whey proteins during storage

Storage conditions lead to alterations in protein conformation at the molecular level, which may influence the protein's functional properties. Covalent and noncovalent interactions occur among WPs, causing protein unfolding and denaturation, resulting in both reversible and irreversible aggregation. As a result, the quantity of native WPs decreases after 3 months of high-temperature storage, with a corresponding rise in denatured proteins. These alterations are clearly dependent on temperature; low temperatures do not exhibit these changes even after 15 months of storage (Norwood et al., 2016). At higher storage temperatures, the flexibility and mobility of WPs increase, leading to a partial loss of their tertiary structure. WPC shows increased surface hydrophobicity and fewer exposed AAs on its particle surfaces after 1 month of storage at 60°C and 22% relative humidity (RH) (Burgain et al., 2016). Besides denaturation, WPs can also undergo lactosylation during storage. Although research on lactosylation in WPC powders is sparse, Le et al. (2012) found that six out of 12 lysine residues in α-lactalbumin (α-LA) lactosylate during the storage of milk protein concentrates for up to 12 weeks at 40°C and 84% RH. This suggests similar changes could occur in WPC, particularly those with higher lactose content. These structural changes in WPs may consequently impact their functional properties. For example, the formation of insoluble aggregates due to high storage temperatures reduces WPC solubility (Zhou et al., 2008), thereby affecting various functional attributes such as emulsification, gelation, and foaming.

Intricate interactions among WPs occur due to storage conditions and compositional changes, impacting secondary structural motifs such as α-helices, β-sheets, β-turns, and random coils in these proteins (Fig. 6.6). In their native state, WPC predominantly feature β-sheets, identifiable by two prominent peaks at 1628–1629 cm^{-1} and 1693 cm^{-1}

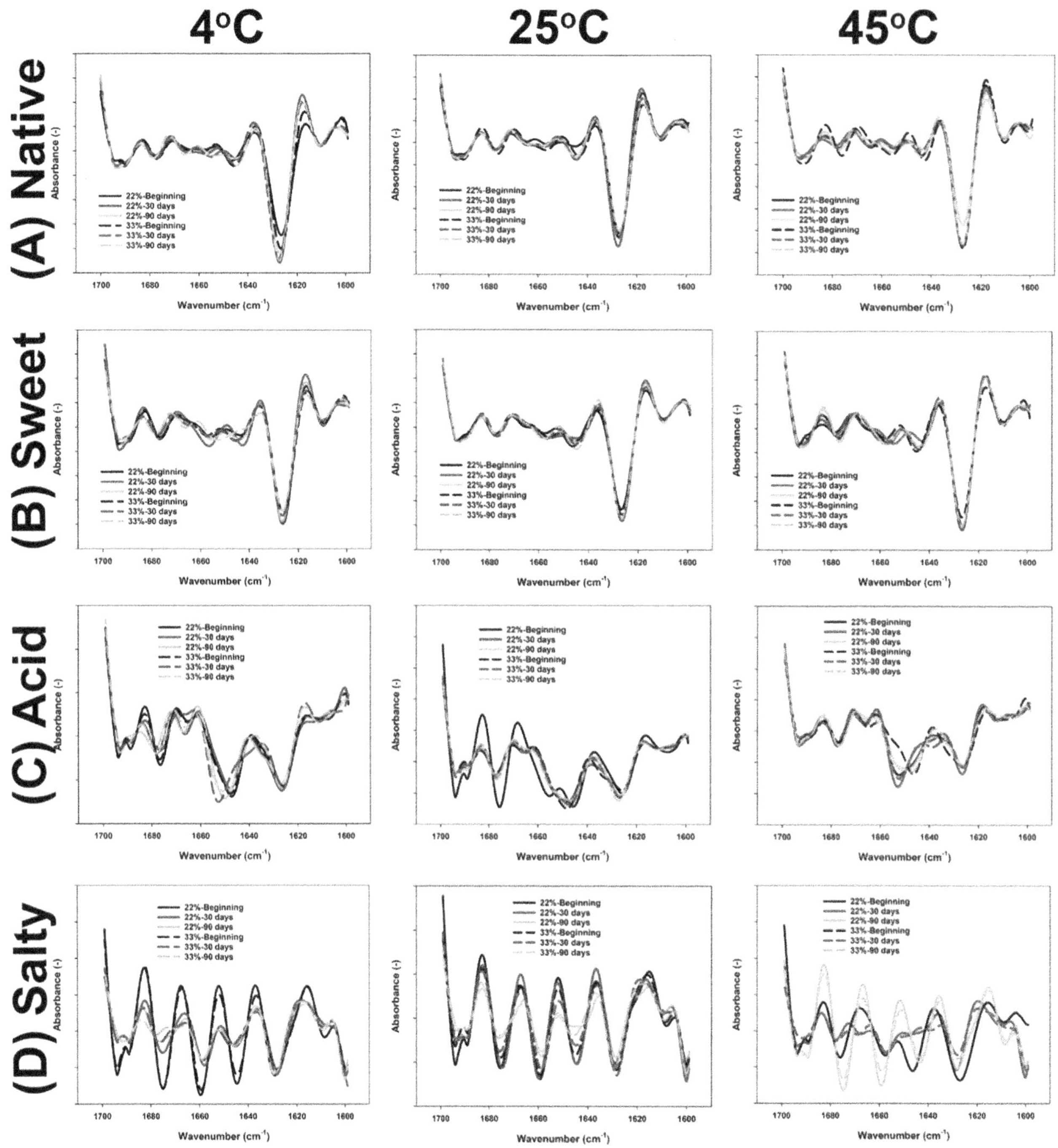

FIGURE 6.6 Second derivative of the amide I region of different whey protein concentrate powders (A—native; B—sweet; C—acid; D—salty) stored at 22 or 33% relative humidity and 4, 25 or 45°C at the beginning and after 30 and 90 days of storage. *Based on the data from Nishanthi et al. (2017c).*

(Fig. 6.6A). Other structures, including α-helices, random coils, and β-turns, are observed at lower intensities regardless of RH. In addition to the native structures, intramolecular β-sheets and cross-linking (at 1682 cm^{-1} and 1615 cm^{-1}) typically arise early in storage, likely as a result of thermal effects during pasteurization and spray drying in WPC processing (Nishanthi et al., 2017c). Over extended storage periods, WPs experience substantial structural changes, particularly at higher RH (33%) compared to 22% RH, and these alterations do not seem to be influenced by storage temperature (Fig. 6.6A). A complete distortion of α-helices leads to an increase in the intensity of randomly ordered peptides (1644 cm^{-1}). As α-helices elongate, the intensity of β-turns (1660 cm^{-1}) also rises, attributed to the widening of the β-turn between β-strand I and H in β-LG. The broadening of β-sheet peaks, especially at 33% RH, suggests potential intermolecular cross-linking due to elevated moisture levels. Concurrently, the intensity of intramolecular antiparallel β-sheet aggregates (Lefevre and Subirade, 1999) increases, likely linked to dimerization or polymerization processes, as numerous β-sheets transition toward the molecular core. Higher RH facilitates increased exposure of AA side chain residues (1610 cm^{-1}) (Lefevre and Subirade, 1999). Although the random coils formed from elongated α-helices may be small, they possess a significant surface area, thus exposing a wide array of AA residues. Furthermore, enhanced protein hydration caused by increased RH promotes the movement of hydrophilic AA residues to the molecule's surface.

All these structural changes are accompanied by the appearance of large aggregates (Mw $\geq$ 250 kDa) that are covalently linked via disulphide bridges, alongside the subsequent decrease of β-LG levels associated with temperature, humidity, and storage time. α-LA remains relatively stable during the first 30 days of storage, but its levels start to decline as it becomes involved in the aggregation process later on (Nishanthi et al., 2017c). It is known that α-LA has only four disulphide bonds located at Cys_6-Cys_{120}, Cys28-Cys11, Cys_{61}-Cys_{77}, and Cys_{73}-Cys_{91} (McSweeney and Fox, 2013) and contains no free thiol groups. Additionally, the holo-form of α-LA, which is abundant in native-WPC, contributes to its high thermal stability (Boye and Alli, 2000). Therefore, the role of α-LA in overall aggregation is limited. Like other proteins, BSA also participates in aggregation; however, its involvement depends on storage time. BSA forms disulphide crosslinks with β-LG (Matsudomi et al., 1994) due to a free thiol group at Cys_{34}. Beyond disulphide bonding, WPs in native-WPC also engage in Maillard reaction-driven aggregation, likely caused by the lactosylation of lysine residues in these proteins.

Sweet WPC powders mainly contain intermolecular and antiparallel aggregated β-sheets (1630 and 1685 cm^{-1}) during the early storage phase (Fig. 6.6B). The appearance of interconnected β-sheets at the beginning of storage highlights the role of thermal treatments used in the production of sweet-WPC (Nishanthi et al., 2017a). Similar to native-WPC powder, significant structural changes take place after prolonged storage. Irrespective of storage conditions, β-sheets experience a loss of conformational stability and partially transition into intramolecular and cross-linked β-sheets (1680 cm^{-1}). At the same time, intermolecular hydrogen bonds are broken, leading to the formation of stronger new intermolecular hydrogen bonds (Ngarize et al., 2004). These changes seem to be influenced by the temperature of storage (Nishanthi et al., 2017c). The balance between monomer and dimer forms of β-LG molecules (Lefevre and Subirade, 1999) shifts toward the dimeric form, establishing strongly linked β-strand structures. This dimerization also leads to a significant increase in deeply buried β-type structures (1691 cm^{-1}). Additionally, dimer formation occurs at the expense of intramolecular α-helices, resulting in a higher number of random coils. The conversion of α-helices into random coils appears to depend on storage temperature but remains unaffected by RH.

Sweet-WPC samples demonstrate behavior similar to native-WPC, characterized by the formation of protein aggregates that are covalently linked through disulphide bridges over the storage period (Nishanthi et al., 2017c). Maximum aggregation occurs at high temperatures and elevated RH. Alongside the formation of aggregates, the rate of disappearance of β-LG is greatest under these conditions, indicating that β-LG is the primary protein involved in aggregation. In contrast, α-LA exhibits only minor participation in aggregation across all storage scenarios, while BSA shows no involvement. Additionally, the creation of Maillard reaction aggregates is also present, with approximate molecular weights of 71–72 kDa and $\geq$250 kDa (Nishanthi et al., 2017c). Their content gradually increases with temperature, humidity, and time. Sweet-WPC generally contains low levels of lactose, potentially leading to the lactosylation of lysine segments in β-LG. Among these, Lys8 at the N-terminus, Lys141 at the C-terminus, and Lys138 in the third turn of the α-helix are readily accessible, thus likely serving as the primary sites for this reaction.

In acid-WPC, β-sheets (1620–1635 cm^{-1}) are mostly absent at the start of storage (Fig. 6.6C). Concurrently, the intensity of intramolecular antiparallel β-sheets (1682 cm-1) becomes more noticeable at low humidity and high temperatures, primarily involving β-strands on the surface, unlike the intact β-strands found deeper within β-LG (1691 cm^{-1}). Other secondary structures, including α-helices, β-turns, and random coils, are present but at low intensity, independent of storage conditions. By day 14 of storage, an increase in exposure of β-sheets (1621 cm^{-1}) indicates partial unfolding of β-LG due to lactic acid (LA)-induced changes in WP (Nishanthi et al., 2017b). After 60 days, three key changes are observed (Fig. 6.6C). First, α-helices completely disappear, resulting in nonnative random structures across all storage conditions. Second, the configuration of β-sheets alters significantly, creating two distinct peaks in the range of 1620–1638 cm^{-1}, suggesting the dimerization of β-LG. Furthermore, the intensity of the deeply located and aggregated β-sheets (1693 cm^{-1}) almost doubles under all conditions. In the presence of LA, β-LG forms dimers, causing a shift of more β-sheets to the molecule's inner side. Third, side chain AA residues (1610 cm^{-1}) are highly exposed, particularly at low humidity, due to the modifications of β-sheets and disruption of α-helices. Toward the end of storage (90 days), a further loss of secondary structures is noted, particularly in samples kept at elevated temperatures and low humidity

(Fig. 6.6C), indicating a significant reduction of random coils and dimeric forms of β-LG and suggesting severe aggregation.

These structural alterations indicate an increase in disulphide-linked aggregates that gradually accumulate during storage, leading to the eventual loss of β-LG A, β-LG B, and α-LA (Nishanthi et al., 2017c). β-LG A is the main proteoform that participates in aggregation (Oldfield et al., 1998), primarily due to disulphide bridging through Val_{118} in β-LG A, as opposed to Ala_{118} in β-LG B. This positioning in β-strand H results in a core that is packed less favorably (McSweeney and Fox, 2013), which enhances the flexibility of β-LG A and exposes the free thiol group. The role of α-LA in protein aggregation is also significant in acid-WPC. Reduced water availability may limit the access of Ca^{2+} to the binding site of α-LA, hindering protein stabilization and leading to a predominance of less stable apo-α-LA. Maillard reaction-induced aggregates are also found in acid-WPC across all storage conditions, which are temperature and RH dependent, predominantly involving α-LA. Out of the 12 lysine residues in α-LA, a maximum of six residues (Lys_{24}, Lys_{112}, Lys_{117}, Lys_{127}, Lys_{133}, and Lys_{141}) can be easily lactosylated (Le et al., 2012).

In comparison to other WPCs, the secondary structures of salty-WPC (Fig. 6.6D) show significant sensitivity to storage conditions during the early days. The amounts of β-sheets (1626, 1636, and 1693 cm^{-1}), random coils (1648 cm^{-1}), and β-turns (1660 cm^{-1}) decrease as temperature and humidity rise. Simultaneously, a subset of deeply located β-sheets (1693 cm^{-1}) diminishes with temperature increases, indicating an exposure of the β-LG core. However, the reactivity of β-LG to temperature does not affect the proportion of crosslinked β-sheets (1616 and 1680 cm^{-1}). Due to osmotic pressure from Na+ in the medium, the β-sheets do not link but instead unfold, creating unordered structures. The dimeric form of β-LG is influenced by temperature, being present at low temperatures but absent at high ones. During continued storage, monomeric β-LG prevailed (Lefevre and Subirade, 1999) under all conditions, indicating that native dimers dissociate over time. As the temperature rises, there is a corresponding reduction in nonnative monomers and an increase in aggregation driven by the crosslinking of β-sheets (1616 cm^{-1}) due to temperature-dependent exposure of reactive sites in β-LG. After 60 days of storage, β-sheets reemerge in the range of 1620–1638 cm^{-1}, with peak intensity increasing alongside storage temperature. Notably, the section indicative of intermolecular β-sheets (1620–1636 cm^{-1}) displayed two peaks at elevated temperatures, representing dimeric forms of β-LG, while a singular peak was observed at low temperatures, indicating monomeric β-LG. The formation of dimeric β-LG also augmented the intensity of intramolecular β-sheets (1680 cm^{-1}), leading to a significantly higher proportion of deeply located β-sheets (1691 cm^{-1}) and a large number of random coils and β-turns. Up to 60 days, the structural changes observed appear unaffected by humidity. However, after 90 days of storage, humidity begins to influence the interactions between β-sheets. A high level of intermolecular hydrogen bonds noted at low RH suggests an abundance of β-LG dimers (1620–1636 cm^{-1}), while the predominance of intramolecular hydrogen bonds at elevated RH indicates a greater presence of monomeric β-LG. Prolonged exposure to high humidity diminishes protein-protein interactions, facilitating protein hydration and the formation of nearly native protein monomers.

Like other WPs, the structural changes in salty-WPC can be linked to disulphide-linked protein aggregates exceeding a Mw of 250 kDa. The level of aggregation increases over time. Both β-LG A and β-LG B are involved in disulphide bonding; however, β-LG A completely disappears after 14 days (Nishanthi et al., 2017c), indicating a rapid reaction rate (Oldfield et al., 1998). The contribution of α-LA to disulphide bonding is also evident, with the extent related to storage temperature and time, without a clear link to humidity. Therefore, thiol/disulphide interchange reactions between β-LG and α-LA likely serve as the primary aggregation pathway. In the absence of a reactive thiol group, thiol/disulphide interchange occurs via the disulphide bonds at Cys_6-Cys_{120} and Cys_{28}-Cys_{111} on the surface of α-LA, which subsequently forms new disulphide bonds with Cys_{66} and Cys_{160} on the surface of β-LG. Additionally, BSA significantly participates as storage temperature, humidity, and time increase. With a free thiol group at Cys_{34} and 17 disulphide bonds in its primary structure, BSA engages in both disulphide bridging and thiol/disulphide interchange reactions with β-LG (Matsudomi et al., 1994) and α-LA (Matsudomi et al., 1993). Similar to sweet-WPC, salty-WPC features aggregates connected via covalent nondisulphide bonds in the BSA-LF region, with an approximate Mw of 71–72 kDa. Furthermore, the quantity of large aggregates (Mw > 250 kDa) formed through lactosylation increases alongside storage temperature, humidity, and time.

8. Conclusions

In commercially sourced whey streams, variations in composition and processing methods influence the properties of WPs. Although the properties of WPs in native and sweet whey are comparable, changes in the pH and ionic strength of acid whey, along with heat treatment, result in conformational changes and aggregation of WPs due to hydrophobic interactions and covalent linkages. In contrast, WPs in salty whey aggregate through robust covalent bonds, as well as weaker van der Waals, hydrogen, and electrostatic interactions, which expose hydrophobic sites. Covalent aggregates, characterized by intramolecular β-sheet aggregation, lose the descriptors of their native secondary structure. The unique properties of WPs in different whey streams are further modified by processing methods, particularly during ultrafiltration and spray drying, depending on the type of whey stream. Acid and salty WPCs form aggregated proteins through covalent crosslinking among WPs (β-LG, α-LA, BSA, LF) and/or caseins, primarily driven by compositional differences (high acidity and salinity) and heat treatments such as pasteurization. During spray drying, the aggregation of proteins and physical characteristics transition from liquid WPC to dried powders; however, this transfer is less pronounced in acid and salty whey. The attributes of WPC powders reflect the combined effects of the original feed stream, concentration, and changes resulting from spray drying. Additionally, storage conditions significantly affect the molecular properties of WPC powders from different whey streams, with protein aggregation resulting from covalent crosslinking that is accelerated at elevated temperatures. The physical properties of acid and salty WPC powders differ from those of native and sweet WPC powders, which may influence their functionality. The functionality of various WPC powders is contingent upon their composition and storage conditions. Changes in protein conformation arise from the interplay of storage and compositional variations, altering the functional properties. The elevated concentration of sodium salts in salty WPC powder causes denaturation and aggregation, leading to reduced solubility and increased heat stability. Denatured WPs and the high-fat coverage on the surface of salty WPC particles result in enhanced emulsification activity but reduced emulsion stability when compared to native WPC powder. Consequently, salty WPC powder may be utilized in heat-processed emulsions, such as beverages, cheese spreads, and dressings, with minor modifications to control its salty flavor. Conversely, high concentrations of lactic acid in acid WPC powder resulted in solubility comparable to that of native WPC powder.

Current research indicates that addressing distinct compositional differences is essential for managing whey streams. The processing results in unique WP products with specific functionalities that could be used in particular applications. At present, the current state of knowledge is robust regarding how processing, storage, and applications impact the behavior of native and sweet proteins; however, this knowledge is lacking for acid and salty WPCs. Considering the environmental impact of these two streams and the difficulties associated with their handling, there is certainly a growing interest in developing this knowledge.

References

Abd El-Salam, M., El-Shibiny, S., Salem, A., 2009. Factors affecting the functional properties of whey protein products: a review. Food Rev. Int. 25, 251–270.

Alsaed, A.K., Ahmad, R., Aldoomy, H., El-Qader, S.A., Saleh, D., Sakejha, H., Mustafa, L., 2013. Characterization, concentration and utilization of sweet and acid whey. Pakistan J. Nutr. 12, 172–177.

American Dairy Products Institute (ADPI), 2016. Whey Protein Concentrate (WPC) Standard.

Bernard, C., Regnault, S., Gendreau, S., Charbonneau, S., Relkin, P., 2011. Enhancement of emulsifying properties of whey proteins by controlling spray-drying parameters. Food Hydrocoll. 25, 758–763.

Blaschek, K., Wendorff, W.L., Rankin, S.A., 2007. Survey of salty and sweet whey composition from various cheese plants in Wisconsin. J. Dairy Sci. 90, 2029–2034.

Bogahawaththa, D., Vasiljevic, T., 2022. Advances in dairy protein science: whey proteins. In: Huppertz, T., Vasiljevic, T. (Eds.), Understanding and Improving the Functional and Nutritional Properties of Milk, first ed. Burleigh Dodds Science Publishing.

Bong, D.D. & Moraru, C.I (2014) Use of micellar casein concentrate for Greek-style yogurt manufacturing: effects on processing and product properties, J. Dairy Sci., 97, 1259-1269.

Boye, J., Alli, I., 2000. Thermal denaturation of mixtures of a-lactalbumin and β-lactoglobulin: a differential scanning calorimetric study. Food Res. Int. 33, 673–682.

Burgain, J., El Zein, R., Scher, J., Petit, J., Norwood, E.A., Francius, G., Gaiani, C., 2016. Local modifications of whey protein isolate powder surface during high temperature storage. J. Food Eng. 178, 39–46.

Chandra, R.K., Singh, G., Shridhara, B., 1989. Effect of feeding whey hydrolysate, soy and conventional cow milk formulas on incidence of atopic disease in high-risk infants. Ann. Allergy 63, 102–106.

Chandrapala, J., Duke, M.C., Gray, S.R., Zisu, B., Weeks, M., Palmer, M., Vasiljevic, T., 2015. Properties of acid whey as a function of pH and temperature. J. Dairy Sci. 98, 4352–4363.

Chandrapala, J., McKinnon, I., Augustin, M.-A., Udabage, P., 2010. The influence of milk composition on pH and calcium activity measured in situ during heat treatment of reconstituted skim milk. J. Dairy Res. 77, 257–264.

Chegini, G., Taheri, M., 2013. Whey powder: process technology and physical properties: a Review. Middle East J. Sci. Res. 13, 1377–1387.

Datta, R., Tsai, S.P., Bonsignore, P., Moon, S.H., Frank, J.R., 1995. Technological and economic potential of poly (lactic acid) and lactic acid derivatives. FEMS (Fed. Eur. Microbiol. Soc.) Microbiol. Rev. 16, 221–231.

de Faria, J.T., Minim, V.P.R., Minim, L.A., 2013. Evaluating the effect of protein composition on gelation and viscoelastic characteristics of acid-induced whey protein gels. Food Hydrocoll. 32, 64–71.
de la Fuente, M.A., Singh, H., Hemar, Y., 2002. Recent advances in the characterisation of heat-induced aggregates and intermediates of whey proteins. Trends Food Sci. Technol. 13, 262–274.
Deeth, H.C., Bansal, N., 2019. Whey Proteins: From Milk to Medicine. Academic Press, London, United Kingdom.
Dissanayake, M., Ramchandran, L., Donkor, O.N., Vasiljevic, T., 2013a. Denaturation of whey proteins as a function of heat, pH and protein concentration. Int. Dairy J. 31, 93–99.
Dissanayake, M., Ramchandran, L., Piyadasa, C., Vasiljevic, T., 2013b. Influence of heat and pH on structure and conformation of whey proteins. Int. Dairy J. 28, 56–61.
Dupont, D., Arnould, C., Rolet-Repecaud, O., Duboz, G., Faurie, F., Martin, B., Beuvier, E., 2006. Determination of bovine lactoferrin concentrations in cheese with specific monoclonal antibodies. Int. Dairy J. 16, 1081–1087.
Faldt, P., Bergenstahl, B., Carlsson, G., 1993. The surface coverage of fat on food powders analyzed by ESCA (electron spectroscopy for chemical analysis). Food Struct. 12, 10.
Gregory, A., 1987. Desalination of sweet-type whey salt drippings for whey solid recovery. Bull. IDF 212, 38–48.
Grummer, J., Bobowski, N., Karalus, M., Vickers, Z., Schoenfuss, T., 2013. Use of potassium chloride and flavor enhancers in low sodium Cheddar cheese. J. Dairy Sci. 96, 1401–1418.
Hussain, R., Gaiani, C., Aberkane, L., Scher, J., 2011. Characterization of high-milk protein powders upon rehydration under various salt concentrations. J. Dairy Sci. 94, 14–23.
Israelachvili, J.N., 2011. Intermolecular and Surface Forces: Revised, third ed. Academic Press.
Jeyarajah, S., Allen, J.C., 1994. Calcium binding and salt-induced structural changes of native and preheated beta-lactoglobulin. J. Agric. Food Chem. 42, 80–85.
Jimenez, X.T., Cuenca, A.A., Jurado, A.T., Corona, A.A., Urista, C.R.M., 2012. Traditional methods for whey protein isolation and concentration: effects on nutritional properties and biological activity. J. Mex. Chem. Soc. 56, 369–377.
Josephson, R.V., Rizvi, S.S.H., Harper, W.J., 1975. Compositional differences in whey systems. J. Food Sci. 40, 479–483.
Kapoor, R., Metzger, L., 2004. Evaluation of salt whey as an ingredient in processed cheese. J. Dairy Sci. 87, 1143–1150.
Kim, E.H.J., Chen, X.D., Pearce, D., 2002. Surface characterization of four industrial spray-dried dairy powders in relation to chemical composition, structure and wetting property. Colloids Surf. B Biointerfaces 26, 197–212.
Kim, E.H.-J., Chen, X.D., Pearce, D., 2003. On the mechanisms of surface formation and the surface compositions of industrial milk powders. Dry. Technol. 21, 265–278.
Kim, E.H.-J., Chen, X.D., Pearce, D., 2009a. Surface composition of industrial spray-dried milk powders. 2. Effects of spray drying conditions on the surface composition. J. Food Eng. 94, 169–181.
Kim, E.H.J., Chen, X.D., Pearce, D., 2009b. Surface composition of industrial spray-dried milk powders. 3. Changes in the surface composition during long-term storage. J. Food Eng. 94, 182–191.
Koh, L.L.A., Nguyen, H.T.H., Chandrapala, J., Zisu, B., Ashokkumar, M., Kentish, S.E., 2014. The use of ultrasonic feed pre-treatment to reduce membrane fouling in whey ultrafiltration. J. Membr. Sci. 453, 230–239.
Kontopidis, G., Holt, C., Sawyer, L., 2004. Invited review: β-lactoglobulin: binding properties, structure, and function. J. Dairy Sci. 87, 785–796.
Kosikowski, F.V., 1979. Whey utilization and whey products1. J. Dairy Sci. 62, 1149–1160.
Lagrange, V., Whitsett, D., Burris, C., 2015. Global market for dairy proteins. J. Food Sci. 80 (S1).
Laiho, S., Ercili-Cura, D., Forssell, P., Myllarinen, P., Partanen, R., 2015. The effect of dynamic heat treatments of native whey protein concentrate on its dispersion characteristics. Int. Dairy J. 49, 139–147.
Le, T.T., Deeth, H.C., Bhandari, B., Alewood, P.F., Holland, J.W., 2012. A proteomic approach to detect lactosylation and other chemical changes in stored milk protein concentrate. Food Chem. 132 (3), 1177–1184.
Lefevre, T., Subirade, M., 1999. Structural and interaction properties of β-lactoglobulin as studied by FTIR spectroscopy. Int. J. Food Sci. Technol. 34, 419–428.
Lievore, P., Simões, D.S., Silva, K., Drunkler, N., Barana, A., Nogueira, A., Demiate, I., 2013. Chemical characterisation and application of acid whey in fermented milk. J. Food Sci. Technol. 1–10.
Lu, Y., McMahon, D., 2015. Effects of sodium chloride salting and substitution with potassium chloride on whey expulsion of Cheddar cheese. J. Dairy Sci. 98, 78–88.
Matsudomi, N., Oshita, T., Kobayashi, K., 1994. Synergistic interaction between β-lactoglobulin and bovine serum albumin in heat-induced gelation. J. Dairy Sci. 77, 1487–1493.
Matsudomi, N., Oshita, T., Kobayashi, K., Kinsella, J.E., 1993. Alpha-lactalbumin enhances the gelation properties of bovine serum albumin. J. Agric. Food Chem. 41, 1053–1057.
McSweeney, P.L.H., Fox, P.F., 2013. Advanced Dairy Chemistry, 1A. Spinger, London.
Ngarize, S., Herman, H., Adams, A., Howell, N., 2004. Comparison of changes in the secondary structure of unheated, heated, and high-pressure-treated β-lactoglobulin and ovalbumin proteins using Fourier transform Raman spectroscopy and self-deconvolution. J. Agric. Food Chem. 52, 6470–6477.
Nijdam, J., Langrish, T., 2006. The effect of surface composition on the functional properties of milk powders. J. Food Eng. 77, 919–925.
Nishanthi, M., Chandrapala, J., Vasiljevic, T., 2017c. Properties of whey protein concentrate powders obtained by spray drying of sweet, salty and acid whey under varying storage conditions. J. Food Eng. 214C, 137–146.

Nishanthi, M., Vasiljevic, T., Chandrapala, J., 2017a. Properties of whey proteins obtained from different whey streams. Int. Dairy J. 66, 76–83.

Nishanthi, M., Vasiljevic, T., Chandrapala, J., 2017b. Compositional and structural properties of whey proteins of sweet, acid and salty whey concentrates and their respective spray dried powders. Int. Dairy J. 74, 49–56.

Norwood, E.A., Chevallier, M., Le Floch-Fouéré, C., Schuck, P., Jeantet, R., Croguennec, T., 2016. Heat-induced aggregation properties of whey proteins as affected by storage conditions of whey protein isolate powders. Food Bioprocess Technol. 9, 993–1001.

Oldfield, D.J., Singh, H., Taylor, M.W., Pearce, K.N., 1998. Kinetics of denaturation and aggregation of whey proteins in skim milk heated in an ultra-high temperature (UHT) pilot plant. Int. Dairy J. 8, 311–318.

Precedence Research, 2024. Whey Protein Market Size to Attain USD 22.95 Bn by 2034. Precedenceresearch.com; Precedence Research. www.precedenceresearch.com/whey-protein-market.

Rahaman, T., Vasiljevic, T., Ramchandran, L., 2015. Conformational changes of beta-lactoglobulin induced by shear, heat, and pH-Effects on antigenicity. J. Dairy Sci. 98, 4255–4265.

Rudloff, S., Lönnerdal, B., 1992. Solubility and digestibility of milk proteins in infant formulas exposed to different heat treatments. J. Pediatr. Gastroenterol. Nutr. 15, 25–33.

Sanderson, W., Brady, A., Whitehead, G., Oldham, I., Brockwell, I., 1996. Recycling Salt Solution in Cheese Processing and Apparatus Therefore. Murray Goulburn Co-Operative Co. Limited, Australia, p. 237. Assignee, US Pat(5).

Sava, N., Van der Plancken, I., Claeys, W., Hendrickx, M., 2005. The kinetics of heat-induced structural changes of β-lactoglobulin. J. Dairy Sci. 88, 1646–1653.

Sousa, G.T., Lira, F.S., Rosa, J.C., de Oliveira, E.P., Oyama, L.M., Santos, R.V., Pimentel, G.D., 2012. Dietary whey protein lessens several risk factors for metabolic diseases: a review. Lipids Health Dis. 11, 67.

Stevenson, R.J., Chen, X.D., 1996. A study of volatile "trapping" in spray dried whey protein concentrate by "crushing" and/or vacuuming, and detection by solid-phase microextraction/gas chromatography/mass spectrometry. Food Res. Int. 29, 495–504.

Uversky, V., Lyubchenko, Y., 2013. Bio-nanoimaging: Protein Misfolding and Aggregation. Academic Press.

Vasbinder, A.J., de Kruif, C.G., 2003. Casein-whey protein interactions in heated milk: the influence of pH. Int. Dairy J. 13, 669–677.

Zhang, Z., Arrighi, V., Campbell, L., Lonchamp, J., Euston, S.R., 2016. Properties of partially denatured whey protein products: formation and characterisation of structure. Food Hydrocoll. 52, 95–105.

Zhou, P., Liu, X., Labuza, T.P., 2008. Effects of moisture-induced whey protein aggregation on protein conformation, the state of water molecules, and the microstructure and texture of high-protein-containing matrix. J. Agric. Food Chem. 56, 4534–4540.

Zhu, H., Damodaran, S., 1994. Effects of calcium and magnesium ions on aggregation of whey protein isolate and its effect on foaming properties. J. Agric. Food Chem. 42, 856–862.

Further reading

Mimouni, A., Schuck, P., Bouhallab, S., 2005. Kinetics of lactose crystallization and crystal size as monitored by refractometry and laser light scattering: effect of proteins. Le Lait 85, 253–260.

Monahan, F.J., McClements, D.J., Kinsella, J.E., 1993. Polymerization of whey proteins in whey protein-stabilized emulsions. J. Agric. Food Chem. 41, 1826–1829.

Chapter 7

Interactions of whey proteins during processing

Dilema Wijegunawardhana[1], Mayumi Silva[1], H.S.T. Hapuarachchi[2], Divya Sandamali Gunasekara[3] and Jayani Chandrapala[1]

[1]*School of Science, STEM College, RMIT University, Bundoora, VIC, Australia;* [2]*Faculty of Science, University of Colombo, Colombo, Sri Lanka;* [3]*SPC Global, Shepparton, VIC, Australia*

1. Whey proteins: Manufacturing, composition, and product variants

Whey proteins are high-quality proteins obtained from milk during the cheese-making process. When milk is turned into cheese, its main protein, caseins are separated to form a solid curd, while the remaining liquid is called whey. This process removes about 80% of the total milk proteins as casein, and the remaining 20%, which stay dissolved in the liquid, are whey proteins (Smithers, 2008). This separation occurs through two main coagulation methods: enzymatic coagulation using rennet or acid-induced coagulation. In enzymatic coagulation, the enzyme chymosin specifically cleaves the Phe105-Met106 bond in κ-casein, which removes the hydrophilic glycomacropeptide from the micelle surface. This cleavage destabilizes the micellar structure and triggers the aggregation of αs- and β-caseins into a curd matrix (Fox et al., 2015). In contrast, acid coagulation involves lowering the milk pH to approximately 4.6, the isoelectric point of caseins, which neutralizes surface charges on the micelles. This loss of electrostatic repulsion causes the micelles to aggregate and precipitate spontaneously, without the enzymatic action (Walstra et al., 2005). The liquid whey obtained after curd removal contains a rich mixture of soluble proteins, lactose, minerals, and water-soluble vitamins. Among these, the primary protein components include β-lactoglobulin (50%–55%), α-lactalbumin (20%–25%), bovine serum albumin (BSA) (5%–10%), immunoglobulins (10%–15%), and glycomacropeptides (GMP; up to 15%) (Patel, 2015; Yiğit et al., 2023). Native liquid whey contains approximately 0.6%–0.8% protein, 4.5% 5.0% lactose, 0.5%–0.6% minerals, and over 93% water, which necessitates further processing to produce protein-enriched ingredients (Fox et al., 2017).

Due to its dilute composition, liquid whey must undergo processing to concentrate and isolate the protein fraction for commercial applications. The initial step involves clarification to remove fat globules and residual casein fines, followed by membrane filtration, predominantly ultrafiltration (UF) and diafiltration (DF). Ultrafiltration uses membranes with a molecular weight cut-off between 10 and 30 kDa, which effectively retain major whey proteins like β-lactoglobulin (18.4 kDa) and α-lactalbumin (14.2 kDa), while allowing lactose (342 Da), salts, and water to pass through. After ultrafiltration and diafiltration, the whey retentate is further concentrated using vacuum evaporation (to ~45%–65% total solids) for low-to medium-protein products, or nanofiltration for high-protein isolates, which preserves protein stability by avoiding heat. The concentrate is then dried by spray drying, where rapid moisture removal (inlet: 170–200°C; outlet: 80–90°C) reduces water content to <5%, producing stable whey protein powders (Guo, 2019).

Whey proteins are commercially available in several product forms, each differing in protein content and degree of processing. Whey Protein Concentrate (WPC) typically contains 35%–80% protein and retains variable amounts of lactose and fat, depending on processing conditions. It is widely used for its nutritional value and functional attributes such as emulsification and water binding (Guo, 2019; Smithers, 2008). Whey Protein Isolate (WPI) contains ≥90% protein, with significantly reduced lactose and fat levels due to additional purification steps such as ion-exchange or microfiltration, making it suitable for lactose-intolerant individuals and high-protein applications (Patel, 2015). Hydrolyzed Whey Protein (WPH) is enzymatically treated to break down peptide bonds, improving digestibility and reducing allergenicity. Although specific protein percentages can vary by the base material used (WPC or WPI), WPH is often

Whey. https://doi.org/10.1016/B978-0-443-27648-4.00001-3

applied in clinical and infant nutrition but may have a bitter taste due to short peptides (Pang et al., 2025). Native Whey Protein is extracted directly from skim milk rather than cheese whey using low-temperature membrane filtration, which preserves native protein structures and yields isolates with ~95% protein, very low fat, and minimal lactose, and is marketed for sports and medical nutrition due to its superior amino acid profile (Carter et al., 2021; Song et al., 2025).

2. Structure of individual whey protein fractions

β-Lg is the predominant whey protein in bovine milk, contributing approximately 50%–55% of the total whey protein fraction. It is a globular protein composed of 162 amino acids with a molecular weight of approximately 18.3 kDa (Broersen, 2020; Simion et al., 2015). Its secondary structure comprises about 50% β-sheet, 15% α-helix, and 15%–20% reverse turns, forming a highly ordered conformation (Chamani et al., 2006). The β-sheet regions organize into nine antiparallel β-strands (A through I), which fold into a tightly packed β-barrel. This barrel defines the core of the protein and encloses a central hydrophobic cavity, known as the calyx, which is responsible for the reversible binding of small hydrophobic ligands such as retinol, vitamin D, and long-chain fatty acids (Kontopidis et al., 2004; Sawyer and Kontopidis, 2000). β-Lg contains 9 β strands, 8 of which fold in to 2 β-sheets that face each other. A 3-turn α-helix located above one of the β sheet. Three Cys residues among five of them such as Cys 106, Cys 119 and Cys 121 located within a strong hydrophobic pocket between one side of the helix G and H strands segments. Through the phenyl ring of Phe 136, Cys 106 - Cys 119 disulfide bond is separated from Cys 121 (Creamer et al., 2004). The tertiary structure of β-Lg is further stabilized by two intramolecular disulfide bonds (Cys66–Cys160 and Cys106-Cys119), along with a free thiol group at Cys121, which is buried under native pH conditions (~6.8). Upon thermal unfolding, typically occurring above 70°C, the thiol group becomes exposed, promoting intermolecular disulfide bond formation and subsequent protein aggregation (Qin et al., 1998).

α-Lactalbumin (α-La) is the second most abundant whey protein in bovine milk, representing approximately 20%–25% of the total whey protein content. It is a small, globular protein consisting of 123 amino acid residues with a molecular weight of approximately 14.2 kDa (Lisak Jakopović, 2016; Permyakov and Berliner, 2000). Structurally, α-La is characterized by a high α-helix content, comprising roughly 26% α-helix, 14% β-sheet, and 60% disordered structures (Spöttel et al., 2021). The protein adopts a compact tertiary structure stabilized by four intramolecular disulfide bridges (Cys6–Cys120, Cys28–Cys111, Cys61–Cys77, and Cys73–Cys91), which provide thermal and structural resilience. A key feature of α-La is its ability to bind a single calcium ion (Ca^{2+}) within a hydrophilic loop region, a binding that is essential for maintaining its native, folded conformation. Under physiological conditions (pH ~6.5–7.0), α-La maintains a stable tertiary structure; however, upon calcium removal—often under acidic or chelating conditions—the protein undergoes a conformational shift into a molten globule state, where much of its secondary structure is retained but the tertiary contacts are lost (Permyakov, 2020; Permyakov and Berliner, 2000; Wang et al., 2011).

BSA is a major globular protein found in bovine blood plasma, accounting for approximately 60% of total serum protein. It is composed of 583 amino acid residues and has a molecular weight of about 66.5 kDa (Holm et al., 2007; Koh et al., 2019). Structurally, BSA is organized into three homologous domains, each predominantly α-helical, resulting in a secondary structure comprising approximately 68% a-helix, 5% b-sheet, 8% b-turn and 19% random structure (Su et al., 2008). The structure of BSA is predominantly α-helical (67%) and rest of the polypeptide consists with turn and extended or flexible regions between subdomains with the absence of β-sheets. The polypeptide is organized in 3 homologous α-helix domains (I, II, III) and 10 helices are contained in each α-helix domain. These 10 helices are again divided in to antiparallel six helix and four subdomains (IA, IB etc.). Two tryptophan residues with intrinsic fluorescence are contained in BSA where one tryptophan residue is positioned in 134 and second tryptophan residue is positioned in 212. Trp-134 is located at molecule surface and Trp-212 is located within a hydrophobic pocket of the protein. These Trp residues are separately located in different domains which are hydrophilic and hydrophobic, within the BSA molecules (Soares et al., 2007; Zhang et al., 2008). The protein adopts a compact tertiary structure that is stabilized by 17 intramolecular disulfide bridges and includes a single free thiol group at Cys34, which contributes to its redox reactivity and ligand-binding capacity (Chu and Letcher, 2021).

Immunoglobulins (Igs) are vital glycoproteins present in bovine milk, playing a crucial role in passive immunity for newborn calves. The predominant class in bovine milk is Immunoglobulin G (IgG), which constitutes approximately 60%–70% of the total immunoglobulin content in mature milk, with IgG1 being the most abundant subclass (Korhonen et al., 2000). Structurally, each IgG molecule is a Y-shaped tetramer composed of two identical heavy (H) chains (~53 kDa each) and two identical light (L) chains (~23 kDa each), linked by disulfide bonds. This configuration results in a complete IgG molecule with a molecular weight of approximately 180 kDa (Korhonen et al., 2000; Su et al., 2008). The secondary structure of bovine IgG is predominantly β-sheet, comprising about 62% β-sheet, 29% β-turns, and 2% α-helix, with the remainder being unordered structures (Ivancic et al., 2022).

3. Interactions of whey proteins during processing

Whey proteins exhibit diverse interactions during food processing, which significantly influence their structural, functional, and nutritional properties. These interactions can be both covalent and noncovalent, depending on the processing conditions and the composition of the food matrix. Covalent interactions include disulfide bond formation and Maillard reaction-driven conjugation with carbohydrates, while noncovalent interactions involve hydrophobic interactions, hydrogen bonding, electrostatic interactions, and van der Waals forces. Such interactions are crucial in determining solubility, thermal stability, emulsifying capacity, foaming properties, and digestibility of whey protein-based systems. In this chapter, the primary focus will be on thermal and nonthermal processing-induced interactions, as these two categories are of dominant industrial relevance.

3.1 Heat induced transition and interaction of whey proteins

Thermal processing causes significant structural changes in whey proteins, triggering the exposure of previously buried reactive groups, particularly thiol (-SH) groups and disulfide (–S–S–) bonds. These reactive sites are responsible for initiating a cascade of protein-protein interactions, including intermolecular disulfide bond formation, thiol-disulfide interchange reactions, and covalent or noncovalent aggregation. Such interactions occur not only within the same protein but also between different whey proteins and other components like casein, significantly altering the protein network in dairy systems. During heating, whey proteins such as β-lactoglobulin (β-Lg), α-lactalbumin (α-La), and BSA undergo substantial conformational rearrangements. These proteins, normally stabilized by intramolecular disulfide bonds, hydrophobic core interactions, and hydrogen bonding, begin to unfold above 65°C. In β-Lg, this thermal denaturation converts its dimeric structure into monomers, exposing its reactive thiol group at Cys121, and destabilizing internal disulfide bonds (Cys66–Cys160 and Cys106–Cys119). These changes enable β-Lg to participate in thiol-disulfide interchange reactions with itself and with other proteins (Invernizzi et al., 2008; Iwaoka et al., 2019; Schokker et al., 2000). Likewise, α-La, although more heat-stable and lacking free thiols in its native state, becomes susceptible to disulfide bond cleavage at temperatures above 70°C, particularly at surface-exposed linkages such as Cys6–Cys120 and Cys28–Cys111 allowing it to function as a disulfide acceptor. BSA also unfolds within the 65–70°C range, exposing Cys34 and hydrophobic patches, thereby engaging in both covalent interactions with β-Lg and α-La and noncovalent interactions through hydrophobic domains (Asaduzzaman et al., 2021; Schokker et al., 2000).

The dominant molecular mechanism underpinning these covalent interactions is the thiol-disulfide interchange, a redox-mediated reaction wherein a thiolate anion (RS^-) formed from a deprotonated thiol-nucleophilically attacks an electrophilic disulfide bond, forming a new disulfide bridge while releasing another thiol. This process is governed by the local pKa, solvent accessibility, and structural flexibility of the thiol-bearing residues, with Cys121 of β-Lg playing a central role due to its heightened reactivity once exposed upon thermal unfolding (Nagy, 2013). In parallel with these covalent processes, hydrophobic interactions intensify as previously buried nonpolar side chains (e.g., leucine, isoleucine, valine) become solvent-accessible. These hydrophobic domains drive protein clustering, thereby increasing the frequency and spatial alignment of reactive groups for further covalent cross-linking. The synergistic effect between thiol-disulfide interchange and hydrophobic association ultimately promotes the formation of large, three-dimensional whey protein aggregates (Li et al., 2022).

Molecular dynamics simulations by Zhang et al. (2023) revealed that β-Lg unfolds through five metastable conformations when heated at 500 K, with the most open structure (QIII) showing a critical spatial arrangement where Cys121 (–SH) and the disulfide bond S-S66-160 are located at opposite ends of the molecule, favoring intermolecular thiol-disulfide interchange reactions involved in aggregation. The unfolding also significantly altered the hydration behavior, as the density of water molecules in the first hydration shell (0–0.24 nm from the protein surface) increased, and more notably, the diffusion coefficients of water rose with unfolding. Specifically, in the native state (QO), the diffusion coefficient of hydration shell 1 water was $1.25 \pm 0.73 \times 10^{-5}$ cm^2/s, which increased to $1.62 \pm 1.04 \times 10^{-5}$ cm^2/s in the unfolded QIII state. Similarly, water in shell 2 (0.24–0.32 nm) diffused faster, rising from $1.39 \pm 0.71 \times 10^{-5}$ cm^2/s (QO) to $1.79 \pm 1.04 \times 10^{-5}$ cm^2/s (QIII). This acceleration in water mobility was linked to increased hydrophobic exposure due to unfolding, which promoted stronger water-water interactions in the hydration shells. Interestingly, despite more hydration water being present, the unfolded β-Lg conformation had enhanced interaction strength with surrounding water, particularly in the first and second shells, as quantified by higher δginter values. This structural flexibility and increased solvent accessibility explain why unfolded β-Lg is more prone to aggregation: even with more hydration, the protein's reactive thiol groups remain exposed and mobile, facilitating intermolecular bonding and aggregate growth.

Jia et al. (2025) employed molecular dynamic simulations to explore how high temperatures (500 K) specifically affect the unfolding behavior of BSA. Their results revealed rapid and extensive structural changes, highlighted by a backbone root-mean-square deviation (RMSD) increase beyond 5 Å within just 5 ns and eventually surpassing 1 nm after 100 ns, indicative of profound protein unfolding. This thermal exposure dramatically reduced the intramolecular hydrogen bonding network from approximately 480 ± 10 at physiological conditions (310 K, pH 7.4) to about 350 ± 19 at 500 K, signifying significant loss of secondary structural stability. Domain-specific analyses further emphasized these disruptions, with Domain III exhibiting the greatest structural deviation (~1.0 nm). Correspondingly, circular dichroism (CD) simulations confirmed a substantial decrease in α-helical content, declining sharply from 74% under native conditions to around 50% at elevated temperatures, thus emphasizing the significant denaturation of secondary structural elements. Additionally, the unfolding increased solvent exposure, particularly in Domain II, which displayed the highest solvent-accessible surface area (SASA) increase. Principal component analysis (PCA) indicated that thermal stress forced BSA out of its native, well-defined energy state into a broader energy landscape, thereby populating multiple intermediate unfolded conformations. Collectively, these findings clearly demonstrate that high-temperature unfolding not only disrupts structural integrity at both the secondary and tertiary levels but also substantially increases solvent exposure, enhancing protein reactivity and propensity for aggregation, analogous to the mechanisms observed for β-Lg (Fig. 7.1).

It is important to note that the extent and nature of whey protein unfolding, interaction, and aggregation during heat treatment are highly dependent on multiple external and intrinsic factors (Table 7.1). These include pH, which influences the ionization state of reactive groups; ionic strength, which modulates electrostatic interactions; temperature, which determines the degree of protein denaturation; whey protein concentration, which affects the likelihood of molecular collisions; and the presence of other ingredients such as other proteins, minerals, polysaccharides, or polyphenols, which can either stabilize or destabilize protein structures. Understanding the interplay of these variables is crucial for tailoring processing conditions in dairy systems to achieve desired textural, functional, and nutritional outcomes.

3.1.1 Effect of pH on heat-induced whey protein interactions

The extent and nature of heat-induced whey protein interactions are highly pH-dependent, as pH influences both the net charge of the proteins and the ionization state of thiol groups, thereby modulating thiol-disulfide interchange and aggregation behavior. Near the isoelectric point (pI) of whey proteins (approximately pH 5.2–5.4 for β-lactoglobulin), the net charge on protein molecules is minimal, reducing electrostatic repulsion and favoring protein aggregation through hydrophobic interactions and covalent disulfide bond formation. At these pH levels, the solubility of denatured whey proteins is also reduced, which promotes the formation of dense aggregates with increased particle size and lower dispersion stability (Lam and Ikeda, 2017; Majhi et al., 2006). In contrast, at neutral or slightly alkaline pH (6.5–7.5) commonly used in milk and yogurt processing thiol groups are more likely to deprotonate, forming reactive thiolate anions (RS^-) that drive thiol-disulfide interchange reactions. This increases the probability of intermolecular disulfide bond formation between β-Lg and other proteins such as κ-casein or BSA (Nagy, 2013). For instance, a study by Anema & Li (2003) investigated the effects of heating reconstituted skim milk at pH 6.55. They observed that upon heating at temperatures ranging from 75 to 100°C for up to 60 min, there was an increase in casein micelle size during the initial stages of heating, which plateaued with prolonged heating. This enlargement correlated with the association of denatured whey proteins with the casein micelles. The study reported that the maximum increase in casein micelle size was about 30–35 nm, and this change was highly correlated with the levels of denatured whey proteins associating with the micelles. The researchers concluded that the heat-induced changes in casein micelle size were a consequence of whey proteins associating with the micelles, which could influence the textural properties of dairy products. In another study, the effect of milk pH at heating on protein complex formation and yogurt texture was examined. Milk was heated at different pH levels, and the resulting protein interactions were analyzed. The study found that heating milk at pH 6.8 resulted in a relatively compact protein network due to a balanced contribution from both soluble protein/κ-casein complexes and whey protein-casein micelle-associated complexes. Yogurt prepared with milk heated at pH 6.8 showed significantly higher storage modulus (G') values, shorter gelation times, higher water-holding capacity, firmness, and a more compact protein network compared to those prepared at pH 6.4, 7.2, and unheated milk. These findings demonstrate that adjusting milk pH before heating can be an important factor in achieving uniform quality in yogurt production (Mahomud et al., 2021).

Conversely, at acidic pH values below 5.0, thiol-disulfide interchange is significantly suppressed due to reduced thiol ionization, and protein aggregation is primarily governed by noncovalent hydrophobic interactions. Under these conditions, whey proteins are more prone to form coarse, less functional aggregates with diminished solubility and stability.

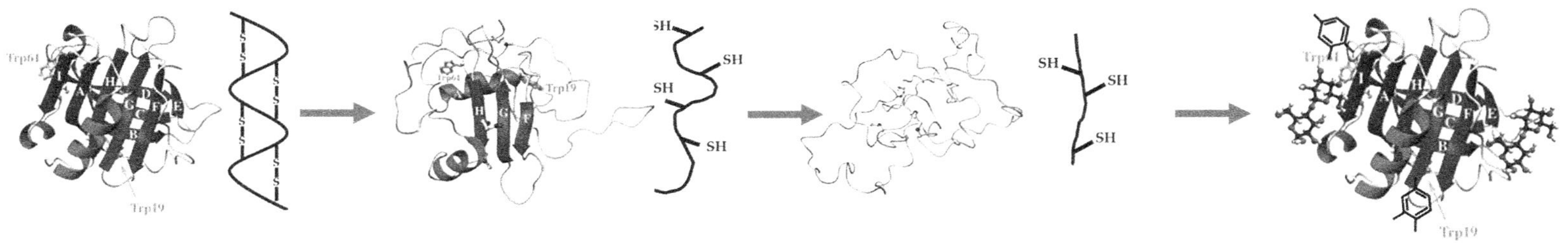

FIGURE 7.1 Schematic representation of heat-induced unfolding and aggregation of whey proteins. Native whey proteins maintain a stabilized, folded globular structure due to intramolecular disulfide bonds. Upon heating, partial unfolding exposes reactive thiol (–SH) groups. Further heat treatment results in complete unfolding, with multiple free thiol groups exposed. These exposed thiol groups facilitate aggregation through the formation of intermolecular disulfide bonds, leading to protein-protein cross-linking.

TABLE 7.1 Summary of key studies investigating the effects of heat treatment conditions (temperature, pH, ionic strength, and duration) on the denaturation, aggregation, solubility, cross-linking, and functional interactions of whey proteins in various systems.

Aspect studied	Thermal treatment and conditions	Heat-induced changes and interactions	References
Denaturation, solubility, turbidity, structural stability, and aggregation as a function of pH and protein concentration	The study aimed to investigate how pH and protein concentration affect the thermal denaturation and aggregation behavior of whey proteins. Whey protein dispersions (10%, 17.5%, and 25% w/w) were adjusted to pH 4.0, 5.0, or 6.0 and thermally treated at 140°C for 5, 10, or 20 s. Measurements included capillary electrophoresis, solubility, turbidity, SDS-PAGE, and DSC.	Denaturation rate of β-lactoglobulin followed first-order kinetics and was highest at pH 5% and 10% concentration. At pH 4, denaturation was slower and solubility remained higher (up to 87%) due to decreased molecular mobility. Turbidity peaked at pH 5 after 20 s, indicating enhanced aggregation. Peak denaturation temperature (Td) reached 85.1°C at pH 4 (10%), while heat stability was lowest at pH 5.	Dissanayake et al. (2013)
Denaturation, aggregation, solubility, and cross-linking of whey proteins as influenced by pH and heating time	Whey protein isolate (WPI) powders were adjusted to pH 2.5, 4.5, and 6.5 and heated at 100°C for up to 24 h with controlled water activity (0.23). Solubility, turbidity, SDS-PAGE, and DLS were used to measure protein solubility, aggregate size, and aggregation kinetics.	Heat treatment induced pH-dependent aggregation of whey proteins. At pH 2.5, proteins remained soluble with small aggregates forming. At pH 4.5 and 6.5, both soluble and insoluble aggregates formed, with the proportion of insoluble aggregates increasing with pH. SDS-PAGE revealed the formation of disulfide bonds and other covalent cross-links (e.g., isopeptide) in aggregates at higher pH, while at pH 2.5, only disulfide bonds predominated. Aggregates increased in size with heating time, reaching 221 ± 86 nm at pH 6.5 after 24 h.	Gulzar et al. (2011)
Coacervation, rheological properties, phase transitions, and electrostatic interactions	The study aimed to evaluate the effects of ionic strength and hydrogen bonding on WPI and flaxseed gum (FG) coacervates. WPI-FG systems (1:1 ratio, 0.05% w/w) were heated at 90°C, with NaCl (0–50 mM) and urea (0–100 mM) added to study the effects of ionic strength and hydrogen bonding. Turbidity, optical density (OD_{600}), and rheological properties (viscosity and G′) were analyzed.	Increasing NaCl concentration (0–50 mM) decreased coacervate formation, lowering turbidity at pHmax and shifting the pH transition points, indicating reduced electrostatic interaction and suppressed coacervate formation. Urea (50–100 mM) shifted pH transition points to more acidic values (pHmax decreased from 3.6 to 3.2) and suppressed hydrogen bonding, resulting in reduced viscosity and lower gel-like properties of WPI-FG coacervates. The coacervates exhibited shear-thinning behavior, with decreased dynamic viscosity and viscoelasticity in the presence of NaCl and urea.	J. Liu et al. (2020)

Protein aggregation, gelation, functional properties (fracture strain, modulus), disulfide and noncovalent interactions	WPC solutions (15% w/w, pH 6.9) were heated at 80°C for 30 min or 85°C for 60 min. Samples were treated in the presence of sodium dodecyl sulfate (SDS), N-ethylmaleimide (NEM), or dithiothreitol (DTT) to selectively block disulfide or noncovalent interactions. Rheology and PAGE were used to measure gel strength and protein interactions.	Heating led to the formation of both disulfide (covalent) and noncovalent interactions. SDS-WPC gels formed only disulfide-linked aggregates and had a low storage modulus (~80 Pa) and low fracture stress. DTT-WPC gels, devoid of disulfide bonds, had higher G′ (~29,000 Pa) and higher fracture strain, indicating a more rubbery texture. NEM-WPC gels had intermediate properties with some disulfide linkages but mainly noncovalent interactions, resulting in moderate stiffness. These results suggest disulfide linkages increase gel rubberiness, while noncovalent interactions increase gel stiffness.	Havea et al. (2009)
Formation of hydrogel, protein interactions, encapsulation efficiency, and rheology	Whey protein isolate fibrils (WPF) and gliadin nanoparticles (GNP) were prepared by heating WPI in a water bath at pH 2.0 (85°C for 3 h). GNP were prepared by antisolvent precipitation at pH 2.0. The ratio of WPF to GNP was varied (1:0 to 4:1), and the effects of pH (2.0–6.0) and NaCl concentration (0–500 mM) on hydrogel formation were investigated.	The addition of GNP significantly improved the gel strength of WPF-based hydrogels. The strongest hydrogel (G′ = 419 Pa) was formed at a 4:1 ratio of WPF:GNP at pH 4.0, exhibiting excellent rheological properties and stability. Higher NaCl concentrations enhanced the hydrogel's strength by promoting electrostatic interactions. Hydrogels formed at pH 2.0 and 3.0 showed weak gel structures, with a decrease in G′ at higher NaCl concentrations. Encapsulation efficiency for curcumin was highest at 89.76% at pH 2.0 and decreased to 79.24% at pH 6.0, with lower curcumin retention observed at higher pH due to gel structural degradation.	Zhu et al. (2022)
Interaction between whey protein and anthocyanin (ACN) during heat treatment	Whey protein was preheated at different temperatures (40, 50, 60, 70, 80°C) for 30 min at pH 3, then mixed with anthocyanin-rich extracts from purple corn, grape, and black carrot (50 μM). Fluorescence spectroscopy was used to analyze the interaction at 25°C, 35°C, and 45°C.	The fluorescence intensity of whey protein decreased with increasing anthocyanin concentration, indicating strong interaction. The quenching mechanism was static, primarily driven by hydrophobic interactions. Preheating increased the binding affinity of whey protein for anthocyanins, with the binding constant (Ks) being higher for preheated whey protein than native whey protein. The quenching efficiency was highest at 45°C.	Ren and Monica Giusti (2021)
Effect of heating on WPC properties (solubility, gel texture, digestibility)	WPC preheated at 70°C, 80°C, and 90°C, then centrifuged and dried. The solubility, gel texture, voluminosity, and digestibility were tested.	Higher heating temperatures reduced solubility, with the lowest solubility at 90°C (16.04 ± 0.74%). Gel hardness increased with lower pH, while acidification rates were faster at 80°C and 90°C. At 80°C, digestibility significantly improved to 30.76 ± 0.07%, compared to native WPC (27.38 ± 0.09%).	Andoyo et al. (2023)

Continued

TABLE 7.1 Summary of key studies investigating the effects of heat treatment conditions (temperature, pH, ionic strength, and duration) on the denaturation, aggregation, solubility, cross-linking, and functional interactions of whey proteins in various systems.—cont'd

Aspect studied	*Thermal treatment and conditions*	*Heat-induced changes and interactions*	*References*
Effect of heat-induced whey proteins/κ-casein complex on acid gelation of yak milk	Yak milk was adjusted to pH 6.6, 7.0, and 7.4 and heated at 85°C for 10 min. The serum phase from heated yak milk was mixed with casein micelles from unheated yak milk and vice versa. Acid gelation was initiated by the addition of glucono-D-lactone (GDL).	Heat-induced whey proteins/κ-casein complexes formed during heating at different pH values, with soluble whey protein/κ-casein complexes predominating at higher pH (7.0). These complexes increased the storage modulus (G′), water-holding capacity (WHC), and firmness of the gel. Micelle-bound complexes were less effective in enhancing gelation compared to the soluble complexes.	Xu et al. (2015)
Heat-induced denaturation and combination of whey proteins with casein	Raw milk was heated at 65°C, 75°C, 85°C, and 100°C for 10, 20, and 30 min. Denaturation of whey proteins (α-LA and β-LG) and their interaction with casein micelles were studied using Native-PAGE and SDS-PAGE.	Denaturation of β-LG increased with temperature. At 100°C, β-LG denaturation was 80%, while α-LA denaturation was 70%. The combination of whey proteins and casein micelles increased significantly at higher temperatures, with casein micelle size increasing by up to 120 nm at 90°C after 30 min of heating. The rate of denaturation was highest at 100°C, with 58% of β-LG forming disulfide bonds with casein.	Qian et al. (2017)
Heat-induced denaturation, aggregation, solubility, and functional properties of whey proteins under varying heat treatments	WPI solutions at concentrations of 1%, 4%, 8%, and 12% (w/w) were prepared at pH values of 6.2, 6.7, and 7.2 and subjected to thermal treatment at 85°C for 30 s using a continuous-flow pilot-scale tubular heat exchanger.	At pH 6.2, whey proteins exhibited the greatest degree of denaturation, with a higher proportion of insoluble aggregates, larger particle sizes, and increased turbidity, particularly at higher concentrations (8%–12%), resulting in markedly reduced heat stability during subsequent processing. In contrast, samples treated at pH 6.7 and 7.2 formed predominantly soluble aggregates (~280–600 kDa), with smaller particle sizes and clearer solutions, indicating better colloidal stability. As protein concentration increased, viscosity also rose, and gel formation occurred especially under lower pH conditions.	Buggy et al. (2018)

Effect of sodium caseinate addition on the formation, structure, and hydration properties of whey protein microparticles (PMs) under dry heating at alkaline pH.	WPI was mixed with sodium caseinate (NaCas) at 0%–100% ratios and dry-heated at 100°C for 36 h at pH 9.5. Micellar casein-enriched milk protein powder (Casmic) was also tested. The browning index (BI), D [4,3] particle diameter, protein microparticle yield, water content, and native whey protein (NWP) content were evaluated.	Incorporating NaCas increased the size and yield of PMs up to ~40% casein, with maximum D [4,3] values of ~270 μm and yield (DY) rising from 0.6 to 0.8 g/g. Water content of PMs decreased from ~30 to 20 g/g with increasing casein. The browning index peaked near 40% casein, indicating Maillard-type crosslinking. PMs formed with micellar casein (Casmic) were smaller (~80 μm), more crosslinked, and retained *less water. Notably, native whey protein denaturation* remained consistent (~80%–90%) across casein levels after DH, confirming crosslinking involved both WPI and casein.	Schong and Famelart (2019)
Effect of pH on structural, textural, and optical properties of heat-induced WPI gels.	WPI solutions (10% w/v; 100 mg/mL) were adjusted to pH 2, 5.6, and 7 and subjected to heating at 80°C for 4 h. Gels were evaluated using AFM, texture analysis, moisture content, and colorimetry.	Gels formed at pH 2 and 7 exhibited clear, consistent, and fine-stranded network structures, with lower force required for compression (~8.5–9.1 N) compared to the isoelectric point at pH 5.6, which showed coarse, amorphous aggregates and required significantly higher force (~12.4 N). AFM confirmed fibrillar structures at pH 2 and 7, but irregular aggregates at pH 5.6. Moisture content was highest at pH 5.6 (93.81%) and lowest at pH 7 (89.61%), with stranded networks at pH 2 and 7 promoting stronger water binding. Color metrics showed higher lightness ($L^* = 68.61$) and opacity (69.51) at pH 5.6, corresponding to larger spherical aggregates, while transparency and chromaticity were greater at acidic and neutral pH due to regular fibrillar aggregation.	Farrokhi et al. (2020)

Cornacchia et al. (2014) investigated WPI systems heated at 90°C for 30 s and observed that at pH 4.0 and 4.5, aggregation occurred primarily through intermolecular hydrophobic interactions, while disulfide bonding contributed only marginally, as confirmed by SDS-PAGE and SDS + DTT analysis. This mechanism coincided with a marked reduction in protein dispersibility, with dispersibility values dropping to ~22% at pH 4.5 in 10% WPI solutions, compared to ~89% at pH 3.5. Additionally, TEM imaging revealed large aggregates (~2000 × 4000 nm) at pH 4.5, in contrast to more dispersed and finer aggregates at pH 3.5.

In summary, pH plays a pivotal role in modulating heat-induced whey protein interactions by influencing protein charge, thiol ionization, and aggregation mechanisms. Near the isoelectric point, reduced electrostatic repulsion favors hydrophobic and covalent aggregation, while at neutral to slightly alkaline pH, enhanced thiol-disulfide interchange promotes covalent cross-linking and improved integration with casein micelles. This enhances gel structure and textural attributes in dairy systems such as yogurt. Conversely, at acidic pH values below 5.0, aggregation is dominated by noncovalent hydrophobic interactions, resulting in larger, less functional protein aggregates with reduced solubility and dispersibility. These findings underscore the importance of precise pH control during thermal processing to tailor whey protein interactions for optimal functional and structural outcomes in dairy applications.

3.1.2 Effect of whey protein concentration on heat-induced interactions

In addition to pH, protein concentration plays a critical role in modulating the structural transitions and interactions of whey proteins during thermal processing. Higher protein concentrations increase the frequency of intermolecular collisions among denatured whey protein molecules, thereby intensifying aggregation through both covalent (e.g., disulfide bonding) and noncovalent (e.g., hydrophobic) interactions. Buggy (2018) demonstrated that when WPI concentration was increased from 1% to 12% (w/w) and heated at 85°C for 30 s, the resulting soluble aggregate size ranged from 280 to 600 kDa, significantly larger than those at lower concentrations. This was accompanied by a notable rise in polydispersity index (PdI) from 0.10 ± 0.03 to 0.25 ± 0.01, indicating a broader and more heterogeneous aggregate population. Furthermore, the denaturation rate of β-Lg increased dramatically from 61% at 1% to over 99% at 12%, reflecting enhanced thermal sensitivity at elevated protein levels. Therefore, careful optimization of whey protein concentration is essential for controlling aggregate size distribution, gel consistency, and functional properties in thermally processed food systems.

3.1.3 Effect of ionic strength and minerals on heat-induced whey protein interactions

The presence of salts and minerals significantly modulates thermal aggregation behavior. Ionic strength influences electrostatic interactions - adding salt (like NaCl) can screen charges on proteins, thereby reducing electrostatic repulsion and promoting aggregation at a given pH. For instance, adding NaCl to a heated whey protein solution is known to increase the rate and extent of aggregation, as evidenced by larger particle size, higher turbidity, and lower protein solubility when salt is present (Perez et al., 2022). Divalent cations (Ca^{2+}, Mg^{2+}) can have even stronger effects: Ca^{2+}can cross-link negatively charged groups on different protein molecules (acting as a bridge), which often leads to stronger gels or precipitates upon heating. In milk, native Ca^{2+} levels cause some whey proteins to associate with casein micelles on heating, whereas in whey protein solutions, adding $CaCl_2$ tends to decrease the temperature needed for gelation and to form firmer gels. Conversely, certain additives can improve heat stability: for example, phosphates or citrates bind Ca^{2+} and mitigate its cross-linking effect, which can increase heat stability of whey proteins (this is a strategy in UHT milk to prevent excessive curdling). Polyphosphate salts like sodium triphosphate not only bind Ca^{2+} but also raise the pH; studies show that adding 0.5% sodium polyphosphate (such as sodium triphosphate) to WPC and heating can either promote large aggregate formation at high levels or, at optimal levels (~0.4%), favor formation of smaller aggregates that remain soluble (Liu et al., 2018). Thus, controlling ionic conditions is a way to steer the aggregation outcome during thermal processing.

Further, Navarra et al. (2007) evaluated the effects of Cu^{2+} and Zn^{2+} ions on the thermal aggregation of β-Lg at 60°C. Their light scattering analysis showed that the presence of Zn^{2+} significantly accelerated aggregation kinetics, with rapid aggregate formation beginning at ~48°C, more than 10°C lower than native β-Lg or Cu-β-Lg controls. At the same 60°C temperature, Zn-β-Lg samples produced large aggregates (diameter ~4500–5000 nm) within 70 min, while Cu-β-Lg and native β-Lg only formed small aggregates (~200 nm) after 11 h. Infrared spectroscopy revealed that Zn^{2+} promoted β-sheet formation (notably at 1623 and 1685 cm^{-1}), indicating interprotein aggregation, whereas Cu^{2+} altered β-sheet structure by shifting the amide I band from 1640 to 1635 cm^{-1} and introducing a unique 1614 cm^{-1} band, attributed to stronger intermolecular hydrogen bonding. These conformational changes correlated with the Cu^{2+}-specific emergence of an additional peak at 1595 cm^{-1}, indicative of carboxylate coordination. Furthermore, the Amide II' region

(1436–1445 cm^{-1}) revealed that Cu^{2+} promoted partial unfolding and subsequent gelation through tight coordination with histidine and carboxyl groups, whereas Zn^{2+} led to rapid precipitation. These findings demonstrate that divalent metals not only influence aggregation rate and size but also modulate the secondary structure trajectory and gelation potential of whey proteins.

Muhammad et al. (2009) showed copper ions (Cu^{2+}) markedly modulate heat-induced aggregation of β-Lg by catalyzing sulfhydryl oxidation reactions over disulfide interchange, resulting in the formation of covalent dimers instead of large disulfide-linked aggregates. At a Cu^{2+}:β-Lg molar ratio of 1, heating at 78°C for 15 min led to conversion of ~80% of β-Lg into covalent dimers, as confirmed by SDS-PAGE and HP-GPC, with no further transformation beyond 30 min. Quantification of free sulfhydryl groups using Ellman's reagent revealed a sharp decrease from 0.92 ± 0.04 mol SH/mol protein (unheated, no Cu^{2+}) to 0.02 ± 0.01 after heating in the presence of copper, confirming near-complete oxidation of Cys121. Mass spectrometry identified three dimer types (AA, BB, AB) with masses of ~36.6 kDa, corresponding to random dimerization of β-Lg variants A and B. CD analysis demonstrated profound structural changes: tertiary structure was nearly lost, and β-sheet content increased, indicative of irreversible unfolding. The aggregation pathway shifted with rising Cu^{2+}: β-Lg ratios, with rate constants (k) increasing from 0.041 to 0.461 $M^{-0.5} \cdot s^{-1}$, while the fraction of native β-Lg (C_t/C_0) dropped below 0.05 within 30 min. Compared to 100 mM NaCl, which promotes non-covalent aggregation by electrostatic screening, Cu^{2+} induced covalent aggregation through oxidative crosslinking, resulting in soluble dimers and oligomers under high ionic strength.

In summary, ionic strength and the presence of specific mineral ions play a pivotal role in determining the nature and extent of heat-induced whey protein interactions. While monovalent salts like NaCl reduce electrostatic repulsion and promote aggregation via charge screening, divalent cations such as Ca^{2+}, Zn^{2+}, and Cu^{2+} exert stronger effects by bridging protein molecules or modifying redox chemistry, thereby influencing aggregation rate, gelation temperature, and structural characteristics. These ions can either enhance or hinder protein stability depending on their concentration, binding affinity, and interaction mechanisms. Furthermore, chelating agents such as phosphates and polyphosphates can modulate the effects of minerals by sequestering cations, improving solubility or controlling aggregate size. As evidenced by studies on Cu^{2+} and Zn^{2+}, divalent metals not only accelerate aggregation but also alter the secondary structure and aggregation pathways of whey proteins. Thus, carefully tuning the ionic environment provides a powerful strategy to control thermal aggregation behavior, gel texture, and solubility in whey protein-based formulations.

4. Heat-induced interactions of whey proteins with other food components: Caseins, carbohydrates, fat and polyphenols

Under thermal conditions, whey proteins not only undergo conformational rearrangements but also engage in complex physicochemical interactions with other food matrix components, including other proteins such as caseins, carbohydrates, polyphenols, and lipids. These interactions are multifaceted and governed by a synergy of molecular mobility, reactive group accessibility, and microenvironmental constraints. Importantly, thermal unfolding exposes not just thiol and hydrophobic domains, but also proton donors/acceptors, glycosylation-prone residues, and amphipathic binding sites, which facilitate unconventional reaction pathways often underrepresented in literature.

4.1 Heat-induced interactions of whey proteins with caseins

Upon heating milk to temperatures between 70°C and 100°C, whey proteins; primarily β-lactoglobulin (β-Lg) and α-lactalbumin (α-La) undergo irreversible structural denaturation. This denaturation exposes hydrophobic regions and reactive thiol groups, enabling the proteins to form inter and intramolecular disulfide bonds and to associate with casein micelles (Hill, 1989; Vasbinder and De Kruif, 2003). Casein micelles themselves undergo minimal structural change during heating (Law et al., 1994), but their surface properties make them central to binding denatured whey proteins via noncovalent and covalent interactions.

As shown in Fig. 7.2, the mechanism by which caseins protect whey proteins (WPs) from irreversible heat-induced aggregation involves both molecular shielding and reversible bonding interactions that resemble chaperone-like behavior (Liyanaarachchi et al., 2015). When whey protein is heated in the absence of caseins, the whey proteins unfold, exposing reactive thiol groups and hydrophobic regions ss temperature increases. These denatured proteins initially form small aggregates through hydrophobic interactions, which are then stabilized by the formation of covalent disulfide bonds. Upon further heating, the aggregation continues via thiol-disulfide interchange and hydrophobic attractions, leading to the formation of large, irreversible protein aggregates. During cooling, these covalent structures do not

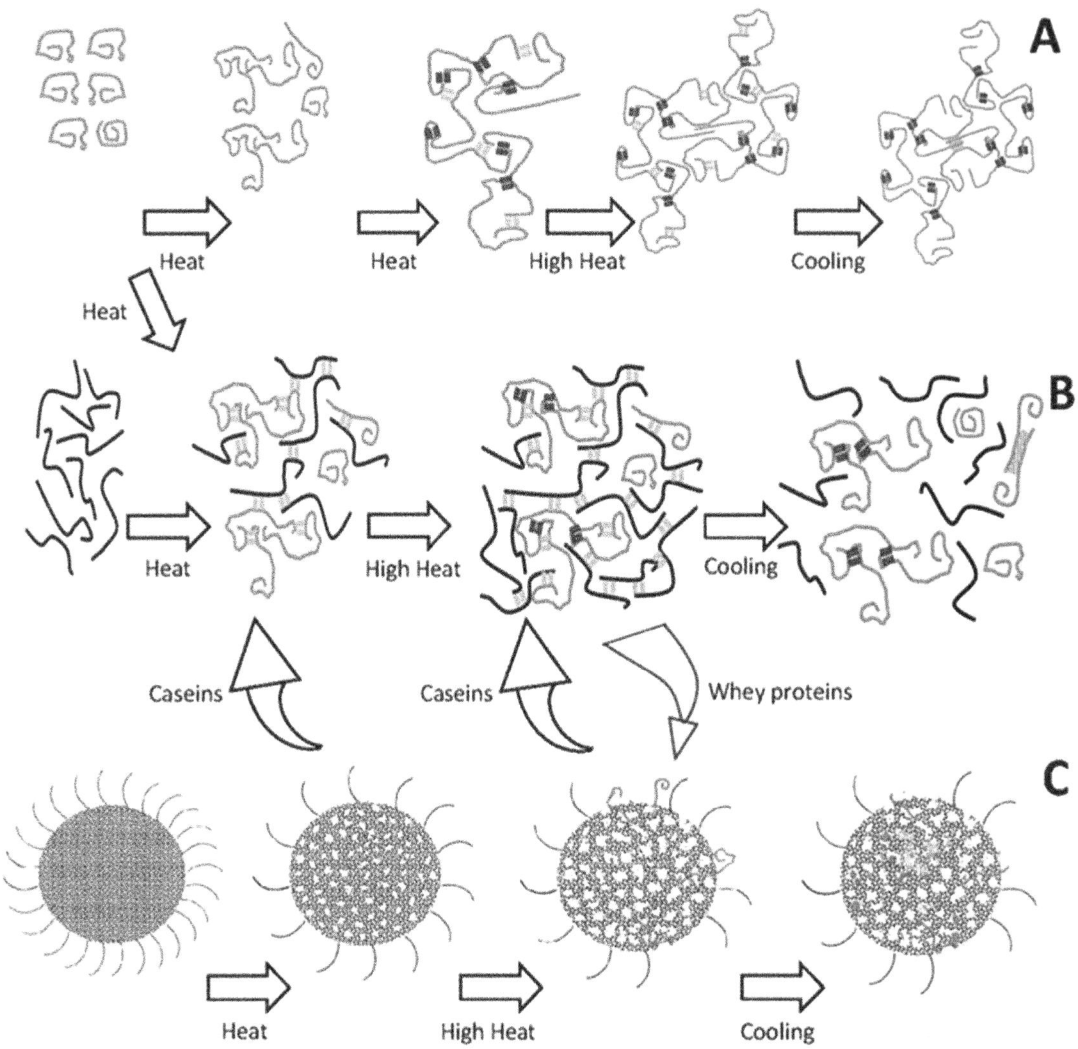

FIGURE 7.2 Proposed mechanism that protects whey proteins (WPs) against heat-denaturation and aggregation, with the chaperone-like action of caseins. A color code is used to mark the caseins (black), WPs (blue [light gray in print version]), covalent disulphide bonds (red [dark gray in print version]), hydrophobic attractions (yellow [very light gray in print version]) and hydrogen bonds (green [gray in print version]). The caseins and WPs are not represented proportional to their size and the figures present only selected types of interactions that take place in the casein/WP mixture (Liyanaarachchi et al., 2015). Panel A: in the absence of caseins, WPs are heat unfolded and small aggregates form via hydrophobic interactions followed by disulphide covalent bond formation. At higher temperatures, more covalent bonds and larger aggregates form while hydrophobic bonds are in place. While cooling down, hydrophobic attractions become weak but hydrogen bonds can be established, leading to larger aggregates; Panel B: with the inclusion of caseins, the serum caseins form hydrophobic bonds with unfolded WPs and their small aggregates during heating, preventing the growth of large WP aggregates. Hydrophobic bonds dissociate when cooling and thus-released WPs may slowly fold back into globular conformations or form noncovalent aggregates (i.e., hydrogen bond formation); Panel C: heat induced dissociation of some k-caseins away from the casein micelle surface and other caseins from the matrix at higher temperatures enhance porosity of the micelle. WPs migrate into the hydrophobic matrix of the micelle and are entrapped inside. Dissociated aS-andb-caseins will help prevent WP denaturation and aggregation via formation of hydrophobic association.

dissociate, resulting in the loss of native structure and solubility. However, when whey proteins is heated with caseins, caseins interact with partially unfolded whey proteins through hydrophobic bonding, acting as a molecular sponge that binds and stabilizes the exposed regions of WPs. These interactions limit further aggregation and the formation of large insoluble complexes. Importantly, these bonds are primarily noncovalent (e.g., hydrophobic and hydrogen bonds), which can dissociate during cooling. This allows the whey proteins to either refold into native-like conformations or form small, reversible aggregates thereby preserving protein functionality. considering this mechanism to the micellar level, where native casein micelles, upon heating, become porous and permit the migration of denatured WPs into their internal structure. Within the micelle, WPs become physically entrapped or stabilized by dissociated αs- and β-caseins, further preventing extensive aggregation. These micelle-trapped WPs are protected from the external aqueous environment, oxidative stress, and irreversible cross-linking. This micelle-centric model underscores the multi-faceted role of caseins in stabilizing milk proteins during high-temperature processing.

The extent and nature of interactions between denatured whey proteins and casein micelles are strongly pH-dependent. At pH < 6.6, denatured whey proteins (especially β-Lg) preferentially associate with casein micelles, whereas at pH > 6.6, they tend to form soluble aggregates (Anema and Klostermeyer, 1997; Corredig and Dalgleish, 1996). Vasbinder & De Kruif (2003) used renneting and electrophoretic techniques to show that the uniformity of the whey protein coating on micelles greatly influences rennet gelation inhomogeneous coatings promote flocculation, while uniform ones inhibit it. Interestingly, β-Lg shows a higher affinity for casein micelles than α-La, likely due to its reactive thiol group. α-La does not associate with micelles unless β-Lg is present, indicating a cooperative mechanism where β-Lg mediates initial binding, and α-La follows by forming soluble β-Lg–α-La complexes (Corredig and Dalgleish, 1999; Jang and Swaisgood, 1990).

Caseins, particularly β-casein and αs1-casein, exhibit chaperone-like activity during thermal processing. Their disordered, flexible, and amphiphilic nature allows them to bind reversibly to exposed hydrophobic patches on denatured whey proteins. This transient interaction stabilizes whey proteins and prevents irreversible aggregation (Bardwell and Jakob, 2012; Bhattacharyya and Das, 1999; O'Kennedy and Mounsey, 2006). Experiments with concentrated milk protein systems showed that increasing casein content reduces the size of heat-induced whey protein aggregates and lowers surface hydrophobicity, especially at higher pH (6.7–7.5), preserving native protein conformation (Liyanaarachchi et al., 2015). The addition of hydrophobic bond blockers (e.g., Tween 20) confirms that hydrophobic interactions are primary drivers of both aggregation and chaperone activity.

κ-casein plays a distinct role by modulating the aggregation behavior of β-Lg. Studies have shown that heating κ-casein with WPI forms stable, mixed aggregates, altering the critical gelation concentration and shifting the gelation pH (Guyomarc'h et al., 2009). While κ-casein affects aggregate composition, it has minimal influence on final gel texture, indicating its role is more structural than functional in gel systems (Baussay et al., 2004; Zúñiga et al., 2010). In contrast, sodium caseinate (containing all casein types) shows weaker suppression of aggregation due to competitive interactions among αs1-, β-, and κ-caseins (Donato et al., 2007).

In conclusion, heat-induced interactions between whey proteins and caseins are complex and highly dependent on environmental factors such as pH, ionic strength, temperature, protein concentration, and the presence of specific casein fractions. Caseins, especially β- and αs1-casein, play a protective, chaperone-like role by reversibly binding to denatured whey proteins and limiting irreversible aggregation. These interactions have critical implications for the textural and functional properties of dairy products, particularly in high-temperature applications such as UHT milk and sterilized beverages. Understanding and manipulating these interactions enables the design of thermally stable, high-protein dairy systems with enhanced functionality and consumer appeal.

4.2 Heat-induced interactions of whey proteins with carbohydrates

Thermal processing promotes both covalent and noncovalent interactions between whey proteins and carbohydrates, significantly impacting the structural and functional properties of dairy systems. β-Lg, the primary heat-sensitive whey protein, undergoes conformational unfolding at temperatures above ~65–75°C, exposing previously buried ε-amino groups of lysine and the thiol group at Cys121. This structural transition increases its reactivity toward carbohydrates such as lactose and maltodextrin, triggering glycation and other heat-induced interactions.

Although both carbohydrates engage with β-Lg during heating, lactose and maltodextrin follow distinct interaction pathways. Lactose, being a reducing disaccharide, readily participates in the Maillard reaction with exposed lysine residues on β-Lg. The reaction begins with the formation of a Schiff base and Amadori product, ultimately progressing to advanced glycation end-products (AGEs) such as Nε-carboxymethyllysine (CML) and Nε-carboxyethyllysine (CEL). In a site-specific analysis using UHPLC-MS/MS, (Meltretter et al. (2014) identified key glycation sites on β-Lg; K14, K47, K60, K75/77/83, K91, and K135 in raw milk heated to 99°C. Notably, the CML modification index reached 1.6 at K60 and 0.03 at K47, with total site modification indices ranging from 0.99 to 56.3, depending on thermal intensity. These findings demonstrate that lactose-induced glycation is both temperature-dependent and site-selective, governed by the degree of protein unfolding and lysine accessibility.

In contrast, maltodextrin, due to its higher molecular weight and reduced availability of free carbonyl groups, exhibits a lower glycation efficiency than lactose but forms noncovalent hydrogen bonds with proteins that stabilize tertiary structure and suppress excessive aggregation. While maltodextrin's participation in the Maillard reaction is more limited, its addition at optimized ratios can reduce whey protein aggregation and improve solubility by maintaining α-helix structure during processing. Furthermore, formulations using maltodextrin as a partial lactose substitute demonstrated enhanced product solubility and emulsification, primarily due to its ability to mitigate Maillard-induced overbrowning and to form a colloidal-stable matrix, especially under spray drying conditions. However, excess maltodextrin (especially at protein-to-carbohydrate ratios exceeding 1:3) can result in preferential sugar–sugar interactions, leading to phase separation and reduced protein dispersibility (Wijegunawardhana et al., 2024).

Beyond covalent glycation, thermal unfolding of whey proteins exposes hydrophilic and hydrophobic domains that facilitate noncovalent interactions with polysaccharides such as maltodextrin, pectin, and gum arabic. These high-molecular-weight carbohydrates form hydration shells around partially unfolded proteins via hydrogen bonding with polar amino acid side chains (e.g., Glu, Ser, Thr). Upon heating, the flexibility of polysaccharide chains is reduced, which results in entropic exclusion a phenomenon where the carbohydrate matrix spatially confines protein molecules and restricts their mobility. This caging effect not only prevents premature aggregation but also promotes proximity-induced covalent crosslinking, particularly in dense dispersions. These noncovalent mechanisms complement the covalent Maillard pathways, working together to modulate the behavior of whey proteins during thermal processing (Fogarty and Laage, 2014; Vijayakumar et al., 1999).

In summary, heat-induced interactions between whey proteins and carbohydrates involve both covalent and noncovalent mechanisms that affect protein functionality and product quality. Lactose readily undergoes Maillard reactions with exposed lysine residues in unfolded β-lactoglobulin, leading to site-specific glycation, while maltodextrin mainly stabilizes proteins through hydrogen bonding, reducing aggregation and preserving solubility. Noncovalent interactions with polysaccharides also contribute to protein stability by forming hydration shells and limiting mobility, thus preventing premature aggregation. These interactions are influenced by processing temperature, protein–carbohydrate ratio, and the structural flexibility of the proteins and carbohydrates involved.

4.3 Heat-induced interactions of whey proteins with polyphenols

Heat-induced structural transitions in β-Lg significantly enhance its ability to interact with dietary polyphenols by exposing hydrophobic and reactive residues that are otherwise buried in its native globular conformation. Al-Shabib et al. (2020) demonstrated that the interaction between catechin and β-Lg occurs through a dynamic quenching mechanism, with Stern-Volmer quenching constants (Ksv) increasing from 2.5×10^4 L/mol at 288 K to 3.5×10^4 L/mol at 310 K, and bimolecular quenching constants (kq) ranging from 2.5×10^{12} to 3.5×10^{12} L/mol·s well above the diffusion-controlled limit, confirming a collision-dominated process. In parallel, binding constants (Ka) also rose with temperature, from 0.132×10^5 to 1.39×10^5 L/mol, indicating that thermal unfolding facilitates stronger catechin binding. Thermodynamic analysis revealed positive enthalpy ($\Delta H = +79.23$ kJ/mol) and entropy ($\Delta S = +354.5$ J/mol·K), confirming that hydrophobic interactions dominate the binding process. Structural characterization using CD and UV–Vis spectroscopy showed enhanced negative ellipticity and a red shift around 218 nm, along with increased absorbance and peak shifts, reflecting exposure of aromatic residues like Trp19 and Trp61 due to conformational changes. Molecular docking confirmed that catechin binds within the β-Lg calyx, involving residues such as Leu58, Ile71, Phe105, and Lys60, with a docking score of -6.82 kcal/mol and a predicted affinity of $\sim 1.01 \times 10^5\ M^{-1}$. Molecular dynamics simulations further supported the formation of a stable complex, showing consistent RMSD (~1.15 Å) and solvent-accessible surface area profiles. These findings confirm that heat not only enhances β-Lg-polyphenol binding via thermally induced exposure of hydrophobic domains but also modulates protein functionality, potentially improving phenolic solubility, antioxidant retention, and colloidal stability in heat-processed dairy-polyphenol systems.

These trends are strongly supported by the findings of Jia et al. (2017) who evaluated β-Lg interactions with three common polyphenols; epigallocatechin-3-gallate (EGCG), chlorogenic acid (CGA), and ferulic acid (FA) and reported that EGCG induced the strongest fluorescence quenching effect (52.17% at 18 μM), followed by CGA (47.12%) and FA (45.91%). Stern-Volmer constants (Ksv) for EGCG decreased from 5.84×10^4 L/mol at 288 K to 3.80×10^4 L/mol at 308 K, with the corresponding binding constants (Ka) increasing from 2.16×10^4 to 5.80×10^4 and 57.89×10^4 L/mol over the same temperature range-indicative of an entropy-driven, endothermic interaction ($\Delta H = +123.05$ kJ/mol, $\Delta S = +0.51$ kJ/mol·K). In contrast, CGA and FA displayed decreasing Ka values with temperature, confirming exothermic interactions driven by van der Waals forces and hydrogen bonding ($\Delta H = -70.48$ kJ/mol for CGA and -187.2 kJ/mol for FA). CD and FTIR analyses showed that EGCG binding shifted β-LG secondary structure from α-helix (16.24%) toward increased β-sheet content (43.77%), suggesting reduced digestibility and enhanced rigidity. Surface hydrophobicity (S_0) declined from 8.51 (native) to 7.56 (EGCG-bound), and particle size decreased from 3777 to 991 nm in the presence of Ca^{2+}, supporting reduced aggregation and enhanced solubility. Molecular docking revealed EGCG formed hydrogen bonds with Glu44, Glu45, Gln59, and Glu157 (bond distances 1.883–2.449 Å), and involved residues Glu15, Thr18, Tyr20, Arg124, Leu156, and Gln159 in stabilizing the complex. CGA interacted with Ser36, Ala37, Lys60,

Trp61, Glu62, and Gly64, forming six hydrogen bonds (e.g., Ser36 at 2.033 and 2.176 Å), while FA primarily involved Gln35, Ser36, and Gly64. FRET analysis confirmed that donor-acceptor distances for all complexes remained <7 nm, verifying static quenching. These findings highlight that thermally induced conformational transitions in β-Lg promote specific, residue-dependent interactions with polyphenols, improving thermal stability, polyphenol solubility, and antioxidant functionality in dairy systems like milk tea formulations.

Recent findings by Khongphakdee et al. (2025) confirmed that heat-induced structural modifications in WPI significantly enhance its capacity to interact with vitexin, a flavone polyphenol through noncovalent interactions such as hydrogen bonding, hydrophobic forces, and van der Waals interactions. Upon sterilization at 121°C for 15 min, WPI underwent conformational unfolding, as evidenced by a marked reduction in α-helix content from 41.0% to 26.2% and a corresponding increase in β-sheet content from 39.4% to 61.6%, indicating greater molecular rigidity and aggregation potential. Intrinsic fluorescence spectroscopy revealed a red shift in emission maximum from 350 to 390 nm and a pronounced intensity reduction, suggesting exposure of aromatic residues such as Trp and Tyr due to tertiary structural disruption. These structural transitions facilitated stronger polyphenol binding, as reflected by enhanced colloidal stability zeta potential shifted from −25.3 mV to −35.1 mV and decreased particle size from 539.5 to 260.8 nm, pointing to reduced aggregation and better dispersion. Although total phenolic content (TPC) decreased before digestion due to thermal degradation, postdigestion analysis revealed up to a 3.2-fold increase in TPC, indicating effective release of vitexin from protein complexes. Additionally, the degree of hydrolysis (%DH) of sterilized WPI-vitexin complexes reached 44.8%, significantly higher than untreated or pasteurized samples, demonstrating improved enzymatic accessibility. In contrast, vitexin-loaded phytosomes (WV-PHN) displayed a higher zeta potential (−39.0 mV) but lower %DH (34.9%), likely due to steric hindrance from the phospholipid matrix. These findings highlight that thermal treatment promotes residue-specific binding and unfolding in whey proteins, enabling improved polyphenol entrapment, colloidal stability, and bioavailability.

Heat-induced conformational changes in whey proteins (WPs), particularly β-Lg, enhance their ability to bind anthocyanins (ACNs) through noncovalent interactions such as hydrophobic forces, hydrogen bonding, and van der Waals interactions. Ren & Monica Giusti (2021) demonstrated that the addition of anthocyanin-rich extracts (100 μM) from purple corn, grape, and black carrot to native whey proteins (3.6 μM) at pH 3.0 resulted in significant fluorescence quenching up to 73% for purple corn and 68% for grape and black carrot indicating strong binding affinity. This interaction induced a red shift in the maximum emission wavelength from 372 to 378 nm, suggesting increased polarity in the local environment of Trp residues due to structural rearrangement. Stern-Volmer and thermodynamic analyses confirmed static quenching behavior with binding constants (Ks) in the range of $1.00\text{–}5.99 \times 10^3\ M^{-1}$, and Gibbs free energy values ($\Delta G°$) between -1.71×10^4 and -2.29×10^4 J/mol, confirming the spontaneity of complex formation. Purple corn and black carrot CAN-WP complexes showed $\Delta H° > 0$ and $\Delta S° > 0$, indicating hydrophobic interactions as the dominant binding force, whereas grape CAN-WP interaction showed $\Delta H° < 0$ and $\Delta S° > 0$, suggesting a mixed mechanism dominated by hydrophobic forces. Preheating WP at 60°C enhanced Trp exposure and increased binding affinity (Ks from 3.33×10^2 to $4.75 \times 10^2\ M^{-1}$), while heating beyond 60°C caused aggregation and reduced binding, likely due to irreversible Tanford-like transitions and disulfide-linked aggregation that shielded binding sites. Collectively, these findings confirm that moderate heat treatment optimizes protein-ACN interactions, improving anthocyanin stability, reducing aggregation, and enhancing compatibility in thermally processed dairy-based systems.

Heat-induced structural modifications in whey proteins, particularly β-lactoglobulin, significantly enhance their interactions with dietary polyphenols by exposing reactive thiol groups and hydrophobic domains. These changes promote both covalent and noncovalent binding mechanisms, dominated by hydrophobic forces, hydrogen bonding, and van der Waals interactions which improve the thermal stability, antioxidant retention, and solubility of polyphenol-protein complexes. Studies using catechin, EGCG, chlorogenic acid, FA, vitexin, and anthocyanins demonstrate that thermal unfolding not only increases polyphenol binding affinity but also leads to structural rearrangements in proteins, including reductions in α-helix content and shifts toward β-sheet-rich, stable configurations. This facilitates enhanced bioavailability, reduced aggregation, and improved colloidal stability in heat-processed dairy systems. Overall, controlled thermal processing can be strategically used to tailor whey protein–polyphenol interactions for the development of functional, phenolic-enriched dairy products.

4.4 Heat-induced interactions of whey proteins with fat

Detailed electrophoretic analyses by Ye et al. (2004a, 2004b) confirmed that heat processing of whole milk at temperatures ranging from 60°C to 95°C facilitates specific covalent interactions between whey proteins and milk fat globule membrane (MFGM) proteins, primarily via disulfide linkages. Notably, β-lactoglobulin (β-Lg) began associating with

MFGM at ≥60°C, with incorporation levels rising up to ~1.12 mg/g fat at 90–95°C, while α-La incorporated at much lower levels (~0.23–0.25 mg/g fat). These interactions are not uniform across the lactation season: early-season milk showed higher β-Lg and α-La associations with MFGM than mid or late-season samples. Importantly, κ-casein was detected only at ≥75°C and was likely involved in β-Lg-MFGM crosslinking through thiol-disulfide interchange. Interestingly, SDS-PAGE under nonreducing conditions revealed that native MFGM proteins such as PAS 6 and PAS 7 progressively disappeared with heating, suggesting their displacement by whey proteins. Two-dimensional electrophoresis further confirmed that β-Lg and α-La integrate into MFGM complexes via covalent bonds, while PAS 7 may either degrade or migrate to the serum phase due to competitive interactions with β-Lg. These binding events occurred prior to complete whey protein denaturation, indicating early-stage monomer dissociation. Moreover, the observed reductions in PAS 6/7 and replacement by β-Lg led to altered interfacial composition, potentially impacting emulsification, heat stability, and aggregation dynamics in thermally processed emulsions.

Hassan et al. (2014) provided compelling evidence that thermally treated WPI nanoparticles form stable complexes with various unsaturated fatty acids, including oleic acid, cis-vaccenic acid, linolenic acid, and their transisomers. Heating WPI with these fatty acids at 60°C for 15 min resulted in substantial structural unfolding and complex formation, confirmed by sharp reductions in surface tension (e.g., from 68.7 to 57.1 mN/m with 0.25 M ratio oleic acid) and far-UV CD signals indicating tertiary structure loss. Notably, cis-configured fatty acids (e.g., oleic and cis-vaccenic acids) induced more pronounced unfolding and binding affinity than their transcounterparts, attributed to their U-shaped configuration which allows stereospecific fitting into exposed hydrophobic pockets of WPI. CD spectra for WPI-cis-vaccenic acid complexes displayed a negative ellipticity at 273 nm and positive peak at 294 nm (near UV), similar to oleic acid-WPI complexes, reflecting aromatic residue interactions and protein unfolding. These conformational changes lowered turbidity in WPI-fatty acid complexes compared to fatty acids alone, indicating enhanced solubility and reduced aggregation-effects beneficial for emulsion stability in heated dairy systems. Furthermore, zeta potential and erythrocyte lysis assays revealed that the cytotoxicity of WPI-fatty acid nanocomplexes increased with fatty acid concentration and was more potent with cis-fatty acids.

These whey protein-fat interactions were further examined in a series of in vivo studies by Carroll et al. (2006), who evaluated the ability of different whey protein matrices WPC80, WPI, and WPCHG to protect unsaturated fatty acids in soybean oil during digestion and promote their incorporation into milk fat. Upon heat-induced emulsion gelation of whey protein-soybean oil systems (WPEG), C18:2 (linoleic acid) and C18:3 (α-linolenic acid) content in milk fat significantly increased across all formulations. Specifically, C18:2 levels rose from 3.29% to 5.88% (WPC80), 2.91%–7.42% (WPI), and 3.57%–6.56% (WPCHG), while C18:3 increased from 0.51% to 0.84%, 0.52%–1.15%, and 0.51%–0.97%, respectively. The transfer efficiency of C18:2 to milk fat reached 30.1%–64.0%, and for C18:3%, 33.2%–33.9%, confirming the protective role of the heat-structured whey protein matrix. The interaction likely involves heat-induced denaturation and surface unfolding of whey proteins such as β-Lg and α-La, exposing hydrophobic and thiol-containing residues that entrap or interface with emulsified oil droplets. This protein-lipid interaction reduces rumen biohydrogenation of polyunsaturated fatty acids, as evidenced by significantly lower trans-9 and trans-11 C18:1 isomers in milk fat of WPEG-fed cows compared to controls. Moreover, plasma C18:2 levels were markedly higher in treated cows (6.13% vs. 3.45%, $P < .01$), reinforcing the matrix's efficacy in shielding PUFA during ruminal transit. These results demonstrate that thermally induced whey protein-fat interactions not only modulate milk lipid profiles but also offer a practical approach to enhancing the nutritional quality of dairy products by increasing PUFA content while minimizing undesirable transfat formation.

In summary, heat-induced structural unfolding of whey proteins facilitates their covalent and noncovalent interactions with lipids, particularly MFGM proteins and unsaturated fatty acids. These interactions, initiated as early as 60°C, involve thiol-disulfide interchange and hydrophobic bonding, leading to the displacement of native MFGM proteins and the formation of stable protein–lipid complexes. Whey proteins such as β-lactoglobulin and α-lactalbumin integrate into the MFGM or encapsulate fatty acids, altering interfacial properties, improving emulsion stability, and reducing protein and lipid aggregation under heat. Cis-unsaturated fatty acids exhibit stronger binding due to their flexible geometry, enhancing solubility and surface activity of protein-fat complexes. In vivo studies further confirm that heat-structured whey protein matrices can protect polyunsaturated fatty acids (PUFAs) like linoleic and α-linolenic acids during digestion, increase their incorporation into milk fat, and reduce biohydrogenation to harmful transisomers. These findings highlight the dual functionality of heat-treated whey proteins as emulsifiers and PUFA carriers, offering a promising strategy for improving the nutritional and functional profile of dairy products.

5. Whey protein interactions during nonthermal processing

Nonthermal technologies such as high-pressure processing (HPP), ultrasound (US), PEF, microwave (MW), and membrane filtration are increasingly used to tailor the structure and functionality of whey proteins without the thermal damage that occurs in traditional heat treatments. These techniques rely on distinct physical mechanisms that selectively disrupt or reorganize protein structures, affecting key interactions such as disulfide bonding, hydrophobic clustering, and electrostatic alignment. Their effects differ across whey proteins primarily β-Lg, α-La, and BSA due to differences in their size, bonding patterns, and stability (Table 7.2).

Compared to thermal processing, nonthermal methods such as HPP, ultrasound (US), and PEF offer distinct advantages, especially in preserving the sensory and nutritional qualities of food. These methods operate at ambient or sublethal temperatures, thereby avoiding thermal degradation of food components and maintaining fresh-like attributes such as flavor, color, and overall nutrient content. Nonthermal technologies are also associated with reduced energy consumption, water usage, and emissions, making them more environmentally friendly and sustainable alternatives to traditional heat treatments (Barros et al., 2023; Pereira and Vicente, 2010). Furthermore, their ability to selectively inactivate microorganisms at low temperatures enhances product safety without compromising quality, aligning with consumer demands for minimally processed, high-quality foods.

5.1 High-pressure processing (HPP): Pressure-induced unfolding and selective denaturation

HPP alters the structure of milk proteins by targeting the noncovalent interactions that stabilize their native conformation. While covalent bonds maintain the primary structure of proteins, secondary and tertiary structures are primarily stabilized by hydrogen bonds, electrostatic interactions, and hydrophobic forces with hydrophobic interactions being the most pressure-sensitive (Ankit Goyal, 2013). When exposed to pressures ranging from 100 to 800 MPa, these noncovalent bonds begin to break down, leading to protein unfolding and the formation of new intermolecular interactions (Fig. 7.3).

Among the whey proteins, β-Lg is the most susceptible to pressure-induced denaturation. It begins to unfold at pressures as low as 100 MPa in mixed systems and is significantly denatured at 200–300 MPa in isolated form. Complete aggregation often occurs by 600 MPa (Ma et al., 2024; Mirarab Razi et al., 2024). This unfolding exposes its reactive thiol group at Cys121, enabling thiol-disulfide exchange reactions that drive the formation of β-Lg-rich aggregates.

In contrast, α-La is far more stable under pressure. Its compact structure, reinforced by four internal disulfide bonds, absence of free thiol groups and a calcium-binding domain, makes it resistant to denaturation. At 600 MPa, α-La shows only ~10% denaturation and retains much of its structure even at 800 MPa, where it enters a molten globule state rather than becoming fully unfolded (Ma et al., 2024; Mirarab Razi et al., 2024; Romo et al., 2023; Siddiqui et al., 2024). These differences in stability allow for selective protein separation β-Lg can be precipitated under pressure while α-La remains in solution, enabling the production of α-La-enriched protein concentrates (Braspaiboon and Laokuldilok, 2024). Similarly Ye et al. (2004a, 2004b) observed that only 10% of the α-LA in milk denatured at 600 MPa when applied for 30 min. The lower susceptibility of this protein to pressure-induced denaturation was attributed to its relatively high conformational stability. Denaturation further increased to 50% when treated at 800 MPa for 30 min.

BSA, due to its larger size and extensive network of 17 disulfide bonds, and a single free thiol (Cys34) is even more resistant to pressure. It remains largely intact up to 400 MPa, and although denaturation begins beyond this point, it proceeds slowly and remains incomplete even at 700 MPa (Ravash et al., 2022; Siddiqui et al., 2024). The extensive disulfide network imposes a strong energetic barrier against pressure-induced unfolding.

Furthermore, pressure-temperature interactions play a critical role in modulating protein responses. Elevated temperatures (50–60°C) enhance pressure-induced denaturation by introducing thermal stress, while lower processing temperatures (~4–10°C) better preserve the native protein structure. For example, α-La retains more of its native structure when treated at 25°C versus 55°C at the same pressure (Rodiles-López et al., 2010; Van der Plancken et al., 2007).

In summary, HPP alters whey protein structure by disrupting noncovalent bonds, with β-lactoglobulin (β-Lg) being highly pressure-sensitive and denaturing at 200–300 MPa. In contrast, α-lactalbumin (α-La) remains largely stable up to 800 MPa due to its compact, disulfide-bonded structure. BSA is even more resistant, unfolding slowly beyond 400 MPa. These differences enable selective protein fractionation. Temperature also modulates HPP effects, higher temperatures promote unfolding, while cooler conditions help preserve native structure. HPP thus offers a targeted, nonthermal method to modify whey proteins for functional dairy applications.

TABLE 7.2 Summary of studies on the effects of nonthermal processing techniques including ultrasound, HPP, pulsed electric field (PEF), and membrane filtration on the physicochemical, structural, and functional properties of whey proteins in dairy systems.

Properties in dairy streams	Nonthermal processing technique	Whey system and treatment conditions	Effect	References
Particle size, solubility, thermal stability, gelation, secondary structure, morphology	Ultrasound	20% amplitude, 19.75 min, 12.75% feed concentration	Ultrasound treatment significantly enhanced whey protein functionality and structure. Mean particle size was reduced from 2.453 ± 0.717 μm to 0.683 ± 0.225 μm due to acoustic cavitation. Solubility increased from 72.22 ± 0.68% to 79.21 ± 1.08%. Thermal resistance improved, with heat coagulation time increasing ~6-fold from 178 ± 10.35 s to 1076.22 ± 8.15 s, and the thermal transition temperature (T_{peak}) shifted from 81.93°C to 93.32°C. The enthalpy of denaturation rose from −592.37 J/g to −3785.17 J/g, indicating increased stability. Rheological analysis showed an increase in G′, pointing to stronger gel formation. SDS-PAGE results revealed aggregation and fragmentation of protein bands. FTIR spectra showed a shift from 1644.4 to 1649.7 cm^{-1}, suggesting increased β-sheet content. SEM and TEM confirmed smaller, rounder, and more uniformly dispersed particles posttreatment.	Khatkar et al. (2018)
Particle size, solubility, zeta potential, turbidity, secondary structure, surface morphology, degree of grafting, polyphenol binding, surface hydrophobicity, free sulfhydryl groups	Ultrasound	A complex of WPI, 40 g/L and rutin (60 mg/mL) was prepared using a pH-driven method. WPI was adjusted to pH 12 and mixed 1:1 with rutin, followed by ultrasound treatment for 20 min at various powers (0, 150, 300, 450, and 600 W) at 0°C. After treatment, pH was neutralized to 7, and samples were dialyzed and freeze-dried. The study aimed to enhance the interaction between WPI and rutin to improve physicochemical stability and functional properties of the complex.	Ultrasound significantly improved the interaction between WPI and rutin. The particle size of the WPI–rutin complex decreased from 457.1 nm (0 W) to 218.5 nm at 600 W due to the disruption of protein aggregates by cavitation. Solubility increased from 72.2% in the untreated sample to a maximum of 84.3% at 300 W, and the absolute zeta potential rose from −22.4 mV to −32.1 mV, indicating improved colloidal stability. FTIR revealed enhanced protein structural order, with β-sheet content increasing from 48.12% to 50.60% and unordered coil decreasing from 10.24% to 7.30%. The degree of polyphenol binding reached 83.58% at 600 W, while the highest grafting degree was observed at 450 W (51.56%). Surface hydrophobicity and free sulfhydryl groups slightly declined due to unfolding and cavitation-induced oxidation. SEM images confirmed the development of a stable, honeycomb-like three-dimensional network structure, reflecting improved microstructural uniformity.	N. Guo et al. (2023)

Zeta potential, particle size, turbidity, secondary structure, surface hydrophobicity, viscosity, emulsifying activity index (EAI), emulsion stability index (ESI)	Ultrasound	This study aimed to investigate the effects of ultrasound on the interactions between WPI and Tremella fuciformis polysaccharide (TPS) at different pH values, and to evaluate the resulting changes in structural and functional properties. Aqueous blends of WPI (0.67%) and TPS (0.17%) at a 4:1 ratio (w/w, 0.84% total solids) were prepared at pH 3.0, 4.5, and 6.5, and subjected to ultrasound treatment at 20 kHz for 0–80 min under pulse mode at 25°C.	The application of ultrasound induced pH-dependent changes in WPI–TPS interactions. At pH 4.5, particle size initially decreased from 4.28 to 2.54 μm after 20 min of sonication, but increased markedly to 21.5 μm at 80 min due to aggregation caused by reduced electrostatic repulsion. Zeta potential measurements confirmed this trend, with values increasing from −24.78 mV to −8.31 mV at pH 4.5, and from −50.28 mV to −28.28 mV at pH 6.5, indicating weaker repulsive forces and enhanced complexation. Turbidity increased in the early stages of sonication at pH 4.5 and 6.5, consistent with complex formation, and then declined with longer exposure. FTIR analysis showed that β-sheet content in the WPI–TPS system at pH 4.5 increased from 29.27% to 38.77%, while random coil content slightly decreased, suggesting increased structural order. Surface hydrophobicity and viscosity were elevated after ultrasound, reflecting greater protein–polysaccharide interaction and network development. The emulsifying activity index (EAI) and emulsion stability index (ESI) both improved significantly after 20 min of treatment, demonstrating functional enhancement of the complex system.	Tian et al. (2024)
Particle size, zeta potential, turbidity, fluorescence, surface hydrophobicity, secondary structure, solubility, foaming, emulsifying, water-holding capacity	Ultrasound	This study aimed to optimize ultrasound treatment of WPI for pickering emulsion formation and assess how structural changes impact its functional properties. WPI (100 g/L) was sonicated using a 40 kHz probe at powers from 320 to 800 W for 0–60 min (optimal: 640 W for 30 min) at 25 ± 5°C in an ice bath.	Ultrasound reduced particle size to a minimum of 1.885 μm at 640 W/30 min, and significantly decreased turbidity due to disaggregation. Zeta potential increased in absolute value, reaching −26.7 mV, indicating enhanced electrostatic stability. Intrinsic fluorescence intensity increased, and surface hydrophobicity was elevated, reflecting partial unfolding and exposure of hydrophobic residues. FTIR analysis showed structural alterations, with increased β-sheet content. Solubility and water-holding capacity improved, while foaming ability increased from 37.89% (untreated) to 62.45%. Emulsifying activity index (EAI) and stability index (ESI) were both significantly enhanced. Pickering emulsions prepared with treated WPI showed excellent thermal and storage stability (no phase separation at $\varphi =$ 80% oil).	N. Li et al. (2024)

Continued

TABLE 7.2 Summary of studies on the effects of nonthermal processing techniques including ultrasound, HPP, pulsed electric field (PEF), and membrane filtration on the physicochemical, structural, and functional properties of whey proteins in dairy systems.—cont'd

Properties in dairy streams	Nonthermal processing technique	Whey system and treatment conditions	Effect	References
Surface hydrophobicity, free sulfhydryl groups, particle size, zeta potential, solubility, secondary structure	Ultrasound	This study aimed to explore the impact of high-intensity ultrasound (HIUS) on the physicochemical and structural properties of β-lactoglobulin for potential food applications. β-Lactoglobulin (1% w/v) solutions were sonicated at 20 kHz with varying amplitudes (20%–40%), durations (10–30 min), and temperatures (40–50°C). A Box–Behnken design was used for optimization.	Ultrasound treatment led to significant structural and functional changes. Surface hydrophobicity increased, reaching a maximum of 5812.08, and free sulfhydryl group content peaked at 5.97 μmol/g, indicating protein unfolding. Particle size increased slightly from 1.21 ± 0.05 nm to 1.66 ± 0.03 nm, while zeta potential increased from 15.47 ± 1.60 mV to 27.63 ± 3.30 mV. Solubility improved from 84.66% to 95.17%. FTIR confirmed an increase in α-helix and β-sheet structures. Fluorescence analysis showed an intensity increase with a blue shift (λ_{max} from 334 to 329 nm), indicating tertiary structure rearrangement. UV absorbance decreased with a shift in λ_{max} from 288 to 285 nm.	Ma et al. (2018)
Particle size, zeta potential, surface hydrophobicity, FTIR structure, foaming, emulsifying, viscosity, coacervation behavior	Ultrasound	This study evaluated the influence of high-intensity ultrasound (HIUS) pretreatment on WPI structure and its subsequent effect on complex coacervation (CC) with κ-carrageenan (KC). WPI (1% w/v) was treated at 24 kHz using amplitudes of 50% and 100% for 2 and 4 min (33–83 W). Coacervation was performed by mixing pretreated WPI with KC (1:1 w/w) under acidic conditions.	HIUS pretreatment significantly improved functional and structural attributes of WPI–KC coacervates. Particle size decreased progressively with increasing intensity and time. Surface hydrophobicity increased at 100% amplitude, indicating enhanced exposure of hydrophobic regions. Zeta potential increased slightly but maintained the cationic character (important for coacervation). FTIR spectra showed major shifts in the amide I region (1600–1700 cm^{-1}), confirming changes in secondary structure. Foaming capacity and stability were improved. Emulsion stability increased with HIUS-treated WPI, and viscosity of the coacervate system was enhanced. Optical microscopy showed more homogeneous microstructures in treated samples.	Vargas et al. (2021)

Particle size, surface hydrophobicity, free sulfhydryl groups, solubility, thermal properties, emulsifying activity, emulsion stability, viscosity	Ultrasound	This study aimed to investigate the effects of high-intensity ultrasound on the physical and functional properties of WPI. A 10% (w/v) WPI dispersion was treated using a 20 kHz ultrasonic probe at 20% and 30% amplitude for 10 or 20 min, corresponding to intensities of approximately 31 W/cm^2 and 69 W/cm^2, respectively. All treatments were performed under controlled temperature conditions (below 40°C).	Ultrasound treatment significantly decreased particle size from 190.4 ± 7.6 nm (untreated) to 138.0 ± 0.6 nm (20 min, 20% amplitude), while increasing span values, indicating broader size distribution. Surface hydrophobicity increased from 5.13×10^5 to 5.77×10^5 under 30% amplitude for 20 min. Free surface sulphydryl content increased from 10.63 to 14.56 μmol/g, and total SH content increased from 39.84 to 44.59 μmol/g, suggesting protein unfolding and exposure of reactive groups. Solubility at pH 7.0 rose from 74.95% (untreated) to 89.70% (30% amplitude, 20 min). DSC analysis showed no significant change in denaturation temperature (Ton: ~65.7°C; Tpeak: ~75.8°C), but enthalpy (ΔH) decreased from 0.54 to 0.28 J/g after treatment at 20% amplitude, indicating altered internal bonding. Emulsifying activity index (EAI) improved from 3.18 to 5.32 m^2/g, and emulsion stability index (ESI) increased from 62.26 to 104.83 min, depending on time and intensity. Apparent viscosity at 1000 s^{-1} shear rate decreased from 5.51 to 4.81 mPa s, while flow behavior shifted toward more Newtonian (n approaching 1). These changes confirmed improved emulsification, solubility, and reduced viscosity after ultrasound.	Shen et al. (2017)
Solubility, particle size, zeta potential, secondary structure, surface morphology, FTIR, XRD, DSC, SEM	High hydrostatic pressure (HHP)	A mixture of WPC, 6% w/v and guar gum (0.25% w/v) was prepared at pH 5 to investigate their interactions under HHP. Samples were treated at 0, 300, and 600 MPa for 0, 10, 20, and 30 min at 25°C, then freeze-dried. The purpose was to explore how pressure and duration affect the physicochemical, structural, and functional properties of the WPC–guar gum system for potential food applications.	HHP significantly affected the WPC–guar gum complex. Solubility decreased with increasing pressure and time due to protein aggregation and increased surface hydrophobicity. Particle size increased significantly, reaching 759.66 nm at 600 MPa for 30 min. Zeta potential became more negative, indicating stronger electrostatic interactions. FTIR showed changes in protein secondary structure (increased α-helix). XRD confirmed higher crystallinity at 600 MPa, and DSC revealed increased denaturation temperature. SEM images displayed larger and more irregular particles with fewer surface cracks, supporting improved microstructural formation. These changes suggest enhanced interactions and potential functional benefits in food formulations.	Mirarab Razi et al. (2024)

Continued

TABLE 7.2 Summary of studies on the effects of nonthermal processing techniques including ultrasound, HPP, pulsed electric field (PEF), and membrane filtration on the physicochemical, structural, and functional properties of whey proteins in dairy systems.—cont'd

Properties in dairy streams	Nonthermal processing technique	Whey system and treatment conditions	Effect	References
Gelation, rheology, microstructure, calcium interaction, protein aggregation, elastic modulus (G′), loss modulus (G″), tan δ	HPP	To explore how pH and calcium levels influence the gelation behavior of high-protein dairy systems under HPP, milk protein concentrate (MPC, 12.5% protein) was treated at 600 MPa for 3 min at 5°C. The pH was adjusted from 6.6 to 5.1 using glucono-δ-lactone, and the calcium content was varied between 24 and 36 mg of Ca/g protein using $CaCl_2$. The treatments were designed to assess changes in rheological and structural properties under varying ionic and acidification conditions.	HPP promoted gelation above pH 5.3, with increasing gel strength at higher pH. G′ increased from ~1400 Pa (pH 5.5) to ~2242 Pa (pH 6.6), while tan δ decreased, indicating more elastic behavior. Increasing calcium from 24 to 36 mg/g protein raised G′ from 2207 to 4109 Pa, suggesting stronger gels due to enhanced Ca^{2+} crosslinking. SEM showed denser, more homogeneous networks with higher pH and calcium.	Wang and Moraru (2021)
Protein denaturation, residual structure, solubility, susceptibility to trypsin hydrolysis, aggregation, particle size	HHP and ultrahigh-pressure homogenization (UHPH)	The study aimed to evaluate how HPP alters the native state and enzymatic digestibility of WPI; 10% w/w, pH 6.5. HHP was applied at 300 MPa for 15 min at 25°C. UHPH was conducted at 250 or 300 MPa, with or without immediate cooling at the HP-valve outlet.	HHP induced partial unfolding and improved susceptibility to trypsin (DH: 6.02%) without major aggregation. UHPH at 300 MPa with cooling preserved structural integrity (residual enthalpy: 95.7%), maintained solubility (90.7%), and enhanced proteolysis (DH: 7.43%). Without cooling, UHPH caused significant aggregation, reduced solubility (66.3%), and lowered hydrolysis efficiency (DH: 5.11%). UHPH at 250 MPa also showed aggregation but no improvement in hydrolysis. Findings highlight the importance of pressure–temperature balance for preserving functionality.	Blayo et al. (2016)
Protein aggregation, turbidity, thiol reactivity, local pH shifts, electrochemical modification	Pulsed electric field	This study aimed to assess how electric field intensity and treatment time influence whey protein aggregation. Reconstituted WPC (0.5% w/w) and filtered native whey (0.4% w/w) were treated using PEF at 1.25–12.5 kV/cm, 10–100 μs pulse length, and 0.38–85 Hz for 5–20 min, maintaining similar energy inputs as thermal treatments.	Protein aggregation was highly dependent on PEF treatment time (pulse length × frequency × duration), rather than EF intensity. Aggregation was attributed to localized deprotonation of thiol groups, cation precipitation, and electrochemical pH shifts. PEF induced aggregation at lower temperatures than thermal treatment. The increase in turbidity (absorbance at 380 nm) correlated with longer treatment time and cysteine concentration, indicating thiol-driven aggregate formation.	Axelrod et al. (2022)

Solubility, surface hydrophobicity, free sulfhydryl content, aggregation/dissociation, emulsifying capacity, microstructure, secondary structure, interfacial tension	Pulsed electric field	Whey protein solution (3 mg/mL) treated with square-wave PEF at 5, 10, 15, and 20 kV/cm for 2–8 ms (pulse width 10 μs, 50 Hz). Structural and functional changes were measured by SDS-PAGE, fluorescence, CD, zeta potential, and emulsion assays.	At 5–10 kV/cm, PEF promoted protein aggregation, reducing solubility by 4.69%–7.14% and increasing surface hydrophobicity. Higher field strengths (15–20 kV/cm) reversed these effects: Solubility increased by up to 10.33%, free amino groups increased by 31.8%, and surface sulfhydryl content increased, indicating partial unfolding and aggregate dissociation. Emulsifying activity index improved from 14.2 to 19.3 m^2/g, and stability index from 67.1 to 88.5 min. Viscosity decreased by 24.7%, and SEM revealed structural loosening and finer surface morphology. CD spectroscopy showed an increase in α-helix and a decrease in random coil, depending on intensity and duration.	Hu et al. (2024)
Intrinsic and extrinsic fluorescence intensity, surface hydrophobicity, protein unfolding	Pulsed electric field	WPI solutions (3% and 5%) were treated at 12, 16, and 20 kV/cm using 10, 20, or 30 exponential decay pulses. Protein unfolding and microenvironmental changes were assessed using tryptophan fluorescence and ANS binding.	PEF treatment induced partial unfolding of whey proteins, as evidenced by an increase in intrinsic fluorescence intensity by up to 26.2% and a red shift in emission peak from 333 to 337 nm, indicating greater exposure of aromatic residues. Extrinsic fluorescence (ANS) increased by up to 21.9%, reflecting an increase in surface hydrophobicity. These changes suggest moderate disruption of tertiary structure without complete denaturation.	Xiang et al. (2011)
Native whey protein content, protein aggregation, digestion profile, peptide bioavailability	Membrane filtration (1.4 μm + 0.2 μm microfiltration)	A split-stream process was used to produce infant milk formula (IMF) from rehydrated skim milk and WPI using sequential ceramic 1.4 μm and PVDF 0.2 μm microfiltration. Filtration was performed at 50°C, TMPs of 0.13 and 0.15 MPa respectively, followed by heat treatment (125°C × 5 s) and spray drying.	Membrane-processed IMF (MEM-IMF) retained significantly higher native (1.10 g/100 g powder) and monomeric whey proteins (1.48 g/100 g) compared to high-heat IMF (HT-IMF) (0.18 and 0.46 g/100 g respectively). MEM-IMF showed reduced protein aggregation and altered digestion profiles with increased release of bioavailable peptides, including enhanced expression of claudin-1 in Caco-2 cells. Carboxymethyl-lysine (CML) levels were not significantly reduced.	Chen et al. (2021)

HPP, high-pressure processing; *WPC*, whey protein concentrate; *WPI*, whey protein isolate.

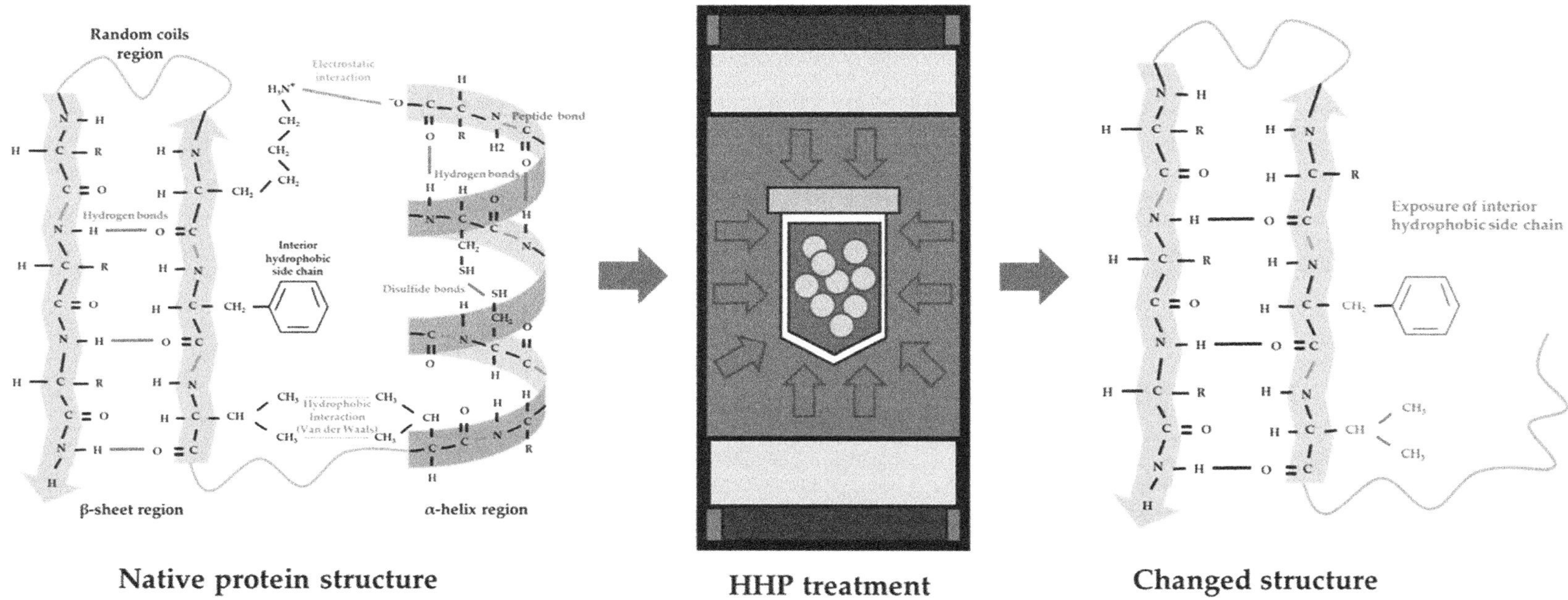

FIGURE 7.3 Schematic representation of high hydrostatic pressure (HHP) induced structural changes in whey proteins. Native protein structure consists of β-sheet and α-helix regions stabilized by hydrogen bonds, disulfide bonds, electrostatic interactions, and interior hydrophobic side chains. HHP treatment disrupts noncovalent interactions, leading to partial unfolding and exposure of interior hydrophobic residues, resulting in altered secondary and tertiary structure. *Adopted by Braspaiboon and Laokuldilok (2024).*

5.2 Ultrasound: Cavitation-induced unfolding and aggregation

Ultrasound (US), particularly high-intensity low-frequency ultrasound (20–100 kHz), modifies the structure of whey proteins primarily through acoustic cavitation-the rapid formation, growth, and collapse of microbubbles in a liquid medium. This process generates extreme but localized physical conditions, including transient temperatures up to ~5000 K and pressures over 1000 atm, accompanied by microjets, turbulence, and high shear forces (Leong et al., 2018; Suslick and Flannigan, 2008; Tinguely et al., 2022). These intense conditions disrupt noncovalent interactions such as hydrogen bonding and hydrophobic associations, leading to partial unfolding of proteins without significantly raising the bulk temperature-thus preserving overall protein functionality (Fig. 7.4).

Among whey proteins, β-Lg is particularly sensitive to ultrasound. Its β-sheet-rich dimeric structure is disrupted by cavitation, exposing its buried thiol group (Cys121), which facilitates thiol-disulfide exchange and promotes the formation of protein aggregates (Li et al., 2024; Wang et al., 2023). The extent of this unfolding is both time- and intensity-dependent. Short treatments (<5 min, <10°C) often result in reversible modifications, while extended or more intense sonication (>15 min, >20°C) can lead to irreversible denaturation and aggregation (Chandrapala et al., 2012) (Fig. 7.1).

Additionally, ultrasound can produce reactive oxygen species (ROS), such as hydroxyl radicals (•OH), during bubble collapse. These radicals further modify proteins by oxidizing susceptible residues like tryptophan and tyrosine and can facilitate covalent interactions between whey proteins and bioactive compounds such as polyphenols and polysaccharides (Ma et al., 2020; Steinhauer et al., 2015). This makes ultrasound a versatile tool not only for unfolding proteins but also for engineering new protein complexes.

Recent findings by Pathak et al. (2022) demonstrated that low-frequency ultrasound (20 kHz, 20 W/cm^3) applied to β-lactoglobulin (β-Lg) solutions (3.3 mg/mL, pH ~7.0) for up to 60 min led to the formation of mesoscopic amyloid crystals, characterized by defined β-sheet stacking and antiparallel arrangement. Despite the solutions appearing visually unchanged, size exclusion chromatography (SEC) revealed that approximately 25% of native β-Lg was converted into high-molecular-weight aggregates after 60 min. Scanning electron microscopy (SEM) analysis showed that mean crystal length increased from 4.7 μm at 15 min to 10.8 μm at 45 min and reached 21.8 μm at 60 min. Thioflavin T fluorescence and FTIR spectra supported these findings, indicating progressive β-sheet enrichment and the formation of antiparallel β-sheet peaks at ~1627 cm^{-1}. SDS-PAGE further confirmed progressive multimerization of β-Lg, with dimer and trimer bands emerging after 30 min, and bands ~20× the monomer mass appearing at 60 min. Notably, amyloid crystal formation initiated earlier than protein unfolding detectable by SEC or PAGE, implying that nucleation at cavitation sites is rapid and potentially driven by localized high shear and thermal effects. These structural transformations suggest that ultrasound enables protein molecules to overcome torsion energy barriers, leading to ordered aggregation and crystal formation without requiring elevated bulk temperatures or chemical initiators.

Further supporting the structural modifications induced by ultrasound, Guo et al. (2023) demonstrated that ultrasonic treatment of WPI-rutin complexes (40 g/L WPI, 60 mg/mL rutin, pH 12, treated for 20 min at 0°C) led to pronounced physicochemical changes. The covalent grafting degree increased from 28.72% (untreated) to 51.56% at 450 W, confirming stronger covalent interactions between rutin and exposed amino groups in WPI due to increased exposure of nucleophilic groups. Concurrently, the polyphenol binding efficiency reached 83.58% at 600 W, suggesting that cavitation and shear forces improved protein-polyphenol affinity. Structural FTIR analysis revealed an increase in β-sheet content from 48.12% to 50.60%, while unordered coil content dropped from 10.24% to 7.30%, indicating a shift to a more ordered secondary structure. This transformation coincided with a decline in free sulfhydryl content, likely due to oxidation by hydroxyl radicals generated during cavitation. Simultaneously, particle size decreased from 457.1 nm (0 W) to 218.5 nm at 600 W, and the absolute zeta potential increased from −22.4 mV to −32.1 mV, indicating better colloidal stability. SEM imaging showed a honeycomb-like three-dimensional network in the 150 W-treated sample, supporting improved microstructural uniformity. These transformations were accompanied by an increase in solubility from 72.2% to 84.3%, attributed to ultrasound-induced unfolding, exposure of hydrophilic residues, and rutin's hydroxyl-mediated hydrogen bonding. Overall, the study confirms that ultrasound can fine-tune WPI–polyphenol complexes by promoting both structural reorganization and enhanced functional performance through cavitation and shear effects.

Recent findings by Ashokkumar et al. (2009) demonstrated that ultrasound treatment significantly enhances the heat stability of whey protein solutions by disrupting preformed aggregates and reducing their viscosity upon subsequent heating. In this study, WPC solutions at 6.4% protein were preheated to 80°C for 1 min, then sonicated at 20 kHz with a calorimetric power of 31 W for 5–20 min. Postheating at 85°C for 20 min showed that nonsonicated samples exhibited a substantial increase in viscosity, indicating irreversible aggregation, whereas sonicated samples retained low viscosity, suggesting disaggregation of heat-induced protein clusters. Laser diffraction analysis confirmed that protein aggregate sizes decreased significantly after sonication, aligning with reduced viscosity data. These effects were reproducible at both

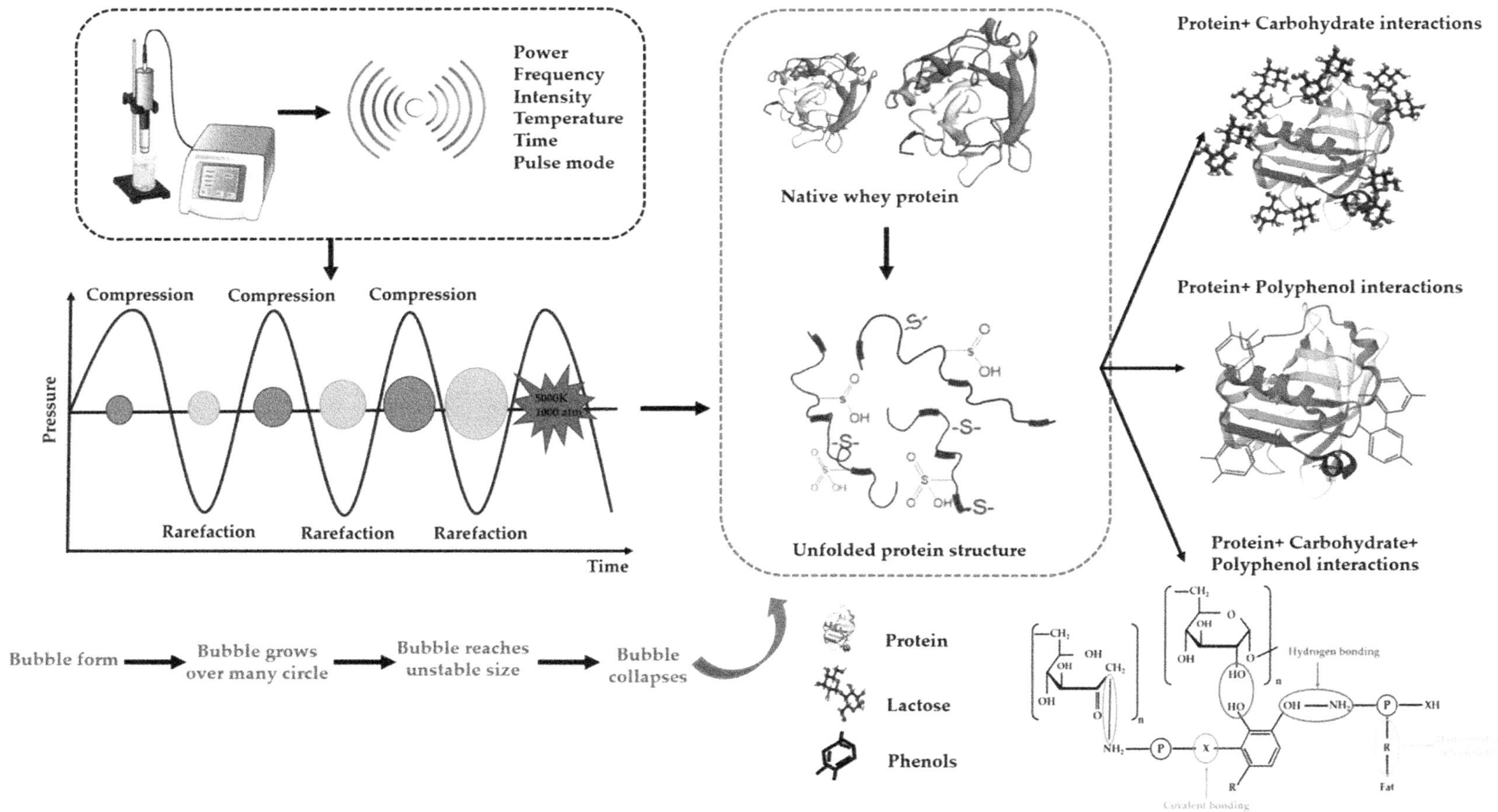

FIGURE 7.4 Mechanism of ultrasound-induced whey protein unfolding and subsequent interactions with carbohydrates and polyphenols. Cavitation generated by ultrasound treatment (controlled by power, frequency, intensity, temperature, and time) leads to alternating compression and rarefaction cycles, resulting in bubble collapse under extreme localized conditions (~5000 K, 1000 atm). This collapse disrupts the native structure of whey proteins, exposing thiol groups and hydrophobic residues. The unfolded proteins exhibit enhanced reactivity, enabling covalent and noncovalent interactions with carbohydrates (via Maillard reactions and hydrogen bonding), polyphenols (through hydrophobic and hydrogen bonding), and mixed protein-carbohydrate-polyphenol complexes, contributing to modified functional properties in food systems.

bench-top and pilot-scale levels (270 W, 300 mL/min flow rate) and were preserved even after freeze- or spray-drying and reconstitution. Notably, sonication without prior heating did not confer the same stability and, in some cases, even promoted gelation during postheating. The authors attributed these changes primarily to physical shear forces generated by acoustic cavitation, as protein solutions sonicated at higher frequencies ($\geq$100 kHz), which generate fewer radicals, did not show similar viscosity reductions. Collectively, these results highlight that ultrasound when applied following thermal pretreatment can prevent reaggregation of whey proteins, thereby improving processability and functional stability in high-protein dairy formulations.

Ultrasound modifies whey proteins by generating cavitation, which disrupts noncovalent bonds and induces partial unfolding, especially in β-lactoglobulin. This exposure of reactive groups promotes aggregation and enhances interactions with polyphenols and other compounds. Time and intensity determine whether changes are reversible or lead to permanent structural rearrangements. Ultrasound also produces ROS that drive covalent bonding. Studies show it enhances β-sheet formation, colloidal stability, and solubility while reducing aggregation. Furthermore, postheat ultrasound treatments improve the thermal stability and viscosity of whey protein systems, supporting its utility in functional dairy formulations.

5.3 Pulsed electric fields (PEF): Dipole disruption and protein realignment

Like ultrasound, PEF technology induces structural changes in whey proteins without significant heat generation. By applying short bursts of high-voltage electricity (20–50 kV/cm) for microto millisecond durations, PEF disrupts noncovalent interactions through dielectric polarization. This process causes reorientation of charged amino acids and dipoles, loosening the protein's tertiary structure and exposing buried hydrophobic regions. The degree of unfolding and aggregation depends on field strength, duration, and conductivity. At higher field intensities or in conductive media, electrochemical side effects such as local pH shifts, ROS generation, and ionization of functional groups (–COOH, $-NH_3^+$) further destabilize protein conformation. These effects can enhance surface hydrophobicity and promote aggregation or functional rearrangement, depending on the processing conditions. PEF is particularly useful for modifying protein solubility, emulsification, or interaction with other molecules in food matrices, and unlike ultrasound, allows precise control over structural modification through electrical parameters (Hu et al., 2024; Taha et al., 2022).

Xiang et al. (2008) investigated the structural responses of WPI to pulsed electric field (PEF) treatment, applying 30 pulses at field strengths of 12, 16, and 20 kV/cm to 3% and 5% WPI solutions. The intrinsic tryptophan fluorescence emission exhibited a red shift from 333 nm in untreated samples to 335–337 nm after PEF treatment, accompanied by increases in fluorescence intensity. For 3% WPI, the intensity rose from 23.11 (control) to 26.91, 28.73, and 29.16 arbitrary units at 12, 16, and 20 kV/cm, respectively. In 5% WPI, a similar trend was observed, with values increasing from 23.97 to 26.57, 28.12, and 29.16 a.u. at the same field strengths. These changes indicate a shift toward a more polar microenvironment surrounding tryptophan residues, suggestive of partial protein unfolding. ANS fluorescence analysis revealed a blue shift in emission maxima from 486 nm (untreated) to 482–484 nm, with intensity increasing from 1.57 to 1.91 a.u. for 3% WPI, and from 1.87 to 2.09 a.u. for 5% WPI at 20 kV/cm, confirming increased surface hydrophobicity. These structural modifications were statistically significant ($P < .05$), and no significant protein degradation or browning was observed, highlighting that PEF is a nonthermal method that alters protein conformation without denaturation. The study confirms that PEF can induce controlled exposure of hydrophobic residues and tertiary structural changes in WPI, potentially enhancing its functional properties in food systems.

Recent investigations by Hu et al. (2024) revealed that pulsed electric field (PEF) treatment significantly modulates the structural, physicochemical, and functional properties of whey proteins (WPs) by promoting either aggregation or dissociation depending on the electric field intensity (EFI) and treatment duration. At lower EFIs (5–10 kV/cm for 4–6 ms), PEF enhanced intermolecular interactions through disulfide bond cross-linking and hydrophobic associations, leading to increased particle sizes (~173 nm), reduced solubility (−6.65% to −7.14%), lower free sulfhydryl content, and higher turbidity indicative of aggregate formation. Conversely, higher EFI levels (15–20 kV/cm for 6–8 ms) triggered dissociation of these aggregates, as evidenced by decreased particle size (−12.75%), elevated zeta potential (−32.53 mV), and improved solubility (↑10.33%), along with enhanced intrinsic fluorescence and α-helix content (↑from 23.53% to 29.68%). SEC and SDS-PAGE analyses confirmed that high EFI promoted disaggregation and increased free amino groups and carbonyl content (up to 4.01 μmol/g), pointing to radical-induced protein oxidation. These modifications translated to improved emulsifying properties, with EAI and ESI increasing by 16.07% and 38.25%, respectively, and a marked reduction in interfacial tension (~10 mN/m). The observed improvements in emulsification and flow behavior were attributed to the generation of smaller, more flexible protein particles with enhanced surface activity and stability.

PEF induce structural changes in whey proteins by disrupting noncovalent bonds and causing dipole reorientation, without significant heat generation. These effects expose hydrophobic residues and modulate protein conformation based on field strength and treatment duration. At lower intensities, PEF promotes aggregation via disulfide cross-linking, while higher intensities favor dissociation, unfolding, and enhanced solubility. Studies confirm increases in surface hydrophobicity, α-helix content, and fluorescence, as well as improved emulsifying activity and interfacial properties. These findings demonstrate PEF's potential to fine-tune whey protein functionality for advanced food applications.

5.4 Microwave (MW) processing: dipolar rotation, volumetric heating, and localized unfolding of whey protein

Microwave (MW) treatment also induces structural changes in whey proteins, but via electromagnetic energy, which acts through both thermal and nonthermal mechanisms. MW radiation, typically at 2.45 GHz, causes dipole rotation and molecular friction, resulting in rapid, uniform, and volumetric heating in water-rich systems like milk. This localized heating can unfold proteins such as β-Lg, α-La, and BSA by destabilizing hydrophobic cores and disrupting hydrogen bonds (Munir et al., 2019). Although MW does not break covalent bonds directly, it can lead to significant tertiary and quaternary structural rearrangement, influencing the solubility and aggregation behavior of whey proteins. In contrast to PEF and US, MW tends to induce faster denaturation due to bulk heating effects, but it can still preserve nutritional properties if optimized carefully.

Microwave (MW) treatment has been shown to significantly alter the structural and immunological properties of whey proteins in milk-based systems through nonthermal denaturation and aggregation mechanisms. Microwave (MW) treatment significantly alters the structural and immunogenic properties of whey proteins in milk-based formulations, primarily through disulfide bond rearrangement and loss of solubility. Kaddouri et al. (2006) demonstrated that continuous MW exposure at 2.45 GHz with power levels ranging from 300 to 700 W for 10–20 min caused a pronounced reduction in β-Lg and α-La band intensities, particularly in whole milk. SDS-PAGE analysis revealed that at 400 W for 15 min, the band intensities of β-Lg and α-La in whole milk dropped drastically compared to those in fresh whey treated under similar conditions. At 700 W for 10 min, band intensity reductions in whole milk were even more substantial, indicating aggregation and molten globule transition. ELISA results showed that IgG binding to β-Lg decreased by over 55% at 400 W and up to 68.7% at 700 W, while α-La binding was reduced by 45%–60%, depending on duration and power. These reductions confirm epitope masking or structural disruption of allergenic sites. Importantly, whole milk samples consistently exhibited stronger reductions in both SDS-PAGE band intensity and immunoreactivity than isolated whey, likely due to the synergistic effects of milk fat and mineral content, which enhance localized microwave heating and protein-lipid interactions. Overall, MW treatment, when optimally controlled, emerges as a promising nonthermal strategy to reduce milk protein allergenicity while preserving nutritional functionality.

Zellal et al. (2011) showed microwave irradiation significantly reduces the allergenicity of B-Lg in whey proteins, with pH playing a critical modulatory role. Zellal et al. (2011) demonstrated in a BALB/c mouse model that microwave treatment at 300 or 700 W for 10–20 min at pH 6.8 (natural milk pH) or pH 4.6 substantially reduced BLG-induced allergenic responses. Jejunal tissues from sensitized mice showed no significant changes in short-circuit current (Isc) or conductance upon challenge with irradiated proteins under these pH conditions, indicating a marked reduction in B-Lg bioactivity. In contrast, treatment at extreme pH values (pH 2 or pH 9) failed to suppress allergenicity, as evidenced by elevated Isc values (e.g., from ~31.6 to 45.5 $\mu A/cm^2$ at 300 W, $P < .01$) and conductance increases (e.g., from 36.6 to 50.3 $mmho/cm^2$ at 700 W, $P < .05$). These effects are attributed to pH-sensitive structural transitions in BLG and the selective removal of conformational epitopes during microwave-induced unfolding. Importantly, allergenicity suppression at pH 6.8 and 4.6 was linked to reduced intestinal ion transport and paracellular permeability, suggesting successful masking of immunogenic sites. The study concludes that microwave treatment especially under mildly acidic or neutral pH can significantly attenuate the antigenic potential of BLG, offering a promising approach for the development of hypoallergenic dairy formulations.

In summary, microwave processing modifies whey proteins by inducing localized unfolding through dipole rotation and rapid volumetric heating, particularly affecting β-Lg, α-La, and BSA. This leads to rearrangement of disulfide bonds, protein aggregation, and reduced solubility without breaking covalent bonds. Studies show MW treatment significantly lowers immunoreactivity by masking allergenic epitopes, especially under neutral or mildly acidic pH conditions, making it an effective strategy for reducing milk allergenicity. Compared to other nonthermal technologies, MW offers rapid structural modification while preserving nutritional functionality when properly optimized (Fig. 7.5).

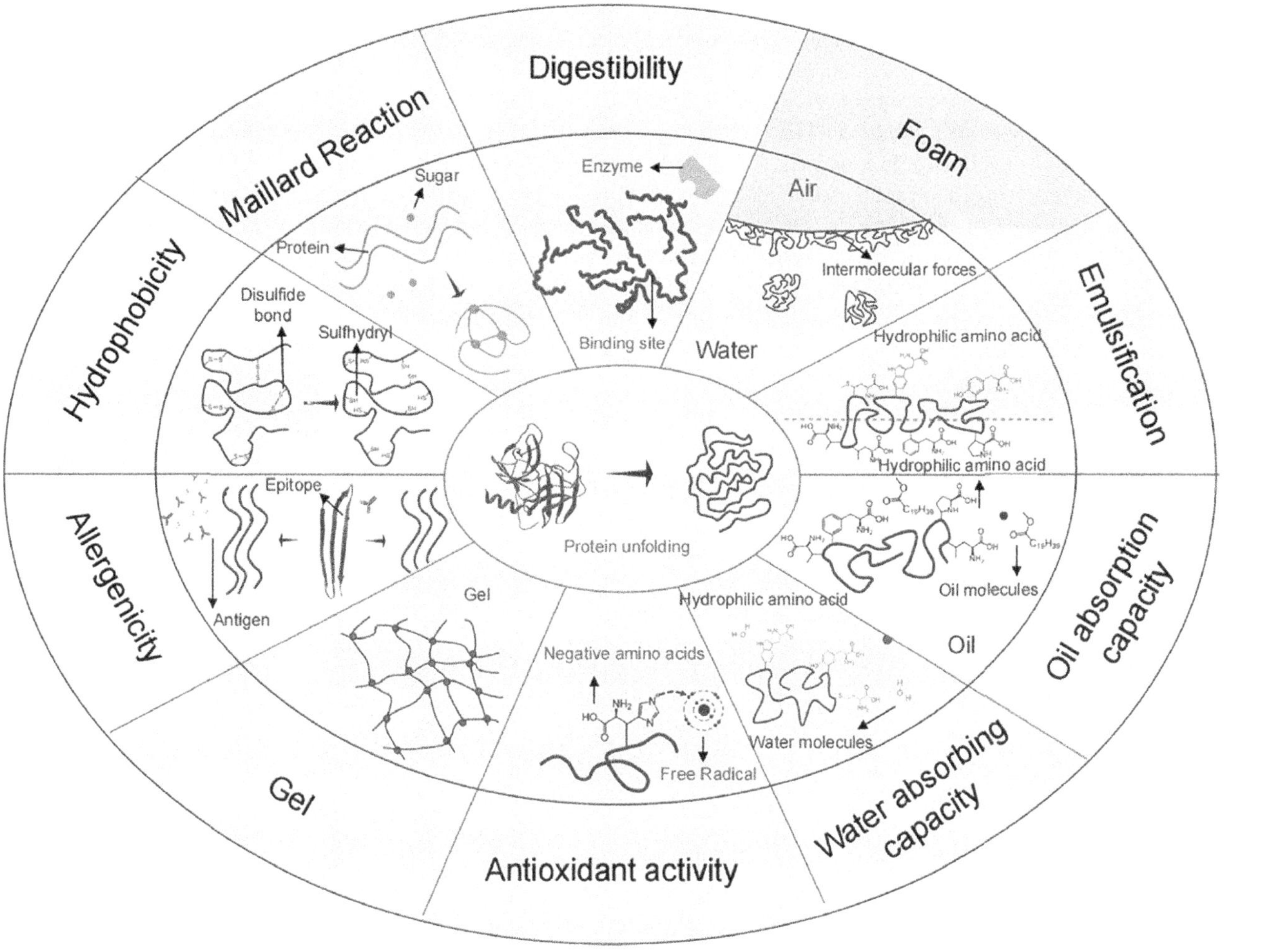

FIGURE 7.5 Effects of protein unfolding on functional and physicochemical properties in food systems. Microwave-induced protein unfolding alters structural features such as disulfide bonding, sulfhydryl group exposure, and epitope presentation, impacting hydrophobicity, allergenicity, and Maillard reactivity. These changes enhance or modify functional attributes including digestibility, gel formation, antioxidant activity, water and oil absorption capacities, foaming, and emulsification. Unfolded proteins exhibit increased interaction with enzymes, free radicals, hydrophilic and hydrophobic molecules, leading to improved bioactivity and altered interaction dynamics in multiphase food matrices. *Adopted by Deng et al. (2022).*

5.5 Filtration (MF/UF): Surface-induced denaturation

Although not traditionally viewed as a denaturing process, membrane filtration such as microfiltration (MF) and ultrafiltration (UF) can lead to surface-induced unfolding of whey proteins. When proteins like β-Lg interact with hydrophobic membrane surfaces, the native conformation can be disrupted, exposing reactive thiol groups that trigger thiol-disulfide interchange and oxidation reactions, leading to aggregate formation (Steinhauer et al., 2015). These aggregates often accumulate on the membrane, contributing to fouling and altering functional behavior. Additionally, shear forces and localized heating during filtration may exacerbate unfolding, increasing surface hydrophobicity and intermolecular interactions. While less dramatic than HPP or US, membrane-induced structural changes are important when proteins are later subjected to drying, heating, or complexation in downstream processes.

During ultrafiltration, ionic environment fluctuations play a crucial role in modulating the structural behavior of whey proteins, contributing significantly to membrane fouling and protein aggregation. demonstrated that dynamic changes in the retention of ions such as K^+, Ca^{2+}, Na^+, Mg^{2+}, and Zn^{2+} alter both the surface characteristics and secondary structure of β-lactoglobulin and other whey proteins. At ~11 min of filtration, elevated concentrations of Ca^{2+} and Na^+ coincided with increased α-helix content and buried hydrophilic groups, triggering greater protein aggregation. Raman spectroscopy confirmed the formation of antiparallel β-sheets (1668 cm^{-1}) and exposure of Trp and Glu residues, indicating structural rearrangements. At 17 min, K^+ concentration peaked, promoting β-sheet formation and enhancing aggregation through hydrophobic interactions and salt-bridge stabilization. Particle size analysis revealed a shift toward larger aggregates (~4500 nm), while zeta potential remained slightly negative, indicating diminished electrostatic repulsion. Surface tension studies showed increased values after 12 min (↑51.5 mN/m), consistent with hydrophobic domain exposure. However, between 21 and 24 min, when ion concentrations returned closer to initial levels, surface hydrophilicity improved, as seen by increased O–H and N–H bands in FTIR spectra and a decrease in surface tension (↓45.8 mN/m) (Wang et al., 2022).

In conclusion, membrane filtration processes like microfiltration (MF) and ultrafiltration (UF), though not typically classified as denaturing, can induce surface-driven unfolding of whey proteins through interactions with hydrophobic membranes and shear forces. This results in exposure of thiol groups, triggering disulfide bond formation and aggregation, particularly in β-Lg. Additionally, dynamic ionic changes during filtration influence protein conformation, leading to secondary structure alterations, hydrophobic exposure, and aggregate formation, which contribute to membrane fouling. These filtration-induced structural changes can impact the proteins' behavior in subsequent thermal or formulation processes, making them critical to understand in dairy processing.

6. Application of whey protein and its interactions in real-world systems

6.1 Infant nutrition: Enhancing digestibility and reducing allergenicity

β-lactoglobulin (β-Lg), are widely used in infant formulas due to their high nutritional value and functional properties. However, β-Lg is also a major allergenic protein absent in human milk and is recognized by specific IgE antibodies in cow's milk-allergic infants. This allergenicity arises from distinct IgE-binding epitopes located at residues 41–60, 102–124, and 149–162 (Hochwallner et al., 2010a, 2010b; Sélo et al., 1999) which trigger immune responses upon ingestion. These epitopes are highly immunogenic and recognized by a majority of cow's milk-allergic individuals, raising concerns about the safety of β-Lg-containing formulas in sensitive populations. Recent studies have explored structural modifications to reduce the allergenicity of β-Lg while maintaining its functional attributes.

A recent study by Jia et al. (2024) demonstrated that combining HHP with glycosylation using galactose oligosaccharides (GOS) effectively reduces the immunoreactivity of β-Lg through structural modification. HHP pretreatment at 200 MPa facilitated the exposure of buried glycosylation sites, particularly at Arg40, an IgE epitope not accessible under native conditions and significantly increased glycosylation at Lys14, Lys101, Lys124, and Arg148. These residues are closely associated with known IgE epitopes. The glycosylated β-Lg produced under optimal conditions (200 MPa, 55°C, 1:4 β-Lg:GOS ratio, 6 h) showed the lowest IgE-binding capacity as measured by indirect competitive ELISA. SDS-PAGE and CD revealed increased molecular weight and a reduction in β-sheet content, indicating epitope masking and partial unfolding. Further, intrinsic fluorescence and UV absorption analysis supported conformational changes that shielded aromatic residues. Functionally, the modified β-Lg displayed enhanced solubility across pH 3-9, improved emulsifying activity by approximately 15%, and significantly elevated DPPH radical scavenging activity. These results confirm that HHP-assisted glycosylation modifies critical structural features of β-Lg, reducing its allergenicity while simultaneously improving its physicochemical functionality thus providing a promising approach for the formulation of safer, hypoallergenic whey-based products for infant nutrition.

6.2 Functional foods and beverages: Improving bioavailability, stability, solubility, and sensory performance

Whey proteins are widely applied in functional food and beverage systems due to their superior solubility, emulsifying properties, and ability to interact with various food components, particularly polyphenols and polysaccharides, which can significantly influence product stability, sensory attributes, and nutritional functionality.

Recent studies have demonstrated that the formation of whey protein-polyphenol (WP-P) complexes not only improves the physical properties of formulations but also enhances the delivery of bioactive compounds. Schneider et al. (2016) developed WP-P mesostructures by combining 20% w/w whey protein isolate (WPI) with juices containing 50–500 μg g^{-1} total polyphenols (TP) at pH 4.5, yielding spray-dried particles with modal sizes ranging from 10–40 μm and protein contents reaching up to 870 mg g^{-1}. At 250 μg g^{-1} TP, protein yield peaked at 56%, and polyphenol retention ranged from 0.47 to 2.00 mg g^{-1}, indicating efficient encapsulation and matrix formation. These interactions strongly affected solubility and rehydration, with WP-P particles retaining over 80% dispersibility at pH 2-3 and 7-8, though stability decreased near the isoelectric point. Zeta potential values shifted from −0.37 mV at pH 4.5 to −34.9 mV at pH 6, suggesting pH-responsive dispersion stability. Moreover, polyphenol conjugation improved functional properties such as foam stability, where WP-MP particles exhibited extended foam drainage half-life compared to nonpolyphenol controls, and antioxidant performance was retained postprocessing. At 200 μg mL^{-1} extract concentration, WP-P particles suppressed iNOS and COX-2 gene expression in LPS-induced RAW 264.7 cells by more than 50% ($P < .001$) and significantly downregulated proinflammatory cytokines IL-1β and IL-6 ($P < .0001$). These findings highlight the dual functionality of WP-P particles not only do they contribute to enhanced product solubility, structure, and stability.

Supporting the functional versatility of whey protein-polyphenol interactions, Abdo et al. (2022) demonstrated that combining 5% whey protein isolate (WPI) with 2.5% beetroot peel extract in a strawberry-based beverage significantly enhanced its nutritional, physicochemical, and sensory properties. This formulation (T2) showed markedly improved TPC (36.12 mg GAE/mL) and total flavonoid content (23.83 mg CE/mL) compared to the control (25.08 and 20.12 mg/mL, respectively), corresponding to a 44% increase in phenolics and a 31% increase in flavonoids. These compositional changes translated into a twofold improvement in antioxidant activity, as measured by DPPH radical scavenging, with the IC_{50} of T2 decreasing to 264.32 μL/mL versus 531.72 μL/mL in the control. Even after 14 days at 4°C, the antioxidant potential of T2 remained high (IC_{50} = 321.92 μL/mL), while the control deteriorated (IC_{50} = 621.88 μL/mL), confirming storage stability. Protein content across formulations remained stable (4.65–4.69 g/100 mL), indicating compatibility of WPI with polyphenol-rich matrices. Additionally, betalain content peaked at 106.44 mg/L in the 5% beetroot extract sample (T3), although storage resulted in 46%–63% degradation, emphasizing the importance of protein-based stabilization systems. Microbial analysis showed strong antimicrobial effects of the beetroot extract, with T2 limiting aerobic mesophilic counts to 3.47 $\log_{10}$ CFU/mL by day 14, well below spoilage thresholds. Sensory evaluation further confirmed the superiority of T2, which received the highest acceptability scores (8.1–8.3 for color, taste, and aroma), outperforming both the control and T3, which exhibited off-flavors. The beverage also maintained stable pH (6.54–6.67) and total soluble solids (12.3–13.6 °Brix) during storage. These results confirm that integrating WPI with antioxidant-rich beetroot extract creates a multifunctional beverage with improved bioactivity, microbial safety, protein stability, and consumer appeal.

7. Conclusion

Thermal and nonthermal processing methods distinctly modulate the structure, functionality, and interaction dynamics of whey proteins, with implications across food formulation, nutritional enhancement, and allergenicity reduction. Thermal treatments, particularly at temperatures above 65°C, induce irreversible denaturation of β-Lg, α-La, and BSA, promoting intermolecular disulfide bond formation, thiol-disulfide interchange, and aggregation. These structural transitions are highly influenced by environmental factors such as pH, protein concentration, and ionic strength. Near-isoelectric conditions favor hydrophobic aggregation, while neutral to slightly alkaline pH enhances thiolate reactivity and disulfide exchange. Increasing protein concentration and divalent ion presence further intensifies aggregation, whereas phosphate additives can moderate it. Unfolded whey proteins also engage in diverse interactions with other food components. Glycation via Maillard reactions with carbohydrates targets exposed lysine residues, while hydrogen bonding with polysaccharides can enhance solubility and reduce excessive aggregation. Polyphenols bind through hydrophobic and van der Waals interactions with newly accessible aromatic residues, improving antioxidant stability. Similarly, interactions with lipids involve incorporation into MFGMs or the formation of protein-lipid nanocomplexes, aiding emulsification and bioactive retention. Moreover, caseins particularly κ-casein, β-casein, and αs1-casein play a protective role during heating. Acting as molecular chaperones, they form reversible hydrophobic and electrostatic associations with denatured whey proteins, limiting irreversible aggregation. Casein micelles can further entrap whey proteins within their porous matrix, enhancing structural stability and preserving functionality in thermally processed dairy systems.

In contrast, nonthermal technologies like HPP, ultrasound (US), PEF, microwave (MW), and membrane filtration offer controlled structural modification without compromising nutritional properties. HPP induces selective denaturation based on protein stability β-Lg is unfolded and aggregated while α-La and BSA resist pressure-induced disruption due to their disulfide-rich structures. Ultrasound promotes partial unfolding and amyloid aggregation via cavitation, exposing thiols and hydrophobic regions, enabling complexation with polyphenols and enhancing thermal stability. PEF reorients dipoles and destabilizes tertiary structures, enhancing surface hydrophobicity and emulsifying properties in a field strength-dependent manner. Microwave treatment modifies whey proteins via rapid, localized heating, reducing allergenicity by disrupting conformational epitopes and facilitating molten globule transitions, particularly under neutral and mildly acidic pH. Filtration processes, though subtle, can also trigger surface-induced unfolding and ion-mediated structural transitions, influencing aggregation, fouling, and downstream processing behavior. Therefore, thermal methods show at promoting covalent restructuring and matrix embedding, while nonthermal techniques offer tunable, residue-specific modifications that preserve functionality and reduce allergenicity.

References

Abdo, E.M., Allam, M.G., Gomaa, M.A.E., Shaltout, O.E., Mansour, H.M.M., 2022. Valorization of whey proteins and beetroot peels to develop a functional beverage high in proteins and antioxidants. Front. Nutr. 9. https://doi.org/10.3389/fnut.2022.984891.

Al-Shabib, N.A., Khan, J.M., Malik, A., Tabish Rehman, M., AlAjmi, M.F., Husain, F.M., Hisamuddin, M., Altwaijry, N., 2020. Molecular interaction of tea catechin with bovine β-lactoglobulin: a spectroscopic and in silico studies. Saudi Pharm. J. 28 (3), 238–245. https://doi.org/10.1016/j.jsps.2020.01.002.

Andoyo, R., Diani, A.Z., Fetriyuna, F., 2023. Effect of heating temperature on physical, functional, and digestibility properties of Whey Protein Concentrate (WPC). IOP Conf. Ser. Earth Environ. Sci. 1230 (1). https://doi.org/10.1088/1755-1315/1230/1/012155.

Anema, S.G., Klostermeyer, H., 1997. Heat-induced, pH-dependent dissociation of casein micelles on heating reconstituted skim milk at temperatures below 100°C. J. Agric. Food Chem. 45 (4), 1108–1115. https://doi.org/10.1021/jf960507m.

Anema, S.G., Li, Y., 2003. Association of denatured whey proteins with casein micelles in heated reconstituted skim milk and its effect on casein micelle size. J. Dairy Res. 70 (1), 73–83. https://doi.org/10.1017/S0022029902005903.

Ankit Goyal, V.S., 2013. High pressure processing and its impact on milk proteins: a review. Res. Rev. J. Dairy Sci. Technol. 2 (1), 12–20. http://www.stmjournals.com/sci/index.php?journal=RRJoDST&page=article&op=view&path%5B%5D=366%5Cn.

Asaduzzaman, M., Mahomud, M.S., Haque, M.E., 2021. Heat-induced interaction of milk proteins: impact on yoghurt structure. Int. J. Food Sci. 2021, 1–10. https://doi.org/10.1155/2021/5569917.

Ashokkumar, M., Lee, J., Zisu, B., Bhaskarcharya, R., Palmer, M., Kentish, S., 2009. Hot topic: sonication increases the heat stability of whey proteins. J. Dairy Sci. 92 (11), 5353–5356. https://doi.org/10.3168/jds.2009-2561.

Axelrod, R., Beyrer, M., Mathys, A., 2022. Impact of the electric field intensity and treatment time on whey protein aggregate formation. J. Dairy Sci. 105 (8), 6589–6600. https://doi.org/10.3168/jds.2021-21395.

Bardwell, J.C.A., Jakob, U., 2012. Conditional disorder in chaperone action. Trends Biochem. Sci. 37 (12), 517–525. https://doi.org/10.1016/j.tibs.2012.08.006.

Barros, J.H.T., de Carvalho Oliveira, L., Cristianini, M., Steel, C.J., 2023. Non-thermal emerging technologies as alternatives to chemical additives to improve the quality of wheat flour for breadmaking: a review. Crit. Rev. Food Sci. Nutr. 63 (11), 1612–1628. https://doi.org/10.1080/10408398.2021.1966380.

Baussay, K., Le Bon, C., Nicolai, T., Durand, D., Busnel, J.P., 2004. Influence of the ionic strength on the heat-induced aggregation of the globular protein β-lactoglobulin at pH 7. Int. J. Biol. Macromol. 34 (1–2), 21–28. https://doi.org/10.1016/j.ijbiomac.2003.11.003.

Bhattacharyya, J., Das, K.P., 1999. Molecular chaperone-like properties of an unfolded protein, αs-casein. J. Biol. Chem. 274 (22), 15505–15509. https://doi.org/10.1074/jbc.274.22.15505.

Blayo, C., Vidcoq, O., Lazennec, F., Dumay, E., 2016. Effects of high pressure processing (hydrostatic high pressure and ultra-high pressure homogenisation) on whey protein native state and susceptibility to tryptic hydrolysis at atmospheric pressure. Food Res. Int. 79, 40–53. https://doi.org/10.1016/j.foodres.2015.11.024.

Braspaiboon, S., Laokuldilok, T., 2024. High hydrostatic pressure: influences on allergenicity, bioactivities, and structural and functional properties of proteins from diverse food sources. Foods 13 (6). https://doi.org/10.3390/foods13060922.

Broersen, K., 2020. Milk processing affects structure, bioavailability and immunogenicity of β-lactoglobulin. Foods 9 (7). https://doi.org/10.3390/foods9070874.

Buggy, A., 2018. Influence of protein concentration on heat-induced aggregation and subsequent functionality of whey proteins. February. https://mural.maynoothuniversity.ie/10047/.

Buggy, A.K., McManus, J.J., Brodkorb, A., Hogan, S.A., Fenelon, M.A., 2018. Pilot-scale formation of whey protein aggregates determine the stability of heat-treated whey protein solutions—effect of pH and protein concentration. J. Dairy Sci. 101 (12), 10819–10830. https://doi.org/10.3168/jds.2017-14177.

Carroll, S.M., DePeters, E.J., Rosenberg, M., 2006. Efficacy of a novel whey protein gel complex to increase the unsaturated fatty acid composition of bovine milk fat. J. Dairy Sci. 89 (2), 640–650. https://doi.org/10.3168/jds.S0022-0302(06)72128-2.

Carter, B.G., Cheng, N., Kapoor, R., Meletharayil, G.H., Drake, M.A., 2021. Invited review: microfiltration-derived casein and whey proteins from milk. J. Dairy Sci. 104 (3), 2465–2479. https://doi.org/10.3168/jds.2020-18811.

Chamani, J., Moosavi-Movahedi, A.A., Rajabi, O., Gharanfoli, M., Momen-Heravi, M., Hakimelahi, G.H., Neamati-Baghsiah, A., Varasteh, A.R., 2006. Cooperative α-helix formation of β-lactoglobulin induced by sodium n-alkyl sulfates. J. Colloid Interface Sci. 293 (1), 52–60. https://doi.org/10.1016/j.jcis.2005.06.015.

Chandrapala, J., Zisu, B., Kentish, S., Ashokkumar, M., 2012. The effects of high-intensity ultrasound on the structural and functional properties of α-Lactalbumin, β-Lactoglobulin and their mixtures. Food Res. Int. 48 (2), 940–943. https://doi.org/10.1016/j.foodres.2012.02.021.

Chen, Y., Callanan, M., Shanahan, C., Tobin, J., Gamon, L.F., Davies, M.J., Giblin, L., Brodkorb, A., 2021. The use of membrane filtration to increase native whey proteins in infant formula. Dairy 2 (4), 515–529. https://doi.org/10.3390/dairy2040041.

Chu, S., Letcher, R.J., 2021. Identification and characterization of serum albumin covalent adduct formed with atrazine by liquid chromatography mass spectrometry. J. Chromatogr., B: Anal. Technol. Biomed. Life Sci. 1163, 122503. https://doi.org/10.1016/j.jchromb.2020.122503. December 2020.

Cornacchia, L., Forquenot De La Fortelle, C., Venema, P., 2014. Heat-induced aggregation of whey proteins in aqueous solutions below their isoelectric point. J. Agric. Food Chem. 62 (3), 733–741. https://doi.org/10.1021/jf404456q.

Corredig, M., Dalgleish, D.G., 1996. Effect of temperature and pH on the interactions of whey proteins with casein micelles in skim milk. Food Res. Int. 29 (1), 49–55. https://doi.org/10.1016/0963-9969(95)00058-5.

Corredig, M., Dalgleish, D.G., 1999. The mechanisms of the heat-induced interaction of whey proteins with casein micelles in milk. Int. Dairy J. 9 (3–6), 233–236. https://doi.org/10.1016/S0958-6946(99)00066-7.

Creamer, L.K., Bienvenue, A., Nilsson, H., Paulsson, M., van Wanroij, M., Lowe, E.K., Anema, S.G., Boland, M.J., Jiménez-Flores, R., 2004. Heat-induced redistribution of disulfide bonds in milk proteins. 1. Bovine β-lactoglobulin. J. Agric. Food Chem. 52 (25), 7660–7668. https://doi.org/10.1021/jf049388y.

Deng, X., Huang, H., Huang, S., Yang, M., Wu, J., Ci, Z., He, Y., Wu, Z., Han, L., Zhang, D., 2022. Insight into the incredible effects of microwave heating: driving changes in the structure, properties and functions of macromolecular nutrients in novel food. Front. Nutr. 9. https://doi.org/10.3389/fnut.2022.941527.

Dissanayake, M., Ramchandran, L., Donkor, O.N., Vasiljevic, T., 2013. Denaturation of whey proteins as a function of heat, pH and protein concentration. Int. Dairy J. 31 (2), 93–99. https://doi.org/10.1016/j.idairyj.2013.02.002.

Donato, L., Guyomarc'h, F., Amiot, S., Dalgleish, D.G., 2007. Formation of whey protein/κ-casein complexes in heated milk: preferential reaction of whey protein with κ-casein in the casein micelles. Int. Dairy J. 17 (10), 1161–1167. https://doi.org/10.1016/j.idairyj.2007.03.011.

Farrokhi, F., Ehsani, M.R., Badii, F., Hashemi, D.A.P., Graduated, M., 2020. Effect of pH on structural properties of heat-induced whey protein gels. J. Food Biosci. Technol. 10 (1), 57–68.

Fogarty, A.C., Laage, D., 2014. Water dynamics in protein hydration shells: the molecular origins of the dynamical perturbation. J. Phys. Chem. B 118 (28), 7715–7729. https://doi.org/10.1021/jp409805p.

Fox, P.F., Guinee, T.P., Cogan, T.M., McSweeney, P.L.H., 2017. Fundamentals of Cheese Science. Springer US. https://doi.org/10.1007/978-1-4899-7681-9.

Fox, P.F., Uniacke-Lowe, T., McSweeney, P.L.H., O'Mahony, J.A., 2015. Dairy chemistry and biochemistry. Nature 267 (5614). https://doi.org/10.1007/978-3-319-14892-2. Springer International Publishing.

Gulzar, M., Bouhallab, S., Jeantet, R., Schuck, P., Croguennec, T., 2011. Influence of pH on the dry heat-induced denaturation/aggregation of whey proteins. Food Chem. 129 (1), 110–116. https://doi.org/10.1016/j.foodchem.2011.04.037.

Guo, M., 2019. Whey Protein Production, Chemistry, Functionality, and Applications. John Wiley and Sons.

Guo, N., Ye, S., Zhou, G., Zhang, Y., Zhang, F., Xu, J., Pan, S., Zhu, G., Wang, Z., 2023. Effect of ultrasound treatment on interactions of whey protein isolate with rutin. Ultrason. Sonochem. 95, 106387. https://doi.org/10.1016/j.ultsonch.2023.106387.

Guyomarc'h, F., Nono, M., Nicolai, T., Durand, D., 2009. Heat-induced aggregation of whey proteins in the presence of κ-casein or sodium caseinate. Food Hydrocoll. 23 (4), 1103–1110. https://doi.org/10.1016/j.foodhyd.2008.07.001.

Hassan, Z., Awad, R., El-sayed, M., Foda, M., Otzen, D., Salama, H., 2014. Interaction between whey protein nanoparticles and fatty acids. Integr. Food Nutr. Metabol. 1 (2), 91–97. https://doi.org/10.15761/IFNM.1000107.

Havea, P., Watkinson, P., Kuhn-Sherlock, B., 2009. Heat-induced whey protein gels: protein-protein interactions and functional properties. J. Agric. Food Chem. 57 (4), 1506–1512. https://doi.org/10.1021/jf802559z.

Hill, A.R., 1989. The β-lactoglobulin-χ-casein complex. Can. Inst. Food Sci. Technol. J. 22 (2), 120–123. https://doi.org/10.1016/S0315-5463(89)70345-X.

Hochwallner, H., Schulmeister, U., Swoboda, I., Balic, N., Geller, B., Nystrand, M., Härlin, A., Thalhamer, J., Scheiblhofer, S., Niggemann, B., Quirce, S., Ebner, C., Mari, A., Pauli, G., Herz, U., Van Tol, E.A.F., Valenta, R., Spitzauer, S., 2010a. Microarray and allergenic activity assessment of milk allergens. Clin. Exp. Allergy 40 (12), 1809–1818. https://doi.org/10.1111/j.1365-2222.2010.03602.x.

Hochwallner, H., Schulmeister, U., Swoboda, I., Focke-Tejkl, M., Civaj, V., Balic, N., Nystrand, M., Härlin, A., Thalhamer, J., Scheiblhofer, S., Keller, W., Pavkov, T., Zafred, D., Niggemann, B., Quirce, S., Mari, A., Pauli, G., Ebner, C., Papadopoulos, N.G., Spitzauer, S., 2010b. Visualization of clustered IgE epitopes on α-lactalbumin. J. Allergy Clin. Immunol. 125 (6), 1279–1285.e9. https://doi.org/10.1016/j.jaci.2010.03.007.

Holm, N.K., Jespersen, S.K., Thomassen, L.V., Wolff, T.Y., Sehgal, P., Thomsen, L.A., Christiansen, G., Andersen, C.B., Knudsen, A.D., Otzen, D.E., 2007. Aggregation and fibrillation of bovine serum albumin. Biochim. Biophys. Acta Proteins Proteom. 1774 (9), 1128–1138. https://doi.org/10.1016/j.bbapap.2007.06.008.

Hu, X., Wang, H., Hu, Y., Tu, Z., 2024. Insight into the effects of pulsed electric field on the structure, aggregation characteristics and functional properties of whey proteins. Food Hydrocoll. 154, 110111. https://doi.org/10.1016/j.foodhyd.2024.110111.

Invernizzi, G., Annoni, E., Natalello, A., Doglia, S.M., Lotti, M., 2008. In vivo aggregation of bovine β-lactoglobulin is affected by Cys at position 121. Protein Expr. Purif. 62 (1), 111–115. https://doi.org/10.1016/j.pep.2008.06.012.

Ivancic, V.A., Lombardo, H.L., Ma, E., Wikström, M., Batabyal, D., 2022. Advancing secondary structure characterization of monoclonal antibodies using microfluidic modulation spectroscopy. Anal. Biochem. 646, 0–5. https://doi.org/10.1016/j.ab.2022.114629.

Iwaoka, M., Mitsuji, T., Shinozaki, R., 2019. Oxidative folding pathways of bovine milk β-lactoglobulin with odd cysteine residues. FEBS Open Bio 9 (8), 1379–1391. https://doi.org/10.1002/2211-5463.12656.

Jang, H.D., Swaisgood, H.E., 1990. Disulfide bond formation between thermally denatured β-lactoglobulin and κ-casein in casein micelles. J. Dairy Sci. 73 (4), 900–904. https://doi.org/10.3168/jds.S0022-0302(90)78746-2.

Jia, C., Qian, X., Zhang, H., Xiao, Y., Sun, A., 2024. Characteristics of IgE/IgG affinity and glycosylation sites in β-lactoglobulin subjected to synergistic HHP and galactose oligogalactose treatment. Lwt 207, 116642. https://doi.org/10.1016/j.lwt.2024.116642.

Jia, J., Gao, X., Hao, M., Tang, L., 2017. Comparison of binding interaction between β-lactoglobulin and three common polyphenols using multi-spectroscopy and modeling methods. Food Chem. 228, 143–151. https://doi.org/10.1016/j.foodchem.2017.01.131.

Jia, Y., Cocker, C., Sampath, J., 2025. Insights into protein unfolding under pH, temperature, and shear using molecular dynamics simulations. Biomacromolecules 26 (4), 2095–2105. https://doi.org/10.1021/acs.biomac.4c00991.

Kaddouri, H., El Mecherfi, K.E., Kheroua, O., Saidi, D., 2006. Microwave treatment modify antigenicity properties of bovine milk proteins. Afr. J. Biotechnol. 5 (13), 1267–1270. https://doi.org/10.4314/ajb.v5i13.43096.

Khatkar, A.B., Kaur, A., Khatkar, S.K., Mehta, N., 2018. Characterization of heat-stable whey protein: impact of ultrasound on rheological, thermal, structural and morphological properties. Ultrason. Sonochem. 49, 333–342. https://doi.org/10.1016/j.ultsonch.2018.08.026.

Khongphakdee, P., Peanparkdee, M., Sae-tan, S., 2025. Dual effects of thermal processing and polyphenol incorporation on bioaccessibility and functionality of soy and whey proteins. Food Chemistry Advances 6, 100905. https://doi.org/10.1016/j.focha.2025.100905.

Koh, B.B., Lee, E.J., Ramachandraiah, K., Hong, G.P., 2019. Characterization of bovine serum albumin hydrolysates prepared by subcritical water processing. Food Chem. 278, 203–207. https://doi.org/10.1016/j.foodchem.2018.11.069.

Kontopidis, G., Holt, C., Sawyer, L., 2004. Invited review: β-lactoglobulin: binding properties, structure, and function. J. Dairy Sci. 87 (4), 785–796. https://doi.org/10.3168/jds.S0022-0302(04)73222-1.

Korhonen, H., Marnila, P., Gill, H.S., 2000. Milk immunoglobulins and complement factors. Br. J. Nutr. 84. https://doi.org/10.1017/s0007114500002282. Suppl. 1.

Lam, C.W.Y., Ikeda, S., 2017. Physical properties of heat-induced whey protein aggregates formed at pH 5.5 and 7.0. Food Sci. Technol. Res. 23 (4), 595–601. https://doi.org/10.3136/fstr.23.595.

Law, A.J., Horne, D.S., Banks, J.M., Leaver, J., 1994. Heat-induced changes in the whey proteins and caseins. Milchwissenschaft-Milk Sci. Int. 49, 125–129.

Leong, T.S.H., Martin, G.J.O., Ashokkumar, M., 2018. Ultrasonic food processing. In: Alternatives to Conventional Food Processing. The Royal Society of Chemistry, pp. 316–354. https://doi.org/10.1039/9781782626596-00316.

Li, N., Zhang, X., Zhu, J., Li, Y., Liu, R., Zhang, P., Wei, S., Fu, X., Peng, X., 2024. Optimization and preparation of ultrasound-treated whey protein isolate pickering emulsions. Foods 13 (20), 3252. https://doi.org/10.3390/foods13203252.

Li, R., Lund, P., Nielsen, S.B., Lund, M.N., 2022. Formation of whey protein aggregates by partial hydrolysis and reduced thermal treatment. Food Hydrocoll. 124, 107206. https://doi.org/10.1016/j.foodhyd.2021.107206.

Lisak Jakopović, K., 2016. Physiological significance, structure and isolation of α-lactalbumin. Mljekarstvo 66 (1), 3–11. https://doi.org/10.15567/mljekarstvo.2016.0101.

Liu, D., Cheng, J., Zhao, C., Guo, M., 2018. Effect of sodium triphosphate on particle size of heat-induced whey protein concentrate aggregates. Food Sci. Nutr. 6 (7), 1940–1949. https://doi.org/10.1002/fsn3.665.

Liu, J., Shim, Y.Y., Reaney, M.J.T., 2020. Ionic strength and hydrogen bonding effects on whey protein isolate–flaxseed gum coacervate rheology. Food Sci. Nutr. 8 (4), 2102–2111. https://doi.org/10.1002/fsn3.1504.

Liyanaarachchi, W.S., Ramchandran, L., Vasiljevic, T., 2015. Controlling heat induced aggregation of whey proteins by casein inclusion in concentrated protein dispersions. Int. Dairy J. 44, 21–30. https://doi.org/10.1016/j.idairyj.2014.12.010.

Ma, S., Gui, S., Hao, R., Dege, X., Zhang, M., 2024. Effect of high pressure processing on the structural and functional properties of the protein in corn gluten meal. CyTA - J. Food 22 (1). https://doi.org/10.1080/19476337.2024.2376227.

Ma, S., Yang, X., Zhao, C., Guo, M., 2018. Ultrasound-induced changes in structural and physicochemical properties of β-lactoglobulin. Food Sci. Nutr. 6 (4), 1053–1064. https://doi.org/10.1002/fsn3.646.

Ma, X., Hou, F., Zhao, H., Wang, D., Chen, W., Miao, S., Liu, D., 2020. Conjugation of soy protein isolate (SPI) with pectin by ultrasound treatment. Food Hydrocoll. 108, 106056. https://doi.org/10.1016/j.foodhyd.2020.106056.

Mahomud, M.S., Haque, M.A., Akhter, N., Asaduzzaman, M., 2021. Effect of milk pH at heating on protein complex formation and ultimate gel properties of free-fat yoghurt. J. Food Sci. Technol. 58 (5), 1969–1978. https://doi.org/10.1007/s13197-020-04708-8.

Majhi, P.R., Ganta, R.R., Vanam, R.P., Seyrek, E., Giger, K., Dubin, P.L., 2006. Electrostatically driven protein aggregation: β-Lctoglobulin at low ionic strength. Langmuir 22 (22), 9150–9159. https://doi.org/10.1021/la053528w.

Meltretter, J., Wüst, J., Pischetsrieder, M., 2014. Modified peptides as indicators for thermal and nonthermal reactions in processed milk. J. Agric. Food Chem. 62 (45), 10903–10915. https://doi.org/10.1021/jf503664y.

Mirarab Razi, S., Mohebbi, M., Mirzababaee, S.M., Hesarinejad, M.A., Khalilian Movahed, M., 2024. The effect of high hydrostatic pressure on the structure of whey proteins-guar gum mixture. Heliyon 10 (1), e24140. https://doi.org/10.1016/j.heliyon.2024.e24140.

Muhammad, G., Croguennec, T., Julien, J., Michel, P., Saïd, B., 2009. Copper modulates the heat-induced sulfhydryl/disulfide interchange reactions of β-Lactoglobulin. Food Chem. 116 (4), 884–891. https://doi.org/10.1016/j.foodchem.2009.03.043.

Munir, M., Nadeem, M., Qureshi, T.M., Leong, T.S.H., Gamlath, C.J., Martin, G.J.O., Ashokkumar, M., 2019. Effects of high pressure, microwave and ultrasound processing on proteins and enzyme activity in dairy systems—a review. Innov. Food Sci. Emerg. Technol. 57, 102192. https://doi.org/10.1016/j.ifset.2019.102192.

Nagy, P., 2013. Kinetics and mechanisms of thiol-disulfide exchange covering direct substitution and thiol oxidation-mediated pathways. Antioxidants Redox Signal. 18 (13), 1623–1641. https://doi.org/10.1089/ars.2012.4973.

Navarra, G., Leone, M., Militello, V., 2007. Thermal aggregation of β-lactoglobulin in presence of metal ions. Biophys. Chem. 131 (1–3), 52–61. https://doi.org/10.1016/j.bpc.2007.09.003.

O'Kennedy, B.T., Mounsey, J.S., 2006. Control of heat-induced aggregation of whey proteins using casein. J. Agric. Food Chem. 54 (15), 5637–5642. https://doi.org/10.1021/jf0607866.

Pang, L., Huang, Z., Li, R., Su, Y., Shi, J., Yang, X., Jiang, Y., 2025. Reducing the allergenicity of whey proteins while improving their functional properties and bioactivity using combined enzymes. Food Chem. 481, 144052. https://doi.org/10.1016/j.foodchem.2025.144052.

Patel, S., 2015. Functional food relevance of whey protein: a review of recent findings and scopes ahead. J. Funct.Foods 19, 308–319. https://doi.org/10.1016/j.jff.2015.09.040.

Pathak, R., Bhangu, S.K., Martin, G.J.O., Separovic, F., Ashokkumar, M., 2022. Ultrasound-induced protein restructuring and ordered aggregation to form amyloid crystals. Eur. Biophys. J. 51 (4–5), 335–352. https://doi.org/10.1007/s00249-022-01601-4.

Pereira, R.N., Vicente, A.A., 2010. Environmental impact of novel thermal and non-thermal technologies in food processing. Food Res. Int. 43 (7), 1936–1943. https://doi.org/10.1016/j.foodres.2009.09.013.

Perez, D., Harte, F., Lopez-Pedemonte, T., 2022. Ionic strength and buffering capacity of emulsifying salts determine denaturation and gelation temperatures of whey proteins. J. Dairy Sci. 105 (9), 7230–7241. https://doi.org/10.3168/jds.2021-21738.

Permyakov, E.A., 2020. α-Lactalbumin, amazing calcium-binding protein. Biomolecules 10 (9), 1210. https://doi.org/10.3390/biom10091210.

Permyakov, E.A., Berliner, L.J., 2000. α-Lactalbumin: structure and function. FEBS (Fed. Eur. Biochem. Soc.) Lett. 473 (3), 269–274. https://doi.org/10.1016/S0014-5793(00)01546-5.

Qian, F., Sun, J., Cao, D., Tuo, Y., & Jiang, S. (2017). <Kosfa-37-44.Pdf>.

Qin, B.Y., Bewley, M.C., Creamer, L.K., Baker, H.M., Baker, E.N., Jameson, G.B., 1998. Structural basis of the tanford transition of bovine β-lactoglobulin. Biochemistry 37 (40), 14014–14023. https://doi.org/10.1021/bi981016t.

Ravash, N., Peighambardoust, S.H., Soltanzadeh, M., Pateiro, M., Lorenzo, J.M., 2022. Impact of high-pressure treatment on casein micelles, whey proteins, fat globules and enzymes activity in dairy products: a review. Crit. Rev. Food Sci. Nutr. 62 (11), 2888–2908. https://doi.org/10.1080/10408398.2020.1860899.

Ren, S., Monica Giusti, M., 2021. Monitoring the interaction between thermally induced whey protein and anthocyanin by fluorescence quenching spectroscopy. Foods 10 (2). https://doi.org/10.3390/foods10020310.

Rodiles-López, J.O., Arroyo-Maya, I.J., Jaramillo-Flores, M.E., Gutiérrez-López, G.F., Hernández-Arana, A., Barbosa-Cánovas, G.V., Niranjan, K., Hernández-Sánchez, H., 2010. Effects of high hydrostatic pressure on the structure of bovine α-lactalbumin. J. Dairy Sci. 93 (4), 1420–1428. https://doi.org/10.3168/jds.2009-2786.

Romo, M., Castellari, M., Bou, R., Gou, P., Felipe, X., 2023. Separation of α-lactalbumin-enriched fractions from caprine and ovine native whey concentrate by combining membrane and high-pressure processing. Foods 12 (14). https://doi.org/10.3390/foods12142688.

Sawyer, L., Kontopidis, G., 2000. The core lipocalin, bovine β-lactoglobulin. Biochim. Biophys. Acta Protein Struct. Mol. Enzymol. 1482 (1–2), 136–148. https://doi.org/10.1016/S0167-4838(00)00160-6.

Schneider, M., Esposito, D., Lila, M.A., Foegeding, E.A., 2016. Formation of whey protein–polyphenol meso-structures as a natural means of creating functional particles. Food Funct. 7 (3), 1306–1318. https://doi.org/10.1039/C5FO01499A.

Schokker, E.P., Singh, H., Creamer, L.K., 2000. Heat-induced aggregation of β-lactoglobulin A and B with α-lactalbumin. Int. Dairy J. 10 (12), 843–853. https://doi.org/10.1016/S0958-6946(01)00022-X.

Schong, E., Famelart, M.H., 2019. Influence of casein on the formation of whey protein microparticles obtained by dry heating at an alkaline pH. Food Res. Int. 122 (April), 96–104. https://doi.org/10.1016/j.foodres.2019.04.001.

Sélo, C., Bernard, C., Créminon, P., Wal, 1999. Allergy to bovine β-lactoglobulin: specificity of human IgE to tryptic peptides. Clin. Exp. Allergy 29 (8), 1055–1063. https://doi.org/10.1046/j.1365-2222.1999.00612.x.

Shen, X., Shao, S., Guo, M., 2017. Ultrasound-induced changes in physical and functional properties of whey proteins. Int. J. Food Sci. Technol. 52 (2), 381–388. https://doi.org/10.1111/ijfs.13292.

Siddiqui, S.A., Khan, S., Bahmid, N.A., Nagdalian, A.A., Jafari, S.M., Castro-Muñoz, R., 2024. Impact of high-pressure processing on the bioactive compounds of milk - a comprehensive review. J. Food Sci. Technol. 61 (9), 1632–1651. https://doi.org/10.1007/s13197-024-05938-w.

Simion, A.M., Aprodu, I., Dumitraşcu, L., Bahrim, G.E., Alexe, P., Stănciuc, N., 2015. Exploring the heat-induced structural changes of β-lactoglobulin -linoleic acid complex by fluorescence spectroscopy and molecular modeling techniques. J. Food Sci. Technol. 52 (12), 8095–8103. https://doi.org/10.1007/s13197-015-1949-2.

Smithers, G.W., 2008. Whey and whey proteins—From "gutter-to-gold." Int. Dairy J. 18 (7), 695–704. https://doi.org/10.1016/j.idairyj.2008.03.008.

Soares, S., Mateus, N., de Freitas, V., 2007. Interaction of different polyphenols with bovine serum albumin (BSA) and human salivary α-amylase (HSA) by fluorescence quenching. J. Agric. Food Chem. 55 (16), 6726–6735. https://doi.org/10.1021/jf070905x.

Song, W., Yun, Y., Lv, Y., Zhang, C., Tang, X., Wang, H., Wang, Z., 2025. Authentication and quality assessment of whey protein-based sports supplements using portable near-infrared spectroscopy and hyperspectral imaging. Food Res. Int. 203, 115807. https://doi.org/10.1016/j.foodres.2025.115807. October 2024.

Spöttel, J., Brockelt, J., Falke, S., Rohn, S., 2021. Characterization of conjugates between α-lactalbumin and benzyl isothiocyanate—effects on molecular structure and proteolytic stability. Molecules 26 (20), 6247. https://doi.org/10.3390/molecules26206247.

Steinhauer, T., Marx, M., Bogendörfer, K., Kulozik, U., 2015. Membrane fouling during ultra- and microfiltration of whey and whey proteins at different environmental conditions: the role of aggregated whey proteins as fouling initiators. J. Membr. Sci. 489, 20–27. https://doi.org/10.1016/j.memsci.2015.04.002.

Su, R., Qi, W., He, Z., Zhang, Y., Jin, F., 2008. Multilevel structural nature and interactions of bovine serum albumin during heat-induced aggregation process. Food Hydrocoll. 22 (6), 995–1005. https://doi.org/10.1016/j.foodhyd.2007.05.021.

Suslick, K.S., Flannigan, D.J., 2008. Inside a collapsing bubble: sonoluminescence and the conditions during cavitation. Annu. Rev. Phys. Chem. 59, 659–683. https://doi.org/10.1146/annurev.physchem.59.032607.093739.

Taha, A., Casanova, F., Šimonis, P., Stankevič, V., Gomaa, M.A.E., Stirkė, A., 2022. Pulsed electric field: fundamentals and effects on the structural and techno-functional properties of dairy and plant proteins. Foods 11 (11). https://doi.org/10.3390/foods11111556.

Tian, L., Roos, Y.H., Biliaderis, C.G., Miao, S., 2024. Effect of ultrasound on the interactions and physicochemical properties of the whey protein isolate-Tremella fuciformis polysaccharide system. Food Hydrocoll. 157 (June). https://doi.org/10.1016/j.foodhyd.2024.110375.

Tinguely, M., Ohtani, K., Farhat, M., Sato, T., 2022. Observation of the formation of multiple shock waves at the collapse of cavitation bubbles for improvement of energy convergence. Energies 15 (7). https://doi.org/10.3390/en15072305.

Van der Plancken, I., Van Loey, A., Hendrickx, M.E., 2007. Kinetic study on the combined effect of high pressure and temperature on the physico-chemical properties of egg white proteins. J. Food Eng. 78 (1), 206–216. https://doi.org/10.1016/j.jfoodeng.2005.09.018.

Vargas, S.A., Delgado-Macuil, R.J., Ruiz-Espinosa, H., Rojas-López, M., Amador-Espejo, G.G., 2021. High-intensity ultrasound pretreatment influence on whey protein isolate and its use on complex coacervation with kappa carrageenan: evaluation of selected functional properties. Ultrason. Sonochem. 70 (August 2020), 105340. https://doi.org/10.1016/j.ultsonch.2020.105340.

Vasbinder, A.J., De Kruif, C.G., 2003. Casein-whey protein interactions in heated milk: the influence of pH. Int. Dairy J. 13 (8), 669–677. https://doi.org/10.1016/S0958-6946(03)00120-1.

Vijayakumar, M., Qian, H., Zhou, H.X., 1999. Hydrogen bonds between short polar side chains and peptide backbone: prevalence in proteins and effects on helix-forming propensities. Proteins Struct. Funct. Genet. 34 (4), 497–507. https://doi.org/10.1002/(SICI)1097-0134(19990301)34:4<497::AID-PROT9>3.0.CO;2-G.

Walstra, P., Walstra, P., Wouters, J.T.M., Geurts, T.J., 2005. Dairy Science and Technology. CRC Press. https://doi.org/10.1201/9781420028010.

Wang, L., Moraru, C.I., 2021. High-pressure structuring of milk protein concentrate: effect of pH and calcium. J. Dairy Sci. 104 (4), 4074–4083. https://doi.org/10.3168/jds.2020-19483.

Wang, S.S.S., Liu, K.N., Wen, W.S., Wang, P., 2011. Fibril Formation of bovine α-lactalbumin is inhibited by glutathione. Food Biophys. 6 (1), 138–151. https://doi.org/10.1007/s11483-010-9199-3.

Wang, W.-Q., Li, J.-J., Zhou, J.-Y., Song, M.-X., Wang, J.-C., Li, X., Tang, C.C., Lu, M.-L., Gu, R.-X., 2022. The effect of ion environment changes on retention protein behavior during whey ultrafiltration process. Food Chem. X 15, 100393. https://doi.org/10.1016/j.fochx.2022.100393.

Wang, Z., Zhao, H., Tao, H., Yu, B., Cui, B., Wang, Y., 2023. Ultrasound improves the physicochemical and foam properties of whey protein microgel. Front. Nutr. 10, 1–11. https://doi.org/10.3389/fnut.2023.1140737.

Wijegunawardhana, D., Wijesekara, I., Liyanage, R., Truong, T., Silva, M., Chandrapala, J., 2024. Process-induced molecular-level protein–carbohydrate–polyphenol interactions in milk–tea blends: a review. Foods 13 (16), 2489. https://doi.org/10.3390/foods13162489.

Xiang, B.Y., Ngadi, M.O., Ochoa-Martinez, L.A., Simpson, M.V., 2011. Pulsed electric field-induced structural modification of whey protein isolate. Food Bioprocess Technol. 4 (8), 1341–1348. https://doi.org/10.1007/s11947-009-0266-z.

Xiang, B.Y., Ngadi, N.M.O., Ochoa-Martinez, L.A., Simpson, M.V., 2008. Pulsed electric field-induced structural modification of whey protein. American Society of Agricultural and Biological Engineers Annual International Meeting 2008, ASABE 2008 9, 5753–5765. https://doi.org/10.13031/2013.25115.

Xu, W., He, S., Ma, Y., Zhang, Y., Wang, R., 2015. Effect of the heat-induced whey proteins/κ-casein complex on the acid gelation of yak milk. RSC Adv. 5 (12), 8952–8956. https://doi.org/10.1039/c4ra14432e.

Ye, A., Anema, S.G., Singh, H., 2004a. High-pressure - induced interactions between milk fat globule membrane proteins and skim milk proteins in whole milk. J. Dairy Sci. 87 (12), 4013–4022. https://doi.org/10.3168/jds.S0022-0302(04)73542-0.

Ye, A., Singh, H., Taylor, M.W., Anema, S., 2004b. Interactions of whey proteins with milk fat globule membrane proteins during heat treatment of whole milk. Le Lait 84 (3), 269–283. https://doi.org/10.1051/lait:2004004.

Yiğit, A., Bielska, P., Cais-Sokolińska, D., Samur, G., 2023. Whey proteins as a functional food: health effects, functional properties, and applications in food. J. Am. Nutr. Assoc. 42 (8), 758–768. https://doi.org/10.1080/27697061.2023.2169208.

Zellal, D., Kaddouri, H., Grar, H., Belarbi, H., Kheroua, O., Saidi, D., 2011. Allergenic changes in β-lactoglobulin induced by microwave irradiation under different pH conditions. Food Agric. Immunol. 22 (4), 355–363. https://doi.org/10.1080/09540105.2011.582094.

Zhang, G., Wang, A., Jiang, T., Guo, J., 2008. Interaction of the irisflorentin with bovine serum albumin: a fluorescence quenching study. J. Mol. Struct. 891 (1–3), 93–97. https://doi.org/10.1016/j.molstruc.2008.03.002.

Zhang, Z., Arrighi, V., Campbell, L., Lonchamp, J., Euston, S.R., 2023. Molecular simulation of partially denatured β-lactoglobulin. Food Hydrocoll. 142 (April), 108811. https://doi.org/10.1016/j.foodhyd.2023.108811.

Zhu, Y., Han, Y., Peng, S., Chen, X., Xie, Y., Liang, R., Zou, L., 2022. Hydrogels assembled from hybrid of whey protein amyloid fibrils and gliadin nanoparticles for curcumin loading: microstructure, tunable viscoelasticity, and stability. Front. Nutr. 9, 1–13. https://doi.org/10.3389/fnut.2022.994740.

Zúñiga, R.N., Tolkach, A., Kulozik, U., Aguilera, J.M., 2010. Kinetics of formation and physicochemical characterization of thermally-induced β-lactoglobulin aggregates. J. Food Sci. 75 (5). https://doi.org/10.1111/j.1750-3841.2010.01617.x.

Chapter 8

Demineralization of whey

Golfo Moatsou and Ekaterini Moschopoulou
Laboratory of Dairy Research, Department of Food Science and Human Nutrition, Agricultural University of Athens, Athens, Greece

1. Introduction

It is well-established that different types of whey streams are generated during the manufacture of different dairy products. Their common feature is that they contain the constituents of milk serum but in different relative concentrations along with traces or small quantities of milk fat. The objective of the present chapter is the treatment of the mineral fraction of various whey streams and of their subfractions, namely deproteinized or/and reduced-lactose whey streams.

The mineral fraction is very important for the exploitation of whey streams as constituents of infant formulas, beverages, sport nutrition or salt substitute. Moreover, it affects the performance of heating, membrane or drying treatments utilized to achieve safety, fractionation or concentration and extension of shelf life of whey fractions or their derivatives (e.g., Barone et al., 2021; Chandrapala and Vasiljevic, 2017). These technological implications can be addressed by demineralization or desalting that involves removal of inorganic salts together with some reduction in the content of organic ions such as lactates and citrates. The partial removal of minerals (25%–30%) results in a whey concentrate that can be used in the manufacture of ice-cream and bakery products or in quark cheese. Highly demineralized (90%–95%) whey concentrate or powder can be a constituent of versatile nutritional products, such as infant formulas. Whey products with removal of 30% or 90% of the initial mineral content are symbolized as D30 or D90, respectively (Dairy Processing Handbook, 2025).

The removal of minerals is essential for infant formulas to simulate the low mineral content of human milk, whereas it increases the relative concentration of lactose and whey protein. The latter effect is very advantageous for these formulations considering the high lactose content (7%) and casein to whey protein ratio (3:2) of human milk (Kelly, 2022). Indicatively, whey powder contains 11%–14.5% protein, 63%–75% lactose, 1%–1.5% fat, 8.2%–8.8% ash and 3.5%–5% moisture. Demineralized whey powder contains 11%–15% protein, 70%–80% lactose, 0.5%–1.8% fat, considerably less ash 1%–7% and 3%–4% moisture (Guo and Wang, 2019).

Demineralization can be also performed before the enzymatic hydrolysis of lactose in whey to improve the taste of the final product or before the evaporation step during the production of lactose from whey permeate or in the "mother liquor" by-product of lactose manufacture from whey. The latter intervention results in demineralized delactosed whey powder (Dairy Processing Handbook, 2025).

The present chapter is based on recent research findings regarding the removal of minerals from various whey streams. The variability and the complexity of the mineral fraction and the related technological implications are discussed. Methods and strategies that have been developed to address the mineral removal from different whey categories and shape the new trends, are presented; a synopsis is depicted in Fig. 8.1.

2. Mineral profile of various whey streams

The two main categories of whey streams are differentiated firstly by their pH. Sweet whey stream, pH > 5.5, is the co-product of the rennet induced coagulation of milk expelled during the syneresis and draining of the cheese curd. The whey expelled during the salting of cheese can be included in this group. Acid whey stream, pH < 5.5, is generated during the straining of acid-coagulated fresh cheeses or Greek-type (strained) yoghurt (yoghurt whey) or during the draining of acid-induced cheese curd (acid cheese whey) or from the production of casein by means of isoelectric precipitation (isoelectric casein whey). In addition, the permeate of the ultrafiltration (UF) treatment of milk can be considered as a type of sweet

Whey. https://doi.org/10.1016/B978-0-443-27648-4.00022-0

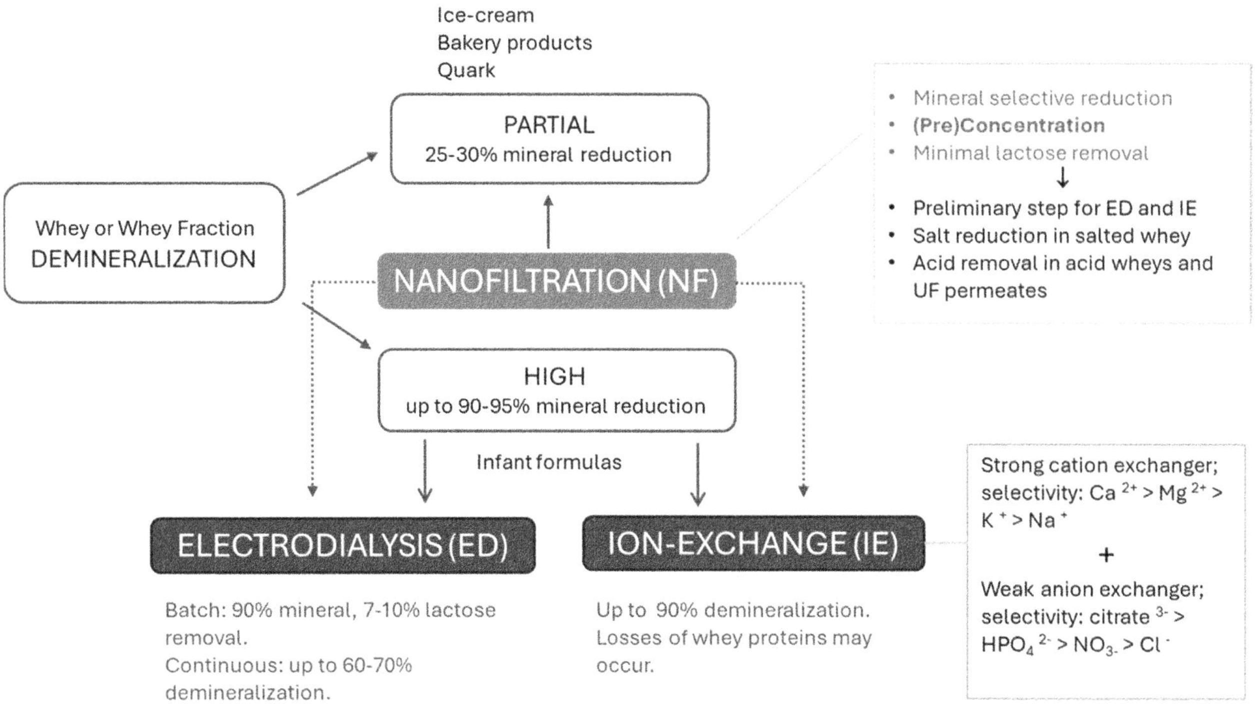

FIGURE 8.1 Synopsis of the demineralization schemes applied in whey or whey fractions based on Kelly (2022) and Dairy Processing Handbook (2025).

whey, whereas the supernatant of the isoelectric preparation of casein is classified as acid whey. Acid whey streams, such as acid whey derived either from the straining of biologically-acidified curd or from the preparation of acid-precipitated casein, are differentiated regarding their detailed composition and mineral fraction. For example, cottage cheese whey contains 0.54% lactic acid, 10 times higher than the content of the whey coming from the preparation of isoelectric casein (Dairy Processing Handbook, 2025).

Recent reports for the mineral fraction of sweet whey derived during the manufacture of different cheese varieties in comparison with ultrafiltration (UF) milk permeate are shown in Table 8.1. The concentrations have been recalculated and are expressed on dry matter in order to be comparable. The same holds true for the minerals of acid whey streams presented in Table 8.2. The contents of proteins or nitrogenous substances are not included in the Tables because occasionally misunderstandings have been observed. The multiplication of all the nitrogen content of a whey stream by 6.38 and the subsequent presentation as protein content, is not always correct. Dairy products that undergo a stage of a rather intense heat treatment are not expected to expel whey containing whey proteins, since they are retained in the main product in their denaturized form. In such cases, the nitrogen content for the most part corresponds to non-protein-nitrogen.

Variable whey mineral contents of different whey streams have been estimated by various research groups (Tables 8.1 and 8.2). Part of these differences can be attributed to analytical practices. On the other hand, there are similarities depicting the production technology of the major dairy product, which can create sub-groups within the two main categories of whey streams. In general, potassium is the most abundant mineral in the whey streams of Tables 8.1 and 8.2 as happens with most milk kinds (Moatsou, 2022). Since potassium is in the soluble phase of milk, it is transferred in the whey in a non-selective way. On the contrary, the transfer in the whey of minerals that are distributed in the colloidal and the soluble phase such as calcium and phosphorus is regulated by the matrix features of the main product.

It is well-established that the cheesemaking technology for the production of rennet-induced curd affects the gross composition of the resultant sweet whey, especially the fat content. Differentiations of the mineral content have been reported. According to the review of Gernigon et al. (2010), the mineral content—expressed as ash content—of Emmental and Gruyere whey fluctuates within 0.48%–0.59% and 0.51%–0.56%, respectively. Cheddar, Gouda and Dutch-type cheese whey contain similar quantities of ash, that is, 0.50%–0.52%. Whey coming from the cheesemaking of hard cheese such as Emmental, Gruyere or Cheddar contain 0.03%–0.04% calcium that is lower than the reported 0.06% for the whey of semi-hard curds of Gouda or Camembert. Acid whey with pH 4.5–4.7 coming from acid-coagulated cheeses

TABLE 8.1 Mineral composition of various sweet whey streams.

	UF milk permate	Cheese whey	Cheese whey	Cheese whey	Cheese whey	Mozzarella whey	Swiss-type cheese whey	Cheddar-type cheese whey	Raclette cheese whey	Soft cheese whey	Cheese whey	Salty cheese whey	Second sweet whey[a]	UF whey permeate[b]
pH	6.37	6.6	6.6	6.1	6.39	6.38	6.52	6.35	6.41	6.42	5.38	5.31	6.12	6–6.3
g/100 g DM[c]														
Lactose	79.1	79	63.3		44.6	68.2	75.7	73.4	71.5	75	49	12.9	79.1	85
Lactic acid	0.9	0.06				1.8		0.88	1.04	1.77	4.01	1.28		
Ash %	8.43	7	7.17	8.12							10.42	67.58		8.26
mg/100 g DM[c]														
Calcium	719	453	563	1770	1676	1073	642	730	745	1008	946	329	2067	540
Phosphorus	746	353.4	431	605	1284	668	356	475	440	563	1090	402	1521	
Sodium	601	529	633	1610	547	705	682	871	618	559	4968	25,114	679	980
Potassium	2687	1017	1933	1210	2257	2652	2669	2555	2394	2699	1923	594	2737	2130
Magnesium	121	632	183	248	176	158	133	146	138	147	272	73	212	120
Chloride		922	1270			1683	1756	1549	1494	2211				210
Citrate		492				3779	2382	2335	2377	2764	103	324		
References	Menchik et al. (2019)	Okawa et al. (2015)	Fischer and Kleinschmidt (2015)	Bebek et al. (2024)	Pasta et al. (2024)	Gernigon et al. (2009)	Gernigon et al. (2009)	Gernigon et al. (2009)	Gernigon et al. (2009)	Gernigon et al. (2009)	Nishanthi et al. (2017b)	Nishanthi et al. (2017b)	Pasta et al. (2024)	Frankowski et al. (2014)

[a]*Whey sub-fraction after being deproteinized by heating.*
[b]*Whey subfraction after being deproteinized by ultrafiltration.*
[c]*DM: dry matter.*

TABLE 8.2 Mineral composition of various acid whey streams.

	Acid whey from iso-electric casein preparation	Yoghurt whey	Yoghurt whey	Yoghurt whey	Yoghurt whey[a]	Yoghurt whey[b]	Cow yoghurt whey	Goat yoghurt whey	Sheep yoghurt whey	Fresh acid coagulated cheese whey	Fresh acid coagulated cheese whey
pH	4.61	4.40	4.52	4.3–4.6	4.47–4.60	4.37–4.42	4.39	4.38	4.11	4.40	
g/100 g DM[c]											
Lactose, %	44	57	57		59.4–61.2	57.5–60.2	61.6	48.2	32.6	53.8	63.3
Lactic acid, %	2.9	7.2	9.9	10	11.6–11.8	12.3–12.8	14.5	17.6	22.1	7.6	13.5
Ash %	11.1	12.5	13.2	9	11.6–12	11.3–11.5	13.7	15.9	15.5	11.4	12.3
mg/100 g DM[c]											
Calcium	1655	2000	2378	2146	1992–2042	1733–1920	1130	1379	1637	1889	1700
Phosphorus	1698	1108	2396	1149	918–981	860–875	1377	1430	1497	1251	910
Sodium	201	645	378	785	1293–1355	1253–1557	1017	1620	1622	624	710
Potassium	662	2817	2396	2452	3677–4235	3798–3874	2835	3530	2580	2573	2945
Magnesium	216	173	306	372	267–275	235–262	206	330	345	183	198
Chloride				2318							2517
Citrate	892		1279	1724							
References	Nishanthi et al. (2017b)	Menchik et al. (2019)	Nishanthi et al. (2017b)	Chen et al. (2016)	Karastamatis et al. (2022)	Karastamatis et al. (2022)	Karela et al. (2025)	Karela et al. (2025)	Karela et al. (2025)	Menchik et al. (2019)	Fischer and Kleinschmidt (2015)

[a] *Immediate straining.*
[b] *Straining after 24 h.*
[c] *DM: dry matter.*

contain high quantities of calcium ranging from 0.09%–0.15% and ash ranging from 0.60%– 0.76% (Gernigon et al., 2010). The ash content of the whey of different rennet coagulated varieties does not differ substantially, if the decisive interventions for the removal of the whey, that is, way of cutting, duration of stirring and cooking, are performed in a similar manner, at similar pH conditions, pH > 6.0. In this respect, the mineral profiles of the UF milk permeate generated at normal milk pH is similar to that of hard-type cheeses (Menchik et al., 2019, Table 8.1). Essentially, UF milk or UF whey permeate contain mineral and lactose up to 85% and they are considered deproteinized sweet whey streams (Table 8.1). It is evident that Soft cheese or Mozzarella cheese that generate whey with high lactic acid content retain less calcium and phosphorus in the cheese mass than Swiss- and Cheddar-type, that is, their whey streams contain approximately 1.8% lactic acid on dry matter versus 0.9% of Cheddar whey (Gernigon et al., 2009, Table 8.1).

Second whey stream is the subfraction emerging after the heat treatment of whey at temperatures up to 90°C that denature whey proteins. The recovered aggregates are the basis of whey cheeses, while the other compounds are transferred in the second whey stream (Pires et al., 2021). Therefore, second whey is essentially a deproteinized whey stream enriched in minerals (Table 8.1). The average ash content reported in the literature review of Fancello et al. (2024) is approximately 0.8%. Macedo et al. (2015) showed that the total solids concentration in ovine cheese whey was 10% and in the respective ovine second cheese whey it was 7.7%. Both lactose and mineral contents expressed on dry matter were concentrated in the second cheese whey. In particular, ovine cheese whey contained 1.5% ash that was 15% of dry matter, while the ash content of the second whey was 1.8% or 23.4% on dry matter (Macedo et al., 2015).

The straining of acid-coagulated cheeses takes place under low pH conditions that induce the solubilization of colloidal calcium phosphate and result in the enrichment of the soluble phase with minerals (Koutina and Skibsted, 2015; Zhao and Corredig, 2016). Accordingly, WPCs manufactured from acid whey exhibit higher moisture and much more higher mineral content compared to sweet whey counterpart. Nishanthi et al. (2017a) report that the moisture contents of acid- and sweet-WPCs were 5.7% and 4.1%, respectively. The respective calcium contents were 150 and 21 mg/100 g dry matter and the respective lactic acid contents were 0.2% and 0.1% on dry matter. These differences along with the high heat treatment applied during the production of acid whey have been considered responsible for the aggregation phenomena observed during the storage of acid WPCs (Nishanthi et al., 2017a, 2017b).

There are different sub-categories of acid-whey streams depending on the production technology of the main product that can influence their mineral profile. Yoghurt whey comes from yoghurt straining, which is prepared from milk treated under high heat conditions that denature whey proteins. The heat treatment conditions of milk used for the production of soft spreadable cheese can be milder compared to those of yoghurt milk, while under the production conditions of isoelectric acid whey coming from casein preparations, denaturation of whey proteins is not expected (Drapala et al., 2018). The sub-types of acid whey streams appeared in Table 8.2 contain much more lactic acid and minerals compared to sweet whey of Table 8.1, mainly due to their enrichment in calcium and phosphorus as previously reported. Moreover, whey streams coming from biological acidification contain a substantial amount of lactic acid opposite the acid whey generated by the isoelectric preparation of casein (Table 8.2).

According to the findings of Karela et al. (2025), the milk kind and the straining method can affect the gross composition and particular minerals of yoghurt acid whey. For example, cow yoghurt whey obtained by means of UF straining contained significantly higher quantities of potassium and sodium compared to those obtained by centrifugation or cloth straining. Accordingly, the ash content on dry matter that expresses the total mineral content was 15.8%, 13.08% and 13.72% for UF, centrifugation and cloth straining, respectively. The same authors report that sheep yoghurt whey exhibits significantly higher contents of ash and major minerals compared to cow and goat counterpart, in accordance to the higher ash content of sheep milk (Moatsou, 2022). The heat treatment of yoghurt milk and the timing of straining had a marginal effect on the ash content of the generated whey streams, whereas the concentrations of calcium and phosphorus were significantly lower when straining was performed after an 24 h stay at 4°C after gel formation (Karastamatis et al., 2022).

2.1 Technological implications

Lactose pre-crystallization after evaporation of whey and before drying is an essential step in the production of various whey powders and concentrates but a high mineral content has an inhibitory effect on this procedure. Demineralization before evaporation is suggested. Ozel et al. (2022) summarized the effect of pH on these phenomena. At low pH, the lactic acid is mainly in the undissociated highly hydroscopic form and the solubility of calcium is elevated. At approximate pH 6.5, the hydroscopic behavior of calcium and lactic acid decreases, while the crystallization rate of lactose increases.

The mineral content of whey streams, impairs the production of whey powders and constituent fractionation. This is especially true for acid whey which contain considerable amounts of calcium and lactic acid. Lactic acid contents of 1% or

0.2% decrease the yield of spray dried lactose powders from 81% -observed in the absence of lactic acid-to 52% or 66%, respectively. Stickiness and formation of large interconnected particles are observed because lactic acid is hydroscopic. Moreover, mineral fouling is developed on the equipment surfaces. In the presence of calcium ions, the hydration layer created by lactic acid is enhanced due to strong ion-dipole interactions between calcium-ion and water and the formation of highly hydroscopic calcium lactate (Chandrapala and Vasiljevic, 2017; Chandrapala et al., 2016a, 2016b; Merkel et al., 2021b). The presence of Ca results in the formation of calcium phosphate, which limits permeation during membrane processing, resulting in scaling, a barrier to filtration, and loss of throughput. Similarly, calcium linked protein aggregates or calcium lactate formation can result in increased viscosity that impairs the drying efficiency (Chandrapala et al., 2015). Corrigan et al. (2023) demonstrated that increasing Ca in formulas containing whey enhanced β-lg aggregation, decreasing thus their heat stability.

The dramatic effect of minerals and in particular of calcium on heat stability, aggregation, gelation, foaming and emulsifying properties is of particular interest for the production and stability of whey beverages or infant formulas that contain whey fractions. The mechanism behind this effect is related to the interaction of the divalent cation with the negative ionizable groups of the whey proteins. The resultant decrease of zeta-potential increases the proximity of whey proteins and promotes attractive forces between proteins. Under these conditions, protein aggregations are formed due to: (i) calcium bridges between the carboxylic groups of aspartic and glutamic amino acids, and (ii) the enhancement of hydrophobic interactions that result from the conformational changes induced by calcium mediated bridges (Bryant and McClements, 2000; Barone et al., 2021, 2022).

Purwanti et al. (2024) summarized the role of calcium in the technological instability of acid whey. The two major mechanisms of heat-induced fouling in dairy processes are (a) protein denaturation/aggregation and (b) precipitation of salts in the form of calcium phosphate (CaP) complexes or CaP-protein/fat complexes. Trisodium citrate (TSC) and ethylenediaminetetraacetic acid disodium salt (Na_2-EDTA) are added to reduce the calcium ions. The findings of Purwanti et al. (2024) indicate that Na_2-EDTA is superior to TSC in reducing the extent of heat-induced changes in reconstituted acid whey due to the presence of larger particle sizes and higher viscosities in TSC-containing suspensions.

3. Technologies to demineralize whey

Demineralization is performed by various ways that remove ions and not undissociated salts. The removal of ions is affected by other compounds and the pH of the whey stream, so a combination of methods is often necessary. The negative charge of proteins that coexist with the minerals in the whey impedes the removal of calcium, that is, counterion, and enhances the removal of chloride, that is, co-ion, whereas the incomplete dissociation of the carboxylic group of organic acids occurred at low pH does not favor their removal (Walstra et al., 2006). Ion exchange, membrane filtration and combinations of them have been used in whey processing aiming to various degrees of minerals removal, which in turn affect the physical and functional properties of the final products. Partial demineralization is mainly based on the utilization of cross-flow membranes specially designed to "leak" particle species that have radii in the nanometer (10^{-9} m) range, that is, nanofiltration. The techniques of electrodialysis and ion-exchange are suitable for high degree demineralization (Dairy Processing Handbook, 2025). Efficacy, effluents, energy consumption and duration of the processing have been considered in several research studies.

3.1 Ion exchange technology

Ion exchange (IE) is mainly used in whey processing to remove inorganic elements. This technique employs resin beads, which are fixed in columns of various designs and absorb minerals from whey in exchange for other ionic species. In general, in this process, multivalent ions are removed more easily than monovalent ions and thus, the selectivity of cations and anions has the increased order $Na^+ < K^+ < Mg^{2+} < Ca^{2+}$ and $Cl^- < NO_3^- < HPO_4^{2-} < citrate^{3-}$ respectively (Dairy Processing Handbook, 2025). Acid whey from productions of acid casein, Quark type cheeses and Greek yoghurt is rich in calcium and phosphorus and hence ion exchange could be more suitable to remove the divalent cations Ca^{2+}. Moreover, according to the overview of Weerasingha et al. (2025), in demineralization of salty whey derived from cheesemaking, ion exchange resins as well as size-exclusion chromatography seem to be the best methods since they achieve as high as 99% salt recovery, compared to membrane filtration and electrodialysis. However, ion exchange in the dairy industry has been replaced by membrane filtration, which is more cost-effective (Kamalesh et al., 2025).

3.2 Membrane technology

Membrane technology employs filtration techniques and is widely applied in whey processing to remove fat, microorganisms and minerals, as well as to fractionate or to concentrate other components, for example, proteins. Membrane processing of milk and dairy streams can be either pressure or electrically driven. In a pressure driven process, the mixture of molecules is driven to the outer layer of membrane with the aid of pressure and separation is achieved either by sieving or by the combination of charge rejection, solubility-diffusion and sieving. Membranes with different diameters of pores are used and the processes are referred to as microfiltration (MF), ultrafiltration (UF), nanofiltration (NF) and reverse osmosis (RO). The most common membrane materials for MF, UF and NF modules are inorganic polymers (ceramics) which are in tubular configuration. Moreover, polymers at spiral-wound and hollow-fiber assemblies are used for NF and RO and UF respectively. Reig et al. (2021) have presented an overview of commercial membranes used in the dairy industry.

More specifically, MF has membrane with pore size 10^{-1}–10^{1} μm and the trans membrane pressure (TMP), which is the pressure difference across the membrane that is, the driving force which pushes the permeate through the membrane, operates at $\leq$1bar (Reig et al., 2021; Kamalesh et al., 2025). It is mainly used in skimmed milk and whey processing for improving microbiological quality and composition. For example, MF processing of sweet whey through membranes with pore size 0.5 μm, at 20°C, results in a permeate with good microbiological quality, almost zero Ca^{2+}, while membrane fouling is minimal (Barukčić et al., 2014). Moreover, in acid whey processing aiming to lactic acid and mineral salts reduction, MF with 1.4 membrane pore size is used prior to NF (Chandrapala et al., 2016a, 2016b; Karela et al., 2025).

In UF processing, the membrane pore size is 10^{-2}–10^{-1} μm, the TMP is 1–10 bar and the molecular weight cut-off 0.2–150 kDa. In sweet whey treatment, UF is widely used to standardize and concentrate the protein content with parallel decrease of mineral content to produce the Whey Proteins Concentrates (WPC) with various levels of concentrations/demineralization. Combined with the diafiltration (DF), UF can produce WPC with 80% protein and low mineral content. In contrast, the modification of pH of acid whey from 4.5 to 6.5 prior to UF seems to increase the mineral retention. Mc Entee et al. (2025), using UF with spiral wound membrane 10 kDa and a volume concentration factor (VCF) 9, produced acid whey powder with total ash content 16.19 ± 1.3 (% w/w of dry matter), compared to the ash content 11.01 ± 0.5 (% w/w of dry matter) of the initial acid whey. This was due to retention of calcium and phosphorous that affected the functional properties of the derived powders, as well.

RO is an energy efficient alternative method for evaporation operating with membranes with pore size 10^{-4}–10^{-3} μm, at 30–60 bar. It is used to dehydrate/concentrate the whey, since all solid compounds are rejected by the membrane, allowing only water and similar size molecules to pass through the membrane. For example, Menchik and Moraru (2019) have proposed a combination of RO with VCF three and forward osmosis (FO), a technique that uses semipermeable membrane at low pressure, with VCF six to concentrate acid whey from Greek yoghurt to 40.2°Brix. On the other hand, it has been demonstrated that in the case of acid whey from Greek yoghurt, RO process is prone to fouling because of the high calcium concentration of the feed. Unfortunately, this problem has not been solved despite the decalcification by precipitation at identified optimal conditions that is, pH > 9 and Ca/P ratio 0.55 (Ortiz Quezada et al., 2021).

3.2.1 Nanofiltration

NF operates with TMP 20 40 bar and separates molecules through the combination of charge rejection, solubility-diffusion and sieving through membranes with pores size 10^{-3}–10^{-2} μm, retaining components with molecular weight 0.2–2 kDa. In whey processing it leads to partial concentration, deacidification and demineralization. The use of NF to improve the composition of acid whey by decreasing lactic acid and minerals contents at both the laboratory and the pilot scale as well as at industrial scale before drying has been investigated by several researchers. It has been demonstrated that lactic acid removal depends on the whey composition and the processing conditions, and it ranges from 30%– 55% at pH 3 with a parallel lactose loss between 1% and 7% (Chandrapala et al., 2016a, 2016b; Bedas et al., 2017; Karela et al., 2025). Moreover, the NF process of acid whey prior to drying either by spray drying or by vacuum evaporation in semi-industrial scale has been proved to decrease the energy cost by 43% (Bedas et al., 2017).

Regarding the demineralization of whey by NF, the dissolved mineral salts are removed inversely proportional to their valence while salt removal is counterbalance of lactose in retentate (Blais et al., 2022). NF partially removes the cations Na^{+} and K^{+} because they have higher content than the Cl^{-} and pass through the membrane in ion pairs with Cl^{-} to keep the electron neutrality (Okawa et al., 2015). In this way, NF can concentrate acid whey to 20%–22% dry matter content and remove 25%–60% of monovalent ions that is, K^{+}, Na^{+}, Cl^{-}, while combined with diafiltration (DF) can decrease the mineral content up to 90% (Barrantes and Morr, 1997, Bedas et al., 2017). By means of nanofiltration in conjunction with diafiltration the composition of acid whey from the preparation of isoelectric casein become similar to that of cheese whey due to the substantial reduction of the high mineral content, in particular that of chloride (Kelly, 2022).

On the other hand, calcium found in acid whey together with lactic acid affects the crystallization of lactose causing stickiness of the produced acid whey powder, because calcium may be binding with lactose. Although it has small binding capacity to lactose, it may be incorporated into lactose crystals because the radius/charge ratio of the lactose is affected by the cations. Moreover, calcium and calcium lactate may affect lactose solubility because the structural behavior of the water molecules that surround the lactose is influenced. Hence, removal of Ca^{2+} as well as of lactic acid is very important. In NF processing of acid whey, which contain more calcium than sweet whey, the pH and the temperature affect the transmission of Ca^{2+} and phosphate ions because calcium complexes with phosphates and lactates (Chandrapala et al., 2016a, 2016b). In this case, a combination of different pHs seems to enhance the removal of Ca^{2+}. Initially, NF at pH three increases lactic acid removal and then NF at pH > 10 leads to precipitation of mineral salts as lactates that can be easily removed from retentate. However, problems of membrane fouling and lactose crystallization can be caused (Bedas et al., 2017).

Table 8.3 presents research studies concerning various NF processing parameters to reduce the mineral content of whey and their findings. In summary, the efficiency of NF for whey demineralization depends on many parameters such as the processing conditions that is, membrane type and pore size, temperature, TMP, flow velocity etc., the pH and composition of whey and subsequent diafiltration (Chandrapala et al., 2016a, 2016b, 2017; Karela et al., 2025).

3.2.2 Electrodialysis

Electrodialysis (ED) is an electrochemical technique process in which electrically charged ions are transferred through an ion-exchange membrane under the influence of application of direct current (DC) that is used as the driving force. In this way, the cations move to cathode and the anions to anode, passing through cation- and anion-permeable membranes separated by spaces respectively. In general, the efficiency of ED depends on the ash content of the feed whey stream, current density, residence time within the membrane and the extent of fouling (Kelly, 2022).

The ion-exchange membranes of ED may be homopolar having negative or positive groups or bipolar having both groups. Moreover, filtration membranes can be used and in this case, the process is called electrodialysis with filtration membranes (EDFM) (Kamalesh et al., 2025). In a study comparing conventional ED having ion-exchange membranes with ED having NF membranes for acid whey processing, a higher demineralization degree (87%) of the total cation concentration was achieved with the IE membranes than with the NF membranes under the same conditions of current application. However, ion-exchange membranes in ED are prone to scaling formation on their surface, especially in the case of acid whey processing.

Except for the membrane type, the processing conditions play a significant role in the efficiency of ED. For example, Dufton et al. (2019) used pulsed electric field (PEF) at a combination of 50s/10s (pulse/pause) or 25s/25s and achieved 10% and 16% better removal of lactic acid from acid whey respectively than using constant current. In contrast, the demineralization rate was similar in both the constant current and the PEF 50s/10s (about 64%), while it was higher in the case of PEF 25s/25s (about 67%). Moreover, the same authors stated that the pause periods favored the rinsing of membranes and hence decreased the scaling and energy consumption.

Regarding the effect of temperature in ED process, it has been demonstrated that increasing it from 15 to 55°C during processing of acid whey, the demineralization degree and especially that of calcium increases because of the low pH that hinders the formation of insoluble calcium salts. In parallel, the time of processing is reduced (Merkel et al., 2021a). The processing time of ED treatment of concentrated sweet whey (total solids 15.8%) under the similar conditions, was decreased by a factor 2.5 to achieve a demineralization degree 62%, while the most effective temperature was shown to be within the range 35–45°C (Merkel et al., 2020).

Composition also affects ED efficiency. Nielsen et al. (2021) have shown that under the same processing conditions (10V for 180 min) less Mg^{2+} and Ca^{2+} are removed from acid whey with high concentration of sugars due to their ability to bind calcium. Moreover, 80% of K^{+}, Mg^{2+} and Ca^{2+} are removed faster from cottage cheese whey than from Quark acid whey, which has higher concentration of minerals. On the other hand, removal of Ca^{2+} from acid whey is affected by the presence of citrate because of the non-charged complex CaHCitr, which is formed at low pH and is not transmitted during ED (Nielsen et al., 2022a).

TABLE 8.3 Processing parameters and main findings for demineralization (% reduction) of microfiltered (MF) acid whey using Nanofiltration.

Whey (feed)	Membrane type	Operating conditions	Approximate % reduction					References
			K	Na	Ca	Mg	P	
MF whey (1.4 μm) from cream cheese, adjusted at pH 3	Thin film HL, 150–300Da	TMP 21 bar, 40°C, VCF 3	83	69	43	32	34	Chandrapala et al. (2016)
	Thin film DL, 150–300Da	TMP 21 bar, 40°C, VCF 3	84	64	42	24	36	
	Thin film DK, 150–300Da	TMP 21 bar, 40°C, VCF 3	80	70	40	28	31	
MF whey (1.4 μm) from cream cheese, adjusted at pH 3	Thin film HL, 150–300Da	NF/DF TMP 21 bar, DF factor 4.2	–	–	45–48	–	49–55	Chandrapala et al. (2017)
Whey from fresh cheese	Polymer spiral wound, cut-off at 200 kg mol-1	TMP 33 bar, VCF 3	53	60	0	0	0	Bedas et al. (2017)
MF whey (1.4 μm) from Greek yoghurt strained by UF	Tubular ceramic, 250Da	TMP 10 bar, 23°C, VCF 1.5	60	47	21	18	23	Karela et al. (2025)
MF whey (1.4 μm) from Greek yoghurt strained by centrifugation	Tubular ceramic, 250Da	TMP 10 bar, 23°C, VCF 1.5	96	72	43	40	46	Karela et al. (2025)
MF whey (1.4 μm) from Greek yoghurt strained by the traditional method	Tubular ceramic, 250Da	TMP 10 bar, 23°C, VCF 1.5	76	69	24	16	20	Karela et al. (2025)
MF ovine whey (1.4 μm) from ovine Greek yoghurt strained by the traditional method	Tubular ceramic, 250Da	TMP 10 bar, 23°C, VCF 1.5	85	64	33	27	33	Karela et al. (2025)

3.3 Combined methods for whey demineralization

As mentioned before, the pressure driven membrane filtration results in partial, about 25%–60%, demineralization of whey, whereas techniques such as IE and ED result in higher demineralization degree, about 90%–95%. However, because of the high energy consumption of IE and ED, the combination of these methods with NF to reduce the processing cost is currently studied. In this case, NF is used as preliminary step before IE or ED to increase the degree of demineralization. A reduction of minerals by >30% has been observed after nanofiltration in both acid and sweet whey which can increase up to approximately 90%, after the addition of ED in the sequence (Talebi et al., 2020; Merkel et al., 2021b).

Table 8.4 presents research findings for the combinations of various methods and the processing parameters utilized to reduce the mineral content of whey.

UF with polymer membranes removed partially Ca^{2+}, total phosphorus and Mg^{2+} from sweet whey and the following treatment of UF permeate by NF removed part of K^{+}, Ca^{2+}, Fe^{2+}, Mg^{+}, Cu^{2+} and Cl^{-}. Further processing of NF retentate by ED resulted in a reduction as low as 93% of ash content in the final product, 89%–94% of Na^{+} and K^{+}, 60%–75% of Ca^{2+} and Mg^{2+}, 75% of total phosphorus and 70% of chlorides, compared to those in unprocessed whey (Melnikova et al., 2023). On the other hand, in salty whey, which contains 4%–10% NaCl, the combination of NF/ED is efficient enough to remove Na^{+}, but not Ca^{2+}. Thus, a process configuration including IE upstream the ED has been proposed to achieve a higher removal degree of Ca^{2+} and Mg^{2+} (El-Okazy et al., 2024). The same research group stated that the inclusion of IE step did not significantly change the ED energy consumption. However, it increased the salt content of effluent because of the NaCl needed for the resin regeneration.

In general, the high levels of calcium and phosphorus in acid whey cause fouling problems during processing. For this reason, a combination process of NF/heating-neutralization/UF for acid whey demineralization, since neutralization at 60°C precipitates calcium phosphate that can be easily removed and used as food ingredient. For this purpose, acid whey was concentrated by NF to 17% total solids, then was heated and neutralized for the precipitation and was classified in hydro-cyclones. Finally, it was processed by means of a UF membrane with a cut off of 150–200 kDa and was dried to produce powder with pH 6.7 that contained mainly ash (95% in DM) and a high proportion of calcium (29% in DM) and phosphorus (15% in DM).

In another study concerning demineralization of acid whey from Skyr production, the application of NF prior to ED increased the separation of Ca^{2+} and Mg^{2+} up to 70% and 67% respectively, but did not affect the K^{+}, Na^{+} and phosphates (Nielsen et al., 2022b). On the other hand, the duration of the combined process was higher than the duration of the process without NF and this was attributed to the increased conductivity of the NF retentate of acid whey because of its higher concentration in divalent ions.

Recently, the combination of NF and ED under pulsed electric field (PEF) for removing lactic acid and minerals from acid whey was studied. The acid whey was pre-treated with NF at a concentration factor 4.33 and after it was processed under two different conditions of ED that is, under two different PEF conditions (5 s/5 s and 15 s/15 s). A demineralization degree of 70% and a removal of lactic acid about 46% were reported. Moreover, it was demonstrated that the rates of calcium and magnesium removal were significantly higher, 25.1% and 47.0% respectively, when PEF 5 s/5 s was used, compared to conventional current. On the other hand, the same process did not affect the reduction level of phosphorus (Poitras et al., 2025).

As mentioned before, NF, followed by DF, increases the demineralization degree. For removing the mineral salts from sweet whey with a parallel concentration up to 24% dry matter, an optimum combination with DF 0.75 step has been suggested regarding the water, the energy and the time consumption (Marx et al., 2019). On the other hand, the same researchers, combining NF with MF and subsequent heat treatment at 80°C/30 s to increase the microbial shelf life of the product, observed that the microbial shelf life decreased as the demineralization degree increased because of the reduced concentration of monovalent ions.

Regarding the combination of RO with ED, Talebi et al. (2021) have stated that it is inefficient for a high degree of lactic acid and salts removal from acid whey. In another study about processing by ED with various current densities, it was demonstrated that the pre-treatment of acid whey by RO increased the processing time needed for 80% removal of the cations, calcium and magnesium, because of their high affinity to the cation exchange membranes that were used (Nielsen et al., 2023). Moreover, the same authors report that the high current density caused lower degree of lactate removal and increased energy consumption by 1.2–2.0 folds.

TABLE 8.4 Processing parameters and main findings for demineralization (% reduction) of various types of whey using combined methods.

Type of whey (feed)	Combined methods	Operating conditions	Approximate % reduction						References
			K	Na	Cl	Ca	Mg	P	
Sweet whey	IE-NF	IE: Column anion -exchange functional group $N^+(CH_3)3$ NF: membrane spiral wound, TMP 12 bar, 10°C, VCF 4	90	91	97	65	65	27	Okawa et al. (2015)
Acid whey from cream cheese	Precipitation at pH 10.5 –NF/DF	NF: Feed pH 3, membrane HL thin film, 150–300 Da, TMP 21 bar, DF factor 4.2	–	–	–	72	–	87	Chandrapala et al. (2017)
Sweet whey	NF-DF-MF-heat treatment	NF: thin film membrane, 10°C, TMP 20 bar, up to DM 24% DF factor 0.75 MF: membrane 1.4 μm, TMP 1bar, 10°C Heating: 80°C/30 s	50	50	86	–	–	–	Marx et al. (2019)
Acid whey and sweet whey	NF-ED	2-step NF: thin film polyamide membrane, 41 bar, 45°C 2 different systems ED: IE membrane, constant current 10V	98–100	77–78	–	80–100	87–100	–	Kaya et al. (2019)
Acid whey from Skyr	NF-ED	NF: Spiral would membrane100-250Da, 50°C, TMP 15 bar, VCF 2.5 ED: 25°C, 180 min, total applied voltage 12.9–14.8 V	100	95	–	84	77	–	Nielsen et al. (2022b)
Acid whey from cream cheese	NF-ED/PEF	NF: Spiral would membrane100-300Da, VCF 4.3 ED under PEF 5 s/5 s	71	56	–	77	58	61	Poitras et al. (2025)
	NF-ED/PEF	NF: Spiral would membrane100-300Da, VCF 4.3 ED under PEF 15 s/15 s	81	67	–	69	46	69	

4. Conclusion

The detailed mineral composition of the whey streams is configurated by the production technology of the main product that generated them and can be very challenging for their further processing and in particular for their demineralization. The selection of a method or a sequence to demineralize whey streams depends on the composition and origin of the streams and the end-users, that is, low or high mineral content is needed for infant nutrition or nutrition supplements, respectively. In this respect, filtration/membrane methods under various conditions have been studied and currently emphasis is given to combined methods, which among others are adequate for highly demineralized products. Fouling and scaling of the membranes, energy consumption, waste generation and the behavior of the demineralized streams during downstream processes such as evaporation and/or spray-drying are important factors for the development of a demineralization protocol. During the last years, there has been a strong research interest with important findings for the production of tailor-made fractions of whey (side) streams, that contribute significantly to the sustainability of the dairy and food sector.

References

Barone, G., Yazdi, S.R., Lillevang, S.K., Ahrne, L., 2021. Calcium: a comprehensive review on quantification, interaction with milk proteins and implications for processing of dairy products. Compr. Rev. Food Sci. Food Saf. 20, 5616–5640. https://doi.org/10.1111/1541-4337.12844.

Barone, G., O'Regan, J., Kelly, A.L., O'Mahony, J.A., 2022. Interactions between whey proteins and calcium salts and implications for the formulation of dairy protein-based nutritional beverage products: a review. Compr. Rev. Food Sci. Food Saf. 21, 1254–1274. https://doi.org/10.1111/1541-4337.12884.

Barrantes, L.D., Morr, C.V., 1997. Partial deacidification and demineralization of cottage cheese whey by nanofiltration. J. Food Sci. 62 (2), 338–341. https://doi.org/10.1111/j.1365-2621.1997.tb03996.x.

Barukčić, I., Božanić, R., Kulozik, U., 2014. Effect of pore size and process temperature on flux, microbial reduction and fouling mechanisms during sweet whey cross-flow microfiltration by ceramic membrane. Int. Dairy J. 39 (1), 8–15. https://doi.org/10.1016/j.idairyj.2014.05.002.

Bebek, M., Sert, D., Mercan, E., 2024. Effects of the demineralisation degree on physicochemical, functional, microstructural and powder flow properties of whey powder. Int. Dairy J. 156, 105982. https://doi.org/10.1016/j.idairyj.2024.105982.

Bedas, M., Tanguy, G., Dolivet, A., Mejean, S., Gaucheron, F., Garric, G., Senard, G., Jeantet, R., Schuck, P., 2017. Nanofiltration of lactic acid whey prior to spray drying: scaling up to a semi-industrial scale. LWT Food Sci. Technol. 79, 355–360. https://doi.org/10.1016/j.lwt.2017.01.061.

Blais, H.N., Schroen, K., Tobin, J.T., 2022. A review of multistage membrane filtration approaches for enhanced efficiency during concentration and fractionation of milk and whey. Int. J. Dairy Technol. 75 (4), 749–760. https://doi.org/10.1111/1471-0307.12884.

Bryant, C.M., McClements, D.J., 2000. Influence of NaCl and $CaCl_2$ on cold-set gelation of heat-denatured whey protein. J. Food Sci. 65, 801–804. https://doi.org/10.1111/j.1365-2621.2000.tb13590.x.

Chandrapala, J., Duke, M.C., Gray, S.R., Zisu, B., Weeks, M., Palmer, M., Vasiljevic, T., 2015. Properties of acid whey as a function of pH and temperature. J. Dairy Sci. 98 (7), 2588–2595. https://doi.org/10.3168/jds.2015-9435.

Chandrapala, J., Vasiljevic, T., 2017. Properties of spray dried lactose powders influenced by presence of lactic acid and calcium. J. Food Eng. 198, 63–71. https://doi.org/10.1016/j.jfoodeng.2016.11.017.

Chandrapala, J., Duke, M.C., Gray, S.R., Weeks, M., Palmer, M., Vasiljevic, T., 2016a. Nanofiltration and nanodiafiltration of acid whey as a function of pH and temperature. Separ. Purif. Technol. 160, 18–27. https://doi.org/10.1016/j.seppur.2015.12.046.

Chandrapala, J., Duke, M.C., Gray, S.R., Weeks, M., Palmer, M., Vasiljevic, T., 2017. Strategies for maximizing removal of lactic acid from acid whey—addressing the un-processability issue. Separ. Purif. Technol. 172, 489–497. https://doi.org/10.1016/j.seppur.2016.09.004.

Chandrapala, J., Wijayasinghe, R., Vasiljevic, T., 2016b. Lactose crystallisation as affected by presence of lactic acid and calcium in model lactose systems. J. Food Eng. 178, 181–189. https://doi.org/10.1016/j.jfoodeng.2016.01.019.

Chen, G.Q., Eschbach, F., Weeks, M., Gras, S.L., Kentish, S.A., 2016. Removal of lactic acid from acid whey using electrodialysis. Separ. Purif. Technol. 158, 230–237. https://doi.org/10.1016/j.seppur.2015.12.016.

Corrigan, B.M., O'Mahony, J.A., Fenelon, M.A., 2023. The effect of whey source on heat-induced aggregation of casein and whey protein mixtures of relevance to infant nutritional product formulation. J. Dairy Sci. 106 (12), 8299–8311. https://doi.org/10.3168/jds.2022-22088.

Dairy Processing Handbook, 2025. Chapter 18. Whey and Milk Fractionation, Revised by Nanne van Dijk. ©Tetra Pak International S.A.

Drapala, K.P., Murphy, K.M., Ho, Q.T., Crowley, S.V., Mulcahy, S., McCarthy, N.A., O'Mahony, J.A., 2018. Multi-component interactions causing solidification during industrial-scale manufacture of pre-crystallized acid whey powders. J. Dairy Sci. 101 (12), 10743–10749. https://doi.org/10.3168/jds.2018-14836.

Dufton, G., Mikhaylin, S., Gaaloul, S., Bazinet, L., 2019. Positive impact of pulsed electric field on lactic acid removal, demineralization and membrane scaling during acid whey electrodialysis. Int. J. Mol. Sci. 20, 797. https://doi.org/10.3390/ijms20040797.

El-Okazy, M.A., Kentish, S.E., Gras, S.L., Chen, G.Q., 2024. Toward a sustainable approach for dairy waste management: the integration of filtration, ion exchange, and electrodialytic processes for whey demineralization. ACS Sustain. Chem. Eng. 12, 13197–13207. https://doi.org/10.1021/acssuschemeng.4c03954.

Fancello, F., Zara, G., Hatami, F., Scano, E.A., Mannazzu, I., 2024. Unlocking the potential of second cheese whey: a comprehensive review on valorisation strategies. Rev. Environ. Sci. Biotechnol. 23, 411–441. https://doi.org/10.1007/s11157-024-09687-2.

Fischer, C., Kleinschmidt, T., 2015. Synthesis of galactooligosaccharides using sweet and acid whey as a substrate. Int. Dairy J. 48, 15–22. https://doi.org/10.1016/j.idairyj.2015.01.003.

Gernigon, G., Piot, M., Beaucher, E., Jeantet, R., Schuck, P., 2009. Physicochemical characterization of Mozzarella cheese wheys and stretchwaters in comparison with several other sweet wheys. J. Dairy Sci. 92 (11), 5371–5377. https://doi.org/10.3168/jds.2009-2359.

Gernigon, G., Schuck, P., Jeantet, R., 2010. Processing of Mozzarella cheese wheys and stretchwaters: a preliminary review. Dairy Sci. Technol. 90 (1), 27–46. https://doi.org/10.1051/dst/2009045.

Guo, M., Wang, G., 2019. History of whey production and whey protein manufacturing. In: Whey Protein Production, Chemistry, Functionality, and Applications. John Wiley & Sons Ltd, pp. 1–9.

Kamalesh, R., Saravanan, A., Yaashikaa, P.R., Vickram, A.S., Thiruvengadam, S., 2025. Recent insights into membrane separation technology for partitioning of nutritional components in dairy products—mechanism and applications. Int. Dairy J. 169, 106340. https://doi.org/10.1016/j.idairyj.2025.106340.

Karastamatis, S., Zoidou, E., Moatsou, G., Moschopoulou, E., 2022. Effect of modified manufacturing conditions on the composition of Greek strained yogurt and the quantity and composition of generated acid whey. Foods 11 (24), 3953. https://doi.org/10.3390/foods11243953.

Karela, M., Sakkas, L., Zoidou, E., Moatsou, G., Milosi, K., Moschopoulou, E., 2025. Acid whey from industrial Greek strained yoghurt: effect of the kind of milk and the way of straining on its composition and processing by nanofiltration. Dairy 6 (21). https://doi.org/10.3390/dairy6030021.

Kaya, N., Altiok, E., Gökkaya, D.S., Kabay, N., Ötleş, S., 2019. Demineralization of cheese whey by sequential nanofiltration (NF) and electrodialysis (ED) processes. J. Membr. Sci. Res. 5, 250–255. https://doi.org/10.22079/JMSR.2019.98013.1230.

Kelly, P.M., 2022. Whey powder, demineralized whey powder and delactosed whey. In: Encyclopedia of Dairy Sciences, third ed. Elsevier Ltd, Amsterdam, pp. 174–185. https://doi.org/10.1016/b978-0-12-818766-1.00162-8.

Koutina, G., Skibsted, L.H., 2015. Calcium and phosphorus equilibria during acidification of skim milk at elevated temperature. Int. Dairy J. 45, 1–7. https://doi.org/10.1016/j.idairyj.2015.01.006.

Macedo, A., Duarte, E., Fragoso, R., 2015. Assessment of the performance of three ultrafiltration membranes for fractionation of ovine second cheese whey. Int. Dairy J. 48, 31–37. https://doi.org/10.1016/j.idairyj.2014.12.003.

Marx, M., Sixta, A., Hofsommer, J., Wörthmann, M., Kulozik, U., 2019. Manufacturing of demineralized whey concentrates with extended shelf life: impact of the degree of demineralization on functional and microbial quality criteria. Food Bioprod. Process. 114, 1–11. https://doi.org/10.1016/j.fbp.2018.10.011.

Mc Entee, S.A., Kelly, A.L., Lawless, F.N., Mccarthy, N.A., Murphy, E.G., 2025. Effect of ultrafiltration of acid whey at pH 4.6 or 6.5 on whey concentrate composition and resultant powder properties. Int. J. Dairy Technol. 78 (1), e13116. https://doi.org/10.1111/1471-0307.13116.

Melnikova, E., Bogdanova, E., Paveleva, D., 2023. The research of whey permeate mineral profile at different stages of membrane filtration. Potr. S. J. F. Sci. 17, 748–758. https://doi.org/10.5219/1904.

Menchik, P., Moraru, C.I., 2019. Nonthermal concentration of liquid foods by a combination of reverse osmosis and forward osmosis. Acid whey: a case study. J. Food Eng. 253, 40–48. https://doi.org/10.1016/j.jfoodeng.2019.02.015.

Menchik, P., Zuber, T., Zuber, A., Moraru, C.I., 2019. Composition of coproduct streams from dairy processing: acid whey and milk permeate. J. Dairy Sci. 102, 3978–3984. https://doi.org/10.3168/jds.2018-15951.

Merkel, A., Farova, H., Voropaeva, D., Yaroslavtse, A., Ahrne, L., Yazdi, S.R., 2021a. The impact of high effective electrodialytic desalination on acid whey stream at high temperature. Int. Dairy J. 114, 104921. https://doi.org/10.1016/j.idairyj.2020.104921.

Merkel, A., Voropaeva, D., Ondrušek, M., 2021b. The impact of integrated nanofiltration and electrodialytic processes on the chemical composition of sweet and acid whey streams. J. Food Eng. 298, 110500. https://doi.org/10.1016/j.jfoodeng.2021.110500.

Merkel, A., Voropaeva, V., Farova, H., Yaroslavtsev, A., 2020. High effective electrodialytic whey desalination at high temperature. Int. Dairy J. 108, 104737. https://doi.org/10.1016/j.idairyj.2020.104737.

Moatsou, G., 2022. Milk composition and properties: interspecies comparison. In: Understanding and Improving the Functional and Nutritional Properties of Milk. Burleigh Dodds Science Publishing Limited, Cambridge. https://doi.org/10.19103/AS.2022.0099.03.

Nielsen, E.N., Gøtke, M., Cordin, U., Leif, H., Skibsted, L.H., Velizarov, S., Crespo, J.G., Ahrné, L.M., 2023. Effect of reverse osmosis pre-processing of acid whey and electrodialysis current density on process performance. Food Bioprod. Process. 140, 72–83. https://doi.org/10.1016/j.fbp.2023.05.002.

Nielsen, E.N., Merkel, A., Yazdi, S.R., Ahrne, L., 2021. The effect of acid whey composition on the removal of calcium and lactate during electrodialysis. Int. Dairy J. 117, 104985. https://doi.org/10.1016/j.idairyj.2021.104985.

Nielsen, E.N., Skibsted, L.H., Yazdi, S.R., Merkel, A., Ahrné, L.M., 2022a. Effect of calcium-binding compounds in acid wheyon calcium removal during electrodialysis. Food Bioprod. Process. 131, 224–234. https://doi.org/10.1016/j.fbp.2021.11.008.

Nielsen, E.N., Skibsted, L.H., YazdiI, S.R., Merkeil, A., Ahrne, L.M., 2022b. Improving electrodialysis separation efficiency of minerals from acid whey by nano-filtration preprocessing. Int. J. Dairy Technol. 75 (4), 820–830. https://doi.org/10.1111/1471-0307.12893.

Nishanthi, M., Chandrapala, J., Vasiljevic, T., 2017a. Properties of whey protein concentrate powders obtained by spray drying of sweet, salty and acid whey under varying storage conditions. J. Food Eng. 214, 137–146. https://doi.org/10.1016/j.jfoodeng.2017.06.032.

Nishanthi, M., Vasiljevic, T., Chandrapala, J., 2017b. Properties of whey proteins obtained from different whey streams. Int. Dairy J. 66, 76–83. https://doi.org/10.1016/j.idairyj.2016.11.009.

Okawa, T., Shimada, M., Ushida, Y., Seki, N., Watai, N., Ohnishi, M., Tamura, Y., Ito, A., 2015. Demineralisation of whey by a combination of nanofiltration and anion-exchange treatment: a preliminary study. Int. J. Dairy Technol. 68 (4), 478–485. https://doi.org/10.1111/1471-0307.12283.

Ortiz Quezada, A.G., Castilla Asaf, A., Sacks, G.L., 2021. Optimization of conditions for Greek style yogurt acid whey demineralization and its effects on filterability. Int. Dairy J. 123, 105163. https://doi.org/10.1016/j.idairyj.2021.105163.

Ozel, B., McClements, D.J., Arikan, C., Kaner, O., Oztop, M.H., 2022. Challenges in dried whey powder production: quality problems. Food Res. Int. 160, 111682. https://doi.org/10.1016/j.foodres.2022.111682.

Pasta, C., Caccamo, M., Petriglieri, R., Difalco, A., Farina, G., Belvedere, G., Marino, G., Marino, V.M., Garavaldi, A., Musi, V., Alcaine, S.D., 2024. Sicilian whey: utilization of ricotta whey in the production of value-added artisanal beers. Fermentation 10, 19. https://doi.org/10.3390/fermentation10010019.

Pires, A.F., Marnotes, N.G., Rubio, O.D., Garcia, A.C., Pereira, C.D., 2021. Dairy by-products: a review on the valorization of whey and second cheese whey. Foods 10, 1067. https://doi.org/10.3390/foods10051067.

Poitras, D., Perreault, V., Gaaloul, S., Schuck, P., Bazinet, L., 2025. Lactic acid removal and demineralization of acid whey by coupling electrodialysis under pulsed electric fields with pre-concentration by nanofiltration: impact on spray drying and powder quality. Desalination 613, 119032. https://doi.org/10.1016/j.desal.2025.119032.

Purwanti, N., Mulcahy, S., Murphy, E.G., 2024. Alteration of physicochemical properties and heating stability of reconstituted acid whey powder by calcium chelating salts. Food Bioprocess Technol. 17, 1278–1293. https://doi.org/10.1007/s11947-023-03198-8.

Reig, M., Vecino, X., Cortina, J.L., 2021. Use of membrane technologies in dairy industry: an overview. Foods 10, 2768. https://doi.org/10.3390/foods10112768.

Talebi, S., Garthe, M., Roghmans, F., Chen, G.Q., Kentish, S.E., 2021. Lactic acid and salt separation using membrane technology. Membranes 11 (2), 107. https://doi.org/10.3390/membranes11020107.

Talebi, S., Suarez, F., Chen, G.Q., Chen, X., Bathurst, K., Kentish, S.E., 2020. Pilot study on the removal of lactic acid and minerals from acid whey using membrane technology. ACS Sustain. Chem. Eng. 8 (7), 2742–2752. https://doi.org/10.1021/acssuschemeng.9b06561.

Walstra, P., Wouters, J.T.M., Guerts, T.J., 2006. Dairy Science and Technology, second ed. CRC Press Taylor & Francis Group, Boca Raton, pp. 354–356.

Weerasingha, V.G., Kelly, A.L., Sheehan, J.J., Alechosseini, A., 2025. Sustainable utilisation of salty whey generated during cheese manufacture: a review. Int. J. Dairy Technol. 78, e70033. https://doi.org/10.1111/1471-0307.70033.

Zhao, Z., Corredig, M., 2016. Serum composition of milk subjected to re-equilibration by dialysis at different temperatures, after pH adjustments. J. Dairy Sci. 99 (4), 2588–2593. https://doi.org/10.3168/jds.2015-9917.

Further reading

Beaulieu, M., Perreault, V., Mikhaylin, S., Bazinet, L., 2020. How overlimiting current condition influences lactic acid recovery and demineralization by electrodialysis with nanofiltration membrane: comparison with conventional electrodialysis. Membranes 10, 113. https://doi.org/10.3390/membranes10060113.

Chandrapala, J., Vasiljevic, T., 2018. Feasibility of spray drying concentrated acid whey after nanofiltration. Food Bioprocess Technol. 11, 1505–1515. https://doi.org/10.1007/s11947-018-2118-1.

Crowley, S.V., Molitor, M.S., Kalscheuer, R., Lu, Y., Kelly, A.L., O' Mahony, J.A., Lucey, J.A., 2019. Size classification of precipitated calcium phosphate using hydrocyclone technology for the recovery of minerals from deproteinised acid whey. Int. J. Dairy Technol. 72 (1), 142–151. https://doi.org/10.1111/1471-0307.12570.

Part 2

Food applications

Chapter 9

Cheeses and ice cream

Tomáš Gryger[1], Eva Samková[2], Itthanan Suttikhana[3], Simona Janoušek Honesová[2], Eva Dadáková[3], Bhavbhuti M. Mehta[4] and Richardos Nikolaos Salek[1]

[1]*Department of Food Technology, Faculty of Technology, Tomas Bata University in Zlín, Zlín, Czech Republic;* [2]*Department of Food Biotechnologies and Agricultural Products' Quality, Faculty of Agriculture and Technology, University of South Bohemia in České Budějovice, České Budějovice, Czech Republic;* [3]*Department of Applied Chemistry, Faculty of Agriculture and Technology, University of South Bohemia in České Budějovice, České Budějovice, Czech Republic;* [4]*Department of Dairy Chemistry, SMC College of Dairy Science, Kamdhenu University, Anand, Gujarat, India*

1. Introduction

1.1 Whey cheeses

The global production of whey (both sweet and acid whey) has been estimated at approximately 200 million tons per year over the past decade, with a continuous upward trend (Ostertag et al., 2023). One important utilization pathway for this dairy by-product is the production of whey cheeses, which represent a unique segment of dairy products, reflecting the nutritional and functional properties of whey. The basic requirements for whey cheese can be found in the current version of the *Standard for Whey Cheeses* (FAO, 1971), defining it as a solid, semisolid, or soft product that is principally obtained through two somewhat distinct approaches.

(1) Coagulation of whey by the application of heat, with or without the addition of acid;
(2) Concentration of whey by heating, followed by stirring and kneading of the resulting product.

In both processes, whey may be preconcentrated, and the addition of milk, cream, or other milk-derived ingredients is also permitted. It is noteworthy that whey cheese obtained by concentration of whey does not involve the coagulation of the proteins; therefore, according to the *General Standard for Cheese*, it would not be classified as cheese (FAO, 1978). Nevertheless, such products are specifically covered by the aforementioned whey cheese standard. Until recently, whey cheeses have not attracted broad consumer interest and have largely been overshadowed by other types of casein cheeses, which benefit from stronger international marketing success. However, in several regions, whey cheeses are among the most traditional and significant food products (Jelen, 2011). In addition, demand for whey cheeses is now gradually increasing, likely reflecting a broader consumer interest in more natural and healthy foods with high nutritional value (Puccio et al., 2022).

Whey cheeses have a strong tradition, especially in Italy and Norway, but they are also widespread in the Mediterranean region and in other Scandinavian countries, while in most other parts of the world, the popularity of these products is considerably more variable. In both Italy and Norway, whey derived from casein cheese production is typically used as the main raw material. However, the two main types of whey cheese differ significantly in their technology and characteristics, with each type representing one of the standard's category (Del Nobile et al., 2009; Modler and Emmons, 2001). Traditional Italian-type whey cheese, largely exemplified by Ricotta, is a typical representative of the "white" acid-heat-coagulated whey cheeses. In contrast, the Norwegian-type cheese known primarily as brown whey cheese or Mysost (Brunost) is produced by evaporating whey, resulting in a product with a high lactose content and distinctive brown color (Bintsis and Papademas, 2023; Skeie and Abrahamsen, 2017). The notable differences between these cheese types are illustrated in the process flowchart (Fig. 9.1), which compares the fundamental technological principles involved in Ricotta and Mysost production.

Whey. https://doi.org/10.1016/B978-0-443-27648-4.00014-1

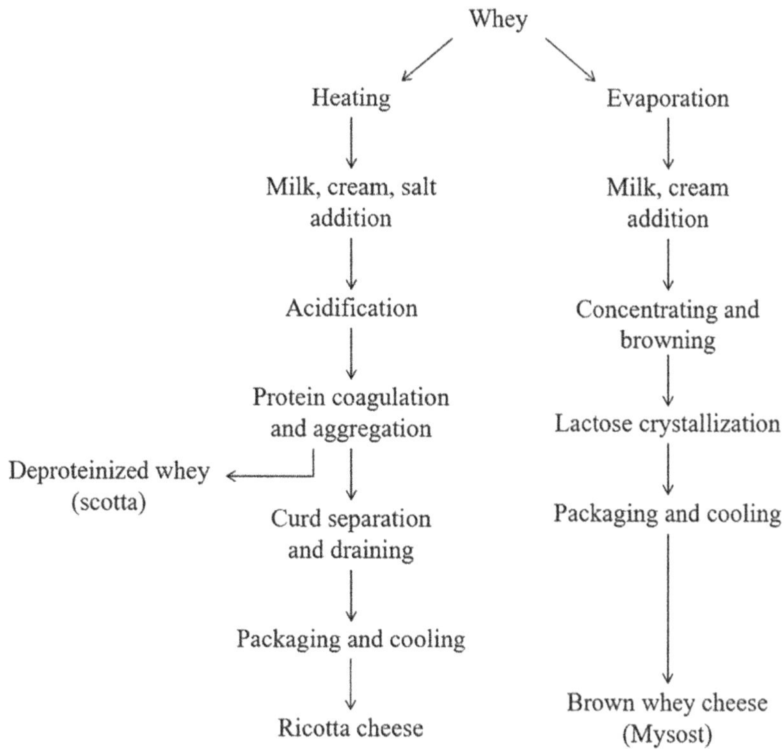

FIGURE 9.1 Alternative processing approaches in whey cheese production.

1.2 Ice cream

The origins of ice cream are surrounded by numerous legends and myths, many of which are not supported by historical evidence. Due to the absence of concrete documentation, it is impossible to determine when or where ice cream IC was first developed definitively; however, its development is closely associated with advances in refrigeration technology (Clarke, 2007; Quinzio, 2009).

The earliest forms of frozen desserts were made by flavoring snow with fruit, fruit juices, or honey. Such practices were recorded across ancient civilizations in Asia (e.g., Mongolia) and Europe (e.g., the Roman Empire). During the 16th and 17th centuries, ice cream began to appear as a luxury commodity in royal courts (France and England) (Goff and Hartel, 2013). A significant milestone came in the 19th century with the invention and patenting of "hand-cranked ice cream freezers." The first patent was granted in the United States in 1843 to Nancy M. Johnson (*U.S. Patent No. 3254—Artificial Freezer*), followed by Thomas Masters in England (*MastersṡPatent Ice-making apparatus*) in 1844. The industrialization of food production in the 20th century, including the development of deep-freezing and automated processing, enabled large-scale manufacturing and global distribution (Goff and Hartel, 2013).

Today, ice cream represents one of the most consumed frozen desserts worldwide, characterized by diverse formulations, textural variations, and flavor profiles. In 2023, the EU produced 3.2 billion liters of ice cream. Among the EU countries, Germany was the leading producer, followed by France and Italy (Eurostat, 2025). Following the EU, the United States produced about 2.7 billion kg of ice cream in 2023 (IDFA, 2025). Revenue in the ice cream market amounts to US$109.24 billion in 2025. The market is expected to grow annually by 5.54% (CAGR, 2025–30) and reach a volume of 24.81 billion kg by 2030. The growth rate is mainly impacted by changing consumer preferences, rising health consciousness, and increasing competition from healthier alternatives (Statista, 2025). Whey, whey protein concentrates (WPC), whey protein isolate (WPI), and whey powder are widely used in the production of ice cream and other frozen desserts to improve their functional properties.

2. Coagulated whey protein cheese—Ricotta cheese

Ricotta cheese is classified as an acid-heat-coagulated whey protein cheese. The name refers to the Latin term *re-cocta* ("recooked"), which is linked to the technology involving the heat treatment of whey. Historically, Ricotta was obtained from the coagulable materials in whey, produced as a by-product during the manufacture of other cheeses. Traditional terminology distinguishes Ricottone as the product derived exclusively from whey proteins, whereas Ricotta may be a blend of milk and whey components (Kindstedt, 2017). Ricotta has a long history dating back to Roman times and is widely consumed in southern Italy. However, similar products can be found in other Mediterranean countries, for example, Anthotyros and Manouri (Greece), Anari (Cyprus), or Urda on the Balkan Peninsula (Del Nobile et al., 2009). In North America and Latin America, it is occasionally available under different names, such as Requesón. Today, Ricotta is produced and consumed worldwide, and its regulation and requirements are determined by individual countries (Paskaš et al., 2019; USDA, 1981). The classification of Ricotta varies depending on factors such as composition, production area, the animal species from which the whey is derived, production scale, and shelf life. The *USDA Specifications for Ricotta Cheeses* (USDA, 1981) classify three main types based on moisture and fat content. Whole milk Ricotta is manufactured from whole milk and contains no more than 80.0% moisture and at least 11.0% milk fat. Part-skim Ricotta is obtained from milk with reduced fat content, with no more than 80.0% moisture and a milk fat content between 6.0% and 11.0%. In contrast, Ricottone is derived from skim milk, whey, or their blends, and is characterized by a moisture content not exceeding 82.5% and less than 1.0% milk fat.

2.1 Sensory and nutritional properties of ricotta cheese

The sensory characteristics of Ricotta cheese (or also called as Ricotta) largely depend on the composition. Fresh Ricotta displays a yellowish-white color and a grainy, crumbly yet compact texture (Fig. 9.2), accompanied by a mild and slightly perceptible flavor (Pintado et al., 2001). Whole milk and part-skim Ricotta varieties typically offer a creamier texture with a slightly sweet or caramel-like taste. In contrast, Ricottone, derived mainly from whey, has a blander flavor profile (Farkye, 2004). Due to its mild taste and gritty texture, Ricotta is rarely consumed on its own; however, these attributes make it highly suitable for pairing with fruits or honey in breakfast dishes, as well as for use as an ingredient in cake fillings, cheesecakes, baked goods, pizzas, and pasta dishes such as lasagna or ravioli. Dry Ricotta, on the other hand, is primarily used for grating (Del Nobile et al., 2009).

The nutritional composition of Ricotta cheese is summarized in Table 9.1. Regardless of the specific draining step, Ricotta has a high moisture content (66%–82%), linked to the water-holding capacity of coagulated whey proteins. The fat content varies significantly (3%–22%), depending on the type of cheese from which the whey was obtained and the amount of milk or cream added to the blend. Due to high water content, Ricotta has a relatively low energy value, making it a recommended component of weight-reduction diets. Furthermore, high protein content (7%–16%) is a notable nutritional advantage, as whey proteins are highly digestible and possess an optimal amino acid profile (Farkye, 2004; Jelen, 2011).

2.2 Technology for the production of Ricotta cheese

Traditional Ricotta production involves heating whey to denature and coagulate the whey proteins, which are then separated from the liquid phase (Modler and Emmons, 2001). However, due to increasing global demand, various

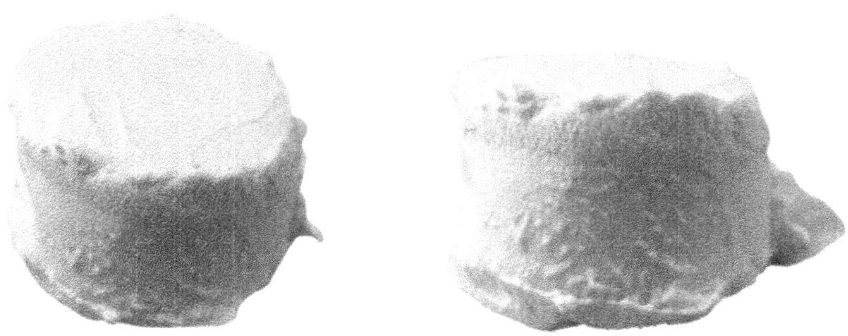

FIGURE 9.2 Ricotta cheese (ITALAT CZ s.r.o., Czech Republic). *Photo: Richardos Nikolaos Salek.*

TABLE 9.1 Compositional data (%) of different ricotta varieties.

Cheese	Country	Source of whey	Moisture	Protein	Fat	Carbohydrate	References
Ricotta di Pecora	Italy	Sheep	70.0	7.8	18.2	3.7	Mucchetti et al. (2002)
Ricotta di Bufala	Italy	Buffalo	75.1	11.7	6.7	4.5	Sameer et al. (2020)
Ricotta di Capra Girgentana	Italy	Goat	67.9	6.7	20.7	3.7	Pizzillo et al. (2005)
Ricotta (5% milk add.)	USA	Cow	81.6	7.0	3.3	NA	Weatherup (1986)
Ricotta (20% milk add.)	USA	Cow	66.5	16.3	11.6	NA	Modler (1988)
Urda	Serbia	Cow	79.6	10.7	5.6	3.2	Paskaš et al. (2019)
Anari (fresh)	Cyprus	Sheep, goat	65.4	11.0	21.7	1.4	Papademas et al. (2017)

NA, not available.

modifications to both ingredients and processing methods have been introduced. These include using different ratios of milk–whey blends, incorporating powdered ingredients, adjusting process parameters (such as temperature and time), employing mechanization, or, conversely, emphasizing traditional production techniques to promote local and artisanal products (Bintsis and Papademas, 2023; Lucey, 2011).

2.2.1 Raw materials for ricotta cheese

The primary raw material is whey obtained from the manufacture of Mozzarella and other types of cheese (Farkye, 2004). Depending on the source, Ricotta can be made from sheep, goat, cow, or buffalo whey, or from a blend of these (Mucchetti et al., 2002). The suitability of whey is determined by its titratable acidity and pH, with optimal whey protein coagulation requiring a pH of at least 6.2 and an ideal titratable acidity ranging between 0.13% and 0.14% lactic acid, with a maximum limit of 0.16% (Chandan, 2003; Farkye, 2004). Acid whey, such as that derived from Quark or Cottage cheese production, is less suitable due to its significantly higher acidity and ash content, which adversely affect the taste and texture of the final product (Jelen, 1992). To improve yield and sensory properties, a certain amount of milk, cream, or other dairy components is often added to the whey (Chandan, 2003).

2.2.2 Manufacturing procedure of ricotta cheese

Ricotta cheese is a typical example of using the principles of agglomeration and precipitation of whey proteins, induced by acidification and heating (Pizzillo et al., 2005). Whey is heated to 75–90°C for 10–30 min, depending on the type of whey and the heating method used, followed by acidification to pH 5.6–6.0 (Lucey, 2011; Mucchetti et al., 2017). To achieve the desired pH that promotes whey protein agglomeration, food-grade acidulants (such as citric, acetic, or lactic acid) are added at concentrations of 1.5%–2.5%. The whey can be enriched with added milk and cream (10%–25%), salt (0.5%–1.5%), and other ingredients (Chandan, 2003). The formed curd is transferred into molds, where the remaining liquid fraction (known as scotta/scotta whey) is drained, followed by packaging and cooling (Mancuso et al., 2014). Scotta is an underutilized whey by-product of ricotta making. The scotta whey (deproteinized whey) can be used for a second cooking of the cheese curd in some traditional Sicilian cheeses (Bonanno et al., 2013), as well as used in the production of beers, organic acids, bioethanol, and polyhydroxyalkanoates. The yield of curd from whey depends on several factors, including the coagulation method, whey composition, and the concentration of added ingredients such as NaCl and acid. For example, Ricotta yield typically ranges from 5% to 6% when whey is used without the addition of milk (Chandan, 2003).

The following processes describe in more detail the essential aspects of Ricotta production, including acidification, heat treatment, and the addition of other ingredients involved in curd formation.

2.2.2.1 Coagulation

The combined application of heat and acidity causes denaturation and aggregation of whey proteins, leading to curd formation (Lucey, 2011). This process is influenced by the balance between the selected temperature, acidity, and salinity of the whey blend, all of which are critical for optimizing Ricotta yield (Anema, 2006). The cheese matrix has a gel-like structure composed of whey proteins (α-lactalbumin and β-lactoglobulin), which interconnect to create a network that traps fat globules and moisture (Stading and Hermansson, 1991). For protein aggregation to occur, changes in processing conditions, particularly an increase in temperature, are essential (Verheul et al., 1998). Denaturation of β-lactoglobulin during heating leads to the exposure of thiol groups, thereby increasing its reactivity (Matsudomi et al., 1991). Additionally, the ionic strength of the whey is an important parameter in protein aggregation. Divalent cations such as calcium and magnesium play a key role by balancing the negative charge of proteins and facilitating bridging between them (Salvatore et al., 2014). Hydrophobic interactions among nonpolar groups also contribute to protein-protein binding (Lucey et al., 2003).

2.2.2.2 Acidification

An important parameter is the pH of the whey, which is closely related to its origin, protein content, and mineral composition. In batch processes, whey (or whey/milk blend) is acidified to pH 5.6–6.0 using a starter culture or acidulant (Lucey, 2011). The pH value significantly affects the final cheese texture, with more acidic pH values resulting in a grainier consistency, while a pH closer to neutrality results in a more elastic cheese texture (Modler, 1988). To promote protein aggregation, various acidulants are used. For sweet whey, the most common acidulants include vinegar, citric acid, or lactic acid solutions. Alternatively, acidification can be achieved by adding acid whey powder to milk (Lucey, 2011). For acid whey, adjusting the pH to values above 6 using food-grade sodium hydroxide or sodium bicarbonate before heating ensures satisfactory protein yields without the need for further pH manipulation (Jelen, 1992). The selection of a suitable acidulant is a critical step that affects the nutritional, organoleptic, and rheological properties of the final product (Araque et al., 2018; El-Sheikh et al., 2010). It is also necessary to consider legislative restrictions regarding permitted food additives. The addition of acidulants can be replaced by using less common substances, mainly in industrial production, such as magnesium sulfate or calcium sulfate, which modify protein-mineral interactions in heated whey and lead to protein coagulation (Mucchetti et al., 2017).

2.2.2.3 Heating

A fundamental condition for the agglomeration of proteins is the heating of the whey blend, which influences the degree of whey protein aggregation. Depending on the species origin of the whey, temperatures between 75 and 90°C are used (e.g., sheep whey 75°C, cow whey 85°C, or buffalo whey 90°C). The high temperature acts on the Ricotta flakes and makes the curd network firmer and more cohesive, thus facilitating separation. Additionally, the size of the agglomerates affects the final structure and the drainage of scotta whey from the cheese (Mucchetti et al., 2017). The formation of aggregates involves denatured α-lactalbumin through free thiol groups as well as hydrophobic or electrostatic interactions (Donato and Guyomarc'h, 2009). However, as temperature increases, the degree of denaturation rises, and the Ricotta yield shows a grainier structure with reduced smoothness and moisture content (Mucchetti et al., 2017). Traditionally, vats with an indirect supply of heat using a double bottom and steam are employed. A more modern method is direct steam injection into the whey, which is more time-efficient (Shelke et al., 2022). There are also systems that allow the use of high pressure, ensuring microbial inactivation while preserving the desired nutritional and sensory properties of Ricotta cheese (Besbes et al., 2002).

2.2.3 Addition of other ingredients

The addition of salt to the whey blend is an important factor in production. Salt affects the interactions between negatively charged whey proteins, improving the aggregation process (Mucchetti et al., 2017). The salt concentration can also impact the structure and porosity of the cheese (Verheul et al., 1998). Moreover, salt significantly enhances the flavor and overall acceptability of the final product (Kamel et al., 2013). Additionally, calcium chloride helps stabilize whey proteins and promotes flocculation, improving the curd's consistency. In industrial production, these additives are used to achieve the desired curd texture, whether for softer or firmer cheese varieties (Lucey, 2011). Furthermore, the character of Ricotta can be influenced by the addition of various dairy ingredients. It is possible to add 10%–25% milk or cream to neutralize acidity, enhance curd cohesiveness, and increase yield, while, for example, in traditional Italian Ricotta, up to 15% raw milk is added (Chandan, 2003; Jelen, 1992). The milk fat content affects the color and firmness of Ricotta, with fat-free

Ricotta being lighter and firmer compared to the full-fat version. In terms of yield, an increased milk solids content is crucial. For this purpose, the addition of skimmed milk powder, condensed milk, or the concentration of whey proteins through whey ultrafiltration has been successfully studied (El-Sheikh et al., 2010; Gomaa et al., 2015).

2.3 Production of Ricotta cheese

The production of Ricotta can follow either traditional small-scale methods or large-scale industrial processing. While the fundamental principles are similar, the differences lie primarily in the raw materials used, processing equipment, and the degree of mechanization.

2.3.1 Small-scale Ricotta cheese production

Traditional Ricotta production is typical of small farms, particularly in Italy and other Mediterranean regions. It relies on natural ingredients, simple tools, and manual techniques. The whey is typically a by-product of local cheese production, such as Pecorino, buffalo mozzarella, or highland sheep cheeses (Bergamaschi et al., 2016; Scatassa et al., 2018; Tripaldi et al., 2020). Raw cow, sheep, or buffalo milk is often added to enrich the whey, along with natural acidifying agents (e.g., fermented whey or vinegar). Heating takes place in open vats, usually over a direct flame or gas burner. Once the target temperature (typically 85–95°C) is reached, a layer of foam and curdled protein forms on the surface. This curd is manually skimmed using traditional wooden or stainless-steel ladles and transferred into conical perforated baskets to allow scotta whey drainage. The Ricotta is first cooled at ambient temperature and then refrigerated. The final product has a delicate texture, sweet milky flavor, and a very short shelf life (Jelen, 1992; Ricciardi et al., 2019).

2.3.2 Large-scale Ricotta cheese production

Modern industrial Ricotta production involves large-scale processing with standardized ingredients and automated systems. Whey, usually derived from pasteurized cheese production, is heated in stainless-steel vats using either low-pressure steam injection or plate heat exchangers. Typically, 10% of cow's milk and 1% of salt are added. Coagulation is induced by food-grade acids, mainly citric or lactic acid (Paskaš et al., 2019; Toppino et al., 2004). In some cases, inorganic salts such as magnesium sulfate may be used instead (Mucchetti et al., 2017). The coagulated curd is mechanically collected and transferred into perforated molds or cloth bags. Light pressing may be applied to promote the drainage of remaining scotta whey. The final step involves thermal treatment of the finished Ricotta to extend its shelf life (Paskaš et al., 2019).

2.4 Shelf life of Ricotta cheese

Ricotta is characterized by high water activity (0.97–0.99) and a rich nutrient profile, which together create favorable conditions for the growth of a wide range of microorganisms, including bacteria, yeasts, and molds (Pala et al., 2016). Traditional artisanal Ricotta has a very limited shelf life, typically less than 1 week, even when stored under refrigeration (Ricciardi et al., 2019). In contrast, industrial production includes a heat treatment step, usually pasteurization of the final product, which extends the shelf life to approximately 20–40 days (Mucchetti and Neviani, 2006). Packaging also plays a key role in preserving freshness and ensuring microbiological safety. Ricotta is commonly vacuum-sealed or packaged under modified atmosphere and stored under controlled refrigeration conditions, which can extend the shelf life up to 4 weeks (Farkye, 2004; Spanu et al., 2017).

2.5 Development pathways of Ricotta cheese

Current research on Ricotta focuses on improving its functional properties as well as developing sustainable strategies to extend its shelf life, including bio-preservation methods, the addition of herbs and plant extracts, and the use of nonthermal processing technologies.

2.5.1 Functional properties and probiotic potential

The growing interest in the functional food sector has led to efforts aimed at enriching dairy products with probiotics and prebiotics, which are known to have beneficial effects on consumer health (Dimidi et al., 2019; Kaur et al., 2022). Ricotta and other whey cheeses have been demonstrated to be suitable matrices for delivering probiotic bacteria into the gastrointestinal tract (Madureira et al., 2011). Several studies have explored the addition of prebiotic strains of lactic acid bacteria (LAB), such as *Lacticaseibacillus paracasei* subsp. *paracasei* and *Lactobacillus delbrueckii* subsp. *bulgaricus*, and have evaluated their viability during refrigerated storage, as well as their influence on textural modifications (e.g., exopolysaccharide production), physicochemical properties (e.g., acidity and moisture content), and sensory acceptability (Carrero-Puentes et al., 2022; Niro et al., 2013; Pires et al., 2021). Another type of functional Ricotta can be produced by incorporating dietary fiber (e.g., at levels of 1%, 3%, or 5%) without significantly altering the expected characteristics of the final product (Zinina et al., 2021). These findings are instrumental in guiding the development and industrial adaptation of functional whey cheeses.

2.5.2 Addition of plant extracts and bio-preservation

The demand for food products made without or with minimal use of synthetic preservatives presents a major challenge for the food industry. Plant-derived extracts contain a wide range of compounds with antioxidant, antibacterial, and antifungal activities that act through various mechanisms (Khorshidian et al., 2018). Numerous studies have investigated the addition of different herbs or plant extracts and their application in combination with packaging materials to prolong the shelf life of whey cheese (Bintsis and Papademas, 2023). In particular, ingredients such as black cumin, thyme, rosemary, and oregano essential oils and extracts have been studied for their antimicrobial efficacy and compatibility with various packaging technologies (Akpinar et al., 2022; Christaki et al., 2022; Ritota and Manzi, 2020). Bio-preservation refers to the use of natural antimicrobial compounds or protective LAB cultures. This approach is generally considered sustainable and consumer-friendly (Bintsis and Papademas, 2023). In whey cheeses, the effectiveness of bacteriocins such as nisin and enterocins has been studied, showing strong inhibitory effects against Gram-positive microorganisms, including *Listeria monocytogenes*, *Bacillus* spp., and *Clostridium* spp. Their combination with other preservation techniques further enhances efficacy (Samelis et al., 2003; Silva et al., 2018; von Staszewski and Jagus, 2008). Promising results have also been reported with the use of *Carnobacterium* spp. cultures applied to the surface of the final product, effectively controlling spoilage microorganisms (Spanu et al., 2017).

2.5.3 Nonthermal treatments

In addition to conventional secondary heat treatments such as pasteurization or thermization followed by refrigerated storage, novel nonthermal technologies are being investigated to extend shelf life while maintaining the authentic texture and sensory attributes of Ricotta-type cheeses (Bintsis and Papademas, 2023). Surface decontamination using continuous UV-C light (1.6–36.8 kJ/m^2) has shown potential for mold growth control; however, excessive exposure may lead to lipid and protein oxidation, adversely affecting quality (Urgu-Ozturk, 2022). Similarly, UV-C and near-UV-visible light treatments have been reported to effectively suppress microbial proliferation and prolong shelf life (Ricciardi et al., 2020). Pulsed light treatment (1.3–15.0 J/cm^2) has also been applied for surface sterilization, effectively inactivating microorganisms through localized photothermal and photophysical effects. Nonetheless, overexposure can promote undesirable changes due to photo-induced protein modifications (Ricciardi et al., 2021). Another promising technique is plasma treatment, either direct or indirect (via plasma effluent exposure) (Ricciardi et al., 2022). High effectiveness has been demonstrated in studies where fresh Ricotta was exposed to X-ray irradiation (0.5–3 kGy), resulting in a shelf life extension of up to 84 days without compromising microbiological safety or sensory characteristics (Ricciardi et al., 2019). As an alternative to refrigeration, hyperbaric storage (100 MPa) at various room temperatures has been proposed for microbial growth inhibition. Furthermore, this method has been estimated to have a carbon footprint approximately 26 times lower than that of conventional cold storage (Duarte et al., 2017).

3. Brown whey cheese

Brown whey cheese is manufactured through the evaporation of whey, combined with milk and cream. It is known under several names, the most common being Mysost (from Norwegian mys = whey and ost = cheese) or Brunost ("brown cheese"). In contrast to Ricotta production, which generates a secondary whey fraction (scotta), Norwegian whey cheese utilizes all whey components, with water vapor as the only by-product (Jelen, 2011). Various types of Brunost have a long tradition in Norway, dating back to the 17th century. The Brunost known today was developed in the second half of the 19th century in the Gudbrandsdalen district in central Norway, when producers began adding goat's milk and butter to the whey to achieve a smoother and milder flavor (Abrahamsen et al., 2008). Therefore, the product may also be referred to as Gjetost ("goat cheese") or Gudbrandsdalsost (Ribeiro and Ribeiro, 2010; Skeie and Abrahamsen, 2017). There are several varieties of brown whey cheese; however, due to localized production and the specificity of the market, the scientific literature does not provide a detailed classification beyond the basic division into sliceable and spreadable types (Skeie and Abrahamsen, 2017). Previously, Norwegian regulations specified the fat content in dry matter and product labeling, which is why local consumers recognize designations such as G35 for Gudbrandsdalsost made from cow's milk with added goat's milk (containing 35% fat in dry matter), or F33 for Fløtemysost made from cow's milk (33% fat in dry matter) (Jelen, 1992; Pintado et al., 2001). However, these regulations have since been abolished, and current product compositions are left to the discretion of the manufacturer (Norwegian Ministry of Agriculture and Food, 2016).

3.1 Sensory and nutritional properties of brown whey cheese

The sensory characteristics reflect the origin of the whey, milk, and cream used, as well as fat content and the degree of browning. Generally, the cheese exhibits a uniform brown color without visible specks. The texture is solid yet smooth and sliceable, without cavities (Fig. 9.3). Upon consumption, the cheese melts readily in the mouth without perceivable lactose crystals. The flavor profile typically includes a sweet, caramel-like aroma with a hint of acidity. A noticeable goaty flavor is also often present. Specific varieties differ, for example, Heidalsost is darker with a more pronounced burnt acidic flavor, Misværost is lighter and yellowish with a very mild taste, and Bestemorost is characterized by a sweeter caramel note due to added sugar. Brown whey cheese is commonly consumed in thin slices with butter on bread, waffles, or pancakes. In culinary applications, it is used in sauces, particularly for venison. Spreadable types are most often served to children on bread (Papademas et al., 2017; Skeie and Abrahamsen, 2017).

Table 9.2 provides the nutritional composition of selected brown whey cheeses. The moisture content of sliceable types generally does not exceed 20%, whereas spreadable variants may contain up to 40% or more. Fat comprises a substantial portion of the product, typically around 30%, although low-fat variants may have roughly half that amount. Swedish products tend to contain less fat than their Norwegian counterparts. Protein levels are generally around 10%, similar to those found in Ricotta. Brown whey cheeses have a significant sugar content. In addition to naturally occurring lactose, some products also contain added sucrose (ranging from 4% to 18% total) (Skeie and Abrahamsen, 2017). Unlike Ricotta, brown whey cheese is a relatively high-energy food. Historically, this was advantageous in Norway as a quick energy source for athletes and laborers with high energy expenditure. However, for modern consumers, lower-energy spreadable variants may be more suitable (Jelen, 1992).

3.2 Technology of brown whey cheese

3.2.1 Raw materials for brown whey cheese

For the production of brown whey cheese, sweet whey with no lactic acid is preferred. Previously, rennet casein was used to obtain such whey, but today it mainly comes from rennet-coagulated cheeses like Norvegia and Jarlsberg. Compared to rennet casein whey, cheese whey has higher levels of lactic acid, minerals, and lactic acid bacteria, requiring immediate pasteurization to prevent further acidification and casein precipitation. The whey blend should have a pH in the range of 6.2–6.5 to avoid excessive acidity and scorching during heating (Skeie and Abrahamsen, 2017). Regional and seasonal differences affect whey composition—protein levels peak in the summer (0.5%) and decline in the autumn (0.46%), while the β-lactoglobulin/α-lactalbumin ratio is highest in the autumn. These variations must be considered in formulation (Johansen et al., 2002). Other ingredients include milk and cream, which adjust the protein and fat content in the whey blend. A higher addition of milk and cream at the expense of whey, in turn, reduces the amount of lactose. Milk and cream also affect the texture of the cheese and reduce the acidic taste, as they increase the buffering capacity of the blend. Swedish variants (Mesost and Messmör) contain butter, which influences the flavor of the cheese, especially due to the concentration of nonvolatile aromatic compounds. Additionally, sucrose is sometimes added to some spreadable variants. Other ingredients may include sweeteners and nondairy components, such as hazelnuts or chocolate (Jelen, 2011; Skeie and Abrahamsen, 2017).

3.2.2 Use of goat milk

In Norway, brown whey cheeses like Brun Geitost and Gudbrandsdalsost traditionally use goat's milk, with Brun Geitost being made entirely from it. Until 1990, over 95% of Norwegian goat milk was used in brown whey cheese production, though rennet- and acid-coagulated goat cheeses have been on the rise. Goat milk often has a rancid flavor due to its high free fatty acid content, which is further correlated with α_{S1}-casein deficiency (Dagnachew et al., 2011; Inglingstad et al., 2017). Since 2007, breeding strategies have improved milk quality, reduced rancidity and enhancing the flavor of goat-based whey cheeses (Inglingstad et al., 2017). This has been positively influenced by an emphasis on feed quality and animal welfare, which between 2001 and 2012 led to an increase in protein and fat content and a reduction in free fatty acids, resulting in a milder flavor profile (Ådnøy, 2014).

FIGURE 9.3 Brown whey cheese (Gudbrandsdalen; TINE SA, Norway). *Photo: Tomáš Gryger.*

3.2.3 Manufacturing procedure of brown whey cheese

The production principle of brown whey cheese is simpler than that of Ricotta and involves gradually heating and concentrating the whey blend, which contains additional ingredients such as milk (20%–30%), milk fat, cream (10%–15%), and other ingredients. A key technological step is the controlled crystallization of lactose in the concentrated whey paste. The thickening occurs in two stages using traditional (or vacuum) dairy evaporators, followed by final evaporation in a special kettle called "gryta" at higher temperatures around 95–115°C. This is necessary to increase the viscosity of the paste and minimize uncontrolled lactose crystallization. In addition, a significant Maillard reaction occurs, providing the typical brown color of the product. The subsequent rapid cooling to around 70°C in scraped-surface agitators/coolers promotes the formation of small lactose crystals (20–40 μm), which is essential to prevent sandiness in the final product. Finally, the cheese is hot-packed into retail plastic containers, parchment paper, or metal tubes (Jelen, 2011; Skeie and Abrahamsen, 2017). Given the production principle, where up to 80% of water is evaporated, the production of brown whey cheese appears to be an efficient and useful method of whey disposal. Skeie and Abrahamsen (2017) reported a yield of 12%–13% for sliceable cheese, Fløtemysost, with milk and cream content (35%). The following technological processes describe in more detail the essential aspects of brown whey cheese production, focusing on protein denaturation, browning, and lactose crystallization during evaporation.

TABLE 9.2 Compositional data (% w/w) of different brown whey cheese varieties.

Cheese	Country	Source of whey	Moisture	Protein	Fat	Carbohydrates	References
Gjestost	Norway	Goat	13.4	9.7	34.0	42.0	Park and Guo (2006)
Primost	Norway	Goat	13.8	10.9	35.0	39.5	Bintsis and Papademas (2023)
Heidalsost	Norway	Goat	33.0	11.0	29.0	27.0	Skeie and Abrahamsen (2017)
Mysost	Norway	Goat, cow	14.9	11.5	24.7	NA	Pintado et al. (2001)
Mesost	Sweden	Goat, cow	20.0	NA	17.0	NA	Bintsis and Papademas (2023)
Prim (spread)	Norway	Goat, cow	31.0	7.5	7.5	46.0	Jelen and Buchheim (1976)
Messmör original	Sweden	Goat	39.8	6.9	5.0	47.0	Skeie and Abrahamsen (2017)

NA, not available.

3.2.3.1 Denaturation of whey proteins

Unlike Ricotta, which is based on the coagulation of whey proteins, the denaturation of whey proteins in brown whey cheese has not been systematically studied, nor has its relationship with product quality been explored. In a more general context, it can be mentioned that in the first evaporation phase in the falling film evaporator, the temperature affecting the whey proteins is <50°C, and the degree of denaturation is still insignificant. However, with increasing thermal load, the structure of the whey proteins begins to unfold while solubility remains high. During the subsequent heating in the kettle, the temperature increases to around 70°C, which is associated with irreversible denaturation of the whey proteins (Skeie and Abrahamsen, 2017). The pH of the whey blend (6.2–6.5) significantly influences the behavior of β-lactoglobulin, with faster aggregation observed at lower pH (de Wit, 2009). The whey proteins aggregate with κ-casein or exist as pure aggregates of β-lactoglobulin (Anema, 2020). Furthermore, it has been found that the presence of free fatty acids increases the denaturation temperature of serum albumin. Therefore, when using goat's milk, which has a relatively high free fatty acid content, a different denaturation process can be expected, potentially affecting the flavor and texture of the brown whey cheese. Heating in the gryta kettle or by scraped surface evaporators increases the temperature to >100°C for several minutes, and extensive denaturation of all whey proteins is expected, which will exist in the final brown whey cheese in the form of aggregates (Edwards and Jameson, 2020).

3.2.3.2 Browning

Browning is generally caused by Maillard reactions, which involve reducing sugars (lactose) and N-terminal amino acids of proteins at temperatures >80°C. Maillard reactions are desirable to achieve the required degree of browning that determines the color and flavor of the cheese. Although consumers describe the flavor of brown whey cheese as caramel-like, the temperature during processing usually does not exceed 110–115°C, and caramelization does not occur. The browning reactions result in a reduction of the lactose content compared to its original content in the whey blend of milk and cream. Additionally, the available lysine content decreases, leading to a lower biological value of brown whey cheese (Skeie and Abrahamsen, 2017). Browning reactions are highly complex and have not yet been fully described. Using gas chromatography–mass spectrometry, the most important aromatic components have been identified, such as 2-furanmethanol, 2-heptanone, butanoic acid methyl ester, 2-furancarboxaldehyde, and 2-nonanone (Abrahamsen et al., 2008). Other aldehydes and melanoids can also be formed during Strecker degradation, which occurs at pH > 6 (Skeie and Abrahamsen, 2017). The degree of browning during production can be influenced by time and temperature, with a darker color resulting from more intense heating. A darker brown color can also develop due to the prolongation of the browning reactions when there is a delay in transferring the cheese from the kettle to the cooling equipment. More acidic whey can also promote Maillard reactions, resulting in a darker color (Steinsholt et al., 1965).

3.2.3.3 Lactose crystallization

Lactose crystals form the main structure of brown whey cheese, accompanied by fat and aggregates of casein and denatured whey proteins. To achieve a smooth texture, small lactose crystals are essential, as larger ones (>100 μm) cause a sandy texture and unpleasant mouthfeel (Jelen and Buchheim, 1976). After evaporation, lactose becomes supersaturated in the moisture phase and starts to crystallize during cooling. Crystallization occurs most effectively under vigorous mixing, which promotes the formation of many small nuclei. Lactose solubility depends on its physical state: stable, metastable, or labile. In the labile state (high supersaturation), spontaneous nucleation can occur; in the metastable state, crystal growth is only possible if nuclei are already present. During cooling from around 110°C to below 70°C, lactose enters a supersaturated state, allowing nucleation and subsequent crystal growth. Cooling is controlled, as too low a temperature increases viscosity, hindering filling and even distribution of nuclei. Industrial processes using scraped-surface heat exchangers achieve small, uniform crystals (<30 μm), while small-scale production often results in coarser textures. Prismatic crystals, typically 20–40 μm in size, are preferred in brown whey cheese. Lactose also exists as α- and β-anomers with differing solubility and crystalline forms, but their specific role in crystallization remains uncertain (Gänzle et al., 2008; Skeie and Abrahamsen, 2017).

3.3 Production of brown whey cheese

Brown whey cheese was originally produced on mountain farms with grazing pastures (Abrahamsen et al., 2008). Industrialization began in the early 20th century, bringing significant advancements in production technology. Given the energy demands of the production process, technologies such as vacuum evaporation and the special kettle LT Gryta, developed by the Norwegian company Landteknikk, were adopted in the mid-20th century. However, in the early 21st century, the widespread use of scraped surface heat exchangers enabled the process to become continuous. Essentially, the production principle remains the same, but the final product characteristics depend on the ingredients used and the technological parameters, which differ between artisanal and industrial production (Skeie and Abrahamsen, 2017).

3.3.1 Small-scale brown whey cheese production

Small-scale production of brown whey cheese remains common, especially in summer at mountain chalets where animals graze. Traditional methods use open kettles, sometimes over wood fires, though most sites today are electrified. Production begins with coagulating skimmed or full-fat milk to obtain whey. While regular cheese is made from the casein fraction, the casein is sometimes discarded and used as animal feed. The whey is then boiled, with milk and cream added in varying ratios. As evaporation proceeds at atmospheric pressure, browning occurs, and viscosity increases, requiring more intense stirring. When the desired color and total solids (TS) are reached, heating stops, and the cheese is cooled with continuous stirring, often aided by sweep agitators, to below 35°C. During cooling, lactose crystallizes into a fine matrix, and avoiding large crystals is crucial for a smooth texture. Once nearly solid, the cheese is molded into blocks, cooled to complete crystallization, and finally waxed or wrapped for storage (Skeie and Abrahamsen, 2017).

3.3.2 Large-scale brown whey cheese production

Industrial brown whey cheese production has traditionally been carried out in batch processes, though continuous systems are now increasingly common. Mechanization is generally used, making large-scale production efficient and ensuring a standard final product. The batch process begins by mixing reverse osmosis–concentrated whey with milk and cream, tailored to the cheese variety. The blend is vacuum-evaporated in falling film evaporators at temperatures of 40–60°C to ~55% TS before being transferred to specially designed closed kettles with two scraped surface mixers for further vacuum concentration to ~75% TS. Heating is intensified under increased pressure (100–110°C), enhancing Maillard browning and lactose solubilization within 10–15 min. The resulting low-viscosity cheese is cooled via scraped surface heat exchangers, initiating crystallization in the form of small lactose crystals. The fluid mass is then filled into molds at ~80°C and cooled to 4°C in a tunnel, enabling complete crystallization and solid texture development. However, most brown whey cheese is now produced on continuous lines using scraped surface heat exchangers and vacuum evaporators in three to four steps with gradually increasing temperature. A holding tube before the final step ensures proper browning. Color and TS are monitored in-line using densitometers and flow meters. A critical point is maintaining optimal TS at the start to prevent protein aggregation and burning (Jelen, 2011; Skeie and Abrahamsen, 2017).

3.4 Shelf life of brown whey cheese

Brown whey cheese can be considered a concentrated product with a high content of dry matter and lactose (Jelen, 1992). Additionally, it undergoes treatment at high temperatures during processing, which ensures a sterile product during hot-pack. The sliceable variants of the cheese typically have a moisture content of around 20% and a low water activity (0.87–0.92), which acts as a barrier limiting the growth of a broad range of microorganisms. The shelf-life, labeled as "best before," is typically about 1 year. In contrast, spreadable variants with higher moisture content and relatively high sugar levels have a shelf-life limited to 4–6 months, with the potential risk primarily being mold growth (Papademas et al., 2017; Skeie and Abrahamsen, 2017). Brown whey cheeses retain their characteristics when stored at 0–5°C (Pintado et al., 2001).

3.5 Development pathways of brown whey cheese

Brown whey cheese is a traditional food primarily in Norway and Sweden, making its production and consumption geographically and culturally specific compared to white Ricotta-type cheeses. As a result, the literature on this topic is limited to more general aspects of production, with a shortage of publications addressing the development of new formulations and technologies and their application to the final product.

3.5.1 Technologies in the 20th century

Skeie and Abrahamsen (2017) noted that the production of brown whey cheese in its current form is complex and highly automated, with know-how mainly concentrated among Norwegian producers. However, it is worth recalling findings published during the rise of industrial production in an effort to increase process efficiency and explore new technological approaches. In addition to vacuum evaporation and specially designed kettles, which became available in the mid-20th century, improvements in lactose crystallization control, the quality effects of raw materials, and aspects of local equipment manufacturing have been documented (Bergum and Nilsen, 1950; Jelen, 1992). During the development of hard and especially spreadable Mysost-like products in the 1970 and 1980s, processes including reverse osmosis concentration were developed to replace vacuum evaporation, alongside attempts to modify formulations and lactose crystallization (Jelen, 1992). Research was conducted on adding dry materials (such as buttermilk) to liquid unprocessed whey to control TS, as well as optimizing mixing, heating, and homogenization to ensure the desired consistency (Guy et al., 1972; Patocka et al., 1990). Additionally, ultrafiltration was explored in Mysost spread technology (Abrahamsen, 1986). This is also related to a successful experimental approach utilizing ultrafiltration retentate from cottage cheese whey mixed with butterfat, resulting in a final product with a higher whey protein content resembling Ricotta-type whey cheeses (Roehl and Jelen, 1986).

3.5.2 Fortification with iron

As an interesting remark, brown whey cheese was historically made in black iron kettles, which, due to the acidic whey, released small amounts of iron into the cheese, making it a notable source of iron in the Norwegian diet. After switching to stainless steel kettles, this natural iron enrichment disappeared, leading producers to fortify the cheese with iron sulfate (10 mg Fe/100 g) near the end of cooking. However, later assessments showed that dietary iron intake was sufficient, and the fortification was largely discontinued in 2001, except for certain spreadable varieties (Abrahamsen et al., 2008).

3.5.3 Advances in the 21st century

Reflecting on the past 2 decades, a marked development has been the introduction of scraped surface heat exchangers in place of kettles, allowing a shift to continuous processes and ensuring a standardized product for both local markets and export. Data from Skeie and Abrahamsen (2017) over the past decade continues to confirm the specific consumption of brown whey cheese, with over 90% of it being consumed in Norway, while the remaining portion is exported to Germany, the USA, and Canada. Additionally, there was historically a Norwegian creamed whey cheese known as Japanprim, intended for the Japanese confectionery industry. Noteworthy production in Sweden is almost exclusively focused on spreadable Messmör (Jelen, 2011; Skeie and Abrahamsen, 2017). Nevertheless, it can be observed that brown whey cheese remains a regional Scandinavian food, and its expansion would require modifications to its sensory profile and efforts in the development and marketing of an innovative product that could be accepted by a broader range of consumers. This can be mainly understood in terms of strategies employed by large industrial producers and their ambitions in other markets.

4. Ice cream

4.1 Composition and microstructure of ice cream

4.1.1 Composition of ice cream

The composition of ice cream and frozen desserts varies significantly by type, such as dairy ice cream, nondairy ice cream, gelato, frozen yogurt, sherbet, and sorbet (Clarke and Cox, 2024). Except for sorbet and nondairy ice cream, these products typically contain a combination of major components, including water, sugar, MSNF (milk solids nonfat), fat, and other ingredients, that is, stabilizers, emulsifiers, flavors, and colors (Fig. 9.4), for example, Harfoush et al. (2024); Hartel et al. (2017).

These components have a unique function, and their ratio plays a crucial role in both the technological processing and the final ice cream products' quality parameters. Water (moisture) is the main part, while the total solid content generally ranges between 28% and 42%, depending on the ice cream type (Goff and Hartel, 2013; Mohammed et al., 2022).

Table 9.3 summarizes how the main components in ice cream formulations influence key physicochemical properties within defined concentration ranges. As the predominant component, water governs freezing behavior and overall phase stability. Sugars contribute by lowering the freezing point and modulating texture. MSNF provides proteins (casein and whey), which help in preventing ice crystal formation and contribute to texture refinement. Fat is crucial for sensory attributes, particularly by influencing mouthfeel and resisting melting. Critically, these components interact synergistically to form colloidal structures essential for microstructure development. Their precise balance enables the subsequent incorporation of air (overrun) and dictates the effectiveness of stabilizers/emulsifiers in maintaining product integrity during storage, for example, Clarke and Cox (2024); Goff et al. (2025); Hartel et al. (2017); and Chang et al. (2025).

The overrun (the incorporation of air during the freezing process) is an important factor in ice cream production, which significantly influences the texture, mouthfeel, and economic yield of the final products. Overrun varies widely, depending on the type of frozen dessert, typically ranging from 60% to 100% in dairy ice cream and below 50% in products like gelato and sherbet (Goff, 2022; Harfoush et al., 2024; VanWees et al., 2020).

Stabilizers and emulsifiers are fundamental components in ice cream formulation due to their impact on the product's texture, structure, and storage stability. Stabilizers, primarily polysaccharides, perform several essential functions, including water binding, viscosity enhancement, and ice and lactose crystallization inhibition. Among these, inhibition of ice recrystallization is considered the most crucial function, as it directly affects the smoothness and creaminess of ice cream during frozen storage (Goff and Hartel, 2013; Li et al., 2025). Stabilizers help prevent shrinkage, control meltdown behavior, and ensure uniform distribution of solid inclusions by reducing their settling tendency (Li et al., 2025).

Emulsifiers play a complementary role by stabilizing fat-water emulsions and promoting partial coalescence of fat globules. This process enhances the whipping ability of the mix, leading to increased overrun and creamier mouthfeel (Warren and Hartel, 2018; Wu et al., 2019). Emulsifiers also influence the microstructure of ice cream by interacting with milk proteins and affecting the air phase, thus contributing to overall product stability and meltdown resistance (Roy et al., 2021).

Emulsifier alternatives have also been explored, including natural proteins and lipids such as whey protein phospholipid concentrate, sodium caseinate, milk polar lipids, casein hydrolyzates, modified wheat gluten, and whipped aquafaba (Balivo et al., 2024; Levin et al., 2016; Park et al., 2024; Pei et al., 2025). These components replace synthetic emulsifiers and enhance emulsion stability and sensory attributes, improving melt resistance, hardness, and foam retention in low-fat or functional ice cream formulations.

Some studies have demonstrated the efficacy of stabilizer-emulsifier blends in improving physical properties. For example, a formulation containing iota-carrageenan and locust bean gum reduced the average ice crystal size after 3 months of frozen storage, highlighting its superior recrystallization inhibition performance (Kaminska-Dwórznicka et al., 2022).

There is a large variety of natural or artificial flavorants and colorants available. The most popular flavoring substances for ice cream are vanilla, chocolate, and fruits. Flavors can be extracted from natural substances or chemically produced (Goff and Hartel, 2013). Recent trends are toward less use of artificial additives, although clean label manufacturing also covers cost and functionality issues (Roobab et al., 2023).

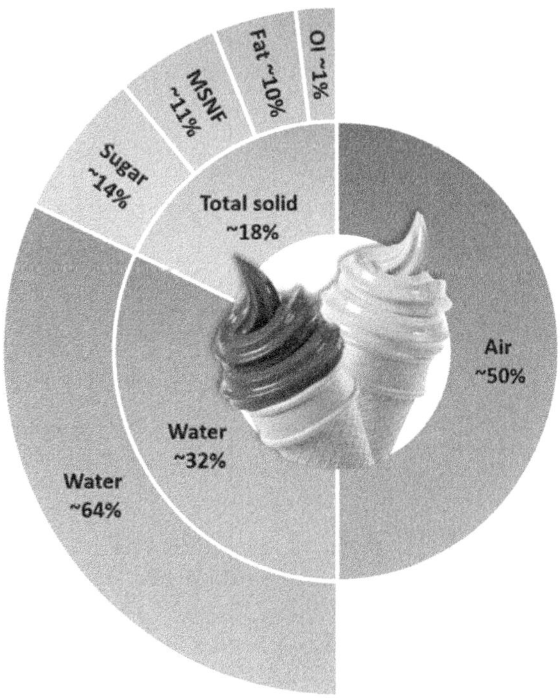

FIGURE 9.4 The overall composition of ice cream (inner circle) and ice cream mixture (outer circle); MSNF = milk solids nonfat; OI = other ingredients, that is, emulsifiers, stabilizers, flavors, and colors.

TABLE 9.3 Functional role of major ice cream (IC) ingredients.

Component	Range (% w/w)	Source	Functional role and consequences
Water	50–64	• Milk ingredients (milk, yogurt, cream, whey) • Fruit components and concentrates • High-quality drinking water	• Solvent: → Key solvent for ice cream ingredients (sugars, proteins, minerals) → Uniform distribution and proper functionality → ↓ Freezing point → Softer texture → Impact on overrun • Ice crystal formation: → Small crystals → Smooth texture → Affects consumer acceptance • Hydration of proteins/stabilizers: → ↑ Viscosity → ↑ Microstructure stability • Melting rate: → Prevents fast dripping → ↑ Scoopability
Sugar	12–16 (up to 20)	• Lactose (milk, yogurt, cream, whey) • Sucrose (sugar beet/cane) • Glucose (corn/wheat starch) • Fructose (fruits/concentrates) • Invert sugar • Honey	• ↑ Total solids: → ↑ Energy value → ↑ Sweetness → Better flavor/acceptability → ↓ Freezing point → Better texture and scoopability • Ice crystal size: → Slows ice crystal growth → Prevents large crystals → Smooth texture • Freezing point: → ↓ Freezing point → ↑ Overrun → ↑ Soft, scoopable consistency
MSNF (milk solids nonfat)	9–12 (optimal 11%)	• Skim milk (fresh, condensed, powder) • Whey powder • Whey protein isolate (WPI) • Buttermilk powder • Milk protein concentrates	• Nutrition: → ↑ Protein and minerals → ↑ Nutritional value • Emulsification: → Stabilizes emulsions → ↑ Overrun • Texture: → Binds water → Controls ice crystals → Prevents sandiness
Fat	10–15 (may vary 8–18)	• Milk fat (yogurt, milk, cream, butter, anhydrous milk fat) • Plant fats (coconut, cocoa, palm)	• Sensory quality: → Creamy mouthfeel → Carrier for flavor compounds • Melting rate: → Interferes with ice crystal formation → ↑ Melt resistance • Air structure: → Stabilizes bubbles → Smoother texture

This table outlines the typical concentration ranges, sources, and functional effects of the primary ingredients used in ice cream production.
Arrows indicate the direction of effect: ↑ = increase, ↓ = decrease, → = leads to/resulting in.
Sources: Information synthesized from Bahram-Parvar (2015); Clarke (2007); Daw and Hartel (2015); Goff and Hartel (2013); Kaminska-Dwórznicka et al. (2022); Kielczewska et al. (2025); Warren and Hartel (2018); Wu et al. (2019).

The composition of ice cream significantly influences its physicochemical, sensory, and rheological properties. In particular, the content of fat, proteins, sugars, and the use of stabilizers and emulsifiers directly impact texture, melting rate, viscosity, and fat phase destabilization.

Notably, fat content shows strong positive correlations with creaminess ($r = 0.80$) and viscosity ($r = 0.65$) but a strong negative correlation with the melting rate ($r = -0.75$), confirming its dual role in enhancing mouthfeel while delaying meltdown (Freire et al., 2020; Warren and Hartel, 2018). Proteins demonstrate positive correlations with both

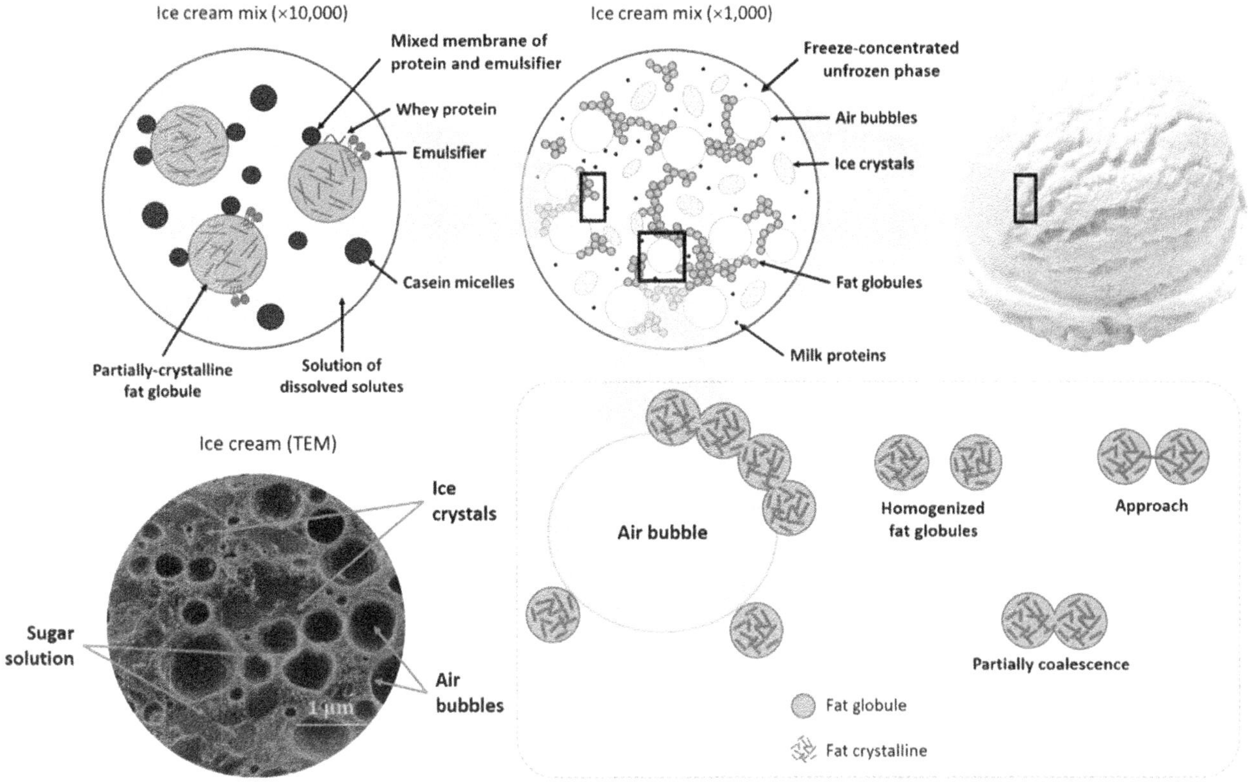

FIGURE 9.5 Microstructural organization of ice cream. *Source for TEM image: © Microfield Scientific LTD/Science Photo Library.*

texture (r = 0.78) and accelerated melting ($r = 0.60$), indicating competitive functionalities in structural support versus thermal stability (Wu and Hartel, 2024; Yan et al., 2021).

Sucrose moderately correlates with foam stability ($r = 0.50$), which is critical for structural integrity. Some stabilizers (e.g., guar gum, carrageenan) exhibit negative correlations with the melting rate ($r = -0.70$) and ice crystal size ($r = -0.60$), directly linking them to structural stabilization through delayed melting and crystal inhibition (Wu and Hartel, 2024). Emulsifiers correlate strongly with fat destabilization ($r = 0.72$), a microstructural process essential for texture development. Finally, overrun (air incorporation) positively correlates with combined texture-melting properties ($r = 0.68$), highlighting its role in moderating heat transfer and sensory perception (Warren and Hartel, 2018).

These empirically established correlations provide a particular assumption for designing novel ice cream formulations, particularly reduced-fat or functionally optimized variants, while preserving desired sensory and structural profiles.

4.1.2 Microstructure of ice cream

Ice cream is a complex dispersion system (Fig. 9.5), often described as a frozen foam (VanWees et al., 2022), comprising multiple phases: gas (air cells/bubbles), solid (ice crystals), fat globules, and a liquid aqueous phase that contains dissolved sugars, proteins, and stabilizers (Clarke and Cox, 2024; Goff and Hartel, 2013; Hartel et al., 2017). The air is represented in an extensive range, from 10% to 60% by volume in most frozen desserts (VanWees et al., 2022).

Air bubbles are a critical component of ice cream, significantly influencing its structure, texture, and sensory acceptance. Their formation depends on the mix composition (fat, proteins, surface-active agents) and processing conditions such as freezer shear, hardening rate, and storage temperature (Pei et al., 2025). The air bubbles are finely dispersed, typically averaging 20–25 μm in diameter, and contribute to the light, smooth texture (Goff and Hartel, 2013). Ice crystals average 35–45 μm in size. According to the protein sources, they may differ from 22 to 41 μm (Daw and Hartel, 2015). About 50% of water freezes during freezing, increasing to 75%–80% after hardening, depending on storage temperature. Fat globules range from 0.5 to 3 μm, averaging ~0.8 μm. During freezing, partial coalescence forms fat globule clusters, typically 5–80 μm, with some exceeding 100 μm. Fat destabilization affects melt resistance and shape maintenance (Daw and Hartel, 2015; Goff and Hartel, 2013).

This figure illustrates the multi-scale structure of ice cream, highlighting the role of whey proteins and fat globules in forming the final frozen matrix. In the unfrozen mix (×10,000), whey proteins and emulsifiers adsorb onto partially crystalline fat globules alongside casein micelles. Upon freezing (×1000), the system develops a multiphase structure composed of ice crystals, air bubbles, fat globules, and a freeze-concentrated serum. Fat globules migrate to the air–water interface during aeration, where partial coalescence creates a stabilizing 3D fat network that enhances meltdown resistance and creaminess. A TEM micrograph (bottom left) shows the distribution of air, sugar solution, and ice crystals within the frozen matrix.

4.2 Whey in ice cream

Common ice cream formulations contain a significant proportion of proteins, primarily derived from dairy ingredients. Within the MSNF component, the ratio of casein to whey proteins varies depending on the type of dairy source used (e. g., whole milk, skim milk), as well as on the specific type of ice cream (Goff and Hartel, 2013).

Whey, a by-product of the dairy industry, is recognized for its high nutritional value, particularly due to its content of high-quality proteins, essential vitamins, and minerals, for example, Paladii et al. (2021). Moreover, with growing interest in sustainable food production, researchers are increasingly exploring the use of food industry by-products. In this context, whey is gaining attention not only as a potential solution to global protein deficiency but also for its beneficial effects on human health, for example, Genovese et al. (2022), Tvorogova et al. (2023).

4.2.1 Role of whey in ice cream

When processing whey, several derivatives are received (Paladii et al., 2021). Whey powder from acid or sweet whey, demineralized whey (25%, 50%, and 90%), and delactosed whey, furthermore WPC with a protein content of 34%, 50%, 60%, 75%, and 80%, WPI, whey protein hydrolyzates (WPH), and protein–mineral concentrates.

In addition to the whey from acid-coagulated cheeses and sweet whey from rennet-coagulated cheeses, Pires et al. (2021) reported the so-called second cheese whey obtained during the production of whey cheeses, such as Ricotta.

Whey and its derivatives are widely used in producing ice cream and other frozen desserts. Among these, whey powder, WPC (34%–80% of protein), and WPI (>90% protein) are the most commonly employed. These ingredients are primarily added to enhance functional properties such as emulsification, water-binding capacity, foam formation, and viscosity (Paladii et al., 2021). According to Kaminska-Dwórznicka et al. (2022), including whey in frozen dairy products reduces overrun. On the other hand, the whey incorporated ice cream was successfully prepared by Sindhu et al. (2018). The ice cream was prepared by substituting milk with whey in various proportions. Whey incorporated ice cream was formulated by substituting milk with whey in the ratio (milk: whey) of 3:1, 1:1, and 1:3, respectively. They have concluded that when whey is incorporated up to 75% in the ice cream, replacing milk could enhance the protein content of the ice cream without affecting sensory qualities, and 80% over run was also obtained.

Compared to conventional sources of MSNF, whey solids are more cost-effective. Conversely, a high amount of lactose in whey can cause freezing point depression. Moreover, the amount of whey added to ice cream is limited in some countries (U.S.) (Goff and Hartel, 2013).

The selection of concrete whey-based ingredients significantly impacts the functional characteristics of the final ice cream. Table 9.4 compares the main compositional characteristics (protein and lactose), functional roles, and implications of commonly used whey-based ingredients in ice cream. The data show that whey derivatives fulfill distinct functional roles, driven by their varying protein, lactose, and mineral contents.

To achieve optimal characteristics of final products, technologists must carefully match the functional capabilities of specific whey ingredients, such as viscosity enhancement, melt resistance, or flavor neutrality, to the desired product characteristics. At the same time, it is important to avoid potential drawbacks, including lactose crystallization, off-flavors, or high mineral content.

4.2.2 Effect of whey ingredient replacement

The use of whey and whey-based ingredients in ice cream production has also become an increasing trend due to their nutritional, technological, and economic advantages (Fig. 9.6). Whey proteins, such WPC or WPI, are often added to ice cream mixes for their high biological value, the ability to form stable foams, and the ability to improve texture and creaminess. Recent research also focuses on the functional use of whey, for example, as a carrier for bioactive compounds, probiotics, or antioxidants. Due to its emulsifying and gelling properties, whey can partially or fully replace other dairy ingredients, helping to reduce production costs. At the same time, whey represents a valuable raw material and an

TABLE 9.4 Comparative properties of whey-based ingredients in ice cream.

Component	Protein (%)	Lactose (%)	Functional properties in ice cream	Formulation implications
Sweet whey	~11	~70	↑ Milk solids nonfat → ↑ Mineral content ↑ Flavor base Risk: Lactose crystallization → Sandiness texture	Monitor crystallization risk during storage
Acid whey	~10	~65	↓ pH → ↑ Tartness ↑ Mineral content Risk: ↑ Syneresis, ↑Off-flavors (salty/metallic)	Requires flavor masking; monitor acidity
WPC-35	~35	~50	↑ Viscosity → ↑ Texture firmness ↑ Foam stability → Creamy mouthfeel	Cost-effective protein enrichment
WPC-80	~80	5–10	↑ Protein → ↑ Melt resistance ↑ Texture firmness ↓ Lactose content	Reduces lactose-related issues
WPI	≥90	<1	↑ Structure → ↑ Thermal stability ↑ Shear resistance → Neutral flavor	Lactose-free; premium cost
Demineralized whey	Varies	High	↓ Ash content → ↑ Flavor clarity → Delicate flavor systems	Reduces mineral bitterness
WPH	Varies	Varies	↑ Digestibility ↑ Bioactive peptides ↓ Allergenicity Risk: ↑ Bitterness	Requires bitterness masking agents

Note: ↑ = Increase; ↓ = Decrease; → = leads to/resulting in.
Abbreviations: *WPC,* whey protein concentrate (% of proteins); *WPI,* whey protein isolate; *WPH,* whey protein hydrolyzate.
Sources: Information synthesized from Gurskiy and Tvorogova (2022); Mykhalevych et al. (2024b); Saentaweesuk and Chaikham (2023); Tekin et al. (2017).

opportunity for waste minimization in the dairy industry. Its utilization allows manufacturers to improve raw material efficiency and support sustainable production principles.

Some studies investigating the substitution of traditional ice cream components with whey-based ingredients reveal significant impacts on the physicochemical, functional, and sensory properties of the final product (Table 9.5). The summary was focused on the specific whey ingredient used, the level of replacement, the observed effects on crucial ice cream properties (e.g., viscosity, overrun, melting rate, texture, color, and sensory attributes), and the product relevance of the substitution. The aim was to follow the options and challenges associated with whey ingredient utilization in frozen dessert formulations.

Measurement conditions: All ice cream formulations were tested under standard storage conditions (−18°C). Sensory data reported where available.

Whey ingredients, such as proteins, phospholipid concentrates, hydrolyzed whey fractions, permeate, and even enzyme-modified whey, offer diverse functionalities as partial or complete replacements for ice cream components, like MSNF, skim milk powder, whole milk, sugar, and synthetic emulsifiers.

The replacement has beneficial effects on ice cream properties, such as increased viscosity and melting rate, along with frequent reductions in overrun and fat destabilization, depending on the whey type and replacement level. While substitutions like hydrolyzed whey for sugar successfully reduce lactose content and calories while enhancing sweetness, others, such as high levels of WPC (80) replacing skim milk powder, can introduce undesirable sensory attributes (sourness and brittleness) despite improving hardness. Additionally, the lactose content in standard whey ingredients may limit their suitability for lactose-intolerant consumers unless prior lactose hydrolysis is performed.

The data underscores the importance of selecting the appropriate whey ingredient and optimizing its inclusion level for the target application (e.g., protein fortification, lactose reduction, clean-label emulsification, or calorie control). Successful implementation depends on balancing technological improvements (e.g., reduced ice crystal size, enhanced

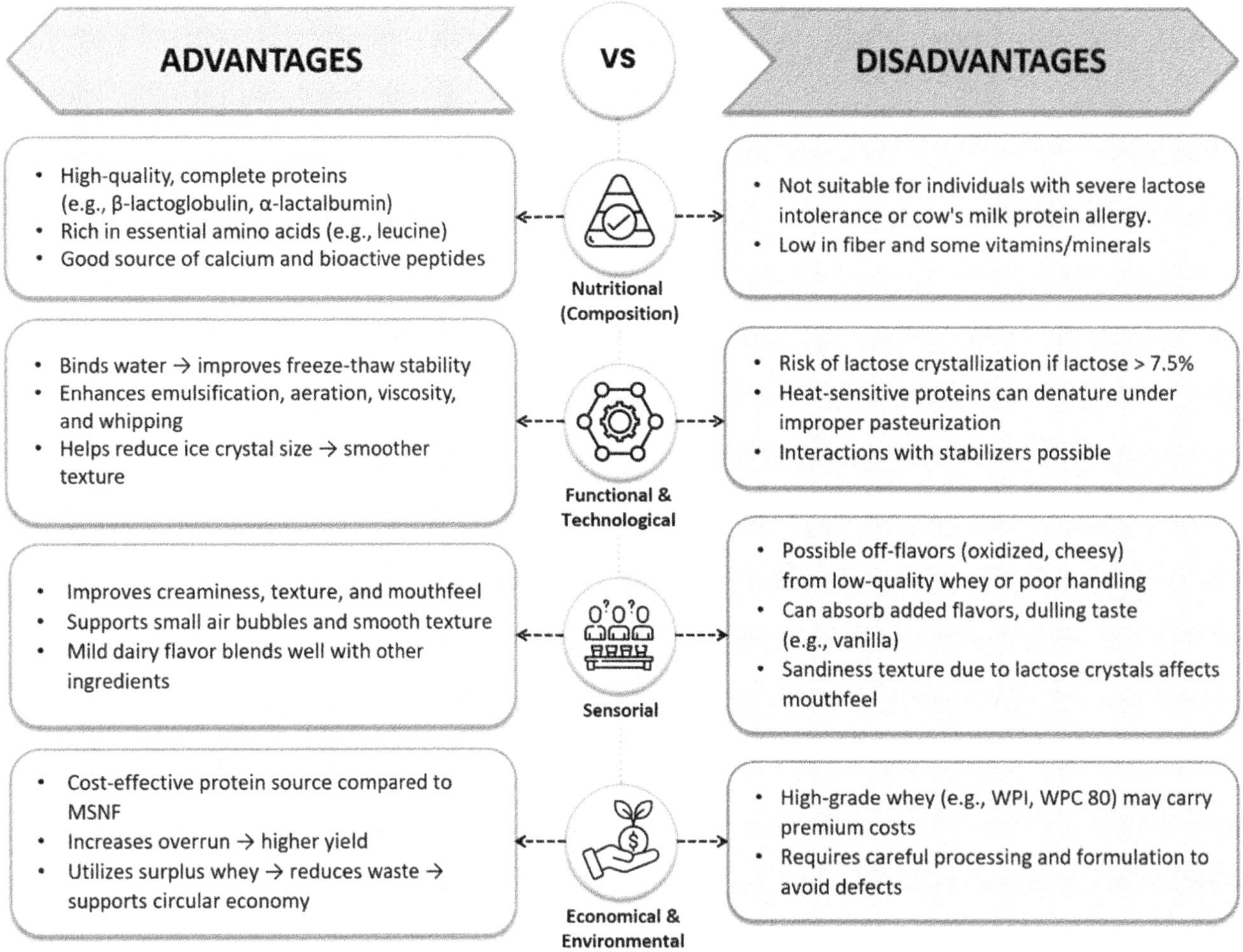

FIGURE 9.6 Advantages and disadvantages of using whey and whey-based products in ice cream. Abbreviations: *MSNF*, milk solids nonfat; *WPI*, whey protein isolate; *WPC 80*, whey protein concentrate (80% protein). *Source: Information synthesized from Paladii et al. (2021); Hartel et al. (2017); Zhao et al. (2023); Gharibzahedi et al. (2018); Genovese et al. (2022).*

creaminess, and specific functional benefits) against potential texture, melt resistance, color, and sensory acceptance drawbacks. Optimal outcomes, such as specific whey protein-to-casein ratios or hydrolyzed whey concentrations, require careful formulation to achieve the desired product characteristics without compromising overall quality.

A coproduct from the production of whey protein isolate and whey protein concentrate is a whey permeate powder, which is one of the most cost-competitive sources of milk solids in the dairy food industry. Due to high lactose content (~80%-82%), whey permeate powder has limitation to use in ice cream, and high lactose content is also responsible for a sandy (gritty) texture in ice cream. Recently, Schweiger et al. (2023) have produced ice cream from whey permeate powder. They have formulated three ice cream mixes containing equivalent total solids, fat, and sugar with varying lactose and added sugar levels. In their formulation, WPC 80 varied from 0.6% to 1.3% (w/w), and whey permeate powder, which varied from 1.7% to 3.3% (w/w) were used. They found that whey permeate powder can be used to produce ice creams of acceptable quality for up to 10 months.

Milk fat improves the body and texture of ice cream. Milk fat contributes textural creaminess to the ice cream, which is one of the desirable sensory properties. The low-fat ice cream has a weak body or a gummy texture, which is not liked by consumers. The carbohydrates or protein-based fat replacers have been tried to produce low-fat diet ice cream. Whey protein is also one of the fat replacers that has similar functionality of fat globules (Ohmes et al., 1998; Lim et al., 2008). Lim et al. (2008) have applied high hydrostatic pressure (HPP) to modify whey protein concentrate for improving body and texture of low-fat ice cream. They used WPC 35 and Washington State University (WSU)-WPC in the experiment. Three batches of low-fat ice cream mix were prepared from WPC 35 without HHP, WSU-WPC with HHP, WSU-WPC without HHP. The 10% WPC 35 or WSU-WPC in all the ice cream mix was used. They observed that high hydrostatic

TABLE 9.5 Summary of research findings on whey substitution and ice cream properties.

Replaced component	Whey ingredient	Replacement level/ratio	Observed effects	Product relevance	References
Nonfat dry milk (NFDM)	MPC, Procream, WPC (blend)	4%–10% total protein (replacing NFDM protein)	↑ Viscosity ↑ Shear-thinning behavior ↑ Melting rate ↓ Fat destabilization	Suitable for protein-enriched bases	Daw and Hartel (2015)
Emulsifiers (synthetic)	WPPC	1%–2% (of mix)	↑ Melting rate ↑ Fat coalescence Slight ↓ Overrun	Clean-label emulsifier substitute	Levin et al. (2016)
Milk powder (MP)	Whey powder + β-galactosidase	0%–100% MP + .5 g/L enzyme (hydrolyzed for 4 h)	↑ Overrun ↔ Lactose hydrolysis (86.6%–97.9%), no negative quality effect	Lactose-free or reduced-lactose options	Tsuchiya et al. (2017)
Whole milk	RW (Ricotta), CW (Cheddar), BUW (Butter)	0%, 25%, 50%, 75%, and100% of milk solids	↑ Melting rate ↑ Titratable acidity ↓ Lightness (L*) ↑ Redness (a*) ↑ Yellowness (b*) ↓ pH ↓ Lipid content BUW: ↑ Hardness RW: ↓ Melting rate	Natural flavor profiles and nutrient recovery	Meneses et al. (2020)
Sugar	Hydrolyzed whey (enzymatic + *L. acidophilus*)	10%–40% solids; 0.1%–0.4% enzyme, 4–8 h hydrolysis	↓ Sugar content ↑ Sweetness ↓ Lactose Improved texture and nutrition	Functional, low-lactose sweeteners	Osmak et al. (2021)
	Hydrolyzed demineralized whey concentrate	0%–30% sugar replacement	↑ Overrun ↑ Melting resistance ↓ Viscosity Sweetness retained	Calorie-reduced, diabetic-friendly desserts	Shevchenko et al. (2022)
Pasteurized milk	Freeze-concentrated whey	0%, 25%, 50%, 75%, and 100% of milk	↑ Total solids ↑ Acidity ↑ Viscosity ↑ Smoothness ↓ Melting rate	Creamy, gelato-like products	Barros et al. (2022)
MSNF	Native WPI	5% protein (of mix)	↑ Air bubble size ↑ Adhesiveness ↓ Ice crystal size	Superior aeration and creaminess; higher cost	Tvorogova et al. (2023)

Continued

TABLE 9.5 Summary of research findings on whey substitution and ice cream properties.—cont'd

Replaced component	Whey ingredient	Replacement level/ratio	Observed effects	Product relevance	References
	WPC (variable casein ratio)	Whey proteins:casein = 13:87 → 54:46	↑ Viscosity ↑ Hardness ↓ Overrun ↓ Ice crystal size	Balanced texture formulation at whey proteins:casein ~20:80	Alvi et al. (2025)
Whey base	WPP	WPC:WPP = 1.3:0.0, 0.6:1.7, 0.0:3.3	✓ Best sensory at 0.6:1.7 ↑ Sandy texture at 10 months (only in 0.0:3.3)	For frozen dairy with <6 months shelf-life	Schweiger et al. (2023)
	β-glucans + vegetable purées	0.75% oat β-glucan, 0.5% yeast β-glucan, 3%–4.5% protein complex, 10% purée	↑ Viscosity ↑ Stability (beet + Mn-enriched whey) ↓ Quality with broccoli/zucchini	Functional and fiber-rich low-fat frozen desserts	Tomczynska-Mleko et al. (2024)
Skim milk powder	WPC 80%	10% of MSNF (iso-nutrient)	↑ Hardness (276 N) ↑ Melting rate (0.24 g/min) ↑ Yellowness (b*) ↓ Overrun (52%) ↓ Lightness (L*) ↓ Sensory fluffiness Sensory notes: sourness, brittleness	High-protein frozen desserts; requires flavor masking	Kielczewska et al. (2025)

Note: ↑ = Increase; ↓ = Decrease; ↔ = No significant change; ✓ = Optimal outcome. All effects are statistically significant at $P < .05$ unless stated otherwise. *WPC 80*, whey protein concentrate (80% protein); *MPC*, milk protein concentrate; *WPI*, whey protein isolate; *WPPC*, whey protein phospholipid concentrate; *WPP*, whey permeate powder; *MSNF*, milk solids nonfat; *RW*, ricotta whey; *CW*, cheese whey; *BUW*, butter whey; L*, a*, b* = CIE color space parameters.

pressure treatment of whey proteins exhibited a positive effect on low-fat ice cream mix overrun and foam stability, as well as the hardness of low-fat ice cream than without treated HPP ice cream mix.

4.3 Processing of ice cream

The production of ice cream, whether formulated with traditional milk ingredients or enriched with whey proteins, follows a series of technological steps. The manufacturing process begins with the preparation and mixing of all ingredients, followed by pasteurization, homogenization, aging, freezing, packaging, and hardening (Fig. 9.7) (Goff and Hartel, 2013).

The initial stage of ice cream production involves formulating and blending dry and liquid ingredients to create a stable emulsion. Recent innovations aim to improve nutritional quality, reduce fat content, and support clean-label products. Notably, multiple emulsions (water-in-oil-in-water) have gained attention for reducing fat content while maintaining the creamy mouthfeel and texture typical of traditional ice cream (Klojdová and Stathopoulos, 2022). These structures also enable the encapsulation of sensitive bioactive compounds, such as vitamins or flavors, improving their stability and controlled release (Soutelino et al., 2024). The application of natural food ingredients, particularly dietary fiber, contributes to the structural stability of these emulsions and enhances their nutritional value (Klojdová and Stathopoulos, 2022).

Similarly, nanoemulsion and microencapsulation technologies have been explored to address formulation challenges, such as sensitive components' stability and uniform distribution (Soutelino et al., 2024). Hossain et al. (2021) have studied the influence of extruded microparticulated whey proteins (eMWPs) and process design in light ice cream processing. They have used inulin, eMWPs, and different combinations of inulin and eMWPs as fat replacements. The two different process designs were employed, namely (i) pasteurization with subsequent homogenization (PH); (ii) homogenization with subsequent pasteurization (HP) for the production of ice cream. The higher overrun of light ice cream treatments of PH was significantly higher. A similar firmness in both the PH and HP groups was observed; however, the products with

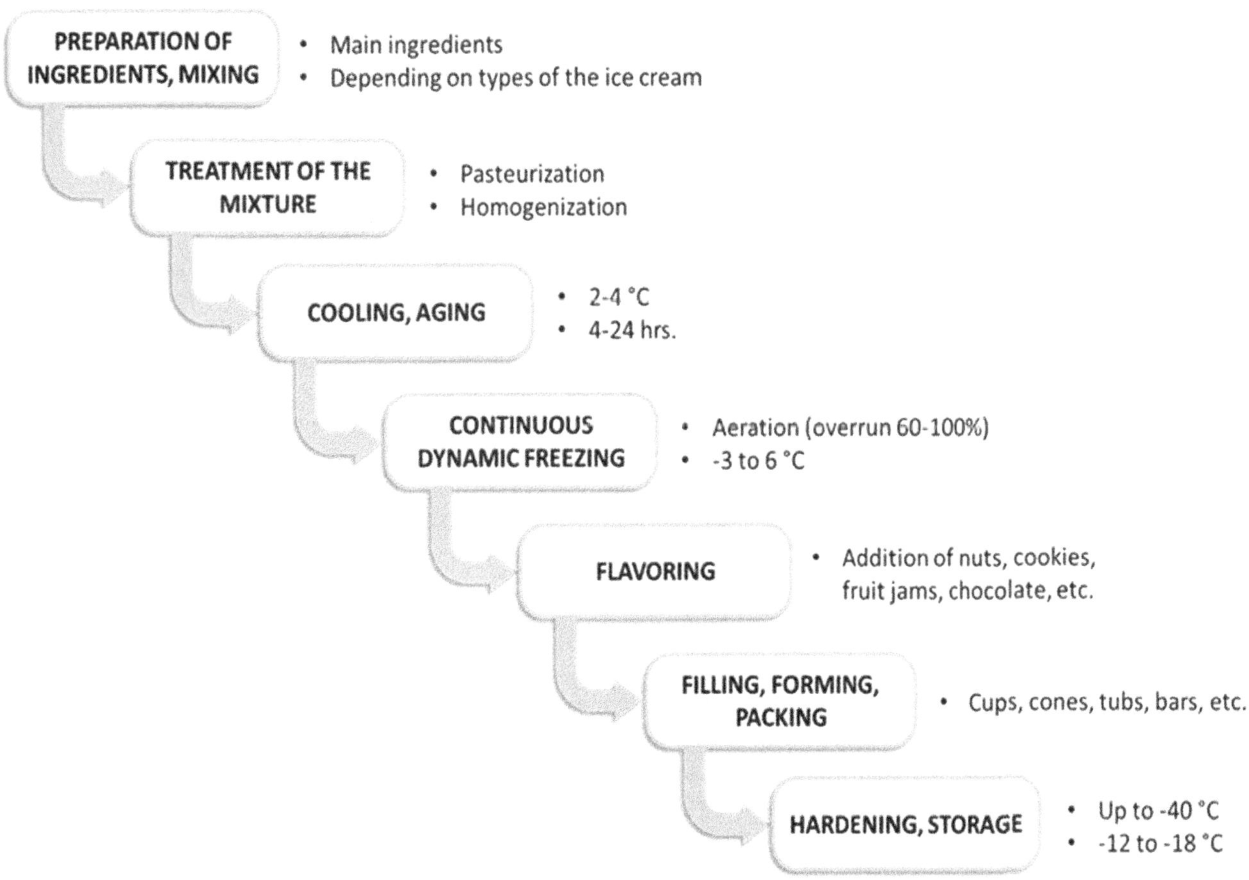

FIGURE 9.7 Schematic flow chart of ice cream processing.

extruded microparticulated whey proteins were firmer compared to other light ice creams. Microbial transglutaminase can be utilized to form new food structures and matrices with high physicochemical stability. This enzyme offers a successful strategy to improve the nutritional and technological characteristics of ice cream through intramolecular cross-linking and reduce production cost by decreasing fat and stabilizer contents (Gharibzahedi et al., 2018).

The ice cream mixture undergoes pasteurization, typically using a plate heat exchanger at 83–85°C for approximately 15 s, followed by homogenization at 73–75°C. Heat treatment ensures food safety by eliminating spoilage and pathogenic microorganisms (Harfoush et al., 2024; Mohammed et al., 2022). Homogenization is carried out at pressures ranging from 14 to 20 MPa to ensure uniform fat dispersion, contributing to a smoother texture and preventing fat separation during storage (Mohammed et al., 2022). Double-stage homogenization at elevated pressures is recommended in some formulations, particularly those involving whey or plant-based fats, to minimize ice crystal formation further and enhance product consistency (Goff et al., 2025). Punsandani et al. (2005) have worked on the physical behavior of dairy ingredients during ice cream processing. They have used low-heat skim milk powder (LHSMP), WPC 35, unheated and heated blends (1:1 ratio) of LHSMP and WPC 35 in ice cream processing. They observed that different proportions of caseins to whey proteins, as well as increased whey protein denaturation, changed the ice cream properties. The use of a high heat-treated blend of LHSMP and WPC 35 resulted in ice creams with the best melt resistance. They concluded that the melting properties of ice cream can be manipulated by the type of MSNF used and the degree of whey protein denaturation.

In addition to conventional thermal processing, nonthermal technologies have acquired attention for their potential to ensure microbial safety. High-pressure-based technologies, such as high-pressure processing, high-pressure homogenization, ultrahigh-pressure homogenization, and high-pressure jet processing, maintain nutritional value without negative impact on flavor or nutrient content (Roobab et al., 2023; Soukoulis and Fisk, 2016).

To ensure microbial safety, any components added to the ice cream mixture after the freezing stage, such as flavorings, colorants, fruits, and nuts, must be pasteurized or otherwise treated against microbial contamination (Goff and Hartel, 2013; Mohammed et al., 2022). The storage life of the final product is influenced by multiple factors, including the type and quality of ice cream, barrier properties of the packaging, and consistently low storage temperature. Under optimal conditions, the shelf-life can range from several weeks to 9 months (Mohammed et al., 2022).

The ice cream mix must be aged for at least 4 hours at 2–5°C with continuous gentle agitation. These steps allow the milk proteins and water to interact and the liquid fat to crystallize, resulting in better air incorporation and improved melting resistance (Mohammed et al., 2022). The most important process in ice cream technology is freezing because it affects the quality characteristics, that is, sensory and textural properties, and efficiency of ice cream production. Some studies have shown that high-intensity ultrasound application in freezing and crystallization processes has promising potential. This technology has been shown to reduce ice crystal size, shorten freezing time, and prevent incrustation on the surface of freezing equipment, leading to improved texture and processing efficiency (Akdeniz and Akalin, 2019). Vetorazi et al. (2024) reported that high-intensity ultrasound, applied at 20 kHz and 500 W for 15 min, has improved ice cream quality without compromising color stability.

One of the major shelf-life issues of ice cream is ice recrystallization, where ice crystals grow larger over time based on thermodynamic ripening principles (Hartel et al., 2017). Incorporating whey-based ingredients into ice cream can expand ice crystal growth and negatively influence textural stability (Li et al., 2023). This is primarily due to the functional properties of whey proteins, which can affect the water-holding capacity and the overall structure of the ice cream matrix (Pehlivanoğlu et al., 2024). Therefore, optimizing protein concentration, incorporating cryoprotectants, and utilizing advanced packaging techniques are used to ensure a comparable shelf-life to conventional ice cream, minimize ice recrystallization, preserve emulsion stability, and prevent undesirable textural changes (Maity et al., 2018; Mykhalevych et al., 2024a).

5. Conclusion

Whey represents a highly versatile raw material with valuable applications across different dairy and frozen products. Ricotta cheese, a traditional whey protein product obtained through acid-heat coagulation, exemplifies the effective upcycling of sweet whey. Its nutritional composition, digestibility, and mild sensory profile make it suitable for both culinary use and functional food development. Technological advances such as optimized acidification, enrichment with milk or cream, and the application of natural antimicrobials or probiotic cultures have further expanded its potential in the context of clean-label innovation. Brown whey cheese, also known as Mysost or Brunost, utilizes the complete whey matrix and offers a distinctly different sensory and nutritional profile. Its manufacture involves extensive evaporation, lactose crystallization, and controlled Maillard browning, yielding a dense, sweet, caramel-colored product with long shelf life. Although its consumption remains mainly regional, modern processing technologies have ensured consistent quality, and reformulated variants could offer new commercial opportunities beyond Scandinavian markets.

In ice cream production, whey and its derivatives, such as WPC and WPI serve as multifunctional ingredients. They improve emulsion stability, viscosity, melting resistance, and aeration, while also contributing to the protein content and allowing for the reduction of milk solids or emulsifiers. These properties support the formulation of high-protein, lactose-reduced, or clean-label frozen desserts in line with current consumer trends.

The studied product categories demonstrate the functional and technological value of whey as more than a by-product. Its targeted use in cheese and frozen applications contributes to nutritional enhancement, process efficiency, and sustainability, and opens new pathways for innovation in modern food systems.

References

Abrahamsen, R.K., 1986. Production of brown whey cheese. Bull. Int. Dairy Feder. 202, 125–130.

Abrahamsen, R.K., Bakkene, G., Gjerde, R., Steinsholt, K., 2008. Brunosten—en norsk historie [Brown whey cheese—a Norwegian story]. TUN Forlag, Oslo.

Ådnøy, T., 2014. The dairy goat industry in Norway: challenges in a historical perspective. Small Rumin. Res. 122, 4–9. https://doi.org/10.1016/j.smallrumres.2014.07.011.

Akdeniz, V., Akalin, A.S., 2019. New approach for yoghurt and ice cream production: high-intensity ultrasound. Trends Food Sci. Technol. 86, 392–398. https://doi.org/10.1016/j.tifs.2019.02.046.

Akpinar, A., Yerlikaya, O., Akan, E., Karagozlu, C., Kinik, O., Uysal, H.R., 2022. The effect of packaging materials on physicochemical, microbiological, and sensorial properties of Turkish whey (Lor) cheese with some plants. J. Food Process. Preserv. 46 (11). https://doi.org/10.1111/jfpp.17060.

Alvi, M.A.U.R., Agbawodike, J.I., Soler-Sánchez, A.G., Martínez-Monteagudo, S.I., 2025. Effect of altering the whey protein/casein ratio on quality properties of ice cream. Int. J. Food Prop. 28 (1), 2459390. https://doi.org/10.1080/10942912.2025.2459390.

Anema, S.G., 2006. Kinetics of the irreversible thermal denaturation and disulfide aggregation of α-lactalbumin in milk samples of various concentrations. J. Food Sci. 66 (1), 2–9. https://doi.org/10.1111/j.1365-2621.2001.tb15573.x.

Anema, S.G., 2020. Chapter 9—the whey proteins in milk: thermal denaturation, physical interactions, and effects on the functional properties of milk. In: Milk Proteins. Elsevier, pp. 325–384. https://doi.org/10.1016/B978-0-12-815251-5.00009-8.

Araque, L.C.O., Darré, M., Ortiz, C.M., Massolo, J.F., Vicente, A.R., 2018. Quality and yield of Ricotta cheese as affected by milk fat content and coagulant type. Int. J. Dairy Technol. 71 (2), 340–346. https://doi.org/10.1111/1471-0307.12431.

Bahram-Parvar, M., 2015. A review of modern instrumental techniques for measurements of ice cream characteristics. Food Chem. 188, 625–631. https://doi.org/10.1016/j.foodchem.2015.05.017.

Balivo, A., d'Errico, G., Genovese, A., 2024. Whipped chickpea aquafaba as a fat replacer in ice cream: effect on sensory and physicochemical properties. J. Food Sci. 89 (12), 8730–8745. https://doi.org/10.1111/1750-3841.17605.

Barros, E.L.D., Silva, C.C., Canella, M.H.M., Verruck, S., Prestes, A.A., Vargas, M.O., Maran, B.M., Esmerino, E.A., Silva, R., Balthazar, C.F., Calado, V.M.D., Prudencio, E.S., 2022. Effect of replacement of milk by block freeze concentrated whey in physicochemical and rheological properties of ice cream. Food Sci. Technol. 42, e12521. https://doi.org/10.1590/fst.12521.

Bergamaschi, M., Cipolat-Gotet, C., Stocco, G., Valorz, C., Bazzoli, I., Sturaro, E., Ramanzin, M., Bittante, G., 2016. Cheesemaking in highland pastures: milk technological properties, cream, cheese and ricotta yields, milk nutrients recovery, and products composition. J. Dairy Sci. 99 (12), 9631–9646. https://doi.org/10.3168/jds.2016-11199.

Bergum, A.T., Nilsen, A., 1950. Experiment on the Evaporation of Milk and Whey under Vacuum, vol 43.

Besbes, S., Blecker, C., Attia, H., Massaux, C., 2002. Comparison of Ricotta cheese made by high pressure treatment with that produced by heat treatment of sweet whey. Sci. Aliment. 22, 601–615. https://doi.org/10.3166/sda.22.601-615.

Bintsis, T., Papademas, P., 2023. Sustainable approaches in whey cheese production: a review. Dairy 4 (2), 249–270. https://doi.org/10.3390/dairy4020018.

Bonanno, A., Tornambè, G., Bellina, V., Mazza, F., Maniaci, G., Di Grigoli, A., 2013. Effect of farming system and cheesemaking technology on the physicochemical characteristics, fatty acid profile, and sensory properties of Caciocavallo Palermitano cheese. J. Dairy Sci. 96 (1), 710–724. https://doi.org/10.3168/jds.2012-5973.

Carrero-Puentes, S., Fuenmayor, C., Jiménez-Pérez, C., Guzmán-Rodríguez, F., Gómez-Ruiz, L., Rodríguez-Serrano, G., Alatorre-Santamaría, S., García-Garibay, M., Cruz-Guerrero, A., 2022. Development and characterization of an exopolysaccharide-functionalized acid whey cheese (requesón) using Lactobacillus delbrueckii ssp. bulgaricus. J. Food Process. Preserv. 46 (6). https://doi.org/10.1111/jfpp.16095.

Chandan, R.C., 2003. Cheeses | soft and special varieties. In: Caballero, B. (Ed.), Encyclopedia of Food Sciences and Nutrition. Academic Press, pp. 1093–1098.

Chang, W.J., Li, K.Y., Qi, X.R., Meng, Z., 2025. Formulation strategies, texture improvement, and sensory perception of healthy ice cream: a review. Food Chem. 481, 144015. https://doi.org/10.1016/j.foodchem.2025.144015.

Christaki, S., Moschakis, T., Hatzikamari, M., Mourtzinos, I., 2022. Nanoemulsions of oregano essential oil and green extracts: characterization and application in whey cheese. Food Control 141, 109190. https://doi.org/10.1016/j.foodcont.2022.109190.

Clarke, C., 2007. The Science of Ice Cream. https://doi.org/10.1039/9781847552150.

Clarke, C., Cox, A., 2024. The Science of Ice Cream. Royal Society of Chemistry. https://doi.org/10.1039/9781837673032.

Dagnachew, B.S., Thaller, G., Lien, S., Ådnøy, T., 2011. Casein SNP in Norwegian goats: additive and dominance effects on milk composition and quality. Genet. Sel. Evol. 43, 31. https://doi.org/10.1186/1297-9686-43-31.

Daw, E., Hartel, R.W., 2015. Fat destabilization and melt-down of ice creams with increased protein content. Int. Dairy J. 43, 33–41. https://doi.org/10.1016/j.idairyj.2014.12.001.

de Wit, J.N., 2009. Thermal behaviour of bovine β-lactoglobulin at temperatures up to 150°C. a review. Trends Food Sci. Technol. 20 (1), 27–34. https://doi.org/10.1016/j.tifs.2008.09.012.

Del Nobile, M.A., Conte, A., Incoronato, A.L., Panza, O., 2009. Modified atmosphere packaging to improve the microbial stability of Ricotta. Afr. J. Microbiol. Res. 3 (4), 137–142.

Dimidi, E., Cox, S., Rossi, M., Whelan, K., 2019. Fermented foods: definitions and characteristics, impact on the gut microbiota and effects on gastrointestinal health and disease. Nutrients 11 (8), 1806. https://doi.org/10.3390/nu11081806.

Donato, L., Guyomarc'h, F., 2009. Formation and properties of the whey protein/κ-casein complexes in heated skim milk—a review. Dairy Sci. Technol. 89 (1), 3–29. https://doi.org/10.1051/dst:2008033.

Duarte, R.V., Moreira, S.A., Fernandes, P.A.R., Santos, D.I., Inácio, R.S., Alves, S.P., Bessa, R.J.B., Saraiva, J.A., 2017. Whey cheese longer shelf-life achievement at variable uncontrolled room temperature and comparison to refrigeration. J. Food Process. Preserv. 41 (6), e13307. https://doi.org/10.1111/jfpp.13307.

Edwards, P.J.B., Jameson, G.B., 2020. Chapter 7—structure and stability of whey proteins. In: Milk Proteins. Elsevier, pp. 251–291. https://doi.org/10.1016/B978-0-12-815251-5.00007-4.

El-Sheikh, M., Farrag, A., Zaghloul, A., 2010. Ricotta cheese from whey protein concentrate. Journal of American Science 6 (8), 321–325.

Eurostat, 2025. EU Ice Cream Production Fell by 1.4% in 2023. Retrieved 30. 6. from. https://ec.europa.eu/eurostat/web/products-eurostat-news/w/ddn-20240807-1.

Farkye, N.Y., 2004. Acid- and acid/rennet-curd cheeses part C: acid-heat coagulated cheeses. In: Fox, P.F., McSweeney, P.L.H., Cogan, T.M., Guinee, T.P. (Eds.), Cheese: Chemistry, Physics and Microbiology. Academic Press, pp. 343–348. https://doi.org/10.1016/S1874-558X(04)80051-4.

Food and Agriculture Organization of the United Nations, 1971. Standard for Whey Cheeses CXS 284-1971. Codex Alimentarius.

Food and Agriculture Organization of the United Nations, 1978. General Standard for Cheese CXS 283-1978. Codex Alimentarius.

Freire, D.O., Wu, B.Q., Hartel, R.W., 2020. Effects of structural attributes on the rheological properties of ice cream and melted ice cream. J. Food Sci. 85 (11), 3885–3898. https://doi.org/10.1111/1750-3841.15486.

Gänzle, M.G., Haase, G., Jelen, P., 2008. Lactose: crystallization, hydrolysis and value-added derivatives. Int. Dairy J. 18 (7), 685–694. https://doi.org/10.1016/j.idairyj.2008.03.003.

Genovese, A., Balivo, A., Salvati, A., Sacchi, R., 2022. Functional ice cream health benefits and sensory implications. Food Res. Int. 161, 111858. https://doi.org/10.1016/j.foodres.2022.111858.

Gharibzahedi, S.M.T., Koubaa, M., Barba, F.J., Greiner, R., George, S., Roohinejad, S., 2018. Recent advances in the application of microbial transglutaminase crosslinking in cheese and ice cream products: a review. Int. J. Biol. Macromol. 107, 2364–2374. https://doi.org/10.1016/j.ijbiomac.2017.10.115.

Goff, H.D., 2022. Ice cream and frozen desserts: product types. In: McSweeney, P.L.H., McNamara, J.P. (Eds.), Encyclopedia of Dairy Sciences, third ed. Academic Press, pp. 498–503. https://doi.org/10.1016/B978-0-12-818766-1.00033-7.

Goff, H.D., Hartel, R.W., 2013. Ice Cream, seventh ed. Springer. https://doi.org/10.1007/978-1-4614-6096-1.

Goff, H.D., Hartel, R.W., Rankin, S.A., 2025. Ice Cream, eighth ed. Springer. https://doi.org/10.1007/978-3-031-77872-8.

Gomaa, M., El-Sokkary, E., Abdel-Aziz, M., Refaey, M., 2015. Effect of reinforcement Ricotta cheese by skim milk powder and milk protein concentrate powder on the quality of the cheese target during storage. J. Food Dairy Sci. 6 (3), 215–225. https://doi.org/10.21608/jfds.2015.48812.

Gurskiy, I.A., Tvorogova, A.A., 2022. The effect of whey protein concentrates on technological and sensory quality indicators of ice cream. Food Process Tech. Technol. 52 (3), 439–448. https://doi.org/10.21603/2074-9414-2022-3-2376.

Guy, E.J., Vettel, H.E., Pallansch, M.J., 1972. Stabilization of milk fat/cheese whey emulsions. Food Technol. 26, 50–55.

Harfoush, A., Fan, Z.Y., Goddik, L., Haapala, K.R., 2024. A review of ice cream manufacturing process and system improvement strategies. Manuf. Lett. 41, 170–181. https://doi.org/10.1016/j.mfglet.2024.09.021.

Hartel, R.W., Rankin, S.A., Bradley, R.L., 2017. A 100-Year Review: milestones in the development of frozen desserts. J. Dairy Sci. 100 (12), 10014–10025. https://doi.org/10.3168/jds.2017-13278.

Hossain, M.K., Petrov, M., Hensel, O., Diakité, M., 2021. Microstructure and physicochemical properties of light ice cream: effects of extruded microparticulated whey proteins and process design. Foods 10, 1433. https://doi.org/10.3390/foods 10061433.

IDFA, 2025. Ice Cream Sales and Trends. Retrieved 30. 6. from. https://www.idfa.org/ice-cream-sales-trends.

Inglingstad, R.A., Skeie, S., Vegarud, G.E., Devold, T.G., Chilliard, Y., Eknæs, M., 2017. Feeding a concentrate rich in rapeseed oil improves fatty acid composition and flavor in Norwegian goat milk. J. Dairy Sci. 100 (9), 7088–7105. https://doi.org/10.3168/jds.2016-12383.

Jelen, P., 1992. Whey cheeses and beverages. In: Whey and Lactose Processing. Springer, Netherlands, Dordrecht, pp. 157–193. https://doi.org/10.1007/978-94-011-2894-0_5.

Jelen, P., 2011. Whey processing: utilization and products. In: Fuquay, J.W. (Ed.), Encyclopedia of Dairy Sciences. Academic Press, Edmonton, pp. 731–737.

Jelen, P., Buchheim, W., 1976. Characteristics of Norwegian whey "cheese." Food Technol. 30 (11).

Johansen, A.G., Vegarud, G.E., Skeie, S., 2002. Seasonal and regional variation in the composition of whey from Norwegian Cheddar-type and Dutch-type cheeses. Int. Dairy J. 12 (7), 621–629. https://doi.org/10.1016/S0958-6946(02)00054-7.

Kamel, B., Boubaker, K., Attia, H., 2013. Implementation of ricotta cheese production process in Tunisia. Int. Food Res. J. 20 (5), 2343–2348.

Kaminska-Dwórznicka, A., Laba, S., Jakubczyk, E., 2022. The effects of selected stabilizers addition on physical properties and changes in crystal structure of whey ice cream. LWT Food Sci. Technol. 154, 112841. https://doi.org/10.1016/j.lwt.2021.112841.

Kaur, H., Kaur, G., Ali, S.A., 2022. Dairy-based probiotic-fermented functional foods: an update on their health-promoting properties. Fermentation 8 (9), 425. https://doi.org/10.3390/fermentation8090425.

Khorshidian, N., Yousefi, M., Khanniri, E., Mortazavian, A.M., 2018. Potential application of essential oils as antimicrobial preservatives in cheese. Innov. Food Sci. Emerg. Technol. 45, 62–72. https://doi.org/10.1016/j.ifset.2017.09.020.

Kielczewska, K., Smoczynski, M., Gutkowska, M., 2025. The use of high-protein preparations in ice cream production. Foods 14 (3), 345. https://doi.org/10.3390/foods14030345.

Kindstedt, P.S., 2017. The history of cheese. In: Papademas, P., Bintsis, T. (Eds.), Global Cheesemaking Technology: Cheese Quality and Characteristics. John Wiley and Sons, pp. 1–19.

Klojdová, I., Stathopoulos, C., 2022. W/o/w multiple emulsions: a novel trend in functional ice cream preparations? Food Chem. X 16, 100451. https://doi.org/10.1016/j.fochx.2022.100451.

Levin, M.A., Burrington, K.J., Hartel, R.W., 2016. Whey protein phospholipid concentrate and delactosed permeate: applications in caramel, ice cream, and cake. J. Dairy Sci. 99 (9), 6948–6960. https://doi.org/10.3168/jds.2016-10975.

Li, A., Zheng, J., Han, X., Yang, S., Cheng, S., Zhao, J., Zhou, W., Lu, Y., 2023. Advances in low-lactose/lactose-free dairy products and their production. Foods 12 (13), 2553. https://www.mdpi.com/2304-8158/12/13/2553.

Li, M., Correa-González, Y.X., Li, T., Wu, T., 2025. Assessing ice recrystallization inhibition effect of stabilizer in ice cream systems: methods and influencing factors. Food Hydrocoll. 160, 110743. https://doi.org/10.1016/j.foodhyd.2024.11074.

Lim, S.–Y., Swanson, B.G., Ross, C.F., Clark, S., 2008. High hydrostatic pressure modification of whey protein concentrate for improved body and texture of low fat ice cream. J. Dairy Sci. 91, 1308–1316. https://doi.org/10.3168/jds.2007-0391.

Lucey, J.A., 2011. Cheese | acid- and acid/heat coagulated cheese. In: Encyclopedia of Dairy Sciences. Elsevier, pp. 698–705. https://doi.org/10.1016/B978-0-12-374407-4.00083-2.

Lucey, J.A., Johnson, M.E., Horne, D.S., 2003. Invited review: perspectives on the basis of the rheology and texture properties of cheese. J. Dairy Sci. 86 (9), 2725–2743. https://doi.org/10.3168/jds.S0022-0302(03)73869-7.

Madureira, A.R., Pintado, M.E., Gomes, A.M.P., Malcata, F.X., 2011. Incorporation of probiotic bacteria in whey cheese: decreasing the risk of microbial contamination. J. Food Protect. 74 (7), 1194–1199. https://doi.org/10.4315/0362-028X.JFP-10-217.

Maity, T., Saxena, A., Raju, P.S., 2018. Use of hydrocolloids as cryoprotectant for frozen foods. Crit. Rev. Food Sci. 58 (3), 420–435. https://doi.org/10.1080/10408398.2016.1182892.

Mancuso, I., Cardamone, C., Fiorenza, G., Macaluso, G., Arcuri, L., Miraglia, V., Scatassa, M.L., 2014. Sensory and microbiological evaluation of traditional ovine ricotta cheese in modified atmosphere packaging. Ital. J. Food Saf. 3 (2). https://doi.org/10.4081/ijfs.2014.1725.

Matsudomi, N., Rector, D., Kinsella, J.E., 1991. Gelation of bovine serum albumin and β-lactoglobulin; effects of pH, salts and thiol reagents. Food Chem. 40 (1), 55–69. https://doi.org/10.1016/0308-8146(91)90019-K.

Meneses, R.B., Silva, M.S., Monteiro, M.L.G., Rocha-Leao, M.H.M., Conte, C.A., 2020. Effect of dairy by-products as milk replacers on quality attributes of ice cream. J. Dairy Sci. 103 (11), 10022–10035. https://doi.org/10.3168/jds.2020-18330.

Modler, H.W., 1988. Development of a continuous process for the production of ricotta cheese. J. Dairy Sci. 71 (8), 2003–2009. https://doi.org/10.3168/jds.S0022-0302(88)79775-1.

Modler, H.W., Emmons, D.B., 2001. The use of continuous ricotta processing to reduce ingredient cost in 'further processed' cheese products. Int. Dairy J. 11 (4–7), 517–523. https://doi.org/10.1016/S0958-6946(01)00082-6.

Mohammed, N.K., Khair, M.F.B., Ahmad, N.H., Hussin, A.S.M., 2022. Ice cream as functional food: a review of health-promoting ingredients in the frozen dairy products. J. Food Process. Eng. 45 (12). https://doi.org/10.1111/jfpe.14171.

Mucchetti, G., Carminati, D., Pirisi, A., 2002. Ricotta fresca vaccina ed ovina:osservazioni sulle tecniche di produzione e sul prodotto. Il Latte 27, 154–166.

Mucchetti, G., Neviani, E., 2006. Microbiologia e tecnologia lattiero—casearia—Qualità e sicurezza. Tecniche Nuove.

Mucchetti, G., Pugliese, A., Paciulli, M., 2017. Characteristics of some important Italian cheeses—Parmigiano Reggiano, Grana Padano, mozzarella, Mascarpone and ricotta. In: Mediterranean Foods. CRC Press.

Mykhalevych, A., Buniowska-Olejnik, M., Polishchuk, G., Puchalski, C., Kaminska-Dwórznicka, A., Berthold-Pluta, A., 2024a. The influence of whey protein isolate on the quality indicators of acidophilic ice cream based on liquid concentrates of demineralized whey. Foods 13 (1), 170. https://doi.org/10.3390/foods13010170.

Mykhalevych, A., Moiseeva, L., Polishchuk, G., Bandura, U., Buniowska-Olejnik, M., 2024b. Comparative analysis of functional and technological properties of β-glucans from oats and yeast in whey ice cream. Ukrain. Food J. 13 (3). https://doi.org/10.24263/2304-974x-2024-13-3-6.

Niro, S., Succi, M., Fratianni, A., Cinquanta, L., Tremonte, P., Sorrentino, E., Panfili, G., 2013. Production of functional ricotta cheese. Agro Food Ind. Hi Tech 24 (6), 56–59.

Norwegian Ministry of Agriculture and Food, 2016. Forskrift Om Kvalitet På Melk Og Melkeprodukter [Regulations on the Quality of Milk and Dairy Products]. https://lovdata.no/dokument/SF/forskrift/2015-06-03-607.

Ohmes, R.L., Marshall, R.T., Heymann, H., 1998. Sensory and physical properties of ice creams containing milk fat or fat replacers. J. Dairy Sci. 81, 1222–1228. https://doi.org/10.3168/jds.S0022-0302(98)75682-6.

Osmak, T., Mieko, S., Bass, O., Mykhalevych, A., Kuzmyk, U., 2021. Enzymatic hydrolysis of lactose in concentrates of reconstituted demineralized whey, intended for ice cream production. Ukrain. Food J. 10 (2), 277–288. https://doi.org/10.24263/2304-974x-2021-10-2-6.

Ostertag, F., Krolitzki, E., Berensmeier, S., Hinrichs, J., 2023. Protein valorisation from acid whey—screening of various micro- and ultrafiltration membranes concerning the filtration performance. Int. Dairy J. 146. https://doi.org/10.1016/j.idairyj.2023.105745.

Pala, C., Scarano, C., Venusti, M., Sardo, D., Casti, D., Cossu, F., Lamon, S., Spanu, V., Ibba, M., Marras, M., Paba, A., Spanu, C., De Santis, E.P.L., 2016. Shelf life evaluation of ricotta fresca sheep cheese in modified atmosphere packaging. Ital. J. Food Saf. 5 (3). https://doi.org/10.4081/ijfs.2016.5502.

Paladii, I.V., Vrabie, E.G., Sprinchan, K.G., Bologa, M.K., 2021. Part 1: classification, composition, properties, derivatives, and application. Surf. Eng. Appl. Electrochem. 57 (5), 579–594. https://doi.org/10.3103/S1068375521050112.

Papademas, P., Bintsis, T., Alichanidis, E., Ardö, Y., 2017. Whey cheeses (heat coagulated). In: Papademas, P., Bintsis, T. (Eds.), Global Cheesemaking Technology: Cheese Quality and Characteristics. John Wiley and Sons. https://doi.org/10.1002/9781119046165.ch13.

Park, J.H., Lee, Y.B., Lee, S.H., Ko, E., Imm, J.Y., 2024. Combination of milk polar lipids and casein hydrolysate as a healthy emulsifier for ice cream. Food Sci. Anim. Resour. 44 (6), 1389–1402. https://doi.org/10.5851/kosfa.2024.e88.

Park, Y.W., Guo, M., 2006. Goat milk products: types of products, manufacturing technology, chemical composition, and marketing. In: Handbook of Milk of Non-Bovine Mammals. Wiley, pp. 59–106. https://doi.org/10.1002/9780470999738.ch4.

Paskaš, S., Miočinović, J., Savić, M., Ješić, G., Rašeta, M., Becskei, Z., 2019. Comparison of the chemical composition of whey cheeses: Urda and ricotta. Maced. Vet. Rev. 42 (2), 151–161. https://doi.org/10.2478/macvetrev-2019-0020.

Patocka, G., Jelen, P., Kalab, M., 1990. Heat induced coagulation in milk with modified casein: whey protein content. In: Brief Communications of the XXIII International Dairy Congress, Montreal, October 8-12, 1990, Vol. II.

Pehlivanoğlu, H., Sunal, Z., Yaman, M., Aksoy, A., 2024. Investigation of the bioaccessibility of functional ice cream with blueberry enriched with whey protein gel [Peynir Altı Suyu protein Jeli ile Zenginleştirilmiş Yaban Mersinli Fonksiyonel Dondurmanın Biyoerişilebilirliğinin Araştırılması]. Tekirdağ Ziraat Fakültesi Dergisi 21 (3), 795–806. https://doi.org/10.33462/jotaf.1399931.

Pei, J.W., Xiong, D.D., Zhang, M.H., Liu, C.H., Zhang, L., Liu, X.B., Duan, X., 2025. Impact of high-soluble modified wheat gluten as an emulsifier on the structure and quality of ice cream. Food Chem. 468, 142473. https://doi.org/10.1016/j.foodchem.2024.142473.

Pintado, M., Macedo, A.C., Malcata, F., 2001. Review: technology, chemistry and microbiology of whey cheeses. Food Sci. Technol. Int. 7 (2), 105–116. https://doi.org/10.1106/TY4X-QX5U-5X5D-TYE8.

Pires, A.F., Marnotes, N.G., Rubio, O.D., Garcia, A.C., Pereira, C.D., 2021. Dairy by-products: a review on the valorization of whey and second cheese whey. Foods 10 (5), 1067. https://doi.org/10.3390/foods10051067.

Pizzillo, M., Claps, S., Cifuni, G.F., Fedele, V., Rubino, R., 2005. Effect of goat breed on the sensory, chemical and nutritional characteristics of ricotta cheese. Livest. Prod. Sci. 94 (1–2), 33–40. https://doi.org/10.1016/j.livprodsci.2004.11.024.

Puccio, V., Moscarelli, A., Messina, P., Francesca, N., Moschetti, G., Gaglio, R., Settanni, L., 2022. Microbiological and physicochemical profile of sugar added ewe's ricotta during storage profilo microbiologico e fisico-chimico della ricotta di pecora zuccherata durante la conservazione. Scienza e Tecnica Lattiero-Casearia 72 (1), 46–52. https://doi.org/10.36138/STLC.01.2022.01.

Punsandani, U., Mary Ann, A., Li Jiang, C., Roderick, P.W.W., 2005. Physical behaviour of dairy ingredients during ice cream processing. Lait 85, 383–394. https://doi.org/10.1051/lait:2005025.

Quinzio, J., 2009. Of Sugar and Snow: A History of Ice Cream Making, 1st ed. University of California Press http://www.jstor.org/stable/10.1525/j.ctt7zw39d.

Ribeiro, A.C., Ribeiro, S.D.A., 2010. Specialty products made from goat milk. Small Rumin. Res. 89 (2–3), 225–233. https://doi.org/10.1016/j.smallrumres.2009.12.048.

Ricciardi, E.F., Del Nobile, M.A., Conte, A., Fracassi, F., Sardella, E., 2022. Effects of plasma treatments applied to fresh ricotta cheese. Innov. Food Sci. Emerg. Technol. 76, 102935. https://doi.org/10.1016/j.ifset.2022.102935.

Ricciardi, E.F., Pedros-Garrido, S., Papoutsis, K., Lyng, J.G., Conte, A., Del Nobile, M.A., 2020. Novel technologies for preserving ricotta cheese: effects of ultraviolet and near-ultraviolet–visible light. Foods 9 (5), 580. https://doi.org/10.3390/foods9050580.

Ricciardi, F.E., Lacivita, V., Conte, A., Chiaravalle, E., Zambrini, A.V., Del Nobile, M.A., 2019. X-ray irradiation as a valid technique to prolong food shelf life: the case of ricotta cheese. Int. Dairy J. 99, 104547. https://doi.org/10.1016/j.idairyj.2019.104547.

Ricciardi, F.E., Plazzotta, S., Conte, A., Manzocco, L., 2021. Effect of pulsed light on microbial inactivation, sensory properties and protein structure of fresh ricotta cheese. Lebensm. Wiss. Technol. 139, 110556. https://doi.org/10.1016/j.lwt.2020.110556.

Ritota, M., Manzi, P., 2020. Natural preservatives from plant in cheese making. Animals 10 (4), 749. https://doi.org/10.3390/ani10040749.

Roehl, D., Jelen, P., 1986. A new whey cheese product from ultrafiltered cottage cheese whey. In: IFT Annual Meeting.

Roobab, U., Inam-Ur-Raheem, M., Khan, A.W., Arshad, R.N., Zeng, X.A., Aadil, R.M., 2023. Innovations in high-pressure technologies for the development of clean label dairy products: a review. Food Rev. Int. 39 (2), 970–991. https://doi.org/10.1080/87559129.2021.1928690.

Roy, S., Hussain, S.A., Prasad, W.G., Khetra, Y., 2021. Effect of emulsifier blend on quality attributes and storage of high protein buffalo milk ice cream. LWT Food Sci. Technol. 150, 111903. https://doi.org/10.1016/j.lwt.2021.111903.

Saentaweesuk, S., Chaikham, P., 2023. Effect of whey protein isolate incorporated with various carbohydrate-based fat replacers on physicochemical and sensorial properties of low-fat chocolate ice cream. Food Res. 7, 167–176. https://doi.org/10.26656/fr.2017.7(1).705.

Salvatore, E., Pes, M., Falchi, G., Pagnozzi, D., Furesi, S., Fiori, M., Roggio, T., Addis, M.F., Pirisi, A., 2014. Effect of whey concentration on protein recovery in fresh ovine ricotta cheese. J. Dairy Sci. 97 (8), 4686–4694. https://doi.org/10.3168/jds.2013-7762.

Sameer, B., Ganguly, S., Khetra, Y., Sabikhi, L., 2020. Development and characterization of probiotic Buffalo milk ricotta cheese. Lebensm. Wiss. Technol. 121. https://doi.org/10.1016/j.lwt.2019.108944.

Samelis, J., Kakouri, A., Rogga, K.J., Savvaidis, I.N., Kontominas, M.G., 2003. Nisin treatments to control Listeria monocytogenes post-processing contamination on Anthotyros, a traditional Greek whey cheese, stored at 4°C in vacuum packages. Food Microbiol. 20 (6), 661–669. https://doi.org/10.1016/S0740-0020(03)00018-2.

Scatassa, M.L., Mancuso, I., Sciortino, S., Macaluso, G., Palmeri, M., Arcuri, L., Todaro, M., Cardamone, C., 2018. Retrospective study on the hygienic quality of fresh ricotta cheeses produced in Sicily, Italy. Ital. J. Food Saf. 7 (1). https://doi.org/10.4081/ijfs.2018.6911.

Schweiger, D., Myers, J., Clark, S., 2023. Whey permeate powder is a suitable ingredient for ice cream. Jds Commun. 4 (6). https://doi.org/10.3168/jdsc.2023-0382.

Shelke, P.A., Sabikhi, L., Khetra, Y., Ganguly, S., Baig, D., 2022. Effect of skim milk addition and heat treatment on characteristics of cow milk Ricotta cheese manufactured from Cheddar cheese whey. Lebensm. Wiss. Technol. 162. https://doi.org/10.1016/j.lwt.2022.113405.

Shevchenko, O., Mykhalevych, A., Polischuk, G., Buniowska-Olejnik, M., Bass, O., Bandura, U., 2022. Technological functions of hydrolyzed whey concentrate in ice cream. Ukrain. Food J. 11 (4). https://doi.org/10.24263/2304-974x-2022-11-4-3.

Sindhu, S., Radhai Sri, S., Vishnupriya, P., 2018. Quality evaluation of whey incorporated ice cream. Int. J. Biomed. Adv. Res. 8 (9), 806–812.

Silva, C.C.G., Silva, S.P.M., Ribeiro, S.C., 2018. Application of bacteriocins and protective cultures in dairy food preservation. Front. Microbiol. 9. https://doi.org/10.3389/fmicb.2018.00594.

Skeie, S., Abrahamsen, R.K., 2017. Chapter 45—Brown whey cheese. In: McSweeney, P.L.H., Patrick, F., Fox, P.F., Cotter, P.D., Everett, D.W. (Eds.), Cheese. Academic Press, Ås, pp. 1117–1132.

Soukoulis, C., Fisk, I., 2016. Innovative ingredients and emerging technologies for controlling ice recrystallization, texture, and structure stability in frozen dairy desserts: a review. Crit. Rev. Food Sci. Nutr. 56 (15), 2543–2559. https://doi.org/10.1080/10408398.2013.876385.

Soutelino, M.E.M., Vieira, G.D., Goulart, M.B., Miranda, K.C., da Conceiçao, R.P., Pimentel, T.C., da Cruz, A.G., Rocha, R.D., 2024. Natural food dyes on dairy products: a critical approach between 2012-2023 literature regarding the technological and functional aspects, health benefits and future trends. Trends Food Sci. Technol. 146, 104370. https://doi.org/10.1016/j.tifs.2024.104370.

Spanu, C., Scarano, C., Piras, F., Spanu, V., Pala, C., Casti, D., Lamon, S., Cossu, F., Ibba, M., Nieddu, G., De Santis, E.P.L., 2017. Testing commercial biopreservative against spoilage microorganisms in MAP packed Ricotta fresca cheese. Food Microbiol. 66, 72–76. https://doi.org/10.1016/j.fm.2017.04.008.

Stading, M., Hermansson, A.-M., 1991. Large deformation properties of β-lactoglobulin gel structures. Food Hydrocoll. 5 (4), 339–352. https://doi.org/10.1016/S0268-005X(09)80046-5.

Statista, 2025. Ice Cream—Worldwide. Retrieved 30. 6. from. https://www.statista.com/outlook/cmo/food/confectionery-snacks/confectionery/ice-cream/worldwide.

Steinsholt, K., Byre, O., Ystgaard, O.M., 1965. Fremstilling Av Fløetemysost F33. I. Virkningen Av Myseslag, Blandingsprosent Og pH I Mysa Med Hensyn På Ostens Kjemiske Og Organoleptiske Kvalitetsegenskaper [Production of Fløtemysost F33. I. The Effect of Whey Source, Amount of Milk and pH in the Whey with Regards to the Chemical and Organoleptic Quality of the Cheese]. Ås.

Tekin, E., Sahin, S., Sumnu, G., 2017. Physicochemical, rheological, and sensory properties of low-fat ice cream designed by double emulsions. Eur. J. Lipid Sci. Technol. 119 (9), 1600505. https://doi.org/10.1002/ejlt.201600505.

Tomczynska-Mleko, M., Mykhalevych, A., Sapiga, V., Polishchuk, G., Terpilowski, K., Mleko, S., Solowiej, B.G., Perez-Huertas, S., 2024. Influence of plant-based structuring ingredients on physicochemical properties of whey ice creams. Appl. Sci. (Basel) 14 (6), 2465. https://doi.org/10.3390/app14062465.

Toppino, P.M., Campagnol, L., Carminati, D., Mucchetti, G., Povolo, M., Benedetti, S., Riva, M., 2004. Shelf life study of packed industrial ricotta cheese. Ital. J. Food Sci. Spec. 252–265.

Tripaldi, C., Rinaldi, S., Palocci, G., Di Giovanni, S., Campagna, M.C., Di Russo, C., Zottola, T., 2020. Chemical and microbiological characteristics of homogenised Ricotta cheese produced from buffalo whey. Ital. J. Food Sci. 32 (2), 292–309.

Tsuchiya, A.C., da Silva, A.D.M., Brandt, D., Kalschne, D.L., Drunkler, D.A., Colla, E., 2017. Lactose-reduced ice cream enriched with whey powder. Semina Cienc. Agrar. 38 (2), 749–758. https://doi.org/10.5433/1679-0359.2017v38n2p749.

Tvorogova, A.A., Gurskiy, I.A., Shobanova, T.V., Smykov, I.T., 2023. Effect of protein concentrates and isolates on the rheological, structural, thermal and sensory properties of ice cream. Curr. Res. Nutr. Food Sci. 11 (1), 294–306. https://doi.org/10.12944/Crnfsj.11.1.22.

United States Department of Agriculture, 1981. USDA Specifications for Ricotta Cheeses.

Urgu-Ozturk, M., 2022. Possibilities of using the continuous type of UV light on the surface of lor (whey) cheese: impacts on mould growth, oxidative stability, sensory and colour attributes during storage. J. Dairy Res. 89 (3), 335–341. https://doi.org/10.1017/S0022029922000590.

VanWees, S.R., Rankin, S.A., Hartel, R.W., 2022. Shrinkage in frozen desserts. Compr. Rev. Food Sci. Food Saf. 21 (1), 780–808. https://doi.org/10.1111/1541-4337.12888.

VanWees, S.R., Rankin, S.A., Hartel, R.W., 2020. The microstructural, melting, rheological, and sensorial properties of high-overrun frozen desserts. J. Texture Stud. 51 (1), 92–100. https://doi.org/10.1111/jtxs.12461.

Verheul, M., Roefs, S.P.E.M., de Kruif, C.G., 1998. Kinetics of heat-induced aggregation of β-lactoglobulin. J. Agric. Food Chem. 46 (3), 896–903. https://doi.org/10.1021/jf970751t.

Vetorazi, M.G., de Oliveira, E.F., Macedo, L.L., Araújo, C.D., Saraiva, S.H., Coelho, J.M., 2024. High-intensity ultrasound improves texture profile and technological characteristics of cupuaçu (Theobroma grandiflorum) ice cream. J. Food Sci. Tech. Mys. 62 (9), 1742–1749. https://doi.org/10.1007/s13197-024-06142-6.

von Staszewski, M., Jagus, R.J., 2008. Natural antimicrobials: effect of MicrogardTM and nisin against Listeria innocua in liquid cheese whey. Int. Dairy J. 18 (3), 255–259. https://doi.org/10.1016/j.idairyj.2007.08.012.

Warren, M.M., Hartel, R.W., 2018. Effects of emulsifier, overrun and dasher speed on ice cream microstructure and melting properties. J. Food Sci. 83 (3), 639–647. https://doi.org/10.1111/1750-3841.13983.

Weatherup, W., 1986. The effect of processing variables on the yield and quality of ricotta cheese. Dairy Ind. Int. 51 (9), 43–45.

Wu, B.Q., Hartel, R.W., 2024. Effect of milk protein-galactomannans interaction on meltdown behavior in the aerated frozen model system. Food Hydrocoll. 155, 110155. https://doi.org/10.1016/j.foodhyd.2024.110155.

Wu, B.Q., Freire, D.O., Hartel, R.W., 2019. The effect of overrun, fat destabilization, and ice cream mix viscosity on entire meltdown behavior. J. Food Sci. 84 (9), 2562–2571. https://doi.org/10.1111/1750-3841.14743.

Yan, L., Yu, D.J., Liu, R., Jia, Y.Y., Zhang, M., Wu, T., Sui, W.J., 2021. Microstructure and meltdown properties of low-fat ice cream: effects of microparticulated soy protein hydrolysate/xanthan gum (MSPH/XG) ratio and freezing time. J. Food Eng. 291, 110291. https://doi.org/10.1016/j.jfoodeng.2020.110291.

Zhao, Y.G., Khalesi, H., He, J., Fang, Y.P., 2023. Application of different hydrocolloids as fat replacer in low-fat dairy products: ice cream, yogurt and cheese. Food Hydrocoll. 138, 108493. https://doi.org/10.1016/j.foodhyd.2023.108493.

Zinina, O., Khitsenko, A., Neverova, O., Stepanov, A., Bykova, O., 2021. Research of the Ricotta made from cheese whey enriched with citrus dietary fiber. IOP Conf. Ser. Earth Environ. Sci. 699, 012006. https://doi.org/10.1088/1755-1315/699/1/012006.

Chapter 10

Whey use in bakery and confectionery products

Alejandro Escobedo[1], Pavel Prieto-Vázquez del Mercado[2] and Ivan Luzardo-Ocampo[1]
[1]*Tecnologico de Monterrey, Institute for Obesity Research, School of Engineering & Sciences, Monterrey, Mexico;* [2]*Centro Universitario de Ciencias de la Salud (CUCS), Universidad de Guadalajara (UdeG), Guadalajara, Mexico*

1. Introduction

Whey is a significant dairy effluent that, when discarded, acts as a major contaminant, but when used in food applications, it becomes a valuable resource for producing food products with enhanced nutritional content and improved technofunctional properties, making it a versatile ingredient (El-Aidie and Khalifa, 2024). It has been estimated that whey production reaches up to 85%–90% of the total milk volume in cheese manufacturing, and in 2022, the global production of 2×10^8 tons annually, with lactose and protein being its main components (4.3% and 0.75%, respectively) (Mabrouki et al., 2022).

Due to its high nutritional content, 50% of the remaining whey is currently valorized into technologically feasible solutions that add value to food and chemical products, justifying its long-standing tradition as one of the most common protein applications in human and animal diets, serving as a techno-functional and health-promoting ingredient (De Almeida et al., 2023). Moreover, whey has also been used as an important ingredient in cooking and baking, offering novel sensory and technological applications in the resulting food products. Yet the production or addition of whey-based proteins still doubles the market price of whey (Sukmana et al., 2025).

Whey is recognized as a rich source of bioactive components, such as milk serum proteins (20% of the total whey composition), primarily α-lactalbumin and β-lactoglobulin, both of which are valuable sources of a wide range of amino acids, particularly tryptophan (Inocente-Alves et al., 2025). Whey proteins promote viscosity, emulsion stability, and gelling properties, due to the way β-lactoglobulin forms aggregates at 78–82°C, acts as a carrier for small hydrophobic ligands, and is stable to proteolytic activity and low pH levels (El-Aidie and Khalifa, 2024). In addition, the enormous diversity of commercially available whey-derived protein products, such as whey protein hydrolyzate, whey protein concentrate, whey protein isolate, and whey protein hydrolyzate, expands whey food applications. In bakery and confections, some of the most popular whey applications, whey can improve dough quality and extensibility, modify pasting properties, and add volume to bakery products, as reflected in changes in textural profiles, which are highly appreciated by consumers (Fluerasu et al., 2025).

Although a plethora of whey applications have been recognized in bakery and confectionery products, further research is needed to develop novel formulations that achieve optimal sensory acceptance among consumers, as improper whey manipulation can negatively impact the technofunctional properties of desired food applications. As long as whey collection and processing are more standardized in countries in the Global South that face environmental problems with whey contamination, whey inclusion in the food industry in these countries promotes the development of more complex infrastructure and the creation of whey applications adapted to local food products (Uribe-Velázquez et al., 2025).

2. Chemical composition

Macronutrient distribution in bakery and confectionery products containing whey ingredients requires careful management to ensure optimal nutritional quality. Although many studies have focused on developing high-protein alternatives,

Whey. https://doi.org/10.1016/B978-0-443-27648-4.01010-0

this is a challenging process due to the complex interactions within the food matrix and considerations of texture and sensory properties in the final products. Table 10.1 presents the effect of whey ingredients on the chemical composition of bakery and confectionery products. A recent study by Himashree et al. (2023) evaluated the nutritional benefits of incorporating liquid whey into bread by replacing water with 25%–100% whey. The results indicated a steady increase in protein and fat levels with higher whey substitution rates. Protein concentration increased from 15.1% in the control bread to 23.4% in the 100% liquid whey bread, while fat content increased from 1.7% to 7.0%. These modifications were accompanied by a reduction in carbohydrate content from 60.5% to 37.8% and a moderate decrease in energy value from 316 to 307 kcal/100 g. Although higher levels of liquid whey increased protein concentration, the elevated fat content and changes in moisture could affect product stability and texture. The addition of WPC further increases the protein content. For instance, replacing 20% of wheat flour with WPC in bread made with either commercial yeast or spontaneous sourdough doubled the protein content regardless of the fermentation method (Ferreyra et al., 2021). Furthermore, the WPC samples had lower moisture content and higher in vitro protein digestibility than the control samples.

Meanwhile, the same expected protein increment was observed in gluten-free bread made from cassava starch and chickpea flour when enriched with 10%–30% WPC (Komeroski et al., 2021). The highest level of replacement resulted in a protein content of 9.7%, compared to only 4.7% in the control sample. Additionally, the fat content increased from 0.3% in the control to 1.9% in the sample containing 30% WPC. These changes affected the energy content, increasing from 236.2 to 292.3 kcal/100 g. Moreover, higher levels of lysine, leucine, isoleucine, and valine, essential amino acids often limited in cereal-based gluten-free products, were shown in WPC-enriched samples. Finally, although the higher levels of WPC increased protein content, the additional fat and energy could affect the nutritional profile of the formulations.

Biscuits and cookies are popular and convenient foods, enjoyed by people of all socioeconomic levels. Their popularity stems from their long shelf life and simple production. WPC and WPI have been used primarily as substitutes for wheat flour up to 100%. For instance, Ranok and Kupradit (2020) replaced all of the wheat flour with either WPC or riceberry flour in the formulation of cookies. They found that the WPC samples had a protein concentration of 38.3%, while the wheat flour and riceberry samples had concentrations of 9.0% and 9.2%, respectively. In addition, the total carbohydrate content was only 27.5%, lower than the 62.1% in the wheat flour sample. Similar replacement levels were evaluated by Borooah et al. (2024), who used WPC to substitute wheat flour at levels ranging from 10%–70%. These biscuits also contained clove basil (*Ocimum gratissimum*), jaggery (a popular South Asian sweetener made from boiling sugarcane juice or palm sap), and ghee. They observed that the 70% WPC level sample increased protein content from 5.8% to 29.8%, while carbohydrates decreased from 72.7% to 52.8%. However, the same sample increased by 62 kcal/100 g, compared to the control. A higher protein content was achieved by adding WPC, increasing it to 26%–54% of the cookie formulation (Marques et al., 2016). The treatment containing 40% WPC showed 45.2% protein and only 11.5% carbohydrates. Furthermore, this sample also had a lower energy content than the control sample, at 2191 kJ/100 g and 2003 kJ/100 g, respectively. WPC has also been used to replace water chestnut (*Trapa nantans*) flour in gluten-free cookies at a 10% substitution level, resulting in an increase in protein content from 8.0% to 14.5% (Sarabhai and Prabhasankar, 2015).

Several studies have evaluated whey ingredients in various cake types, including sponge cake, banana cake, pancakes, cupcakes, and muffins. Norton et al. (2020) incorporated WPI to achieve a cake composition of 6.1%, thereby doubling the protein content from 6% to 12%. In the same study, they added WPI to cupcakes, bringing it to 11.7% of the formulation, and observed a similar increase in protein, from 6% to 12%, compared to the control.

Whey protein can be a valuable addition to banana cakes, as it provides essential amino acids that are typically lacking in bananas. For instance, Camargo et al. (2018) developed banana cakes supplemented with a blend of WPI and WPH, with protein content ranging from 10%–30%. In the sample containing 30% WPI/WPH, the protein concentration increased from 8.4% to 20.9% compared to the control. The amino acid composition was improved, particularly lysine, which increased from 0 mg/g to 32.6 mg/g. In addition to WPC and WPI, other whey ingredients, such as yogurt acid whey, have been used to replace all the water in pancake and pizza crust formulations (Flinois et al., 2019). Acid whey is a byproduct of Greek yogurt and cream cheese production, containing approximately 95% water and 4% lactose and galactose, which is currently used as fertilizer and cattle feed. The authors found increased protein levels in both pancake and pizza crusts, with levels rising from 6% to 8% for pancakes and from 7% to 9% for pizza crusts compared to their control samples. Additionally, there was a slight increase in sodium, calcium, and potassium levels in both products compared to the controls.

For confectionery products, Lazarev and Ershova (2023) developed a marshmallow formulation using sweeteners and WPC to enhance the product's nutritional profile. The formula consisted of adding WPC to achieve a 5%–30% content of the marshmallow, along with Jerusalem artichoke syrup and isomalt. The resulting sample with 30% WPC had a protein content of 8.1%, while the control had only 1.8%.

TABLE 10.1 Effects of the incorporation of whey ingredients on the nutritional profile (protein, carbohydrates, fat, and ash) of bakery and confectionery products.

Product	Whey incorporation	Sample	Protein	CHO	Fat	Ash	Moisture	References
Bread	LW as water substitute at 25%–100%	Control	15.1	60.5	1.7	1.8	21.1	Himashree et al. (2023)
		100% LW	23.4	37.8	7.0	1.9	29.8	
	LW as water substitute at 25%	Control	–	–	–	2.3	37.6	Çiçek et al. (2025)
		25% LW	–	–	–	2.3	37.6	
	Whey coacervate as milk, butter, and salt substitute at 100%	Control	8.8	52.7	2.9	2.6	33	Ingrassia et al. (2022)
		100% Whey coacervate	7.7	47.8	0.3	2.9	42	
	WPC as cassava starch and chickpea flour blend substitute at 10%–30%	Control	4.7	–	0.3	20.5	14.9	Komeroski et al. (2021)
		30% WPC	9.7	–	1.9	14.2	14.9	
	WPC as wheat flour substitute at 20%	Control	10.8	–	–	–	34.7	Ferreyra et al. (2021)
		20% WPC	22.1	–	–	–	27	
Sourdough bread	WPC as wheat flour substitute at 20%	Control	10.2	–	–	–	32.5	Ferreyra et al. (2021)
		20% WPC	21.9	–	–	–	28	
Flat bread	LW as water substitute at 25%–100%	Control	12.6	–	1.7	2.05	37.1	Satouf (2023)
		100% LW	15.1	–	1.8	3.01	38.9	
Biscuits	WPC as wheat flour substitute at 10%–70% All biscuits contained clove basil, jaggery, and ghee	Control	5.8	72.7	20	1.4	3.6	Borooah et al. (2024)
		70% WPC	29.8	52.8	23.6	2.5	3.9	
	W-LPH as wheat flour substitute at 5%–20%	Control	4	79.9	18.5	–	–	Li et al. (2024)
		15% W-LPH	9.1	68.6	19.4	–	–	
	WPC as wheat flour substitute at 7%–20%. Some biscuits contained lupin flour	Control	8.1	78.9	11.7	1.1	5.3	Beshir et al. (2021)
		20% WPC	23.9	58.9	13.7	3.2	5.4	
	WPI added to achieve 7.5% of biscuit	Control	5.6	61	36	–	3.1	Norton et al. (2020)
		7.5% WPI	14	56	34	–	2.3	
	WPC added to achieve 11%–15% protein in biscuits	Control	9	–	–	–	–	Andoyo et al. (2023)
		15% WPC	15	–	–	–	–	
	WPI added to achieve 24.1% of biscuits	Control	8.2	70.6	15	–	–	Hassanzadeh-Rostami et al. (2020)
		24.1% WPI	29.6	50.8	13.5	–	–	
Cookies	WPC added to achieve 6%–24% of cookie. Some cookies contained soy protein isolate	Control	27.7	40	21.3	1.9	9.1	Park et al. (2021)
		24% WPC	25.7	41.3	23.2	1.4	8.5	
	WPC as wheat flour substitute at 100%	Control	9	62.1	22.7	1.1	5	Ranok and Kupradit (2020)

Continued

TABLE 10.1 Effects of the incorporation of whey ingredients on the nutritional profile (protein, carbohydrates, fat, and ash) of bakery and confectionery products.—cont'd

Product	Whey incorporation	Sample	Protein	CHO	Fat	Ash	Moisture	References
		100% WPC	38.3	27.5	26.3	2.7	5.2	
	WPC as wheat flour substitute at 2%–6%	Control	9.1	59.7	22.3	1	6.1	Wani et al. (2015)
		6% WPC	13.2	50.1	23.1	2	11.3	
	WPC as wheat flour substitute at 5%–10%	Control	15.8	–	1.6	1.4	4.1	Rababah et al. (2022)
		10% WPC	21.7	–	2	1.5	4.8	
	WPI and blackcurrant concentrate as wheat flour substitute at 5%–15%	Control	7.6[a]	–	–	1.3[a]	8.4	Wu et al. (2021)
		15% WPI	12[a]	–	–	1.1[a]	9.8	
	WPC added to achieve 26%–54% of cookie	Control	17.1	53.1	27	1.5	1.3	Marques et al. (2016)
		40% WPC	45.2	11.5	28	7.5	7.8	
	WPC as water chestnut flour substitute at 5%–10%	Control	8.0	–	1.0	2.1	11.5	Sarabhai and Prabhasankar (2015)
		10% WPC	14.5	–	0.5	2.4	11.8	
Crackers	WPI as wheat flour substitute at 5%–20%	Control	9.6	–	–	–	4.3	En Tay et al. (2022)
		20% WPI	21.7	–	–	–	2.5	
Cake	WPI added to achieve 6.1% of cake	Control	6	44	23	–	28.8	Norton et al. (2020)
		6.1% WPI	12	41	22	–	25.3	
Sponge cake	WPC as wheat flour substitute at 20% Some cakes contained sourdough	Control	9.1	59.8	3.9	1.3	23.4	Maravić et al. (2022)
		20% WPC	14.8	54.6	2.9	1.4	24.4	
Banana cake	Blend of WPI and WPH added to achieve 10%–30% of cake	Control	8.4	41.6	12.2	2.2	35.6	Camargo et al. (2018)
		30% WPI/WPH	20.9	20.9	14.6	2.8	40.8	
Pancake	YAW as water substitute at 100%	Control	6	31	1.5	–	53.3	Flinois et al. (2019)
		100% YAW	8	25	1	–	–	
Cupcake	WPI added to achieve 11.7% of cupcake	Control	6	51	24	–	23.5	Norton et al. (2020)
		11.7% WPI	12	45	24	–	24.2	
Muffin	WPI and WPH as wheat flour substitute at 5%	Control	10.2	–	14	–	21.5	Bekiroglu et al. (2024)
		5% WPI	11.7	–	13.5	–	21.9	
		5% WPH	12.5	–	14.1	–	21.4	
Pizza crust	YAW as water substitute at 100%	Control	7	42	4.5	–	33.1	Flinois et al. (2019)
		100% YAW	9	44	4.5	–	31.8	
Marshmallow	WPC added to achieve 5%–30% of marshmallow	Control	1.8	–	–	–	14.2	Lazarev and Ershova (2023)
		30% WPC	8.1	–	–	–	15.1	

Note: Values are expressed as percentages.
CHO, carbohydrates; *LW*, liquid whey; *W-LPH*, low-phenylalanine whey protein hydrolysate; *WPC*, whey protein concentrate; *WPH*, whey protein hydrolysate; *WPI*, whey protein isolate; *YAW*, yogurt acid whey.
[a]*Dry weight basis.*

The incorporation of whey ingredients into bakery and confectionery formulations is primarily intended to increase protein content, but it has also been shown to reduce carbohydrate content and improve the amino acid profile. Moreover, whey supplementation can ensure that products lacking essential amino acids, such as those that are egg- and dairy-free, have an adequate nutritional profile. Moreover, whey contributes to the novel field of "functional confections", considering all those confectionery products delivering health benefits due to their composition (Herrera-Cazares et al., 2021). Nevertheless, these improvements must be evaluated in the context of overall product quality, consumer acceptance, and regulatory frameworks.

3. Physicochemical properties

3.1 Dough quality

Dough is a viscoelastic system characterized by a cohesive structure formed by gluten proteins, such as glutenin and gliadin, in wheat-flour dough. During mixing, gluten proteins are mechanically disrupted as the hydrated dough is mixed. Then, gluten proteins form disulfide bonds, generating the characteristic gluten network of the dough (Prieto-Vázquez del Mercado et al., 2022).

The properties of dough can significantly influence the quality of the final baked product (Nogueira et al., 2022). A farinograph is a tool used to determine the amount of water required to achieve the desired dough consistency during mixing, to analyze mixing characteristics, and to predict baking performance (Gani et al., 2015b). Key farinograph parameters include water absorption (WA), dough development time (DDT), dough stability time (DST), and mixing tolerance index (MTI), which assess dough resistance to deformation under constant temperature and mixing speed (Prieto-Vázquez del Mercado et al., 2022; Srikanlaya et al., 2018).

WA indicates the amount of water required to reach maximum dough consistency (500 Brabender Units, BU) (Prieto-Vázquez del Mercado et al., 2022). Table 10.2 shows how water absorption decreased when whey protein was incorporated. Nogueira et al. (2022) replaced 5%–30% of wheat flour in a bread formulation with whey, soy, and pea protein isolates and observed a 21.7% decrease in WA at 30% WPI substitution compared with the control. This reduction in WA was not observed in soy and pea protein products, due to the amino acid composition of these proteins, in which WPI has a higher content of hydrophobic amino acids than pea and soy protein isolate (Nogueira et al., 2022). Pořízka et al. (2023) replaced 5% of high-gluten wheat flour with WPI, resulting in a 5.7% reduction in WA. Similar results were observed by Satouf (2023), who formulated a flat bread replacing 100% of the water with liquid whey, leading to a 6% decrease in WA. These reductions in WA could be attributed to whey's poor water solubility, due to its high hydrophobic amino acid content, and to the formation of complex systems that hinder hydration (Nogueira et al., 2022; Pořízka et al., 2023).

DDT indicates the time from the initial addition of water to the point at which maximum dough consistency is reached. DST measures the dough's ability to maintain the structure and resist breakdown during mixing at 500 BU. MTI refers to the difference in BU from the top of the curve measured at 5 min after the peak is reached (Gani et al., 2015b; Pořízka et al., 2023). Table 10.2 illustrates how whey incorporation modifies farinograph parameters of different bread formulations. Gani et al. (2015b) replaced wheat flour with whey protein concentrate (WPC) or whey protein hydrolyzate (WPH) (5%–15%) and observed that WPC at 15% reduced DDT by 2.2 min, DST by 0.9 min, and MTI increased by 123.2% compared to the control. WPH at 15% also reduced DDT and ST by 0.8 and 0.6 min, respectively, and increased MTI by 102%. In contrast, Nogueira et al. (2022) reported that replacing 30% of wheat flour with whey protein isolated (WPI) increased DDT and MTI by 5.7 min and 160.7%, respectively, while reducing DST by 9.9 min. The increase in DDT and the decrease in DST were consequences of a delay in gluten network formation, due to whey protein restricting optimal hydration of gluten proteins and interfering with gluten network formation (Nogueira et al., 2022).

Dough extensibility (E) is associated with baking performance and refers to the force required to stretch and break the dough structure. It can be measured using an extensograph, alveograph, or texture analyzer (Gani et al., 2015b). Table 10.2 shows how extensograph parameters change with the addition of whey. Pořízka et al. (2023) reported a 5.9% increase in dough extensibility when 5% of wheat flour was replaced with WPI. Conversely, Satouf (2023) observed a 13.2% decrease in extensibility in flatbread when 100% of the water was replaced with liquid whey. A decrease in extensibility values could be due to the interference of sulfhydryl groups of the whey proteins in the normal sulfhydryl/disulfide interchange reactions during wheat flour dough development (Nogueira et al., 2022; Satouf, 2023). Some authors reported that substituting more than 5% of wheat flour with WPI results in inadequate dough consistency, hindering extensograph analysis (Nogueira et al., 2022; Pořízka et al., 2023).

Pasting properties involve different phenomena of starch, including starch gelatinization, pasting, and starch retrogradation, and can be measured using a Rapid Visco Analyzer (RVA) or a rheometer. The RVA provides key parameters

TABLE 10.2 Effects of the incorporation of whey ingredients on dough/blend characteristics and pasting properties of bakery and confectionery products.

Product	Whey incorporation	Dough/blend characteristics	Pasting properties	References
Bread	WPI as wheat flour substitute at 5%	*Farinograph parameters:* Water absorption ↓5.7% Dough development time ↑5 min Doug stability time ↑6 min *Extensograph parameters:* Resistance to extension ↑34% Extensibility ↑6% Ratio R/E ↑25%	–	Pořízka et al. (2023)
	LW as water substitute at 25%–100%	*Texture parameters:* Firmness ↑59% Hardness ↔ Springiness ↔	Peak viscosity ↑28% Breakdown viscosity ↑43% Final viscosity ↑9% Setback viscosity ↑491% Pasting temperature ↑4°C	Himashree et al. (2023)
	WPI as wheat flour substitute at 5%–30%	*Farinograph parameters:* Water absorption ↓22% Dough development time ↑5.7 min Mixing tolerance index ↑161% Doug stability time ↓9.9 min *Extensograph parameters:* Resistance to extension ↑44% Ratio R/E ↑52% Extensibility ↔	Peak viscosity ↓59% Breakdown viscosity ↓68% Final viscosity ↓54% Setback viscosity ↓55% Peak time ↓0.6 min Pasting temperature ↑1.6°C	Nogueira et al. (2022)
	WPC as wheat flour substitute at 5%–15%	*Farinograph parameters:* Water absorption ↓12% Dough development time ↓2.2 min Doug stability time ↓0.9 min Mixing tolerance index ↑123%	Peak viscosity ↓55% Breakdown viscosity ↓85% Trough/hold viscosity breakdown ↓42% Final viscosity ↓41% Setback viscosity ↓39%	Gani et al. (2015b)
	WPH as wheat flour substitute at 5%–15%	*Farinograph parameters:* Water absorption ↓15% Dough development time ↓0.8 min Doug stability time ↓0.6 min Mixing tolerance index ↑102%	Peak viscosity ↓64% Breakdown viscosity ↓78% Trough/hold viscosity breakdown ↓59% Final viscosity ↓57% Setback viscosity ↓54%	
Multigrain bread	LW containing 25% solids as water substitute at 100%	*Texture parameters:* Hardness ↑205% Gumminess ↑104% Cohesiveness ↓18% Springiness ↓15% Resilience ↓47%	–	Paul et al. (2016)

Gluten-free bread	WPC as rice flour substitute at 9%	–	*Before fermentation:* Peak viscosity ↓38% Breakdown viscosity ↓46% Trough/hold viscosity breakdown ↓33% Final viscosity ↓32% Setback viscosity ↓32% Peak time ↓0.2 min Pasting temperature ↔ *After fermentation:* Peak viscosity ↓35% Breakdown viscosity ↓43% Trough/hold viscosity breakdown ↓31% Final viscosity ↓20% Setback viscosity ↓10% Peak time ↓0.2 min Pasting temperature ↔	Yang et al. (2024)
	WPC as rice flour substitute at 2%–6% All breads contained hydroxy-propyl methylcellulose	*Farinograph parameters:* Water absorption ↔ Dough development time ↔ Doug stability time ↓0.73 min	–	Srikanlaya et al. (2018)
Flat bread	LW as water substitute at 25%–100%	*Farinograph parameters:* Water absorption ↓6% Dough development time ↓1.5 min Doug stability time ↑0.9 min Softening degree ↑85% *Alveogram properties:* Pressure ↓19% Extensibility ↓13%	–	Satouf (2023)
Gluten-free muffin	WPI added to achieve 6.2% of muffin	Water absorption ↓31%	Peak viscosity ↓48% Breakdown viscosity ↓56% Final viscosity ↓60% Setback viscosity ↓63% Pasting temperature ↑3.3°C	Bala et al. (2019)

Note: Comparisons were performed between the sample with the highest whey ingredient incorporation and its corresponding control sample. Arrows indicate, ↑ = increase, ↓ = decrease, ↔ = no significant difference.
LW, liquid whey; *WPC*, whey protein concentrate; *WPh*, whey protein hydrolyzate; *WPI*, whey protein isolate.

such as peak viscosity (PV), trough or hold viscosity (HV), breakdown (BV), final viscosity (FV), setback viscosity (SBV), pasting temperature (PTEM), and peak time (PT) (Balet et al., 2019). The analysis begins with sample hydration at temperatures below 50°C. Then, it is heated (typically to 95°C), where gelatinization occurs, simulating the cooking step. Pasting occurs immediately after gelatinization and refers to changes in viscosity before, during, and after the process. In the holding stage, a decrease in viscosity is observed at maximum temperature, and the curve starts to flatten. This is followed by the cooling stage or setback, where viscosity increases again as the starch granules are cooled to 50°C and retrogradation occurs. In the final holding stage, temperature remains constant while viscosity continues to increase until a plateau is reached, known as the final viscosity (Balet et al., 2019; Himashree et al., 2023).

Table 10.2 demonstrates the impact of whey addition on pasting properties. Himashree et al. (2023) replaced 25%, 50%, 75% and 100% of the water in the bread formulation with liquid whey and observed increases in PV, BV, FV, SBV, and PTEM by 27.9%, 42.9%, 9%, 491.2% and 4°C, respectively, at 100% replacement. These improvements in pasting properties were attributed to the gelatinization of whey proteins during heating, modifying the dough viscosity (Himashree et al., 2023). In contrast, Nogueira et al. (2022) found that replacing 30% of wheat flour with WPI decreased PV, BV, FV, SBV, and PT by 59.2%, 68.4%, 53.8%, 54.6% and 0.6 min, while PTEM increased by 1.6°C. These reductions in viscosity parameters could be due to the dilution of starch in the mixture (Nogueira et al., 2022). Yang et al. (2024) evaluated gluten-free bread with 9% WPC as a rice flour replacement and observed reductions in PV, HV, BV, FV, SBV, and PT before and after fermentation compared to a control rice dough. This was attributed to WPC competing with starch for water during gelatinization, thereby reducing water absorption and the dough's swelling capacity (Yang et al., 2024).

Whey can be incorporated into bakery products using different strategies, such as in powder form (replacing flour) or in liquid form (replacing water). Its incorporation significantly alters dough properties, such as water absorption and pasting behavior, and, depending on the amount and form of whey used, can impact the quality and functionality of the final product.

3.2 Volume

The volume modifications of bakery and confectionery products containing whey ingredients are summarized in Table 10.3. Volume is one of the most important characteristics of bread, as it serves as a key measurement for baking performance. Consumers are particularly interested in loaf volume because they prefer breads that appear light rather than dense. However, an excessively high loaf volume is not always desirable (Gani et al., 2015b). For instance, Zhou et al. (2018) examined the effect of WPC on wheat bread volume over a substitution range of 0%–30%. At 10% WPC, the specific volume decreased by 15%, likely due to gluten dilution and water competition. However, at a 30% substitution level, the specific volume increased by 18% compared to the control. This improvement was attributed to whey protein gelation, which reinforced the gaseous cell walls during baking. Despite the increase in volume, crumb hardness increased from 173 to 1291 g at 30% WPC. These findings suggest that while high levels of whey protein could improve structural expansion, they also produce a denser crumb. On the contrary, adding WPC into rice-based doughs enriched with hydroxypropyl methylcellulose (HPMC) at 4% could not increase the specific volume (Srikanlaya et al., 2018). The specific volume was consistent with the low gas retention. During fermentation, the maximum dough height was not higher than that of dough containing only HPMC, but the incorporation of whey protein increased CO_2 retention. Bread volume is influenced by the available sugars, which serve as substrates for yeast fermentation. A high supply of these substrates could improve CO_2 production. However, the bread volume also depends on the dough's CO_2 retention capacity.

Apparently, incorporating whey protein into cookies and biscuits has been shown to increase their volume. Borooah et al. (2024) replaced 70% of the wheat flour with WPC in a biscuit formulation. They observed a 38% increase in volume and a 16% increase in thickness, while the spread ratio decreased by 13%. The authors attributed these changes to the foaming and emulsifying properties of WPC, which facilitated greater air entrapment during mixing, leading to expanded air during baking and increased volume.

Similar results were reported by Hajas et al. (2022), who substituted red and green lentil flours for WPI at 25% in gluten-free cookies. The volume increased by 15% and 22%, and the spread ratio decreased by 7% and 12% for red and green lentil-flour cookies supplemented with WPI, compared with their controls, respectively. They attributed this to hydrogen bonds between whey protein and water, forming a protein network that increases the cookie's height.

In contrast, another study found different results regarding the spread ratio in wheat-flour-based crackers with the addition of WPI (En Tay et al., 2022). It showed that increasing WPI levels led to a higher spread ratio. Crackers containing 20% WPI exhibited a significantly higher spread ratio, 11% more, compared to the control. This differing

TABLE 10.3 Effects of the incorporation of whey ingredients on the physicochemical properties (volume; texture profile: hardness, fracturability, adhesiveness, cohesiveness, resilience, chewiness, gumminess, elasticity, firmness, and springiness; and color: luminosity or L*) of bakery and confectionery products.

Product	Whey incorporation	Volume	Texture profile	Color (L*)	Best formulation	References
Bread	LW as water substitute at 25%–100%	–	Hardness ↑53% Fracturability ↑31% Adhesiveness ↑16% Resilience ↑23% Chewiness ↑84% Gumminess ↑24% Springiness ↓27%	*Crust:* ↑12.1% *Crumb:* ↑7.9%	50% LW	Himashree et al. (2023)
	LW as water substitute at 25%	Specific volume ↑3%	Hardness ↓24% Chewiness ↓23% Gumminess ↔ Elasticity ↔	*Crust:* ↓4%	25% LW	Çiçek et al. (2025)
	Whey coacervate as milk, butter, and salt substitute at 100%	–	Springiness ↑9% Cohesiveness ↑48% Hardness ↔	*Crust:* ↔ *Crumb:* ↔	100% Whey coacervate	Ingrassia et al. (2022)
	WPC as wheat flour substitute at 5%–30%	Specific volume ↑18%	Hardness ↑646% Cohesiveness ↑ 16% Gumminess ↑744% Chewiness ↑770% Resilience ↑31% Springiness ↔	*Crumb:* ↓5%	15% WPC	Zhou et al. (2018)
	WPC as wheat flour substitute at 5%–15%	Volume ↑ 44%	Firmness ↑138%	–	5% WPC	Gani et al. (2015b)
	WPH as wheat flour substitute at 5%–15%	Volume ↑3%	Firmness ↑88%	–	5% WPH	
	WPC as wheat flour substitute at 20%	Specific volume ↓11%	Hardness ↑103% Chewiness ↑20% Cohesiveness ↓20% Springiness ↔	*Crust:* ↓46% *Crumb:* ↓11%	20% WPC	Ferreyra et al. (2021)
Sourdough bread	WPC as wheat flour substitute at 20%	Specific volume ↔	Hardness ↑186% Chewiness ↑147% Cohesiveness ↓14% Springiness ↔	*Crust:* ↓30% *Crumb:* ↓4%	20% WPC	Ferreyra et al. (2021)
Multigrain bread	LW containing 15% solids as water substitute at 100%	–	Hardness ↑34% Cohesiveness ↑44% Springiness ↑ 4% Gumminess ↑91% Chewiness ↑96%	–	15% solids	Paul et al. (2016)
Gluten-free bread	WPC as cassava starch and chickpea flour blend substitute at 10%–30%	Specific volume ↑136%	Hardness ↑223% Chewiness ↑109% Cohesiveness ↓10%	*Crust:* ↓21% *Crumb:* ↔	10% WPC	Komeroski et al. (2021)
Gluten-free bread	WPC as rice flour substitute at 4%. All breads contained hydroxypropyl methylcellulose.	Specific volume ↔	Hardness ↔ Springiness ↔ Cohesiveness ↔ Chewiness ↔	–	4% WPC	Srikanlaya et al. (2018)

Continued

TABLE 10.3 Effects of the incorporation of whey ingredients on the physicochemical properties (volume; texture profile: hardness, fracturability, adhesiveness, cohesiveness, resilience, chewiness, gumminess, elasticity, firmness, and springiness; and color: luminosity or L*) of bakery and confectionery products.—cont'd

Product	Whey incorporation	Volume	Texture profile	Color (L*)	Best formulation	References
Biscuit	WPC as wheat flour substitute at 10%–70%. All biscuits contained clove basil, jaggery, and ghee	Volume ↑38%	–	–	50% WPC	Borooah et al. (2024)
	W-LPH as wheat flour substitute at 5%–20%	–	Hardness ↓86% Fracturability ↔	↓8%	15% W-LPH	Li et al. (2024)
	WPC as wheat flour substitute at 7%–20%. Some biscuits contained lupin flour	Volume ↔	Hardness ↑171% Fracturability ↑446% Adhesiveness ↑100% Resilience ↔	–	7% WPC	Beshir et al. (2021)
	WPI added to achieve 7.5% of biscuit	–	Hardness ↑34% Fracturability ↑95%	↓5%	Control	Norton et al. (2020)
	WPC added to achieve 11%–15% protein in biscuits	–	Hardness ↑134%	–	11% preheated WPC	Andoyo et al. (2023)
	Preheated WPC added to achieve 11%–15% protein in biscuits	–	Hardness ↑79%	–		
Cookies	WPC as sorghum flour substitute at 8%–12%	D/T ratio ↑19%	Hardness ↑≈64%	↓25%	10% WPC	Zhu et al. (2023)
	WPC as corn flour substitute at 8%–12%	D/T ratio ↑26%	Hardness ↔	↓35%		
	WPI as green lentil flour substitute at 25% Some cookies contained inulin and xylitol	Volume ↑22%	Hardness ↓43% Adhesiveness ↔ Cohesiveness ↔	↔	25% WPI +6% inulin	Hajas et al. (2022)
	WPI as red lentil flour substitute at 25% Some cookies contained inulin and xylitol	Volume ↑15%	Hardness ↓48% Adhesiveness ↓77% Cohesiveness ↔	↔		
	WPC added to achieve 6%–24% of cookies Some cookies contained soy protein isolate	–	Hardness ↑43%	↔	24% WPC +24% soy protein isolate	Park et al. (2021)
	WPC as wheat flour substitute at 100%	–	Hardness ↑110% Chewiness ↑360%	↓5%	100% WPC	Ranok and Kupradit (2020)
	WPC as wheat flour substitute at 5%–30%	D/T ratio ↓37%	Hardness ↑154%	↓29%	10% WPC	Tang and Liu (2017)
	WPC as wheat flour substitute at 2%–6%	D/T ratio ↑49%	Hardness ↔	↓14%	4% WPC	Wani et al. (2015)

TABLE 10.3 Effects of the incorporation of whey ingredients on the physicochemical properties (volume; texture profile: hardness, fracturability, adhesiveness, cohesiveness, resilience, chewiness, gumminess, elasticity, firmness, and springiness; and color: luminosity or L*) of bakery and confectionery products.—cont'd

Product	Whey incorporation	Volume	Texture profile	Color (L*)	Best formulation	References
Cracker	WPI as wheat flour substitute at 5%–20%	D/T ratio ↑11%	Hardness ↓68 Fracturability ↑5%	↓5%	20% WPI	En Tay et al. (2022)
Cake	WPI added to achieve 6.1% of cake	–	Hardness ↑106% Chewiness ↑118% Cohesiveness ↑5% Springiness ↔	↔	Control	Norton et al. (2020)
Gluten-free cake	WPC as rice flour substitute at 5% and 10%. Some cakes contained xanthan gum	Specific volume ↑7%	Springiness ↑98% Firmness ↔	↓16%	10% WPC +0.3% xanthan Gum	Ghaemi et al. (2024)
Sponge cake	WPC as wheat flour substitute at 20%. Some cakes contained sourdough	Specific volume ↓15%	Hardness ↑162% Resilience ↑13% Chewiness ↑184% Springiness ↑4% Cohesiveness ↑3% Gumminess ↑172%	*Crust:* ↓19% *Crumb:* ↑8%	20% WPC +30% sourdough	Maravić et al. (2022)
	WPC as wheat flour substitute at 5.6% and 6.5% in a mixture of maize and rice flour	–	–	*Crust:* ↓4% *Crumb:* ↑9%	6.5% WPC	Ammar et al. (2021)
	WPI as egg white protein substitute at 12.5%–100%	Specific volume ↓7%	Resilience ↓28% Cohesiveness ↓17% Firmness ↔ Chewiness ↔ Springiness ↔	–	25% WPI	Díaz-Ramírez et al. (2016)
Banana cake	Blend of WPI and WPH added to achieve 10%–30% of cake	Volume ↑34%	Hardness ↔	↔	20% WPI/WPH	Camargo et al. (2018)
Cupcake	WPI added to achieve 11.7% of cupcake	–	Hardness ↑44% Adhesiveness ↑97% Chewiness ↑62% Cohesiveness ↑14% Gumminess ↑63% Resilience ↔ Springiness ↔	↔	Control	Norton et al. (2020)
Muffin	WPI as wheat flour substitute at 5%	Specific volume ↓14%	*Crust:* Hardness ↑17% Chewiness ↑26% Gumminess ↑24% Resilience ↔ Springiness ↔ Cohesiveness ↔ *Crumb:* Hardness ↔ Resilience ↔ Chewiness ↔ Springiness ↔ Cohesiveness ↔ Gumminess ↔	*Crust:* ↔ *Crumb:* ↔	5% WPH	Bekiroglu et al. (2024)

Continued

TABLE 10.3 Effects of the incorporation of whey ingredients on the physicochemical properties (volume; texture profile: hardness, fracturability, adhesiveness, cohesiveness, resilience, chewiness, gumminess, elasticity, firmness, and springiness; and color: luminosity or L*) of bakery and confectionery products.—cont'd

Product	Whey incorporation	Volume	Texture profile	Color (L*)	Best formulation	References
	WPH as wheat flour substitute at 5%	Specific volume ↓4%	*Crust:* Chewiness ↓21% Gumminess ↓22% Hardness ↔ Resilience ↔ Springiness ↔ Cohesiveness ↔	*Crust:* ↓8%		
			Crumb: Hardness ↔ Resilience ↔ Chewiness ↔ Springiness ↔ Cohesiveness ↔ Gumminess ↔	*Crumb:* ↓2%		
Gluten-free muffin	WPI added to achieve 6.2% of muffin	Specific volume ↑4.3%	Hardness ↓47% Chewiness ↓28% Springiness ↓11% Cohesiveness ↓40% Resilience ↓39%	*Crumb:* ↑5%	16.2% WPI	Bala et al. (2019)
Chocolate	WPI added to achieve 16%–24% of chocolate	–	Hardness ↑214% Fracturability ↑136%	↑13%	24% WPI	Nastaj et al. (2022)
	WPI added to achieve 12%–24% of chocolate	–	Hardness ↑99% Fracturability ↑101%	↑1%	24% WPI	Nastaj et al. (2024)
	WPH added to achieve 6% of chocolate	–	Hardness ↔	–	6% WPH	Rakin et al. (2023)
	WPC added to achieve 2.5%–10% of chocolate	–	Hardness ↓29%	–	10% WPC	Lapčíková et al. (2022)
Gelatin gummy	WPC added to achieve 0.5%–1% of gelatin gummy	–	Hardness ↑82% Springiness ↑22% Gumminess ↑148% Chewiness ↓81% Cohesiveness ↔	–	0.75% WPC	Kumkong et al. (2020)

Note: Comparisons were performed between the sample with the highest whey ingredient incorporation and its corresponding control sample. L* represents the brightness of the sample. Arrows indicate: ↑ = increase, ↓ = decrease, ↔ = no significant difference.
LW, liquid whey; *W-LPH*, low-phenylalanine whey protein hydrolysate; *WPC*, whey protein concentrate; *WPH*, whey protein hydrolysate; *WPI*, whey protein isolate.

outcome may be attributed to variations in the formulation, which could create a dough with lower viscosity, allowing the crackers to spread more rapidly during baking.

Interestingly, whey-protein-supplemented cakes tend to decrease in volume, whereas gluten-free cakes tend to increase in volume. For example, sponge cakes prepared with 20% WPC as a substitute for wheat flour had a 15% lower specific volume compared to the control sample (Maravić et al., 2022). The limited availability of water led to competition among ingredients, making the quantity and composition of the aqueous phase crucial for determining batter density. Surface-active proteins, such as whey proteins, could form films around air bubbles and create hydrogen bonds with water molecules. This process could redistribute water from gluten to other components, resulting in partial dehydration of the gluten network. Consequently, this restricts air expansion and reduces the specific volume of the final product. Díaz-

Ramírez et al. (2016) observed similar outcomes in their study on sponge cakes, where they used WPI to replace egg white protein at substitution levels ranging from 12.5%–100%. The authors found that at 12.5% and 25% WPI substitution, the batter viscosity was adequate to allow bubbles to expand before escaping, leading to a decrease in cell density and an increase in specific volume. However, at 50% and 100% WPI replacement levels, the viscosity decreased, allowing air to escape and resulting in a lower specific volume.

Meanwhile, specific volume increased in gluten-free cake formulated with rice flour and a 10% addition of WPC compared to those without WPC (Ghaemi et al., 2024). The increase in volume could be associated with whey proteins' ability to form a thermal gel during baking. At this stage, protein denaturation occurs, resulting in new interactions between proteins. This process increases the batter's viscosity and traps air, increasing its specific volume. A comparable outcome was observed with banana cakes containing 30% of a blend of WPI and WPH, resulting in a 34% increase in volume compared to the control (Camargo et al., 2018). According to the authors, this was related to the foam and emulsification properties of the whey protein.

Whey protein supplementation can enhance the volume of breads, cookies, and biscuits, primarily due to its foaming and emulsifying properties. However, in gluten-containing cakes, effectiveness is limited by water competition and structural interference. In contrast, in gluten-free cakes, it can improve volume through thermal gelation and air entrapment. These outcomes underscore the need for specific formulations for each type of bakery product to leverage the techno-functional properties of whey proteins.

3.3 Texture

Texture encompasses the physical properties that define the quality and consumer acceptance of bakery and confectionery products. Achieving the desired texture depends on the interactions between ingredients and processing conditions. Therefore, it is essential to understand the functional roles of key ingredients to predict and control textural outcomes (Voong and Westpha, 2023). Table 10.3 shows the effect of whey ingredients on the texture properties of bakery and confectionery products.

Himashree et al. (2023) conducted a study in which they replaced water with liquid whey in wheat bread at substitution levels ranging from 25%–100%. They found that as the level of whey increased, hardness, adhesiveness, resilience, fracturability, gumminess, and chewiness increased progressively. For example, at 100% whey, hardness increased by 53%, gumminess by 24%, and chewiness by 84% compared to the control. Hardness, a primary mechanical attribute, refers to the force required to compress food between the teeth. At the same time, gumminess is the effort, and chewiness is the work required to break up the food for swallowing (Rosenthal and Chen, 2023). These increments were associated with the addition of whey and its interaction with the gluten network during fermentation. Notably, springiness decreased with higher whey concentrations due to increased protein interactions that limit elasticity. Springiness indicates how well a product returns to its original shape after deformation. Bread containing 50% whey exhibited a balanced texture, providing acceptable levels of hardness, adhesiveness, resilience, chewiness, cohesiveness, and gumminess.

A recent study reported that wheat bread made with 25% liquid whey was softer and had lower chewiness than the control (Çiçek et al., 2025). This softening effect was attributed to the incorporation of whey, which contains organic acids and calcium that could disrupt the gluten network and increase protease activity. Additionally, elasticity and gumminess remained unchanged. The incorporation of WPC into sourdough-fermented wheat bread affected its texture. According to Ferreyra et al. (2021), replacing 20% of wheat flour with WPC resulted in a 186% increase in hardness and a 147% increase in chewiness, while maintaining the same springiness as the control sourdough bread that did not contain WPC. The authors attributed these increments to the gel structure formed by gluten and to the aggregation of whey proteins at pH values near their isoelectric points. However, the cohesiveness, the extent to which a material can be deformed before it ruptures, of the bread significantly decreased due to disruptions in the gluten network caused by the high concentration of whey protein.

Whey proteins play a significant role in increasing the hardness and fracturability of most cookie and biscuit formulations. Hardness, as mentioned, is the force required to compress the sample, while fracturability is the force at which the product crumbles or breaks (Rosenthal and Chen, 2023). For instance, adding 7.5% to biscuits increased hardness by 34% and fracturability by 95% (Beshir et al., 2021). Similarly, in a study by Zhu et al. (2023), gluten-free cookies made with sorghum flour containing 8%–11% WPC were 64% harder than those without WPC. These changes can be attributed to the gelation and thermosetting properties of whey proteins, as well as their larger spread ratio. The increased hardness of the cookies could also be due to interactions between proteins and starch via hydrogen bonding.

The hardness of cookies also increased when the WPC was added at levels ranging from 6%–24% (Park et al., 2021). Cookies made with 24% WPC were 43% harder than those made with soy protein isolates. This difference may be due to the higher density of soy protein cookie dough compared to the WPC sample, which tends to have lower dough density and results in greater hardness. The authors noted that the dough made with soy protein was the most challenging to knead, primarily due to its high density, which made it easier to crumble. Even after baking, these cookies crumbled more readily than those made with WPC, which were the easiest to knead. Different outcomes were observed in crackers in which wheat flour was replaced with 20% (En Tay et al., 2022). Fracturability increased by 5%, while hardness decreased by 68%. In this case, the gluten network may have been affected by the incorporation of WPI, which reduced dough strength, resulting in decreased cracker thickness and increased susceptibility to breakage.

Whey protein has notable effects on cake texture, particularly hardness and springiness. For example, sponge cakes that used WPC at 20% to replace wheat flour showed a 162% increase in hardness and a 4% increase in springiness compared with the control (Maravić et al., 2022). Authors attributed this hardness to the high solubility of whey protein, which reduces the available water in the system for sugar solubilization. Therefore, the insolubilized sugar crystallizes during baking, thereby increasing its hardness. Probably, the increased hardness is occurring only in the crust and not in the crumb. In a separate study by Bekiroglu et al. (2024), muffins made with 20% WPC as a wheat-flour substitute showed a 17% increase in crust hardness, while the crumb texture remained unchanged compared to the control. In addition to sugar crystallization, the authors noted that during baking, whey proteins aggregate, forming a protein network that could increase hardness.

The springiness was also higher in gluten-free cakes containing rice flour and 10% WPC (Ghaemi et al., 2024). Springiness increased by 98% because whey protein holds more water, preventing excessive evaporation and contributing to crumb flexibility. In comparison, the rice flour control sample lacks a protein network, leading to increased water evaporation during baking and reduced springiness.

The incorporation of whey protein ingredients into chocolate formulations significantly impacts their textural properties, particularly hardness. Chocolate hardness is a critical quality attribute influenced by the interaction between the crystallized continuous fat phase and the dispersed solid particles. Lapčíková et al. (2022) supplemented milk chocolate with whey powder at concentrations ranging from 2.5%–10% and observed a clear softening effect, correlated with whey concentration. The hardness values decreased from 73 N in the control sample to 52 N at the highest whey concentration. This reduction was attributed to the plasticizing effect of whey particles, which disrupt highly ordered cocoa fat crystals, resulting in a more ductile fracture pattern rather than a brittle one.

Conversely, Rakin et al. (2023) reported no significant differences in hardness for milk chocolates supplemented with either 6% WPC or WPH compared to the control sample. The hardness of the control, WPC, and WPH samples was 572, 549, and 570 g, respectively. These results suggest that moderate whey protein supplementation does not necessarily compromise chocolate hardness. However, the study indicated that WPH significantly increased chocolate viscosity, which could indirectly affect textural perception during consumption. The effect of WPI on chocolate hardness was also investigated by Nastaj et al. (2022). They formulated high-protein (16%–24%), sugar-free dark chocolates without tempering. Hardness increased with the whey protein concentration, reaching 260 N at the highest supplementation level, compared with 83 N in the control sample containing sucrose. This increase was primarily due to the formation of a denser protein network and enhanced interactions between the protein and other chocolate components, resulting in greater structural resistance to mechanical deformation.

In addition to chocolate, the effect of whey protein on the textural profile of gelatin-based gummy confections has also been studied. Kumkong et al. (2020) evaluated the addition of 1% WPC to a gelatin gummy containing gelatin, sucrose, and glucose syrup. The addition of WPC increased hardness by 82%, springiness by 22% and gumminess by 148%, compared to the control. The authors explained that WPC disrupts the triple helix structure of gelatin, thereby altering inter- and intramolecular crosslinking, particularly involving lysine and hydroxylysine residues.

The impact of whey protein on texture varies depending on the product type and formulation. In bread, cookies, biscuits, and cakes, whey protein increases hardness through gelation, protein aggregation, and water-holding capacity. It also improves springiness in gluten-free cakes by enhancing water retention and network formation. On the other hand, chocolate formulations respond differently. Low to moderate levels of whey protein can soften chocolate by disrupting fat crystals. Still, when high levels of protein are added to chocolate, it tends to harden due to denser protein networks and increased interactions among components. These findings highlight the importance of adjusting whey protein incorporation to the specific structural and functional requirements of each product.

3.4 Color

Color is an essential factor to consider when using liquid whey or whey protein as a partial or total replacement, as it significantly influences consumer acceptability and marketability. Table 10.3 illustrates the changes in color, particularly the L* values, of bakery and confectionery products formulated with whey ingredients. For instance, bread samples in which water was replaced with liquid whey exhibited an increasing trend in both crust and crumb lightness (L*) as the concentration of whey increased, indicating a lighter appearance (Himashree et al., 2023). The control bread had a crust L* value of 56.0, which increased to 62.8 with 100% whey substitution. Similarly, crumb L* values increased from 72.9 to 78.6. This trend contrasts with the darkening often observed in whey-enriched cookies and cakes. Although redness (a*) values of the crust increased with the addition of whey, yellowness (b*) values decreased, possibly due to caramelization of sugars. Furthermore, the color difference (ΔE) remained minimal at 50% substitution, suggesting no significant visual differences relative to the control.

Different outcomes were observed in wheat bread supplemented with WPC. Zhou et al. (2018) assessed the effect of adding WPC in concentrations ranging from 5%–30% on the color parameters of breadcrumbs. They found that as WPC levels increased, L* values decreased, with a 5% decrease in the sample containing 30% WPC compared to the control. This darkening effect was attributed to the Maillard reactions, which were favored by the presence of amino acids in whey protein and reducing sugars in the formulation. The authors noted that while WPC contributed to bread browning, its impact was less intense than that of soy protein in other samples, due to differences in crust hardness, which modulate heat transfer during baking.

In wheat-flour-based biscuits, the addition of WPI at 7.5% influenced the color parameters of the biscuits (Norton et al., 2020). The biscuits showed a decrease in L* value, indicating a darker appearance than the control samples. This effect may be due to the increased presence of reducing sugars and amino acids found in whey proteins, which promote Maillard reactions during baking. In addition, whey proteins provide less available water than wheat flour, creating conditions that favor the Maillard reaction.

Similar findings were reported by Zhu et al. (2023), who investigated the impact of WPC incorporation (8%–12%) on the color of gluten-free sorghum and corn cookies. As WPC content increased, the cookies exhibited a decline in L* values, indicating darker cookies than the non-supplemented cookies. Corn cookies formulated with 12% WPC had the lowest L* value, representing a 35% reduction, while sorghum cookies with 12% WPC showed a 25% reduction compared to their control. The color changes were mainly attributed to the Maillard reaction and caramelization during baking. The whey protein ingredients contain lactose residues and free amino groups, particularly lysine, which facilitate the Maillard reaction. Furthermore, whey proteins have higher solubility, allowing greater interaction with reducing sugars, thereby further enhancing browning. Thus, the higher the whey protein concentration, the darker the color developed in the cookies. Despite the trend of decreasing L* values with the incorporation of whey protein in biscuits and cookies, Park et al. (2021) reported no significant differences between L* values of cookies containing 24% WPC and the control sample. The authors related this finding to the fact that all treatments had the same sugar content and were exposed to heat under similar conditions.

Comparable results were observed in sponge cakes and muffins regarding color change. Maravić et al. (2022) reported a significant decrease in crust L* values when 20% of wheat flour was substituted with WPC in sponge cake. The control sample exhibited an L* value of 77.6, while the 20% WPC sample had an L* value of 62.6, indicating darkening. The authors attributed this change to an increase in free amino groups, particularly lysine, and reducing sugars, both of which promote Maillard reactions. Interestingly, while crusts darkened with protein addition, the crumb lightened. This inverse behavior was explained by the intrinsic brightness of whey protein, which can exceed that of wheat flour. For instance, the crumb L* values increased from 77.4 in the control sample to 83.3 in the sponge cake with 20% WPC. These results suggest that the initial color of WPC and the redistribution of water within the crumb matrix contribute to a lighter internal structure.

In muffins, the inclusion of WPH resulted in lower L* values in the crust, particularly at higher degrees of hydrolysis (Bekiroglu et al., 2024). The L* value of the crust decreased from 61.3 in the control sample to 56.4 in the muffin with 5% WPH at a 15% hydrolysis degree. This change was accompanied by an increase in a* values, indicating a more intense red-brown coloration. Conversely, the crumb L* value showed an opposite trend, rising from 76.2 to 78.5 in the 5% WPI sample. The authors noted that enzymatic hydrolysis increased the availability of amino groups, promoting browning reactions in the crust and enhancing brightness in the crumb due to improved protein dispersion and solubility.

The incorporation of whey protein into chocolate formulations affects color parameters, particularly lightness (L*), which indicates the degree of surface luminescence and is often associated with fat bloom or surface crystallization. In untempered white chocolates supplemented with 24% WPI, the L* value was 90.4, higher than the 86.4 of the control sample (Nastaj et al., 2024). Although there were slight variations, the differences in L* across samples were relatively minor, with all values within the expected range for white chocolates. The increase in luminescence with WPI content contrasts with trends observed in bakery products, where higher whey protein levels promote browning through Maillard reactions. Nevertheless, in chocolate, the absence of high-temperature processing, combined with the reduced lactose concentration in WPI and low water activity, limits non-enzymatic browning reactions. Similarly, sugar-free dark chocolates formulated with 24% WPI exhibited L* values that were 13% higher than those of the control chocolate (Nastaj et al., 2022). However, when larger amounts of whey protein were added, the resulting surfaces showed less luminescence. This decrease in L* values was attributed to the greater density of protein networks, which scatter light.

The effect of whey protein on color depends heavily on the type of product. In bakery products, the addition of whey protein consistently darkens the crust due to Maillard reactions. In contrast, it lightens the crumb by redistributing water and the intrinsic brightness of whey proteins. Meanwhile, chocolate processed at lower temperatures and under low-moisture conditions did not exhibit darkening. Instead, whey proteins maintain or enhance chocolate brightness. This suggests that temperature conditions and water activity have a greater influence on the impact of whey protein on color than protein concentration alone.

4. Sensory acceptability

The addition of whey protein and hydrolysates can influence sensory characteristics, including appearance, texture, flavor, and overall acceptability. Table 10.4 summarizes the overall acceptance of bakery and confectionery products supplemented with whey ingredients and the methods used to evaluate consumer acceptance. This includes details on the number of judges, whether they were trained, and the hedonic scale employed.

Several studies used a limited number of judges for hedonic assays. However, it is recommended that these evaluations include at least 25 untrained judges during the development phase to achieve more reliable results (O'Sullivan, 2017). While using trained judges might seem appropriate, it can lead to misleading outcomes, as they often evaluate the product from a different perspective than the average consumer (Lawless and Heymann, 2010).

In a study conducted by Komeroski et al. (2021), gluten-free bread formulated with a combination of cassava starch, chickpea flour, and WPC exhibited improved sensory acceptability, particularly at moderate WPC levels. The study compared the sensory attributes of breads containing 10%, 20%, and 30% WPC with those of control samples made with a blend of cassava and chickpea flours. Among non-celiac panelists, breads with 10% and 20% WPC achieved favorable sensory scores, particularly in appearance, texture, and overall acceptance, with scores above 6.5 on a 9-point hedonic scale. Similarly, among celiac judges, the 10% WPC bread obtained the highest overall acceptability score of 7.5, the same as the control. However, higher supplementation levels led to greater crumb hardness, which may have contributed to lower overall acceptability. These findings suggest that moderate incorporation of WPC improves both nutritional and sensory profiles of gluten-free bread. In contrast, higher concentrations could affect texture and consumer acceptance due to increased hardness and darker coloration.

Comparable outcomes were observed in multigrain bread supplemented with whey that contained 15% total solids, which served as a water substitute (Paul et al., 2016). Although the addition of whey had minimal impact on the dough preparation process, it significantly influenced the sensory and structural characteristics of the final product. Bread made with 15% whey solids, along with optimized proofing conditions (6% yeast, 45°C), received higher ratings in sensory evaluations than the multigrain control. Specifically, scores for both flavor and overall acceptability increased from 7.1 to 7.4 on a 9-point hedonic scale. Judges also noted improvements in texture and taste due to enhanced Maillard reactions and the higher protein content, which contributed to better flavor and structure.

Moreover, biscuits formulated with WPC at substitution levels ranging from 10%–70% showed varying sensory profiles and consumer preferences (Borooah et al., 2024). Biscuits prepared with a 30% WPC replacement for wheat flour were perceived as the most acceptable, attributed to the fluffy texture produced by the emulsifying properties of whey protein. Those with higher substitution levels, up to 70%, received lower scores due to a lack of flavor-pairing from the Maillard reaction, despite increased nutritional value. In contrast, the study by Gani et al. (2015a) found that cookies supplemented with 5%, 10%, and 15% WPH received progressively lower scores for appearance, color, aroma, taste, texture, and overall acceptability. While 5% WPH cookies had an overall acceptability score above 6, formulations containing 10% and 15% WPH had lower scores of 5.4 and 4.2, respectively. This decline was attributed to increased hardness, darker coloration due to Maillard browning, and bitterness from hydrolyzed peptides.

Further evidence of sensory limitations at high whey protein concentrations is provided by Norton et al. (2020), who evaluated consumer perception of biscuits fortified with WPI. They reported lower overall acceptance, moistness, and ease of eating in biscuits supplemented with 7.5% of WPI. This was particularly pronounced in older adults, who described the whey protein biscuits as drier and harder. Mouth-drying was specifically identified as a significant negative sensory characteristic.

Regarding cakes, Camargo et al. (2018) studied the sensory effects of adding a blend of WPI and WPH into banana cakes. Four formulations were developed with 0%, 10%, 20%, and 30% of whey protein. They found that while

TABLE 10.4 Effects of the incorporation of whey ingredients on the overall acceptance of bakery and confectionery products.

Product	Whey incorporation	*n*	Training	Sample	Overall acceptance	Scale	References
Bread	WPC or WPH as wheat flour substitute at 5%–15%	9	Yes	Control	6.7	7	Gani et al. (2015b)
				5% WPC	6.1		
				5% WPH	6.3		
	WPI as wheat flour substitute at 5%–15%	20	No	Control	4[a]	5	Pořízka et al. (2023)
				5% WPI	4[a]		
Multigrain bread	LW containing 15% solids as water substitute at 100%	15	No	Control	7.1	9	Paul et al. (2016)
				15% solids	7.4		
Gluten-free bread	WPC as cassava starch and chickpea flour blend substitute at 10%–30%	11	No (Non-celiac)	Control	5.8	9	Komeroski et al. (2021)
				20% WPC	6.7		
		11	No (Celiac)	Control	7.5		
				10% WPC	7.5		
Flat bread	LW as water substitute at 25%–100%	10	No	Control	79.4	100	Satouf (2023)
				50% LW	82.8		
Biscuits	WPC as wheat flour substitute at 10%–70%	20	Yes	Control	6.8	9	Borooah et al. (2024)
				30% WPC	8.2		
	W-LPH as wheat flour substitute at 5%–20%	15	No	Control	≈8.6	9	Li et al. (2024)
				15% W-LPH	≈8.8		
	WPC as wheat flour substitute at 7%–20%	10	Yes	Control	83.0	100	Beshir et al. (2021)
				7% WPC	93.7		
	WPI added to achieve 7.5% of biscuit	84	No	Control	6.2	9	Norton et al. (2020)
				7.5% WPI	5.3		
Cookies	WPC added to achieve 6%–24% of cookies	120	No	Control	4.9	9	Park et al. (2021)
				15% WPC	5.1		
	WPC as wheat flour substitute at 100%	30	No	Control	6.7	9	Ranok and Kupradit (2020)
				100% WPC	6.4		
	WPC as wheat flour substitute at 5%–30%	50	No	Control	6.9	9	Tang and Liu (2017)
				10% WPC	6.2		
	WPC as wheat flour substitute at 2%–6%	–	No	Control	4.6	5	Wani et al. (2015)
				4% WPC	4.8		
	WPC as wheat flour substitute at 5%–10%	60	No	Control	6.9	9	Rababah et al. (2022)
				10% WPC	7.3		
	WPC or WPH as wheat flour substitute at 5%–15%	9	Yes	Control	6.7	7	Gani et al. (2015a)
				5% WPC	6.1		
				5% WPH	6.3		

Continued

TABLE 10.4 Effects of the incorporation of whey ingredients on the overall acceptance of bakery and confectionery products.—cont'd

Product	Whey incorporation	*n*	Training	Sample	Overall acceptance	Scale	References
Crackers	WPI as wheat flour substitute at 5%–20%	20	No	Control	≈3.8	5	En Tay et al. (2022)
				20% WPI	≈4.3		
Cake	WPI added to achieve 6.1% of cake.	84	No	Control	6.6	9	Norton et al. (2020)
				6.1% WPI	5.0		
Gluten-free cake	WPC as rice flour substitute at 5% and 10%	10	No	Control	2.5	5	Ghaemi et al. (2024)
				10% WPC	4.2		
Sponge cake	WPC as wheat flour substitute at 5.6% and 6.5% in a mixture of maize and rice flour	60	No	Control	3.9	6	Ammar et al. (2021)
				6.5% WPC	3.8		
	WPI as egg white protein substitute at 12.5%–100%	125	No	Control	21.3	28	Díaz-Ramírez et al. (2016)
				25% WPI	21.2		
Banana cake	Blend of WPI and WPH added to achieve 10%–30% of cake	61	No	Control	7.4	9	Camargo et al. (2018)
				20% WPI/WPH	7.4		
Cupcake	WPI added to achieve 11.7% of cupcake	70	No	Control	65.4	100	Norton et al. (2020)
				11.7% WPI	51.3		
Muffin	WPC as wheat flour substitute at 33.3%	60	No	Control	7.4	9	Machado et al. (2021)
				33.3% WPC	7.2		
Marshmallow	WPC added to achieve 5%–30% of marshmallow	11	No	Control	4.8	5	Lazarev and Ershova (2023)
				15% WPC	4.7		

Note: Overall acceptance is expressed as mean values, and for each study, the sample with the highest overall acceptance score is presented, along with its corresponding control sample.
LW, liquid whey; *W-LPH*, low-phenylalanine whey protein hydrolysate; *WPC*, whey protein concentrate; *WPH*, whey protein hydrolysate; *WPI*, whey protein isolate.
[a]*Values expressed as median.*

appearance, color, flavor, and overall acceptance remained similar across all treatments, the highest incorporation level exhibited lower texture acceptance, with a score of 6.69, compared to 7.41–7.62 for the other treatments. Despite this, all formulations received favorable sensory scores, with global acceptance values above 7 on the 9-point hedonic scale. Similarly, muffins developed for hemodialysis patients, enriched with 33% WPC, showed comparable appearance, texture, taste, and overall acceptance compared to the control, as assessed by 60 untrained judges (Machado et al., 2021).

Sugar-free marshmallows enriched with 15% WPC received scores comparable to the control sample across several attributes, including taste, smell, texture, color, consistency, structure, surface, and form (Lazarev and Ershova, 2023). In contrast, the 5% WPC sample was described as having a dry, dense texture due to its low moisture content, while the 30% WPC sample exhibited a strong protein odor and an unpleasant taste.

Whey protein supplementation can improve the sensory acceptance of bakery and confectionery products when used in moderation. It has the potential to improve the flavor, texture, and overall acceptability of bread, biscuits, cakes, and marshmallows when whey protein is incorporated at levels that do not compromise product structure or moisture retention. However, excessive whey protein can lead to undesirable effects, such as increased hardness, dryness, and off-flavors. Therefore, it is essential to optimize formulations to balance an enhanced nutritional profile with consumer acceptance.

5. Potential biological effects

Bakery and confectionery products have been assessed for their potential biological effects, with a primary focus on antioxidant capacity. These assays include the 1,1-diphenyl-2-picrylhydrazyl (DPPH) free radical scavenging assay, the 2,2-azino-bis(3-ethylbenzthiazoline-6-sulfonic acid) (ABTS) free radical scavenging assay, the ferric reducing antioxidant power (FRAP) assay, and Fe^{2+}–chelating activity tests. Additionally, a clinical trial is investigating changes in weight and appetite associated with these products. For example, Li et al. (2024) demonstrated that substituting 20% of the wheat flour with W-LPH significantly enhanced antioxidant activity, with DPPH and ABTS radical-scavenging values increasing by approximately 40% and 71%, respectively, compared to the control. This effect was attributed to the bioactive peptides generated during enzymatic hydrolysis, which are known to exhibit strong electron-donating capacity and radical-scavenging activity.

In biscuits enriched with clove basil, jaggery, and ghee, replacing 70% of the wheat flour with WPC increased the DPPH radical-scavenging capacity by 106% compared to the control (Borooah et al., 2024). This improvement was primarily attributed to the phenolic compounds present in clove basil. Consistently, when used as a complete substitute for wheat flour in gluten-free cookies, WPC significantly enhanced antioxidant activity (Ranok and Kupradit, 2020). Compared to the control, the ABTS radical scavenging activity increased by approximately 33%, the FRAP by 54%, and the Fe^{2+}–chelating activity by 123%. These improvements were even greater than those observed in samples made with a combination of 50% WPC and 50% riceberry flour, which is recognized for its anthocyanin content.

According to Rakin et al. (2023), chocolates enriched with WPH showed a 17% increase in DPPH radical-scavenging activity and a 37% increase in ABTS radical-scavenging activity compared with the control. The authors noted that the presence of bioactive peptides produced through enzymatic hydrolysis was responsible for these outcomes. Importantly, there were no significant negative changes in sensory characteristics, indicating that moderate levels of WPH can enhance chocolate's oxidative stability without compromising its quality.

On the other hand, biscuits fortified with WPI at 24.1% of the formulation were evaluated in an 8-week clinical trial involving participants with overweight and obesity (Hassanzadeh-Rostami et al., 2020). Those who consumed these biscuits showed a mean reduction of 1.86 kg in body weight and a decrease of 2.6 cm in waist circumference compared to the control group. These outcomes were attributed to the appetite-suppressing effects of whey protein, including increased GLP-1 secretion and reduced energy intake, with no effects on lean body mass. The study highlighted the potential of whey protein-enriched bakery products to support weight management when combined with a meal plan.

6. Conclusions

Incorporating whey ingredients into bakery and confectionery products offers numerous opportunities for improvement. One of the main advantages is enhancing nutritional profiles by increasing protein content, lowering carbohydrate levels, and providing essential amino acids. These nutritional gains are accompanied by notable changes in physicochemical properties, including volume, texture, and color.

Whey proteins, especially, increase the volume of breads, cookies, and gluten-free cakes due to their effective foaming, emulsifying, and thermal gelling abilities. Conversely, in gluten-containing cakes, adding whey reduces volume due to water competition and structural interference. Whey proteins significantly influence textural properties such as hardness and springiness and can vary with product type, protein amount, and specific formulation.

Although moderate use of whey generally enhances sensory acceptance, excessive amounts can negatively impact texture, moisture retention, and taste. This underscores the importance of careful formulation optimization. Ultimately, a strategic approach to using whey ingredients can balance increased nutritional value with favorable consumer acceptance, leading to functional and commercially successful bakery and confectionery products.

References

Ammar, I., Gharsallah, H., Ben Brahim, A., Attia, H., Ayadi, M.A., Hadrich, B., Felfoul, I., 2021. Optimization of gluten-free sponge cake fortified with whey protein concentrate using mixture design methodology. Food Chem. 343. https://doi.org/10.1016/j.foodchem.2020.128457.

Andoyo, R., Rahmasari, S.D., Moody, S.D., Nurhasanah, S., 2023. The effect of preheated whey protein concentrate addition on high protein biscuit. IOP Conf. Ser. Earth Environ. Sci. 1230 (1), 012167. https://doi.org/10.1088/1755-1315/1230/1/012167.

Bala, M., Arun Kumar, T.V., Tushir, S., Nanda, S.K., Gupta, R.K., 2019. Quality protein maize based muffins: influence of non-gluten proteins on batter and muffin characteristics. J. Food Sci. Technol. 56 (2), 713–723. https://doi.org/10.1007/s13197-018-3529-8.

Balet, S., Guelpa, A., Fox, G., Manley, M., 2019. Rapid visco analyser (RVA) as a tool for measuring starch-related physiochemical properties in cereals: a review. Food Anal. Methods 12 (19), 2344–2360. https://doi.org/10.1007/s12161-019-01581-w.

Bekiroglu, H., Karaman, S., Bozkurt, F., Sagdic, O., 2024. Characterization of some physicochemical, textural, and antioxidant properties of muffins fortified with hydrolyzed whey protein. Food Sci. Nutr. 12, 8105–8117. https://doi.org/10.1002/fsn3.4422.

Beshir, E.O., Khallaf, M.F., El-Daim, Y.A.A., Awad, R.A., Kamel, M.A.A., 2021. Sweet lupin and whey protein concentrate as supplementants for utilizing in semi-hard biscuit and its chemical properties. Egypt. J. Chem. 64, 7417–7427. https://doi.org/10.21608/ejchem.2021.98412.4583.

Borooah, R., Das, B., Dutta Baruah, L., Alam, S., Saikia, A., Das Purkayastha, M., 2024. Novel biscuits using whey protein and *Ocimum gratissimum* to mitigate protein burden: quality parameters, *in vitro* protein digestibility, SWOT analysis and implication on public health. Food Human. 3, 100318. https://doi.org/10.1016/j.foohum.2024.100318.

Camargo, L.R., Silva, L.M., Komeroski, M.R., Kist, T.B.L., Rodrigues, C.E., Rios, A. de O., Silva, M.M., Doneda, D., Schmidt, H. de O., Oliveira, V. R., 2018. Effect of whey protein addition on the nutritional, technological and sensory quality of banana cake. Int. J. Food Sci. Technol. 53 (11), 2617–2623. https://doi.org/10.1111/ijfs.13857.

Çiçek, S., Aydin, F., Bozoğlu, F.T., 2025. Enrichment of bread with whey treated with high hydrostatic pressure. Res. Agric. Sci. 56 (2), 102–111. https://doi.org/10.17097/agricultureatauni.1500607.

De Almeida, M.P.G., Mockaitis, G., Weissbrodt, D.G., 2023. Got whey? Sustainability endpoints for the dairy industry through resource biorecovery. Fermentation 9 (10), 897. https://doi.org/10.3390/fermentation9100897.

Díaz-Ramírez, M., Calderón-Domínguez, G., García-Garibay, M., Jiménez-Guzmán, J., Villanueva-Carvajal, A., Salgado-Cruz, M. de la P., Arizmendi-Cotero, D., Del Moral-Ramírez, E., 2016. Effect of whey protein isolate addition on physical, structural and sensory properties of sponge cake. Food Hydrocoll. 61, 633–639. https://doi.org/10.1016/j.foodhyd.2016.06.020.

El-Aidie, S.A.M., Khalifa, G.S.A., 2024. Innovative applications of whey protein for sustainable dairy industry: environmental and technological perspectives-a comprehensive review. Compr. Rev. Food Sci. Food Saf. 23 (2), 13319. https://doi.org/10.1111/1541-4337.13319.

En Tay, R.R., Agatha, T., Somang, G., Yuliarti, O., Lin Tan, E.L., 2022. Structuring wheat flour-based crackers using whey protein isolate. Int. Dairy J. 128, 105314. https://doi.org/10.1016/j.idairyj.2021.105314.

Ferreyra, L.S., Verdini, R.A., Soazo, M., Piccirilli, G.N., 2021. Impact of whey protein addition on wheat bread fermented with a spontaneous sourdough. Int. J. Food Sci. Technol. 56 (9), 4738–4745. https://doi.org/10.1111/ijfs.15052.

Flinois, J.C., Dando, R., Padilla-Zakour, O.I., 2019. Yogurt acid whey utilization for production of baked goods: pancakes and pizza crust. Foods 8 (12), 615. https://doi.org/10.3390/foods8120615.

Fluerasu, D., Neagu, C., Dossa, S., Negrea, M., Jianu, C., Berbecea, A., Stoin, D., Lalescu, D., Brezovan, D., Cseh, L., 2025. The use of whey powder to improve bread quality: a sustainable solution for utilizing dary by-products. Foods 14 (16), 2911. https://doi.org/10.3390/foods14162911.

Gani, A., Broadway, A.A., Ahmad, M., Ashwar, B.A., Wani, A.A., Wani, S.M., Masoodi, F.A., Khatkar, B.S., 2015a. Effect of whey and casein protein hydrolysates on rheological, textural and sensory properties of cookies. J. Food Sci. Technol. 52 (9), 5718–5726. https://doi.org/10.1007/s13197-014-1649-3.

Gani, A., Broadway, A.A., Masoodi, F.A., Wani, A.A., Maqsood, S., Ashwar, B.A., Shah, A., Rather, S.A., Gani, A., 2015b. Enzymatic hydrolysis of whey and casein protein- effect on functional, rheological, textural and sensory properties of breads. J. Food Sci. Technol. 52 (12), 7697–7709. https://doi.org/10.1007/s13197-015-1840-1.

Ghaemi, P., Arabshahi-Delouee, S., Aalami, M., Hosseini Ghaboos, S.H., 2024. The effect of whey protein concentrate, soy protein isolate, and xanthan gum on textural and rheological characteristics of gluten-free batter and cake. J. Food Process. Preserv. 2024, e5571107. https://doi.org/10.1155/2024/5571107.

Hajas, L., Benedek, C., Csajbókné Csobod, É., Juhász, R., 2022. Development of protein- and fiber-enriched, sugar-free lentil cookies: impact of whey protein, inulin, and xylitol on physical, textural, and sensory characteristics. Foods 11 (23), 3819. https://doi.org/10.3390/foods11233819.

Hassanzadeh-Rostami, Z., Abbasi, A., Faghih, S., 2020. Effects of biscuit fortified with whey protein isolate and wheat bran on weight loss, energy intake, appetite score, and appetite regulating hormones among overweight or obese adults. J. Funct.Foods 70, 103743. https://doi.org/10.1016/j.jff.2019.103743.

Herrera-Cazares, L.A., Luzardo-Ocampo, I., Ramírez-Jiménez, A.K., Gutiérrez-Uribe, J., Campos-Vega, R., Gaytán-Martínez, M., 2021. Influence of the extrusion process on the release of phenolic compounds from mango (*Mangifera indica* L.) bagasse-added confections and evaluation of their bioaccessibility, intestinal permeability, and antioxidant capacity. Food Res. Int. 148, 110591. https://doi.org/10.1016/j.foodres.2021.110591.

Himashree, P., Gokul Devi, E., Hariharan, K.C., Sunil, C.K., 2023. Impact of whey incorporation on acrylamide formation and the quality of bread. J. Food Meas. Char. 17 (6), 5748–5758. https://doi.org/10.1007/s11694-023-02078-1.

Ingrassia, R., Torres, P., Bojanich, L., Ratti, J., Baldor, S., Ramunno, C., Dotta, G., Vidal Tesón, A., Forastieri, P., Soazo, M., Spelzini, D., Narambuena, C., Boeris, V., 2022. Concentration of proteins and fat from whey by coacervation: evaluation of its incorporation in bread. J. Food Process. Preserv. 46 (2). https://doi.org/10.1111/jfpp.16297.

Innocente-Alves, C., Sulzbach, S.L., Marques, A.L., Faustino, A.M., Calegari-Alves, Y.P., Verli, H., Santi, L., Beys-da-Silva, W.O., 2025. Whey as a biotechnological resource: current trends and future directions. ACS Food Sci. Technol. 5 (8), 2892–2906. https://doi.org/10.1021/acsfoodscitech.5c00508.

Komeroski, M.R., Homem, R.V., Schmidt, H. de O., Rockett, F.C., de Lira, L., Vitória da Farias, D., Kist, T.L., Doneda, D., Rios, A. de O., Ruffo de Oliveira, V., 2021. Effect of whey protein and mixed flours on the quality parameters of gluten-free breads. Int. J. Gastron. Food Sci. 24, 100361. https://doi.org/10.1016/j.ijgfs.2021.100361.

Kumkong, N., Banjongsinsiri, P., Laohakunjit, N., Vatanyoopaisarn, S., Thumthanaruk, B., 2020. Influence of natural colour blends of freeze-dried Gac aril and pulp on the quality of whey protein-mixed gelatin-based chewables. Heliyon 6 (12), e05817. https://doi.org/10.1016/j.heliyon.2020.e05817.

Lapčíková, B., Lapčík, L., Salek, R., Valenta, T., Lorencová, E., Vašina, M., 2022. Physical characterization of the milk chocolate using whey powder. LWT 154, 112669. https://doi.org/10.1016/j.lwt.2021.112669.

Lawless, H.T., Heymann, H., 2010. Acceptance testing. In: Lawless, H.T., Heymann, H. (Eds.), Sensory Evaluation of Food: Principles and Practices. Springer Science+Business Media, New York, pp. 325–348.

Lazarev, V., Ershova, A., 2023. Functional confectionery product using whey protein concentrate development. E3S Web of Conferences 451 (1), 04008. https://doi.org/10.1051/e3sconf/202345104008.

Li, H., Zhang, S., Chen, H., Yang, L., Li, J., Liu, X., Luo, Z., Li, H., Yu, J., 2024. Preparation and characterisation of low phenylalanine biscuits with whey protein hydrolysates. Int. J. Dairy Technol. 77 (1), 262–268. https://doi.org/10.1111/1471-0307.13028.

Mabrouki, J., Abbassi, M.A., Khiari, B., Jellali, S., Zorpas, A.A., Jeguirim, M., 2022. The dairy biorefinery: integrating treatment process for Tunisian cheese whey valorization. Chemosphere 293, 133567. https://doi.org/10.1016/j.chemosphere.2022.133567.

Machado, J., Miyahira, R.F., Marques, M., Moura-Nunes, N., Guimarães, R.R., Zago, L., Santana, I., Leite Junior, M., Avesani, C.M., 2021. Development of muffins as dialysis snacks for patients undergoing hemodialysis: results of chemical composition and sensory analysis. J. Nephrol. 34 (4), 1281–1289. https://doi.org/10.1007/s40620-020-00831-z.

Maravić, N., Škrobot, D., Dapčević-Hadnađev, T., Pajin, B., Tomić, J., Hadnađev, M., 2022. Effect of sourdough and whey protein addition on the technological and nutritive characteristics of sponge cake. Foods 11 (14), 1992. https://doi.org/10.3390/foods11141992.

Marques, G. de A., São José, J.F.B. de, Silva, D.A., Silva, E.M.M. da, 2016. Whey protein as a substitute for wheat in the development of no added sugar cookies. LWT 67, 118–126. https://doi.org/10.1016/j.lwt.2015.11.044.

Nastaj, M., Sołowiej, B.G., Stasiak, D.M., Mleko, S., Terpiłowski, K., Łyszczek, R.J., Tomasevic, I.B., Tomczyńska-Mleko, M., 2022. Development and physicochemical properties of reformulated, high-protein, untempered sugar-free dark chocolates with addition of whey protein isolate and erythritol. Int. Dairy J. 134, 105450. https://doi.org/10.1016/j.idairyj.2022.105450.

Nastaj, M., Sołowiej, B.G., Terpiłowski, K., Kucia, W., Tomasevic, I.B., Podkościelna, J., 2024. Effects of whey proteins and erythritol combination on the physical and quality characteristics of untempered, high-protein white chocolates without the saccharose addition. Int. Dairy J. 157 (7). https://doi.org/10.1016/j.idairyj.2024.106007.

Nogueira, A. de C., Aquino, R.K. do N. de, Steel, C.J., 2022. Empirical rheology of wheat flour doughs with pea, soybean and whey protein isolates. Food Sci. Technol. 42. https://doi.org/10.1590/fst.23921.

Norton, V., Lignou, S., Bull, S.P., Gosney, M.A., Methven, L., 2020. Consistent effects of whey protein fortification on consumer perception and liking of solid food matrices (cakes and biscuits) regardless of age and saliva flow. Foods 9 (9), 1328. https://doi.org/10.3390/foods9091328.

O'Sullivan, M.G., 2017. Sensory affective (hedonic) testing. In: A Handbook for Sensory and Consumer-Driven New Product Development. Elsevier, pp. 39–57. https://doi.org/10.1016/b978-0-08-100352-7.00003-8.

Park, H.R., Kim, G.H., Na, Y., Oh, J.E., Cho, M.S., 2021. Physicochemical and sensory properties of protein-fortified cookies according to the ratio of isolated soy protein to whey protein. Food Sci. Biotechnol. 30 (5), 653–661. https://doi.org/10.1007/s10068-021-00909-9.

Paul, S., Kulkarni, S., Rao, K.J., 2016. Effect of Indian cottage cheese (paneer)-whey on rheological and proofing characteristics of multigrain bread dough. J. Texture Stud. 47 (2), 142–151. https://doi.org/10.1111/jtxs.12168.

Pořízka, J., Slavíková, Z., Bidmonová, K., Vymětalová, M., Diviš, P., 2023. Physiochemical and sensory properties of bread fortified with wheat bran and whey protein isolates. Foods 12 (13), 2635. https://doi.org/10.3390/foods12132635.

Prieto-Vázquez del Mercado, P., Mojica, L., Morales-Hernández, N., 2022. Protein ingredients in bread: technological, textural and health implications. Foods 11 (16), 2399. https://doi.org/10.3390/foods11162399.

Rababah, T., Al-U'Datt, M., Al-Mahasneh, M., Alsaad, A., Gammoh, S., Mahili, H., Abdhamead, K.B., Almajwal, A., Ajouly, T., Bartkute-Norkuniene, V., 2022. Improving the functional and sensory properties of cookies by ultrasonic treatment of whey proteins. J. Food Qual. 2022, 1–8. https://doi.org/10.1155/2022/6902592.

Rakin, M., Bulatović, M., Zarić, D., Krunić, T., Lončarević, I., Petrović, J., Pajin, B., 2023. Influence of added whey proteins and hydrolysates on rheological and textural characteristics of milk chocolate. Food Sci. Technol. Int. 29 (6), 610–618. https://doi.org/10.1177/10820132221103758.

Ranok, A., Kupradit, C., 2020. Effect of whey protein and riceberry flour on quality and antioxidant activity under gastrointestinal transit of gluten-free cookies. AIMS Agric. Food 5 (3), 434–448. https://doi.org/10.3934/AGRFOOD.2020.3.434.

Rosenthal, A., Chen, J., 2023. Introduction to the measurement of sensory and instrumental food texture. In: Food Texturology: Measurement and Perception of Food Textural Properties. Springer International Publishing, pp. 3–16. https://doi.org/10.1007/978-3-031-41900-3_1.

Sarabhai, S., Prabhasankar, P., 2015. Influence of whey protein concentrate and potato starch on rheological properties and baking performance of Indian water chestnut flour based gluten free cookie dough. LWT 63 (2), 1301–1308. https://doi.org/10.1016/j.lwt.2015.03.111.

Satouf, M., 2023. The effect of adding concentrated cheese whey prepared by ultrafiltration on the quality of Syrian bread. J. Food Sci. Technol. 60 (12), 3102–3110. https://doi.org/10.1007/s13197-023-05823-y.

Srikanlaya, C., Therdthai, N., Ritthiruangdej, P., Zhou, W., 2018. Effect of hydroxypropyl methylcellulose, whey protein concentrate and soy protein isolate enrichment on characteristics of gluten-free rice dough and bread. Int. J. Food Sci. Technol. 53 (7), 1760–1770. https://doi.org/10.1111/ijfs.13761.

Sukmana, H., Rizwana, I., Dobozi, R., Al-Tayawi, A.N., Szabó, B.P., Özer, B., Kertész, S., 2025. A review of various protein separation techniques and valorization of cheese whey and buttermilk. Eur. J. Food Res. Technol. 251 (2025), 4263–4279. https://doi.org/10.1007/s00217-025-04877-w.

Tang, X., Liu, J., 2017. A comparative study of partial replacement of wheat flour with whey and soy protein on rheological properties of dough and cookie quality. J. Food Qual. 2017, 1–10. https://doi.org/10.1155/2017/2618020.

Uribe-Velázquez, T., Díaz-Vázquez, D., Barajas-Álvarez, P., González-López, M.E., Gradilla-Hernández, M.S., Garcia-Amezquita, L.E., Carrillo-Nieves, D., García-Cayuela, T., 2025. From waste to value: mitigating the environmental impact of whey in Jalisco, Mexico. J. Clean. Prod. 501, 145334. https://doi.org/10.1016/j.jclepro.2025.145334.

Voong, A.K., Westpha, L., 2023. Texture of bakery products. In: Food Texturology: Measurement and Perception of Food Textural Properties. Springer International Publishing, pp. 291–310. https://doi.org/10.1007/978-3-031-41900-3_14.

Wani, S.H., Gull, A., Allaie, F., Safapuri, T.A., 2015. Effects of incorporation of whey protein concentrate on physicochemical, texture, and microbial evaluation of developed cookies. Cogent Food Agric. 1 (1), 10922406. https://doi.org/10.1080/23311932.2015.1092406.

Wu, G., Hui, X., Stipkovits, L., Rachman, A., Tu, J., Brennan, M.A., Brennan, C.S., 2021. Whey protein-blackcurrant concentrate particles obtained by spray-drying and freeze-drying for delivering structural and health benefits of cookies. Innov. Food Sci. Emerg. Technol. 68, 102606. https://doi.org/10.1016/j.ifset.2021.102606.

Yang, Y., Zhang, S. Di, Ma, C.M., Bian, X., Zhang, G., Liu, X.F., Guo, X.X., Zhang, N., 2024. Effects of four kinds of improver on fermentation characteristics of gluten-free rice dough. J. Food Sci. 89 (11), 7680–7692. https://doi.org/10.1111/1750-3841.17371.

Zhou, J., Liu, J., Tang, X., 2018. Effects of whey and soy protein addition on bread rheological property of wheat flour. J. Texture Stud. 49 (1), 38–46. https://doi.org/10.1111/jtxs.12275.

Zhu, L., Snider, L., Vu, T.H., Desam, G.P., Herald, T.J., Dogan, H., Khaled, A.Y., Adedeji, A.A., Alavi, S., 2023. Effect of whey protein concentrate on rheological properties of gluten-free doughs and their performance in cookie applications. Sustainability 15 (13), 10170. https://doi.org/10.3390/su151310170.

Chapter 11

Infant formulas

Cristine Couto de Almeida[1], Vania M. Flosi Paschoalin[1] and Carlos Adam Conte Junior[1, 2]
[1]Federal University of Rio de Janeiro (UFRJ), Department of Biochemistry, Chemistry Institute, Graduate Studies in Food Science, Rio de Janeiro, Brazil; [2]Center for Food Analysis (NAL), Technological Development Support Laboratory (LADETEC), Federal University of Rio de Janeiro (UFRJ), Cidade Universitária, Rio de Janeiro, Brazil

1. Introduction

The global production of whey-based products has increased in recent years, underscoring its growing significance in the food industry. Whey, a by-product of cheese production, represents a rich source of nutrients, including soluble proteins, lactose, vitamins, minerals, and a small percentage of fat. This by-product is valued for its nutritional qualities and the potential health advantages associated with its bioactive compounds. As a result, whey has been used as an ingredient in many foods and beverages, such as those for clinical nutrition, sports nutrition, infant nutrition, and health foods (Timby et al., 2017; Almeida et al., 2021a; Zhao et al., 2022; Saadi et al., 2024).

In the context of infant nutrition, breastmilk is regarded as the gold standard because it meets all of an infant's nutritional needs. Additionally, it contains a variety of bioactive compounds that are essential for healthy development and immune protection. For this reason, it is recommended that infants be exclusively breastfed for the first 6 months. After this period, safe and nutritious complementary foods should be introduced, while continuing breastfeeding until at least 2 years of age (WHO, 2003; Meek and Noble, 2022).

Nevertheless, there are situations where breastfeeding may not be practical or recommended due to specific maternal or infant health conditions. In such cases, alternative feeding options should be considered to ensure that the infant receives adequate nutrition. These alternatives include pasteurized donor human milk, which retains many of the beneficial properties of human milk, and commercially prepared infant formulas that are designed to meet the nutritional needs of infants at different stages of development. Infant formulas can offer a balanced proportion of nutrients that resemble those found in human milk, although they are not identical. They can be categorized into standard milk-based, which includes three different phases based on the infant's age, and specialized formulas designed for infants with specific medical conditions or dietary requirements, such as allergies, lactose intolerance, or metabolic disorders (Jiang and Guo, 2021; Zhong et al., 2022; Luo et al., 2023; WHO, 2023; Li and Zhu, 2024).

Infant formula is a product formulated with plant or animal milk, or a combination of both, along with other ingredients proven to be suitable for feeding infants. Bovine milk is the most commonly used raw material due to its easy availability and lower cost than milk from other species. While plant-based formulas, typically formulated with soy or rice, are generally recommended for bovine milk protein allergy, lactose intolerance, and vegetarian or vegan diets (Martorell-Aragonés et al., 2015; Maryniak et al., 2022; Luo et al., 2023; WHO, 2023). Infant formulas are regulated by several health agencies worldwide, guided by the standards established by the Codex Alimentarius Commission (Codex), which defines the minimum and maximum nutrient levels along with quality and safety requirements. Additionally, each country may implement specific regulations to address local policies, nutritional needs, and safety standards (Codex Ailmentarius Commission, 2007).

Although bovine milk is the primary ingredient in most infant formulas, it must undergo significant modifications to ensure it is safe and nutritious for infants. One of the main distinctions between human and bovine milk and milk from other animal species is their protein profiles. Human milk predominantly contains whey proteins, with a whey-to-casein ratio of about 60:40, while bovine is mainly composed of caseins, with a ratio of about 20:80. Additionally, the composition of whey proteins varies not only in ratio but also in quality and quantity between human milk and that of bovine or other species. Whey contains several protein fractions, each with bioactive functions crucial for infant nutrition

Whey. https://doi.org/10.1016/B978-0-443-27648-4.00017-7

and development. Human milk is particularly rich in proteins such as α-lactalbumin (α-La), lactoferrin (Lf), immunoglobulins (Igs), lysozyme (Lz) and human serum albumin (HSA), and other minor proteins. In contrast, while bovine milk contains similar whey protein fractions, its overall content is significantly lower than human milk. Human milk primarily contains α-La, while bovine milk mostly consists of β-lactoglobulin (β-Lg), which makes up about half of its whey protein. Unlike human milk, β-Lg is absent in human milk and can be a potential allergen for infants (Martorell-Aragonés et al., 2015; Lönnerdal, 2016; Almeida et al., 2021b; Felice et al., 2021; Maryniak et al., 2022; Luo et al., 2023).

One of the essential steps in developing bovine milk-based infant formulas is to make their composition similar to human milk to meet infants' nutritional and physiological demands. For this, the adjusting of profile and total protein content is needed to improve digestibility and avoid renal overload due to excess proteins. The incorporation of whey proteins enriches the formulas, reducing the total protein content and increasing solubility, which increases the bioavailability of essential amino acids and bioactive peptides and also reduces allergenic potential. Additionally, demineralization of whey is necessary to reduce the mineral content, align it more closely with human milk, and prevent renal overload in infants, as bovine milk contains high levels of macrominerals. In addition, demineralized whey also retains lactose, an essential carbohydrate for infant energy needs, contributing to the nutritional balance of the final product (Thompson et al., 2017; Okawa et al., 2015; Patro-Gołąb et al., 2016).

In addition to adjusting the protein profile and demineralizing whey, further challenges related to industrial processing must be addressed. During production, key steps such as heat treatment, evaporation, and spray drying are employed to ensure product safety, nutritional quality, and stability. However, these processes can lead to the denaturation or degradation of heat-sensitive nutrients, adversely affecting the formula's nutritional quality. Thus, to mitigate these losses, in addition to adjusting the whey-to-casein protein ratio, common strategies include fortifying formulas with vitamins and minerals to compensate for nutrient depletion during heat treatment (Thompson et al., 2017; Masum et al., 2021; Jiang and Guo, 2021).

Advances in scientific knowledge and processing technologies have significantly enhanced infant formulas by combining adjustments to protein profile, mitigation of nutrient losses during processing, and the inclusion of bioactive compounds to closely approximate the complexity and benefits of human milk. The incorporation of bioactive components, such as nucleotides, probiotics, prebiotics, synthetic sources of human milk oligosaccharides (HMOs), long-chain polyunsaturated fatty acids (LC-PUFAs), milk fat globule membrane fraction (MFGM) and bioactive proteins like α-La and Lf has been shown to replicate some of the functional benefits found in human milk (Lien et al., 2018; Lu et al., 2019; Vandenplas et al., 2019; Luo et al., 2023). Ongoing research continues to explore and validate the efficacy of these bioactive compounds in infant nutrition, striving to bridge the gap between infant formula and human milk.

2. Nutritional composition of whey fractions in human milk and bovine milk: Essential components for infant formulas

Human milk is considered the gold standard for infant nutrition and serves as the benchmark for optimizing the composition of infant formulas. The following sections will discuss the components—proteins, carbohydrates, vitamins, and minerals—found in whey from several mammalian species, with a particular emphasis on bovine milk, which is the primary ingredient in most infant formulas.

2.1 Proteins

Neonates and infants experience intense anabolic activity during the first 6 months of life, characterized by rapid growth, physical development, and the maturation of the brain and immune system, impacting long-term health results. As part of this metabolic requirement, adequate protein intake is crucial for healthy development. This is why exclusive breastfeeding is particularly important during this period, as human milk provides high-quality proteins in adequate proportions of 0.9–1.2 g per 100 mL of mature human milk (Table 11.1). The quantity and quality of proteins consumed during early life stages support physical and neurological growth and impact long-term health results (Nayak et al., 2020; Meek and Noble, 2022; Milani et al., 2023).

The protein content in milk from other species is generally higher than in human milk. Bovine milk contains about 3.4 g/100 mL, and goat milk has a similar protein concentration, with roughly 3.2 g/100 mL. Sheep milk, which has one of the highest protein contents among these species, contains approximately 5.7 g/100 mL (Table 11.1) (Nayak et al., 2020; Felice et al., 2021; Prosser, 2021; Luo et al., 2023). Considering that the protein requirement for a full-term neonate ranges from 2.0 to 2.5 g/kg/day during the initial weeks and that this requirement decreases to about 1.15 g/kg/day after

TABLE 11.1 Nutrient composition of mature human milk and bovine milk.

Nutrients	Unit	Human milk	Bovine milk
Energy	*kcal*	65–70	60–70
Protein	*g*/100 mL	1.20	3.4
Whey	*g*/100 mL	0.72	0.6
Casein	*g*/100 mL	0.31	2.6
α_{s1}-Casein	*g*/100 mL	–	1.2–1.5
α_{s2}-Casein	*g*/100 mL	–	0.3–0.4
β-Casein	*g*/100 mL	0.26–0.44	0.9–1.1
κ-Casein	*g*/100 mL	0.09–0.13	0.2–0.4
Whey: Casein	%	60:40	20:80
α-Lactalbumin	*g*/100 mL	0.19–0.34	0.12–0.13
β-Lactoglobulin	*g*/100 mL	Absent	0.32–0.33
Lactoferrin	*g*/100 mL	0.2–0.4	0.002–0.035
Lysozyme	*mg*/100 mL	5.5–7.5	1.0–1.5
Immunoglobulin A	*mg*/100 mL	100	13.0
Immunoglobulin G	*mg*/100 mL	4.0	72.0
Immunoglobulin M	*mg*/100 mL	10.0	40.0
Carbohydrates	*g*/100 mL	7.5	4.5
Lactose	*g*/100 mL	7–7.5	4.4–5.6
Human milk oligosaccharides	*g*/100 mL	0.5–1.0	–
Bovine milk oligosaccharides	*g*/100 mL	–	0.003–0.009
Lipids: Total	*g*/100 mL	3.90	3.40
Micronutrients			
Sodium	*mg*	15	42.0
Calcium	*mg*	33	112.0
Copper	*μg*	39	20
Iodine	*μg*	7	–
Iron	*mg*	70	0.1
Magnesium	*mg*	3	11.0
Manganese	*μg*	0.003	6.0
Phosphorus	*mg*	95	91.0
Potassium	*mg*	55	145.0
Selenium	*μg*	0.016	1.8
Zinc	*mg*	0.16	0.4
Fluoride	*mg*	0.01	–
Vitamin A	*μg*-RE[1]	0.53	0.37
Vitamin D	*μg*	0.001	0.0008
Vitamin E	*mg*	5.4	1.1
Vitamin K	*mg*	0.015	0.03
Thiamin (B1)	*μg*	0.15	0.42

Continued

TABLE 11.1 Nutrient composition of mature human milk and bovine milk.—cont'd

Nutrients	Unit	Human milk	Bovine milk
Riboflavin (B_2)	*μg*	0.37	1.72
Pyridoxine (B_6)	*μg*	0.10	0.48
Cobalamin (B_{12})	*μg*	0.0003	0.0045
Niacin (B_3)	*μg*	1.7	0.92
Folic Acid (B_9)	*μg*	0.043	0.053
Ascorbic acid (C)	*mg*	47	18
Pantothenic Acid (B_5)	*mg*	2.1	3.6
Inositol	*mg*	300	160

Source: Gapper et al. (2007), Allen and Hampel (2019), Francisquini et al. (2020), Byrne et al. (2021), Milani et al. (2023).

the fourth month, it is clear that the protein content in milk from other species significantly exceeds that of human milk (Milani et al., 2023). High protein intake in infants can overload their immature kidneys, requiring more water to dilute and excretion. Additionally, when protein intake surpasses their needs, excess amino acids turn into glucose and fatty acids, promoting fat storage. This process is linked to rapid weight gain, resulting in a higher body mass index (BMI), which increases the risk of obesity later in life (Patro-Gołąb et al., 2016; Rossiter et al., 2021; Wallingford, 2023).

In addition to the optimal protein content, human milk contains specific proteins with physiological functions that go beyond basic nutrition. These proteins can be categorized into two primary groups: (i) soluble proteins found in the whey (whey proteins) and (ii) insoluble proteins (caseins). One of the main differences between human and bovine milk is the proportion of caseins to whey proteins. Human milk is whey-dominant, comprising about 80% whey proteins in colostrum, which decreases to approximately 60% in mature milk. In contrast, bovine milk, frequently used in infant formulas, has an inverse ratio of 80% caseins to 20% whey proteins in mature milk (Fig. 11.1) (Lönnerdal, 2016; Liao et al., 2017; Almeida et al., 2021b).

Whey proteins are a high-quality protein source compared to casein due to their high content of essential amino acids and their ability to release bioactive peptides with beneficial physiological effects. The main proteins in human whey include α-La, Lf, Igs, Lz, and HSA. These proteins play a key role in promoting infant well-being by providing a variety of beneficial effects, including support for physiological functions and protective roles (Table 11.2). Beyond their bioactive functional properties, whey proteins are digested and absorbed more efficiently than casein due to their structural characteristic. Their high solubility allows digestive enzymes to break them down more easily, enabling rapid and efficient absorption. In contrast, caseins form less soluble micellar aggregates, which slow down digestion and result in a more gradual release of amino acids (Lönnerdal, 2016; Liao et al., 2017; Almeida et al., 2021a; Zhao et al., 2022).

Bovine milk, on the other hand, has a higher proportion of casein than whey proteins. About 95% of the casein in milk is present as colloidal particles, known as micelles. Variation in the structure of bovine milk casein micelles influences digestion and absorption rates, which may result in discomfort or reduced efficiency of nutrient absorption. Although bovine milk contains many of the same protein fractions as the whey found in human milk, they are present in significantly lower concentrations. Additionally, bovine milk contains high levels of β-Lg, a protein that is absent from human milk and is known for its allergenic potential, raising concerns about its use for infants (Martorell-Aragonés et al., 2015; Liao et al., 2017; Almeida et al., 2021b; Maryniak et al., 2022).

Considering these differences, an important step to enhance infant formulas is the addition of whey proteins. This adjustment not only brings the protein profile closer to that of human milk but also reduces overall protein content while improving solubility, digestibility, and the bioavailability of essential amino acids and bioactive peptides. This strategy is supported by several clinical trials, which have shown that formulas with reduced protein content, between 1.77 and 2.2 g/100 kcal, promote healthier weight gain during the first 2 years without overloading the infant's metabolism, while higher protein levels (2.9–4.4 g/100 kcal) are linked to a heightened risk of obesity by the age of six (Kouwenhoven et al., 2020; Luo et al., 2023; Milani et al., 2023; Tinghäll Nilsson et al., 2024).

FIGURE 11.1 Concentrations of whey proteins and caseins in mature human and bovine milk. These values can vary slightly depending on factors like the stage of lactation and the diet. *n.d.*, not detectable.

The main whey proteins will be discussed, highlighting their structural characteristics, bioactive functions, and applications in infant formulas. This approach will allow us to understand how each bioactive protein contributes to infant development and how technological advances can bring formulas closer to the properties of human milk.

2.1.1 α-Lactalbumin (α-La)

α-Lactalbumin is a globular protein with a molecular mass of ~14.2 kDa composed of 123 amino acids, organized into α-helices and β-sheets. Its structure is highly stable due to four disulfide bonds, a high-affinity calcium-binding site, and an isoelectric point of 4.5. α-La exists in three different forms based on its calcium-binding state: apo-α-La (calcium-depleted), native-α-La (bound to one calcium ion), and holo-α-La (bound to two calcium ions) (Jakopović et al., 2016; Permyakov, 2020).

α-Lactalbumin is found in all mammal milk, with variations in concentration, amino acid sequence, and bioactivity compared to human α-La. Despite these variations, α-La maintains the same essential physiological function in lactose biosynthesis that occurs within the epithelial cells of the lactating mammary gland, where it binds to β-1,4-galactosyltransferase (GT-1) to form the lactose synthase complex. This complex combines UDP-galactose and glucose to produce lactose, the main milk carbohydrate that provides energy for infants. In addition, lactose production creates an osmotic gradient that boosts milk volume, reflecting the protein's dual role in nutrition and milk production (Jakopović et al., 2016; Layman et al., 2018; Zhang et al., 2021).

In human milk, α-La comprises about 36% of whey proteins, with concentrations peaking at approximately 0.42 g/100 mL in the first 2 weeks postpartum, and stabilizes over the first month, ranging from 0.19 to 0.34 g/100 mL (Zhang et al., 2021). In contrast, milk from other mammalian species generally contains lower concentrations of α-La. In bovine milk, α-La is the second most important whey protein, comprising about 17% of total whey proteins, with a typical

TABLE 11.2 Whey proteins and their known potential benefits for infant health.

Whey proteins	Potential benefits to infant health
α-La	Source of EAAs, especially BCAAs (i.e., tryptophan) Antimicrobial Immunostimulatory Increase brain serotonin level (regulating sleep)
Lf	Immunomodulation Antibacterial activity Antiviral activity Modulates inflammation Promotes growth of beneficial bacteria Inhibits pathogenic bacterial growth
Igs	Modulation of the immune response Antimicrobial Promotes growth of beneficial bacteria
Lz	Antioxidant activity Antibacterial activity Antiviral activity Iron-binding glycoprotein
OPN	Immunomodulatory Supports cognitive and intestinal development Influences cell proliferation and repair
β-Lg	Source of EAAs, especially BCAAs Act as a carrier molecule (vitamins A and D, palmitic acid, fatty acids, phospholipids, and aromatic compounds)
GMP	Source of BCAAs Antibacterial activity Antiviral activity Promoting the growth of beneficial bacteria Immunomodulatory
Serum albumin	Source of EAAs Binds fatty acids
LPO	Antibacterial activity Antiviral activity Catalyzes oxidation of molecules in the presence of hydrogen peroxide

α-Lactalbumin, α-La; Lactoferrin, Lf; Immunoglobulins, Igs; Lysozyme, Lz; Osteopontin, OPN; β-Lactoglobulin, β-Lg; Glycomacropeptide, GMP; Lactoperoxidase, LPO; EAAs, essential amino acids; BCAAs, branched-chain amino acids.
Source: Lönnerdal (2016), Almeida et al. (2021a), Zhao et al. (2022), Ge et al. (2024a), Szyller et al. (2024).

concentration of 0.12–0.13 g/100 mL (Table 11.1). Goat milk has α-La levels between 0.07 and 0.23 g/100 mL and sheep milk ranges from 0.10 to 0.19 g/100 mL (Felice et al., 2021; Ge et al., 2024a).

α-Lactalbumin is high in essential amino acids, especially branched-chain amino acids (BCAAs) such as isoleucine, leucine, and valine, which are crucial for muscle metabolism and growth. The amino acid sequences of human and bovine α-La share about 72% similarity, but key differences significantly affect digestibility and the release of bioactive peptides. Human α-La has a higher cysteine content in its N-terminal region, facilitating disulfide bond formation, which enhances structural stability and resistance to denaturation. Conversely, bovine α-La is less stable and more prone to enzymatic degradation, increasing the exposure to allergenic epitopes and the risk of immune reactions mediated by IgE and IgG antibodies (Martorell-Aragonés et al., 2015; Almeida et al., 2021a; Maryniak et al., 2022; Ge et al., 2024a).

A common approach employed to reduce the risk of triggering immune reactions in infants is the production of hydrolyzed infant formulas. These formulas involve the predigestion of proteins into smaller fragments, making them less likely to cause allergic reactions. Another approach consists of attaching calcium ions to proteins, stabilizing the protein's structure by masking allergenic epitopes. This reduces their antibody recognition, lowering the likelihood of immune responses. Importantly, this method preserves the protein's bioactivity and nutritional value, ensuring its suitability for infant nutrition (Jiménez-Saiz et al., 2015; Permyakov, 2020).

α-Lactalbumin is widely recognized as a bioactive protein with multifaceted benefits beyond its nutritional value. During digestion, its peptides exhibit immunomodulatory functions, support intestinal development, and confer antimicrobial and antioxidant properties. Moreover, α-La plays a pivotal role in infant growth, brain development, and sleep regulation, which are critical components of overall development. All these effects have been substantiated by clinical evidence demonstrating the positive impact of α-La-enriched formulas on infant growth and metabolic health (Table 11.3) (Layman et al., 2018; Nielsen et al., 2020; Petersen et al., 2020; Tinghäll Nilsson et al., 2023, 2024).

α-Lactalbumin has been shown to enhance growth and metabolic health in infants. Clinical trials indicate that formulas enriched with α-La support growth metrics, such as weight and BMI, comparable to those of breastfed infants while maintaining lower insulin and IGF-1 levels than standard formulas (Tinghäll Nilsson et al., 2024). Chetta et al. (2022) demonstrated that α-La-enriched formulas adequately met essential amino acid requirements in infants and improved energy efficiency. However, excessive α-La levels above those found in human milk did not provide additional benefits and could induce metabolic stress (Nielsen et al., 2020).

α-Lactalbumin enhances the innate immune response by boosting immune cell activity. The tripeptide Gly-Leu-Phe, found in human and bovine milk, strengthens phagocytic function and reduces inflammation. These peptides also promote a healthy gut microbiome by inhibiting pathogenic bacteria and supporting the growth of *bifidobacteria*, essential for gastrointestinal health. By fostering a healthy gut environment, α-La indirectly reduces intestinal inflammation, which is particularly important for preventing inflammatory diseases of the gut, such as necrotizing enterocolitis (NEC) in infants. Additionally, α-La interacts with soluble CD14, which recognizes lipopolysaccharides (LPS) from Gram-negative bacteria, facilitating a quick immune response and supporting a healthy gut microbiota (Ward, 2014; Layman et al., 2018; Beverly et al., 2020; He et al., 2022).

α-Lactalbumin has strong antioxidant properties that protect cells from oxidative damage, especially in infants with developing antioxidant defenses. Its benefits are mainly due to its amino acid profile, particularly high levels of cysteine, which helps produce glutathione, a key antioxidant that neutralizes reactive oxygen species (ROS) and maintains redox balance. These properties are further enhanced when α-La interacts with Lf, which binds free iron and provides additional protection against oxidative stress. α-La also has metal-chelating properties, binding transition metals like iron and copper to curb ROS formation (Joubran et al., 2015; Permyakov, 2020).

α-Lactalbumin improves mineral absorption by forming soluble complexes with calcium, which helps prevent its precipitation during digestion and increases bioavailability. Research shows that formulas enriched with α-La improve zinc absorption in infants, but their impact on iron bioavailability remains uncertain. In vitro studies suggest that peptide-iron complexes can improve iron absorption in intestinal Caco-2 cells, but further research is needed to confirm these effects in vivo. Short-term research involving humans has not reported significant improvements in iron absorption, underscoring the necessity for longer studies to evaluate the full potential of α-La in enhancing iron bioavailability (Layman et al., 2018; Permyakov, 2020).

α-Lactalbumin supports infant neurodevelopment due to its high tryptophan content, which is metabolized into serotonin, essential for functions like appetite, sleep, memory, and mood. In infants, serotonin aids neuronal maturation and cognitive development and serves as a precursor to melatonin, promoting healthy sleep patterns vital for brain growth (Shoff et al., 2025). Research indicates that tryptophan-rich formulas, especially those fortified with α-La, lead to shorter sleep onset and improved sleep efficiency by maintaining elevated tryptophan levels during nighttime, aligning with natural melatonin peaks for better circadian rhythm regulation. Additionally, peptides from α-La digestion can cross the blood-brain barrier, potentially offering neuroprotective effects (Brück, 2017).

Research suggests that α-La supplementation may support infant growth, enhance immune function, and improve gastrointestinal health, but caution is advised. Although bovine α-La offers benefits, it also has weaknesses, including lesser stability and a decreased arginine content compared to human α-La. If α-La were used as the sole whey protein source in infant formula with a typical whey-to-casein ratio of 60:40, it would result in a significant arginine deficit (Ge et al., 2024a). To address this limitation, a balanced formulation strategy is essential. This approach involves increasing the proportion of α-La while reducing β-Lg and selectively retaining other bioactive proteins. While this strategy shows promise, ensuring that α-La concentrations remain within physiologically appropriate levels is equally important. Animal studies suggest that α-La concentrations exceeding those naturally found in human milk do not provide additional benefits and may even induce metabolic stress. Therefore, α-La enrichment, though valuable, must be implemented judiciously to optimize infant health outcomes and avoid potential metabolic imbalances (Fleddermann et al., 2014; Nielsen et al., 2020).

TABLE 11.3 Clinical studies evaluating the impact of protein supplementation in infant formulas on growth and health outcomes.

Bioactive protein	Biological function	Population/ Species	Experimental group	Control group	Time of treatment	Main outcomes	References
α-La	Effects on growth, protein metabolism and nutrient intake	84 healthy newborn infants	Low-protein IF enriched with α-La (1.89 g ptn/100 kcal) or SF	BF	12 months	There were no significant differences in Z-scores for weight, length, head circumference, weight-for-length, body mass index, or fat percentage between the study groups at the end of the intervention.	Petersen et al. (2020)
		328 healthy-term infants	SF (2.2 g protein/100 kca) or Low-protein IF with α-La (1.75 g ptn/100 kcal) or CGMP reduced (1.76 g ptn/100 kcal)	BF	6 months	Lower-protein IF that had more α-La or less GMP supported growth and metabolic health similar to the BF group. In the α-La group, weight gain from 4 to 6 months was equal to that of the BF group. All IF types were well tolerated, with no major side effects. The experimental IF also showed better protein use than the SF.	Tinghäll Nilsson et al. (2023)
	Improvement of obesity and related metabolic diseases	245 healthy-term infants (2–6 months)	Low-protein IF enriched with α-La (1.75 g ptn/100 kcal) or SF (2.2 g ptn/100 kcal)	BF	12 months	Weight gain and BMI at 12 months were higher in SF than BF infants, but low-protein IF groups did not differ significantly from the BF group. S-insulin and C-peptide levels were higher in the SF group compared to the BF group, while they were more similar between the low-protein IF groups and the BF group. Serum IGF-1 levels at 12 months were comparable across all study groups.	Tinghäll Nilsson et al. (2024)
	Infant growth and development	216 full-term infants (≤14 days)	α-La enriched IF (α-LA 2.2 g/L)	Standard IF (α-La 1.3 g/L)	8 weeks	α-LA-enriched IF met EAA requirements of the infants, and significantly	Chetta et al. (2022)

		398 full-term infants	α-LA enriched IF +LC-PUFA (protein, 13 g/L)	Control IF (protein 15 g/L); HM (protein 1.2 ± 0.2 g/L)	28–120 days after birth	Improved IF with α-LA met the protein requirements for full growth and significantly improved energy efficiency	Fleddermann et al. (2014)
		137 full-term infants (≤40 days)	α-La-enriched IF (Ptn 1.43 g/100 kcal, α-La 26%; II. Ptn 1.9 g/100 kcal, α-La 26%; III. Ptn 2.18 g/100 kcal, α-La 11%)	Human milk (Ptn 1.28 g/100 kcal, α-La 33%)	0–4 months	The weight, height and head circumference growth of group I (low protein, high α-La) fed infants were most similar to breastfed infants	Oropeza-Ceja et al. (2018)
		40 newborn premature pigs	α-La enriched diet (α-La 18 or 6.3 g/L)	Standard diet (α-La 0.7 g/L)	19 days	The α-La-enriched IF had beneficial effects on the physical and organ (growth, gut, immune system and brain) development of preterm infants, but excess α-La (above human milk levels) did not provide additional beneficial effects.	Nielsen et al. (2020)
	Gut health and development	145 healthy-term infants (5–14 days)	IF enriched with α-La + GOS (3.0 g/L) or IF enriched with α-La without GOS	BF	8 weeks	IF enriched with α-La and GOS raised fecal *bifidobacteria* levels to those similar to BF, surpassing levels in IF without GOS. It also softened stools to resemble those of BF infants without increasing stool frequency or affecting hydration.	Wernimont et al. (2014)
Lf	Prevent infection and Supports immune system maturation	50 VLBW neonates (<1500 g)	200 mg/d bLf	Placebo (saline)	During the hospitalization period	Infants treated with Lf had fewer sepsis episodes, and none developed NEC.	Akin et al. (2014)
		375 neonates (<2500g)	200 mg/kg/d bLf	Placebo (maltodextrin)	4 weeks	Sepsis occurred in 12.6% of the Lf group, while the placebo group had a rate of 22.1%.	Ochoa et al. (2015)
		743 VLBW neonates (<1500 g)	100 mg/d bLf	Placebo (5% glucose solution)	1 month	The group receiving bLf (2.0%) had a significantly lower occurrence of NEC compared to the control group (5.4%).	Manzoni et al. (2014)
		260 infants (4–6 months)	Lf-fortified formula (38 mg/100 g Lf)	No Lf-fortified formula	3 month	The Lf-fortified formula group showed fewer cases of respiratory and diarrhea-related diseases compared to the control group.	Chen et al. (2016)

Continued

TABLE 11.3 Clinical studies evaluating the impact of protein supplementation in infant formulas on growth and health outcomes.—cont'd

Bioactive protein	Biological function	Population/ Species	Experimental group	Control group	Time of treatment	Main outcomes	References
		108 infants with Anemia (6–9 months)	Lf-fortified formula (38 mg/100 g, 76 mg/100 g Lf)	No Lf-fortified formula	3 month	The group receiving Lf-fortified formula had fewer cases of diarrhea and respiratory tract infections, with the 76 mg/100 g bLf-fortified formula showing a stronger effect.	Chen et al. (2021)
	Intestinal development	59 preterm infants	100 mg bLf + probiotic	Probiotics	1 month	Infants with bLf supplementation showed greater microbiota species richness, but the impact on microbiota development was minimal compared to controls.	Grzywacz et al. (2020)
		202 healthy term infants	Lf (1 g/L) + *Bifidobacterium animalis* (3.7 ± 2.1 × 104 CFU/g)	Infant formula	8 weeks	Supplementing with bLf and probiotics may improve gut maturation markers in infant stools at 4 weeks. At 1 week, *Bifidobacterium* levels were higher in the bLf group, but no significant difference was observed at 4 weeks when compared to the group control.	Castanet et al. (2020)
		62 healthy-term infants	Infant formula with Lf (0.6 g/L) + bovine MFGM	Bovine milk-based infant formula	12 months	*Bacteroides uniformis* and *Bacteroides plebeius* were more abundant in the Lf + MFGM group at 4 months.	Chichlowski et al. (2021)
	Neurodevelopment	Healthy-term infants ages 10–14 days	Infant formula with Lf (0.6 g/L) + bovine MFGM (5 g/L)	Bovine milk-based infant formula	365 days of age	Lf + MFGM enhanced the neurodevelopmental profile by day 365 and improved language scores by 545 days, along with a reduction in the occurrence of respiratory-related adverse events and diarrhea	Li et al. (2019)
		414 neonates (500–2000 g)	200 mg/kg/day bLf	Placebo	8 weeks	bLf supplementation in low birth weight infants had no significant effect on neurodevelopment, sepsis rates, or growth outcomes.	Ochoa et al. (2021)

	Ameliorate iron deficiency anemia	260 healthy infants (4–6 months)	Fortified formula (38 mg/100 g bLf and 4 mg/100 g iron)	Formula (4 mg/100 g iron)	3 months	In the group fortified with bLf, higher amounts of total body iron, hemoglobin, serum ferritin, and serum transferrin receptor were found, along with a lower occurrence of anemia and iron deficiency.	Chen et al. (2015)
		108 anemic infants (6–9 months)	Lf-fortified formula (38 mg/100 g, 76 mg/100 g Lf)	Formula without Lf	3 months	The group receiving 76 mg/100 g of bLf showed higher hemoglobin levels than both the 38 mg/100 g bLf group and the control group.	Chen et al. (2020)
		120 children with iron deficiency anemia	100 mg Lf with 30 mg iron	Iron polymaltose complex (6 mg/kg/d)	1 month	Serum ferritin levels showed greater improvement in the Lf with iron group compared to the iron polymaltose complex group.	El-Hawy et al. (2021)
		180 healthy full-term infants (6 ± 2 weeks)	Low iron formula with fortified bLf (1.0 g/L)	Low iron formula with unfortified bLf	Up to 6 months of age	Incorporating Lf did not influence iron levels.	Björmsjö et al. (2021)

IF, infant formula; *IGF-1*, Insulin-like growth factor-1; *α-La*, α-Lactoalbumin; *Ptn*, protein; *SF*, standard formula; *BF*, breastfed; *GOS*, oligofructose; *Lf*, Lactoferrin; *bLf*, bovine lactoferrin; *VLBW*, very low birth weight; *NEC*, necrotizing enterocolitis; *MFGM*, milk fat globule membrane.

2.1.2 Lactoferrin (Lf)

Lactoferrin is an iron-binding glycoprotein with a molecular weight of approximately 80 kDa. Structurally, Lf consists of two homologous globular lobes (N-lobe and C-lobe), each containing an active site for binding ferric iron ions (Fe^{3+}), connected by an α-helical bridge stabilized by 17 disulfide bonds. Each iron-binding site includes two tyrosine residues, one aspartate, and one histidine, with carbonate acting as a counterion that facilitates iron coordination (Wang et al., 2019; Ianiro et al., 2023; Zhang et al., 2024).

Iron binding is cooperative, starting at the C-lobe and facilitating subsequent binding at the N-lobe, allowing each Lf molecule to carry up to two iron ions. As a result, Lf can exist as iron-saturated (holo) or unsaturated (apo), with naturally low iron saturation (15%–20%) resembling levels in human and bovine milk. This natural low saturation allows Lf to exhibit distinct bioactivities depending on its form. The holo form is more resistant to enzymatic and thermal denaturation, tolerating temperatures up to 90°C, while the apo form is stable between 60.5°C and 71.0°C. The stability variations highlight their relevance in industrial processes and functional applications. The holo form is advantageous in pasteurization due to its resistance to high temperatures, while the apo form retains enhanced antimicrobial and antioxidant properties owing to its iron-binding capacity. Most commercially available Lf has a naturally low iron saturation level. However, techniques like iron supplementation or the use of magnetic particles can boost this saturation. In this way, as the saturation level of iron directly impacts Lf's biological functions, adjusting it can optimize the stability and efficacy of Lf for various nutritional and industrial applications, tailoring its functionality to specific needs (Liu et al., 2018; Wang et al., 2019; Cao et al., 2023; Ostertag et al., 2025).

A comparative analysis of Lf across mammalian species reveals distinctions in composition, concentration, and functionality. Human Lf (hLf) is the second most abundant whey protein, with concentrations peaking in colostrum at approximately 0.6–0.8 g/100 mL and gradually decreasing to around 0.2–0.4 g/100 mL in mature milk. In comparison, milk from other species contains significantly lower levels of Lf. Bovine colostrum has approximately 0.15–0.5 g/100 mL, while mature bovine milk contains only 0.002–0.035 g/100 mL. In goat colostrum, the levels range from 0.0455 to 0.2 g/100 mL, and mature goat milk has levels between 0.002 and 0.02 g/100 mL. For sheep colostrum, the concentration is about 0.07–0.08 g/100 mL. Human lactoferrin is composed of 710 amino acids, while bovine Lf (bLf) consists of 708, sharing approximately 77% sequence similarity (Lin et al., 2021; Zhang et al., 2024).

While both share a common structure, the glycan compositions vary between species and throughout lactation, influencing their bioactivity. hLf consists of complex glycans that are sialylated and fucosylated, whereas bLf has simpler glycans, such as N-acetyllactosamine and mannose. Despite its simpler structure, bLf effectively supports immune function and gut health, highlighting its significance as a bioactive component in infant formulas (Kumar and Mattad, 2021; Lin et al., 2021; Ianiro et al., 2023; Zhang et al., 2024). Its structure allows interaction with biomolecules like Lz and LPS, enhancing Lz antibacterial activity by disrupting bacterial membranes through LPS binding, increasing membrane permeability, and aiding pathogen destruction. Lf can also bind to the free form of LPS, neutralizing its ability to trigger an inflammatory response (Drago-Serrano et al., 2017).

In recent years, there has been a growing interest in the inclusion of Lf, specifically bLf, in infant nutrition due to its numerous health benefits. Widely incorporated into commercial infant formulas, bLf is recognized for its safety profile, as it is categorized as Generally Recognized as Safe (GRAS) by the FDA (GRAS Notice, 2016; Superti, 2020). The health benefits associated with bLf are numerous and significant. It exhibits bacteriostatic activity, immunomodulatory and antiinflammatory action, intestinal health support, neurodevelopment promotion, and iron absorption improvement (Table 11.3) (Manzoni et al., 2014; Chen et al., 2016; Castanet et al., 2020; Sienkiewicz et al., 2022; Cao et al., 2023).

Moreover, the hydrolysis of Lf by pepsin in the gastrointestinal tract generates bioactive peptides, such as lactoferricins, which have even more potent bacteriostatic activity than intact Lf. Cao et al. (2023) further highlighted lactoferrampin, a peptide exclusive to bovine Lf, as a key contributor to its enhanced antimicrobial properties. These characteristics make bLf a valuable bioactive compound for application in infant formulas. Its supplementation has shown great potential to promote the general health of newborns, strengthening the immune system and contributing to healthy development (Donovan, 2016; Castanet et al., 2020). Clinical trials demonstrate Lf potential in reducing infections and promoting immune health in neonates. For instance, Manzoni et al. (2014) found that supplementation with bLf significantly reduced the incidence of NEC and late-onset sepsis in very low birth weight (VLBW) infants compared to a placebo group. Similarly, Akin et al. (2014) reported fewer sepsis episodes in VLBW neonates receiving bLf, with no cases of NEC. Lf-fortified formulas have also shown reductions in respiratory and gastrointestinal infections in infants, alongside alleviating iron deficiency anemia (Chen et al., 2016). Chen et al. (2020) demonstrated that infants receiving formulas fortified with 76 mg/100 g bLf exhibited higher hemoglobin levels and fewer anemia-related symptoms than those on lower doses or control formulas.

Lactoferrin exhibits strong antimicrobial properties against bacteria, viruses, and fungi, mainly due to its ability to bind iron, which restricts a vital element for pathogen growth. Its antimicrobial mechanisms also include direct interaction with bacterial membranes, creating holes in the outer layers of Gram-negative bacteria. These breaches allow Lz to penetrate and degrade structural components, further intensifying the antimicrobial effects. In addition, the digestion of Lf in the gastrointestinal tract, which generates the peptides lactoferricins and lactoferrampins, reinforces gastrointestinal defences and enhances infant protection against infections. These antimicrobial peptides, exclusive to bLf, strengthen the role of the protein as a functional component in infant formulas (Manzoni et al., 2014; Donovan, 2016; Ianiro et al., 2023).

In addition to its antimicrobial action, Lf has immunomodulatory and antiinflammatory properties that help balance immune responses. By interacting with intestinal receptors, Lf modulates immune responses through endocytosis and nuclear transport. Once bound, Lf is internalized into cells via endocytosis and transported to the nucleus, where it binds to DNA at specific sites, acting as a transcription factor. This process regulates the expression of cytokine genes, such as interleukin-1β and transforming growth factor-β, which are critical for immune modulation. In vitro studies indicate that the hLf and bLf forms of the protein can modulate cytokine production and support balanced T-cell responses (Donovan, 2016; Cao et al., 2023).

Another important aspect of Lf is its contribution to gastrointestinal health. Research indicates that this protein promotes cell proliferation, strengthening the intestinal barrier function and contributing to a beneficial intestinal microbiota. In vivo and in vitro experiments show that Lf supplementation increases the depth of intestinal crypts and villi, favoring nutrient absorption and intestinal health in animal models. The partial resistance of Lf to digestion in the gastrointestinal tract, demonstrated by the presence of intact bLf in feces, allows its direct interaction with intestinal cells, enhancing its beneficial effects (Manzoni et al., 2014; Donovan, 2016; Dyrda-Terniuk and Pomastowski, 2023).

In addition, Lf has been associated with neurodevelopment in infants, especially for its contribution to improving iron absorption, an essential element for brain growth and cognitive function during the first years of life. Although the exact mechanisms are still being investigated, studies suggest that adequate iron levels may positively influence neuronal development, indicating a potential role for Lf in supporting better cognitive outcomes (Manzoni et al., 2014; Donovan, 2016).

While the main mechanism is still being explored, Lf has been associated with better cognitive outcomes in infants. Some research indicates that by improving iron absorption, Lf might aid neurodevelopment, as sufficient iron levels are associated with brain growth and function during early development. Nevertheless, additional studies are required to confirm a direct connection between Lf supplementation in formulas and cognitive improvements (Donovan, 2016; Manzoni et al., 2014; Li et al., 2019).

To ensure that bLf retains its structural integrity and bioactivity during industrial processing, some strategies must be implemented. This aspect is particularly important given that lactoferrin exhibits partial resistance to gastrointestinal digestion and must remain functional throughout processing and digestion to provide its intended health benefits. Micro and nanoencapsulation techniques offer a viable solution by creating protective coatings, such as proteins, polysaccharides, or liposomes, that enable controlled release in the intestine, enhance bioavailability, and mitigate gastric degradation. Another approach, known as PEGylation, involves the covalent attachment of polyethylene glycol (PEG) to bLf. This technique increases bLf resistance to proteases, thereby extending its bioavailability. Other techniques, like iron saturation and chitosan-based absorption enhancers, have also been explored, but their effectiveness is limited due to instability in acidic conditions (Liu et al., 2018; Superti, 2020; Li et al., 2022).

Recombinant human lactoferrin (rhLf) holds significant promise for enriching infant formulas due to its structural similarity to native hLf and its potential to provide comparable health benefits. Produced through advanced biotechnological methods, rhLf is synthesized using transgenic host expression systems, including transgenic rice, fungi, and cows. These systems enable the large-scale production of rhLf with an amino acid sequence and protein folding closely mirroring that of natural hLf. This structural congruence ensures that rhLf retains the functional properties of hLf, making it a valuable ingredient in infant formula. By combining rhLf with innovative protection strategies like encapsulation or PEGylation, infant formulas can provide bioactive benefits more closely resembling those of human milk. However, further research is needed to determine the long-term safety, potential immunogenicity, and overall efficacy of rhLf in supporting infant health outcomes (Li et al., 2014; Aly et al., 2016).

2.1.3 Immunoglobulins (Igs)

Immunoglobulins (Igs) are globular proteins with antimicrobial properties and protective functions essential for immune defense. They are in varying concentrations in blood serum and milk. Mammalian milk contains five isotypes of Igs, categorized based on their structure: IgA, IgG, IgM, IgD, and IgE. Among these, IgA, IgG, and IgM are particularly significant for newborns' immune systems. The highest concentrations of these three Igs are found in colostrum, which provides critical immunological support during the neonatal period when the immune system is still developing (Czosnykowska-Łukacka et al., 2020).

Human and bovine milk have different immunological profiles, particularly regarding the types and concentrations of Igs. In human milk, secretory IgA (sIgA) is the most abundant, accounting for over 90% of its antibody content, with approximately 100 mg/100 mL concentrations in mature milk. IgG and IgM are present in much lower concentrations, around 4.0 mg/100 mL and 10.0 mg/100 mL, respectively. In contrast, bovine milk is predominantly rich in IgG, constituting 72.0 mg/100 mL, while IgA is present at much lower levels, approximately 13.0 mg/100 mL, and IgM at 40.0 mg/100 mL. Furthermore, in bovine milk and colostrum, IgG occurs as monomeric (~160 kDa), while IgA (~370 kDa) and IgM (~1000 kDa) form polymeric Ig. IgG can also be present in two subclasses, IgG1 and IgG2, with IgG1 comprising approximately 80% of the total Ig content (Bogahawaththa et al., 2017; Czosnykowska-Łukacka et al., 2020).

Secretory IgA plays a key role in shaping the infant's gut microbiota and immune system by supporting the growth of beneficial bacteria and facilitating the adhesion of commensal strains, such as *Bifidobacterium* and *Lactobacillus*, to the intestinal epithelium. Additionally, sIgA contributes to the development of immune cells, promoting tolerant host-microbe interactions within the gastrointestinal tract. IgG activates phagocytes, modulates excessive inflammatory responses to allergens, and aids in immune regulation through glycosylated structures similar to those in sIgA. Meanwhile, IgM contributes to immune defense by opsonizing Gram-negative bacteria and, together with sIgA, facilitates the immune exclusion of pathogens (Rogier et al., 2014; Donaldson et al., 2018; Donald et al., 2022).

Although there is a difference between the two sources, the Igs present in bovine colostrum, especially IgG and IgA, can potentially promote beneficial effects on the human immune system. The bovine IgG1 subtype supports passive immunity by neutralizing pathogens, reducing inflammation, and enhancing intestinal barrier integrity. Their ability to bind human pathogens and allergens highlights their potential role in cross-species immune modulation (Gazi et al., 2023).

Bovine IgG immunoprotective properties make it highly relevant in dietary supplements and infant formulas aimed at improving immune health. Studies have demonstrated its efficacy, such as the reduction of diarrhea incidence and duration in infants consuming colostral IgG-enriched formulas and decreased viral and microbial infections in IgG-fortified formulas (Tawfeek et al., 2003). These findings underscore bovine IgG promising role in infant nutrition. However, its use in commercial infant formulas remains under clinical investigation, with further studies needed to confirm its safety and efficacy for widespread application (Gapper et al., 2007; Bogahawaththa et al., 2017).

2.1.4 Lysozyme (Lz)

Lysozyme is a protein of approximately 14 kDa, composed of a polypeptide chain with 130 amino acids. It is widely distributed in bodily fluids such as breast milk, tears, saliva, blood, and other secretions, where it plays a crucial role in the body's nonspecific defenses. In human milk, the concentration of Lz is significantly higher than in other species, ranging between 5.5 and 7.5 mg/100 mL, whereas in bovine milk, it is only 1.0–1.5 mg/100 mL (Table 11.1). Goat milk contains approximately 2.5 mg/100 mL of Lz, while its amount in sheep milk is insignificant. Interestingly, unlike other human milk proteins, the concentration of Lz progressively increases with prolonged lactation, enhancing its benefits throughout the breastfeeding period (Wu et al., 2019; Leśnierowski and Yang, 2020).

Lysozyme contributes to developing a healthy gut microbiota and combating a wide range of pathogens, including Gram-positive bacteria, some Gram-negative bacteria, viruses, fungi, yeasts, and parasites. This enzyme is well-known for its potent antimicrobial activity, as it can break down the β(1,4) glycosidic bonds between N-acetylmuramic acid and N-acetyl-D-glucosamine in peptidoglycan. Its effectiveness is particularly effective against Gram-positive bacteria. In its native monomeric form, Lz is highly efficient at destroying Gram-positive bacteria by disrupting their cell walls, which consist of a single layer of peptidoglycan. This simpler structure makes Gram-positive bacteria more vulnerable to Lz, as opposed to Gram-negative bacteria, which have a more complex cellular structure and are protected by an additional outer membrane, making them more resistant to monomeric Lz form. However, when Lz is modified to its dimeric form, it gains enhanced antimicrobial capabilities, allowing it to overcome the outer membrane of Gram-negative bacteria and broaden its antibacterial spectrum, effectively targeting both Gram-positive and Gram-negative microorganisms (Leśnierowski, 2015; Vilcacundo et al., 2018; Leśnierowski and Yang, 2020).

In addition to its antimicrobial properties, Lz exhibits antiinflammatory, immunomodulatory, and antiviral functions. Moreover, in synergy with other immune proteins, such as Lf and Igs, Lz amplifies its role in maintaining the digestive health of infants. The interaction between Lz and Lf enhances effectiveness against Gram-negative bacteria by disrupting their outer membrane, which allows other antimicrobial agents to act more effectively. Lz and these other proteins have been shown to limit neutrophil migration to damaged tissues, acting as an antiinflammatory agent (Wu et al., 2019; Ferraboschi et al., 2021).

Recent studies have achieved notable progress in altering Lz by employing different thermal and thermo-chemical techniques. These methods enable the transformation of Lz into a new oligomeric form of the enzyme, with varying degrees of oligomerization and hydrophobicities, enhancing Lz interaction with cellular membrane components and effectiveness against Gram-negative bacteria, viruses, and other microorganisms (Leśnierowski and Yang, 2020). However, the low natural concentration of Lz in bovine milk poses challenges for obtaining adequate quantities for infant formulas, making extraction costly and inefficient for large-scale applications.

A widely used alternative source is chicken egg white (HEWL), which contains much higher Lz concentrations, ranging between 250–350 mg/100 mL (Leśnierowski, 2015). Nevertheless, despite its industrial application as a food additive, Lz derived from eggs presents significant structural and functional differences compared to human milk Lz (hLz). These differences may pose an obstacle to its safe use in infant nutrition, mainly due to the potential allergenic risk of the egg-derived protein (EFSA et al., 2023).

Recombinant human lysozyme (rhLz) production has been investigated as a safer and more effective alternative to traditional sources. This method enables the creation of a protein that is functionally similar to that found in human milk, ensuring high safety and compatibility for use in infant formulas. This strategy addresses the limitations associated with traditional Lz sources (Zhao et al., 2023).

2.1.5 β-Lactoglobulin (β-Lg)

β-Lactoglobulin is the most abundant whey protein in bovine whey, comprising approximately 10%–15% of total milk proteins and 58% of whey protein, with a concentration of 0.32–0.33 g/100 mL (Table 11.1). Structurally, β-Lg is characterized as a compact globular protein composed of 162 amino acid residues, with a molecular weight of about 18.4 kDa. The structure of β-La contains two disulfide bonds at Cys66-Cys160 and Cys106-Cys119 and one free sulfhydryl group on Cys121 buried within the protein structure (Nayak et al., 2020; Felice et al., 2021; Chun and Lee, 2022).

β-lactoglobulin is absent in human milk, and its exact biological function is unclear, although it serves as an excellent source of essential amino acids, particularly leucine. Its nutritional significance in infant formulas lies in its ability to complement amino acid profiles, addressing the high demand for leucine and other essential amino acids during infancy. Due to its structure, β-Lg is believed to bind to and transport small hydrophobic ligands. With multiple binding sites, it can interact with compounds such as vitamins A and D, palmitic acid, and various other hydrophobic substances. Additionally, β-Lg demonstrates a strong binding affinity for fatty acids, phospholipids, and aromatic compounds (Almeida et al., 2021b; Felice et al., 2021).

Although β-Lg has beneficial properties, it can also cause allergic reactions in sensitive individuals, particularly infants. The allergenic nature of β-Lg is influenced by the presence of specific epitopes that can bind to IgE or IgG antibodies. This binding triggers the release of inflammatory mediators. In the context of infant formulas, these allergic reactions pose a challenge, especially for nonbreastfed infants with cow's milk protein allergy (CMPA). Partially and extensively hydrolyzed formulas are now the primary choice in the infant formula market for addressing CMPA. These formulations aim to reduce β-Lg allergenicity while maintaining the protein's nutritional value (Jiménez-Saiz et al., 2015; Martorell-Aragonés et al., 2015; Maryniak et al., 2022; Xiong et al., 2023; Sun et al., 2025).

Several studies have investigated methods to hydrolyze or remove β-Lg in order to create hypoallergenic formulas. The main methods currently used to reduce allergic reactions to bovine milk include heat treatment, glycation, and enzymatic hydrolysis. Heat treatment primarily impacts the secondary and tertiary structures of antigens, which disrupts their conformational epitopes. However, this method has not achieved the desired reduction in protein antigenicity. Glycation works by masking allergenic epitopes, but it is an expensive process that involves complex procedures. In contrast, enzymatic hydrolysis is widely applied in the production of infant formulas due to its practicality and high specificity. This process disrupts the spatial structure of antigenic proteins, breaking them down into peptides or amino acids. As a result, it effectively reduces allergic reactions triggered by β-Lg (Xiong et al., 2023; Sun et al., 2025).

Recent studies have focused on optimizing enzymatic hydrolysis to target antigenic epitopes in β-Lg more effectively. For instance, research has demonstrated that specific proteases, such as pepsin, chymotrypsin A, papain, and stem bromelain, can significantly reduce β-Lg antigenicity under optimal conditions. These enzymes degrade β-Lg into shorter peptides, altering its structural properties and cleaving linear epitopes, thereby diminishing its allergenic potential. This approach provides a theoretical basis for developing targeted enzymatic hydrolysis technologies aimed at producing low-allergenicity dairy-based materials (Sun et al., 2025).

Additionally, combining enzymatic hydrolysis with other processing techniques has been explored to further reduce allergenicity. Methods such as heat treatment, pressure processing, and fermentation, when used alongside enzymatic hydrolysis, can modify protein structures more extensively, potentially decreasing their allergenic properties while maintaining nutritional quality and sensory attributes. This integrated approach offers promising avenues for creating hypoallergenic food products, including infant formulas (Ge et al., 2024b).

2.2 Carbohydrates

Carbohydrates in human milk possess distinct characteristics that set them apart from those in other species. The predominant carbohydrate is lactose, a disaccharide composed of galactose and glucose, which accounts for about 87% of the total carbohydrate content. HMOs account for around 12%, while traces of free monosaccharides, such as glucose and galactose, make up about 1%. Additionally, glycoconjugates, such as glycoproteins, glycopeptides, and glycolipids, are found in specific fractions of the milk. Lactose, HMOs, and free monosaccharides are predominantly present in the whey portion, whereas glycolipids are associated with the milk fat globule membrane (Martin et al., 2016; Amissah et al., 2018; Almeida et al., 2021a, 2025).

Carbohydrates in infant formulas are broadly categorized into glycemic and nonglycaemic types. Glycaemic carbohydrates, such as lactose, are digested and absorbed in the small intestine, providing an essential energy source. Nonglycaemic carbohydrates, like oligosaccharides, remain undigested, reaching the large intestine to exert prebiotic effects. Lactose and free monosaccharides serve as the primary energy source during the first months of life while also regulating milk osmotic pressure and facilitating the absorption of some minerals, like calcium and magnesium, essential for bone and neuromuscular development. Additionally, lactose metabolism promotes the growth of probiotic microorganisms, such as *Bifidobacterium* and *Lactobacillus*, contributing to a balanced microbiota (Hettinga, 2019; Wicinski et al., 2020; Felice et al., 2021).

On the other hand, HMOs are highly complex structural compounds that are indigestible by the host but exert functions in the large intestine, where they selectively modulate the growth of beneficial bacteria, inhibit pathogen adhesion to the intestinal epithelium, and promote immunomodulation through interactions with cellular receptors. These functions are essential for maintaining intestinal health and supporting the immune system during early life. HMOs, with concentrations ranging from 0.5 to 1.0 g/100 mL in human milk, include over 200 structurally diverse types classified into fucosylated, nonfucosylated, and sialylated categories. This diversity is influenced by maternal genotype, particularly the expression of α1,2-fucosyltransferase 2 (FUT2) and α1,3/4-fucosyltransferase 3 (FUT3), which determine fucosylation patterns and variations during different lactation phases, with higher abundance in colostrum. Moreover, HMOs serve as a source of sialic acid for brain development and cognitive functions, highlighting their multifunctionality (Ackerman et al., 2018; Vandenplas et al., 2019; Wicinski et al., 2020; Alliet et al., 2022; Almeida et al., 2025).

Although glycoconjugates are less abundant than lactose and HMOs, they enhance human milk immunological and antimicrobial properties. By mimicking glycan receptors on epithelial cells and interacting with mucosal surfaces, they help prevent microbial adhesion, inhibit pathogen colonization, and promote the maturation of the infant's immune system, bolstering both innate and adaptive defenses (Plaza-Díaz et al., 2018; Almeida et al., 2025).

In contrast to human milk, in bovine milk, lactose also predominates, though in slightly lower concentrations (~5%), corresponding to approximately 4.4–5.6 g/100 mL. Bovine milk oligosaccharides (BMOs), are present in much lower concentrations (0.003–0.009 g/100 mL) and lack the structural diversity of HMOs, limiting their bioactive potential. Conversely, caprine milk contains higher oligosaccharide levels (0.025–0.03 g/100 mL), with structures more similar to HMOs than BMOs (Robinson, 2019; Almeida et al., 2025).

BMOs exhibit structural similarities to some HMOs but lack equivalent functional attributes, particularly concerning immunomodulation and interaction with human microbiota. A distinctive feature is the absence of fucose in most BMOs, which limits their effectiveness in modulating microbiota or serving as functional barriers against pathogens. This limitation highlights the need for changes in infant formulas to better replicate the benefits of HMOs. Research indicates that combining BMOs with probiotics can improve their functional properties, thereby increasing their potential for inclusion in infant formulas (Reverri et al., 2018; Robinson, 2019; Almeida et al., 2025).

Lactose is often retained as the primary carbohydrate source in infant formulas. However, in cases of lactose intolerance, its concentration is reduced or replaced with alternative carbohydrates, such as maltodextrin and starch. To replicate the effects of HMOs, galactooligosaccharides (GOS) and fructooligosaccharides (FOS) are added, which act as functional prebiotics, stimulating the growth of beneficial bacteria. However, these alternatives face limitations. Their simplified structures cannot fully replicate the biological complexity of HMOs, particularly in immunomodulation and pathogen defense. Advances in biotechnology have facilitated the synthesis of HMOs, including 2′-fucosyllactose (2′-FL) and lacto-N-tetraose (LNT), which are incorporated into some commercial infant formulas. Nevertheless, these products still face challenges related to cost and large-scale production (Puccio et al., 2017; Reverri et al., 2018; Robinson, 2019; Vandenplas et al., 2019).

2.3 Vitamins and minerals

Whey is a natural source of essential vitamins and minerals that support several biological processes. However, due to differences in the natural composition of bovine whey and the impacts of thermal processing, fortification with these nutrients is required in infant formulas. This fortification ensures that infants receive adequate levels of key micronutrients, compensating for losses that occur during heat treatment and addressing the demineralization processes required to make whey suitable for infant consumption (Yakoob and Lo, 2017; Guo, 2020; Almeida et al., 2022; Moraes et al., 2022).

Vitamins are essential micronutrients that support infant growth and cognitive development. They play a crucial role in cellular metabolism and help prevent various health issues. Their content in milk can also reflect the nutritional status and physiological reserves of the mother. Among them, water-soluble vitamins found in the whey fraction of milk, such as the B-complex vitamins (thiamine, riboflavin, pyridoxine, cobalamin, and folic acid) and vitamin C, stand out for their function as coenzymes in cellular metabolism and their rapid excretion, requiring constant replenishment. In contrast, fat-soluble vitamins (A, D, E, and K), present in greater amounts in the fatty fraction of milk, predominantly act in cellular structure and are stored in the body for use in processes related to cellular differentiation, immune system function, and calcium metabolism (Yakoob and Lo, 2017; Qiao et al., 2022).

In human milk, these vitamins' content reflects the infant's specific needs. Compared to bovine milk, human milk contains higher amounts of vitamins A, E, C, niacin, and inositol, all crucial for neonatal development. In contrast, bovine milk has higher concentrations of vitamins B1, B2, B6, B12, K, and pantothenic acid (Table 11.1). Despite these differences, the bioavailability of specific vitamins, such as B12, is more efficient in human milk due to associations with proteins like haptocorrin, which are in higher proportions (Hampel et al., 2017; Allen and Hampel, 2019; Magan et al., 2020).

In addition to compositional differences, the impact of heat processing on vitamins is a critical point in infant formulas derived from bovine milk. Water-soluble vitamins, such as C and folic acid, are particularly vulnerable to heating, while fat-soluble vitamins, such as A and E, are more stable. To address these challenges, advanced technologies such as microencapsulation have been implemented to protect sensitive vitamins during processing. Additionally, postprocessing fortification, widely used in infant formulas, ensures adequate levels of essential vitamins after heating. This dual approach helps maintain the nutritional quality of formulas, aligning them more closely with the composition of human milk (Allen and Hampel, 2019; Guo, 2020; Jiang and Guo, 2021).

Minerals are essential for the body, forming tissues and key components of enzymes and other molecules. Bovine milk contains significantly higher levels of macrominerals (Na, K, Cl, Ca, Mg, and P) compared to human milk, which explains its considerably greater total mineral content (~5.8 g/L) compared to that of mature human milk (~1.6 g/L) (Table 11.1). However, the chemical form and interactions of these minerals with other milk components influence their bioavailability. In human milk, minerals such as Fe, Ca, and Zn are associated with proteins or soluble compounds that promote their absorption. In contrast, insoluble forms like phosphates, more common in bovine milk, reduce absorption efficiency. The lactose in human milk facilitates the absorption of Ca and Mg, while Lf increases the utilization of Fe (Yakoob and Lo, 2017; Francisquini et al., 2020; Almeida et al., 2022).

In contrast, caseins in bovine milk may limit the bioavailability of these nutrients during digestion by forming casein complexes, which sequester Ca, Mg, Fe, and Zn. Among them, Ca is one of the most affected, as it can be internalized in casein micelles when milk is subjected to heat treatment and becomes unavailable for absorption. This underscores the importance of demineralization and fortification strategies in the preparation of infant formulas (Liao et al., 2017; Francisquini et al., 2020; Almeida et al., 2022).

The high levels of electrolytes (Na, Cl, K, and P) in bovine milk result in a potential renal solute load, which can be harmful to the immature infant kidney, increasing the risk of hypertonic dehydration. Therefore, infant formulas derived from bovine milk should undergo demineralization to reduce these excessive macrominerals. This process removes significant Na and Ca while retaining beneficial minerals like Mg and Zn. Combined ion exchange and electrodialysis techniques are used to remove these minerals, allowing first-stage infant formulas to better meet the nutritional needs of infants, avoiding risks associated with excessive mineral intake during this critical developmental phase. In follow-up formulas for older infants, these processes are generally unnecessary, as their more mature kidneys can handle higher mineral levels without adverse effects (Okawa et al., 2015; Alfaris et al., 2022; Almeida et al., 2022).

3. Infant formulas

Infant formulas are carefully designed to meet the diverse nutritional needs of infants who cannot be exclusively breastfed. Initially developed in the early 20th century as modifications of bovine milk, modern infant formulas have evolved substantially, with advances and discoveries in infant nutrition science and improvements in food processing technologies. Today, many formulations closely resemble the composition of human milk, striving to achieve similar physiological responses (Timby et al., 2017; Ventura et al., 2018).

Recent innovations in infant formula development have been driven by new insights into essential nutrients, their required quantities, and potential novel components that can enhance infant health. These include adjustments in the macronutrient profile to address digestive deficiencies, specialized protein hydrolysates for allergy management, and enhanced additives to better replicate human milk bioactivity. These innovations now offer viable alternatives to breastfeeding, particularly for infants with specific medical or dietary needs (Thompson et al., 2017; Lien et al., 2018; Vandenplas et al., 2019; Lu et al., 2019; Masum et al., 2021).

3.1 Classification of infant formula

The composition of infant formulas must be carefully tailored to meet the diverse needs of infants, following current scientific knowledge and international standards. Factors such as age, allergies, lactose intolerance, reflux, and various health conditions are important in determining the suitable formula composition.

To align with the developmental stages of infants, formulas are broadly categorized into three types based on age. Infant formula (Phase 1) is designed for newborns and babies up to 6 months old, providing essential nutrients to support early growth and development. Follow-on formula (Phase 2) is formulated for infants aged 6–18 months, offering additional vitamins and minerals to sustain the rapid growth and developmental milestones typical of this stage. Lastly, the Growing-up formula (Phase 3) is tailored for toddlers between 1 and 3 years of age, addressing their evolving nutritional needs while transitioning to a more varied diet (Codex Ailmentarius Commission, 2007; Brazil, 2011a, 2011b, 2014; EU, 2016).

For infants with specific medical or dietary needs, specialized formulas are available. Hypoallergenic formulas, which include extensively hydrolyzed or amino acid-based formulas, cater to infants with bovine milk protein allergies or other severe intolerances. Soy-based formulas, prepared using soy protein isolates, are suitable for infants with lactose intolerance or galactosemia. For those with severe allergies or malabsorption issues, elemental formulas, which use free amino acids instead of whole proteins, provide an alternative. Antiregurgitation formulas, thickened with ingredients like starches or carob seed flour, are intended for infants experiencing gastroesophageal reflux. Additionally, lactose-free formulas cater to infants with congenital or temporary lactase deficiency, ensuring they receive adequate nutrition without the complications of lactose intolerance (ESPGHAN, 2009; Brazil, 2014; Thompson et al., 2017; Almeida et al., 2021a, 2021b; Masum et al., 2021).

3.2 Regulatory guidelines for infant formula

The manufacturing and marketing of infant formulas are subject to strict regulations to ensure their safety and nutritional adequacy. The Codex Commission, a joint food standards program managed by the Food and Agriculture Organization (FAO) and the World Health Organization (WHO), has established standards detailing the minimum and maximum levels of each nutrient, as well as the quality and safety requirements for infant formulas (Table 11.4). These standards ensure consistency in nutritional adequacy and address the use of innovative additives. Although the Codex provides a broad framework, each country develops its specific regulations, which typically align with international guidelines but may include additional requirements to address local policies, nutritional concerns, and safety standards (Codex Alimentarius Commission, 2007; Brazil, 2011a, 2011b, 2014; EU, 2016; WHO, 2023).

In addition to the essential nutrient composition, bioactive compounds such as prebiotics (GOS, FOS, and inulin), synthetic sources of HMOs (2′-FL and LNT), LC-PUFAs such as DHA and ARA, can be optionally added to infant formulas as novel ingredients to improve infant health. These additives align with GRAS determinations, which consider rigorous safety standards, including toxicological studies, for the vulnerable population of infants. In addition, α-la, Lf, and MFGM compounds have received attention for their immunological and nutritional properties. Despite GRAS recognition, existing regulations do not specify optimal addition levels of these compounds, requiring further clinical studies to confirm their efficacy and determine doses that maximize benefits without exceeding the recommended protein

TABLE 11.4 Minimum and maximum nutritional parameters per 100 kcal established by the Codex Alimentarius for infant formulas.

		Codex Alimentarius	
Nutrients	**Unit**	**Min.**	**Max.**
Energetic value	*kcal*	60	70
Carbohydrates	*g*	9.0	14.9
Lactose	*g*	–	–
Protein	*g*	1.8	3.0
Linoleic acid	*mg*	0.3	1.2
α-linolenic acid	*mg*	50	–
Sodium	*mg*	20	60
Calcium	*mg*	50	140
Iron	*mg*	0.3	1.3
Potassium	*mg*	60	160
Chloride	*mg*	50	160
Phosphor	*mg*	25	90
Magnesium	*mg*	5.0	15
Iodine	*μg*	10	50
Copper	*μg*	35	80
Zinc	*mg*	0.5	1.5
Selenium	*μg*	1.0	9.0
Manganese	*μg*	1.0	50
Vitamin A	*μg*-RE[1]	60	180
Vitamin D	*μg*	1.0	2.5
Vitamin E	*mg*	0.5	5.0
Vitamin C	*mg*	8.0	30
Vitamin K	*μg*	4.0	25
Niacin	*μg*	300	1500
Vitamin B6	*μg*	35	175
Pantothenic acid	*μg*	400	2000
Vitamin B12	*μg*	0.1	0.5
Biotin	*μg*	1.5	7.5
Choline	*mg*	7.0	50
Taurine	*mg*	–	12
L-Carnitine	*mg*	1.2	–

Source: Codex Alimentarius Commission (2007).

levels, and therefore avoid metabolic overload in infants (Masum et al., 2021; Vandenplas et al., 2019; Vargas et al., 2023).

3.3 Formulation of infant formula

The production of infant formulas begins with the separation and processing of whey. Whey is obtained during the coagulation of milk using rennet or acidifying agents, resulting in two types of whey: sweet whey and acid whey, depending on the technological process employed. With its neutral pH, sweet whey retains higher concentrations of essential proteins such as α-La and Lf, making it the better choice for infant formula. This initial processing step ensures that the raw material aligns with the specific nutritional and functional requirements of the final product (Okawa et al., 2015; Jiang and Guo, 2021).

Technological advancements in whey demineralization, such as combining nanofiltration with ion-exchange resins, have further optimized whey processing by effectively reducing mineral content while preserving key components like proteins and lactose. This method significantly lowers excessive Na, K, and Cl levels, better aligning the mineral profile with human milk. Moreover, this process preserves the sensory characteristics of whey, ensuring acceptable flavor and odor, while maintaining its thermal stability during later processing stages (Okawa et al., 2015; Guo, 2020; Salaria et al., 2023).

Infant formulas are manufactured in various forms, including powdered, ready-to-feed, and concentrated liquid formulas. In some countries, such as Brazil, only powdered formulas are regulated and widely used due to their longer shelf life and ease of transportation and storage. Among the manufacturing processes, the wet mixing-spray drying method is the most common due to its ability to ensure uniform nutrient distribution (Masum et al., 2021; Jiang and Guo, 2021; Salaria et al., 2023).

The wet mixing-spray drying process involves several stages. Ingredients are first mixed in a liquid state to create a consistent emulsion. This emulsion undergoes pasteurization to eliminate harmful bacteria and homogenization to create a stable and uniform mixture. For hypoallergenic formulas, fats are incorporated early in the process, while heat-sensitive ingredients, such as vitamins and fats, are added after pasteurization. High-viscosity components, like starches or carob seed flour, are directly added to the spray dryer. Alternatively, to optimize costs, specific maltodextrins, and other ingredients may be blended with the powder postdrying (Masum et al., 2021; Jiang and Guo, 2021; Salaria et al., 2023).

Evaporation is often used as a preliminary step to concentrate the solids in the mixture, reducing drying costs. During this stage, heat treatment is applied to eliminate pathogenic organisms, ensuring microbiological safety. The concentrated liquid is then spray-dried using hot air to produce a fine powder. Multistage spray dryers are commonly employed, as they enhance product quality regarding agglomeration, solubility, dispersibility, and wettability. Additional techniques, such as ultrafiltration and enzymatic hydrolysis, may also improve digestibility or cater to specific dietary requirements. Managing the inlet and outlet temperatures during spray drying is an essential step for maintaining protein solubility and preventing Maillard reactions. Once the powdered formula meets quality standards, it is packaged in metal cans under a nitrogen atmosphere. This packaging technique minimizes oxidation of milk fats and LC-PUFAs, preserving the nutritional integrity of the product (de Oliveira Lino et al., 2016; Masum et al., 2021; Jiang and Guo, 2021; Salaria et al., 2023) (Fig. 11.2).

Maintaining stability during transportation and storage is also crucial for the quality and shelf life of infant formulas. High humidity can lead to lactose crystallization, Maillard reactions, and decreased solubility, compromising the finished product. To address these risks, manufacturers must rigorously monitor and control environmental factors, such as relative humidity and temperature, during production, handling, storage, and transportation. Thus, ensuring consistent physicochemical properties through optimized processing parameters, including heating, homogenization, evaporation, and spray drying, is essential for delivering a stable and safe product to consumers (Condurso et al., 2020; Masum et al., 2021).

4. Innovative approaches to align infant formulas closely with human milk

The quest to create infant formulas more closely replicating human milk has driven significant advancements in processing technologies and biotechnological approaches. However, replicating its complex composition in formulas requires innovative methods for isolating and utilizing bioactive compounds from bovine milk, such as α-La and Lf. These proteins are integral to promoting growth, boosting immunity, and enhancing cognitive development in infants (Lönnerdal, 2016; Layman et al., 2018).

Efforts to improve protein purification processes have centered on ensuring high purity, scalability, and cost-efficiency. Key techniques include ultrafiltration (UF), enzymatic hydrolysis, selective precipitation, and

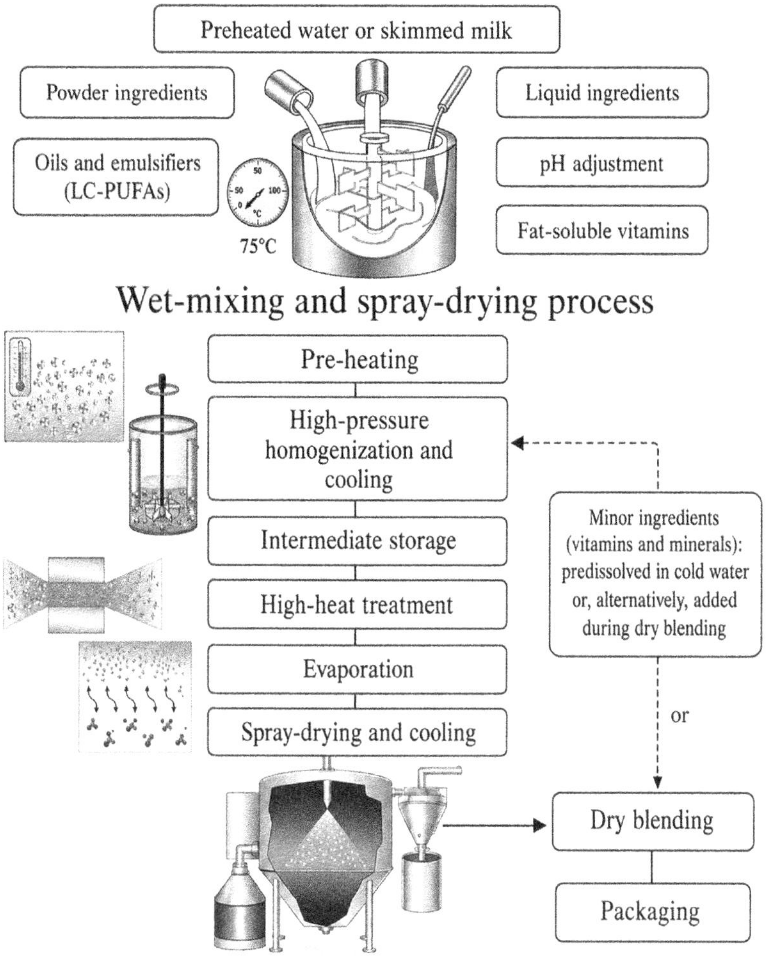

FIGURE 11.2 Manufacturing flow chart for powdered infant formula.

chromatography (Masum et al., 2021). UF membranes with specific molecular weight cutoffs enable the separation of α-La from larger proteins like β-Lg and Igs, while optimized conditions such as controlled pH and temperature enhance the efficiency of this process. Enzymatic hydrolysis, employing enzymes like trypsin, further refines the purity by selectively degrading β-Lg while preserving α-La. When combined with diafiltration, this process achieves purities of up to 95% and ensures industrial scalability (Ventura et al., 2018; Ge et al., 2024b).

Selective precipitation and thermal aggregation methods also play a vital role in protein isolation. For instance, adjusting ionic concentration and pH or applying thermal precipitation at specific conditions can separate unwanted proteins like β-Lg while retaining α-La. Chromatography, though highly precise, is generally reserved for specialized applications due to its high cost and complexity (Liao et al., 2017; Ventura et al., 2018; Zhao et al., 2022).

Recent innovations have extended beyond traditional methods, exploring the potential of fermentation to generate bioactive peptides from milk proteins. By utilizing microorganisms like *Lactobacillus helveticus*, specific proteins like α-La are hydrolyzed into smaller peptides with immunological and gut health benefits. These bioactive peptides mimic some of the functional benefits of human milk, offering promising applications in enriched formulas (Ge et al., 2024a).

Recent advancements in biotechnological methods, particularly through the use of transgenic approaches, have further transformed the landscape of infant formula development. An achievement in this field is the successful production of human-like proteins, such as α-La and Lf, utilizing microbial systems and alternative biological hosts. These innovations have led to a good yield, suggesting a viable pathway to enhance the nutritional profile of infant formulas (Li and Zhu, 2024).

5. Conclusion

This chapter explored whey compounds in bovine milk, the main ingredient in infant formula, and compared them to human milk, highlighting its role as the gold standard in infant nutrition. Despite technological and scientific advancements in infant formula development, significant challenges remain in replicating the whole composition and functional benefits of human milk. Human milk contains a wide array of bioactive compounds that go beyond basic nutrients, including immunomodulatory proteins, growth hormones, and specific oligosaccharides, playing crucial roles in gut health, immune and cognitive development, and infection protection. Bovine whey serves as an important base for the production of infant formulas due to its high biological value, containing essential proteins such as α-La and Lf, albeit at lower concentrations than in human milk. These proteins have demonstrated significant benefits in supporting infant development, including antimicrobial, antiinflammatory, and neurodevelopmental functions. However, fundamental differences in protein composition, such as the presence of β-Lg, absent in human milk and potentially allergenic, remain major limitations.

Technological adaptations, such as enzymatic hydrolysis, membrane filtration, and the incorporation of natural and synthesized bioactive compounds, have contributed to bringing infant formulas closer to the nutritional profile of human milk. Furthermore, novel approaches, including the use of nonthermal technologies and biotechnology for the production of recombinant human proteins, show significant potential to overcome challenges associated with traditional heat treatments that often compromise the nutritional integrity of milk components. Despite these advancements, much must be done to create an infant formula that fully matches human milk benefits. The development of novel technologies and processing strategies will continue to be essential to minimizing negative impacts on the nutritional and functional value of formulas, ensuring food safety, and promoting infant well-being.

References

Ackerman, D.L., Craft, K.M., Doster, R.S., Weitkamp, J.H., Aronoff, D.M., Gaddy, J.A., Townsend, S.D., 2018. Antimicrobial and antibiofilm activity of human milk oligosaccharides against *Streptococcus agalactiae, Staphylococcus aureus*, and *Acinetobacter baumannii*. ACS Infect. Dis. 4 (3), 315–324. https://doi.org/10.1021/acsinfecdis.7b00183.

Akin, I.M., Atasay, B., Dogu, F., Okulu, E., Arsan, S., Karatas, H.D., Ikinciogullari, A., Turmen, T., 2014. Oral lactoferrin to prevent nosocomial sepsis and necrotizing enterocolitis of premature neonates and effect on T-regulatory cells. Am. J. Perinatol. 31 (12), 1111–1120. https://doi.org/10.1055/s-0034-1371704.

Alfaris, N.A., Alothman, Z.A., Aldayel, T.S., Wabaidur, S.M., Altamimi, J.Z., 2022. Evaluation and comparison of the nutritional and mineral content of milk formula in the Saudi Arabia market. Front. Nutr. 9, 851229. https://doi.org/10.3389/fnut.2022.851229.

Allen, L.H., Hampel, D., 2019. Water-soluble vitamins in human milk: factors affecting their concentration and their physiological significance. Nestle Nutr. Inst. Workshop Ser. 90, 69–81. https://doi.org/10.1159/000490296.

Alliet, P., Vandenplas, Y., Roggero, P., Jespers, S.N.J., Peeters, S., Stalens, J.P., et al., 2022. Safety and efficacy of a probiotic-containing infant formula supplemented with 2'-fucosyllactose: a double-blind randomized controlled trial. Nutr. J. 21 (1), 11. https://doi.org/10.1186/s12937-022-00764-2.

Almeida, C.C., Mendonça Pereira, B.F., Leandro, K.C., Costa, M.P., Spisso, B.F., Conte-Junior, C.A., 2021a. Bioactive compounds in infant formula and their effects on infant nutrition and health: a systematic literature review. Int. J. Food Sci. 2021. https://doi.org/10.1155/2021/8850080. Article 8850080.

Almeida, C.C., Baião, D.d.S., Leandro, K.C., Paschoalin, V.M.F., Costa, M.P.d., Conte-Junior, C.A., 2021b. Protein quality in infant formulas marketed in Brazil: assessments on biodigestibility, essential amino acid content and proteins of biological importance. Nutrients 13 (11). https://doi.org/10.3390/nu13113933. Article 3933.

Almeida, C.C., Baião, D.D.S., Rodrigues, P.A., Saint'Pierre, T.D., Hauser-Davis, R.A., Leandro, K.C., Paschoalin, V.M.F., da Costa, M.P., Conte-Junior, C.A., 2022. Macrominerals and trace minerals in commercial infant formulas marketed in Brazil: compliance with established minimum and maximum requirements, label statements, and estimated daily intake. Front. Nutr. 9, 857698. https://doi.org/10.3389/fnut.2022.857698.

Almeida, C.C., Paschoalin, V.M.F., Conte-Junior, C.A., 2025. Compostos bioativos e probióticos em fórmulas infantis. In: Faintuch, J. (Ed.), Microbioma, disbiose, probióticos e bacterioterapia, 2ª ed. Manole.

Aly, E., López-Nicolás, R., Darwish, A.A., Frontela-Saseta, C., Ros-Berruezo, G., 2016. Supplementation of infant formulas with recombinant human lactoferrin and/or galactooligosaccharides increases iron bioaccessibility as measured by ferritin formed in Caco-2 cell model. Food Res. Int. 89, 1048–1055. https://doi.org/10.1016/j.foodres.2016.08.030.

Amissah, E.A., Brown, J., Harding, J.E., 2018. Carbohydrate supplementation of human milk to promote growth in preterm infants. Cochrane Database Syst. Rev. https://doi.org/10.1002/14651858.CD000280.pub2.

Beverly, R.L., Huston, R.K., Markell, A.M., McCulley, E.A., Martin, R.L., Dallas, D.C., 2020. 'Milk peptides survive in vivo gastrointestinal digestion and are excreted in the stool of infants. J. Nutr. 150 (4), 712–721.

Björmsjö, M., Hernell, O., Lönnerdal, B., Berglund, S.K., 2021. Reducing iron content in infant formula from 8 to 2 mg/L does not increase the risk of iron deficiency at 4 or 6 months of age: a randomized controlled trial. Nutrients 13 (3), 3. https://doi.org/10.3390/nu13010003.

Bogahawaththa, D., Chandrapala, J., Vasiljevic, T., 2017. Thermal denaturation of bovine immunoglobulin G and its association with other whey proteins. Food Hydrocoll. 72, 350–357.

BRAZIL. Agência Nacional de Vigilância Sanitária, 2011a. RDC n° 43, de 19 de setembro de 2011. Dispõe sobre o regulamento técnico para fórmulas infantis para lactentes. Disponível em. http://www.ibfan.org.br/site/wp-content/uploads/2014/06/Resolucao_RDC_n_43_de_19_de_setembro_de_2011.pdf.

BRAZIL. Agência Nacional de Vigilância Sanitária, 2011b. RDC n° 44, de 19 de setembro de 2011. Dispõe sobre o regulamento técnico para fórmulas infantis de seguimento para lactentes e crianças de primeira infância. Disponível em. http://www.ibfan.org.br/site/wp-content/uploads/2014/06/Resolucao_RDC_n_44_de_19_de_setembro_de_2011.pdf.

BRAZIL. Agência Nacional de Vigilância Sanitária, 2014. RDC n° 48, de 25 de setembro de 2014. Dispõe sobre o regulamento técnico para fórmulas infantis para lactentes destinadas a necessidades dietoterápicas específicas e fórmulas infantis de seguimento para lactentes e crianças de primeira infância destinadas a necessidades dietoterápicas específicas. Disponível em. https://bvsms.saude.gov.br/bvs/saudelegis/anvisa/2014/rdc0048_25_09_2014.html.

Brück, W.M., 2017. The benefits of alpha-lactalbumin in early childhood. In: Dairy in Human Health and Disease across the Lifespan, pp. 117–126. https://doi.org/10.1016/b978-0-12-809868-4.00008-x.

Byrne, M.E., O'Mahony, J.A., O'Callaghan, T.F., 2021. Compositional and functional considerations for bovine-, caprine- and plant-based infant formulas. Dairy 2 (4), 695–715. https://doi.org/10.3390/dairy2040054.

Cao, X., Ren, Y., Lu, Q., Wang, K., Wu, Y., Wang, Y., Zhang, Y., Cui, X., Yang, Z., Chen, Z., 2023. Lactoferrin: a glycoprotein that plays an active role in human health. Front. Nutr. 9. https://doi.org/10.3389/fnut.2022.1018336. Article 1018336.

Castanet, M., Costalos, C., Haiden, N., Hascoet, J.-M., Berger, B., Sprenger, N., et al., 2020. Early effect of supplemented infant formulae on intestinal biomarkers and microbiota: a randomized clinical trial. Nutrients 12, 1481. https://doi.org/10.3390/nu12051481.

Chen, K., Chai, L., Li, H., Zhang, Y., Xie, H.-M., Shang, J., et al., 2016. Effect of bovine lactoferrin from iron-fortified formulas on diarrhea and respiratory tract infections of weaned infants in a randomized controlled trial. Nutrition 32 (2), 222–227. https://doi.org/10.1016/j.nut.2015.08.010.

Chen, K., Jin, S., Chen, H., Cao, Y., Dong, X., Li, H., Zhou, Z., Liu, C., 2021. Dose effect of bovine lactoferrin fortification on diarrhea and respiratory tract infections in weaned infants with anemia: a randomized, controlled trial. Nutrition 90, 111288. https://doi.org/10.1016/j.nut.2021.111288.

Chen, K., Zhang, G., Chen, H., Cao, Y., Dong, X., Li, H., et al., 2020. Dose effect of bovine lactoferrin fortification on iron metabolism of anemic infants. J. Nutr. Sci. Vitaminol. 66 (1), 24–31.

Chen, K., Zhang, L., Li, H., Zhang, Y., Xie, H., Shang, J., Tian, W., Yang, P., Chai, L., Mao, M., 2015. Iron metabolism in infants: influence of bovine lactoferrin from iron-fortified formula. Nutrition 31 (2), 304–309. https://doi.org/10.1016/j.nut.2014.07.006.

Chetta, K.E., Newton, D.A., Wagner, C.L., Baatz, J.E., 2022. Free fatty acid and α-lactalbumin-oleic acid complexes in preterm human milk are cytotoxic to fetal intestinal cells in vitro. Front. Nutr. 9, 918872.

Chichlowski, M., Bokulich, N., Harris, C.L., Wampler, J.L., Li, F., Berseth, C.L., et al., 2021. Effect of bovine milk fat globule membrane and lactoferrin in infant formula on gut microbiome and metabolome at 4 months of age. Curr. Dev. Nutr. 5 (5). https://doi.org/10.1093/cdn/nzab027.

Chun, S.H., Lee, K.W., 2022. Immune-enhancing effects of β-lactoglobulin glycated with lactose following in vitro digestion on cyclophosphamide-induced immunosuppressed mice. J. Dairy Sci. 105 (1), 623–636.

Codex Alimentarius Commission, 2007. Standard for infant formula and formulas for special medical purposes intended for infants. CODEX STAN 72–108. Available at: https://www.fao.org/fao-who-codexalimentarius.

Condurso, C., Cincotta, F., Merlino, M., Stanton, C., Verzera, A., 2020. Stability of powdered infant formula during secondary shelf-life and domestic practices. Int. Dairy J. https://doi.org/10.1016/j.idairyj.2020.104761.

Czosnykowska-Łukacka, M., Lis-Kuberka, J., Królak-Olejnik, B., Orczyk-Pawiłowicz, M., 2020. Changes in human milk immunoglobulin profile during prolonged lactation. Front. Pediatr. 8. Article 428.

de Oliveira Lino, M., da Silva Rodrigues, J.B., de Souza Oliveira, R.P., de Carvalho, A.F., do Prado, G.M., 2016. Physical-chemical and microbiological stability of whole milk powder packed in aluminum cans under modified atmosphere. Food Sci. Technol. 36 (2), 274–278.

Donald, K., Petersen, C., Turvey, S.E., Finlay, B.B., Azad, M.B., 2022. Secretory IgA: linking microbes, maternal health, and infant health through human milk. Cell Host Microbe 30 (5), 650–659. https://doi.org/10.1016/j.chom.2022.02.005.

Donaldson, G.P., Ladinsky, M.S., Yu, K.B., Sanders, J.G., Yoo, B.B., Chou, W.C., et al., 2018. Gut microbiota utilize immunoglobulin A for mucosal colonization. Science 360 (6390), 795–800.

Donovan, S.M., 2016. The role of lactoferrin in gastrointestinal and immune development and function: a preclinical perspective. J. Pediatr. 173, S16–S28. https://doi.org/10.1016/j.jpeds.2016.02.072.

Drago-Serrano, M.E., Campos-Rodríguez, R., Carrero, J.C., De la Garza, M., 2017. Lactoferrin: balancing ups and downs of inflammation due to microbial infections. Int. J. Mol. Sci. 18 (3), 501. https://doi.org/10.3390/ijms18030501.

Dyrda-Terniuk, T., Pomastowski, P., 2023. The multifaceted roles of bovine lactoferrin: molecular structure, isolation methods, analytical characteristics, and biological properties. J. Agric. Food Chem. 71 (51), 20500–20531. https://doi.org/10.1021/acs.jafc.3c06887.

EFSA Panel on Food Contact Materials, Enzymes and Processing Aids (CEP), Lambré, C., Barat Baviera, J.M., Bolognesi, C., Cocconcelli, P.S., Crebelli, R., et al., 2023. Safety evaluation of the food enzyme lysozyme from hens' eggs. EFSA J. 21 (3), e07916. https://doi.org/10.2903/j.efsa.2023.7916.

El-Hawy, M.A., Abd Al-Salam, S.A., Bahbah, W.A., 2021. Comparing oral iron bisglycinate chelate, lactoferrin, lactoferrin with iron and iron polymaltose complex in the treatment of children with iron deficiency anemia. Clinical Nutrition ESPEN 46, 367–371. https://doi.org/10.1016/j.clnesp.2021.08.040.

ESPGHAN Committee on Nutrition, Agostoni, C., Braegger, C., et al., 2009. Breastfeeding: a commentary by the ESPGHAN committee on nutrition. J. Pediatr. Gastroenterol. Nutr. 49, 112–125. https://doi.org/10.1097/01.mpg.0000304464.60788.bd.

European Union, 2016. Off. J. Eur. Union L 25 (2 February 2016). Available at: https://eur-lex.europa.eu/legal-content/PT/TXT/PDF/?uri=OJ:L:2016:025:FULL.

Felice, V.D., Owens, R.A., Kennedy, D., Hogan, S.A., Lane, J.A., 2021. Comparative structural and compositional analyses of cow, buffalo, goat and sheep cream. Foods 10 (11). https://doi.org/10.3390/foods10112643. Article 2643.

Ferraboschi, P., Ciceri, S., Grisenti, P., 2021. Applications of lysozyme, an innate immune defense factor, as an alternative antibiotic. Antibiotics (Basel, Switzerland) 10 (12), 1534. https://doi.org/10.3390/antibiotics10121534.

Fleddermann, M., Demmelmair, H., Grote, V., Nikolic, T., Trisic, B., Koletzko, B., 2014. Infant formula composition affects energetic efficiency for growth: the BeMIM study, a randomized controlled trial. Clin. Nutr. 33 (4), 588–595.

Francisquini, J.d.A., Nunes, L., Martins, E., Stephani, R., Perrone, Í.T., Carvalho, A. F. de, 2020. How the Heat Treatment Affects the Constituents of Infant Formulas: A Review. https://doi.org/10.1590/1981-6723.27219.

Gapper, L.W., Copestake, D.E.J., Otter, D.E., Indyk, H.E., 2007. Analysis of bovine immunoglobulin G in milk, colostrum and dietary supplements: a review. Anal. Bioanal. Chem. 389 (1), 93–109. https://doi.org/10.1007/s00216-007-1391-z.

Gazi, I., Reiding, K.R., Groeneveld, A., Bastiaans, J., Huppertz, T., Heck, A.J., 2023. Key changes in bovine milk immunoglobulin G during lactation: NeuAc sialylation is a hallmark of colostrum immunoglobulin GN-glycosylation. Glycobiology 33 (2), 115–125.

Ge, X., Zhang, J., Regenstein, J.M., Liu, D., Huang, Y., Qiao, Y., Zhou, P., 2024a. 'α-Lactalbumin: functional properties and potential health benefits. Food Biosci. 60, 104371. https://doi.org/10.1016/j.fbio.2024.104371.

Ge, X., Qu, X., Xie, C., Zang, J., Wu, W., Lv, L., 2024b. The influence on the structure and allergenicity of milk β-lactoglobulin by methylglyoxal during thermal processing. Food Res. Int. 196, 115043. https://doi.org/10.1016/j.foodres.2024.115043.

GRAS Notice (GRN) No. 669, 2016. Bovine Milk-Derived Lactoferrin in Term Infant Formulas. Assessed June 24, 2022. Available at: https://www.fda.gov/media/124472/download.

Grzywacz, K., Butcher, J., Li, J., Barrington, K., Mohamed, I., Stintzi, A., 2020. 'Bovine lactoferrin supplementation does not disrupt microbiota development in preterm infants receiving probiotics. J. Pediatr. Gastroenterol. Nutr. 71 (2), 216–222. https://doi.org/10.1097/MPG.0000000000002734.

Guo, M., 2020. Human Milk Biochemistry and Infant Formula Manufacturing Technology, second ed. Woodhead Publishing. Available at: https://www.sciencedirect.com/book/9780081028988.

Hampel, D., Shahab-Ferdows, S., Islam, M.M., Peerson, J.M., Allen, L.H., 2017. Vitamin concentrations in human milk vary with time within feed, circadian rhythm, and single-dose supplementation. J. Nutr. 147 (4), 603–611. https://doi.org/10.3945/jn.116.242941.

He, T., Rombouts, W., Einerhand, A.W.C., Hotrum, N., van de Velde, F., 2022. Gastric protein digestion of goat and cow milk infant formula and human milk under simulated infant conditions. Int. J. Food Sci. Nutr. 73 (1), 28–38.

Hettinga, K.A., 2019. Lactose in the dairy production chain. In: Lactose, pp. 231–266. https://doi.org/10.1016/b978-0-12-811720-0.00006-4.

Ianiro, G., Rosa, L., Bonaccorsi di Patti, M.C., et al., 2023. Lactoferrin: from the structure to the functional orchestration of iron homeostasis. Biometals 36, 391–416. https://doi.org/10.1007/s10534-022-00453-x.

Jakopović, K.L., Barukčić, I., Božanić, R., 2016. Physiological significance, structure and isolation of α-lactalbumin. Mljekarstvo: Časopis za Unaprjeđenje Proizvodnje i Prerade Mlijeka 66 (1), 3–11.

Jiang, S.L., Guo, M.R., 2021. Processing technology for infant formula. In: Guo, M. (Ed.), Human Milk Biochemistry and Infant Formula Manufacturing Technology, second ed. Woodhead Publishing, pp. 223–240. https://doi.org/10.1016/B978-0-08-102898-8.00008-8.

Jiménez-Saiz, R., Benedé, S., Molina, E., López-Expósito, I., 2015. Effect of processing technologies on the allergenicity of food products. Crit. Rev. Food Sci. Nutr. 55 (13), 1902–1917.

Joubran, Y., Moscovici, A., Lesmes, U., 2015. 'Antioxidant activity of bovine alpha-lactalbumin Maillard products and evaluation of their in vitro gastro-duodenal digestive proteolysis. Food Funct. 6 (4), 1229–1240. https://doi.org/10.1039/c4fo01165a.

Kouwenhoven, S.M.P., Antl, N., Finken, M.J.J., Twisk, J.W.R., van der Beek, E.M., Abrahamse-Berkeveld, M., et al., 2020. A modified low-protein infant formula supports adequate growth in healthy, term infants: a randomized, double-blind, equivalence trial. Am. J. Clin. Nutr. 111 (5), 962–974. https://doi.org/10.1093/ajcn/nqz308.

Kumar, B.S.G., Mattad, S., 2021. Comprehensive analysis of lactoferrin N-glycans with site-specificity from bovine colostrum using specific proteases and RP-UHPLC-MS/MS. Int. Dairy J. 119, 104999. https://doi.org/10.1016/j.idairyj.2021.104999.

Layman, D.K., Lönnerdal, B., Fernstrom, J.D., 2018. Applications for α-lactalbumin in human nutrition. Nutr. Rev. 76 (6), 444–460. https://doi.org/10.1093/nutrit/nuy004.

Leśnierowski, G., 2015. Lysozyme and its modified forms: properties, potential for its production and application. In: Watson, R.R., de Meester, F. (Eds.), Handbook of Eggs in Human Function. Wageningen Academic Publishers, Wageningen, pp. 483–495, 978-90-8686-254-2.

Leśnierowski, G., Yang, T., 2020. Lysozyme and its modified forms: a critical appraisal of selected properties and potential. Trends Food Sci. Technol. https://doi.org/10.1016/j.tifs.2020.11.004.

Li, J., Zhu, F., 2024. Whey protein hydrolysates and infant formulas: effects on physicochemical and biological properties. Compr. Rev. Food Sci. Food Saf. 23 (3). https://doi.org/10.1111/1541-4337.13337. Article e13337.

Li, Q., Hu, W., Zhao, J., Wang, J., Dai, Y., Zhao, Y., et al., 2014. Supplementation transgenic cow's milk containing recombinant human lactoferrin enhances systematic and intestinal immune responses in piglets. Mol. Biol. Rep. 41, 2119–2128.

Li, F., Wu, S.S., Berseth, C.L., Harris, C.L., Richards, J.D., Wampler, J.L., et al., 2019. Improved neurodevelopmental outcomes associated with bovine milk fat globule membrane and lactoferrin in infant formula: a randomized, controlled trial. J. Pediatr. https://doi.org/10.1016/j.jpeds.2019.08.030.

Li, W., Liu, B., Lin, Y., et al., 2022. The application of lactoferrin in infant formula: the past, present and future. Crit. Rev. Food Sci. Nutr. https://doi.org/10.1080/10408398.2022.2157792.

Liao, Y., Weber, D., Xu, W., Durbin-Johnson, B.P., Phinney, B.S., Lönnerdal, B., 2017. Absolute quantification of human milk caseins and the whey/casein ratio during the first year of lactation. J. Proteome Res. https://doi.org/10.1021/acs.jproteome.7b00486.

Lien, E., Richard, C., Hoffman, D., 2018. DHA and ARA addition to infant formula: current status and future research directions. Prostagl. Leukot. Essent. Fat. Acids 128, 26–40. https://doi.org/10.1016/j.plefa.2017.09.005.

Lin, T., Meletharayil, G., Kapoor, R., Abbaspourrad, A., 2021. Bioactives in bovine milk: chemistry, technology, and applications. Nutr. Rev. 79, 48–69.

Liu, F., Zhang, S., Li, J., McClements, D.J., Liu, X., 2018. Recent development of lactoferrin-based vehicles for the delivery of bioactive compounds: complexes, emulsions, and nanoparticles. Trends Food Sci. Technol. 79, 67–77. https://doi.org/10.1016/j.tifs.2018.06.013.

Lönnerdal, B., 2016. Bioactive proteins in human milk: health, nutrition, and implications for infant formulas. J. Pediatr. 173 (Suppl. 1), S4–S9. https://doi.org/10.1016/j.jpeds.2016.02.070.

Lu, Y., Ryan, S.L., Zhao Avena-Woods, C., 2019. Prebiotics and probiotics in dietetics practice. J. Acad. Nutr. Diet. 119 (9), 1433–1443.

Luo, G., Zhu, Y., Ni, D., Chen, J., Zhang, W., Mu, W., 2023. Infant formulae: key components, nutritional value, and new perspectives. Food Chem. 424. https://doi.org/10.1016/j.foodchem.2023.136393. Article 136393.

Magan, J.B., Carboy, T.F., Zheng, J., Zhang, L., Mandal, R., Hennessy, D., et al., 2020. Effect of diet on the vitamin B profile of bovine milk-based protein ingredients. Foods 9 (5). Article 578.

Manzoni, P., Meyer, M., Stolfi, I., Rinaldi, M., Cattani, S., Pugni, L., et al., 2014. 'Bovine lactoferrin supplementation for prevention of necrotizing enterocolitis in very-low-birth-weight neonates: a randomized clinical trial. Early Hum. Dev. 90 (Suppl. 1), S60–S65.

Martin, C.R., Ling, P.-R., Blackburn, G.L., 2016. Review of infant feeding: key features of breast milk and infant formula. Nutrients 8 (5), 279.

Martorell-Aragonés, A., Echeverría-Zudaire, L., Alonso-Lebrero, E., Boné-Calvo, J., Martín-Muñoz, M.F., Nevot-Falcó, S., et al., 2015. Position document: IgE-mediated cow's milk allergy. Allergol. Immunopathol. 43 (5), 507–526. https://doi.org/10.1016/j.aller.2015.01.003.

Maryniak, N.Z., Sancho, A.I., Egon, B.H., Bøgh, K.L., 2022. Alternatives to cow's milk-based infant formulas in the prevention and management of cow's milk allergy. Foods 11 (7). https://doi.org/10.3390/foods11070926. Article 926.

Masum, A.K.M., Chandrapala, J., Huppertz, T., Adhikari, B., Zisu, B., 2021. Production and characterization of infant milk formula powders: a review. Dry. Technol. https://doi.org/10.1080/07373937.2020.1767645.

Meek, J.Y., Noble, L., Section on Breastfeeding, 2022. Policy statement: breastfeeding and the use of human milk. Pediatrics 150 (1), e2022057988.

Milani, G.P., Edefonti, V., De Cosmi, V., Bettocchi, S., Mazzocchi, A., Silano, M., et al., 2023. Protein and growth during the first year of life: a systematic review and meta-analysis. Pediatr. Res. 94 (3), 878–891. https://doi.org/10.1038/s41390-023-02531-3.

Moraes, M.R., do Nascimento da Silva, E., Sanches, V.L., et al., 2022. Bioaccessibility of some minerals in infant formulas. J. Food Sci. Technol. 59, 2004–2012. https://doi.org/10.1007/s13197-021-05215-0.

Nayak, C., Ramachandra, C.T., Kumar, G.M., 2020. A comprehensive review on composition of donkey milk in comparison to human, cow, buffalo, sheep, goat, camel and horse milk. Mysore J. Agric. Sci. 54 (3), 42–50.

Nielsen, C.H., Hui, Y., Nguyen, D.N., Ahnfeldt, A.M., Burrin, D.G., Hartmann, B., Bering, S.B., 2020. Alpha-lactalbumin enriched whey protein concentrate to improve gut, immunity and brain development in preterm pigs. Nutrients 12 (1), 245.

Ochoa, T., Loli, S., Mendoza, K., Carcamo, C., Bellomo, S., Cam, L., et al., 2021. 'Effect of bovine lactoferrin on prevention of late-onset sepsis in infants <1500 g: A pooled analysis of individual patient data from two randomized controlled trials. Biochem. Cell. Biol. 99 (1), 14–19. https://doi.org/10.1139/bcb-2020-0046.

Ochoa, T.J., Zegarra, J., Cam, L., Llanos, R., Pezo, A., Cruz, K., Zea-Vera, A., Cárcamo, C., Campos, M., Bellomo, S., NEOLACTO Research Group, 2015. Randomized controlled trial of lactoferrin for prevention of sepsis in Peruvian neonates less than 2500 g. Pediatr. Infect. Dis. J. 34 (6), 571–576. https://doi.org/10.1097/INF.0000000000000593.

Okawa, T., Shimada, M., Ushida, Y., Seki, N., Watai, N., Ohnishi, M., Ito, A., 2015. 'Demineralisation of whey by a combination of nanofiltration and anion-exchange treatment: a preliminary study. Int. J. Dairy Technol. 68, 478–485.

Oropeza-Ceja, L.G., Rosado, J.L., Ronquillo, D., García, O.P., Caamaño, M.C.D., García-Ugalde, C., Viveros-Contreras, R., Duarte-Vázquez, M.A., 2018. Lower protein intake supports normal growth of full-term infants fed formula: a randomized controlled trial. Nutrients 10 (7), 886. https://doi.org/10.3390/nu10070886.

Ostertag, F., Grimm, V.J., Hinrichs, J., 2025. Iron saturation and binding capacity of lactoferrin—development and validation of a colorimetric protocol for quality control. Food Chem. 463. https://doi.org/10.1016/j.foodchem.2024.141365. Article 141365.

Patro-Gołąb, B., Zalewski, B.M., Kouwenhoven, S.M., Karaś, J., Koletzko, B., van Goudoever, J.B., Szajewska, H., 2016. Protein concentration in milk formula, growth, and later risk of obesity: a systematic review. J. Nutr. 146 (3), 551–564. https://doi.org/10.3945/jn.115.223651.

Permyakov, E.A., 2020. 'α-Lactalbumin, amazing calcium-binding protein. Biomolecules 10 (9), 1210. https://doi.org/10.3390/biom10091210.

Petersen, H., Nomayo, A., Zelenka, R., Foster, J., Tvrdík, J., Jochum, F., 2020. Adequacy and safety of α-lactalbumin–enriched low-protein infant formula: a randomized controlled trial. Nutrition 74, 110728.

Plaza-Díaz, J., Fontana, L., Gil, A., 2018. Human milk oligosaccharides and immune system development. Nutrients 10 (8), 1038. https://doi.org/10.3390/nu10081038.

Prosser, C.G., 2021. Compositional and functional characteristics of goat milk and relevance as a base for infant formula. J. Food Sci. 86 (2), 257–265. https://doi.org/10.1111/1750-3841.15574.

Puccio, G., Alliet, P., Cajozzo, C., et al., 2017. Effects of infant formula with human milk oligosaccharides on growth and morbidity: a randomized multicenter trial. J. Pediatr. Gastroenterol. Nutr. 64, 624–631.

Qiao, W., Chen, J., Zhang, M., Wang, Y., Yang, B., Zhao, J., et al., 2022. A cohort study of vitamins contents in human milk from maternal-infant factors. Front. Nutr. 9. https://doi.org/10.3389/fnut.2022.993066. Article 993066.

Reverri, E.J., Devitt, A., Kajzer, J., et al., 2018. Review of the clinical experiences of feeding infants formula containing the human milk oligosaccharide 2′-fucosyllactose. Nutrients 10 (10), 1346.

Robinson, R.C., 2019. Structures and metabolic properties of bovine milk oligosaccharides and their potential in the development of novel therapeutics. Front. Nutr. 6. https://doi.org/10.3389/fnut.2019.00050.

Rogier, E.W., Frantz, A.L., Bruno, M.E., Wedlund, L., Cohen, D.A., Stromberg, A.J., Kaetzel, C.S., 2014. Secretory antibodies in breast milk promote long-term intestinal homeostasis by regulating the gut microbiota and host gene expression. Proc. Natl. Acad. Sci. U. S. A 111 (8), 3074–3079. https://doi.org/10.1073/pnas.1315792111.

Rossiter, C., Cheng, H., Appleton, J., Campbell, K.J., Denney-Wilson, E., 2021. Addressing obesity in the first 1000 days in high-risk infants: systematic review. Matern. Child Nutr. 17 (3), e13178. https://doi.org/10.1111/mcn.13178.

Saadi, S., Makhlouf, C., Nacer, N.E., Halima, B., Faiza, A., Kahina, H., et al., 2024. Whey proteins as multifunctional food materials: recent advancements in hydrolysis, separation, and peptidomimetic approaches. Compr. Rev. Food Sci. Food Saf. 23 (1). https://doi.org/10.1111/1541-4337.13288. Article e13288.

Salaria, A., Arora, S., Mehla, R., Singh, T.P., Anuj, 2023. Recent developments for formulation of infant foods. In: Malik, J.A., Goyal, M.R., Kumari, A. (Eds.), Food Process Engineering and Technology: Safety, Packaging, Nanotechnologies and Human Health. Springer Nature Singapore, pp. 363–395. https://doi.org/10.1007/978-981-99-6831-2_19.

Shoff, S., He, X., Lee, H., Zhang, Z., Mishchuk, D.O., Ozturk, G., Barile, D., Hartvigsen, M.L., Kvistgaard, A.S., Slupsky, C.M., 2025. The role of alpha-lactalbumin in modulating tryptophan metabolism and serotonin synthesis. NPJ Sci. Food 9, 120. https://doi.org/10.1038/s41538-025-00497-6.

Sienkiewicz, M., Jaśkiewicz, A., Tarasiuk, A., Fichna, J., 2022. Lactoferrin: an overview of its main functions, immunomodulatory and antimicrobial role, and clinical significance. Crit. Rev. Food Sci. Nutr. 62 (22), 6016–6033. https://doi.org/10.1080/10408398.2021.1895063.

Sun, P., Wu, X., Sun, Q., Zhao, Q., Mu, G., Kong, F., 2025. Optimizing β-Lactoglobulin antigenicity through single enzyme hydrolysis: exploring structural changes and effects on linear epitopes. Food Chem. 464. https://doi.org/10.1016/j.foodchem.2024.141770. Article 141770.

Superti, F., 2020. Lactoferrin from bovine milk: a protective companion for life. Nutrients 12 (9). https://doi.org/10.3390/nu12092562. Article 2562.

Szyller, H., Antosz, K., Batko, J., Mytych, A., Dziedziak, M., Wrześniewska, M., et al., 2024. Bioactive components of human milk and their impact on child's health and development: literature review. Nutrients 16 (10). https://doi.org/10.3390/nu16101487. Article 1487.

Tawfeek, H.I., Najim, N.H., Al-Mashikhi, S., 2003. Efficacy of an infant formula containing anti-*Escherichia coli* colostral antibodies from hyperimmunized cows in preventing diarrhea in infants and children: a field trial. Int. J. Infect. Dis. 7 (2), 120–128.

Thompson, C.A., Habanek-Kuleta, A., Vonderwell, M., Walker, M., 2017. Changing infant feeding guidelines. Am. Fam. Physician 95 (2), 79–89.

Timby, N., Hernell, O., Vaarala, O., Melin, M., Lönnerdal, B., 2017. Whey-based infant formula: current status and future directions. Nutrients 9 (7). Article 743.

Tinghäll Nilsson, U., Hernell, O., Lönnerdal, B., Hartvigsen, M.L., Jacobsen, L.N., Kvistgaard, A.S., Karlsland Åkeson, P., 2023. Low-protein formulas with alpha-lactalbumin-enriched or glycomacropeptide-reduced whey: effects on growth, nutrient intake and protein metabolism during early infancy: a randomized, double-blinded controlled trial. Nutrients 15 (4). https://doi.org/10.3390/nu15041010. Article 1010.

Tinghäll Nilsson, U., Lönnerdal, B., Hernell, O., Kvistgaard, A.S., Jacobsen, L.N., Karlsland Åkeson, P., 2024. Low-protein infant formula enriched with alpha-lactalbumin during early infancy may reduce insulin resistance at 12 months: a follow-up of a randomized controlled trial. Nutrients 16 (7). https://doi.org/10.3390/nu16071026. Article 1026.

Vandenplas, Y., Berger, B., Carnielli, V.P., Ksiazyk, J., Lagström, H., Sanchez Luna, M., et al., 2019. Human milk oligosaccharides: 2'-fucosyllactose (2'FL) and lacto-N-neotetraose (LNnT) in infant formula. Nutrients 11 (4). Article 848.

Vargas, A.J., Assar, C., Bremer, A.A., Carlson, S.J., Fasano, J., Gahche, J., et al., 2023. Science surrounding the safe use of bioactive ingredients in infant formula: federal comment. Pediatr. Res. 94 (2), 420–422. https://doi.org/10.1038/s41390-023-02512-6.

Ventura, A.L.M., Melo, T.M.S., Moraes, J., Pires, A.C.M., Silveira, N.F.A., 2018. Process optimization for powdered infant formula: a review. Food Bioprocess Technol. 11 (11), 1970–1990.

Vilcacundo, R., Méndez, P., Reyes, W., Romero, H., Pinto, A., Carrillo, W., 2018. Antibacterial activity of hen egg white lysozyme denatured by thermal and chemical treatments. Sci. Pharm. 86 (4), 48. https://doi.org/10.3390/scipharm86040048.

Wallingford, J.C., 2023. Perspective: assuring the quality of protein in infant formula. Adv. Nutr. 14 (4), 585–591. https://doi.org/10.1016/j.advnut.2023.04.008.

Wang, B., Timilsena, Y.P., Blanch, E., Adhikari, B., 2019. Lactoferrin: structure, function, denaturation, and digestion. Crit. Rev. Food Sci. Nutr. 59 (4), 580–596.

Ward, T.L., 2014. Characterizing Immune-Modulatory Components of Human Milk: The Fate and Function of Soluble CD14 and the Human Milk Metagenome. Doctoral dissertation, Université d'Ottawa/University of Ottawa.

Wernimont, S., Northington, R., Kullen, M.J., Yao, M., Bettler, J., 2014. Effect of an α-lactalbumin-enriched infant formula supplemented with oligofructose on fecal microbiota, stool characteristics, and hydration status: a randomized, double-blind, controlled trial. Clin. Pediatr. 54 (4), 359–370. https://doi.org/10.1177/0009922814553433.

Wicinski, M., Sawicka, E., Gebalski, J., et al., 2020. Human milk oligosaccharides: health benefits, potential applications in infant formulas, and pharmacology. Nutrients 12 (1), 266.

World Health Organization (WHO), 2003. Global Strategy for Infant and Young Child Feeding. WHO, Geneva, Switzerland. Available at: https://apps.who.int/iris/bitstream/handle/10665/42590/9241562218.pdf?sequence=1.

World Health Organization (WHO), 2023. WHO Guideline for Complementary Feeding of Infants and Young Children 6–23 Months of Age. Available at: https://www.who.int/publications/i/item/9789240081864. (Accessed 18 January 2024).

Wu, T., Jiang, Q., Wu, D., Hu, Y., Chen, S., Ding, T., Ye, X., Liu, D., Chen, J., 2019. What is new in lysozyme research and its application in the food industry? A review. Food Chem. 274, 698–709.

Xiong, Z., Wang, W., Ma, X., Zhang, X., Wu, Z., Yang, A., Li, X., 2023. Development of a two-step hydrolysis hypoallergenic cow's milk formula and evaluation of residue allergenicity by peptidomics and immunoreactivity analysis. J. Agric. Food Chem. 71 (32), 12237–12249.

Yakoob, M.Y., Lo, C.W., 2017. Nutrition (Micronutrients) in child growth and development. J. Dev. Behav. Pediatr. 38 (8), 665–679. https://doi.org/10.1097/dbp.0000000000000482.

Zhang, J., Zhao, A., Lai, S., Yuan, Q., Jia, X., Wang, P., Zhang, Y., 2021. Longitudinal changes in the concentration of major human milk proteins in the first six months of lactation and their effects on infant growth. Nutrients 13 (5). Article 1476.

Zhang, Y., Zhang, L., Xia, S., Hou, Y., Wu, T., Zhou, P., 2024. Site-specific glycoproteomic analysis of purified lactoferrin from human and animal milk. Int. J. Biol. Macromol. 254. https://doi.org/10.1016/j.ijbiomac.2023.127766. Article 127766.

Zhao, C., Chen, N., Ashaolu, T.J., 2022. Whey proteins and peptides in health-promoting functions – a review. Int. Dairy J. 126, 105269. https://doi.org/10.1016/j.idairyj.2021.105269.

Zhao, M., Huang, M., Li, Z., 2023. Exploring the therapeutic potential of recombinant human lysozyme: a review on wound management system with antibacterial. Front. Bioeng. Biotechnol. 11. https://doi.org/10.3389/fbioe.2023.1292149. Article 1292149.

Zhong, Q., Lu, Q., Peng, N., Liang, X.H., 2022. Amino acid-based formula vs. extensively hydrolyzed formula in the treatment of feeding intolerance in preterm infants: study protocol for a randomized controlled trial. Front. Nutr. 9, 854121.

Further reading

American Academy of Pediatrics, 2012. Breastfeeding and the use of human milk. Pediatrics 129 (3), e827–e841.

Brandelli, A., Daroit, D.J., Corrêa, A.P.F., 2015. Whey as a source of peptides with remarkable biological activities. Food Res. Int. https://doi.org/10.1016/j.foodres.2015.01.016.

de Souza Oliveira, R.P., de Morais, M.B., 2018. The evolution of infant formulas: from the origin to the present. Food Res. Int. 109, 307–314.

European Society for Paediatric Gastroenterology, Hepatology and Nutrition, European Academy of Paediatrics, European Society for Paediatric Research, et al, 2024. World Health Organization (WHO) guideline on the complementary feeding of infants and young children aged 6–23 months 2023: a multisociety response. J. Pediatr. Gastroenterol. Nutr. 79 (1), 181–188. https://doi.org/10.1002/jpn3.12248.

Fewtrell, M., Bronsky, J., Campoy, C., Domellöf, M., Embleton, N., Fidler Mis, N., et al., 2017. Complementary feeding: a position paper by the European society for paediatric gastroenterology, hepatology, and nutrition (ESPGHAN) committee on nutrition. J. Pediatr. Gastroenterol. Nutr. 64 (1), 119–132.

Peila, C., Riboldi, L., Coscia, A., 2024. Role of the biological active components of human milk on long-term growth and neurodevelopmental outcome. Ital. J. Pediatr. 50. https://doi.org/10.1186/s13052-024-01773-z. Article 201.

Rêgo, C., Pereira-Da-Silva, L., Ferreira, R., 2018. Consenso sobre fórmulas infantis: A opinião dos especialistas portugueses sobre sua composição e suas indicações. Acta Pediátrica Port. 31, 754–765. https://doi.org/10.20344/amp.10620.

Sánchez, C., Franco, L., Regal, P., Lamas, A., Cepeda, A., Fente, C., 2021. Breast milk: a source of functional compounds with potential application in nutrition and therapy. Nutrients 13 (3). Article 1026.

Victora, C.G., Bahl, R., Barros, A.J.D., França, G.V.A., Horton, S., Krasevec, J., et al., 2016. Breastfeeding in the 21st century: epidemiology, mechanisms, and lifelong effect. Lancet 387 (10017), 475–490.

Chapter 12

Beverages

Anita Rejdlová, Eva Lorencová and Richardos Nikolaos Salek
Department of Food Technology, Faculty of Technology, Tomas Bata University in Zlín, Zlín, Czech Republic

1. Introduction

As the dairy industry continues to expand production volume, the generation of whey as a valuable by-product has also increased significantly. However, improper handling and disposal of whey can pose serious environmental risks due to its composition. For instance, when whey is released into wastewater systems, it can lower dissolved oxygen levels, thereby threatening aquatic ecosystems. Similarly, its uncontrolled disposal on land may alter the physicochemical properties of the soil, potentially leading to reduced agricultural productivity. Therefore, the effective processing and management of whey are essential to mitigate these environmental impacts (Zandona et al., 2021; Yadav et al., 2015).

Bovine milk is highly valued for its versatility, as it can be consumed directly or used as a fundamental ingredient in a wide range of dairy products. Today, the dairy industry produces a variety of milk-based goods, including liquid milk, cheese, yoghurt, and powdered milk, as well as enriched products such as protein-rich foods and lactose. Furthermore, cream-based products, butter, and fermented milk are also widely available. According to Eurostat, Europe produced over 148 million tonnes of milk in 2020, with nearly half of this amount allocated to cheese production (Tsermoula et al., 2021; Eurostat, 2020). In the cheese-making process, as much as 90% of the milk volume is transformed into whey. Moreover, this by-product is categorized into two primary types: sweet whey and acid whey. Sweet whey, the most prevalent type, is generally produced during cheese manufacture, while acid whey is predominantly derived from the preparation of quark cheese and yoghurt. Sweet whey, abundant in proteins, lactose, and minerals, exhibits significant nutritional and technological attributes. Advancements in processing technologies have increased its importance, broadening its applications in the food business (Fox, 2008; Risner et al., 2019; Tsermoula et al., 2021). At the same time, the dairy industry produces salted whey, which is obtained in the production of Cheddar and Colby cheese (Kapoor and Metzger, 2004; Nishanthi et al., 2017).

Recognized as a significant source of highly nutritious and biologically active proteins, along with other chemical substances, whey protein constitutes roughly 20% of the overall protein content of cow's milk. These proteins possess remarkable functional attributes, such as foaming, emulsification, water-binding, and gelation capabilities, which have led to an increasing demand for whey protein products as food additives (Svanborg et al., 2015; Celeghin et al., 2016; Ghanimah, 2018). Recent advancements in whey processing technology, including ultrafiltration, microfiltration, and enzymatic treatments, have shown new opportunities for whey utilization. These developments facilitate the synthesis of high-purity whey protein isolates (WPI) and other specialized products, addressing an increasing market demand for whey-based ingredients (Lappa et al., 2019).

Whey is frequently considered a byproduct of production and is commonly mixed with water from cleaning procedures prior to being released into wastewater treatment facilities, potentially leading to significant environmental issues (Besediuk et al., 2024). According to available data, only slightly more than 50% of the produced sweet whey undergoes further processing for food-related applications. Of this, approximately half is utilized in its liquid form, particularly in the production of ricotta cheese and fermented whey-based beverages. Additionally, around 30% of the processed whey is transformed into powder, then utilized in the manufacturing of infant formula. Fifteen percent is processed and sold as lactose and its derivatives. The residual fraction is utilized for functions such as thickening or the extraction of whey proteins (Carvalho et al., 2013; Rama et al., 2019; Besediuk et al., 2024).

Whey. https://doi.org/10.1016/B978-0-443-27648-4.00010-4

2. History and type of beverages

The use of whey as a health-promoting beverage has deep historical roots and continues to evolve in modern food science. Its therapeutic potential was recognized as early as 460 BC by Hippocrates, the father of medicine, who recommended whey for treating a range of ailments including tuberculosis, jaundice, liver and digestive disorders, and skin diseases (Mitrovic et al., 2023). In contemporary food technology, whey-based beverages have gained prominence due to their excellent nutritional profile and functional versatility. These beverages can be broadly classified into four main categories:

- fruit or vegetable juice-based drinks,
- dairy-type beverages,
- carbonated "Rivella-type" drinks,
- and alcoholic formulations (Chavan et al., 2015).

Among the various types of whey-based beverages, those formulated with fruit or vegetable juices are the most commonly developed and widely accepted. These beverages typically consist of unprocessed or deproteinized whey, or ultrafiltration (UF) permeates, blended with fruit juices to enhance sensory appeal and mask the naturally salty or sour taste of whey. Citrus fruits, particularly orange and lemon, are preferred due to their strong flavor profiles and acidity, which effectively neutralize undesirable whey notes. Additionally, fruits such as mango, passion fruit, apple, strawberry, and raspberry are often used, either individually or in mixed formulations, to improve palatability and provide nutritional benefits (Chavan et al., 2015; Joshi et al., 2020). Vegetable-based variants, while less common, also exist; examples include formulations with carrot or tomato juice, which provide unique flavor profiles and additional vitamins (Nursiwi et al., 2017; Arsić et al., 2018; Naik et al., 2023).

Dairy-type whey beverages include both fermented and unfermented products, such as drinkable yogurts and flavored milk drinks. Unfermented versions typically have a neutral pH, while fermented types, like kefir or sour milk, are more acidic and often contain probiotic cultures. To improve texture and prevent sedimentation, stabilizers such as pectin or xanthan gum are commonly added. These beverages can be enriched with whey protein concentrates or isolates, offering a nutritious and cost-effective alternative to traditional dairy drinks (Chavan et al., 2015).

Carbonated whey-based beverages offer a refreshing alternative to traditional soft drinks and are typically made using deproteinized whey to reduce foaming and improve clarity. The most notable example is Rivella, introduced in Switzerland in 1952, produced by fermenting whey. The drink is clear, mildly acidic (pH ~3.7), and available in variants like Rivella Green with green tea extracts. Other examples include Bodrost (Russia), a beer-like whey drink, and Tai (Brazil), a soft drink fortified with whey protein. Products such as Frusighurt, Big MR, and Taksi also use fruit flavors to improve taste and consumer appeal. Additionally, hydrolyzed whey is used in drinks like Lactofruit to enhance sweetness and reduce lactose intolerance (Chavan et al., 2015).

Additionally, whey can be fermented to produce mildly alcoholic beverages, including beers and wines, through the action of specific yeast strains (Ali et al., 2015; Chavan et al., 2015; Yang et al., 2024). Study Ali et al. (2015) developed a method for producing whey beer by integrating cheese whey with malt wort, aiming to create a value-added fermented beverage that reduces whey waste while offering potential nutritional and commercial benefits. The process involved pasteurizing sweet whey, blending it with a prepared malt wort (at a 60:40 whey-to-wort ratio), and fermenting the mixture using *Saccharomyces cerevisiae* yeast. Fermentation was carried out under controlled conditions to optimize alcohol production and flavor development. The resulting whey beer exhibited acceptable sensory characteristics, including taste, appearance, and overall palatability, making it a feasible alternative to traditional beer.

3. Environmental impact of whey

Sweet whey, typically regarded as a waste by-product, has recently been transformed into a valuable raw resource with considerable market potential owing to technological advancements and research. Innovations in processing have facilitated the development of high-value components, thereby accelerating market growth for whey-derived ingredients compared to other dairy-based options (Fox, 2008). Cheese production exerts significant environmental pressure, mostly due to the substantial quantity of whey, frequently regarded as a by-product. The whey is combined with wash water from the facility and thereafter directed to the wastewater treatment plants. The composition of this wastewater may differ based on the final product, the system employed, and the operational techniques utilized. It often consists of milk or dairy residues mixed with wash water containing alkaline and acidic cleaning agents from processing equipment (Carvalho et al., 2013).

Wastewater generated from cheese and casein manufacturing is characterized by elevated concentrations of various indicators, as dairy wastewater comprises significant amounts of organic compounds, including serum proteins, lactose, mineral salts (calcium, magnesium, phosphorus, and sodium), and milk fat (Besediuk et al., 2024). Key indicators of wastewater comprise the values of Biological Oxygen Demand (BOD) and Chemical Oxygen Demand (COD) (Chatzipaschali and Stamatis, 2012). The Chemical Oxygen Demand (COD) of cheese whey generally varies from 50,000 to 80,000 mg/L, whereas its Biological Oxygen Demand (BOD) normally ranges from 40,000 to 60,000 mg/L (Kapoor and Metzger, 2004). Lactose specifically contributes to elevated BOD and COD values; therefore, prioritizing the recovery of lactose from whey is essential for mitigating environmental impact while enhancing the economic viability of whey recovery (Pires et al., 2021). Salty whey poses more significant processing difficulties than sweet whey due to its higher salinity and elevated biochemical oxygen demand and chemical oxygen demand (COD) levels (Kapoor and Metzger, 2004). Whey generates a waste load that is 100–175 times greater than the equivalent volume of household wastewater (Barukčić et al., 2019). The rapid depletion of oxygen in soil, caused by the breakdown of proteins and carbohydrates in whey, presents a considerable disposal difficulty, especially given the substantial volumes involved (Chatzipaschali and Stamatis, 2012; Domingos et al., 2018).

Mawson (1994) delineates three approaches for managing acid whey: direct utilization or disposal (e.g., as animal feed or a food ingredient), direct stabilization (physical or chemical treatments to enhance stability against microbial degradation), and conversion processes (the transformation of lactose into alternative compounds via biotransformation or chemical reactions).

4. Composition, properties and health benefits of whey: Technological advancements and nutritional insights

Whey was formerly regarded predominantly as waste and largely utilized in animal feed. Nonetheless, owing to advancements in dairy technology and contemporary research, whey is currently regarded as a value-added raw material (Fox, 2008; Fernández-Gutiérrez, 2017).

Two distinct kinds of whey may be produced depending on the coagulation process employed. Sweet whey is generated when rennet enzymes, primarily chymosin, combine with milk caseins. The enzymatic coagulation process is notably common in cheese production, with the pH of sweet whey typically varying from 5.9 to 6.6. Acid whey often occurs as a by-product of acid-coagulated cheeses, such as quark, ricotta, and Greek yoghurt. In contrast, acid whey is produced either through fermentation or by adding organic or mineral acids resulting in a lower pH of 4.3–4.6 (Venetsaneas et al., 2009; Rocha-Mendoza et al., 2021; Bandara et al., 2023). In addition to the two primary types of whey, the production of Cheddar and Colby cheeses results in salty whey. These cheeses are produced using the dry salting technique, where salt is directly applied to the curd. The moisture in the curd causes the dissolution of salt on its surface and subsequent diffusion into the curd. This process results in the shrinking of curds and the release of whey. Excess whey is released during the subsequent pressing phase. The whey produced by this procedure has a pH value of around 5.2. It is further distinguished by its elevated salt concentration of around 4%–10%.

Whey is a yellow-green liquid, tinted by riboflavin (vitamin B2), that remains after the extraction of caseins and fat from milk, comprising lactose, soluble salts, and globular proteins (Rocha-Mendoza et al., 2021). The composition and characteristics of whey differ based on the milk source, the animal's diet, the processing technology employed, and several environmental conditions (Park et al., 2007). Table 12.1 presents basic characteristics of three types of whey.

Typically, the water content of whey is 93% (w/w). The acidic coagulation of casein takes place when pH value is around 4.6, respectively around the isoelectric point of milk. resulting in a greater degree of milk protein precipitation than in sweet whey. Additionally, the reduced pH of acid whey promotes the solubilization of colloidal calcium phosphate from casein micelles, causing a calcium concentration nearly twice as high as that of other whey types. Furthermore, acid whey exhibits a higher ash concentration; however, it contains less lactose, leading to an increased lactic acid content. This occurs because lactic acid bacteria (LAB) metabolize lactose into lactic acid during curd formation. In contrast, the effectiveness of lactic acid bacteria in sweet coagulation is reduced compared to rennet-induced coagulation, resulting in a lower concentration of lactic acid in sweet whey (Panesar et al., 2007; Carvalho et al., 2013; Fernández-Gutiérrez, 2017; Rocha-Mendoza et al., 2021; Bandara et al., 2023).

Whey comprises constituents of significant nutritional value and biological efficacy. Primarily, these are whey proteins, which play a crucial role in promoting physiological development and overall well-being. Protein deficiency in nutrition presents a significant global health concern; therefore, incorporating foods high in protein into human diets is essential. Whey proteins are characterized by a globular structure, possess a significant number of α-helix structures. Hydrophilic and hydrophobic amino acids, together with acidic and basic amino acids, are uniformly distributed

TABLE 12.1 Comparative composition of sweet, acid, and salty whey ($g \cdot L^{-1}$) (Rejdlová et al., 2024).

Composition ($g \cdot L^{-1}$)	Sweet whey	Acid whey	Salty whey
Total solids	5.72–8.50	5.20–5.90	8.60–22.40
Ash	0.50–0.65	0.60–0.80	14.80–15.00
Lactose	3.06–5.02	3.00–4.60	2.60–3.00
Total protein	0.60–1.20	0.20–0.50	0.30–0.85
Fat	0.05–0.40	0.01–0.32	0.60–0.80
Lactic acid	0.14–0.28	0.50–0.60	0.28–0.72
Ca	0.03–0.06	0.09–0.14	0.13–0.21
K	0.08–0.16	0.10–0.14	0.10–0.21
Mg	0.01–0.02	0.01–0.02	0.01–0.02
Na	0.02–0.32	0.02–0.05	0.10–0.52
Citrate	0.06–0.07	0.07–0.09	0.07–0.08
Phosphate	0.06–0.07	0.06–0.10	0.08–0.09

throughout the polypeptide chain (Jeličić et al., 2008; Minj and Anand, 2020). Whey proteins are thermally unstable and degrade at temperatures of 60 °C. During pasteurization, they undergo denaturation and subsequent precipitation (Barukčić et al., 2015).

The Food and Agriculture Organization of the United Nations (FAO) introduced the Digestible Indispensable Amino Acid Score (DIAAS), a method designed to emphasize the importance of protein quality. This approach evaluates food protein quality based on three key criteria: overall protein content, the availability of essential amino acids, and their digestibility (Bailey et al., 2023). Based on DIAAS values, the FAO established a scale for classifying protein quality: <75 (no quality claim); 75–99 (high-quality protein); and ≥ 100 (good quality protein) (Herreman et al., 2020). Whey proteins are classified as high-quality proteins, emphasizing their superior nutritional value. However, the DIAAS value may be influenced by purity and the production technique. The DIAAS value of liquid whey is approximately 90, while WPC ranges between 110 and 134, and WPI has a DIAAS value of 125 (Fox, 2008; Herreman et al., 2020).

The concentration of whey protein is influenced by several factors, including the variety of whey, type of milk, stage of lactation, nutrition the processing technology during cheese manufacturing and the processes involved in the production of whey powder, WPC or WPI (Pelegrine and Gasparetto, 2005; Szafrańska et al., 2024). The primary constituents of whey proteins are β-lactoglobulin (β-LG), α-lactalbumin (α-LA), bovine serum albumin (BSA), immunoglobulins (IG), lactoferrin (LF), lactoperoxidase (LP), and minor amounts of glycomacropeptide (GMP) (Pgntado et al., 2001). Furthermore, sweet whey comprises casein macropeptide (CMP), generated after the breakdown of κ-casein by rennet (Barukčić et al., 2019).

Whey proteins exhibit an amino acid profile that closely aligns with human muscle tissue, making them one of the most nutritionally valuable protein sources. They are particularly rich in essential amino acids, with a notably high concentration of branched-chain amino acids (BCAAs), including valine, leucine, and isoleucine (Bilyi et al., 2021). Beyond their role in muscle health, BCAAs function as metabolic regulators, influencing protein synthesis, glucose homeostasis, and lipid metabolism, thereby playing a crucial role in weight management. Additionally, whey proteins contain sulfur-containing amino acids that support metabolic regulation and serve as precursors to glutathione, a potent intracellular antioxidant. When compared to protein sources such as egg, meat, and soy, whey proteins offer a more balanced and bioavailable supply of essential amino acids, further enhancing their nutritional significance (Pires et al., 2021). Additionally, approximately 14 percent of whey protein comprises hydrolysates (including single amino acids, di-, tri-, and polypeptides), which enhance digestion and are crucial for the production of several beneficial compounds (Gorissen et al., 2018).

A preliminary study suggested that salty whey from Cheddar cheese may exhibit a substantial increase in LF concentration compared to sweet whey (Rao and Wendorff, 2003). This finding was confirmed by Blaschek et al. (2007), who reported that salty whey contained higher levels of LF than sweet whey; however, the concentrations of α-LA and β-LG were lower. The higher salt concentration in whey is likely responsible for the increased solubility of LF.

Whey proteins can be modified through various methods to generate bioactive peptides that elicit a variety of physiological effects. A bioactive peptide is defined as a well-characterized fragment of a protein that exerts a beneficial influence on bodily functions and, consequently, contributes to enhanced health outcomes. The biological activity of such peptides is primarily determined by their specific amino acid sequence and molecular size, which typically ranges from 2 to 20 amino acid residues (Korhonen, 2009).

Bioactive peptides and other compounds formed during whey fermentation may exhibit anti-inflammatory, antioxidant, and antibacterial properties, thereby enhancing the overall health benefits of whey-based products (Egger and Ménard, 2017; Jakopović et al., 2019). Among these bioactive peptides, lactoferrin and β-lactoglobulin are particularly notable due to their diverse functional properties, including antibacterial and antioxidant activities. Lactoferrin and its derivatives, such as lactoferricin, demonstrate potent antimicrobial effects by disrupting bacterial membranes and supporting immune function. In contrast, β-lactoglobulin is recognized for its antibacterial and hypotensive properties, further underscoring its significance in functional foods (Darewicz et al., 2011; Bartoń et al., 2025).

The release of bioactive peptides occurs not only during fermentation but also through the enzymatic hydrolysis of whey proteins in the human digestive system or via the action of plant and microbial proteases. These peptides act as signaling molecules, influencing various physiological processes, including cardiovascular, immune, neurological and gastrointestinal functions. A growing body of scientific evidence highlights the beneficial effects of whey proteins on human health, demonstrating their antibacterial, antioxidant, antihypertensive, antidiabetic, and immunomodulatory properties. Additionally, whey proteins have been associated with weight management and the prevention or alleviation of allergic reactions, further reinforcing their nutritional and functional importance (Egger and Ménard, 2017; Jakopović et al., 2019; Bartoń et al., 2025).

Lactose accounts for approximately 75% of whey solids, making it the predominant component of whey (Pires et al., 2021). Furthermore, lactose possesses a low caloric value, low glycemic index, and minimal cariogenic potential in comparison to other carbohydrates (de Souza et al., 2010). Lactose in whey is crucial for sustaining sufficient amount of magnesium and enhancing the digestion of dairy fats and other nutrients in the human digestive system, besides not contributing to the formation of dental plaque. Moreover, the thermal processing of whey transforms a portion of lactose into lactulose, which serves as a growth enhancer for bifidobacteria (Tratnik, 2003; Jeličić et al., 2008).

Nearly all minerals present in milk are transferred into whey, along with substances added during technological processing or introduced through contact with equipment. Minerals are typically classified into macroelements (K, Ca, Mg, Na, P) and microelements (Fe, Mn, Cu, Se, Zn) (Hoac et al., 2007; Bleoussi et al., 2020; Besediuk et al., 2024). Among the most abundant macroelements in whey are potassium, calcium, phosphorus, and sodium. Their high concentrations result both from the natural composition of the milk and from the addition of calcium chloride during cheesemaking processes. Among microelements, iron and zinc are present in significant amounts, which also reflects the original composition of the milk. Selenium is found in two forms—bound to protein fractions (immunoglobulins, BSA, β-LG, and α-LA) and as amino acid derivatives (selenomethionine, selenocysteine)—with approximately 65% bound to proteins and 35% present as free amino acids (Hoac et al., 2007; Besediuk et al., 2024).

Water-soluble vitamins present in milk also migrate into whey, although their concentrations vary significantly and are greatly influenced by the storage conditions of the whey. Whey contains substantial amounts of riboflavin, folic acid, and cobalamin, with the latter two binding to whey proteins and often remaining in the whey after cheese production. Whey may have higher levels of riboflavin than milk, due to the activity of certain lactic acid bacteria employed in cheese production (Jeličić et al., 2008).

The individual vitamins present in whey contribute to its biological value through their specific physiological functions. For instance, β-carotene acts as a provitamin A, which is converted into the active form of vitamin A according to the body's needs. Additionally, it can be safely stored in subcutaneous adipose tissue without adverse effects (Tang, 2010). Vitamins B1 and B2 support energy metabolism by participating in the conversion of nutrients into usable energy. Whey is naturally rich in B-complex vitamins, which are commonly found in the daily diet (Hanna et al., 2022). Although deficiencies of these vitamins are rare in developed countries, certain physiological or metabolic conditions may increase the body's demand for them. Moreover, research indicates that a higher intake of B vitamins may yield beneficial effects even in healthy individuals. Similarly, the consumption of whey in its unprocessed form may also offer health advantages (Besediuk et al., 2024).

5. Microbiology of whey-based beverages

Fermentation represents a fundamental process in the production of fermented whey-based beverages. This process involves not only LAB but also cultures derived from milk and water kefir, and even the symbiotic culture of bacteria and yeast (SCOBY) (Tu et al., 2019; Bandara et al., 2023; Rejdlová et al., 2023, 2024). The enzymes produced by these microorganisms during fermentation catalyze the breakdown of lactose, proteins, and lipids into various metabolites, including lactic acid, alcohols, peptides, and volatile fatty acids. These compounds contribute significantly to the development of characteristic flavors and aromas, enhance the nutritional profile, and extend the shelf life of whey. Nevertheless, the specific biochemical transformations that occur during fermentation vary considerably depending on the microbial cultures employed (Pescuma et al., 2015; Bandara et al., 2023).

5.1 Lactic acid bacteria

Lactic acid bacteria are Gram-positive, non-spore-forming microorganisms exhibiting either rod-shaped or coccoid morphology. They thrive under anaerobic or microaerophilic conditions and are naturally present in milk, on plant surfaces, and within the gastrointestinal tracts of humans and animals. A unifying characteristic of these bacteria is their ability to ferment lactose. They generally grow optimally at temperatures ranging from 25 to 40°C and at a pH between 4.0 and 6.0. Notably, certain strains are recognized as probiotics due to their beneficial health effects (Bishop et al., 2022).

5.1.1 Lactose fermentation

Among the microbial groups utilized in fermentation, LAB are the most prevalent. During the conversion of lactose into lactic acid, the acidity of the substrate increases, creating increasingly inhospitable conditions for non-LAB microorganisms (Shiby and Mishra, 2013). Based on their primary metabolic pathways, LAB are classified into two main groups: homofermentative and heterofermentative. Homofermentative LAB predominantly produce lactic acid as the main metabolic product, whereas heterofermentative strains generate additional by-products such as carbon dioxide, ethanol, and/or acetic acid alongside lactic acid (Bandara et al., 2023).

The pathway of lactose fermentation is provided in Fig. 12.1. Lactose is transported across the cell membrane via two distinct systems: the lactose permease system or the phosphoenolpyruvate-dependent phosphotransferase system. Depending on the transport mechanism, lactose is either hydrolyzed in its non-phosphorylated form into glucose and galactose by β-galactosidase, or in its phosphorylated form into glucose and galactose-6-phosphate by phosphorylated β-galactosidase. These monosaccharides subsequently enter various metabolic routes, including the glycolytic Embden–Meyerhof–Parnas (EMP) pathway, the phosphoketolase pathway (for glucose), the tagatose-6-phosphate pathway (for galactose-6-phosphate), and the Leloir pathway (for galactose). LAB species that possess aldolase—such as *Streptococcus*, *Pediococcus*, and both obligately homofermentative and facultatively heterofermentative *Lactobacillus*—undergo homolactic fermentation, resulting in lactate as the exclusive end product. Conversely, LAB with phosphoketolase activity can be divided into two categories. The first group, which includes *Leuconostoc* spp. and obligately heterofermentative lactobacilli, utilizes the "6P-gluconate pathway," producing equimolar amounts of CO_2, lactate, and acetate. The second group follows the "bifidus pathway," which yields acetate and lactate in a molar ratio of 3:2 (Shiby and Mishra, 2013).

5.2 Kefir starter cultures

Kefir starter cultures can also be used in the production of fermented whey-based beverages (Athanasiadis et al., 2004; Mazaheri Assadi et al., 2008; Londero et al., 2012; M'hir et al., 2019). There are two types of kefir grains—milk kefir grains and water kefir grains—which differ primarily in their microbial composition. Both types generally contain LAB and yeasts, and in some cases, also a small proportion of acetic acid bacteria (Plessas et al., 2017).

The microorganisms in milk kefir are embedded in a complex network of proteins and carbohydrates, commonly referred to as kefiran. Kefiran is a polysaccharide composed of glucose and galactose, and its main producer is *Lactobacillus kefiranofaciens*. The resulting fermented product, kefir, exhibits numerous health-promoting properties, including anticancer, anti-inflammatory, and antibacterial effects, along with notable antioxidant activity. Typical LAB species associated with milk kefir include *Lentilactobacillus kefiri*, *Lactobacillus kefiranofaciens*, *Lactobacillus kefiranofaciens* subsp. *kefirgranum*, and *Lentilactobacillus parakefiri* (Guzel-Seydim et al., 2021).

In contrast, the polysaccharide matrix that forms the structural basis of water kefir grains consists mainly of α-1,6-glucan, with a smaller proportion of levan present in the network. This glucan is completely insoluble in aqueous environments, and its primary producer is *Lentilactobacillus hilgardii*. Notably, water kefir grains do not contain proteins or lipids (Gulitz et al., 2011; Coma et al., 2019; Moretti et al., 2022). LAB species found in water kefir grains include

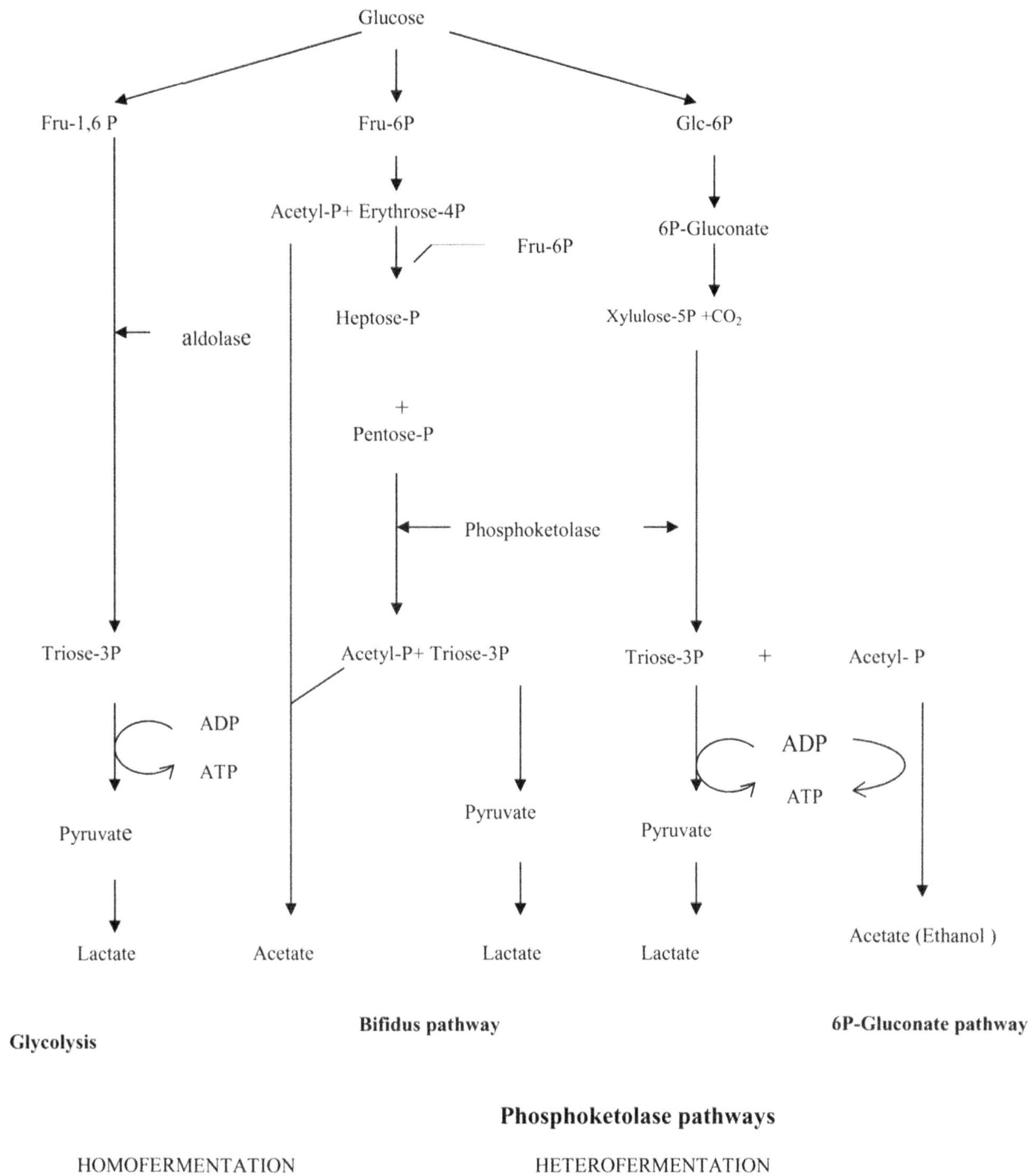

FIGURE 12.1 Fermentation of lactose by lactic acid bacteria. *Adapted from Kandler (1983).*

Lacticaseibacillus paracasei, *Lentilactobacillus hilgardii* and *Liquorilactobacillus nagelii*. Additionally, the presence of yeast, particularly *Saccharomyces cerevisiae*, is essential for the fermentation process (Guzel-Seydim et al., 2021). Bacteria of the genera *Acetobacter* and *Gluconobacter* are also commonly encountered in the microbial community of water kefir (Lynch et al., 2021).

5.3 Symbiotic culture of bacteria and yeasts

Symbiotic Culture of Bacteria and Yeasts (SCOBY) is commonly used in the production of kombucha, a sweetened fermented beverage typically made from black tea, although green tea, oolong, or other varieties may also be used. Originally from China, kombucha has gained global popularity in recent years. During fermentation, a cellulose-based pellicle forms on the surface of the medium, produced by bacteria present in the SCOBY culture. The resulting beverage has a pleasant, fruity, acidic, and effervescent flavor due to the presence of various organic acids and carbon dioxide released during the fermentation process (Coelho et al., 2020; Laavanya et al., 2021).

Kombucha contains several organic acids, such as acetic, gluconic, and glucuronic acids. Additionally, it includes various sugars, water-soluble vitamins, amino acids, biogenic amines, purines, pigments, lipids, proteins, hydrolytic enzymes, ethanol, carbon dioxide, polyphenols, and minerals (e.g., manganese, iron, nickel, copper, zinc, lead, cobalt, chromium, cadmium), along with anions (fluoride, chloride, bromide, iodide, nitrate, phosphate, sulfate), and numerous metabolic by-products of yeasts and bacteria (Laavanya et al., 2021).

The SCOBY microbiome consists primarily of acetic acid bacteria (AAB), yeasts, and LAB. LAB may be intentionally added to kombucha to enhance its antimicrobial and antioxidant capacity. The microbial composition is influenced by fermentation conditions such as temperature, duration, type of tea used, and source of sugars. In the early stages of fermentation, yeasts dominate, being better adapted to high sugar concentrations. In later phases, acid-tolerant microorganisms become more prevalent (Morales, 2021).

The fermentation process involves several metabolic pathways. Yeasts hydrolyze sucrose into glucose and fructose via invertase, followed by ethanol production through glycolysis. AAB subsequently oxidize glucose to gluconic and glucuronic acids, and ethanol is converted into acetic acid via the sequential action of alcohol dehydrogenase and aldehyde dehydrogenase. Furthermore, AAB synthesize and polymerize cellulose chains, forming a dense fibrous matrix on the surface (Morales, 2021; Liu et al., 2022).

The dominant bacterial genera in kombucha include Komagataeibacter and Acetobacter, as well as *Lactobacillus* and *Lactococcus* (LAB). Common yeast genera found in kombucha include *Saccharomyces, Zygosaccharomyces, Brettanomyces, Candida, Torulospora, Kloeckera, Pichia*, and *Schizosaccharomyces* (Harrison and Curtin, 2021).

Acetic acid bacteria (AAB) are Gram-negative, non-spore-forming, motile microorganisms belonging to the family Acetobacteraceae. They are typically ellipsoidal to rod-shaped, measuring approximately 1–4 μm in length. Optimal growth occurs at temperatures between 25 and 30 °C and a pH range of 5.0–6.5, although they can tolerate pH values as low as 3.0–4.0 (Laureys et al., 2020; Bishop et al., 2022). Notably, *Komagataeibacter* species can accumulate up to 10%–20% acetic acid in the medium, whereas *Acetobacter* species typically accumulate no more than 8% (Gomes et al., 2018).

6. Fermented and non-fermented whey-based beverages, carbonated whey-based beverages

Whey, a by-product of cheese production, has gained increasing application in the development of both non-fermented and fermented whey-based beverages over the past decade. These products offer a novel approach to utilizing whey for human consumption and can be enriched with nutraceuticals, probiotics, and prebiotics. Additionally, they may be blended with milk or fruit juices to enhance their nutritional and sensory properties. A key limitation of conventional processing, particularly heat treatment, is the occurrence of heat-induced sedimentation. Therefore, the integration of non-thermal processing techniques can help preserve the nutritional quality of whey-based beverages (Özer and Evrendilek, 2022).

Fermented whey beverages appeal to a broader consumer base, primarily due to their reduced lactose content and lower allergenic properties. In addition, compared to non-fermented whey drinks, fermented whey beverages offer an extended shelf life, higher concentrations of phenolic compounds, and enhanced antioxidant activity (Szafrańska et al., 2024). Table 12.2 depicts the production process for fermented whey beverages. Nevertheless, whey can exhibit undesirable organoleptic properties, particularly the flavor of acid whey, and may also present an unpleasant astringency, defined as "complex sensations resulting from the contraction, drawing, or puckering of the epithelium due to exposure to substances such as alums or tannins" (ASTM, 1989; Djurić et al., 2004; Rocha-Mendoza et al., 2021).

Different combinations of flavoring ingredients improve or obscure the organoleptic properties of whey-based beverages. The most commonly used components are fruit or, less frequently, vegetable derivatives, including concentrates, juices, syrups, pulps, or nectars (Djurić, 2004; Ahmed, 2023). Additionally, chocolate, cocoa, vanilla, and several cereals (including rice, oats, or barley) were employed to augment the flavor of whey beverages (Jeličić, 2008). Various types of whey, including liquid whey, whey powder, WPC and WPI can be utilized to produce whey beverages (Chavan et al., 2015).

TABLE 12.2 Review of studies on whey-based beverage production and development.

Type of whey-based beverage	References
Nonfermented whey-based beverages	Shekilango et al. (1997)
	Djurić et al. (2004)
	Cruz et al. (2009)
	Yadav et al. (2010)
	Ismail et al. (2011)
	Sady et al. (2013)
	Chatterjee et al. (2015)
	Goudarzi et al. (2015)
	Chavda et al. (2016)
	Gimhani and Liyanage (2018)
	Singh et al. (2021)
	Nedanovska et al. (2022)
	Nkurunziza et al. (2022)
	Zhakupova et al. (2022)
	Ahmed et al. (2023)
Fermented whey-based beverages	Athanasiadis et al. (2004)
	Mazaheri Assadi et al. (2008)
	Magalhães et al. (2010)
	Londero et al. (2012)
	Castro et al. (2013)
	Akpinar et al. (2015)
	da Silveira et al. (2015)
	Pogoń et al. (2015)
	Sabokbar et al. (2015)
	Nursiwi et al. (2017)
	Abdul-Alim et al. (2018)
	Aly et al. (2019)
	Kaur et al. (2019)
	M'hir et al. (2019)
	Skryplonek et al. (2019)
	Legarová and Kouřimská (2010)
	León-López et al. (2020)
	Priyanka et al. (2022)
	Rodríguez-Hernández et al. (2020)
	Dinkçi et al. (2023)
	Rejdlová et al. (2023)

Continued

TABLE 12.2 Review of studies on whey-based beverage production and development.—cont'd

Type of whey-based beverage	References
	Elkot et al. (2024)
	Rejdlová et al. (2024)
	Szafrańska et al. (2024)
	Bartoń et al. (2025)
Carbonated whey-based beverages	Sameen et al. (2013)
	Pareek et al. (2014)
	Silva e Alves et al. (2018)

6.1 Nonfermented whey-based beverages

A study conducted by Djurić et al. (2004) investigated the formulation of whey beverages with diverse fruit components. This research focused on the formulation of whey beverages utilizing orange, pear, peach, and apple juices at concentrations of 2%, 4%, and 6%. The research investigated the impact of varying amounts of fruit juice solids on the organoleptic properties of the beverages. A higher concentration of fruit juice improved the overall acceptance of the whey beverages. The drink containing 6% orange juice garnered positive feedback for its refreshing nature; however, lower juice concentrations resulted in undesirable flavor profiles. The amalgamation of whey and peach juice earned the highest overall evaluation. The incorporation of orange juice into the whey beverages failed to mask the whey flavor, and the whey drinks had an increased sediment level. Whey drinks flavored with pear juice exhibited increased sediment level, adversely impacting the visual appeal of the whey beverages. The flavor enhanced with the augmentation of pear juice concentration. The whey beverages infused with apple juice exhibited undesirable taste, odor, and appearance. Peach-whey beverages have superior overall quality, characterized by favorable color, appearance, stability, and flavor, especially in formulations containing 6% dry matter. The peach nectar successfully conceals the undesirable flavor of whey, enhancing the beverage's sensory attractiveness.

Cruz et al. (2009) explored an unconventional combination of whey and acerola juice, creating three distinct formulations with varying ratios: 50% v/v whey and 50% v/v acerola juice, 70% v/v whey and 30% v/v acerola juice, and 30% v/v whey and 70% v/v acerola juice. The resulting whey-based beverages exhibited favorable sensory characteristics and, due to the addition of acerola, served as a significant source of vitamin C. Among the formulations, the beverage containing 70% acerola juice had the highest concentration of vitamin C, though it also exhibited the most sediment. Despite this, the increased sedimentation did not impact consumer acceptability, making the beverage with 70% v/v acerola juice the most preferred.

The combination of whey-based beverages with orange juice is prevalent in production. Sady et al. (2013) conducted a comparison between a regular orange beverage and an orange beverage fortified with whey. The findings gathered indicated that orange beverages containing whey displayed elevated amounts of protein, glucose, ash content, and vitamin B12. Nevertheless, they exhibited diminished amounts of sugar, fructose, and vitamin C, along with decreased antioxidant activity. Conversely, increased whey incorporation adversely affected the organoleptic characteristics of the beverage.

Chatterjee et al. (2015) explored beverages with varying concentrations of concentrated whey, orange juice (40%, 50%, and 60%), and sugar (7% w/v, 8% w/v, and 9% w/v). The study also assessed the influence of sodium benzoate as a preservative on the shelf life of whey-based drinks stored at ambient temperature ($30 \pm 2°C$) and refrigerated conditions ($7 \pm 1°C$). Among the tested formulations, the beverage containing 40% orange juice and 8% sugar exhibited the most favorable sensory qualities. The incorporation of preservatives was shown to extend the shelf life of the whey-based beverages. At room temperature, the preservative-free drink lasted for 5 days, while the version with preservative remained stable for 11 days. Under refrigeration, the shelf life increased to 49 days for the preservative-free variant and 91 days for the preserved beverage.

Goudarzi et al. (2015) investigated the development of beverages incorporating WPI and whey protein hydrolyzate (**WPH**) in combination with apple juice. The WPI concentration in the formulations ranged from 1% to 5% w/v, while WPH was standardized at 6% w/v. Apple juice was added at levels of 4%–8% w/v. A higher WPI content adversely influenced the sensory characteristics of the samples. Conversely, increasing the proportion of apple juice enhanced the sensory attributes of WPI-containing beverages but had no notable impact on WPH-based drinks. Additionally, the findings indicated that samples with pH values of 3.15 and 3.75 exhibited the lowest sedimentation levels.

The study by Gimhani and Liyanage (2018) focused on the development of ready-to-drink whey beverages incorporating passion fruit, watermelon, and nelli juices to enhance both the sensory and nutritional properties of whey. Notably, significant differences were observed in the organoleptic properties of the three formulations. Among the tested options, the passion fruit-flavored beverage was the most preferred, as it effectively masked whey's undesirable flavor and odor. Furthermore, the beverage demonstrated stability during storage, with no significant sedimentation observed.

In a study by Nedanovska et al. (2022), pasteurized sheep whey was used to create whey-based beverages flavored with mango, pineapple, and tropical fruits. The physicochemical, microbiological, and sensory properties of the model samples were evaluated over a 15-day storage period. During this time, the beverages displayed a notable amount of sediment, which decreased as storage progressed. The aggregation of whey proteins was primarily influenced by heat treatment and the sample's pH. The reduced pH during storage likely enhanced the surface charge of the denatured whey proteins, facilitating the dissolution of the sediment. Among the different combinations, the blend of whey and tropical fruit received the highest sensory rating. Microbiological analysis indicated a significant presence of bacteria, potentially due to contamination occurring during the storage period.

A variety of fruits is employed in the production of whey-based beverages. These primarily include citrus fruits such as oranges, lemons, and grapefruits. Traditional fruits like apples and pears are often employed, among additional varieties such as mangoes, bananas, and papayas. An assortment of berries, including raspberries, strawberries, blueberries, mulberries, and blackberries, is commonly included in various beverages (Shekilango et al., 1997; Yadav et al., 2010; Ismail et al., 2011; Singh et al., 2021; Nkurunziza et al., 2022; Zhakupova et al., 2022). In addition, some beverages also feature vegetable juices, incorporating ingredients such as carrot, pumpkin, or parsley juice (Purkiewicz and Pietrzak-Fiećko, 2021).

6.2 Fermented whey-based beverages

The manufacture of fermented whey beverages represents a notably appealing approach to using all whey. Whey, owing to its elevated lactose concentration, provides an appropriate environment for the propagation of lactic acid bacteria, yeasts, or kefir cultures. During fermentation, microbial enzymes catalyze the degradation of lactose, peptides, and lipids, resulting in the production of lactic acid, alcohols, and other sensory-active chemicals. These chemicals can enhance the distinct organoleptic characteristics of the beverages while prolonging their shelf life. Examples encompass acetaldehyde, ethyl acetate, isoamyl alcohol, isobutanol, 1-propanol, isopentyl alcohol, and 1-hexanol, which have been found by high-performance liquid chromatography (Pescuma et al., 2008; Magalhães et al., 2011; Macwan et al., 2016; Bandara, 2023). The manufacture of fermented whey-based beverages utilizes lactic acid bacteria (*Pediococcus pentosaceus, Lactiplantibacillus plantarum*), which are essential for suppressing pathogenic germs. Consequently, these beverages improve antioxidant activity and concurrently assist in lowering blood cholesterol levels (Jitpakdee et al., 2022).

The sensory and rheological attributes of probiotic whey beverages, emphasizing the influence of additives, fermentation duration, and whey concentration on these aspects, were examined in the study by Castro et al. (2013). The research entailed the creation of beverages with differing whey concentrations (0%, 20%, 35%, 50%, 65%, and 80% v/v), infused with 2% v/v strawberry juice, and fermented using *Lactobacillus acidophilus*. Of the evaluated formulations, the sample with 65% v/v whey attained the highest sensory rating score. The findings further demonstrated that substituting casein with whey protein enhanced gel brittleness.

The research conducted by da Silveira et al. (2015) sought to produce a probiotic chocolate goat milk beverage that includes *Bifidobacterium lactis*, goat cheese whey, and prebiotics (inulin and oligofructose). It assessed the impact of these substances on the physicochemical characteristics and sensory acceptance of the beverage, so responding to the growing demand for functional dairy products with probiotic advantages. Seven formulations were created utilizing different quantities of whey (15–45 mL/100 mL) and prebiotics (0–6 g/100 mL). The beverage underwent fermentation with *B. lactis*, and the physicochemical properties, including pH, acidity, total solids, viscosity, and syneresis were assessed over a 28-day storage duration. The results indicate that incorporating goat cheese whey and prebiotics into probiotic dairy beverages enhances their physicochemical stability and sensory appeal, while maintaining probiotic viability. The formulation comprising 45 mL/100 mL of whey and 6 g/100 mL of prebiotics demonstrated outstanding efficacy, especially regarding texture, viscosity, and consumer preference.

The research by Nursiwi et al. (2017) examined a distinctive amalgamation of probiotic whey beverage and tomato juice. Fermentation was performed using the probiotic bacteria *Lactobacillus acidophilus* and *Lactiplantibacillus plantarum*. Tomato juice was used in proportions of 5%, 10%, and 15%. The physicochemical and organoleptic characteristics of fermented whey-based beverages were assessed over an 18-hour fermentation period. The data revealed that the beverage containing 18% tomato juice had the maximum antioxidant activity, whereas the whey beverage with 5% tomato juice was deemed the most acceptable in sensory evaluation.

A probiotic beverage based on whey and flavored with black mulberry juice was created in a study conducted by Abdul-Alim et al. (2018). The concentrations of the fruit juice were 25%, 50%, and 75%. Furthermore, beverages composed purely of 100% whey and 100% black mulberry juice were also produced. A bacterial starter culture comprising *Lactobacillus delbrueckii* subsp. *bulgaricus*, *Streptococcus thermophilus*, and *Lacticaseibacillus rhamnosus* was utilized for fermentation, with the addition of *Bifidobacterium animalis* subsp. *lactis*. The physicochemical, microbiological, and sensory aspects of the generated beverages were assessed over a period of 21 days. The vitality of the probiotic bacteria *Lacticaseibacillus rhamnosus* and *Bifidobacterium animalis* ssp. *lactis* was sustained for 14 days, but diminished after 21 days. Moreover, a greater concentration of components in the beverage resulted in enhanced antioxidant activity. Nevertheless, throughout the experiment, both antioxidant activity and phenolic content diminished. The sensory analysis results indicated that all beverages earned favorable ratings, with the drink comprising 75% black mulberry juice achieving the highest score.

Aly et al. (2019) conducted a study on the formulation of whey beverages that included fruit components, namely cactus pear juice at percentages of 10%, 20%, and 30%, and kiwi juice at concentrations of 10% and 20%. The fermented beverages were integrated with *Lactiplantibacillus plantarum*. During a 20-day interval, specific qualities of the model beverages were assessed, encompassing drink quality, microbe viability, and organoleptic characteristics. The findings indicated that samples with elevated fruit juice content demonstrated enhanced antioxidant activity and greater concentrations of phenolic compounds. Moreover, the addition of fruit components augmented the fiber content and increased the levels of solids and soluble solids in the beverages. The type and concentration of the fruit juices did not substantially influence the viability of the bacteria. The sensory analysis indicated that the beverage with a 30% fruit component was evaluated as possessing the most favorable attributes. The study conclusively showed including fruit juices enhanced the quality of whey-based beverages.

Legarová and Kouřimská (2010) investigated the commercial viability of whey beverages supplemented with 25% and 50% semi-skimmed milk and fermented with yogurt cultures (*Lactobacillus delbrueckii* subsp. *bulgaricus* and *Streptococcus thermophilus*). The aim of the research was to evaluate the commercial potential of these whey-based dairy beverages and to improve the whey flavor through fermentation using commercial yogurt starter cultures. The course of fermentation was monitored by changes in pH and titratable acidity. The sensory profile of non-fermented and fermented drinks was assessed using unstructured graphical scales. No significant differences in acidity were found between samples fermented for 3 or 4 h; however, a substantial disparity was observed between whey drink samples with and without milk supplementation. The results indicated that fermentation using yogurt culture did not produce statistically significant improvements in the organoleptic properties of the whey drink. In contrast, the addition of milk emerged as the principal factor influencing not only the total sensory quality of the whey drinks but also their flavor, appearance, color, viscosity, and homogeneity.

León-López et al. (2020) examined the addition of hydrolyzed collagen at concentrations of 0.30%, 0.50%, 0.75%, and 1% into fermented whey beverages, utilizing lactic acid bacteria (*Lacticaseibacillus rhamnosus*, *Lactobacillus delbrueckii* subsp. *bulgaricus*, *Lactobacillus delbrueckii*, and *Streptococcus thermophilus*) for the fermentation process. The 1% collagen treatment demonstrated significant radical inhibition, with values of 48.30% for ATBS and 30.06% for DPPH, along with the highest protein content (9.75 ± 0.20 g/L). The viscosity of the beverages augmented with elevated collagen concentrations. The use of hydrolyzed collagen improved the bioavailability, nutritional value, and antioxidant activity of the beverage. It also exhibited antibacterial characteristics, as no pathogenic microbes were identified in the treated beverages. No significant variations in fat and lactose levels were observed across the treatments; however, the pH of the control was lower than that of the beverages treated with hydrolyzed collagen. The Fourier transform-infrared spectroscopy exhibited spectra indicative of lactose and collagen amides. The study demonstrated that hydrolyzed collagen enhanced the nutritional quality and functional characteristics of whey-based beverages, resulting in greater absorption rates attributed to its low molecular weight.

Dinkçi et al. (2023) concentrated on the production of probiotic beverages employing whey protein concentrate sourced from sheep, cow and goat milk. Moreover, the beverages were enhanced with kiwi powder. The beverages were fermented utilizing *Streptococcus thermophilus*, alongside the probiotic bacteria *Bifidobacterium animalis* subsp. *lactis* and *Lactobacillus acidophilus*. The incorporation of WPC increased the protein concentration of the beverages. The beverages contained a substantial number of phenolic compounds and exhibited heightened antioxidant activity. The kind and source of WPC significantly influenced the quality of the probiotic beverages, as they contained high concentrations of free amino acids, including essential and branched-chain amino acids.

A study by Elkot et al. (2024) focused on manufacturing probiotic whey beverage fortified with spirulina at concentrations of 0.25%, 0.50%, and 0.75% (w/w). Furthermore, the beverages were infused with lemon and mint juice. Spirulina (*Arthrospira platensis*), a microalga, is recognized for its high fiber and protein content. It serves as a preservative that maintains the textural quality of food by enhancing sensory attributes and improving rheological properties. Additionally, Spirulina offers several beneficial effects, including its potential application in the prevention and management of hypercholesterolemia, cardiovascular diseases, and diabetes. The probiotic culture used for beverage production consisted of *Streptococcus thermophilus*, *Lactobacillus acidophilus*, and *Bifidobacterium* spp. Subsequently, the probiotic beverages were stored at a refrigerator temperature of 4 ± 1°C for a duration of 21 days. Throughout storage, the concentrations of vitamins, minerals, antioxidants, and phenolic components in the beverages rose, while their probiotic qualities were effectively preserved. The addition of 0.50% (w/w) spirulina significantly improved the structural characteristics and sensory attributes of the final product.

A different approach to producing fermented whey beverages entails the use of kefir starting culture (Athanasiadis et al., 2004; Mazaheri Assadi et al., 2008; Londero, 2012; M'hir et al., 2019). The bacteria in kefir starting culture comprise *Lactobacillus kefiranofaciens*, *Lentilactobacillus kefiri*, *Lentilactobacillus parakefiri*, *Lactococcus lactis*, *Kluyveromyces marxianus*, *Saccharomyces unisporus*, and *Saccharomyces cerevisiae*.

The study by Magalhães et al. (2010) investigated the structural integrity and microbial composition of Brazilian kefir grains and beverages produced using milk and whole or deproteinized whey as fermentation substrates. The primary objective was to assess the stability of kefir grains and determine whether probiotic bacteria transferred to the resulting beverages. Confocal Laser Scanning Microscopy combined with fluorescence staining revealed that yeasts formed macro-clusters within the grain's polysaccharide-rich matrix, which was primarily composed of kefiran and bacteria. Furthermore, molecular profiling through Denaturing Gradient Gel Electrophoresis (PCR-DGGE) identified key microbial species, including yeasts such as *Kluyveromyces marxianus*, *Saccharomyces cerevisiae*, and *Kazachstania unispora*, alongside bacterial strains affiliated with *Lactobacillus kefiranofaciens* subsp. *kefirgranum*, *Lactobacillus kefiranofaciens* subsp. *kefiranofaciens*, and an uncultured *Lactobacillus* species. The findings demonstrated that kefir grains maintained a stable microbial structure when whey was used as a fermentation medium instead of milk. Notably, the core bacterial community present in the kefir grains was also detected in the fermented whey-based beverages, confirming their probiotic potential.

The study by Londero et al. (2012) compared the growth, acidification capacity, and microbiological composition of kefir grains fermented in whey at 20°C, 30°C, and 37°C with those derived from milk. The results showed that higher temperatures not only altered grain structure but also reduced yeast diversity and increased the protein-to-polysaccharide ratio. In contrast, at 20°C, kefir grains remained stable with minimal changes. Furthermore, fermented whey at 20°C contained fewer lactic acid bacteria but a higher yeast count compared to fermented milk. Nevertheless, both exhibited similar microbiological profiles. Therefore, these findings suggest that 20°C is the optimal temperature for preserving kefir grains as whey fermentation starters while maintaining their probiotic potential.

Sabokbar et al. (2015) explored the effects of fermentation temperature (20–30°C) and kefir grain concentration (2%–8% w/v) on the antioxidant properties and phenolic content of whey-based beverages flavored with apple juice. Through response surface methodology, they identified the optimal conditions for maximizing these attributes: a fermentation temperature of 24.82°C and kefir grain concentration of 7.56% (w/v).

The study by Rejdlová et al. (2023) focused on the production of fermented whey-based beverages using starter cultures of milk and water kefir, which were additionally flavored with carrot juice at varying concentrations (65%, 75%, 85%, and 95% w/w). The impact of different types of kefir grains and carrot juice concentration on the properties of the beverages was assessed using physicochemical, rheological, and sensory analyses. The results from the physicochemical and rheological analyses indicated that the type of kefir grains did not significantly affect the properties of the fermented beverages. However, rheological analysis demonstrated a potential effect of whey concentration on the flow curves of the Power-Law model, with all beverages exhibiting non-Newtonian behavior. In the sensory analysis, beverages containing 75% (w/w) carrot juice were rated the most favorable, regardless of the starter culture used.

Rejdlová et al. (2024) conducted a following study examining the utilization of water kefir grains in the manufacturing of fermented whey-based beverages, which were flavored with different concentrations of black currant juice (10% and 20% w/w). The impact of the 0.25% (w/w) and 0.50% (w/w) hydrocolloid (furcellaran) incorporation on the beverage's qualities was assessed by rheological and mechanical vibration damping methods. Samples containing 0.50% (w/w) furcellaran demonstrated a greater quantity of sediment than those with 0.25% (w/w). Following 48 h of fermentation, the samples exhibited ethanol concentrations ranging from 0.92% to 4.86% (v/v). The investigation established that all model samples demonstrated non-Newtonian behavior, with the sample containing 20% (w/w) black currant juice receiving the highest evaluation.

The study by Bartoń et al. (2025) investigated the production of fermented whey-based beverages using goat or cow whey, a by-product of curd and aged cheese production, as the main raw material. To enhance flavor, the beverages were infused with organic sea buckthorn and rosehip juice in a 1:1 ratio (whey to juice). Fermentation was carried out using a lyophilized LAB culture containing *Streptococcus salivarius* subsp. *thermophilus*, *Bifidobacterium bifidum*, *Lactobacillus acidophilus*, and *Lactobacillus delbrueckii* subsp. *bulgaricus*. A comprehensive physico-chemical, rheological, and microbiological analysis was conducted on the beverages. The results indicated that the addition of fruit juices not only improved their physicochemical and rheological properties but also ensured microbiological safety. Notably, the beverages exhibited lower viscosity, a desirable characteristic for this type of product.

Recently, SCOBY culture has also been used in the production of fermented whey beverages. A study by Belloso-Morales and Hernández-Sánchez (2003) investigated the use of SCOBY culture for the fermentation of two types of sweet whey (fresh and reconstituted) and fresh acid whey. Based on the results of the study, it was possible to say that fermentation proceeded well in both types of sweet whey. However, based on sensory analysis, the drinks were too sour with salty taste.

In a study by Sfoglia et al. (2024), they produced two types of fermented whey drinks, respectively without and with the addition of sucrose (70 g/L). Fermentation with SCOBY culture was carried out for 7 days at 20°C. Both model samples had an alcohol content of more than 0.5% (v/v) at the end of fermentation, which led to the classification of the beverages as alcoholic. After 7 days of whey fermentation, the population proportions of *Komagataeibacter hansenii*, *Komagataeibacter rhaeticus* and *Gluconobacter oxydans* decreased, while *Klebsiella oxytoca* and *Chromohalobacter beijerinckii* increased. The addition of sugar at the beginning of fermentation also led to a decrease in the relative abundance of *K. hansenii*, *K. rhaeticus* and *G. oxydans*, although the reduction was lower for *K. hansenii*. When whey was treated with sugar, the ratio of *K. oxytoca* and *C. beijerinckii* increased compared to the original consortia.

Fermentation of acid and sweet whey by SCOBY culture was investigated by Thibodeau et al. (2025). The microbial community successfully adapted to whey, with different types of whey (Gouda, queso fresco, Greek yoghurt) undergoing kombucha-like fermentation. During fermentation, the pH was significantly reduced below 4, ensuring the microbial safety of the drink. However, the use of lactose during fermentation remained mostly incomplete. The main products of fermentation were acetic acid and ethanol, the concentration of which varied according to the type of whey. The highest concentration of acetic acid was observed in Gouda whey, while ethanol was predominant in queso fresco. Smaller amounts of butyric acid and lactic acid were also detected, with their occurrence varying between whey types. During fermentation, there were changes in the relative abundance of bacterial and yeast species between fermentation cycles. Significant differences in bacterial composition were observed in whey fermentations, with *Acetobacter* and *Komagataeibacter* species occurring in different proportions depending on the whey type. The study shows that it is possible to produce a kombucha-like beverage from whey using an adapted SCOBY. This process has the potential to utilize whey as a by-product of the dairy industry and thus contribute to more sustainable processing, although further improvements in lactose utilization are needed.

6.3 Carbonated whey-based beverages

The production of carbonated whey beverages has become the focus of scientific research, driven by both the functional potential of whey proteins and the growing consumer demand for healthier beverage alternatives. These studies have explored various aspects of formulation, including the incorporation of probiotics, the use of different stabilizers, flavoring agents, and carbonation techniques, all aimed at improving sensory acceptability, nutritional quality, and shelf stability. Research findings have demonstrated that whey serves as an effective base for the development of refreshing, low-fat, and protein-rich drinks that can be enhanced with bioactive components to provide additional health benefits (Sameen et al., 2013; Pareek et al., 2014; Silva e Alves et al., 2018).

A study by Sameen et al. (2013) investigated the development of a carbonated flavored whey drink (CFWD) using mozzarella whey and evaluated the effect of different stabilizers—carrageenan, carboxymethylcellulose (CMC), and their combination—on its quality during 30 days of refrigerated storage. The beverages were formulated with 8% sugar, citric acid, flavor, and color (orange red), then manually carbonated and stored at 4°C. Over time, lactose, total solids, fat, and viscosity declined, while pH and microbial counts increased. Among the tested formulations, the sample containing carrageenan alone demonstrated the highest sensory acceptability and maintained better viscosity and appearance. The results indicated that carrageenan was the most suitable stabilizer for preserving the sensory and physico-chemical quality of CFWD, supporting its use in the commercial development of functional whey-based beverages.

Pareek et al. (2014) conducted a study to develop a healthy, fruit-based carbonated whey beverage by blending orange juice with paneer whey in three different ratios: 50:50, 60:40, and 70:30. Sensory evaluation revealed that the 70:30 ratio (orange juice to whey), referred to as OJCW-C, received the highest scores for color, flavor, appearance, and overall acceptability. Nutritional analysis of the most preferred formulation showed an increase in key nutrients—such as protein, vitamin C, total sugars, and essential minerals like sodium and potassium—compared to standard orange juice. The study demonstrated that the incorporation of whey into fruit juices not only enhances the nutritional profile but also offers a palatable and refreshing alternative to conventional carbonated soft drinks. Moreover, the simple and economical production process suggests strong potential for commercial application, while simultaneously addressing the environmental concern of whey disposal.

Silva e Alves et al. (2018) developed a probiotic functional carbonated beverage using sweet cheese whey and evaluated its microbiological and physicochemical properties during 28 days of refrigerated storage. The beverage was formulated with sugar, pectin, sodium citrate, natural flavoring, and the probiotic strain *Bifidobacterium animalis* subsp. *lactis* BB-12. The probiotic was selected based on its superior viability and low acidification compared to *Lactobacillus acidophilus* La-5. The final beverage maintained stable pH and titratable acidity throughout storage and exhibited minimal sedimentation (1%). Microbiological analyses confirmed the absence of pathogens and spoilage organisms, while probiotic counts remained above 10^6 CFU/mL, meeting the recommended threshold for health benefits. The results demonstrated that the beverage maintained good microbiological stability and probiotic viability, indicating its potential as a functional, health-promoting alternative to traditional soft drinks.

7. Technological processes employed in the production of whey-based beverages

Various technological methods are employed in the production of dairy beverages, each affecting the final product's properties. However, these techniques can negatively impact the quality of whey proteins by causing denaturation and aggregation, which may reduce their nutritional value, digestibility, and bioavailability. Therefore, selecting processing methods that minimize these adverse effects is essential for preserving the functional and nutritional integrity of whey proteins (Pellegrino et al., 2013).

In parallel, ongoing research is exploring advanced technologies, such as intensive ultrasound, cold plasma, microwave, and ohmic heating, which have shown promise in enhancing product quality. Among these, high-intensity ultrasound is particularly effective in preserving the bioactivity of the product. However, sonication at temperatures below 55°C may not be sufficiently effective in controlling bacterial contamination (Coimbra et al., 2020; Silva, 2020; Oliveira et al., 2022).

Table 12.3 presents a comprehensive summary of the technological methods employed in the production of whey-based beverages.

Kelleher et al. (2018) conducted a study to examine the effects of direct and indirect heat processing on whey drinks. Specifically, the study compared pre-heating at 70°C with final heating at 121°C, and pre-heating at 80°C with final heating at 135°C, with corresponding final hold times of 30 and 2 s, respectively. The drinks contained whey protein

TABLE 12.3 **A comprehensive summary of the technological processes employed in the production of whey-based beverages.**

Technological process	References
Direct and indirect heat processing	Kelleher et al. (2018)
Thermosonication	Barukčić et al. (2015) Oliveira et al. (2022)
Ultrasound and thermosonication	Jeličić et al. (2012)
High-intensity ultrasound	Herrera-Ponce et al. (2022)
Supercritical carbon dioxide	Iuan Chen et al. (2010) Yuk et al. (2014) Ceni et al. (2016) Amaral et al. (2018)
Microfiltration	Castro-Muñoz and Fíla (2018) Nazir et al. (2019) Vieira et al. (2020)

isolate (WPI) at concentrations of 4%, 6%, and 8% (w/w). As a result, significant differences were observed between the beverages subjected to direct and indirect heating, particularly in terms of their appearance and sensory attributes. Notably, beverages that underwent direct heating exhibited less protein denaturation, reduced viscosity, and fewer changes in their volatile component profiles. These findings suggest that direct thermal processing is especially effective for beverages with higher protein content.

Oliveira et al. (2022) examined the impacts of short-term thermal treatment and thermosonication on the characteristics of whey drinks. Thermosonification involved a synergistic interaction between temperature and cavitation, resulting in beverages with enhanced properties. This method offered several advantages, as it positively influences the characteristics of the treated beverages. Compared to those subjected to conventional pasteurization, thermosonified beverages exhibited improved rheological properties, greater microbial stability, and a more favorable volatile profile. Moreover, there was a reduction in the breakdown of ascorbic acid in the beverage. Furthermore, beverages treated through thermosonification also demonstrated increased antioxidant activity, along with enhanced antihypertensive and antidiabetic effects.

Research by Jeličić et al. (2012) investigated the influence of different levels of ultrasound power (240 W, 320 W, 400 W), applied alone or in combination with thermal pretreatment, on microbial inactivation and sensory properties of whey from rennet cheese, compared to traditional pasteurization processes. While ultrasonication alone did not significantly reduce microbial counts, combining it with preheating to 35°C or 45°C enhanced microbial reduction, likely due to cavitation. The most effective treatment was 400 W with preheating to 55°C, which achieved a greater microbial reduction than pasteurization, possibly due to the added heat. Sensory analysis revealed significant improvements in texture, with smoother mouthfeel, no sediment, and consistent color, outpacing pasteurized samples.

As part of the study conducted by Barukčić et al. (2015), researchers examined the impact of ultrasound on the fermentation process using both pasteurized and thermosonicated whey. The study also focused on monitoring sensory properties and microbial activity. Fermentation was carried out with the bacterium *Lactobacillus acidophilus* as well as a yogurt culture consisting of *Streptococcus thermophilus* and *Lactobacillus delbrueckii* subsp. *bulgaricus*. Notably, whey treated with thermosonication (480 W for 10 min at 55°C) demonstrated more favorable microbiological and organoleptic properties compared to pasteurized whey, particularly exhibiting a lower sediment content. Furthermore, the highest increase in viable microorganisms was observed in whey treated with ultrasound (84 W for 150 s). Additionally, based on the obtained results, it can be concluded that whey fermented with *Lactobacillus acidophilus* contained a higher number of viable microorganisms.

In a study conducted by Herrera-Ponce et al. (2022), whey-oat fermented beverages were produced using high-intensity ultrasound treatment (frequency of 40 kHz and 1 W/cm^2) for varying durations of 0, 3, and 10 min. Following ultrasound treatment, the samples were inoculated with *Lacticaseibacillus casei* to initiate fermentation. The results indicated that the beverage containing an equal ratio of whey and oats (50:50) and treated with high-intensity ultrasound for 3 min exhibited the highest antioxidant activity, the greatest number of viable *Lacticaseibacillus casei* bacteria, and received the highest scores in the sensory evaluation. These findings suggest that ultrasound treatment not only enhanced the fermentation process but also improved the overall functionality of the resulting beverages.

The use of supercritical carbon dioxide represents an advanced and innovative technique in beverage processing. This non-thermal, gentle method helps preserve the physicochemical, nutritional, and sensory properties of food, making it a valuable alternative to conventional treatments. Supercritical carbon dioxide processing has been successfully applied to the treatment of fruit juices, demonstrating its effectiveness in maintaining product quality. Unlike traditional pasteurization methods, this technique does not alter the key physicochemical characteristics of whey beverages, including pH, soluble solids, titratable acidity, and bioactive compounds such as phenolics and anthocyanins. As a result, supercritical carbon dioxide offers a promising approach to enhancing the quality and stability of whey-based beverages while retaining their nutritional and sensory integrity (Iuan Chen et al., 2010; Yuk et al., 2014; Ceni et al., 2016; Amaral et al., 2018).

Finally, it is crucial to emphasize the growing utilization of microfiltration technology in dairy processing. This membrane filtration technique uses pressure to facilitate the separation process. It provides multiple benefits, such as decreased energy expenses, exceptional separation efficiency, operational simplicity, and elevated productivity. Microfiltration effectively separates high-value components, including proteins, carbohydrates, pectin and phenolic compounds. Furthermore, in the dairy sector, microfiltration facilitates bacterial elimination, allowing for the manufacture of milk with a prolonged shelf life (Castro-Muñoz and Fíla, 2018; Nazir et al., 2019; Vieira et al., 2020). Choosing the right technological method for producing whey-based beverages is vital to preserve its technological properties and prevent degradation. An improper selection of processing techniques may negatively affect the quality of the beverage, underscoring the importance of selecting an appropriate manufacturing approach to maintain its integrity (Oliveira et al., 2022).

8. Conclusion

Whey is an important by-product of the dairy industry due to its composition and nutritional profile. However, its use is still limited, which has prompted efforts to incorporate whey into various food products. Traditionally, whey was either used as livestock feed or discharged into waste water, contributing to environmental pollution. In recent years, however, one of the important uses of whey processing is the production of non-fermented whey-based beverages.

Flavored whey beverages have gained significant popularity due to the positive impact of added flavoring agents (fruit/vegetable juices) on its taste, aroma and appearance. Among these, fermented whey-based beverages distinguish themselves through their elevated levels of phenolic compounds and enhanced antioxidant properties. Recent innovations in food processing technologies, such as thermosonication, supercritical carbon dioxide treatment, ultrasound, and microfiltration, have played a pivotal role in further improving the quality of whey-based beverages. These methods contribute to higher concentrations of ascorbic acid, better microbial balance, and enhanced organoleptic and rheological qualities. By leveraging whey for beverage production, this approach not only increases the versatility of whey but also offers an eco-friendly solution to mitigate the environmental impact of improper whey disposal. Ongoing advancements in whey processing techniques hold great promise for expanding its applications within the food industry.

References

Abdul-Alim, T.S., Zayan, A.F., Campelo, P.H., Bakry, A.M., 2018. Development of new functional fermented product: mulberry-whey beverage. J. Nutr. Food Technol. 1 (3), 64–69.

Ahmed, T., Sabuz, A.A., Mohaldar, A., Fardows, H.S., Inbaraj, B.S., Sharma, M., Sridhar, K., 2023. Development of novel whey-mango based mixed beverage: effect of storage on physicochemical, microbiological, and sensory analysis. Foods 12 (2), 237.

Akpinar, A.S.L.I., Torunoglu, F.A., Yerlikaya, O.K.T.A.Y., Kinik, O.Z.E.R., Akbulut, N.E.C.A.T.I., Uysal, H.R., 2015. Fermented probiotic beverages produced with reconstituted whey and cow milk. Agro Food Ind. Hi Tech. 26(4).

Ali, M.N., Singh, A., Broadway, A.A., Singh, M., 2015. Technology for manufacturing of whey beer by incorporating malt wort. The Allahabad Farmer 71 (1).

Aly, E., Darwish, A.A., Tawfek, M., 2019. Quality characteristics of sweet whey-based fruits beverages fermented with Lactobacillus plantarum. Egypt. J. Food Sci. 47 (2), 141–254.

Amaral, G.V., Silva, E.K., Cavalcanti, R.N., Martins, C.P., Andrade, L.G.Z., Moraes, J., Cruz, A.G., 2018. Whey-grape juice drink processed by supercritical carbon dioxide technology: physicochemical characteristics, bioactive compounds and volatile profile. Food Chem. 239, 697–703.

Arsić, S., Bulatović, M., Zarić, D., Kokeza, G., Subić, J., Rakin, M., 2018. Functional fermented whey carrot beverage-qualitative, nutritive and techno-economic analysis. Rom. Biotechnol. Lett. 23 (2), 13496–13504.

ASTM, A., 1989. Standard definitions of terms relating to sensory evaluation of materials and products. Annu. Book ASTM Stand. 2.

Athanasiadis, I., Paraskevopoulou, A., Blekas, G., Kiosseoglou, V., 2004. Development of a novel whey beverage by fermentation with kefir granules. Effect of various treatments. Biotechnol. Prog. 20 (4), 1091–1095.

Bailey, H.M., Fanelli, N.S., Stein, H.H., 2023. Effect of heat treatment on protein quality of rapeseed protein isolate compared with non-heated rapeseed isolate, soy and whey protein isolates, and rice and pea protein concentrates. J. Sci. Food Agric. 103 (14), 7251–7259.

Bandara, T.A., Munasinghe-Arachchige, S.P., Gamlath, C.J., 2023. Fermented whey beverages: a review of process fundamentals, recent developments and nutritional potential. Int. J. Dairy Technol. 76 (4), 737–757.

Bartoń, M., Waraczewski, R., Sołowiej, B.G., 2025. Organic sea buckthorn or rosehip juices on the physicochemical, rheological, and microbial properties of organic goat or cow fermented whey beverages. Appl. Sci. 15, 1240.

Barukčić, I., Jakopović, K.L., Herceg, Z., Karlović, S., Božanić, R., 2015. Influence of high intensity ultrasound on microbial reduction, physico-chemical characteristics and fermentation of sweet whey. Innov. Food Sci. Emerg. Technol. 27, 94–101.

Barukčić, I., Lisak Jakopović, K., Božanić, R., 2019. Valorisation of whey and buttermilk for production of functional beverages–an overview of current possibilities. Food Technol. Biotechnol. 57 (4), 448–460.

Belloso-Morales, G., Hernández-Sánchez, H., 2003. Manufacture of a beverage from cheese whey using a “tea fungus” fermentation. REVISTA LATINOAMERICANA DE MICROBIOLOGIA-MEXICO 45 (1/2), 5.

Besediuk, V., Yatskov, M., Korchyk, N., Kucherova, A., Maletskyi, Z., 2024. Whey-From waste to a valuable resource. J. Agric. Food Res., 101280

Bilyi, V., Merzlov, S., Narizhniy, S., Mashkin, Y., Merzlova, H., 2021. Amino acid composition of whey and cottage cheese under various rennet enzymes. Sci. Horiz. 24 (9), 19–25.

Bishop, P., Pitts, E.R., Budner, D., Thompson-Witrick, K.A., 2022. Kombucha: biochemical and microbiological impacts on the chemical and flavor profile. Food Chem. Adv. 1, 100025.

Blaschek, K.M., Wendorff, W.L., Rankin, S.A., 2007. Survey of salty and sweet whey composition from various cheese plants in Wisconsin. J. Dairy Sci. 90 (4), 2029–2034.

Bleoussi, R., Konfo, C.T., Tchekessi, C.C., Sachi, P., Banon, J.S.B., Djogbe, A.A., Bokossa, Y.I., 2020. Nutritional quality and use of whey in human food for its valorization. World J. Adv. Res. Rev. 8 (1), 284.

Carvalho, F., Prazeres, A.R., Rivas, J., 2013. Cheese whey wastewater: characterization and treatment. Sci. Total Environ. 445, 385–396.

Castro, W.F., Cruz, A.G., Bisinotto, M.S., Guerreiro, L.M.R., Faria, J.A.F., Bolini, H.M.A., Cunha, R.L., Deliza, R., 2013. Development of probiotic dairy beverages: rheological properties and application of mathematical models in sensory evaluation. J. Dairy Sci. 96 (1), 16–25. https://doi.org/10.3168/jds.2012-5590.

Castro-Muñoz, R., Fíla, V., 2018. Membrane-based technologies as an emerging tool for separating high-added-value compounds from natural products. Trends Food Sci. Technol. 82, 8–20.

Celeghin, A.G., Minetti, F., Contini, L.E., Miccolo, M.E., Rubiolo, A.C., Olivares, M.L., 2016. Syneresis and sensory acceptability of desserts based on whey proteins concentrates. J. Food Nutr. Res. 4, 478–482.

Ceni, G., Silva, M.F., Valério Jr, C., Cansian, R.L., Oliveira, J.V., Dalla Rosa, C., Mazutti, M.A., 2016. Continuous inactivation of alkaline phosphatase and Escherichia coli in milk using compressed carbon dioxide as inactivating agent. J. CO2 Util. 13, 24–28.

Chatterjee, G., De Neve, J., Dutta, A., Das, S., 2015. Formulation and statistical evaluation of a ready-to-drink whey-based orange beverage and its storage stability. Rev. Mex. Ing. Quim. 4 (2), 253–264.

Chatzipaschali, A.A., Stamatis, A.G., 2012. Biotechnological utilization with a focus on anaerobic treatment of cheese whey: current status and prospects. Energies 5, 3492–3525.

Chavan, R.S., Shraddha, R.C., Kumar, A., Nalawade, T., 2015. Whey based beverage: its functionality, formulations, health benefits and applications. J. Food Process. Technol. 6 (10), 1000495.

Chavda, V.R., Sangwan, R.B., Solanki, A.K., Modi, R.B., Parmar, M.P., 2016. Development of whey-protein enriched strawberry whey beverage using response surface methodology. Indian J. Dairy Sci. 69.

Coelho, R.M.D., de Almeida, A.L., do Amaral, R.Q.G., da Mota, R.N., de Sousa, P.H.M., 2020. Kombucha. Int. J. Gastron. Food Sci. 22, 100272.

Coimbra, L.O., Vidal, V.A., Silva, R., Rocha, R.S., Guimarães, J.T., Balthazar, C.F., Cruz, A.G., 2020. Are ohmic heating-treated whey dairy beverages an innovation? Insights of the Q methodology. LWT 134, 110052.

Coma, M.E., Peltzer, M.A., Delgado, J.F., Salvay, A.G., 2019. Water kefir grains as an innovative source of materials: study of plasticiser content on film properties. Eur. Polym. J. 120, 109234.

Cruz, A.G., Sant'Ana, A.D.S., Macchione, M.M., Teixeira, Â.M., Schmidt, F.L., 2009. Milk drink using whey butter cheese (queijo manteiga) and acerola juice as a potential source of vitamin C. Food Bioprocess Technol. 2, 368–373.

da Silveira, E.O., Neto, J.H.L., da Silva, L.A., Raposo, A.E., Magnani, M., Cardarelli, H.R., 2015. The effects of inulin combined with oligofructose and goat cheese whey on the physicochemical properties and sensory acceptance of a probiotic chocolate goat dairy beverage. LWT-Food Sci. Technol. 62 (1), 445–451.

Darewicz, M., Dziuba, B., Minkiewicz, P., Dziuba, J., 2011. The preventive potential of milk and colostrum proteins and protein fragments. Crit. Rev. Food Sci. Nutr. 51, 375–389.

de Souza, R.R., Bergamasco, R., da Costa, S.C., Feng, X., Faria, S.H.B., Gimenes, M.L., 2010. Recovery and purification of lactose from whey. Chem. Eng. Process. Process Intensif. 49 (11), 1137–1143.

Dinkçi, N., Akdeniz, V., Akalın, A.S., 2023. Probiotic whey-based beverages from cow, sheep and goat milk: antioxidant activity, culture viability, amino acid contents. Foods 12 (3), 610.

Djurić, M., Carić, M., Milanović, S., Tekić, M., Panić, M., 2004. Development of whey-based beverages. Eur. Food Res. Technol. 219, 321–328.

Domingos, J.M., Puccio, S., Martinez, G.A., Amaral, N., Reis, M.A., Bandini, S., Bertin, L., 2018. Cheese whey integrated valorisation: production, concentration and exploitation of carboxylic acids for the production of polyhydroxyalkanoates by a fed-batch culture. Chem. Eng. J. 336, 47–53.

Egger, L., Ménard, O., 2017. Update on bioactive peptides after milk and cheese digestion. Curr. Opin. Food Sci. 14, 116–121.

Elkot, W.F., Elmahdy, A., Talaat, H., Alghamdia, O.A., Alhag, S.K., Al-Shahari, E.A., Ismail, H.A., 2024. Development and characterization of a novel flavored functional fermented whey-based sports beverage fortified with Spirulina platensis. Int. J. Biol. Macromol. 258, 128999.

Eurostat, 2020. Agriculture, Forestry and Fishery Statistics—Milk and Milk Products Statistics. Publications Office of the European Union, Luxembourg.

Fernández-Gutiérrez, D., Veillette, M., Giroir-Fendler, A., Ramirez, A.A., Faucheux, N., Heitz, M., 2017. Biovalorization of saccharides derived from industrial wastes such as whey: a review. Rev. Environ. Sci. Biotechnol. 16, 147–174.

Fox, P.F., 2008. Milk: an overview. Milk Proteins 1–54.

Ghanimah, M.A., 2018. Functional and technological aspects of whey powder and whey protein products. Int. J. Dairy Technol. 71 (2), 454–459.

Gimhani, K.H.I., Liyanage, A.L.C.J., 2018. Development and quality evaluation of ready to drink fruit flavoured whey beverage. Int. J. Sci. Res. Publ. 9 (8), 779–783.

Gomes, R.J., de Fatima Borges, M., de Freitas Rosa, M., Castro-Gómez, R.J.H., Spinosa, W.A., 2018. Acetic acid bacteria in the food industry: systematics, characteristics and applications. Food Technol. Biotechnol. 56 (2), 139.

Gorissen, S.H., Crombag, J.J., Senden, J.M., Waterval, W.H., Bierau, J., Verdijk, L.B., van Loon, L.J., 2018. Protein content and amino acid composition of commercially available plant-based protein isolates. Amino Acids 50, 1685–1695.

Goudarzi, M., Madadlou, A., Mousavi, M.E., Emam-Djomeh, Z., 2015. Formulation of apple juice beverages containing whey protein isolate or whey protein hydrolysate based on sensory and physicochemical analysis. Int. J. Dairy Technol. 68 (1), 70–78.

Gulitz, A., Stadie, J., Wenning, M., Ehrmann, M.A., Vogel, R.F., 2011. The microbial diversity of water kefir. Int. J. Food Microbiol. 151 (3), 284–288.

Guzel-Seydim, Z.B., Gökırmaklı, Ç., Greene, A.K., 2021. A comparison of milk kefir and water kefir: physical, chemical, microbiological and functional properties. Trends Food Sci. Technol. 113, 42–53.

Hanna, M., Jaqua, E., Nguyen, V., Clay, J.B., 2022. Vitamins: functions and uses in medicine. Perm. J. 26 (2), 89–97.

Harrison, K., Curtin, C., 2021. Microbial composition of SCOBY starter cultures used by commercial kombucha brewers in North America. Microorganisms 9 (5), 1060.

Herreman, L., Nommensen, P., Pennings, B., Laus, M.C., 2020. Comprehensive overview of the quality of plant-And animal-sourced proteins based on the digestible indispensable amino acid score. Food Sci. Nutr. 8 (10), 5379–5391.

Herrera-Ponce, A.L., Salmeron-Ochoa, I., Rodriguez-Figueroa, J.C., Santellano-Estrada, E., Garcia-Galicia, I.A., Alarcon-Rojo, A.D., 2022. High-intensity ultrasound as pre-treatment in the development of fermented whey and oat beverages: effect on the fermentation, antioxidant activity and consumer acceptance. J. Food Sci. Technol. 1–9.

Hoac, T., Lundh, T., Purup, S., Önning, G., Sejrsen, K., Åkesson, B., 2007. Separation of selenium, zinc, and copper compounds in bovine whey using size exclusion chromatography linked to inductively coupled plasma mass spectrometry. J. Agric. Food Chem. 55 (10), 4237–4243.

Ismail, A.E., Abdelgader, M.O., Ali, A.A., 2011. Microbial and chemical evaluation of whey-based mango beverage. Adv. J. Food Sci. Technol. 3 (4), 250–253.

Jakopović, K.L., Barukčić, I., Božanić, R., 2019. Bioactive components derived from bovine milk. Mljekarstvo 69 (3), 151–161.

Jeličić, I., Božanić, R., Brnčić, M., Tripalo, B., 2012. Influence and comparison of thermal, ultrasonic and thermo-sonic treatments on microbiological quality and sensory properties of rennet cheese whey. Mljekarstvo: Časopis Za Unaprjeđenje Proizvodnje I Prerade Mlijeka 62 (3), 165–178.

Jeličić, I., Božanić, R., Tratnik, L., 2008. Whey based beverages-new generation of dairy products. Mljekarstvo 58 (3), 257–274.

Jitpakdee, J., Kantachote, D., Kanzaki, H., Nitoda, T., 2022. Potential of lactic acid bacteria to produce functional fermented whey beverage with putative health promoting attributes. LWT 160, 113269.

Joshi, J., Gururani, P., Vishnoi, S., Srivastava, A., 2020. Whey based beverages: a review. Octa J. Biosci. 8, 30–37.

Kandler, O., 1983. Carbohydrate metabolism in lactic acid bacteria. Antonie Leeuwenhoek 49, 209–224.

Kapoor, R., Metzger, L.E., 2004. Evaluation of salt whey as an ingredient in processed cheese. J. Dairy Sci. 87 (5), 1143–1150.

Kaur, S., Bhise, S.R., Kaur, A., Minhas, K.S., 2019. Development of naturally carbonated paneer whey fermented beverage blended with pineapple and strawberry juice. Nutr. Food Sci. 49 (4), 528–547.

Kelleher, C.M., O'Mahony, J.A., Kelly, A.L., O'Callaghan, D.J., Kilcawley, K.N., McCarthy, N.A., 2018. The effect of direct and indirect heat treatment on the attributes of whey protein beverages. Int. Dairy J. 85, 144–152.

Korhonen, H., 2009. Milk-derived bioactive peptides: from science to applications. J. Funct.Foods 1 (2), 177–187.

Laavanya, D., Shirkole, S., Balasubramanian, P., 2021. Current challenges, applications and future perspectives of SCOBY cellulose of Kombucha fermentation. J. Clean. Prod. 295, 126454.

Lappa, I.K., Papadaki, A., Kachrimanidou, V., Terpou, A., Koulougliotis, D., Eriotou, E., Kopsahelis, N., 2019. Cheese whey processing: integrated biorefinery concepts and emerging food applications. Foods 8, 347.

Laureys, D., Britton, S.J., De Clippeleer, J., 2020. Kombucha tea fermentation: a review. J. Am. Soc. Brew. Chem. 78 (3), 165–174.

Legarová, V., Kouřimská, L., 2010. Sensory quality evaluation of whey-based beverages. Mljekarstvo 60 (4), 280.

León-López, A., Pérez-Marroquín, X.A., Campos-Lozada, G., Campos-Montiel, R.G., Aguirre-Álvarez, G., 2020. Characterization of whey-based fermented beverages supplemented with hydrolyzed collagen: antioxidant activity and bioavailability. Foods 9 (8), 1106.

Liu, Y., Zheng, Y., Yang, T., Mac Regenstein, J., Zhou, P., 2022. Functional properties and sensory characteristics of kombucha analogs prepared with alternative materials. Trends Food Sci. Technol. 129, 608–616.

Londero, A., Hamet, M.F., De Antoni, G.L., Garrote, G.L., Abraham, A.G., 2012. Kefir grains as a starter for whey fermentation at different temperatures: chemical and microbiological characterisation. J. Dairy Res. 79 (3), 262–271.

luan Chen, J., Zhang, J., Song, L., Jiang, Y., Wu, J., Hu, X.S., 2010. Changes in microorganism, enzyme, aroma of hami melon (Cucumis melo L.) juice treated with dense phase carbon dioxide and stored at 4 C. Innov. Food Sci. Emerg. Technol. 11 (4), 623–629.

Lynch, K.M., Wilkinson, S., Daenen, L., Arendt, E.K., 2021. An update on water kefir: microbiology, composition and production. Int. J. Food Microbiol. 345, 109128.

M'hir, S., Rtibi, K., Mejri, A., Ziadi, M., Aloui, H., Hamdi, M., Ayed, L., 2019. Development of a novel whey date beverage fermented with kefir grains using response surface methodology. J. Chem. 2019 (1), 1218058.

Macwan, S.R., Dabhi, B.K., Parmar, S.C., Aparnathi, K.D., 2016. Whey and its utilization. Int. J. Curr. Microbiol. Appl. Sci. 5 (8), 134–155.

Magalhães, K.T., Dias, D.R., de Melo Pereira, G.V., Oliveira, J.M., Domingues, L., Teixeira, J.A., Schwan, R.F., 2011. Chemical composition and sensory analysis of cheese whey-based beverages using kefir grains as starter culture. Int. J. Food Sci. Technol. 46 (4), 871–878.

Magalhães, K.T., Pereira, M.A., Nicolau, A., Dragone, G., Domingues, L., Teixeira, J.A., Schwan, R.F., 2010. Production of fermented cheese whey-based beverage using kefir grains as starter culture: evaluation of morphological and microbial variations. Bioresour. Technol. 101 (22), 8843–8850.

Mawson, A.J., 1994. Biokonverze pro využití syrovátky a snižování odpadu. Bioresour. Technol. 47, 195–203.

Mazaheri Assadi, M., Abdolmaleki, F., Mokarrame, R.R., 2008. Application of whey in fermented beverage production using kefir starter culture. Nutr. Food Sci. 38 (2), 121–127.

Minj, S., Anand, S., 2020. Whey proteins and its derivatives: bioactivity, functionality, and current applications. Dairy 1 (3), 233–258.

Mitrovic, N.M., Mitrovic, M.D., Tasic, N., Milanovic, N.D., 2023. Whey proteins-facts and possibilities in modern medicine. Med. J./Medicinski Časopis 57 (2).

Morales, D., 2021. Biological activities of kombucha beverages: the need of clinical evidence. Trends Food Sci. Technol. 105, 323–333.

Moretti, A.F., Moure, M.C., Quiñoy, F., Esposito, F., Simonelli, N., Medrano, M., León-Peláez, Á., 2022. Water kefir, a fermented beverage containing probiotic microorganisms: from ancient and artisanal manufacture to industrialized and regulated commercialization. Future Foods 5, 100123.

Naik, B., Kohli, D., Walter, N., Gupta, A.K., Mishra, S., Khan, J.M., Kumar, V., 2023. Whey-carrot based functional beverage: development and storage study. J. King Saud Univ. Sci. 35 (6), 102775.

Nazir, A., Khan, K., Maan, A., Zia, R., Giorno, L., Schroën, K., 2019. Membrane separation technology for the recovery of nutraceuticals from food industrial streams. Trends Food Sci. Technol. 86, 426–438.

Nedanovska, E., Jakopović, K.L., Daniloski, D., Vaskoska, R., Vasiljevic, T., Barukčić, I., 2022. Effect of storage time on the microbial, physico-chemical and sensory characteristics of ovine whey-based fruit beverages. Int. J. Food Sci. Technol. 57 (8), 5388–5398.

Nishanthi, M., Chandrapala, J., Vasiljevic, T., 2017. Compositional and structural properties of whey proteins of sweet, acid and salty whey concentrates and their respective spray dried powders. Int. Dairy J. 74, 49–56.

Nkurunziza, F., Theoneste, T., Sandrine, U., Clement, B., Upadhyay, S.K., Sharma, A.K., 2022. Processing and development of papaya (Carica papaya L.) juice with incorporation of whey. Indian J. Appl. Pure Bio 37 (3), 653–659.

Nursiwi, A., Nurhartadi, E., Utami, R., Sari, A.M., Laksono, P.W., Aprilia, E.N., 2017. Characteristic of fermented whey beverage with addition of tomato juice (Lycopersicum esculentum). IOP Conf. Ser. Mater. Sci. Eng. 193 (1), 012009.

Oliveira, G.A., Guimarães, J.T., Ramos, G.L.P., Esmerino, E.A., Pimentel, T.C., Neto, R.P., Cruz, A.G., 2022. Benefits of thermosonication in orange juice whey drink processing. Innov. Food Sci. Emerg. Technol. 75, 102876.

Özer, B., Evrendilek, G.A., 2022. Whey beverages. In: Dairy Foods. Woodhead Publishing, pp. 117–137.

Panesar, P.S., Kennedy, J.F., Gandhi, D.N., Bunko, K., 2007. Bioutilisation of whey for lactic acid production. Food Chem. 105 (1), 1–14.

Pareek, N., Gupta, A., Sengar, R., 2014. Preparation of healthy fruit based carbonated whey beverages using whey and orange juice. Asian J. Dairy Food Res. 33 (1), 5–8.

Park, Y.W., Juárez, M., Ramos, M.G.F.W., Haenlein, G.F.W., 2007. Physico-chemical characteristics of goat and sheep milk. Small Rumin. Res. 68 (1–2), 88–113.

Pelegrine, D., Gasparetto, C., 2005. Whey proteins solubility as function of temperature and pH. LWT Food Sci. Technol. 38, 77–80.

Pellegrino, L., Masotti, F., Cattaneo, S., Hogenboom, J.A., De Noni, I., 2013. Nutritional quality of milk proteins. In: Advanced Dairy Chemistry, fourth ed. Publisher: Springer, Boston, pp. 515–538.

Pescuma, M., de Valdez, G.F., Mozzi, F., 2015. Whey-derived valuable products obtained by microbial fermentation. Appl. Microbiol. Biotechnol. 99, 6183–6196.

Pescuma, M., Hébert, E.M., Mozzi, F., de Valdez, G.F., 2008. Whey fermentation by thermophilic lactic acid bacteria: evolution of carbohydrates and protein content. Food Microbiol. 25 (3), 442–451.

Pgntado, M., Macedo, A.C., Malcata, F.X., 2001. Review: technology, chemistry and microbiology of whey cheese. Food Sci. Tech. Int. 7, 105–116.

Pires, A.F., Marnotes, N.G., Rubio, O.D., Garcia, A.C., Pereira, C.D., 2021. Dairy by-products: a review on the valorization of whey and second cheese whey. Foods 10, 1067.

Plessas, S., Nouska, C., Mantzourani, I., Kourkoutas, Y., Alexopoulos, A., Bezirtzoglou, E., 2017. Microbiological exploration of different types of kefir grains. Fermentation 3 (1), 1.

Pogoń, K., Sady, M., Jaworska, G., Grega, T., 2015. Characteristics of orange-whey fermented beverages. Acta Aliment. 44 (2), 170–177.

Priyanka, Bhushan, K., Kocher, G.S., 2022. Valorisation of whey for fermented beverage production using functional starter yeast. Acta Aliment.

Purkiewicz, A., Pietrzak-Fiećko, R., 2021. Antioxidant properties of fruit and vegetable whey beverages and fruit and vegetable mousses. Molecules 26 (11), 3126.

Rama, G.R., Kuhn, D., Beux, S., Maciel, M.J., de Souza, C.F.V., 2019. Potential applications of dairy whey for the production of lactic acid bacteria cultures. Int. Dairy J. 98, 25–37.

Rao, R.D., Wendorff, W.L., 2003. Compositional differences between whey, salty whey. and press whey from commercial manufacture of Cheddar cheese. J. Dairy Sci. 86 (Suppl. 1), 365.

Rejdlová, A., Salek, R.N., Mříšková, Z., Lorencová, E., Kůrová, V., Adámek, R., Sumczynski, D., 2023. Physical characterization of a novel carrot juice whey-enriched beverage fermented with milk or water kefir starter cultures. Foods 12, 3368.

Rejdlová, A., Vašina, M., Lorencová, E., Hružík, L., Salek, R.N., 2024. Assessment of different levels of blackcurrant juice and furcellaran on the quality of fermented whey-based beverages using rheological and mechanical vibration damping techniques. Foods 13, 1855.

Risner, D., Tomasino, E., Hughes, P., Meunier-Goddik, L., 2019. Volatile aroma composition of distillates produced from fermented sweet and acid whey. J. Dairy Sci. 102 (1), 202–210.

Rocha-Mendoza, D., Kosmerl, E., Krentz, A., Zhang, L., Badiger, S., Miyagusuku-Cruzado, G., García-Cano, I., 2021. Invited review: acid whey trends and health benefits. J. Dairy Sci. 104 (2), 1262–1275.

Rodríguez-Hernández, G., Rentería-Monterrubio, A.L., Gutiérrez-Méndez, N., Ortega-Gutiérrez, J.A., Santellano-Estrada, E., Rojas-González, S., Chávez-Martínez, A., 2020. Effect of probiotic cultures on the angiotensin converting enzyme inhibitory activity of whey-based fermented beverages. J. Food Sci. Technol. 57, 3731–3738.

Sabokbar, N., Khodaiyan, F., Moosavi-Nasab, M., 2015. Optimization of processing conditions to improve antioxidant activities of apple juice and whey based novel beverage fermented by kefir grains. J. Food Sci. Technol. 52, 3422–3432.

Sady, M., Jaworska, G., Grega, T., Bernas, E., Domagala, J., 2013. Application of acid whey in orange drink production. Food Technol. Biotechnol. 51 (2), 266.

Sameen, A., Tariq, M.R., Huma, N., Khan, M.I., 2013. Effect of stabilizers on the quality of carbonated flavoured whey drink. Afr. J. Agric. Res. 8 (5), 445–448.

Sfoglia, N.M., Rossi, D.M., Sant'Anna, V., Fabricio, M.F., Ayub, M.A.Z., Hickert, L.R., 2024. Milk whey fermented by kombucha SCOBY: effect of sugar on the physicochemical and microbiological properties of the new beverage. Waste Biomass Valoriz. 1–9.

Shekilango, S.A., Jelen, P., Bagdan, G.C., 1997. Production of whey-banana beverages from acid whey and overripe bananas. Milchwissenschaft 52 (4), 209–212.

Shiby, V.K., Mishra, H.N., 2013. Fermented milks and milk products as functional foods—a review. Crit. Rev. Food Sci. Nutr. 53 (5), 482–496.

Silva e Alves, A.T., Spadoti, L.M., Zacarchenco, P.B., Trento, F.K., 2018. Probiotic functional carbonated whey beverages: development and quality evaluation. Beverages 4 (3), 49.

Silva, F.V., 2020. Ultrasound assisted thermal inactivation of spores in foods: pathogenic and spoilage bacteria, molds and yeasts. Trends Food Sci. Technol. 105, 402–415.

Singh, H.K., Puranik, D.B., Patwadi, P.S., Sain, M., 2021. Process optimisation for the production of papaya leaf extract based therapeutic whey beverage. Environ. Conserv. J. 22 (1&2), 153–158.

Skryplonek, K., Dmytrów, I., Mituniewicz-Małek, A., 2019. Probiotic fermented beverages based on acid whey. J. Dairy Sci. 102 (9), 7773–7780.

Svanborg, S., Johansen, A., Abrahamsenn, R., Skeie, S., 2015. The composition and functional properties of whey protein concentrates produced from buttermilk are comparable with those of whey protein concentrates produced from skimmed milk. J. Dairy Sci. 98, 5829–5840.

Szafrańska, J.O., Waraczewski, R., Bartoń, M., Wesołowska-Trojanowska, M., Maziejuk, W., Nowak, P., Sołowiej, B.G., 2024. Effects of organic fruit juices on physicochemical, microbiological, and antioxidative aspects of fermented whey beverages from organic goat and cow milk, produced at laboratory and industrial scales. J. Dairy Sci. 107 (12), 10481–10496.

Tang, G., 2010. Bioconversion of dietary provitamin A carotenoids to vitamin A in humans. Am. J. Clin. Nutr. 91 (5), 1468S–1473S.

Thibodeau, A., Reardon, P., Bartlett, B., Curtin, C., 2025. Changes in microbial community structure during adaptation of kombucha symbiotic culture of bacteria and yeast to fermentation of sweet and acid whey. J. Dairy Sci. 108 (5), 4761–4784.

Tratnik, L., 2003. Uloga sirutke u proizvodnji funkcionalne mliječne hrane. Mljekarstvo: časopis za unaprjeđenje proizvodnje i prerade mlijeka 53 (4), 325–352.

Tsermoula, P., Khakimov, B., Nielsen, J.H., Engelsen, S.B., 2021. WHEY-The waste-stream that became more valuable than the food product. Trends Food Sci. Technol. 118, 230–241.

Tu, C., Tang, S., Azi, F., Hu, W., Dong, M., 2019. Use of kombucha consortium to transform soy whey into a novel functional beverage. J. Funct.Foods 52, 81–89.

Venetsaneas, N., Antonopoulou, G., Stamatelatou, K., Kornaros, M., Lyberatos, G., 2009. Using cheese whey for hydrogen and methane generation in a two-stage continuous process with alternative pH controlling approaches. Bioresour. Technol. 100 (15), 3713–3717.

Vieira, A.H., Balthazar, C.F., Guimaraes, J.T., Rocha, R.S., Pagani, M.M., Esmerino, E.A., Cruz, A.G., 2020. Advantages of microfiltration processing of goat whey orange juice beverage. Food Res. Int. 132, 109060.

Yadav, J.S.S., Yan, S., Pilli, S., Kumar, L., Tyagi, R.D., Surampalli, R.Y., 2015. Cheese whey: a potential resource to transform into bioprotein, functional/nutritional proteins and bioactive peptides. Biotechnol. Adv. 33 (6), 756–774.

Yadav, R.B., Yadav, B.S., Kalia, N., 2010. Development and storage studies on whey-based banana herbal (Mentha arvensis) beverage. Am. J. Food Technol. 5 (2), 121–129.

Yang, Y., Wang, H., Shuang, Q., Xia, Y., 2024. Novel insights into flavor formation in whey fermented wine: a study of microbial metabolic networks. LWT 197, 115911.

Yuk, H.G., Sampedro, F., Fan, X., Geveke, D.J., 2014. Nonthermal processing of orange juice using a pilot-plant scale supercritical carbon dioxide system with a gas–liquid metal contactor. J. Food Process. Preserv. 38 (1), 630–638.

Zandona, E., Blažić, M., Režek Jambrak, A., 2021. Whey utilization: sustainable uses and environmental approach. Food Technol. Biotechnol. 59 (2), 147–161.

Zhakupova, G.N., Tultabayeva, T.C., Nurtayeva, A.B., Kundyzbayeva, N.D., Sagandyk, A.T., 2022. Research and development of the refreshing whey drinks technology. J. Almat. Technol. Univ. 4, 99–105.

Chapter 13

Whey protein and muscle synthesis: Implications for healthy aging and sports performance

Isabel Gigli
Agronomy College, National University of La Pampa (UNLPam), La Pampa, Argentina

1. Introduction

As the global population is projected to reach 9.1 billion by 2050, the Food and Agriculture Organization (FAO) estimates that food production must increase by 70%, with developing countries needing to nearly double their output (FAO, 2009). Simultaneously, approximately 30%–40% of all food produced worldwide is lost or wasted each year, revealing significant inefficiencies in the global food system. Whey—the liquid by-product of cheese production—is a striking example of this paradox. Despite an estimated global production exceeding 10 billion tons annually, only about half is processed into food products, while the rest is often discarded or discharged into wastewater systems, leading to environmental pollution. Yet, whey is far from mere waste: it is a rich source of high quality protein and other bioactive components beneficial to human health (Besediuk et al., 2024).

The global demographic shift toward an aging population presents unprecedented challenges to healthcare systems, economies, and public health. One of the most critical physiological changes associated with aging is the progressive loss of skeletal muscle mass and function, a condition known as sarcopenia, which significantly contributes to frailty, reduced mobility, and loss of independence. Simultaneously, interest in optimizing sports performance and muscle recovery in both elite and recreational athletes continues to grow, with a strong focus on nutritional interventions that support muscle adaptation and repair. In this context, whey protein emerges as a superior source due to its rapid digestibility, efficient amino acid absorption, and bioactive peptide content. Beyond its physiological advantages, whey also represents a strategic opportunity within circular economy frameworks, offering a sustainable pathway to revalorize dairy industry byproducts and reduce environmental impact.

The relevance of leucine amino acid as a potent activator of the mTOR signaling pathway has placed it at the center of strategies aimed at stimulating muscle protein synthesis (MPS). Whey protein, due to its high leucine concentration, has been shown to induce stronger anabolic responses compared to other protein sources such as casein and soy. These findings are particularly important for older adults, who experience anabolic resistance—a reduced sensitivity to anabolic stimuli such as dietary amino acids and resistance exercise. This is also relevant for athletes, who frequently enter a catabolic state following exercise, thereby increasing the need for readily available amino acids to shift the balance toward synthesis.

Furthermore, whey protein's bioactive compounds may exert additional health-promoting effects beyond muscle metabolism. These effects include modulation of immune function, antioxidant activity, and potential benefits in metabolic disorders, which are relevant to both aging populations and physically active individuals. Whey protein is a rich source of cysteine, a sulfur-containing amino acid that serves as a precursor for glutathione synthesis, one of the body's most important intracellular antioxidants. Elevated glutathione levels contribute to enhanced cellular redox status and protection against oxidative stress, which is linked to aging and chronic disease progression (Sadowska-Bartosz and Bartosz, 2014; Wu et al., 2004). However, most of these effects have been studied in vitro, and the physiological evidence from whey dietary intervention remains inconsistent (for a review, see Corrochano et al., 2018). Additionally, digestion of

Whey. https://doi.org/10.1016/B978-0-443-27648-4.00011-6

whey protein releases bioactive peptides with immunomodulatory properties, such as lactoferrin, alpha-lactalbumin, and beta-lactoglobulin derivatives, which can influence the innate and adaptive immune responses (Nagpal et al., 2011). In metabolic health, whey protein supplementation has demonstrated beneficial effects on glycemic control and lipid metabolism. Studies indicate that whey protein intake improves postprandial glycemia through mechanisms including increased insulin secretion and delayed gastric emptying (Chiang et al., 2022). These properties make whey protein a promising adjunct in managing metabolic syndrome components, type 2 diabetes, and obesity, conditions that disproportionately affect older adults. Together, these functional attributes position whey protein not only as an effective nutritional supplement for muscle maintenance but also as a multifunctional food ingredient that supports systemic health, particularly in populations experiencing aging-related physiological decline or heightened metabolic demands due to physical activity.

This chapter explores the dual potential of whey as a sustainable food resource and a valuable dietary protein, focusing on its applications in healthy aging and sports performance—two key areas where nutritional strategies can have profound impacts on quality of life and functional capacity.

2. Whey protein supplementation

Whey protein has received much attention for its role in muscle recovery and the prevention of muscle mass loss, particularly in older adults and athletes. Despite the widespread use of protein supplements, research findings on protein supplementation are often contradictory. Several factors contribute to these inconsistencies, including variations in exercise protocols, treatment durations, dietary patterns, measured response variables, and statistical analyses across studies. Moreover, a major limitation in the literature is that most studies focus on young athletic men, with relatively fewer investigations conducted in women and older populations.

The International Society of Sports Nutrition (ISSN) states:

> *An acute exercise stimulus, particularly resistance exercise, and protein ingestion both stimulate muscle protein synthesis (MPS) and are synergistic when protein consumption occurs before or after resistance exercise*
>
> (Jäger et al., 2017).

The Recommended Dietary Allowance (RDA) for protein is 0.8 g per kilogram of body weight. However, for individuals aiming to build muscle mass, research suggests a higher intake of approximately 0.25 g of high quality protein per kilogram of body weight per meal. Protein supplementation offers a practical way to increase protein intake without consuming full meals, commonly using powders, ready-to-drink shakes, and protein bars.

The amino acid profile is a critical factor in protein supplementation. In particular, leucine plays a central role in stimulating MPS (Wilkinson et al., 2013). Among the branched-chain amino acids (BCAAs), leucine has been shown to be particularly effective at stimulating protein synthesis, as evidenced by increased incorporation of radioactive phenylalanine into muscle proteins (Anthony et al., 2000). The effect is mediated by the mTOR signaling pathway, as treatment with rapamycin—an mTOR inhibitor—abolishes the leucine-induced stimulation of protein synthesis in rats. However, this response is independent of insulin, as leucine administration does not elevate insulin levels. Additionally, in healthy young men, both leucine and its metabolite β-hydroxy-β-methylbutyrate (HMB) enhance anabolic signaling via mTOR in an insulin-independent manner. The signaling effect of leucine appears to be more potent than that observed with HMB (Wilkinson et al., 2013).

Insulin and exercise interact to balance two opposing metabolic regulatory conditions. Insulin is secreted after food intake to promote the storage of glucose and fatty acids—glucose as glycogen and fatty acids as triglycerides. During exercise, fuel stores need to be mobilized, therefore, insulin secretion is suppressed. In the postexercise period, there is a need to restore glycogen stores. This process is regulated by an increase in muscle insulin sensitivity (for an extensive review see Richter et al., 2021). Whey protein has been shown to temporarily reduce postprandial insulin sensitivity—a response that aligns with the anabolic demands of protein synthesis—not only in young male athletes but also in postmenopausal women (Smith et al., 2015). In other words, nutritional anabolic responses are greater when protein intake is combined with exercise, facilitating muscle adaptation and repair.

Given leucine's potent role in stimulating muscle protein synthesis, identifying protein sources rich in leucine is essential. Whey protein contains a higher leucine concentration than casein and soy, highlighting its superior effectiveness in supporting muscle health (Rasmussen, 2008).

The high leucine content of whey protein plays a pivotal role in activating the mTOR signaling pathway, a key regulator of muscle protein synthesis, thereby reinforcing whey's effectiveness as a nutritional strategy to promote muscle maintenance and growth (Fig. 13.1).

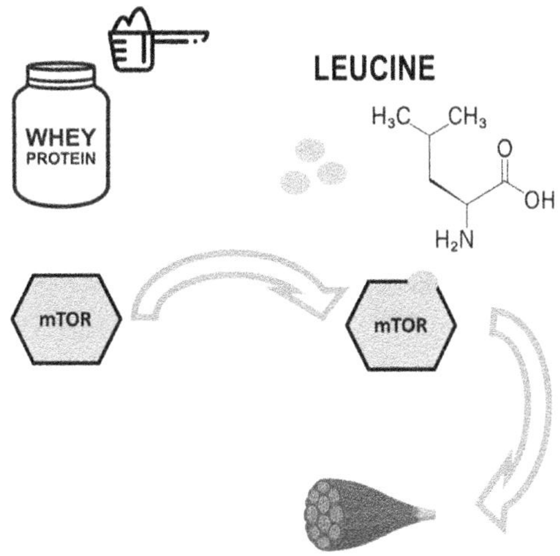

FIGURE 13.1 Graphical abstract: Whey protein is rich in leucine, a branched-chain amino acid that plays a key role in stimulating muscle protein synthesis (MPS) by activating the mTOR signaling pathway.

3. Gastric dynamics of whey, casein, and soy proteins

The physiological effects of dietary protein are strongly influenced not only by their amino acid composition but also by their digestion and absorption kinetics. This aspect, often referred to as gastric dynamics, plays a critical role in determining how quickly amino acids become available in the bloodstream, thereby influencing muscle protein synthesis (MPS), metabolic signaling, and recovery processes. Understanding these dynamics is particularly important in two physiologically distinct yet nutritionally demanding scenarios: athletic performance, where the goal is rapid muscle repair and growth, and in aging, where slowed digestion and anabolic resistance hinder the maintenance of muscle mass.

Whey and soy proteins are classified as fast-digesting proteins due to their high solubility in the acidic environment of the stomach. This property facilitates their rapid transit through the gastric system and prompt release into the small intestine, where proteolytic enzymes break them down into absorbable amino acids and small peptides. In contrast, casein protein behaves quite differently due to its unique physicochemical properties. At the stomach's acidic pH (approximately 1.5–3.5), casein is far below its isoelectric point (pH ~4.6), which promotes coagulation and gel formation, which slow gastric emptying (Boirie et al., 1997). Table 13.1 provides a summary of the comparative properties of whey, casein, and soy proteins.

The rate at which amino acids appear in the bloodstream significantly influences the effectiveness of muscle protein synthesis. Rapidly digested proteins, such as whey, produce a sharp and transient spike in plasma amino acids. Among all amino acids, leucine is a key trigger through activation of the mTOR signaling pathway. Slow proteins such as casein, on the other hand, result in more sustained aminoacidemia, potentially reducing muscle protein breakdown over a longer period but may not be as effective in acutely stimulating MPS (Boirie et al., 1997; Dangin et al., 2003). This dynamic becomes especially relevant postexercise, when the body is most responsive to anabolic stimuli. In studies comparing postexercise ingestion of whey, soy, and casein, both whey and soy proteins increased plasma insulin concentrations, enhancing nutrient uptake and recovery (Tang et al., 2009). However, only whey induced a robust increase in circulating essential amino acids (EAAs). The superior anabolic response observed with whey protein may thus stem from both its faster gastric emptying and its higher leucine content (constituting approximately 10%–11% of its total amino acids), compared to soy and casein.

In the context of aging, digestion kinetics are particularly critical. Older adults experience a phenomenon known as anabolic resistance—characterized by a reduced sensitivity of skeletal muscle to dietary amino acids. In this setting, the rapid and substantial increase in plasma leucine provided by fast proteins like whey may help overcome this resistance and more effectively stimulate MPS. Additionally, older individuals may have altered gastric emptying and digestive efficiency, further supporting the use of easily digestible proteins with high bioavailability.

In summary, the gastric dynamics of whey protein confer unique advantages for both athletes and older adults. Its rapid digestion promotes quick amino acid delivery and potent stimulation of muscle protein synthesis—key factors for optimizing recovery in sport and preserving muscle health with aging. These characteristics—combined with its high leucine content and digestibility—position whey as a superior dietary protein in a wide range of physiological contexts.

TABLE 13.1 Comparative properties of whey, casein, and soy proteins in relation to muscle health and supplementation strategies.

Characteristic	Whey protein	Casein protein	Soy protein	References
Digestion speed	Fast	Slow	Intermediate	Driskell (2007)
DIAAS^{+}	High	Excellent	High	Herreman et al. (2020)
BCAA^{++t}	High	Moderate	Lower	Driskell (2007)
Insulin response	High	Low	Moderate	Butteiger et al. (2013)
Glutamine	Moderate	High	Moderate	Butteiger et al. (2013)
Phytoestrogens^{+++}	None	None	Present ()	Cederroth et al. (2012)
Lactose	May contain	May contain	None	Driskell (2007)

DIAAS^{+}, digestible indispensable amino acid score, a protein quality measure that accounts for amino acid digestibility at the ileal level; *BCAA*$^{++t}$, branched-chain amino acids; *Phytoestrogens*$^{+++}$, in form of isoflavones.

4. Whey protein and sarcopenia

Age-related muscle loss, known as sarcopenia, is a progressive condition that significantly compromises physical function, reduces independence, and increases the risk of falls, frailty, and metabolic disorders. It results from a disruption in muscle protein turnover, where muscle protein breakdown (MPB) exceeds muscle protein synthesis (MPS)—a process further exacerbated by physical inactivity, hormonal changes, and reduced anabolic sensitivity to dietary protein, commonly referred to as anabolic resistance.

Among dietary interventions, protein supplementation has emerged as a key strategy to slow the progression of sarcopenia. Not all proteins are equally effective, however. Whey protein has demonstrated superior efficacy in stimulating MPS compared to other protein sources such as casein and soy. This is primarily attributed to its rapid digestion rate, high bioavailability, and elevated leucine content, a branched-chain amino acid (BCAA) that serves as a potent trigger for anabolic signaling pathways (Tang et al., 2009).

In older men aged 60–75 years, the dose-dependent effects of whey protein following resistance exercise have been explored. Doses of 30–40 g of whey protein were found to enhance p70S6K phosphorylation, a marker of mTORC1 pathway activation, which plays a central role in initiating muscle protein synthesis during postexercise recovery (D'Souza et al., 2014). Although 30 g of soy protein combined with carbohydrates produced a comparable anabolic signal at 2 h postexercise, it failed to sustain this response at 4 h, unlike whey (Mitchell et al., 2015). This finding suggests that whey protein supports more prolonged anabolic signaling, a critical factor in overcoming anabolic resistance in aging muscle.

Beyond molecular markers, the functional benefits of whey supplementation are equally compelling. Combined with resistance training, whey protein supplementation has been shown to improve both muscle mass and strength in older adults (Mertz et al., 2021). In a study of older women, consuming 35 g of whey protein either before or after resistance training over a 12-week period led to significant gains in skeletal muscle mass (SMM), strength, and functional performance, regardless of timing (Nabuco et al., 2018). These findings underscore the practical relevance of whey supplementation across sexes and different training regimens.

Further evidence supports whey's role in managing clinical sarcopenia. In individuals diagnosed with sarcopenia, daily supplementation with 22 g of whey protein alongside regular physical activity led to significant improvements in fat free mass, strength, and physical function, as well as reductions in inflammatory markers (Rondanelli et al., 2016). These antiinflammatory effects observed are particularly relevant given the role of chronic low-grade inflammation in the pathophysiology of sarcopenia.

While cross-study comparisons remain difficult due to variability in sample size, protein dose, study duration, and statistical methodology, the overall body of evidence supports the conclusion that whey protein supplementation enhances MPS and muscle function in older adults. Importantly, combining whey protein supplementation with resistance exercise is essential to effectively counteract anabolic resistance.

Implementing targeted whey protein supplementation strategies offers a practical and evidence based approach to preserve muscle mass, maintain physical function, and improve quality of life in aging populations. As the global demographic shift toward older age continues, such interventions will become increasingly important in reducing healthcare burdens and supporting independent living in later life.

5. Whey protein supplementation in sport

Resistance exercise, even in previously untrained individuals, induces a robust anabolic response, improving net muscle protein balance by increasing fractional synthesis rates (FSR) and suppressing protein breakdown (Phillips et al., 1997). This effect is amplified by the availability of dietary amino acids, which enhances muscle protein synthesis (MPS) through both systemic and intracellular mechanisms. On a systemic level, amino acid ingestion promotes increased blood flow to skeletal muscle, improving the delivery of amino acids to the target tissue (Biolo et al., 1997). At the cellular level, elevated plasma amino acid concentrations enhance muscle uptake and intracellular availability, facilitating the initiation of anabolic signaling (Mitchell et al., 2016). Together, these processes underpin the critical role of protein ingestion in optimizing the postexercise anabolic response.

However, resistance and endurance exercise induce a catabolic state, characterized by increased muscle protein breakdown (MPB). For athletes, particularly those engaged in high-frequency or high-intensity training, nutritional strategies that favor rapid amino acid availability are essential to shifting the net balance toward protein synthesis and recovery. Whey protein is uniquely suited to meet this demand due to its rapid gastric emptying, high digestibility, and rich leucine content—factors that enable a swift and potent stimulation of MPS following exercise.

The concept of nutrient timing, specifically the "anabolic window" following exercise, has been widely discussed in sports nutrition. Some guidelines recommend consuming protein within 2 h postexercise, arguing that this period represents a heightened anabolic environment in which muscle is more responsive to amino acid provision (Jäger et al., 2017). During this time, muscle tissue is prepared for repair and adaptation, due to increased insulin sensitivity, elevated translational activity, and greater membrane permeability for amino acid transporters. Consuming whey protein, which provides a substantial leucine bolus, has been shown to amplify the postexercise anabolic response (Rustad et al., 2016; Sollie et al., 2018).

Nevertheless, recent studies have questioned the exclusive importance of this narrow time frame. A growing body of evidence indicates that total daily protein intake, as well as even distribution of protein across meals, may have a greater impact on muscle recovery, hypertrophy, and performance outcomes than strict adherence to a postexercise protein window (Antonio et al., 2024; Jäger et al., 2017).

Importantly, the effectiveness of protein timing may vary by population. For instance, a study in postmenopausal women found that the timing of whey protein ingestion—either before or after resistance exercise—had no significant influence on changes in lean body mass, strength, or functional capacity over 12 weeks (de Branco et al., 2020). This suggests that longer-term nutritional strategies may be more important than the precise timing of protein intake in certain populations.

In summary, whey protein supplementation supports exercise adaptation and recovery through rapid amino acid delivery and potent stimulation of anabolic signaling, particularly via leucine mediated mTORC1 activation. While postexercise protein consumption remains a practical and evidence based strategy for enhancing muscle recovery, the emphasis is shifting toward ensuring adequate total protein intake and regular high quality protein dosing throughout the day. Whey protein, due to its digestibility, amino acid profile, and palatability, remains a cornerstone supplement in sports nutrition, suitable for athletes across training modalities and age groups.

6. Anabolic resistance

A critical challenge in aging and intensive training contexts is anabolic resistance, a diminished muscle sensitivity to dietary protein that impairs stimulation of muscle protein synthesis (MPS). This phenomenon arises from several factors, including mitochondrial dysfunction, chronic low-grade inflammation, and altered intracellular signaling pathways such as mTORC1 (Breen and Phillips, 2011). As a consequence, older adults and some athletes may require higher doses of high quality protein to elicit maximal anabolic responses.

Combining whey protein supplementation with resistance exercise synergistically improves muscle sensitivity to amino acids, thus helping to overcome anabolic resistance (Breen and Phillips, 2011).

Overall, addressing anabolic resistance through optimized whey protein dosing—both in quantity per meal and distribution across the day—alongside tailored exercise interventions is fundamental to maximizing muscle maintenance and performance in both healthy aging and athletic populations.

7. Whey protein products

Whey protein products vary significantly depending on the type of processing and the specific nutritional or functional needs of consumers. One technical challenge in whey processing, particularly with acid whey—produced when curd is formed by acidification—is its tendency to agglomerate and form lumps during drying, complicating handling and powder uniformity. To mitigate this, acid whey is often neutralized, and additives such as skim milk powder or cereal ingredients are incorporated to facilitate drying and improve powder flowability.

Advanced separation technologies such as nanofiltration and electrodialysis have been employed to remove lactic acid and salts effectively from whey. These processes not only reduce acidity but also improve the flavor profile and nutritional quality of the whey, enabling the production of higher grade protein ingredients suitable for specialized applications. The removal of salt and lactic acid also enhances drying efficiency and powder stability, which are critical for large-scale production.

Whey protein products are commonly classified into three major types—whey protein concentrate (WPC), whey protein isolate (WPI), and whey protein hydrolyzate (WPH)—each differing in processing intensity and nutritional composition (see Table 13.2). WPC is typically produced by ultrafiltration and contains 30%–80% protein by weight, retaining a notable amount of fats and lactose, which can impact both digestibility and allergenicity. WPI undergoes more rigorous processing such as cross-flow microfiltration or ion-exchange chromatography to remove most of the fat and lactose, resulting in protein concentrations of 90%–95%. This form is preferred for individuals seeking high protein purity with minimal lactose content. WPH is enzymatically hydrolyzed to break down peptide bonds, producing smaller peptides that can be absorbed more rapidly and may be less allergenic, making it attractive for clinical nutrition and sports supplementation.

The functional properties of whey proteins—such as solubility, emulsification, foaming, and gelation—are heavily influenced by the source of whey (sweet vs. acid), cheese production methods, transportation conditions, and processing parameters. Variations in the content of insoluble or denatured proteins can affect these functional properties and, consequently, the performance of whey proteins in food formulations or supplements (Onwulata et al., 2004). For instance, excessive heat treatment during processing can denature whey proteins, reducing their bioactivity and solubility.

Recent trends in commercial whey products include the development of native whey protein, which is extracted directly from skim milk rather than cheese whey, preserving the native protein structures and potentially offering superior functional and nutritional properties. Additionally, there is growing interest in whey protein blends that combine fast absorbing whey with slower-digesting proteins like casein, aiming to optimize muscle protein synthesis over longer periods.

Overall, the diversity of whey protein products allows them to meet a wide spectrum of dietary and functional needs—from athletes requiring rapid amino acid delivery to support muscle recovery, to older adults seeking to counteract sarcopenia, and individuals with lactose intolerance or milk allergies looking for suitable alternatives.

TABLE 13.2 Whey protein products.

Whey protein type	Processing method	Protein content	Other components	Functional properties	Typical uses
Whey Protein Concentrate (WPC)	Ultrafiltration	30%–80% by weight	Contains fats, lactose, minerals	Moderate solubility; nutritional profile preserved	General sports nutrition, food formulations
Whey Protein Isolate (WPI)	Cross-flow microfiltration or ion-exchange chromatography	90%–95% by weight	Low fat, low lactose	High purity; better solubility; reduced allergenicity	Lactose intolerant consumers, clinical nutrition, athletes
Whey Protein Hydrolyzate (WPH)	Enzymatic hydrolysis (predigested peptides)	80%–90% by weight	Very low fat and lactose	Rapid absorption; lower allergenicity; improved digestibility	Medical nutrition, infant formula, postexercise recovery

8. Circular economy: Valorizing whey for health and sustainability

In the global dairy industry, whey is often treated as a waste product rather than recognized as a valuable nutritional resource. It is estimated that 180–190 million tonnes of whey are produced annually worldwide, mostly as a byproduct of cheese manufacturing (Sambashivaiah et al., 2023). Alarmingly, up to 50% of this whey is discarded or underutilized, especially in small- and medium-scale dairy facilities. This practice not only reflects a lost economic opportunity, but also contributes to serious environmental contamination.

Whey has a high organic load, with Biological Oxygen Demand (BOD) levels of 35–45 kg/m^3 and Chemical Oxygen Demand (COD) of 60–80 kg/m^3, making it a significant pollutant if discharged untreated into aquatic systems. These values indicate that whey can rapidly deplete oxygen in water bodies, disrupt ecosystems, and contribute to eutrophication.

Beyond environmental impact, discarding whey represents a missed opportunity to address major public health concerns. Incorporating whey into the human food chain supports two critical needs. By redirecting whey from waste to nutritional use, the dairy sector can actively contribute to the circular bioeconomy, where byproducts are valorized rather than discarded. This approach reduces environmental pollution, creates new economic value from existing production streams, promotes functional food development aligned with health and wellness, supports the United Nations Sustainable Development Goals (SDGs), including:

SDG 2: End hunger and promote nutrition.
SDG 3: Ensure healthy lives at all ages.
SDG 12: Ensure sustainable production and consumption.

Valorizing whey protein bridges environmental sustainability with human health advancement. Its targeted application in the contexts of aging and athletic performance not only transforms a waste stream into a health-supporting asset, but also reflects a broader shift toward sustainable, circular food systems (Fig. 13.2).

9. Conclusion

The role of whey protein in muscle recovery, strength development, and the prevention of muscle loss is well supported by scientific evidence. Across diverse populations—from young athletes to older adults—whey protein supplementation has been shown to enhance muscle protein synthesis (MPS), improve recovery, and mitigate age-related muscle decline. In older adults, whey protein has demonstrated significant benefits in counteracting sarcopenia, improving skeletal muscle mass, strength, and functional capacity, particularly when combined with resistance exercise. Given the challenges of anabolic resistance in aging populations, a higher protein intake is necessary to effectively stimulate MPS and support muscle preservation. In athletic populations, whey protein stands up to other protein sources due to its rapid digestion,

FIGURE 13.2 Visual representation of whey in a circular economy framework. The illustration highlights whey as a by-product of cheese production, emphasizing its recovery and transformation into high quality protein supplements.

high leucine content, and ability to stimulate anabolic signaling. Research highlights its effectiveness in enhancing postexercise recovery, reducing muscle soreness, and maximizing muscle adaptations. Additionally, protein timing—particularly pre and postexercise ingestion—plays a crucial role in optimizing performance and recovery. Contrary to common concerns, high-protein diets do not pose a risk to kidney health in healthy individuals, reinforcing the safety of protein supplementation in physically active populations.

The commercial production of whey has also evolved to optimize its nutritional value and functionality. Various whey protein products are available, ranging from concentrates and isolates to hydrolysates, each with unique processing methods and applications. Innovations in acid whey processing, including nano-filtration and electro-dialysis, have improved flavor, drying efficiency, and mineral balance, making whey-derived products more versatile in both dietary and industrial applications. Valorizing whey, rather than discarding it, is crucial for promoting sustainability and reducing waste. As a nutrient-rich byproduct, whey offers diverse applications that can enhance the nutritional value of dairy systems. Its use in sports nutrition and healthy aging, particularly in supporting muscle health and protein synthesis, highlights its potential to improve both athletic performance and age-related muscle maintenance.

References

Anthony, J.C., Yoshizawa, F., Anthony, T.G., Vary, T.C., Jefferson, L.S., Kimball, S.R., 2000. Leucine stimulates translation initiation in skeletal muscle of postabsorptive rats via a rapamycin-sensitive pathway. J. Nutr. 130 (10), 2413–2419. https://doi.org/10.1093/jn/130.10.2413.

Antonio, J., Evans, C., Ferrando, A.A., Stout, J.R., Antonio, B., Cintineo, H.P., Harty, P., Arent, S.M., Candow, D.G., Forbes, S.C., Kerksick, C.M., Pereira, F., Gonzalez, D., Kreider, R.B., 2024. Common questions and misconceptions about protein supplementation: what does the scientific evidence really show? J. Int. Soc. Sports Nutr. 21 (1). https://doi.org/10.1080/15502783.2024.2341903.

Besediuk, V., Yatskov, M., Korchyk, N., Kucherova, A., Maletskyi, Z., 2024. Whey—from waste to a valuable resource. J. Agric. Food Res. 18, 101280. https://doi.org/10.1016/J.JAFR.2024.101280.

Biolo, G., Tipton, K.D., Klein, S., Wolfe, R.R., 1997. An abundant supply of amino acids enhances the metabolic effect of exercise on muscle protein. Am. J. Physiol. Endocrinol. Metabol. 273 (1), E122–E129. https://doi.org/10.1152/ajpendo.1997.273.1.E122.

Boirie, Y., Dangin, M., Gachon, P., Vasson, M.-P., Maubois, J.-L., Beaufrère, B., 1997. Slow and fast dietary proteins differently modulate postprandial protein accretion. Proc. Natl. Acad. Sci. 94 (26), 14930–14935. https://doi.org/10.1073/pnas.94.26.14930.

Breen, L., Phillips, S.M., 2011. Skeletal muscle protein metabolism in the elderly: interventions to counteract the "anabolic resistance" of ageing. Nutr. Metabol. 8 (1), 68. https://doi.org/10.1186/1743-7075-8-68.

Butteiger, D.N., Cope, M., Liu, P., Mukherjea, R., Volpi, E., Rasmussen, B.B., Krul, E.S., 2013. A soy, whey and caseinate blend extends postprandial skeletal muscle protein synthesis in rats. Clin. Nutr. 32 (4), 585–591. https://doi.org/10.1016/J.CLNU.2012.10.001.

Cederroth, C.R., Zimmermann, C., Nef, S., 2012. Soy, phytoestrogens and their impact on reproductive health. Mol. Cell. Endocrinol. 355 (2), 192–200. https://doi.org/10.1016/J.MCE.2011.05.049.

Chiang, S.-W., Liu, H.-W., Loh, E.-W., Tam, K.-W., Wang, J.-Y., Huang, W.-L., Kuan, Y.-C., 2022. Whey protein supplementation improves postprandial glycemia in persons with type 2 diabetes mellitus: a systematic review and meta-analysis of randomized controlled trials. Nutr. Res. 104, 44–54. https://doi.org/10.1016/j.nutres.2022.04.002.

Corrochano, A.R., Buckin, V., Kelly, P.M., Giblin, L., 2018. Invited review: whey proteins as antioxidants and promoters of cellular antioxidant pathways. J. Dairy Sci. 101 (6), 4747–4761. https://doi.org/10.3168/jds.2017-13618.

D'Souza, R.F., Markworth, J.F., Figueiredo, V.C., Della Gatta, P.A., Petersen, A.C., Mitchell, C.J., Cameron-Smith, D., 2014. Dose-dependent increases in p70S6K phosphorylation and intramuscular branched-chain amino acids in older men following resistance exercise and protein intake. Physiol. Rep. 2 (8), e12112. https://doi.org/10.14814/phy2.12112.

Dangin, M., Guillet, C., Garcia-Rodenas, C., Gachon, P., Bouteloup-Demange, C., Reiffers-Magnani, K., Fauquant, J., Ballèvre, O., Beaufrère, B., 2003. The rate of protein digestion affects protein gain differently during aging in humans. J. Physiol. 549 (2), 635–644. https://doi.org/10.1113/jphysiol.2002.036897.

de Branco, F.M.S., Carneiro, M.A.S., Rossato, L.T., Nahas, P.C., Teixeira, K.R.C., de Oliveira, G.N., Orsatti, F.L., de Oliveira, E.P., 2020. Protein timing has no effect on lean mass, strength and functional capacity gains induced by resistance exercise in postmenopausal women: a randomized clinical trial. Clin. Nutr. 39 (1), 57–66. https://doi.org/10.1016/j.clnu.2019.01.008.

Driskell, J.A. (Ed.), 2007. Sports Nutrition. CRC Press. https://doi.org/10.1201/9781420008500.

Food and Agriculture Organization of the United Nations, 2009. How to Feed the World in 2050. FAO. https://www.fao.org/fileadmin/templates/wsfs/docs/expert_paper/How_to_Feed_the_World_in_2050.pdf.

Herreman, L., Nommensen, P., Pennings, B., Laus, M.C., 2020. Comprehensive overview of the quality of plant- and animal-sourced proteins based on the digestible indispensable amino acid score. Food Sci. Nutr. 8 (10), 5379–5391. https://doi.org/10.1002/fsn3.1809.

Jäger, R., Kerksick, C.M., Campbell, B.I., Cribb, P.J., Wells, S.D., Skwiat, T.M., Purpura, M., Ziegenfuss, T.N., Ferrando, A.A., Arent, S.M., Smith-Ryan, A.E., Stout, J.R., Arciero, P.J., Ormsbee, M.J., Taylor, L.W., Wilborn, C.D., Kalman, D.S., Kreider, R.B., Willoughby, D.S., Antonio, J., 2017. International society of sports nutrition position stand: protein and exercise. J. Int. Soc. Sports Nutr. 14 (1). https://doi.org/10.1186/s12970-017-0177-8.

Mertz, K.H., Reitelseder, S., Bechshoeft, R., Bulow, J., Højfeldt, G., Jensen, M., Schacht, S.R., Lind, M.V., Rasmussen, M.A., Mikkelsen, U.R., Tetens, I., Engelsen, S.B., Nielsen, D.S., Jespersen, A.P., Holm, L., 2021. The effect of daily protein supplementation, with or without resistance training for 1 year, on muscle size, strength, and function in healthy older adults: a randomized controlled trial. Am. J. Clin. Nutr. 113 (4), 790–800. https://doi.org/10.1093/ajcn/nqaa372.

Mitchell, C.J., Della Gatta, P.A., Petersen, A.C., Cameron-Smith, D., Markworth, J.F., 2015. Soy protein ingestion results in less prolonged p70S6 kinase phosphorylation compared to whey protein after resistance exercise in older men. J. Int. Soc. Sports Nutr. 12 (1). https://doi.org/10.1186/s12970-015-0070-2.

Mitchell, W.K., Wilkinson, D.J., Phillips, B.E., Lund, J.N., Smith, K., Atherton, P.J., 2016. Human skeletal muscle protein metabolism responses to amino acid nutrition. Adv. Nutr. 7 (4), 828S–838S. https://doi.org/10.3945/AN.115.011650.

Nabuco, H.C.G., Tomeleri, C.M., Sugihara Junior, P., Fernandes, R.R., Cavalcante, E.F., Antunes, M., Ribeiro, A.S., Teixeira, D.C., Silva, A.M., Sardinha, L.B., Cyrino, E.S., 2018. Effects of whey protein supplementation pre- or post-resistance training on muscle mass, muscular strength, and functional capacity in pre-conditioned older women: a randomized clinical trial. Nutrients 10 (5), 563. https://doi.org/10.3390/nu10050563.

Nagpal, R., Behare, P., Rana, R., Kumar, A., Kumar, M., Arora, S., Morotta, F., Jain, S., Yadav, H., 2011. Bioactive peptides derived from milk proteins and their health beneficial potentials: an update. Food Funct. 2 (1), 18–27. https://doi.org/10.1039/C0FO00016G.

Onwulata, C.I., Konstance, R.P., Tomasula, P.M., 2004. Minimizing variations in functionality of whey protein concentrates from different sources. J. Dairy Sci. 87 (3), 749–756. https://doi.org/10.3168/jds.S0022-0302(04)73218-X.

Phillips, S.M., Tipton, K.D., Aarsland, A., Wolf, S.E., Wolfe, R.R., 1997. Mixed muscle protein synthesis and breakdown after resistance exercise in humans. Am. J. Physiol. Endocrinol. Metabol. 273 (1), E99–E107. https://doi.org/10.1152/ajpendo.1997.273.1.E99.

Rasmussen, C.J., 2008. Nutritional supplements for endurance athletes nutritional supplements. In: Greenwood, M., Kalman, D., Antonio, J. (Eds.), Sports and Exercise. Humana Press, pp. 369–407.

Richter, E.A., Sylow, L., Hargreaves, M., 2021. Interactions between insulin and exercise. Biochem. J. 478 (21), 3827–3846. https://doi.org/10.1042/BCJ20210185.

Rondanelli, M., Klersy, C., Terracol, G., Talluri, J., Maugeri, R., Guido, D., Faliva, M.A., Solerte, B.S., Fioravanti, M., Lukaski, H., Perna, S., 2016. Whey protein, amino acids, and vitamin D supplementation with physical activity increases fat-free mass and strength, functionality, and quality of life and decreases inflammation in sarcopenic elderly. Am. J. Clin. Nutr. 103 (3), 830–840. https://doi.org/10.3945/ajcn.115.113357.

Rustad, P.I., Sailer, M., Cumming, K.T., Jeppesen, P.B., Kolnes, K.J., Sollie, O., Franch, J., Ivy, J.L., Daniel, H., Jensen, J., 2016. Intake of protein plus carbohydrate during the first two hours after exhaustive cycling improves performance the following day. PLoS One 11 (4), e0153229. https://doi.org/10.1371/journal.pone.0153229.

Sadowska-Bartosz, I., Bartosz, G., 2014. Effect of antioxidants supplementation on aging and longevity. BioMed Res. Int. 2014, 1–17. https://doi.org/10.1155/2014/404680.

Sambashivaiah, S., Cope, M., Mukherjea, R., Selvam, S., George, N., Kuriyan, R., Kurpad, A.V., 2023. The effect of soy and whey protein supplementation on glucose homeostasis in healthy normal weight Asian Indians. J. Nutr. Metab. 2023, 1–9. https://doi.org/10.1155/2023/2622057.

Smith, G.I., Yoshino, J., Stromsdorfer, K.L., Klein, S.J., Magkos, F., Reeds, D.N., Klein, S., Mittendorfer, B., 2015. Protein ingestion induces muscle insulin resistance independent of leucine-mediated mTOR activation. Diabetes 64 (5), 1555–1563. https://doi.org/10.2337/db14-1279.

Sollie, O., Jeppesen, P.B., Tangen, D.S., Jernerén, F., Nellemann, B., Valsdottir, D., Madsen, K., Turner, C., Refsum, H., Skålhegg, B.S., Ivy, J.L., Jensen, J., 2018. Protein intake in the early recovery period after exhaustive exercise improves performance the following day. J. Appl. Physiol. 125 (6), 1731–1742. https://doi.org/10.1152/japplphysiol.01132.2017.

Tang, J.E., Moore, D.R., Kujbida, G.W., Tarnopolsky, M.A., Phillips, S.M., 2009. Ingestion of whey hydrolysate, casein, or soy protein isolate: effects on mixed muscle protein synthesis at rest and following resistance exercise in young men. J. Appl. Physiol. 107 (3), 987–992. https://doi.org/10.1152/japplphysiol.00076.2009.

Wilkinson, D.J., Hossain, T., Hill, D.S., Phillips, B.E., Crossland, H., Williams, J., Loughna, P., Churchward-Venne, T.A., Breen, L., Phillips, S.M., Etheridge, T., Rathmacher, J.A., Smith, K., Szewczyk, N.J., Atherton, P.J., 2013. Effects of leucine and its metabolite β-hydroxy-β-methylbutyrate on human skeletal muscle protein metabolism. J. Physiol. 591 (11), 2911–2923. https://doi.org/10.1113/jphysiol.2013.253203.

Wu, G., Lupton, J.R., Turner, N.D., Fang, Y.-Z., Yang, S., 2004. Glutathione metabolism and its implications for health. J. Nutr. 134 (3), 489–492. https://doi.org/10.1093/jn/134.3.489.

Chapter 14

Edible films and coatings

Essam Hebishy[1], Suresh Gurusamy[2] and Ayten A. Tas[2, 3]

[1]*Centre d'Innovació, Recerca i Transferència en Tecnologia dels Aliments (CIRTTA), TECNIO, Departament de Ciència Animal i del Aliments, Facultat de Veterinària, Universitat Autònoma de Barcelona, Bellaterra, Spain;* [2]*Centre of Excellence in Agri-food Technologies, National Centre for Food Manufacturing, College of Sciences, University of Lincoln, Spalding, United Kingdom;* [3]*Department of Health Professions, Faculty of Health and Education, Manchester Metropolitan University, Manchester, United Kingdom*

Abbreviations

ACE Angiotensin-converting enzyme
BSA Bovine serum albumin
CFU Colony-forming unit
CMC Carboxy methyl cellulose
EAB Elongation at break
EDTA Ethylene diamine tetraacetic acid
EFC Edible films and coatings
EO Essential oil
GABA Gamma-amino butyric acid
GWP Global warming potential
HPP Hydrostatic pressure processing
HT Heat treatment
LAB Lactic acid bacteria
LDL Low-density lipoprotein
NK Natural killer (cells)
OP Oxygen permeability
PEG Polyethylene glycol
PET Polyethylene
RH Relative humidity
SDGs Sustainable Development Goals
TBARS Thiobarbituric Acid Reactive Species
TS Tensile strength
UF Ultrafiltration
UV Ultraviolet
WP Whey protein
WPC Whey protein concentrate
WPI Whey protein isolate
WVP Water vapor permeability

1. Introduction

Modern consumption patterns have led to an increase in solid waste generation, with packaging being a significant contributor to this trend. The efforts to reduce solid waste focus on novel waste disposal methods and sustainable packaging alternatives (Gagliarini et al., 2019; Rodriguez-Turienzo et al., 2013). Whey, the liquid drained from cheese curd after milk is coagulated, poses environmental and economic challenges due to its high organic matter content. Treatment for subsequent disposal is time-consuming and laborious (Amaral and Silva, 2021). The global whey protein

Whey. https://doi.org/10.1016/B978-0-443-27648-4.00013-X

(WP) market is growing rapidly, with significant efforts being made to valorize whey to meet Sustainable Development Goals (SDGs) (Buchanan et al., 2023; Macwan et al., 2016).

Food supply chains are often unsustainable, with dairy products having a substantial environmental impact due to resource depletion and greenhouse gas emissions. The dairy industry also generates wastewater and pollutants (Kim et al., 2013; Mirabella et al., 2014; Prazeres et al., 2012; Roma et al., 2015; Sala et al., 2012). Cheese whey is one of the most significant dairy streams and has a large environmental impact.

Packaging is essential for extending the shelf life of products and ensuring safe transportation. There is a growing interest in using biodegradable and environmentally friendly materials derived from sustainable sources for food packaging (Han, 2014). Edible films and coatings made from lipids, polysaccharides, and proteins provide excellent barrier properties, enhancing food preservation and extending the shelf life of foods. Milk proteins, including WPs, are particularly notable for their performance in edible films and coatings, replacing fossil-based plastics. Utilizing milk proteins in edible films and coatings can address food packaging needs and waste management concerns (Guimarães et al., 2018).

WP coatings have shown potential for improving food safety while meeting the growing consumer demand for environmentally friendly packaging alternatives. Additionally, WP films are highly transparent, biodegradable, and suitable for various food applications. Despite extensive research on the use of WP in edible coatings and films, significant gaps remain in understanding and optimizing its properties for widespread industrial use. More comprehensive research is needed to integrate the latest advancements in additives and film formation methods. Moreover, limited research quantifies the biodegradability of WP films and their ecological footprint compared to conventional packaging materials.

This chapter provides a comprehensive overview of the current research landscape regarding whey and whey proteins in food packaging. The chapter comprehensively covers the state-of-the-art developments and related research on WP-based edible films, their processing techniques, and potential uses in the food industry. It discusses recent advancements in the formulation and use of WP-based films, focusing on incorporating antimicrobial agents, plasticizers, and active compounds such as essential oils (EOs) into their formulation and the impact this would have on quality, shelf life and safety. Various studies highlight the effectiveness of WP films in inhibiting microbial growth, enhancing the mechanical strength of the packaging, and offering excellent oxygen and water vapor barrier properties. Future innovations are expected to focus on multifunctional films simultaneously delivering several active compounds, such as antioxidants and antimicrobials.

2. Whey and whey proteins

2.1 Overview of whey and whey proteins

Whey proteins (WPs) have emerged as a promising resource for sustainable food packaging. Whey is the serum fraction of milk that remains after the curd separates due to casein coagulation from milk by acidic agents or proteolytic enzymes. The casein removal process determines the type and composition of the whey produced. Table 14.1 shows the classification of whey based on pH and titratable acidity (Macwan et al., 2016).

WPs can be available in three forms such as whey protein concentrate (WPC, containing 29%–89% protein, fat and lactose), whey protein isolate (WPI, containing 90%–95% protein, virtually no fat and minimal lactose (<1%)) and whey protein hydrolyzate (WPH, this is the semi-digested form of the protein to reduce allergenicity) (Gangurde et al., 2011; Patel, 2015).

2.2 Protein recovery methods from whey

WP production involves selectively recovering and concentrating proteins from liquid whey by removing lactose, lipids, and mineral salts to achieve high purity and functional integrity. Traditional and emerging separation technologies, such

TABLE 14.1 Types of whey based on acidity level and pH (Macwan et al., 2016).

Type of whey	pH	Titratable acidity (%)
Acid whey	4.0–5.0	0.40–0.60
Medium acid whey	5.0–5.8	0.20–0.40
Sweet whey	5.5–6.6	0.10–0.20

as ultrafiltration, nanofiltration, ion-exchange chromatography, electrodialysis, and selective precipitation, remain central to this process.

Recent advancements have introduced innovative nonthermal and hybrid processing methods, including high hydrostatic pressure (HHP), ultrasonication, and extrusion, demonstrating significant improvements in protein functional properties, stability and nutritional value (Kheto et al., 2024). These techniques, while effective, require careful control of the processing conditions to maintain the integrity of protein structure, stability and bioactivity (Wang and Guo, 2019; Kheto et al., 2024).

Separation methods predominantly rely on membrane-based technologies, with ultrafiltration (UF) representing the primary industrial method for separating WPs. In UF, hydrostatic pressure drives the liquid phase against a semi-permeable membrane, resulting in the selective retention of proteins while allowing water and small molecules to permeate through the membrane. Nanofiltration, microfiltration and reverse osmosis are the other membrane separation techniques (Ganju and Gogate, 2017).

Numerous novel techniques have been developed for whey protein recovery, aiming to enhance the efficiency and yield of the extraction process. These include acid precipitation, salting-out using high salt concentrations, membrane distillation utilizing a temperature gradient, electrostatic interactions between the functional groups of protein molecules and anion exchange membranes, and surfactant-coated bubbles that aid in separating whey proteins by attracting the proteins to their surfaces. Affinity chromatography and ion-exchange techniques offer high selectivity for isolating specific protein fractions (Aslam et al., 2021).

After isolation, spray drying and freeze drying are the primary methods to produce stable WP powders for commercial use (Ganju and Gogate, 2017; Wang and Guo, 2019).

2.3 Whey protein fractions and their composition

WPs have a diverse range of protein fractions. β-lactoglobulin is the primary constituent of whey, produced in the mammary glands and released into milk (Chen et al., 2023). α-Lactalbumin comprises 123 amino acids in one peptide chain. Bovine serum albumin (BSA), a single-chain globular non-glycoprotein, is one of the most abundant elements in plasma. Serum albumins, the most abundant protein soluble in the circulatory system, serve a variety of physiological roles. Lactoferrin is found in dairy products and, to a smaller extent, in exocrine fluids, including bile and tears. Bovine immunoglobulin G is a significant component of the immune activity in colostrum (Aslam et al., 2021). WP also contains several proteins in low quantities but with high activity as growth regulators, such as lactoperoxidase, milk fat globular membrane proteins, and vitamin-binding proteins (Minj and Anand, 2020). The composition of each fraction is shown in Table 14.2.

2.4 Health benefits of whey proteins

Whey proteins (WPs) are rich in branched-chain amino acids (20%–25%), particularly leucine, isoleucine, and valine, which are essential for muscle metabolism and recovery during intense exercise (Brestenský et al., 2015; Lam et al.,

TABLE 14.2 Whey protein fractions and their composition, concentration and molecular weight (Brew et al., 1970; Madureira et al., 2007).

WP constituent	Number of amino acid residues	Concentration (g/L)	Molecular weight (kDa)
β-Lactoglobulin	162	1.3	18,277
α-Lactalbumin	123	1.2	14,175
Glycomacropeptide	64	1.2	6700
Bovine serum albumin	582	0.4	66,267
Bovine lactoferrin	700	0.1	80,000
Bovine lactoperoxidase	612	0.03	70,000
Milk fat globular membrane proteins	200	0.02	25,000
Vitamin-binding proteins	300	0.01	45,000

2019). These amino acids are rapidly absorbed and support muscle protein synthesis, making WPs valuable in sports and clinical nutrition.

Whey proteins are highly bioavailable due to their excellent amino acid profile, high solubility and ease of digestion. They also enhanced glutathione synthesis, a potent intracellular antioxidant that supports immune function and cellular detoxification, thereby reducing oxidative stress and inflammation associated with aging and chronic disease (Frestedt et al., 2008). Additionally, WPs modulate immune responses by promoting the production of immunoglobulin G4 and enhancing the activity of natural killer cells (Aslam et al., 2021).

WPs exhibit cardioprotective effects by lowering LDL cholesterol and reducing the risk of atherosclerosis (Gangurde et al., 2011). Its bioactive peptides inhibit angiotensin-converting enzyme (ACE), aiding in blood pressure regulation (Patel, 2015). Components such as lactoferrin and immunoglobulins exhibit antibacterial and antiviral activity, helping to inhibit pathogens in the gastrointestinal tract (Aslam et al., 2021). They also support liver protection, enhance immune and digestive functions, reduce blood pressure, and protect against gastric mucosal injury (Gangurde et al., 2011).

3. Whey as a waste and related environmental challenges

Historically, whey was regarded as having little usage, and it was mainly used as animal feed (Aslam et al., 2021). In 2024, the global WP market size was estimated at 12.64 billion USD, projected to increase to 22.63 billion USD by 2032 (Fortune Business Insights, 2025). Efforts have been made to find sustainable uses of whey beyond animal feed. Whey valorization supports economic strategies and environmental goals, which align with the United Nations' Sustainable Development Goals (SDGs) for clean water, industry innovation, and responsible consumption (United Nations, 2024). Modern lifestyles, consumption patterns, and technologies have led to an increase in solid waste generation (Rodriguez-Turienzo et al., 2013). Packaging accounts for 15%–20% of solid waste collected in some countries and up to 30% in the more industrialized ones. Efforts to reduce solid waste have focused on two main areas: (1) novel waste disposal methods and (2) sustainable packaging alternatives (Gagliarini et al., 2019). Humans consume more resources than our planet can replenish (Sala et al., 2012). Many food supply chains, especially in dairy production, are unsustainable due to the negative environmental impact (Roma et al., 2015). While dairy products are nutritious, they contribute to resource depletion, greenhouse gas emissions, and pollution (Kim et al., 2013).

Milk production is particularly damaging and can lead to acidification, eutrophication, and a high global warming potential (GWP) (González-García et al., 2013; Mirabella et al., 2014; Prazeres et al., 2012). Cheese production and packaging have a significant environmental impact and energy demand, consuming 8%–12% of energy and contributing around 3%–4% of the GWP in the industry. The carbon footprint of whey powder is higher than that of commonly consumed meats and plant-based food products (such as chicken, pork, and eggs), although it is lower than that of beef and lamb (van Middelaar et al., 2011). Cheese manufacturing generates large quantities of whey, which requires proper disposal treatment to minimize environmental impact (Prazeres et al., 2012; Rajeshwari et al., 2000; Rivas et al., 2011). Regulations in many countries have prompted the dairy industry to seek sustainable waste management methods for whey, including its use as animal feed (Prazeres et al., 2012).

4. Sustainable food packaging

Packaging is mainly used to extend the shelf life of food products. This becomes particularly important when foods must be shipped over long distances, as their nutritional content, safety, and quality can be compromised during transit. Food packaging has undergone significant evolution in recent years, incorporating new features and adapting to market needs, particularly for perishable food products. Packaging materials must be reliable in terms of non-toxicity, microbiological stability, and physical and barrier capabilities; they should have acceptable sensory qualities and product compatibility and be manufactured cost-effectively.

There has been increased interest in employing edible dietary components and biopolymers derived from sustainable sources or commercial by-products as packaging materials. These biopolymers are entirely biodegradable, environmentally friendly, and superior to petroleum-based compounds routinely used in the food industry (Han, 2014). Studies on environmentally friendly packaging materials have paved the way for studies on bio-based packaging.

5. Edible films and coatings

Edible films and coatings (EFCs) are thin layers of edible material used to pack or cover food, preserving it during distribution and marketing. These layers help protect the product against mechanical harm and biological, chemical, and microbial activity (Wu et al., 2024). Price, functional attributes, accessibility, optical properties, gas barrier effect, water resistance, mechanical and antimicrobial properties, and sensory appeal are the factors that influence the use of EFCs in food-related applications, particularly for extremely delicate items such as horticultural products (Kumar et al., 2022).

EFCs are environmentally friendly alternatives to traditional polymers that help prevent ecological harm. They are sustainable, biodegradable, and edible. They have the same functions as artificial packaging. However, selecting ingredients for EFCs should be customized to particular food applications, considering the principal drivers of quality degradation (Kumar et al., 2022).

A perfect edible film must have the following characteristics: (1) Be free of allergens, and non-digestible and hazardous components (2) Offer rigidity to avoid mechanical harm during shipment, handling, and display (3) Stick firmly to the food surface, to ensure consistent coverage (4) Regulate the movement of water to keep the desired water content of the preserved food (5) Be partially permeable to maintain the proper ratio of both anaerobic and aerobic gases, avoiding aging (6) Keep the aroma, nutritional, and biological aspects of the food without affecting its flavor or appearance (7) Limit deterioration, pesticide invasion, microbal growth, and various other damage on the food surface (8) Improve the product's esthetic and sensory qualities (9) Ideally function as vehicle to deliver vitamins, minerals, taste, and colors; (10) Limit antioxidant substances and antimicrobial agents to the surface, reducing cost and invasive taste, and lastly (11) Be easy to manufacture and cost-effective (Umaraw et al., 2020).

Edible packaging is made from lipids, polysaccharides and proteins, alone or in combination. Edible film (Assefa and Admassu, 2013) acids (Assefa and Admassu, 2013).

Proteins are ideal for biodegradable films and coatings due to their favorable physical and optical properties. They are excellent barriers to oxygen, smells, and chemical substances. They, alone or in combination with other substances, can be added to impart new properties or enhance existing ones, such as antibacterial, antioxidant, improved taste, and color. Several protein compounds, including animal proteins (fish proteins, ovalbumen, casein, gelatine, collagen), and plant sources (wheat, zein, soya and soyflour, gluten, peanuts, corn, rice, and cottonseed), have been studied with EFCs (Chen et al., 2019). Milk proteins, including casein and WPs, are especially notable for their outstanding water solubility and emulsifying capacity, enhancing their performance in EFCs while increasing nutritional value (Kandasamy et al., 2021). The protein-based films are well-suited for hydrophilic meat surfaces, providing oxygen and carbon dioxide barriers while allowing water vapor to permeate; however, incorporating hydrophobic compounds and plasticizers can significantly enhance their moisture resistance (Chen et al., 2019; Wagh et al., 2014).

Polysaccharide-based films provide limited moisture protection but are selectively permeable to oxygen and carbon dioxide, while remaining impermeable to fats and oils. Cellulose, pectin, converted starch, alginate, carrageenan, agar, gums, chitosan, and pullulan can all be used to create these films. Starch is especially promising because of its ability to degrade, accessibility, cost-effectiveness, sustainability, and low toxicity, making it a viable alternative to plastic packaging (Sanyang et al., 2015).

Lipids serving as a biological film base include oils and fats from animals and plants, waxes, organic resins, and surface-active substances. Lipid based films are frequently mixed with hydrocolloids, including proteins, cellulose, starch, and derivatives, to improve film characteristics (Galus and Kadzińska, 2015).

Compared to synthetic polymers, EFCs offer superior biodegradability and oxygen barrier properties, but they often exhibit poor water vapor resistance due to their hydrophilic nature (Abdulla et al., 2024; Saklani et al., 2019). A contemporary technique for developing edible and recyclable films involves creating composite films that combine the advantages of various materials, including lipids, polysaccharides, and proteins, thereby overcoming these limitations. This results in improved mechanical strength, flexibility, barrier properties and protection against moisture migration, gases, and microbial activity. As with artificial polymers, the structural and barrier capabilities of composite biological films strongly rely on the features and compatibility of their constituent polymers (Wu et al., 2024). Composite films also allow for the customization of properties to suit specific applications, offering better performance than single-component films. Additionally, they can be optimized for cost-effectiveness and functionality, making them ideal for diverse packaging and preservation needs (Chen et al., 2019; Rubilar et al., 2015).

Recent advancements in EFCs include smart coatings and nanostructured films (NSFs). NSFs incorporate nanoparticles (NPs), such as silver NPs, to enhance the mechanical, barrier, and antimicrobial properties of the films (Wang et al., 2025). Smart coatings may include pH-sensitive or temperature-responsive probes or indicators for real-time monitoring of a food product's freshness (Filho et al., 2021).

The production of EFCs at a lab scale has shown great promise; however, scaling up to an industrial scale is challenging. This stems from high energy requirements and costs, as well as the high prices of biopolymers. Additionally, active additives such as essential oils and nanoparticles may require strict regulatory considerations before use, which further complicates the process. Finally, from a sensory point of view, EFCs must offer transparency, maintain the taste and aroma, and provide clear information about what is included and the benefits they bring to the product (Jeya Jeevahan et al., 2020; Weng et al., 2024).

Future directions in the evolution of EFCs may bring multifunctionality, combining the enhancement of barrier properties with the delivery of active additives. A strong emphasis on using food waste as a starting material in sustainable EFCs production would also integrate circular economy principles in the process. Environmental impact is expected to become the central focus, and life cycle assessment (LCA) will be employed to quantify the ecological benefits of EFCs, compared to plastic packaging. Future research on smart and active packaging is also on the horizon, involving the implementation of new sensors to measure the release of active compounds and monitor food safety and quality in real-time. These advancements will collectively redefine the landscape of food preservation, making it more adaptive, sustainable, and consumer-friendly (Gupta et al., 2024; Kumar et al., 2023).

6. Whey protein edible films and coatings

6.1 Preparation methods of whey protein edible films and coatings

Traditional methods for creating edible films are categorized into two main processes: wet and dry. The wet process involves dissolving the polymer and spreading it on a smooth, flat surface using solvents, with controlled drying to remove the solvent. This approach is highly energy-intensive (due to oven drying) and better suited to laboratory settings with limited commercial applications.

Injection, blow molding, extrusion, and heat pressing are popular dry processes for making edible films (Fig. 14.1) (Mellinas et al., 2016). One of the dry methods, extrusion, is a faster and more efficient approach that reduces the cost of creating biopolymer films compared to artificial polymers. Extrusion and compression molding are well-known commercial processes for producing films. Transparent, elastic WP edible films are subjected to a thermal-compression molding stage before extrusion, which is an important step that is both cost and time-efficient. Extruded WP films can be mass-made into pouches to store and serve various foods and dry materials. Thermal-plastic extrusion is a technique for creating protein casings and films that do not require the addition or removal of solvents. The bulk of the raw material is melted and forced across the extruder die into the desired shape during continuous extrusion. The heat energy is provided by the heat source (barrel/die heaters) and the mechanical work generated by friction between the screws (Abeykoon et al., 2014; Kandasamy et al., 2021).

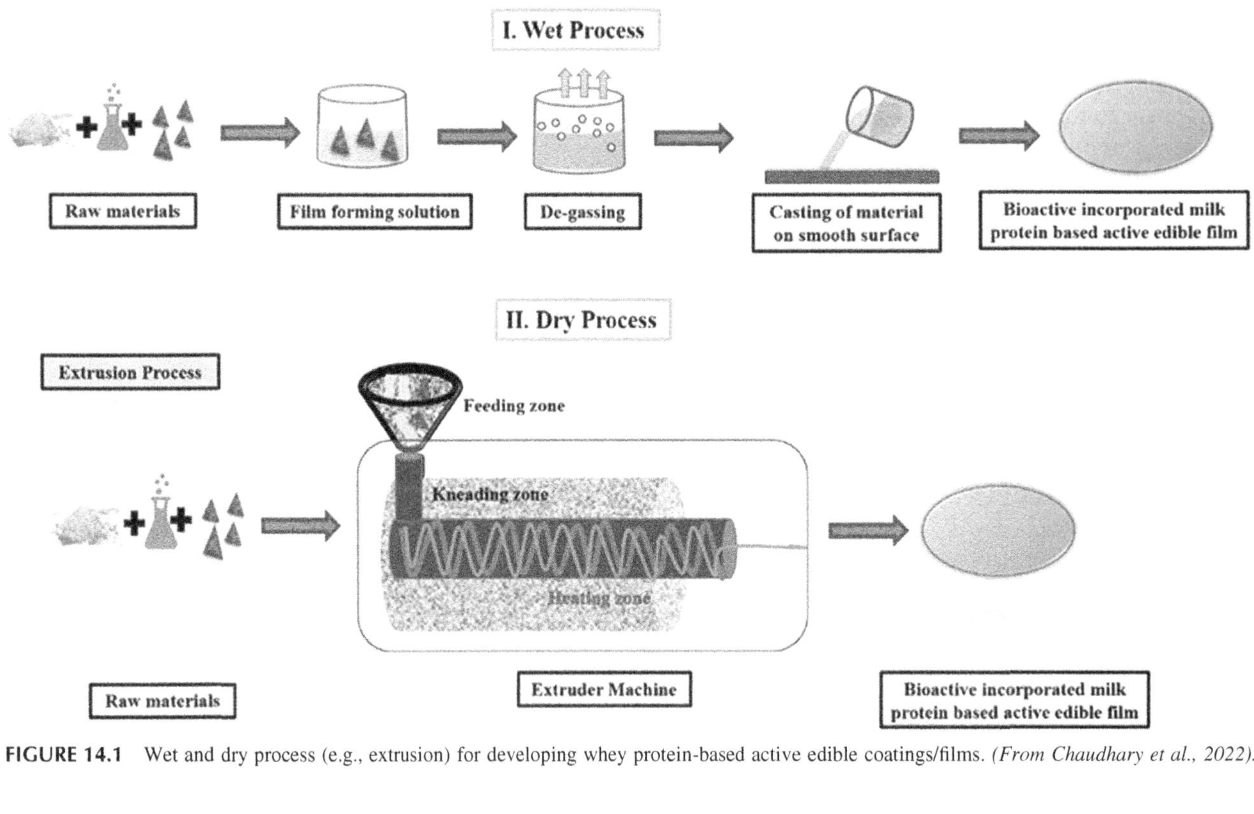

FIGURE 14.1 Wet and dry process (e.g., extrusion) for developing whey protein-based active edible coatings/films. *(From Chaudhary et al., 2022).*

WP films are primarily produced via the casting method, in which a film solution is dropped onto flat, smooth surfaces to form a solidified gel that is subsequently utilized for wrapping foods. WP films produce a highly interacting polymer matrix with a three-dimensional gel network, which can incorporate other film-forming chemicals. WP-based EFCs are used to package various foods, including peanuts, walnuts, frozen fish, fruits, and breakfast cereals, to increase moisture, scent, gas and fat protection (Henriques et al., 2015). WP-based films are often more transparent than artificial films, and their quality attributes are determined by their thickness and the inclusion of plasticizers.

The production of WP edible films has been widely studied due to their potential to enhance food preservation while being environmentally friendly. They are produced by extracting WPs from cheese whey, followed by heat precipitation or UF methods to concentrate the proteins. The protein solution is then cast into films, which are dried to form thin, flexible sheets that serve as protective layers for food products. Although film production has not extensively explored the thermoplastic behavior of specific proteins, the extrusion process relies on the thermoplastic behavior of raw materials, including plasticizing agents and chemical additives, which have been thoroughly studied for WP. Some studies used glycerol as a plasticizer to create extruded, clear, versatile, and uniform WP sheets (Pan et al., 2025; Weng et al., 2024).

The extruder configuration and operating conditions enable the heat denaturation and cross-linking of WP sheets, resulting in mechanical properties equivalent to or improved over those of heat-denatured films in solvent casting. A study found that the quantity of plasticizer, which was greater than that of solvent-cast films, had no impact on the stretching of extruded films. Increased tensile strength (TS) was attained (Weng et al., 2024).

6.2 Whey protein in edible films and coatings: technological aspects

Despite extensive research on the use of WP in EFCs, significant gaps remain in understanding and optimizing their properties for widespread industrial use. More comprehensive evaluations are needed that integrate the latest advancements in additives, film formation methods, and the influence of environmental conditions. Furthermore, limited research has quantified the biodegradability of WP films and their ecological footprint compared to conventional packaging materials, although the environmental benefits are acknowledged. This section addresses these gaps by providing an updated analysis of WP-based films and coatings innovations, focusing on their mechanical properties, environmental impact, and potential for incorporating active ingredients (Pan et al., 2025; Wu et al., 2024).

6.2.1 Whey protein properties relevant to edible films and coatings

The attempts to reuse and recycle agricultural and industrial waste have spurred scientific interest in using WP to produce EFCs. WP possesses good barrier properties against oxygen, aromas and oil; however, it is a poor moisture barrier due to the hydrophilic nature of its amino acids (Wang et al., 2019). WP films exhibit several mechanical properties that make them suitable for food packaging applications. These properties include flexibility and tensile properties. Flexibility is crucial, allowing the films to conform to various shapes and surfaces without breaking (Wu et al., 2024). Tensile properties, such as TS and elongation, indicate the film's ability to withstand mechanical stress and deformation during handling and storage (Martins et al., 2024; Ramos et al., 2013). These properties are influenced by the film's composition, processing techniques, and the presence of plasticizers and other additives (Pires et al., 2024).

WP is the ideal packaging material due to its outstanding biodegradability, high heat stability, non-toxicity, ability to create strong inter-cross links, and micelle generation. Because milk proteins possess unique structural and functional properties, they are well-suited for creating innovative, active packaging solutions that cater to the demands of the food and nutraceutical industries (Chaudhary et al., 2022). For example, the WHEYLAYER project produced films and laminates coated with WP, which exhibit strong oxygen-barrier effectiveness and improved recyclability, as they are based on biodegradable whey protein (Pires et al., 2024; Tharanathan, 2003).

Furthermore, WP can generate transparent films with superior physical and barrier properties compared to polysaccharide-based films. They are essential for consumer appeal, and they may provide surface sterility. Apart from the improved barrier properties, these films decompose rapidly. However, to develop environmentally friendly food packaging, which includes components with an improved barrier to moisture transfer and increased flexibility, WPs must be coupled with suitable plasticizers (Basiak et al., 2017). Whey protein isolate (WPI) and plasticizers were used to develop novel bio-based coating solutions to avoid brittle films. A polyethylene terephthalate coating was applied at a semi-industrial scale using optimum processing conditions to minimize the energy consumption associated with this production stage. This proprietary approach also provided optimal denaturation of the protein-based coating on polyethylene (PET) substrate films, improving barrier qualities (Schmid et al., 2013).

6.2.2 Additives in whey protein films and coatings

Active packaging is an innovative approach that preserves or extends the shelf life of foods while maintaining their freshness, safety, and quality. According to European Regulation (EC) No. 450/2009 (European Union, 2009). Active packaging systems purposefully contain components that discharge or absorb materials into or from the packaged food or its surroundings. Typically, active packaging is created by adding bioactive substances to a film-forming solution before casting the film. An alternative is to add the active agent to the film-forming fluid after it has been encapsulated in colloidal particles. New opportunities in the food processing industry are presented by developments in the movement of active chemicals from the micro-to the nanoscale (Muhammad et al., 2020).

Milk proteins have a more extended release; they are quickly dispersed and are biocompatible. That is why they are ideal vehicles for micro- or nano-encapsulated bioactive compounds, such as antioxidants, nutraceuticals, and antimicrobials. Because of their diverse properties, milk proteins can be utilized alone or in combination with other biopolymers to create various materials, including hydrogels, nanoparticles, nanocomposites, nanocapsules, and more (Fig. 14.2) (Poonia, 2019).

In addition to being ecologically beneficial, WP-based coatings can offer further advantages by incorporating active/functional agents. WP has gained popularity recently for its ability to deliver and protect nutraceuticals, including probiotics, vitamins, bioactive peptides, essential oils, and antioxidants (Chaudhary et al., 2022). Using WP to encapsulate nutritional supplements enables food scientists to create nutritious foods with added health benefits. WP as nanocarriers is especially significant in nanobiotechnology, which uses nanoparticles smaller than 1050 nm, providing a substantial advantage over typical packaging technologies (Shendurse et al., 2018)The following sections focus on the different additives in WP films and coatings that improve the films' functional and antimicrobial properties.

6.2.2.1 Antimicrobial agents

WP-based EFCs have been examined for their ability to deliver antimicrobial agents to various microbiological environments and food surfaces. They can transport and release antimicrobials such as sodium lactate, lysozyme, nisin, potassium sorbate, lactoperoxidase system, *p*-aminobenzoic acid, sorbic acid, natamycin, e-polylysine, and essential oils (Bryaskova et al., 2025; Lainioti and Druvari, 2024). Antimicrobial coatings/films are, in fact, highly novel in the global context of active packaging, as they slow down, reduce, or even prevent the spread of infectious and deteriorating microorganisms on food packaging. The selection of appropriate coatings to maintain antibacterial agents on the outermost layer may lower the level of ingestion of antibacterial substances by consumers, while also preventing the need to compensate for the amounts conveyed into the products (Henriques et al., 2015; Wang and Guo, 2019). The most common antimicrobial agents in edible coatings include inorganic gases, bacteriocins, organic acids, enzymes, polysaccharides, spices, fungicides, plant extracts, and fatty acids (Ramos et al., 2012a).

WP solutions can be combined with antimicrobial agents such as propionic acid, benzoic acid, sorbic acid, sodium benzoate, or sodium potassium to enhance their effectiveness. Additionally, ethylenediaminetetraacetic acid (EDTA) can be incorporated into edible WP films to inhibit bacterial and fungal growth, along with natural antagonists (Bryaskova et al., 2025). Incorporating sorbic acid or *p*-aminobenzoic acid into WP isolate films (pH 5.2) can inhibit the growth of *Listeria monocytogenes*, *Salmonella typhimurium*, and *Escherichia coli* O157:H7 (Lainioti and Druvari, 2024). Sorbic acid is commonly used as a preservative due to its safety and minimal impact on organoleptic properties (Chen et al., 2025; Hauser and Wunderlich, 2011). Sorbic acid is a GRAS (Generally Recognized as Safe) additive, and it is effective against yeasts and molds, although it only partially inhibits bacterial growth (Wang et al., 2024). This antimicrobial agent has been utilized in antimicrobial films, demonstrating effectiveness in controlling microbial growth (Silveira et al., 2007), including *Escherichia coli*, *Penicillium* spp., and *Lactobacillus plantarum*. Cell-free supernatants from probiotic bacteria contain antimicrobial compounds, making them a promising candidate for incorporation into edible films (Beristain-Bauza et al., 2016). *Lactobacillus sakei* is a lactic acid bacterium used as a bio-preservative and has been added to meat products due to its ability to inhibit *E. coli* O157:H753. WPI films incorporated with the cell-free supernatant of *Lb. sakei* can inhibit *L. monocytogenes* and *E. coli* growth during refrigerated storage (Beristain-Bauza et al., 2017).

6.2.2.2 Other functional compounds

WP films are also used to carry additives and antioxidants. For instance, vitamin E is incorporated with WPs to shield products from oxidation and extend their shelf life. Incorporating ascorbic acid into WP films reduces food oxidation and color changes during storage (Bryaskova et al., 2025). Furthermore, WP films can carry flavors and colors, enhancing the performance of packaging materials (Jooyandeh, 2011).

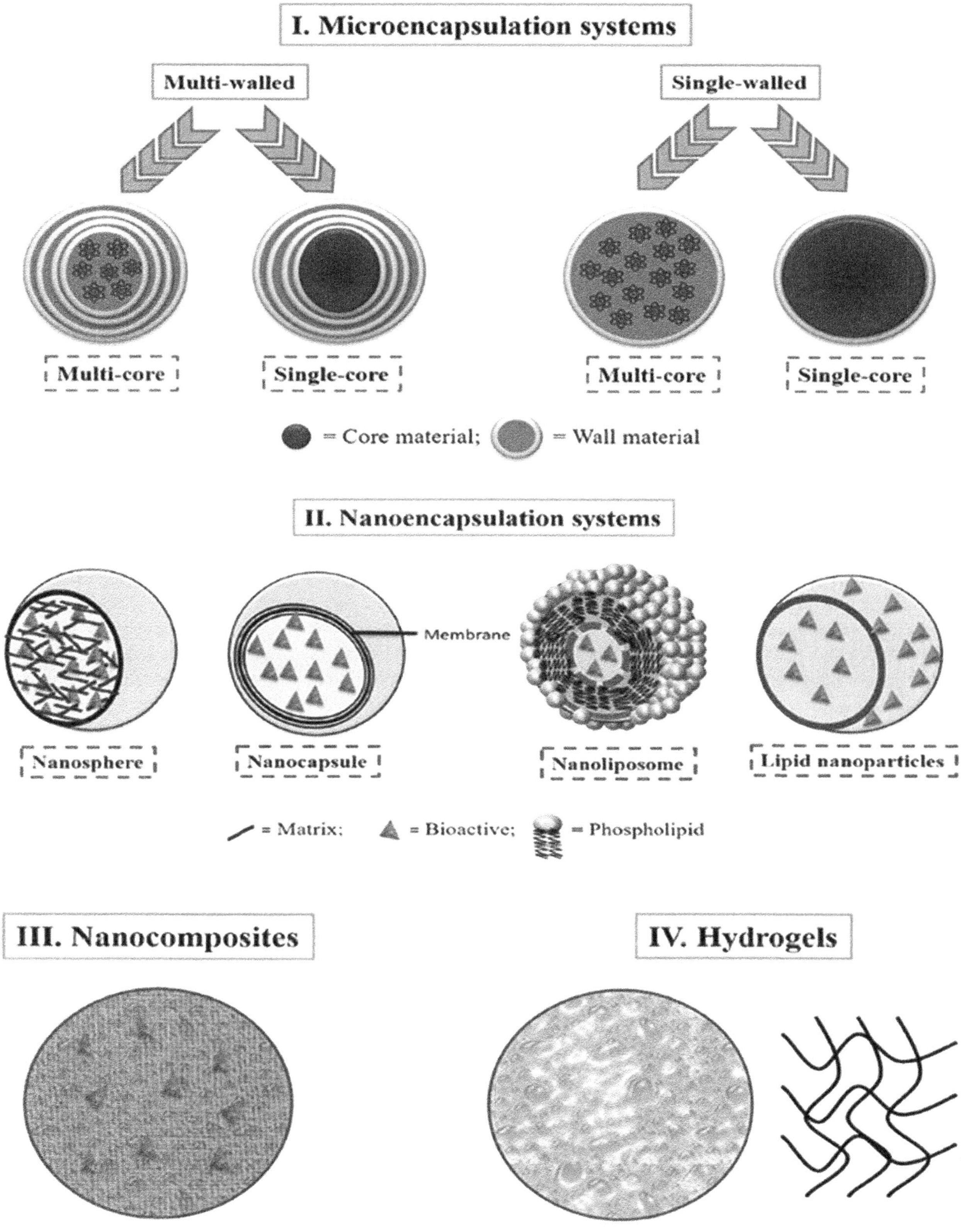

FIGURE 14.2 Milk protein-based hydrogels, nanoparticles and nanocomposites. *(From Chaudhary et al., 2022).*

WP comprises durable polymers due to the strong bonding between amino acid chains and their interactions with hydrogen atoms. When no plasticizing agents are incorporated to diminish the strength and bonding of hydrogen, the films become brittle and useless (Hammam, 2019). Cohesion refers to the internal molecular bonding within the polymer, which is vital for the film's mechanical strength. The stronger the molecular interaction between adjacent polymer chains, the better the cohesion. Key factors affecting this include the molecular structure, mass distribution, and functional groups of the polymer. Films with poor cohesion are likely to have reduced mechanical integrity, especially if there is an incompatibility between the WP and other additives, resulting in weaker films. Surfactants, such as Tween or lecithin, are recommended to improve surface adhesion, particularly in hydrophilic edible films, such as WP coatings. These reduce surface tension and enhance wettability, helping the film adhere better to the surface of food products (Xue et al., 2019).

6.2.2.3 Plasticizers

WP edible films must possess excellent moisture barrier qualities to package fruits and vegetables effectively and prevent microbiological decomposition. However, WP's hydrophilic nature can affect these barrier properties. Water vapor permeability (WVP) is crucial for controlling food moisture levels and interactions, effectively minimizing water vapor condensation. Differences in molecular weight, chemical structure, and the presence of many hydroxyl groups all contribute to changes in film solubility, mechanical characteristics, and barrier properties. Factors influencing WVP include the hydrophilic nature of WP, the relative humidity (RH) of the environment, the type of plasticizer, and film thickness. The plasticizer selection and RH control are critical for improving WVP (Gökkaya Erdem et al., 2021).

The permeability of gases, flavor, and fragrance are essential variables in the overall effectiveness of WP edible films in preserving food quality (Wang et al., 2019). Oxygen permeability (OP) impacts food quality by preventing oxidative damage, with WP films having higher OP than synthetic polymers. Factors influencing OP include the polar nature of WP, WPI/plasticizer ratio, RH, temperature, and film thickness (Kaneto and Uto, 2020). Table 14.3 summarizes the factors affecting the characteristics of WP films.

A three-dimensional protein network is formed during film processing in pure WP films. A highly cross-linked network leads to stronger and more rigid materials, enhancing modulus and TS. Additionally, the material becomes less flexible, reducing elongation (Schmid and Müller, 2019). Plasticizers are often used with WP as a raw ingredient for films and coatings to overcome their moisture barrier and mechanical restrictions (Ramos et al., 2012b) by reducing intermolecular interactions between protein chains and increasing flexibility, stretching, strength, and tear resistance (Tan et al., 2022).

One study assessed the efficiency of plasticizers (propylene glycol, glycerol, sucrose, sorbitol, polyethylene glycol (PEG)) in films made from β-lactoglobulin (Sothornvit and Krochta, 2001). The plasticizers varied in size, shape, and composition. Glycerol was the most effective plasticizer due to its small size and hygroscopic nature, which attracted

TABLE 14.3 Factors affecting the characteristics of whey protein films.

Property	Measurement technique	Key findings	Influencing factors	References
Water vapor permeability (WVP)	Cup method	Crucial for controlling food moisture levels and interactions, minimizing water vapor condensation	Hydrophilic nature of WP, RH, plasticizer type, and film thickness	Wang et al. (2019)
Oxygen permeability (OP)	Gas permeability measurement	Impacts food quality by preventing oxidative damage; WP films have higher OP than synthetic polymers	Polar nature of WP, WPI/plasticizer ratio, RH, temperature, and film thickness	Khwaldia et al. (2004)
Mechanical properties	Tensile strength and elongation tests	Stronger and less flexible with higher modulus and tensile strength; reduced elongation	Cross-linked WP network, plasticizer selection and ratio, intermolecular interactions	Schmid and Müller (2019)
Moisture content and solubility	Equilibrium moisture content measurement	Enhanced moisture barrier and solubility with glycerol; higher solubility with sorbitol	Glycerol levels, lactose content, and internal polymer network	Kim and Ustunol (2001)
Thermal stability	Differential scanning calorimetry (DSC)	Greater thermal stability with lower water vapor and gas permeability	WPI versus WPC content, glycerol ratio, internal molecular structure	Ramos et al. (2013)

more water and enhanced the plasticizing effect. Sucrose, which has a bulky ring structure, was less efficient than the linear plasticizers. The mechanical performance is impacted by the ratio and type of plasticizer used. Glycerol-plasticized β-lactoglobulin films showed increased elongation and reduced TS.

Adding glycerol to WP films enhanced the equilibrium moisture content and solubility (Henriques et al., 2015). Higher glycerol levels increased WP films' solubility and water vapor permeability, but decreased mechanical resistance, apparent young modulus, and glass transition temperature (Sothornvit et al., 2003). Similar investigations found that incorporating sorbitol into WPI films improved moisture and oxygen barriers compared to glycerol-plasticized films (Krochta and Perez-Gago, 2002). Cross-linking agents, such as formaldehyde, can significantly enhance the physical strength of glycerol-plasticized WPI films. Nonetheless, these compounds do not have food safety certification and are often unsuitable for use in food packaging (Ustunol and Mert, 2006).

WPI and WPC films were observed to be transparent, flexible, and uniform. WPI films exhibited lower moisture content and solubility, reduced water vapor and gas permeability, and greater intensity, TS, elongation, and thermal stability than WPC films with the same glycerol ratio as a plasticizer (Ramos et al., 2013). These differences arise from variations in lactose, fat, minerals, and total solids content, which influence the internal molecular structure of the films. A decrease in lactose content reduces the spacing within the polymer network, thus enhancing the film matrix. In contrast, an increase in lactose raises film permeability due to its role as a plasticizer.

6.2.2.4 Cross-linking agents

WP films are generated by cross-linking, which involves forming strong molecular bonds in amino acids and restricting the movement of polymer chains. Several approaches have been tested to enhance protein films, making them stronger and more impermeable to water. Some examples include protein modification (denaturation), the use of other macromolecules, bioactive chemicals, and plasticizers. Heating, acid/base treatments, the addition of aldehydes, phenolic compounds, $CaCl_2$, enzymes, and using methods such as irradiation, and ultrasonic/microwave-aided treatment can be used to achieve cross-linking in WP films (Calva-Estrada et al., 2019). Various cross-linking procedures are used to improve the quality of WP edible films. Ionizing radiation, such as γ-irradiation, alters protein structure irreversibly by oxidizing amino acids, forming free radicals, breaking covalent bonds, and causing recombination and polymerization reactions. γ-irradiation produces hydroxyl radicals from water, which react with protein residues, favoring aromatic peptides over aliphatic ones (Ibáñez et al., 2024; Sabato et al., 2001). When applied to WP, γ-irradiation creates dityrosine linkages between protein chains, resulting in intractable and sterile films (Schmid and Müller, 2019). UV rays can also create films by covalently bonding with WP. UV irradiation converts double-bonded molecules and aromatic rings into free radicals, which generate new cross-links and form a protein film network. Larger radiation doses result in better interactions; UV-irradiated WP films exhibit increased rigidity, yet their barrier characteristics are not significantly affected (Di Pierro et al., 2018).

Proteins have been cross-linked using enzymatic methods that employ transglutaminase, lipoxygenase, peroxidase, and polyphenol oxidase. Transglutaminase is capable of catalyzing covalent cross-linking activities between protein molecules, generating extremely heavy biopolymers (Purewal et al., 2023). Chemical cross-linking agents such as carbonyldiimidazole, glutaraldehyde and dialdehyde starch significantly improved the insoluble nature and TS of WPI film-forming solutions (Infurna et al., 2021).

6.3 Recent advancements in the development of whey protein edible films and coatings

Recent studies on developing WP EFCs are summarized in Table 14.4, which compiles data from studies conducted between 2020 and 2025. Any literature older than these dates can be found in two extensive review papers (Pires et al., 2024; Yoo et al., 2023). Recent developments have focused on enhancing the functional properties of the films, including their antimicrobial activity and suitability for various food matrices. Innovations in formulations have also shown a clear trend in sustainable and active packaging.

Current trends in the preparation of WP-EFCs include the use of bioactive compounds with antimicrobial and antioxidant properties to enhance the quality and safety of food products. Examples include the use of antimicrobial compounds in the WP films, either by direct addition, such as essential oils (EOs), or through fermentation to yield metabolites such as bacteriocins.

Probiotics, such as bacteriocin-producing *Lactobacillus curvatus* 54M16, were used to inhibit *Listeria innocua* in ham slices (La Storia et al., 2020). In another study, the incorporation of *Lactobacillus acidophilus* and *Lb. plantarum*, and a mixed culture of both in WPI-carrageenan films, improved WVP and color but decreased mechanical properties (TS and

TABLE 14.4 Recent studies on developing whey protein-based films and coatings.

Film type	Antimicrobial/ Functional agent	Food product	Amount	Results	References
WP-inulin-gelatin	Bacteriocin-producing LAB (*Lactobacillus curvatus* 54M16)	Only film	Final concentration of 8.54 and 8.56 CFU/mL	LAB decreased the viscosity of the film-forming solution and improved the film's plasticity, but it did not affect barrier properties. Growth of *Lb. innocua* C6 was inhibited in a model system (cooked ham discs).	La Storia et al. (2020)
WPI	Lemon and bergamot EO	Only film	2%–5%	Bergamot EO was more effective than lemon in inhibiting microbial growth. The optimal concentration of bergamot EO was 4.5%. Films produced desirable WVP and OP.	Çakmak et al. (2020)
WPI	Curcumin	Only film	0.5%–2.5% w/w based on WPI content	Good antioxidant activity and lower WVP in samples containing curcumin. The strength of the films was improved with lower concentrations (0.2–0.5 mL/mL).	Taghavi Kevij et al. (2020)
Liquid whey	Cassava starch	Only film	7.50%–11.25%	The addition of cassava starch resulted in a decrease in solubility; films with a higher proportion of whey were more resistant to thermal degradation. The optimal composition was 67.5% of whey and 7.5% of cassava starch.	Martins et al. (2020)
Heat-treated (HT) WPC film	Tarragon EO	Only film	0.5%–2.5%	Heat treatment resulted in thicker films with better barrier properties to moisture and light. HT films also exhibited higher antimicrobial effectiveness and enhanced physical properties, making them more suitable for edible packaging applications.	Socaciu et al. (2020)
WPI-psyllium seed gum		Only film		WPI-psyllium seed gum composite film exhibited better overall properties than films made solely from psyllium seed gum or WPI. Among the tested ratios, the 1:1 WPI to psyllium seed gum blend demonstrated the most favorable combination of physicochemical and mechanical characteristics.	Zhang et al. (2020)
WPI	Nettle (*Urtica dioica*) leaf extract	Only film	10%–50%	Increased total phenolic content and antioxidant activity, as well as antimicrobial activity against S. aureus and E. coli, were also observed.	Mohammadian et al. (2021)

TABLE 14.4 Recent studies on developing whey protein-based films and coatings.—cont'd

Film type	Antimicrobial/ Functional agent	Food product	Amount	Results	References
WP	Clove EO	Only film	5%–15%	Sample containing 10% EO produced the best results regarding thickness, elongation, TS and WVP; the growth of *S. aureus* and *E. coli* was inhibited.	Fahrullah and Rosyidi (2021)
WPI	Chitosan	Only film	3% (w/w/)	Edible films obtained by electrostatic interactions between negatively charged WPI and positively charged chitosan (complex coacervation); stronger network structure, lower WVP and solubility and higher TS were observed when compared to mono-component WPI and chitosan films.	Tavares et al. (2021)
WP and candelilla wax film	Tarbush (*Flourensia cernua*) polyphenols	Only film	1.15–1.35	Polyphenolic extract had a significant effect on the solubility and transparency of the films without having an impact on thickness and WVP.	Alvarez-Perez et al. (2022)
Gelatin-WP-chitosan		Only film		The optimal formulation consisted of 6 g of gelatin, 2.5 g of WP, and 1.48 g of chitosan; this sample exhibited lower solubility and moisture content, but better mechanical properties and greater transparency.	Herrera-vázquez et al. (2022)
WPI-carrageenan	*Lactobacillus acidophilus, Lb. plantarum,* and mixed culture	Only film	Final concentration of 9 log cfu/g whey solution	The incorporation of probiotic bacteria influenced WVP and color and decreased mechanical properties (TS and EAB) over storage at 4 and 25°C for 30 days.	Sogut et al. (2022)
Fermented cheese whey–cassava peel starch	*Whey fermented with Candida tropicalis for 24 h*	Only film	Varying whey to starch ratios (1:3 to 3:1 v/v)	Good mechanical and barrier properties to extend the shelf-life of particularly vegetables with antimicrobial activity toward *Pseudomonas auruginosa.*	Utama et al. (2022)
WPC	Spent coffee grounds	Only film	To achieve a phenolic content of 0.5–2.5 mg/g film	Antioxidant activity was maintained during storage at 5°C for 12 months. Film properties can be controlled by adjusting the pH of the WP solution. WVP was significantly decreased at higher pH values, indicating a potential for preventing moisture loss.	Papadaki et al. (2022)

Continued

TABLE 14.4 Recent studies on developing whey protein-based films and coatings.—cont'd

Film type	Antimicrobial/ Functional agent	Food product	Amount	Results	References
WPI	Gamma-amino butyric acid (GABA)	Only film	0.1%–1% w/v	All films were colorless and transparent; hydration and thermal properties were improved at low concentrations of GABA (0.1% and 0.5%), but impaired at 1%.	He et al. (2022)
WPC—*Lepidium perfoliatum* gum and transglutaminase	Oregano EO	Only film	0.5%–8%	Composite films loaded with oregano EO had antibacterial activity and can be used for biodegradable packaging.	Alipour et al. (2023)
LAB-fermented whey solution-potato starch		Only film	Whey solution: potato starch 1:1 v/v	Formulated bioactive edible films had adequate characteristics to be applied to food products; moisture, color, WVP, mechanical properties and microbial load were measured at the beginning and after 14 days of storage (4 ± 1 °C) using plastic packaging materials as controls.	Hernández-Carranza et al. (2023)
WPC and tapioca starch	Ginger EO	Only film	0.1%–0.5%	Optimal results were obtained with 0.5% EO, considering factors such as thickness, solubility, elongation, tensile strength, and inhibition of *E. coli* and *S. aureus*.	Pratiwi et al. (2024)

(covering recent research papers published from 2020 to 2025; plasticizers in the formulations were not included) [*EAB*, Elongation at break; *EO*, Essential oil; *LAB*, Lactic acid bacteria; *OP*, Oxygen permeability; *TS*, Tensile strength; *WP*, Whey protein; *WPC*, Whey protein concentrate; *WPI*, Whey protein isolate; *WVP*, Water vapor permeability].

EAB) for 30 days (Sogut et al., 2022). Other antimicrobials from plant extracts, such as nettle leaf extract (10%–50%), were used to inhibit *S. aureus* and *E. coli* (Mohammadian et al., 2021). Cheese whey subjected to 24 h fermentation using *Candida tropicalis* was mixed with cassava peel starch (1:3 to 3:1 v/v), and it yielded good mechanical and barrier properties, as well as an extended shelf life for vegetables, with antimicrobial activity against *Pseudomonas auruginosa* (Utama et al., 2022).

Essential oils (EOs) have also been applied, such as lemon or bergamot EOs at 4.5% to reduce microbial growth (Çakmak et al., 2020), and clove EO (5%–15%), which demonstrated the best mechanical properties and antimicrobial performance at 10% EO (Fahrullah and Rosyidi, 2021). Heat treatment of WPC films before incorporating tarragon EO enhanced the moisture and light barrier properties of the films, as well as their antimicrobial activities (Socaciu et al., 2020). Other innovations in EO-added WP films include the addition of gums (*Lepidium perfoliatum* gum) and cross-linking enzymes (transglutaminase), which, when combined with oregano EO, exhibit enhanced antimicrobial activity and biodegradability (Alipour et al., 2023). A recent study obtained optimal results (film properties and inhibition of microbial growth) with ginger EO (0.5%) in films made from WPC and tapioca starch (Pratiwi et al., 2024).

Antioxidants have also been used to enhance the properties of WP EFCs and food quality. Curcumin (0.5%–2.5%) resulted in good antioxidant activity and lower WVP, especially at low concentrations (Taghavi Kevij et al., 2020). WPC mixed with spent coffee grounds increased the phenolic content, which maintained antioxidant activity for 12 months and reduced WVP at higher pH values (Papadaki et al., 2022).

Several innovations have been recently reported to enhance the physicochemical, barrier, thermal, and optical properties of WP-based EFCs. Blending WP with other polysaccharides is an example of this. The addition of cassava starch to liquid whey resulted in films with reduced solubility and enhanced thermal stability (Martins et al., 2020). WPI-psyllium seed gum composite film (1:1) exhibited better overall properties than films made solely from psyllium seed gum or WPI (Zhang et al., 2020). Complex coacervates of WPI and chitosan yielded a stronger network, lower WVP and solubility and a higher tensile strength (Tavares et al., 2021). Similarly, WP and candelilla wax films, with added tarbush (*Flourensia cernua*) polyphenols, exhibited a significant effect on the solubility and transparency of the films, without affecting their thickness and WVP (Alvarez-Perez et al., 2022). WPI films enriched in gamma-aminobutyric acid (GABA) improved solubility and thermal properties, mainly when used at low concentrations (He et al., 2022). Finally, a multicomponent approach, including WP, chitosan, and gelatin blends, was optimized to enhance mechanical properties while reducing solubility and moisture content (Herrera-vázquez et al., 2022).

These studies collectively highlight the versatility of WP-based films in delivering functional agents, improving mechanical and barrier properties, and adapting to various food systems. The integration of natural extracts, probiotics, and fermentation-derived compounds reflects a shift toward multifunctional and sustainable packaging. Future research is expected to focus on scalability, regulatory compliance, and consumer acceptance to facilitate broader industrial adoption.

6.4 Application of whey protein films and coatings as active edible packaging of food products

EFCs preserve food from harmful environmental conditions during storage and transit while preventing mass and water transfer, permitting them to extend the shelf life of food products (Avramescu et al., 2020). Table 14.5 presents the current food applications for WP-based films and coatings, encompassing the literature from 2000 to 2025. Any literature older than these dates can be found in two extensive review papers (Kandasamy et al., 2021; Pires et al., 2024). WP-based films and coatings exhibit notable antibacterial activity against various bacteria, including aerobic mesophilic bacteria, *Escherichia coli*, lactic acid bacteria (LAB), *Staphylococcus aureus*, psychrotrophic bacteria, and *Bacillus cereus*. They also enhance the preservation of physical parameters such as organoleptic quality, storage life, color, and texture.

6.4.1 Dairy products

Dairy applications dominate the use of WP-based films, particularly in cheese preservation. Various forms of whey protein (such as WPI, WPC, liquid acid WPC, and liquid whey) were used in diverse types of cheeses (including soft rennet-curd cheese, sliced Kasar cheese, and soft cheese).

WP coatings incorporating *Lactobacillus buchneri* (Guimarães et al., 2020) effectively inhibited fungal contamination and ochratoxin A production in semi-hard cheese during storage at 25°C, demonstrating the potential of probiotic-loaded films in enhancing microbial safety. A more complex formulation involved liquid acid WPC combined with apple pectin and sunflower oil, enriched with indigenous *Lactobacillus helveticus* MI-LH13 (Vasiliauskaite et al., 2022). This coating was applied to artisanal fresh acid-curd cheese at a concentration of 0.2 g/100 g. The coating supported the viability of *Lb. helveticus* for 14 days at both 4°C and 23°C, significantly reduced yeast counts by day 14, and suppressed mold growth until day 21. Furthermore, the coating helped preserve moisture, which slowed down discoloration in cheese.

WPI-based antimicrobial films incorporating oregano essential oil (EO), garlic EO, nisin, and natamycin were tested on sliced Kasar cheese (a semi-hard Turkish cheese) inoculated with pathogens (Seydim et al., 2020). These films, applied at a concentration of 2% (w/v), successfully reduced surface contamination during 15 days of storage at 4°C, thereby offering a promising strategy to minimize food waste and enhance food safety. WPC films containing cumin EO were applied to Iranian white cheese at concentrations of 1% and 2% (Nemati et al., 2023). These edible films not only extended the product's shelf life but also contributed to a more sustainable manufacturing process and enhanced the cheese's functional value. WP films incorporating milk-derived antimicrobial peptides, specifically α_{s2}-casein$_{151\text{-}181}$ and α_{s2}-casein$_{182\text{-}207}$, were applied to soft cheese Hernande (Zhang et al., 2024). At a final concentration of 100 μg/mL, the peptide α_{s2}-casein$_{151\text{-}181}$ proved to be a strong alternative to nisin, effectively controlling surface contamination and enhancing microbial stability by targeting both native microbiota and post-processing contaminants.

An innovative composite film was made from furcellaran (a polysaccharide derived from red algae) and WPI, and enriched with yerba mate (*Ilex paraguariensis*, a plant species of the holly genus native to South America) and white tea extracts (Pluta-Kubica et al., 2020). When applied to soft rennet-curd cheese, this film significantly extended shelf life by reducing yeast and mold counts compared to conventional plastic packaging. It also inhibited the growth of coliform bacteria during 3 weeks of refrigerated storage. However, the appearance and smell of the cheese were found to be negatively affected after the film was removed.

Several studies have investigated the incorporation of herbal extracts and spirulina into WP-based films to improve their functional properties and enhance their effectiveness. One study produced WPC films supplemented with aqueous extracts of rosemary and sage and applied them to soft cheese (Kontogianni et al., 2022). The addition of these extracts

TABLE 14.5 Uses of whey protein-based films and coatings in food products.

Category	Film type	Antimicrobial/ Functional agent	Food product	Amount	Results	References
Dairy products	WP	*Lactobacillus buchneri* UTAD104	Cheese		*Lb. buchneri* application prevented fungal contamination by *Penicillium nordicum* in cheese for at least 25 days of storage at 25°C; ochratoxin A was not found in cheeses treated with the coating containing *Lb. buchneri*.	Guimarães et al. (2020)
	Furcellaran (genus of red algae)—WPI	Yerba mate and white tea extract	Soft rennet-curd cheese		Film application prolonged the shelf life of soft cheese, resulting in lower yeast and mold counts compared to plastic packaging, and inhibited coliform bacteria during 3 weeks of refrigerated storage. There was a small negative impact on the appearance and smell of the cheese samples after removing the film.	Pluta-Kubica et al. (2020)
	WPI	Oregano EO, garlic EO, nisin and natamycin	Sliced Kasar cheese (semi-hard Turkish cheese) inoculated with pathogens	2% (w/v)	WPI-based antimicrobial layers reduced surface contamination during storage at 4°C for 15 days and, consequently, can prevent food waste.	Seydim et al. (2020)
	WPC	Aqueous rosemary and sage extracts	Soft cheese	2%	The mechanical properties of the films improved with the addition of extracts, with sage extract having a more pronounced effect, possibly due to its higher polyphenol content.	Kontogianni et al. (2022)
	WPC	Green tea (*Camellia sinensis* L.) extract	Fresh cheese (Latin-style)	2.5%	The films effectively delayed lipid oxidation and inhibited the growth of microorganisms, including *E. coli*, at 5°C for 1 week.	Robalo et al. (2022)
	Liquid acid WPC-apple pectin-sunflower oil	Indigenous *Lactobacillus helveticus* MI-LH13	Artisanal fresh acid-curd cheese	0.2 g/100 g	The formulation supported the viability of *Lb. helveticus* for 14 days of storage at both 4 °C and 23 °C; however, the yeast count was significantly reduced by day 14, and mold growth was suppressed until day 21. Discoloration of the cheese was slowed down since moisture was preserved.	Vasiliauskaite et al. (2022)

TABLE 14.5 Uses of whey protein-based films and coatings in food products.—cont'd

Category	Film type	Antimicrobial/ Functional agent	Food product	Amount	Results	References
	WPC film	Cumin EO	Iranian white cheese	1% and 2%	The edible coating improved the shelf life, contributed to the development of a more sustainable manufacturing process, and increased the functional value of the product.	Nemati et al. (2023)
	WPC	Spirulina	Kefalotyri (Greek) cheese	0.5%–4%	The film containing 2% spirulina had the highest total phenolic content and desirable color parameters; it was easily digestible and prevented the growth of spoilage bacteria at refrigerated storage for 60 days.	Kontogianni et al. (2023)
	Whey (liquid)	Konjac flour	Dangke vheese (Indonesian hard cheese)	1.5%	The quality of hard cheese was preserved for 20 days of storage at 5°C.	Malaka et al. (2024)
	WP	Milk derived antimicrobial peptides (α_{s2}-casein$_{151-181}$ and α_{s2}-casein$_{182-207}$)	Soft cheese Hernande	Final concentration of 100 μg/mL film solution	Antimicrobial peptide αs2-casein 151–181 was an excellent alternative to nisin for controlling contamination on the cheese surface, thereby enhancing microbial stability associated with both native microbiota and post-processing contamination.	Zhang et al. (2024)
Red meat	Cassava starch and WPI film	Cinnamon oil and rambutan peel extract	Salami		The release mechanisms controlled antioxidant activity, and the antibacterial effect was strongly dependent on food matrix components. The redness of the coated salami was maintained during storage.	Chollakup et al. (2020)
	WPI	*Laurel (Laurus nobilis* L.) and sage (*Salvia officinalis*) extract	Cooked meatballs		The antioxidant activity of the extracts enabled the control of lipid oxidation on the product surface during refrigerated storage for 1–7 days.	Akcan and Serdaroglu (2022)
Poultry	WPC film	Ethanolic extract of brown edible seaweed (*Fucus vesiculosus*)	Chicken breast		Reduced malonaldehyde, decreased lipid oxidation, and extended shelf life	Andrade et al. (2021)
	WP	Fish gelatin hydrolyzates	Chicken breast fillets		Coating with WP and hydrolyzates resulted in minor changes in cooking loss, pH and oxidation parameters when compared to samples coated with either WP or hydrolyzate (samples stored for 16 days at 4°C).	Sabzipour-Hafshejani et al. (2022)

Continued

TABLE 14.5 Uses of whey protein-based films and coatings in food products.—cont'd

Category	Film type	Antimicrobial/ Functional agent	Food product	Amount	Results	References
Fish and seafood	WPI film	Tarragon EO (2.5%)	Brook trout		Reduced TBARS values suggest antioxidant characteristics. A reduction in the total surviving microorganism count, lactic acid bacteria, psychrophilic count, and H2S-producing microorganisms indicated antibacterial action, resulting in a 3-day increase in storage life.	Socaciu et al. (2021)
	WP film	Aqueous clove (*Syzygium aromaticum* L.) extract	Brown trout fillets		An edible coating containing whey protein-nanoliposome extract was more effective in maintaining quality at 4°C compared to a combination of whey protein-free-formed extract and whey protein.	kosh goftar et al. (2025)
	Chitosan-WPI-Fe^{3+} composite nanoparticles	Lemon EO	Shrimps	10% w/w	Quality was preserved through reduced moisture loss, delayed pH changes, and maintained textural properties during refrigerated storage.	Li et al. (2025)
Fruits and vegetables	WPI	Chitosan	Strawberries		The film extended the shelf life of the fruit to 8 and 5 days when stored at 5°C and 20°C, respectively.	Muley and Singhal (2020)
	WPI	Lemon and lemon grass EO	Fresh-cut pears	Lemon oil at 1.0% or lemon grass oil at 0.5%	The coating resulted in a reduction in color change, and firmness was also maintained. The films exhibited excellent oxygen and carbon dioxide permeability, along with a good level of transparency.	Galus et al. (2021)
	WPC	Rosemary EO	Fresh spinach	0.3% and 0.6%	The coating prevented weight loss and reduced oxygen permeability, exhibiting a significant antimicrobial effect against coliforms.	Abedi et al. (2021)
	Tamarind starch and WPC coating	Thyme EO nano-emulsion	Tomatoes	5%–20%	The shelf life was extended to 14 days at room temperature, and the film retained its antioxidant and antimicrobial properties.	Ghoshal and Shivani (2022)
	WP-based emulsion	Transglutaminase and sunflower oil	Fresh-cut apples	Transglutaminase 60 units/g; sunflower oil 1.5%–3.0%	Weight loss and browning index were reduced significantly.	Xin et al. (2023)
	WPI-alginate films	Curcumin	Apples		Whey protein increased the antioxidant capacity and hydrophobicity of the film, while reducing its transparency. Curcumin had similar effects in addition to improving the UV-blocking efficiency.	Botalo et al. (2024)

TABLE 14.5 Uses of whey protein-based films and coatings in food products.—cont'd

Category	Film type	Antimicrobial/ Functional agent	Food product	Amount	Results	References
	WPC coating	Chitosan	Apples (golden delicious)		Coatings effectively reduced weight loss, acidity, and enhanced firmness during cold storage.	Tarawneh et al. (2025)
Other foods	WPC and carboxy methyl cellulose (CMC)	Rosemary extract	Sunflower seed kernels	800 and 1500 ppm	Preferred mechanical properties were achieved with an 8% WPC and 0.6% CMC composition; better color characteristics were obtained with this optimized composite film. However, rosemary extract enhanced oxidation in the kernels.	Hosseini et al. (2020)
	WPI	Sodium montmorillonite/sodium metabisulphite	Fresh eggs	Montmorillonite nanoparticles: whey (1:100); sodium metabisulphite 0.055 g/ 100 mL	Coated eggs maintained their integrity for up to 8 weeks of storage at 25 ± 3 °C, exhibiting high foam stability for up to 5 weeks, which partially eliminated the need for refrigeration.	de Araújo Soares et al. (2021)
	WPI/pectin	Microbial transglutaminase	Roasted peanuts		Water uptake, oxygen permeability and oxidation were reduced in the coated peanuts, preserving crunchy texture and improving consumer acceptability.	Rossi-Márquez et al. (2021)
	Bi-layer edible films based on starch (primary layer)	LAB-fermented whey and/or mango pulp powder solutions (secondary layer)	Sushi	Whey and mango 5%, *Lb. rhamnosus* 10^9 CFU/ mL	Storage at 4 and 20°C for 28 days; at low storage temperatures, edible films containing LAB-fermented whey and whey-mango solutions maintained a sufficient probiotic count (10^6 CFU/g) to deliver their health benefits.	Hernández-Carranza et al. (2024)
	Potato starch or sweet potato flour	Fermented (*Lb. rhamnosus*) and non-fermented whey solution	Surimi		The fermented whey solution maintained *Lb. rhamnosus* at a concentration greater than 10^8 CFU/g, even at 20°C. However, surimi covered with films had a concentration of less than 10^6 CFU/g, which was below the minimum recommended level.	Macías-González et al. (2024)

(covering recent research papers published from 2020 to 2025; plasticizers in the formulations were not included) [*EO*, Essential oil; *TBARS*, Thiobarbituric acid reactive substances; *WP*, Whey protein; *WPC*, Whey protein concentrate; *WPI*, Whey protein isolate].

improved the mechanical properties of the films, with sage extract exhibiting a more pronounced effect, likely due to its higher polyphenol content. Similarly, WPC films containing green tea (*Camellia sinensis* L.) extract were used on Latin-style fresh cheese (Robalo et al., 2022). These films, applied at a concentration of 2.5%, effectively delayed lipid oxidation and inhibited microbial growth, including *E. coli*, during a 1-week storage period at 5°C. Another study investigated the addition of spirulina (in WPC films for the preservation of Kefalotyri, a traditional hard Greek cheese made from a mixture of sheep's and goat's milk (Kontogianni et al., 2023). The film containing 2% spirulina exhibited the highest total phenolic content and desirable color characteristics. The film effectively prevented the growth of spoilage bacteria during 60 days of refrigerated storage. The antimicrobial activity of the film was attributed to a wide diversity of bioactive compounds in spirulina.

A recent study combined liquid whey with konjac flour (at concentrations of 1%, 1.5% and 2%), and measured their physical characteristics (Malaka et al., 2024). The coating with the most desirable characteristics (containing 1.5% konjac flour) was used to coat Dangke cheese (a traditional hard cheese from Indonesia). This coating preserved the quality of the cheese for 20 days at 5°C as measured by sensory evaluation.

6.4.2 Meat and meat products

6.4.2.1 Red meat

A composite film containing cassava starch and WPI, enriched with cinnamon bark oil and rambutan (*Nephelium lappaceum*) peel extract, was developed as an active packaging. The antimicrobial activity of this film was investigated in vitro and in a real food system (salami). The film exhibited the highest efficacy in delaying microbial growth in salami. The sample packaged with composite film retained its color for up to 5 days as measured by a colourimeter (Chollakup et al., 2020).

Another study investigated the use of WPI films containing extracts from laurel (*Laurus nobilis* L.) and sage (*Salvia officinalis*) on cooked meatballs during frozen storage (−18°C for 60 days). The higher total phenolic content in the meatballs treated with these natural antioxidant extracts contributed to the oxidative stability of the frozen meatballs (Akcan and Serdaroglu, 2022). It was noted that there was a need to improve the acceptability and sensory properties of the coated meatballs.

6.4.2.2 Poultry

One study developed a WPC-based active film using the ethanolic extract of brown edible seaweed (*Fucus vesciculosus*) and investigated its properties and application on chicken breast meat (Andrade et al., 2021). This active edible film (including the extract obtained from freeze-dried seaweed and 75% ethanol (v/v)) was able to inhibit lipid oxidation in chicken breast samples during a 25-day storage period. Another study using chicken breast fillets employed a WP-based coating containing fish gelatin hydrolyzates (Sabzipour-Hafshejani et al., 2022). Coated samples exhibited minor changes in cooking loss, pH, total volatile base nitrogen, and oxidation parameters compared to samples coated with either WP or fish gelatin hydrolyzate. These parameters increased in uncoated samples during 16 days of cold storage (4°C).

6.4.2.3 Fish and seafood

The use of heat-denatured WPI films enriched with tarragon EO (2.5%) was studied to preserve brook trout during storage at 4°C (Socaciu et al., 2021). Tarragon EO conferred antioxidant and antimicrobial properties to the film. The use of the film significantly reduced thiobarbituric acid reactive substances (TBARS), suggesting the antioxidant potential of tarragon EO. Additionally, the film demonstrated antibacterial properties by lowering the number of total viable microorganisms, psychrophilic bacteria, lactic acid bacteria, and hydrogen sulfide-producing microorganisms. As a result, the quality of the fish was preserved during storage. Due to the low level of incorporation of EO, the organoleptic properties were not significantly affected.

Another study (kosh goftar et al., 2025) explored the application of WP films combined with aqueous clove extract encapsulated in nanoliposomes on the quality (chemical, physical, microbial and sensory) of brown trout fillets stored at 4°C for 16 days. The edible coating (with nanoliposome extract of clove) was more effective in maintaining quality compared to the combination of WP and extract (in the free form) or WP only.

An interesting study integrated a novel nanoparticle system composed of chitosan, WPI, and Fe^{3+} using film-forming technology. Lemon EO was gradually incorporated into this nanoparticle system (10% w/w), and a Pickering emulsion was formed. The base for this nano-edible film consisted of polyvinyl alcohol and glycerin. This film preserved shrimp quality by reducing moisture loss, delaying pH changes and maintaining the textural properties of shrimps during refrigerated storage for 6 days (Li et al., 2025).

6.4.3 Fruits and vegetables

Numerous studies have been conducted on apples using WP-based films. One study developed a WP-based emulsion incorporating transglutaminase (60 units/g) and sunflower oil (1.5 w/w) for the preservation of fresh-cut apples (Xin et al., 2023). Both ingredients were added to increase the water resistance of the protein-based film. The film reduced the weight loss and browning index (a value obtained from L*, a* and b* values, which were measured by a colourimeter) of fresh-cut apples at 8°C for 8 days. One approach in producing edible films is to include polysaccharides to act as gas barriers and proteins to enhance the mechanical properties of the film. The incorporation of hydrophobic additives, such as lipids, helps reduce the water permeability. One study employed alginate (a polysaccharide), WPI, and curcumin (a hydrophobic polyphenolic compound) to improve the overall properties of an edible film for use with whole apples (stored at 25°C for 21 days) (Botalo et al., 2024). WP increased the antioxidant capacity and hydrophobicity, while reducing transparency. Curcumin had similar effects, but it also improved the UV-blocking efficiency. There were also studies using apples that employed simpler ingredients to create edible coatings. One study utilized WPC and chitosan to monitor the quality of Golden Delicious apples at both room and refrigerated temperatures (Tarawneh et al., 2025). This edible coating effectively reduced weight loss and total titratable acidity, while enhancing the firmness of the apples during cold storage.

WPI and chitosan conjugate films were used to extend the post-harvest shelf life of strawberries. The film extended the shelf life of the fruit to 8 and 5 days when stored at 5°C and 20°C, respectively (Muley and Singhal, 2020). Several studies have incorporated EO with WPs for the preservation of fruits and vegetables. Such examples include WPI-lemon EO (1%) and WPI-lemon grass EO (0.5%) applied to fresh-cut pears (Galus et al., 2021), WPC and rosemary EO applied to fresh spinach (Abedi et al., 2021), and tamarind starch-WPC with thyme EO nanoemulsion applied to tomatoes (Ghoshal and Shivani, 2022). These studies confirm the shelf life extension of these products, with the retention of quality parameters and the antimicrobial and/or antioxidant properties of the films (Table 14.5).

6.4.4 Other foods

WPC and carboxy methyl cellulose (CMC) composite film, which incorporated rosemary extracts (at 800 and 1500 ppm), was applied to sunflower seed kernels (Hosseini et al., 2020). Desirable mechanical properties and better color characteristics were obtained with the film containing 8% WPC and 0.6% CMC. However, the use of rosemary extract enhanced oxidation in the kernels. Fresh eggs coated with WPI-sodium montmorillonite nanoparticles and sodium metabisulphite maintained their internal quality for up to 8 weeks (at 25°C and 70% RH) (de Araújo Soares et al., 2021). This partially eliminated the need for refrigeration. A WPI and pectin-based edible film cross-linked with microbial transglutaminase was developed to improve the shelf life of roasted peanuts (Rossi-Márquez et al., 2021). The reduction in water uptake and oxygen permeability, and subsequently a reduced rate of oxidation, enabled the preservation of a crunchy texture, improving consumer acceptability.

Another study explored bi-layer edible films for sushi, using starch as the primary layer and a secondary layer comprising LAB-fermented whey and/or mango pulp powder (Hernández-Carranza et al., 2024). These films containing *Lactobacillus rhamnosus* at 10^9 CFU/mL and 5% whey or whey-mango solutions were tested under storage at 4°C and 20°C for 28 days. At lower temperatures, the films successfully maintained a sufficient probiotic count (10^6 CFU/mL) to deliver expected health benefits. LAB-fermented whey was used in another study in combination with purple sweet potato flour or potato starch for the preservation of surimi stored at 4°C and 20°C for up to 28 days (Macías-González et al., 2024). Coated surimi samples had good acceptability scores; however, the probiotic count (*Lb. rhamnosus)* was lower than 10^6, which was below the threshold to exert its beneficial effects.

7. Conclusions and future trends

Whey protein-based edible films and coatings significantly advance sustainable packaging solutions. As an outcome of the dairy sector, whey protein presents a unique opportunity to valorize an otherwise environmentally problematic waste product while addressing the growing demand for biodegradable, renewable packaging materials. Whey protein films' physical and barrier characteristics make them particularly suited for preserving food quality, increasing storage life, and protecting against oxygen and microbial contamination. Whey protein films demonstrate excellent transparency and flexibility, which enhances their appeal for food packaging applications, especially for highly perishable goods. Incorporating various additives, such as antimicrobial and antioxidant agents, into whey protein films can enhance food preservation and safety by inhibiting microbial growth and spoilage. This approach aligns well with the goals of modern food packaging, which seeks to increase storage life while ensuring the reliability and safety of the product. Moreover, whey protein-based films reflect the broader trend toward circular economy principles, where waste materials are repurposed for new applications, reducing reliance on non-renewable resources. The potential scalability of these films, combined with continuous improvements in processing and formulation, suggests that whey protein coatings could become an integral part of future food packaging strategies. Overall, whey protein-based films are a breakthrough in creating functional, environmentally conscious packaging solutions, paving the way for future innovations in the food packaging industry.

Recent trends highlight the increasing use of whey protein, which has excellent barrier properties against oxygen, making it an ideal candidate for increasing the storage life of perishable foods. Looking ahead, the future of edible whey protein films/coatings lies in the development of multifunctional films that not only serve as protective barriers but also deliver active substances such as antibacterials to enhance food safety and quality. As the field evolves, regulatory frameworks and consumer acceptance will play critical roles in the widespread adoption of these innovative solutions.

References

Abdulla, S.F., Shams, R., Dash, K.K., 2024. Edible packaging as sustainable alternative to synthetic plastic: a comprehensive review. Environ. Sci. Pollut. Control Ser. https://doi.org/10.1007/S11356-024-32806-Z.

Abedi, A., Lakzadeh, L., Amouheydari, M., 2021. Effect of an edible coating composed of whey protein concentrate and rosemary essential oil on the shelf life of fresh spinach. J. Food Process. Preserv. 45 (4), e15284. https://doi.org/10.1111/JFPP.15284.

Abeykoon, C., Kelly, A.L., Brown, E.C., Vera-Sorroche, J., Coates, P.D., Harkin-Jones, E., Howell, K.B., Deng, J., Li, K., Price, M., 2014. Investigation of the process energy demand in polymer extrusion: a brief review and an experimental study. Appl. Energy 136, 726–737. https://doi.org/10.1016/J.APENERGY.2014.09.024.

Akcan, T., Serdaroglu, M., 2022. Edible protein films containing bay leaves (Laurus Nobilis L.) and sage (Salvia Officinalis) extracts: preventing oxidation and application on refrigerated cooked meatballs. Dokuz Eylül Üniversitesi Mühendislik Fakültesi Fen ve Mühendislik Dergisi 24 (72), 927–938. https://doi.org/10.21205/DEUFMD.2022247221.

Alipour, A., Rahaiee, S., Rajaei Litkohi, H., Jamali, S.N., Jafari, S.M., 2023. Development and optimization of whey protein-Lepidium perfoliatum gum packaging films: an approach towards antimicrobial and biodegradable films. Ind. Crop. Prod. 196, 116447. https://doi.org/10.1016/J.INDCROP.2023.116447.

Alvarez-Perez, O.B., Ventura-Sobrevilla, J.M., Torres-León, C., Rojas-Molina, R., Rodríguez-Herrera, R., Aguilar-González, M.A., Aguilar, C.N., 2022. Development and characterization of whey protein films incorporated with tarbush polyphenols and candelilla wax. Food Biosci. 45, 101505. https://doi.org/10.1016/J.FBIO.2021.101505.

Amaral, D. de A., Silva, J.A. F. da., 2021. Whey in the industry: environmental and valorization impacts. Revista Científica Multidisciplinar Núcleo Do Conhecimento 41–57.

Andrade, M.A., Barbosa, C.H., Souza, V.G.L., Coelhoso, I.M., Reboleira, J., Bernardino, S., Ganhão, R., Mendes, S., Fernando, A.L., Vilarinho, F., Silva, A.S., Ramos, F., 2021. Novel active food packaging films based on whey protein incorporated with seaweed extract: development, characterization, and application in fresh poultry meat. Coatings 11 (2), 229. https://doi.org/10.3390/COATINGS11020229.

Aslam, M., Khalid, A., Tahir, G., Mukhtar, H., 2021. Recent developments in purification techniques for whey valorization. J. Biomed. Res. Environ. Sci. 2 (9), 876–887. https://doi.org/10.37871/JBRES1326.

Assefa, Z., Admassu, S., 2013. In: Development and Characterization of Antimicrobial Packaging Films. https://doi.org/10.4172/2157-7110.1000235.

Avramescu, S.M., Butean, C., Popa, C.V., Ortan, A., Moraru, I., Temocico, G., 2020. Edible and Functionalized films/coatings—performances and perspectives. Coatings 10 (7), 687. https://doi.org/10.3390/COATINGS10070687.

Basiak, E., Lenart, A., Debeaufort, F., 2017. Effects of carbohydrate/protein ratio on the microstructure and the barrier and sorption properties of wheat starch-whey protein blend edible films. J. Sci. Food Agric. 97 (3), 858–867. https://doi.org/10.1002/JSFA.7807.

Beristain-Bauza, S., Del Carmen, S., Mani-López, E., Palou, E., López-Malo, A., 2017. Antimicrobial activity of whey protein films supplemented with Lactobacillus sakei cell-free supernatant on fresh beef. Food Microbiol. 62, 207–211. https://doi.org/10.1016/j.fm.2016.10.024.

Beristain-Bauza, S., Mani-López, E., Palou, E., López-Malo, A., 2016. Antimicrobial activity and physical properties of protein films added with cell-free supernatant of Lactobacillus rhamnosus. Food Control 62, 44–51. https://doi.org/10.1016/J.FOODCONT.2015.10.007.

Botalo, A., Inprasit, T., Ummartyotin, S., Chainok, K., Vatthanakul, S., Pisitsak, P., 2024. Smart and UV-resistant edible coating and films based on alginate, whey protein, and curcumin. Polymers 16 (4), 447. https://doi.org/10.3390/POLYM16040447.

Brestenský, M., Nitrayová, S., Patráš, P., Heger, J., Nitray, J., 2015. Branched chain amino acids and their importance in nutrition. J. Microbiol. Biotechnol. Food Sci. 5 (2), 197–202. https://doi.org/10.15414/JMBFS.2015.5.2.197-202.

Brew, K., Castellino, F.J., Vanaman, T.C., Hill, R.L., 1970. The complete amino acid sequence of bovine α-lactalbumin. J. Biol. Chem. 245 (17), 4570–4582. https://doi.org/10.1016/S0021-9258(19)63827-5.

Bryaskova, R., Philipova, N., Bakov, V., Georgiev, N., 2025. Innovative antibacterial polymer coatings. Appl. Sci. 15 (4), 1780. https://doi.org/10.3390/APP15041780.

Buchanan, D., Martindale, W., Romeih, E., Hebishy, E., 2023. Recent advances in whey processing and valorisation: technological and environmental perspectives. Int. J. Dairy Technol. 76 (2), 291–312. https://doi.org/10.1111/1471-0307.12935.

Çakmak, H., Özselek, Y., Turan, O.Y., Fıratlıgil, E., Karbancioğlu-Güler, F., 2020. Whey protein isolate edible films incorporated with essential oils: antimicrobial activity and barrier properties. Polym. Degrad. Stabil. 179, 109285. https://doi.org/10.1016/J.POLYMDEGRADSTAB.2020.109285.

Calva-Estrada, S.J., Jiménez-Fernández, M., Lugo-Cervantes, E., 2019. Protein-based films: advances in the development of biomaterials applicable to food packaging. Food Eng. Rev. 11 (2), 78–92. https://doi.org/10.1007/S12393-019-09189-W.

Chaudhary, V., Kajla, P., Kumari, P., Bangar, S.P., Rusu, A., Trif, M., Lorenzo, J.M., 2022. Milk protein-based active edible packaging for food applications: an eco-friendly approach. Front. Nutr. 9, 942524. https://doi.org/10.3389/FNUT.2022.942524/XML/NLM.

Chen, G.Q., Qu, Y., Gras, S.L., Kentish, S.E., 2023. Separation technologies for whey protein fractionation. Food Eng. Rev. 15 (3), 438–465. https://doi.org/10.1007/S12393-022-09330-2.

Chen, H., Wang, J., Cheng, Y., Wang, C., Liu, H., Bian, H., Pan, Y., Sun, J., Han, W., 2019. Application of protein-based films and coatings for food packaging: a review. Polymers 11 (12), 2039. https://doi.org/10.3390/POLYM11122039.

Chen, X., Ding, X., Huang, Y., Zhao, Y., Chen, G., Xu, X., Xu, D., Jiao, B., Zhao, X., Liu, G., 2025. Recent advances in polysaccharide-based nanocomposite films for fruit preservation: construction, applications, and challenges. Foods 14 (6), 1012. https://doi.org/10.3390/FOODS14061012.

Chollakup, R., Pongburoos, S., Boonsong, W., Khanoonkon, N., Kongsin, K., Sothornvit, R., Sukyai, P., Sukatta, U., Harnkarnsujarit, N., 2020. Antioxidant and antibacterial activities of cassava starch and whey protein blend films containing rambutan peel extract and cinnamon oil for active packaging. Lebensm. Wiss. Technol. 130. https://doi.org/10.1016/J.LWT.2020.109573.

de Araújo Soares, R., Borges, S.V., Dias, M.V., Piccoli, R.H., Fassani, E.J., Silva, E.M. C. da., 2021. Impact of whey protein isolate/sodium montmorillonite/sodium metabisulfite coating on the shelf life of fresh eggs during storage. Lebensm. Wiss. Technol. 139, 110611. https://doi.org/10.1016/J.LWT.2020.110611.

Di Pierro, P., Mariniello, L., Giosafatto, V.L., Esposito, M., Sabbah, M., Porta, R., 2018. Dairy whey protein-based edible films and coatings for food preservation. Food Packag. Preserv. 439–456. https://doi.org/10.1016/B978-0-12-811516-9.00013-0.

European Union, 2009. In: Commission Regulation (EC) No 450/2009 of 29 May 2009 on Active and Intelligent Materials and Articles Intended to Come into Contact with Food. https://eur-lex.europa.eu/legal-content/EN/ALL/?uri=CELEX:32009R0450.

Fahrullah, M.E., Rosyidi, D., 2021. Characterization and antimicrobial activity of whey edible film composite enriched with clove essential oil. Trop. Anim. Sci. J. 44 (3), 369–376. https://doi.org/10.5398/TASJ.2021.44.3.369.

Filho, J.G.O., Miranda, M., Ferreira, M.D., Plotto, A., 2021. Nanoemulsions as edible coatings: a potential strategy for fresh fruits and vegetables preservation. Foods 10, 2438. https://doi.org/10.3390/FOODS10102438.

Fortune Business Insights, 2025. In: Whey Protein Market Size, Share and Trends Report. www.grandviewresearch.com/industry-analysis/whey-protein-market#:~:text=Report%20Overview,10.48%25%20from%202022%20to%202030.

Frestedt, J.L., Zenk, J.L., Kuskowski, M.A., Ward, L.S., Bastian, E.D., 2008. A whey-protein supplement increases fat loss and spares lean muscle in obese subjects: a randomized human clinical study. Nutr. Metab. 5 (1). https://doi.org/10.1186/1743-7075-5-8.

Gagliarini, N., Diosma, G., Garrote, G.L., Abraham, A.G., Piermaria, J., 2019. Whey protein-kefiran films as driver of probiotics to the gut. Lebensm. Wiss. Technol. 105, 321–328. https://doi.org/10.1016/J.LWT.2019.02.023.

Galus, S., Kadzińska, J., 2015. Food applications of emulsion-based edible films and coatings. Trends Food Sci. Technol. 45 (2), 273–283. https://doi.org/10.1016/J.TIFS.2015.07.011.

Galus, S., Mikus, M., Ciurzyńska, A., Domian, E., Kowalska, J., Marzec, A., Kowalska, H., 2021. The effect of whey protein-based edible coatings incorporated with lemon and lemongrass essential oils on the quality attributes of fresh-cut pears during storage. Coatings 11 (7), 745. https://doi.org/10.3390/COATINGS11070745/S1.

Gangurde, H.H., Chordiya, M.A., Patil, P.S., Baste, N.S., 2011. Whey protein. Scholars' Res. J. 1 (2), 69–78. https://doi.org/10.4103/2249-5975.99663.

Ganju, S., Gogate, P.R., 2017. A review on approaches for efficient recovery of whey proteins from dairy industry effluents. J. Food Eng. 215, 84–96. https://doi.org/10.1016/J.JFOODENG.2017.07.021.

Ghoshal, G., Shivani, 2022. Thyme essential oil nano-emulsion/tamarind starch/whey protein concentrate novel edible films for tomato packaging. Food Control 138, 108990. https://doi.org/10.1016/J.FOODCONT.2022.108990.

Gökkaya Erdem, B., Dıblan, S., Kaya, S., 2021. A comprehensive study on sorption, water barrier, and physicochemical properties of some protein- and carbohydrate-based edible films. Food Bioprocess Technol. 14 (11), 2161–2179. https://doi.org/10.1007/S11947-021-02712-0.

González-García, S., Hospido, A., Moreira, M.T., Feijoo, G., Arroja, L., 2013. Environmental life cycle assessment of a galician cheese: San Simon da Costa. J. Clean. Prod. 52, 253–262. https://doi.org/10.1016/j.jclepro.2013.03.006.

Guimarães, A., Abrunhosa, L., Pastrana, L.M., Cerqueira, M.A., 2018. Edible films and coatings as carriers of living microorganisms: a new strategy towards biopreservation and healthier foods. Compr. Rev. Food Sci. Food Saf. 17 (3), 594–614. https://doi.org/10.1111/1541-4337.12345.

Guimarães, A., Ramos, Ó., Cerqueira, M., Venâncio, A., Abrunhosa, L., 2020. Active whey protein edible films and coatings incorporating lactobacillus buchneri for penicillium nordicum control in cheese. Food Bioprocess Technol. 13 (6), 1074–1086. https://doi.org/10.1007/S11947-020-02465-2/TABLES/4.

Gupta, D., Lall, A., Kumar, S., Patil, T.D., Gaikwad, K.K., 2024. Plant-based edible films and coatings for food-packaging applications: recent advances, applications, and trends. Sustain. Food Technol. 2 (5), 1428–1455. https://doi.org/10.1039/D4FB00110A.

Hammam, A.R.A., 2019. Technological, applications, and characteristics of edible films and coatings: a review. SN Appl. Sci. 1 (6), 1–11. https://doi.org/10.1007/S42452-019-0660-8/FIGURES/2.

Han, J.H., 2014. Edible films and coatings: a review. In: Innovations in Food Packaging, second ed., pp. 213–255. https://doi.org/10.1016/B978-0-12-394601-0.00009-6.

Hauser, C., Wunderlich, J., 2011. Antimicrobial packaging films with a sorbic acid based coating. Procedia Food Sci. 1, 197–202. https://doi.org/10.1016/J.PROFOO.2011.09.031.

He, Z., Zhao, J., Liu, C., Li, W., Wang, Y., 2022. Ameliorating effect of γ-aminobutyric acid on the physical performance of whey protein films. Food Hydrocoll. 124, 107207. https://doi.org/10.1016/J.FOODHYD.2021.107207.

Henriques, M., Gomes, D., Pereira, C., 2015. Whey protein edible coatings: recent developments and applications. In: Nedovic, V., Raspor, P., Levic, J., Daponjac, V.T., V Barbosa-Canovas, G. (Eds.), Emerging and Traditional Technologies for Safe, Healthy and Quality Food. Springer, pp. 177–196. https://doi.org/10.1007/978-3-319-24040-4_10.

Hernández-Carranza, P., Fierro-Corona, G., Tapia-Maruri, D., Ruiz-Martínez, I., Ávila-Reyes, S.V., Ruiz-López, I.I., Ochoa-Velasco, C.E., 2023. Bioactive edible films based on LAB-fermented whey solution and potato starch: characterization and storage behavior. Food Bioprocess Technol. 16 (12), 3045–3056. https://doi.org/10.1007/S11947-023-03115-Z/FIGURES/4.

Hernández-Carranza, P., Mendoza-Gutiérrez, B.A., Estévez-Sánchez, K.H., Ramírez-López, C., Beristain-Bauza, S. del C., Avila-Reyes, S.V., Ruíz-López, I.I., Ochoa-Velasco, C.E., 2024. Development and application of bioactive bi-layer edible films based on starch and LAB-fermented whey and/or mango solution. Fermentation 10 (2), 105. https://doi.org/10.3390/FERMENTATION10020105.

Herrera-vázquez, S.E., Dublán-garcía, O., Arizmendi-cotero, D., Gómez-oliván, L.M., Islas-flores, H., Hernández-navarro, M.D., Ramírez-durán, N., 2022. Optimization of the physical, optical and mechanical properties of composite edible films of gelatin, whey protein and chitosan. Molecules 27 (3), 869. https://doi.org/10.3390/MOLECULES27030869.

Hosseini, H., Yaghoubi Hamgini, E., Jafari, S.M., Bolourian, S., 2020. Improving the oxidative stability of sunflower seed kernels by edible biopolymeric coatings loaded with rosemary extract. J. Stored Prod. Res. 89, 101729. https://doi.org/10.1016/J.JSPR.2020.101729.

Ibáñez, B., Melero, A., Montoro, A., San Onofre, N., Soriano, J.M., 2024. Molecular insights into radiation effects and protective mechanisms: a focus on cellular damage and radioprotectors. Curr. Issues Mol. Biol. 46 (11), 12718–12732. https://doi.org/10.3390/CIMB46110755.

Infurna, G., Cavallaro, G., Lazzara, G., Milioto, S., Dintcheva, N.T., 2021. Understanding the effects of crosslinking and reinforcement agents on the performance and durability of biopolymer films for cultural heritage protection. Molecules 26 (11), 3468. https://doi.org/10.3390/MOLECULES26113468.

Jeya Jeevahan, J., Chandrasekaran, M., Venkatesan, S.P., Sriram, V., Britto Joseph, G., Mageshwaran, G., Durairaj, R.B., 2020. Scaling up difficulties and commercial aspects of edible films for food packaging: a review. Trends Food Sci. Technol. 100, 210–222. https://doi.org/10.1016/J.TIFS.2020.04.014.

Jooyandeh, H., 2011. Whey protein films and coatings: a review. Pakistan J. Nutr. 10 (3), 293–301. https://doi.org/10.3923/PJN.2011.296.301.

Kandasamy, S., Yoo, J., Yun, J., Kang, H.-B., Seol, K.-H., Kim, H.-W., Ham, J.-S., 2021. Application of whey protein-based edible films and coatings in food industries: an updated overview. Coatings 11 (9), 1056. https://doi.org/10.3390/COATINGS11091056.

Kaneto, K., Uto, S., 2020. A new method to evaluate oxygen permeability and biofuel diffusion in barrier films using fuel cells. Trans. Mater. Res. Soc. Jpn 45 (4), 131–134. https://doi.org/10.14723/TMRSJ.45.131.

Kheto, A., Adhikary, U., Dhua, S., Sarkar, A., Kumar, Y., Vashishth, R., Shrestha, B.B., Saxena, D.C., 2024. A review on advancements in emerging processing of whey protein: enhancing functional and nutritional properties for functional food applications. Food Safe. Health 3 (1), 23–45. https://doi.org/10.1002/FSH3.12067.

Khwaldia, K., Banon, S., Desobry, S., Hardy, J., 2004. Mechanical and barrier properties of sodium caseinate-anhydrous milk fat edible films. Int. J. Food Sci. Technol. 39 (4), 403–411. https://doi.org/10.1111/J.1365-2621.2004.00797.X.

Kim, D., Thoma, G., Nutter, D., Milani, F., Ulrich, R., Norris, G., 2013. Life cycle assessment of cheese and whey production in the USA. Int. J. Life Cycle Assess. 18 (5), 1019–1035. https://doi.org/10.1007/S11367-013-0553-9/FIGURES/9.

Kim, S.J., Ustunol, Z., 2001. Solubility and moisture sorption isotherms of whey-protein-based edible films as influenced by lipid and plasticizer incorporation. J. Agric. Food Chem. 49 (9), 4388–4391. https://doi.org/10.1021/JF010122Q/ASSET/IMAGES/LARGE/JF010122QF00002.JPEG.

Kontogianni, V.G., Kasapidou, E., Mitlianga, P., Mataragas, M., Pappa, E., Kondyli, E., Bosnea, L., 2022. Production, characteristics and application of whey protein films activated with rosemary and sage extract in preserving soft cheese. Lebensm. Wiss. Technol. 155, 112996. https://doi.org/10.1016/J.LWT.2021.112996.

Kontogianni, V.G., Kosma, I., Mataragas, M., Pappa, E., Badeka, A.V., Bosnea, L., 2023. Innovative intelligent cheese packaging with whey protein-based edible films containing spirulina. Sustainability 15 (18), 13909. https://doi.org/10.3390/SU151813909/S1.

kosh goftar, M., Bolandi, M., Shariatifar, N., Ziaolhagh, S., 2025. Incorporation of whey protein with aqueous clove (Syzygium aromaticum L.) extract (nanoliposome and free forms) as an edible coating to prevent bacterial growth and oxidative spoilage of brown trout fillets. Food Control 176, 111386. https://doi.org/10.1016/J.FOODCONT.2025.111386.

Krochta, J., Perez-Gago, M., 2002. Formation and properties of whey protein films and coatings. In: Protein-Based Films and Coatings. CRC Press. https://doi.org/10.1201/9781420031980.CH6.

Kumar, L., Ramakanth, D., Akhila, K., Gaikwad, K.K., Kumar, L., Ramakanth, D., Akhila, K., Gaikwad, K.K., 2022. Edible films and coatings for food packaging applications: a review. EnvCL 20 (1), 875–900. https://doi.org/10.1007/S10311-021-01339-Z.

Kumar, N., Pratibha, P.J., Yadav, A., Upadhyay, A., Neeraj, S.S., Petkoska, A.T., Heena, S.S., Gniewosz, M., Kieliszek, M., 2023. Recent trends in edible packaging for food applications — perspective for the future. Food Eng. Rev. 15 (4), 718–747. https://doi.org/10.1007/S12393-023-09358-Y.

La Storia, A., Di Giuseppe, F.A., Volpe, S., Oliviero, V., Villani, F., Torrieri, E., 2020. Physical properties and antimicrobial activity of bioactive film based on whey protein and Lactobacillus curvatus 54M16 producer of bacteriocins. Food Hydrocoll. 108, 105959. https://doi.org/10.1016/J.FOODHYD.2020.105959.

Lainioti, G.C., Druvari, D., 2024. Designing antibacterial-based quaternary ammonium coatings (surfaces) or films for biomedical applications: recent advances. Int. J. Mol. Sci. 25 (22), 12264. https://doi.org/10.3390/IJMS252212264.

Lam, F.-C., Khan, T.M., Faidah, H., Haseeb, A., Khan, A.H., 2019. Effectiveness of whey protein supplements on the serum levels of amino acid, creatinine kinase and myoglobin of athletes: a systematic review and meta-analysis. Syst. Rev. 8 (1), 130. https://doi.org/10.1186/S13643-019-1039-Z.

Li, Y., Sun, J., Ai, C., Song, S., Yang, J., 2025. Edible nano-film incorporated with lemon essential oil-loaded pickering emulsion for cold storage improvement of Penaeus vannameia. Food Biosci. 69, 106900. https://doi.org/10.1016/J.FBIO.2025.106900.

Macías-González, A., Hernández-Carranza, P., Ruiz-López, I.I., Ochoa-Velasco, C.E., 2024. Using LAB-fermented whey for developing bioactive edible films based on purple sweet potato flour/potato starch. Int. J. Food Sci. Technol. 59 (4), 2632–2646. https://doi.org/10.1111/IJFS.17007.

Macwan, S.R., Dabhi, B.K., Parmar, S.C., Aparnathi, K.D., 2016. Whey and its utilization. Int. J. Curr. Microbiol. Appl. Sci. 5 (8), 134–155. https://doi.org/10.20546/ijcmas.2016.508.016.

Madureira, A.R., Pereira, C.I., Gomes, A.M.P., Pintado, M.E., Xavier Malcata, F., 2007. Bovine whey proteins – overview on their main biological properties. Food Res. Int. 40 (10), 1197. https://doi.org/10.1016/J.FOODRES.2007.07.005.

Malaka, R., Maruddin, F., Arief, F.A., Hakim, W., Irwansyah, K., Putranto, W.S., Astawan, M., Kadir, R.W., 2024. Evaluation dangke cheese processing by edible film coating made from whey combined with konjac flour. J. Food Process. Preserv. 2024 (1), 4676923. https://doi.org/10.1155/2024/4676923.

Martins, V.F.R., Pintado, M.E., Morais, R.M.S.C., Morais, A.M.M.B., 2024. Recent highlights in sustainable bio-based edible films and coatings for fruit and vegetable applications. Foods 13 (2), 318. https://doi.org/10.3390/FOODS13020318.

Martins, Y.A.A., Ferreira, S.V., Silva, N.M., Sandre, M.F.B., Filho, J.G.O., Leão, P.V.T., Leão, K.M., Nicolau, E.S., Plácido, G.R., Egea, M.B., da Silva, M.A.P., 2020. Edible films of whey and cassava starch: physical, thermal, and microstructural characterization. Coatings 10 (11), 1059. https://doi.org/10.3390/COATINGS10111059.

Mellinas, C., Valdés, A., Ramos, M., Burgos, N., Del Carmen Garrigós, M., Jiménez, A., 2016. Active edible films: current state and future trends. J. Appl. Polym. Sci. 133 (2). https://doi.org/10.1002/APP.42631.

Minj, S., Anand, S., 2020. Whey proteins and its derivatives: bioactivity, functionality, and current applications. Dairy 1 (3), 233–258. https://doi.org/10.3390/DAIRY1030016.

Mirabella, N., Castellani, V., Sala, S., 2014. Current options for the valorization of food manufacturing waste: a review. J. Clean. Prod. 65, 28–41. https://doi.org/10.1016/J.JCLEPRO.2013.10.051.

Mohammadian, M., Moghaddam, A.D., Sharifan, A., Dabaghi, P., Hadi, S., 2021. Structural, physico-mechanical, and bio-functional properties of whey protein isolate-based edible films as affected by enriching with nettle (Urtica dioica L.) leaf extract. J. Food Meas. Char. 15 (5), 4051–4060. https://doi.org/10.1007/S11694-021-00988-6/TABLES/3.

Muhammad, D.R.A., Sedaghat Doost, A., Gupta, V., bin Sintang, M.D., Van de Walle, D., Van der Meeren, P., Dewettinck, K., 2020. Stability and functionality of xanthan gum–shellac nanoparticles for the encapsulation of cinnamon bark extract. Food Hydrocoll. 100. https://doi.org/10.1016/J.FOODHYD.2019.105377.

Muley, A.B., Singhal, R.S., 2020. Extension of postharvest shelf life of strawberries (Fragaria ananassa) using a coating of chitosan-whey protein isolate conjugate. Food Chem. 329. https://doi.org/10.1016/j.foodchem.2020.127213.

Nemati, V., Hashempour-Baltork, F., Sadat Gharavi-Nakhjavani, M., Feizollahi, E., Marangoni Júnior, L., Mirza Alizadeh, A., 2023. Application of a whey protein edible film incorporated with cumin essential oil in cheese preservation. Coatings 13 (8), 1470. https://doi.org/10.3390/COATINGS13081470.

Pan, J.-N., Sun, J., Shen, Q.-J., Zheng, X., Zhou, W.-W., Pan, J.-N., Sun, J., Shen, Q.-J., Zheng, X., Zhou, W.-W., 2025. Fabrication, properties, and improvement strategies of edible films for fruits and vegetables preservation a comprehensive review. Food Innov. Adv. 4 (1), 43–52. https://doi.org/10.48130/FIA-0025-0003.

Papadaki, A., Kachrimanidou, V., K. Lappa, I., Andriotis, H., Eriotou, E., Mandala, I., Kopsahelis, N., 2022. Tuning the physical and functional properties of whey protein edible films: effect of pH and inclusion of antioxidants from spent coffee grounds. Sustain. Chem. Pharm. 27, 100700. https://doi.org/10.1016/J.SCP.2022.100700.

Patel, S., 2015. Functional food relevance of whey protein: a review of recent findings and scopes ahead. J. Funct.Foods 19, 308–319. https://doi.org/10.1016/J.JFF.2015.09.040.

Pires, A.F., Díaz, O., Cobos, A., Pereira, C.D., 2024. A review of recent developments in edible films and coatings-focus on whey-based materials. Foods 13 (16), 2638. https://doi.org/10.3390/FOODS13162638.

Pluta-Kubica, A., Jamróz, E., Kawecka, A., Juszczak, L., Krzyściak, P., 2020. Active edible furcellaran/whey protein films with yerba mate and white tea extracts: preparation, characterization and its application to fresh soft rennet-curd cheese. Int. J. Biol. Macromol. 155, 1307–1316. https://doi.org/10.1016/J.IJBIOMAC.2019.11.102.

Poonia, A., 2019. Recent trends in nanomaterials used in dairy industry. Nanosci. Sustain. Agricult. 375–396. https://doi.org/10.1007/978-3-319-97852-9_17.

Pratiwi, D., Saputri, R.A., Setiyawan, A.I., 2024. Whey edible film enhanced with ginger essential oil: characterization and antimicrobial activity. IOP Conf. Ser. Earth Environ. Sci. 1377 (1), 012036. https://doi.org/10.1088/1755-1315/1377/1/012036.

Prazeres, A.R., Carvalho, F., Rivas, J., 2012. Cheese whey management: a review. J. Environ. Manag. 110, 48–68. https://doi.org/10.1016/J.JENVMAN.2012.05.018.

Purewal, S.S., Kaur, A., Bangar, S.P., Singh, P., Singh, H., 2023. Protein-based films and coatings: an innovative approach. Coatings 14 (1), 32. https://doi.org/10.3390/COATINGS14010032.

Rajeshwari, K.V., Balakrishnan, M., Kansal, A., Lata, K., Kishore, V.V.N., 2000. State-of-the-art of anaerobic digestion technology for industrial wastewater treatment. Renew. Sustain. Energy Rev. 4 (2), 135–156. https://doi.org/10.1016/S1364-0321(99)00014-3.

Ramos, O.L., Fernandes, J.C., Silva, S.I., Pintado, M.E., Malcata, F.X., 2012a. Edible films and coatings from whey proteins: a review on formulation, and on mechanical and bioactive properties. Crit. Rev. Food Sci. Nutr. 52 (6), 533–552. https://doi.org/10.1080/10408398.2010.500528.

Ramos, O.L., Pereira, J.O., Silva, S.I., Fernandes, J.C., Franco, M.I., Lopes-da-Silva, J.A., Pintado, M.E., Malcata, F.X., 2012b. Evaluation of antimicrobial edible coatings from a whey protein isolate base to improve the shelf life of cheese. J. Dairy Sci. 95 (11), 6282–6292. https://doi.org/10.3168/JDS.2012-5478.

Ramos, O.L., Reinas, I., Silva, S.I., Fernandes, J.C., Cerqueira, M.A., Pereira, R.N., Vicente, A.A., Fátima Poças, M., Pintado, M.E., Malcata, F.X., 2013. In: Effect of Whey Protein Purity and Glycerol Content upon Physical Properties of Edible Films Manufactured Therefrom. https://doi.org/10.1016/j.foodhyd.2012.05.001.

Rivas, J., Prazeres, A.R., Carvalho, F., 2011. Aerobic biodegradation of precoagulated cheese whey wastewater. J. Agric. Food Chem. 59 (6), 2511–2517. https://doi.org/10.1021/JF104252W.

Robalo, J., Lopes, M., Cardoso, O., Silva, A.S., Ramos, F., 2022. Efficacy of whey protein film incorporated with Portuguese green tea (Camellia sinensis L.) extract for the preservation of Latin-style fresh cheese. Foods 11 (8), 1158. https://doi.org/10.3390/FOODS11081158.

Rodriguez-Turienzo, L., Cobos, A., Diaz, O., 2013. Effects of microbial transglutaminase added edible coatings based on heated or ultrasound-treated whey proteins in physical and chemical parameters of frozen Atlantic salmon (Salmo salar). J. Food Eng. 119 (3), 433–438. https://doi.org/10.1016/J.JFOODENG.2013.06.015.

Roma, R., Corrado, S., De Boni, A., Forleo, M.B., Fantin, V., Moretti, M., Palmieri, N., Vitali, A., Camillo, D.C., 2015. Life cycle assessment in the livestock and derived edible products sector. In: Life Cycle Assessment in the Agri-Food Sector, pp. 251–332. https://doi.org/10.1007/978-3-319-11940-3_5.

Rossi-Márquez, G., Helguera, M., Briones, M., Dávalos-Saucedo, C.A., Di Pierro, P., 2021. Edible coating from enzymatically reticulated whey protein-pectin to improve shelf life of roasted peanuts. Coatings 11 (3), 329. https://doi.org/10.3390/COATINGS11030329.

Rubilar, J.F., Zúñiga, R.N., Osorio, F., Pedreschi, F., 2015. Physical properties of emulsion-based hydroxypropyl methylcellulose/whey protein isolate (HPMC/WPI) edible films. Carbohydr. Polym. 123, 27–38. https://doi.org/10.1016/j.carbpol.2015.01.010.

Sabato, S.F., Ouattara, B., Yu, H., D'Aprano, G., Le Tien, G., Mateescu, M.A., Lacroix, M., 2001. Mechanical and barrier-properties of cross-linked soy and whey protein based films. J. Agric. Food Chem. 49 (3), 1397–1403. https://doi.org/10.1021/JF0005925/ASSET/IMAGES/LARGE/JF0005925F00003.JPEG.

Sabzipour-Hafshejani, F., Mirzapour-Kouhdasht, A., Khodaei, D., Taghizadeh, M.S., Garcia-Vaquero, M., 2022. Impact of whey protein edible coating containing fish gelatin hydrolysates on physicochemical, microbial, and sensory properties of chicken breast fillets. Polymers 14 (16), 3371. https://doi.org/10.3390/POLYM14163371.

Saklani, P., Kishor Das, S., Mohan Singh, S., 2019. A review of edible packaging for foods. Int. J. Curr. Microbiol. Appl. Sci. 8 (7), 2885–2895. https://doi.org/10.20546/ijcmas.2019.807.359.

Sala, S., Bianchi, A., Wolf, M.-A., Bligny, J.-C., Mubareka, S., Vandecasteele, I., Bouraoui, F., Castellani, V., De Camillis, C., 2012. In: Water Footprint in the Context of Sustainability Assessment. Report on the Application of Life Cycle Based Indicators of Water Consumption in the Context of Integrated Sustainability Impact Analysis. https://doi.org/10.2788/81274.

Sanyang, M.L., Sapuan, S.M., Jawaid, M., Ishak, M.R., Sahari, J., 2015. Effect of plasticizer type and concentration on tensile, thermal and barrier properties of biodegradable films based on sugar palm (Arenga pinnata) starch. Polymers 7 (6), 1106–1124. https://doi.org/10.3390/POLYM7061106.

Schmid, M., Hinz, L.V., Wild, F., Noller, K., 2013. Effects of hydrolysed whey proteins on the techno-functional characteristics of whey protein-based films. Materials 6 (3), 927–940. https://doi.org/10.3390/MA6030927.

Schmid, M., Müller, K., 2019. Whey protein-based packaging films and coatings. In: Whey Proteins: From Milk to Medicine, pp. 407–437. https://doi.org/10.1016/B978-0-12-812124-5.00012-6.

Seydim, A.C., Sarikus-Tutal, G., Sogut, E., 2020. Effect of whey protein edible films containing plant essential oils on microbial inactivation of sliced Kasar cheese. Food Packag. Shelf Life 26, 100567. https://doi.org/10.1016/J.FPSL.2020.100567.

Shendurse, A., Gopikrishna, G., Patel, A., Pandya, A., 2018. Milk protein based edible films and coatings–preparation, properties and food applications. J. Nutrit. Health Food Eng. 8 (2). https://doi.org/10.15406/JNHFE.2018.08.00273.

Silveira, M.F.A., Soares, N.F.F., Geraldine, R.M., Andrade, N.J., Botrel, D.A., Gonçalves, M.P.J., 2007. Active film incorporated with sorbic acid on pastry dough conservation. Food Control 18 (9), 1063–1067. https://doi.org/10.1016/J.FOODCONT.2006.07.004.

Socaciu, M.I., Fogarasi, M., Semeniuc, C.A., Socaci, S.A., Rotar, M.A., Mureşan, V., Pop, O.L., Vodnar, D.C., 2020. Formulation and characterization of antimicrobial edible films based on whey protein isolate and tarragon essential oil. Polymers 12 (8), 1748. https://doi.org/10.3390/POLYM12081748.

Socaciu, M.I., Fogarasi, M., Simon, E.L., Semeniuc, C.A., Socaci, S.A., Podar, A.S., Vodnar, D.C., 2021. Effects of whey protein isolate-based film incorporated with tarragon essential oil on the quality and shelf-life of refrigerated brook trout. Foods 10 (2), 401. https://doi.org/10.3390/FOODS10020401.

Sogut, E., Filiz, B.E., Seydim, A.C., 2022. Whey protein isolate- and carrageenan-based edible films as carriers of different probiotic bacteria. J. Dairy Sci. 105 (6), 4829–4842. https://doi.org/10.3168/JDS.2021-21245.

Sothornvit, R., Krochta, J.M., 2001. Plasticizer effect on mechanical properties of β-lactoglobulin films. J. Food Eng. 50 (3), 149–155. https://doi.org/10.1016/S0260-8774(00)00237-5.

Sothornvit, R., Olsen, C.W., McHugh, T.H., Krochta, J.M., 2003. Formation conditions, water-vapor permeability, and solubility of compression-molded whey protein films. J. Food Sci. 68 (6), 1985–1999. https://doi.org/10.1111/J.1365-2621.2003.TB07006.X.
Taghavi Kevij, H., Salami, M., Mohammadian, M., Khodadadi, M., 2020. Fabrication and investigation of physicochemical, food simulant release, and antioxidant properties of whey protein isolate-based films activated by loading with curcumin through the pH-driven method. Food Hydrocoll. 108, 106026. https://doi.org/10.1016/J.FOODHYD.2020.106026.
Tan, S.X., Andriyana, A., Ong, H.C., Lim, S., Pang, Y.L., Ngoh, G.C., 2022. A comprehensive review on the emerging roles of nanofillers and plasticizers towards sustainable starch-based bioplastic fabrication. Polymers 14 (4), 664. https://doi.org/10.3390/POLYM14040664.
Tarawneh, H., Obeidat, H., Shawaqfeh, S., Al-massad, M., Alrabadi, N., Hiary, M., Talhouni, M., Alrosan, M., 2025. Effectiveness of edible coating in extending shelf life and enhancing quality properties of golden delicious apple. Disc. Food 5 (1), 1–8. https://doi.org/10.1007/S44187-025-00425-3.
Tavares, L., Souza, H.K.S., Gonçalves, M.P., Rocha, C.M.R., 2021. Physicochemical and microstructural properties of composite edible film obtained by complex coacervation between chitosan and whey protein isolate. Food Hydrocoll. 113, 106471. https://doi.org/10.1016/J.FOODHYD.2020.106471.
Tharanathan, R.N., 2003. Biodegradable films and composite coatings: past, present and future. Trends Food Sci. Technol. 14 (3), 71–78. https://doi.org/10.1016/S0924-2244(02)00280-7.
Umaraw, P., Munekata, P.E.S., Verma, A.K., Barba, F.J., Singh, V.P., Kumar, P., Lorenzo, J.M., 2020. Edible films/coating with tailored properties for active packaging of meat, fish and derived products. Trends Food Sci. Technol. 98, 10–24. https://doi.org/10.1016/J.TIFS.2020.01.032.
United Nations, 2024. Progress towards the Sustainable Development Goals. https://unstats.un.org/sdgs/files/report/2024/SG-SDG-Progress-Report-2024-advanced-unedited-version.pdf.
Ustunol, Z., Mert, B., 2006. Water solubility, mechanical, barrier, and thermal properties of cross-linked whey protein isolate-based films. J. Food Sci. 69 (3). https://doi.org/10.1111/J.1365-2621.2004.TB13365.X.
Utama, G.L., Dinika, I., Nurmilah, S., Masruchin, N., Nurhadi, B., Balia, R.L., 2022. Characterization of antimicrobial composite edible film formulated from fermented cheese whey and cassava peel starch. Membranes 12 (6), 636. https://doi.org/10.3390/MEMBRANES12060636.
van Middelaar, C.E., Berentsen, P.B.M., Dolman, M.A., de Boer, I.J.M., 2011. Eco-efficiency in the production chain of Dutch semi-hard cheese. Livest. Sci. 139 (1–2), 91–99. https://doi.org/10.1016/J.LIVSCI.2011.03.013.
Vasiliauskaite, A., Mileriene, J., Songisepp, E., Rud, I., Muizniece-Brasava, S., Ciprovica, I., Axelsson, L., Lutter, L., Aleksandrovas, E., Tammsaar, E., Salomskiene, J., Serniene, L., Malakauskas, M., 2022. Application of edible coating based on liquid acid whey protein concentrate with indigenous lactobacillus helveticus for acid-curd cheese quality improvement. Foods 11 (21), 3353. https://doi.org/10.3390/FOODS11213353.
Wagh, Y.R., Pushpadass, H.A., Emerald, F.M.E., Nath, B.S., 2014. Preparation and characterization of milk protein films and their application for packaging of cheddar cheese. J. Food Sci. Technol. 51 (12), 3767–3775. https://doi.org/10.1007/S13197-012-0916-4.
Wang, C., Killpatrick, A., Humphrey, A., Guo, M., 2019. Whey protein functional properties and applications in food formulation. In: Whey Protein Production, Chemistry, Functionality, and Applications, pp. 157–204. https://doi.org/10.1002/9781119256052.CH7.
Wang, G., Guo, M., 2019. Manufacturing technologies of whey protein products. In: Guo, M. (Ed.), Whey Protein Production, Chemistry, Functionality and Applications. John Wiley and Sons, pp. 27–49.
Wang, M., Wei, Z., Zhang, Z., 2024. Antimicrobial edible films for food preservation: recent advances and future trends. Food Bioprocess Technol. 17 (6), 1391–1411. https://doi.org/10.1007/S11947-023-03178-Y.
Wang, X., Wang, Y., Zhu, H., Chen, S., Xue, J., Shen, C., Xiao, Z., Luo, Y., 2025. Emerging strategies for food preservation: recent advances and challenges in nanotechnology for edible coatings. Agricult. Prod. Process. Stor. 1, 13. https://doi.org/10.1007/s44462-025-00015-7.
Weng, S., Marcet, I., Rendueles, M., Díaz, M., 2024. Edible films from the laboratory to industry: a review of the different production methods. Food Bioprocess Technol. 18 (4), 3245–3271. https://doi.org/10.1007/S11947-024-03641-4.
Wu, Y., Wu, H., Hu, L., 2024. Recent advances of proteins, polysaccharides and lipids-based edible films/coatings for food packaging applications: a review. Food Biophys. 19 (1), 29–45. https://doi.org/10.1007/S11483-023-09794-7.
Xin, Y., Yang, C., Zhang, J., Xiong, L., 2023. Application of whey protein-based emulsion coating treatment in fresh-cut apple preservation. Foods 12 (6), 1140. https://doi.org/10.3390/FOODS12061140.
Xue, M., Ji, Y., Ou, J., Wang, F., Li, C., Lei, S., Li, W., 2019. Surface wettability and strong adhesion of medical polyurethane elastomer porous films by microphase separation. AIP Adv. 9 (7), 75309. https://doi.org/10.1063/1.5107459/22569.
Yoo, J., Kandasamy, S., Kim, H.W., Bae, H.-H., Ham, J.-S., 2023. Whey protein-based edible films and coatings in food industry. J. Dairy Sci. Biotechnol. 41 (4), 219–229. https://doi.org/10.22424/JDSB.2023.41.4.219.
Zhang, R., Wang, B., Zhang, F., Zheng, K., Liu, Y., 2024. Milk-derived antimicrobial peptides incorporated whey protein film as active coating to improve microbial stability of refrigerated soft cheese. Int. J. Food Microbiol. 419, 110751. https://doi.org/10.1016/J.IJFOODMICRO.2024.110751.
Zhang, X., Zhao, Y., Li, Y., Zhu, L., Fang, Z., Shi, Q., 2020. Physicochemical, mechanical and structural properties of composite edible films based on whey protein isolate/psyllium seed gum. Int. J. Biol. Macromol. 153, 892–901. https://doi.org/10.1016/J.IJBIOMAC.2020.03.018.

Part 3

Functional properties and health effects

Chapter 15

Functional properties of whey proteins in food formulation

Qihui Wu[1,2], Usman Amin[1,2] and Haotian Zheng[1,2]

[1]*Food Rheology Laboratory, Department of Food, Bioprocessing and Nutrition Sciences, North Carolina State University, Raleigh, NC, United States;* [2]*Southeast Dairy Foods Research Center, North Carolina State University, Raleigh, NC, United States*

1. Introduction

Depending on the cow's breed, bovine milk contains ~3.0%–3.7% (w/w) protein, which makes up ~25%–26% (w/w) milk dry matter, in which milk fat is included (Horne, 2017). There are two major types of proteins present in the milk protein sol as part of the aqueous phase of the Oil-in-Water (O/W) emulsion of fresh milk; they are caseins and whey proteins (WPs), making up approximately 80% and 20% (w/v) of total milk proteins, respectively (Horne, 2017). In general, milk proteins are functional components, which are isolated and processed, yielding different ingredients for food formulations (Patel et al., 2022). Milk proteins can construct different model structures, for example, gels, emulsions, and foams with diversified rheological and textural properties, which determine the physical stability and mouthfeel (Amin et al., 2024, 2025; An and Zheng, 2025a; Chen et al., 2020; Ondiek et al., 2021; Shi and Zheng, 2025; Zheng and Meletharayil, 2025). In general, upon different processing treatments, proteins derived either from milk or from plant sources may self-interact and may interact with nonmilk components, such as polysaccharides and phenolic compounds (PCs), impacting physical stability and texture of different consumer foods (Kumar et al., 2017, 2018, 2021a, 2021b, 2023; Lin et al., 2024a, 2024b; Rivero et al., 2024; Shao et al., 2025; Wu et al., 2025; Zheng, 2018). Milk protein ingredients available in the market are often in powdered form; therefore, protein powder functionalities, such as flowability and rehydration properties, are of important concern to the quality of processed foods; approaches for characterizing protein powder functionalities have been investigated and discussed in the literature (An and Zheng, 2025b). Knowledge and approaches regarding characterization of rheological and textural properties relevant to semisolid foods are available in different book chapters (Zheng, 2019, 2022). The chemistry and properties of milk caseins are elaborated in the literature (Horne, 2020; Zheng, 2024). The WP properties and whey products processing methods are discussed in "Chapter 5" of the present book.

Concentrated, isolated, and fractionated WPs derived from cheese whey and skim milk are widely used in different consumer foods for their high nutritional value and techno-functionalities. The active techno-functionalities of WPs allow them to be used as the primary building blocks for constructing diversified food structures, such as protein sol, gel, emulsion, foam, and/or a complex structure in the form of a combination of the referred model structures. In general, the acid- and sweet-whey streams contain five major proteins/polypeptides, including β-lactoglobulin (β-Lg), α-lactalbumin (α-La), immunoglobulins (e.g., IgG), bovine serum albumin (BSA), and proteose peptone (PP); glycomacropeptide (GMP) is found in cheese whey. These different WPs enable diversified health benefits or potential in terms of enhancing immune responses, contributing positive effects on body weight management and mental health; these benefits have been summarized in a book chapter elsewhere in review articles, in which example research outcomes were briefly introduced (Minj and Anand, 2020; Sharma, 2019; Yiğit et al., 2023). Therefore, to apply WP in processed foods in different forms and physical states, for example, fluids (beverages), semisolids (soft gels) and solids (protein bars, powders, and frozen desserts), studies on protein chemistry, food structure, and texture, have been conducted on mining and modifying WP functionalities (Amin et al., 2024, 2025; Imtiaz et al., 2012; Ipsen, 2017; Kumar et al., 2018, 2021a; Lian et al., 2026;

Whey. https://doi.org/10.1016/B978-0-443-27648-4.00021-9

Nicolai, 2016; Rivero et al., 2024; Sarkar et al., 2017; Shi and Zheng, 2025; Wagoner and Foegeding, 2017). Kelly (2019) presented comprehensive examples regarding the application of WP ingredients in different processed foods, such as beverages, fermented foods, protein bars, desserts, cheeses, etc.

WPs are heat-labile and can be denatured and aggregated. Undesirable and uncontrolled aggregation of WPs might result in sedimentation in fluid protein sols and syneresis or coarse textural attributes in protein gels. Self-assembling WPs into colloidal particles is an effective approach for improving WP ingredient functionalities for stabilizing different model systems, for example, gels, emulsions, and foams (Amin et al., 2024, 2025; Nicolai, 2016; Nicolai and Durand, 2013; Shi and Zheng, 2025). Besides self-assembling WPs, on the other hand, "chaperone-protein" effects on protein functionality modification have been introduced and elaborated (Kelly, 2019). "Chaperone-protein" effects are an intervention of thermal WP aggregation behavior by using caseins and other nonprotein ligands, for example, carbohydrates, to form glycated and/or conjugated particles (Chen et al., 2015; O'Kennedy and Mounsey, 2006; Wagoner and Foegeding, 2017). These restricted particle sizes of the resulting protein aggregates may be beneficial to the development of ultra high temperature (UHT) beverages with fortified amounts of WPs. In our opinion, "chaperone-protein" effects need to be investigated further for tackling issues of heat stability and coarse texture of high-protein food systems with significantly fortified amounts of WPs. The potential chaperone candidates may be casein micelles, fractionated caseins, vegan proteins, polysaccharides, and polyphenols. The scope of the present chapter focuses on (1) elaborating generic WP ingredient functionalities using three basic model systems, such as sols, interfaces, and gels, (2) self-assembling WPs as a modification approach for improving WP ingredient functionalities, and (3) chaperone–protein and ligand–protein interactions for modifications on WP ingredient functionalities for diversified applications.

2. Whey protein ingredient functionalities relevant to consumer foods

2.1 Whey protein dispersion

WP ingredients, such as whey protein isolate (WPI) and whey protein concentrate (WPC), are typically produced as spray-dried powders and redispersed in aqueous systems to utilize their ingredient functionalities, such as emulsifying, foaming, thickening, and gelling ability. In food formulations, WPs are typically part of aqueous dispersion systems, often in environments with varying pH levels, salt ions, and other interacting molecules. Their ability to disperse effectively and interact with these components is critical to their ingredient functionalities across a wide range of applications in the food and beverage industry. As globular proteins, WPs are compact, three-dimensional, nearly spherical particles stabilized by the net effect of covalent (i.e., disulfide linkage) and noncovalent (i.e., electrostatic, hydrophobic, hydrogen bonding, and van der Waals) interactions (Damodaran and Parkin, 2017; Nicolai, 2019). In the aqueous model system, the presence of salts (added or inherently present in WPI or WPC) and/or lowering pH toward their isoelectric point (IEP or pI) disturbed the electrostatic charge balance, resulting in the partial unfolding of the native structure of WPs (Barone et al., 2022; Nicolai, 2019). This can increase the turbidity of the WP dispersion, which may be detrimental to the appearance of WP-based beverage systems (Barone et al., 2022; Ryan et al., 2012). However, unfolding of protein can expose reactive groups, including the hydrophobic and free sulfhydryl (–SH) groups, which may be beneficial for their interfacial and gelation behaviors (refer to Sections 2.2 and 2.3). WP dispersions are also exposed to elevated temperatures under various processing conditions, causing their aggregation and gelation, which further influence their functionality for structuring and texturizing applications (refer to Section 3.1). Additionally, the processing of WP dispersions that include additional components like carbohydrates, polyphenols, and other proteins can lead to the formation of aggregates, conjugates, complexes, and complex-coacervate systems involving WPs. The relevant functionalities of such systems have been explored in the subsequent sections (refer to Sections 4–7).

2.2 Interfacial behaviors: Emulsification and foaming agent

Various food and beverage products, such as ice cream, salad dressings, spreads, toppings, etc., are structured as emulsions and/or foams. These systems rely on the interfacial phenomena to deliver appealing texture and distinctive sensory attributes. WPs are widely recognized as effective surface-active agents for these applications due to their ability to form a thin, dense, two-dimensional (2D) viscoelastic interfacial film that stabilizes the dispersed phase in emulsion and foam systems (Cao et al., 2018; Fan et al., 2021; Zhou et al., 2020). The manufacturing strategies and stabilization/destabilization mechanisms for emulsion and foam systems have been extensively discussed by McClements (2015) in Food Emulsions: Principles, Practices, and Techniques.

Interfacial adsorption kinetics of WPs, studied using drop tensiometry for oil–water/air–water interfaces, is typically described as a three-stage process consisting of, (i) diffusion-controlled adsorption, (ii) unfolding/penetration, and (iii) their rearrangement, to form a dense viscoelastic film at the oil–water/air–water interface, as shown in Fig. 15.1 (Zhou et al., 2021a, 2021b).

During homogenization and whipping processes, WPs diffuse through the bulk dispersion system and adsorb at the newly created oil–water/air–water interfaces, thereby reducing the interfacial tension and promoting the emulsion droplet and foam bubble formations. The initial adsorption of WP at the interface is a rapid process and is characterized as diffusion-controlled (Güzey et al., 2006; Tamm et al., 2012; Zhou et al., 2020, 2021b). At a given protein concentration, the diffusion-controlled transport of various WP subunits, such as α-La, β-Lg, BSA, etc., to the interface does not necessarily depend on their molecular size, as described by the Stokes–Einstein diffusion model (Schröder et al., 2017; Zhou et al., 2021a). Their diffusion and the subsequent adsorption also depend on their interfacial affinity, which is governed by a combination of electrostatic, hydration, hydrophobic, and conformational forces. For instance, among the major WPs, β-Lg has shown preferential adsorption at the oil–water interface, as compared to the α-La and BSA (Paulsson and Dejmek, 1992; Schröder et al., 2017). However, under acidic conditions, α-La showed higher interfacial activity, particularly at the air–water interface, due to its molten globule state. These behaviors have been attributed to their higher surface hydrophobicity in their respective states (Cornec et al., 2001). Additionally, WPs showed a faster adsorption rate at the oil–water interface relative to the air–water interface (Jung et al., 2010). It was suggested that hydrophobic patches may have a greater affinity for an oil phase compared to the air phase. After the diffusion-controlled adsorption, WPs undergo extensive conformational rearrangements at the interface during longer adsorption times. Among WPs, α-La has shown a much faster rate of unfolding as compared to β-Lg (Cornec et al., 1999). The unfolding of protein increases the number of interfacial contacts and allows the formation of inter/intramolecular interactions (covalent and noncovalent) with neighboring molecules, which result in a dense, 2D viscoelastic interfacial film (Sagis and Scholten, 2014; Zhou et al., 2020). β-Lg has been shown to form disulfide interactions in the interfacial film (Schröder et al., 2017). Such interactions can impart significant elasticity and mechanical strength in the 2D interfacial gel, resulting in improved

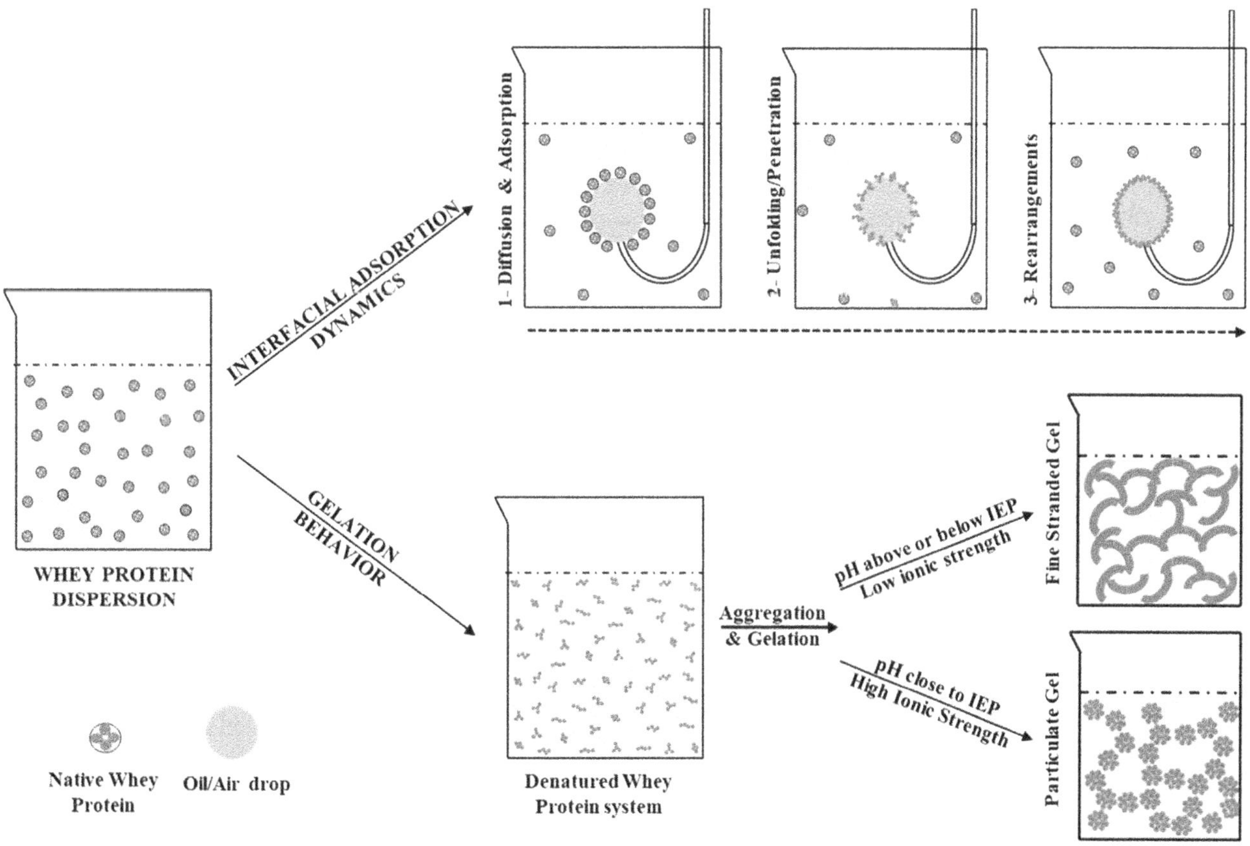

FIGURE 15.1 Schematic representation of interfacial adsorption behavior (top row) of whey proteins at oil–water/air–water interface and the gelation (bottom row) behavior in surrounding environments of varying pH and ionic strength.

stability against coalescence (Sagis and Scholten, 2014; Schröder et al., 2017). WP concentration is another important factor determining the stability of the interfacial film. The critical interfacial concentration (CIC) of WPs was found to be between 0.20% and 0.31% (w/w) for oil–water interfaces and 0.10%–0.13% (w/w) for air–water, respectively (Tamm et al., 2012; Zhou et al., 2021a). At CIC, WP fully covers to form a dense, monolayer interface, and more protein may destabilize or collapse the 2D interfacial arrangement.

Both in-process and complex structural modifications of WP have been made for the improvement of their interfacial adsorption dynamics and stability of the resulting interfacial film (Cornec et al., 2001; Schröder et al., 2017; Zhou et al., 2021a; Zhu, 2024). In-process parameters, such as protein concentration, pH, ionic strength, and application of heat treatment, can reduce electrochemical and conformational energy barriers in the interfacial film formation (Lam and Nickerson, 2015; Waniska and Kinsella, 1985; Zhou et al., 2021a). For instance, a decrease in pH and/or an increase in ionic strength reduces the electrostatic repulsions and helps unfold the native protein structure, facilitating the adsorption and protein–protein interactions at the interface. Similarly, heat treatment disrupts the native structure of WPs, exposing the hydrophobic patches, which improves their adsorption at the interface (Lam and Nickerson, 2015; Zhou et al., 2020). These aspects have been thoroughly reviewed elsewhere (Zhou et al., 2021b). The structural modifications of WPs include a combination of chemical processes, such as succinylation, complexation, and conjugation, physical processes, such as heat treatment and sonication, and enzymatic processes, such as hydrolysis and cross-linking (Zhu, 2024). These modifications disrupt the native conformation of WP, which mainly impacts its solubility and surface hydrophobicity (Zhu, 2024). In addition, some methods, such as WP aggregation and glycosylation, increase the viscosity of the WP dispersion systems, which have additional advantages against various emulsion/foam destabilization mechanisms. The impact of various modification approaches on the structure and functionality of WPs has been discussed in the subsequent sections.

2.3 Gelation

WP ingredients, including WPC and WPI, can form gels and gel-like structures under appropriate conditions, which have been widely utilized in food formulations to modify texture, improve water-holding capacity, and enhance structural integrity in a variety of food and beverage applications (Havea et al., 2004; Uribe-Alvarez et al., 2023; Verheul and Roefs, 1998). In general, gelation of WPs occurs through a sequential process of denaturation, aggregation, and network formation (gelation), as shown in Fig. 15.1 (Havea et al., 2004; Uribe-Alvarez et al., 2023; Verheul and Roefs, 1998). Denaturation of WP disrupts the native protein structure, exposing the initially buried hydrophobic amino acids, free sulfhydryl groups, and disulfide bonds between the cysteine residues, thereby facilitating irreversible aggregation (Nicolai et al., 2011; Nicolai and Durand, 2013). When the WP dispersion system is heated (temperature >65°C) above the critical gelation concentration (C_g), the denaturation, aggregation, and network formation occur simultaneously in a single-step process (Alting et al., 2003; Verheul and Roefs, 1998). The resulting gels are commonly referred to as heat-induced or heat-set gels. Alternatively, when the WP dispersions are heated below their C_g, the proteins denature and form aggregates in the initial step. These protein aggregate dispersion systems are made away from the IEP pH and at low ionic strength to maintain sufficient electrostatic repulsion among the protein molecules. In the second step, whey protein aggregate (WPA) dispersions can be cooled and induced into a gel structure by changes in pH, ionic strength, or the addition of cross-linking agents (Alting et al., 2003; Amin et al., 2024). The gelation, in this case, does not necessarily involve heat treatment; these gels are referred to as cold-induced or cold-set WP gels (Amin et al., 2024). C_g depends on the WP composition and the environment around (Nicolai et al., 2011). In general, C_g will reduce when the electrostatic repulsions between the WP molecules decrease. The properties of both heat-set and cold-set WP gels can be tailored for specific applications by manipulating the intrinsic and extrinsic properties of WP dispersions and have been discussed separately below:

Heat-set gels: Heat-set WP gels are of great interest in processed meat, bakery, and dairy products, as well as meat and dairy analogs, due to their ability to impart distinctive textural properties and enhance water-holding capacity within complex food matrices. In the heat-set gelation process, WP dispersions are heated to induce protein denaturation and irreversible aggregation that subsequently grow into a three-dimensional gel network through secondary aggregation (Brodkorb et al., 2016; Havea et al., 2009; Yan et al., 2023). The primary spatial structure of heat-set WP gels is formed at the gel point and is determined by the amount of aggregates at that point. Protein incorporated in the gel structures in the later stages (secondary aggregation) thickens the gel strands, resulting in a coarse gel structure (Verheul and Roefs, 1998). The aggregation and gelation process of heat-set WP gels can be modulated for desired attributes by manipulating intrinsic properties of WPs, for example, type/proportion of different protein subunits in their composition, and extrinsic factors, such as ionic strength, pH, presence of salt ions, heat treatment conditions, and mechanical shear (Havea et al., 2009; Nicolai and Durand, 2013; Verheul and Roefs, 1998). For instance, major WPs, such as β-Lg and bovine serum albumin, each contain a free cysteine residue with a reactive sulfhydryl (–SH) group, enabling inter-/intramolecular disulfide–thiol interchange reactions, contributing to the increase in gel strength (storage modulus, G') (Gezimati et al., 1997; Hines and Foegeding, 1993). In contrast, α-La cannot form a gel itself (up to 20% protein concentration) due to the lack of a free sulfhydryl group (Hines and Foegeding, 1993); however, it is seemingly involved in the network formation through hydrophobic interactions in WP gel formulation. It is expected that the gels formed by noncovalent interactions are relatively stiff and brittle, while the gels formed by disulfide interactions are relatively more rubbery, and the rubberiness increases with an increase in the degree of disulfide linkage (Havea et al., 2004; Verheul and Roefs, 1998). Additionally, the presence of disulfide in WP gel has been found to reduce the syneresis (Alting et al., 2003). The formation of intermolecular disulfide bonds slows down due to the presence of high electrostatic repulsion between charged protein molecules and due to the gel network formation during the early heating of WP dispersion, preventing the cysteine residues from colliding more frequently (Wagner et al., 2020). Electrostatic repulsions also prevent gel network formation through noncovalent interactions such as hydrophobic, hydrogen bonding, etc., by preventing the protein molecules from coming close to each other. Electrostatic repulsions can be screened by adjusting the ionic strength of the WP dispersions through the addition of salts, for example, NaCl and $CaCl_2$, and controlling the pH (Verheul and Roefs, 1998; Ju and Kilara, 1998). Controlling the ionic strength along with pH eventually impacts the gel structure and rigidity. For instance, at high ionic strength and pH values near the IEP, WP typically forms turbid, milk-white gels with a particulate structure that exhibits poor water-holding capacity. In contrast, gels formed at low ionic strength and pH values far from the IEP are transparent with a fine-stranded structure and exhibit good water-holding capacity. The mixed gels with both fine-stranded and particulate structures can be formed by adjusting the ionic strength and pH to moderate levels (Havea et al., 2004; Langton and Hermansson, 1992; Nicolai and Durand, 2013). Heating conditions, for example, heating rate, temperature–time combination, etc., also impact the gel structure and strength of WP gels. Heating at a high temperature (>75°C) simply increases the gel strength due to an increase in the amount of protein aggregates (Verheul and Roefs, 1998). However, high temperature quickly increases the denaturation rate while the aggregation process becomes rate-limiting, impacting the aggregate structure and eventually resulting in a coarser gel structure (Nicolai and Durand, 2013; Verheul and Roefs, 1998).

Cold-set gels: Cold-set WP gels have a significant advantage for texturizing clean-label and heat-labile food matrices, as well as for encapsulation of heat-sensitive nutrients, drugs, and pharmaceuticals. In the cold-set gelation, the denaturation and aggregation steps are separated from the gelation step (Alting et al., 2003). First, WPAs are manufactured through the heat treatment of WP dispersion at an appropriate ionic strength and pH. The morphology of the resulting protein aggregates has been discussed in the subsequent section. Initial aggregation of protein allows the formation of cold-set WP gel at very low protein concentrations (2%, w/w), significantly lower than those of heat-set WP gels (Amin et al., 2024, 2025). Gelation in protein aggregate dispersion is induced in a separate step by adding salt ions, acidification, or through the application of cross-linking agents (Alting et al., 2003; Amin et al., 2024; Liang et al., 2020). The gelation step in cold-set gelation is understood as a two-stage process: the first stage involves the formation of a space-filling protein network driven by physical interactions, while the second stage involves the increase in gel stiffness and hardness through covalent interactions between the gel structural elements (Alting et al., 2003). Consequently, the gelation kinetics, as well as structural and mechanical properties of resulting cold-set gels, are influenced by the specific cold gelation method employed. For instance, salts, such as NaCl and $CaCl_2$, primarily promote gelation by screening the electrostatic interactions (salt-induced). It was reported that increasing the salt concentration accelerated the gelation rate and improved the gel strength of cold-set WP gels. This effect was more pronounced for $CaCl_2$ as it forms additional salt bridges between the protein molecules (Alting et al., 2003). Lowering the pH close to the IEP (acid-induced) of protein is another method to induce gelation in the aggregate dispersions and is usually achieved by using glucono-δ-lactone (GDL) or an acidic solution, for example, hydrochloric acid (HCl), citric acid, etc. (Amin et al., 2024; Kharlamova et al., 2018a; Li et al., 2023). GDL progressively induces acidification without the need for stirring or agitation. It was reported that both HCl and GDL produce similar effects on gel formation, provided that flocculation caused by rapid acid addition is avoided (Kharlamova et al., 2018a). Additionally, slower acidification allows more time for gel structure development, leading to a stiffer gel at a given pH (Kharlamova et al., 2018a). Lastly, enzymes, such as transglutaminase (TGase), have been used to induce cold gelation in heat-denatured WP dispersion (Li et al., 2023). The resulting gels had relatively higher physical stability and gel strength than acid-induced gels. TGase can catalyze the formation of covalent bonds between the lysine residues and glutamine. These bonds are nearly 20 times stronger than hydrophobic and hydrogen bonds typically formed during acidification (Akamittath and Ball Jr., 1992; Li et al., 2023). Like heat-set gels, gelation kinetics, structural, and mechanical properties of cold-set WP gels are impacted by the protein composition, aggregate size, gelation temperature, along with the method of gelation used (Alting et al., 2003; Bryant and McClements, 2000a; Kharlamova et al., 2018a, 2018b; Liang et al., 2020; Rabiey and Britten, 2009). The formation of disulfiebonds was observed during cold gelation, largely contributing to the gel strength and gel hardness and influenced by WP composition (Alting et al., 2000). The larger aggregate size and higher temperature cause a faster gelation rate (Kharlamova et al., 2018b). However, contradictory findings have been reported regarding the impact of aggregate size on the gel strength (G') (Amin et al., 2024; Kharlamova et al., 2018b).

3. Whey protein self-assemblies as a functional protein ingredient

Variation in the physical and/or chemical environment of the WPs lead to the unfolding of their molecular structure. This exposes the initially buried reactive amino acid residues, which interact and self-assemble into various morphologies, referred to as WPAs. This section only discusses the heat- and solvent-induced aggregation processes.

3.1 Formation and structure of whey protein aggregates

3.1.1 Heat-induced aggregates

Heat-induced WPAs are formed through irreversible aggregation in a complex self-assembly pathway when an aqueous WP system in a hermetically sealed container or in a heat exchanger is subject to elevated temperature under controlled environmental conditions, such as pH, ionic strength, shear, etc. In an aqueous environment, the controlled thermal treatment modifies the structure of WPs. The dimers are changed into monomers, the native protein structure is unfolded, and initially buried reactive sites become more accessible. These reactive groups promote intermolecular interactions through hydrogen bonding, hydrophobic, and/or disulfide interactions, thus forming finite-sized WPAs (Nicolai et al., 2011; Nicolai and Durand, 2013). The concentration of WP dispersion must be below the C_g to avoid the formation of a continuous gel network. The self-assembly pathway of heat-induced WPA can be altered by changing the pH and/or ionic strength of WP dispersion, resulting in distinct aggregate morphologies, such as fibrils, spheres, and irregular/fractal shapes (Mudgal et al., 2011; Nicolai, 2016; Nicolai and Durand, 2013; Schmitt et al., 2009; Gélébart et al., 2019). For instance, heat-induced WPAs formed below the IEP (e.g., pH <3.35) exhibited a linear structure that varied from long, rigid, semiflexible fibrils to shorter, soft, and flexible strands with a high aspect ratio (Nicolai and Durand, 2013; Nicolai, 2016; Jiang et al., 2022). Their manufacturing involves extensive heating of the aqueous WP system under acidic conditions, which results in its hydrolysis. These individual peptides from acid hydrolysis, along with monomeric protein units, act as nuclei that grow and refold into cross-β-sheets. Sometimes, enzymatically hydrolyzed peptides are used to accelerate nucleation and growth and eventually fibril formation (Gao et al., 2013). During prolonged heating, these cross-β-sheets form protofilaments and, finally, mature fibrils (Adamcik and Mezzenga, 2012; Bolder et al., 2007; Mudgal et al., 2011). These fibril structures also coexist as spherulites, which are radially oriented fibrils with a core (Bolder et al., 2006). In contrast, heat-induced WPAs near or above their pI (pH 4.6–7.0) exhibited a range of morphologies, including spherical and curved strands or fractal structures; these aggregate systems are considered as soft and porous structures (Amin et al., 2024; Inthavong et al., 2016; Nicolai and Durand, 2013). Under these conditions, thermally induced protein–protein covalent and noncovalent interactions between protein monomers lead to the formation of dimers or oligomers. Above their critical association concentration, these dimers and oligomers further interact to form primary aggregates or soluble aggregates, such as microgels (near their pI) or curved strands (around neutral pH) (Andoyo et al., 2018; Mahmoudi et al., 2011; Nicolai et al., 2011; Ryan et al., 2013). At a higher protein concentration, the curved strands aggregate into fractal clusters.

Among WPs, β-Lg is shown to be mainly involved in the formation of fibril aggregates, although other WPs, such as α-La and BSA, may also form fibrillar structures (Bolder et al., 2007; Goers et al., 2002; Veerman et al., 2003). In contrast, all major WP units, for example, β-Lg, α-La, and BSA, are seemingly involved in the aggregate formation through disulfide exchange, disulfide bonds, as well as noncovalent interactions (Havea et al., 2001). Disulfide interactions are mainly involved in heat-induced aggregation under neutral conditions; however, aggregation does not necessarily depend on these interactions around or below their pI (Mudgal et al., 2011; Nicolai, 2016; Schokker et al., 1999). The extent, size, and morphology of heat-induced aggregates can be further regulated by varying the protein concentration, pH, ionic strength, mechanical shear, and thermal treatment to fine-tune their functionality for diverse applications (Nicolai et al., 2011; Nicolai and Durand, 2013).

3.1.2 Antisolvent-induced aggregates

Solvent-induced aggregation of WPs is achieved by treating the aqueous WP dispersions with a water–organic solvent mixture, such as water–alcohols (i.e., ethanol, isopropanol), under controlled environment (Feng et al., 2021; Nikolaidis et al., 2017; Nikolaidis and Moschakis, 2018; Pal et al., 2016). Alcohols, particularly ethanol, have commonly been utilized for manufacturing solvent-induced WPAs for food-grade applications. Ethanol can induce cold denaturation in WPs due to its lower polarity and reduced dielectric coefficient relative to water (Wagner et al., 2020). However, the effect of ethanol on WP conformation depends on its concentration in the medium. For instance, ethanol may have minimal effect on the structure of WPs at lower solvent concentrations (i.e., <20%). However, at moderate solvent concentrations (i.e., 20%–50%), ethanol promotes intramolecular hydrogen bonding, thereby enhancing the α-helical structure while reducing the presence of β-sheets. It also weakens the intermolecular noncovalent interactions, resulting in the unfolding of the tertiary protein structure. As the solvent concentration increases further (i.e., >50%), the local hydrogen bonding becomes destabilized due to enhanced intermolecular protein–protein interactions. These modifications promote the transition of α-helices into an intermolecular β-sheet structure, ultimately leading to protein aggregation (Feng et al., 2021; Nikolaidis et al., 2017; Wagner et al., 2020). Lastly, a very high ethanol concentration can reduce the intermolecular hydrophobic interactions, leading to a decrease in protein aggregation (Kayser et al., 2020; Wagner et al., 2020). After the incubation period, ethanol can be removed through evaporation and drying techniques. Both reversible and irreversible WP aggregation were observed in the solvent-induced aggregation process, depending on the environmental conditions, for example, solvent concentration, pH of the WP system, treatment temperature, etc. (Gülseren et al., 2012; Nikolaidis and Moschakis, 2018). Nikolaidis and Moschakis (2018) reported that β-Lg retained about 34% of its denaturation after ethanol removal. Heat treatment has also been used to improve the ethanol-induced aggregation behavior of WPs (Nikolaidis and Moschakis, 2018). Distinct morphologies have been observed for ethanol-induced WPAs, depending on the ethanol concentration, pH, and ionic strength of the WP system. Kayser et al. (2020) reported short and worm-like aggregates at pH 2.0 after 5 h of incubation (30°C), whereas Sentoukas et al. (2023) reported hard-sphere morphology at higher pH (3.0–7.0). These aggregates can further build up to form supramolecular structures of fractal appearance (Wagner et al., 2020).

3.2 Functionality of whey protein aggregates

WPAs have been explored for food structuring and texturizing applications through their thickening, gelling, emulsifying, and foaming abilities in food systems. Interfacial behavior of WPs has been discussed in Section 2.2. WPAs usually have a larger particle size than native proteins; therefore, WPAs have shown a slower diffusion rate at oil–water and air–water interfaces (Mahmoudi et al., 2011; Zhou et al., 2020). Despite their slower diffusion rate, interfacial tension in both WPI and WPA systems has been shown to decrease to a similar extent (Mahmoudi et al., 2011; Zhou et al., 2020). In some cases, WPA systems showed faster adsorption rate at air–water or oil–water interface, presumably due to the presence of monomeric or hydrolyzed proteins (in case of fibril WPA system) (Jung et al., 2010; Mahmoudi et al., 2011). In foam systems, particularly, these monomeric and/or hydrolyzed proteins in the WPA system improve their foamability (Farrokhi et al., 2019; Jung et al., 2010; Mahmoudi et al., 2011; Schmitt et al., 2007). Zhou et al. (2020) proposed a denser and more brittle 2D interfacial structure for WPI-stabilized interfaces, while a coarser and thicker three-dimensional interfacial structure for WPA-stabilized interfaces. Therefore, WPA-stabilized systems showed high emulsion and foam stabilities during storage; however, the thicker and coarser structure of interfacial films made them more susceptible to deformation under dynamic conditions (Xu et al., 2024; Zhou et al., 2020). The distinct morphologies of WPAs influence how they spread, rearrange, and interact at the interface to form an interfacial film. For instance, fibril WPAs, due to their rigid and semiflexible structures with high aspect ratio, proved to be more effective emulsion and foam stabilizers through the Pickering effect (Noskov et al., 2020; Xu et al., 2024). The long fibers form stronger interfacial networks with a higher elastic modulus and provide additional stabilization against flocculation and coalescence due to their thickening effect (Jung et al., 2010; Xu et al., 2024). In contrast, microgel WPAs, being soft particles, formed a Mickering emulsion system with smooth interfacial films (Amin et al., 2025). The stability of Mickering emulsions depends on the extent of swelling and deformation of microgel particles at the interface (Schmidt et al., 2011). The WPAs also function as structural building blocks in the structural development of cold-set gels. In the WPA dispersion system, the cold gelation can be induced under ambient conditions by adding salt or through acidification; the detailed cold gelation mechanism has been discussed in Section 2.3. WPAs proved to be more effective than WPI in forming phase-stable cold-set gels at low protein concentration (2%) under neutral, slightly acidic conditions (Amin et al., 2025). In addition to cold-set gels, WPAs also impact the heat-induced gelation process by accelerating the gelation in the WP system (Kharlamova et al., 2018b). The morphology of WPAs also influences the gelation dynamics and the characteristics of WP gels. For instance, Kharlamova et al. (2018a, 2018b) reported that microgel WPAs formed weaker gels than fractal WPAs due to their dense structure, and more proteins were required to form the gel. Whereas, Khalesi et al. (2021) suggested that adding fibril WPAs in WPI gels can assist in improving the intermolecular interactions and increasing the formation of junction points, affecting the size, structure, and distribution of pores in the WPI gel network.

Additionally, soluble WPAs have shown enhanced thermal stability in food and beverage formulations. Heat treatment during the manufacturing of WPAs can expose the charged amino acid residues, increasing the net surface charge of the resulting WPAs. These highly charged soluble WPAs may have higher thermal stability, even in the presence of salt (Ryan et al., 2012).

4. Whey protein–starch systems

Starch is a plant-sourced polysaccharide and the main carbohydrate in the human diet. It exists naturally in the form of granules and consists of two major components, which are amylose and amylopectin (Wang and Copeland, 2013). Amylose accounts for ~20%–30% of the granules and has a linear structure, primarily made up of α-1,4-linked D-glucose units with a degree of polymerization (DF) typically ranging from 100 to 10,000 (Pérez and Bertoft, 2010; Seung, 2020). In contrast, amylopectin is a highly branched biopolymer, with α-1,4-linked D-glucose chains (DF < 100) as the backbones joined by many α-1,6-linked branches (Pérez and Bertoft, 2010; Seung, 2020). When heated in excess water, granular starch changes from a highly ordered structure to an amorphous disordered state, a process known as gelatinization (Wang and Copeland, 2013). This process is accompanied by the solubilization of starch granules and an increase in solution viscosity, making gelatinized starch an ideal ingredient as a thickening or stabilizing agent in various food applications. The WP–starch system has also attracted considerable interest in recent years, as combining the two food materials with distinct functional properties has shown potential in developing more functional and nutritionally balanced food products.

4.1 Formation and structure of WP–starch complexes

There are different ways to prepare WP–starch mixtures, from simple blending to cogelatinization and to codry heating. Simple blending refers to direct mixing of WP and starch without additional heat treatment, with starch possibly added in a pregelatinized or modified form (Niu et al., 2023; Ye et al., 2004; Zhang et al., 2022). For example, in the study of WP–starch-stabilized emulsion, the starch was gelatinized at 90°C for 1 h and cooled to 60°C prior to mixing with whey protein hydrolysate (WPH) and oil for emulsion preparation (Ye et al., 2004). This method minimizes the interactions between WP and starch and could avoid heat-induced gelation from WP that may affect downstream applications (Zhang et al., 2022). Cogelatinization is a more common approach to preparing WP–starch complexes, where WP and starch are mixed together following a heat treatment at ~85–100°C for 15–30 min. During the heat treatment, both WP and starch undergo extensive structural changes, which facilitate the intermolecular interactions between the two components. The major interactions have been found to be hydrogen bonding and hydrophobic interactions, whereas ionic bonding was weak, and no covalent bonds were formed between WP and starch (Fan et al., 2023; Zhang et al., 2022). The strong interactions between the two components and the physical barrier formed by denatured WP contributed to slower digestion of starch, which is considered beneficial for certain health-vulnerable populations (e.g., individuals with diabetes) (Yang et al., 2019; Zhang et al., 2022).

Another useful method to combine WP with starch is known as codry heating, where the mixture of the two components is heated at low moisture conditions (e.g., <10%) for a certain period of time. The temperature may range from 60°C to 130°C depending on the specific application requirements. Under the heating conditions, both WP and starch undergo substantial structural modifications that can result in improved functionality and a wider range of applications. WPs alone could form both soluble and insoluble aggregates linked by disulfide bonds and other covalent bonds, accompanied by partial denaturation (Gulzar et al., 2011). Dry heating also affects the structure of starch, leading to modified surface and decreased relative crystallinity (Gou et al., 2019). It is expected that codry heating of WP and starch could enhance the interactions between the two components, leading to the formation of WP–starch agglomerates that may benefit downstream applications. For instance, one study on WP–rice starch mixture showed larger aggregates formed after a dry heat treatment, and the extent of agglomeration increased with the heating time (Zhu et al., 2020). The study also reported that the internal ordered structure of starch was partially disrupted, as indicated by decreased crystallinity and altered spectral patterns, resulting from heat treatment and WP incorporation.

4.2 Functionality of WP–starch complexes

Combining WP and starch (native or modified) can substantially influence the functionality of food complexes in ways that are of great applied significance for the food industry. WP–starch complexes are considered to have better wettability compared with starch or protein alone, therefore allowing better adsorption onto the oil–water surface during the emulsifying process (Wang et al., 2023). In addition, due to the presence of abundant hydroxyl groups, the gelatinized starch can adsorb and bind a high amount of water through hydrogen bonding. This enables them to function as stabilizers, reducing the flocculation and coalescence of oil droplets by increasing the solution viscosity and providing physical hindrance between particles. For example, in one study analyzing the impact of starch type on the emulsifying properties of WP–starch (amorphous) complexes, the complexes showed better emulsifying ability than amorphous starch or WPI alone by forming a stable oil emulsion. The emulsions stabilized with potato starch–WP, pea starch–WP, corn starch–WP, and wheat starch–WP mixtures had smaller particle sizes and no creaming after 30-day storage, whereas the emulsions prepared from Japonica rice starch–WP, waxy rice starch–WP, and tapioca starch–WP complexes had severe phase separation (Wang et al., 2023). In another work, WPs were complexed with taro starch through cogelatinization, with WP contents ranging from 0% to 13% based on a total solid content of 5% (w/w) (Fan et al., 2023). It was found that the particle size of the WP–taro starch stabilized emulsion decreased with the increase in WP concentration. The complex system with 13% WP exhibited the smallest emulsion droplets, significantly outperforming the emulsions prepared from pure amorphous starch or WPI, indicating the presence of a synergistic effect of combining the two ingredients to improve emulsifying performance. Similar enhanced emulsifying ability and emulsion stability were achieved when WP–starch complexes were used to produce high internal phase emulsion (HIPE) (Guo et al., 2021; Zheng et al., 2024). As an example, Guo et al. (2021) reported that the WP–starch complex stabilized HIPE showed a smaller particle size than the WPI alone, and the particle size decreased with the increase of starch concentration (0%–7.5% w/w). In addition, the HIPE formed with 7.5% (w/w) starch and WP (2%, w/w) was stable at high ionic strength (NaCl 0–500 mM) and under thermal conditions (50–90°C for 30 min), with minimum changes in oil droplet size.

The interactions between WP and starch can also influence gelling behavior and gel properties. By controlling the WP and starch ratio, the combined system has the potential to form viscoelastic gels with desired rheological behavior. When starch is heated in water above the critical concentration and gelatinization temperature, the granules will adsorb water and swell, and eventually release dissolved starch molecules, forming a viscoelastic mass, commonly known as a paste (with certain fluidity) or a starch gel (a defined shape without fluidity) (Ai and Jane, 2015; BeMiller, 2011). Different from WP-based gels (see Section 2.3), the formation of starch gel is driven primarily by noncovalent hydrogen bonding and physical entanglement. As a result, the gel properties and the intermolecular interactions for the mixed gels will be highly dependent on the dominating phase but will also be affected by the minor component.

In a starch-dominating system (e.g., starch >75% of the total mass), the presence of WP can increase gelatinization temperature, restrict starch swelling, reduce gel viscosity and firmness, and retard retrogradation (Fan et al., 2025; Hu et al., 2020; Kumar et al., 2018, 2023). WPs can bind with starch granules as a result of hydrogen bonding and hydrophobic interactions, which could inhibit amylose leaching out and water binding, leading to reduced swelling and a higher gelatinization temperature (Hu et al., 2020; Liu et al., 2021). At low concentrations, WPs remain as a dispersed phase and cannot form a cross-linked network that contributes to the enhancement of gels during the hydrothermal treatment; thus, substitution of starch with WPs results in a weaker starch gel (Kumar et al., 2018, 2023). In addition, after the gelatinization, the dispersed WP aggregates could hinder the molecular associations and hydrogen formation among the starch polymer chains, therefore leading to reduced recrystallization or retrogradation of gelatinized mixed gels (Fan et al., 2025; Hu et al., 2020). In a WP-dominating system (e.g., WP>50% of the total mass), increasing the starch fraction could have contrasting effects on the gel formation. Studies on the rheological properties of WP–cassava starch complexes showed that heat-induced mixed gels with a starch fraction between 0 and 0.25 exhibited higher gel firmness than pure WP gels, while a further increase in starch fraction led to weaker gels (Aguilera and Baffico, 1997; Aguilera and Rojas, 1996). At low concentrations, a synergistic effect was observed between WP and gelatinized cassava starch, which is thought to result from the concentrating effect of the substituted starch, which reduced free water and therefore increased the effective concentration of WP in the system (Aguilera and Baffico, 1997). This synergism makes it economically beneficial to partially replace WP with starch while maintaining the gel properties. In other cases, this phenomenon was not observed, and replacing WP with 10%–50% starch (e.g., corn starch, cross-linked waxy maize starch) led to a continued decrease in gel strength (Aguilera and Rojas, 1996; Ravindra et al., 2004). When WP and starch are structurally or thermodynamically incompatible, the dispersed starch may disrupt the formation and continuity of the WP network, and this impact is stronger than the concentrating effect, therefore, leading to weaker gels (Aguilera and Baffico, 1997; Ravindra et al., 2004).

5. Whey protein–nonstarch polysaccharide complexes and conjugates

Nonstarch polysaccharides (NSPs), such as pectin and carrageenan, are widely used as stabilizers or gelling agents in the food industry. They are mostly water-soluble and can form viscous fluids or gels at a very low concentration with or without heat treatment, a feature that makes them distinctly different from starch. Common NSPs used as food texture modifiers include, but are not limited to, pectin, tree gums (e.g., gum Arabic, gum karaya, gum tragacanth), seed gums (e.g., locust bean gum, guar gum, tara gum), xanthan gum, carrageenan, agar, alginate, gellan gum and carboxymethyl cellulose (CMC) (Damodaran and Parkin, 2017; McClements, 2015). Many of these polysaccharides carry charged groups, different from those of neutral starch, and can effectively interact with proteins to form functional complexes or conjugates under certain conditions. For example, pectin and gum Arabic are anionic polysaccharides containing carboxylic groups, which are negatively charged at a pH higher than the pKa (~2.8–4.1 for pectin and ~2.2 for gum Arabic, respectively) (Celus et al., 2018; Sabet et al., 2021), whereas chitosan is a cationic polysaccharide, which possesses positively charged amine groups ($-NH_3^+$) when the pH is lower than ~6.5 (Ding et al., 2019). When combined with WPs, NSPs can offer colloidal stability through the steric hindrance and their thickening effect, leading to improved emulsion and foam stability (O'Mahony et al., 2019; Turgeon et al., 2007). In addition, most NSPs are resistant to digestion by human enzymes but could be moderately fermented by gut bacteria and generate health-promoting short-chain fatty acids (Kumar et al., 2012).

5.1 Formation and structure of WP–NSP complexes and conjugates

NSPs can form noncovalent complexes with WPs using simple physical mixing or form covalent conjugates with WPs with dry heating, wet heating, chemical, or enzymatic cross-linking (Fig. 15.2). Physical mixing is the most commonly used method in practical applications due to its simplicity and convenience. Some polysaccharides, such as locust bean gum and carrageenan, are not soluble in cold water; thus, it is common to dissolve them in hot water prior to mixing with WP. For the charged polysaccharides, the pH of the solution is a critical factor influencing the formation of the WP–polysaccharide complexes. At neutral pH, the anionic polysaccharides and WPs are cosoluble and stable, and limited electrostatic interactions exist among the biopolymers (Wagoner et al., 2016). On lowering the pH, soluble complexes form between WPs and polysaccharides due to molecular interactions between positively charged local domains exposed by WPs and negatively charged polysaccharides (Wijaya et al., 2017). For example, for pectin and its derivatives, an acidic pH at ~4–5 is often used to form soluble complexes with WPs (Kazmierski et al., 2003; Wagoner and Foegeding, 2017; Wijaya et al., 2017). As pH further decreases until WPs and polysaccharides become oppositely charged, complex coacervation or precipitation would occur due to electrostatic attraction among the biopolymers (De Kruif and Tuinier, 2001). Coacervation can be useful for encapsulation of bioactive compounds; however, the resulting phase separation may not be desirable for various applications (e.g., protein beverage). Apart from pH and ionic strength, the ratio and concentration of the biopolymers are also important factors affecting the formation of WP–NSP complexes. More detailed information on protein–polysaccharide interactions can be found in earlier review works (De Kruif et al., 2004; De Kruif and Tuinier, 2001; McClements, 2006; Schmitt and Turgeon, 2011; Turgeon et al., 2007).

For covalent interaction, dry heating is widely used to trigger Maillard reactions between WPs and NSPs. WPs and polysaccharides are mixed in the solution, spray- or freeze-dried, and then heated at a temperature ranging from 60°C to 100°C for a certain period of time (e.g., a few hours to several days) (Du et al., 2022). During this process, the amino groups of WPs, mostly the ε-amino groups of lysine, react with the free carbonyl groups of polysaccharides, resulting in the formation of glycoproteins (Setiowati et al., 2020). This process also induces tertiary and secondary structural changes in the WPs, with previous studies confirming spectral differences between heat-treated conjugates and unheated WP control when analyzed by Fourier-transform infrared spectroscopy (FT-IR) and intrinsic fluorescence spectroscopy (Kan et al., 2024; Spotti et al., 2014). Key reaction parameters include temperature, treatment time, pH, relative humidity, and type and ratio of polysaccharide (Sedaghat Doost et al., 2019; Setiowati et al., 2020). For example, increasing the temperature and incubation time often results in a greater degree of conjugation, but this may lower the solubility and increase the brownish color of the conjugates. Compared to dry heating, wet heat treatment is a simpler alternative, in which WPs and polysaccharides are directly dissolved and heated in aqueous solution. It avoids the spray- or freeze-drying step as required by dry heating, and takes less time to make the conjugates. However, protein aggregation and

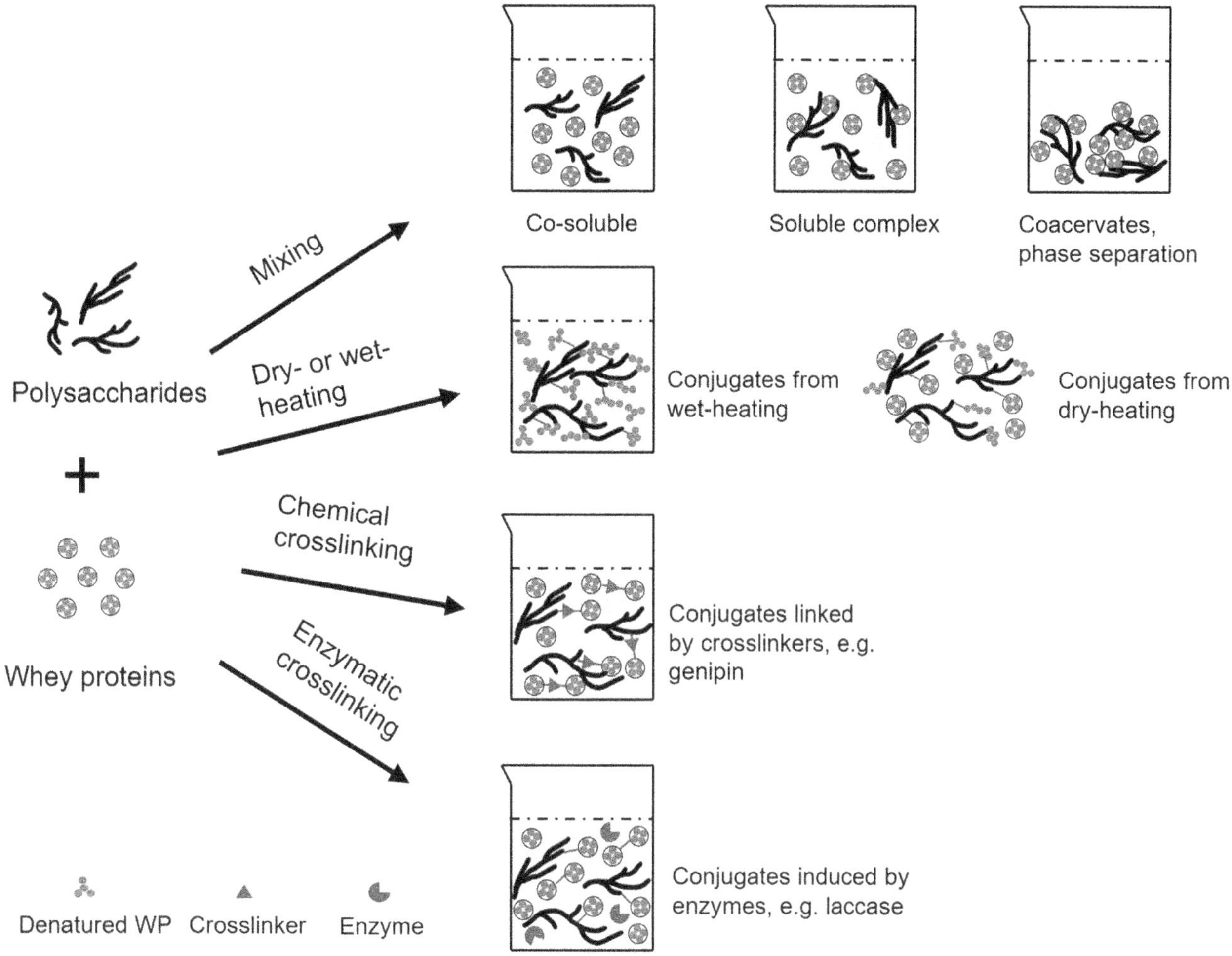

FIGURE 15.2 Schematic representation of different methods in preparing whey protein–nonstarch polysaccharide complexes and conjugates.

polymerization may occur using this method under a high temperature or a long heating time (Sedaghat Doost et al., 2019).

For chemical cross-linking, cross-linkers, such as glutaraldehyde and carbodiimide, have been reported to produce protein–polysaccharide conjugates; however, these methods are not commonly used for food production due to safety concerns (Hattori et al., 1994; Liu et al., 2012). More recently, genipin, a compound found in the plant Gardenia Jasminoides, has emerged as a natural cross-linking agent. By reacting with free amine groups (–NH_2) on polysaccharides and proteins, genipin can cross-link the two components to form conjugates. This approach has been successfully applied to cross-link sugar beet pectin with WPs (Lin et al., 2020). In 2023, genipin was approved by the FDA as a color additive, and its potential applications as a cross-linker are expected to expand in the future (Natallia et al., 2024).

Enzymatic cross-linking is another useful approach to producing WP–NSP conjugates. Some common enzymes used for this purpose include TGase (EC 2.3.2.13), laccase (EC 1.10.3.2), and peroxidase (EC 1.11.1.7). TGase is widely used in the production of plant- and animal-based gels as it can catalyze acyl transfer reactions to form isopeptide bonds among proteins. Some protein-containing polysaccharides, such as gum Arabic, guar gum, and locust bean gum, can potentially act as a TGase amino donor substrate and form conjugates with WPs through TGase. It has been confirmed that gum Arabic and sodium caseinate formed covalent conjugates using the enzyme (Flanagan and Singh, 2006); however, similar applications for WPs are not documented yet. Laccase is a copper-containing oxidase that catalyzes the oxidation of PCs and aromatic and aliphatic amines, with oxygen concurrently reduced to water (Kudanga et al., 2011). The resulting reactive radicals can then undergo covalent coupling reactions to form dimers or polymers within the protein–polysaccharide mixture (Kudanga et al., 2011). Past studies showed that WP–pectin conjugates can be successfully fabricated with this method, resulting in modified physical and functional properties compared with the pure WPs (Chen et al., 2012, 2018; Gazme and Madadlou, 2014; Jung and Wicker, 2012). Peroxidase is another enzyme that has been reported in the literature to produce protein–polysaccharide conjugates (Liu et al., 2015, 2017). This enzyme oxidizes hydrogen donors (e.g., aromatic phenols, amines) in the presence of hydrogen peroxide (H_2O_2). The reaction leads to the formation of quinones, which further react with other food components to form covalent conjugates (Zeeb et al., 2014). Studies by Liu et al. (2017) showed that BSA formed covalent bonds with corn fiber gum (CFG) using peroxidase as the cross-linker. The surface morphology of the conjugates was found to be more compact and rougher than that of the mixture without enzyme treatment.

5.2 Functionality of WP–NSP complexes and conjugates

NSPs play a critical role in the modern food industry due to their versatile functionalities. They are highly hydrophilic and can form viscous fluids or gels with particular rheological properties. When combined with WPs, the formed complexes or conjugates can exhibit superior physical and functional properties that are highly desirable for various food applications. The stability and solubility of WP under different pH and temperature conditions are of significant importance in practical applications. Natural WPs are generally not stable under heat treatment at low pH conditions and tend to form aggregates, which negatively affect downstream applications (e.g., ready-to-drink acidic beverages). This undesirable property can be modified by cross-linking WPs with polysaccharides. Previous studies reported that the solubility of WPI–dextran conjugates (5% w/v protein) obtained by dry heating was maintained above 90% at pH 4.5 and 5.5 after a thermal treatment (70°C, 75°C, 80°C for 30 min), significantly higher than that of pure WPI (~7%–58%) (Wang and Ismail, 2012). Similarly, WPI (1 mg mL^{-1}) cross-linked with gum Arabic (GA) by dry heating showed higher solubility than WPI alone through pH ranges of 4–10 (Chen et al., 2019a, 2019b). WPI–GA conjugates also showed good stability against heat treatment (50–100°C for 1 h), with protein solubility maintained at ~70%–100% at pH 7, whereas the native WPI had lower values at ~60%–75% (Chen et al., 2019a, 2019b). The improved solubility and stability of WPs are primarily ascribed to increased hydrophilicity and the higher physical and steric hindrance from cross-linked polysaccharides. On the other hand, WP–gellan conjugates prepared from wet heating showed decreased solubility at pH 4 and 5 compared to that of untreated WPs (Nooshkam and Varidi, 2020). In this study, WP–gellan mixture was heated at 90°C for 30–90 min, which resulted in protein denaturation and increased surface hydrophobicity. This higher hydrophobicity of WPs likely facilitated the protein aggregation and precipitation under low pH conditions near the pI.

The influence of polysaccharides on the emulsifying and foaming properties of WP highly depends on the type and concentration of polysaccharides, as well as the complexation/conjugation method. WPs conjugated with polysaccharides, such as CFG, GA, and CMC through Maillard reactions, generally showed improved emulsifying activity and stability compared to WP alone (Chen et al., 2019a; Jiang et al., 2023; Kan et al., 2024; Yadav et al., 2010). The presence of polysaccharides can increase the viscosity of the aqueous phase as well as the steric and physical hindrance between the particles, which could inhibit the coalescence of oil droplets, thus leading to improved emulsifying properties. WP–CMC and WP–GA conjugated systems also exhibited better emulsion stability than the mixed systems (no heat treatment applied), suggesting the conjugated WP–polysaccharides possibly formed stronger interfacial films at the oil–water interface or provided greater spatial resistance to inhibit oil droplet aggregation (Chen et al., 2019a; Jiang et al., 2023; Kan et al., 2024). In addition, the emulsification properties are affected by Maillard reaction time. Chen et al. (2019a) reported that the emulsifying ability and stability of WP–GA conjugates prepared from dry heating (60°C) continued to increase with glycation reactions up to day 5 but decreased at day 7. Extended incubation at high temperatures resulted in decreased protein solubility and lower surface hydrophobicity, which possibly led to decreased emulsifying properties (Chen et al., 2019a). Different findings were reported for the WP–gellan system, where the WP–gellan conjugates had similar emulsifying properties to pure WPI, but their emulsion stability was lower than that of the mixed system (not heat-treated) (Nooshkam and Varidi, 2020). Interestingly, in this study, the foaming capacity and stability of WP–gellan conjugates were found to be superior compared with those of the mixed system and pure WPI. A similar observation was also observed in WP–carrageenan conjugates, which showed improved foaming properties (higher overrun values and drainage time) than the mixed WP–carrageenan system and pure WPI (Wang et al., 2015). In addition, the foaming stability increased with the concentration of carrageenan for all systems. High concentration of polysaccharides corresponds to increased viscosity values and less mobility of the aqueous phase around air bubbles, which could reduce foam drainage (Wang et al., 2015).

The interactions between WP and NSPs can be either noncovalent bonding, which includes weak attractive or repulsive forces between molecules, or covalent bonding formed through the Maillard reaction or chemical-induced cross-linking. These interactions can lead to the formation of mixed gels with greater flexibility in structural and textural properties compared with pure WP gels. In addition, extrinsic conditions, such as pH, ionic strength, temperature, and protein and polysaccharide concentrations, can also influence the gelling behavior and gel properties. WP–polysaccharide mixed gels can be mainly formed through heat treatment, cold-set, and acid gelation.

Heat-induced gelation occurs when the WP–polysaccharide mixture undergoes intensive heat treatment, often in the temperature range of 80–90°C for 30 min. The subsequent interactions among the protein and polysaccharide molecules lead to the formation of a cross-linked gel network. Studies of heat-induced WP–xantham gum mixture showed that the gelation temperature increased with increasing heating rate (0.1, 1, 5, 10, 20°C min^{-1}) and was lower than that of pure WPI solution under the same heating rate (Li et al., 2006). In addition, the firmness of mixed gels was strongly affected by both the concentration of xanthan gum and heating rate. When the concentration of xanthan was at 0.1% and 0.2%, the mixed gels showed similar gel firmness to the WPI gels, and the firmness decreased with increasing heating rate. As the concentration of xanthan increased to 0.5% and 1.0%, the firmness of the mixed gel became higher at higher heating rates but was generally lower than that of the WPI control, indicating an antagonistic effect on gel structure formation. At high concentrations, the anionic xanthan possibly became less compatible with WPI due to increased steric repulsion with WPs, which further led to phase separation and weaker gels (Li et al., 2006; McClements, 2006). The pH of WP–xanthan mixture is another factor that influences gel formation. Bertrand and Turgeon (2007) reported that the gel strength increased as the pH decreased from 6.5 to 5.5, due to stronger protein–protein interactions at pH near the pI. The impact of xanthan on gel formation was pH-dependent. At pH 6.5 and 6.0, increasing gum concentration (0.01%, 0.03%, 0.06%) led to firmer gels, whereas at pH 5.5, this led to softer gels. Previous studies also investigated the influence of dry heating on the gel properties of the WP–dextran system (Spotti et al., 2013). It was found that the conjugate gels had lower gel firmness than the mixed gels, and a longer dry-heating time (2, 5, and 9 days at 60°C) led to a decrease in gel strength. The higher level of glycated proteins due to longer Maillard reactions possibly increased the steric repulsion among the protein molecules and thus affected gel network formation.

WP–polysaccharide mixtures can form cold-set gels after a heat-cool step under particular pH, mineral, protein, or polysaccharide conditions. The initial heat treatment unfolds the proteins and enhances the binding of denatured proteins to polysaccharides. The structure of WP–polysaccharide complex is further stabilized by the intermolecular disulfide (S–S) bonds formed by WPs during the heat treatment (Krzeminski et al., 2014). Upon cooling, the intermolecular interactions (e.g., hydrogen bonding or hydrophobic interactions) among the biopolymers increase, especially with the addition of salts, leading to the formation of ordered gel networks. Bryant and McClements (2000b) reported that heat-denatured WPI solution (8.5% w/w) formed a cold-set gel with the addition of 200 mM sodium chloride, and the gel firmness further increased with the addition of xanthan gum (0.05%, 0.1%, and 0.2% w/w). Controlling the initial pH of WP–polysaccharide mixture is also critical for the formation of cold-set gel. It was found that WP–low-methoxyl pectin (LMP) mixture at pH 2 prior to heating formed gel upon cooling, while those at higher pH conditions (pH 5.5, 3.5, and 2.5) did not (Wijaya et al., 2017). The lower charge density of LMP at pH 2 possibly contributed to the aggregation between WP–pectin molecules through hydrophobic interactions and the formation of a self-supporting gel (Wijaya et al., 2017). Acid gelation is, to some extent, similar to cold gelation. After heat treatment, the pH of WP–polysaccharide mixture is reduced to or near the pI of the protein. For example, in the study by Zhang et al. (2014), WP–pectin mixture (5% w/w protein) was heated at 85°C for 30 min and then acidified to pH ~4.7 using GDL. The gel strength of acid-induced gels increased with pectin concentration up to 0.25% (w/w) and then dropped at higher concentrations (0.375%, 0.5% w/w). Pectin with a higher charge density (or degree of esterification) led to enhanced gel strength, which was attributed to stronger pectin–protein interactions during the acidification process (Zhang and Vardhanabhuti, 2014).

6. Whey protein–phenolic compound complexes and conjugates

PCs, which primarily exist in plants, are known to have diverse health benefits by acting as antioxidants or antiinflammatory agents, which can protect against chronic diseases (Samtiya et al., 2021). They also reduce the risk of diet-induced obesity as observed in animal models (Anhê et al., 2015; Roopchand et al., 2015). The addition of PCs to WPs to form protein–PC complexes and conjugates can therefore provide substantial health benefits to consumers when used as a food ingredient.

6.1 Formation and structure of WP–PC complexes

The formation of WP–PC complexes or conjugates mainly depends on the preparation methods, which affect the bonding between the WPs and PCs (Fig. 15.3). WP–PC complexes are typically prepared by mixing WPs and PCs in the dispersion to form noncovalent bondings, which include hydrophobic attraction, hydrogen bonding, electrostatic interactions, and van der Waals forces (Baba et al., 2021). WP–PC conjugates are formed through covalent bonding between WPs and

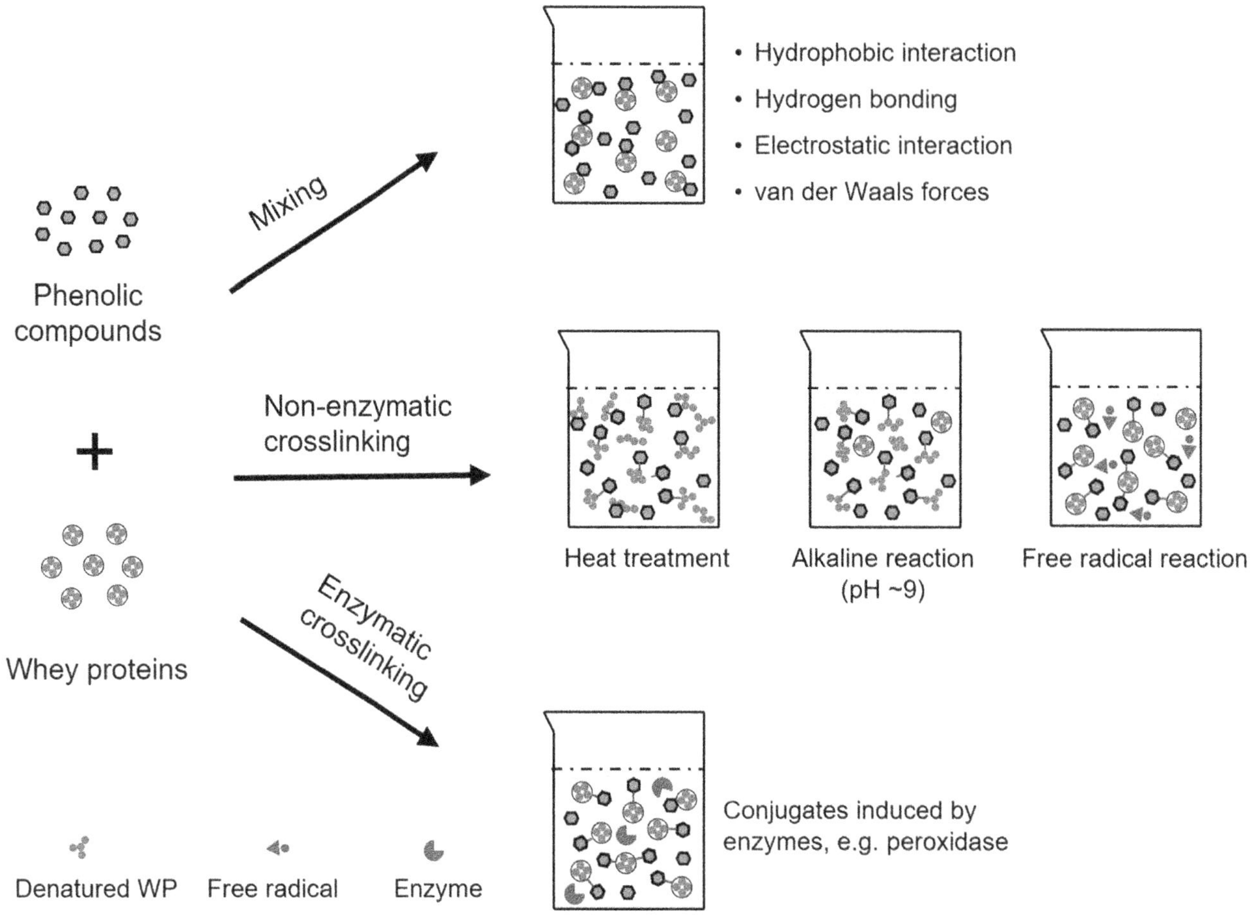

FIGURE 15.3 Schematic representation of different methods in preparing whey protein–phenolic compound complexes and conjugates.

polyphenols. The nucleophilic groups, such as $–NH_2$ and –SH groups on the protein, react with the hydroxyl groups on the polyphenols driven by nonenzymatic (or chemical) or enzymatic reactions (Baba et al., 2021; Quan et al., 2019).

The nonenzymatic approaches include heat treatment, free radicals, alkaline, and chemical coupling methods. During the heat treatment, polyphenols can undergo autoxidation to form semiquinone or quinone structures, which can react with nucleophilic residues of the proteins to form conjugates (Zhang et al., 2024). Alkaline reaction is another common method to fabricate protein–polyphenol conjugates. Under alkaline conditions (e.g., pH 9), polyphenols are oxidized to *o*-quinones in the presence of oxygen, which further react with proteins (Manzoor et al., 2024; Quan et al., 2019). Free radical-induced conjugation typically involves the use of ascorbic acid and H_2O_2 to generate hydroxyl free radicals to induce the formation of covalent bonds (Baba et al., 2021; Quan et al., 2019). Both free radical grafting and alkaline treatment are considered effective, safe, and scalable strategies to obtain the target products. Regarding chemical coupling, it is generally not recommended for food applications due to the reagents (e.g., glutaraldehyde) used for cross-linking being harmful to humans (Liu et al., 2019). The enzymes used for the conjugation reactions include peroxidase, laccase, and tyrosinase (EC 1.14.18.1) (Zeeb et al., 2014). These enzymes catalyze the oxidation of PCs to highly reactive *o*-quinones in the presence of oxygen, which then interact with nucleophilic residues of the proteins to form conjugates (Zeeb et al., 2014). More detailed information on the mechanisms of protein–PC interactions can be found in the reviews by Quan et al. (2019) and Ebrahimi et al. (2025).

The binding of PCs to WPs affects the structure of the proteins, leading to conformational changes and partial unfolding of the structure. The structure of proteins is stabilized by several forces, such as hydrogen bonding and hydrophobic interactions between the amino acids. The external forces from noncovalent or covalent bonding between PCs and WPs can therefore destabilize the protein structure, causing partial unfolding. These structural changes further affect protein hydration, intermolecular interactions, and eventually the functional performance of the products. Analytical techniques, such as fluorescence spectroscopy, FT-IR, and SAXS (small-angle X-ray scattering), have been widely utilized to explore the structural changes of WPs after the addition of PCs. For example, protein intrinsic fluorescence spectroscopy is a useful tool to identify the changes in the tertiary structure of proteins. Past studies showed that the fluorescence intensity decreased as WPs complexed with PCs (e.g., chlorogenic acid, quercetin, gallic acid), with a red shift for the maximum emission peak (Cheng et al., 2023; Jiang et al., 2018; Meng and Li, 2021). These observations suggest the binding caused exposure of the tryptophan-containing hydrophobic residues to a more hydrophilic or polar environment through protein unfolding (Jiang et al., 2018). The binding also affects the secondary structure of WPs. By using FT-IR spectroscopy, it was found that the β-sheet of WPs decreased significantly after the addition of the PCs (e.g., quercetin, gallic acid, chlorogenic acid, and epigallocatechin gallate [EGCG], etc.), while the contents of β-turn or random coil increased, indicating a looser protein structure (Cheng et al., 2023; Meng and Li, 2021; Tian et al., 2023). However, an increase in β-sheet was also observed for WPs complexed with tea polyphenols (catechin, epicatechin, epicatechin gallate [ECG], epigallocatechin gallate [EGCG]) (Kanakis et al., 2011). The opposite changes in the secondary structure could be related to the specific nature of the WPs and PCs.

6.2 Impact of PC on the functionality of WP

The complexation or conjugation of PCs with WPs has been reported to have various effects on the functionality of WPs, which extends the applicability of WPs as a health-promoting functional ingredient. Water solubility is a basic property of proteins that is highly related to their functional performance and application in food and beverages. In general, WPs have a high water solubility ranging from approximately 85%–90% at neutral pH conditions (Jiang et al., 2018; Pelegrine and Gasparetto, 2005; Vasava et al., 2022). Past studies have shown different results regarding the changes in solubility of WPs after they bind with PCs. For instance, Jiang et al. (2018) reported that the addition of chlorogenic acid (240 μmol g^{-1} protein) increased the solubility of WPI from ~86% to ~99% at pH 7, leading to a ~15% rise. In another study, it was found that the solubility of WPI at pH 7 increased from ~83% to ~90%–93% upon noncovalent binding of gallic acid, chlorogenic acid, and epigallocatechin gallate (EGCG) (Meng and Li, 2021). The increase in protein solubility could be related to partial unfolding of protein structure with the addition of PCs, which results in exposure of internal hydrophilic and hydrophobic residues, leading to improved protein hydration and solubility. Meanwhile, the binding of PCs to the hydrophobic groups on the protein surface and the introduction of polar phenolic hydroxyl and carboxyl groups could also contribute to an increase in protein hydrophilicity or solubility (Jiang et al., 2018). On the other hand, some studies showed that the solubility of WPs decreased with the interaction of PCs. For instance, one study by Prigent et al. (2009) showed that the majority of α-La remained insoluble (<10% solubility) in the presence of procyanidins (degree of polymerization of 7.4) between pH 3 and 6. At pH 7, the solubility of α-La resumed to ~80%, yet it remained considerably lower than that of the untreated control (~100%). Vasava et al. (2022) reported that the solubility of WPC slightly decreased from 85%–92% to 74%–87% when added with ferulic acid and caffeic acid. The solubility could also be affected by the interaction mode between WPs and PCs. It was reported that WPC noncovalently linked with quercetin showed a decrease in solubility, while the protein solubility increased for covalent interactions (Cheng et al., 2023). Some PCs may act as bridges or cross-linkers to link multiple protein molecules together through hydrogen bonding, causing protein aggregation or precipitation (Cheng et al., 2023).

Emulsifying and foaming properties of WP are also influenced by the presence of PC. Higher emulsifying and foaming properties have been observed in the WP–PC complexes or conjugates compared with WP alone (Jiang et al., 2018; Meng and Li, 2021; Prigent et al., 2009; Tian et al., 2023; Vasava et al., 2022). It is considered that the binding of PCs to proteins could cause conformational changes, such as partial unfolding of the proteins, which facilitates protein adsorption at the oil–water interface and reduces phase separation (Ribeiro et al., 2025). The greater extent of protein unfolding and the presence of PCs at the air–water interface could strengthen intermolecular interactions (e.g., protein cross-linking), leading to improved film stability (Jiang et al., 2018; Prigent et al., 2009). In other studies, the emulsifying or foaming properties of WPs were negatively impacted by the addition of PCs (Cheng et al., 2023; Li and Girard, 2023). For instance, Cheng et al. (2023) reported that WPC–quercetin complexes showed decreased emulsion and foaming stability compared to WP alone. The opposite findings reported in the literature could again be attributed to the unique properties of each type of PC, which cause different responses in the functional performance when they interact with WPs.

7. Whey protein and plant protein mixtures

Plant proteins have distinct physical and functional properties compared with animal-derived proteins, such as WPs, which attract a growing interest in designing food systems consisting of both plant and animal proteins. The increasing research interest also stems from a growing consumer demand for alternative proteins, driven by a greater awareness of health and environmental sustainability. The addition of plant proteins to WPs not only improves the nutritional value of plant proteins but also opens up new avenues for developing novel food products with unique techno-functional properties.

7.1 Formation and structure of WP–PP mixture

There are three major methods to prepare the WP–PP mixture, which are direct mixing, coprecipitation, and enzymatic cross-linking. In many cases, whey and plant protein dispersions are directly mixed in a specific ratio depending on the application needs. Treatments like pH shifting, ultrasound, or heating may be applied to the samples prior to or after the blending to modify the protein properties. For example, Chihi et al. (2016) applied a heat treatment after the mixing of β-Lg and pea globulin dispersions at 85°C for 60 min. The heating step induced the formation of soluble whey–pea protein aggregates that could be used as ingredients for novel applications. For simple blending without additional treatments, the interactions between the heteroproteins are mainly hydrogen bonding and electrostatic interactions, and van der Waals forces, which have minimal effect on the tertiary or secondary structure of the protein molecules. More intensive treatments on the protein blends, such as pH shifting or heating, can lead to partial unfolding or extensive denaturation of whey and plant proteins. For instance, in the study by Zhao et al. (2024), the pH of WP and pea protein dispersions was adjusted to 12, respectively, followed by an ultrasound treatment or no treatment. The samples were then adjusted back to pH 7 and blended at a mass ratio of 1:1. The WP–pea protein mixtures showed increased surface hydrophobicity and higher intrinsic fluorescence intensity than the untreated control, indicating unfolding of protein structure and exposure of hydrophobic residues (Zhao et al., 2024).

Another approach to preparing a WP–PP mixture is coprecipitation. This method was first mentioned in the study by Thompson (1977), in which a cheese whey and rapeseed protein mixture was heated to 95°C, acidified to pH 4.6, and maintained at 95°C for another 15 min prior to centrifugation; the precipitates were washed and recovered for use. The method was later adopted for the preparation of whey–soy and whey–pea coprecipitates by other researchers (Alu'datt et al., 2012; Kristensen et al., 2020, 2021). Enzymatic cross-linking using TGase is another route to prepare WP–PP conjugates. This technique relies on the catalytic activity of the TGase, which catalyzes acyl-transfer reactions between the ε-amino groups of lysine residues and the γ-carboxamide groups of glutamine residues, leading to the formation of isopeptide bonds (or covalent cross-links) between proteins (Anuradha and Prakash, 2009). Relevant examples include WP–soy protein, WP–sesame protein, and WP–pea protein conjugates as reported in the previous literature (Anuradha and Prakash, 2009; Kim et al., 2023). Analysis of the whey–pea protein conjugates showed that the tryptophan fluorescence intensity was higher than that of the untreated protein mixtures, which again was attributed to the opening of hydrophobic regions of proteins (Kim et al., 2023). Taken together, the direct mixing of whey and plant proteins is a more common method for preparing protein complexes, as it requires fewer processing steps or avoids the use of enzymes compared with the other two approaches, that is, coprecipitation and enzymatic cross-linking.

7.2 Functionality of whey–plant protein mixtures

The functional properties of the whey–plant protein mixture are modulated by the ratio of each protein and the way they interact with each other. As whey and plant proteins behave distinctly differently from each other, it is necessary to understand whether their combination has a synergistic or antagonistic effect on the functional performance of the food system.

It is known that WPs have excellent water solubility (90%–100%, pH 7) while plant proteins, for example, soybean, pea, and rice proteins, typically have lower water solubility that could limit their application in certain food products, such as protein-based beverages. Studies have shown that the combination of WP with plant protein contributes to the improvement of the water solubility of plant proteins, though the overall water solubility is still lower than that of WPs. By complexation with WP (1:1 ratio) at alkaline conditions, the solubility of rice protein increased to ~50%, significantly higher than ~1.7% of the untreated control (Wang et al., 2019). Increasing the WP-to-rice-protein ratio to 2 led to a slight increase in the solubility (~55%), but the effect was insignificant (Wang et al., 2019). It is considered that WP could be incorporated into the partially unfolded rice protein structures under alkaline conditions, leading to increased surface charge, improved protein stability, and better solvation in water upon neutralization (Wang et al., 2018, 2019). Similar improvement was also observed for pea protein, where the solubility increased from ~10% in the untreated control to an estimated level of ~30% when blended with WP (1:1 ratio) (Zhao et al., 2024).

Partial replacement of WPs with native plant proteins, in general, leads to decreased emulsifying properties. Compared with WPs, plant proteins exhibit reduced water solubility and a greater molecular size, which results in their lower efficiency for adsorption onto oil-water interfaces during homogenization. This is evidenced in earlier studies, where plant proteins, such as soybean, pea, faba bean, and lentil, have been reported to generate larger oil droplets than WPs (Gumus et al., 2017; Tan et al., 2022). Regarding WP–PP mixtures, recent studies have shown that WP blended with pea protein or soy protein isolate (SPI) at various ratios had larger oil droplets or lower emulsifying index (emulsifying activity index, emulsifying stability index) values than WP alone (Li et al., 2025; Zhang et al., 2021; Zhao et al., 2024). Moreover, the emulsifying ability of the dual protein system decreased with the increasing ratio of plant proteins (Kim et al., 2023; Li et al., 2025; Zhang et al., 2021). Foaming ability is closely related to emulsifying properties, as both are driven by the action of the surface-active proteins that stabilize the interfaces. Similar to emulsifying properties, the presence of plant proteins could negatively affect the formation and stability of foam in the WP–PP system. For instance, whey–soy protein mixtures (25%, 50%, and 75% SPI) displayed similar foaming capacity compared with the WPI dispersion; however, their foam stability was significantly lower than that of the WPI control (Alves et al., 2022). Whey–pea protein mixtures displayed the same trend as above, as shown in the study by Kristensen et al. (2021), in which the whey–pea mixtures prepared from coprecipitate and direct blending both showed decreasing foam overrun with the increase of pea protein ratio in the dual protein systems.

WP–PP complex is also an ideal ingredient for producing high-protein food gels. As detailed in Section 2.3, WPs can form heat- and cold-set gels under specific conditions. Similarly, plant proteins are also widely used as a gelling material, though their gelling ability is generally considered weaker compared with WPs. This means a higher protein concentration may be required to form a plant protein-based gel, or the gel formed has a lower gel strength compared with WP gel formed under the same conditions. As mentioned earlier, plant proteins possess a high molecular weight and are prone to creating water-insoluble aggregates during the production process, resulting in a reduced efficiency for gel formation. Many products also contain a significant amount of impurities, such as fiber and starch, that could negatively affect the gelling process. By simply changing the ratio of WP to plant protein, it is feasible to modulate the gel properties without adding extra gelling agents (e.g., $CaCl_2$). For example, increasing soy protein or pea protein fraction ranging from 0% to 90% resulted in weaker gels that were prepared from WP–plant protein mixtures and heat treatment (Jose et al., 2016; McCann et al., 2018; Wong et al., 2013; Xia et al., 2022). In other studies, partial substitution of WPs with plant proteins (soy, pea) did not always compromise gel formation. Roesch and Corredig (2005) reported that the gel strength remained similar between the WP–soy protein mixed gel (10%–50% soy protein) and the pure WP gel. In another study, the gel firmness of WP–pea protein gels remained fairly constant even though the WPI was substituted up to 75%, suggesting that pea protein isolate was an ideal substitute for WPI in certain applications (Kornet et al., 2021). The contrasting findings in the literature are likely due to plant proteins, whose gelling properties strongly depend on the purification process and degree of denaturation.

8. Conclusion

As outlined above, the functional properties of WPs are highly dependent on the intrinsic nature of the proteins, which inherently have a high water solubility, desirable emulsifying and foaming ability, as well as good gelling properties. Processing conditioning, such as heat treatment or pH adjustment, can strongly affect the functional properties of WPs, leading to improved dispersion and emulsion stability, better foaming or gel properties. The formulation of WPs with other food materials (e.g., plant proteins) or nutraceuticals (e.g., polyphenols) opens new avenues for modulating the functionality and nutritional values of WP ingredients and products, and the development of relevant products has attracted increasing interest on the market. It is expected that WP ingredients and their derived products will continue to increase on the market in the foreseeable future.

References

Adamcik, J., Mezzenga, R., 2012. Proteins fibrils from a polymer physics perspective. Macromolecules 45 (3), 1137–1150. https://doi.org/10.1021/ma202157h.

Aguilera, J.M., Baffico, P., 1997. Structure-mechanical properties of heat-induced whey protein/cassava starch gels. J. Food Sci. 62 (5), 1048–1066. https://doi.org/10.1111/j.1365-2621.1997.tb15035.x.

Aguilera, J.M., Rojas, E., 1996. Rheological, thermal and microstructural properties of whey protein-cassava starch gels. J. Food Sci. 61 (5), 962–966. https://doi.org/10.1111/j.1365-2621.1996.tb10911.x.

Ai, Y., Jane, J., 2015. Gelatinization and rheological properties of starch. Starch—Stärke 67 (3–4), 213–224. https://doi.org/10.1002/star.201400201.

Akamittath, J.G., Ball, J.,H.R., 1992. Transglutaminase mediated polymerization of crude actomyosin refined from mechanically deboned poultry meat. J. Muscle Foods 3 (1), 1–14. https://doi.org/10.1111/j.1745-4573.1992.tb00666.x.

Alting, A.C., Hamer, R.J., de Kruif, C.G., Visschers, R.W., 2000. Formation of disulfide bonds in acid-induced gels of preheated whey protein isolate. J. Agric. Food Chem. 48 (10), 5001–5007.

Alting, A.C., Hamer, R.J., de Kruif, C.G., Visschers, R.W., 2003. Cold-set globular protein gels: interactions, structure and rheology as a function of protein concentration. J. Agric. Food Chem. 51 (10), 3150–3156. https://doi.org/10.1021/jf0209342.

Alu'datt, M.H., Alli, I., Nagadi, M., 2012. Preparation, characterization and properties of whey-soy proteins co-precipitates. Food Chem. 134 (1), 294–300. https://doi.org/10.1016/j.foodchem.2012.02.142.

Alves, A.C., Martha, L., Casanova, F., Tavares, G.M., 2022. Structural and foaming properties of whey and soy protein isolates in mixed systems before and after heat treatment. Food Sci. Technol. Int. 28 (6), 545–553. https://doi.org/10.1177/10820132211031756.

Amin, U., Lin, Y., Zuo, X., Zheng, H., 2024. Soft matter approach for creating novel protein hydrogels using fractal whey protein assemblies as building blocks. Food Hydrocoll. 151, 109828. https://doi.org/10.1016/J.FOODHYD.2024.109828.

Amin, U., Yeung, C.K., Zheng, H., 2025. Construction of cold-set Mickering emulsion gel using whey protein assembly particles as oil-water interfacial stabilizer and gelling agent: phase stability, nonlinear rheology, and tribology. J. Dairy Sci. 108 (6), 5580–5597. https://doi.org/10.3168/JDS.2024-26001.

An, H., Zheng, H., 2025a. Investigation of the relation between interfacial properties of milk protein ingredients and their emulsifying properties. J. Dairy Sci. 108 (10), 10533–10547. https://doi.org/10.3168/JDS.2025-26514.

An, H., Zheng, H., 2025b. Investigation of correlations between powder functionalities and powder surface properties for milk protein ingredients. J. Dairy Sci. 108 (1), 190–205. https://doi.org/10.3168/JDS.2024-25289.

Andoyo, R., Dianti Lestari, V., Mardawati, E., Nurhadi, B., 2018. Fractal dimension analysis of texture formation of whey protein-based foods. Int. J. Food Sci. 2018, 7673259. https://doi.org/10.1155/2018/7673259.

Anhê, F.F., Roy, D., Pilon, G., Dudonné, S., Matamoros, S., Varin, T.V., Garofalo, C., Moine, Q., Desjardins, Y., Levy, E., Marette, A., 2015. A polyphenol-rich cranberry extract protects from diet-induced obesity, insulin resistance and intestinal inflammation in association with increased *Akkermansia* spp. population in the gut microbiota of mice. Gut 64 (6), 872–883. https://doi.org/10.1136/gutjnl-2014-307142.

Anuradha, S.N., Prakash, V., 2009. Altering functional attributes of proteins through cross-linking by transglutaminase—a case study with whey and seed proteins. Food Res. Int. 42 (9), 1259–1265. https://doi.org/10.1016/j.foodres.2009.06.012.

Baba, W.N., McClements, D.J., Maqsood, S., 2021. Whey protein–polyphenol conjugates and complexes: production, characterization, and applications. Food Chem. 365, 130455. https://doi.org/10.1016/j.foodchem.2021.130455.

Barone, G., O'Regan, J., Kelly, A.L., O'Mahony, J.A., 2022. Interactions between whey proteins and calcium salts and implications for the formulation of dairy protein-based nutritional beverage products: a review. Compr. Rev. Food Sci. Food Saf. 21 (2), 1254–1274. https://doi.org/10.1111/1541-4337.12884.

BeMiller, J.N., 2011. Pasting, paste, and gel properties of starch–hydrocolloid combinations. Carbohydr. Polym. 86 (2), 386–423. https://doi.org/10.1016/j.carbpol.2011.05.064.

Bertrand, M.-E., Turgeon, S.L., 2007. Improved gelling properties of whey protein isolate by addition of xanthan gum. Food Hydrocoll. 21 (2), 159–166. https://doi.org/10.1016/j.foodhyd.2006.03.005.

Bolder, S.G., Hendrickx, H., Sagis, L.M.C., van der Linden, E., 2006. Fibril assemblies in aqueous whey protein mixtures. J. Agric. Food Chem. 54 (12), 4229–4234. https://doi.org/10.1021/jf060606s.

Bolder, S.G., Vasbinder, A.J., Sagis, L.M., van der Linden, E., 2007. Heat-induced whey protein isolate fibrils: conversion, hydrolysis, and disulphide bond formation. Int. Dairy J. 17 (7), 846–853.

Brodkorb, A., Croguennec, T., Bouhallab, S., Kehoe, J.J., 2016. Heat-induced denaturation, aggregation and gelation of whey proteins. In: McSweeney, P.L.H., O'Mahony, J.A. (Eds.), Advanced Dairy Chemistry: Volume 1B: Proteins: Applied Aspects. Springer, pp. 155–178. https://doi.org/10.1007/978-1-4939-2800-2_6.

Bryant, C.M., McClements, D.J., 2000a. Influence of NaCl and CaCl2 on cold-set gelation of heat-denatured whey protein. J. Food Sci. 65 (5), 801–804. https://doi.org/10.1111/j.1365-2621.2000.tb13590.x.

Bryant, C.M., McClements, D.J., 2000b. Influence of xanthan gum on physical characteristics of heat-denatured whey protein solutions and gels. Food Hydrocoll. 14 (4), 383–390. https://doi.org/10.1016/S0268-005X(00)00018-7.

Cao, Y., Xiong, Y.L., Cao, Y., True, A.D., 2018. Interfacial properties of whey protein foams as influenced by preheating and phenolic binding at neutral pH. Food Hydrocoll. 82, 379–387. https://doi.org/10.1016/j.foodhyd.2018.04.020.

Celus, M., Kyomugasho, C., Van Loey, A.M., Grauwet, T., Hendrickx, M.E., 2018. Influence of pectin structural properties on interactions with divalent cations and its associated functionalities. Compr. Rev. Food Sci. Food Saf. 17 (6), 1576–1594. https://doi.org/10.1111/1541-4337.12394.

Chen, B., Li, H., Ding, Y., Suo, H., 2012. Formation and microstructural characterization of whey protein isolate/beet pectin coacervations by laccase catalyzed cross-linking. LWT Food Sci. Technol. 47 (1), 31–38. https://doi.org/10.1016/j.lwt.2012.01.006.

Chen, H., Ji, A., Qiu, S., Liu, Y., Zhu, Q., Yin, L., 2018. Covalent conjugation of bovine serum album and sugar beet pectin through Maillard reaction/laccase catalysis to improve the emulsifying properties. Food Hydrocoll. 76, 173–183. https://doi.org/10.1016/j.foodhyd.2016.12.004.

Chen, H., Zheng, H., Anne Brennan, M., Chen, W., Guo, X., Brennan, C.S., 2020. Effect of black tea infusion on physicochemical properties, anti-oxidant capacity and microstructure of acidified dairy gel during cold storage. Foods 9 (6), 831. https://doi.org/10.3390/foods9060831.

Chen, W., Lv, R., Wang, W., Ma, X., Muhammad, A.I., Guo, M., Ye, X., Liu, D., 2019a. Time effect on structural and functional properties of whey protein isolate-gum acacia conjugates prepared via Maillard reaction. J. Sci. Food Agric. 99 (10), 4801–4807. https://doi.org/10.1002/jsfa.9735.

Chen, W., Ma, X., Wang, W., Lv, R., Guo, M., Ding, T., Ye, X., Miao, S., Liu, D., 2019b. Preparation of modified whey protein isolate with gum acacia by ultrasound Maillard reaction. Food Hydrocoll. 95, 298–307. https://doi.org/10.1016/j.foodhyd.2018.10.030.

Chen, Y., Chen, X., Guo, T.L., Zhou, P., 2015. Improving the thermostability of β-lactoglobulin via glycation: the effect of sugar structures. Food Res. Int. 69, 106–113. https://doi.org/10.1016/j.foodres.2014.12.017.

Cheng, J., Dudu, O.E., Zhang, J., Wang, Y., Meng, L., Wei, W., Li, X., Yan, T., 2023. Impact of binding interaction modes between whey protein concentrate and quercetin on protein structural and functional characteristics. Food Hydrocoll. 142, 108787. https://doi.org/10.1016/j.foodhyd.2023.108787.

Chihi, M.-L., Mession, J., Sok, N., Saurel, R., 2016. Heat-induced soluble protein aggregates from mixed pea globulins and β-lactoglobulin. J. Agric. Food Chem. 64 (13), 2780–2791. https://doi.org/10.1021/acs.jafc.6b00087.

Cornec, M., Cho, D., Narsimhan, G., 1999. Adsorption dynamics of α-lactalbumin and β-lactoglobulin at air–water interfaces. J. Colloid Interface Sci. 214 (2), 129–142. https://doi.org/10.1006/jcis.1999.6230.

Cornec, M., Kim, D.A., Narsimhan, G., 2001. Adsorption dynamics and interfacial properties of α-lactalbumin in native and molten globule state conformation at air–water interface. Food Hydrocoll. 15 (3), 303–313. https://doi.org/10.1016/S0268-005X(01)00031-5.

Damodaran, S., Parkin, K.L., 2017. Fennema's Food Chemistry, Fifth Edition. CRC Press. https://doi.org/10.1201/9781315372914.

De Kruif, C.G., Tuinier, R., 2001. Polysaccharide protein interactions. Food Hydrocoll. 15 (4–6), 555–563. https://doi.org/10.1016/S0268-005X(01)00076-5.

De Kruif, C.G., Weinbreck, F., De Vries, R., 2004. Complex coacervation of proteins and anionic polysaccharides. Curr. Opin. Colloid Interface Sci. 9 (5), 340–349. https://doi.org/10.1016/j.cocis.2004.09.006.

Ding, L., Huang, Y., Cai, X., Wang, S., 2019. Impact of pH, ionic strength and chitosan charge density on chitosan/casein complexation and phase behavior. Carbohydr. Polym. 208, 133–141. https://doi.org/10.1016/j.carbpol.2018.12.015.

Du, Q., Zhou, L., Lyu, F., Liu, J., Ding, Y., 2022. The complex of whey protein and pectin: interactions, functional properties and applications in food colloidal systems—a review. Colloids Surf. B Biointerfaces 210, 112253. https://doi.org/10.1016/j.colsurfb.2021.112253.

Ebrahimi, P., Lante, A., Grossmann, L., 2025. Protein-polyphenol complexation vs. conjugation: a review on mechanisms, functional differences, and antioxidant-emulsifier roles. Food Hydrocoll. 169, 111590. https://doi.org/10.1016/j.foodhyd.2025.111590.

Fan, H., Zhu, P., Hui, G., Shen, Y., Yong, Z., Xie, Q., Wang, M., 2023. Mechanism of synergistic stabilization of emulsions by amorphous taro starch and protein and emulsion stability. Food Chem. 424, 136342. https://doi.org/10.1016/j.foodchem.2023.136342.

Fan, Y., Peng, G., Pang, X., Wen, Z., Yi, J., 2021. Physicochemical, emulsifying, and interfacial properties of different whey protein aggregates obtained by thermal treatment. Lebensm. Wiss. Technol. 149, 111904. https://doi.org/10.1016/j.lwt.2021.111904.

Fan, Z., Chai, J., Zhu, L., Yang, C., Wang, J., Lu, X., Shan, Y., 2025. Effects of whey protein aggregates with different sizes on the gelatinization and retrogradation behavior of wheat starch. J. Sci. Food Agric. 105 (8), 4271–4279. https://doi.org/10.1002/jsfa.14165.

Farrokhi, F., Badii, F., Ehsani, M.R., Hashemi, M., 2019. Functional and thermal properties of nanofibrillated whey protein isolate as functions of denaturation temperature and solution pH. Colloids Surf. A Physicochem. Eng. Asp. 583, 124002. https://doi.org/10.1016/j.colsurfa.2019.124002.

Feng, Y., Ma, X., Kong, B., Chen, Q., Liu, Q., 2021. Ethanol induced changes in structural, morphological, and functional properties of whey proteins isolates: influence of ethanol concentration. Food Hydrocoll. 111, 106379. https://doi.org/10.1016/J.FOODHYD.2020.106379.

Flanagan, J., Singh, H., 2006. Conjugation of sodium caseinate and gum Arabic catalyzed by transglutaminase. J. Agric. Food Chem. 54 (19), 7305–7310. https://doi.org/10.1021/jf061220k.

Gao, Y.-Z., Xu, H.-H., Ju, T.-T., Zhao, X.-H., 2013. The effect of limited proteolysis by different proteases on the formation of whey protein fibrils. J. Dairy Sci. 96 (12), 7383–7392. https://doi.org/10.3168/jds.2013-6843.

Gazme, B., Madadlou, A., 2014. Fabrication of whey protein–pectin conjugate particles through laccase-induced gelation of microemulsified nanodroplets. Food Hydrocoll. 40, 189–195. https://doi.org/10.1016/j.foodhyd.2014.02.017.

Gélébart, P., Riaublanc, A., Famelart, M.-H., Jonchère, C., Beaumal, V., Anton, M., Garnier, C., 2019. Protein aggregates modulate the texture of emulsified and acidified acid milk gels. Food Hydrocoll. 93, 176–188. https://doi.org/10.1016/j.foodhyd.2019.02.007.

Gezimati, J., Creamer, L.K., Singh, H., 1997. Heat-induced interactions and gelation of mixtures of β-lactoglobulin and α-lactalbumin. J. Agric. Food Chem. 45 (4), 1130–1136. https://doi.org/10.1021/jf960564f.

Goers, J., Permyakov, S.E., Permyakov, E.A., Uversky, V.N., Fink, A.L., 2002. Conformational prerequisites for α-lactalbumin fibrillation. Biochemistry 41 (41), 12546–12551. https://doi.org/10.1021/bi0262698.

Gou, M., Wu, H., Saleh, A.S.M., Jing, L., Liu, Y., Zhao, K., Su, C., Zhang, B., Jiang, H., Li, W., 2019. Effects of repeated and continuous dry heat treatments on properties of sweet potato starch. Int. J. Biol. Macromol. 129, 869–877. https://doi.org/10.1016/j.ijbiomac.2019.01.225.

Gülseren, İ., Fang, Y., Corredig, M., 2012. Whey protein nanoparticles prepared with desolvation with ethanol: characterization, thermal stability and interfacial behavior. Food Hydrocoll. 29 (2), 258–264. https://doi.org/10.1016/j.foodhyd.2012.03.015.

Gulzar, M., Bouhallab, S., Jeantet, R., Schuck, P., Croguennec, T., 2011. Influence of pH on the dry heat-induced denaturation/aggregation of whey proteins. Food Chem. 129 (1), 110–116. https://doi.org/10.1016/j.foodchem.2011.04.037.

Gumus, C.E., Decker, E.A., McClements, D.J., 2017. Formation and stability of ω-3 oil emulsion-based delivery systems using plant proteins as emulsifiers: lentil, pea, and faba bean proteins. Food Biophys. 12 (2), 186–197. https://doi.org/10.1007/s11483-017-9475-6.

Guo, B., Hu, X., Wu, J., Chen, R., Dai, T., Liu, Y., Luo, S., Liu, C., 2021. Soluble starch/whey protein isolate complex-stabilized high internal phase emulsion: interaction and stability. Food Hydrocoll. 111, 106377. https://doi.org/10.1016/j.foodhyd.2020.106377.

Güzey, D., Gülseren, İ., Bruce, B., Weiss, J., 2006. Interfacial properties and structural conformation of thermosonicated bovine serum albumin. Food Hydrocoll. 20 (5), 669–677. https://doi.org/10.1016/j.foodhyd.2005.06.008.

Hattori, M., Nagasawa, K., Ametani, A., Kaminogawa, S., Takahashi, K., 1994. Functional changes in beta-lactoglobulin by conjugation with carboxymethyl dextran. J. Agric. Food Chem. 42 (10), 2120–2125. https://doi.org/10.1021/jf00046a009.

Havea, P., Carr, A.J., Creamer, L.K., 2004. The roles of disulphide and non-covalent bonding in the functional properties of heat-induced whey protein gels. J. Dairy Res. 71 (3), 330–339.

Havea, P., Singh, H., Creamer, L.K., 2001. Characterization of heat-induced aggregates of β-lactoglobulin, α-lactalbumin and bovine serum albumin in a whey protein concentrate environment. J. Dairy Res. 68 (3), 483–497. https://doi.org/10.1017/S0022029901004964.

Havea, P., Watkinson, P., Kuhn-Sherlock, B., 2009. Heat-induced whey protein gels: protein–protein interactions and functional properties. J. Agric. Food Chem. 57 (4), 1506–1512.

Hines, M.E., Foegeding, E.A., 1993. Interactions of .alpha.-lactalbumin and bovine serum albumin with beta-lactoglobulin in thermally induced gelation. J. Agric. Food Chem. 41 (3), 341–346. https://doi.org/10.1021/jf00027a001.

Horne, D.S., 2017. Characteristics of milk. In: Fennema's Food Chemistry, fifth ed. CRC Press.

Horne, D.S., 2020. Chapter 6—casein micelle structure and stability. In: Boland, M., Singh, H. (Eds.), Milk Proteins, third ed. Academic Press, pp. 213–250 https://www.sciencedirect.com/science/article/pii/B9780128152515000062.

Hu, Y., He, C., Zhang, M., Zhang, L., Xiong, H., Zhao, Q., 2020. Inhibition from whey protein hydrolysate on the retrogradation of gelatinized rice starch. Food Hydrocoll. 108, 105840. https://doi.org/10.1016/j.foodhyd.2020.105840.

Imtiaz, S.R., Kuhn-Sherlock, B., Campbell, M., 2012. Effect of dairy protein blends on texture of high protein bars. J. Texture Stud. 43 (4), 275–286. https://doi.org/10.1111/j.1745-4603.2011.00337.x.

Inthavong, W., Kharlamova, A., Chassenieux, C., Nicolai, T., 2016. Structure and flow of dense suspensions of protein fractal aggregates in comparison with microgels. Soft Matter 12 (10), 2785–2793. https://doi.org/10.1039/C5SM02893K.

Ipsen, R., 2017. Microparticulated whey proteins for improving dairy product texture. Int. Dairy J. 67, 73–79. https://doi.org/10.1016/j.idairyj.2016.08.009.

Jiang, F., Pan, Y., Peng, D., Huang, W., Shen, W., Jin, W., Huang, Q., 2022. Tunable self-assemblies of whey protein isolate fibrils for Pickering emulsions structure regulation. Food Hydrocoll. 124, 107264. https://doi.org/10.1016/j.foodhyd.2021.107264.

Jiang, J., Zhang, Z., Zhao, J., Liu, Y., 2018. The effect of non-covalent interaction of chlorogenic acid with whey protein and casein on physicochemical and radical-scavenging activity of in vitro protein digests. Food Chem. 268, 334–341. https://doi.org/10.1016/j.foodchem.2018.06.015.

Jiang, Z., Huangfu, Y., Jiang, L., Wang, T., Bao, Y., Ma, W., 2023. Structure and functional properties of whey protein conjugated with carboxymethyl cellulose through maillard reaction. Lebensm. Wiss. Technol. 174, 114406. https://doi.org/10.1016/j.lwt.2022.114406.

Jose, J., Pouvreau, L., Martin, A.H., 2016. Mixing whey and soy proteins: consequences for the gel mechanical response and water holding. Food Hydrocoll. 60, 216–224. https://doi.org/10.1016/j.foodhyd.2016.03.031.

Ju, Z.Y., Kilara, A., 1998. Aggregation induced by calcium chloride and subsequent thermal gelation of whey protein isolate. J. Dairy Sci. 81 (4), 925–931. https://doi.org/10.3168/jds.S0022-0302(98)75652-8.

Jung, J., Wicker, L., 2012. Laccase mediated conjugation of heat treated β-lactoglobulin and sugar beet pectin. Carbohydr. Polym. 89 (4), 1244–1249. https://doi.org/10.1016/j.carbpol.2012.04.028.

Jung, J.-M., Gunes, D.Z., Mezzenga, R., 2010. Interfacial activity and interfacial shear rheology of native β-lactoglobulin monomers and their heat-induced fibers. Langmuir 26 (19), 15366–15375. https://doi.org/10.1021/la102721m.

Kan, X., Zhang, K., Fan, X., Chen, L., Zeng, X., 2024. Maillard conjugates of whey protein isolate and gum Arabic: enhanced functional properties and unique gut microbiota regulation. Food Hydrocoll. 152, 109911. https://doi.org/10.1016/j.foodhyd.2024.109911.

Kanakis, C.D., Hasni, I., Bourassa, P., Tarantilis, P.A., Polissiou, M.G., Tajmir-Riahi, H.-A., 2011. Milk β-lactoglobulin complexes with tea polyphenols. Food Chem. 127 (3), 1046–1055. https://doi.org/10.1016/j.foodchem.2011.01.079.

Kayser, J.J., Arnold, P., Steffen-Heins, A., Schwarz, K., Keppler, J.K., 2020. Functional ethanol-induced fibrils: influence of solvents and temperature on amyloid-like aggregation of beta-lactoglobulin. J. Food Eng. 270, 109764. https://doi.org/10.1016/j.jfoodeng.2019.109764.

Kazmierski, M., Wicker, L., Corredig, M., 2003. Interactions of (β-Lactoglobulin and high-methoxyl pectins in acidified systems. J. Food Sci. 68 (5), 1673–1679. https://doi.org/10.1111/j.1365-2621.2003.tb12312.x.

Kelly, P., 2019. Chapter 9—whey protein ingredient applications. In: Deeth, H.C., Bansal, N. (Eds.), Whey Proteins. Academic Press, pp. 335–375. https://www.sciencedirect.com/science/article/pii/B9780128121245000102.

Khalesi, H., Sun, C., He, J., Lu, W., Fang, Y., 2021. The role of amyloid fibrils in the modification of whey protein isolate gels with the form of stranded and particulate microstructures. Food Res. Int. 140, 109856. https://doi.org/10.1016/j.foodres.2020.109856.

Kharlamova, A., Chassenieux, C., Nicolai, T., 2018a. Acid-induced gelation of whey protein aggregates: kinetics, gel structure and rheological properties. Food Hydrocoll. 81, 263–272. https://doi.org/10.1016/j.foodhyd.2018.02.043.

Kharlamova, A., Nicolai, T., Chassenieux, C., 2018b. Calcium-induced gelation of whey protein aggregates: kinetics, structure and rheological properties. Food Hydrocoll. 79, 145–157. https://doi.org/10.1016/j.foodhyd.2017.11.049.

Kim, W., Wang, Y., Ye, Q., Yao, Y., Selomulya, C., 2023. Enzymatic cross-linking of pea and whey proteins to enhance emulsifying and encapsulation properties. Food Bioprod. Process. 139, 204–215. https://doi.org/10.1016/j.fbp.2023.03.011.

Kornet, R., Shek, C., Venema, P., Jan van der Goot, A., Meinders, M., van der Linden, E., 2021. Substitution of whey protein by pea protein is facilitated by specific fractionation routes. Food Hydrocoll. 117, 106691. https://doi.org/10.1016/j.foodhyd.2021.106691.

Kristensen, H.T., Denon, Q., Tavernier, I., Gregersen, S.B., Hammershøj, M., Van der Meeren, P., Dewettinck, K., Dalsgaard, T.K., 2021. Improved food functional properties of pea protein isolate in blends and co-precipitates with whey protein isolate. Food Hydrocoll. 113, 106556. https://doi.org/10.1016/j.foodhyd.2020.106556.

Kristensen, H.T., Møller, A.H., Christensen, M., Hansen, M.S., Hammershøj, M., Dalsgaard, T.K., 2020. Co-precipitation of whey and pea protein—indication of interactions. Int. J. Food Sci. Technol. 55 (8), 2920–2930. https://doi.org/10.1111/ijfs.14553.

Krzeminski, A., Prell, K.A., Weiss, J., Hinrichs, J., 2014. Environmental response of pectin-stabilized whey protein aggregates. Food Hydrocoll. 35, 332–340. https://doi.org/10.1016/j.foodhyd.2013.06.014.

Kudanga, T., Nyanhongo, G.S., Guebitz, G.M., Burton, S., 2011. Potential applications of laccase-mediated coupling and grafting reactions: a review. Enzym. Microb. Technol. 48 (3), 195–208. https://doi.org/10.1016/j.enzmictec.2010.11.007.

Kumar, L., Brennan, M.A., Mason, S.L., Zheng, H., Brennan, C.S., 2017. Rheological, pasting and microstructural studies of dairy protein–starch interactions and their application in extrusion-based products: a review. Starch—Stärke 69 (1–2), 1600273. https://doi.org/10.1002/star.201600273.

Kumar, L., Brennan, M., Brennan, C., Zheng, H., 2022a. Influence of whey protein isolate on pasting, thermal, and structural characteristics of oat starch. J. Dairy Sci. 105 (1), 56–71. https://doi.org/10.3168/jds.2021-20711.

Kumar, L., Brennan, M., Brennan, C., Zheng, H., 2022b. Thermal, pasting and structural studies of oat starch-caseinate interactions. Food Chem. 373, 131433. https://doi.org/10.1016/j.foodchem.2021.131433.

Kumar, L., Brennan, M., Zheng, H., Brennan, C., 2018. The effects of dairy ingredients on the pasting, textural, rheological, freeze-thaw properties and swelling behaviour of oat starch. Food Chem. 245, 518–524. https://doi.org/10.1016/J.FOODCHEM.2017.10.125.

Kumar, L., Brennan, M., Zheng, H., Kumar, G., Brennan, C., 2023. Protein fortification in oat flour gel using various dairy protein ingredients: an evaluation of textural and pasting properties. Int. J. Dairy Technol. 77 (1), 71–80. https://doi.org/10.1111/1471-0307.13018.

Kumar, V., Sinha, A.K., Makkar, H.P.S., De Boeck, G., Becker, K., 2012. Dietary roles of non-starch polysaccharides in human nutrition: a review. Crit. Rev. Food Sci. Nutr. 52 (10), 899–935. https://doi.org/10.1080/10408398.2010.512671.

Lam, R.S.H., Nickerson, M.T., 2015. The effect of pH and temperature pre-treatments on the physicochemical and emulsifying properties of whey protein isolate. LWT—Food Sci. Technol. 60 (1), 427–434. https://doi.org/10.1016/j.lwt.2014.07.031.

Langton, M., Hermansson, A.-M., 1992. Fine-stranded and particulate gels of β-lactoglobulin and whey protein at varying pH. Food Hydrocoll. 5 (6), 523–539. https://doi.org/10.1016/S0268-005X(09)80122-7.

Li, D., Li, S., Zhao, R., Shen, Y., Yin, F., Wang, C., 2025. Fabrication and characterisation of whey–pea protein-based emulsion gels induced by transglutaminase cross-linking. Int. J. Dairy Technol. 78 (1), e13143. https://doi.org/10.1111/1471-0307.13143.

Li, H., Zhang, Y., Liu, T., Zhang, L., Li, M., Li, H., Li, D., Wang, X., Yu, J., 2023. Transglutaminase, glucono-δ-lactone, and citric acid–induced whey protein isolation–milk fat emulsion gel embedding lutein and its application in processed cheese. J. Dairy Sci. 106 (10), 6635–6645. https://doi.org/10.3168/jds.2022-23097.

Li, J., Ouldeleya, M., Gunasekaran, S., 2006. Gelation of whey protein and xanthan mixture: effect of heating rate on rheological properties. Food Hydrocoll. 20 (5), 678–686. https://doi.org/10.1016/j.foodhyd.2005.07.001.

Li, N., Girard, A.L., 2023. Impact of pH and temperature on whey protein-proanthocyanidin interactions and foaming properties. Food Hydrocoll. 134, 108100. https://doi.org/10.1016/j.foodhyd.2022.108100.

Lian, J., Yang, K., Chen, J., Zhu, Y., 2026. Tribological properties and interfacial adsorption behavior of milk protein-based formulations: roles of whey protein type and thermal processing. Food Hydrocoll. 170, 111695. https://doi.org/10.1016/j.foodhyd.2025.111695.

Liang, X., Ma, C., Yan, X., Zeng, H., McClements, D.J., Liu, X., Liu, F., 2020. Structure, rheology and functionality of whey protein emulsion gels: effects of double cross-linking with transglutaminase and calcium ions. Food Hydrocoll. 102, 105569. https://doi.org/10.1016/j.foodhyd.2019.105569.

Lin, J., Guo, X., Ai, C., Zhang, T., Yu, S., 2020. Genipin crosslinked sugar beet pectin-whey protein isolate/bovine serum albumin conjugates with enhanced emulsifying properties. Food Hydrocoll. 105, 105802. https://doi.org/10.1016/j.foodhyd.2020.105802.

Lin, Y., Cheng, N., Jiang, Y., Grace, M.H., Lila, M.A., Hoskin, R.T., Zheng, H., 2024a. Colloidal and interfacial properties of spray dried pulse protein-blueberry polyphenol particles in model dispersion systems. Food Chem. 457, 140073. https://doi.org/10.1016/J.FOODCHEM.2024.140073.

Lin, Y., Maloney, K., Drake, M., Zheng, H., 2024b. Synergistic functionality of transglutaminase and protease on modulating texture of pea protein based yogurt alternative: from rheological and tribological characterizations to sensory perception. Food Hydrocoll. 150, 109652. https://doi.org/10.1016/j.foodhyd.2023.109652.

Liu, J., Liu, Q., Yang, Y., Zhao, S., Jin, Z., Zhu, K., Xu, L., Jiao, A., 2021. Effects of whey protein on the in vitro digestibility and physicochemical properties of potato starch. Int. J. Biol. Macromol. 193, 1744–1751. https://doi.org/10.1016/j.ijbiomac.2021.11.011.

Liu, J., Yong, H., Yao, X., Hu, H., Yun, D., Xiao, L., 2019. Recent advances in phenolic–protein conjugates: synthesis, characterization, biological activities and potential applications. RSC Adv. 9 (61), 35825–35840. https://doi.org/10.1039/C9RA07808H.

Liu, Y., Ma, L., Gao, C., 2012. Facile fabrication of the glutaraldehyde cross-linked collagen/chitosan porous scaffold for skin tissue engineering. Mater. Sci. Eng. C 32 (8), 2361–2366. https://doi.org/10.1016/j.msec.2012.07.008.

Liu, Y., Qiu, S., Li, J., Chen, H., Tatsumi, E., Yadav, M., Yin, L., 2015. Peroxidase-mediated conjugation of corn fiber gum and bovine serum albumin to improve emulsifying properties. Carbohydr. Polym. 118, 70–78. https://doi.org/10.1016/j.carbpol.2014.10.059.

Liu, Y., Yadav, M.P., Chau, H.K., Qiu, S., Zhang, H., Yin, L., 2017. Peroxidase-mediated formation of corn fiber gum-bovine serum albumin conjugates: molecular and structural characterization. Carbohydr. Polym. 166, 114–122. https://doi.org/10.1016/j.carbpol.2017.02.069.

Mahmoudi, N., Axelos, M.A.V., Riaublanc, A., 2011. Interfacial properties of fractal and spherical whey protein aggregates. Soft Matter 7 (17), 7643–7654. https://doi.org/10.1039/C1SM05262D.

Manzoor, M., Tchameni, Z.F.N., Bhat, Z.F., Jaiswal, A.K., Jaglan, S., 2024. Recent insights on the conformational changes, functionality, and physiological properties of plant-based protein–polyphenol conjugates. Food Bioprocess Technol. 17 (8), 2131–2154. https://doi.org/10.1007/s11947-023-03212-z.

McCann, T.H., Guyon, L., Fischer, P., Day, L., 2018. Rheological properties and microstructure of soy-whey protein. Food Hydrocoll. 82, 434–441. https://doi.org/10.1016/j.foodhyd.2018.04.023.

McClements, D.J., 2006. Non-covalent interactions between proteins and polysaccharides. Biotechnol. Adv. 24 (6), 621–625. https://doi.org/10.1016/j.biotechadv.2006.07.003.

McClements, D.J., 2015. Food Emulsions: Principles, Practices, and Techniques, Third Edition, 0 ed. CRC Press. https://doi.org/10.1201/b18868.

Meng, Y., Li, C., 2021. Conformational changes and functional properties of whey protein isolate-polyphenol complexes formed by non-covalent interaction. Food Chem. 364, 129622. https://doi.org/10.1016/j.foodchem.2021.129622.

Minj, S., Anand, S., 2020. Whey proteins and its derivatives: bioactivity, functionality, and current applications. Dairy 1 (3), 233–258. https://doi.org/10.3390/dairy1030016.

Mudgal, P., Daubert, C.R., Clare, D.A., Foegeding, E.A., 2011. Effect of disulfide interactions and hydrolysis on the thermal aggregation of β-lactoglobulin. J. Agric. Food Chem. 59 (5), 1491–1497. https://doi.org/10.1021/jf101893v.

Natallia, L., Dama, A., Gorica, E., Darya, K., Peña-Corona, S.I., Cortés, H., Santini, A., Büsselberg, D., Leyva-Gómez, G., Sharifi-Rad, J., 2024. Genipin's potential as an anti-cancer agent: from phytochemical origins to clinical prospects. Med. Oncol. 41 (8), 186. https://doi.org/10.1007/s12032-024-02429-y.

Nicolai, T., 2016. Formation and functionality of self-assembled whey protein microgels. Colloids Surf. B Biointerfaces 137, 32–38. https://doi.org/10.1016/j.colsurfb.2015.05.055.

Nicolai, T., 2019. Gelation of food protein-protein mixtures. Adv. Colloid Interface Sci. 270, 147–164. https://doi.org/10.1016/j.cis.2019.06.006.

Nicolai, T., Britten, M., Schmitt, C., 2011. β-Lactoglobulin and WPI aggregates: formation, structure and applications. Food Hydrocoll. 25 (8), 1945–1962. https://doi.org/10.1016/j.foodhyd.2011.02.006.

Nicolai, T., Durand, D., 2013. Controlled food protein aggregation for new functionality. Curr. Opin. Colloid Interface Sci. 18 (4), 249–256. https://doi.org/10.1016/j.cocis.2013.03.001.

Nikolaidis, A., Andreadis, M., Moschakis, T., 2017. Effect of heat, pH, ultrasonication and ethanol on the denaturation of whey protein isolate using a newly developed approach in the analysis of difference-UV spectra. Food Chem. 232, 425–433. https://doi.org/10.1016/j.foodchem.2017.04.022.

Nikolaidis, A., Moschakis, T., 2018. On the reversibility of ethanol-induced whey protein denaturation. Food Hydrocoll. 84, 389–395. https://doi.org/10.1016/j.foodhyd.2018.05.051.

Niu, W., Wu, L., Gong, W., Kang, X., Zhang, J., Nishinari, K., Zhao, M., 2023. Properties of complexes of whey protein isolate fibrils (WPIF) and octenyl succinate starch (OSS) and their applications in emulsion stabilization and custard cream. Food Hydrocoll. 142, 108822. https://doi.org/10.1016/j.foodhyd.2023.108822.

Nooshkam, M., Varidi, M., 2020. Whey protein isolate-low acyl gellan gum Maillard-based conjugates with tailored technological functionality and antioxidant activity. Int. Dairy J. 109, 104783. https://doi.org/10.1016/j.idairyj.2020.104783.

Noskov, B.A., Bykov, A.G., Gochev, G., Lin, S.-Y., Loglio, G., Miller, R., Milyaeva, O.Y., 2020. Adsorption layer formation in dispersions of protein aggregates. Adv. Colloid Interface Sci. 276, 102086. https://doi.org/10.1016/j.cis.2019.102086.

O'Kennedy, B.T., Mounsey, J.S., 2006. Control of heat-induced aggregation of whey proteins using casein. J. Agric. Food Chem. 54 (15), 5637–5642. https://doi.org/10.1021/jf0607866.

O'Mahony, J.A., Drapala, K.P., Mulcahy, E.M., Mulvihill, D.M., 2019. Chapter 7—Whey protein–carbohydrate conjugates. In: Deeth, H.C., Bansal, N. (Eds.), Whey Proteins. Academic Press, pp. 249–280. https://www.sciencedirect.com/science/article/pii/B9780128121245000084.

Ondiek, W., Wang, Y., Sun, L., Zhou, L., On, S.L.W., Zheng, H., Ravi, G., 2021. Removal of aflatoxin b1 and t-2 toxin by bacteria isolated from commercially available probiotic dairy foods. Food Sci. Technol. Int., 1082013220987916 https://doi.org/10.1177/1082013220987916.

Pal, S., Maity, S., Sardar, S., Chakraborty, J., Halder, U.C., 2016. Insight into the co-solvent induced conformational changes and aggregation of bovine β-lactoglobulin. Int. J. Biol. Macromol. 84, 121–134. https://doi.org/10.1016/j.ijbiomac.2015.11.055.

Patel, A.S., Yeung, C.K., Brennan, C., Zheng, H., 2022. Additives in dairy foods: ingredients and formulation. In: McSweeney, P.L.H., McNamara, J.P. (Eds.), Encyclopedia of Dairy Sciences, third ed. Academic Press, pp. 663–673 https://www.sciencedirect.com/science/article/pii/B9780128187661001616.

Paulsson, M., Dejmek, P., 1992. Surface film pressure of β-lactoglobulin, α-lactalbumin and bovine serum albumin at the air/water interface studied by wilhelmy plate and drop volume. J. Colloid Interface Sci. 150 (2), 394–403. https://doi.org/10.1016/0021-9797(92)90209-5.

Pelegrine, D.H.G., Gasparetto, C.A., 2005. Whey proteins solubility as function of temperature and pH. LWT Food Sci. Technol. 38 (1), 77–80. https://doi.org/10.1016/j.lwt.2004.03.013.

Pérez, S., Bertoft, E., 2010. The molecular structures of starch components and their contribution to the architecture of starch granules: a comprehensive review. Starch—Stärke 62 (8), 389–420. https://doi.org/10.1002/star.201000013.

Prigent, S.V.E., Voragen, A.G.J., Van Koningsveld, G.A., Baron, A., Renard, C.M.G.C., Gruppen, H., 2009. Interactions between globular proteins and procyanidins of different degrees of polymerization. J. Dairy Sci. 92 (12), 5843–5853. https://doi.org/10.3168/jds.2009-2261.

Quan, T.H., Benjakul, S., Sae-leaw, T., Balange, A.K., Maqsood, S., 2019. Protein–polyphenol conjugates: antioxidant property, functionalities and their applications. Trends Food Sci. Technol. 91, 507–517. https://doi.org/10.1016/j.tifs.2019.07.049.

Rabiey, L., Britten, M., 2009. Effect of protein composition on the rheological properties of acid-induced whey protein gels. Food Hydrocoll. 23 (3), 973–979. https://doi.org/10.1016/j.foodhyd.2008.07.009.

Ravindra, P., Genovese, D.B., Foegeding, E.A., Rao, M.A., 2004. Rheology of heated mixed whey protein isolate/cross-linked waxy maize starch dispersions. Food Hydrocoll. 18 (5), 775–781. https://doi.org/10.1016/j.foodhyd.2003.12.004.

Ribeiro, A.C., Bravo, C., Ramos, R.M., Dias, R., Mateus, N., De Freitas, V., Pérez-Gregorio, R., Soares, S., 2025. Polyphenols as enhancers of emulsifying and antioxidant properties of egg and yeast protein emulsions: understanding the molecular interactions behind their technological and functional potential. Food Hydrocoll. 167, 111399. https://doi.org/10.1016/j.foodhyd.2025.111399.

Rivero, W.C., Shah, U., Wang, Q., Zheng, H., Salvi, D., 2024. Exploration of plasma-activated water (PAW) as a cleaning-in-place (CIP) solution for fouling removal and microbial reduction. J. Food Process Eng. 47 (7), e14669. https://doi.org/10.1111/JFPE.14669.

Roesch, R.R., Corredig, M., 2005. Heat-induced soy–whey proteins interactions: formation of soluble and insoluble protein complexes. J. Agric. Food Chem. 53 (9), 3476–3482. https://doi.org/10.1021/jf048870d.

Roopchand, D.E., Carmody, R.N., Kuhn, P., Moskal, K., Rojas-Silva, P., Turnbaugh, P.J., Raskin, I., 2015. Dietary polyphenols promote growth of the gut bacterium *Akkermansia muciniphila* and attenuate high-fat diet–induced metabolic syndrome. Diabetes 64 (8), 2847–2858. https://doi.org/10.2337/db14-1916.

Ryan, K.N., Vardhanabhuti, B., Jaramillo, D.P., van Zanten, J.H., Coupland, J.N., Foegeding, E.A., 2012. Stability and mechanism of whey protein soluble aggregates thermally treated with salts. Food Hydrocoll. 27 (2), 411–420. https://doi.org/10.1016/j.foodhyd.2011.11.006.

Ryan, K.N., Zhong, Q., Foegeding, E.A., 2013. Use of whey protein soluble aggregates for thermal stability—a hypothesis paper. J. Food Sci. 78 (8), R1105–R1115. https://doi.org/10.1111/1750-3841.12207.

Sabet, S., Rashidinejad, A., Melton, L.D., Zujovic, Z., Akbarinejad, A., Nieuwoudt, M., Seal, C.K., McGillivray, D.J., 2021. The interactions between the two negatively charged polysaccharides: gum Arabic and alginate. Food Hydrocoll. 112, 106343. https://doi.org/10.1016/j.foodhyd.2020.106343.

Sagis, L.M.C., Scholten, E., 2014. Complex interfaces in food: structure and mechanical properties. Trends Food Sci. Technol. 37 (1), 59–71. https://doi.org/10.1016/j.tifs.2014.02.009.

Samtiya, M., Aluko, R.E., Dhewa, T., Moreno-Rojas, J.M., 2021. Potential health benefits of plant food-derived bioactive components: an overview. Foods 10 (4), 839. https://doi.org/10.3390/foods10040839.

Sarkar, A., Kanti, F., Gulotta, A., Murray, B.S., Zhang, S., 2017. Aqueous lubrication, structure and rheological properties of whey protein microgel particles. Langmuir 33 (51), 14699–14708. https://doi.org/10.1021/acs.langmuir.7b03627.

Schmidt, S., Liu, T., Rütten, S., Phan, K.-H., Möller, M., Richtering, W., 2011. Influence of microgel architecture and oil polarity on stabilization of emulsions by stimuli-sensitive core–shell poly(N-isopropylacrylamide-co-methacrylic acid) microgels: mickering versus Pickering behavior? Langmuir 27 (16), 9801–9806. https://doi.org/10.1021/la201823b.

Schmitt, C., Bovay, C., Rouvet, M., Shojaei-Rami, S., Kolodziejczyk, E., 2007. Whey protein soluble aggregates from heating with NaCl: physico-chemical, interfacial, and foaming properties. Langmuir 23 (8), 4155–4166. https://doi.org/10.1021/la0632575.

Schmitt, C., Bovay, C., Vuilliomenet, A.-M., Rouvet, M., Bovetto, L., Barbar, R., Sanchez, C., 2009. Multiscale characterization of individualized β-lactoglobulin microgels formed upon heat treatment under narrow pH range conditions. Langmuir 25 (14), 7899–7909. https://doi.org/10.1021/la900501n.

Schmitt, C., Turgeon, S.L., 2011. Protein/polysaccharide complexes and coacervates in food systems. Adv. Colloid Interface Sci. 167 (1–2), 63–70. https://doi.org/10.1016/j.cis.2010.10.001.

Schokker, E.P., Singh, H., Pinder, D.N., Norris, G.E., Creamer, L.K., 1999. Characterization of intermediates formed during heat-induced aggregation of β-lactoglobulin AB at neutral pH. Int. Dairy J. 9 (11), 791–800. https://doi.org/10.1016/S0958-6946(99)00148-X.

Schröder, A., Berton-Carabin, C., Venema, P., Cornacchia, L., 2017. Interfacial properties of whey protein and whey protein hydrolysates and their influence on O/W emulsion stability. Food Hydrocoll. 73, 129–140. https://doi.org/10.1016/j.foodhyd.2017.06.001.

Sedaghat Doost, A., Nikbakht Nasrabadi, M., Wu, J., A'yun, Q., Van der Meeren, P., 2019. Maillard conjugation as an approach to improve whey proteins functionality: a review of conventional and novel preparation techniques. Trends Food Sci. Technol. 91, 1–11. https://doi.org/10.1016/j.tifs.2019.06.011.

Sentoukas, T., Charitou, G., Wagner, J., Wagemans, A.M., Moschakis, T., Papagiannopoulos, A., 2023. Formation of nanoparticles from ethanol-denatured whey proteins. Food Struct. 37, 100337. https://doi.org/10.1016/j.foostr.2023.100337.

Setiowati, A.D., Wijaya, W., Van Der Meeren, P., 2020. Whey protein-polysaccharide conjugates obtained via dry heat treatment to improve the heat stability of whey protein stabilized emulsions. Trends Food Sci. Technol. 98, 150–161. https://doi.org/10.1016/j.tifs.2020.02.011.

Seung, D., 2020. Amylose in starch: towards an understanding of biosynthesis, structure and function. New Phytol. 228 (5), 1490–1504. https://doi.org/10.1111/nph.16858.

Shao, Y., He, L., Zheng, H., 2025. Prediction of phase stability of UHT reconstituted skim milk after ambient storage using Raman spectroscopy. Int. Dairy J. 161, 106142. https://doi.org/10.1016/J.IDAIRYJ.2024.106142.

Sharma, R., 2019. Chapter 17—whey proteins in functional foods. In: Deeth, H.C., Bansal, N. (Eds.), Whey Proteins. Academic Press, pp. 637–663. https://www.sciencedirect.com/science/article/pii/B9780128121245000187.

Shi, H., Zheng, H., 2025. Mesoscale fractal whey protein particles derived from microscale linear-shaped protein assemblies (Part 2): foaming properties and heat stability. J. Dairy Sci. 109 (3), 2150–2164. https://doi.org/10.3168/jds.2025-27190.

Spotti, M.J., Martinez, M.J., Pilosof, A.M.R., Candioti, M., Rubiolo, A.C., Carrara, C.R., 2014. Influence of Maillard conjugation on structural characteristics and rheological properties of whey protein/dextran systems. Food Hydrocoll. 39, 223–230. https://doi.org/10.1016/j.foodhyd.2014.01.014.

Spotti, M.J., Perduca, M.J., Piagentini, A., Santiago, L.G., Rubiolo, A.C., Carrara, C.R., 2013. Gel mechanical properties of milk whey protein–dextran conjugates obtained by Maillard reaction. Food Hydrocoll. 31 (1), 26–32. https://doi.org/10.1016/j.foodhyd.2012.08.009.

Tamm, F., Sauer, G., Scampicchio, M., Drusch, S., 2012. Pendant drop tensiometry for the evaluation of the foaming properties of milk-derived proteins. Food Hydrocoll. 27 (2), 371–377. https://doi.org/10.1016/j.foodhyd.2011.10.013.

Tan, Y., Lee, P.W., Martens, T.D., McClements, D.J., 2022. Comparison of emulsifying properties of plant and animal proteins in oil-in-water emulsions: whey, soy, and RuBisCo proteins. Food Biophys. 17 (3), 409–421. https://doi.org/10.1007/s11483-022-09730-1.

Thompson, L.U., 1977. Coprecipitation of rapeseed and cheese whey proteins using acid and heat treatment. Canad. Inst. Food Sci. Technol. J. 10 (1), 43–48. https://doi.org/10.1016/S0315-5463(77)73435-2.

Tian, R., Han, X., Tian, B., Li, G., Sun, L., Tian, S., Qin, L., Wang, S., 2023. Effects of covalent binding of different polyphenols on structure, rheology and functional properties of whey protein isolate. LWT 184, 114968. https://doi.org/10.1016/j.lwt.2023.114968.

Turgeon, S.L., Schmitt, C., Sanchez, C., 2007. Protein–polysaccharide complexes and coacervates. Curr. Opin. Colloid Interface Sci. 12 (4–5), 166–178. https://doi.org/10.1016/j.cocis.2007.07.007.

Uribe-Alvarez, R., Murphy, C.P., Coleman-Vaughan, C., O'Shea, N., 2023. Evaluation of ionic calcium and protein concentration on heat-and cold-induced gelation of whey protein isolate gels as a potential food formulation for 3D food printing. Food Hydrocoll. 142, 108777.

Vasava, H., Singh, R., Yadav, T., 2022. Characterisation of whey protein–polyphenol conjugates prepared by the noncovalent and covalent methods for their effect on the functional properties of whey proteins. Int. J. Dairy Technol. 75 (3), 563–574. https://doi.org/10.1111/1471-0307.12874.

Veerman, C., Sagis, L.M.C., Heck, J., van der Linden, E., 2003. Mesostructure of fibrillar bovine serum albumin gels. Int. J. Biol. Macromol. 31 (4), 139–146. https://doi.org/10.1016/S0141-8130(02)00074-0.

Verheul, M., Roefs, S.P., 1998. Structure of particulate whey protein gels: effect of NaCl concentration, pH, heating temperature, and protein composition. J. Agric. Food Chem. 46 (12), 4909–4916.

Wagner, J., Biliaderis, C.G., Moschakis, T., 2020. Whey proteins: musings on denaturation, aggregate formation and gelation. Crit. Rev. Food Sci. Nutr. 60 (22), 3793–3806. https://doi.org/10.1080/10408398.2019.1708263.

Wagoner, T.B., Foegeding, E.A., 2017. Whey protein–pectin soluble complexes for beverage applications. Food Hydrocoll. 63, 130–138. https://doi.org/10.1016/j.foodhyd.2016.08.027.

Wagoner, T.B., Vardhanabhuti, B., Foegeding, E.A., 2016. Designing whey protein–polysaccharide particles for colloidal stability. Annu. Rev. Food Sci. Technol. 7 (1), 93–116. https://doi.org/10.1146/annurev-food-041715-033315.

Wang, M., Shen, Y., Wang, B., Liu, S., Zhu, P., 2023. Effect of starch type on the physicochemical and emulsifying properties of amorphous starch–whey protein isolate mixtures. LWT 185, 115134. https://doi.org/10.1016/j.lwt.2023.115134.

Wang, Q., Ismail, B., 2012. Effect of Maillard-induced glycosylation on the nutritional quality, solubility, thermal stability and molecular configuration of whey protein. Int. Dairy J. 25 (2), 112–122. https://doi.org/10.1016/j.idairyj.2012.02.009.

Wang, R., Xu, P., Chen, Z., Zhou, X., Wang, T., 2019. Complexation of rice proteins and whey protein isolates by structural interactions to prepare soluble protein composites. LWT 101, 207–213. https://doi.org/10.1016/j.lwt.2018.11.006.

Wang, S., Copeland, L., 2013. Molecular disassembly of starch granules during gelatinization and its effect on starch digestibility: a review. Food Funct. 4 (11), 1564. https://doi.org/10.1039/c3fo60258c.

Wang, T., Yue, M., Xu, P., Wang, R., Chen, Z., 2018. Toward water-solvation of rice proteins via backbone hybridization by casein. Food Chem. 258, 278–283. https://doi.org/10.1016/j.foodchem.2018.03.084.

Wang, Z., Zhang, S., Vardhanabhuti, B., 2015. Foaming properties of whey protein isolate and λ-carrageenan mixed systems. Journal of Food Science 80 (8). https://doi.org/10.1111/1750-3841.12940.

Waniska, R.D., Kinsella, J.E., 1985. Surface properties of beta-lactoglobulin: adsorption and rearrangement during film formation. J. Agric. Food Chem. 33 (6), 1143–1148.

Wijaya, W., Van Der Meeren, P., Patel, A.R., 2017. Cold-set gelation of whey protein isolate and low-methoxyl pectin at low pH. Food Hydrocoll. 65, 35–45. https://doi.org/10.1016/j.foodhyd.2016.10.037.

Wong, D., Vasanthan, T., Ozimek, L., 2013. Synergistic enhancement in the co-gelation of salt-soluble pea proteins and whey proteins. Food Chem. 141 (4), 3913–3919. https://doi.org/10.1016/j.foodchem.2013.05.082.

Wu, Q., Moncada, M., Zheng, H., 2025. Improving the texture and structure of high protein yogurt alternative using pea protein hydrolysate as functional ingredient. Food Struct., 100458 https://doi.org/10.1016/j.foostr.2025.100458.

Xia, W., Zhu, L., Delahaije, R.J.B.M., Cheng, Z., Zhou, X., Sagis, L.M.C., 2022. Acid-induced gels from soy and whey protein thermally-induced mixed aggregates: rheology and microstructure. Food Hydrocoll. 125, 107376. https://doi.org/10.1016/j.foodhyd.2021.107376.

Xu, X., Zhang, Y., Han, M., Guo, Q., 2024. Whey protein fibrils enhance fat-related texture of emulsion systems: translating structural changes to textural perception. Food Hydrocoll. 146, 109208. https://doi.org/10.1016/j.foodhyd.2023.109208.

Yadav, M.P., Parris, N., Johnston, D.B., Onwulata, C.I., Hicks, K.B., 2010. Corn fiber gum and milk protein conjugates with improved emulsion stability. Carbohydr. Polym. 81 (2), 476–483. https://doi.org/10.1016/j.carbpol.2010.03.003.

Yan, J., Li, S., Chen, G., Ma, C., McClements, D.J., Liu, X., Liu, F., 2023. Formation, physicochemical properties, and comparison of heat- and enzyme-induced whey protein-gelatin composite hydrogels. Food Hydrocoll. 137, 108384. https://doi.org/10.1016/j.foodhyd.2022.108384.

Yang, C., Zhong, F., Douglas Goff, H., Li, Y., 2019. Study on starch-protein interactions and their effects on physicochemical and digestible properties of the blends. Food Chem. 280, 51–58. https://doi.org/10.1016/j.foodchem.2018.12.028.

Ye, A., Hemar, Y., Singh, H., 2004. Flocculation and coalescence of droplets in oil-in-water emulsions formed with highly hydrolysed whey proteins as influenced by starch. Colloids Surf. B Biointerfaces 38 (1–2), 1–9. https://doi.org/10.1016/j.colsurfb.2004.07.011.

Yiğit, A., Bielska, P., Cais-Sokolińska, D., Samur, G.U., 2023. Whey proteins as a functional food: health effects, functional properties, and applications in food. J. Am. Nutr. Assoc. 42 (8), 758–768. https://doi.org/10.1080/27697061.2023.2169208.

Zeeb, B., Fischer, L., Weiss, J., 2014. Stabilization of food dispersions by enzymes. Food Funct. 5 (2), 198. https://doi.org/10.1039/c3fo60499c.

Zhang, K., Huang, J., Wang, D., Wan, X., Wang, Y., 2024. Covalent polyphenols-proteins interactions in food processing: formation mechanisms, quantification methods, bioactive effects, and applications. Front. Nutr. 11, 1371401. https://doi.org/10.3389/fnut.2024.1371401.

Zhang, S., Hsieh, F.-H., Vardhanabhuti, B., 2014. Acid-induced gelation properties of heated whey protein–pectin soluble complex (Part I): effect of initial pH. Food Hydrocoll. 36, 76–84. https://doi.org/10.1016/j.foodhyd.2013.07.029.

Zhang, S., Vardhanabhuti, B., 2014. Acid-induced gelation properties of heated whey protein–pectin soluble complex (Part II): effect of charge density of pectin. Food Hydrocoll. 39, 95–103. https://doi.org/10.1016/j.foodhyd.2013.12.020.

Zhang, S., Yang, C., Zhu, S., Zhong, F., Huang, D., Li, Y., 2022. Understanding the mechanisms of whey protein isolate mitigating the digestibility of corn starch by *in vitro* simulated digestion. Food Hydrocoll. 124, 107211. https://doi.org/10.1016/j.foodhyd.2021.107211.

Zhang, X., Zhang, S., Xie, F., Han, L., Li, L., Jiang, L., Qi, B., Li, Y., 2021. Soy/whey protein isolates: interfacial properties and effects on the stability of oil-in-water emulsions. J. Sci. Food Agric. 101 (1), 262–271. https://doi.org/10.1002/jsfa.10638.

Zhao, R., Fu, W., Li, D., Dong, C., Bao, Z., Wang, C., 2024. Structure and functionality of whey protein, pea protein, and mixed whey and pea proteins treated by pH shift or high-intensity ultrasound. J. Dairy Sci. 107 (2), 726–741. https://doi.org/10.3168/jds.2023-23742.

Zheng, H., 2018. Polymers for structure design of dairy foods. In: Gutiérrez, T.J. (Ed.), Polymers for Food Applications. Springer International Publishing, pp. 509–528. https://doi.org/10.1007/978-3-319-94625-2_19.

Zheng, H., 2019. Introduction: measuring rheological properties of foods. In: Joyner, H.S. (Ed.), Rheology of Semisolid Foods. Springer International Publishing, pp. 3–30. https://doi.org/10.1007/978-3-030-27134-3_1.

Zheng, H., 2022. Texture. Encyclopedia of Dairy Sciences. https://doi.org/10.1016/B978-0-12-818766-1.00379-2 vols. 616–625.

Zheng, H., 2024. Chemistry and properties of the caseins. Casein: Structural Properties, Uses, Health Benefits and Nutraceutical Applications, pp. 23–36. https://doi.org/10.1016/B978-0-443-15836-0.00022-6.

Zheng, H., Meletharayil, H., 2025. Constructing aqueous foams from milk components: structure and interfaces. Curr. Opin. Food Sci. 63, 101290. https://doi.org/10.1016/J.COFS.2025.101290.

Zheng, X.-Q., Wang, D.-D., Xue, S., Cui, Z.-Y., Yu, H.-Y., Wei, J.-T., Chen, H.-H., Mu, H.-Y., Chen, R., 2024. Composite formation of whey protein isolate and OSA starch for fabricating high internal phase emulsion: a comparative study at different pH and their application in biscuits. Int. J. Biol. Macromol. 259, 129094. https://doi.org/10.1016/j.ijbiomac.2023.129094.

Zhou, B., Tobin, J.T., Drusch, S., Hogan, S.A., 2021a. Dynamic adsorption and interfacial rheology of whey protein isolate at oil-water interfaces: effects of protein concentration, pH and heat treatment. Food Hydrocoll. 116, 106640. https://doi.org/10.1016/j.foodhyd.2021.106640.

Zhou, B., Tobin, J.T., Drusch, S., Hogan, S.A., 2021b. Interfacial properties of milk proteins: a review. Adv. Colloid Interface Sci. 295, 102347. https://doi.org/10.1016/j.cis.2020.102347.

Zhou, X., Sala, G., Sagis, L.M.C., 2020. Bulk and interfacial properties of milk fat emulsions stabilized by whey protein isolate and whey protein aggregates. Food Hydrocoll. 109, 106100. https://doi.org/10.1016/j.foodhyd.2020.106100.

Zhu, F., 2024. Modifications of whey proteins for emulsion based applications: current status, issues and prospectives. Food Res. Int. 178, 113935.

Zhu, P., Wang, M., Du, X., Chen, Z., Liu, C., Zhao, H., 2020. Morphological and physicochemical properties of rice starch dry heated with whey protein isolate. Food Hydrocoll. 109, 106091. https://doi.org/10.1016/j.foodhyd.2020.106091.

Chapter 16

Whey protein foams: Preparation, types, and rheology

Alberto Tecante[1] and Ricky Frank López-Santiago[1, 2]

[1]*Departamento de Alimentos y Biotecnología, Facultad de Química, Universidad Nacional Autónoma de México, Cd. Universitaria, Mexico City, Mexico;* [2]*Institute of Physical Chemistry, RWTH Aachen University, Aachen, Germany*

1. Introduction

Whey proteins have long been recognized as functional ingredients in food products. This recognition has led to abundant studies on their functional properties, particularly their physicochemical behavior in aqueous environments. Properties such as foam formation, emulsification, gelation, and water retention have been systematically examined over the past 50 years. Whey protein foaming depends on factors such as extraction, purification, and processing, which also influence its functional qualities, including foam formation. Consequently, exploring these aspects remains important. Different commercial preparations produce foams with distinct physical properties (Morr and Foegeding, 1990). The industrial spray drying conditions used to produce whey powder affect its foaming properties (Osorio et al., 2014). Therefore, it is crucial to continue researching this functional characteristic to gain a deeper understanding of its behavior.

2. Whey proteins

2.1 Structure, amino acid profile, and molar mass

In this document, whey refers to the liquid by-product resulting from the precipitation of cow's milk proteins (Deeth and Bansal, 2019). Precipitation can occur through the activity of microorganisms, the addition of acid, or enzymatic action, such as the addition of rennet. Whey can be either acid or sweet, depending on the process used to produce it. Acidic whey is a by-product of cheese making during casein separation (Kilara and Vaghela, 2018). The commercial product is typically a powder composed of a high proportion of protein, lactose, lipids, minerals, and water (Ramos et al., 2016). The ratios of these components can vary depending on the production process and the purification method used to create specific commercial forms for different applications. Whey concentrate contains at least 25 g of protein per 100 g (usually around 80 g), while whey isolate has 90 g or more per 100 g. β-Lactoglobulin makes up about 60% of whey protein, with α-lactalbumin at 20%, and bovine serum albumin at 3%. The remaining components include immunoglobulins (approximately 10 g/100 g) and minor components such as lactoferrin, osteopontin, lactophorin, glycomacropeptide, and enzymes (Deeth and Bansal, 2019; Edwards and Jameson, 2020).

β-Lactoglobulin consists of 162 amino acid residues and has a molar mass of 18,623 g/mol α-Lactalbumin contains 123 amino acids with a molar mass of 14,178 g/mol. Albumin is the largest, with 583 amino acids and a molar mass of 66,399 g/mol. However, other components, such as immunoglobulin G and lactoferrin, which account for a smaller portion of the total mass, contain more amino acids and have higher molar masses than the primary proteins (Edwards and Jameson, 2020).

Fig. 16.1 shows the primary sequence of β-lactoglobulin A (Kilara and Vaghela, 2018). Cysteine residues form two intrachain disulfide bonds, specifically between positions 66–160 and 119–121. However, Cys 106 may bond with either Cys 121 or Cys 119 in a 50:50 ratio, leaving one free sulfhydryl group between these residues. The secondary structure includes nine beta strands that create a hydrophobic core, enabling interactions with other hydrophobic ligands. Additionally, the molecule's surface features one α-helix.

Whey. https://doi.org/10.1016/B978-0-443-27648-4.00018-9

1				5					10					15					20
Leu	Ile	Val	Thr	Gln	Thr	Met	Lys	Gly	Leu	Asp	Ile	Gln	Lys	Val	Ala	Gly	Thr	Thr	Trp
21				25					30					35					40
Ser	Leu	Ala	Met	Ala	Ala	Ser	Asp	Ile	Ser	Leu	Leu	Asp	Ala	Gln	Ser	Ala	Pro	Leu	Arg
41				45					50					55					60
Val	Tyr	Val	Glu	**Glu**	Leu	Lys	Pro	Thr	Pro	Glu	Gly	Asp	Leu	Glu	**Ile**	Leu	Leu	Gln	Lys
61				65					70					75					80
Asp	Glu	Asn	**Asp**	Glu	Cys	Ala	Gln	Lys	Lys	Ile	Ile	Ala	Glu	Lys	Thr	Lys	Ile	Pro	Ala
81				85					90					95					100
Val	Phe	Lys	Ile	Asp	Ala	Leu	Asn	Glu	Asn	Lys	Val	Leu	Val	Leu	Asp	Thr	Asp	Tyr	Lys
101				105					110					115					120
Lys	Thr	Leu	Leu	Phe	Cys	Met	Glu	Asn	Ser	Ala	Glu	Pro	**Glu**	Gln	Ser	Leu	Val	Cys	Gln
121				125					130					135					140
Cys	Leu	Val	Arg	Thr	Pro	Glu	Val	Asp	Asp	Glu	Ala	Leu	Glu	Lys	Phe	Asp	Lys	Ala	Leu
141				145					150					155					160
Lys	Ala	Leu	Pro	Met	His	Ile	Arg	Leu	Ser	Phe	Asn	Pro	Thr	Gln	Leu	Glu	Glu	Gln	Cys
161	162																		
His	Ile	OH																	

Glu(45) is Gln in variant D, **Ile**(56) is His in variant C, **Asp**(64) is Gly in variants B and C, and **Glu**(114) is Ala in variants B and C.

FIGURE 16.1 The amino acid sequence of β-lactoglobulin A.

The protein's superstructure depends on pH. Under acidic conditions (pH < 3.0) and alkaline conditions (pH > 8.0), the monomeric form dominates. At pH levels between 3.1 and 5.1, along with low temperatures and high protein concentrations, an octamer is present. A dimer occurs at other pH levels (Kilara and Vaghela, 2018). The properties of whey are those of β-lactoglobulin because it is the major component.

Fig. 16.2 shows the amino acid sequence of α-lactalbumin, which has 123 residues and a molar mass of 14,146 g/mol. This protein forms four intrachain disulfide bonds between cysteine residues 6–60, 28–111, 61–77, and 73–91 (Kilara and Vaghela, 2018). Its secondary structure includes three antiparallel beta sheets between residues 41–44, 47–50, and 55–56, three long alpha helices between residues 5–11, 23–24, and 86–98, and three short helices between residues 18–20, 77–80, and 115–118 (Chrysina et al., 2000). This protein exhibits a high affinity for calcium ions.

β-Lactoglobulin and α-lactalbumin are proteins produced in the mammary gland, while bovine serum albumin originates from the bloodstream. Albumin has 17 disulfide bonds, one free sulfhydryl group, and binds fatty acids as well as mono- and divalent cations (Kilara and Vaghela, 2018).

3. Liquid foams

3.1 Foaming capacity and foam stability

Whey proteins are water-soluble and foam when stirred vigorously. Foaming capacity, or overrun—the volume increase caused by trapped air—is calculated as described (Damodaran, 2008).

1				5					10					15					20
Glu	Gln	Leu	Thr	Lys	Cys	Glu	Val	Phe	**Gln**	Glu	Leu	Lys	Asp	Leu	Lys	Gly	Tyr	Gly	Gly
21				25					30					35					40
Val	Ser	Leu	Pro	Glu	Trp	Val	Cys	Thr	Thr	Phe	His	Thr	Ser	Gly	Tyr	Asp	Thr	Glu	Ala
41				45					50					55					60
Ile	Val	Glu	Asn	Asn	Gln	Ser	Thr	Asp	Tyr	Gly	Leu	Phe	Gln	Ile	Asn	Asn	Lys	Ile	Trp
61				65					70					75					80
Cys	Lys	Asn	Asp	Gln	Asp	Pro	His	Ser	Ser	Asn	Ile	Cys	Asn	Ile	Ser	Cys	Asp	Lys	Thr
81				85					90					95					100
Leu	Asn	Asn	Asp	Leu	Thr	Asn	Asn	Ile	Met	Cys	Var	Lys	Lys	Ile	Leu	Asp	Lys	Val	Gly
101				105					110					115					120
Ile	Asn	Tyr	Trp	Leu	Ala	His	Lys	Ala	Leu	Cys	Ser	Glu	Lys	Leu	Asp	Gln	Trp	Leu	Cys
121		123		125															
Glu	Lys	Leu	OH																

Gln(10) is Arg in variant B.

FIGURE 16.2 The amino acid sequence of α-lactalbumin.

$$\text{Overrun}\ (\%) = \left(\frac{V_F - V_0}{V_0}\right)100 \quad (16.1)$$

V_F represents the foam volume, while V_0 indicates the initial liquid volume.

Foaming happens when nitrogen gas is bubbled into a liquid. This process assesses the dynamic foaming capacity. The formation of foam is constantly monitored to determine the rate of foaming.

Foam stability refers to the foam's ability to resist gravity and mechanical stress, thereby preserving the integrity of the liquid film surrounding the gas bubbles. It is measured by half-life—the time it takes for foam to shrink to half of its original volume (Damodaran, 2008). Stability can also be expressed by the ratio of the volume of liquid drained at time t (V_{liq}) to the total volume drained at the endpoint (V_f).

$$\text{Stability} = \frac{V_{liq}}{V_f} \quad (16.2)$$

3.2 Factors affecting foaming

The functional properties of whey include its ability to bind water, form foam and gel, and stabilize oil-in-water emulsions. Whey also offers important health benefits to humans (Ramos et al., 2016). The interaction between whey proteins and various common food components demonstrates these properties. Researchers often study these compounds in aqueous solutions as model systems rather than in actual foods. The main interacting components are water, hydrocolloids, lipids, and carbohydrates (Kilara and Vaghela, 2018). Each application aims for the ideal combination of ingredients and conditions. Key factors affecting foaming and foam stability include pH, lipids, carbohydrates like sugar, protein concentration and modification, and the rheological properties of both the liquid phase and the foam.

The earliest reports on the foaming ability of whey date back at least 50 years (Devilbiss et al., 1975). Abundant studies have examined this topic since then. However, this chapter only reviews research from the past 15 years. The main aspects studied are the overrun and the stability of the foam. Applications require long-lasting stability that whey alone cannot provide. A thickening agent is often added to the aqueous phase to improve the stability of native whey foam. Polysaccharides are preferred for this purpose because of their strong hydrophilicity. Xanthan and pectin are the most commonly used polysaccharides to stabilize whey foams because they can increase the viscosity of the aqueous phase at low concentrations. Other polysaccharides, such as carrageenans, have also been added for this reason.

Table 16.1 summarizes key observations from representative reports on the foaming capacity and stability of whey protein concentrate (WPC) and whey protein isolate (WPI) polysaccharide mixtures over the past 15 years. Foaming was produced by mechanical agitation, while Perez et al. (2010) used nitrogen gas injection into the liquid phase.

The current trend involves altering whey proteins through physical methods to improve their foaming ability and foam stability, without using polysaccharides. These alterations include airgel dehydration (Chen et al., 2013), fibrillization (Oboroceanu et al., 2014), spray drying conditioning (Osorio et al., 2014), ultrasonication (Tan et al., 2015, 2016), gelation (Lazidis et al., 2016), aging (Norwood et al., 2016), heating (Lajnaf et al., 2018; Audebert et al., 2019; Vidotto and Tavares, 2020), and pH conditioning (Lajnaf et al., 2018; Nastaj and Solowiej, 2020).

3.3 Whey foams, mixtures with polysaccharides, and surfactants

This section presents our experimental results along with a detailed discussion of the observed effects. The materials include whey protein isolate (WPI: Lactalis, France), food-grade guar (Gomas Naturales, Mexico), xanthan (Keltrol, CP Kelco, San Diego, USA), λ-carrageenan (MSC Co., Korea), sodium stearoyl lactylate (Palsguard, Mexico), sorbitan monolaurate (Span 20) (Sigma, USA), anhydrous glycerol (Baker, Mexico), sorbitol (Sigma, USA), sodium chloride (Baker, Mexico), and sodium hydroxide (Baker, Mexico). Distilled water was used as the solvent.

A thermobalance (MB45-2A, OHAUS, USA) was used to measure the moisture content of WPI and polysaccharides. The moisture contents (g/100 g) of WPI, guar, xanthan, and carrageenan were 10, 12, 13, 14, and 15, respectively.

Solutions with various compositions and components were prepared using a WPI concentration of 100 g/kg. The WPI solution was made by dissolving the powder in 100 mM NaCl, followed by stirring for 2 hours. The pH was adjusted to 7.0 with 2.5 M NaOH and stored at 4°C overnight. Guar (G), xanthan (X), and λ-carrageenan (λC) solutions were prepared at room temperature and left overnight to ensure complete dissolution. The surfactants, sodium stearoyl lactylate (SSL) or Span 20, were dissolved in distilled water with stirring for 30 min at 70°C, then left overnight at room temperature. The WPI-surfactant and surfactant-polysaccharide solutions were prepared by sequentially adding the WPI or

TABLE 16.1 Representative reports on the foaming capacity and foam stability of mixtures of whey proteins and polysaccharides.

Liquid phase	Polysaccharides	Highlight	References
WPC, WPI, 10 g/kg in 0.05 mol/L Trizma buffer, pH 7.0	Sodium alginate (SA), λ-carrageenan (λC), 0–10 g/L	SA and λC improve foaming capacity and stability, but λC is better.	Perez et al. (2010)
WPI-Durian seed gum (DSG) in water	Arabic gum (AG), SA, κ-carrageenan (κC), guar gum (GG); 20 g/kg	WPI-DSG shows higher overrun and stability. AG enhances overrun and stability more than the other polysaccharides.	Amid et al. (2013)
WPC, 80 g/kg in water	Xanthan gum (XG), AG, a mixture of carrageenans, 50–200 g/kg	Gum-free WPC does not foam at neutral pH. Viscoelastic gums contribute to stabilizing foams.	Liszka-Skoczylas et al. (2014)
WPI, 100 g/kg in water	XG, 0.5, 1.0, and 1.5 g/kg. GG, 1.5, 2.7, and 3.0 g/kg	Foams show solid-like rheological behavior. XG decreases overrun but stabilizes foams more efficiently than GG.	García-García et al. (2015)
WPC, 100, 150, 200, 250 g/kg in water	XG, 0.5 and 1.5 g/kg	XG viscoelasticity plays a crucial role in stabilizing foam. Low concentration yields better results.	Martinez-Padilla et al. (2015)
WPI, 50 g/kg in water, pH 6.2, 6.5, 7.0	λC, 0–2.5 g/kg	A higher λC concentration results in higher stability. Higher elasticity and viscosity correspond to better stability.	Wang et al. (2015)
WPC in water, pH 4, 16.5 g/kg	Pectin (P), 3.3 g/kg	WPI-pectin electrostatic complexation enhances the foaming capacity and stability of WPI compared to native WPC. Heating WPI-P complexes before foaming allows for the control of structure and viscosity.	Oduse et al. (2017)
WPI in water, pH 5, 6, 7, 1.0 g/kg total biopolymer concentration	High-acyl gellan (HAG), WPI-HAGG ratios; 2:1, 3:1, 5:1, 10:1, 15:1, 20:1	Heat-induced WPI-HAGG complexes produce stable foams. A 2:1 ratio yields the highest foaming and stability at pH 6 and 7.	Zhu and Griffith Jones (2024)

polysaccharide, as previously described. The WPI-polysaccharide and WPI-polysaccharide-surfactant solutions were made by mixing the components separately in a 6:4 ratio. Liquid foams were created by stirring 100 mL of each solution with a rotor-stator (Ultra Turrax T25 digital, IKA, Germany) at 15,400 rpm for 5 min at room temperature (García-García et al., 2015). The overrun was calculated using Eq. (16.1). Foam stability was determined by measuring the liquid drained every 10 minutes from 100 mL of foam in a 100 mL graduated cylinder and calculating it using Eq. (16.2).

3.4 Foam stability

Experimental data were fitted to an adapted Gompertz-type equation (Zwietering et al., 1990), as shown.

$$\frac{V_{liq}}{V_f} = \phi_e \cdot \exp\left\{ -\exp\left[1 - \frac{\mu_t \cdot e \cdot t}{\phi_e}\right]\right\} \tag{16.3}$$

In this equation, $e = 2.718$, μ_t is the slope of the linear part of the exponential increase of V_{liq}/V_f over time. The parameter μ_t is a kinetic constant indicating foam stability: the lower its value, the more stable the foam. Parameter ϕ_e represents the value of V_{liq}/V_f at long periods when the liquid has fully drained, and the foam no longer exists; ideally, ϕ_e approaches one as time approaches infinity. Foam stability is characterized by the collapse time, t_C, the experimental half-life, $(t_{1/2})_{exp}$, μ_t, ϕ_e, and $t_{1/2}$ predicted by Eq. (16.3).

3.5 Bubble size

The size of the liquid foam bubbles was measured using an optical microscope (Olympus BX-45, Olympus Corporation, Japan). About 0.5 mL of foam was placed on a slide, and at least 350 bubbles were measured (García-García et al., 2015).

Image analysis was performed with Image-Pro Plus version 4.5 software. Bubble size data were grouped into categories, with the number of categories denoted as N_c and the class interval as W.

$$N_c = \sqrt{N_b} \tag{16.4}$$

$$W = \frac{d_{max} - d_{min}}{N_c} \tag{16.5}$$

N_b indicates the number of bubbles measured, while d_{max} and d_{min} are the largest and smallest recorded diameters.

3.6 Rheological behavior

Rheological tests were performed at 25°C with at least two repetitions using a rotational rheometer (ARES-RFS III, TA Instruments, USA). Solutions were examined in a double-walled concentric cylinder setup, with an inner diameter of 27.94 mm for the cup, an inner diameter of 29.5 mm for the inner cylinder, an outer diameter of 32.0 mm for the outer cylinder, and an outer diameter of 34 mm for the cup. Liquid foams were tested with a straight six-blade vane measuring 33.5 mm in length and 18 mm in diameter. The foam was placed into the rotating cup, and the vane was immersed in the sample.

3.6.1 Steady-flow

The variations in viscosity with shear rate were examined over ranges from 1 to 300 s^{-1} for solutions and 0.1 to 100 s^{-1} for liquid foams. The flow behavior was characterized using the power-law equation.

$$\eta(\dot{\gamma}) = K\dot{\gamma}^{n-1} \tag{16.6}$$

K is the consistency coefficient, and n is the flow behavior index.

3.6.2 Unsteady oscillatory flow

The zone of linear viscoelasticity (ZLV) was identified through strain sweeps ranging from 10% to 100% for solutions and 0.10%–10.0% for liquid foams at a frequency of 6.28 rad/s. Frequency sweeps within the ZLV were performed over ranges of 1–100 rad/s for solutions and 0.1–100 rad/s for foams. The viscoelastic behavior was analyzed based on how the storage modulus, $G'(\omega)$, the loss modulus, $G''(\omega)$, and their ratio, $\tan\delta = G''(\omega)/G'(\omega)$, vary with frequency.

3.7 Statistical analysis

The results show the average of repeated measurements along with their standard deviations. Significant differences were detected using a one-way analysis of variance (ANOVA). When a significant difference was found, a multiple comparison test was conducted with the Tukey method at a 95% confidence level.

3.8 WPI-λ-carrageenan

3.8.1 Overrun and stability of liquid foams

Table 16.2 displays the overrun of WPI and WPI-λC liquid foams. WPI foams exhibit the highest overrun. Adding λC decreases the overrun as the polysaccharide concentration increases.

Adding 1.0 and 1.5 g/kg λC results in similar overrun levels. Beyond 1.5 g/kg λC, the overrun decreases significantly compared to WPI foams. As the polysaccharide concentration increases, the liquid phase becomes more viscous, which limits the incorporation of air. Consequently, the viscosity of the liquid batch affects the interfacial adsorption of WPI proteins. Adding this polysaccharide reduces the proteins' foaming ability. However, λC concentrations below 2.0 g/kg still produce overruns similar to those of WPI alone. Foam stability, indicated by the change in V_{liq}/V_f over time, is shown in Fig. 16.3.

Higher concentrations result in more stable foams. Table 16.3 presents the data-fitting results based on Eq. (16.3).

The fitting capacity of Eq. (16.3) is excellent, and its predictions are meaningful because of the physical significance of its two parameters. Notice that ϕ_e converged to one as expected, and μ_t decreased steadily at higher polysaccharide concentrations. Therefore, the Gompertz-type model accurately and consistently describes the behavior of these mixtures. Table 16.4 shows the collapse time, t_C, the experimental half-life $(t_{1/2})_{exp}$, and the half-life predicted by Eq. (16.3).

WPI foams collapsed quickly, even at the highest overrun. However, the collapse time remains relatively long, indicating some degree of stability. This behavior depends on WPI concentration, neutral pH, and ionic strength from NaCl. Under these conditions, WPI exhibits a high foaming ability, resulting in a stable foam. Increasing the λC concentration to 1.0 and 1.5 g/kg produces more stable foams, but with 9.0% less overrun than WPI. This suggests slightly less foam but greater stability. The stability of these two λC concentrations is similar. Collapsing times are twice as long for 2.0 and 3.0 g/kg λC, but they produce 32% and 38% less foam than WPI. These foams are much more stable but generate less foam overall.

Between 0 and 10 min, the liquid draining rate (LDR) was highest for WPI and decreased steadily with λC concentration. The volume of liquid draining increased linearly with λC concentration from 10 to 30 min. This linear trend continued up to 40 min at a λC concentration of 3.0 g/kg. Afterward, the LDR steadily decreased until reaching the collapse time, then remained in a linear region. Fig. 16.1 shows how this behavior affects foam stability. The half-life increases proportionally with λC concentration, confirming foam stability. Overrun and stability results are consistent, indicating that various WPI-λC combinations can provide a range of techno-functional properties, depending on the application—whether short-term or long-term stability is needed—along with good foaming characteristics. Therefore, both low and high λC concentrations can be used depending on the goal, allowing a balance between overrun and stability.

3.8.2 Bubble size

The bubble size distribution of the 100 g/kg WPI foam (Fig. 16.4a) shows that the bubbles are larger than those in the λC foam (Fig. 16.4b). The size difference is more noticeable compared to foams containing 1.5, 2.0, and 3.0 g/kg λC; therefore, the foam without λC experiences greater pressure differences between bubbles, leading to coalescence and subsequent liquid drainage. The disparity decreases as λC concentration increases, indicating that the coalescence rate slows with higher λC levels, which in turn reduces the drainage rate. This behavior is consistent with the stability of foam.

3.8.3 Flow and viscoelastic behavior of solutions and liquid foams

3.8.3.1 λ-carrageenan

Fig. 16.5a shows the steady-flow behavior of solutions with 1.0, 1.5, 2.0, and 3.0 g/kg λC. Incipient shear-thinning observed, with η equal to 2.80×10^{-3} at 10 s^{-1}. The 3.0 g/kg λC solution had a zero-shear viscosity of about 8.7 mPa s at shear rates below 10 s^{-1}.

Table 16.5 shows the power-law parameters for carrageenan solutions. The flow behavior index, n, is close to one, indicating only a slight deviation from Newtonian behavior. The consistency index, K, shows that lambda carrageenan has a limited thickening ability. However, these viscosity levels might still be adequate to stabilize liquid foams when using this polysaccharide.

Fig. 16.5b displays the typical viscoelastic response of polysaccharide solutions, where at low frequencies, the loss modulus $G''(\omega)$ exceeds the storage modulus $G'(\omega)$. The crossover points shift to higher frequencies as the λC concentration increases: 14.0, 17.7, 25.0, and 36 rad/s for concentrations of 1.0, 1.5, 2.0, and 3.0 g/kg, respectively. Below 3 rad/s, $G'(\omega)$ shows more significant deviations from its average value, but beyond this frequency, differences among all λC concentrations become negligible. Increasing the polysaccharide concentration has little effect on elastic properties, although $G''(\omega)$ varies with concentration and increases with λC across the entire frequency range.

3.8.3.2 WPI-λ-carrageenan

Fig. 16.6a shows the flow behavior of WPI and its mixtures with different λC concentrations.

TABLE 16.2 Overrun (%) of 100 g/kg WPI and its mixtures with different λC concentrations (g/kg).

Foam	Overrun (%)	Foam	Overrun (%)
WPI	220 ± 0.00[(a)]	WPI-2.0λC	150 ± 0.00[(c)]
WPI-1.0λC	205 ± 7.07[(b)]	WPI-4.0λC	135 ± 7.07[(d)]
WPI-1.5λC	200 ± 0.00[(b)]		

Different letters in the same column indicate statistically significant differences ($P < 0.05$).

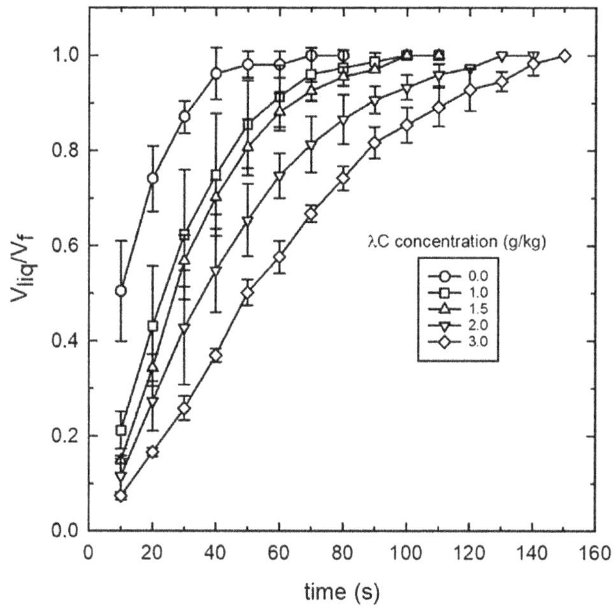

FIGURE 16.3 Stability of 100 g/kg WPI-λC liquid foams. Lines are visual guides.

TABLE 16.3 Fitting parameters for Eq. (16.3) used with the data in Fig. 16.3.

Foam	ϕ_e	μ_t (min^{-1})	r^2
WPI	0.993[(a)]	0.0437[(b)]	0.9667
WPI-1.0λC	1.00[(a)]	0.0211[(c)]	0.9995
WPI-1.5λC	1.00[(a)]	0.0185[(d)]	0.9967
WPI-2.0λC	1.00[(a)]	0.0136[(e)]	0.9980
WPI-3.0λC	1.07[(a)]	0.00971[(f)]	0.9950

Different letters in the same column indicate statistically significant differences ($P < 0.05$).

The apparent viscosity of the 100 g/kg WPI solution is lower than that of its mixtures with λC. WPI exhibits Newtonian behavior with a viscosity of about 2.3 mPa s. Therefore, WPI lacks thickening properties, and its liquid foams are unlikely to remain stable over time. A noticeable synergistic effect occurs between WPI and carrageenan, as their mixtures display significantly higher viscosities than the individual components.

Table 16.6 reveals that solutions without λC and those with 1.0 and 1.5 g/kg λC exhibit similar n values, indicating comparable shear-thinning behavior. However, the K value at 1.0 g/kg λC is 28% lower. The behavior shifts notably in

TABLE 16.4 Characteristic times for foam stability with 100 g/kg WPI and various λC concentrations (g/kg).

Foam	t_C (min)	$(t_{1/2})_{exp}$ (min)	$t_{1/2}$ (min)
WPI	70[(a)]	10.7 ± 3.67[(d)]	11.6
WPI-1.0λC	100[(a)]	23.7 ± 6.37[(e)]	23.8
WPI-1.5λC	100[(a)]	27.0 ± 1.83[(e)]	27.2
WPI-2.0λC	130[(b)]	35.1 ± 8.01[(f)]	37.0
WPI-3.0λC	150[(c)]	51.0 ± 3.60[(g)]	48.2

Different letters in the same column indicate statistically significant differences ($P < 0.05$).

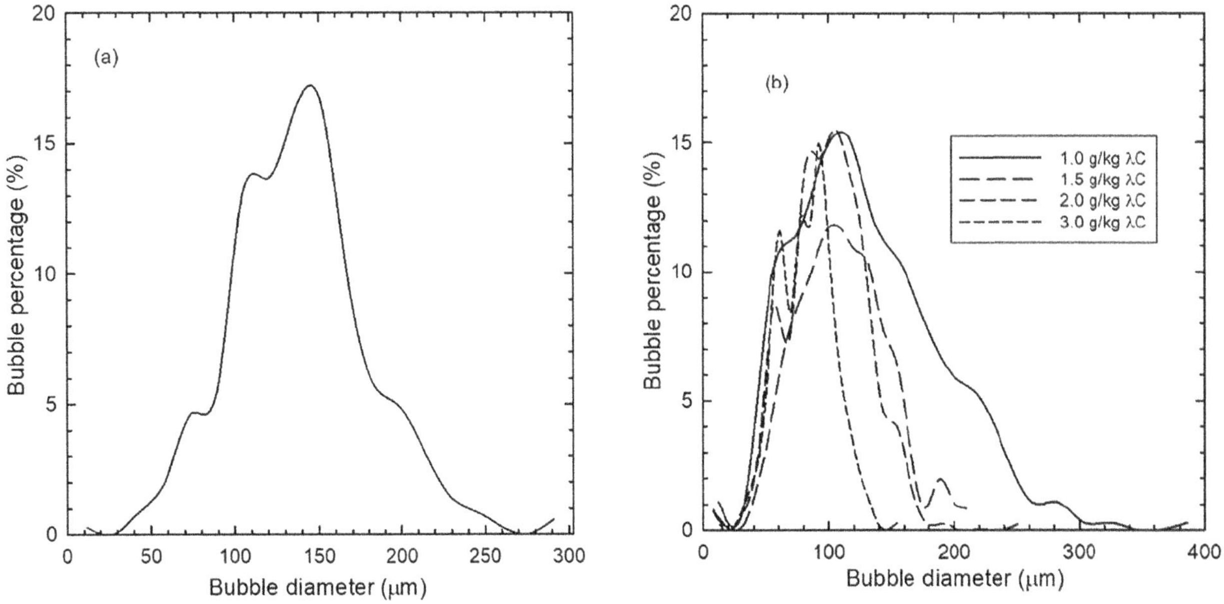

FIGURE 16.4 Frequency distribution: (a) WPI, (b) WPI with λC.

precursor solutions containing 100 g/kg WPI with 2.0 or 3.0 g/kg λC. The degree of shear-thinning increases, and the viscosity of the 3.0 g/kg λC solution decreases from 852 to 67 mPa s over the shear rate range of 1–250 s^{-1}, implying this solution effectively stabilizes foam. The change in thickening strength is particularly significant for λC at a concentration of 3.0 g/kg.

Fig. 16.6b shows that liquid foams also exhibit shear-thinning behavior, but their viscosities are higher than those of their precursor solutions. The drop in apparent viscosity indicates an irreversible breakdown of the foam structure caused by shear (Liszka-Skoczylas et al., 2014). However, the higher viscosities compared to the precursor solutions can result from increased resistance to flow, promoted by the gas dispersed in the foam, which acts as a filler and reinforces the liquid phase surrounding the bubbles. The flow behavior index, n, is lower than in the precursor solutions, and the consistency index, K, is significantly higher. Therefore, liquid foams are structured materials with greater resistance to

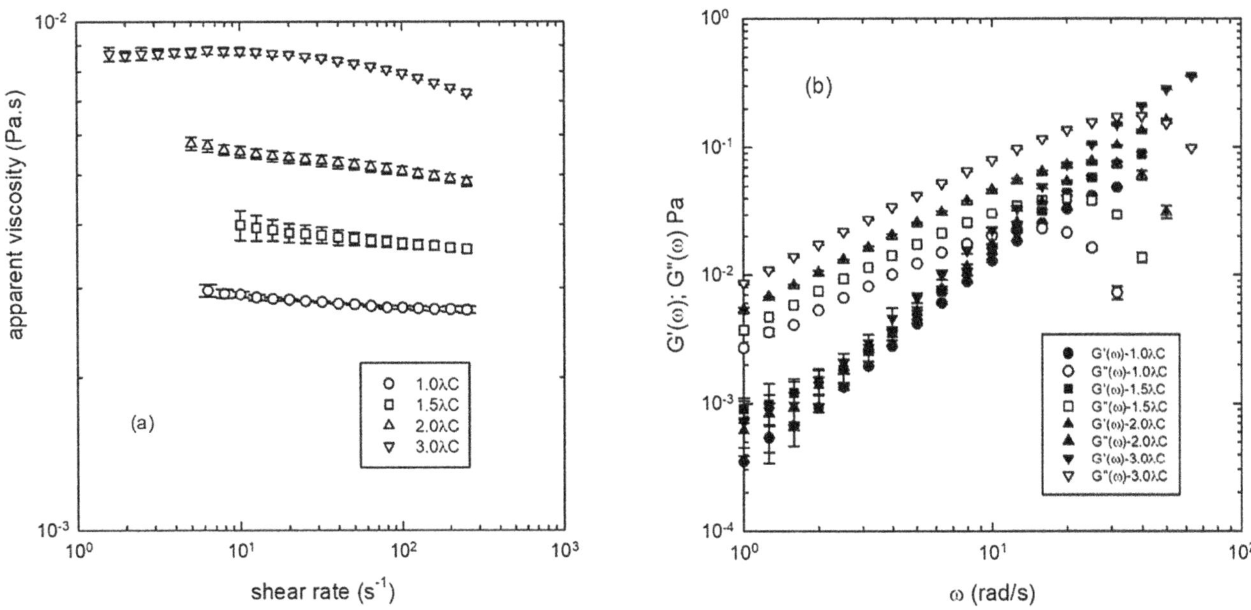

FIGURE 16.5 (a) Shear rate sweep and (b) frequency sweep (strain range: 30%–50%) for 1.0, 1.5, 2.0, and 3.0 g/kg λC solutions.

TABLE 16.5 Power-law parameters for λ-carrageenan solutions.

Concentration (g/kg)	K (Pa·s^n)	n	r^2	Shear rate range (s^{-1})
1.0	0.00305	0.98	0.9596	6–250
1.5	0.00426	0.97	0.9893	10–250
2.0	0.00614	0.96	0.9947	5–250
3.0	0.0105	0.94	0.9630	15–250

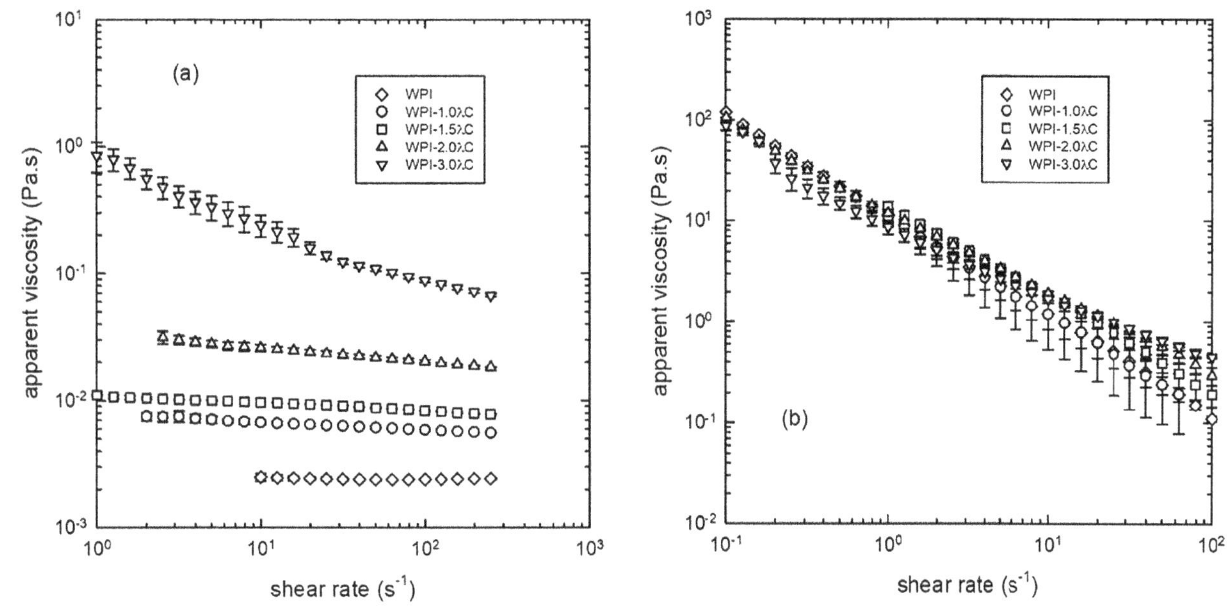

FIGURE 16.6 Shear rate sweep of precursor solutions (a) and their corresponding liquid foams (b) prepared with 100 g/kg WPI and 0, 1.0, 1.5, 2.0, and 3.0 g/kg λC.

flow; yet they are irreversibly destroyed under high shear rates. The liquid foams are remarkably stable when mixed with 100 g/kg WPI and 3.0 g/kg λC. The viscosity of this mixture decreases more slowly, meaning foam destruction by shear occurs at a slower rate, which explains its improved stability.

Fig. 16.7 shows the linear viscoelastic response of the precursor solutions, as indicated by changes in the frequency of the dynamic moduli and the loss angle.

The loss modulus, G"(ω), exceeds the storage modulus, G′(ω), at frequencies below 10 rad/s for WPI and its mixtures with carrageenan. The points where the moduli intersect depend on the polysaccharide concentration, occurring at 12.0,

TABLE 16.6 Power-law parameters for WPI-λC mixtures.

	Precursor solutions			Liquid foams		
Mixtures	K (Pa·s^n)	n	r^2	K (Pa·s^n)	n	r^2
WPI	2.30[a]	1.00	1.000	11.4	0.017	0.9996
WPI-1.0λC	0.00784	0.94	0.9945	10.7	0.036	0.9996
WPI-1.5λC	0.0110	0.94	0.9871	14.6	0.081	0.9988
WPI-2.0λC	0.0335	0.89	0.9960	12.9	0.19	0.9983
WPI-3.0λC	0.737	0.53	0.9820	10.3	0.26	0.9842

[a]*η in mPa·s.*

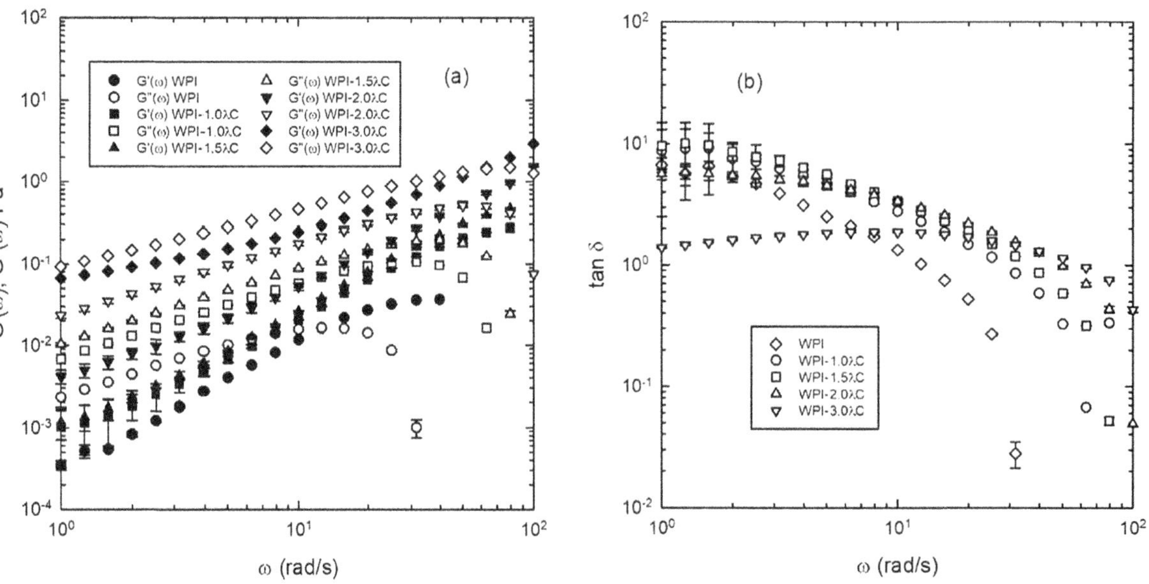

FIGURE 16.7 Precursor solutions: (a) Frequency sweep and (b) tan δ, 100 g/kg WPI with 0, 1.0, 1.5, 2.0, and 3.0 g/kg λC. Strain range: 30%–50%.

28.0, 38.0, 48.2, and 63.5 rad/s for 0.0, 1.0, 1.5, 2.0, and 3.0 g/kg λC, respectively. Adding λC enhances the viscous response and extends the effective frequency range. In WPI combined with 3.0 g/kg carrageenan, the moduli are near the crossover point. Increasing the λC level to 3.0 g/kg causes only slight changes in the moduli. This proximity suggests a conformational transition, with tan δ approaching one, as shown in Fig. 16.7b. Such a transition results from the rearrangement of the polymer chains, supported by the known affinity between carrageenans and milk proteins, indicating that this change occurs at this specific WPI-to-λC ratio.

Fig. 16.7b shows how the loss angle varies with frequency. In solutions containing 100 g/kg WPI and its mixtures with 1.0, 1.5, and 2.0 g/kg λC, the tan δ steadily decreases as frequency increases. In contrast, WPI without λC declines more quickly. The behavior of the solution with 100 g/kg WPI and 3.0 g/kg λC is different: tan δ slightly increases from 1.5 to 2.0, then drops to about 0.5, corresponding to angles from 60 degrees to 27 degrees, which indicates its viscous nature. The changes in viscoelastic properties caused by higher λC concentrations are clearly visible.

Fig. 16.8 shows the viscoelastic behavior of liquid foams. G′(ω) remains higher than G''(ω) at all frequencies, indicating a solid-like response related to interfacial energy density (Cohen-Addad and Höhler, 2014). In Fig. 16.8b, G′(ω), G″(ω), and tan δ mostly show no significant differences among the liquid foams, except at 3.0 g/kg λC, where they have lower moduli and higher loss angles. The tan δ values range from 0.25 to 1.0, corresponding to loss angles from 14 degrees to 45 degrees. After 10 rad/s, tan δ gradually increases with frequency but remains below 1.0.

3.9 Sodium stearoyl lactylate (SSL) and SSL-polysaccharide

3.9.1 Overrun and stability of liquid foams

SSL is an anionic surfactant widely used across various industrial sectors and is among the most common surface-active agents in processed foods. Its critical micellar concentration (CMC) is 0.402 g/kg (Ortiz-Tafoya and Tecante, 2018). Given SSL's surface activity, it is important to study its foaming behavior and how polysaccharides influence it. The review includes xanthan, guar, neutral hydrocolloids, and λC. Xanthan, like λC, is anionic and known for its strong thickening ability even at concentrations below 1.0 g/kg, so significant viscosity increases are unnecessary. Guar, a neutral hydrocolloid, has moderate to low thickening power. Exploring the differences in thickening capability and ionic nature of these polysaccharides can help characterize the foaming properties of SSL before it is mixed with WPI. This approach aims to improve understanding of how surface-active molecules with different molar masses behave.

Table 16.7 displays the overrun (%) for SSL foam prepared with 2.0, 4.0, and 6.0 g/kg SSL, as well as mixtures containing 2.0 g/kg SSL and polysaccharides. The SSL concentrations are 5, 10, and 15 times higher than the surfactant's CMC. This guarantees complete interfacial adsorption, although micellization also occurs.

SSL has a higher foaming capacity than WPI; however, foam production tends to be inconsistent, with large deviations from the average overrun. When SSL concentration reaches 6.0 g/kg, the overrun decreases, but the fluctuations from the

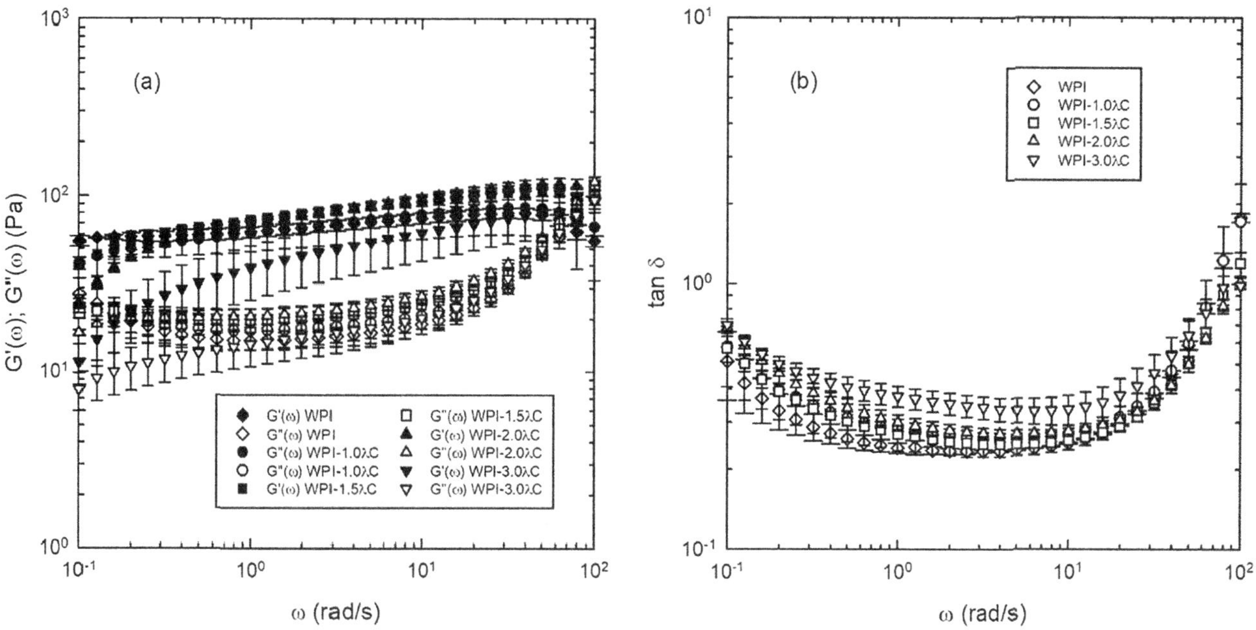

FIGURE 16.8 Liquid foams: (a) Frequency sweep and (b) tan δ, 100 g/kg of WPI with 0, 1.0, 1.5, 2.0, and 3.0 g/kg λC. Strain range: 2%–5%.

average disappear, likely because the high concentration quickly saturates the interface. The stabilizing effect of hydrocolloids was examined at an SSL level of 2.0 g/kg. All three polysaccharides reduce the overrun threefold, but they all produce the same value, indicating they similarly influence SSL's foaming ability. This suggests that, under the tested conditions, ionic properties and thickening abilities do not affect the overrun.

Fig. 16.9 shows the foam stability of SSL and its mixes with guar, xanthan, and λC. The stability patterns match those observed during foaming. For example, liquid drainage in foam with 2 g/kg SSL remains steady from 30 to 50 min after an initial quick drainage, then suddenly drains completely. SSL-polysaccharide mixtures display similar step-like stability patterns, with clear differences in their stability levels.

Table 16.8 displays the regression parameters from Eq. (16.3). Despite some foams having low correlation coefficients and step patterns, clear trends can still be observed.

Further research is necessary to understand how foam interacts with SSL and guar gum. This foam shows a 30-minute lag phase during which no liquid drainage occurs. Eq. (16.3) does not apply when a lag occurs, but it can be adjusted by introducing a new regression parameter, λ, which indicates where the tangent at the inflection point intersects the x-axis. Although this parameter does not always match the actual lag time, it sometimes does. The updated Eq. (16.7) now incorporates this parameter.

$$\frac{V_{liq}}{V_f} = \phi_e \cdot \exp\left\{ - \exp\left[1 + \frac{\mu_t \cdot e \cdot (\lambda - t)}{\phi_e} \right] \right\} \tag{16.7}$$

TABLE 16.7 Overrun (%) of SSL and its mixtures with λC, xanthan (X), and guar (G).

Code	Overrun (%)	Code	Overrun (%)
2.0SSL	370 ± 42.4[(a)]	2.0SSL-2.0λC	135 ± 7.07[(c)]
4.0SSL	205 ± 49.5[(a)]	2.0SSL-1.5X	130 ± 14.1[(c)]
6.0SSL	300 ± 0.00[(b)]	2.0SSL-1.5G	130 ± 0.00[(c)]

Different letters in the same column indicate statistically significant differences ($P < 0.05$).

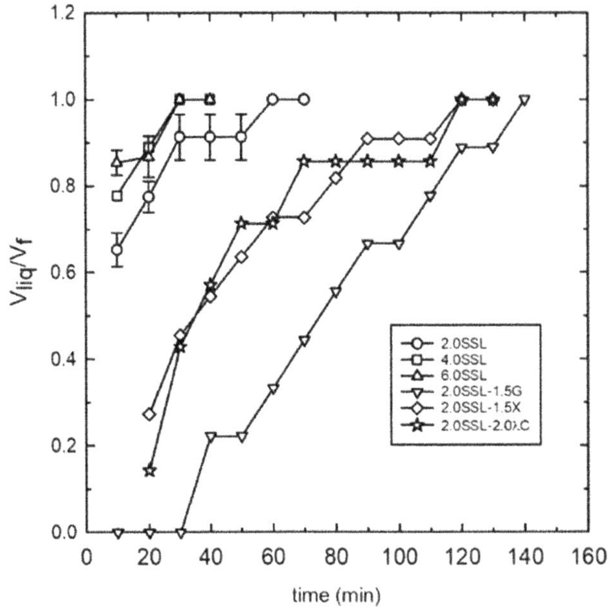

FIGURE 16.9 Stability of SSL and SSL-polysaccharide foams. Lines serve as visual guides.

TABLE 16.8 Fitting parameters for Eq. (16.3) based on the data in Fig. 16.7.

Foam	ϕ_e	μ_t (min^{-1})	r^2	Foam	ϕ_e	μ_t (min^{-1})	r^2
2.0SSL	0.942[(a)]	0.0618[(b)]	0.7731	2.0SSL-1.5G*	1.12[(a)]	0.0107[(e)]	0.9865
4.0SSL	0.990[(a)]	0.0848[(c)]	0.8588	2.0SSL-1.5X	1.00[(a)]	0.0131[(f)]	0.9775
6.0SSL	0.984[(a)]	0.103[(d)]	0.5420	2.0SSL-2.0λC	0.984[(a)]	0.0136[(f)]	0.9456

Different letters in the same column indicate statistically significant differences ($P < 0.05$).
Data fitted to Eq. (16.2) with $\lambda = 29.3$ min.

Table 16.8 shows $\lambda = 29.3$ min for the 2.0SSL-1.5G foam. This value is 97.6%, close to the experimental lag time. According to μ_t, SSL foam stability follows the trend 2.0SSL > 4.0SSL > 6.0SSL. The stability of mixtures of SSL with xanthan or λC is comparable but lower than 2.0SSL-1.5G. This combination is the most stable and has the best regression coefficient. The poor regression coefficients of SSL foams agree with the large deviations and overrun. Therefore, SSL has a foaming capacity higher than WPI, but the foams are unstable and exhibit irregular behavior.

Table 16.9 shows the collapse time, experimental half-life, and predictions based on Eqs. (16.3) and (16.7).

TABLE 16.9 Stability parameters of SSL and SSL-polysaccharide foams.

Code	t_C (min)	$(t_{1/2})_{exp}$ (min)	$t_{1/2}$ (min)
2.0SSL	60	5.94 ± 1.65[(a)]	8.67
4.0SSL	30	5.34 ± 0.00[(a)]	5.99
6.0SSL	30	4.69 ± 0.12[(a,b)]	4.97
2.0SSL-1.5G	140	84.5 ± 0.00[(c)]	76.1
2.0SSL-1.5X	120	35.0 ± 0.00[(d)]	38.4
2.0SSL-2.0λC	120	35.0 ± 0.00[(d)]	37.6

Different letters in the same column indicate statistically significant differences ($P < 0.05$).

SSL foams disintegrate in less than 60 min at both 4.0 and 6.0 g/kg SSL. They drain quickly within the first 10 min after forming, regardless of surfactant concentration. Adding polysaccharides more than doubled the foam stability for guar and exactly doubled it for xanthan and carrageenan. The half-life of SSL foams was less than 10 min across all surfactant levels, indicating rapid initial drainage followed by slower drainage afterward. The half-life only represents a small part of the total collapse time—around one-10th at 2.0 g/kg SSL and one-sixth at 4.0 and 6.0 g/kg SSL. Calculating the liquid draining rate (LDR) is difficult because SSL drains quickly below 10 min, while polysaccharide mixtures drain more slowly.

Over 10–30 min, the liquid drains steadily at an average rate of 1.3 mL/min across all three surfactant concentrations. It remains unclear why 2.0 g/kg SSL stayed stationary for 20 min at 90% of the total drained volume before finally completing drainage. Unlike overrun, adding a polysaccharide improves foam stability. The foam fully drains in 120–140 min because of increased viscosity caused by the polysaccharides. Incorporating guar gum boosts stability and delays initial drainage by 30 min. After this, the foam destabilizes at a consistent linear rate. Xanthan and λC provide similar stability but are less effective than guar gum. This indicates that guar-SSL interactions differ from those with xanthan and λC. According to the flow behavior shown in Fig. 16.10a, a foam with xanthan is expected to exhibit greater stability because of its thickening properties.

Xanthan, λC, and SSL are anionic, so surfactant-polysaccharide interactions are unlikely due to electrostatic repulsion. No counterions were added to the liquid phase, meaning the charge densities of the polysaccharide and SSL were not screened. In contrast, SSL-guar systems could experience van der Waals attractions between molecules. Consequently, electrostatic effects might oppose viscosity effects in xanthan and λC foams, leading to lower stability than expected compared to guar. This is important because viscosity is not always the key factor in stabilization. Additionally, a low molar mass surfactant behaves quite differently from a high molar mass surfactant.

3.9.2 Flow and viscoelastic properties of solutions and liquid foams

Fig. 16.10a and Table 16.10 illustrate the shear-thinning behavior of the solutions.

The apparent viscosity of 2.0 SSL with 1.5 g/kg xanthan clearly varies with shear rate. Fig. 16.10b shows that liquid foams behave similarly to those of WPI-λC in Fig. 16.6b. However, they exhibit noticeable dispersion at high shear rates, caused by uneven foaming that leads to inconsistent shear disruption.

Fig. 16.11 illustrates the viscoelastic behavior of SSL-polysaccharide solutions. The loss modulus exceeds the storage modulus between approximately 1 and 20 rad/s, indicating that viscous behavior dominates, regardless of the polysaccharide used. The crossover points occur at different frequencies, reflecting the distinct thickening abilities and viscoelastic responses of each polysaccharide: guar at 79 rad/s (extrapolated), xanthan at 25.2 rad/s, and λC at 49.5 rad/s. The shift from viscous to elastic dominance occurs when $G'(\omega) > G''(\omega)$. As shown in Fig. 16.9a, the similar dynamic

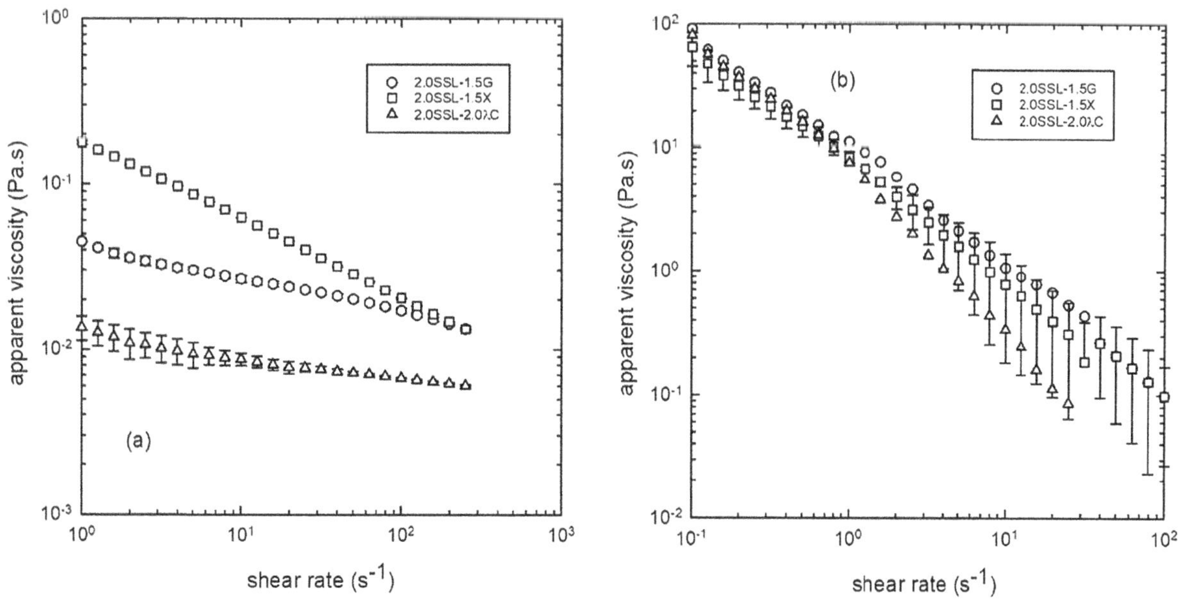

FIGURE 16.10 Shear rate sweep of precursor solutions (a) and foams (b) for 2.0 SSL with 1.5 g/kg guar, 1.5 g/kg xanthan, and 2.0 g/kg λC.

TABLE 16.10 Power-law parameters.

	Precursor solutions			Liquid foams		
Mixtures	K (Pa·s^n)	n	r^2	K (Pa·s^n)	n	r^2
2.0SSL-1.5G	0.0423	0.80	0.9928	9.83	0.046	0.9968
2.0SSL-1.5X	0.184	0.52	0.9997	7.36	0.053	0.9968
2.0SSL-2.0λC	0.0121	0.87	0.9714	5.81	−0.24	0.9939

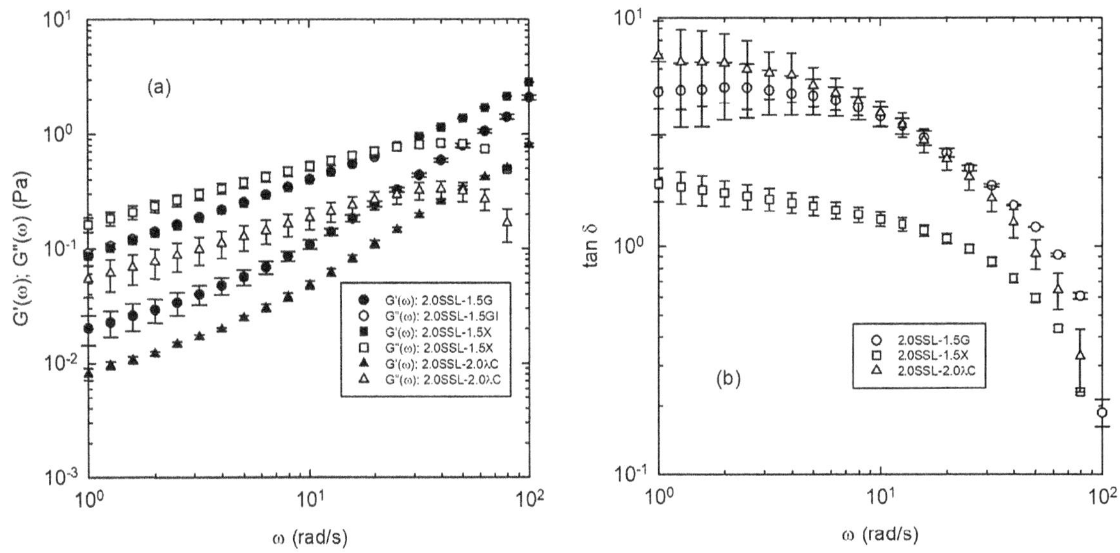

FIGURE 16.11 Precursor solutions: (a) frequency sweep and (b) tan δ of 2.0 SSL with 1.5 g/kg guar, 1.5 g/kg xanthan, and 2.0 g/kg λC. Strain range: 10%–50%.

moduli for xanthan-based SSL solutions result in a notably lower tan δ compared to surfactant mixtures with guar and carrageenan. The decreasing loss angle with increasing frequency confirms the dominance of the elastic response.

Fig. 16.12a shows the viscoelastic properties of foams made from the solutions in Figs. 16.10a and 16.11. Notably, their dynamic moduli are higher than those of the original solutions. Although G′(ω) stays above G''(ω), the moduli crossover happens at 22.2 and 11.2 rad/s for foams with guar and xanthan, respectively. In contrast, this crossover does not occur for carrageenan. The dynamic moduli of the three polysaccharide-based foams vary significantly. However, their loss angles, tan δ, are pretty similar, ranging from 0.4 to 0.8—equivalent to 22–39 degrees at frequencies below 20 rad/s, as shown in Fig. 16.12b. Above this frequency, the loss angles for guar and xanthan foams increase to 51 and 56 degrees, with tan δ rising from 1.25 to 1.50. These moderate angles indicate a viscous nature, though the foams are more stable because they can exhibit some solid-like behavior. This trait is important for foam stability.

3.10 WPI-SSL and WPI-SSL-polysaccharide

3.10.1 Overrun and stability of liquid foams

After evaluating the foaming capacity and stability of WPI and SSL individually, their combined effects were also studied to identify potential cooperative interactions. Additionally, polysaccharides were added to assess their stabilizing effects when used with WPI and SSL. The hypothesis suggests that foams without polysaccharides will be more stable, indicating that the stabilizing role of these hydrocolloids may be less significant when WPI and SSL are combined.

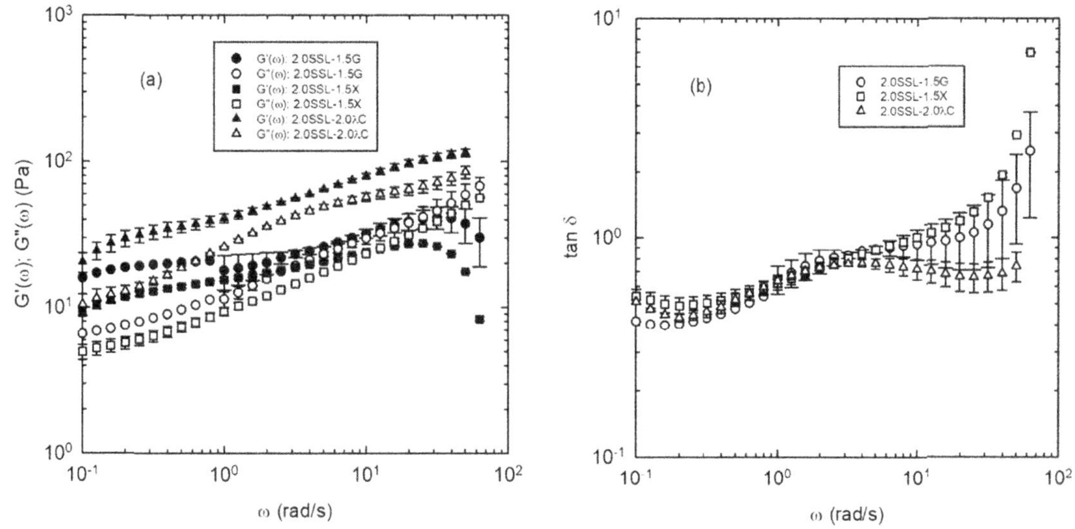

FIGURE 16.12 Liquid foams: (a) Frequency sweep, and (b) tan δ of 2.0 SSL with 1.5 g/kg guar, 1.5 g/kg xanthan, and 2.0 g/kg λC. Strain range: 0.6%–5%.

Table 16.11 shows the overrun (%) of WPI-SSL and WPI-SSL-polysaccharide foams with 100 g/kg WPI, 2.0 and 4.0 g/kg SSL, 1.5 g/kg guar, 1.5 g/kg xanthan, and 2.0 g/kg λC.

The overruns of WPI-2.0SSL and WPI-4.0SSL foams are 32% and 54% lower, respectively, compared to WPI alone (Table 16.2). When compared with SSL alone (Table 16.9), they are 59.5% and 72.7% lower, respectively. These significant reductions in overrun indicate that WPI and SSL do not work well together, especially at an SSL concentration of 4.0 g/kg. Less overrun means less interfacial adsorption, which suggests that electrostatic attraction between the negatively charged SSL and positively charged areas of WPI proteins may lead to less protein and surfactant adsorbing at the interface. This effect is more noticeable at 4.0 g/kg SSL. Adding polysaccharides increased the overrun for WPI-4.0SSL, though it still remained lower than WPI-2.0SSL. The decrease in overrun with WPI alone can be explained by SSL's interfacial adsorption, which may block WPI protein adsorption. This phenomenon occurs when two surfactants are used to make foam (Lexis and Willenbacher, 2014). These authors found that an anionic surfactant adsorbs more strongly at the interface than WPI. Foam with 2.0 SSL and 1.5 g/kg guar, 1.5 g/kg xanthan, or 2.0 g/kg λC reduced foaming capacity by 13%, 13%, and 10%, respectively, compared to foam without polysaccharides. This reduction is likely due to increased viscosity. Compared to foams with the same polysaccharide and WPI amounts but no SSL, the overrun decreased by 13%, 19%, and 10%, respectively. These results suggest that adding SSL reduces foaming capacity because SSL's interfacial adsorption blocks the protein-polysaccharide complex from adsorbing, thereby stabilizing the foam by slowing liquid drainage. This effect has been reported in protein-polysaccharide systems (Narchi et al., 2009).

Fig. 16.13 shows the stability of these foams. Combining WPI with 4.0 g/kg SSL produces more stable foams than with 2.0 g/kg SSL, likely because a larger excess of SSL surpasses its CMC. Although this surfactant can interact with protein regions through electrostatic attraction, enough free SSL remains for interfacial adsorption. Foam stability with guar and λC is similar to that of intermediate WPI-SSL foams. The decreased draining rate is linked to increased viscosity (Fig. 16.14a). Notably, foam stability with xanthan improves significantly, with viscosity more than 10 times higher. Xanthan carries a negative charge, which allows electrostatic interactions with proteins (García-García et al., 2015).

TABLE 16.11 Overrun (%) of WPI-SSL and its mixtures with λC, xanthan (X), and guar (G).

Code	Overrun (%)	Code	Overrun (%)
WPI-2.0SSL	150 ± 0.00[(a)]	WPI-2.0SSL-1.5G	130 ± 0.00[(c)]
WPI-4.0SSL	101 ± 1.41[(b)]	WPI-2.0SSL-2.0λC	135 ± 7.07[(c)]
		WPI-2.0SSL-1.5X	130 ± 14.1[(c)]

Different letters in the same column indicate statistically significant differences ($P < 0.05$).

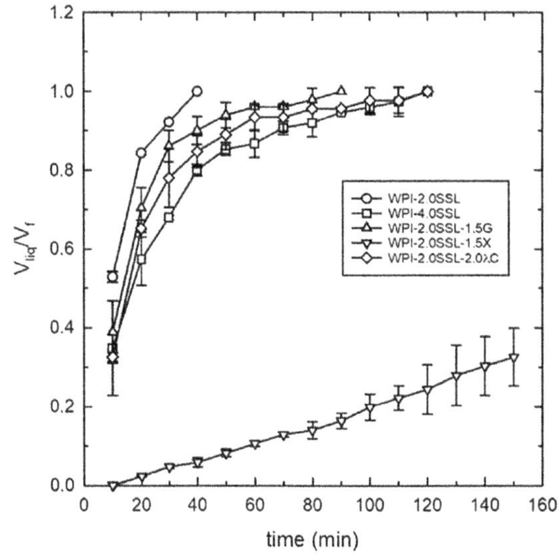

FIGURE 16.13 Displays the stability of WPI-SSL and WPI-SSL-polysaccharide foams. Lines are visual guides.

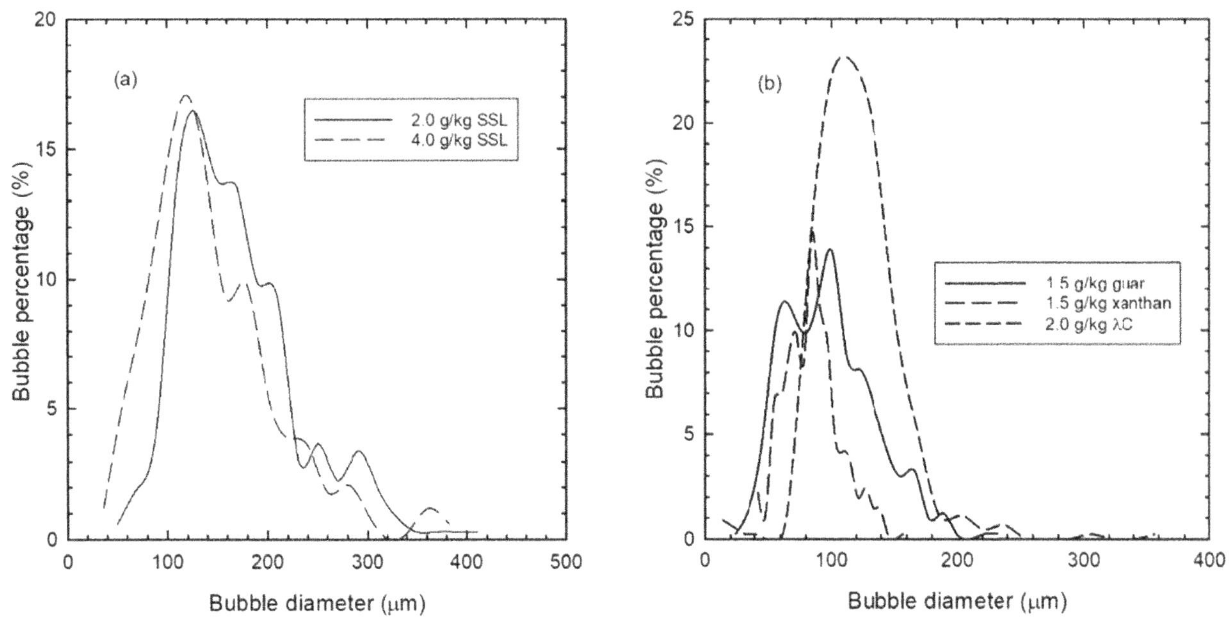

FIGURE 16.14 Frequency distribution of bubbles: (a) WPI-SSL, and (b) WPI-2.0 SSL-polysaccharides.

Table 16.12 shows the regression parameters for Eqs. (16.3) and (16.7). The stability hierarchy from most to least stable is: WPI-2.0SSL-1.5X, followed by WPI-4.0SSL, then WPI-2.0SSL-1.5λC, WPI-2.0SSL-1.5G, and finally WPI-2.0SSL. Foam containing xanthan does not reach $V_{liq}/V_f = 1$ within the observation period, with initial liquid drainage slowing down over time. According to Eq. (16.7), it is not possible to determine the time required for complete drainage ($\phi_e = 1$).

Table 16.13 shows the half-lives of protein-surfactant-polysaccharide foams, which correspond to the results in Fig. 16.13. Statistically, only the foam with xanthan has a significantly different $t_{1//2}$ compared to the other formulations in Table 16.13, which include 2.0 g/kg SSL and an additional polysaccharide.

3.10.2 Bubble size

The frequency distribution (Fig. 16.14a) shows that bubble size increases as SSL concentration rises. At 4.0 g/kg SSL, there is less variation in bubble size, which improves stability. The formulation with λC does not significantly change

TABLE 16.12 Stability parameters of WPI-SSL and WPI-SSL-polysaccharide foams.

Foam	ϕ_e	μ_t (min^{-1})	r^2
WPI-2.0SSL	0.997[a]	0.0518[c]	0.9908
WPI-4.0SSL	0.948[a]	0.0265[d]	0.9463
WPI-2.0SSL-1.5G	0.973[a]	0.0376[e]	0.9895
WPI-2.0SSL-1.5X*	0.543[b]	0.00267[f]	0.9950
WPI-2.0SSL-2.0λC	0.962[a]	0.0316[g]	0.9766

Different letters in the same column indicate statistically significant differences ($P < 0.05$).
*Data fitted to Eq. (16.2) with $\lambda = 25.7$ min.

TABLE 16.13 Stability parameters of WPI-SSL and WPI-SSL-polysaccharide foams.

Foam	t_C (min)	$(t_{1/2})_{exp}$ (min)	$t_{1/2}$ (min)
WPI-2.0SSL	40[a]	9.32 ± 0.30[d]	9.74
WPI-4.0SSL	120[b]	15.7 ± 4.62[e]	20.1
WPI-2.0SSL-1.5G	120[b]	13.0 ± 1.99[e]	13.8
WPI-2.0SSL-1.5X	>150[c]	>150[f]	287
WPI-2.0SSL-2.0λC	120[b]	14.6 ± 0.34[e]	16.6

Different letters in the same column indicate statistically significant differences ($P < 0.05$).

bubble size, unlike guar and xanthan (Fig. 16.14b). However, a reduction in disparity occurs, further enhancing foam stability. The average bubble size decreases by 33% with guar and xanthan compared to other foams. This results in greater stability because the Archimedean force has a more significant influence on larger bubbles, causing them to reach the surface more quickly.

3.10.3 *Flow and viscoelastic properties of solutions and liquid foams*

Fig. 16.15a illustrates the steady flow characteristics of the precursor solutions. WPI and SSL mixtures display Newtonian behavior, with viscosities ranging from 2.3 to 2.8 mPa s. SSL slightly increases WPI's viscosity, mainly enhancing foaming rather than viscosity itself. Fig. 16.15b illustrates the flow behavior of liquid foams, exhibiting shear-thinning properties and a noticeable increase in apparent viscosity in the precursor solutions. Adding 4.0 g/kg SSL improves foam stability and causes a significant increase in apparent viscosity compared to WPI alone. In contrast, the mixture with 2.0 g/kg SSL does not differ significantly from WPI.

Fig. 16.16 shows the viscoelastic properties of WPI and its mixtures with SSL. The similarity in the dynamic moduli of SSL-free WPI and the WPI-SSL mixtures indicates comparable viscoelastic behavior. However, adding SSL shifts the moduli crossover point from 12.5 rad/s in pure WPI to 14.2 and 15.7 rad/s for mixtures with 2.0 and 4.0 g/kg SSL, respectively. This change suggests that the surfactant reduces WPI's relaxation time by about 25%. The loss angles, shown in Fig. 16.16b, remain consistent across the range from 84 to 6 degrees, confirming the viscous nature observed in Fig. 16.16a at low and moderate frequencies.

Fig. 16.17 illustrates the viscoelastic properties of liquid foams containing WPI and SSL.

WPI foams exhibit solid-like behavior, with $G'(\omega)$ higher than $G''(\omega)$, and maintain a consistent storage modulus across a wide frequency range. The loss modulus typically has a concave shape, crossing at about 70.7 rad/s. Adding 2.0 g/kg of SSL reduces the dynamic moduli, but it preserves the solid-like behavior, shifting the crossover point to 52.7 rad/s, which indicates a slight increase in relaxation time due to the addition of SSL. Increasing SSL to 4.0 g/kg causes the moduli to cross above 80 rad/s, with the viscoelastic response changing notably; the dynamic moduli become frequency-dependent and intersect. Fig. 16.17b shows that tan δ values are similar for WPI foam and the mixture with 2.0 g/kg SSL. Below 10 rad/s, loss angles are around 11 degrees, signaling elastic behavior, while above this, viscous effects prevail, with loss

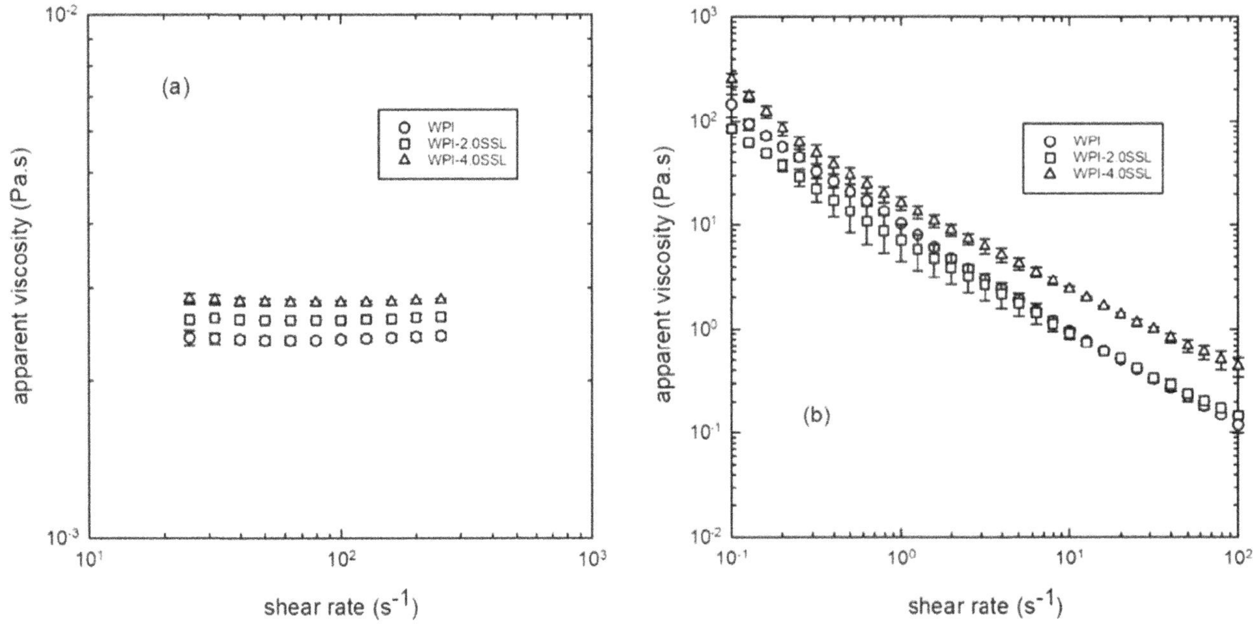

FIGURE 16.15 Shear rate sweep: (a) Solutions and (b) foams. 100 g/kg WPI with 0, 2.0, and 4.0 g/kg SSL.

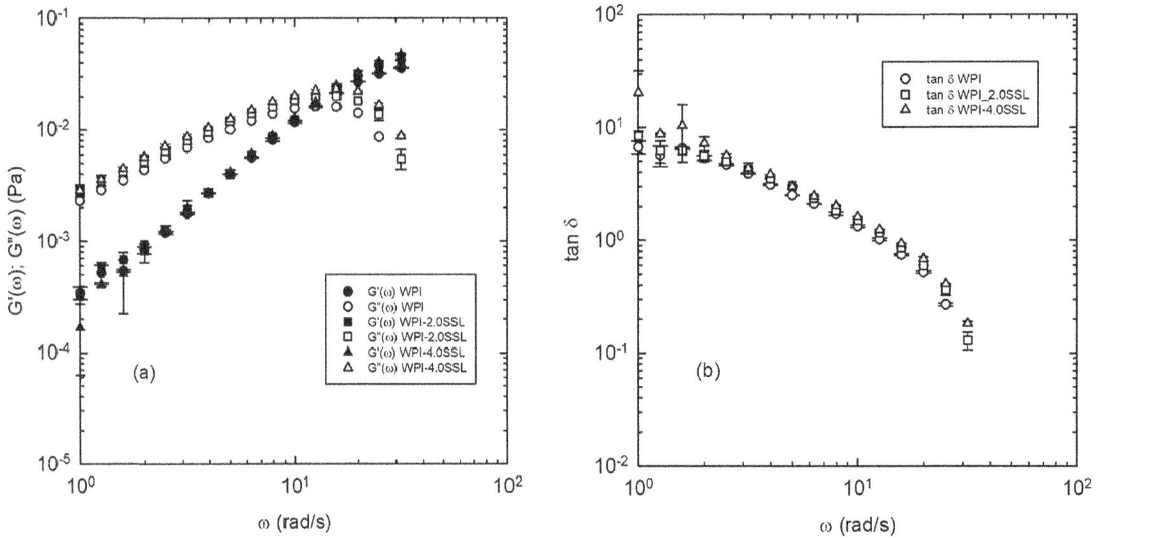

FIGURE 16.16 Precursor solutions: (a) Frequency sweep, and (b) tan δ of 100 g/kg WPI with 0, 2.0, and 4.0 g/kg SSL. Strain: 50%.

angles ranging from 45 to 72 degrees at 100 rad/s for both samples. The similar dynamic moduli at 4.0 g/kg SSL produce a tan δ of 0.6, corresponding to a phase angle of 30.9 degrees. Because the dynamic moduli are comparable, tan δ remains frequency-independent. Overall, the addition of SSL alters the viscoelastic properties and causes significant changes at higher concentrations.

Fig. 16.18 illustrates the steady-flow behavior of precursor solutions and the associated liquid foams in WPI-SSL-polysaccharide mixtures.

Table 16.14 presents the power-law parameters for solutions and foams. The impact of xanthan is clear in both systems. Guar and carrageenan exhibit similar shear-thinning behavior, but the latter yields thicker solutions. Conversely, in foams, the opposite trend is observed. The shear-thinning and thickening properties of guar exceed those of carrageenan and xanthan.

Fig. 16.19 displays the frequency sweep of the precursor solutions along with their corresponding tan δ values. These solutions show typical behavior, with G″(ω) being larger than G′(ω) at low and moderate frequencies, where both moduli

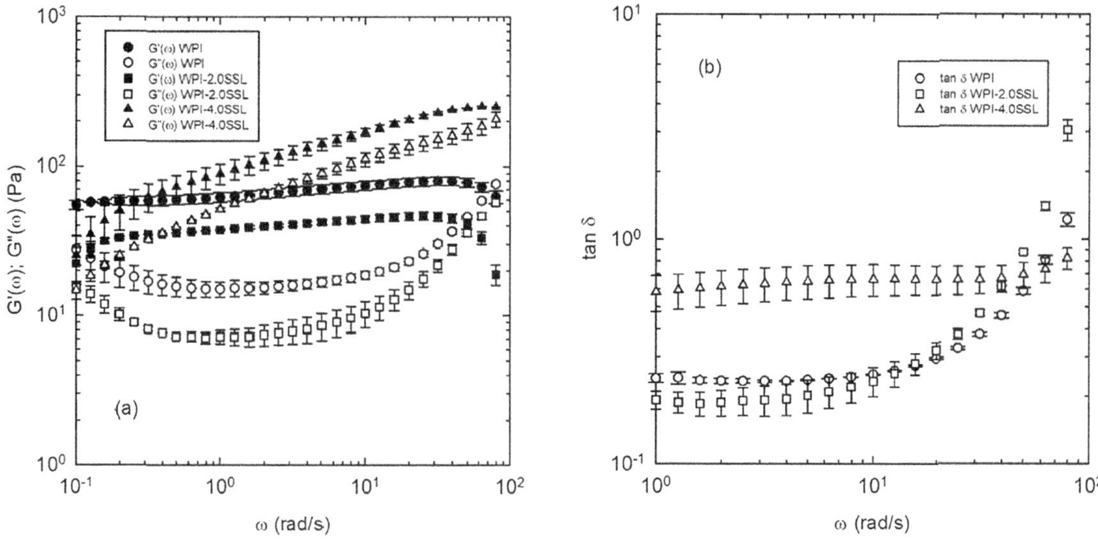

FIGURE 16.17 Liquid foams: (a) Frequency sweep, and (b) tan δ of 100 g/kg WPI with 0, 2.0, and 4.0 g/kg SSL. Strain range: 3%–5%.

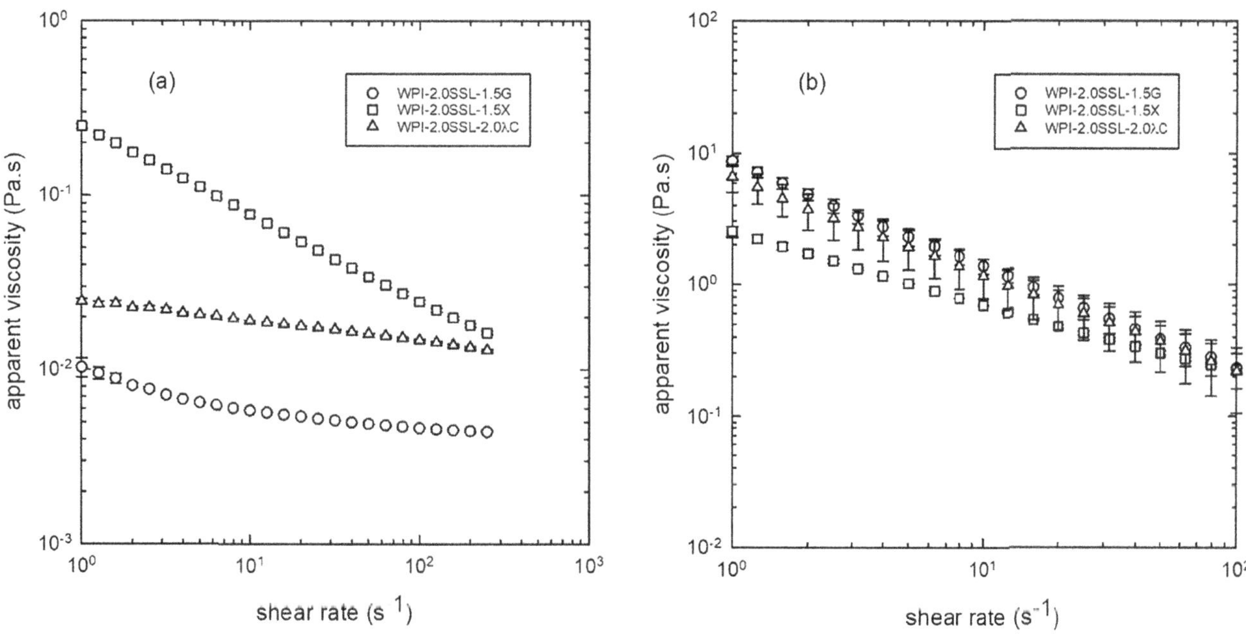

FIGURE 16.18 Rate sweep: (a) solutions and (b) foams of 100 g/kg WPI with 2.0 g/kg SSL and 1.5 g/kg guar, 1.5 g/kg xanthan, and 2.0 g/kg λC. *Based on the previous results, the SSL concentration was set at 2.0 g/kg. Both solutions and foams exhibit shear-thinning behavior. Xanthan significantly changes the viscosity and shear-thinning properties of the solutions. Foams containing guar and λC behave similarly, with no notable differences in their apparent viscosities. This pattern also appears in xanthan-containing foams, but only at shear rates above 10 s^{-1}. Below this threshold, the apparent viscosities of xanthan foams are lower than those of guar and carrageenan (Fig. 16.18b).*

are highly dependent on frequency and cross over at 23.7, 25.4, and 47.5 rad/s for guar, xanthan, and carrageenan, respectively. The relative positions of the dynamic moduli highlight the differences in the viscoelastic properties of these polysaccharides. Guar has lower moduli than carrageenan, while xanthan's are higher than both. Notably, the dynamic moduli in WPI-xanthan mixtures are remarkably similar, due to xanthan concentration and its rod-like structure, which results in more elastic than viscous behavior. Consequently, $G'(\omega) > G''(\omega)$ can be expected.

Fig. 16.19b shows that the guar and λC solutions have similar tan δ behaviors at low frequencies, with λC showing higher values above 2.5 rad/s. The decrease in tan δ for these two polymers indicates loss angles of 83.0 degrees–9.4 degrees for guar and 34 degrees for carrageenan. The elastic part of xanthan in the foams is greater than that of guar and

TABLE 16.14 Power law parameters.

	Solutions			Foams		
Mixtures	K (Pa·s^n)	n	r^2	K (Pa·s^n)	n	r^2
WPI-2.0SSL	2.65[a]	1.0	n.a.	7.97	0.090	0.9972
WPI-4.0SSl	2.86[a]	1.0	n.a.	19.2	0.12	0.9907
WPI-2.0SSL-1.5G	0.00871	0.86	0.9230	9.79	0.15	0.9959
WPI-2.0SSL-1.5X	0.248	0.50	0.9996	2.48	0.46	0.9995
WPI-2.0SSL-2.0λC	0.0250	0.89	0.9973	6.82	0.24	0.9989

[a]*viscosity in mPa·s;* n.a. = *not applicable.*

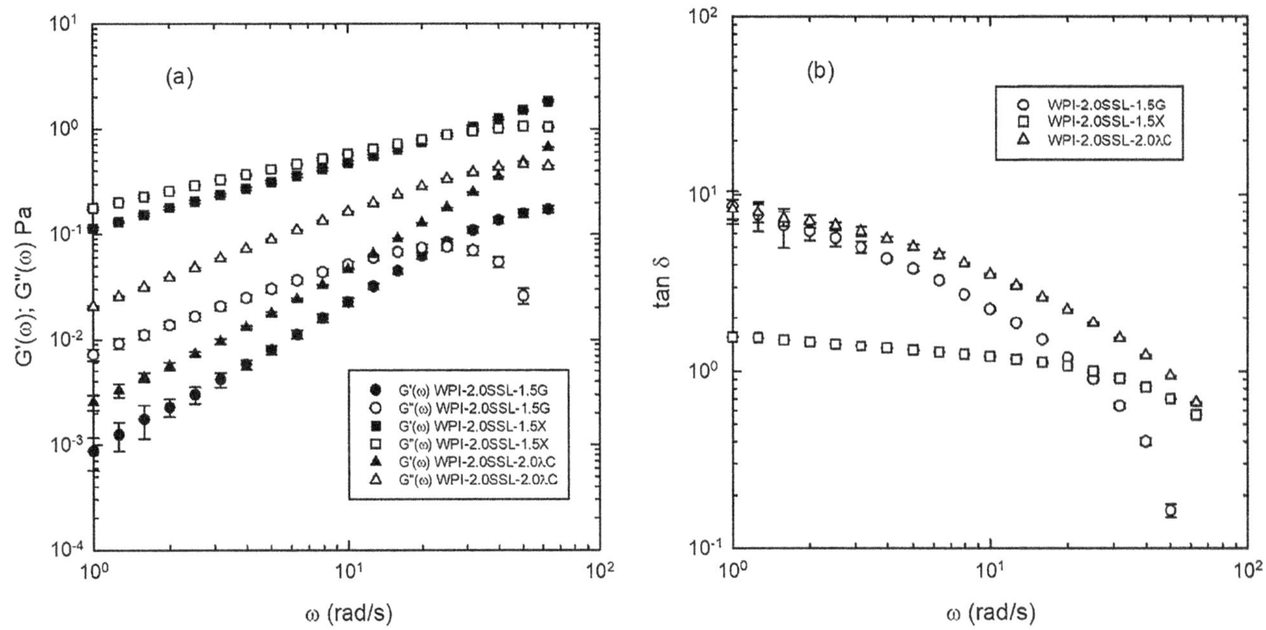

FIGURE 16.19 Precursor solutions: (a) frequency sweep and (b) tan δ of 2.0 g/kg SSL with 1.5 g/kg guar, 1.5 g/kg xanthan, and 2.0 g/kg λC. Strain at 50%.

carrageenan. The tangent of the loss angle ranges from 0.6 to 1.6, corresponding to angles of 31.0 degrees and 58.0 degrees, respectively. However, these angles are still far from the solid-like region.

Fig. 16.20 shows the viscoelastic response of foams made from SSL-polysaccharides with 2.0 g/kg surfactant. Several differences are observed in the precursor solutions. In the foams with guar and carrageenan, $G'(\omega) > G''(\omega)$, indicating that both moduli slightly depend on frequency. Guar's moduli are higher than those of carrageenan, and crossover occurs at 64.0 and 44.8 rad/s, respectively. This means that the elastic component dominates a significant portion of the frequency range. The moduli of foams with xanthan are lower, more dependent on frequency, and are similar over a large part of the spectrum. However, a distinct moduli crossover occurs at 7.8 rad/s. Adding xanthan changes the viscoelastic properties of the foams. Fig. 16.20b shows the behavior of tan δ. In the range of 0.4–10 rad/s, the loss angles of foams with guar, carrageenan, and xanthan are about 14.0, 21.0, and 45 degrees, respectively. Higher angles are observed at both low and high frequencies; however, the elastic contribution to the foam's viscoelasticity decreases in the order guar > carrageenan > xanthan. The increased shear-thinning properties of xanthan may explain this, as this effect counterbalances the elastic contribution in the liquid film around the air bubbles in the foam. This matches the foamability and stability seen when xanthan was added.

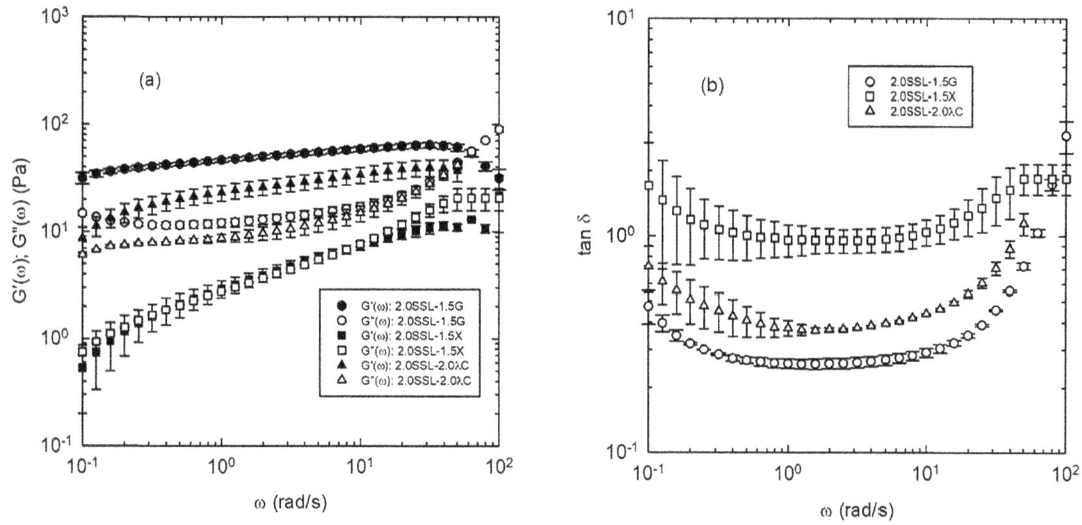

FIGURE 16.20 Liquid foams: (a) Frequency sweep and (b) tan δ of 2.0 g/kg SSL with 1.5 g/kg guar, 1.5 g/kg xanthan, and 2.0 g/kg λC. Strain at 5%.

3.11 Span 20 and Span 20-polysaccharide

3.11.1 Overrun and stability of liquid foams

The initial examination also investigated how a nonionic, low-molar-mass surfactant affects foaming in the absence of polysaccharides. The surfactant's CMC is 2.4×10^{-5} mol/g (Wu et al., 2010), which equals 8.32 g/kg. Therefore, foams were prepared using concentrations several times above the CMC—specifically, 8, 10, 12, 14, and 16 times higher—corresponding to 64, 80, 96, 128, and 160 g/kg, respectively, to reach the maximum possible overrun within this range. Table 16.15 shows the corresponding overruns.

The foam overrun with Span 20 at a concentration of 14 times the CMC is the only one that significantly differs from all other series. The increase in overrun to a maximum of 450% occurs because of the increased interfacial area during foam formation; thus, more surfactant is needed to cover the water-air interface. All other foams have similar overruns comparable to those of previous systems. The overrun data show significant deviations from the average, indicating inconsistent foaming between repetitions. The maximum overrun of this series demonstrates the foam's capacity when surfactant concentration greatly exceeds the CMC. However, further increases in CMC lead to poor and inconsistent foaming, likely due to surfactant micellization in the lamella. Adding polysaccharides decreases the overrun, and although micellization might influence this, increasing the Span 20 concentration to levels similar to those without polysaccharides results in overrun values close to the original. These concentrations were 20, 28, 36, and 42 times above the CMC, corresponding to 160, 224, 288, and 336 g/kg, respectively. Table 16.16 displays the results. Again, notable deviations from the average overrun are observed, indicating that differences among various combinations are not substantial, even though the average suggests otherwise. However, the foam with 1.5 g/kg guar and a Span 20 concentration 28 times above the CMC stands out from the rest. Foam stability measurements could not be performed because their lifespan was too short.

Adding the surfactant did not change the rheological behavior of polysaccharide solutions. The foams stayed unstable, so bubble size and rheological properties could not be measured.

TABLE 16.15 Foaming capacity of Span 20 at or above its CMC.

Code	Overrun (%)	Code	Overrun (%)
CMC	$277 \pm 36^{(a)}$	14CMC	$450 \pm 75^{(b)}$
10CMC	$300 \pm 75^{(a)}$	16CMC	$175 \pm 100^{(a)}$
12CMC	$300 \pm 0.00^{(a)}$		

Different letters in the same column indicate statistically significant differences ($P < 0.05$).

TABLE 16.16 The foaming capacity of Span 20-polysaccharides greatly surpasses the surfactant's CMC.

Code	Overrun (%)	Code	Overrun (%)
20CMC-1.5G	300 ± 75[(a)]	36CMC-1.5G	225 ± 35[(a) (b)]
20CMC-1.5X	250 ± 25[(a)]	36CMC-1.5X	150 ± 74[(b)]
20CMC-2.0λC	170 ± 42[(b)]	36CMC-2.0λC	125 ± 25[(b) (d)]
28CMC-1.5G	400 ± 75[(c)]	42CMC-1.5G	200 ± 0.00[(b)]
28CMC-1.5X	250 ± 75[(a) (b)]	42CMC-1.5X	125 ± 50[(b) (d)]
28CMC-2.0λC	250 ± 0.0[(a)]	42CMC-2.0λC	150 ± 74[(b)]

Different letters in the same column indicate statistically significant differences ($P < 0.05$).

3.12 WPI-Span 20 and WPI-Span 20-polysaccharide

3.12.1 Overrun and stability of liquid foams

Table 16.17 presents the foam overrun obtained by mixing 100 g/kg WPI with Span 20 and polysaccharides.

More consistent foaming was achieved with Span 20-WPI mixtures. The large deviations observed with Span 20 alone were no longer present. Additionally, the same overrun was achieved for 20 and 28CMC, which are 160 and 224 g/kg Span 20, respectively, and a lower overrun was reached at 36CMC (288 g/kg). Since Span 20 is moderately hydrophilic, it can interact with the hydrophobic regions of WPI. This could explain the reduction in overrun at the highest surfactant concentration. The foaming capacity of WPI-28CMC decreased to roughly the same level as WPI-36CMC when polysaccharides were added. This capacity showed no significant difference regardless of the polysaccharide.

Fig. 16.21 displays the foam stability of the combinations listed in Table 16.17.

The stability of the foams with Span 20 (Fig. 16.21) does not show a significant difference at various surfactant concentrations above its CMC. Concerning the superior foaming observed for the Span 20-Polysaccharide and WPI-Span 20 combinations, foams were prepared with WPI, 28CMC Span 20, and 1.5 g/kg guar, 1.5 g/kg xanthan, and 2.0 g/kg λC. The foams created when guar is added display the same stability as the polysaccharide-free foams. Only foams made with λC and xanthan exhibit different behavior. The stabilizing effect of xanthan is notable, as indicated by its low draining rate and the fact that visible liquid drainage started 20 min after foam formation. This effect is supported by the half-life data of the foams shown in Table 16.19.

Table 16.18 shows the regression parameters of Eqs. (16.3) and (16.7) for the foams, as illustrated in Fig. 16.21. The stability ranking from highest to lowest is as follows: WPI-28CMC-1.5X > WPI-28CMC-2.0λC > WPI-28CMC-1.5G > WPI-28CMC = WPI-20CMC > WPI-36CMC. In this case, foams with polysaccharides demonstrate better stability. The least stable foam is the mixture of proteins and Span 20, with a concentration 36 times higher than its CMC.

Table 16.19 shows the collapse time, experimental half-life, and predicted half-life calculated using Eqs. (16.3) and (16.7). The latter was used for foams containing xanthan. The collapse time relates to the stability parameters in Table 16.19. Foams with xanthan had a collapse time longer than the observation period, while foams with protein, 36 times Span 20 CMC, had the shortest time. The predicted half-life values closely match the experimental half-lives. The half-life for foams with xanthan could not be determined because Eq. (16.7) is undefined when $\phi_e < 0.5$.

TABLE 16.17 Foaming capacity of WPI with Span 20 and polysaccharides.

Code	Overrun (%)	Code	Overrun (%)
WPI-20CMC	185 ± 7.07[(a)]	WPI-28CMC-1.5G	135 ± 7.07[(b)]
WPI-28CMC	185 ± 7.07[(a)]	WPI-28CMC-1.5X	135 ± 7.07[(b)]
WPI-36CMC	140 ± 14.1[(b)]	WPI-28CMC-2.0λC	150 ± 0.00[(b)]

Different letters in the same column indicate statistically significant differences ($P < 0.05$).

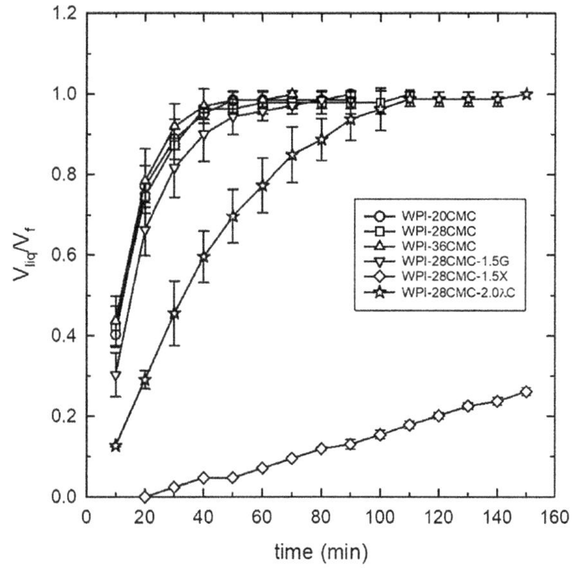

FIGURE 16.21 Shows the stability of the combinations in Table 16.19. Lines are visual guides.

TABLE 16.18 Stability parameters of WPI-Span 20 and WPI-Span 20-polysaccharide foams.

Foam	ϕ_e	μ_t (min^{-1})	r^2
WPI-20CMC	0.992[(a)]	0.0413[(c)]	0.9956
WPI-28CMC	0.983[(a)]	0.0411[(c)]	0.9448
WPI-36CMC	0.997[(a)]	0.0439[(d)]	0.9990
WPI-28CMC-1.5G	0.998[(a)]	0.0326[(e)]	0.9829
WPI-28CMC-1.5X*	0.369[(b)]	0.00231[(f)]	0.9948
WPI-28CMC-2.0λC	1.00[(a)]	0.0148[(g)]	0.9978

Different letters in the same column indicate statistically significant differences ($P < 0.05$).
Data fitted to Eq. (16.2) with $\lambda = 31.2$ min.

TABLE 16.19 Half-life stability of WPI-Span 20 polysaccharide foams.

Foams	t_C (min)	$(t_{1/2})_{exp}$ (min)	$t_{1/2}$ (min)
WPI-20CMC	90[(a)]	12.1 ± 0.81[(f)]	12.3
WPI-28CMC	110[(b)]	11.9 ± 1.02[(f)]	12.5
WPI-36CMC	70[(c)]	11.5 ± 1.58[(f)]	11.5
WPI-28CMC-1.5G	100[(b)]	14.8 ± 1.66[(g)]	15.5
WPI-28CMC-1.5X	>150[(d)]	<150[(c)]	Not Defined
WPI -28CMC-2.0λC	150[(e)]	32.7 ± 5.13[(h)]	34.0

Different letters in the same column indicate statistically significant differences ($P < 0.05$).

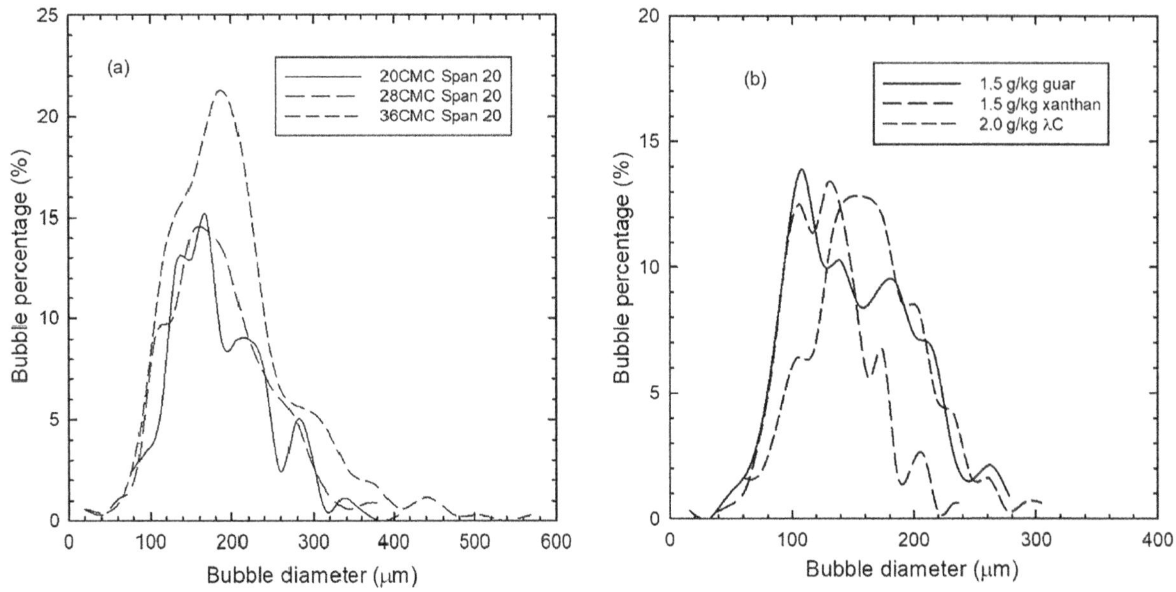

FIGURE 16.22 Frequency distribution of bubbles: (a) WPI-Span 20, (b) WPI-28CMC-Span 20-polysaccharides.

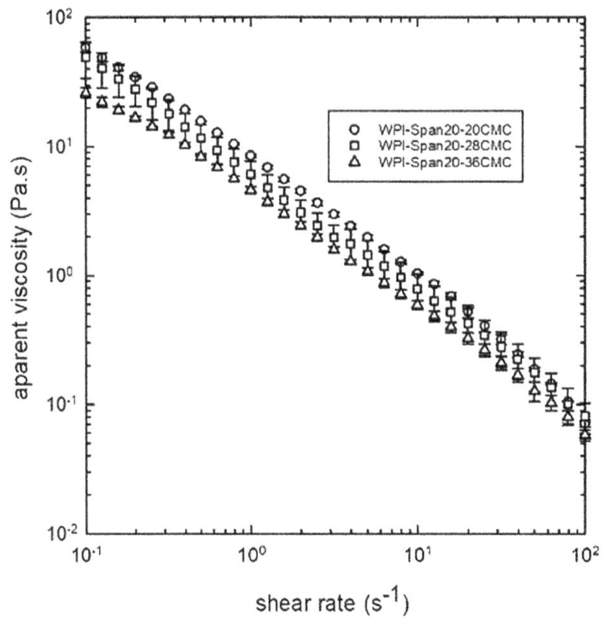

FIGURE 16.23 Flow curve of WPI-Span 20 foams.

3.12.2 Bubble size

The frequency distribution for the foams with 20CMC and 28CMC Span 20 (Fig. 16.22a) shows no difference in size, while there is less dispersion at 36CMC Span 20. However, the average bubble size remains constant, which explains why the drainage rate is similar at any Span 20 concentration. Fig. 16.22b indicates no significant change in size dispersion or size, suggesting that bubble size is not a key factor in stability compared to the WPI-SSL-Polysaccharide system.

3.12.3 Flow and viscoelastic properties of precursor solutions and their foams

WPI solutions containing 20, 28, or 36 times the CMC of Span 20 exhibit Newtonian behavior with a viscosity of approximately 3 mPa s. Fig. 16.23 shows the flow curves for concentrations above the CMC, which are not significantly different and exhibit nonNewtonian shear-thinning behavior. The foams with λC have slightly lower apparent viscosities.

TABLE 16.20 Power law parameters of WPI-Span 20-polysaccharide foams.

	Foams		
Mixtures	K (Pa·s^n)	n	r^2
WPI-28CMC-1.5G	8.11	0.056	0.9963
WPI-28CMC-1.5X	6.14	0.087	0.9993
WPI-28CMC-2.0λC	4.27	0.12	0.9977

Table 16.20 shows the power-law parameters of the foams. The low flow index for guar and xanthan confirms the shear-thinning behavior of these foams, while the n value for carrageenan foams is low enough to help stabilize them. This trend matches the observed stability of these foams, as they drain within 50 min. Despite this, a concentration 28 times higher than the CMC (28CMC) was chosen for further experiments.

Fig. 16.24 shows the viscoelastic behavior of WPI-Span 20 solutions at different surfactant concentrations above their CMC. Only minor variations are observed in the dynamic moduli across various concentrations. Similar to the flow behavior, adding Span 20 did not significantly change the viscoelastic response (Fig. 16.24a).

The moduli crossing points were also close to each other: 13.8 rad/s for 20CMC, 14.2 rad/s for 28CMC, and 36CMC. The tangent of the loss angle, as shown in Fig. 16.24b, was not significantly different, decreasing from 3.9 to 0.11, which corresponds to a change from 76 to 6.3 degrees. These systems transition from a highly viscous contribution to behavior that is much closer to a purely elastic response within a narrow frequency range.

Fig. 16.25 illustrates the viscoelastic behavior of the foams formed from the solutions shown in Fig. 16.24.

The storage modulus was consistently higher than the loss modulus across the entire frequency range. This difference indicates the dominance of elastic behavior. The challenge of producing uniform batch-to-batch foams results in significant variations in the dynamic moduli (Fig. 16.25a). Consequently, no apparent differences are observed among the three Span 20 concentrations. The same trend is observed for tan δ, as shown in Fig. 16.25b, which varies from 0.6 to 1.5, corresponding to angles of 31.0 degrees–58 degrees. These angles represent the midpoint of the viscoelastic spectrum, aligning with 0 degrees for purely elastic behavior and 90 degrees for purely viscous behavior. Therefore, when analyzing the dynamic response, the foams display solid-like properties but are not completely elastic, as indicated by the loss angles, which suggest a viscous component. As a result, adding Span 20 does not significantly alter the viscoelastic properties of the foams.

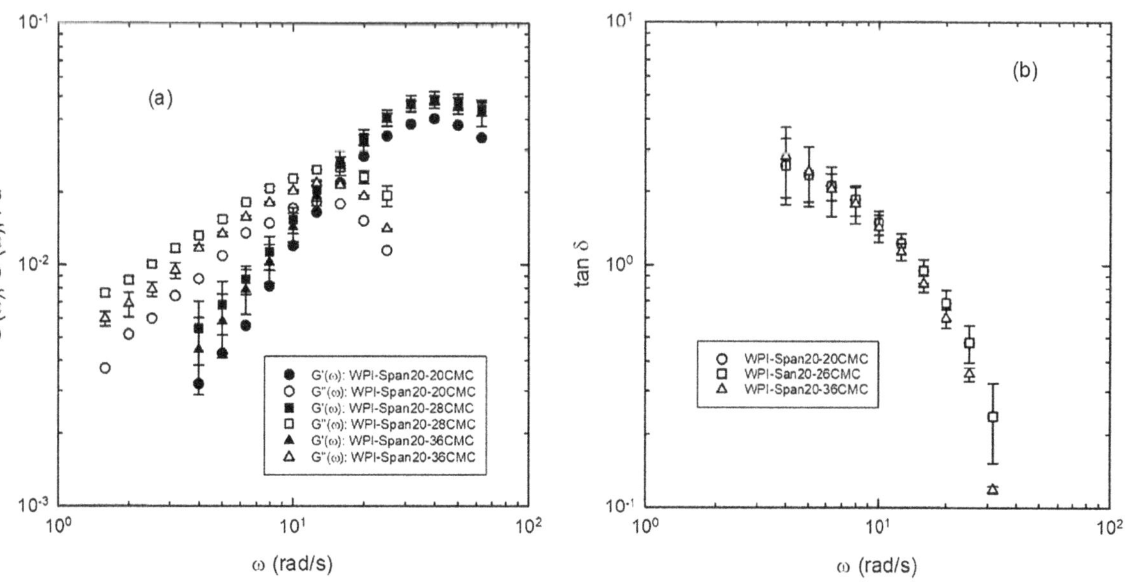

FIGURE 16.24 Frequency sweep (a) and tan δ (b) of WPI-Span 20 precursor solutions with varying surfactant concentrations above their critical micellar concentration.

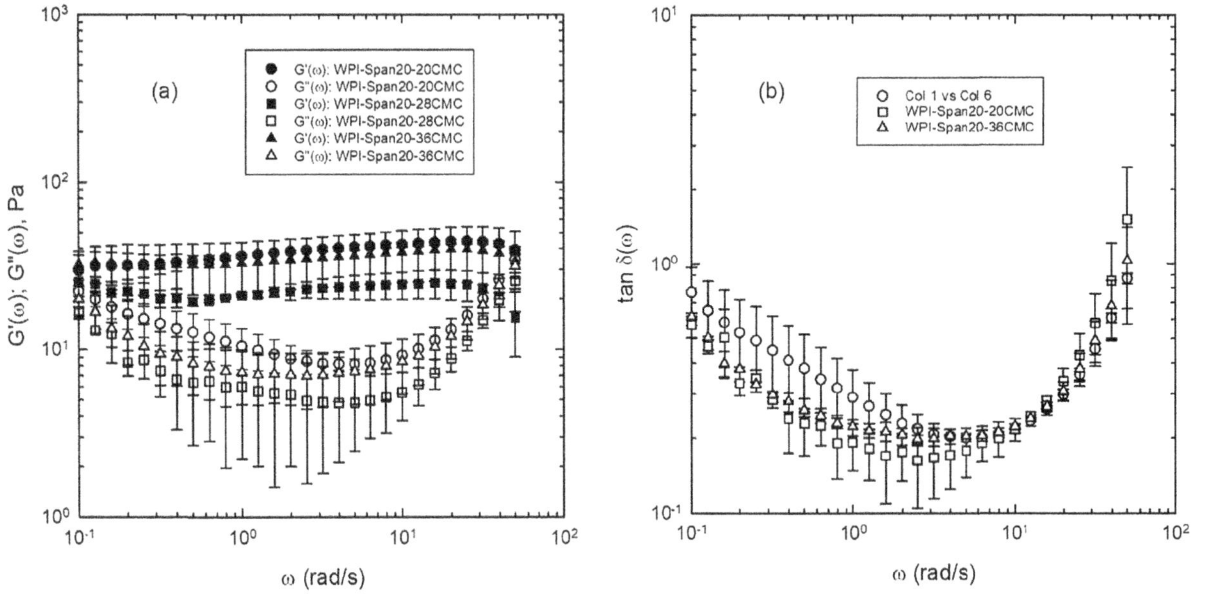

FIGURE 16.25 Frequency sweep (a) and tan δ (b) of WPI-Span 20 foams with different surfactant concentrations above their CMC. Strain range: 3%–6%.

Fig. 16.26 shows the flow behavior of solutions (Fig. 16.26a) and foams (Fig. 16.26b) of WPI mixtures with a Span 20 concentration 28 times above their critical micellar concentration (28CMC). Fig. 16.26a illustrates the thickening ability and shear-thinning properties of xanthan compared to guar and carrageenan when mixed with WPI. The solutions below 30 s^{-1} with guar exhibit erratic signals due to the difficulty in measuring low viscosity at lower shear rates. This behavior is typical of its low thickening ability. Lambda carrageenan thickens WPI solutions but only produces moderate shear thinning. Table 16.21 shows the power-law parameters of these solutions and foams. Fig. 16.26b demonstrates a significant increase in apparent viscosity of the foams generated using the solutions shown in Fig. 16.26a.

The nonNewtonian behavior and thickening ability of guar and carrageenan are similar, supported by the data in Table 16.21. The foam with xanthan shows less shear-thinning behavior and has lower apparent viscosities, although these values match those of guar and carrageenan at a shear rate of 100 s^{-1}. Therefore, the surfactant in the mixture does not influence the flow properties of WPI. In contrast, the polysaccharides modify these properties to different degrees, depending on whether they are in solution or foam.

Fig. 16.27 illustrates the viscoelastic response of WPI solutions containing 28CMC Span 20 and polysaccharides. The effect of Span 20 is similar to that of SSL. The loss modulus exceeds the storage modulus at low and moderate frequencies. However, a distinct separation exists between the $G'(\omega)$ and $G''(\omega)$ signals for each polymer. As a result, the degree of viscoelasticity varies, with guar being less viscoelastic than carrageenan and xanthan. The dynamic moduli crossover occurs at 27.0, 35.9, and 44.4 rad/s for guar, xanthan, and carrageenan, respectively. Between 1 and 5 rad/s, the tan δ values for guar and carrageenan are equal. Beyond this frequency, the tan δ of guar decreases more rapidly than that of carrageenan. These angles still range between 83.0 and 28.3 degrees, indicating that the response is far from purely elastic. The tan δ of mixtures with xanthan is lower and decreases from 67.0 to 17.0 degrees, due to the higher contribution of this polysaccharide to viscoelasticity. However, the viscous characteristic always remains evident.

The dynamic moduli of the foams with WPI, Span 20, and xanthan are the only ones showing significant differences (Fig. 16.28a). The dominance of $G'(\omega)$ over $G''(\omega)$ across the entire frequency range for guar and λC confirms the structured nature of the foams. The dynamic moduli of the foams with xanthan remain consistent from 0.32 to 25.0 rad/s, similar to the foam with SSL. The tan δ reaches a minimum for all three polymers, as shown in Fig. 16.28b, but remains nearly the same for the foams with guar and carrageenan, consistent with the mechanical spectrum in Fig. 16.28a. Loss angles range from 38 to 58 degrees, passing through 14.7 degrees, indicating a transition from elastic to viscous behavior, although they still exhibit noticeable viscous properties. Xanthan-containing foams have a minimum at tan $\delta = 1$ ($\delta = 45$ degrees) and vary from 66.4 to 46.7 degrees.

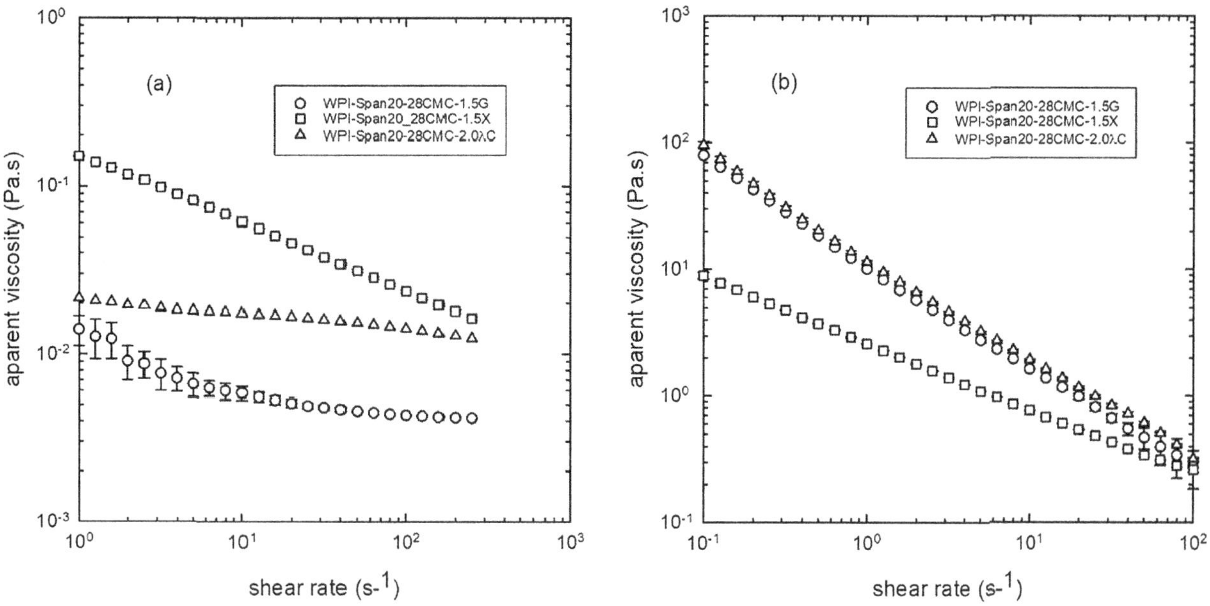

FIGURE 16.26 Flow curves for WPI with Span 20 (28CMC) and polysaccharide solutions (a) and foams (b).

TABLE 16.21 Power law parameters.

	Solutions			Foams		
Mixtures	K (Pa·s^n)	n	r^2	K (Pa·s^n)	n	r^2
WPI-20CMC	0.015	0.538	0.978	8.869	0.172	0.999
WPI-28CMC	0.018	0.731	0.513	6.275	0.098	0.999
WPI-36CMC	0.004	0.934	0.790	4.860	0.258	0.997
WPI-28CMC-1.5G	0.00599	0.93	0.9629[a]	10.8	0.19	0.9987
WPI-28CMC-1.5X	0.156	0.59	0.9997	2.59	0.48	0.9994
WPI-28CMC-2.0λC	0.0212	0.91	0.9817	12.4	0.21	0.9978

[a]*From 30 s^{-1}.*

4. Solid foams

Solid foams were created from the precursor solutions listed in Table 16.22. A 50 mL sample of solution was stirred with a rotor-stator (Ultra Turrax, T25 digital, IKA, Germany) at 15,400 rpm for 5 min at room temperature to optimize foam, overrun, and stability (García-García et al., 2015). The resulting liquid foam was poured into either a glass container measuring 14 × 10 × 4 cm or a 600 mL beaker with an internal diameter of 8.5 cm and a height of 12 cm. The foam was heated in a microwave oven at 1000 W for 6 minutes, reaching a final temperature of 55 ± 5°C. Immediately after heating, the foam was air-dried in a convection oven (Precision, Thermo Electric Corporation, USA) at 45 ± 1°C for about 2.5 h. Then, the foam was removed from the container, placed on a waxed flat surface, and dried further in the oven for 24 h. The plasticizer concentration used for forming the solid foam was 70 g/kg glycerol and 120 g/kg sorbitol. Images of the solid foams were taken with a digital camera (Nikon Coolpix P50, Japan).

The WPI-plasticizer and WPI-polysaccharide-plasticizer solutions were prepared by adding the plasticizer, either glycerol or sorbitol, to the WPI and WPI-polysaccharide solutions, then stirring for 1 hour at room temperature. Plasticizers were used to help produce solid foams after the liquid foams dried. The moisture content of solid foams was measured using a thermobalance (MB45-2A, OHAUS, USA). Samples were heated for 5 or 10 min at 100°C, with at least three repetitions.

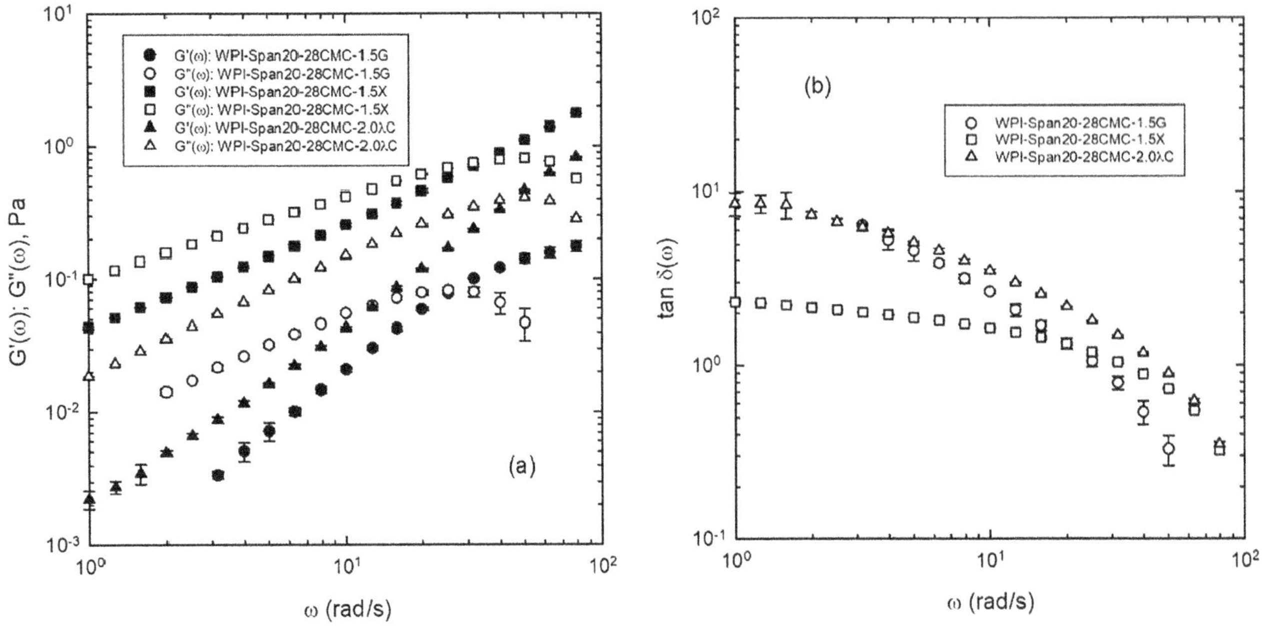

FIGURE 16.27 Viscoelastic behavior of WPI with Span 20 (28CMC) and polysaccharide solutions. Frequency sweep (a) and loss angle (b).

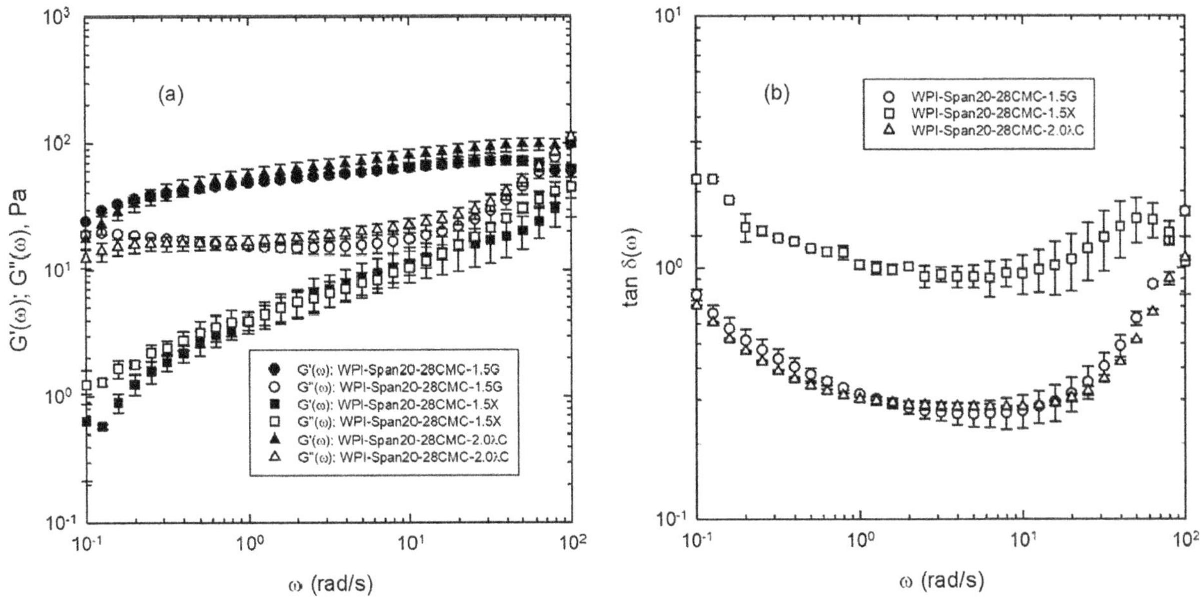

FIGURE 16.28 Liquid foams: (a) Frequency sweep and (b) tan δ of WPI with 28CMC Span 20 and polysaccharides. Strain range: 3%–5%.

4.1 Pore size

The pore size of solid foams was measured using the same optical microscope employed for liquid foams. A thin slice, approximately 0.5 mm thick, of solid foam placed parallel to the foam surface was examined on a slide, and at least 350 pores were measured.

TABLE 16.22 Solid foams from precursor solutions with the composition shown.

Foam	Code
WPI and glycerol (control)	WPI-70GLYC
WPI-GLYC and 1.5 g/kg guar	WPI-70GLYC-1.5G
WPI- GLYC and 2.0 g/kg λ-carrageenan	WPI-70GLYC-2.0-λC
WPI-GLYC and 1.5 g/kg xanthan	WPI-70 GLYC-1.5X

TABLE 16.23 Combinations used to produce solid foams.

Combination	Overrun (%)	Combination	Overrun (%)
WPI-70GLY	$210 \pm 14.1^{(a)}$	WPI-120SRB	$190 \pm 0.00^{(b)}$
WPI-70GLY-1.5G	$180 \pm 7.07^{(b)}$	WPI-120SRB-1.5G	$180 \pm 14.1^{(b)}$
WPI-70GLY-1.5X	$130 \pm 14.1^{(c)}$	WPI-120SRB-1.5X	$120 \pm 14.1^{(c)}$
WPI-70GLY-2.0λC	$170 \pm 7.07^{(b)}$	WPI-120SRB-2.0λC	$180 \pm 7.07^{(b)}$

Different letters in the same column indicate statistically significant differences ($P < 0.05$).

4.2 Mechanical properties

The solid foams were conditioned in a desiccator containing a supersaturated solution of magnesium nitrate hexahydrate (Baker, Mexico) for 24 h at a relative humidity of $50 \pm 5\%$. Uniaxial tensile tests were performed on foam strips measuring 7×1 cm x approximately 0.5 cm, at a crosshead speed of 60 mm/min, using a 100 N load cell on a mechanical testing machine (Sintech 1/S, MTS, USA). Young's modulus, maximum stress, and Hencky's maximum strain were calculated from the stress-Hencky strain data. At least eight samples were tested for each type of solid foam.

4.3 WPI-Plasticizer-Polysaccharide precursor to solid foams

4.3.1 Overrun and stability of liquid foams

Glycerol and sorbitol are two common plasticizers used to produce plastic films from WPI. It is expected that, similar to plastic films, a solid material will form when the precursor liquid foam is dehydrated in a mixture with a plasticizer. The main conclusions from the previous information for preparing a solid foam with the highest possible overrun are: 1) Adding a surfactant to a WPI solution decreases the overrun, probably due to competitive adsorption between the surfactant and whey proteins. 2) For the WPI-Surfactant-Polysaccharide system, the type of polysaccharide added affects the stability, flow, and viscoelastic properties, allowing the formation of stable foams even though they produce less foam overall. Therefore, the mixtures used to create solid foam contain both plasticizers and polysaccharides. The polysaccharides help maintain foam stability and reduce liquid drainage during dehydration. The combinations used are shown in Table 16.23, with 70 g/kg glycerol and 120 g/kg sorbitol, resulting in more stable and solid foams.

Table 16.23 shows that the highest overrun occurs in foams without polysaccharides. WG has an overrun of 200%, and WS has 190%, which decreases with the addition of a polysaccharide, similar to the patterns seen in the WPI-surfactant-polysaccharide systems. The foams with the lowest overrun are those containing λ-carrageenan and xanthan.

Fig. 16.29 shows the stability of WPI-Plasticizer-Polysaccharide combinations.

Polysaccharide-free foams are the least stable, with WG and WS completely draining by 70 min. The most stable foams contain xanthan. WGX drains 0.19 at 150 min, and WSX drains a fraction of 0.18. Foams with guar are less stable than those with λC, with WGLC and WSLC draining at 150 min, and WGG and WSG at 100 and 110 min, respectively. This pattern matches the stability order based on the type of polysaccharide added when compared to foams with surfactant. Table 16.24 shows the regression parameters of Eqs. (16.3) and (16.7) for liquid foams with glycerol and sorbitol.

Table 16.25 shows that, based on the analysis of variance, the half-life of all pairs except WGX-WSX is not significantly different. The half-life of this pair remains above 150 min regardless of the plasticizer.

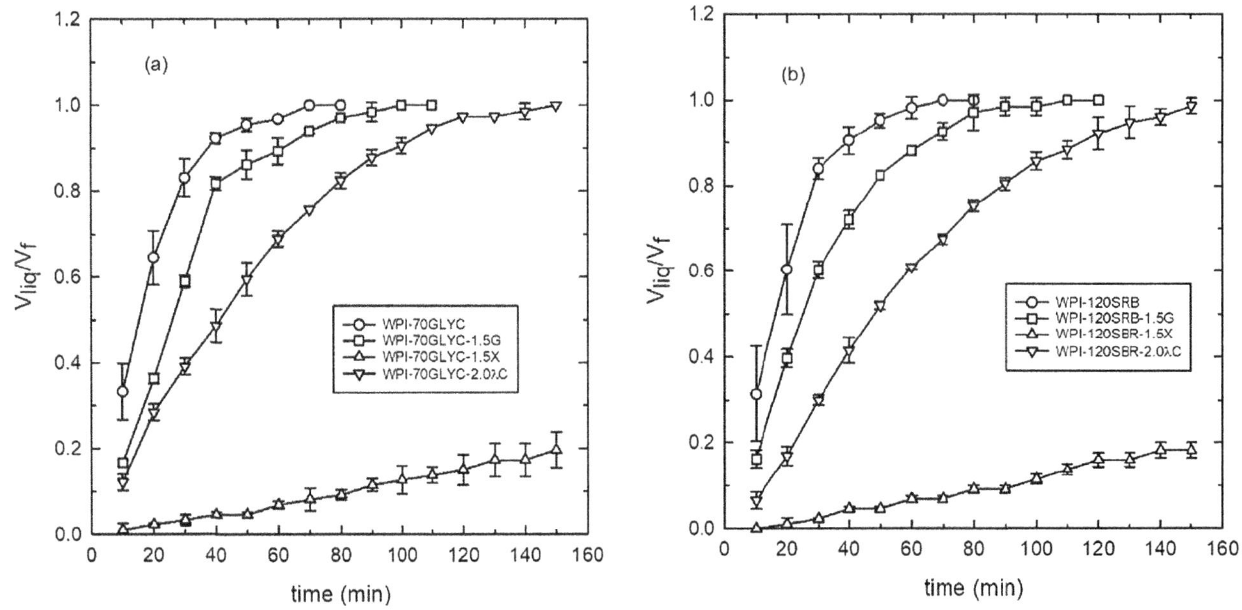

FIGURE 16.29 Stability of WPI-Plasticizer-Polysaccharide foams. (a) Glycerol, (b) sorbitol. Lines are visual guides.

TABLE 16.24 Stability parameters of foams with glycerol and sorbitol from Eqs. (16.3) and (16.7).

Combination	ϕ_e	μ_t (min^{-1})	r^2
WPI-70GLY	0.997[(a)]	0.0332[(a)]	0.9960
WPI-70GLY-1.5G	1.00[(a)]	0.0205[(b)]	0.9910
WPI-70GLY-1.5X*	0.279[(b)]	0.00152[(c)]	0.9942
WPI-70GLY-2.0λC	1.01[(a)]	0.0123[(d)]	0.9954
WPI-120SRB	1.00[(a)]	0.0317[(e)]	0.9958
WPI-120SRB-1.5G	1.00[(a)]	0.0196[(f)]	0.9954
WPI-120SRB-1.5X**	0.270[(b)]	0.00158[(c)]	0.9867
WPI-120SRB-2.0λC	1.02[(a)]	0.0102[(g)]	0.9935

Different letters in the same column indicate statistically significant differences ($P < 0.05$).
*Eq. (16.2) $\lambda = 17.3$ min.
**Eq. (16.2) $\lambda = 23.7$ min.

Half-life stability decreases when surfactants are added to the mixtures instead of plasticizers. This change in $t_{1/2}$ for mixtures with glycerol or sorbitol results from an increase in viscosity caused by the plasticizer (Fig. 16.28) compared to those with the surfactant. Additionally, there is no competitive adsorption between the plasticizer and the protein. This action improves the stability of the liquid foam.

In summary, the stability of the WPI-Surfactant-Polysaccharide and WPI-Plasticizer-Polysaccharide systems mainly depends on the type of polysaccharide added because the thickening effects differ. Additionally, the interface visco-elasticity varies depending on the type of hydrocolloid used. The properties of these systems are primarily influenced by the type of polysaccharide and its interaction with the protein.

4.3.2 Bubble size

The frequency distributions show that foams with the largest bubbles are those without polysaccharides. The bubble size decreases when polysaccharides are added (Fig. 16.30). This pattern also occurs in foams with surfactants instead of plasticizers. The WG and WS foams have identical frequency distributions, with an average bubble diameter of about

TABLE 16.25 Stability parameters of liquid foams, precursors to solid foams.

Combination	$(t_{1/2})_{exp}$ (min)	$t_{1/2}$ (min)
WPI-70GLY	14.9 ± 2.12[(a)]	15.2[(f)]
WPI-70GLY-1.5G	25.7 ± 0.97[(b)]	24.5[(g)]
WPI-70GLY-1.5X*	>150[(c)]	Not Defined
WPI-70GLY-2.0λC	41.7 ± 2.88[(d)]	40.4[(h)]
WPI-120SRB	16.2 ± 4.00[(a)]	15.9[(f)]
WPI-120SRB-1.5G	24.5 ± 0.99[(b)]	25.7[(g)]
WPI-120SRB-1.5X**	>150[(c)]	Not Defined
WPI-120SRB-2.0λC	47.9 ± 1.11[(e)]	48.3[(i)]

Different letters in the same column indicate statistically significant differences ($P < 0.05$).
*From Eq. (16.2) with $\lambda = 17.3$ min.
**From Eq. (16.2) with $\lambda = 23.7$ min.

110 μm. In contrast, the WGG and WGLC foams measure nearly 105 and 97 μm, respectively. The last two distributions are similar, except that WGG shifts toward larger sizes. This suggests that WGG is less stable than WGLC, since a larger bubble size indicates lower stability. For WGX, the average diameter is around 75 μm, with a more uniform distribution compared to the other foams. The smaller bubble size indicates higher stability.

WSG and WSLC foams have average bubble sizes of 78 and 90 μm, respectively, with similar bubble distributions, while WSX has an average size of 75 μm. The sizes for WSG and WSLC are smaller compared to their Glycerol counterparts, aligning with their higher stability, as indicated by longer $t_{1/2}$. The WSX foam exhibits the same frequency distribution as WGX, suggesting that they would have similar stability curves. However, this is not the case, as the parameters k and $t_{1/2}$ differ significantly, highlighting interactions between the plasticizer and xanthan, which is a polysaccharide with a strong negative character.

4.3.3 Flow and viscoelastic characteristics of solutions and foams

Fig. 16.31 demonstrates that the WPI-plasticizer mixtures without polysaccharides exhibit Newtonian behavior and have the lowest viscosity, approximately 2.8 mPa s.

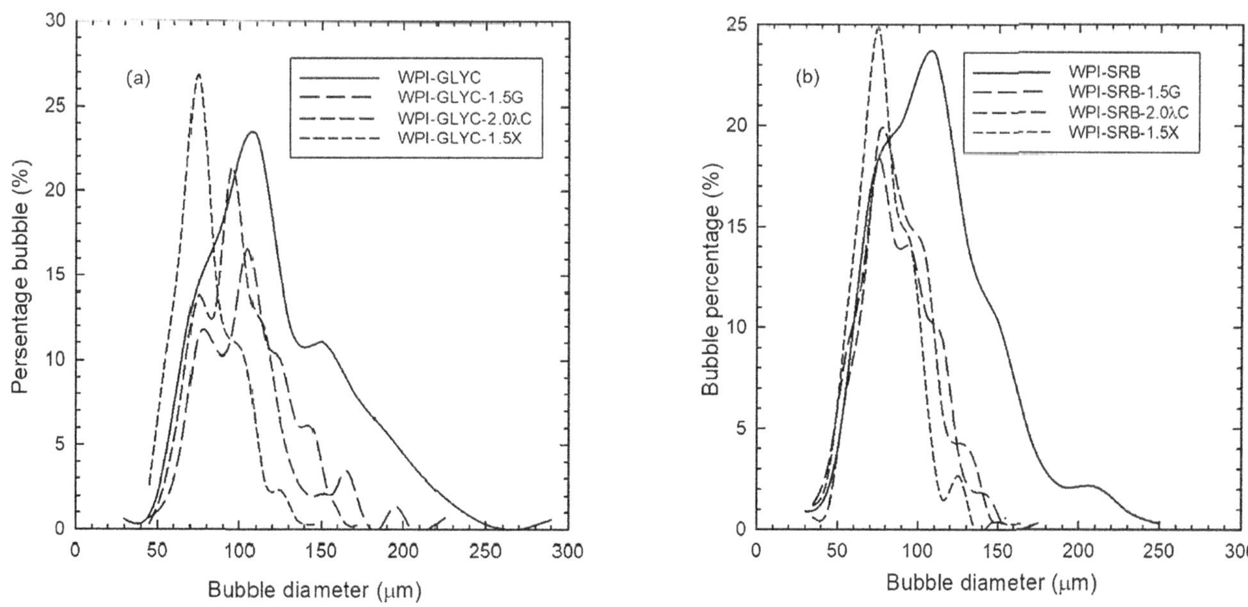

FIGURE 16.30 Bubble frequency distribution: (a) WPI-GLYC-polysaccharide, and (b) WPI-SRB-polysaccharide.

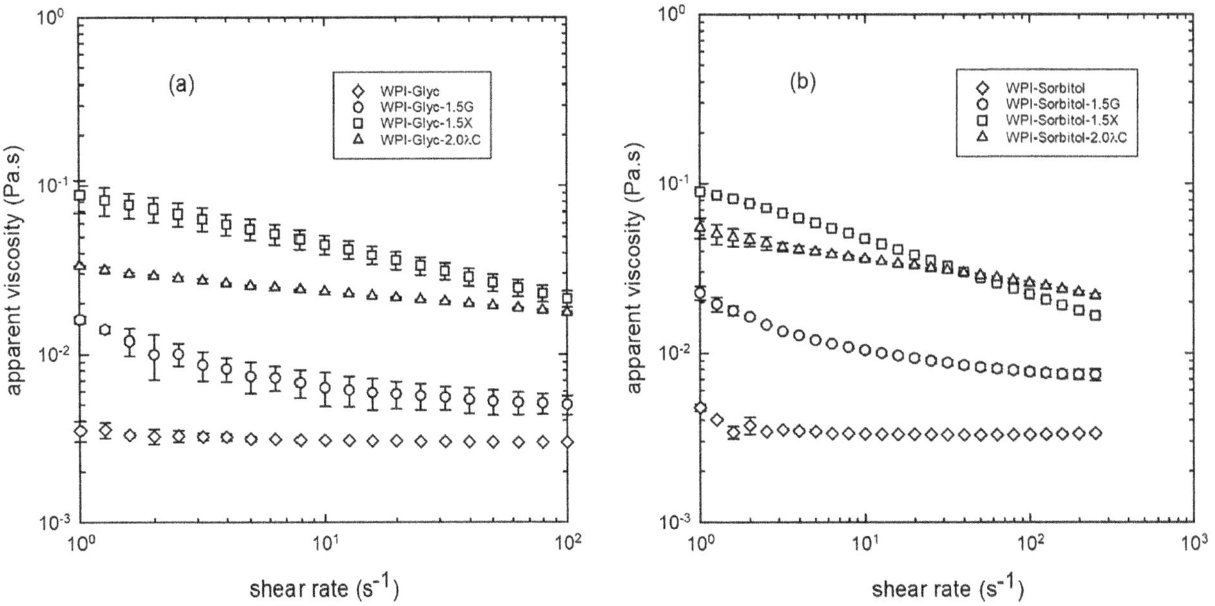

FIGURE 16.31 Shear rate sweep of WPI-Plasticizer precursor solutions with and without polysaccharides: (a) Glycerol, (b) sorbitol.

Adding a polysaccharide increases their viscosity and alters their steady-shear behavior. Mixtures with polysaccharides exhibit shear-thinning properties. Replacing glycerol with sorbitol slightly raises the viscosity of the precursor solutions, but they respond similarly to changes in shear rate.

Fig. 16.32a shows that for all precursor solutions, G"(ω) exceeds G′(ω). In mixtures with glycerol, the moduli crossing points occur at 15.9, 31.6, 50.1, and 31.6 rad/s for those without polysaccharides and for those with guar, xanthan, and carrageenan, respectively. Therefore, the solution without polysaccharides relaxes more slowly than those with polysaccharides.

The elastic contribution of each polysaccharide likely influences this behavior. Solutions containing polysaccharides exhibit similar behaviors since their moduli are not significantly different. This is supported by the tan δ data shown in Fig. 16.33a. Only the solution without polysaccharides displays a different behavior. Similarly, in the mixtures with

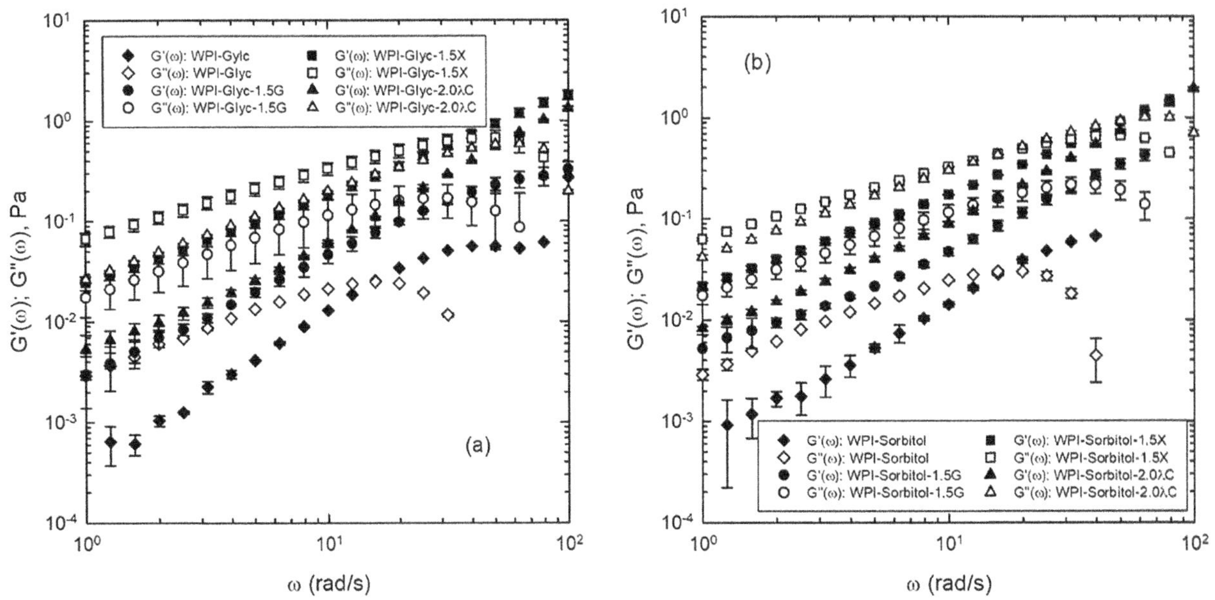

FIGURE 16.32 Frequency sweep of WPI-Plasticizer precursor solutions with and without polysaccharides: (a) Glycerol, (b) sorbitol. Strain range: 20%–50% for glycerol and 30%–50% for sorbitol.

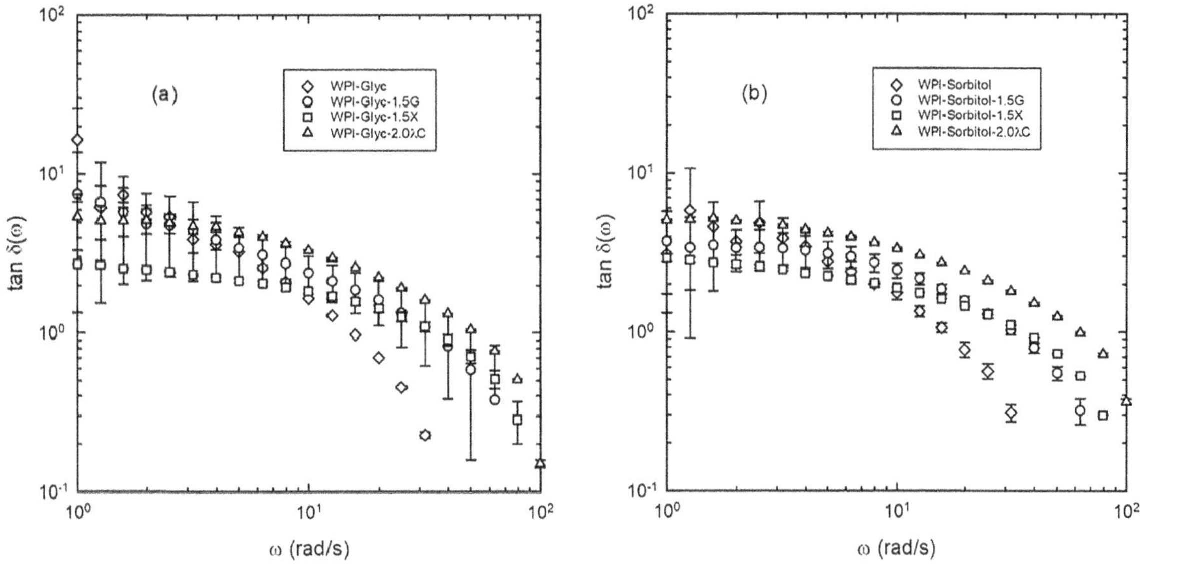

FIGURE 16.33 tan δ of WPI-plasticizer precursor solutions with and without polysaccharides: (a) Glycerol, (b) sorbitol. Strain range: 20%–50% for glycerol and 30%–50% for sorbitol.

sorbitol, G′(ω) equals G″(ω) at 15.9, 31.6, 50.1, and 39.8 rad/s for those without polysaccharide, guar, xanthan, and carrageenan, respectively. Fig. 16.33b shows the same dependence and tan δ values as those for glycerol in Fig. 16.33a. The type of plasticizer does not affect the viscoelastic behavior.

The steady-flow behavior shown in Fig. 16.34 for the WPI/Plasticizer/Polysaccharide foams, with and without polysaccharides, demonstrates the same shear dependence and trend regardless of the type of plasticizer. The foams with the lowest resistance to shear deformation and an initial apparent viscosity of about 10 Pa s are those that do not contain polysaccharides. Foams containing polysaccharides show similar initial apparent viscosities (around 80 Pa s). Xanthan-containing foams, whether with glycerol or sorbitol, are more resistant to shear deformation, meaning they experience smaller viscosity drops compared to the other foams. The same behavior is seen in the foams with surfactants.

Table 16.26 shows the flow parameters. Polysaccharide-free WPI-plasticizer solutions exhibit similar flow behavior with a slight deviation from Newtonian behavior. Adding polysaccharides enhances the shear-thinning properties, with

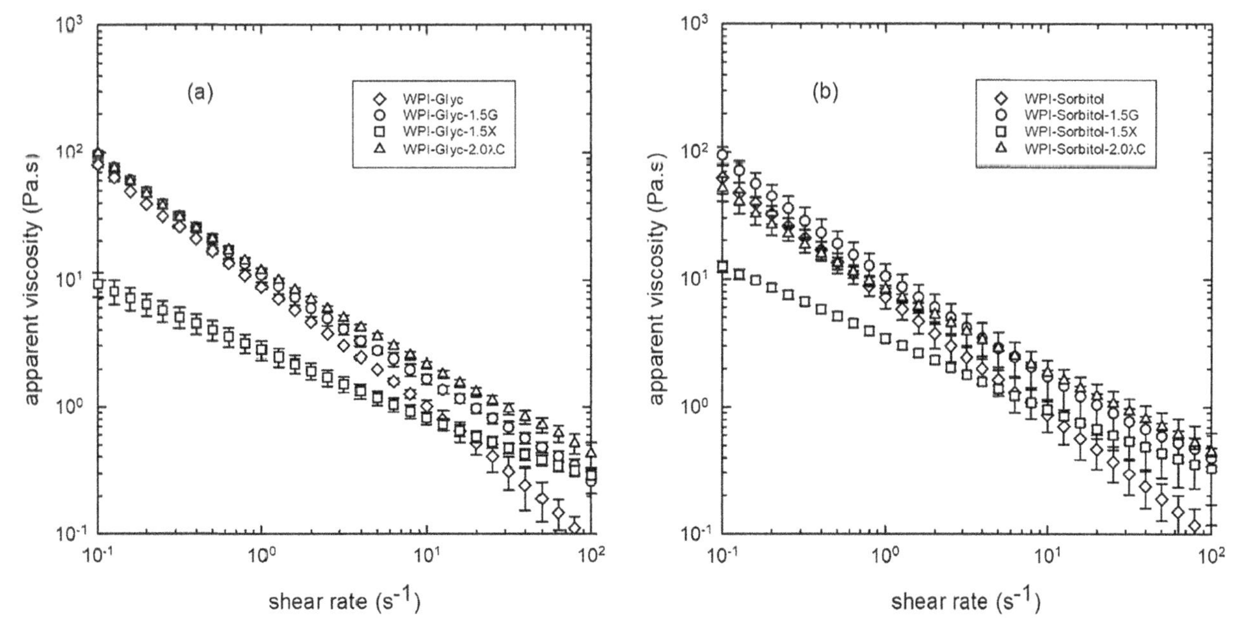

FIGURE 16.34 Flow curve of WPI-Plasticizer foams with and without polysaccharides: (a) Glycerol, (b) sorbitol.

TABLE 16.26 Flow parameters.

	Solution			Foam		
Mixture	K (Pa·s^n)	n	r^2	K (Pa·s^n)	n	r^2
WPI-70GLYC	0.00331	0.97	n.a.	8.76	0.035	0.9993
WPI-70GLYC-1.5G	0.0115	0.81	0.8628	11.8	0.16	0.9978
WPI-70GLYC-1.5X	0.0909	0.69	0.9990	2.78	0.49	0.9991
WPI-70GLYC-2.0λC	0.0319	0.87	0.9965	13.0	0.24	0.9966
WPI-120SRB	0.00373	0.97	n.a.	7.17	0.069	0.9997
WPI-120SRB-1.5G	0.0175	0.82	0.9099	11.5	0.22	0.9942
WPI-120SRB-1.5X	0.0953	0.69	0.9985	3.48	0.46	0.9986
WPI-120SRB-2.0λC	0.0517	0.85	0.9951	8.85	0.34	0.9974

n.a., not applicable.

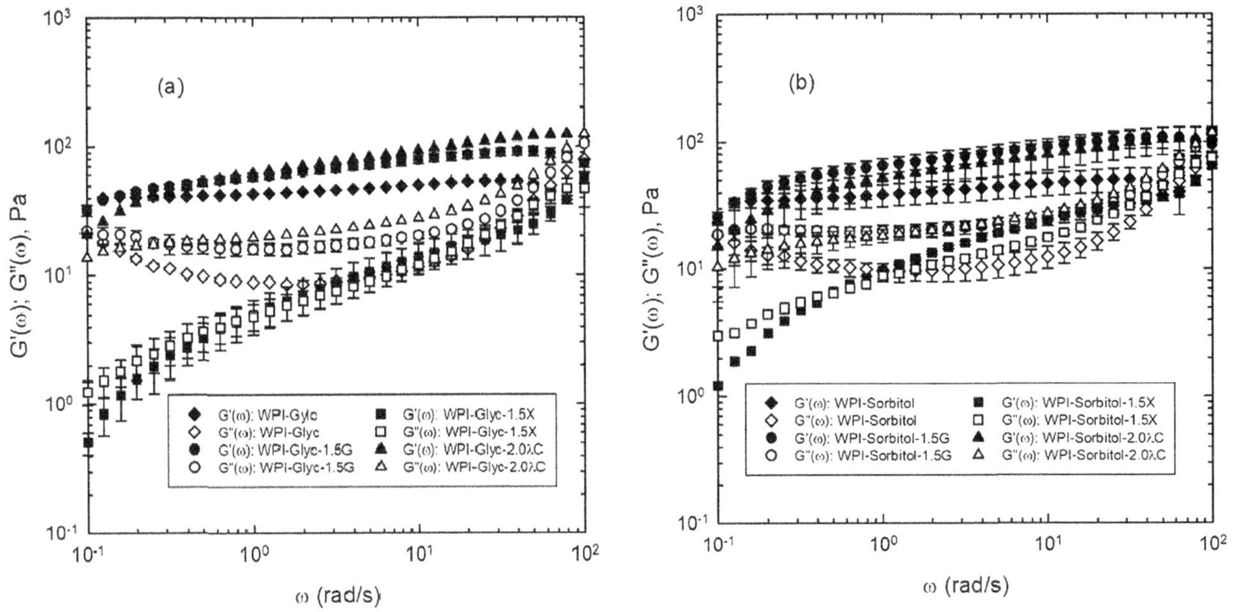

FIGURE 16.35 Dynamic spectra of WPI-plasticizer foams with and without polysaccharide (g/kg): (a) Glycerol, (b) sorbitol. Strain range: 0.4%–3%.

xanthan being the most effective thickener regardless of the plasticizer used. Incorporating glycerol or sorbitol raises both the apparent viscosity and the degree of shear-thinning.

Fig. 16.35a displays the frequency sweep of the foams, showing a clear dominance of G′(ω) over G″(ω), except for the mixture with glycerol and xanthan, where, within a frequency range of 2.6–100 rad/s, the moduli have similar values, indicating equal viscoelasticity. The moduli crossing occurs at 63.1, 79.4, and 98.1 rad/s for WPI without polysaccharides, WPI with guar, and carrageenan, respectively. These crossings indicate elastic behavior over substantial frequency ranges when guar or λC is added.

The foams with sorbitol, as shown in Fig. 16.35b, display the same viscoelastic behavior as their counterparts with glycerol. Therefore, polysaccharides play a key role in determining the viscoelastic properties of liquid foams. Fig. 16.35b shows that the gap between G′(ω) and G’’(ω) is slightly smaller than in the mixtures with glycerol. This occurs because of the addition of sorbitol and its interaction with polysaccharides and proteins. The WPI-sorbitol-xanthan mixture exhibits two crossing points. The viscous behavior dominates from 0 to 5.50 rad/s, followed by elastic behavior from 5.50 to 32.02 rad/s, and then the viscous behavior again from 32.02 to 100 rad/s. These different behaviors at various frequencies

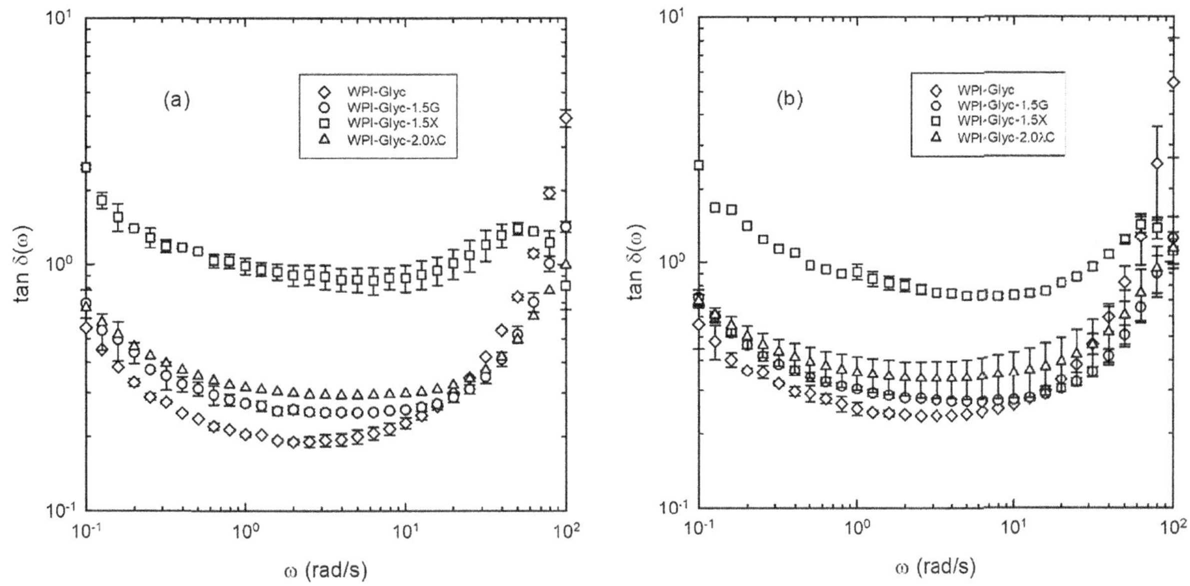

FIGURE 16.36 tan δ for WPI-plasticizer liquid foams with and without polysaccharide (g/kg): (a) Glycerol, (b) sorbitol. Strain range: 0.4%–3%.

suggest some interaction between xanthan and the plasticizer, likely due to the negative charges of xanthan and the plasticizer, which also respond differently to shear deformation (Fig. 16.36).

4.4 WPI-Plasticizer-Polysaccharide

4.4.1 Morphology

Fig. 16.37 displays the morphology and microscope images of different solid foams. Differences in appearance, color, and porosity are evident. Adding the polysaccharide results in a surface that is less porous and produces denser foams. Some foams also have a rougher surface, indicating observable rugosity. These images demonstrate that solid foam can be made using the method described here. WPI-GLYC-2.0λC, WPI-GLYC-1.5X, WPI-SRB-2.0λC, and WPI-SRB-2.5X reached maximum volumes of 29.9, 34.1, 47.3, and 39.1 cm^3, respectively. The foam volume was estimated by displacing spherical particles, 1 mm in diameter, which approximates the volume of a disc. These results suggest that adding λ-carrageenan and xanthan increases the final volume.

4.4.2 Pore size

Fig. 16.38 shows that WPI-70GLYC foam has larger pores, while the polysaccharide-containing foams have smaller pores. Foams with a higher final volume, such as WPI-70GLYC-2.0λC and WPI-70GLYC-1.5X, exhibit smaller pore sizes. Foams with λ-carrageenan and xanthan have more uniform pore sizes that match their surface appearance. Therefore, adding λ-carrageenan and xanthan results in smaller pore sizes.

Smaller pore sizes are seen in foams with larger final volumes, such as WPI-70GLYC-2.0λC, WPI-70GLYC-1.5X, WPI-120SRB-2.0λC, and WPI-120SRBS-1.5X. The difference in pore size is less evident for foams containing λC and xanthan, as shown by their uniform surfaces (Fig. 16.37, left). Therefore, adding λC and xanthan encourages smaller pore sizes, which matches the increased stability of the precursor liquid foams because they are more stable. Less coalescence results in smaller final pores.

Furthermore, foams with sorbitol show less variation in pore size compared to those with glycerol, which correlates with the higher final volume for foams with sorbitol. This suggests that adding sorbitol reduces pore size variation, resulting in higher final volumes. WPI-120SRB and WPI-SRB-1.5G foams have similar pore sizes, consistent with their comparable final volume and surface area (Fig. 16.38). WPI-120SRB-2.0λC foam exhibits the largest pores and the most uniform surface.

4.4.3 Mechanical properties of solid foams

Table 16.27 shows the moisture content of the solid foams.

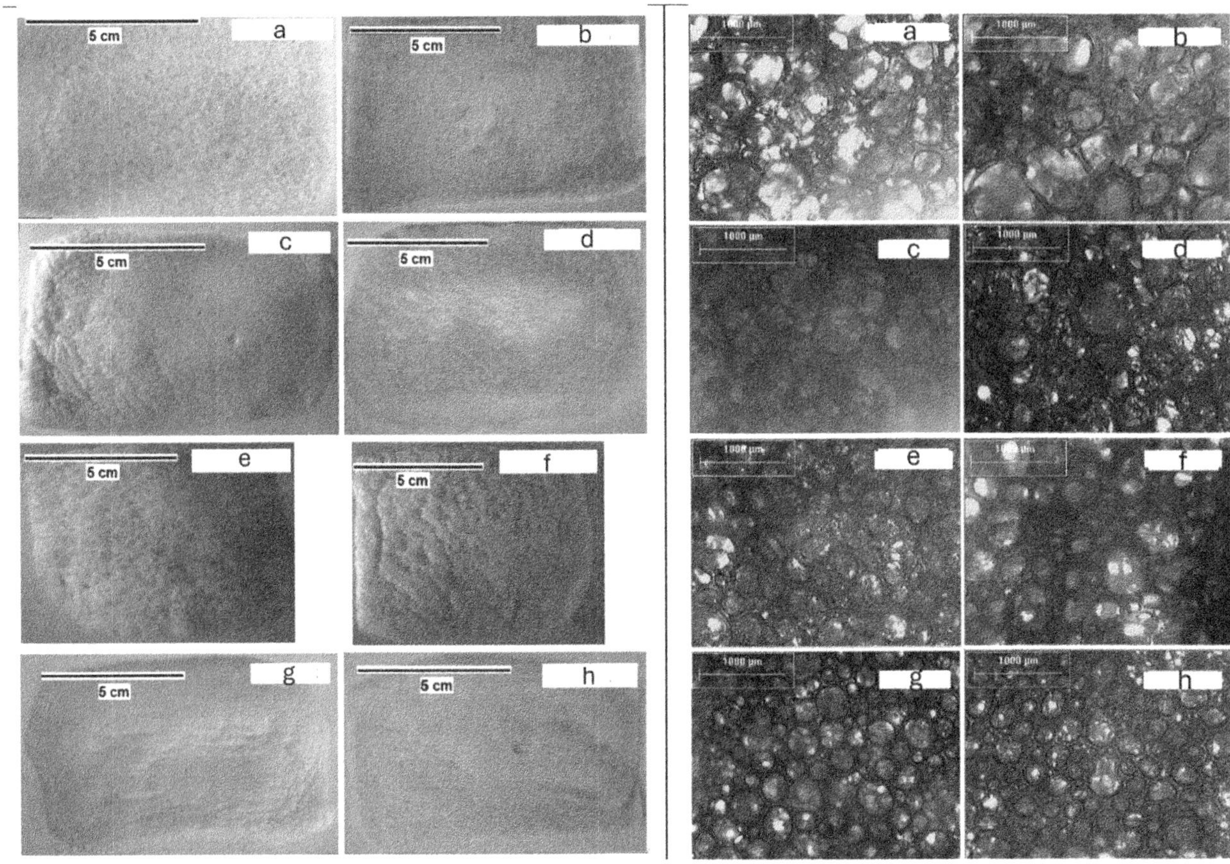

FIGURE 16.37 Left: Morphology of solid foams; right: Microscope images of solid foams. WPI-70GLYC (a), WPI-70GLYC-1.5G (b), WPI-70GLYC-2.0λC (c), WPI-70GLYC-1.5X (d), WPI-120SRB (e), WPI-120SRB-1.5G (f), WPI-120SRB-2.0λC (g), WPI-120SRB-1.5X (h).

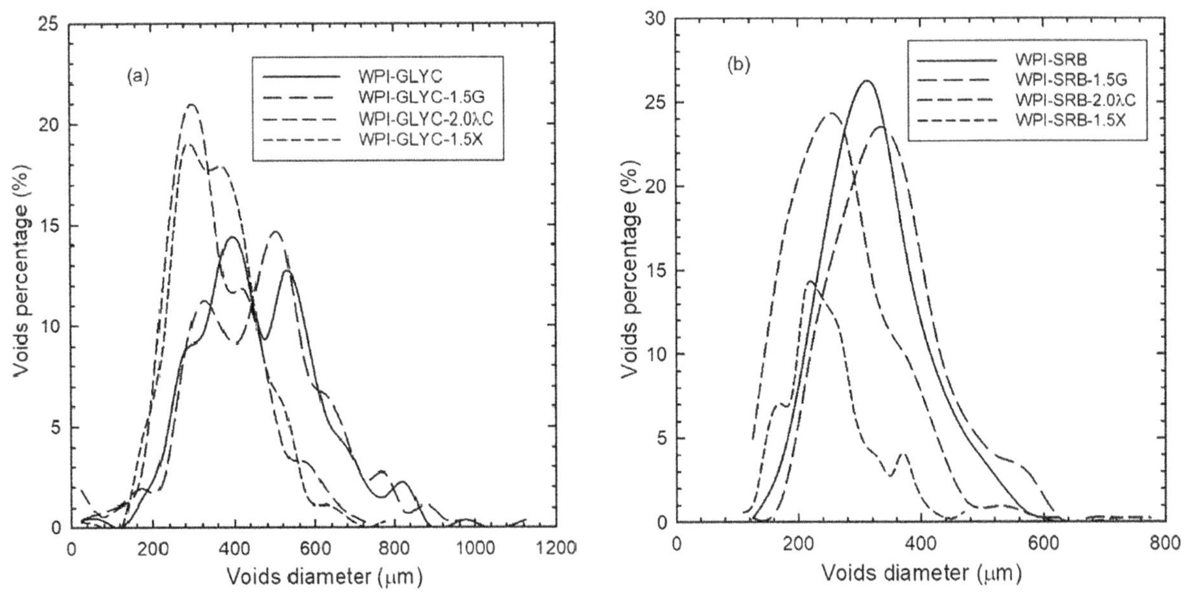

FIGURE 16.38 Pore frequency distribution: (a) WPI-70GLYC-polysaccharides, (b) WPI-120SRB-polysaccharides.

TABLE 16.27 Moisture and mechanical properties of solid foam.

Mixture	Moisture (g/100 g)	σ_{max} (Pa)	Young's modulus (Pa)
WPI-70GLYC	$6.03 \pm 0.84^{(a)}$	$93.3 \pm 13.0^{(e)}$	$1449 \pm 214.7^{(i)}$
WPI-70GLYC-1.5G	$7.05 \pm 0.59^{(a)}$	$126 \pm 17.0^{(f)}$	$2214 \pm 473.5^{(j)}$
WPI-70GLYC-2.0λC	$9.12 \pm 0.88^{(b)}$	$60.1 \pm 8.24^{(g)}$	$816.2 \pm 150.3^{(k)}$
WPI-70GLYC-1.5X	$11.0 \pm 0.72^{(c)}$	$38.8 \pm 5.57^{(h)}$	$529.8 \pm 104.7^{(m)}$
WPI-120SRB	$5.31 \pm 0.01^{(a)}$	$86.9 \pm 8.02^{(e)}$	$2128 \pm 346.1^{(j)}$
WPI-120SRB-1.5G	$4.25 \pm 0.26^{(d)}$	$97.2 \pm 12.6^{(e)}$	$2618 \pm 232.7^{(j)}$
WPI-120SRB-2.0λC	$5.09 \pm 0.28^{(a)}$	$118 \pm 15.6^{(f)}$	$2025 \pm 406.5^{(j)}$
WPI-120SRB-1.5X	$4.08 \pm 0.33^{(d)}$	$96.0 \pm 24.7^{(e)}$	$1649 \pm 484.8^{(j)}$

Different letters in the same column indicate statistically different values ($P < 0.05$).

In glycerol-containing foams, moisture content ranges from 6.03 to 11.0 g/100 g, but only those with carrageenan and xanthan show different moisture levels. In sorbitol-containing foams, the moisture range is 4.08–5.31 g/100 g; however, only those with guar and xanthan have humidity levels that differ from the rest, not among each other. Generally, foams with glycerol have higher moisture than those with sorbitol due to glycerol's hydrophilic nature. These differences are linked to the type of plasticizer, as they are visible even in mixtures with the same polysaccharide. Solid foams with glycerol display significantly different maximum stresses, with guar being more resistant and xanthan being weaker. Foams with sorbitol show a more uniform stress distribution, similar to those with glycerol. The Young's moduli of glycerol-based foams vary widely, with guar showing a higher modulus. Xanthan-based foams have the lowest modulus, indicating a more rigid structure with less deformability. Sorbitol-based foams are more flexible and homogeneous than those with glycerol.

The standard deviations are relatively high, showing that solid foams are heterogeneous materials. This heterogeneity is visible on the surface of the foams (Fig. 16.37, left) and inside (Fig. 16.37, right). As a result, when strains are small, the structure collapses because the cells compress until the material reaches its breaking point. Foams with sorbitol have higher maximum stress and Young's moduli, except for WPI-70GLYC-1.5G and WPI-120SRB-1.5G. Smaller pore sizes, as shown in Fig. 16.37, are observed in foams with sorbitol, resulting in better packing within the aerated structure, which helps resist rupture and enables these foams to withstand higher stresses. A dense, tightly packed structure better resists deformation of the original cell framework (Corradini and Peleg, 2008). These foams also show greater resistance to uniaxial deformation in the linear stress-strain zone, corresponding to higher Young's moduli. Adding polysaccharides reduces the resistance to deformation in this linear elastic zone, except for foams containing glycerol with carrageenan or xanthan. Additionally, foams with glycerol are softer than those with sorbitol. This softness indicates they contain more moisture, as lower hardness usually correlates with lower water content in the material.

5. Conclusions

In a WPI-polysaccharide solution, increasing the concentration of the same polysaccharide reduces the foam's overrun and improves its stability. This occurs due to higher apparent viscosity and the range of frequencies where elastic behavior dominates.

Adding a surfactant to a WPI solution decreases the overrun, likely because of competitive absorption between the surfactant and the protein. The overrun changes depending on the amount of surfactant added.

In the WPI-Surfactant-Polysaccharide system, the type of polysaccharide affects stability, flow, and rheological properties. The most stable foam forms with xanthan because of the viscoelasticity this polysaccharide provides, where the elastic and viscous behaviors are similar. The Gompertz-type equation offers excellent predictions of stability for all foams.

The nature of the surfactant does not change the viscoelastic and flow properties of the precursor solutions. However, it does influence these properties in the liquid foams. In the WPI-Surfactant-Polysaccharide system, the change in these properties is mainly caused by the polysaccharide.

Using a plasticizer instead of a surfactant increases the apparent viscosity of the precursor solutions, their foams, and the system's stability while keeping the same polysaccharide. Foams with the same polysaccharide but different plasticizers can be considered equivalent.

Solid foams with sorbitol show less variation in pore size, and adding λC and xanthan encourages smaller pores and reduces disparity. The stability of the liquid foam also helps form smaller pores and minimizes disparity.

Foams containing sorbitol show higher maximum stress and Young's moduli compared to those with glycerol. The foam with glycerol and guar has the highest maximum stress and Young's modulus. The foams with the greatest maximum stress and Young's moduli are those with smaller pore sizes and lower disproportion.

The mechanical response to stress in solid foams does not depend on the viscoelastic behavior of their precursor liquid foams because the foam pairs contain the same polysaccharide but different plasticizers. Similar behaviors are observed in the frequency range, but the maximum stress and Young's modulus differ in magnitude.

Foams containing glycerol show a higher water swelling capacity and a slower breakdown of their structure when hydrated. WPI-70GLYC-1.5X, WPI-GLYC-2.0λC, and WPI-120SRB-1.5X foams dissolve faster, while foams with guar dissolve more slowly.

References

Amid, B.T., Mirhosseini, H., Poorazrang, H., Mortazavi, S.A., 2013. Implications of partial conjugation of whey protein isolate to durian seed gum through Maillard reactions: foaming properties, water holding capacity, and interfacial activity. Molecules 18, 15110–15125. https://doi.org/10.3390/molecules181215110.

Audebert, A., Beaufils, S., Lechevalier, V., Le Floch-Fouéré, C., Saint-Jalmes, A., Pezennec, S., 2019. How foam stability against drainage is affected by the conditions of prior whey protein powder storage and dry heating. A multidimensional experimental approach. J. Food Eng. 242, 153–162. https://doi.org/10.1016/j.foodeng.2018.08.029.

Chen, H.-B., Wang, Y.-Z., Schiraldi, D.A., 2013. Foam-like materials based on whey protein isolate. Eur. Polym. J. 49, 3387–3391. https://doi.org/10.1016/j.europolymj.2013.07.019.

Chrysina, E.D., Brew, K., Acharya, K.R., 2000. Crystal structures of Apo- and holo-bovine α-lactalbumin at 2.2-Å resolution reveal an effect of calcium on inter-lobe interactions. J. Biol. Chem. 275 (47), 37021–37029. https://doi.org/10.1074/jbc.M004752200.

Cohen-Addad, S., Höhler, R., 2014. Rheology of foams and highly concentrated solutions. Curr. Opin. Colloid Interface Sci. 19, 536–548. https://doi.org/10.1016/j.cocis.2014.11.003.

Corradini, M.G., Peleg, M., 2008. Solid food foams. In: Aguilera, J.M., Lillford, P.J. (Eds.), Food Materials Science: Principles and Practice. Springer, New York. https://doi.org/10.1007/978-0-387-71947-4.

Damodaran, S., 2008. Amino acids, peptides, and proteins. In: Fennema, O.R., Parkin, L., Damodaran, S. (Eds.), Food Chemistry, fourth ed. CRC Press, USA, pp. 217–331.

Deeth, H., Bansal, N., 2019. Whey proteins: an overview. In: Deeth, H.C., Bansal, N. (Eds.), Whey Proteins. From Milk to Medicine. Academic Press, Elsevier, UK, pp. 1–50. https://doi.org/10.1016/B978-0-12-812124-5.00001-1.

Devilbiss, E.D., Holsinger, V.H., Posati, L.P., Pallansch, M.J., 1975. Properties of whey protein concentrate foams. Food Technol. 28, 40–48.

Edwards, P.J.B., Jameson, G.B., 2020. Structure and stability of whey proteins. In: Boland, M., Singh, H. (Eds.), Milk Proteins: From Expression to Food, third ed. Academic Press, Elsevier, UK, pp. 251–291. https://doi.org/10.1016/B978-0-12-815251-5.00007-4.

García-García, E.M., Ramírez-Gilly, M., Tecante, A., 2015. Formation, stability, flow and viscoelastic properties of foams prepared with whey protein isolate and its individual mixtures with xanthan, guar and refined sugar. In: Wyatt, M. (Ed.), Whey Proteins. Nova Science Publishers, USA, pp. 117–147.

Kilara, A., Vaghela, M.N., 2018. Whey proteins. In: Yada, R.K. (Ed.), Proteins in Food Processing, second ed. Woodhead Publishing, Elsevier, UK, pp. 93–126. https://doi.org/10.1016/B978-0-08-100722-8.00005-X.

Lajnaf, R., Picart-Palmade, L., Cases, E., Attia, H., Marchesseau, S., Ayadi, M.A., 2018. The foaming properties of camel and bovine whey: the impact of pH and heat treatment. Food Chem. 240, 295–303. https://doi.org/10.1016/foodchem.2017.07.064.

Lazidis, A., Hancocks, R.D., Spyropoulos, F., Kreuß, M., Berrocal, R., Norton, I.T., 2016. Whey protein fluid gels for the stabilization of foams. Food Hydrocoll. 53, 209–217. https://doi.org/10.1016/foodhyd.2015.02.022.

Lexis, M., Willenbacher, N., 2014. Yield stress and elasticity of aqueous foams from protein and surfactant solution. The role of continuous phase viscosity and interfacial properties. Colloids Surf. A Physicochem. Eng. Asp. 459, 177–185. https://doi.org/10.1016/j.colsurfa.2014.06.030.

Liszka-Skoczylas, M., Ptaszek, A., Zmudzinski, D., 2014. The effect of hydrocolloids on producing stable foams based on whey protein concentrate (WPC). J. Food Eng. 129, 1–11. https://doi.org/10.1016//j.foodeng.2014.01.002.

Martinez-Padilla, L.P., García-Rivera, J.L., Romero-Arreola, V., Casas-Alencáster, N.B., 2015. Effects of xanthan gum rheology on the foaming properties of whey protein concentrate. J. Food Eng. 156, 22–30. https://doi.org/10.1016//j.foodeng.2015.01.018.

Morr, C.V., Foegeding, E.A., 1990. Composition and functionality of commercial whey and milk protein concentrates: a status report. Food Technol. 44 (4), 100–112.

Narchi, I., Vial, Ch., Djelveh, G., 2009. Effect of protein-polysaccharide mixtures on the continuous manufacturing of foamed food products. Food Hydrocoll. 23, 188–201. https://doi.org/10.1016/j.foodhyd.2007.12.010.

Nastaj, M., Solowiej, B.G., 2020. Effect of various pH values on foaming properties of whey protein preparations. Int. J. Dairy Technol. 73 (4), 683–694. https://doi.org/10.1111/1471-0307.12705.

Norwood, E.-A., Le Floch-Fouéré, C., Briar-Bion, V., Schuck, P., Croguennec, T., Jeante, R., 2016. Structural markers of the evolution of whey protein isolate powder during aging and effects on foaming properties. J. Dairy Sci. 99, 5265–5272. https://doi.org/10.3168/jds.2015-10788.

Oboroceanu, D., Wang, L., Magner, E., Auty, M.A.E., 2014. Fibrillization of whey protein improves foaming capacity and foam stability at low protein concentrations. J. Food Eng. 121, 102–111. https://doi.org/10.1016//j.foodeng.2013.08.023.

Oduse, K., Campbell, L., Longchamp, J., Euston, S.R., 2017. Electrostatic complexes of whey protein and pectin as foaming and emulsifying agents. Int. J. Food Prop. 20 (Suppl. 3), S3027–S3041. https://doi.org/10.1080/10942912.2017.1396478.

Ortiz-Tafoya, M.C., Tecante, A., 2018. Physicochemical characterization of sodium stearoyl lactylate(SSL), polyoxyethylene sorbitan monolaurate (Tween20) and κ-carrageenan. Data Brief 19, 642–650. https://doi.org/10.1016/j.dib.2018.05.064.

Osorio, J., Monjes, J., Pinto, M., Ramírez, C., Simpson, R., Vega, O., 2014. Effects of spray drying conditions and the addition of surfactants on the foaming properties of a whey protein concentrate. LWT Food Sci. Technol. 58, 109–115. https://doi.org/10.1016/j.lwt.2014.02.016.

Perez, A.A., Carrara, C.R., Carrera-Sanchez, C., Santiago, L.G., Rodriguez-Patino, J.M., 2010. Interfacial and foaming characteristics of milk whey protein and polysaccharide mixed systems. AIChE J. 56 (4), 1107–1117. https://doi.org/10.1002/aic.

Ramos, O.L., Pereira, R.N., Rodrigues, R.M., Teixeira, J.A., Vicente, A.A., Malcata, F.X., 2016. Whey and whey powders: production and uses. In: Caballero, B., Finglas, P.M., Toldrá, F. (Eds.), Encyclopedia of Food and Health. Elsevier Inc, pp. 498–505. https://doi.org/10.1016/b978-0-12-384947-2.00747-9.

Tan, M.C., Chin, N.L., Yusof, Y.A., Taip, F.S., Abdullah, J., 2015. Characterization of improved foam aeration and rheological properties of ultrasonically treated whey protein suspension. Int. Dairy J. 43, 7–14. https://doi.org/10.1016/j.dairyj.2014.09.013.

Tan, M.C., Chin, N.L., Yusof, Y.A., Abdullah, J., 2016. Effect of high power ultrasonic treatment on whey protein foaming quality. Int. J. Food Sci. Technol. 51, 617–624. https://doi.org/10.1111/ijfs.13013.

Vidotto, D.C., Tavares, G.M., 2020. Impact of dry heating in an alkaline environment on the structure and foaming properties of whey proteins. Food Bioprocess Technol. 13, 1755–1764. https://doi.org/10.1007/s11947-020-02519-5.

Wang, Z., Zhang, S., Vardhanabhuti, B., 2015. Foaming properties of whey protein isolate and λ-carrageenan mixed systems. J. Food Sci. 80 (8), N1893–N1902. https://doi.org/10.1111/1750-3841.12940.

Wu, Y., Iglauer, S., Shuler, P., Tang, Y., Goddard, W.A., 2010. Alkylpolyglucoside surfactant-sorbitan ester cosurfactant formulations for improved oil recovery. Tenside Surfactants Deterg. 47 (5), 280–287. https://doi.org/10.3139/113.110078.

Zhu, R., Griffith Jones, O., 2024. Effect of high acyl gellan gum and pH on the structural and foaming properties of heated whey protein suspensions. Food Chem. 449, 139955. https://doi.org/10.1016/foodchem.2024.139255.

Zwietering, M.H., Jongenburger, I., Rombouts, F.M., van't Riet, K., 1990. Modeling of the bacterial growth curve. Appl. Environ. Microbiol. 56 (6), 1875–1881.

Chapter 17

Digestion of whey and its products

Juan Camilo Charris-Barreto[1], María Ximena Quintanilla-Carvajal[1] and Bhavbhuti M. Mehta[2]
[1]*Universidad de La Sabana, Facultad de Ingeniería, Bogotá, Colombia;* [2]*Department of Dairy Chemistry, SMC College of Dairy Science, Kamdhenu University, Anand, Gujarat, India*

1. Introduction

Digestion is a complex biological process. It involves a series of interdependent stages that allow the release of nutrients from the food matrix for eventual absorption and transport via the circulatory system to the various tissues and organs where they will be used, either for growth and maintenance or as a source of energy. This absorption is assessed under the concept of bioavailability, which is the fraction of an ingested component that is available for utilization at the site of action in normal physiological function, and is comprised by two subconcepts: Bioaccessibility, defined as the fraction of a nutrient that is released from the food matrix within the gastrointestinal tract and is therefore available for absorption through the intestinal walls; and bioactivity, which includes the events from the absorption of the bioactive, its transport via the circulation to the site of action and its interaction with biomolecules, the transformations it may undergo, and the physiological responses that may be induced (Sandoval-Cuella and Quintanilla-Carvajal, 2021).

Previously considered as a waste-stream from cheese production, whey protein (the second big set of proteins in bovine milk, ~20%; ~80% for casein) is now widely studied for its multiple health benefits given for its richness in both sulphur-containing and branched-chain essential amino acids. Its benefits include: Beneficial effect on body recovery and performance, and on lean mass maintenance during physical exercise; stronger satiety effect, appetite suppression, and mood control; vascular function improvement, glucose tolerance, lipid metabolism, and regulation of adipokines, among others (Boscaini et al., 2023; Cava et al., 2024).

Considering the above, focusing on the digestion of whey is a key factor to be enlightened to generate viable solutions, advocate for its correct use, and optimize its effectiveness.

2. Structure and composition of whey and its products and its relation with digestion

2.1 Major whey proteins

- β-lactoglobulin (18.4 kDa): Being the most important whey protein, it possesses a globular structure that consists of 162 amino acid residues in its primary structure that is stable against acids and enzymes in the stomach (Boscaini et al., 2023). Plays a crucial role in the overall nutritional value of whey due to its abundance and high content of branched-chain amino acids (Tang et al., 2025). It is responsible for the mitigation of hypocholesterolemia, antiviral activity, and the suppression of pathogen adhesion. Nonetheless, it is the main component of cow's milk that causes milk allergy (Pillai et al., 2024). It also possesses excellent emulsifying and foaming properties (Delgado-Macuil et al., 2025). Its 3D structure is presented in Fig. 17.1.
- α-lactalbumin (14.4 kDa): As the second most prevalent protein in whey, present in both human's and cow's milk, this globular structured protein of 123 amino acid residues serves as a rich source of branched-chain essential amino acids like Tryptophan and Cysteine, which are building blocks for serotonin, hormone that regulates the mood (Cava et al., 2024; Pillai et al., 2024). Also, it possesses excellent emulsifying and foaming properties (Delgado-Macuil et al., 2025), plays a significant role in lactose synthesis in mammary glands (Tang et al., 2025), and protects against gastric mucosal injuries (Boscaini et al., 2023). Its crystal structure is shown in Fig. 17.2.

Whey. https://doi.org/10.1016/B978-0-443-27648-4.00008-6

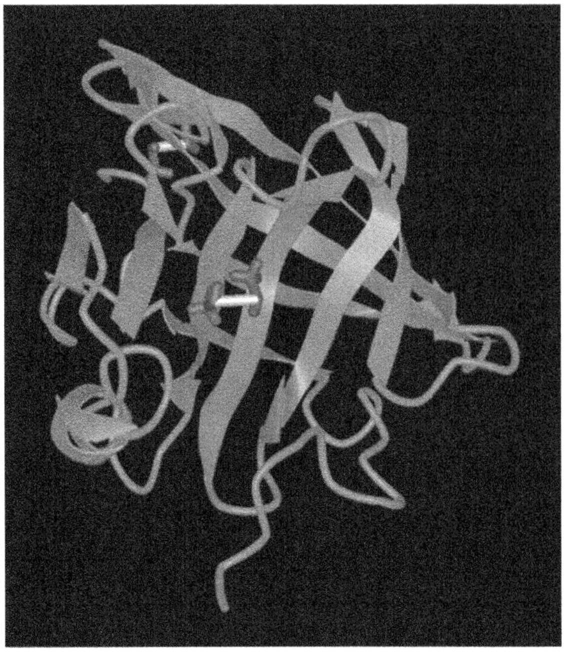

FIGURE 17.1 3D structure of bovine β-lactoglobulin A (Kuwata et al., 1999).

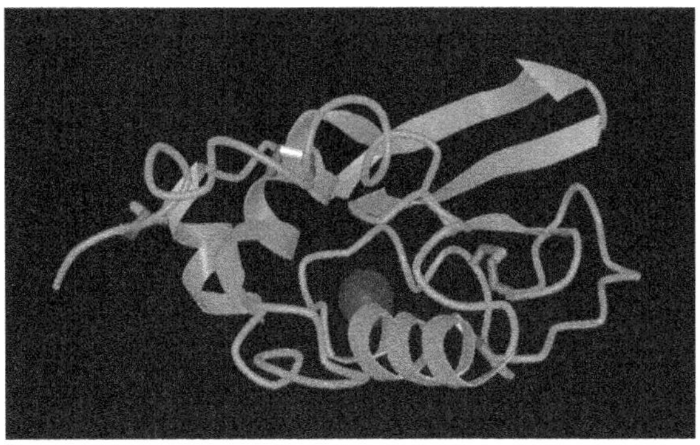

FIGURE 17.2 Crystal structure of bovine α-lactalbumin (Chrysina et al., 2000).

- Immunoglobulins (Heavy chain—55 kDa, Light chain—22 kDa): These antibodies, which are postnatal transferred via colostrum, provide passive immunity and potentially contribute to the immunomodulatory effects associated with whey protein consumption (Madureira et al., 2007). The majority of immunoglobulins are made up of a total of four polypeptide chains, with each one containing two identical heavy chains joined together by covalent and noncovalent disulfide bridges (Pillai et al., 2024). An example of a human immunoglobulin is illustrated in Fig. 17.3.
- Bovine serum albumin (66.5 kDa): This 582 amino acid residues protein that forms part of cell inner and outer membranes is known for its ability to bind and transport various molecules including fatty acids, hormones, and minerals (Madureira et al., 2007); contributes to whey's overall nutritional profile and may enhance the bioavailability of certain nutrients (Tang et al., 2025). Its 3D structure is illustrated in Fig. 17.4.
- Lactoferrin (80 kDa): This single-chain iron-binding protein is a member of the transferrin protein family, which comprises two homologous lobes, each of which can bind one ferric ion. It is produced by mucosal epithelial cells in several mammalian species and is present in tears, saliva, and gastrointestinal tract fluids. Its properties include stimulation of beneficial intestinal bacteria, antiviral, immunomodulatory, anticancer, antiinflammatory and antibacterial activities (effective against both gram-positive and gram-negative), exerting both bacteriostatic (sequestrating iron away from the pathogen, limiting its use and thus, down-regulating the expression of virulence factors) and bactericidal (direct interaction with bacterial surface proteins) activity (Boscaini et al., 2023). Its crystal structure is presented in Fig. 17.5.

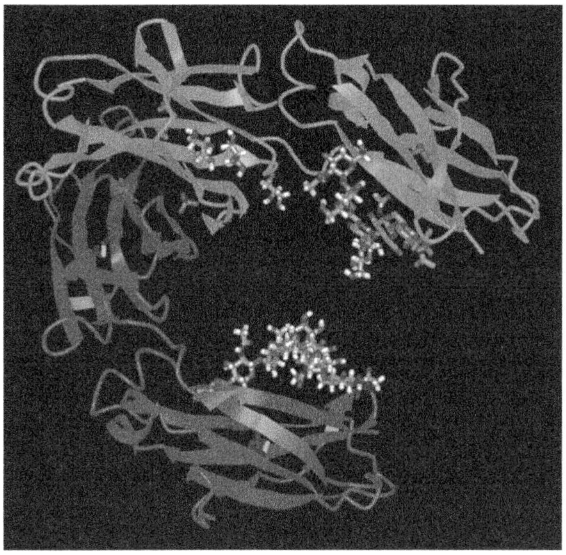

FIGURE 17.3 Crystal structure of human immunoglobulin G1-crystallizable fragment (Lobner et al., 2017).

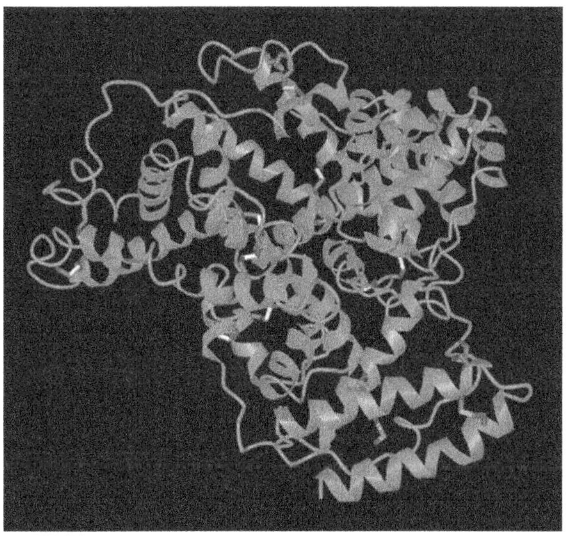

FIGURE 17.4 Bovine serum albumin crystal structure (Bujacz, 2012).

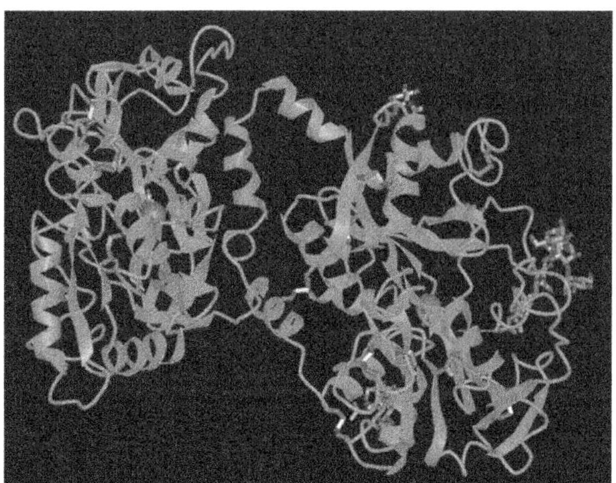

FIGURE 17.5 Differic bovine lactoferrin 3D structure (Moore et al., 1997).

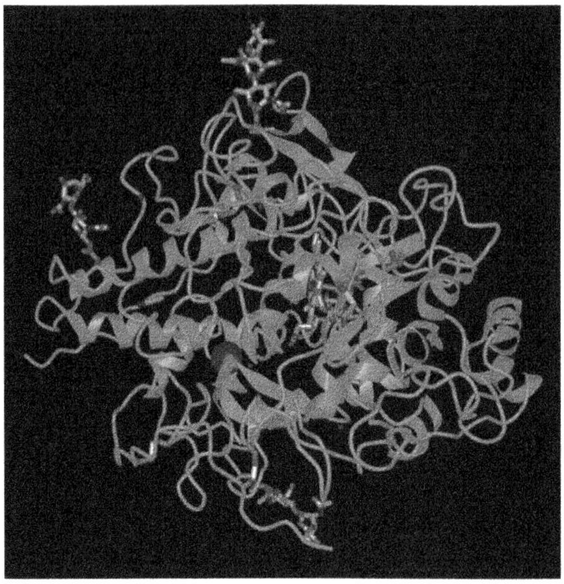

FIGURE 17.6 Crystal structure of bovine lactoperoxidase (Singh et al., 2010).

- Lactoperoxidase (78 kDa): It is a heme-containing chain glycoprotein enzyme that plays an important role in oral health and binds a covalently linked heme prosthetic group, a derivative of protoporphyrin IX, in its catalytic center. Its antibacterial activity (against both gram positive and gram negative) relies on the presence of both the thiocyanate anion and hydrogen peroxide, which reaction is catalyzed by this enzyme to form short-life intermediary compounds (hypothiocyanite and hypothiocyanous acid) which are thought to invoke its antibacterial activity by glycolysis (and other essential metabolic pathways) inhibition through oxidation of sulfhydryl groups of bacterial proteins (Boscaini et al., 2023). Its 3D structure is shown in Fig. 17.6.

2.2 Types of whey products

The whey protein concentrate (WPC) is the most common and economical form of whey obtained by 1) ultrafiltration of liquid whey to remove water, lactose and minerals, 2) diafiltration with water to wash out residual nonprotein components, and finally, and 3) spray-drying the washed concentrate to obtain the powder form (35%–55% w/w protein content). Nonetheless, this type is limited to nonlactose-intolerant consumers but is widely used to isolate β-lactoglobulin and α-lactalbumin for further analysis. However, this concentrate can go another step further in its refining through ion exchange chromatography or membrane filtration to obtain a whey protein isolate (WPI) with >90% w/w protein contents, making it a choice for lactose-intolerant or low-carbohydrate diet consumers (Madureira et al., 2007; Delgado-Macuil et al., 2025).

Following the previous steps, whey protein can be upgraded into a hydrolyzate (Whey Protein Hydrolyzate—WPH) by (1) breaking down the WPH using proteolytic enzymes, (2) heat-deactivate the enzymes used, (3) filtrate the matrix to retain intact proteins, and (4) spray-drying it to produce the final powder. This is to enhance digestibility and absorption rates in the gastrointestinal tract but may bring a bitter taste to the product (Tang et al., 2025).

Nevertheless, along with their typical use, other applications stand out for their low-cost and sustainable nature since it is an abundant by-product which eases the scalability of new processes. For instance, whey fermentation in pH-controlled and microencapsulated cultures to produce lactic acid, polysaccharides, poly-hydroxyalkanoates, polyhydroxybutyrate, bioethanol, butanol, algae to produce antioxidant compounds, natural pigments, or absorb excessive nutrients to reduce the environmental impact of dairy waste, and microalgae for bioremediation. All this by using either natural or metabolic-engineered strains (Delgado-Macuil et al., 2025).

2.3 Physicochemical properties relevant to digestion

When it comes to digestion, physicochemical properties such as solubility, heat stability, and aggregation behavior play a key role in the quality of functional food. For whey proteins, pH-dependent behavior significantly influences their solubility, stability, and interactions with each other. Here is where the isoelectric point of each protein starts the scene, since we can use techniques such as acylation, phosphorylation, succinylation, or glycation, to manipulate the molecules and increase or decrease effects according to the desired application (i.e., whey proteins typically exhibit negative zeta potentials at ~7 pH, contributing to their stability in solution through electrostatic repulsion). Also, even though heat treatments cause denaturation and hence a reduction in functionality, they can also expose hidden reactive groups, potentially enhancing digestibility and the proteins' ability to form gels or emulsions, but cautiousness is mandatory because high temperatures can cause Millard Reactions, spoiling the product quality (Tang et al., 2025).

3. Mechanisms of whey protein digestion

3.1 Oral, gastric, intestinal, and liver phases

Digestion process starts immediately in the mouth, where mechanical breakdown (chewing/mastication) is done by the teeth to make large pieces of food into smaller ones that are covered with saliva to ease its passing through the esophagus (swallowing) toward the stomach through the esophageal sphincter, where parallelly in the gastric mucosa, the presence of food leads pepsinogen to be secreted by the chief cells and HCl by parietal cells, which activates the pepsin (endopeptidase, active form of pepsinogen) to initiate the chemical digestion of proteins helped by peristalsis (muscular contractions of the stomach), which converts the partially digested protein into the chyme (more uniformed mixture), where the protein that was not broken down during cooking or chewing denatures by acid and mechanical stresses induced, unfolding the 3D structure which gets rid of the protein functionality to reveal just the polypeptide chain, where the pepsin begins hydrolyzing the peptide bonds (proteolysis) between hydrophobic amino acids, especially those involving aromatic residues, creating shorter polypeptides (Sorensen, 2025; Cherian, 2019; Callahan et al., 2020).

Then, the chyme is emptied by the stomach into the small intestine, where the great part of protein digestion occurs: The hormone secretin, in the duodenum, stimulates the secretion of trypsinogen, chymotrypsinogen, and procarboxypeptidase (inactive forms) from the pancreas. Trypsinogen is then converted into its active form, trypsin, by the enterokinase (secreted at the duodenum), which selectively cleaves bonds containing Lys and Arg, activating the other two proteases-chymotrypsin (selectively cleaves large hydrophobic residues in the carboxyl side) and carboxypeptidase (selectively cleaves C-terminal peptide or ester bond), respectively (Cherian, 2019). These enzymes break proteins down to tripeptides, dipeptides and single amino acids that then, helped by muscle contractions of the small intestine, are mixed and propelled through the lumen to the absorption sites in the lower parts of the small intestine entering the enterocytes (intestinal cells) using active transport (requires ATP) where are completely broken by cellular machinery and absorbed into the bloodstream (Sorensen, 2025). Some amino acids may compete against each other for the absorption site because specific carrier proteins based on the nature of the amino acid (e.g., neutral, basic, acid, large, small) are involved in their transport (Cherian, 2019).

Once in the blood, amino acids are transported to the liver (the "checkpoint" for distribution or further breakdown, if necessary) along with other macronutrients, where they become part of the body's amino acid pool and can be sent to different metabolic pathways depending on the need and availability of energy (Callahan et al., 2020).

Considering that amino acids contain N, their further catabolism releases ammonia, which, since it is toxic, is converted into urea by the liver and excreted in the urine thanks to its high solubility in water, which makes it an appropriate choice for transporting excess nitrogen out of the body (Sorensen, 2025).

Enough energy	Not enough energy
❖ Protein synthesis in cells around the body occurs as either essential or nonessential amino acids. ❖ Making other nitrogen compounds. ❖ Rearranged and stored as fat.	❖ Rearranged into glucose for fuel for the brain and erythrocytes. ❖ Metabolized as fuel for an immediate source of ATP.

3.2 Factors affecting digestibility

The digestibility of whey protein is a central factor determining its nutritional value and functional efficacy in food systems. Understanding the factors that affect whey protein digestibility is crucial not only for optimizing the nutritional outcomes for consumers but also for innovating processing methods and product formulations. Digestibility is influenced by the intrinsic properties of the protein, the methods employed during processing, as well as the dietary matrix within

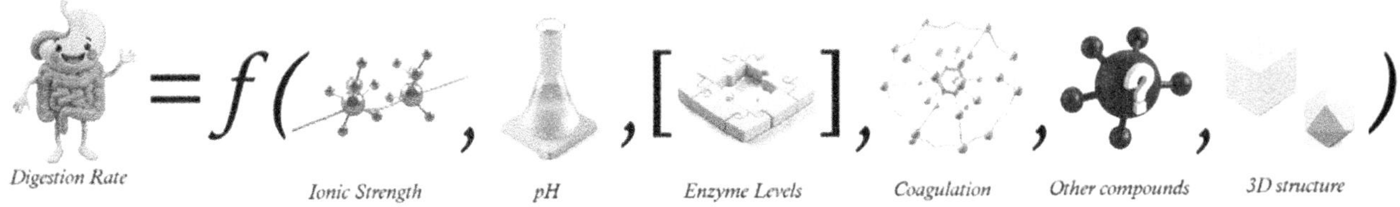

FIGURE 17.7 Factors affecting whey protein digestion.

which the protein is consumed. This section delves into the scientific principles governing whey protein digestibility, addressing how protein structure, processing techniques, enzymatic activity, and interactions with other dietary components play interdependent roles in modulating its bioavailability (Fig. 17.7).

Whey proteins, such as β-lactoglobulin, α-lactalbumin, and bovine serum albumin, possess inherently globular conformations stabilized by disulfide bonds and hydrophobic interactions. These native conformations are key determinants of how susceptible the proteins are to proteolytic enzymes in the gastrointestinal tract. The compact tertiary and quaternary structures of whey proteins facilitate precise enzyme-substrate interactions under physiological conditions. Enzymes, including pepsin in the stomach and pancreatic trypsin and chymotrypsin in the small intestine, preferentially cleave specific peptide bonds exposed on the protein surface. However, the folding state can either expose or shield these sites, influencing the rate of digestion. Native whey proteins typically present a readily digestible structure, enabling efficient enzymatic hydrolysis. Thus, for whey proteins, a balance between protein unfolding and steric hindrance, depending on the precise processing conditions, will result in an increase or a decrease in digestibility (Sabari et al., 2023).

3.2.1 Protein aggregation

Processing fields such as heating, pH adjustment, or mechanical shear can induce aggregation or crosslinking of whey proteins. These aggregations form gel-like networks with limited porosity, thus restricting the access of digestive enzymes. Crosslinking, whether enzymatic or chemical, results in more robust gels that slow proteolysis owing to the "one-by-one" enzymatic mechanism and reduced surface area relative to volume. Such physical barriers lower the rate and extent of digestion compared to dispersed native proteins (Liang et al., 2025).

3.2.2 Denaturation and hydrolysis

Contrastingly, controlled protein denaturation by heat, pressure, or enzymatic predigestion unfolds protein tertiary structures. These unfolding exposes previously hidden cleavage sites, significantly improving enzymatic access and hydrolysis rates. Protein hydrolysates derived from whey often exhibit superior digestibility and faster absorption kinetics, advantageous for clinical and sport nutrition applications (Zhang et al., 2022).

3.2.3 Processing techniques

The methods employed during manufacturing greatly affect whey protein structure and digestibility as follows:

Thermal Processing: Heat treatment modifies whey protein structure through unfolding, aggregation, and Maillard reaction initiation. Moderate heating at acidic or neutral pH promotes partial unfolding that enhances digestibility by exposing protease cleavage sites. Excessive heating, particularly at alkaline pH, leads to irreversible complex modifications, including amino acid degradation and crosslinking that diminish enzyme accessibility. Heat-induced unfolding must be carefully balanced to optimize digestibility without compromising protein integrity (Accardo et al., 2022).

pH and Ionic Strength Effects: The pH and salt concentrations during processing influence protein solubility and aggregation behavior. Acidic conditions tend to produce looser, more digestible protein aggregates, while neutral to alkaline conditions promote denser, more heat-stable aggregates that resist enzymatic breakdown. Adjusting these parameters allows modulation of digestive kinetics and protein bioavailability in final products (Accardo et al., 2022). For instance, Kaspchak et al. (2018) studied the effect of heating and ionic strength on the interaction of bovine serum albumin and the antinutrients tannic and phytic acids, and their influence on in vitro protein digestibility, finding that high ionic strength increases affinity between tannins and the serum protein due to exposure of hydrophobic protein groups, decreasing protein digestibility. Likewise, phytic acid and bovine serum albumin binding occurs due to electrostatic interactions; thus, as ionic strength rises, interaction decreases, and protein digestibility increases. Also, De Wit and van Kessel (1996) have reported that ionic strength, in different measures, helps to solubilize protein products near their isoionic point.

Dehydration and Storage: Powdered whey proteins may be subjected to spray drying or freeze drying. These processes impact particle morphology and protein conformation. Improper drying or long-term storage may induce oxidative damage, aggregation, or alterations in secondary structure that negatively affect digestibility. As was mentioned above, thermal treatment, such as spray drying, causes denaturation of whey globular proteins, leading to the unfolding of their tertiary structures and disruption of hydrophobic and covalent interactions within the protein. This structural alteration increases the accessibility of amino acids, making them more available for utilization by probiotics (Bauer Estrada et al., 2022).

3.2.4 Interaction with other dietary components

Whey protein consumed within complex food matrices interacts with various biochemical constituents that influence digestion. Some examples are:

Polyphenols and Flavonoids: Dietary flavonoids and polyphenols may interact with whey proteins, forming protein-polyphenol complexes and influencing proteolytic enzyme activity. Such interactions can alter protein conformation, reduce enzyme accessibility, or inhibit enzymes like trypsin and pepsin, impairing digestibility. Contradictory findings exist whereby some studies report negligible effects on the hydrolysis kinetics of whey proteins in the presence of catechin-rich compounds, suggesting that these interactions are complex and matrix-dependent. Flavonoids can bind both the protein substrates in food matrices and the proteolytic enzymes involved in the digestive process, which may alter the bioavailability of the proteins (Cirkovic Velickovic and Stanic-Vucinic, 2018). It has been reported that flavonoids can affect trypsin, chymotrypsin, pepsin, and pancreatin, reducing protein digestibility (Świeca et al., 2013). In contrast, when whey protein is mixed with catechin simulating a chocolate milk matrix, it was found that although its presence does affect the whey protein's secondary structure by slightly increasing the amount of α-helix and β-sheets, it does not significantly affect ($P > .05$) the amount of solubilized protein, the degree of hydrolysis, nor the peptides profile (Accardo et al., 2023).

Despite the need of further research in the effect of flavonoids presence in whey due to the contradictory information, the interaction mechanisms of this molecules with WPI where studied by Li et al. (2021), showing that the binding process of flavonoids with WPI proteins is a spontaneous process mainly driven by hydrophobic force and is favored by higher temperatures and concentrations of the flavonoid, concluding that the skeleton of WPI is not affected during the interaction, but the microenvironmental hydrophobicity around tryptophan residues may change; that flavonoids have better binding affinity with α-lactalbumin than β-lactoglobulin, and that all flavonoids enter in the same hydrophobic cavity of both proteins, locating in the binding site by binding with similar amino residues. However, their molecular orientations in the same cavity were different, which may be due to the torsion of the 3D structure of flavonoids during the interaction.

Considering the presence of casein, the other important set of proteins present in milk, the κ-casein-whey protein complex shows better digestibility by trypsin, pepsin, and chymotrypsin than that without the complex. However, Sánchez-Rivera et al. (2015) found resistance of casein to gastric hydrolysis because of aggregates formed between whey protein and caseins in heated skimmed milk. Similarly, Ye et al. (2016) also studied the formation and breakdown of structured clots from heated whole milk during gastric digestion. They have reported that the denatured whey can associate with casein micelles and fat globules. This leads to reduced casein-casein and casein-fat globule interactions, leading to a looser structure (Guo et al., 2020).

Reducing Sugars and Maillard Products: The presence of reducing sugars in whey formulations can initiate Maillard reactions under heat, producing conjugates that modify protein structure. The shorter the carbonic chain of the sugar, the greater the number of open chain forms, and the greater the sugar's reactivity to the amino groups of protein, modulating digestibility by either exposing hidden cleavage sites or inducing steric hindrance to enzyme action (Sabari et al., 2023). Optimizing conditions to mitigate adverse Maillard reaction effects is essential for preserving digestibility.

Lipids and Emulsion Interfaces: Whey proteins frequently act as emulsifiers, binding to lipid droplets in dairy and beverage systems. The structural rearrangement at the oil-water interface and subsequent interactions with digestive emulsifiers like bile salts influence digestion rates. Emulsified protein systems can undergo variable gastric breakdown rates contingent on gel strength and particle size, affecting overall protein bioavailability. Lipids serve as a vital energy source during prolonged endurance exercise, and WPI is often used in food emulsions to influence lipid digestibility. Recent research has shown that stabilizing emulsions with short WPI fibrils—created by controlled heating and shear treatment—enhances both the stability and digestion of lipids. These short fibrils, smaller and more hydrophobic than native proteins or long fibrils, form emulsions with excellent dispersibility during digestion. This structural modification improves in vitro lipid digestion rates and accelerates lipid absorption in vivo. In the context of whey protein digestion, these findings highlight how the protein's interaction with lipid droplets at emulsion interfaces can significantly affect overall lipid digestibility and absorption. By tailoring whey protein fibril structure around oil droplets, it is possible to optimize lipid digestion and enhance nutrient bioavailability, especially in functional foods designed for the energy demands of endurance exercise (Leng et al., 2025a).

Fibers and Food Matrix Complexity: Dietary fibers and other macronutrients present in complex meals can delay gastric emptying, alter enzyme diffusion rates, or sequester proteins, modifying the bioaccessibility and digestion of whey proteins. The holistic food matrix must be considered during formulation to ensure favorable digestibility profiles. The incorporation of dietary fibers, such as fructooligosaccharides and xylooligosaccharides, into whey protein formulations can significantly influence whey digestion through the formation of Maillard conjugates. These conjugates, created by binding WPC to fibers via the Maillard reaction, act as effective emulsifiers that impact the physical characteristics of emulsions, including larger oil droplet sizes and high encapsulation efficiency. The structural modifications from fiber-protein conjugation affect the stability and digestion of whey-based emulsions, with fiber type influencing storage stability. This highlights how dietary fibers and complex food matrices modulate whey protein digestibility by altering its structural interactions and matrix behavior, ultimately affecting the bioaccessibility and release of nutrients during digestion (Urango et al., 2024).

3.2.5 Role of digestive enzymes and their modulators

Enzyme Concentrations and Activity Levels: Digestibility is intrinsically linked to the availability and activity of digestive proteases, which can vary physiologically or pathologically. Enzymatic capacity can be a limiting factor, especially in aging populations or those with compromised gastrointestinal function. Melchior et al. (2023) adapted an in vitro digestion model simulating elderly gastrointestinal conditions—adjusting pH, enzyme and bile salt concentrations, and digestive phase durations—to assess protein digestibility from various sources, including whey. Compared to an adult digestion model, the elderly simulation showed a marked reduction in proteolysis, with whey protein digestibility decreasing by about 20%. This highlights how age-related changes in digestive enzyme levels and gastrointestinal environment can significantly impair whey protein hydrolysis and absorption. Understanding these enzyme-related modulators in elderly digestion is crucial for developing whey-based nutritional products tailored to support optimal protein utilization in aging populations.

Enzyme Inhibitors in Food Systems: As was mentioned above, certain components, such as polyphenols, possess inhibitory effects on protease enzymes. These interactions reduce hydrolytic rates and must be accounted for when optimizing formulations for protein bioavailability. In the same way, it is possible to accomplish the enhancement by exogenous enzymes in whey digestion; supplementing with exogenous proteases has been shown to enhance whey protein breakdown. Such strategies are implemented in clinical nutrition to overcome limitations of endogenous enzyme insufficiency (Matthews et al., 2025).

3.2.6 Physical characteristics and food structure effects

The macro- and microstructural organization of whey protein in foods affects digestion kinetics:

Gel Formation: Whey protein gels vary in hardness and network density. Softer gels disintegrate rapidly during gastric digestion, whereas dense, particulate gels afford physical resistance that slows enzyme activity and peptide release, but gels can overall release a more diverse peptide profile thanks to a more complete digestion (Leng et al., 2025b; Luo et al., 2021) demonstrated that the structure of heat-set whey protein emulsion gels significantly affects whey digestion and nutrient bioaccessibility. Hard gels formed under high ionic strength (200 mM NaCl) had a compact, particulate network, resulting in slower gel disintegration and reduced protein hydrolysis during gastric digestion compared to softer gels formed at low ionic strength (10 mM NaCl). The soft gels showed earlier oil droplet coalescence and faster digestion, whereas hard gels maintained gel integrity longer, delaying breakdown. During intestinal digestion, gastric digesta from hard gels also disintegrated more slowly. Importantly, lipid digestion extent correlated positively with the bioaccessibility of capsaicinoids (CAP) embedded in the gels, as lipolytic products aided CAP solubilization in micelles, enhancing its release. This highlights that gel hardness and microstructure control protein digestion kinetics and nutrient release, with softer gels promoting faster whey proteolysis and improved bioactive compound accessibility.

Particle Size and Surface Area Considerations: Larger aggregated particles exhibit reduced surface-area-to-volume ratios, limiting enzyme interaction compared to finely dispersed proteins. Processing approaches that modulate particle size can therefore manipulate digestions pace. Mao et al. (2021) showed that variations in protein composition influence the size of emulsified droplets within whey protein gels. Larger droplets and aggregates decrease the surface-area-to-volume ratio, limiting enzyme accessibility and slowing digestion. Modifying particle size through processing alters the gel microstructure and thereby regulates digestion rates. Also, even though whey proteins possess a globular structure, which means a good protein-enzyme area of contact, when they are crosslinked, this state displays slower in vitro digestion rates due to the stability of the gel structure which impedes the contact between the protein and the enzyme, following the "one-by-one" mechanism, and the lower surface-area-to-volume ratio compared to its noncrosslinked homolog decreases the contact area between digestive enzymes and proteins (Leng et al., 2025b).

Another example of the effect of microstructure is given by Dong et al. (2022), where WPI hydrogels made up with different NaCl concentrations were tested during in vitro gastrointestinal digestion, concluding that at higher NaCl concentrations, coarser gels are obtained, hence, slower loss of dry matter, lower rates and extent of proteolysis were measured due to restricted moisture absorption, limited swelling and coarse microstructure with more severe protein aggregation.

3.2.7 Gut microbial interactions

Emerging studies demonstrate that undigested whey proteins reaching the colon undergo fermentation by gut microbiota, influencing metabolic outputs and gut health.

Prebiotic-Like Effects: Whey protein modulates microbial populations favoring beneficial genera capable of generating metabolites that influence host physiology and potentially feed back on protein utilization. Whey protein modulates gut microbial populations by selectively stimulating beneficial bacteria such as Bifidobacterium and Lactobacillus. These bacteria produce metabolites like short-chain fatty acids that positively influence host physiology, including improved gut health and enhanced immune responses. Whey protein's ability to promote such probiotic genera suggests its potential as a prebiotic dietary component that can indirectly support protein utilization and overall metabolic health (Sánchez-Moya et al., 2017).

Microbiota-Dependent Utilization: Interindividual variability in microbial composition results in differential fermentation patterns affecting amino acid availability and overall protein nutritional outcomes. The composition of an individual's gut microbiota varies widely, leading to differential fermentation capacities of undigested whey protein fractions in the colon. This variability affects amino acid availability and overall protein nutritional outcomes, as different microbial communities produce distinct metabolite profiles. Personalized differences in microbial fermentation can thus modulate how effectively whey protein is utilized in postintestinal digestion (Boscaini et al., 2020).

There are additional influential factors such as host physiological status, including age, health condition, and gastrointestinal transit time, which markedly influence whey's digestibility. Moreover, the presence of multiple food components within mixed meals adds complexity in real-world digestive scenarios.

4. Recent advances in whey digestion research (2015–25)

4.1 In Vitro and in vivo models

On the one hand, in vitro digestion models are divided into two groups: static, which start from initial conditions with average and constant values throughout the experiment; and dynamic, which include parameter variation in time, looking to simulate more accurately the conditions in vivo, simulating the flow of digestive fluids, gastric emptying times, and peristaltic movements, among others. Thus, static models are most widely used due to economic, practical, and simplicity issues, since they help to screen the effects of composition, structure, and processing over food matrices, but since there is no consensus in the physiological conditions used, even though there are some widely used models (i.e., COST INFOGEST protocol), interlaboratory comparability poses a huge challenge. An overall algorithm of a sample preparation for an in vitro assessment is shown as an example in Fig. 17.8 (Sandoval-Cuella and Quintanilla-Carvajal, 2021).

On the other hand, in vivo experiments are very tedious since they involve high costs and involve ethical considerations when performing assays due to the involvement of animals, which would be subsequently sacrificed, and if assessed in humans, involve compounds whose dangerousness has not been established yet (Guerra et al., 2012).

After digestion experiments are conducted, it is important to evaluate the digesta obtained in both in vivo and in vitro assays to know the impact of the digestion method on the peptides profile for certain processing steps and matrix features. For instance, evaluating the models through the peptidomics approach, Sutantawong et al. (2024) found differences in the peptides profile between in vitro intestinal digestion (porcine pepsin with pH adjusted to 3.0 with HCl and after digestion, chyme was neutralized with NaOH and mixed with bovine bile mixture and porcine pancreatin) and in vivo intestinal digestion (3 adults aged 20–25 years, each consumed 40 g of WPI powder dissolved in water and 5 mg of polyethylene glycol-28 as a marker): Even though both models liberated 200 identical peptide sequences from 23 protein common precursors, this accounted the 40% of unique peptide counts identified in vitro and 19% of peptides identified in vivo, which indicates that the peptide profile generated in vivo is more diverse than the generated in vitro, possibly because of the presence of a broader array of proteases in the gut. But considering the proteomics approach with a similar setup assay, Sutantawong et al. (2025) found that even though β-lactoglobulin appeared more intact in the human jejunum compared with in vitro intestinal digesta, where no intact protein bands were visually detectable, the high Spearman's correlation coefficient (0.92) indicated similar intact survival profiles of the proteins analyzed between in vivo and in vitro intestinal digestion.

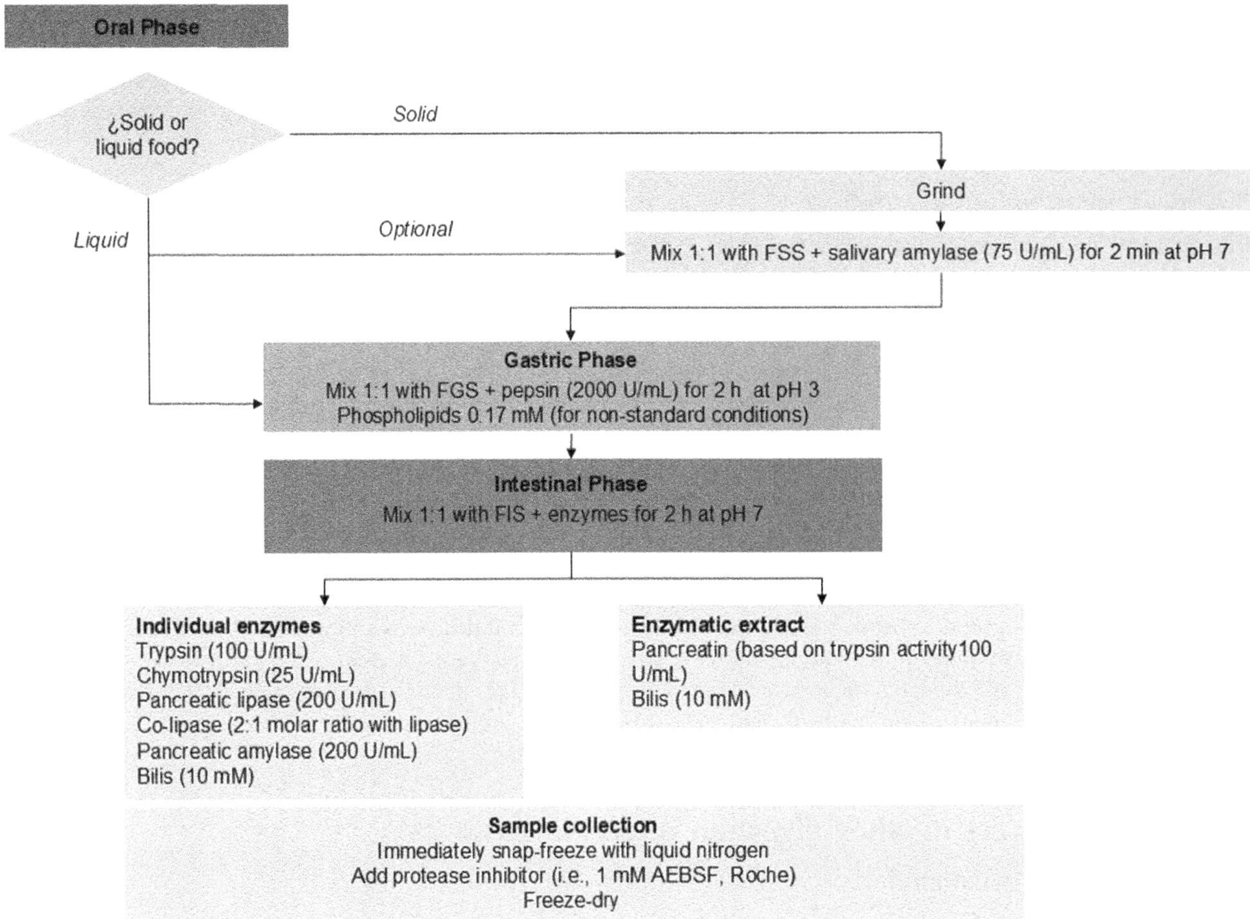

FIGURE 17.8 In vitro sample preparation algorithm. *Taken from Sandoval-Cuella and Quintanilla-Carvajal (2021).*

4.2 Bioactive peptide release during digestion

Bioactive peptides are specific protein fragments that exert beneficial physiological effects beyond their nutritional value. Whey proteins, primarily β-lactoglobulin, α-lactalbumin, bovine serum albumin, and minor fractions, serve as rich sources of these peptides. During gastrointestinal digestion, whey proteins undergo enzymatic hydrolysis, resulting in the liberation of a variety of bioactive peptides that are commonly identified and quantified by LC-MS/MS, UPLC-MS/MS, and SDS-PAGE (Sutantawong et al., 2024, 2025; Czelej et al., 2025; Leng et al., 2025b), and the prediction of their bioactive activity can be done using databases (i.e., BIOPEP-UWM database) (Leng et al., 2025b). These smaller molecules exert a variety of functionalities like antibacterial, antihypertensive, antioxidant, anticarcinogenic, hypocholesterolemia, opioid-like effects, immunomodulatory, among others (Pillai et al., 2024). Naturally, whey proteins present functionalities like (Madureira et al., 2007).

- Whole whey proteins: Prevention of breast, intestinal and chemically induced cancer, increment glutathione levels, which increase tumor cell vulnerability and help in HIV patients' treatment; antimicrobial activities and increase satiety response.
- β-Lactoglobulin: Enhancement of pregrastic esterase activity, transfer of passive immunity, and regulation of mammary gland phosphorus metabolism.
- α-Lactalbumin: Prevention of cancer and treatment of chronic stress-induced disease.
- Bovine serum albumin: Antimutagenic function and prevention of cancer.
- Whey immunomodulation: Antibacterial activity (i.e., HIV), antifungal activity, opioid activity.

But this section comprehensively explores the mechanisms of bioactive peptide release from whey proteins, their biological activities, influencing factors, and implications for human health.

The release of bioactive peptides from whey protein is a multi-step process mediated by gastrointestinal proteases. Digestion begins in the stomach, where pepsin initiates hydrolysis by cleaving peptide bonds preferentially at aromatic amino acids such as phenylalanine, tyrosine, and tryptophan. However, whey proteins, especially β-lactoglobulin, exhibit some resistance to gastric pepsin, maintaining partial native structure under acidic conditions. As the chyme moves to the small intestine, pancreatic enzymes—including trypsin, chymotrypsin, elastase, and carboxypeptidases—further digest the proteins into smaller peptides (Sutantawong et al., 2024).

Emerging peptidomic analyses employing mass spectrometry have identified hundreds to thousands of unique peptides derived from whey proteins after in vivo digestion. These peptides span a range of lengths and amino acid sequences, some corresponding to known bioactive motifs. For instance, certain peptides inhibit the angiotensin-converting enzyme (ACE), thereby aiding blood pressure regulation. Others act as antioxidants, neutralizing reactive oxygen species, or exhibit antimicrobial activity by disrupting pathogenic bacterial membranes. Immunomodulatory peptides influence cytokine secretion and immune cell activation, while opioid peptides modulate nociception and satiety (Amigo and Hernández-Ledesma, 2020; Sutantawong et al., 2024).

Indeed, beta-casomorphins, opioid-like peptides derived from casein but also found in minor amounts from whey matrices, exemplify multifunctional peptides capable of crossing the intestinal barrier and engaging peripheral opioid receptors, influencing gastrointestinal motility and pain perception. This layered enzymatic breakdown and peptide release highlights whey protein's capacity to produce physiologically impactful molecules during normal digestion (Olvera-Rosales et al., 2023).

❖ Biological Activities of Whey-Derived Bioactive Peptides

The bioactivities attributed to whey-derived peptides encompass multiple physiological systems.

- Antihypertensive Activity: Being the most extensively studied activity, these peptides inhibit ACE, preventing the conversion of angiotensin I to angiotensin II, a potent vasoconstrictor. This reduces blood pressure and improves cardiovascular health (Korhonen and Pihlanto, 2006).
- Antioxidant Effects: Peptides scavenge free radicals and chelate prooxidative metal ions, thereby protecting tissues from oxidative damage linked to aging and chronic diseases (Amigo and Hernández-Ledesma, 2020).
- Antimicrobial Properties: Certain whey peptides disrupt bacterial cell membranes or inhibit enzyme systems of pathogens, contributing to intestinal health and immune defense (Olvera-Rosales et al., 2023).
- Immunomodulation: Some peptides stimulate immune cells, modulate cytokine profiles, and enhance barrier function, potentially aiding in infection resistance and inflammatory disease management (Korhonen and Pihlanto, 2006).
- Opioid-Like Effects: Peptides such as beta-casomorphins can bind opioid receptors, influencing gastrointestinal transit, appetite regulation, and stress responses (Olvera-Rosales et al., 2023).

These biological effects result from peptide interaction with receptors, enzymes, or signaling molecules, and their efficacy depends on peptide availability after digestion and absorption.

❖ Factors Influencing Peptide Release and Bioavailability

The generation and activity of bioactive peptides from whey protein are influenced by intrinsic protein structure, processing conditions, digestion dynamics, and host factors.

- Protein Conformation and Processing: Native whey proteins have tightly folded structures. Processing techniques such as heating, enzymatic hydrolysis, or fermentation can unfold proteins, rendering peptide bonds more accessible to proteases and enhancing the release of bioactive peptides. For example, mild heat treatment partially denatures whey, increasing digestion rates and peptide yield without losing functional properties (Amigo and Hernández-Ledesma, 2020).
- Enzymatic Specificity and Digestive Environment: Different gastrointestinal enzymes target specific peptide bonds, producing unique peptide profiles depending on enzyme activity and sequence context. Variations in pH, enzyme concentration, and transit times affect the extent and rate of hydrolysis (Sutantawong et al., 2024).
- Food Matrix Effects: The presence of lipids, carbohydrates, and fibers shapes the physical environment of digestion, influencing enzyme accessibility to whey proteins and stabilizing or limiting peptide release. Emulsions or gels may delay breakdown but also protect bioactive peptides from premature degradation (Olvera-Rosales et al., 2023).
- Peptide Stability: Once released, peptides may undergo further hydrolysis by brush border enzymes or gut microbes. Stability to these enzymes and resistance to gastric acidity are crucial for peptides to retain activity and be absorbed intact (Amigo and Hernández-Ledesma, 2020).

- Absorption and Transport: Bioactive peptides enter systemic circulation primarily via paracellular routes, carrier-mediated transport, or transcytosis in the intestinal epithelium. Efficient absorption determines their capacity to exert systemic health effects (Olvera-Rosales et al., 2023).
- Host Factors: Individual variability in digestive enzyme levels, gut microbiota composition, gastrointestinal health, and genetics modulate peptide release, stability, and absorption, influencing the net physiological impact (Amigo and Hernández-Ledesma, 2020).

5. Health implications and applications

Whey-derived bioactive peptides have promising applications in preventive health and therapeutic nutrition. Their antihypertensive effects suggest use in managing mild hypertension or as adjuncts to pharmaceutical agents. Antioxidant and immunomodulatory peptides may support healthy aging, reduce inflammation, and help prevent chronic diseases such as cardiovascular disorders and metabolic syndrome. Antimicrobial peptides contribute to gut microbiota balance and infection resistance.

Clinical trials examining purified whey peptides or whey hydrolysates enriched in bioactive sequences indicate improvements in blood pressure, oxidative stress markers, and immune parameters. However, variability in peptide profiles and bioavailability underscores the importance of optimizing whey processing and delivery to maximize health benefits (Korhonen and Pihlanto, 2006; Olvera-Rosales et al., 2023).

Moreover, formulations incorporating whey-derived peptides are being developed as functional foods, nutraceuticals, and medical nutrition therapies tailored for specific populations such as the elderly, athletes, or individuals with metabolic disorders.

Finally, advancements in peptidomics, in vitro digestion models, and controlled clinical trials are refining our understanding of whey-protein peptide release and action in humans. Future research aims to.

- Map comprehensive peptide profiles generated during digestion under physiological conditions.
- Explore the synergistic effects of peptide mixtures versus isolated peptides.
- Elucidate molecular mechanisms of peptide-receptor interactions.
- Develop enhanced whey processing methods to tailor peptide generation.
- Investigate personalized nutrition approaches considering host variability in peptide utilization.

5.1 Survival of intact proteins and peptides

One of the big challenges to overcome is the high resistance of β-Lactoglobulin to proteolysis (digested in the intestinal phase) in the gastrointestinal tract due to the presence of phosphatidylcholine and the existence of two intramolecular disulfide bridges in the molecule, which can trigger allergenicity mostly in children below 2 years of age, because gastric proteolysis in the infant stage is much lower compared to that of an adult. Gastric acidic fluid, protein denatured, and low gastric pH will lead to activation of pepsin; in contrast, high gastric pH will decrease pepsin output, inhibiting the digestion of protein in infants (Sabari et al., 2023). Also, a predigested product like WPH can reduce the allergenic potential of whey proteins by breaking down epitopes responsible for allergic reactions (Tang et al., 2025).

6. Physiological and health implications

6.1 Protein quality and bioavailability

There is limited data about the quality of whey protein supplements available on the market. Hence, Pehlivanoğlu et al. (2022) evaluated 14 commercial whey protein supplements available in Istanbul, Turkey, finding that some of them have discrepancies of over a 10% between the measured and the declared total amino acids content. Besides, the in vitro digestibility measured by PDCAAS ranged from 50.4% to 79.6% among the supplements. Also, individual amino acid contents vary significantly among samples: while high levels of glycine were detected in some samples, very low levels of cysteine were detected compared to the reference whey protein, suggesting that collagen or casein may be present in these products, bringing back the relevance in the presence of nonprotein compounds discussed previously and the importance of having good sources of whey protein as raw materials to manufacture a high quality product.

6.2 Effects on muscle, bone, and metabolic health

The process of muscle growth (hypertrophy) is a complex physiological phenomenon that involves multiple interconnected mechanisms and the capability of muscle cells to synthesize new proteins, leading to muscle repair and growth when the muscle protein synthesis (MPS) exceeds protein breakdown over time, making it a constant balance process. Consequently, the proven efficacy of whey protein for stimulating MPS is due to factors like its rapid digestion and absorption that quickly increases amino acid levels in blood, creating an anabolic environment conducive to protein synthesis; the fact that whey protein contain all essential amino acids in amounts that meet human intake requirement; and the high leucine content in whey, which triggers the mTOR protein that acts as a master switch for protein synthesis, integrating signals from nutrients, growth factors, and mechanical stress to regulate cell growth and proliferation (Tang et al., 2025). These effects have been studied, indicating that consumption of whey protein following resistance exercise in the presence of carbohydrates (Tang et al., 2007) and a higher dose of protein (Macnaughton et al., 2016) can boost MPS, because carbohydrates and proteins share features in their digestive pathway and active transport through the enterocytes (Sitrin, 2014). Moreover, consuming whey protein alongside carbohydrates can enhance the insulin response, potentially improving nutrient uptake and glycogen resynthesis (Cava et al., 2024).

Moving on to the effects on bone mass, there is still lack of comparable information, but some of the insights obtained so far are: High protein intake in murine models led to decreased bone density along with changes in urinary pH and calcium concentration; from individuals consuming whey protein was observed an increased urinary calcium excretion and decreased urinary pH, modifications in serum calcidiol and calcium levels, along with a decrease in parathyroid hormone levels in older adults, and no significant impact on bone mineral density was observed on menopausal women (Cava et al., 2024).

In another vein, obese is nowadays a very common disease, which made researchers turned to look for quick and effective solutions, and protein supplements, often rich in whey protein, are commonly part of weight-loss strategies based on the rationale that protein is more satiating than the other macronutrients, but may not be as effective in older as in younger adults nor in solid as in liquid protein, because protein is the most satiating of the macronutrients in this population and solid protein has been found to be more satiating than liquid, respectively. Also, failing to achieve substantial long-term weight loss with a high-protein-intake diet may be because protein is not as satiating in obese individuals as in nonobese individuals (Oberoi et al., 2020).

7. Recent trends in whey digestion

7.1 Hydrolyzed whey proteins

Understanding the effects of hydrolysis on whey proteins is important to achieve the goals of the developing product. Overall processing methods to obtain the basic whey protein products were described in section 2.2, but we are going to delve more into the impact of the hydrolyzed product (WPH) in digestion against the concentrated and isolated versions.

On one hand, WPH generally exhibits the fastest absorption rate due to its predigested nature, followed by WPI thanks to its higher solubility across a wide range of pH given by its high purity, and then WPC, which exhibits the lowest rate due to the presence of other nonprotein components, specifically fat, and its weakened solubility at low pH levels (Tang et al., 2025).

On the other hand, since microstructure may alter the digestion rate, as explained in section 3.2, emulsification, foaming, and gelation may affect this rate. Given that, WPC performs the best in these areas due to the intact nature of its proteins and the presence of phospholipids, followed by the WPI and WPH, with their ability reduced or almost null (Tang et al., 2025).

All these characteristics must be considered, for instance, in applications to sports nutrition. Since athletes seek the fastest recuperation and muscle reparation possible, a fast digestion and high bioavailability can enhance amino acid transport to muscles for protein synthesis, resulting in muscle growth and repair (Pillai et al., 2024).

7.2 Interactions with gut microbiota

The microbiota is essentially a set of bacterial colonies, which start right after birth, that live in the gut and even communicate with the central nervous system to cooperate with the host body in its appropriate function. Thus, gut microbiota has been proven to be affected by whey proteins. For instance, a study of whey protein consumption coupled with exercise in professional rugby players showed a higher microbial diversity than the controls; and another study with sedentary adults that underwent short-term exercise and/or dietary whey protein supplementation resulted in changes in the gut virome, while the microbial composition and function were modest. However, whey protein appears to play important roles in the development of the infant's gut microbiota by stimulating beneficial bacteria species and providing protection against opportunistic pathogens (Boscaini et al., 2023).

Based on the aforementioned, colonic gut microbiota responsible for the fermentation of carbohydrates and proteins can be easily changed by diet, altering the production of short-chain fatty acids (butyrate, propionate, and acetate in major proportions) under anaerobic conditions, changing the energy intake patterns and exerting antiinflammatory effects on the host (van der Hee and Wells, 2021).

7.3 Fermented whey-based products

A recent systematic review of health-promoting effects of consumption of whey-based fermented products demonstrates the significant potential of fermented whey as a functional food due to its rich nutritional content and the bioactive compounds generated during fermentation. A systematic review of 12 human intervention studies found that consuming fermented whey improves several health parameters in adults, including muscle mass, glycemic control, lipid profiles (especially triglycerides and LDL cholesterol), immune function (e.g., increased natural killer cell activity), and reductions in oxidative stress and inflammation. Additional benefits for gastrointestinal and urinary tract health were also reported. These effects are linked to enhanced bioavailability of branched-chain amino acids, bioactive peptides, and microbial metabolites such as exopolysaccharides and short-chain fatty acids produced during fermentation. The reviewed studies showed fermented whey to be safe and well-tolerated. However, more extensive, long-term clinical research is needed to confirm these findings and clarify the mechanisms behind the health benefits. Overall, fermented whey holds promise as a safe and effective functional food for improving adult health and nutrition (Sar et al., 2025).

8. Comparison of the digestibility of whey protein and casein

Comparing caseins and whey proteins structures, caseins have relatively open, flexible structures, whereas whey proteins have a globular and compact structure that makes them heat labile, with caseins presenting an isoelectric point at 4.6 pH, but whey proteins remain soluble in acidic conditions, being predictable that caseins coagulate while whey proteins remain soluble in gastric conditions, characterized by low pH levels. Egger et al. (2017) had consistent conclusions with the above during in vivo pig digestion, finding that the casein clot formed was slowly disintegrated from the surface and hence degraded in the duodenal, though the in vitro sample showed a faster hydrolysis rate; but found that β-lactoglobulin was resistant to pepsin, different than casein, which possessing an open structure it shows high hydrolysis by pepsin (van Lieshout et al., 2020). Thus, due to coagulation of casein micelles in the stomach, casein is generally considered as slow emptying milk protein. However, whey proteins are considered as fast-emptying milk proteins due to their solubility, rapidly emptying from the stomach and rapidly passing through to the duodenum. Hence, whey proteins offer a fast postprandial rate of release of amino acids into the bloodstream in humans (Boirie et al., 1997). The comparison of the nutrient digestion, nutrient absorption, and physiological responses of casein and whey proteins is shown in Table 17.1.

In another study, Tunick et al. (2016) subjected milk to heating and homogenization, then subjected it to in vitro digestion, finding that native whey proteins remained intact after gastric digestion, whereas casein was almost completely digested.

TABLE 17.1 Comparison of the nutrient digestion, nutrient absorption, and physiological responses of casein and whey proteins.

Parameters	Molecular level (caseins vs whey proteins)
Bioaccesibility	*In vitro:* Caseins were more easily digested than whey proteins during the static gastric digestion. The digestibility of the latter increased by the partial unfolding of the whey proteins. *In vivo:* Rapid appearance of mostly intact protein form in the jejunum after β-lactoglobulin digestion, whereas caseins were more slowly recovered in a more degraded form.
Bioavailability	*In vivo:* According to the plasma amino acids appearance, whey proteins and caseins are considered as "fast" and "slow" digested proteins, respectively.
Physiological responses	• Casein consumption inhibited whole protein breakdown but only a slight increase in whole protein synthesis compared to whey proteins. • Whey proteins stimulated muscle protein synthesis. • Whey proteins seemed to have a greater satiety-promoting power in the short term, in contrast to the long-term effect shown by the slow protein digestion of caseins.

Modified from Mulet-Cabero et al. (2020) with permission.

9. Conclusion

Recent research has significantly advanced our understanding of whey protein digestion, highlighting the complexity of its breakdown, the release of bioactive peptides, and the influence of product structure and processing. Whey protein is considered a fast-emptying protein, while casein is known to digest slowly. Various processing-induced changes in the microstructure of different dairy products may impact their overall digestive behavior. The validation of in vitro digestion results for whey proteins and casein requires animal and/or human trials, which are considered time-consuming and expensive. While whey remains a superior protein source in terms of digestibility and health benefits, ongoing research is needed to fully elucidate the physiological roles of its digestive products and to optimize its use in functional foods and medical nutrition.

References

Accardo, F., et al., 2022. Structural and chemical changes induced by temperature and pH hinder the digestibility of whey proteins. Food Chem. 387, 132884. https://doi.org/10.1016/j.foodchem.2022.132884.

Accardo, F., et al., 2023. Evaluation of in vitro whey protein digestibility in a protein-catechins model system mimicking milk chocolate: interaction with flavonoids does not hinder protein bioaccessibility. Food Res. Int. 169, 112888. https://doi.org/10.1016/j.foodres.2023.112888.

Amigo, L., Hernández-Ledesma, B., 2020. Current evidence on the bioavailability of food bioactive peptides. Molecules 25 (19), 4479. https://doi.org/10.3390/molecules25194479.

Bauer Estrada, K., et al., 2022. Whey as food-grade culture medium on an industrial scale that protects probiotics during in vitro digestion. Front. Food Sci. Technol. 2, 894761. https://doi.org/10.3389/frfst.2022.894761.

Boirie, Y., et al., 1997. Slow and fast dietary proteins differently modulate postprandial protein accretion. Proc. Natl. Acad. Sci. U. S. A 94 (26), 14930–14935. https://doi.org/10.1073/pnas.94.26.14930.

Boscaini, S., et al., 2020. Age- and duration-dependent effects of whey protein on high-fat diet-induced changes in body weight, lipid metabolism, and gut microbiota in mice. Physiol. Rep. 8 (15), 19. https://doi.org/10.14814/phy2.14523.

Boscaini, S., et al., 2023. The "Whey" to good health: whey protein and its beneficial effect on metabolism, gut microbiota and mental health. Trends Food Sci. Technol. 133, 1–14. https://doi.org/10.1016/j.tifs.2022.12.009.

Bujacz, A., 2012. Structures of bovine, equine and leporine serum albumin. Acta Crystallogr., Sect. D: Biol. Crystallogr. 68 (10), 1278–1289. https://doi.org/10.1107/S0907444912027047.

Callahan, A., Leonard, H., Powell, T., 2020. Nutrition: science and everyday application. Eugene, Oregon: Open Oregon Educational Resources, first ed. Available at: https://openoregon.pressbooks.pub/nutritionscience/.

Cava, E., et al., 2024. Investigating the health implications of whey protein consumption: a narrative review of risks, adverse effects, and associated health issues. Healthcare 12 (2), 246. https://doi.org/10.3390/healthcare12020246.

Cherian, G., 2019. A Guide to the Principles of Animal Nutrition. Oregon State University, Corvallis, OR. Available at: https://open.oregonstate.education/animalnutrition/.

Chrysina, E.D., Brew, K., Acharya, K.R., 2000. Crystal structures of apo- and holo-bovine α-lactalbumin at 2.2-Å resolution reveal an effect of calcium on inter-lobe interactions. J. Biol. Chem. 275 (47), 37021–37029. https://doi.org/10.1074/jbc.M004752200.

Cirkovic Velickovic, T.D., Stanic-Vucinic, D.J., 2018. The role of dietary phenolic compounds in protein digestion and processing technologies to improve their antinutritive properties. Compr. Rev. Food Sci. Food Saf. 17 (1), 82–103. https://doi.org/10.1111/1541-4337.12320.

Czelej, M., et al., 2025. Whey protein enzymatic breakdown: synthesis, analysis, and discovery of new biologically active peptides in papain-derived hydrolysates. Molecules 30 (7), 1451. https://doi.org/10.3390/molecules30071451.

De Wit, J.N., van Kessel, T., 1996. Effects of ionic strength on the solubility of whey protein products. A colloid chemical approach. Food Hydrocoll. 10 (2), 143–149. https://doi.org/10.1016/S0268-005X(96)80028-2.

Delgado-Macuil, R.J., et al., 2025. Recent biotechnological applications of whey: review and perspectives. Fermentation 11 (4), 217. https://doi.org/10.3390/fermentation11040217.

Dong, X., et al., 2022. Mechanistic study on in vitro disintegration and proteolysis of whey protein isolate gels: effect of the strength of sodium ions. Food Hydrocoll. 132, 107862. https://doi.org/10.1016/j.foodhyd.2022.107862.

Egger, L., et al., 2017. Physiological comparability of the harmonized INFOGEST in vitro digestion method to in vivo pig digestion. Food Res. Int. 102, 567–574. https://doi.org/10.1016/j.foodres.2017.09.047.

Guerra, A., et al., 2012. Relevance and challenges in modeling human gastric and small intestinal digestion. Trends Biotechnol. 30 (11), 591–600. https://doi.org/10.1016/j.tibtech.2012.08.001.

Guo, Q., et al., 2020. Destructuring and restructuring of foods during gastric digestion. Compr. Rev. Food Sci. Food Saf. 19 (4), 1658–1679. https://doi.org/10.1111/1541-4337.12558.

Kaspchak, E., Mafra, L.I., Mafra, M.R., 2018. Effect of heating and ionic strength on the interaction of bovine serum albumin and the antinutrients tannic and phytic acids, and its influence on in vitro protein digestibility. Food Chem. 252, 1–8. https://doi.org/10.1016/j.foodchem.2018.01.089.

Korhonen, H., Pihlanto, A., 2006. Bioactive peptides: production and functionality. Int. Dairy J. 16 (9), 945–960. https://doi.org/10.1016/j.idairyj.2005.10.012.

Kuwata, K., et al., 1999. Solution structure and dynamics of bovine β-lactoglobulin A. Protein Sci. 8 (11), 2541–2545. https://doi.org/10.1110/ps.8.11.2541.

Leng, J., Li, L., et al., 2025a. Enhancing lipid digestion and absorption rate: the role of whey protein isolate short fibrils in emulsions. Int. J. Biol. Macromol. 285, 138351. https://doi.org/10.1016/j.ijbiomac.2024.138351.

Leng, J., Jiang, Y., et al., 2025b. Unveiling the slow digestion and peptide profiles of polymerised whey gel via heat and TGase crosslinking: an in vitro/vivo perspective. Food Chem. 464 (Pt 3), 141829. https://doi.org/10.1016/j.foodchem.2024.141829.

Li, J., et al., 2021. Interaction mechanism of flavonoids with whey protein isolate: a spectrofluorometric and theoretical investigation. Food Chem. 355, 129617. https://doi.org/10.1016/j.foodchem.2021.129617.

Liang, Y., et al., 2025. Formation of non-enzymatic protein modifications (nePTMs) on caseins and whey proteins during dry heating and the impact on their in vitro infant digestion profile. Food Chem., 145896 https://doi.org/10.1016/j.foodchem.2025.145896.

Lobner, E., et al., 2017. Fcab-HER2 Interaction: a Ménage à Trois. Lessons from X-Ray and Solution Studies. Structure 25 (6), 878–889. https://doi.org/10.1016/j.str.2017.04.014.

Luo, N., et al., 2021. Effect of gel structure on the in vitro gastrointestinal digestion behaviour of whey protein emulsion gels and the bioaccessibility of capsaicinoids. Molecules 26 (5), 1379. https://doi.org/10.3390/molecules26051379.

Macnaughton, L.S., et al., 2016. The response of muscle protein synthesis following whole-body resistance exercise is greater following 40 g than 20 g of ingested whey protein. Physiol. Rep. 4 (15), e12893. https://doi.org/10.14814/phy2.12893.

Madureira, A.R., et al., 2007. Bovine whey proteins—overview on their main biological properties. Food Res. Int. 40 (10), 1197–1211. https://doi.org/10.1016/j.foodres.2007.07.005.

Mao, C., et al., 2021. Physicochemical properties and digestive kinetics of whey protein gels filled with potato and whey protein mixture emulsified oil droplets: effect of protein ratios. Food Funct. 12 (13), 5927–5939. https://doi.org/10.1039/d1fo00653c.

Matthews, J.J., et al., 2025. Understanding dietary protein quality: digestible indispensable amino acid score and beyond. J. Nutr. https://doi.org/10.1016/j.tjnut.2025.07.005.

Melchior, S., et al., 2023. Comparison of protein in vitro digestibility under adult and elderly conditions: the case study of wheat, pea, rice, and whey proteins. Food Res. Int. 163, 112147. https://doi.org/10.1016/j.foodres.2022.112147.

Moore, S.A., et al., 1997. Three-dimensional structure of diferric bovine Lactoferrin at 2.8 Å resolution. J. Mol. Biol. 274 (2), 222–236. https://doi.org/10.1006/jmbi.1997.1386.

Mulet-Cabero, A.-I., et al., 2020. Dairy structures and physiological responses: a matter of gastric digestion. Crit. Rev. Food Sci. Nutr. 60 (22), 3737–3752. https://doi.org/10.1080/10408398.2019.1707159.

Oberoi, A., et al., 2020. Acute effects of whey protein on energy intake, appetite and gastric emptying in younger and older, obese men. Nutr. Diabetes 10 (1), 37. https://doi.org/10.1038/s41387-020-00139-8.

Olvera-Rosales, L.B., et al., 2023. Bioactive peptides of whey: obtaining, activity, mechanism of action, and further applications. Crit. Rev. Food Sci. Nutr. 63 (30), 10351–10381. https://doi.org/10.1080/10408398.2022.2079113.

Pehlivanoğlu, H., Bardakçi, H.F., Yaman, M., 2022. Protein quality assessment of commercial whey protein supplements commonly consumed in Turkey by in vitro protein digestibility-corrected amino acid score (PDCAAS). Food Sci. Technol. 42, e64720. https://doi.org/10.1590/fst.64720.

Pillai, A.T., Morya, S., Kasankala, L.M., 2024. Emerging trends in bioavailability and pharma-nutraceutical potential of whey bioactives. J. Nutr. Metab. 2024, 8455666. https://doi.org/10.1155/2024/8455666.

Sabari, S., et al., 2023. In vitro infant digestion of whey proteins isolate–lactose. Foods 12 (3), 667. https://doi.org/10.3390/foods12030667.

Sánchez-Moya, T., et al., 2017. In vitro modulation of gut microbiota by whey protein to preserve intestinal health. Food Funct. 8 (9), 3053–3063. https://doi.org/10.1039/c7fo00197e.

Sánchez-Rivera, L., et al., 2015. Peptide mapping during dynamic gastric digestion of heated and unheated skimmed milk powder. Food Res. Int. 77, 132–139. https://doi.org/10.1016/j.foodres.2015.08.001.

Sandoval-Cuella, C.E., Quintanilla-Carvajal, M.X., 2021. Evaluación de la bioaccesibilidad de compuestos bioactivos en alimentos funcionales: Modelos de digestión in vitro. In: Alamilla-Beltrán, L., et al. (Eds.), Subproductos agroindustriales y recursos autóctonos. Procesamiento y técnicas de análisis. INTA Ediciones, p. 283.

Sar, T., et al., 2025. A systematic review of health promoting effects of consumption of whey-based fermented products on adults. Front. Nutr. 12, 1651365. https://doi.org/10.3389/fnut.2025.1651365.

Singh, A.K., et al., 2010. Mode of binding of the tuberculosis prodrug isoniazid to heme peroxidases: binding studies and crystal structure of bovine lactoperoxidase with isoniazid at 2:7 å resolution. J. Biol. Chem. 285 (2), 1569–1576. https://doi.org/10.1074/jbc.M109.060327.

Sitrin, M.D., 2014. Digestion and absorption of carbohydrates and proteins. In: Leung, P.S. (Ed.), The Gastrointestinal System: Gastrointestinal, Nutritional and Hepatobiliary Physiology, first ed. Springer Dordrecht, Buffalo, NY, USA, pp. 137–158. https://doi.org/10.1007/978-94-017-8771-0_6.

Sorensen, J., 2025. Human Nutrition. Langara College, Nutrition and Food Service Management Program, Vancouver, British Columbia. Available at: https://pressbooks.bccampus.ca/nutr1100/.

Sutantawong, S., et al., 2024. Release of bioactive peptides from whey proteins across digestion in adult humans and a static in vitro digestion model. J. Funct.Foods 122, 106540. https://doi.org/10.1016/j.jff.2024.106540.

Sutantawong, S., et al., 2025. Survival of intact bovine whey proteins across in vivo and simulated static in vitro models of the adult gastrointestinal tract. Food Chem. 465 (Pt 1), 142013. https://doi.org/10.1016/j.foodchem.2024.142013.

Świeca, M., et al., 2013. The influence of protein-flavonoid interactions on protein digestibility in vitro and the antioxidant quality of breads enriched with onion skin. Food Chem. 141 (1), 451–458. https://doi.org/10.1016/j.foodchem.2013.03.048.

Tang, C., et al., 2025. Chemical properties of whey protein in protein powders and its impact on muscle growth in athletes: a review. Nat. Prod. Commun. 20 (3), 1–17. https://doi.org/10.1177/1934578X251326124.

Tang, J.E., et al., 2007. Minimal whey protein with carbohydrate stimulates muscle protein synthesis following resistance exercise in trained young men. Appl. Physiol. Nutr. Metabol. 32 (6), 1132–1138. https://doi.org/10.1139/H07-076.

Tunick, M.H., et al., 2016. Effect of heat and homogenization on in vitro digestion of milk. J. Dairy Sci. 99 (6), 4124–4139. https://doi.org/10.3168/jds.2015-10474.

Urango, A.C.M., et al., 2024. Maillard conjugates of whey protein and prebiotic dietary fibers: emulsion properties, encapsulation efficiency and chemical stability after in vitro gastrointestinal digestion. Colloids Surf. A Physicochem. Eng. Asp. 703 (Part 1), 135193. https://doi.org/10.1016/j.colsurfa.2024.135193.

van Lieshout, G.A.A., et al., 2020. How processing may affect milk protein digestion and overall physiological outcomes: a systematic review. Crit. Rev. Food Sci. Nutr. 60 (14), 2422–2445. https://doi.org/10.1080/10408398.2019.1646703.

van der Hee, B., Wells, J.M., 2021. Microbial regulation of host physiology by short-chain fatty acids. Trends Microbiol. 29 (8), 700–712. https://doi.org/10.1016/j.tim.2021.02.001.

Ye, A., et al., 2016. The formation and breakdown of structured clots from whole milk during gastric digestion. Food Funct. 7 (10), 4259–4266. https://doi.org/10.1039/c6fo00228e.

Zhang, L., et al., 2022. Effects of sequential enzymatic hydrolysis and transglutaminase crosslinking on functional, rheological, and structural properties of whey protein isolate. LWT Food Sci. Technol. (Lebensm.-Wiss.-Technol.) 153, 112415. https://doi.org/10.1016/j.lwt.2021.112415.

Chapter 18

Nutritional applications of whey

Jeferson Silva Cunha, Flaviana Coelho Pacheco, Viviane Lopes Pereira, Isabela Soares Magalhães, Danielly Aparecida de Souza, Irene Andressa, Fabio Ribeiro dos Santos and Bruno Ricardo de Castro Leite Júnior

Department of Food Technology, Federal University of Viçosa, University Campus, Viçosa, Brazil

1. Introduction

Whey is one of the most abundant by-products of the dairy industry, generated during the enzymatic or acidic coagulation of caseins in cheese production (Mestawet et al., 2024). Historically, it was considered waste with no commercial value and was often discarded directly into water bodies or soil. This disposal practice has led to significant environmental issues, primarily due to whey's high organic matter content and elevated biochemical oxygen demand (Djouab and Aïder, 2019). Over time, the necessity of repurposing whey in the food industry has become increasingly apparent, driven not only by environmental concerns but also by the rising demand for natural and sustainable ingredients. Today, whey, especially whey proteins, is recognized as a valuable coproduct with substantial economic and nutritional importance.

Whey protein is a highly versatile ingredient in the food industry, with applications spanning a wide range of products, including yogurts, fermented and nonfermented beverages, bakery and meat products. Beyond its industrial applications, whey protein plays a key role in products designed for specific populations, such as nutritional supplements for athletes, children, and the elderly, primarily due to its contribution to muscle mass development and overall nutritional value.

In this context, this chapter aims to provide a comprehensive overview of whey, particularly whey proteins, emphasizing their health benefits and their significance across various market segments.

2. Nutritional composition of whey

Whey is highly valued for its nutritional profile and is primarily composed of water (~93%), lactose (4.5%–5.0%), soluble proteins (0.60%–0.85%), minerals (~0.53%), and fat (~0.36%), with the protein fraction being the most extensively utilized in the food industry (Karim and Aider, 2022). Whey proteins are renowned as functional ingredients, as their regular consumption offers numerous health benefits, including antioxidant activity (Wang et al., 2024a, 2024b), anti-inflammatory effects (Zhao et al., 2023), enhanced satiety (Ho et al., 2023), and the prevention and treatment of certain cancers (Ramani et al., 2024; Thampy et al., 2024a, 2024b).

Whey proteins account for approximately 20% of the total protein content in cow milk, with β-lactoglobulin (β-LG, 50%–55%) and α-lactalbumin (α-LA, 20%–25%) as the predominant fractions. β-LG is the most abundant whey protein, characterized by its globular structure consisting of 162 amino acid residues and a molecular weight of 18.3 kDa, contributing roughly 3.2 g/L in bovine milk. Due to its size, β-LG exhibits resistance to the acidic and enzymatic conditions of the stomach, with digestion primarily occurring in the small intestine (Haraguchi et al., 2006).

β-LG is highly valued in both the food and pharmaceutical industries due to its ability to form complexes with hydrophobic bioactive compounds of interest that are sensitive to the gastrointestinal environment, thereby enhancing their bioavailability (Grinberg et al., 2023). Several bioactive compounds, such as curcumin (Grinberg et al., 2023), astaxanthin (Liu et al., 2018), and resveratrol (do Vale et al., 2022), have been reported to form complexes with these proteins.

α-Lactalbumin (α-LA) is primarily characterized by its high tryptophan content, a key precursor of serotonin, which may help alleviate symptoms associated with anxiety and depression (Sharma, 2019). It consists of 123 amino acid residues and has a molecular weight of 14.2 kDa, exhibiting a simple chain structure that is easily digested. This makes

Whey. https://doi.org/10.1016/B978-0-443-27648-4.00009-8

α-LA particularly attractive for formulating dietary supplements, especially for elderly and pediatric populations (Jakopović et al., 2016).

One of the key features of α-LA is its interaction with minerals, particularly calcium. Calcium enhances the thermal stability and improves the techno-functional properties of α-LA, making it highly valuable for food industry applications, especially under various processing conditions (Wang et al., 2024a, 2024b). Additionally, α-LA interacts with fatty acids, with in vitro studies suggesting potential anticancer properties and their role in promoting the destruction of tumor cells (Fang et al., 2018; Uversky et al., 2017).

Recent in vivo studies have highlighted the bioactive properties of peptides derived from β-LG hydrolysis, such as reductions in body weight (Ge et al., 2024), improvements in insulin sensitivity (Chen et al., 2022), and regulatory effects on lipid metabolism (Chen et al., 2021). Furthermore, the composition, techno-functional, and biological properties of whey protein are influenced by both the processing conditions and the source of raw materials. Advanced technologies, such as ultrafiltration, spray drying, and membrane filtration, are commonly employed for the extraction and concentration of whey proteins (Mestawet et al., 2024).

Whey protein concentrate (WPC) is marketed at different concentrations, typically ranging from 35% to 80%, preserving both macro and micronutrients present in whey, though it tends to be high in calories due to the presence of carbohydrates. Whey protein isolate (WPI) typically contains protein concentrations of 90% or higher and is commonly used in the formulation of dietary supplements (Mestawet et al., 2024). A primary disadvantage of WPI is the loss of micronutrients and bioactive protein fractions, such as lactoferrin, during the purification process. However, the protein composition of isolates/concentrates varies according to their intended applications and the target population (Minj and Anand, 2020).

Proteins found in both isolates and concentrates remain in their native forms, necessitating enzymatic hydrolysis by the digestive system for absorption. In the case of hydrolyzed proteins, they undergo a prehydrolysis process, which can be acid, alkaline, or enzymatic, resulting in more digestible and efficiently absorbed proteins in the intestine (Minj and Anand, 2020). The composition of hydrolyzates varies according to the degree and conditions of hydrolysis as well as their intended application.

3. Nutritional benefits of whey

Whey protein (WP) is widely recognized as a high-quality protein source with significant health and nutritional benefits. WP possesses characteristics such as (i) rapid digestion, (ii) high bioavailability, and (iii) a complete essential amino acid (EA) profile. These attributes make WP a preferred choice in various food formulations and nutritional supplements. Historically, whey was considered a waste product in the dairy industry due to its susceptibility to spoilage. However, technological advances and a deeper understanding of its properties have transformed this byproduct into a highly valued functional ingredient (Alrosan et al., 2021, 2022a; de Castro et al., 2017). In addition to its nutritional value, WP has important technological properties. WP exhibits over 95% solubility, which facilitates its application in liquid and semisolid products. It also demonstrates emulsification capacity, water retention, and emulsion stability, with values ranging from 45% to 66.05% (Wen-Qiong et al., 2019). These characteristics make WP a versatile choice for industries seeking functional solutions in their food products.

Compared to other protein sources (Table 18.1), whey protein stands out for its superior protein quality. It contains approximately 43% of essential amino acids (EAs), including leucine, valine, and cysteine, which play a fundamental role in protein synthesis and support various metabolic processes. In contrast, plant-based proteins, such as those from wheat, lupin, and oats, have significantly lower EA levels, ranging from 21% to 40.64% (Gorissen et al., 2018; Kheto et al., 2024a, 2024b). Among animal proteins, such as those from milk, eggs, and casein, EA levels range from 32% to 39%, still lower than those of WP (Floret et al., 2023).

Furthermore, WP is highly digestible, ensuring greater efficiency in the absorption and metabolic utilization of amino acids (Fig. 18.1). Another important benefit of WP is its role in weight control and appetite regulation. Studies suggest that WP stimulates the release of anorectic hormones, such as peptide YY and cholecystokinin, in the gastrointestinal tract. These hormones help reduce caloric intake, promote satiety, and contribute to glycemic control (Rigamonti et al., 2019). This capacity to stabilize blood glucose levels makes WP particularly beneficial for managing metabolic conditions, such as obesity and insulin resistance, while also helping to prevent chronic diseases related to metabolism (Patel, 2015).

WP's immunological and antioxidant benefits are also extensively documented (Castejon et al., 2021; Ha et al., 2021). The presence of bioactive components, such as immunoglobulins and lactoferrin, supports the immune system and helps protect against oxidative stress. Lactoferrin, for instance, exhibits antimicrobial and immunomodulatory properties, while cysteine, an amino acid found in WP, serves as a precursor to glutathione, a crucial antioxidant for cellular defense

TABLE 18.1 Composition of essential amino acids in animal and plant protein sources.

Protein source	Essential amino acids	References
Whey protein	Leucine, Isoleucine, Valine, Lysine, Methionine, Threonine, Phenylalanine, Histidine, Tryptophan	Naidoo et al. (2018)
Casein	Leucine, Isoleucine, Valine, Lysine, Methionine, Threonine, Phenylalanine, Histidine	McGregor and Poppitt (2013)
Egg protein	Leucine, Isoleucine, Valine, Lysine, Methionine, Threonine, Phenylalanine, Histidine	Matsuoka and Sugano (2022)
Rice protein	Leucine, Isoleucine, Valine, Lysine, Methionine, Threonine, Phenylalanine, Histidine, Arginine	Jayaprakash et al. (2022)
Pea protein	Leucine, Isoleucine, Valine, Lysine, Methionine, Threonine, Phenylalanine, Histidine	Lu et al. (2020)
Hemp protein	Leucine, Isoleucine, Valine, Lysine, Methionine, Threonine, Phenylalanine, Histidine, Arginine	Xu et al. (2022)
Wheat protein	Leucine, Isoleucine, Valine, Lysine, Methionine, Threonine, Phenylalanine, Histidine	Khan et al. (2014)
Lupin protein	Leucine, Isoleucine, Valine, Lysine, Methionine, Threonine, Phenylalanine, Histidine	Domínguez et al. (2023)
Oat protein	Leucine, Isoleucine, Valine, Lysine, Methionine, Threonine, Phenylalanine, Histidine	Kumar et al. (2021), Mel and Malalgoda (2022)

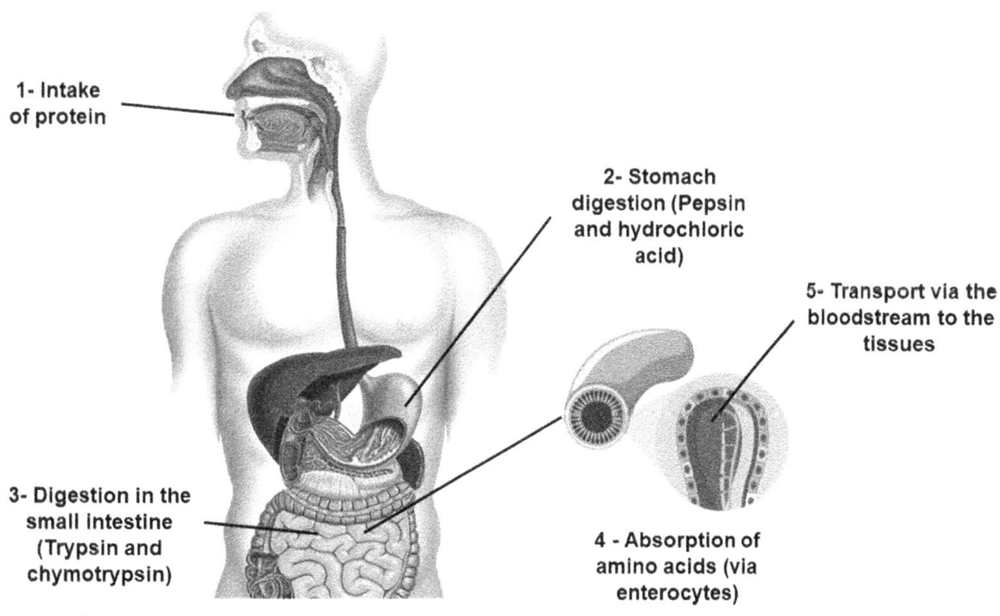

FIGURE 18.1 Whey protein digestion and absorption: From the gastrointestinal tract to amino acid absorption.

(Marshall, 2004; Krissansen, 2007). Glutathione plays a vital role in preventing cellular damage caused by reactive oxygen species, thus contributing to the protection against chronic diseases like obesity, cardiovascular disease, and neuropsychiatric disorders (Rocha-Mendoza et al., 2021).

Additionally, WP consumption has been extensively studied in sports, particularly regarding muscle recovery and protein synthesis following physical exercise (Fabre et al., 2022; Hilkens et al., 2021). Branched-chain amino acids (BCAAs), such as leucine, isoleucine, and valine, are essential in this process, acting as metabolic fuel and stimulating muscle protein synthesis (de Castro et al., 2017). WP also helps reduce postprandial lipemia, as studies have shown that its consumption leads to lower exposure of arteries to triglyceride-rich lipoproteins, suggesting a potential protective effect against cardiovascular diseases (Pal et al., 2010).

Despite some limitations reported in certain studies (Davies et al., 2020), WP supplementation continues to offer distinct advantages. WP's unique attributes—such as efficiency, affordability, and broad availability—solidify its status as one of the most widely used proteins in dietary supplements. WPC is particularly favored among WP derivatives, owing to its low production cost and versatility. Additionally, WP supplementation has been linked to reduced oxidative stress, enhanced hepatic glutathione levels, and prevention of nonalcoholic fatty liver disease (Hamad et al., 2011). WP is also beneficial for specific populations, including older adults, who often face sarcopenia, and athletes seeking to optimize muscle recovery. Its remarkable combination of EAs, high digestibility, and bioavailability offers numerous benefits, from muscle growth to improved metabolic balance.

Therefore, whey protein is distinguished by its exceptional profile of essential amino acids, metabolic advantages, and immunomodulatory properties. Whether used in sports nutrition, public health initiatives, or weight management, WP's versatility and unique benefits reinforce its status as a vital ingredient in both food and nutritional supplements, offering a wealth of insights for those working in the field.

4. Nutritional applications in specific populations

In recent years, the consumption of WPC has gained increasing popularity, primarily due to their ability to enhance nutritional adequacy through high protein content. Beyond their nutritional value, numerous studies have highlighted the bioactive properties of whey proteins, offering significant benefits to specific populations. These benefits extend to the prevention and/or adjunctive treatment of various diseases associated with lifestyle factors and aging, such as diabetes, cancer, malnutrition, sarcopenia, and cardiovascular disease (Zhao et al., 2022). Table 18.2 provides an overview of the nutritional applications of whey protein in different clinical conditions and across various age groups, demonstrating its broad potential to support health and well-being.

4.1 Clinical nutrition: Support for postsurgical recovery and treatment of malnutrition

Effective clinical nutrition plays a crucial role in the recovery of patients, regardless of their clinical condition. In general, hospitalized patients, especially those recovering from surgery, are at risk of protein catabolism and malnutrition due to inflammation, increased energy expenditure, and reduced food intake following surgery (Hao et al., 2024; Niitsu et al., 2016).

TABLE 18.2 Nutritional applications of whey protein in specific populations.

Clinical condition	Age range	Intervention	Main effects	References
Moderate acute malnutrition	Children	Dairy-based supplement with whey protein	Weight gain, improvement in growth markers and recovery rates from malnutrition	Stobaugh et al. (2016)
Cancer and malnutrition	Adults	Whey protein isolate	Improved body composition, muscle strength, body weight and reduced chemotherapy toxicity	Cereda et al. (2019)
Chronic kidney disease and malnutrition	Adults	Whey protein isolate	Adequacy of protein intake, improvements in weight, BMI, skin fold measurements, serum urea and CRP level.	Sahathevan et al. (2018)
Sarcopenia	Elderly	Whey protein enriched with vitamin D and leucine	Gain muscle mass and improve lower limb function	Bauer et al. (2015)
Limited mobility	Elderly	Whey protein and physical activity	Increased muscle density and greater loss of intermuscular fat	Englund et al. (2018)
Type 2 diabetes mellitus	Elderly	Whey protein	Reduction of postprandial blood glucose	Watson et al. (2019)

In this context, Niitsu et al. (2016) investigated the effect of whey protein intake combined with rehabilitation exercises in the immediate postoperative period for patients with hip fractures. Participants who received whey protein supplementation showed significantly greater improvement in knee extension strength compared to the control group, in addition to better walking and toileting skills.

Several constituents may be associated with the beneficial effects of whey protein on postsurgical recovery. Lactoferrin, for example, has antiinflammatory and immunomodulatory properties. Leucine contributes to muscle repair and recovery, while cysteine plays a key role in the synthesis of glutathione, an essential antioxidant and mediator of immune and liver function (Zhao et al., 2022). Other constituents are also linked to a range of bioactive functions, such as promoting tissue repair, eliminating pathogens, and detoxifying the body (Cereda et al., 2019).

Whey protein can also be used in other clinical conditions, such as malnutrition. Azimi et al. (2020) evaluated the effect of administering ready-to-use food supplements on growth indicators in children with mild or moderate malnutrition. Among the protein sources used, the authors highlighted animal proteins, such as whey protein, as essential for improving the biological protein value and amino acid profile of the supplement, noting that these proteins have a significant positive impact on lean mass and growth factors in malnourished children. As a result, children who received the supplement showed a significant increase in weight, height, and body mass index, in addition to a lower prevalence of diarrhea and fever compared to the control group.

The effectiveness of whey proteins in treating malnutrition is further supported by a study that compared a plant-based food supplement with soy protein to a supplement based on dairy ingredients containing whey protein concentrate, both used in the treatment of children with moderate acute malnutrition. The study found that despite the higher total protein and energy content of the plant-based supplement, the dairy-based supplement led to higher recovery rates from malnutrition, greater weight gain, and better growth markers (Stobaugh et al., 2016).

Several factors explain these effects. The high content of BCAAs in animal-based foods can enhance muscle metabolism. Specifically, whey protein contains bioactive factors that promote muscle synthesis and has been shown to help reduce adipose tissue while preserving lean mass (Azimi et al., 2020; Bahwere et al., 2014). Furthermore, its BCAAs and sulfur-containing amino acids play a significant role in regulating various metabolic and cardiovascular functions, as well as helping to maintain blood glucose homeostasis (Zhao et al., 2022).

Additionally, it has been reported that the lactose content of whey protein can play a prebiotic role, increasing the production of short-chain fatty acids in the gut, improving the composition of the intestinal microbiota, and enhancing the immune system. This may lead to faster recovery from illnesses and better overall health maintenance (Azimi et al., 2020).

4.2 Nutrition of the elderly: Aging and sarcopenia

The quality of life in the elderly is directly related to their physical performance, which can be significantly compromised by age-related losses in muscle mass and function. This clinical condition, known as sarcopenia, is a risk factor for mobility in older adults, increasing the likelihood of falls and fractures (Bauer et al., 2015). The progression of sarcopenia is multifaceted and involves, among other factors, neuromuscular deterioration and reduced intake of nutrients, such as proteins and essential amino acids, which are crucial for managing sarcopenia (Camargo et al., 2020; Bauer et al., 2015).

In addition to inadequate consumption, it is well-documented that aging leads to changes in digestive tract functionality, which can affect nutrient absorption. In this context, Lavoisier et al. (2024) studied the impact of age on the digestive capacity of caseins and whey protein. The authors found that protein hydrolysis was significantly reduced in older individuals, attributing this reduction to lower pepsin concentrations, higher pH levels in the gastric phase, and reduced concentrations of pancreatin and bile salts in the intestinal phase (Lavoisier et al., 2024).

These physiological changes associated with aging underscore the importance of employing strategies to increase protein intake in the diets of elderly individuals. Among the available options, whey protein supplementation stands out as a promising approach to aid nutritional adequacy and enhance the health of older adults (Camargo et al., 2020).

The relationship between whey protein intake in the diets of elderly individuals and physical and clinical outcomes demonstrates that supplementation boosts protein synthesis in the body, improving muscle performance and aerobic capacity. This helps protect against sarcopenia and reduces the risk of falls. Moreover, studies suggest that whey protein consumption aids in recovery from illnesses and helps prevent cardiovascular and metabolic risks, as well as complications from hepatic steatosis (Camargo et al., 2020).

Other observed benefits include a significant reduction in interleukin-6 (an inflammatory marker), increased insulin-like growth factor-1, adequate protein intake, and improvements in the ability to perform daily activities (Li et al., 2024). Therefore, adequate supplementation can be a valuable ally in the aging process, as it not only helps maintain muscle mass but also plays a critical role in improving quality of life and overall health.

4.3 Whey protein in special diets: Individuals who are lactose intolerant or have other restrictions

Whey protein is highly digestible, with intestinal hydrolysis occurring faster than that of casein. However, lactose, the most abundant component in whey, is not well-digested by a portion of the population, leading to intolerance to this disaccharide. During the production of whey concentrate, ultrafiltration is used to concentrate proteins, but some lactose remains in the product. To address this, lactose hydrolysis has been implemented to expand the possibilities of using whey in specific situations (Argenta et al., 2021).

Another increasingly common dietary restriction in the population is food allergies. The main whey protein responsible for allergic reactions is β-Lactoglobulin (β-LG), which can trigger IgE-mediated hypersensitivity, leading to degranulation and the release of mediators (Zhao et al., 2022). For individuals with allergies to these proteins, hydrolyzed whey protein is an alternative, as in vitro studies suggest that whey peptides with molecular masses below 3 kDa have negligible antigenicity (Knipping et al., 2014).

Lastly, dietary restrictions for cancer patients are particularly noteworthy, as treatments often generate side effects that include symptoms with nutritional impacts, such as loss of appetite. Additionally, protein consumption recommendations for cancer patients are set at higher levels compared to the general population, making it even more challenging to meet these needs through food alone. These factors highlight the critical need for nutritional interventions, especially to increase protein intake, with whey protein being an excellent supplement option during treatment (Yi et al., 2020; Faccio et al., 2021).

5. Impact of whey on metabolic health

In the general population, metabolic health is more strongly influenced by factors such as diet and physical activity than by genetics (Nouri et al., 2022). Diet plays a crucial role in the prevention of diabetes and cardiovascular disease, particularly through reducing the consumption of added sugars and increasing the intake of proteins, unprocessed fats, and vegetables (Lesgards, 2023). Whey protein promotes satiety due to its high content of BCAAs, such as leucine, isoleucine, tryptophan, and valine (Ho et al., 2022). These amino acids stimulate the production of satiety-related hormones, including cholecystokinin (CCK), glucagon-like peptide-1 (GLP-1), peptide YY (PYY), and ghrelin (Lueders et al., 2022). Additionally, protein intake reduces energy consumption by influencing appetite control mechanisms. Leucine, in particular, plays a prominent role by rapidly reaching the brain and contributing to appetite suppression (Sperringer et al., 2017).

Moreover, whey protein supports vascular health, including reductions in inflammation and cholesterol (Price et al., 2022). According to Price et al. (2022), whey protein can lower proinflammatory cytokines and boost endogenous antioxidant enzymes, making it a promising dietary supplement for treating metabolic conditions such as obesity and type II diabetes (Prokopidis et al., 2023). Figure 18.2 illustrates the key mechanisms by which whey protein helps reduce postprandial glycemia.

5.1 Effect of whey on glycemic control and diabetes

Whey protein contains specific amino acids, especially BCAAs, which are essential for stimulating insulin secretion (Lueders et al., 2022). Because it is rapidly digested, whey protein promotes a quick increase in BCAAs in the bloodstream, favoring insulin release and improving postprandial hyperglycemia. Akhavan et al. (2010) observed that consuming portions greater than 20 g of whey protein resulted in a significant reduction in blood glucose, accompanied by a marked increase in insulin levels.

The reduction in blood glucose and whey protein's insulinotropic effect is observed in both healthy individuals and patients with type 2 diabetes (Frid et al., 2005). Although the exact mechanism by which whey protein stimulates insulin secretion is not yet fully understood, it is believed that the high content of essential amino acids released during digestion is one of the main factors responsible for this response (Jakubowicz and Froy, 2013). Among these amino acids, leucine stands out for promoting insulin secretion by pancreatic β cells, either through its deaminated metabolite, alpha-ketoisocaproic acid (KIC), or by allosteric activation of glutamate dehydrogenase, which increases glutamate oxidation (Nouri et al., 2022; Lesgards, 2023).

Another mechanism by which whey protein may help reduce postprandial glycemia is through appetite suppression and decreased caloric intake (Mignone et al., 2015). This effect appears to be mediated by the action of intestinal hormones, particularly the stimulation of cholecystokinin (CCK), peptide YY (PYY), and GLP-1 (Giezenaar et al., 2018; Nouri et al., 2022).

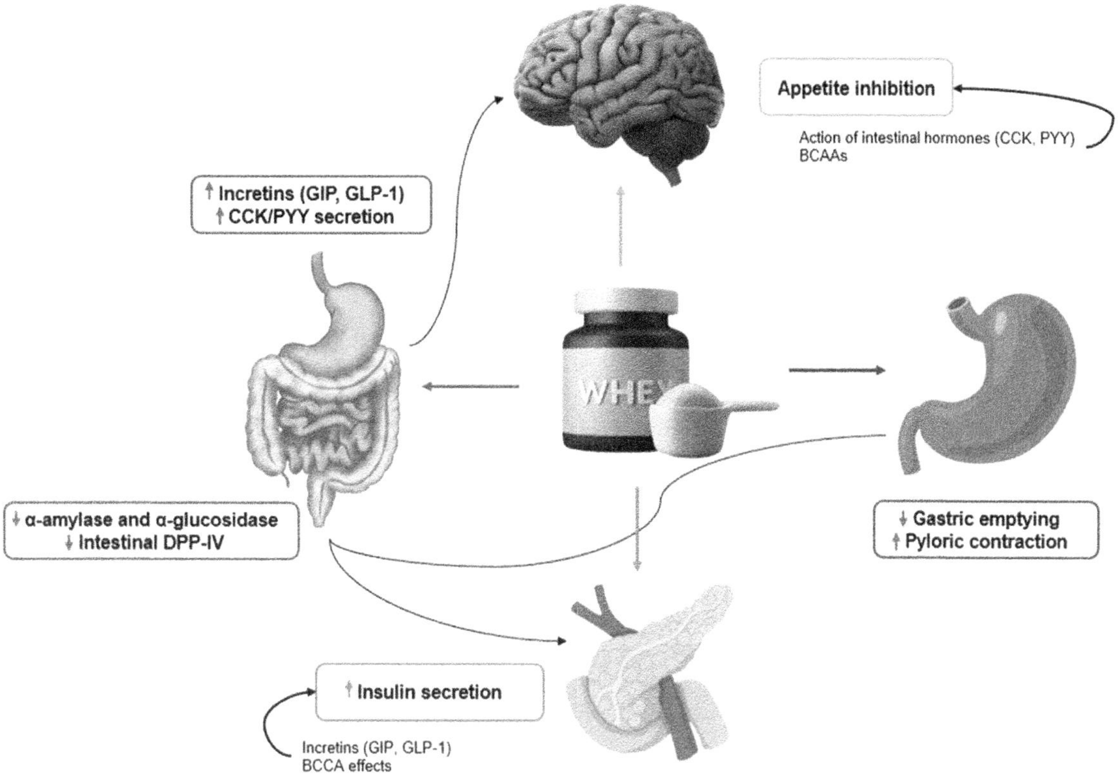

FIGURE 18.2 Mechanisms of whey protein in reducing postprandial glycemia. *GIP*, glucose-dependent insulinotropic polypeptide; *GLP-1*, glucagon-like peptide-1; *CCK*, cholecystokinin; *PYY*, peptide YY; *DPP-IV*, dipeptidyl peptidase-IV; *BCAAs*, branched-chain amino acids.

5.2 Effect of whey on cardiovascular health

Hypertension is a condition characterized by an abnormal increase in blood pressure, which can cause significant damage to various organs in the body (Fisher and Curfman, 2018). This condition is associated with serious complications, such as cardiovascular disease and kidney failure (Ettehad et al., 2016). Natural products have gained prominence in recent years due to their nutritional and therapeutic properties, demonstrating efficacy in several health conditions (Rouhi-Boroujeni et al., 2017). Among these natural compounds, whey protein stands out and has been associated with reducing the risk of cardiovascular disease and mortality in individuals who consume dairy products (Guo et al., 2017). These foods are rich in bioactive substances with important properties, including reducing inflammation, combating oxidative stress, and regulating blood pressure (Vajdi et al., 2023). Evidence suggests that whey protein acts by inhibiting angiotensin-converting enzyme (ACE), which contributes to blood pressure control and improved vascular reactivity (Vajdi et al., 2023). For example, in a study by Oberoi et al. (2022), the authors reported that the ingestion of beverages containing whey protein resulted in a reduction in blood pressure in healthy older men, with a significant decrease observed between 120 and 180 min after consumption.

5.3 Effect of whey on gut health

WP benefits gut health by stimulating the growth of beneficial bacteria, such as *Bifidobacterium* and *Lactobacillus*, and providing protection against pathogens like pathogenic strains of *Escherichia coli* (Subramanian et al., 2014). While its effects are most evident in the development of the infant microbiota, its impact on adults remains poorly understood (Boscaini et al., 2023). Studies in animal models suggest that WP, compared to other protein sources such as meat and casein, reduces colonocyte DNA damage and the production of toxic metabolites, benefiting the gut microbiota (Toden et al., 2007).

Similarly, WP-enriched diets have been shown to increase populations of *Lactobacillus* and *Bifidobacterium* while reducing *Clostridium* (McAllan et al., 2014). Additionally, WP consumption has been linked to increased production of short-chain fatty acids (SCFAs), which contribute to colonocyte energy and intestinal barrier integrity (Sanchez-Moya et al., 2017). However, divergent results have been reported in studies using different methods and populations. For example, in mice fed high-fat diets, the effects of WP varied according to age and duration of intervention (Boscaini et al., 2020). Specific changes in the microbiota were observed, including an increase in *Allobaculum* and a reduction in genera such as *Clostridium* sensu stricto and *Streptococcus*, alongside improvements in lipid metabolism and reduced weight gain on high-fat diets (Boscaini et al., 2020). Despite these promising findings, the underlying mechanisms and the long-term impact on stable microbiotas still require further investigation.

6. Other nutritional applications

The main products derived from whey include whey powder (12%–14% protein), whey protein concentrates (WPC, with 35%–80% protein), and whey protein isolates (WPI, ≥90% protein). These products have distinct compositions, with variations in the contents of macronutrients (proteins, lipids, and lactose) and micronutrients (minerals) (Brodkorb et al., 2016). As described previously, WPC is the most widely used among the products due to its wide availability and low production cost (Kheto et al., 2024a, 2024b). In addition, WP (WPC and WPI) has several benefits, such as a composition rich in essential amino acids, high bioavailability, and rapid digestibility. These factors make it an excellent ingredient for the development of fortified foods aimed at diverse nutritional needs, and it is of great importance in sports and medicine (Soltani et al., 2017).

Athletes, the elderly, children, and individuals with specific dietary restrictions are among the groups that benefit most from whey. Athletes commonly use protein bars, shakes, and snacks enriched with whey, as this protein is essential in optimizing muscle metabolism. Foods rich in leucine (the main branched-chain amino acid present in WP) serve as energy fuels and contribute to increased protein synthesis, positively affecting muscle recovery and growth (Giles et al., 2024). The use of WP by those engaged in physical activity should be properly guided, considering each person's individual needs and specific goals. For example, in cases of muscle hypertrophy, WP is indicated when the individual's protein needs are not met solely through food, necessitating supplementation (Burke et al., 2019).

For the elderly population, the intake of foods fortified with WP is linked to combating sarcopenia, a geriatric syndrome characterized by significant loss of muscle mass, strength, and physical function, which is increasingly recognized as a public health issue (Nasimi et al., 2023). Sarcopenia is associated with the worsening of metabolic disorders, including type 2 diabetes mellitus (Camargo et al., 2020). Hernández-Lepe et al. (2023) found that WP supplementation combined with resistance exercises led to improvements in muscle mass, total lean mass, strength, and, most importantly, the quality of life in elderly individuals with sarcopenia. Similarly, Kamińska et al. (2023) observed that the synergistic effect of WP supplementation and physical exercise improved muscle mass and lower limb functionality in elderly individuals. In addition to these positive effects, WP supplementation has also been associated with a reduction in complications following delicate surgeries, contributing to the healing of wounds and pressure ulcers, with fewer infectious cases (Cawood et al., 2012).

WP has proven to be a promising ingredient in baby food since adequate protein intake is essential for the growth and development of children aged 2 to 5 (Zhang et al., 2023). For children, appropriate use of WP can contribute to a significant reduction in the incidence of cardiovascular diseases, enhancing metabolic health and cardiorespiratory capacity (Camargo et al., 2020).

Another important application of whey protein is in products designed for consumers with lactose intolerance or those seeking reduced-fat foods. WP contains small amounts of lactose, but the isolated versions (e.g. in WPI) are practically free of this carbohydrate (Mehra et al., 2021). This allows for the development of lactose-free products, such as shakes and desserts, which provide all the nutritional benefits of whey without causing digestive discomfort. Fat-free products are also a growing trend, especially among consumers seeking healthy, low-calorie options (Kim et al., 2019). In addition to being a lean protein source, whey has functional properties that aid in the texture and flavor of fat-free foods, such as yogurts and ice cream (Xavier et al., 2024; Roy et al., 2022).

WP's versatility enables its use in a range of new functional products. It is widely utilized to prepare products such as noodles, pasta, bread, biscuits, cakes, and cookies to improve their nutritional value and sensory characteristics while maintaining their original flavor and texture (de Castro et al., 2017; Narmada et al., 2020). Table 18.3 illustrates the products prepared from WP supplementation that have functional appeal.

In addition to fortification, WP can also be utilized in the development of edible films, coatings, fat replacements, hydrogels, and nanoparticles, owing to its rheological and morphological properties (de Castro et al., 2017).

TABLE 18.3 Functional products developed with whey protein supplementation.

Product	Benefits	References
Cassava flour muffins and cookies fortified with WP	Higher protein content; Nutraceutical carriers	Jisha and Padmaja (2011)
WP-fortified wheat bread	Better nutritional quality, with high protein content; Higher calcium levels	Erben and Osella (2017)
Cookie type cookies with WP	Better sensory acceptance with up to 25% WP, delivering a protein-rich product	Parate et al. (2011)
Sheep ice cream with WP	More nutritious ice cream that caters to consumers who avoid beef products	Moschopoulou et al. (2021).
Camel milk yogurt with WPI	After simulated in vitro digestion, WP hydrolyzates provided inhibition of α-amylase and α-glucosidase, inhibition of angiotensin-converting enzyme	Ali et al. (2023)

7. Challenges and opportunities

The nutritional application of whey faces important challenges, such as lactose intolerance and allergies to dairy proteins, limiting its use for certain individuals. The high concentration of lactose in whey limits its use in formulations aimed at individuals with lactose intolerance, a condition characterized by low activity of the enzyme lactase in the gut (da Silva et al., 2022). Faced with this challenge, it is crucial to develop and implement lactose-free dairy protein alternatives to enable safe and effective consumption by people with hypolactasia. Enzymatic hydrolysis and membrane separation technologies aim at the efficient removal of lactose from whey while preserving its functional properties and nutritional profile (Li et al., 2023).

Enzymatic hydrolysis, performed through the application of the enzyme lactase, converts lactose into glucose and galactose, monosaccharides that are easier to absorb by the gastrointestinal tract. This approach not only improves product digestibility but also contributes to sensory enhancement by imparting natural sweetness that can reduce or eliminate the need for additional sweeteners (Kaushal et al., 2022). An alternative strategy is enzyme encapsulation, which is designed to promote the controlled release of lactase along the gastrointestinal tract. This approach allows lactose hydrolysis to be performed directly in the gut, reducing the risk of enzyme degradation during the storage period and increasing the enzyme's effectiveness in the body (Zhang et al., 2017). This method seeks to optimize the bioavailability of lactase and ensure that its activity is preserved until the moment of its action, thus enhancing the digestion of dairy products by individuals with lactose intolerance.

Membrane separation technologies, including ultrafiltration and nanofiltration, have been shown to be effective for lactose removal while preserving whey protein fractions, such as immunoglobulins and lactoferrin, recognized for their bioactive activities and health benefits. The combined application of ultrafiltration and nanofiltration allows the reduction of lactose to undetectable levels, ensuring the maintenance of the integrity and quality of the whey protein profile (Chen et al., 2023). This approach offers an advanced solution for producing dairy products that meet the needs of individuals with lactose intolerance, without compromising the functional value of proteins.

Allergic reactions to whey proteins, such as beta-lactoglobulin and alpha-lactalbumin, constitute another considerable challenge for the food industry, given the high nutritional value of whey in food formulations and supplements. These proteins can induce intense immune responses, affecting both children and adults, which requires the development of safe alternatives with an equivalent nutritional profile (Zhang et al., 2022). The need for solutions that reduce the allergenic potential of whey proteins is essential to increase the accessibility and safety of protein products intended for individuals with hypersensitivity. To this end, techniques such as enzymatic hydrolysis, microbial fermentation, and thermal processing have been the subject of extensive research.

Enzymatic hydrolysis works by fragmenting proteins into smaller peptides, reducing their allergic reactivity. While extensive hydrolysis can significantly decrease the antigenicity of whey proteins, making them safer for individuals with mild allergies, it may not fully eliminate the risk for those with severe hypersensitivity (Saubenova et al., 2024). Controlled fermentation using specific bacterial strains, such as *Lactobacillus*, can induce structural modifications to proteins, making them less recognizable by the immune system and reducing the likelihood of allergic reactions (Shi et al., 2014). However, this method has limitations in completely removing allergenic epitopes and is more suitable for

TABLE 18.4 Technological innovations and functional applications of whey.

Technology/Strategy	Application	Advantages	References
Microencapsulation	Functional foods and supplements	Improved bioavailability and stability of bioactive compounds	Mehta et al. (2022)
Fermentation with probiotics	Fermented dairy products and functional beverages	Allergen reduction, generation of bioactive peptides	Rosa et al. (2023)
Enzymatic hydrolysis	Infant formula and sports nutrition	Facilitates digestibility and increases amino acid absorption	Li and Zhu (2024)
Ultrafiltration and nanofiltration	Whey protein isolate and concentrate	Increased concentration of bioactive proteins and lactose removal	Brião et al. (2024)
Fortification with minerals and vitamins	Enriched products for specific audiences (elderly, pregnant women)	Addition of essential nutrients, increased nutritional value	Traylor et al. (2024)
Enzymatic synthesis of galacto-oligosaccharides (GOS)	Functional foods and beverages	Improved gut health and synergistic effect with whey proteins	Eskandarloo and Abbaspourrad (2018)
Lactose recovery for lactose-free formulations	Lactose-free products for intolerants	Market expansion for lactose intolerant consumers	Dekker et al. (2019)
Freeze-drying of bioactive peptides	Infant formula and functional foods	Maintenance of functional activity and stability during storage	Haas et al. (2024)

individuals with lower sensitivity. Alternatively, thermal processing can reduce allergenic potential by causing protein denaturation and conformational changes, which decreases the exposure of specific epitopes (Zhang et al., 2021). Nevertheless, the effectiveness of this approach depends on the applied processing conditions and may not be sufficient for those with severe hypersensitivity. Therefore, both fermentation and thermal processing should be combined with other technologies to enhance the safety of whey-based protein products.

In addition to these technologies, plant-based proteins, such as pea and rice, have emerged as promising alternatives in the formulation of allergen-free dairy products (Tang et al., 2024). However, replicating the nutritional and functional profile of whey presents significant challenges. To address these challenges, research has focused on combining different plant proteins to optimize the content of essential amino acids while also enhancing characteristics such as solubility and texture, thereby improving the sensory experience to more closely resemble that of dairy products. Therefore, strategies that combine protein modification techniques with plant protein sources can expand access to nutritionally complete and safe products for individuals with whey protein allergies.

Strategies to improve the nutritional value and utilization of whey have been the focus of extensive research, aiming to enhance its functional properties and broaden its application in various food formulations. Table 18.4 presents the most promising approaches that not only improve the nutritional profile of whey but also enhance its functionality in products designed to meet diverse dietary needs, such as improving digestibility, preserving bioactive compounds, enriching with essential nutrients, and optimizing solubility and stability.

8. Conclusion

Whey protein has emerged as a valuable ingredient in human nutrition due to its high-quality protein content, excellent digestibility, and rich essential amino acid profile. Its versatility makes it a key component in supporting muscle, immune, and metabolic health. Widely used by athletes and specific groups, such as the elderly and individuals with specialized dietary needs, whey protein offers numerous benefits. For athletes, it plays a vital role in muscle recovery and protein synthesis, largely due to its high leucine content. For older adults, incorporating whey protein into the diet has shown potential in maintaining muscle mass and improving overall quality of life, helping to prevent age-related muscle loss.

Beyond its impact on muscle health, whey protein also demonstrates positive effects on metabolic health, including supporting glycemic control and cardiovascular well-being, making it a promising option for managing conditions like type 2 diabetes and hypertension. However, certain challenges, such as lactose intolerance and milk protein allergies, limit its use in some populations. Technological advances, such as enzymatic hydrolysis and innovative filtration methods, are helping to address these limitations, thereby increasing its accessibility to a broader range of consumers.

In conclusion, while whey protein stands out as a highly beneficial nutritional ingredient, overcoming existing limitations through innovative strategies is essential for its broader application. This will ensure its continued expansion as a vital component in food products and clinical nutrition solutions.

References

Akhavan, T., Luhovyy, B.L., Brown, P.H., Cho, C.E., Anderson, G.H., 2010. Effect of premeal consumption of whey protein and its hydrolysate on food intake and postmeal glycemia and insulin responses in young adults. Am. J. Clin. Nutr. 91 (4), 966–975. https://doi.org/10.3945/ajcn.2009.28406.

Ali, A.H., Alsalmi, M., Alshamsi, R., Tarique, M., Bamigbade, G., Zahid, I., Ayyash, M., 2023. Effect of whey protein isolate addition on set-type camel milk yogurt: rheological properties and biological activities of the bioaccessible fraction. J. Dairy Sci. 106 (12), 8221–8238.

Alrosan, M., Tan, T.C., Easa, A.M., Gammoh, S., Alu'datt, M.H., 2021. Mechanism of the structural interaction between whey and lentil proteins in the unique creation of a protein structure. J. Food Sci. 86 (12), 5282–5294.

Alrosan, M., Tan, T.C., Easa, A.M., Gammoh, S., Alu'datt, M.H., 2022. Recent updates on lentil and quinoa protein-based dairy protein alternatives: nutrition, technologies, and challenges. Food Chem. 383, 132386.

Argenta, A.B., Nogueira, A., Scheer, A.D.P., 2021. Hydrolysis of whey lactose: kluyveromyces lactis β-galactosidase immobilisation and integrated process hydrolysis-ultrafiltration. Int. Dairy J. 117, 105007. https://doi.org/10.1016/j.idairyj.2021.105007.

Azimi, F., Esmaillzadeh, A., Alipoor, E., Moslemi, M., Yaseri, M., Hosseinzadeh-Attar, M.J., 2020. Effect of a newly developed ready-to-use supplementary food on growth indicators in children with mild to moderate malnutrition. Public Health 185, 290–297. https://doi.org/10.1016/j.puhe.2020.06.025.

Bahwere, P., Banda, T., Sadler, K., Nyirenda, G., Owino, V., Shaba, B., Collins, S., 2014. Effectiveness of milk whey protein-based ready-to-use therapeutic food in treatment of severe acute malnutrition in Malawian under-5 children: a randomised, double-blind, controlled non-inferiority clinical trial. Matern. Child Nutr. 10 (3), 436–451. https://doi.org/10.1111/mcn.12112.

Bauer, J.M., Verlaan, S., Bautmans, I., Brandt, K., Donini, L.M., Maggio, M., Cederholm, T., 2015. Effects of a vitamin D and leucine-enriched whey protein nutritional supplement on measures of sarcopenia in older adults, the provide study: a randomized, double-blind, placebo-controlled trial. J. Am. Med. Dir. Assoc. 16 (9), 740–747. https://doi.org/10.1016/j.jamda.2015.05.021.

Boscaini, S., Cabrera-Rubio, R., Nychyk, O., Speakman, J.R., Cryan, J.F., Cotter, P.D., Nilaweera, K.N., 2020. Age-and duration-dependent effects of whey protein on high-fat diet-induced changes in body weight, lipid metabolism, and gut microbiota in mice. Physiol. Rep. 8 (15), e14523.

Boscaini, S., Skuse, P., Nilaweera, K.N., Cryan, J.F., Cotter, P.D., 2023. The 'Whey' to good health: whey protein and its beneficial effect on metabolism, gut microbiota and mental health. Trends Food Sci. Technol. 133, 1–14.

Brião, V.B., Mossmann, J., Seguenka, B., Graciola, S., Piccin, J.S., 2024. Integrating whey processing: ultrafiltration, nanofiltration, and water reuse from diafiltration. Membranes 14 (9), 191. https://doi.org/10.3390/membranes14090191.

Brodkorb, A., Croguennec, T., Bouhallab, S., Kehoe, J.J., 2016. Heat-induced denaturation, aggregation and gelation of whey proteins. In: Advanced Dairy Chemistry: Volume 1B: Proteins: Applied Aspects. Springer, New York, NY, pp. 155–178. https://doi.org/10.1007/978-1-4939-2800-2_6.

Burke, L.M., Castell, L.M., Casa, D.J., Close, G.L., Costa, R.J., Desbrow, B., Stellingwerff, T., 2019. International association of athletics federations consensus statement 2019: nutrition for athletics. Int. J. Sport Nutr. Exerc. Metabol. 29 (2), 73–84.

Camargo, L.R., Doneda, D., Oliveira, V.R., 2020. Whey protein ingestion in elderly diet and the association with physical, performance and clinical outcomes. Exp. Gerontol. 137, 110936. https://doi.org/10.1016/j.exger.2020.110936.

Castejon, A.M., Spaw, J.A., Rozenfeld, I., Sheinberg, N., Kabot, S., Shaw, A., Packer, E.E., 2021. Improving antioxidant capacity in children with autism: a randomized, double-blind controlled study with cysteine-rich whey protein. Front. Psychiatr. 12, 669089.

Cawood, A.L., Elia, M., Stratton, R.J., 2012. Systematic review and meta-analysis of the effects of high protein oral nutritional supplements. Ageing Res. Rev. 11 (2), 278–296.

Cereda, E., Turri, A., Klersy, C., Cappello, S., Ferrari, A., Filippi, A.R., Caccialanza, R., 2019. Whey protein isolate supplementation improves body composition, muscle strength, and treatment tolerance in malnourished advanced cancer patients undergoing chemotherapy. Cancer Med. 8 (16), 6923–6932. https://doi.org/10.1002/cam4.2517.

Chen, H., Qi, X., Guan, K., Gu, Y., Wang, R., Li, Q., Ma, Y., 2021. Peptides released from bovine α-lactalbumin by simulated digestion alleviated lipid accumulation induced by free fatty acids in HepG2 cells. J. Funct.Foods 85, 104618. https://doi.org/10.1016/j.jff.2020.103954.

Chen, H., Sun, Y., Zhao, H., Qi, X., Cui, H., Li, Q., Ma, Y., 2022. α-Lactalbumin peptide Asp-Gln-Trp alleviates hepatic insulin resistance and modulates gut microbiota dysbiosis in high-fat diet-induced NAFLD mice. Food Funct. 13 (19), 9878–9892. https://doi.org/10.1039/d1fo04157a.

Chen, G.Q., Qu, Y., Gras, S.L., Kentish, S.E., 2023. Separation technologies for whey protein fractionation. Food Eng. Rev. 15, 438–465. https://doi.org/10.1007/s12393-022-09330-2.

da Silva, A.T., de Lima, J.J., Reis, P., Passos, M., Baumgartner, C.G., Sereno, A.B., Krüger, C.C.H., Cândido, L.M.B., 2022. Application of lactose-free whey protein to Greek yogurts: potential health benefits and impact on rheological aspects and sensory attributes. Foods 11 (23), 3861. https://doi.org/10.3390/foods11233861.

Davies, R.W., Bass, J.J., Carson, B.P., Norton, C., Kozior, M., Wilkinson, D.J., Brook, M.S., Atherton, P.J., Smith, K., Jakeman, P.M., 2020. The effect of whey protein supplementation on myofibrillar protein synthesis and performance recovery in resistance-trained men. Nutrients 12 (3), 845.

de Castro, R.J.S., Domingues, M.A.F., Ohara, A., Okuro, P.K., dos Santos, J.G., Brexó, R.P., Sato, H.H., 2017. Whey protein as a key component in food systems: physicochemical properties, production technologies and applications. Food Struct. 14, 17–29. https://doi.org/10.1016/j.foostr.2017.05.004.

Dekker, P.J.T., Koenders, D., Bruins, M.J., 2019. Lactose-free dairy products: market developments, production, nutrition and health benefits. Nutrients 11 (3), 551. https://doi.org/10.3390/nu11030551.

Djouab, A., Aïder, M., 2019. Whey permeate integral valorisation via in situ conversion of lactose into lactulose in an electro-activation reactor modulated by anion and cation exchange membranes. Int. Dairy J. 89, 6–20. https://doi.org/10.1016/j.idairyj.2018.07.019.

do Vale, R.T.R., de Paula, H.M.C., Coelho, Y.L., Rezende, J.D.P., Vidigal, M.C.T.R., da Silva, L.H.M., Pires, A.C.D.S., 2022. β-Lactoglobulin and resveratrol nanocomplex formation is driven by solvation water release. Food Res. Int. 158, 111567. https://doi.org/10.1016/j.foodres.2022.111567.

Domínguez, R., Bermúdez, R., Pateiro, M., Lucas-González, R., Lorenzo, J.M., 2023. Optimization and characterization of lupin protein isolate obtained using alkaline solubilization-isoelectric precipitation. Foods 12 (20), 3875. https://doi.org/10.3390/foods12203875.

Englund, D.A., Kirn, D.R., Koochek, A., Zhu, H., Travison, T.G., Reid, K.F., Fielding, R.A., 2018. Nutritional supplementation with physical activity improves muscle composition in older adults with limited mobility: the VIVE2 study. J. Gerontol. A 73 (1), 95–101. https://doi.org/10.1093/gerona/glx141.

Erben, M., Osella, C.A., 2017. Optimization of mold wheat bread fortified with soy flour, pea flour and whey protein concentrate. Food Sci. Technol. Int. 23 (5), 457–468.

Eskandarloo, H., Abbaspourrad, A., 2018. Production of galacto-oligosaccharides from whey permeate using β-galactosidase immobilized on functionalized glass beads. Food Chem. 251, 115–124. https://doi.org/10.1016/j.foodchem.2018.01.068.

Ettehad, D., et al., 2016. Blood pressure lowering for prevention of cardiovascular disease and death: a systematic review and meta-analysis. Lancet 387 (10022), 957–967.

Fabre, M., et al., 2022. Effects of native whey protein and carbohydrate supplement on physical performance and plasma markers of muscle damage and inflammation during a simulated rugby sevens tournament: a double-blind, randomized, placebo-controlled, crossover study. Nutrients 14 (22), 4780.

Faccio, A.A., et al., 2021. Oral nutritional supplementation in cancer patients who were receiving chemo/chemoradiation therapy: a multicenter, randomized phase II study. Nutr. Cancer 73 (3), 442–449. https://doi.org/10.1080/01635581.2020.1758170.

Fang, B., et al., 2018. α-Lactalbumin-oleic acid complex kills tumor cells by inducing excess energy metabolism but inhibiting mRNA expression of the related enzymes. J. Dairy Sci. 101, 4853–4863. https://doi.org/10.3168/JDS.2017-13731.

Fisher, N.D., Curfman, G., 2018. Hypertension—a public health challenge of global proportions. JAMA 320 (17), 1757–1759.

Floret, C., et al., 2023. Replacement of animal proteins in food: how to take advantage of nutritional and gelling properties of alternative protein sources. Crit. Rev. Food Sci. Nutr. 63 (7), 920–946.

Frid, A.H., Nilsson, M., Holst, J.J., Björck, I.M., 2005. Effect of whey on blood glucose and insulin responses to composite breakfast and lunch meals in type 2 diabetic subjects. Am. J. Clin. Nutr. 82 (1), 69–75. https://doi.org/10.1093/ajcn/82.1.69.

Ge, X., et al., 2024. α-Lactalbumin: functional properties and potential health benefits. Food Biosci. 60, 104371. https://doi.org/10.1016/J.FBIO.2024.104371.

Giezenaar, C., et al., 2018. Effect of gender on the acute effects of whey protein ingestion on energy intake, appetite, gastric emptying and gut hormone responses in healthy young adults. Nutr. Diabetes. 8 (1), 40. https://doi.org/10.1038/s41387-018-0048-7.

Giles, H., et al., 2024. A narrative review investigating the potential effect of lubrication as a mitigation strategy for whey protein-associated mouthdrying. Food Chem. 436, 137603.

Gorissen, S.H., et al., 2018. Protein content and amino acid composition of commercially available plant-based protein isolates. Amino Acids 50 (12), 1685–1695.

Grinberg, V.Y., et al., 2023. Energetics and mechanism of complexation between β-lactoglobulin and oligochitosan. Food Hydrocoll. 134, 108021. https://doi.org/10.1016/J.FOODHYD.2022.108021.

Guo, J., et al., 2017. Milk and dairy consumption and risk of cardiovascular diseases and all-cause mortality: dose–response meta-analysis of prospective cohort studies. Eur. J. Epidemiol. 32 (4), 269–287.

Ha, D.J., et al., 2021. Dietary whey protein supplementation increases immunoglobulin G production by affecting helper T cell populations after antigen exposure. Foods 10 (1), 194.

Haas, J., et al., 2024. Effects of spray drying and freeze drying on the protein profile of whey protein concentrate. J. Food Sci. 89 (11), 7477–7493. https://doi.org/10.1111/1750-3841.17349.

Hamad, E.M., et al., 2011. Protective effect of whey proteins against nonalcoholic fatty liver in rats. Lipids Health Dis. 10, 1–7.

Hao, X., et al., 2024. Risk factors for hospitalization-associated disability among older patients: a systematic review and meta-analysis. Ageing Res. Rev., 102516 https://doi.org/10.1016/j.arr.2024.102516.

Haraguchi, F.K., De Abreu, W.C., De Paula, H., 2006. Proteínas do soro do leite: composição, propriedades nutricionais, aplicações no esporte e benefícios para a saúde humana. Rev. Nutr. 19, 479–488. https://doi.org/10.1590/S1415-52732006000400007.

Hernández-Lepe, M.A., et al., 2023. Exercise programs combined with diet supplementation improve body composition and physical function in older adults with sarcopenia: a systematic review. Nutrients 15 (8), 1998.

Hilkens, L., et al., 2021. Whey protein supplementation does not accelerate recovery from a single bout of eccentric exercise. J. Sports Sci. 39 (3), 322–331.

Ho, T.M., Bhandari, B.R., Bansal, N., 2022. Functionality of bovine milk proteins and other factors in foaming properties of milk: a review. Crit. Rev. Food Sci. Nutr. 62 (17), 4800–4820.

Ho, C.Y., et al., 2023. Pre-habilitation with exercise and high whey-protein-based meal replacement therapy promotes weight loss and preserves muscle mass before bariatric surgery. Asian J. Surg. 46, 3716–3721. https://doi.org/10.1016/J.ASJSUR.2023.03.026.

Jakopović, K.L., Barukčić, I., Božanić, R., 2016. Physiological significance, structure and isolation of α-lactalbumin. Mljekarstvo: Časopis Za Unaprjeđenje Proizvodnje I Prerade Mlijeka 66, 3–11. https://doi.org/10.15567/MLJEKARSTVO.2016.0101.

Jakubowicz, D., Froy, O., 2013. Biochemical and metabolic mechanisms by which dietary whey protein may combat obesity and Type 2 diabetes. J. Nutr. Biochem. 24 (1), 1–5. https://doi.org/10.1016/j.jnutbio.2012.07.008.

Jayaprakash, G., Bains, A., Chawla, P., Fogarasi, M., Fogarasi, S., 2022. A narrative review on rice proteins: Current scenario and food industrial application. Polymers 14 (15), 3003. https://doi.org/10.3390/polym14153003.

Jisha, S., Padmaja, G., 2011. Whey protein concentrate fortified baked goods from cassava-based composite flours: nutritional and functional properties. Food Bioprocess Technol. 4. https://doi.org/10.1007/s11947-010-0432-5.

Kamińska, M.S., et al., 2023. The impact of whey protein supplementation on sarcopenia progression among the elderly: a systematic review and meta-analysis. Nutrients 15 (9), 2039.

Karim, A., Aider, M., 2022. Comprehensive utilisation of electro-activated whey-based media in cell growth, metabolite production and aroma compounds synthesis using a starter culture originated from kefir grains. Int. Dairy J. 126, 105276. https://doi.org/10.1016/J.IDAIRYJ.2021.105276.

Kaushal, J., Singh, G., Arya, S.K., 2022. Emerging trends and future prospective in enzyme technology. In: Value-Addition in Food Products and Processing through Enzyme Technology. Elsevier, pp. 491–503. https://doi.org/10.1016/b978-0-323-89929-1.00036-6.

Khan, M.S., Ali, E., Ali, S., Khan, W.M., Sajjad, M.A., Hussain, F., 2014. Assessment of essential amino acids in wheat proteins: a case study. J. Biodivers. Environ. Sci. 4, 185–189.

Kheto, A., Adhikary, U., Dhua, S., Sarkar, A., Kumar, Y., Vashishth, R., Shrestha, B.B., Saxenan, D.C., 2024a. A review on advancements in emerging processing of whey protein: enhancing functional and nutritional properties for functional food applications. Food Saf. Health 3 (1), 23–45. https://doi.org/10.1002/fsh3.12067.

Kheto, A., Sehrawat, R., Gul, K., Kumar, L., 2024b. Effect of extraction pH on amino acids, nutritional, in-vitro protein digestibility, intermolecular interactions, and functional properties of guar germ proteins. Food Chem. 444, 138628.

Kim, S.J., et al., 2019. Dietary fat intake and age modulate the composition of the gut microbiota and colonic inflammation in C57BL/6J mice. BMC Microbiol. 19, 1–11.

Knipping, K., et al., 2014. Development of β-lactoglobulin-specific chimeric human IgEκ monoclonal antibodies for in vitro safety assessment of whey hydrolysates. PLoS One 9 (8), e106025. https://doi.org/10.1371/journal.pone.0106025.

Krissansen, G.W., 2007. Emerging health properties of whey proteins and their clinical implications. J. Am. Coll. Nutr. 26 (6), 713S–723S.

Kumar, L., Sehrawat, R., Kong, Y., 2021. Oat proteins: a perspective on functional properties. LWT Food Sci. Technol. (Lebensm.-Wiss.-Technol.) 152, 112307.

Lavoisier, A., Chevalier, S., Henry, G., Ossemond, J., Harel-Oger, M., Garric, G., Morzel, M., 2024. Impact of age on the digestion of cream cheese formulated with opposite caseins to whey proteins ratios: an in vitro study. Food Res. Int., 114621 https://doi.org/10.1016/j.foodres.2024.114621.

Lesgards, J.F., 2023. Benefits of whey proteins on type 2 diabetes mellitus parameters and prevention of cardiovascular diseases. Nutrients 15 (5), 1294.

Li, J., Zhu, F., 2024. Whey protein isolates in novel food formulations. Food Res. Int. 165, 112316.

Li, A., Zheng, J., Han, X., Yang, S., Cheng, S., Zhao, J., Zhou, W., Lu, Y., 2023. Advances in low-lactose/lactose-free dairy products and their production. Foods 12 (13), 2553. https://doi.org/10.3390/foods12132553.

Li, M.L., Zhang, F., Luo, H.Y., Quan, Z.W., Wang, Y.F., Huang, L.T., Wang, J.H., 2024. Improving sarcopenia in older adults: a systematic review and meta-analysis of randomized controlled trials of whey protein supplementation with or without resistance training. J. Nutr. Health Aging 28 (4), 100184. https://doi.org/10.1016/j.jnha.2024.100184.

Liu, C., Liu, Z., Sun, X., Zhang, S., Wang, S., Feng, F., Wang, D., Xu, Y., 2018. Fabrication and characterization of β-lactoglobulin-based nanocomplexes composed of chitosan oligosaccharides as vehicles for delivery of astaxanthin. J. Agric. Food Chem. 66, 6717–6726. https://doi.org/10.1021/ACS.JAFC.8B00834/ASSET/IMAGES/MEDIUM/JF-2018-00834B_0009.

Lu, Z.X., He, J.F., Zhang, Y.C., Bing, D.J., 2020. Composition, physicochemical properties of pea protein and its application in functional foods. Crit. Rev. Food Sci. Nutr. 60 (15), 2593–2605. https://doi.org/10.1080/10408398.2019.1651248.

Lueders, B., Kanney, B.C., Krone, M.J., Gannon, N.P., Vaughan, R.A., 2022. Effect of branched-chain amino acids on food intake and indicators of hunger and satiety: a narrative summary. Hum. Nutr. Metab. 30, 200168. https://doi.org/10.1016/j.hnm.2022.200168.

Marshall, K.N., 2004. Therapeutic applications of whey protein. Alternative Med. Rev. 9 (2), 136–156.

Matsuoka, R., Sugano, M., 2022. Health functions of egg protein. Foods 11 (15), 2309. https://doi.org/10.3390/foods11152309.

McAllan, L., Skuse, P., Cotter, P.D., Connor, P.O., Cryan, J.F., Ross, R.P., Nilaweera, K.N., 2014. Protein quality and the protein to carbohydrate ratio within a high fat diet influences energy balance and the gut microbiota in C57BL/6J mice. PLoS One 9 (2), e88904.

McGregor, R.A., Poppitt, S.D., 2013. Milk protein for improved metabolic health: a review of the evidence. Nutr. Metab. 10 (1), 46. https://doi.org/10.1186/1743-7075-10-46.

Mehra, R., Kumar, H., Kumar, N., Ranvir, S., Jana, A., Buttar, H.S., Guiné, R.P., 2021. Whey proteins processing and emergent derivatives: an insight perspective from constituents, bioactivities, functionalities to therapeutic applications. J. Funct.Foods 87, 104760.

Mehta, N., et al., 2022. Microencapsulation as a noble technique for the application of bioactive compounds in the food industry: a comprehensive review. Appl. Sci. 12 (3), 1424. https://doi.org/10.3390/app12031424.

Mel, R., Malalgoda, M., 2022. Oat protein as a novel protein ingredient: Structure, functionality, and factors impacting utilization. Cereal Chem. 99 (1), 21–36. https://doi.org/10.1002/cche.10488.

Mestawet, A.T., France, T.C., Mulcahy, P.G.J., O'Mahony, J.A., 2024. Microfiltration retentate co-product from whey protein isolate production - composition, processing, applications and potential for value addition. Trends Food Sci. Technol. 153, 104739. https://doi.org/10.1016/J.TIFS.2024.104739.

Mignone, L.E., Wu, T., Horowitz, M., Rayner, C.K., 2015. Whey protein: The "whey" forward for treatment of type 2 diabetes? World J. Diabetes. 6 (14), 1274–1284. https://doi.org/10.4239/wjd.v6.i14.1274.

Minj, S., Anand, S., 2020. Whey proteins and its derivatives: bioactivity, functionality, and current applications. Dairy 2020 1, 233–258. https://doi.org/10.3390/DAIRY1030016.

Moschopoulou, E., Dernikos, D., Zoidou, E., 2021. Ovine ice cream made with addition of whey protein concentrates of ovine-caprine origin. Int. Dairy J. 122, 105146.

Naidoo, K., Naidoo, R., Bangalee, V., 2018. Understanding the amino acid profile of whey protein products. Glob. J. Health Sci. 10 (9), 45.

Narmada, M., Lakshmi, V.V., Hymavathi, T.V., Kameswari, S.L., 2020. Sensory and nutritional properties of protein enriched biscuits. Int. Res. J. Pure Appl. Chem. 21 (10), 1–8.

Nasimi, N., Sohrabi, Z., Nunes, E.A., Sadeghi, E., Jamshidi, S., Gholami, Z., Phillips, S.M., 2023. Whey protein supplementation with or without vitamin D on sarcopenia-related measures: a systematic review and meta-analysis. Adv. Nutr. 14 (4), 762–773.

Niitsu, M., Ichinose, D., Hirooka, T., Mitsutomi, K., Morimoto, Y., Sarukawa, J., Yamazaki, K., 2016. Effects of combination of whey protein intake and rehabilitation on muscle strength and daily movements in patients with hip fracture in the early postoperative period. Clin. Nutr. 35 (4), 943–949. https://doi.org/10.1016/j.clnu.2015.07.006.

Nouri, M., Gargari, B.P., Tajfar, P., Tarighat-Esfanjani, A., 2022. A systematic review of whey protein supplementation effects on human glycemic control: a mechanistic insight'. Diabet. Metabol. Syndr. Clin. Res. Rev. 16 (7), 102540.

Oberoi, A., Giezenaar, C., Lange, K., Jones, K.L., Horowitz, M., Chapman, I., Soenen, S., 2022. Blood pressure and heart rate responses following dietary protein intake in older men. Nutrients 14 (9), 1913.

Pal, S., Ellis, V., Ho, S., 2010. Acute effects of whey protein isolate on cardiovascular risk factors in overweight, post-menopausal women. Atherosclerosis 212 (1), 339–344. https://doi.org/10.1016/j.atherosclerosis.2010.05.032.

Parate, V.R., Kawadkar, D.K., Sonawane, S.S., 2011. Study of whey protein concentrate fortification in cookies variety biscuits. Int. J. Food Eng. 7 (2).

Patel, S., 2015. Emerging trends in nutraceutical applications of whey protein and its derivatives. J. Food Sci. Technol. 52 (11), 6847–6858. https://doi.org/10.1007/s13197-015-1894-0.

Price, D., Jackson, K.G., Lovegrove, J.A., Givens, D.I., 2022. The effects of whey proteins, their peptides and amino acids on vascular function. Nutr. Bull. 47 (1), 9–26.

Prokopidis, K., Mazidi, M., Sankaranarayanan, R., Tajik, B., McArdle, A., Isanejad, M., 2023. Effects of whey and soy protein supplementation on inflammatory cytokines in older adults: a systematic review and meta-analysis. Br. J. Nutr. 129 (5), 759–770.

Ramani, A., Hazra, T., Mudgil, S., Mudgil, D., 2024. Emerging potential of whey proteins in prevention of cancer. Food Human. 2, 100199. https://doi.org/10.1016/j.foohum.2023.12.007.

Rigamonti, A.E., et al., 2019. Whey proteins reduce appetite, stimulate anorexigenic gastrointestinal peptides and improve glucometabolic homeostasis in young obese women. Nutrients 11 (2), 247. https://doi.org/10.3390/nu11020247.

Rocha-Mendoza, D., et al., 2021. Invited review: Acid whey trends and health benefits. J. Dairy Sci. 104 (2), 1262–1275. https://doi.org/10.3168/jds.2020-19038.

Rosa, L.S., et al., 2023. Probiotic fermented whey-milk beverages: effect of different probiotic strains on the physicochemical characteristics, biological activity, and bioactive peptides. Food Res. Int. 164, 112396. https://doi.org/10.1016/j.foodres.2022.112396.

Rouhi-Boroujeni, H., Heidarian, E., Rouhi-Boroujeni, H., Deris, F., Rafieian-Kopaei, M., 2017. Medicinal plants with multiple effects on cardiovascular diseases: a systematic review. Curr. Pharm. Des. 23 (7), 999–1015.

Roy, S., Hussain, S.A., Prasad, W.G., Khetra, Y., 2022. Quality attributes of high protein ice cream prepared by incorporation of whey protein isolate. Appl. Food Res. 2 (1), 100029.

Sahathevan, S., Se, C.H., Ng, S., Khor, B.H., Chinna, K., Goh, B.L., Karupaiah, T., 2018. Clinical efficacy and feasibility of whey protein isolates supplementation in malnourished peritoneal dialysis patients: a multicenter, parallel, open-label randomized controlled trial. Clin. Nutr. ESPEN 25, 68–77. https://doi.org/10.1016/j.clnesp.2018.04.002.

Sanchez-Moya, T., López-Nicolás, R., Planes, D., González-Bermúdez, C.A., Ros-Berruezo, G., Frontela-Saseta, C., 2017. In vitro modulation of gut microbiota by whey protein to preserve intestinal health. Food Funct. 8 (9), 3053–3063.

Saubenova, M., Oleinikova, Y., Rapoport, A., Maksimovich, S., Yermekbay, Z., Khamedova, E., 2024. Whey as an optimal substrate for the development of functional products enriched with bioactive peptides. Preprints. https://doi.org/10.20944/preprints202406.0540.v1.

Sharma, R., 2019. Whey Proteins in Functional Foods. Whey Proteins: From Milk to Medicine, pp. 637–663. https://doi.org/10.1016/B978-0-12-812124-5.00018-7.

Shi, J., Luo, Y., Xiao, Y., Li, Z., Xu, Q., Yao, M., 2014. Effects of fermentation by Lactobacillus casei on the antigenicity and allergenicity of four bovine milk proteins. Int. Dairy J. 35 (1), 75–80. https://doi.org/10.1016/j.idairyj.2013.10.010.

Soltani, M., Say, D., Guzeler, N., 2017. Functional properties and nutritional quality of whey proteins. J. Int. Environ. Appl. Sci. 12 (4), 334–338.

Sperringer, J.E., Addington, A., Hutson, S.M., 2017. Branched-chain amino acids and brain metabolism. Neurochem. Res. 42, 1697–1709.

Stobaugh, H.C., Ryan, K.N., Kennedy, J.A., Grise, J.B., Crocker, A.H., Thakwalakwa, C., Trehan, I., 2016. Including whey protein and whey permeate in ready-to-use supplementary food improves recovery rates in children with moderate acute malnutrition: a randomized, double-blind clinical trial. Am. J. Clin. Nutr. 103 (3), 926–933. https://doi.org/10.3945/ajcn.115.124636.

Subramanian, S., Huq, S., Yatsunenko, T., Haque, R., Mahfuz, M., Alam, M.A., Gordon, J.I., 2014. Persistent gut microbiota immaturity in malnourished Bangladeshi children. Nature 510 (7505), 417–421.

Tang, Q., Roos, Y.H., Miao, S., 2024. Comparative studies of structural and thermal gelation behaviours of soy, lentil and whey protein: a pH-dependency evaluation. Food Hydrocoll. 146, 109240. https://doi.org/10.1016/j.foodhyd.2023.109240.

Thampy, A., Palani Kumar, M.K., Serva Peddha, M., Reddy, M., 2024a. The effectiveness of whey proteins in prevention and treatment of cancer: a review. Crit. Rev. Food Sci. Nutr. 64, 2088–2104. https://doi.org/10.1080/10408398.2022.2121256.

Thampy, A., Palani Kumar, M.K., Serva Peddha, M., Reddy, M., 2024b. A eficácia das proteínas do soro de leite na prevenção e tratamento do câncer: Uma revisão. Crit. Rev. Food Sci. Nutr. 64, 2088.

Toden, S., Bird, A.R., Topping, D.L., Conlon, M.A., 2007. Differential effects of dietary whey, casein and soya on colonic DNA damage and large bowel SCFA in rats fed diets low and high in resistant starch. Br. J. Nutr. 97 (3), 535–543.

Traylor, D.A., Nunes, E.A., Lees, M., Plucinski, M.J., Vondrasek, M., Phillips, S.M., 2024. Unveiling the potential of a whey protein- and leucine-enriched multi-nutrient formula high in vitamin D3: a scoping review on its role in enhancing body composition, strength, and physical function in sarcopenia. Arch. Gerontol. Geriatr. Plus 1 (4), 100058. https://doi.org/10.1016/j.aggp.2024.100058.

Uversky, V.N., El-Fakharany, E.M., Abu-Serie, M.M., Almehdar, H.A., Redwan, E.M., 2017. Divergent anticancer activity of free and formulated camel milk α-lactalbumin. Cancer Invest. 35, 610–623. https://doi.org/10.1080/07357907.2017.1373783.

Vajdi, M., Musazadeh, V., Zareei, M., Adeli, S., Karimi, A., Hojjati, A., Farhangi, M.A., 2023. The effects of whey protein on blood pressure: a systematic review and dose-response meta-analysis of randomized controlled trials. Nutr. Metabol. Cardiovasc. Dis. 33 (9), 1633–1646.

Wang, C., Wen, H. xing, Yang, S., Ma, C. yue, Wang, X. mei, Tu, Z.C., Shao, Y.H., Liu, J., 2024a. Effect of dynamic high-pressure microfluidization on the structural, emulsifying properties, in vitro digestion and antioxidant activity of whey protein isolate. Int. J. Biol. Macromol. 283, 137720. https://doi.org/10.1016/J.IJBIOMAC.2024.137720.

Wang, Z., Liu, D., Hong, X., Tao, X., Zhang, J., Zhang, J., Hou, Y., Wu, T., Liu, X., Zhou, P., 2024b. A ligação do cálcio afeta a digestão gastrointestinal in vitro da α-lactalbumina bovina em condições de lactentes, adultos e idosos. Int. Dairy J. 15.

Watson, L.E., Phillips, L.K., Wu, T., Bound, M.J., Checklin, H.L., Grivell, J., Rayner, C.K., 2019. A whey/guar "preload" improves postprandial glycaemia and glycated haemoglobin levels in type 2 diabetes: a 12-week, single-blind, randomized, placebo-controlled trial. Diabetes Obes. Metabol. 21 (4), 930–938. https://doi.org/10.1111/dom.13604.

Wen-Qiong, W., Yun-Chao, W., Xiao-Feng, Z., Rui-Xia, G., Mao-Lin, L., 2019. Whey protein membrane processing methods and membrane fouling mechanism analysis. Food Chem. 289, 468–481.

Xavier, L.S., Pacheco, F.C., Nalon, G.A., Cunha, J.S., Santos, F.R.D., Pacheco, A.F.C., Leite Júnior, B.R.D.C., 2024. Strategies to improve the quality of goat yogurt: whey protein supplementation and milk pre-treatment with high shear dispersion assisted by ultrasound. Foods 13 (10), 1558.

Xu, J., et al., 2022. Effect of hemp protein on the physicochemical properties and flavor components of plant-based yogurt. LWT 172, 114145. https://doi.org/10.1016/j.lwt.2022.114145.

Yi, H.C., Ibrahim, Z., Abu Zaid, Z., Mat Daud, Z.A., Md Yusop, N.B., Omar, J., Jamhuri, N., 2020. Impact of enhanced recovery after surgery with preoperative whey protein-infused carbohydrate loading and postoperative early oral feeding among surgical gynecologic cancer patients: an open-labelled randomized controlled trial. Nutrients 12 (1), 264. https://doi.org/10.3390/nu12010264.

Zhang, H., Duan, S., Yu, Y., Wang, J., Chen, X.D., Szeto, I.M.Y., Jin, Y., 2023. Impact of casein-to-whey protein ratio on gastric emptying, proteolysis, and peptidome profile of fermented milk during in vitro dynamic gastrointestinal digestion in preschool children. Food Chem. 405, 134840.

Zhang, Z., Ma, R., Xu, Y., Chi, L., Li, Y., Mu, G., Zhu, X., 2022. Investigation of the structure and allergic potential of whey protein by both heating sterilization and simulation with molecular dynamics. Foods 11 (24), 4050. https://doi.org/10.3390/foods11244050.

Zhang, Y., Min, L., Zhang, S., Zheng, N., Li, D., Sun, Z., Wang, J., 2021. Proteomics analysis reveals altered nutrients in the whey proteins of dairy cow milk with different thermal treatments. Molecules 26 (15), 4628. https://doi.org/10.3390/molecules26154628.

Zhang, Z., Zhang, R., McClements, D.J., 2017. Lactase (β-galactosidase) encapsulation in hydrogel beads with controlled internal pH microenvironments: Impact of bead characteristics on enzyme activity. Food Hydrocoll. 67, 85–93. https://doi.org/10.1016/j.foodhyd.2017.01.005.

Zhao, C., Chen, N., Ashaolu, T.J., 2022. Whey proteins and peptides in health-promoting functions–A review. Int. Dairy J. 126, 105269.

Zhao, Q., Zheng, W., Yuan, Z., Wang, X., Huang, A., 2023. Anti-inflammatory effect of two novel peptides derived from Binglangjiang buffalo whey protein in lipopolysaccharide-stimulated RAW264.7 macrophages. Food Chem. 429, 136804. https://doi.org/10.1016/J.FOODCHEM.2023.136804.

Further reading

FAOSTAT (Food and Agriculture Organization Corporate Statistical Database), 2021. Crops and Livestock Products. Available at: (Accessed 19 December 2024).

Giesenaar, C., et al., 2018. Effects of substitution, and adding of carbohydrate and fat to whey-protein on energy intake, appetite, gastric emptying, glucose, insulin, ghrelin, CCK and GLP-1 in healthy older men—a randomized controlled trial. Nutrients 10 (2), 113.

Liu, S., Lim, S.L., 2020. Whey protein isolate and hydrolysates in functional foods. Crit. Rev. Food Sci. Nutr. 60 (11), 1–17.

Magar, K.D., Dholakia, D.P., 2021. Whey protein hydrolysates in human health applications: impacts of digestion, bioavailability, and health outcomes. Crit. Rev. Food Sci. Nutr. 61 (9), 1525–1535.

Oliveira, D., Mazzutti, S., Martins, S.F., Lima, A., 2021. Bioprocessing and applications of whey protein: a review. Food Rev. Int. 37 (4), 429–448.

Radhakrishnan, M., Premavalli, K., 2014. Processing and nutritional quality of whey protein. J. Food Sci. Technol. 51 (10), 2782–2793. https://doi.org/10.1007/s11483-014-0672-x.

Wang, Z., Zhang, B., 2023. Effects of whey protein on cognitive function and neuroplasticity. Food Funct. 14 (5), 1968–1980.

Zambrowicz, A., Swiderska, S., Kaczmarek, Z., 2022. Whey proteins as functional ingredients in food: recent advances and emerging trends. Trends Food Sci. Technol. 115, 133–144. https://doi.org/10.1016/j.tifs.2021.11.015.

Chapter 19

Whey: A source of bioactive peptides

Carla Paulo Vieira[1, 2], Italo Rennan Sousa Vieira[1, 2], Abel Alberto Massingue Júnior[3], Katherine Gutiérrez-Alzate[4, 5], Joselene Conceição Nunes Nascimento[4], Arlen Carvalho de Oliveira Almeida[1, 2, 5], Hannay Crystynah Almeida de Souza[1, 2], Anamaria M.P. Dos Santos[1, 2], Thiago da Silveira Álvares[6] and Carlos Adam Conte Junior[1, 2]

[1]*Center for Food Analysis (NAL), Technological Development Support Laboratory (LADETEC), Federal University of Rio de Janeiro (UFRJ), Cidade Universitária, Rio de Janeiro, Brazil;* [2]*Laboratory of Advanced Analysis in Biochemistry and Molecular Biology (LAABBM), Department of Biochemistry, Federal University of Rio de Janeiro (UFRJ), Cidade Universitária, Rio de Janeiro, Brazil;* [3]*Department of Fish Processing and Quality Control Engineering, Higher Institute Polytechnic of Quissico, Division of Fisheries and Aquaculture, Inhambane, Mozambique;* [4]*Program in Food Science (PGAli), Faculty of Pharmacy, Federal University of Bahia (UFBA), Salvador, Brazil;* [5]*Laboratorio de Inspeção e Tecnologia de Leite e Derivados (LAITLACTEOS), Federal University of Bahia (UFBA), Salvador, Brazil;* [6]*Multidisciplinary Center UFRJ-Macaé, Federal University of Rio de Janeiro, Rio de Janeiro, Brazil*

1. Introduction

Whey is a by-product generated during dairy processing. However, its high perishability and lower nutritional density compared with milk limit its commercial use. In this context, hydrolysis represents a viable strategy to enhance the commercial value of whey by releasing bioactive peptides (Irazoqui et al., 2024). These peptides are amino acid sequences encrypted within the protein structure, and various technologies have emerged to liberate them from the native protein matrix (Luparelli et al., 2025).

In the gastrointestinal tract, the sequential action of digestive enzymes plays a central role. Pepsin initiates the hydrolysis of α-lactalbumin and β-lactoglobulin, generating precursor fragments that are subsequently cleaved by trypsin and chymotrypsin, releasing bioactive sequences such as YGLF (α-lactorphin) and YLLF (β-lactorphin) (Zhao and Ashaolu, 2020). These peptides are mainly produced through enzymatic hydrolysis, and their physiological relevance depends on factors such as protease resistance, stability, and efficient intestinal absorption (Buey et al., 2023).

From a technological standpoint, controlled enzymatic hydrolysis is the most widely used strategy for obtaining whey-derived peptides, allowing precise modulation of protein breakdown and optimization of the resulting peptide profile. Additional approaches, such as microbial fermentation, chemical synthesis, and in silico enzymatic simulation, have also been employed to identify and produce bioactive fragments from whey proteins (Olvera-Rosales et al., 2022). In fermented dairy products, strains of *Lactobacillus* and *Lactococcus* are particularly relevant for naturally generating bioactive peptides (Buey et al., 2023; Singh and Gaur, 2024).

More recently, emerging processing techniques have expanded the possibilities for producing whey-derived peptides. Ultrasound and high hydrostatic pressure treatments induce conformational changes that expose hydrophobic regions, improving enzyme accessibility and enhancing hydrolysis efficiency (Singh and Gaur, 2024). These innovative methods, combined with advancements in biotechnology, have enabled the development of hydrolyzates enriched in active peptides, broadening their applications in functional foods and nutraceuticals.

Additionally, experimental evidence has deepened the understanding of the bioactivity and mechanisms of action of whey protein hydrolyzates (WPH). Reported biological activities include antihypertensive, antidiabetic, antioxidant, immunomodulatory, and gut microbiota–modulating effects (Brandelli et al., 2015). In this context, this chapter aims to provide an updated overview of current methods for obtaining bioactive peptides from whey, their reported bioactivities and mechanisms of action, safety and regulatory considerations, and the main challenges and future perspectives in the field.

Whey. https://doi.org/10.1016/B978-0-443-27648-4.00006-2

2. Digestion models for obtaining of bioactive peptides

In recent years, interest in whey-derived bioactive peptides (BAPs) has increased markedly due to their potential antihypertensive, antioxidant, antimicrobial, and immunomodulatory effects (Saubenova et al., 2024). These peptides are released during digestion or obtained through targeted extraction methods. This section reviews common in vitro and in vivo digestion models, as well as the principal approaches used to extract whey BAPs.

2.1 From *in vitro* and *in vivo* digestion

Whey protein digestion can be investigated using both in vitro and in vivo models. Each approach offers complementary perspectives on protein cleavage and BAPs production. Understanding these pathways is crucial for guiding the use of BAPs in biological and technological applications.

2.1.1 *In vitro digestion*

In vitro digestion is a fundamental approach for investigating the release of BPAs from whey proteins, allowing for the precise simulation of the gastrointestinal process under controlled conditions (Luparelli et al., 2025). This process employs digestive enzymes, such as pepsin, trypsin, and pancreatin, to hydrolyze proteins into smaller peptides and free amino acids. Furthermore, several experimental variables, such as pH, temperature, enzyme concentration, and incubation time, can be standardized to facilitate a more precise evaluation of the factors influencing the formation and stability of BPAs (Atma et al., 2024).

Recent investigations have demonstrated the potential of in vitro digestion to generate BAPs from whey and other milk proteins. In the study by D'Opazo et al. (2023), a hydrolysis degree greater than 90% was achieved for pancreatin after 24 h of digestion of bovine casein with pancreatin and trypsin. One hundred and twenty peptides (<5 kDa) were obtained and identified in the gastric, duodenal, and colonic fractions. Of these, five peptides were able to cross Caco-2 cell monolayers, indicating their potential for bioavailability. Trypsin hydrolyzates exhibited a higher antioxidant capacity (47% free-radical reduction; 32,112 μM Trolox equivalent) than pancreatin hydrolyzates (15%; 2862 μM of TE), highlighting the enzyme's specificity as a determinant of the yield of functional peptides. Similarly, Li et al. (2025) investigated simulated gastrointestinal digestion of donkey whey protein hydrolyzates (DWHs) and separated peptides by molecular-weight fractions (<1 kDa, 1–3 kDa, 3–10 kDa, >10 kDa). The low-molecular-weight fractions (DWH-I and DWH-II) displayed the highest antioxidant capacity, with DPPH-scavenging rates up to 54.8% at 2.5 mg/mL. Amino-acid analysis revealed significant enrichment of Glu, Leu, and Lys after digestion, suggesting enhanced nutritional quality and digestibility.

The effect of processing and structural modification prior to in vitro digestion was also investigated. Shao et al. (2024) evaluated the effect of ultrasonic treatment prior to sham digestion on the biological potential of β-lactoglobulin (BLG). It was observed that sonicated samples produced a higher proportion of short peptides (<3 kDa, approximately 42% of the total) and exhibited superior $ABTS^{\bullet+}$ scavenging capacity ($IC_{50} = 0.19$ mg/mL) compared to untreated controls ($IC_{50} = 0.44$ mg/mL). The peptides also increased the activities of antioxidant enzymes (SOD, CAT, GSH-Px) in HepG2 cells, demonstrating that ultrasonic pretreatment can promote proteolysis efficiency and increase BPAs release.

In the study by Ren et al. (2025), the digestion of goat whey proteins with different casein ratios (80:20 and 40:60) and heat treatments (wet vs. dry) were evaluated. The results revealed that wet heating improved digestibility and increased peptide intensity during in vitro digestion in infants, whereas dry heating and glycation hindered protein cleavage, resulting in lower levels of BAPs. Peptide transport assays through Caco-2 cells demonstrated that shorter and more hydrophilic peptides were preferentially absorbed, highlighting the importance of processing conditions in the kinetics of digestion and bioavailability of peptides.

Furthermore, studies have shown that controlled enzymatic digestion, often combined with analytical peptidomics using liquid chromatography-tandem mass spectrometry (LC-MS/MS), produces peptides with angiotensin-converting enzyme (ACE) inhibitory, antioxidant, and antimicrobial properties. The variability among different enzymatic systems (pancreatin, trypsin, and flavorzyme) and milk sources (bovine, caprine, and donkey) underscores how substrate composition and enzyme specificity determine the molecular profile and bioactivity of the released peptides (Campanhon et al., 2024; Grigorean et al., 2024).

Although the in vitro digestion model does not reproduce the dynamic and complex environment of the human gastrointestinal tract, including interactions with mucus, microbiota metabolism, and peptide transport, it remains a promising tool for screening and characterizing BAPs derived from whey. Furthermore, it offers low cost, good reproducibility, and ethical viability. Integrating this method with advanced analytical techniques and subsequently validating it into cellular models can reduce the gap between laboratory simulations and physiological relevance.

2.1.2 *In vivo digestion*

In vivo digestion represents the most physiologically relevant approach for studying the release, transformation, and absorption of BAPs derived from whey proteins (Sutantawong et al., 2025). Unlike simulated systems, in vivo models reproduce the complex physiological environment of the gastrointestinal tract, encompassing factors such as enzyme secretion, bile salt dynamics, intestinal transit, and host–microbiota interactions (Sumny and Kempka, 2025). This allows a more accurate assessment of peptide bioavailability, metabolic fate, and biological activity under realistic conditions.

Although in vivo digestion is crucial for understanding the physiological behavior of whey peptides, studies conducted directly in humans remain limited, primarily due to the technical and ethical challenges associated with collecting digestive fluids. Thus, a more precise evaluation of whey-derived peptidome should be performed under physiological conditions to provide insights into how the digestion patterns observed in vitro translate to real human systems.

In the study by Xue et al. (2021), peptide profiles obtained from in vivo human digestion were compared with a standardized in vitro model. In the study, adult volunteers consumed whey protein isolate, and duodenal effluent was collected via nasoduodenal intubation to monitor peptide release during digestion. LC-MS/MS analysis identified hundreds of peptides in human duodenal samples, 61 of which showed known biological activities, including ACE inhibitory effects, DPP-IV inhibitory effects, antioxidant, antimicrobial, immunomodulatory, and opioid effects. Among them, β-casein fragments YPFPGPI (75–81) and EMPFPK (123–128) were highlighted for their potential antihypertensive and antimicrobial properties. Despite the substantial qualitative similarity between the in vitro and in vivo peptide profiles, significant quantitative differences were observed, emphasizing the influence of physiological conditions, such as pH gradients, bile salt secretion, and digestive transit, on peptide abundance and persistence.

Another recent study evaluated the peptidomic profile of in vivo digestion in humans, combined with a static in vitro comparison. The authors reported that in vivo digestion 61 previously known BAPs and seven newly identified sequences (Sutantawong et al., 2024). These included peptides with ACE-inhibitory, antimicrobial, antioxidant, and immunomodulatory potential. Furthermore, the study also confirmed that the human gastrointestinal tract generates a highly complex peptide network that cannot be fully replicated in vitro, even under standardized INFOGEST conditions. Despite its critical importance for understanding the translation of in vitro bioactivity into actual physiological function, in vivo digestion remains an underexplored model due to limitations in ethics, reproducibility, and scalability.

In vivo studies using animal models have also been conducted aiming to mimic human gastrointestinal physiology. Barbé et al. (2014) investigated the digestion of dairy matrices in minipigs, identifying more than 16,000 distinct peptides in duodenal effluents collected over 5 hours after ingestion of different dairy matrices (liquid milk, acid gels, and rennet gels). While the specific cleavage sites within protein sequences remained highly conserved across all matrices, the kinetics and quantity of peptide release varied significantly depending on the food structure. Acid gels, for instance, favored rapid proteolysis due to their low pH, which optimized pepsin activity. In contrast, rennet gels generated approximately three times fewer peptides because of their denser network and slower gastric emptying. The study provided a comprehensive in vivo peptide map, illustrating how the microstructure of food modulates peptide accessibility and digestion rate without altering the enzymatic cleavage mechanism.

Despite recent advances, in vivo digestion studies of BAPs from whey are limited, and the field still relies predominantly on in vitro data. Research opportunities may explore in vivo approaches combining advanced analytical tools, such as peptidomics, stable isotope screening, and metabolomic profiling, to validate the physiological relevance of BAPs identified in vitro. These approaches may be crucial for understanding how whey BAPs are released, absorbed, and exert beneficial health effects, thus strengthening their application in functional foods and nutraceuticals.

2.2 Extraction of bioactive peptides: Conventional and emerging technologies methods

Extracting BAPs from whey is a crucial step in enhancing the functional properties of milk proteins. Traditionally, processes such as enzymatic hydrolysis, microbial fermentation, and chemical or thermal methods are employed. However, with the advancement of non-thermal technologies, new approaches such as ultrasound-assisted extraction, microwaves, high hydrostatic pressure, and pulsed electric field have shown great potential for increasing the yield and efficiency of BAP release. Table 19.1 summarizes the main studies on traditional extraction methods and emerging techniques for BAP production from whey. This section discusses both conventional methods and emerging technologies that are standing out for improving the efficiency, yield, and bioactivity of whey-derived BAPs.

TABLE 19.1 Comparative overview of conventional and emerging extraction methods for obtaining bioactive peptides (BAPs) from whey proteins.

Whey type	Extraction method	Enzyme/reagent	Key conditions	Peptide yield	Evaluated bioactivity	Peptides identified	References
Yak whey protein concentrate (YWPC)	Enzymatic hydrolysis	Alkaline protease (compared with trypsin and papain)	Optimized by RSM: E/S = 7500 U/g, 62°C, pH 8, 2.5 h	17.21 mg/mL (total peptides); degree of hydrolysis (DH) = 35–41%	α-Amylase inhibition (22.06%), XOD inhibition (17.15%), ABTS scavenging (69.55%)	<1 kDa fraction = highest activity	Hao et al. (2024)
Whey protein concentrate (78%)	Enzymatic hydrolysis	Chicken pepsin extract (CPE) *vs.* porcine pepsin (PP)	WPC 2% (w/v), phosphate buffer pH 2; E/S = 0.5–1% (w/w); 50°C for CPE, 37°C for PP; inactivation 95°C/10 min	DH = 4–13%; complete α-La degradation, partial β-Lg hydrolysis (62%–73%)	Antioxidant (ABTS, ORAC); solubility, emulsifying and foaming activity	<11 kDa peptides by SDS-PAGE	Kheroufi et al. (2022)
Bovine whey protein	Enzymatic hydrolysis	Cardosins A and B (from *Cynara cardunculus*)	Optimized by RSM; E/S = 1:150 (w/w); 24 h; pH 5.3, 60°C; complete α-La hydrolysis, partial β-Lg (40%)	DH α-La = 95–100%, DH β-Lg = 25–40%	ACE-inhibitory (IC_{50} 6.47 μg mL^{-1})	28 peptides (e. g., VSLPEW, IAFSQYL, SLACQCLVR)	Estévez et al. (2020)
Whey protein	Microbial fermentation	*Lactobacillus rhamnosus* B2-1	Fermentation 48 h at 37°C (anaerobic) in desalted whey-MRS medium; peptides purified by macroporous resin D101 and Sephadex G-15	Fraction B11 = 51.4% AA content; Glu and Asp major AAs (11.18%, 6.01%)	Antioxidant (ABTS 84.36%, ·OH 62.43%, ORAC 1726 μM TE/g); antitumor on LoVo, KB and Cal-27 cells	B11 fraction (<1 kDa); rich in acidic and hydrophobic AAs	Guo et al. (2024)
Whey protein concentrate (WPC80)	Microbial fermentation	*Streptococcus thermophilus* RBC06, RBC20, RBN16	72 h at 37°C; 1% v/v inoculum (2×10^6 CFU/mL); aerobic conditions; peptides (<3 kDa) analyzed by Q Exactive Orbitrap	Peptide conc. 9.93–5.82 mmol Leu eq/L; DH ≈ 7–12%	ACE- (IC_{50} 5–10 μmol/mL), DPP-IV- and antioxidant (ABTS) inhibitory activity; strongest in RBC06	241–357 peptides; 32–47 bioactive (ACE, DPP-IV, antioxidant); VPP (12.8 mg/L), IPP (0.56 mg/L)	Solieri et al. (2022)
Sweet whey (reconstituted 11% w/v)	Ohmic heating (physical-assisted release of bioactive peptides)	None (thermal–electrical treatment)	Electric fields: 2, 4, 5, 7, 9 V/cm at 60 Hz; 72–75°C for 15 s (HTST); compared with conventional heating	Maximum ACE inhibition: 93.2% (4 V/cm) versus 33.7% (conventional); antioxidant capacity (DPPH) = 48.2% (2 V/cm)	ACE-inhibitory and antioxidant activities; release and preservation of bioactive peptides at low–moderate field intensity	Not identified individually (low-molecular-weight peptides inferred by microscopy and rheology)	Costa et al. (2018)
Whey protein isolate (WPI, 90%)	Ultrasound-assisted enzymatic hydrolysis	Bromelain + ProteAXH and Papain W-40 + ProteAXH (1:1 w/w)	Ultrasound pretreatment: 10 min, 25 kHz, 650 W (ice bath); enzymatic hydrolysis: 55°C, pH 7.0, 1 h, E/S = 1:50 (w/w)	DH increased by 1.5%–3.3%; 456 peptides identified; 14 protein sources	Antioxidant (ABTS, FRAP, DPPH), α-glucosidase inhibition, reduced IgE binding (~70%), improved foam and emulsion capacity	456 peptides (<10 kDa) from β-lactoglobulin, α-lactalbumin, and lactotransferrin	Pang et al. (2024a)

Whey protein isolate (WPI, 91.7%)	Ultrasound-assisted enzymatic hydrolysis	Bromelain (vegetable cysteine protease, E/S = 1:20 w/w)	Ultrasound pretreatment: 20 kHz, 500 W, 25%–50% amplitude, 10 min (≤25°C); hydrolysis at 55°C, pH 7.0 for 1 h; inactivation 90°C/15 min	Fractions <5 kDa increased after ultrasound; F3 (3–5 kDa) and F4 (1–3 kDa) showed highest ACE inhibition ($IC_{50} \approx$ 34–38 μg mL^{-1})	ACE-inhibitory activity (up to 42% reduction in IC_{50} vs. control); stable after simulated digestion	Peptide fractions 1–5 kDa most active; structural unfolding (↓ α-helix, ↑ β-sheet) confirmed by FTIR	Abadía-García et al. (2021)
Whey protein concentrate (WPC 77.5%)	Ultrasound- and ohmic-heating-assisted enzymatic hydrolysis	Pepsin (E/S = 2%, w/w)	Pretreatments: ultrasound (24 kHz, 400 W, 5–15 min, ice bath) and ohmic heating (150 V/cm, 50 Hz, 5–15 s); hydrolysis at pH 2, 37°C, 3 h	DH up to 37.7% (OH 15 s + US 15 min); F1 fraction (>10 kDa) showed highest antioxidant and antibacterial activity	Antioxidant (ABTS, DPPH, reducing power) and antibacterial (B. cereus, S. aureus, L. monocytogenes, S. Typhimurium, E. coli)	F1 fraction (>10 kDa) most active; combined OH + US produced less hydrophobic peptides (RT 4–6 min by RP-HPLC)	Alizadeh and Aliakbarlu (2020)
Whey protein isolate (WPI, 90%)	Ultrasound-assisted enzymatic hydrolysis (endo/exopeptidases combination)	PROTIN AY50C (*B. licheniformis* endopeptidase) + Protease A "Amano" 2SD (*A. oryzae* endo/exopeptidase)	Ultrasound: 25 kHz, 650 W, 10 min (ice bath); hydrolysis: 55°C, pH 8.0/7.0, 1 h, E/S = 1:50 (w/w); enzyme inactivation 95°C/5 min	DH = 38.65%; 176 peptides, 5 master proteins (mainly β-Lg and α-La); 78–207 allergenic peptides identified	Antioxidant (ABTS, DPPH, FRAP), α-glucosidase inhibition, reduced IgE-binding capacity (↓ to 55.6% vs. control), improved solubility, emulsifying and foaming activity	176 total peptides; 4 emulsifying peptides; several ACE- and DPP-IV-inhibitory peptides predicted by BIOPEP and MBPDB databases	Pang et al. (2024b)
Whey protein concentrate (WPC 80%)	Microwave-assisted enzymatic hydrolysis	Pepsin (27 U mg^{-1} protein)	Continuous microwave heating at 37–70°C (optimum 65°C); 1.25% WPC (w/v) in glycine buffer pH 2; residence time = 2 s/cycle; total 30 min; compared to conventional heating	DH = 14.1% (microwave) versus 7.1% (thermal); β-Lg completely degraded at 65°C (microwave), partially under conventional heating	Reduced allergenicity: β-hexosaminidase release ↓ from 46% → 6.5%; improved pepsinolysis efficiency; potential hypoallergenic hydrolyzate	Mainly low-MW peptides from β-lactoglobulin and α-lactalbumin (confirmed by SDS-PAGE and RP-HPLC)	El Mecherfi et al. (2019)
Whey protein isolate (WPI ≥92%)	High hydrostatic pressure–assisted enzymatic digestion	Pepsin (1:200, pH 1.9, 15 min) + Trypsin (1:200), Chymotrypsin (1:87), and Aminopeptidase (1:120), pH 7.4, 60 min	WPI pressurized at 550 MPa, 20°C, 1 min; digestion at 37°C; ultrafiltration MWCO 10 kDa	DH: pWPI = 99%, nWPI = 95%; faster peptide release; >80% peptides <1 kDa	Antioxidant (FRAP ↑ 21% vs. control), anti-inflammatory (↓ IL-8 secretion by 25%–28% in LPS-stimulated cells)	31 peptide peaks in pWPI (<1 kDa), including 9 new peptides absent in nWPI	Iskandar et al. (2015)

2.2.1 Conventional extraction methods

2.2.1.1 Enzymatic hydrolysis

Among conventional methods, enzymatic hydrolysis is the most widely used approach for producing and extracting BAPs from whey proteins, allowing precise control over peptide size, amino acid composition, and biological activity (Luparelli et al., 2025). Recent studies have demonstrated that both conventional and emerging proteolytic systems can generate whey-derived peptides with high biological potential, including antioxidant, antihypertensive, and anti-inflammatory activities (Hao et al., 2024; Kheroufi et al., 2022).

Hydrolysis using classical enzymes, such as pepsin, trypsin, and chymotrypsin, is well established for releasing low-molecular-weight peptides with radical-scavenging and ACE-inhibitory activity. Hao et al. (2024) optimized the enzymatic hydrolysis of yak whey protein concentrate (YWPC) using alkaline protease, revealing that conditions of 62°C, pH 8, and enzyme–substrate ratio of 7500 U/g yielded hydrolyzates with a degree of hydrolysis (DH) of 35%–41% and a total peptide concentration of 17.2 mg/mL. The ultrafiltered fraction (<1 kDa) showed the highest antioxidant activity (ABTS = 69.55%), α-amylase inhibition (22.06%), and xanthine oxidase inhibition (17.15%), suggesting the release of multifunctional peptides with hypoglycemic and uric acid–lowering potential. Therefore, it can be inferred that enzymatic parameters and substrate origin significantly influence the peptide profile and bioactivity spectrum.

Similarly, Kheroufi et al. (2022) demonstrated that both chicken- and porcine-derived pepsins efficiently hydrolyzed α-lactalbumin in whey protein concentrate, producing hydrolyzates with enhanced ABTS and ORAC antioxidant activities. Although β-lactoglobulin remained partially resistant to cleavage, the enzymatic treatment improved solubility (up to 93%), emulsifying, and foaming properties, indicating a direct relationship between the DH and technological improvement. These findings support the use of specific proteases for tailoring functional peptides with simultaneous physicochemical enhancement.

In addition to mammalian enzymes, plant proteases have also gained significant relevance due to their sustainability and distinct cleavage selectivity. In the study by Estévez et al. (2020), the use of *Cynara cardunculus* cardosins A and B, optimized by a central composite rotatable design, resulted in up to 40% hydrolysis of β-lactoglobulin at pH 6.3 and 50°C, as well as complete degradation of α-lactalbumin at pH levels ranging from 5 to 7. The hydrolyzates exhibited potent ACE inhibitory activity (IC_{50} = 6.47 μg/mL) and contained well-characterized peptides, such as VSLPEW and IAFSQYL, known for their antihypertensive potential. Furthermore, the β-lactoglobulin fraction obtained after partial hydrolysis proved suitable for forming gels, nanofibrils, and spray-dried microparticles, demonstrating a link between the generation of bioactive peptides and their structural applications.

Thus, alternative biocatalysts, such as plant or avian proteases, expand the functional diversity of whey-derived peptides. Although classical enzymes offer reproducible yields, challenges related to catalytic selectivity and process intensification still exist. Overcoming these limitations is fundamental to achieving greater control over hydrolysis kinetics and subsequent fractionation, enabling the production of peptide ingredients with well-defined molecular profiles and more predictable biological effects.

2.2.1.2 Microbial fermentation

Fermentation has emerged as an eco-friendly and cost-effective alternative to enzymatic hydrolysis for generating BAPs from whey proteins, harnessing the proteolytic activity of lactic acid bacteria and other microbial strains (Singh, 2025). During fermentation, proteases secreted by microorganisms hydrolyze whey proteins, releasing peptides with various biological functions, such as antioxidant, antihypertensive, and anti-diabetic effects (Irazoqui et al., 2024)). The biochemical outcome depends on strain specificity, substrate composition, and process parameters, including pH, temperature, and fermentation time (Haj Mustafa et al., 2024).

In recent years, several studies have explored the ability of lactic acid bacteria to release peptides from whey protein substrates. Solieri et al. (2022) investigated *Streptococcus thermophilus* strains isolated from natural whey starters of Parmigiano Reggiano for their ability to ferment whey protein concentrate (WPC). Among the seven strains tested, *S. thermophilus* RBC06, RBC20, and RBN16 exhibited the most significant proteolytic activity, generating low-molecular-weight peptides (<3 kDa) with potent antioxidant, ACE-inhibitory, and DPP-IV-inhibitory activities. Peptidomic analysis revealed over 350 distinct peptides, primarily derived from β- and α-caseins, with known bioactivities, including the antihypertensive tripeptides Val-Pro-Pro (VPP) and Ile-Pro-Pro (IPP). Notably, the RBC06 strain produced the highest VPP levels (12.8 mg/L), highlighting its potential for the industrial production of bioactive peptide-enriched whey beverages.

Complementarily, Guo et al. (2024) focused on the fermentation of demineralized whey by *Lactobacillus rhamnosus* B2-1, isolating novel antioxidant peptides after a multistep purification process (D101 macroporous resin and Sephadex

G-50/G-15 chromatography) and subsequent LC-MS/MS identification. Three peptides (KYPVEPF, LEASPEVI, and YPFPGPIHNS) exhibited exceptional free radical scavenging capacity (ABTS $IC_{50} = 0.031–0.042$ mg/mL) and high ORAC values (>6400 μM Trolox eq/g). These peptides demonstrated cytoprotective effects in H_2O_2-induced HepG2 cells, enhancing catalase and superoxide dismutase activities while reducing reactive oxygen species by over 60%. Molecular docking studies revealed that these peptides bind to the Keap1-Nrf2 complex, thereby preventing oxidative damage at the cellular level. In summary, the study provided a mechanistic basis for the functionality of the peptides, connecting fermentation-derived hydrolyzates to specific antioxidant pathways.

Therefore, microbial fermentation can be considered to offer a double benefit: adding value to dairy by-products and producing functional peptides without the need for commercial enzymes. However, variability between strains and limited control over peptide profiles remain significant challenges. Future work should integrate in silico prediction with peptidomics and cell-based assays to optimize microbial consortia, aiming for consistent peptide yields and enhanced bioactivities.

2.2.1.3 Chemical/thermal methods

Traditional chemical and thermal methods are based on the cleavage of peptide bonds by acid or alkaline hydrolysis, or by heat-induced denaturation. Although these strategies are less specific than enzymatic hydrolysis, they have been commonly employed to produce peptide-rich hydrolyzates at low cost and on an industrial scale (Saadi et al., 2024).

Acid hydrolysis has been widely employed for the complete cleavage of proteins and the quantification of amino acids. However, for BAPs production, this approach presents serious drawbacks: it destroys sensitive residues such as tryptophan, serine, and threonine, and produces nonspecific fragmentation that compromises bioactivity. Similarly, alkaline hydrolysis can generate extensive racemization and oxidative modification of sulfur-containing amino acids. These limitations make purely chemical hydrolysis unsuitable for obtaining functional peptides, despite its effectiveness in depolymerizing whey proteins (Ouyang et al., 2023).

On the other hand, thermal treatments have been explored as supportive processes that modify protein conformation and enhance enzymatic hydrolysis rather than directly generating peptides. Moderate heat (60–80°C) induces partial denaturation of β-lactoglobulin and α-lactalbumin, exposing reactive groups and improving digestibility. Costa et al. (2018) demonstrated that ohmic heating of sweet whey (72–75°C, 15 s, 4 V/cm) significantly increased ACE-inhibitory activity (93.2% inhibition *vs.* 33.7% in conventional heating) and antioxidant potential, suggesting that electrically assisted thermal treatments can trigger limited peptide release and structural rearrangements beneficial to functionality. Likewise, El Mecherfi et al. (2019) demonstrated that microwave-assisted heating accelerated the pepsin hydrolysis of whey proteins, doubling the DH and completely degrading β-lactoglobulin, while reducing allergenicity by nearly 85%. These findings confirm that controlled heating, especially when combined with electromagnetic or ohmic fields, can enhance peptide liberation efficiency and reduce immunoreactivity.

While chemical and purely thermal methods are effective for modifying the structure and denaturing whey proteins, their lack of selectivity and potential for degrading amino acids limit their use in producing bioactive peptides. Modern adaptations, such as microwave- or ohmic-assisted thermal processing, have revitalized these approaches, providing better control over reaction kinetics and molecular integrity. However, compared to enzymatic and biotechnological routes, chemical and thermal hydrolysis remain less precise and are best suited as pretreatments to enhance subsequent enzymatic or microbial hydrolysis rather than as standalone extraction methods.

2.2.2 *Emerging technologies for extraction*

In recent years, emerging technologies have gained prominence due to their ability to improve efficiency, reduce processing time, and minimize extraction costs for BAPs. Among these technologies, extraction assisted by ultrasound, microwaves, high hydrostatic pressure, and pulsed electric fields is the most promising (Luparelli et al., 2025).

2.2.2.1 Ultrasound-assisted extraction

Ultrasound-assisted extraction has emerged as one of the most efficient physical approaches to enhance enzymatic hydrolysis and release of BAPs from whey proteins (Kadam et al., 2015). High-intensity ultrasound (typically 20–40 kHz) promotes protein unfolding through cavitation, microstreaming, and shear forces that disrupt noncovalent interactions (Qian et al., 2023). This increases the accessibility of enzymatic cleavage sites, accelerates proteolysis, and reduces processing time and enzyme dosage (Pang et al., 2024a). Recent studies have demonstrated that ultrasound, either as a standalone pretreatment or in combination with other technologies, can significantly improve the peptide yield, functional properties, and biological activities of whey hydrolyzates (Abadía-García et al., 2021; Alizadeh and Aliakbarlu, 2020; Pang et al., 2024b).

In a systematic evaluation of whey protein concentrate (77.5%), Alizadeh and Aliakbarlu (2020) demonstrated that ultrasonic pretreatment (24 kHz, 400 W, 5–15 min) before pepsin hydrolysis enhanced the DH > 25% compared to the untreated control. When ultrasound was combined with ohmic heating (150 V cm^{-1}, 15 s), the DH reached 37.7%, producing fractions with superior ABTS, DPPH, and reducing power activities. The high-molecular-weight fraction (>10 kDa) exhibited the most pronounced antioxidant and antibacterial effects, indicating that cavitation not only accelerates proteolysis but also modulates peptide aggregation and functionality. The synergy between ultrasound and electric field treatment disrupted intramolecular hydrogen bonds and hydrophobic domains, facilitating enzyme–substrate interactions and generating less hydrophobic peptide profiles detectable by RP-HPLC.

On the other hand, the study by Pang et al. (2024a) found that ultrasound pretreatment (25 kHz, 650 W, 10 min) improved enzyme accessibility, reduced the average peptide size, and altered the zeta potential of the hydrolyzates, confirming the structural loosening of β-lactoglobulin and α-lactalbumin. The hydrolyzates exhibited enhanced antioxidant capacity (65.9%), α-glucosidase inhibition, and improved emulsifying and foaming properties, while allergenicity (IgE-binding capacity) was reduced to 55.6%. Peptidomic analysis identified >170 peptides, including dipeptidyl-peptidase IV (DPP-IV) and ACE-inhibitory sequences, underscoring ultrasound's role in tailoring the bioactive peptide spectrum.

Other investigations have also established ultrasound as a useful tool for enhancing enzymes of plant and animal origin. Abadía-García et al. (2021) reported that sonication of whey proteins prior to hydrolysis by papain and bromelain significantly increased proteolytic efficiency and antioxidant potential. The study highlighted how ultrasound breaks down tertiary and quaternary structures without causing extensive thermal degradation, thus allowing controlled exposure of hydrophobic amino acids that are critical for peptide bioactivity.

Despite its advantages, ultrasound-assisted extraction is a method that requires the optimization and standardization of extraction parameters to enable the adequate production of BAPs. Excessive sonication can lead to oxidative side reactions and peptide fragmentation, potentially reducing the biological efficacy of the peptide. Furthermore, scalability and energy efficiency remain challenges for industrial applications. However, when properly integrated with enzymatic or electrically assisted hydrolysis, ultrasound provides a reproducible, low-cost, and sustainable route for generating BAPs whey with multifunctional properties, as well as improved emulsification and reduced allergenicity.

2.2.2.2 Microwave-assisted extraction

Microwave-assisted extraction is an emerging technology that can intensify protein hydrolysis and modify peptide functionality through rapid, volumetric heating (Luparelli et al., 2025). In whey systems, microwaves promote protein unfolding and enhance enzymatic access, leading to faster hydrolysis and potentially improved bioactivity. However, despite its theoretical advantages, studies applying microwaves for the extraction of BAPs from whey are still limited.

In the study by El Mecherfi et al. (2019), continuous microwave irradiation during pepsinolysis of whey protein concentrate doubled the DH compared to conventional heating (14.1% vs. 7.1%) and completely degraded β-lactoglobulin at 65°C. The process also reduced allergenicity, lowering basophil degranulation from 46% to 6.5%, confirming its potential for producing hypoallergenic hydrolyzates. Meanwhile, Nooshkam and Madadlou (2016) utilized microwaves to induce Maillard conjugation between pre-hydrolyzed whey peptides and lactose, thereby enhancing antioxidant activity (approximately 50%) and solubility. This discovery demonstrates that microwaves can also make existing peptides functional.

These results indicate that microwave-assisted processes can accelerate proteolysis, reduce allergenicity, and improve peptide functionality. Yet, the lack of recent investigations highlights a research gap. Current evidence largely stems from combined or indirect applications, underscoring the need for modern studies focused on controlled, scalable microwave-assisted hydrolysis and its molecular effects on whey-derived peptides.

2.2.2.3 High hydrostatic pressure

High hydrostatic pressure (HHP) processing is a non-thermal technology that modifies the structural conformation of whey proteins, increasing enzyme accessibility and enhancing peptide release (Blayo et al., 2016). When combined with enzymatic hydrolysis, HHP enhances proteolysis efficiency, improves bioactivity, and reduces allergenicity compared to conventional treatments.

In the study by Iskandar et al. (2015), HHP pretreatment of whey protein isolate (600 MPa, 25°C, 15 min) before pancreatin digestion enhanced digestibility by 28% and doubled antioxidant activity (DPPH, ORAC) relative to untreated samples. The pressurized whey showed increased β-sheet and random coil content, indicating controlled unfolding that facilitated hydrolysis without compromising nutritional quality. Similarly, Landim et al. (2022) applied HHP (400 MPa, 5–35 min) as a pretreatment to whey protein concentrate hydrolyzed by Novo Pro-DVR or ficin. The treatment enhanced hydrolysis efficiency, resulting in a higher antioxidant capacity (ORAC >1100 μmol TE g^{-1}) and complete elimination of β-lactoglobulin immunoreactivity when combined with Novo Pro-DVR. Ficin-derived hydrolyzates also induced vascular relaxation in isolated rat aorta, demonstrating biological relevance beyond chemical assays.

Jiang et al. (2023) confirmed the capacity of HHP (500 MPa, 10 min) to alter the secondary and tertiary structure of whey proteins, reducing allergenicity by 76% and increasing the release of low-molecular-weight peptides (<10 kDa) during simulated infant digestion. The structural unfolding, evidenced by a 1.4-fold increase in surface hydrophobicity and a transition from α-helix to β-sheet conformation, enhanced the susceptibility of pepsin and trypsin and facilitated epitope disruption. Consequently, antigenicity retention after digestion decreased to 39%, reinforcing HHP as a tool for producing hypoallergenic and more digestible whey ingredients.

In general, HHP acts as an efficient non-thermal intensification tool, promoting protein unfolding, increased enzyme accessibility, and higher yields of BAPs. Its benefits include enhanced antioxidant activity, digestibility, and reduced allergenicity. However, excessive pressures may cause aggregation and loss of enzymatic accessibility. Despite promising outcomes, studies on HHP-assisted extraction of whey BAPs remain limited, particularly on an industrial scale. Further research combining peptidomics and process modeling is needed to clarify the relationships between structure and function, and to optimize large-scale applications.

2.2.2.4 Pulsed electric field

Pulsed electric field (PEF) is a non-thermal processing technology that applies short, high-voltage electric pulses to biological matrices, inducing electroporation and structural modification of macromolecules (Roobab et al., 2022). In dairy systems, PEF has been explored primarily for microbial inactivation, emulsion stability, and structural modification of milk and whey proteins, offering a promising tool for protein valorization beyond traditional thermal processing (Vashisht et al., 2024).

Experimental evidence demonstrates that PEF can induce partial unfolding of globular milk proteins, thereby improving their solubility, emulsification capacity, and enzymatic susceptibility. Hu et al. (2024) reported that treating whey protein isolate at 20–40 kV cm^{-1} increased surface hydrophobicity and β-sheet content by approximately 15%, enhancing its digestibility and foaming stability. Similarly, Axelrod et al. (2022) demonstrated that PEF-assisted microfiltration of skim milk altered the protein conformation and particle size distribution, thereby improving membrane permeability and protein recovery. Moreover, studies in microbial and plant matrices demonstrated that PEF pretreatment can enhance enzymatic hydrolysis and peptide yield by disrupting cell membranes and facilitating enzyme access, an outcome that is theoretically translatable to whey protein systems.

Despite these advances, there is a notable lack of experimental studies that apply PEF directly to the extraction or production of BAPs from whey. Current reports primarily focus on the structural and functional modification of whey proteins, rather than on controlled proteolysis or peptide generation. This gap likely arises from technical challenges, such as optimizing electric field intensity, pulse duration, and conductivity of protein solutions to prevent local heating and aggregation. Additionally, the mechanisms through which PEF might enhance enzymatic cleavage, via molecular dipole alignment, charge redistribution, or disruption of hydrophobic cores, remain poorly characterized for dairy proteins.

3. Bioactivities of whey-derived peptides

Several studies have reported the bioactivities of whey-derived peptides, mainly in vitro, in vivo, and in situ (Table 19.2).

3.1 Antihypertensive activity

Hypertension is a long-term condition characterized by persistently elevated blood pressure. The angiotensin-converting enzyme (ACE) is a central component of the renin–angiotensin system, helping to regulate blood pressure and maintain

TABLE 19.2 Effects and conditions of intervention with bioactive peptides from milk whey for different health benefits.

Health effect	Model/ Participants	Identified bioactive peptide	Source matrix	Enzyme or MO responsible for the production	Concentration Tested	Intervention Time	Effect tested/ condition	Effect compared to control	Action mechanism	References
Antidiabetic	In vitro: Enzyme inhibition	FVVAPFPEVFR IAMNPWHHVTPFR SPAQTLQWQVLPNAVPAK	Sheep milk whey (Panchali breed)	*Lactiplantibacillus plantarum* KGL3A (MG722814)	9.38 mg/mL	48h	α-amylase (67.14%), α-glucosidase (71.32%), and lipase (64.15%) inhibition	Increase	Direct inhibition of target enzymes; peptides <3 kDa are more active	Pipaliya et al. (2025)
Antidiabetic	In vitro: Enzyme inhibition	DDQNPHSSNICNISCDK, FCMENSAEPEQSLACQCLVR, AASDISLLDAQSAPLR, and LDQWLCEK.	cheddar cheese whey	*Lactiplantibacillus plantarum* KGL3A	7.64 mg/mL	48h	α-amylase (65.88%), α-glucosidase (63.55%), and lipase (61.80%) inhibition	Increase	α-amylase, α-glucosidase, and lipase inhibition	Das et al. (2025)
Antidiabetic	In vitro: Enzyme inhibition	LIVTQTMKGLD (f17–27)	β-lactoglobulin isolated from Gir cow milk	Pepsin	2.5 mg/mL	12h	DPP-IV inhibitory potential (IC50 value = 6.64 mmolL-1).	Increase	DPP-IV inhibition of the pepsin-treated hydrolyzate was 68.5%,	Kumar et al., (2024a,b)
Hypoglycemic	in vitro: α-glucosidase inhibitory activity	YPVEPF, VPYPQ, LPYPY	Whey from Rubing cheese by-product	Papain	0.5, 1, 1.5, 2.5, and 5 mg/mL	24h	α-glucosidase inhibition; Glucose consumption in HepG2 cells	Increase	Inhibition of α-glucosidase via hydrogen bonding and hydrophobic interactions	Li et al. (2023)
Antidiabetic	In vitro: Human Salivary Gland (HSG) cell line	Not individually identified (Hydrolyzate mixture)	Whey Protein Isolate (WPI), β-Lactoglobulin (β-LG), α-Lactalbumin (α-LA)	Pepsin Pancreatin	0.313 mg/mL	24 h	DPP-IV inhibition and GLP-1 secretion	Increase	Bioactive peptides in hydrolyzates stimulate GLP-1 secretion from enteroendocrine L-cells (simulated by HSG cells).	Bunsroem et al. (2022)
Antidiabetic	In vitro *(semi-industrial peptide fraction recovery)*	LIVTQTMK, VAGTWY	Whey protein concentrate hydrolyzate	Enzyme mixture (trypsin and chymotrypsin)	2.86 mg/mL	6h	DPP-IV inhibitory activity	Increase	Competitive inhibition: peptides interfere with DPP-IV (slowing incretin degradation)	Faucher et al. (2022)
Antidiabetic	In vitro: enzymatic assays (DPP-IV inhibition, α-glucosidase, α-amylase, inhibition)	APFPE, IPP, LPPT, FPP, VVPPF, VVVPPF	Parmigiano-Reggiano Cheese	Natural proteolysis during 24-month ripening	1.2–2 mg/mL	30–60 min	enzymatic activity of α-glucosidase, α-amylase, and dipeptidyl peptidase-IV (DPP-IV)	Increase	Inhibits multiple enzymes, including α-glucosidase, α-amylase, and DPP-IV, which reduces glucose absorption and glycemic spikes.	Martini et al. (2021)
Antidiabetic	In vitro Inhibition of α-amylase and α-glucosidase (antidiabetic potential)	PAGNFLMNGLMHR; PAVACCLPPLPCHM; MLPLMLPFTMGY; PAGNFLPPVAAAPVM; CCGM; MFE; FCCLGPVPP	camel whey proteins	Pepsin	3.8 mg/mL	120–360 min	Inhibitory activity of α-amylase (AAM) and α-glucosidase (AG)	Increase	Binding to digestive enzymes, inhibiting their activity	Baba et al. (2021)

Antidiabetic	In vitro (enzymatic hydrolyzates)	LAHKPL, ILDKEGIDY (α-lactalbumin-derived); VPV, YPI, VPF (β–CN–derived)	Camel whey protein concentrate (WPC, 44.7 ± 3.4% protein)	Trypsin	Hydrolyzates: 2.8×10^{-2}–2.8 mg/mL (protein eq.); Synthetic peptides: 1.3–2000 μM; Positive control (IPI): 36.6×10^{-2}–36.6 μM	60 min	DPP-IV inhibitory activity	Increase	VPV, VPF, LPVPQ, YPI, and VL inhibit the DPP-IV enzyme ($IC_{50} < 100$ μM), with VPV standing out ($IC_{50} = 6.6 \pm 0.5$ μM), improving glycemic control.	Nongonierma et al. (2019)
Antioxidant	PKYPVEPF, LEASPEVI, YPFPGPIHNS	NR	Fermented bovine whey	Lactobacillus rhamnosus B2-1	100 μg/mL	NR	Protection of HepG2 cells against H_2O_2	Significant	Increase in SOD/catalase; Keap1/Nrf2 binding	Guo et al. (2024)
Antioxidant	Bovine serum before/after digestion	<3 kDa sweet whey fraction	Cow milk	INFOGEST + ultrafiltration	NR	NR	Increase in ORAC, FRAP, ABTS assays; ↑ SOD & CAT	NR	Free score + induction of antioxidant enzymes	Dalaka et al. (2023)
Antioxidant	Aged mice and scopolamine	Whey protein hydrolyzate	Cow milk	NR	NR	In vivo intervention - 10 days	Cognitive improvement, ↓ $A\beta_{1-42}$	NR	Reduction of oxidative stress; modulation of microbiota	Ding et al. (2023)
Antioxidant	Peptide VLPVPQK	casein-derived	Cow milk	osteoblastic cells	50–200 ng/mL	24 h	excess hydrogen peroxide (H_2O_2)	No significant	reduced ROS production	Ding et al. (2023)
Antioxidant	In vitro: DPPH and ABTS assays	α-lactalbumin-derived peptides	Cow milk whey	Trypsin	100–500 μg/mL	NR	Radical scavenging capacity	Increase	Hydrogen atom or electron donation	Rival et al. (2001)
Antioxidant	Mixed serum peptides (various proteins)	NR	Cow milk	Enzymes/LAB and fermentation	NR	NR	Antioxidant activities	NR	Electron donation, membrane action	Saubenova et al. (2024)
Antioxidant	Peptides extracted by advanced methodologies	NR	Cow milk	Ultrasound, microwaves, fermentation, enzymes	NR	NR	NR	NR	Higher degree of hydrolysis = higher activity	KiliÇ (2024)
Antimicrobial	Various whey peptides against food pathogens	*Enterococcus faecium* in vivo and ex vivo	Cow milk	Bovine lactoferrin-derived AMPs	NR	NR	Inhibition of food-borne pathogens	Designed AMPs showed high antimicrobial activity on free cells and biofilm, with low mammalian cytotoxicity	membrane-activating mechanisms.	Corrêa et al. (2023)
Antimicrobial	In vitro: Agar diffusion test against E. coli, L. monocytogenes	Lactoferricin B	Cow milk	Pepsin (gastric digestion)	10–100 μg/mL	NR	Inhibition zone diameter	Increase	Disruption of the bacterial membrane	Pellegrini et al. (2001)
Antimicrobial	In vitro: MRSA and food bacteria	fermented βcasein (novel polypeptide)	Cow milk	Extended fermentation by LAB	NR	6 days of fermentation	Inhibition zone/inhibitory activity	Increase	Cell rupture	Santos et al. (2021)

Continued

TABLE 19.2 Effects and conditions of intervention with bioactive peptides from milk whey for different health benefits.—cont'd

Health effect	Model/ Participants	Identified bioactive peptide	Source matrix	Enzyme or MO responsible for the production	Concentration Tested	Intervention Time	Effect tested/ condition	Effect compared to control	Action mechanism	References
Antioxidant & antimicrobial	Review/meta-analysis	Various peptides (βlactoglobulin, casein)	Cow milk	Various hydrolysis/ extraction methods	NR	NR	Antioxidant capacity and microbiocide	Obvious in comparison	Various mechanisms (in vitro)	Olvera-Rosales et al. (2023)
Neuroprotection	In vivo: D-galactose-induced aging model/male C57BL/6N mice	Not individually identified (Hydrolysated mixture)	Cow milk whey	Commercial whey hydrolyzed peptides obtained via enzymatic hydrolysis (Hydrolyzate mixture with protein peptide content of 67.2%)	0.3, 1.5, and 3.0 g/kg bw	30 days	*Serum* MDA SOD activity GSH-Px *Brain tissue* IL-1β and TNF-α MDA *Liver* SOD activity GSH-Px *Space Exploration Capability (Open-Field Test)* Number of squares crossed and number of rearing *Learning and Memory Capability (Morris Water Maze)* Escape latency in the place navigation test Swimming velocity Time spent in the target quadrant and distance in the target quadrant *Novel object Recognition test* New object index	Decreased Increased Increased Decreased Decreased Decreased Increased Increased Decreased Increased Increased Increased	Reduction of the degeneration of hippocampal nerve cells, reducing the apoptosis of nerve cells, improving the activity of AChE in brain tissue, and improving the expression of p-CaMKII and BDNF synaptic plasticity proteins	Xiao-Chen Yu et al. (2021)
Neuroprotection	In vitro: H_2O_2-induced oxidative stress in HT22 cells	Whey protein hydrolyzate (WPH) containing the pentapeptide leucine-aspartate-isoleucineglutamine-lysine (LDIQK)	Cow milk whey	Hydrolyzed whey protein concentrate by Alcalase, Protamex, and Flavourzyme	50, 100, and 200 μg/ mL	Pre-treatment for 2h followed by exposure to 300 μM H_2O_2 for 24h.	Cell viability ROS Calcium ion (Ca^{2+}) influx p-tau BAX ChAT BDNF BCL2	Increased Decreased Decreased Decreased Decreased Increased Increased Increased	Activation of the Nrf2 pathway, while inhibiting the MAPK signaling pathway	Chang et al. (2024)
Neuroprotection	In vitro: H_2O_2-injured HT-22 cells	Peptide LYLKPR	Fermented yak milk	Peptide LYLKPR derived from fermented yak milk	25, 50, 100, 150, and 200 μM	24h	*Activity* SOD CAT	Increased Increased Decreased	Inhibition of the NLRP3 inflammasome via	Jiang et al. (2022)

				by *Lactiplantibacillus plantarum* JLAU103			ROS MDA *Intracellular level* Mature IL-18 *Expression of synapse-related proteins* BDNF Synaptophysin PSD95	Decreased Decreased Increased Increased Increased	modulation of the Nrf2/Keap-1/HO-1 pathway. In addition to indirectly inhibiting the NLRP3 inflammasome via mitigation of oxidative stress, the peptide LYLKPR can bind to the Keap-1 protein and directly inhibit the inflammasome. Inhibition of the NLRP3 inflammasome prevented the cleavage of pro-IL-18 and pro-IL-1β by caspase-1.	
Neuroprotection	In vivo*:* Alzheimer's mice model/5×FAD transgenic mice and PS19 tauopathy mice	Whey-Derived β-Lactolin (GTWY) tetrapeptide	Cow milk whey	Cow milk whey via enzymatic hydrolysis	5% w/w whey digestion (0.28 mg of β-lactolin)	3.5 months	*Brain cortex and hippocampus* *Levels of inflammatory cytokines* IL-1α IL-1β MIP-1α IL-12p70 *Brain cortex* Infiltration of activated microglia Amyloid-β level Synaptophysin Dopamine BDNF Insulin-like growth factor 1 Phosphorylated tau to total tau ratio *Behavioral analysis* Long-term object memory	Decreased Decreased Decreased Decreased Decreased Decreased Decreased Attenuated decrease Attenuated decrease Attenuated decrease Attenuated decrease Decreased Improved	NR	Ano et al. (2020)
Neuroprotection	In vivo*:* Transgenic *C.elegans* strain CL4176 with Aβ peptide induction at 25°C	Not individually identified (Hydrolysated mixture)	Cheese whey	Whey protein concentrate samples hydrolyzed with commercial enzyme	200, 250, 300, 350, and 400 μg mL^{-1}	30h	Percentage of paralyzed worms	Decreased	NR	Giraldo et al. (2024)
Neuroprotection	In vivo*:* Scopolamine-induced cognitive impairment model (ICR mice) and aged C57BL/6J mice	Not individually identified (Hydrolysated mixture)	Cheese whey	NR	10 and 100 mg/kg bw per day	10 days	Spatial Memory and Cognitive Ability *Brain tissue* Aβ1-42 level *Serum* Aβ1-42 level *Histopathological analysis of hippocampus* Neuronal damage	Improved Decreased Decreased Decreased	Upregulation of TTR, Ppp1r1b, CAR-1, and Rbp1 proteins in the hippocampus.	Ding et al. (2023)

Continued

TABLE 19.2 Effects and conditions of intervention with bioactive peptides from milk whey for different health benefits.—cont'd

Health effect	Model/ Participants	Identified bioactive peptide	Source matrix	Enzyme or MO responsible for the production	Concentration Tested	Intervention Time	Effect tested/ condition	Effect compared to control	Action mechanism	References
Neuroprotection	In vivo: Scopolamine-induced amnesia mouse model (C57BL/6J mice)	Trp-Tyr (WY)–containing peptides from whey protein	Cow's milk whey protein	Whey protein digested with 10 different enzymes, including protease P [amano] 3SD from *A. melleus* or Thermoase C100 from *B. stearothermophilus*	10 mg/kg bw daily	1h	Memory function *Brain tissue* Dopamine levels	Improved Increased	Direct inhibition of monoamine oxidase-B activity in brain tissue by whey peptides; pretreatment with a dopamine receptor antagonist abolished the memory function improvement induced by whey peptides.	Ano et al. (2018)
Neuroprotection	In vivo: Regulation of response speed in CD-1 mice. Response speed is the individual's ability to perceive and react to harmful stimuli	Not individually identified (Hydrolysated mixture)	Whey Protein Hydrolyzate	NR	2000 mg/kg bw daily	6 consecutive weeks	*Response speed against* Pain Mechanical strength stimuli Prepulse inhibition	Increased Increased Increased	Proteomic analysis of the hippocampus revealed changes in proteins related to arginine and proline metabolism, as well as neuroactive ligand–receptor interactions	Ren et al. (2025)
Neuroprotection	In vitro: LPS-induced BV-2 cell line and primary microglia cultures	Metal-binding peptides from whey protein	Cow's milk whey protein isolate	Whey protein isolate hydrolyzed with alcalase, followed by isolation of metal-binding peptides.	100 or 1000 $\mu g\ mL^{-1}$	Pretreatment for 1h followed by exposure to LPS for 24h.	*mRNA and protein levels* Nitric oxide TNF-α	Decreased Decreased	Suppression of the nuclear factor kappa B (NF-κB) pathway due to the nuclear translocation of NF-κB p65 being decreased by roughly 30% upon treatment with the copper-binding peptide.	Caetano-Silva et al. (2021)
Antihypertensive	In vitro: cell-free assay. In vitro enzymatic assay using HHL (Hippuryl-His-Leu) as a substrate	Not individually identified (Hydrolysated mixture)	NR	Whey protein concentrate hydrolyzate with pepsin	5% WPC solution at 65°C after 8 h of pepsin hydrolysis	NR	ACE-inhibitory activity	Increased by up to 58.37%	NR	Mansinhbhai et al. (2023)

Antihypertensive	In vivo: *Male* spontaneously hypertensive rats	Not individually identified (Hydrolysated mixture)	NR	Whey protein fermentation with Aspergillus oryzae, *Aspergillus mellus,* and *Bacillus licheniformis.*	500 mg/kg and 1000 mg/kg bw daily	6 weeks	Systolic blood pressure Tissue injuries caused by hypertension	Decreased Decreased	Inhibition of ACE activity (20.81%), with a consequent reduction in renin concentration. Moreover, the increased abundance of *Akkermansia, Bacteroides,* and *Lactobacillus* promoted higher short-chain fatty acid levels	Dou et al. (2024)
Antihypertensive	In vivo: Spontaneously Hypertensive Wistar Rats	Peptide Arg-Leu-Ser-Asn-Pro	NR	Whey protein treated with *Lactobacillus helveticus* LB10 crude protease	60, 30, and 15 mg/kg bw daily	Short- and long-term periods	Systolic blood pressure Endothelial cytokines Malondialdehyde HDL-C LDL-C Total cholesterol	Decreased Decreased Decreased Decreased Decreased Decreased	Rebalancing of renin–angiotensin system (RAS) signaling in serum and tissues. Moreover, the peptide modulated gut microbiota homeostasis, renal tissue damage, and lung injury.	Zhao et al. (2025)
Antihypertensive	In vivo: Spontaneously hypertensive rats	Not individually identified (Hydrolysated mixture)	NR	Enzymatic Hydrolysis of whey protein concentrate using alcalase	400 mg/kg bw daily	24 h	Blood pressure	Dose-dependent decrease	In vitro ECA-1 inhibition	Hussein et al. (2020)
Antihypertensive	In vitro enzymatic assay using HHL (Hippuryl-His-Leu) as a substrate	Not individually identified (Hydrolysated mixture)	NR	Ultrafiltration protein-rich retentate obtained from ricotta cheese exhausted whey, followed by fermentation with a *Lactobacillus helveticus* strain	60 μL of the peptides mixture	1 h	ACE-inhibitory activity	Increased	NR	Pontonio et al. (2021)
Antihypertensive	In vitro enzymatic assay using HHL (Hippuryl-His-Leu) as a substrate	Not individually identified (Hydrolysated mixture)	Cow's milk	Whey protein concentrate was hydrolyzed using the purified *Sporisorium reilianum* aspartyl protease and the commercial enzymes chymotrypsin and trypsin individually, as well as in different enzymatic combinations	80 μL of the peptides mixture	30 min	ACE-inhibitory activity	Increased by up to 92.35%	NR	Tovar-Jiménez et al. (2021)

Continued

TABLE 19.2 Effects and conditions of intervention with bioactive peptides from milk whey for different health benefits.—cont'd

Health effect	Model/ Participants	Identified bioactive peptide	Source matrix	Enzyme or MO responsible for the production	Concentration Tested	Intervention Time	Effect tested/ condition	Effect compared to control	Action mechanism	References
Antihypertensive	In vivo: Spontaneously hypertensive rats	VGINYW peptide	NR	Hydrolysis of α-lactalbumin present in whey	5 mg per kg bw	Short-term effect: 8 h; Long-term effect: once-daily dosing for 4 weeks	Systolic blood pressure Oxidative stress	Decreased Decreased	Regulating the renin–angiotensin system by inhibiting ACE activity, reducing angiotensin II, upregulating the expression levels of ACE2 and AT2 receptor, and downregulating AT1 receptor. Alleviate hypertension-associated intestinal microbiota dysbiosis by restoring microbiota diversity and increasing SCFA-producing microorganisms.	Xie et al. (2022)
Antihypertensive	In situ: Fermented cheese whey	Not individually identified (Hydrolysated mixture)	Cow's milk whey	Fermentation of cheese whey by its native microbiota	Alfa-amino groups (360–456 μg Gly/mL)	120 h of fermentation	ACE-inhibitory activity	Increased by 60%–70%	NR	Mazorra-Manzano et al. (2020)
Antihypertensive	In situ: Whey protein hydrolyzed by trypsin	Not individually identified (Hydrolysated mixture)	Sheep/goat whey protein concentrate	Trypsin-hydrolyzed whey protein concentrate without pH control	Tryptic hydrolyzates of 5% (w/w) solutions	1h	ACE-inhibitory activity	Increased >80%	NR	Sakkas et al. (2022)
Antihypertensive	Clinical trial: healthy volunteers. A single-blind, placebo-controlled pilot study.	Whey protein hydrolyzate containing dipeptides isoleucine–tryptophan and tryptophan–leucine	NR	Whey protein solution (5% w/v) was first hydrolyzed with chymotrypsin followed by adding thermolysin	5, 10 or 50g	Single dose	Plasma ACE activity Systolic and diastolic blood pressure	Decreased Without effect	NR	Martin et al. (2020)
Antihypertensive	In situ: Sweet whey obtained from miniature fresh, Chanco and Gouda-style model cheeses	Not individually identified (Hydrolysated mixture)	Cow's milk	Cheese processing through starter culture and coagulase	IC50 = 4.8–6.3 mg/mL	Ripening time of 30 minute	ACE-inhibitory activity	Increased	NR	Godoy et al. (2020)

Antihypertensive	*In situ:* addition of whey protein hydrolyzate in pork frankfurters	Not individually identified (Hydrolysated mixture)	NR	Whey protein concentrate hydrolyzed with Alcalase	IC50 = 258.78 μg/mL	5h	ACE-inhibitory activity	Increased	NR	Barrón-Ayala et al. (2020)
Antitumoral	*In vitro:* human MCF-7 breast cancer cells	QYKIPDWFLNR and ALPMHIR peptides	Goat's milk	Goat milk whey hidrolised with pepsin or trypsin	5, 10, 25, 50, and 100 μg mL $^{-1}$	24 and 48h	Viability	Decreased	Cell death by apoptosis and necrosis. In the molecular docking analysis, QYKIPDWFLNR peptide interacted with the GSK 3-β, PKM2, RIPK1 and MUC 1, ALPMHIR peptide interacted with the LDH B with the highest docking scores	Sahna et al. (2022)
Antitumoral	*In vitro:* human MCF-7 breast cancer cells	Not individually identified (Hydrolysated mixture)	Camel's milk	Whey proteins hydrolyzed using pepsin, trypsin, and both enzymes.	400 μg/mL	24h	Viability	Decreased	NR	Taghipour et al. (2023)
Antitumoral	*In vitro:* B16F10 melanoma cells	Not individually identified (Hydrolysated mixture)	Cow's milk	Whey-derived peptides from *Eupenicillium javanicum* metalloprotease hydrolysis	4 mg mL^{-1}	48h	Viability	Decreased	Modulation of proteins involved in proliferation, energy metabolism, apoptosis, and metastatic and malignancy rates	Neto et al. (2022)
Antitumoral	*In vitro:* human colon carcinoma cells (HCT116)	Peptides CWPH P-4.3 and CWPH P-5.2	Camel's milk	Whey protein hydrolyzates using gastric enzyme pepsin	CWPH P-4.3 IC_{50} = 231 μg/mL CWPH P-5.2 IC_{50} = 221 μg/mL	72h	Viability	Decreased	Induction of G2/M cell cycle arrest through the modulation of Cdk1, p-Cdk1, Cyclin B1, p-histone H3, p21, and p53 expression	Murali et al. (2021)
Antitumoral	*In vitro:* human liver cancer cells (HepG2)	Not individually identified (Hydrolysated mixture)	Camel's milk	Whey protein hydrolyzates produced by gastric and pancreatic enzymes	400 to 1000 μg/mL	24h	Viability	Decreased	NR	Kamal et al. (2018)

NR: not reported; Bw: Body weight; p-CaMKII (phosphorylated form of Calcium/calmodulin-dependent protein kinase II); BDNF: Brain-Derived Neurotrophic Factor; AChE: Acetylcholinesterase; SOD: Superoxide dismutase; MDA: Malondialdehyde; GSH-Px: glutathione peroxidase; ROS: reactive oxygen species; p-tau: phosphorylated tau; BAX: Bcl-2-associated X protein; ChAT: choline acetyltransferase; BCL2: B-cell lymphoma-2; Nrf2: nuclear factor erythroid 2-related factor 2; MAPK: mitogen-activated protein kinase; MIP: macrophage inflammatory protein; Ppp1r1b: Protein Phosphatase 1 Regulatory Inhibitor Subunit 1B; TTR: transthyretin; CAR-1: carbonic anhydrase 1; Rbp1: Retinol Binding Protein 1; ECA: Angiotensin-Converting Enzyme; PKM2: pyruvate kinase M2; GSK 3-β: Glycogen synthase kinase 3-β; RIPK-1: receptor-interacting serine/threonine-protein kinase 1; LDH B: L-lactate dehydrogenase B physiological balance in mammals. The conversion of ACE 1 to ACE 2—an enzyme with vasoconstrictive effects—is a key mechanism for controlling vascular tone (Coates, 2003). Under pathological conditions, however, excessive vasoconstriction can lead to elevated blood pressure and increased cardiac workload (Sweitzer, 2003).

In this context, cell-free in vitro assays have reported that hydrolysis of whey protein concentrate (WPC) or whey protein isolate (WPI) by enzymes such as pepsin can increase ACE-inhibitory activity by up to 58% (Mansinhbhai et al., 2023). Similarly, microbial fermentation has demonstrated this potential: Lactobacillus helveticus strains applied for 1 h significantly enhanced the ACE-inhibitory activity of ricotta cheese exhausted whey (Pontonio et al., 2021). Moreover, proteases isolated from microorganisms, including the fungus Sporisorium reilianum, have shown antihypertensive potential (Tovar-Jiménez et al., 2021). These findings demonstrate that enzymatic hydrolysis can release peptides with antihypertensive activity from the native protein matrix of whey.

This inhibitory effect has been supported by in situ studies, where processing fresh cheese whey via its native microbiota (Mazorra-Manzano et al., 2020) or through starter cultures and coagulase inhibited ACE by up to 70% (Godoy et al., 2020). The enhancement of dietary antihypertensive potential through hydrolysis during processing suggests that consumption of such peptides or peptide-enriched foods could be a strategy for developing biofunctional foods. Indeed, adding whey protein hydrolyzed for 5 h with Alcalase increased the ACE-inhibitory potential of pork frankfurters (Barrón-Ayala et al., 2020).

To evaluate the potential of these peptides against hypertension, in vivo studies have been conducted (Table 19.3). Recent investigations have focused on spontaneously hypertensive rat models (Dou et al., 2024; Zhao et al., 2025; Hussein et al., 2020; Xie et al., 2022). Administration of 500 and 1000 mg/kg body weight daily for 6 weeks of whey protein fermented by microorganisms reduced systolic blood pressure and tissue injuries caused by hypertension in male rats (Dou et al., 2024). In another study, broader parameters were assessed: daily administration of 15–60 mg/kg body weight decreased systolic blood pressure, endothelial cytokines, malondialdehyde, HDL-C, LDL-C, and total cholesterol over long-term administration (Zhao et al., 2025). This evaluation is particularly relevant, as vascular dysfunction is influenced by proinflammation, oxidative stress, and cholesterol levels.

Oxidative stress can damage macromolecules, particularly through plasma lipoprotein oxidation. Accumulation of LDL in plasma increases susceptibility to oxidation, leading to atherogenic dyslipidemia. Lipid peroxidation of LDL particles contributes to activation of various pathophysiological mechanisms, including receptor-mediated activation of pro-inflammatory signaling pathways (Vekic et al., 2023). Collectively, these findings suggest that whey peptides may mitigate hypertension through homeostasis across multiple pathways.

In preclinical models, blood pressure reduction by whey peptides has been reported to be dose-dependent (Hussein et al., 2020) and effective in both short- and long-term interventions on systolic blood pressure and oxidative stress (Xie et al., 2022). The mechanisms of action are diverse: ACE-inhibitory activity observed in in vitro and in situ studies has been corroborated in in vivo hypertensive models. Whey peptides classically regulate the renin–angiotensin system by inhibiting ACE activity, reducing angiotensin II, and modulating renin in tissues and serum (Xie et al., 2022; Hussein et al., 2020; Zhao et al., 2025; Dou et al., 2024).

Interestingly, whey peptides have also been shown to alleviate hypertension-associated intestinal microbiota dysbiosis by restoring microbial diversity and increasing SCFA-producing microorganisms (Xie et al., 2022; Zhao et al., 2025; Dou et al., 2024). Administration of these peptides increased the abundance of Akkermansia, Bacteroides, and Lactobacillus, consequently raising short-chain fatty acid levels (Dou et al., 2024). This modulation of gut microbiota was accompanied by reduced renal tissue damage and lung injury (Zhao et al., 2025). The link between microbiota and hypertension is complex; through prevention of dysbiosis, whey peptides positively influence postbiotic and gut metabolite production, including SCFAs, which in turn modulate the renin–angiotensin system, gut barrier, endothelium, inflammation, and oxidative stress, collectively exerting antihypertensive effects (Tsafack et al., 2022).

Despite promising preclinical results, single doses of 5–50 g of whey peptides did not significantly affect systolic or diastolic blood pressure in healthy volunteers, even though plasma ACE activity was reduced. This highlights a gap in understanding the inhibitory action of these peptides in humans and underscores the need for better characterization of their mechanisms in clinical trials (Martin et al., 2020).

3.2 Antitumor activity

There is emerging evidence supporting the anticarcinogenic role of whey-derived peptides. Although studies remain relatively limited, recent research suggests that these peptides can modulate signaling pathways involved in cancer development, as well as regulate inflammatory responses, which play a key role in several stages of tumor progression (Quintieri et al., 2025).

TABLE 19.3 Overview of whey-derived opioid peptides, their precursor proteins, sequences, target receptors, and reported physiological effects.

Peptide/ Fragment	Precursor protein	Sequence (example)	Target receptor	Reported physiological effect	Type of evidence	Critical comment	References
α-lactorphin	α-lactalbumin (α-LA)	Tyr–Gly–Leu–Phe (YGLF)	μ (weak agonist)	Inhibition of guinea pig ileum contractions; naloxone-reversible	Experimental (in vitro)	Weak but consistent opioid effect; released after pepsin action	Antila et al. (1991)
β-lactorphin	β-lactoglobulin (β-LG)	Tyr–Leu–Leu–Phe (YLLF)	μ (weak binding); receptor-independent	Contractile effect in ileum; not reversed by naloxone	Experimental (in vitro)	Mechanism unclear; receptor binding documented	Antila et al. (1991)
β-lactorphin	β-lactoglobulin (β-LG)	Tyr–X–X–Phe	μ opioid receptor	Analgesic and behavioral modulation in animals	In vivo (rats, neonates)	Limited bioavailability; CNS access uncertain	Pihlanto-Leppälä (2001)
β-lactorphin/ α-LA-derived variant)	α-lactalbumin (α-LA)	Tyr–Gly–Leu–Phe	μ opioid receptor	Analgesia/pain modulation	Early experimental (in vitro)	Early evidence; human relevance unknown	Antila et al. (1991)
β-lactotensin	β-lactoglobulin (β-LG)	His–Ile–Ala–Leu (HIAL)	Unclear/non-classical	Contraction of longitudinal ileum muscle	In vitro	Effect not clearly opioid; mechanism not defined	Pihlanto-Leppälä (1997); Antila et al. (1991)

Legend/Notes: -μ and δ receptors: Classic opioid receptors (mu and delta subtypes) involved in analgesia, gut motility, and central nervous system effects.-In vitro: Laboratory studies using isolated tissues or cell models; -In vivo: Animal or human studies examining physiological responses;-Pepsin proteolysis: Enzymatic digestion in the stomach that releases bioactive peptides from milk proteins; –CNS/GI: Central nervous system/Gastrointestinal tract.

Breast and colorectal cancers are among the most frequently investigated models for assessing the efficacy of whey-derived peptides. Whey protein hydrolyzates (WPH) obtained through hydrolysis with gastrointestinal peptidases markedly suppressed TNF-α and IL-6 production in HT-29 human colon cancer cells, indicating reduced pro-inflammatory signaling in these tumor cells. In line with these findings, camel whey protein hydrolyzates (CWPHs) showed enhanced anti-inflammatory activity, ranging from 732.50 to 3779.16 μg Diclofenac Sodium Equivalent (DSE)/mg protein. Consequently, CWPHs significantly inhibited the growth of human colon carcinoma HCT-116 cells (Murali et al., 2021). Indeed, camel whey peptides induced G2/M cell cycle arrest through the modulation of several cell-cycle regulatory proteins, including Cdk1, Cyclin B1, p21, and p53 (Murali et al., 2021). Beyond anti-inflammatory and cell-cycle regulatory effects, the iron-binding capacity of whey also prevents oxidative damage, contributing to the reduction of colon cancer risk (Ramani et al., 2024).

Regarding preclinical studies in breast cancer, doses ranging from 5 to 400 μg/mL of whey protein hydrolyzate mixtures or isolated whey-derived peptides reduced the viability of carcinogenic cells after 24 h of treatment (Sahna et al., 2022; Taghipour et al., 2023). The primary mechanism involved was cell death by apoptosis and necrosis. In MCF-7 cells, the enzymes glycogen synthase kinase 3-β (GSK-3β), pyruvate kinase M2 (PKM2), and L-lactate dehydrogenase (LDH-B), along with the mucin 1 (MUC1) protein, were downregulated, whereas receptor-interacting serine/threonine-protein kinase 1 (RIPK1) was upregulated following treatment with whey hydrolyzates.

GSK-3β is a key enzyme involved in Wnt and Hedgehog signaling pathways, insulin signaling, neuronal development, transcription, cell division, and cell death. PKM2 activity is frequently increased in benign and malignant proliferative tissues, LDH-B is overexpressed in breast and lung cancers and is considered a potential therapeutic target, while RIPK1 activation promotes cell death in radioresistant breast cancers. Therefore, the regulatory effects of whey hydrolyzates on these pathways support the potential role of whey-derived peptides in breast cancer therapy (Sahna et al., 2022).

Beyond these cancer types, melanoma (Neto et al., 2022) and liver cancer (Kamal et al., 2018) models have also been investigated, with reduced cancer cell viability reported in both cases. In melanoma, the proposed mechanisms included the modulation of proteins involved in proliferation, energy metabolism, apoptosis, and metastatic and malignancy rates (Neto et al., 2022).

Additional anticarcinogenic mechanisms have been proposed, such as enhancing antioxidant defenses and influencing signaling pathways including PI3K/Akt, mTOR, and Wnt/β-catenin. Similarly, α-lactalbumin has been suggested to interact with cell-surface modulators, altering calcium transport, intracellular calcium levels, and cell growth rate (Elmas et al., 2025).

It is also important to highlight the selectivity reported for some peptides toward cancer cells, with minimal effects on normal cells. In this context, lactoferricin B exhibited in vitro cytotoxicity against rat and human cancer cell lines—including leukemia, fibrosarcoma, various carcinomas, and neuroblastoma—while not significantly affecting the viability of normal fibroblasts, lymphocytes, epithelial cells, endothelial cells, or erythrocytes. This selectivity has been attributed to its strongly cationic nature, which enables preferential binding to negatively charged cancer cells, whereas healthy untransformed cells are largely spared due to their net neutral charge (Pepe et al., 2013).

Whey protein supplementation is also a viable and cost-effective strategy for managing cancer cachexia syndrome. This may be attributed to whey protein's high leucine content and its intrinsic capacity to modulate insulin-like growth factor I (IGF-I) concentrations, thereby helping to mitigate muscle atrophy and anorexia in cancer patients (Ramani et al., 2024). However, human intervention studies with whey protein are still in their infancy and must be expanded across different tumor types to provide solid evidence for its potential role as an adjuvant in cancer therapy (Thampy et al., 2024).

3.3 Antidiabetic activity

The World Health Organization defines diabetes as a chronic metabolic disease characterized by persistently high blood glucose concentrations. Over time, this metabolic disorder promotes the development of severe complications that compromise cardiovascular function (ischemic heart disease, cerebrovascular disease), renal function (diabetic nephropathy), ocular function (retinopathy), and neurological function (peripheral and autonomic neuropathy). The most recent epidemiological data indicate that approximately 830 million people worldwide have diabetes, representing a growing challenge for global public health (World Health Organization, 2025). Type 2 diabetes mellitus (T2DM) is the most prevalent form, accounting for approximately 90% of diagnosed cases, while type 1 diabetes mellitus (T1DM) and gestational diabetes have a lower incidence. T2DM is characterized by a combination of two pathological processes: insulin resistance in peripheral tissues, such as skeletal muscle, liver, and adipose tissue, and progressive dysfunction of pancreatic β cells, leading to insufficient insulin production to meet metabolic demand (DeFronzo et al., 2015; Tenenbaum et al., 2023).

In recent years, scientific research has highlighted the remarkable potential of bioactive peptides derived from milk proteins, due to their functional relevance and potential impact on health. Whey protein concentrate and, to a greater extent, whey protein isolate stand out for their high protein content, with values of 65%–80% and over 90% on a dry basis, respectively. Among their main fractions are β-lactoglobulin and α-lactalbumin, recognized as key sources of bioactive peptides (Andrade et al., 2019; Bunsroem et al., 2022). These compounds can be released through controlled enzymatic hydrolysis or microbial fermentation processes, especially using lactic acid bacteria (LAB) (Nongonierma et al., 2019; Guha et al., 2021; Saubenova et al., 2024). These peptides not only represent an innovative way of valorizing whey but have also been shown to have multiple beneficial physiological effects, among which their antidiabetic properties stand out, with particular relevance for blood glucose control and the prevention of T2D (Mignone et al., 2015; Tulipano et al., 2015).

Peptides derived from milk proteins, such as β-lactoglobulin and α-lactalbumin, exert their antidiabetic effects through various mechanisms, mainly related to the modulation of glucose homeostasis. These include the inhibition of digestive enzymes responsible for the breakdown of carbohydrates, such as α-amylase and α-glucosidase, whose action releases monosaccharides from complex carbohydrates (Mignone et al., 2025; Baba et al., 2021a,b). Inhibiting these enzymes slows down the digestion of carbohydrates, preventing the rapid release of large amounts of glucose into the intestine. Therefore, intestinal glucose absorption is significantly delayed and distributed more gradually throughout the distal intestinal tract, helping to reduce postprandial blood glucose spikes after meals, a critical factor in the management of DM2 (Kahn et al., 2014; Martini et al., 2021). Various studies indicate that nutritional strategies are becoming increasingly important, and the oral intake of whey proteins and their hydrolyzates has demonstrated a remarkable inhibitory capacity on α-amylase and α-glucosidase, showing a considerable inhibitory capacity on these enzymes responsible for raising glucose, which points to an efficacy similar to that of conventional oral medications such as acarbose, miglitol, voglibose, and emiglitate, commonly used to control DM2 and attenuate postprandial hyperglycemia. These medications can cause adverse reactions, a disadvantage that these protein alternatives could help mitigate (Kahn et al., 2014; Patil et al., 2015).

The study by Kamal et al. (2018) showed that hydrolysis time and the type of digestive enzyme used are important factors in the release of bioactive sequences, given that camel milk whey hydrolyzates generated with pepsin reached the highest inhibition of α-glucosidase (78.59%) 3 hours after hydrolysis, followed by those produced with chymotrypsin at 3 and 6 h (77.88% and 77.45%, respectively), while those produced with chymotrypsin achieved up to 93.6% α-amylase inhibition after 6 hours of hydrolysis, surpassing those produced with pepsin (85.0%) and trypsin (57.5%) in the same time. These findings were corroborated and expanded upon by Baba et al. (2021a,b), who optimized the conditions for hydrolysis with pepsin (time, temperature, and enzyme: substrate ratio). Their study identified hydrolyzates with high inhibitory potency. For α-glucosidase, hydrolyzate H19 obtained at 45°C, 120 min, and an enzyme: substrate ratio of 2% showed the lowest 50% inhibitory concentration (IC_{50}) (0.64 mg/mL), with 97.62% inhibition compared to the intact protein ($IC_{50} = 26.9$ mg/mL). On the other hand, for α-amylase, hydrolyzate H26 (obtained at 45°C, 240 min, and an enzyme: substrate ratio of 0.5%) exhibited the highest inhibitory potency, with a CI_{50} of 0.29 mg/mL and a 95.48% increase in inhibition. These findings show that mild hydrolysis conditions (low temperature and low enzyme concentration) can favor the generation of peptides with high antidiabetic bioactivity.

The identification of specific peptides has confirmed this potential, with various studies characterizing sequences with marked hypoglycemic or antidiabetic activity. In hydrolyzates obtained from Rubing cheese by-products, the YPVEPF peptide showed the highest α-glucosidase inhibition ($CI_{50} = 3.52$ mg/mL), standing out as a candidate for application in functional foods (Li et al., 2023). Similarly, the hydrolysis of camel milk whey also generated peptides such as PAGNFLMNGLMHR, PAVACCLPPLPCHM, MLPLMLPFTMGY, and PAGNFLPPVAAAPVM, all with antidiabetic potential (Baba et al., 2021a,b). In mature cheeses such as Parmigiano-Reggiano, the IPP peptide generated during natural proteolysis over 24 months of maturation proved to be the most active against α-glucosidase (Martini et al., 2021). This evidence confirms that peptides derived from whey, both bovine and camel, represent a promising strategy for the prevention and management of T2D, either as ingredients in functional foods or as candidates for the development of new-generation nutraceuticals.

The inhibition of α-glucosidase and α-amylase enzymes has also been attributed to the generation of bioactive peptides derived from fermentation with LAB. Strains such as *Lactiplantibacillus plantarum* have proteolytic capacity and can release peptides with an inhibitory effect, the intensity of which depends on both the strain used and the fermentation time, with more extended periods favoring greater enzyme inhibition (Das et al., 2025; Pipaliya et al., 2025). This antidiabetic potential has been associated with the proteolytic activity of the cultures, as demonstrated by Rosa et al. (2023), who reported that *Lactobacillus acidophilus* La-05 and *Bifidobacterium* Bb-12 exhibited a more marked inhibition of α-glucosidase, reinforcing their role in the dietary management of diabetes.

Another widely studied mechanism is the inhibition of dipeptidyl peptidase-IV (DPP-IV), an enzyme that rapidly degrades incretin hormones such as glucagon-like peptide-1 (GLP-1) and glucose-dependent insulinotropic peptide (GIP) by removing their N-terminal dipeptides. These incretins, GLP-1 and GIP, are peptides derived from the intestine that are released within minutes after food intake and stimulate glucose-dependent insulin secretion (Campbell and Drucker, 2013; Tulipano et al., 2015). Under normal conditions, the amount of insulin produced is sufficient to maintain glucose balance; however, in individuals with prediabetes or diabetes, greater production is required to control hyperglycemia.

In this context, DPP-IV inhibition has been identified as an effective strategy for improving insulin secretion. Peptides from whey protein hydrolyzates can modulate DPP-IV activity, stimulating incretin hormone secretion and delaying enzymatic degradation. Various studies have shown that both whey protein concentrate hydrolyzates and whey protein isolate hydrolyzates have significant inhibitory activity on DPP-IV, reinforcing their potential as a nutritional strategy in diabetes control. Bioactive compounds such as APFPE, IPP, and LPPT have been identified in Parmigiano-Reggiano matured for 12 months, demonstrating possible antidiabetic activity due to the application of an integrated approach that combined peptidomics, in silico analysis, and a structure-activity relationship (Martini et al., 2021).

Also, peptides derived from α-lactalbumin (LAHKPL and ILDKEGIDY) have been detected in hydrolyzed whey protein concentrate hydrolyzates from camel whey protein. In addition, three potent peptides derived from β-casein were identified: VPV, YPI, and VPF, with DPP-IV IC50 values of 6.6 ± 0.5, 35.0 ± 2.0, and 55.1 ± 5.8 μM, respectively. After IPI, VPV is the second most potent DPP-IV inhibitory peptide identified to date, supporting the role of camel milk as an antidiabetic agent (Nongonierma et al., 2019). Kumar et al., 2024a,b demonstrated that β-lactoglobulin hydrolyzates obtained from pepsin were more effective in inhibiting DPP-IV, reaching an IC50 value of 3.78 mg/mL, which indicates superior activity compared to preparations with trypsin and Flavourzyme. Furthermore, among the peptides released, LIVTQTMKGLD (f17–27) was identified as the most active, with an IC50 of 6.64 mmol/L, which could be associated with an increase in GLP-1 levels and an antihyperglycemic effect.

Both in silico and in vitro analyses have shown that bovine β-lactoglobulin is a more efficient source of DPP-IV inhibitors than α-lactalbumin (Tulipano et al., 2015). However, other studies, such as that by Bunsroem et al. (2022), have shown opposite results, with α-lactalbumin exhibiting greater inhibitory potential than β-lactoglobulin, bovine serum albumin, and lactoferrin, attributed to its higher content of short peptide sequences. Overall, the evidence supports that different fractions of milk proteins possess peptides with the ability to inhibit DPP-IV, prolonging the action of incretins, promoting insulin secretion, and contributing to the reduction of postprandial blood glucose (Lambeir et al., 2003; Patil et al., 2015; Bunsroem et al., 2022).

Technological advances, such as the use of ultrafiltration membrane electrodialysis (EDUF), have been shown to significantly improve the inhibitory capacity of whey peptides. Faucher et al. (2022) reported that the recovery fraction obtained by EDUF showed up to four times higher in vitro inhibition of DPP-IV in the peptide recovery fraction compared to the initial hydrolyzate, highlighting its antidiabetic potential. However, when evaluating the antidiabetic effect of this fraction in obese mice fed a diet rich in fat and sucrose, the results were less promising. After 6 weeks of supplementation, insulin resistance was not reversed, and at week eight, glucose intolerance was confirmed, with no modulatory effects on the glycemic or insulinemic response observed. This discrepancy is related to the digestive process and, in particular, to the degradation and limited bioavailability of bioactive peptides. After ingestion, these compounds undergo partial degradation in the stomach and more intense degradation in the intestine, where various peptidases reduce their stability and hinder their transport into the bloodstream, thus limiting their antidiabetic effect in vivo (Renaud et al., 2023).

3.4 Antioxidant activity

The antioxidant characteristics of bioactive peptides are associated with their composition, formation, and hydrophobicity. Histidine, glutamic acid, proline, tyrosine, cysteine, methionine, and phenylalanine are all amino acids with antioxidant properties (Ibrahim et al., 2018; Kashung and Karuthapandian, 2025; Zaky et al., 2022).

Antioxidants are used as nutraceuticals and functional additives in food and many other products (Du and Li, 2022). Antioxidants protect the body by eliminating free radicals and inhibiting lipid peroxidation reactions, thus preventing oxidative damage (Alzaydi et al., 2023).

Peptides with antioxidant activity can be obtained from the enzymatic hydrolysis of proteins contained in food. Peptides derived from milk proteins have been shown to have antioxidant properties that prevent the peroxidation of essential fatty acids (Sánchez and Vázquez, 2017).

Peptides involved in antioxidant activity must be able to induce or participate in single-electron transfer reactions (Bhandari et al., 2020). Many proteins, protein hydrolyzates, and specific amino acids have demonstrated significant antioxidant activity (Kumari and Shahidi, 2024).

Oxidation is a necessary metabolism in the human and vertebrate body. However, it produces reactive oxygen species (ROS) and free radicals, which disrupt homeostasis and lead to oxidative stress (Jakubczyk et al., 2020; Purohit et al., 2024; Shahidi and Saeid, 2025). Extensive research has been conducted on antioxidant peptides derived from casein and whey proteins (Besharati and Lackner, 2023).

The oxidative process is induced by free radicals, such as reactive oxygen species (ROS): superoxide anion radical (O2), hydroxyl radical (OH), hydrogen peroxide (H2O2), lipid radical (ROO) and reactive nitrogen species, such as nitrogen oxide (NO), and generates damage to lipids, proteins, DNA and carbohydrates (Jakubczyk et al., 2020; Tadesse and Emire, 2020; González-Osuna et al., 2024).

As a result of the excessive production of free radicals, cellular proteins, membrane lipids, DNA, and enzymes are damaged, which can lead to various diseases such as rheumatoid arthritis, diabetes, atherosclerosis, and oxidative damage to DNA, which can also cause cancer (Punia et al., 2020).

Amino acids bind to pro-oxidant metal ions to exert activity, eliminate the OH radical, and/or inhibit lipid peroxidation. Therefore, each amino acid contributes as an antioxidant in a unique way, depending on its type (Zaky et al., 2022).

Reactive oxygen species (ROS) such as superoxide anion free radicals, hydroxyl free radicals, lipoxyl free radicals, nitrogen dioxide, nitric oxide free radicals, and hydrogen peroxide are found in biological systems (Ye et al., 2022).

3.5 Antimicrobial activity

Antimicrobials act directly to protect against bacteria, fungi, viruses, and protozoa, with antimicrobial activity. This cationic part of the peptides is responsible for the electrostatic interaction of pores in the cytoplasmic membrane with anionic phospholipids. Due to the formation of pores, an ionic imbalance occurs in the membrane and, consequently, cell lysis. The functions of antimicrobial peptides are carried out in three phases: (i) initial attraction by the bacterial cell wall; (ii) subsequent binding to the cell membrane and (iii) the eventual insertion of the peptides into the host cell membrane, causing membrane permeabilization (Bhandari et al., 2020; Wijesekara et al., 2024).

Antimicrobial peptides are classified into four categories based on their origin: microbial, animal, plant, and other, and their antibacterial effects vary depending on their source (Gong et al., 2021). Antimicrobial peptides can also be categorized into anionic and cationic peptides based on their net charge (Purohit et al., 2024). These peptides, also known as host defense peptides, act as an important barrier against pathogenic bacterial invasion of the body (Wijesekara et al., 2024).

Most peptides have a chain of 10–50 or 10–55 amino acids, with an overall positive charge and structures containing hydrophobic and hydrophilic regions (Alzaydi et al., 2023; Kumari and Shahidi, 2024). In studies on the antimicrobial activity of milk-derived peptides, peptide fragments from casein, β-lactoglobulin, and α-lactalbumin were mainly identified (Piotto et al., 2012).

Antimicrobial peptides mainly contain an α-helical structure, cationic and amphipathic, while a smaller number of α-helical peptides are hydrophobic. These are mostly short-chain peptides, with 10–55 amino acids, an overall positive charge, and structures containing hydrophobic and hydrophilic regions (Kumari and Shahidi, 2024).

Antimicrobial peptides can be classified into three families according to structural characteristics: linear α-helical peptides; cyclic peptides with disulphide bridges, and open cyclic peptides (Sánchez and Vázquez, 2017).

Antimicrobial peptides directly affect bacteria, yeasts, and viruses. They can be categorized differently depending on their structure, amino acid sequence, or biological function. Peptides and protein hydrolyzates with antibacterial activity are important in preventing food spoilage and the transmission of food-borne pathogens. Interestingly, many antimicrobial peptides exhibit additional bioactivities, such as antioxidant activities, immunomodulation, and wound healing (Daliri et al., 2017).

The main factors influencing their antibacterial activity include electrostatic interactions between the peptide and anionic and positively charged lipids present on the surface of the target microorganism. Some peptides target microbial cells, causing cell death through other mechanisms, such as a) binding to DNA or RNA to disrupt the genetic material and prevent replication and the transcription process, b) inhibiting protein synthesis, and c) inhibiting cell wall synthesis (González-Osuna et al., 2024).

The therapeutic approach based on antibacterial peptides has significant potential for treating infections caused by antibiotic-resistant bacteria, which pose a growing threat to human health (Saubenova et al., 2024).

3.6 Action on the gastrointestinal system

Bioactive peptides derived from whey proteins exert a broad spectrum of effects on the gastrointestinal system, from modulation of intestinal motility and the gut microbiota to protection against oxidative stress, inflammation, and even chemopreventive outcomes.

3.6.1 Modulation of the gut microbiota

Whey-derived peptides contribute to the modulation of the intestinal microbiota by promoting the growth of beneficial bacterial genera such as *Lactobacillus* and *Bifidobacterium* (Yu et al., 2016). The glycomacropeptide (GMP) is a key bioactive component in this effect. Although not classified as a conventional whey protein, GMP is released from κ-casein during the initial stages of cheese manufacture through the enzymatic activity of chymosin or pepsin, and also during gastrointestinal digestion (Rackerby et al., 2024). GMP is extensively glycosylated, sialylated, and phosphorylated due to post-translational modifications (Rackerby et al., 2024). Its sialic acid content is central to its biological functionality, particularly in shaping microbial composition. Particular *Bifidobacterium* species harbor sialidase enzymes capable of cleaving sialic acid from glycoconjugates such as GMP, enabling their utilization as carbon and energy sources (Egan et al., 2014).

Experimental evidence demonstrates that increased GMP concentrations selectively enhance the growth of beneficial microbes, conferring a competitive advantage over potentially harmful species such as *Escherichia coli* and *Clostridium perfringens* (Nishiyama and Mukai, 2022). Lactoferrin (LF), another major whey protein, also exerts prebiotic-like effects by sequestering free iron, thereby stimulating the proliferation of *Lactobacillus* and *Bifidobacterium* while restricting the growth of iron-dependent pathogens such as *E. coli* (Zhao et al., 2023; Vega-Bautista et al., 2019). Additional peptides display antioxidant activity, restoring intestinal homeostasis by reducing reactive oxygen species (ROS) levels and attenuating oxidative damage. Such effects preserve epithelial barrier integrity and promote a stable ecological niche for commensal microorganisms (Riaz Rajoka et al., 2021; Sun et al., 2024).

3.6.2 Regulation of gastrointestinal motility

Beyond microbial modulation, GMP has been shown to regulate gastrointestinal function via induction of cholecystokinin (CCK), a hormone pivotal in the control of appetite, food intake, gallbladder contraction, intestinal motility, gastric emptying, and pancreatic enzyme secretion (Beucher et al., 1994). This downregulation of gastric secretions is closely associated with satiety. Opioid peptides derived from whey, including α-lactorphin (YGLF) and β-lactorphin (YLLF), act as agonists of opioid receptors (Teschemacher et al., 1997). Their activation suppresses acetylcholine release within the enteric plexus, thereby reducing peristalsis (Holzer, 2009). This mechanism prolongs nutrient–mucosa contact time, enhancing absorption and nutrient bioavailability (Zhao et al., 2021), ultimately improving nutritional efficiency. Such regulation is critical for maintaining gastrointestinal health, as it prevents the accumulation of undigested substrates in the lumen, which may otherwise favor the proliferation of undesirable microorganisms and promote inflammation (Mosele et al., 2015).

3.6.3 Anti-inflammatory properties in the digestive system

Whey-derived bioactive peptides also contribute to immunomodulation, as will be further detailed in subsequent sections. Concerning gastrointestinal inflammation, specific peptides such as β-lactoglobulin have demonstrated anti-inflammatory activity in Caco-2 intestinal epithelial cells by attenuating TNF-α-induced IL-8 expression and enhancing phosphorylation of proteins that suppress pro-inflammatory signaling pathways (Olvera-Rosales et al., 2023). Similarly, studies employing goat whey in murine models of colitis have reported improved intestinal inflammation, decreased leukocyte infiltration, downregulation of pro-inflammatory cytokines (IL-1β, IL-6, IL-17, TNF-α), and upregulation of mucin expression, which collectively strengthen the intestinal mucosal barrier (Van der Sluis et al., 2006; Olvera-Rosales et al., 2023). These findings underscore the relevance of whey peptides as modulators of gastrointestinal inflammation.

3.6.4 Chemopreventive potential in the digestive system

Whey proteins and their derived peptides exhibit promising chemopreventive properties within the gastrointestinal tract. Notably, peptides such as lactoferrin and lactoferricin are recognized for their antitumor activities, including induction of apoptosis and inhibition of angiogenesis (Fernández-Tomé and Hernández-Ledesma, 2020). In rodent models of intestinal tumorigenesis, dietary supplementation with whey protein hydrolyzates inhibited the development of aberrant crypt foci—microscopic lesions in the colonic mucosa regarded as preneoplastic indicators (Xiao et al., 2006). Such early lesions are precursors of adenomatous polyps and, eventually, colorectal carcinoma (Bird, 1995).

Further studies have demonstrated that higher doses and frequent administration of whey hydrolyzates significantly reduced the incidence of microscopic tumors, reinforcing their efficacy in colon cancer prevention (Attaallah et al., 2012). These chemopreventive attributes highlight the broader importance of whey-derived peptides in gastrointestinal health and disease prevention.

3.7 Immune system modulation

The immune system represents the body's primary line of defense against pathogenic microorganisms and plays a central role in maintaining overall health. Approximately 90% of human diseases are estimated to be associated with abnormalities in this system (Pulendran and Davis, 2020). Whey is composed of a wide variety of native proteins that act as precursors to bioactive peptides with immunomodulatory properties, including β-lactoglobulin (β-LG), α-lactalbumin (α-LA), lactoferrin (LF), and bovine serum albumin (BSA). In addition to these, glycomacropeptide (GMP), although not a native whey protein, is a bioactive peptide released from κ-casein during the enzymatic coagulation or by the action of pepsin in digestion, which has also been widely recognized for its functional properties in immunological modulation (Córdova-Dávalos et al., 2019).

3.7.1 β-lactoglobulin (β-LG)

The immunomodulatory properties of β-LG-derived peptides have been evidenced by studies demonstrating their capacity to regulate spleen cell proliferation, modulate immunoglobulin synthesis, and influence the balance between pro- and anti-inflammatory cytokine responses (Bamdad et al., 2017; Tai et al., 2016).

Classical studies have indicated that commercial variants of β-LG (A and B) stimulate the proliferation of murine splenic lymphocytes in vitro, and that this effect is attenuated after enzymatic hydrolysis with trypsin (Wong et al., 1998). However, subsequent investigations suggested that proteolytic digestion of β-LG could, in certain contexts, potentiate its mitogenic activity, implying that peptides derived from the hydrolytic process might contribute to the modulation of the immune response (Mahmud, Matn, and Otani et al., 2004). Indeed, peptides extracted after simulated gastrointestinal digestion of β-LG promoted dendritic cell maturation and directed the T cell response toward a Th1 phenotype (Ma et al., 2022).

In recent years, studies involving β-LG and its derived peptides have focused not only on characterizing their ability to modulate immune cell activity and cytokine production, but also on elucidating their allergenic potential. Several studies have explored the functionality of these peptides, both in their free form and in ligand-bound form, primarily in relation to their potential to influence the allergic immune response (Mayerhofer et al., 2022; Roth-Walter et al., 2021; Złotkowska et al., 2021).

Ligand-bound forms (holo-β-LG) often exhibit attenuation of cell activation associated with allergic reactions by promoting the proliferation of regulatory T cells or Th1-type responses, which are generally associated with protection against allergies. In contrast, ligand-free forms (apo-β-LG) tend to favor the activation of pro-allergic pathways and the induction of Th2-type T cell responses (Afify et al., 2021; Roth-Walter et al., 2021). This difference in immune function is due to the ability of holo-β-LG to transport small bioactive ligands that can influence the immune microenvironment, reducing inflammation and promoting tolerance.

For example, pretreatment with holo-β-LG has shown potential for protection against pollen allergy, mainly by attenuating the Th2 response in splenocytes of mice sensitized with β-LG. This effect was evidenced by a reduction in the production of pro-allergic cytokines, a lower frequency of $CD4^{+}CRTh2^{+}$ cells, decreased antigen presentation, and reduced levels of specific antibodies (Afify et al., 2021). In contrast, apo-β-LG has been linked to the induction of allergic responses by promoting increased activation of $CD4^{+}$ T cells and the overexpression of Th2-type cytokines (Hufnagl et al., 2018).

3.7.2 α-lactalbumin (α-LA)

Recent reviews have highlighted the bioactive potential attributed to peptides derived from whey proteins, including α-LA, in modulating immunity and exerting anti-inflammatory action (Saubenova et al., 2024). Bounous and Kongshavn (1985) pioneered the demonstration that mice fed a lactalbumin-based diet had a significantly higher number of antibody-forming cells in the spleen compared to those fed protein diets containing casein, soy, or wheat (Bounous and Kongshavn, 1985). The hydrolysis of α-LA by digestive proteases demonstrated that the released peptides can stimulate splenocyte proliferation and cytokine secretion (Jacquot et al., 2010). Furthermore, the anti-inflammatory potential of α-LA was demonstrated in a study employing the combination of α-lactalbumin and the phage KPP22, which significantly reduced the secretion of interleukins IL-1β and IL-6 in a model of inflammation induced by *Pseudomonas aeruginosa* PAO1 (Ghadimi et al., 2024).

3.7.3 Lactoferricin (LFcin)

Lactoferricin (LFcin) is one of the main peptides derived from lactoferrin (LF), responsible for its immunomodulatory effects. Lactoferrin, in turn, is capable of exerting pro-inflammatory properties by inducing human M2-type macrophages to acquire an M1-like phenotype, characterized by increased expression of pro-inflammatory cytokines, such as IL-1β (Gao et al., 2018). Interestingly, although previous studies suggest pro-inflammatory effects, lactoferrin has been shown in ocular surface cell models to reduce TLR2 expression and IL-β secretion, indicating that its immunomodulatory effects are highly dependent on the cell type, activation context, and structural form of the protein (Regueiro et al., 2022).

The study conducted by Yami et al. (2023) suggests that lactoferrin can reduce pro-inflammatory cytokines, including IL-6 and TNF-α, and stimulate the production of IL-10. Similarly, lactoferricin has demonstrated the ability to attenuate the inflammatory response mediated by human and murine macrophages, promoting a reduction in the production of lipopolysaccharide (LPS)-induced pro-inflammatory cytokines, such as tumor necrosis factor-alpha (TNF-α) and interleukin-6 (Malone et al., 2022). More specifically, it is believed that LF may differentially regulate the production of inflammatory cytokines, depending on the physiological or pathological context, thereby ensuring immune homeostasis (Kim et al., 2025).

3.7.4 Bovine serum albumin (BSA)

The immunomodulatory effects of bovine serum albumin (BSA) and its peptides are attributed to their ability to serve as precursors for glutathione (GSH) synthesis, as well as the direct impact of GSH on the immune system and the body's antioxidant defenses (Iyer et al., 2009). Additionally, the oral intake of BSA in the diet increases antioxidant capacity and enhances innate immunity, helping to counteract thioacetamide (TAA)-induced stress (Abasubong et al., 2023).

3.7.5 Glycomacropeptide (GMP)

The regulatory effect of glycomacropeptide on adaptive immunity is primarily attributed to its ability to modulate lymphocyte proliferation and cytokine expression. Interestingly, undigested GMP stimulated the secretion of the pro-inflammatory cytokines TNF-α and IL-1β in macrophage-like cells, suggesting a possible role in the induction of inflammatory responses (Olsen et al., 2023). However, the digested forms of GMP showed lower expression of these cytokines, suggesting that digestion alters their effects (Olsen et al., 2023).

In the context of innate immunity, GMP has been shown to stimulate the proliferation and activity of cells with a phenotype similar to that of human macrophages, an effect that intensifies after digestion with pepsin (Córdova-Dávalos et al., 2019). Furthermore, GMP reduces the production of pro-inflammatory mediators, such as nitrites, interleukin (IL)-1β, and tumor necrosis factor (TNF)-α, while promoting increased expression of proteins and enzymes associated with the anti-inflammatory and antioxidant response, including SOCS3 and heme oxygenase 1 (HMOX1) (Córdova-Dávalos et al., 2023). Additionally, studies have demonstrated that GMP and its hydrolyzates exhibit anti-inflammatory effects, attenuating inflammatory responses and oxidative stress in LPS-stimulated macrophages (Cheng et al., 2015; Li et al., 2018). Therefore, most evidence indicates that GMP predominantly exerts anti-inflammatory effects, modulating cytokine production and oxidative stress pathways in a cell- and context-dependent manner (Jauregui-Rincón et al., 2019).

3.8 Antithrombotic and antilipidemic activity

Interest in milk-derived antithrombotic peptides is based on the observation that specific regions of milk proteins exhibit structural homology with key hemostatic proteins such as fibrinogen and von Willebrand factor. Studies have demonstrated that sequences derived from κ-casein and lactoferrin can mimic motifs present in fibrinogen, thereby blocking its binding to platelets and reducing platelet aggregation (Jollès et al., 1986; Mazoyer et al., 1990; Qian et al., 1995). This type of activity is particularly relevant for the prevention of arterial thrombosis, a pathophysiological process that plays a central role in events such as myocardial infarction and stroke (Bal dit Sollier et al., 1996). The potential of these peptides lies not only in their antiplatelet activity but also in their ability to act in a more modulated manner compared to classical drugs such as aspirin or clopidogrel, which are associated with a risk of hemorrhagic effects (Chabance et al., 1997).

3.8.1 Antithrombotic peptides

Lactoferrin represents a relevant source of whey proteins from which antithrombotic peptides have been characterized (Xu et al., 2020; Fernández-Musoles et al., 2013). The nonapeptide LRPVAAEIY, derived from bovine lactoferrin, demonstrates potent anticoagulant activity through dual-binding mechanisms to thrombin (Xu et al., 2020). At a 5 mM concentration, LRPVAAEIY prolonged activated partial thromboplastin time by 13.4 s, prothrombin time by 1.7 s, and thrombin time by 5.1 s, with fibrinogen clotting time extended from 15.3 ± 0.4 to 20.2 ± 0.5 s (Xu et al., 2020). Mechanistic characterization using molecular docking, circular dichroism spectroscopy, and fluorescence quenching revealed that LRPVAAEIY binds simultaneously to thrombin's active site (engaging ASP189, LYS60F, and TRP60D residues) and exosite-I (interacting with LYS36 and ARG77A), inducing conformational changes that increase α-helix content from 25.6% to 56.7% while eliminating β-sheet structures (Xu et al., 2020). This dual-binding mode distinguishes LRPVAAEIY from many synthetic anticoagulants, suggesting a unique mechanism of action (Xu et al., 2020).

Additional lactoferrin-derived peptides with cardiovascular protective properties include GILRPY and REPYFGY, which were identified through proteinase K hydrolysis of lactoferrin to generate fractions with a molecular weight below 3 kDa (Fernández-Musoles et al., 2013). These peptides exhibit dual vasopeptidase inhibitory activity, targeting both angiotensin-converting enzyme (ACE) and endothelin-converting enzyme (ECE) (Fernández-Musoles et al., 2013). In spontaneously hypertensive rats, the oral administration of lactoferrin hydrolyzates containing these peptides resulted in significant reductions in blood pressure, suggesting antihypertensive effects that may indirectly support antithrombotic function through improved endothelial tone (Fernández-Musoles et al., 2013).

Lactic acid bacteria can also generate antithrombotic peptides during fermentation processes (Rendon-Rosales et al., 2019). *Lactococcus lactis* strains Lc-571 and Lc-572 produce thrombin-inhibitory compounds, with water-soluble fractions below 3 kDa exhibiting IC_{50} values of 0.045 and 0.049 mg/mL, respectively, for thrombin inhibition (Rendon-Rosales et al., 2019). This bioactivity was maintained following simulated gastrointestinal digestion, suggesting oral bioavailability potential (Rendon-Rosales et al., 2019).

Buffalo β-lactoglobulin exhibits bioactivity through mechanisms distinct from those of lactoferrin-derived peptides (Arpitha et al., 2017). A peptide fraction from buffalo colostrum β-lactoglobulin exhibits anti-hemorrhagic activity by inhibiting zinc-dependent metalloproteinases, chelating zinc ions, and preventing fibrinogenolysis and fibrinolysis in a model of *Echis carinatus* venom-induced hemorrhage (Arpitha et al., 2017). Molecular docking studies confirmed zinc chelation as a key mechanism (Arpitha et al., 2017). Although not evaluated in conventional cardiovascular thrombosis models, these findings indicate protective effects on hemostasis in venom-induced coagulopathy.

From a production standpoint, proteinase K hydrolysis has been employed to generate lactoferrin-derived peptides with ACE and ECE inhibitory activities, resulting in the production of the GILRPY and REPYFGY peptides (Fernández-Musoles et al., 2013). Pepsin hydrolysis generates lactoferrin fractions below 3 kDa that demonstrate strong ACE-inhibitory and antivasoconstrictor activities (Fernández-Musoles et al., 2013). Emerging green technologies, including ultrasound-assisted processing, high-pressure processing, pulsed electric field treatment, and microwave-assisted extraction, are being explored as pre-treatments to enhance enzymatic accessibility and improve the yields of bioactive peptides (Landim et al., 2021; Shi et al., 2025).

The LRPVAAEIY sequence contains hydrophobic residues (leucine, proline, valine, alanine) that facilitate interactions with hydrophobic pockets in thrombin, charged residues (arginine, glutamic acid) enabling electrostatic interactions with the active site, and an aromatic tyrosine residue at the C-terminus contributing π-π stacking and hydrogen bonding for binding specificity (Xu et al., 2020). Proline residues introduce structural bends that allow peptides to conform to enzyme binding pockets, enhancing affinity and specificity (Xu et al., 2020; Fernández-Musoles et al., 2013). Mechanistically, direct thrombin inhibition through simultaneous binding to the active site and exosite-I prevents conversion of fibrinogen to fibrin and reduces thrombin-mediated platelet aggregation (Xu et al., 2020). Vasopeptidase inhibition through dual ACE/ECE targeting provides indirect antithrombotic effects by promoting vasodilation and improving endothelial function (Fernández-Musoles et al., 2013), while metalloproteinase inhibition and zinc chelation represent alternative mechanisms for preventing hemorrhage and fibrinogen degradation (Arpitha et al., 2017).

3.8.2 Antilipidemic peptides

Whey protein consumption reduces LDL-cholesterol by 9.96%–11.23%, total cholesterol by 6.63%–10.25%, significantly decreases triglycerides, and increases HDL-cholesterol by 4.12%–4.49% based on a meta-analysis of 37 studies encompassing 2344 individuals (Badely et al., 2019). These effects were accompanied by significant reductions in both systolic and diastolic blood pressure (Badely et al., 2019). The most rigorous human clinical evidence comes from a randomized controlled trial examining 65 adults with abdominal obesity over 12 weeks, where consumption of 60 g per day of whey protein hydrolyzate combined with low-fiber products significantly reduced both postprandial and fasting triglycerides, total cholesterol, apolipoprotein B-100, and apolipoprotein B-48 (Rakvaag et al., 2019). The reduction in apolipoprotein B-48 indicates modulation of intestinal lipid absorption and chylomicron metabolism (Rakvaag et al., 2019).

Animal model studies have elucidated multiple pathways through which whey proteins exert lipid-lowering effects. Whey protein hydrolyzate alleviated atherosclerosis in apolipoprotein E knockout (apoE−/−) mice fed a Western diet by regulating hepatic lipid metabolism (Wang et al., 2022). The hydrolyzate down-regulated genes involved in cholesterol biosynthesis, up-regulated genes promoting cholesterol excretion, inhibited de novo lipogenesis, and activated fatty acid β-oxidation pathways (Wang et al., 2022). Remarkably, the intervention increased the n-3/n-6 polyunsaturated fatty acid ratio by 2.69-fold in serum and 3.64-fold in liver (Wang et al., 2022).

Lactoferrin emerges as a key mediator of cholesterol-lowering effects. Lactoferrin administration in apoE−/− mice downregulated HMG-CoA reductase, upregulated cholesterol 7-alpha hydroxylase, and downregulated NPC1L1 (Ling et al., 2019). The optimal dose of 10 mg/mL reduced serum and hepatic cholesterol while increasing fecal cholesterol excretion and concurrently reduced atherosclerotic lesions and hepatic steatosis (Ling et al., 2019). Beyond these direct effects, lactoferrin binds to the LDL receptor–related protein 1 (LRP1) through its arginine-rich N-terminus, which structurally resembles apolipoprotein E, thereby facilitating the cellular uptake and metabolic processing of lipoproteins (Levy et al., 2023). Inverse relationships exist between circulating lactoferrin levels and fasting triglycerides and glucose, along with direct correlations to HDL-cholesterol (Levy et al., 2023). Lactoferrin also preserves intestinal barrier integrity and modulates gut microbiota composition, providing indirect benefits for metabolic health through the gut–liver axis (Levy et al., 2023).

Whey protein hydrolyzates influence the secretion of incretin hormones, which, in turn, affect lipid metabolism. Intraduodenal whey protein hydrolyzate administration at 1.5–3 kcal/min produced dose-dependent effects on gut hormone secretion, with pronounced increases in GLP-1, GIP, and CCK (Hutchison et al., 2015). Co-ingestion of whey protein hydrolyzate with milk minerals rich in calcium potently stimulated GLP-1 secretion (Chen et al., 2019). These incretin hormones regulate glucose homeostasis and influence lipid metabolism and satiety signaling (Hutchison et al., 2015; Chen et al., 2019).

A crucial cautionary finding emerged from studies showing that whey peptide supplementation in high-fat–fed mice increased body weight gain, worsened glucose intolerance, and promoted ectopic fat accumulation with mitochondrial dysfunction, indicating that dietary context and baseline metabolic status influence outcomes (D'Souza et al., 2021). This context-dependent effect highlights the complexity of metabolic responses to bioactive peptides and underscores the need for careful clinical trial design and population selection.

The molecular mechanisms underlying cholesterol-lowering effects involve both direct and indirect pathways. HMG-CoA reductase inhibition by lactoferrin reduces cholesterol biosynthesis through mechanisms partly analogous to those of statin drugs (Ling et al., 2019). NPC1L1 downregulation reduces intestinal cholesterol absorption, complementing increased excretion to create a negative cholesterol balance (Ling et al., 2019). Activation of SREBP transcription factors coordinates the expression of lipogenic and oxidative enzymes, with whey peptides promoting oxidative pathways over synthetic ones (Wang et al., 2022). Enhanced insulin sensitivity, mediated partly through GLP-1 and other incretins, improves overall glucose and lipid homeostasis in insulin-resistant states (Hutchison et al., 2015; Chen et al., 2019).

3.9 Neuroprotective activity

Aging is the main risk factor for most neurodegenerative disorders, including Alzheimer's and Parkinson's disease. Mitochondrial dysfunction and the resulting oxidative stress that accompany aging are well-established contributors to the progression of these conditions (Hou et al., 2019). In aged mitochondria, the production of reactive oxygen species (ROS) increases, mitochondrial membrane potential decreases, ATP synthesis becomes impaired, and the activity of respiratory enzyme complexes declines. As a consequence, proteins undergo oxidation and aggregation, while mitochondrial DNA accumulates mutations and deletions, further compromising cellular function (Cenini et al., 2019).

In this context, recent evidence has reported a neuroprotective role for whey-derived peptides, shown in both in vitro and in vivo studies, particularly in murine aging models (Table 19.3). HT22 cells, a murine hippocampal neuronal line, displayed increased cell viability after pretreatment with 200 μg/mL of hydrolyzed whey protein (WPH) (Chang et al., 2024). This effect was associated with attenuation of H_2O_2-induced oxidative stress: WPH and the peptide LYLKPR reduced ROS and malondialdehyde (MDA) levels while increasing the activity of antioxidant enzymes such as catalase (CAT) and superoxide dismutase (SOD). Consequently, BCL2, an anti-apoptotic protein, increased, whereas BAX, a pro-apoptotic marker, decreased. These changes in apoptotic signaling were accompanied by reduced calcium influx and improved cell viability (Jiang et al., 2022; Chang et al., 2024).

Another noteworthy finding was the increased expression of synaptic proteins—BDNF, synaptophysin, and PSD95—indicating enhanced synaptic plasticity. Moreover, choline acetyltransferase, an enzyme involved in acetylcholine synthesis, was upregulated, demonstrating a modulation of neuronal function–related factors after whey peptide administration (Jiang et al., 2022; Chang et al., 2024).

The antioxidant mechanism described in vitro was associated with activation of the Nrf2/Keap1/HO-1 pathway by the peptides. Modulation of this pathway suppresses the NLRP3 inflammasome. In addition to the direct inactivation of the NLRP3 inflammasome by the binding of the LYLKPR peptide to Keap1, there was an indirect inhibition mediated by oxidative-stress mitigation (Jiang et al., 2022; Chang et al., 2024). Importantly, inhibition of the NLRP3 inflammasome prevented the cleavage of pro-IL-18 into mature IL-18 by caspase-1, thereby reducing pro-inflammatory signaling (Jiang et al., 2022).

Metal-binding whey peptides were also shown to inhibit transcription and expression of NO and TNF-α in LPS-stimulated primary microglia, demonstrating anti-neuroinflammatory potential not only via NLRP3 suppression but also by inhibiting the NF-κB pathway through reduced nuclear translocation of NF-κB p65 (Caetano-Silva et al., 2021). This anti-neuroinflammatory activity is relevant because neuroinflammation induced by genetic alterations in CNS cells or peripheral immune infiltration can trigger protein aggregation, a hallmark of neurodegeneration (Zhang et al., 2023).

These in vitro findings have been supported by murine aging models. Administration of whey protein hydrolyzate at 3.0 g/kg bw for 30 days reduced oxidative stress and inflammation in a D-galactose–induced aging model. The peptides lowered MDA, IL-1β and TNF-α levels while increasing SOD and glutathione peroxidase (GSH-Px) activity in serum and brain tissue (Xiao-Chen Yu et al., 2021). Consequently, mice exhibited improved space exploration behavior as well as enhanced learning and memory.

Similarly, promising results were reported in a scopolamine-induced cognitive-impairment model. Daily administration of a whey hydrolyzate mixture (10–100 mg/kg bw) for up to 10 days improved spatial memory, cognitive ability, and memory performance. This improvement may be attributed to the reduction of Aβ1–42 levels in serum and brain, since neurotoxic amyloid-β aggregates originate from amyloid precursor protein. As these protein levels decreased, hippocampal neuronal damage was attenuated and dopamine levels increased (Ano et al., 2018; Ding et al., 2023).

A decrease in amyloid-β levels together with increased dopamine in the cerebral cortex was also observed in an Alzheimer's murine model. In this study, 3.5 months of treatment with 0.28 mg β-lactolin attenuated the decline in synaptophysin and BDNF, indicating improved synaptic plasticity. Pro-inflammatory cytokines (IL-1α, IL-1β, MIP-1α, and IL-12p70) were reduced in the cortex and hippocampus, and infiltration of activated microglia was prevented. Long-term object memory was significantly enhanced (Ano et al., 2020).

Additional neuroprotective effects include modulation of response speed in mice. Response speed—the ability to detect and respond to harmful stimuli—was increased by administration of whey-derived peptides, although mechanisms remain unclear (Ren et al., 2025). Whey peptides also prevented the age-related decline in insulin-like growth factor 1, suggesting a regulatory effect on glucose metabolism, which is impaired in neurodegenerative diseases such as Alzheimer's (Ano et al., 2020).

Regarding in vivo mechanisms of action, whey peptides improve acetylcholinesterase (AChE) activity in brain tissue and increase expression of synaptic plasticity related proteins such as p-CaMKII and BDNF. Choline acetyltransferase (ChAT), which promotes acetylcholine synthesis, was increased, counteracting the cholinergic dysfunction observed in aging mice, who typically show elevated AChE activity and reduced ChAT (Xiao-Chen Yu et al., 2021). Additionally, whey peptides directly inhibit monoamine oxidase-B, the enzyme responsible for dopamine degradation. Thus, inhibition by the GTWY peptide increased hippocampal dopamine levels (Ano et al., 2018).

Another mechanism reported was the upregulation of transthyretin (TTR), Ppp1r1b, carbonic anhydrase 1 (CAR-1), and Rbp1 in the hippocampus following whey hydrolyzate administration. TTR confers neuroprotection by binding Aβ and preventing its aggregation; Ppp1r1b encodes DARPP-32, a regulator of dopamine signaling that affects learning; CAR-1 is essential for signal processing and long-term synaptic modification and may regulate GABAergic output; and Rbp1 contributes to memory via RXR-mediated signaling (Ding et al., 2023).

Despite promising results in in vivo experimental models, there is still a lack of clinical studies investigating the neuroprotective potential of whey-derived peptides. Human trials are therefore essential to determine the translational relevance of these preclinical findings.

3.10 Opioid activity

Milk proteins, in addition to their well-recognized nutritional value, represent an important source of bioactive peptides. Among the various fragments identified, opioid peptides stand out. These can be produced endogenously by the organism (endogenous peptides) or obtained through the diet (exogenous peptides) (Liu and Udenigwe, 2018; Kaur et al., 2019). Exogenous opioid peptides are generated during the enzymatic digestion of dietary proteins, controlled enzymatic hydrolysis, or through technological processes applied in the food industry (Olvera-Rosales et al., 2022). They exhibit a broad range of biological functions, including antimicrobial, antioxidant, immunomodulatory, and antihypertensive activities (Kaur et al., 2019; Liu and Udenigwe, 2018; Olvera-Rosales et al., 2022).

Within this context, particular attention has been given to opioid-active peptides derived from whey proteins, especially fragments released from α-lactalbumin and β-lactoglobulin, which are capable of interacting with specific receptors and modulating relevant physiological responses.

3.10.1 Structure and classification of opioid peptides

From a structural perspective, opioid peptides typically consist of short amino-acid sequences—most often four to eight residues—and share a characteristic N-terminal motif, commonly Tyr-Gly-Gly-Phe, which is essential for receptor interaction. Endogenous opioid peptides include endorphins, enkephalins, dynorphins, and endomorphins. Because dietary proteins may contain similar structural motifs, they represent important precursors of exogenous opioid peptides (also known as exorphins) (Kaur et al., 2019; Liu and Udenigwe, 2018). These dietary peptides can act as agonists or antagonists of opioid receptors and influence physiological processes related to nociception, mood regulation, memory, and gastrointestinal motility (Antila et al., 1991; Liu and Udenigwe, 2018).

Opioid receptors, the main targets of milk-derived peptides, are widely distributed throughout the central nervous system and peripheral tissues. They belong to the G protein-coupled receptor family, characterized by seven transmembrane domains, and nclude four major subtypes: μ (mu), δ (delta), κ(kappa), and the nociceptin/orphanin receptor, the latter recognized for its structural and functional similarity to classical opioid receptors (Liu and Udenigwe, 2018; Palmer et al., 2022). Each subtype mediates distinct physiological responses, such as analgesia (Fernández-Tomé et al., 2016; Kaur et al., 2019; Nyberg, 2006; Pihlanto-Leppälä, 2001), modulation of gastrointestinal motility (Nyberg, 2006), and regulation of emotional and behavioral states (Brandelli et al., 2015; Kaur et al., 2019; Olvera-Rosales et al., 2022; Zhao and Ashaolu, 2020). In the context of milk, whey proteins are among the primary dietary sources capable of generating peptides that interact with these receptors. Fig. 19.1 illustrates the general pathway of peptide release, receptor engagement, and the subsequent physiological effects.

In milk, the generation of opioid peptides is directly influenced by the type of protein serving as precursor. Within the whey protein fraction, two major proteins, α-lactalbumin and β-lactoglobulin, are recognized as the primary sources of opioid-active fragments. α-Lactalbumin gives rise to α-lactorphin, while β-lactoglobulin produces β-lactorphin and β-lactotensin, all of which display agonistic activity toward opioid receptors (Liu and Udenigwe, 2018). Other minor whey proteins, such as lactoferrin, can generate bioactive peptides such as lactoferroxins, although these have been less extensively characterized in terms of opioid functionality.

The interaction of these peptides with opioid receptors is strongly dependent on specific structural features, particularly the presence of an N-terminal tyrosine (Tyr) residue, often accompanied by an aromatic amino acid (phenylalanine or tyrosine) at the third or fourth position. This motif represents a key determinant of binding affinity and receptor activation (Fernández-Tomé et al., 2016; Liu and Udenigwe, 2018).

For exogenous opioid peptides to exert functional activity within the central nervous system, several criteria must be met: (i) they must be released during gastrointestinal digestion; (ii) they must exhibit partial resistance to proteolytic degradation; (iii) they must reach systemic circulation; and (iv) they must be capable of crossing the blood–brain barrier to access central opioid receptors (Zioudrou et al., 1979).

3.10.2 Main groups of whey-derived opioid peptides

The study of milk-derived opioid peptides has evolved substantially over the past decades, and within this context, whey proteins have emerged as major precursors of physiologically active fragments. Among the whey fraction, α-lactalbumin (α-LA) and β-lactoglobulin (β-LG) stand out for their capacity to release peptides with opioid-like properties during gastrointestinal digestion or controlled enzymatic hydrolysis (Buey et al., 2023). These peptides—primarily α-lactorphin, β-lactorphin, and β-lactotensin—possess structural motifs compatible with opioid receptor binding, enabling them to modulate diverse physiological processes.

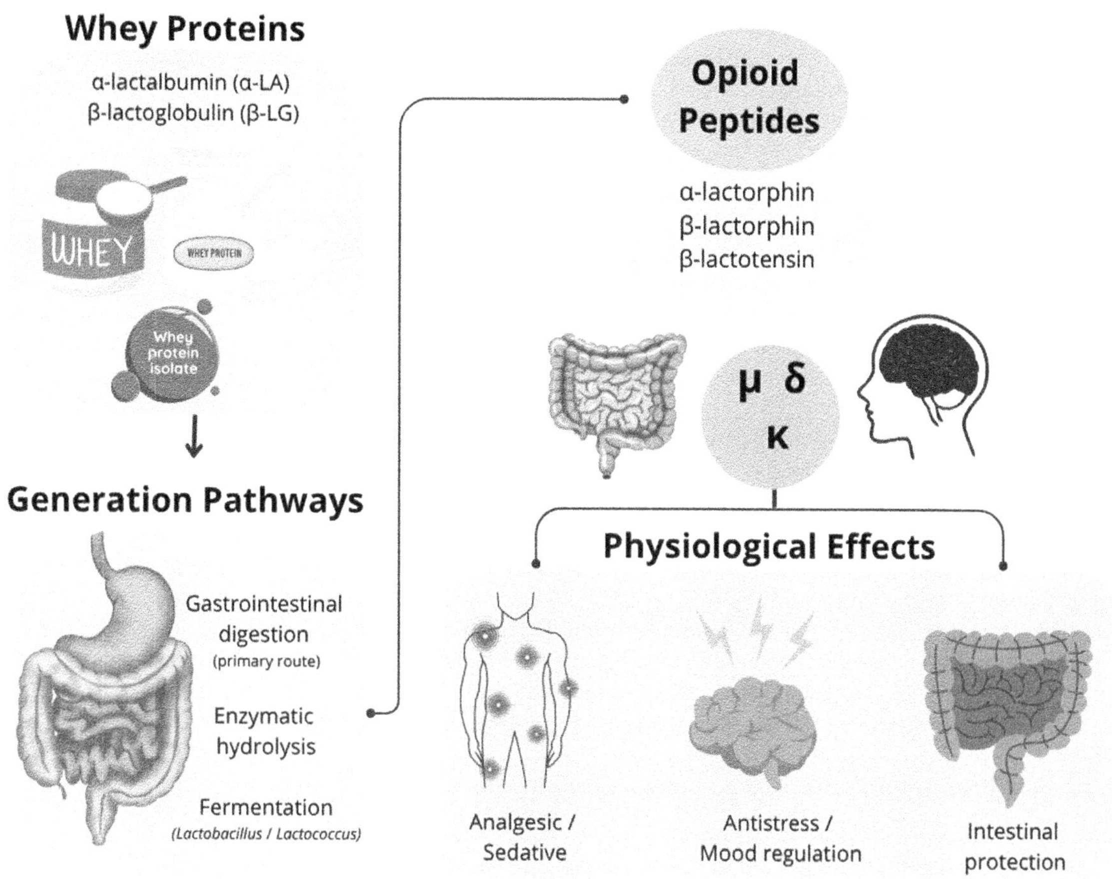

FIGURE 19.1 Potential physiological functionality of whey-derived bioactive opioid peptides. Whey proteins, particularly α-lactalbumin and β-lactoglobulin, serve as precursors for opioid-active peptides generated through enzymatic hydrolysis, gastrointestinal digestion, or fermentation. These peptides interact with μ, δ, and κ opioid receptors, with the potential to contribute to physiological effects such as analgesia, stress modulation, and intestinal protection.

According to current literature, whey-derived opioid peptides exert effects that extend beyond local gastrointestinal activity. Evidence suggests that these fragments may influence central nervous system pathways, modulate cardiovascular tone, and participate in satiety and feeding behavior regulation, contributing to communication along the gut–brain axis (Kaur et al., 2019; Olvera-Rosales et al., 2022). Their multifunctionality highlights the importance of whey proteins as a rich source of bioactive compounds capable of acting on opioid and non-opioid pathways.

Although whey-derived peptides generally exhibit milder potency compared with synthetic opioids, this characteristic may represent an advantage, as it reduces the likelihood of adverse effects while maintaining functional biological activity. Moreover, their antihypertensive, neuromodulatory, and appetite-regulating properties make them attractive candidates for applications in functional nutrition and health-promoting dairy products (Kaur et al., 2019). However, more comprehensive investigations into bioavailability, metabolic stability, absorption mechanisms, and controlled clinical trials are necessary to substantiate their potential clinical significance.

3.10.3 Whey proteins (α-lactorphin, β-lactorphin, β-lactotensin)

Milk whey proteins, particularly α-lactalbumin (α-LA) and β-lactoglobulin (β-LG), are recognized as important precursors of bioactive peptides released during gastrointestinal digestion or controlled enzymatic hydrolysis. Among the different peptides generated from these proteins, those exhibiting opioid-like activity have drawn significant scientific interest due to their structural compatibility with opioid receptors and their multifunctional physiological effects.

α-Lactorphin is a tetrapeptide (Tyr–Gly–Leu–Phe) derived from α-lactalbumin through the action of pepsin and other digestive proteases. Its opioid-related activity has been demonstrated in classical pharmacological models, where it inhibits contractions in the guinea pig ileum—an effect completely reversed by naloxone, confirming its interaction with opioid receptors (Zhao and Ashaolu, 2020). In addition to its effects on intestinal motility, α-lactorphin exhibits endothelium-dependent vasodilatory properties, suggesting a role in vascular tone regulation and potential contributions to antihypertensive activity (Brandelli et al., 2015). Further studies attribute immunomodulatory, antimicrobial, antioxidant, and antithrombotic properties to this peptide, underscoring its relevance as a multifunctional compound within the whey protein fraction (Buey et al., 2023).

β-Lactorphin (YLLF), derived from β-lactoglobulin, exhibits moderate affinity for opioid receptors and has been the subject of several pharmacological investigations. In binding assays, it demonstrated the ability to displace ligands in rat brain homogenates, confirming its interaction with opioid binding sites (Zhao and Ashaolu, 2020). However, in isolated ileum preparations, it induced a contractile response that was not reversed by the addition of naloxone, suggesting that its effects are not exclusively mediated through opioid receptor activation and may involve alternative or complementary physiological pathways. In animal models, β-lactorphin has been associated with anxiolytic and hypotensive effects (Brandelli et al., 2015), reinforcing its importance not only within the opioid context but also in the regulation of behavior and cardiovascular function.

Another relevant fragment is β-lactotensin (HIAL), also derived from β-lactoglobulin. This tetrapeptide has been described as capable of inducing contractions in the longitudinal muscle of the guinea pig ileum, even in the absence of electrical stimulation or external agonists, suggesting a distinct mechanism of action that has not yet been fully elucidated (Pihlanto-Leppälä, 1997). The identification of β-lactotensin further reinforces the diversity of bioactive fragments derived from whey, whose activity is not limited to classical modulation of opioid receptors but may involve different signaling pathways.

Overall, peptides derived from α-lactalbumin and β-lactoglobulin expand the understanding of whey proteins as a rich source of physiologically active compounds. Their ability to modulate gastrointestinal, cardiovascular, and neuro-behavioral functions highlights the relevance of whey not only as a nutritional component but also as a matrix with potential applications in functional foods, nutraceuticals, and health-promoting formulations.

3.10.4 Bioavailability and physiological relevance

The bioavailability of whey-derived opioid peptides is determined largely by their stability during digestion and by their capacity to reach systemic circulation in an intact or partially preserved form. As low-molecular-weight molecules, peptides released from α-lactalbumin and β-lactoglobulin remain susceptible to degradation by gastric and intestinal proteases, which can reduce the fraction available for absorption (Singh and Gaur, 2024). After surviving gastrointestinal hydrolysis, absorbed peptides undergo first-pass hepatic metabolism, which may further modify or degrade their structure, limiting their systemic bioactive potential (Zhao and Ashaolu, 2020).

Transport across the intestinal epithelium occurs mainly through the PepT1 transporter, which facilitates the uptake of di- and tripeptides, although paracellular diffusion and other minor pathways may allow the passage of intact peptide sequences (Liu and Udenigwe, 2018). The efficiency of absorption depends on several physicochemical properties, such as peptide size, hydrophobicity, and charge, which influence both permeability and resistance to enzymatic inactivation. Despite the relatively low systemic concentrations detected in vivo, whey-derived peptides can exert local effects on the intestinal mucosa as well as systemic actions on distant target tissues.

To enhance their stability and functional activity, several strategies have been explored, including encapsulation in lipid or polymer matrices, gastroprotective formulations, and targeted structural modifications designed to increase resistance to proteolysis (Singh and Gaur, 2024). Abordagens in silico complementares também têm sido utilizadas para prever sítios de clivagem enzimática e otimizar sequências peptídicas com maior probabilidade de sobreviver à digestão e exercer efeitos biológicos (Olvera-Rosales et al., 2022). Despite these advances, the literature highlights that many peptides, such as αs1-casein exorphins, still face challenges related to degradation resistance and lack sufficient absorption and transport studies to confirm effects beyond the gastrointestinal tract (Fernández-Tomé et al., 2016).

3.10.5 Physiological effects

The physiological effects of whey-derived opioid peptides arise mainly from their ability to interact with μ-, δ-, and κ-opioid receptors, which are widely distributed throughout the central and peripheral nervous systems (Pihlanto-Leppälä, 2001; Garg et al., 2016). Although these peptides exhibit lower potency than pharmacological opioids, their mild and modulatory activity is considered advantageous, offering potential health benefits with a substantially lower risk of adverse effects.

Table 19.3 summarizes the principal opioid-active peptides derived from whey proteins, compiling key information on α-lactorphin, β-lactorphin, and β-lactotensin—fragments released from α-lactalbumin and β-lactoglobulin through gastrointestinal digestion or enzymatic hydrolysis. The table integrates data from experimental assays, review articles, and biochemical analyses, allowing comparison of peptide sequences, precursor proteins, receptor interactions, and reported physiological effects.

This approach allows for an integrated visualization of consolidated findings as well as existing knowledge gaps, highlighting both the current state of research and areas requiring further investigation.

Activation of opioid receptors by whey-derived peptides triggers the classical intracellular signaling cascade associated with G protein–coupled opioid receptors (Fig. 19.1). This pathway involves the inhibition of adenylyl cyclase, reducing intracellular cAMP levels. Therefore, potassium (K^+) channels open, leading to postsynaptic membrane hyperpolarization and reduced neuronal excitability. Simultaneously, the inhibition of voltage-gated calcium (Ca^{2+}) channels decreases neurotransmitter release at the synaptic cleft. Together, these mechanisms contribute to analgesic, neuromodulatory, and gastrointestinal effects (Olvera-Rosales et al., 2022).

Among whey-derived peptides, α-lactorphin, released from α-lactalbumin (α-LA), is one of the most studied fragments. It inhibits intestinal contractions in classical ileum assays, and this response is completely reversed by naloxone, confirming interaction with μ-opioid receptors (Zhao and Ashaolu, 2020). Beyond its intestinal activity, α-lactorphin demonstrates endothelium-dependent vasodilation, contributing to acute reductions in blood pressure in experimental models (Brandelli et al., 2015; Olvera-Rosales et al., 2022). β-Lactorphin, derived from β-lactoglobulin (β-LG), exhibits moderate affinity for μ-opioid receptors and has been associated with anxiolytic and hypotensive effects in vivo (Brandelli et al., 2015; Zhao and Ashaolu, 2020). Interestingly, its contractile effects on intestinal tissue are not reversed by naloxone, suggesting the involvement of both opioid and non-opioid physiological pathways (Zhao and Ashaolu, 2020).

Another relevant whey-derived peptide is β-lactotensin, which induces contractions in the longitudinal muscle of the guinea pig ileum even without external stimuli. Its activity appears to follow a non-classical mechanism, not fully explained by opioid receptor activation, indicating that whey proteins generate a diverse array of peptides with distinct modes of action (Pihlanto-Leppälä, 1997).

Collectively, these whey-derived opioid peptides modulate not only gastrointestinal motility but also cardiovascular responses, neurobehavioral effects, and potentially the gut–brain axis, influencing the release of gastrointestinal hormones involved in satiety and energy balance (Kaur et al., 2019). Despite substantial progress, most evidence arises from in vitro or animal studies, and further research, particularly pharmacokinetic and clinical investigations, is required to clarify their bioavailability, physiological impact, and relevance in the context of dietary intake.

4. Factors affecting peptide bioactivity

The factors that affect the peptide bioactivity are illustrated in Fig. 19.2. The bioactivity of peptides is affected by the polypeptides degraded by cellular peptidases, and the molecular weight can remain intact and exert its activity. The bioavailability of isolated peptides depends mainly on the degree of hydrolysis during isolation (Dalaka et al., 2023).

Bioactive peptides are released during enzymatic proteolysis (gastrointestinal digestion, in vitro hydrolysis with proteolytic enzymes) of proteins and also during food processing (cooking, fermentation, maturation) (Daliri et al., 2017).

Depending on the exact processing conditions, whey and/or casein modifications can include denaturation, aggregation, glycation, oxidation, cross-linking, racemization, and dephosphorylation (Auestad and Layman, 2021).

Different processing treatments are applied to the raw materials according to the requirements of the product, depending on which, the proteins present are subject to changes about their functional and biological properties and these changes (Fig. 19.2) are caused mainly due to pH or acid or alkaline treatments, and application of chemical treatments, as well as processing steps (Bhandari et al., 2020).

The bioactivity of peptides is directly related to their primary structure and the physicochemical properties derived from this composition. However, short peptides (2–10 amino acids) tend to have greater steric accessibility and better interaction with enzymes and cell receptors (Korhonen and Pihlanto, 2006).

Aromatic and hydrophobic amino acids, such as Trp, Tyr, and Phe, are often responsible for antioxidant, antimicrobial, and antihypertensive activities, as they contribute to proton donation capacity or hydrophobic interaction with the biological target (Udenigwe and Aluko, 2012). The three-dimensional conformation and charge distribution influence the affinity of the peptide for enzymes such as ACE, DPP-IV, and renin (Daliri et al., 2017).

The bioavailability of peptides is crucial for them to exert systemic effects after ingestion. Low-molecular-weight peptides, especially di- and tripeptides, have a higher rate of intestinal absorption and can cross the mucosa through active transport mediated by PEPT1, passive transcellular diffusion, or paracellular routes (Shen and Matsui, 2017). Resistance to degradation by gastrointestinal peptidases is crucial, as rapid hydrolysis significantly reduces bioactive potential (Shahidi and Zhong, 2008).

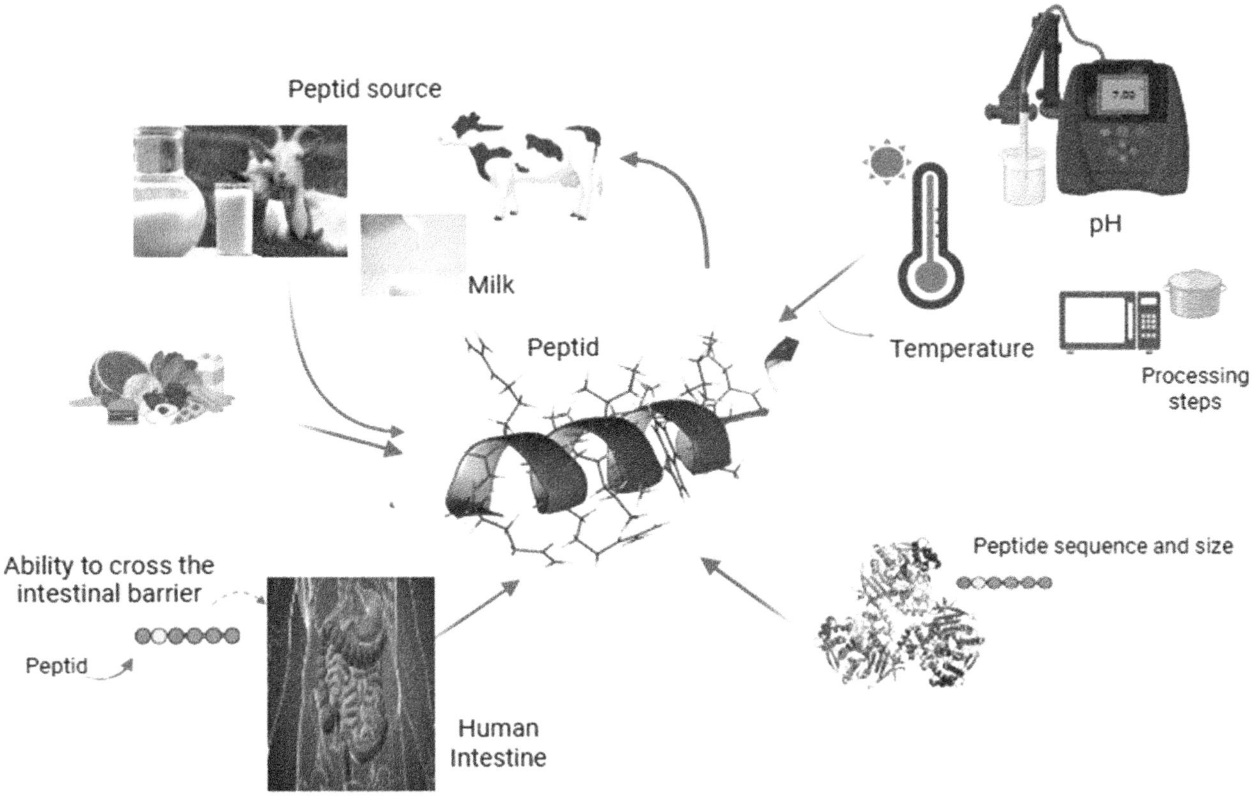

FIGURE 19.2 Factors affecting the bioavailability of peptides.

Food processing has a direct influence on the generation and stability of bioactive peptides. Heat-induced protein denaturation can facilitate protease access to bioactive sequences, increasing the release of functionally relevant peptides (Phelan et al., 2009). However, excessive heat treatment can cause Maillard reactions or structural degradation, reducing bioactivity (Elias et al., 2008).

The pH alters protein ionization and enzyme activity, modulating hydrolysis efficiency and the stability of the peptides produced (Segura-Campos et al., 2011). The pH also modulates protein solubility and conformation, affecting the efficiency of enzymatic hydrolysis (Kilara and Panyam, 2003). Processes such as fermentation have been shown to significantly increase the generation of peptides with antioxidant and antihypertensive activities (Hernández-Ledesma et al., 2014a,b).

Temperature can induce protein denaturation, promote the action of proteases, and increase the generation of bioactive peptides; however, excessive heating can promote degradation, oxidation, and loss of bioactivity (Korhonen and Pihlanto, 2006).

5. Applications of bioactive peptides

Bioactive peptides derived from whey, released from whey proteins through enzymatic hydrolysis, chemical processes, or microbial fermentation, have attracted scientific and technological interest, as they are considered functional ingredients with high added value and specific health benefits (Park et al., 2015). This is due to their composition and amino acid sequence, which usually contain 2 to 20 amino acid residues, presenting a wide range of biological activities, including anticancer, antihypertensive, anti-inflammatory, antimicrobial, antioxidant, immunomodulatory, antiproliferative, antithrombotic, opioid modulators, mineral binders, growth stimulants, and appetite suppressants (Hernández-Ledesma et al., 2014a,b; Saubenova et al., 2024; Das et al., 2025).

These functions depend on the composition and order of amino acid residues in the peptide sequence and, in most cases, are inactive in precursor proteins until released by proteolysis. These functions contribute to strengthening the immune system and managing various chronic diseases. Therefore, once isolated and characterized, these peptides are mainly used in the development of nutraceuticals, functional foods, and pharmaceutical products. The choice of application depends on the specific bioactivity of the peptide fraction, its stability under gastrointestinal and thermal conditions, its bioavailability, and its organoleptic properties.

The application of bioactive peptides from whey in nutraceuticals and functional foods is the most common and commercially viable use developed for the industry, given that these peptides can be used as natural food preservatives due to their antimicrobial and antioxidant activities and, at the same time, enrich the functional and healthy profile of conventional foods (Saubenova et al., 2024). Bioactive peptides in whey have techno-functional properties that are highly relevant to food formulation. These include improving the solubility of protein fractions, stabilizing emulsions, promoting foam formation and stability, and increasing water and oil retention capacity (Madadlou and Abbaspourrad, 2018; Aguilar-Toalá et al., 2022; Singh et al., 2022). These characteristics facilitate their incorporation into various food matrices and provide key technological advantages, such as texture optimization, increased yield, and extended product shelf life.

Nowadays, it is common to find protein hydrolyzates rich in bioactive whey peptides marketed in dietary supplements mainly for high-performance athletes (Grando et al., 2024), physically active individuals, or specific formulations for metabolic diseases or older adults. According to Klaewkla et al. (2022), most commercial supplements combine whey protein isolate and hydrolysate with whey protein concentrate to improve flavor, although the latter contains a lower proportion of protein. Likewise, whey protein hydrolyzates are characterized by lower foaming and foam stabilization capabilities, lower viscosity, better emulsifying properties, and higher solubility than protein concentrates (Power et al., 2009).

Bioactive peptides derived from whey have also been used as encapsulation matrices for nutraceutical compounds, taking advantage not only of their ability to form capsules, but also of their intrinsic bioactive potential. This strategy allows the development of delivery systems with high resistance to gastric digestion, improving the protection of encapsulated compounds against degradation in the gastrointestinal tract. In addition, the surface functionalization of these capsules with lectins (natural cell recognition molecules) could enhance intestinal absorption (Madadlou and Abbaspourrad, 2018).

In addition, bioactive peptides derived from whey contribute to improving the mechanical properties of capsules, promoting fermentation activity and the survival of probiotics in functional beverages, such as isotonic sports drinks based on whey permeate containing probiotic bacteria encapsulated in alginate and bioactive whey peptides, demonstrating that the hydrolysis of these proteins has a positive effect on antioxidant activity and maintains the viability of probiotics for at least 28 days of storage (Soltani et al., 2025). Therefore, these types of beverages represent a natural and functional alternative to conventional commercial options, demonstrating the potential of bioactive whey peptides in both nutraceutical applications and bioactive compound delivery systems.

Whey protein hydrolyzates have been extensively studied for their bioactive properties with potential impact on human health. In particular, whey-derived peptides have been shown to possess antihypertensive activity by inhibiting angiotensin-converting enzyme (ACE), positioning them as a promising alternative to pharmacological or conventional supplements used in the control of hypertension (Dou et al., 2024).

The ingestion of whey protein hydrolyzates rich in bioactive peptides has shown superior effects compared to mixtures of free amino acids. These hydrolyzates have been shown to raise plasma peptide levels and more effectively stimulate phosphorylation of the mammalian target of rapamycin (mTOR) after resistance exercise, thereby promoting muscle protein synthesis (Luiking et al., 2016). In athletes, this characteristic is particularly relevant for physical performance and recovery, as it contributes to reducing exercise-induced damage and inducing beneficial adaptations in connective tissue (Fassina et al., 2019; König et al., 2021).

In particular, consumption of whey hydrolyzates, a source of bioactive peptides, is associated with increased muscle torque and maximum hypertrophy derived from strength training, as well as contributing to reduced muscle damage by lowering serum levels of creatine kinase and lactate dehydrogenase during aerobic exercise. It can also improve maximum torque without adverse effects on creatine kinase or tumor necrosis factor alpha (TNF-α) levels during strength activities. In contrast, the consumption of whey protein concentrate has been mainly associated with weight loss, increased satiety, and improved body composition, without compromising lean mass (Fassina et al., 2019).

Food science and technology have also sought to expand the application of these bioactive whey peptides in functional foods, due to their potential to improve health beyond basic nutrition (Dullius et al., 2018). Therefore, bioactive peptides derived from whey protein hydrolysis have emerged as ingredients of great interest in the food industry due to their superior nutritional and technological properties compared to intact proteins. Enzymatic hydrolysis and microbial fermentation allow low molecular weight peptide fractions to be obtained with better digestibility, greater heat stability, and outstanding technological functions, such as improved foaming and emulsifying properties (Grando et al., 2024).

These characteristics favor their incorporation into various food matrices, even in complex systems such as confectionery products, where they interact with fatty and sugary phases.

Stožinić et al. (2024) demonstrated that the addition of whey peptides, obtained both by enzymatic hydrolysis and by fermentation with *Lactobacillus rhamnosus* ATCC 7469, to a dairy cream used as a confectionery filling produced differentiated effects: enzymatic hydrolyzates exhibited greater bioactivity, antioxidant activity, structural firmness, and sensory acceptance, establishing themselves as the most effective strategy for enriching value-added products; while fermented hydrolyzates, although less potent in oxidative inhibition and with a lower elastic limit, provided a more flexible rheological profile, potentially advantageous in formulations requiring plasticity and manageability.

As for snacks, bioactive peptides obtained from the fractionation of whey protein hydrolyzates have been used as functional ingredients to reduce salt content in formulations designed for healthier consumption. One example is the cashew nut (*Anacardium occidentale* L.), tropical fruit rich in ascorbic acid. Its coating with peptides allowed it to maintain its angiotensin-converting enzyme (ACE) inhibitory activity. It was observed that incorporating a larger peptide fraction into the formulation could enhance this antihypertensive activity. In fact, at low concentrations such as 2%, the extract was able to replace up to 30% of conventional salt, directly reducing the sodium content of the snack and ensuring a positive impact on health, without compromising its appearance, taste, overall acceptance, or consumers' intention to purchase (Amorim et al., 2018).

Bioactive peptides from whey have been used as antifreeze agents in fishery products such as fish surimi. Their antioxidant, preservative, safe, and sustainable properties improve product quality and shelf life by protecting the structural and functional integrity of myofibrillar proteins during freeze-thaw cycles. In addition, they contribute to improving rheological properties and emulsifying stability, establishing themselves as effective natural additives for the frozen food industry (Peng et al., 2024). Similarly, the addition of hydrolyzed bioactive whey proteins to hamburgers subjected to freeze-thaw cycles reduced lipid and protein oxidation, as well as protein aggregation, increasing water retention and preserving the texture of the products (Liu et al., 2022).

Kong et al. (2023) confirmed that hydrolyzed bioactive whey proteins improve water retention and antioxidant capacity in pork burgers throughout successive freezing and thawing cycles, helping to maintain their quality and functional stability. Furthermore, coating chicken breasts stored at 4°C for 12 days with hydrolyzed bioactive whey proteins combined with innovative pretreatments, such as ultrasound, lemon verbena (Aloysia citrodora) essential oil, and selenium nanoparticles, can mitigate the chemical and microbiological deterioration of this product. In addition, these coatings have antioxidant activity and help to preserve the organoleptic properties of the product (Karimi-Dehkordi et al., 2021).

Dairy beverages represent one of the main food matrices for the incorporation of bioactive peptides derived from whey or protein hydrolyzates. This relevance is due to their wide acceptance by consumers, their high nutritional value, and the physicochemical conditions they offer, which favor the protection of peptides against degradation during storage and gastrointestinal transit, thus increasing the effectiveness of their physiological effects (Abd El-Fattah et al., 2018). The addition of bioactive peptides obtained through tryptic hydrolysis of yogurt proteins has demonstrated antioxidant properties and natural antimicrobial effects against pathogens such as *Staphylococcus aureus* and *Escherichia coli* in fermented dairy products, such as doogh (traditional Iranian fermented milk), contributing to the maintenance of product quality and sensory acceptability over time (Karimi et al., 2022).

Likewise, several studies have reported the application of bioactive serum peptides in the development of dairy beverages, whether enriched or through fermented products in which lactic acid bacteria (LAB) generate bioactive peptides, which are capable of exerting hypoallergenic effects (Yukalo et al., 2022), antihypertensive (Chen et al., 2007; Rosa et al., 2023), hypocholesterolemic (Rosa et al., 2023), and antioxidant (Dinkçi et al., 2023; Rosa et al., 2023) effects. The combination of peptides with probiotic strains of BAL amplifies the functional profile of the product, promoting both intestinal and cardiovascular health (Abd El-Fattah et al., 2018).

6. Safety of bioactive peptides

The use of bioactive peptides derived from whey proteins in functional foods and dietary supplements has increased substantially, driving scientific interest not only in their efficacy but also in their safety profile (Cao et al., 2025; Pang et al., 2025; Zhao and Ashaolu, 2020). The close similarity between human gastrointestinal digestion and the industrial processes employed to generate peptides and free amino acids further supports the perception of safety associated with the consumption of these compounds in the human diet (Zhao and Ashaolu, 2020). Although whey-derived peptides are widely recognized for their beneficial biological activities—including antioxidant, antimicrobial, antihypertensive, and immunomodulatory effects (Beltrán-Barrientos et al., 2016; Hussein et al., 2020; Kaur et al., 2019; Liu and Udenigwe,

2018; Olvera-Rosales et al., 2022)—as well as analgesic properties (Fernández-Tomé et al., 2016; Kaur et al., 2019; Nyberg, 2006; Pihlanto-Leppälä, 2001) and the capacity to modulate gastrointestinal function (Liu and Udenigwe, 2018), toxicological, allergenic, and regulatory aspects still require deeper investigation, particularly considering the broad heterogeneity of products generated through enzymatic hydrolysis.

The recognition of whey protein concentrate as a safe ingredient by regulatory authorities—most notably the United States Food and Drug Administration (FDA)—has substantially supported its widespread incorporation into functional foods. This regulatory acknowledgment is formalized through its designation as Generally Recognized as Safe (GRAS) (Boland, 2011). Products derived from whey proteins are considered safe, particularly when manufactured under controlled conditions and in accordance with good manufacturing practices, including the use of enzymatic or fermentation-based processes conducted under moderate operational parameters (Pang et al., 2025).

Nevertheless, important caveats must be considered. During the production and processing of peptide ingredients, undesirable reactions may occur, including the formation of biogenic amines, Maillard reaction products, and peptides retaining residual allergenic potential, all of which may pose health risks (Patil et al., 2022). Furthermore, excessive intake, prolonged use, or exposure by individuals belonging to sensitive population groups may lead to adverse effects such as toxicity or allergic responses (Liu et al., 2020).

Experimental studies using animal models have significantly advanced the safety assessment of whey-derived bioactive peptides, consistently demonstrating an absence of toxic effects even at high doses and under prolonged exposure scenarios. Table 19.4 provides a synthesized overview of the principal findings reported in the literature to date.

Despite these promising findings, the primary contemporary concern centers on the allergenic potential of these compounds, particularly given their structural complexity and the potential chemical alterations that may arise during production processes (Hussein et al., 2020; Cao et al., 2025; Pang et al., 2025; Zhao and Ashaolu, 2020). The occurrence of structural modifications and the formation of potentially harmful secondary compounds—such as Maillard reaction by-products or biogenic amines—underscore the need for continuous and rigorous safety evaluations regarding the consumption of these peptides.

The following sections address the key aspects related to the safety of whey-derived bioactive peptides, including the origin and nature of the precursor protein, the enzymatic hydrolysis processes involved, potential allergenicity and toxicity risks, and the regulatory criteria that govern their use in foods and dietary supplements.

6.1 Origin of the protein and enzymatic hydrolysis

Safety evaluation of whey-derived peptides begins with an assessment of the origin and intrinsic characteristics of the protein raw material. Whey proteins are widely incorporated into the food industry due to their high nutritional value and versatile techno-functional properties (Cao et al., 2025; Pang et al., 2025; Zhao and Ashaolu, 2020). However, these proteins also contain allergenic epitopes capable of triggering adverse immune reactions in individuals sensitized to cow's milk proteins (Pang et al., 2025; Cao et al., 2025), making it essential to understand how different processing strategies may alter their structural integrity and, consequently, their safety profile.

Enzymatic hydrolysis is one of the most commonly employed techniques to generate bioactive peptides from whey proteins and is traditionally regarded as a mild process because it operates under moderate temperature and pH conditions. Nevertheless, hydrolysis can induce substantial structural modifications. These include the loss of sulfhydryl (–SH) groups, often resulting from reactions with reactive carbonyl compounds formed during processing, as well as additional chemical transformations that may yield peptide derivatives whose toxicological relevance is not yet fully understood (Liu et al., 2020).

Beyond these changes, specific thermal or chemical conditions may promote the formation of iso-peptides, characterized by covalent linkages alternative to classical α-peptide bonds, such as ε-(γ-glutamyl)-lysine crosslinks, generated through glycation, dehydration, cyclization, or oxidative degradation. These structural rearrangements can reduce protein extractability, impair digestibility, and alter the functional properties of the resulting peptides (Patil et al., 2022). Collectively, these phenomena demonstrate that although hydrolysis is essential for producing bioactive peptides, it may also introduce structural alterations that directly affect the safety and biological attributes of the final products.

During hydrolysis, additional chemical reactions may occur, including deamination, racemization, cyclic formation of susceptible residues (Asp, Glu, Thr), β-elimination, and oxidation of labile amino acids such as Trp, Lys, Arg, Asn, Cys, and Met. These reactions can modify the sequence, size, and conformation of the peptides produced, thereby influencing their biological activity as well as their allergenic or toxic potential (Liu et al., 2020). Understanding these modifications is therefore critical for establishing robust safety assessments for whey-derived peptides intended for food and nutraceutical applications.

TABLE 19.4 Summary of safety assessments of whey-derived bioactive peptides and whey protein hydrolyzates.

Assessment category	Experimental model	Peptides or hydrolyzates evaluated	Analytical methods used	Key safety findings	Authors' conclusion	References
Cytotoxicity	Cell lines (Caco-2, HT29, THP-1, HeLa)	Lactorfins, lactoferricin, lactoferrampin, WPI hydrolyzates	MTT, LDH, apoptosis assays, microscopy	Selective cytotoxicity toward tumor cells; normal cells preserved	Whey peptides considered safe; potential antitumor activity	Beltrán-Barrientos et al. (2016)
Genotoxicity/ Mutagenicity	In vitro assays (Ames, comet), cell culture	Whey-derived peptides and whey hydrolyzates	Chromosomal aberration test, comet assay	No DNA damage at food-relevant concentrations	No evidence of mutagenicity; safe for dietary use	Beltrán-Barrientos et al. (2016)
Acute toxicity	Wistar rats	Alcalase-generated WPC hydrolyzate	OECD 425, LD_{50}, clinical evaluation	No mortality; absence of toxic signs	$LD_{50} > 2500$ mg/kg; classified as safe	Hussein et al. (2020)
Subchronic toxicity	SHR and Wistar rats	Hydrolyzed WPC	Histopathology, serum biochemistry	No hepatic, renal or hematological alteration	Safety confirmed with repeated dosing	Hussein et al. (2020)
Allergenicity (IgE-mediated)	ELISA; IgE from milk-allergic patients; in vitro models	β-LG, α-LA and their hydrolyzates	IgE-binding assay, epitope mapping	Hydrolysis or UHP/PEF/CP processing reduces IgE binding by 30%–50%	Hydrolyzates show reduced allergenicity	Cao et al. (2025); Pang et al. (2025)
Impact of processing technologies	Physicochemical & immunological analyses	Non-thermally processed whey proteins	Conformational epitope mapping; LC-MS/MS	UHP and PEF reduce conformational epitopes without loss of function	Promising technologies to lower allergenic risk	Cao et al. (2025)
Functional safety (bioactivity–risk balance)	Systematic reviews; in vivo studies	Lactorfins, lactoferricin, β-lactotensin	Integrated evaluation of bioactivity and toxicology	No adverse effects; cardiovascular and gastrointestinal benefits	Safe at dietary levels; purified peptides require individual assessment	Zhao and Ashaolu (2020)

5-dimethylthiazol-2-yl)-2, 5-diphenyl tetrazolium bromide assay; *CP,* Cold Plasma; *ELISA,* Enzyme-Linked Immunosorbent Assay; *LC-MS/MS,* Liquid Chromatography–Tandem Mass Spectrometry; *LD50,* Median Lethal Dose; *LDH,* Lactate Dehydrogenase Release Assay; *MTT,* 3-(4; *OECD,* Organisation for Economic Co-operation and Development; *PEF,* Pulsed Electric Fields; *SHR,* Spontaneously Hypertensive Rats.; *UHP,* Ultra-High Pressure; *WPC,* Whey Protein Concentrate; *WPI,* Whey Protein Isolate

6.2 Allergenicity

Allergenicity remains one of the major safety concerns associated with whey-derived bioactive peptides. Although enzymatic hydrolysis is widely employed to reduce the immunoreactivity of intact proteins, evidence indicates that partially hydrolyzed fragments may retain allergenic epitopes capable of being recognized by IgE antibodies, particularly in individuals who are already sensitized, such as infants with cow's milk protein allergy (Liu et al., 2020; Pang et al., 2025).

The degree of hydrolysis is a critical determinant of immunogenic reduction: extensively hydrolyzed products generally exhibit lower allergenic potential compared with partially hydrolyzed proteins or native whey proteins. Among whey components, β-lactoglobulin (β-LG) and α-lactalbumin (α-LA) stand out as the principal allergens responsible for IgE-mediated reactions in infants and young children sensitized to cow's milk proteins (Cao et al., 2025). These findings underscore that, even in hypoallergenic formulations, residual peptide fragments may retain immunologically active structures, necessitating rigorous allergenicity assessment—especially for hydrolyzates intended for vulnerable populations.

Classical and contemporary studies consistently demonstrate that whey proteins—predominantly β-lactoglobulin (β-LG) and α-lactalbumin (α-LA), are among the major cow's milk allergens, responsible for allergic manifestations particularly in infants and young children (Beltrán-Barrientos et al., 2016). Although most patients develop tolerance by approximately 4 years of age, the persistence of immunological reactivity in sensitized individuals reinforces the need for stringent evaluation of whey-derived ingredients used in functional foods and infant formulas. The literature indicates that, even after hydrolysis, certain peptides may retain linear or conformational epitopes capable of activating mast cells and basophils or binding to specific IgE, thereby perpetuating the risk of adverse reactions (Liu et al., 2020). These residual epitopes are especially critical in highly sensitive populations and have direct implications for the formulation of hypoallergenic products.

Recent evidence further expands this understanding. Patil et al. (2022) demonstrated that during enzymatic hydrolysis, whey proteins can generate peptides that retain, and, in rare cases, even potentiate, allergenicity, depending on their sequence, molecular size, and the exposure of specific residues. This phenomenon arises because the partial disruption of the protein's three-dimensional structure may expose previously hidden antigenic determinants or modify the stability of existing epitopes, thereby influencing their interaction with the immune system. Consequently, even when a substantial reduction in allergenic potential is achieved, complete elimination of immunogenicity cannot always be guaranteed, particularly in individuals with a prior diagnosis of cow's milk protein allergy.

On the other hand, recent technological advances have shown promise in mitigating these risks. In a recent study, Pang et al. (2025) demonstrated that combining endopeptidases and exopeptidases—such as bromelain, papain, and proteases like ProteAXH, promotes a more complete degradation of allergenic epitopes present in whey protein isolate (WPI). This integrated enzymatic approach yielded hydrolyzates with approximately 47%–52% reductions in IgE-binding capacity, alongside improvements in structural stability and key functional properties, including emulsifying capacity. These findings reinforce that strategically optimized enzyme combinations can substantially reduce allergenicity without compromising essential technological attributes required for food applications.

Given this scenario, allergenicity assessment of whey-derived peptides should rely on a suite of complementary methodologies, including in vitro assays (e.g., IgE-binding tests, competitive inhibition assays, basophil activation tests, and epitope-specific analyses), in addition to controlled clinical studies when the products are intended for vulnerable populations. In summary, allergenicity remains one of the most critical components of the safety profile of whey-derived bioactive peptides, necessitating multidisciplinary approaches and highly sensitive analytical strategies to ensure their safe use in conventional foods, functional formulations, and products designed for high-risk groups.

6.3 Toxic compounds

Toxicity assessment is a critical component of the safety evaluation of whey-derived bioactive peptides. Overall, in vitro and in vivo studies indicate that these peptides, at concentrations typically consumed in foods or supplements, do not elicit relevant toxic effects (Beltrán-Barrientos et al., 2016; Hussein et al., 2020). Cytotoxicity assays conducted in various cell lines, as well as acute and subchronic toxicity tests in animal models—including standardized protocols such as the OECD Guideline 425—consistently report the absence of adverse clinical signs and a lack of significant hepatic, renal, or hematological alterations (Beltrán-Barrientos et al., 2016; Hussein et al., 2020).

However, protein hydrolyzates are inherently complex mixtures and may contain peptide fractions or secondary by-products that remain poorly characterized. For this reason, each newly produced hydrolyzate or peptide fraction should be subjected to a comprehensive battery of validated toxicological studies to ensure the absence of cytotoxic, genotoxic, or mutagenic effects (Liu et al., 2020).

In addition to the peptides themselves, increasing attention has been directed toward the formation of toxic compounds generated through chemical reactions that occur during the processing of food proteins. Processes such as intense heating, alkaline treatments, and advanced Maillard reactions can lead to the formation of potentially harmful by-products, including Amadori compounds, furans, biogenic amines, nitrosamines, and acrylamide some of which have been associated with mutagenic and carcinogenic effects depending on time–temperature conditions (Patil et al., 2022; Liu et al., 2020). These compounds may induce oxidative stress, chromosomal instability, and DNA damage, representing a critical consideration in the safety assessment of processed protein derivatives.

Another relevant aspect concerns the formation of chemically modified or racemized amino acids during thermal or alkaline processing. Racemization, which generates D-amino acids, may alter biological recognition, reduce digestibility, and in some cases be associated with genotoxic or mutagenic effects—particularly when incorporated into partially degraded peptide structures (Liu et al., 2020). These alterations underscore the importance of evaluating not only the resulting peptides but also the structural by-products generated throughout processing.

In contrast to these potential risks, classical milk-derived bioactive peptides, such as VPP (Val-Pro-Pro) and IPP (Ile-Pro-Pro), have a well-established history of safe use. Short-term studies in rodents and dogs, including repeated-dose assays, demonstrated the absence of adverse effects even at concentrations substantially higher than those expected for human consumption (Beltrán-Barrientos et al., 2016). Genotoxicity assays, including the micronucleus test, have also confirmed that these peptides do not induce chromosomal damage, further supporting their safety for food applications (Beltrán-Barrientos et al., 2016).

In summary, the toxicity of whey-derived peptides depends both on their intrinsic structural characteristics and on the processing conditions applied. Consequently, systematic monitoring of reactive compounds, detailed characterization of peptide fractions, and the application of validated toxicological assays are essential to ensure the safety of hydrolyzates intended for food, nutraceutical, or clinical use.

6.4 Regulatory aspects of whey-derived peptides

The regulation of whey-derived bioactive peptides encompasses the demonstration of safety, quality, and efficacy—requirements common to all functional ingredients obtained from food sources. Although these peptides are generally regarded as safe when manufactured under good manufacturing practices , their approval depends on scientific evidence demonstrating the absence of toxicity, reduced allergenic potential, and adequate product stability (Beltrán-Barrientos et al., 2016; Liu et al., 2020).

In the United States, a major regulatory milestone is the designation of whey protein concentrate as Generally Recognized as Safe (GRAS) by the Food and Drug Administration (FDA), under regulation 21 CFR 184.1979c (Boland, 2011). This classification reflects the long-standing record of safe use of whey proteins and the absence of significant toxic effects reported in standardized studies. However, this status does not automatically extend to newly developed hydrolyzates or structurally modified peptide fractions, which must undergo individual evaluation to demonstrate safety, stability, and scientific substantiation for any proposed functional claims (Hussein et al., 2020).

Whey-derived bioactive peptides are among the most widely approved functional ingredients in markets such as the European Union, Japan, China, and the United States. Their authorization depends on meeting regulatory requirements that, while varying across jurisdictions, follow shared fundamental principles. These include detailed chemical characterization of the ingredient, toxicological and allergenic safety assessment, demonstration of stability, and physiological efficacy supported by clinical evidence. Despite a strong safety record, regulatory challenges persist—particularly concerning industrial process variability, long-term effects, and jurisdiction-specific standards. Therefore, alignment between scientific evidence and legal requirements is essential to ensure the safe acceptance of novel whey-derived peptides in the global market (Chalamaiah et al., 2019).

In summary, although substantial evidence supports the favorable safety profile of whey-derived peptides, significant regulatory challenges remain, especially those related to long-term safety, manufacturing variability, and compliance with country-specific requirements. Harmonizing scientific data with regulatory expectations is critical to ensuring that new whey-derived peptide ingredients are safely accepted and utilized in international markets.

7. Current trends and future perspectives

The field of whey bioactive peptides has experienced remarkable growth in technological capabilities and scientific understanding. Mass spectrometry-based peptidomics, particularly when combined with liquid chromatography and tandem mass spectrometry (LC-MS/MS), now allows researchers to identify and characterize thousands of peptides within complex whey hydrolyzates (Luparelli et al., 2025). Ion mobility mass spectrometry adds another dimension by separating peptides based on their molecular shape, which is especially valuable for distinguishing isomeric peptides that share the same mass but differ in sequence or three-dimensional architecture (Österlund et al., 2022). A significant advancement has been the integration of computational tools into the peptide discovery process. The AlphaFold2 algorithm can now model peptide–protein interactions, including how β-lactoglobulin binds amyloidogenic targets, revealing recognition mechanisms without the need for crystallography (Österlund et al., 2022). Researchers are further developing quantitative structure–activity relationship (QSAR) models that correlate peptide structural features with biological outcomes, enabling the prediction of bioactive sequences before synthesis (Guan et al., 2025).

On the production side, there has been a clear shift toward greener extraction technologies that reduce environmental impact while improving hydrolysis efficiency. Ultrasound-assisted processing generates localized high-pressure and high-temperature microenvironments through acoustic cavitation, facilitating protein unfolding and enhancing enzymatic accessibility (Shi et al., 2025). This approach has been shown to improve the calcium and thermal stability of whey hydrolyzates through glycosylation reactions (Shi et al., 2025). High-pressure processing applies pressures of 100–1000 MPa to unfold proteins without heat, creating more hydrolysis-susceptible substrates while preserving thermolabile bioactives (Landim et al., 2021). This technology simultaneously improves bioactivity and reduces allergenicity—an advantage for sensitive applications such as infant formula (Landim et al., 2021). Other emerging technologies include pulsed electric fields that modify protein structure non-thermally, microwave-assisted processing for rapid and energy-efficient denaturation, and subcritical water hydrolysis, which enables protein breakdown without the need for added enzymes (Irazoqui et al., 2024)). The enzymatic toolbox has similarly expanded, with novel fungal, bacterial, and recombinant proteases designed for enhanced stability and specificity (Irazoqui et al., 2024)). Enzyme immobilization systems also facilitate catalyst recovery and reuse, thereby significantly reducing industrial costs (Irazoqui et al., 2024)).

Delivering bioactive peptides from manufacturing to physiological sites of action remains a significant challenge. The gastrointestinal tract rapidly degrades peptides via acid hydrolysis and proteolytic enzymes, limiting their bioavailability. Nanoencapsulation technologies offer a promising strategy. For example, phosphatidylcholine nanoliposomes encapsulating peptides such as nisin reduced toxicity while maintaining bioactivity, increasing survival to 73.6% in *Caenorhabditis elegans* compared to lower rates with free peptide (Boelter et al., 2023). Liposomes, nanoemulsions, and nanovesicles can shield peptides from degradation and enhance intestinal transport (Gharibzahedi et al., 2023; Shadrack et al., 2025). Structural modifications, such as PEGylation and glycosylation, also extend the peptide circulation time (Shadrack et al., 2025). Spray-drying is the most economical industrial encapsulation method, and advances in low-temperature drying and the use of protective carriers, such as maltodextrin and cyclodextrins, improve peptide stability during processing (Gharibzahedi et al., 2023).

Scaling from laboratory to industry involves economic and technical constraints. Membrane filtration technologies, such as ultrafiltration and nanofiltration, can separate peptides based on molecular weight, thereby enriching fractions containing bioactive sequences (Luparelli et al., 2025). Chromatographic purification achieves superior precision but remains cost-prohibitive for large-scale food applications (Luparelli et al., 2025). Integrated continuous systems that couple enzymatic hydrolysis with in-line monitoring and downstream purification are emerging to improve productivity (Irazoqui et al., 2024). Enzyme expenses, which often account for 30%–50% of production costs, represent a significant economic barrier; however, immobilization technologies can reduce these costs by up to 70%–80% through multiple reuses (Irazoqui et al., 2024). A growing emphasis on whey valorization within circular economy frameworks has transformed whey from an industrial waste stream into a high-value raw material, effectively eliminating substrate costs (Dimassi et al., 2025; Silva et al., 2025). In the commercial landscape, an analysis of 1142 approved health foods in China revealed that whey protein ranks among the most frequently included functional ingredients, with claims spanning immune enhancement, bone health, skin hydration, and fatigue reduction (Duan et al., 2024).

The translation of promising laboratory findings into human health products is influenced by regulatory complexities that vary across jurisdictions. In the United States, bioactive peptide products can be marketed as dietary supplements without requiring premarket approval, although specific health claims must be substantiated by regulatory authorities (Kumar et al., 2024a,b). The European Union imposes stricter requirements, with health claims evaluated and authorized by EFSA (Kumar et al., 2024a,b). In the U.S., a GRAS designation requires comprehensive safety dossiers that include toxicology, allergenicity assessments, and determination of appropriate intake levels (Kumar et al., 2024a,b). Clinical translation faces additional challenges: manufacturers must establish standardized characterization methods to ensure batch consistency, generate robust toxicology data, and define dose–response relationships (Kumar et al., 2024a,b). Clinical trial design is further complicated by the modest effect sizes typically observed with nutraceuticals, often necessitating large sample sizes (Kumar et al., 2024a,b).

One promising direction lies in identifying peptides with multiple complementary bioactivities. Fermented whey products exemplify this concept, demonstrating improvements in muscle mass, glycemic control, lipid metabolism, immune function, oxidative stress, inflammation, and gastrointestinal health (Sar et al., 2025). These broad effects appear to arise from synergistic interactions among bioactive peptides, branched-chain amino acids, and microbial metabolites such as exopolysaccharides and short-chain fatty acids (Sar et al., 2025). This suggests that whole fermented matrices may outperform isolated peptide fractions by delivering multifaceted physiological benefits (Sar et al., 2025). However, interindividual variability poses challenges: differences in genetics, peptide absorption capacity, receptor sensitivity, and gut microbiota composition influence who benefits most from specific peptides (Elmas et al., 2025). This creates a compelling future for personalized nutrition, integrating nutrigenomics, metabolomics, and microbiome profiling to tailor whey peptide interventions (Elmas et al., 2025).

Sustainability considerations increasingly shape research and industry priorities. Global whey production exceeds 180–190 million tons annually, and valorizing this by-product stream into functional peptides simultaneously addresses waste management, environmental sustainability, and public health opportunities (Taesuwan et al., 2024; Marwan-Abdelbaset et al., 2025). Life cycle assessments consistently demonstrate that whey-derived bioactive peptides have a more favorable environmental profile than many synthetic pharmaceuticals, characterized by lower carbon emissions, reduced water usage, and minimal hazardous waste generation (Silva et al., 2025).

Looking forward, several research directions hold particular promise. Systematic screening of all major whey proteins using standardized hydrolysis conditions, combined with peptidomics, could generate comprehensive databases that link peptide sequences to biological activities (Luparelli et al., 2025). Structure-guided computational design may yield optimized peptides with improved potency, stability, or bioavailability relative to naturally occurring sequences (Guan et al., 2025). Comprehensive pharmacokinetic studies that track intact peptides and metabolites following oral administration would clarify absorption, tissue distribution, and the bioactive forms (Kumar et al., 2024a,b). Large-scale, long-term clinical trials evaluating cardiovascular endpoints are essential for establishing definitive efficacy and safety (Kumar et al., 2024a,b). Combination strategies that pair whey peptides with other bioactives, established therapies, or specific dietary patterns could uncover synergistic approaches (Taesuwan et al., 2024). Ultimately, rapid and cost-effective bioassays for industrial quality control would ensure that commercial products consistently deliver their intended bioactivities (Irazoqui et al., 2024)). The convergence of advanced analytical technologies, innovative production methods, improved delivery systems, and computational design positions the field for substantial progress in translating whey bioactive peptides from promising biochemical entities into practical tools for cardiometabolic health.

8. Conclusion

Bioactive peptides derived from whey through hydrolysis have gained substantial interest as functional ingredients in the food industry, enhancing the value of whey due to their combined nutritional and technological properties. These peptides may be released during digestion or obtained through targeted extraction processes. Emerging processing technologies, such as ultrasound-assisted extraction, microwave treatment, high hydrostatic pressure, and pulsed electric fields, have become prominent because they increase efficiency, reduce processing time, and lower extraction costs.

The most widely described bioactivities include antihypertensive, anti-inflammatory, antimicrobial, antioxidant, immunomodulatory, antithrombotic, opioid-modulating, and mineral-binding effects, whereas anticancer and neuroprotective activities are more recently explored. Biofunctionality, however, depends strongly on peptide composition and amino acid sequence, being directly influenced by the precursor protein. Processing conditions, such as pH and temperature, also play a decisive role in shaping bioactivity. In this context, more hydrophilic, low–molecular weight peptides, particularly di- and tripeptides, tend to be more bioavailable due to enhanced intestinal absorption.

Beyond their biological effects, whey-derived peptides exhibit techno-functional properties relevant to food formulation, including improved solubility of protein fractions, emulsion stabilization, promotion of foam formation and stability, increased water retention, and the enhancement of fermentation activity and probiotic viability in functional beverages. As a result, these peptides have been widely used in the development of nutraceuticals, functional foods, and pharmaceutical products.

Despite their promising health effects in preclinical models, clinical translation faces toxicological and regulatory challenges, particularly due to the broad heterogeneity of products generated through enzymatic hydrolysis. Additionally, undesirable reactions, including the formation of biogenic amines and Maillard reaction products, may occur during production and processing of these bioactive ingredients. A promising direction involves integrating computational tools into the peptide discovery and design pipeline, enabling the selection of candidates through structure-guided computational design. One key advantage of these tools is the ability to identify peptides with multiple complementary bioactivities, allowing synergistic interactions and potentially enhancing beneficial bioactive effects in future applications.

References

Abadía-García, L., Castaño-tostado, E., Cardador-Martínez, A., Martín-del-Campo, S.T., Amaya-llano, S.L., 2021. Production of ACE inhibitory peptides from whey proteins modified by high intensity ultrasound using bromelain. Foods 10, 2099. https://doi.org/10.3390/foods10092099.

Abasubong, K.P., Jiang, G.-Z., Guo, H., Wang, X., Huang, Y., Dai, Y.-J., Li, X.-F., Dong, Y., Gabriel, N.N., Liu, W., 2023. Oral bovine serum albumin administration alleviates inflammatory signals and improves antioxidant capacity and immune response under thioacetamide stress in blunt snout bream fed a high-calorie diet. Fish Shellfish Immunol. 141, 108996.

Abd El-Fattah, A., Sakr, S., El-Dieb, S., Elkashef, H., 2018. Developing functional yogurt rich in bioactive peptides and gamma-aminobutyric acid related to cardiovascular health. Lwt 98, 390–397. https://doi.org/10.1016/j.lwt.2018.09.022.

Afify, S.M., Pali-Schöll, I., Hufnagl, K., Hofstetter, G., El-Bassuoni, M.A.-R., Roth-Walter, F., Jensen-Jarolim, E., 2021. Bovine holo-beta-lactoglobulin cross-protects against pollen allergies in an innate manner in BALB/c mice: potential model for the farm effect. Front. Immunol. 12, 611474.

Aguilar-Toalá, J.E., Quintanar-Guerrero, D., Liceaga, A.M., Zambrano-Zaragoza, M.L., 2022. Encapsulation of bioactive peptides: a strategy to improve the stability, protect the nutraceutical bioactivity and support their food applications. RSC Adv. 12 (11), 6449–6458. https://doi.org/10.1039/D1RA08590E.

Alizadeh, O., Aliakbarlu, J., 2020. Effects of ultrasound and ohmic heating pretreatments on hydrolysis, antioxidant and antibacterial activities of whey protein concentrate and its fractions. Lwt 131, 109913. https://doi.org/10.1016/j.lwt.2020.109913.

Alzaydi, A., Barbhuiya, R.I., Routray, W., Elsayed, A., Singh, A., 2023. Bioactive peptides: synthesis, applications, and associated challenges. In: Food Bioengineering, pp. 273–290. https://doi.org/10.1002/fbe2.12057.

Amorim, M., Pereira, J.O., Silva, L.B., Ormenese, R.C.S.C., Pacheco, M.T.B., Pintado, M., 2018. Use of whey peptide fraction in coated cashew nut as functional ingredient and salt replacer. LWT 92, 204–211. https://doi.org/10.1016/j.lwt.2017.12.075.

Andrade, J., Pereira, C.G., de Almeida Junior, J.C., Viana, C.C.R., de Oliveira Neves, L.N., da Silva, P.H.F., dos Anjos, V.D.C., 2019. FTIR-ATR determination of protein content to evaluate whey protein concentrate adulteration. Lwt 99, 166–172. https://doi.org/10.1016/j.lwt.2018.09.079.

Ano, Y., Ayabe, T., Kutsukake, T., Ohya, R., Takaichi, Y., Uchida, S., Yamada, K., Uchida, K., Takashima, A., Nakayama, H., 2018. Novel lactopeptides in fermented dairy products improve memory function and cognitive decline. Neurobiol. Aging 72, 23–31. https://doi.org/10.1016/j.neurobiolaging.2018.07.016.

Ano, Y., Ohya, R., Takaichi, Y., Washinuma, T., Uchida, K., Takashima, A., Nakayama, H., 2020. L-Lactolin, a whey-derived lacto-tetrapeptide, prevents Alzheimer's disease pathologies and cognitive decline. J. Alzheimers Dis. 73, 1331–1342. https://doi.org/10.3233/JAD-190997.

Antila, P., Paakkari, I., Järvinen, A., Mattila, M.J., Laukkanen, M., Pihlanto-Leppälä, A., Mäntsälä, P., Hellman, J., 1991. Opioid peptides derived from in-vitro proteolysis of bovine whey proteins. Int. Dairy J. 1 (4), 215–229. https://doi.org/10.1016/0958-6946(91)90015-z.

Arpitha, A., Sebastin Santhosh, M., Rohit, A.C., Girish, K.S., Vinod, D., Aparna, H.S., 2017. Inhibition of snake venom metalloproteinase by β-lactoglobulin peptide from Buffalo (Bubalus bubalis) colostrum. Appl. Biochem. Biotechnol. 182 (4), 1415–1432. https://doi.org/10.1007/s12010-017-2407-6.

Atma, Y., Murray, B.S., Sadeghpour, A., Goycoolea, F.M., 2024. Encapsulation of short-chain bioactive peptides (BAPs) for gastrointestinal delivery: a review. Food Funct. 15, 3959–3979. https://doi.org/10.1039/D3FO04195F.

Attaallah, W., Yılmaz, A.M., Erdoğan, N., Yalçın, A.S., Aktan, A.Ö., 2012. Whey protein versus whey protein hydrolyzate for the protection of azoxymethane and dextran sodium sulfate induced colonic tumors in rats. Pathol. Oncol. Res. 18, 817–822.

Auestad, N., Layman, D.K., 2021. Dairy bioactive proteins and peptides: a narrative review. Nutr. Rev. 79, 36–47. https://doi.org/10.1093/nutrit/nuab097.

Axelrod, R., Beyrer, M., Mathys, A., 2022. Impact of the electric field intensity and treatment time on whey protein aggregate formation. J. Dairy Sci. 105, 6589–6600. https://doi.org/10.3168/jds.2021-21395.

Baba, W.N., Mudgil, P., Kamal, H., Kilari, B.P., Gan, C.Y., Maqsood, S., 2021a. Identification and characterization of novel α-amylase and α-glucosidase inhibitory peptides from camel whey proteins. J. Dairy Sci. 104 (2), 1364–1377. https://doi.org/10.3168/jds.2020-19271.

Baba, W.N., Baby, B., Mudgil, P., Gan, C.-Y., Vijayan, R., Maqsood, S., 2021b. Pepsin generated camel whey protein hydrolysates with potential antihypertensive properties: identification and molecular docking of antihypertensive peptides. LWT 143, 111135. https://doi.org/10.1016/j.lwt.2021.111135.

Badely, M., Sepandi, M., Samadi, M., Parastouei, K., Taghdir, M., Akbari, A., Ghasemi, M., Samavat, S., Rezaei, F., 2019. The effect of whey protein on the components of metabolic syndrome in overweight and obese individuals; a systematic review and meta-analysis. Diabetes Metabol. Syndr.: Clin. Res. Rev. 13 (6), 3121–3131. https://doi.org/10.1016/j.dsx.2019.11.001.

Bal dit Sollier, C., Drouet, L., Pignaud, G., Chevallier, C., Caen, J., Fiat, A.M., Izquierdo, C., Jollès, P., 1996. Effect of κ-casein split peptides on platelet aggregation and on thrombus formation in the guinea-pig. Thromb. Res. 81 (4), 427–437. https://doi.org/10.1016/0049-3848(96)00015-1.

Bamdad, F., Bark, S., Kwon, C.H., Suh, J.-W., Sunwoo, H., 2017. Anti-inflammatory and antioxidant properties of peptides released from β-lactoglobulin by high hydrostatic pressure-assisted enzymatic hydrolysis. Molecules 22, 949.

Barbé, F., Le Feunteun, S., Rémond, D., Ménard, O., Jardin, J., Henry, G., Laroche, B., Dupont, D., 2014. Tracking the in vivo release of bioactive peptides in the gut during digestion: mass spectrometry peptidomic characterization of effluents collected in the gut of dairy matrix fed mini-pigs. Food Res. Int. 63, 147–156. https://doi.org/10.1016/J.FOODRES.2014.02.015.

Barrón-Ayala, C.G., Valenzuela-Melendres, M., Camou, J.P., Sebranek, J.G., Dávila-Ramírez, J.L., Cumplido-Barbeitia, G., 2020. Pork frankfurters prepared with hydrolyzed whey: preliminary product quality aspects and inhibitory activity of the resulting peptides on angiotensin-converting enzyme. Meat Sci. 166, 108111. https://doi.org/10.1016/j.meatsci.2020.108111.

Beltrán-Barrientos, L.M., García, H.S., Torres-Llanez, M.J., González-Córdova, A.F., Hernández-Mendoza, A., Vallejo-Cordoba, B., 2016. Safety of milk-derived bioactive peptides. Int. J. Dairy Technol. 70 (1), 16–22. https://doi.org/10.1111/1471-0307.12338.

Besharati, M., Lackner, M., 2023. Bioactive peptides: a review. The EuroBiotech J. 7, 176–188. https://doi.org/10.2478/ebtj-2023-0013.

Beucher, S., Levenez, F., Yvon, M., Corring, T., 1994. Effects of gastric digestive products from casein on CCK release by intestinal cells in rat. J. Nutr. Biochem. 5, 578–584.

Bhandari, D., Rafiq, S., Gat, Y., Gat, P., Waghmare, R., Kumar, V., 2020. A review on bioactive peptides: physiological functions, bioavailability and safety. Int. J. Pept. Res. Therapeut. 26, 139–150. https://doi.org/10.1007/s10989-019-09823-5.

Bird, R.P., 1995. Role of aberrant crypt foci in understanding the pathogenesis of colon cancer. Cancer Lett. 93, 55–71.

Blayo, C., Vidcoq, O., Lazennec, F., Dumay, E., 2016. Effects of high pressure processing (hydrostatic high pressure and ultra-high pressure homogenisation) on whey protein native state and susceptibility to tryptic hydrolysis at atmospheric pressure. Food Res. Int. 79, 40–53. https://doi.org/10.1016/J.FOODRES.2015.11.024.

Boelter, J.F., Garcia, S.C., Göethel, G., Charão, M.F., de Melo, L.M., Brandelli, A., 2023. Acute toxicity evaluation of phosphatidylcholine nanoliposomes containing nisin in *Caenorhabditis elegans*. Molecules 28 (2), 563. https://doi.org/10.3390/molecules28020563.

Boland, M., 2011. 'Whey Proteins', Handbook of Food Proteins. Elsevier, pp. 30–55. https://doi.org/10.1533/9780857093639.30.

Bounous, G., Kongshavn, P.A.L., 1985. Differential effect of dietary protein type on the B-cell and T-cell immune responses in mice. J. Nutr. 115, 1403–1408.

Brandelli, A., Daroit, D.J., Corrêa, A.P.F., 2015. Whey as a source of peptides with remarkable biological activities. Food Res. Int. 73, 149–161. https://doi.org/10.1016/j.foodres.2015.01.016.

Buey, B., Layunta, E., Latorre, E., Mesonero, J.E., 2023. Potential role of milk bioactive peptides on the serotonergic system and the gut-brain axis. Int. Dairy J. 137, 105534. https://doi.org/10.1016/j.idairyj.2022.105534.

Bunsroem, K., Prinyawiwatkul, W., Thaiudom, S., 2022. The influence of whey protein heating parameters on their susceptibility to digestive enzymes and the antidiabetic activity of hydrolysates. Foods 11 (6), 829. https://doi.org/10.3390/foods11060829.

Caetano-Silva, M.E., Rund, L.A., Vailati-Riboni, M., Bertoldo Pacheco, M.T., Johnson, R.W., 2021. Copper-binding peptides attenuate microglia inflammation through suppression of NF-κB pathway. Mol. Nutr. Food Res. 65 (22), e2100153. https://doi.org/10.1002/mnfr.202100153.

Campanhon, I.B., De Aguiar, P.F., Bezerra, F.F., Soares, M.R., Torres, A.G., 2024. Human colostrum in vitro protein digestion: peptidomics by liquid chromatography-Orbitrap-high-resolution MS and prospection for bioactive peptides via bioinformatics. Br. J. Nutr. 131, 17–26. https://doi.org/10.1017/S0007114523001459.

Campbell, J.E., Drucker, D.J., 2013. Pharmacology, physiology, and mechanisms of incretin hormone action. Cell Metabol. 17 (6), 819–837. https://doi.org/10.1016/j.cmet.2013.04.008.

Cao, Z., Chang, Z., Gao, Y., Xue, B., Yang, Y., Zhao, L., Ge, W., 2025. Effects of non-thermal treatments on goat whey protein allergenicity: a study based on conformational and linear epitopes. LWT 219, 117517. https://doi.org/10.1016/j.lwt.2025.117517.

Cenini, G., Lloret, A., Cascella, R., 2019. Oxidative stress in neurodegenerative diseases: from a mitochondrial point of view. Oxidat. Med. Cell. Longev. 2019, 18. https://doi.org/10.1155/2019/2105607.2105607.

Chalamaiah, M., Keskin Ulug, S., Hong, H., Wu, J, 2019. Regulatory requirements of bioactive peptides (protein hydrolysates) from food proteins. J. Funct. Foods 58, 123–129. https://doi.org/10.1016/j.jff.2019.04.050.

Chang, Y.B., Jung, E.-J., Jo, K., Suh, H.J., Choi, H.-S., 2024. Neuroprotective effect of whey protein hydrolysate containing leucine-aspartate-isoleucine-glutamine-lysine on HT22 cells in hydrogen peroxide–induced oxidative stress. J. Dairy Sci. 107 (5), 2620–2632. https://doi.org/10.3168/jds.2023-24284.

Chen, G.W., Tsai, J.S., Pan, B.S., 2007. Purification of angiotensin I-converting enzyme inhibitory peptides and antihypertensive effect of milk produced by protease-facilitated lactic fermentation. Int. Dairy J. 17 (6), 641–647. https://doi.org/10.1016/j.idairyj.2006.07.004.

Chen, Y.C., Smith, H.A., Hengist, A., Chrzanowski-Smith, O.J., Manoukian, L., Allerton, D.M., Sayer, R.D., Walhin, J.P., Gonzalez, J.T., Thompson, D., Betts, J.A., Afman, L.A., 2019. Co-ingestion of whey protein hydrolysate with milk minerals rich in calcium potently stimulates glucagon-like peptide-1 secretion: an RCT in healthy adults. Eur. J. Nutr. 59 (6), 2449–2462. https://doi.org/10.1007/s00394-019-02092-4.

Cheng, X., Gao, D., Chen, B., Mao, X., 2015. Endotoxin-binding peptides derived from casein glycomacropeptide inhibit lipopolysaccharide-stimulated inflammatory responses via blockade of NF-κB activation in macrophages. Nutrients 7, 3119–3137.

Córdova-Dávalos, L.E., Jiménez, M., Salinas, E., 2019. Glycomacropeptide bioactivity and health: a review highlighting action mechanisms and signaling pathways. Nutrients 11, 598.

Coates, D., 2003. The angiotensin converting enzyme (ACE). Int. J. Biochem. Cell Biol. 35, 769–773. https://doi.org/10.1016/S1357-2725(02)00309-6.

Córdova-Dávalos, L.E., Cervantes-García, D., Ballona-Alba, M.F., Santos-López, A., Esquivel-Basaldúa, A.S., Gallegos-Alcalá, P., Jiménez, M., Salinas, E., 2023. Protective effect of glycomacropeptide on the inflammatory response of U937 macrophages. Foods 12, 1528.

Corrêa, T., Vilela, C., Ribeiro, S., 2023. Bioactive antimicrobial peptides from food proteins: perspectives and challenges for controlling foodborne pathogens. Pathogens 12 (3), 477. https://doi.org/10.3390/pathogens12030477.

Costa, N.R., Cappato, L.P., Pereira, M.V.S., Pires, R.P.S., Moraes, J., Esmerino, E.A., Silva, R., Neto, R.P.C., Tavares, M.I.B., Freitas, M.Q., Silveira Júnior, R.N., Rodrigues, F.N., Bisaggio, R.C., Cavalcanti, R.N., Raices, R.S.L., Silva, M.C., Cruz, A.G., 2018. Ohmic Heating: a potential technology for sweet whey processing. Food Res. Int. 106, 771–779. https://doi.org/10.1016/j.foodres.2018.01.046.

D'Souza, K., Acquah, C., Mercer, A., Pyle, W.G., Chabowski, A., Aukema, H.M., 2021. Whey peptides exacerbate body weight gain and perturb systemic glucose and tissue lipid metabolism in male high-fat fed mice. Food Funct. 12 (8), 3552–3561. https://doi.org/10.1039/d0fo02610g.

Dalaka, E., Politis, I., Theodorou, G., 2023. Antioxidant activity of sweet whey derived from bovine, ovine, and caprine milk obtained from various small-scale cheese plants in Greece before and after in vitro simulated gastrointestinal digestion. Antioxidants 12, 1676. https://doi.org/10.3390/antiox12091676.

Daliri, E., Oh, D., Lee, B., 2017. Bioactive peptides. Foods 6, 32. https://doi.org/10.3390/foods6050032.

Das, B., Basaiawmoit, B., Sakure, A.A., Maurya, R., Bishnoi, M., Kondepudi, K.K., Hati, S., 2025. Production and identification of antidiabetic and antioxidative peptides from fermented cheddar cheese whey with anti-inflammatory activity: a valorisation of dairy waste. J. Food Sci. Technol. 1–16. https://doi.org/10.2174/0113816128350545241231161241.

DeFronzo, R.A., Ferrannini, E., Groop, L., Henry, R.R., Herman, W.H., Holst, J.J., Weiss, R., 2015. Type 2 diabetes mellitus. Nat. Rev. Dis. Primers 1 (1), 1–22. https://doi.org/10.1038/nrdp.2015.19.

Dimassi, O., Jaber, L., Fleyfel, L., Hamadeh, S., 2025. Milk supply in Lebanon: economic challenges and the role of traditional dairy products. Foods 14 (17), 3115. https://doi.org/10.3390/foods14173115.

Ding, N., Meng, H., Wu, C., Yokoyama, W., Hong, H., Luo, Y., Tan, Y., 2023. Whey protein hydrolysate renovates age-related and scopolamine-induced cognitive impairment. Nutrients 15 (5), 1228. https://doi.org/10.3390/nu15051228.

Dinkçi, N., Akdeniz, V., Akalın, A.S., 2023. Probiotic whey-based beverages from cow, sheep and goat milk: antioxidant activity, culture viability, amino acid contents. Foods 12 (3), 610. https://doi.org/10.3390/foods12030610.

Dou, P., Li, X., Zou, X., Wang, K., Yao, L., Sun, Z., Hong, H., Luo, Y., Tan, Y., 2024. Antihypertensive effects of whey protein hydrolysate involve reshaping the gut microbiome in spontaneously hypertensive rats. Food Sci. Hum. Wellness 13 (4), 1974–1986. https://doi.org/10.26599/FSHW.2022.9250164.

Du, Z., Li, Y., 2022. Review and perspective on bioactive peptides: a roadmap for research, development, and future opportunities. J. Agric. Food Res. 9, 100353. https://doi.org/10.1016/j.jafr.2022.100353.

Duan, H., Liu, G., Liu, J., Wang, Z., Bao, S., Chang, X., Yan, W., 2024. Review: application of protein-based raw materials in health foods in China. Foods 14 (1), 20. https://doi.org/10.3390/foods14010020.

Dullius, A., Goettert, M.I., de Souza, C.F.V., 2018. Whey protein hydrolysates as a source of bioactive peptides for functional foods–Biotechnological facilitation of industrial scale-up. J. Funct.Foods 42, 58–74. https://doi.org/10.1016/j.jff.2017.12.063.

D'Opazo, V., Calpe, J., Escrivá, L., Melo Nazareth, T., Meca, G., Luz, C., 2023. Identification and evaluation of antioxidant activity of bioactive casein-derived peptides during in vitro digestion and transepithelial transport in Caco-2 cells. Food Biosci. 53, 102763. https://doi.org/10.1016/J.FBIO.2023.102763.

Egan, M., O'Connell Motherway, M., Ventura, M., van Sinderen, D., 2014. Metabolism of sialic acid by Bifidobacterium breve UCC2003. Appl. Environ. Microbiol. 80, 4414–4426.

El Mecherfi, K.E., Curet, S., Lupi, R., Larré, C., Rouaud, O., Choiset, Y., Rabesona, H., Haertlé, T., 2019. Combined microwave processing and enzymatic proteolysis of bovine whey proteins: the impact on bovine β-lactoglobulin allergenicity. J. Food Sci. Technol. 56, 177–186. https://doi.org/10.1007/s13197-018-3471-9.

Elias, R.J., Kellerby, S.S., Decker, E.A., 2008. Antioxidant activity of proteins and peptides. Crit. Rev. Food Sci. Nutr. 48 (5), 430–441. https://doi.org/10.1080/10408390701425615.

Elmas, S., Fındık, M., Kıyak, R., Taşkın, G., Cîrţînă, D., Dîrnu, R., Guţă, N., Mecu, R.-M., Bîcă, M.-D., 2025. Anticancer potential of whey proteins: a systematic review of bioactivity and functional mechanisms. Int. J. Mol. Sci. 26 (21), 10406. https://doi.org/10.3390/ijms262110406.

Estévez, N., Fuciños, P., Fuciños, C., Jauregi, P., Tovar, C.A., Rúa, M.L., 2020. Hydrolysis of whey protein as a useful approach to obtain bioactive peptides and a β-Lg fraction with different biotechnological applications. Food Hydrocoll. 109, 106095. https://doi.org/10.1016/J.FOODHYD.2020.106095.

Fassina, P., Nunes, G.Q., Adami, F.S., Goettert, M.I., Volken de Souza, C.F., 2019. Importance of cheese whey processing: supplements for sports activities-a review. Pol. J. Food Nutr. Sci. 69 (1). https://doi.org/10.31883/pjfns-2019-0008.

Faucher, M., Geoffroy, T.R., Thibodeau, J., Gaaloul, S., Bazinet, L., 2022. Semi-industrial production of a DPP-IV and ACE inhibitory peptide fraction from whey protein concentrate hydrolysate by electrodialysis with ultrafiltration membrane. Membranes 12 (4), 409. https://doi.org/10.3390/membranes12040409.

Fernández-Musoles, R., Salom, J.B., Martínez-Maqueda, D., López-Díez, J.J., Recio, I., Manzanares, P., 2013. Antihypertensive effects of lactoferrin hydrolyzates: inhibition of angiotensin- and endothelin-converting enzymes. Food Chem. 139 (1–4), 994–1000. https://doi.org/10.1016/j.foodchem.2012.12.049.

Fernández-Tomé, S., Hernández-Ledesma, B., 2020. Gastrointestinal digestion of food proteins under the effects of released bioactive peptides on digestive health. Mol. Nutr. Food Res. 64, 2000401.

Fernández-Tomé, S., Martínez-Maqueda, D., Girón, R., Goicoechea, C., Miralles, B., Recio, I., 2016. Novel peptides derived from αs1-casein with opioid activity and mucin stimulatory effect on HT29-MTX cells. J. Funct.Foods 25, 466–476. https://doi.org/10.1016/j.jff.2016.06.023.

Gao, C.-H., Dong, H.-L., Tai, L., Gao, X.-M., 2018. Lactoferrin-containing immunocomplexes drive the conversion of human macrophages from M2- into M1-like phenotype. Front. Immunol. 9, 37.

Garg, S., Nurgali, K., Kumar Mishra, V., 2016. Food proteins as source of opioid peptides – a review. Curr. Med. Chem. 23 (9), 893–910. https://doi.org/10.2174/0929867323666160219115226.

Ghadimi, D., Kaya, A.Ş., Krüger, S., Röcken, C., Schäfer, H., Uchiyama, J., Matsuzaki, S., Bockelmann, W., 2024. Exploring interplay between bovine milk-derived α-lactalbumin, pathogenic bacteria, and bacteriophages at the molecular interface of inflammation. Comp. Immunol. Microbiol. Infect. Dis. 115, 102271.

Gharibzahedi, S.M.T., Moghadam, M., Amft, J., Tolun, A., Hasabnis, G., Altintas, Z., 2023. Recent advances in dietary sources, health benefits, emerging encapsulation methods, food fortification, and new sensor-based monitoring of Vitamin B12: a critical review. Molecules 28 (22), 7469. https://doi.org/10.3390/molecules28227469.

Giraldo, N.D., Sánchez, J.D., López, A., González, N., Balaguer, F., Redondo, J., Llopis, S., Barrena, M., Klotz, B., 2024. Functional assessment of a whey protein hydrolysate: in vivo responses in lipid regulation, oxidant defense, and neural protection in *Caenorhabditis elegans*. Int. Dairy J. 154, 105921. https://doi.org/10.1016/j.idairyj.2024.105921.

Godoy, J., Peslerbes, M., Vyhmeister, S., Vargas-Bello-Pérez, E., Fellenberg, M.A., Ibáñez, R.A., 2020. Changes in the chemical and in-vitro anti-hypertensive properties of sweet whey obtained from miniature fresh, Chanco and Gouda-style model cheeses. J. Dairy Res. 87 (4), 488–492. https://doi.org/10.1017/S0022029920001041.

Gong, T., Fu, J., Shi, L., Chen, X., Zong, X., 2021. Antimicrobial peptides in gut health: a review. Front. Nutr. 8, 751010. https://doi.org/10.3389/fnut.2021.751010.

González-Osuna, M.F., Bernal-Mercado, A.T., Wong-Corral, F.J., Ezquerra-Brauer, J.M., Soto-Valdez, H., Castillo, A., Rodríguez-Figueroa, J.C., Del-Toro-Sánchez, C.L., 2024. Bioactive peptides and protein hydrolysates used in meat and meat products' preservation: a review. ACS Food Sci. Technol. 4, 1003–1016. https://doi.org/10.1021/acsfoodscitech.3c00605.

Grando, L.M., Fassina, P., Leonhardt, F., Ethur, E.M., Neutzling Lehn, D., Goettert, M.I., Volken de Souza, C.F., 2024. Protein hydrolysates from buffalo cheese whey: production and characterisation for food supplementation. Int. J. Food Sci. Technol. 59 (4), 2816–2822. https://doi.org/10.1111/ijfs.17047.

Grigorean, G., Du, X., Kuhfeld, R., Haberl, E.M., Lönnerdal, B., 2024. Effect of in vitro digestion on bioactive peptides related to immune and gut health in intact cow's milk and hydrolyzed protein-based infant formulas. Nutr. 2024 16, 3268. https://doi.org/10.3390/NU16193268.

Guan, H., Yan, C., Zhu, X., Wang, X., Liu, M., Huang, M., Zhou, L., Wang, Z., Liu, H., 2025. 'Protein-Derived chelating peptides: preparation, conformational relationships, and biological activities. J. Agric. Food Chem. 73 (38), 23809–23823. https://doi.org/10.1021/acs.jafc.5c09028.

Guha, S., Sharma, H., Deshwal, G.K., Rao, P.S., 2021. A comprehensive review on bioactive peptides derived from milk and milk products of minor dairy species. Food Prod. Process. Nutr. 3 (1), 2. https://doi.org/10.1186/S43014-020-00045-7.

Guo, H., Zang, C., Zheng, L., Ding, L., Yang, W., Shan Ren, N., Guan, H., 2024. Novel antioxidant peptides from fermented whey protein by Lactobacillus rhamnosus B2-1: separation and identification by in vitro and in silico approaches. J. Agric. Food Chem. 72, 23306–23319. https://doi.org/10.1021/ACS.JAFC.4C07531.

Haj Mustafa, M., Soleimanian-Zad, S., Albukhaty, S., 2024. Whey protein concentrate hydrolyzed by microbial protease: process optimization and evaluation of its dipeptidyl peptidase inhibitory activity. Waste Biomass Valoriz. 15, 2259–2271. https://doi.org/10.1007/S12649-023-02306-1/TABLES/7.

Hao, L., Li, X., Zhao, B., Song, X., Zhang, Y., Liang, Q., 2024. Enzymatic hydrolysis optimization of yak whey protein concentrates and bioactivity evaluation of the ultrafiltered peptide fractions. Molecules 29, 1403. https://doi.org/10.3390/MOLECULES29061403.

Hernández-Ledesma, B., García-Nebot, M.J., Fernández-Tomé, S., Amigo, L., Recio, I., 2014a. Dairy protein hydrolysates: peptides for health benefits. Int. Dairy J. 38 (2), 82–100. https://doi.org/10.1016/j.idairyj.2013.11.004.

Hernández-Ledesma, B., del Mar Contreras, M., Recio, I., 2014b. Bioactive peptides produced by food-grade proteolytic enzymes. Compr. Rev. Food Sci. Food Saf. 13 (4), 712–728. https://doi.org/10.1111/1541-4337.12084.

Holzer, P., 2009. Opioid receptors in the gastrointestinal tract. Regul. Pept. 155, 11–17.

Hou, Y., Dan, X., Babbar, M., et al., 2019. Ageing as a risk factor for neurodegenerative disease. Nat. Rev. Neurol. 15 (10). https://doi.org/10.1038/s41582-019-0244-7.

Hu, X., Wang, H., Hu, Y., Tu, Z., 2024. Insight into the effects of pulsed electric field on the structure, aggregation characteristics and functional properties of whey proteins. Food Hydrocoll. 154, 110111. https://doi.org/10.1016/j.foodhyd.2024.110111.

Hufnagl, K., Ghosh, D., Wagner, S., Fiocchi, A., Dahdah, L., Bianchini, R., Braun, N., Steinborn, R., Hofer, M., Blaschitz, M., Roth, G.A., Hofstetter, G., Roth-Walter, F., Pacios, L.F., Jensen-Jarolim, E., 2018. Retinoic acid prevents immunogenicity of milk lipocalin Bos d 5 through binding to its immunodominant T-cell epitope. Sci. Rep. 8, 1598.

Hussein, F.A., Chay, S.Y., Ghanisma, S.B.M., Zarei, M., Auwal, S.M., Hamid, A.A., Wan Ibadullah, W.Z., Saari, N., 2020. Toxicity study and blood pressure–lowering efficacy of whey protein concentrate hydrolysate in rat models, plus peptide characterization. J. Dairy Sci. 103 (3), 2053–2064. https://doi.org/10.3168/jds.2019-17462.

Hutchison, A.T., Feinle-Bisset, C., Fitzgerald, P.C., Standfield, S., Horowitz, M., Clifton, P.M., Luscombe-Marsh, N.D., 2015. Comparative effects of intraduodenal whey protein hydrolysate on antropyloroduodenal motility, gut hormones, glycemia, appetite, and energy intake in lean and obese men. Am. J. Clin. Nutr. 102 (6), 1323–1331. https://doi.org/10.3945/ajcn.115.114538.

Ibrahim, H.R., Isono, H., Miyata, T., 2018. Potential antioxidant bioactive peptides from camel milk proteins. Animal Nutrition 4, 273–280. https://doi.org/10.1016/j.aninu.2018.05.004.

Irazoqui, J.M., Santiago, G.M., Mainez, M.E., Amadio, A.F., Eberhardt, M.F., 2024. Enzymes for production of whey protein hydrolysates and other value-added products. Appl. Microbiol. Biotechnol. 108 (1), 354. https://doi.org/10.1007/S00253-024-13117-2.

Iskandar, M.M., Lands, L.C., Sabally, K., Azadi, B., Meehan, B., Mawji, N., Skinner, C.D., Kubow, S., 2015. High hydrostatic pressure pretreatment of whey protein isolates improves their digestibility and antioxidant capacity. Foods 4, 184–207. https://doi.org/10.3390/foods4020184.

Iyer, S.S., Accardi, C.J., Ziegler, T.R., Blanco, R.A., Ritzenthaler, J.D., Rojas, M., Roman, J., Jones, D.P., 2009. Cysteine redox potential determines pro-inflammatory IL-1β levels. PLoS One 4, e5017.

Jacquot, A., Gauthier, S.F., Drouin, R., Boutin, Y., 2010. Proliferative effects of synthetic peptides from β-lactoglobulin and α-lactalbumin on murine splenocytes. Int. Dairy J. 20, 514–521.

Jakubczyk, A., Karaś, M., Rybczyńska-Tkaczyk, K., Zielińska, E., Zieliński, D., 2020. Current trends of bioactive peptides—new sources and therapeutic effect. Foods 9, 846. https://doi.org/10.3390/foods9070846.

Jauregui-Rincón, J., Salinas-Miralles, E., Chávez-Vela, N., Jiménez-Vargas, M., 2019. Glycomacropeptide: biological activities and uses. In: Gigli, I. (Ed.), Whey - Biological Properties and Alternative Uses. IntechOpen.

Jiang, Y., Qi, Y., Liu, X., Fang, L., Gao, Y., Liu, C., Wu, D., Wang, X., Zhao, F., Wang, J., Min, W., 2022. Neuroprotective effects of fermented yak milk-derived peptide LYLKPR on H2O2-injured HT-22 cells. Food Funct. 13 (23), 12021–12038. https://doi.org/10.1039/d2fo02131e.

Jiang, H., Zhang, Z., Wang, Y., Gao, J., Yuan, Q., Mao, X., 2023. Effects of high hydrostatic pressure treatment on the antigenicity, structural and digestive properties of whey protein. Lwt 178, 114628. https://doi.org/10.1016/j.lwt.2023.114628.

Jollès, P., Lévy-Toledano, S., Fiat, A.M., Soria, C., Gillessen, D., Thomaidis, A., Dunn, F.W., Caen, J.P., 1986. Analogy between fibrinogen and casein: effect of an undecapeptide isolated from κ-casein on platelet function. Eur. J. Biochem. 158 (2), 379–382. https://doi.org/10.1111/j.1432-1033.1986.tb09764.x.

Kadam, S.U., Tiwari, B.K., Álvarez, C., O'Donnell, C.P., 2015. Ultrasound applications for the extraction, identification and delivery of food proteins and bioactive peptides. Trends Food Sci. Technol. 46, 60–67. https://doi.org/10.1016/J.TIFS.2015.07.012.

Kahn, S.E., Cooper, M.E., Del Prato, S., 2014. Pathophysiology and treatment of type 2 diabetes: perspectives on the past, present, and future. Lancet 383 (9922), 1068–1083. https://doi.org/10.1016/S0140-6736(13)62154-6.

Kamal, H., Jafar, S., Mudgil, P., Murali, C., Amin, A., Maqsood, S., 2018. Inhibitory properties of camel whey protein hydrolysates toward liver cancer cells, dipeptidyl peptidase-IV, and inflammation. J. Dairy Sci. 101 (10), 8711–8720. https://doi.org/10.3168/jds.2018-14586.

Karimi, N., Pourahmad, R., Taheri, S., Eyvazzadeh, O., 2022. Effect of addition of bioactive peptide obtained from enzymatic hydrolysis of yogurt whey on qualitative properties of doogh. Iran. Food Sci. Technol. Res. J. 18 (2). https://doi.org/10.22067/ifstrj.2021.70916.1059.

Karimi-Dehkordi, M., Ghaffarnezhad, M., Mohammadi, F., Ghirati, M., Rezaeifar, M., Rajabi, N., Alizadeh, O., 2021. Whey protein coating incorporated with essential oil, bioactive peptides and nanoparticle extends shelf-life of chicken breast slices. J. Food Meas. Char. 15 (6), 5266–5276. https://doi.org/10.1007/s11694-021-01088-1.

Kashung, P., Karuthapandian, D., 2025. Milk-derived bioactive peptides. Food Prod. Process Nutr. 7, 6. https://doi.org/10.1186/s43014-024-00280-2.

Kaur, J., Kumar, V., Sharma, K., Kaur, S., Gat, Y., Goyal, A., Tanwar, B., 2019. Opioid peptides: an overview of functional significance. Int. J. Pept. Res. Therapeut. 26 (1), 33–41. https://doi.org/10.1007/s10989-019-09813-7.

Kheroufi, A., Brassesco, M.E., Campos, D.A., Mouzai, A., Boughellouta, H., Pintado, M.E., 2022. Whey protein-derived peptides: the impact of chicken pepsin hydrolysis upon whey proteins concentrate on their biological and technological properties. Int. Dairy J. 134, 105442. https://doi.org/10.1016/J.IDAIRYJ.2022.105442.

Kilara, A., Panyam, D., 2003. Peptides from milk proteins and their biological activity. Food Sci. Technol. 14 (12), 475–482. https://doi.org/10.1016/j.tifs.2003.09.002.

KiliÇ, G., 2024. Extraction of bioactive peptides from whey proteins by conventional and novel technologies. In: INTECHOPEN. https://doi.org/10.5772/intechopen.1005645.

Kim, J.W., Lee, J.S., Choi, Y.J., Kim, C., 2025. The multifaceted functions of lactoferrin in antimicrobial defense and inflammation. Biomolecules 15, 1174.

Klaewkla, J., Hudthagosol, C., Chaijenkij, K., Panya, A., Sang-Ngoen, D., Phonsatta, N., Kaewkul, K., 2022. Whey protein concentrate mixed beverages and plasma amino acid response in young males. Food Res. 6 (3), 62–68. https://doi.org/10.26656/fr.2017.

Kong, L., Liu, C., Tang, H., Yu, P., Wen, R., Peng, X., Yu, X., 2023. Hygroscopicity and antioxidant activity of whey protein hydrolysate and its ability to improve the water holding capacity of pork patties during freeze– thaw cycles. LWT 182, 114784. https://doi.org/10.1016/j.lwt.2023.114784.

König, D., Kohl, J., Jerger, S., Centner, C., 2021. Potential relevance of bioactive peptides in sports nutrition. Nutrients 13 (11), 3997. https://doi.org/10.3390/nu13113997.

Korhonen, H., Pihlanto, A., 2006. Bioactive peptides: production and functionality. Int. Dairy J. 16 (9), 945–960. https://doi.org/10.1016/j.idairyj.2005.10.012.

Kumar, A., Kumar, M.S., Champalli, S.R., Sabikhi, L., Sonarthi, H., 2024. Inhibition of dipeptidyl peptidase-IV by hydrolysates of beta-lactoglobulin isolated from Gir cow milk. Food Chem. Adv. 5, 100842. https://doi.org/10.1016/j.focha.2024.100842.

Kumar, M., Kumar, D., Kumar, D., Garg, Y., Chopra, S., Bhatia, A., 2024. Therapeutic potential of nanocarrier mediated delivery of peptides for wound healing: current status, challenges and future prospective. AAPS PharmSciTech 25 (5), 108. https://doi.org/10.1208/s12249-024-02827-5.

Kumari, S., Shahidi, F., 2024. Bioactive peptides as antioxidants and antimicrobials: fundamentals and applications. J. Food Bioact. 2–12. https://doi.org/10.31665/JFB.2024.18367.

Lambeir, A.M., Durinx, C., Scharpé, S., De Meester, I., 2003. Dipeptidyl-peptidase IV from bench to bedside: an update on structural properties, functions, and clinical aspects of the enzyme DPP IV. Crit. Rev. Clin. Lab Sci. 40 (3), 209–294. https://doi.org/10.1080/713609354.

Landim, A.P.M., Matsubara, N.K., da Silva-Santos, J.E., Mellinger-Silva, C., Rosenthal, A., 2021. Application of preliminary high-pressure processing for improving bioactive characteristics and reducing antigenicity of whey protein hydrolysates. Food Sci. Technol. Int. 28 (6), 489–501. https://doi.org/10.1177/10820132211022106.

Landim, A.P.M., Matsubara, N.K., da Silva-Santos, J.E., Mellinger-Silva, C., Rosenthal, A., 2022. Application of preliminary high-pressure processing for improving bioactive characteristics and reducing antigenicity of whey protein hydrolysates. Food Sci. Technol. Int. 28, 489–501. https://doi.org/10.1177/10820132211022106.

Levy, E., Marcil, V., Tagharist Ép Baumel, S., Dahan, N., Delvin, E., Spahis, S., 2023. Lactoferrin, osteopontin and lactoferrin-osteopontin complex: a critical look on their role in perinatal period and cardiometabolic disorders. Nutrients 15 (6), 1394. https://doi.org/10.3390/nu15061394.

Li, T., Gao, D., Du, M., Cheng, X., Mao, X., 2018. Casein glycomacropeptide hydrolysates inhibit PGE2 production and COX2 expression in LPS-stimulated RAW 264.7 macrophage cells via Akt mediated NF-κB and MAPK pathways. Food Funct. 9, 2524–2532.

Li, Y., Fan, Y., Liu, J., Meng, Z., Huang, A., Xu, F., Wang, X., 2023. Identification, characterization and in vitro activity of hypoglycemic peptides in whey hydrolysates from rubing cheese by-product. Food Res. Int. 164, 112382. https://doi.org/10.1016/j.foodres.2022.112382.

Li, W., Ren, Y., Wei, M., Liu, Q., Liu, Y., Zhang, C., Tan, C., Guo, C., Wang, S., Ning, C., 2025. Effect of in vitro simulated gastrointestinal digestion on the structural properties and antioxidant activity of donkey whey protein hydrolysates. J. Food Meas. Char. 19, 7284–7296. https://doi.org/10.1007/s11694-025-03461-w.

Ling, C.J., Min, Q.Q., Yang, J.R., Li, N.N., Luo, L.P., Pan, W., Xiong, Y.Y., Liu, J., Zhao, A.Z., Wang, Y.J., Ma, L.T., Wu, Z.H., 2019. Lactoferrin alleviates the progression of atherosclerosis in mice fed with high-fat/cholesterol diet through cholesterol homeostasis. J. Med. Food 22 (10), 1000–1008. https://doi.org/10.1089/jmf.2018.4389.

Liu, Z., Udenigwe, C.C., 2018. Role of food-derived opioid peptides in the central nervous and gastrointestinal systems. J. Food Biochem. 43 (1), e12629. https://doi.org/10.1111/jfbc.12629.

Liu, L., Li, S., Zheng, J., Bu, T., He, G., Wu, J., 2020. Safety considerations on food protein-derived bioactive peptides. Trends Food Sci. Technol. 96, 199–207. https://doi.org/10.1016/j.tifs.2019.12.022.

Liu, C., Kong, L., Yu, P., Wen, R., Yu, X., Xu, X., Peng, X., 2022. Whey protein hydrolysates improved the oxidative stability and water-holding capacity of pork patties by reducing protein aggregation during repeated freeze–thaw cycles. Foods 11 (14), 2133. https://doi.org/10.3390/foods11142133.

Luiking, Y.C., Abrahamse, E., Ludwig, T., Boirie, Y., Verlaan, S., 2016. Protein type and caloric density of protein supplements modulate postprandial amino acid profile through changes in gastrointestinal behaviour: a randomized trial. Clin. Nutr. 35 (1). https://doi.org/10.1016/j.clnu.2015.02.013.

Luparelli, A., Trisciuzzi, D., Schirinzi, W.M., Caputo, L., Smiriglia, L., Quintieri, L., Nicolotti, O., Monaci, L., 2025. Whey proteins and bioactive peptides: advances in production, selection and bioactivity profiling. Biomedicines 13, 1311. https://doi.org/10.3390/BIOMEDICINES13061311.

Ma, X., Yang, F., Meng, X., Wu, Y., Tong, P., Gao, J., Chen, H., Li, X., 2022. Immunomodulatory role of BLG-derived peptides based on simulated gastrointestinal digestion and DC-T cell from mice allergic to cow's milk. Foods 11, 1450.

Madadlou, A., Abbaspourrad, A., 2018. Bioactive whey peptide particles: an emerging class of nutraceutical carriers. Crit. Rev. Food Sci. Nutr. 58 (9), 1468–1477.

Malone, A., Clark, R.F., Hoskin, D.W., Power Coombs, M.R., 2022. Regulation of macrophage-associated inflammatory responses by species-specific lactoferricin peptides. Front. Biosci. (Landmark Ed) 27, 43.

Mansinhbhai, C.H., Sakure, A., Liu, Z., Maurya, R., Das, S., Basaiawmoit, B., Bishnoi, M., Kondepudi, K.K., Padhi, S., Rai, A.K., Mishra, B.K., Hati, S., 2023. Anti-inflammatory, ACE inhibitory, antioxidative activities and release of novel antihypertensive and antioxidative peptides from whey protein hydrolysate with molecular interactions. J. Am. Nutr. Assoc. 42 (4), 371–385. https://doi.org/10.1080/07315724.2022.2052201.

Martin, M., Hagemann, D., Nguyen, T.T., Schwarz, L., Khedr, S., Moskopp, M.L., Henle, T., Deussen, A., 2020. Plasma concentrations and ACE-inhibitory effects of tryptophan-containing peptides from whey protein hydrolysate in healthy volunteers. Eur. J. Nutr. 59, 1135–1147. https://doi.org/10.1007/s00394-019-01974-x.

Martini, S., Solieri, L., Cattivelli, A., Pizzamiglio, V., Tagliazucchi, D., 2021. An integrated peptidomics and in silico approach to identify novel anti-diabetic peptides in Parmigiano-Reggiano cheese. Biology 10 (6), 563. https://doi.org/10.3390/biology10060563.

Marwan-Abdelbaset, E., Lu, X., Samy-Kamal, M., Tan, D., 2025. From waste to value: microbial valorization of low-cost renewable resources for hyaluronic acid production. Waste Manag. 209, 115179. https://doi.org/10.1016/j.wasman.2025.115179.

Mayerhofer, H., Zednik, K., Pali-Schöll, I., 2022. The extended farm effect: the milk protein β-lactoglobulin in stable dust protects against allergies. ALS 6, 111–117.

Mazorra-Manzano, M.A., Robles-Porchas, G.R., González-Velázquez, D.A., Torres-Llanez, M.J., Martínez-Porchas, M., García-Sifuentes, C.O., González-Córdova, A.F., Vallejo-Córdoba, B., 2020. Cheese whey fermentation by its native microbiota: proteolysis and bioactive peptides release with ACE-inhibitory activity. Fermentation 6 (1), 19. https://doi.org/10.3390/fermentation6010019.

Mazoyer, E., Lévy-Toledano, S., Rendu, F., Hermant, L., Lu, H., Fiat, A.M., Jollès, P., Caen, J., 1990. KRDS, a new peptide derived from human lactotransferrin, inhibits platelet aggregation and release reaction. Eur. J. Biochem. 194 (1), 43–49. https://doi.org/10.1111/j.1432-1033.1990.tb19424.x.

Mignone, L.E., Wu, T., Horowitz, M., Rayner, C.K., 2015. Whey protein: the "whey" forward for treatment of type 2 diabetes? World J. Diabetes 6 (14), 1274–1284. https://doi.org/10.4239/wjd.v6.i14.1274.

Mignone, E., Ephraums, L.A., Routley, A., Judith, M., Emily, H., Chapman, S., Gagliardi, L., Mahesh, U., 2025. MON-535 glucagon-like-peptide-1 receptor agonists are effective in patients with cystic fibrosis. J. Endocrine Soc. 9 (Supplement_1), 2472–2972. https://doi.org/10.1210/jendso/bvaf149.902.

Mosele, J.I., Macià, A., Motilva, M.-J., 2015. Metabolic and microbial modulation of the large intestine ecosystem by non-absorbed diet phenolic compounds: a review. Molecules 20, 17429–17468.
Murali, C., Mudgil, P., Gan, C.-Y., Tarazi, H., El-Awady, R., Abdalla, Y., Amin, A., Maqsood, S., 2021. Camel whey protein hydrolysates induced G2/M cell cycle arrest in human colorectal carcinoma. Sci. Rep. 11, 7062. https://doi.org/10.1038/s41598-021-86391-z.
Neto, Y.A.A.H., Garzon, N.G.R., Coitinho, L.B., Sobral, L.M., Leopoldino, A.M., Cataldi, T.R., Labate, C.A., Cabral, H., 2022. Fungal metalloprotease generate whey-derived peptides that may be involved in apoptosis in B16F10 melanoma cells. Food Sci. Technol. 42. https://doi.org/10.1590/fst.43022.
Nishiyama, K., Mukai, T., 2022. Sialic acid impact on the gut microbiome and function. Glycoforum 25, A3.
Nongonierma, A.B., Cadamuro, C., Le Gouic, A., Mudgil, P., Maqsood, S., FitzGerald, R.J., 2019. Dipeptidyl peptidase IV (DPP-IV) inhibitory properties of a camel whey protein enriched hydrolysate preparation. Food Chem. 279, 70–79. https://doi.org/10.1016/j.foodchem.2018.11.142.
Nooshkam, M., Madadlou, A., 2016. Microwave-assisted isomerisation of lactose to lactulose and Maillard conjugation of lactulose and lactose with whey proteins and peptides. Food Chem. 200, 1–9. https://doi.org/10.1016/j.foodchem.2015.12.094.
Nyberg, F., 2006. Casomorphins and hemorphins—opioid active peptides released by partial hydrolysis of structural proteins. In: Handbook of Biologically Active Peptides. Elsevier, pp. 1339–1344. https://doi.org/10.1016/b978-012369442-3/50189-6.
Olsen, W., Liang, N., Dallas, D.C., 2023. Macrophage-immunomodulatory actions of bovine whey protein isolate, glycomacropeptide, and their in vitro and in vivo digests. Nutrients 15, 4942.
Olvera-Rosales, L.B., Cruz-Guerrero, A.E., García-Garibay, J.M., Gómez-Ruíz, L.C., Contreras-López, E., Guzmán-Rodríguez, F., González-Olivares, L.G., 2022. Bioactive peptides of whey: obtaining, activity, mechanism of action, and further applications. Crit. Rev. Food Sci. Nutr. 63 (30), 10351–10381. https://doi.org/10.1080/10408398.2022.2079113.
Olvera-Rosales, L.B., Cruz-Guerrero, A.E., García-Garibay, J.M., Gómez-Ruíz, L.C., Contreras-López, E., Guzmán-Rodríguez, F., González-Olivares, L.G., 2023. Bioactive peptides of whey: obtaining, activity, mechanism of action, and further applications. Crit. Rev. Food Sci. Nutr. 63 (30), 10351–10381. https://doi.org/10.1080/10408398.2022.2079113.
Österlund, N., Vosselman, T., Leppert, A., Gräslund, A., Jörnvall, H., Ilag, L.L., Marklund, E.G., Elofsson, A., Johansson, J., Sahin, C., Landreh, M., 2022. Mass spectrometry and machine learning reveal determinants of client recognition by antiamyloid chaperones. Mol. Cell. Proteomics 21 (10), 100413. https://doi.org/10.1016/j.mcpro.2022.100413.
Ouyang, K., Xie, H., Wang, Y., Woo, M.W., Chen, Q., Lai, S., Xiong, H., Zhao, Q., 2023. Whey protein isolate nanofibrils formed with phosphoric acid: formation, structural characteristics, and emulsion stability. Food Hydrocoll. 135, 108170. https://doi.org/10.1016/J.FOODHYD.2022.108170.
Palmer, C.B., Meyrath, M., Canals, M., Kostenis, E., Chevigné, A., Szpakowska, M., 2022. Atypical opioid receptors: unconventional biology and therapeutic opportunities. Pharmacol. Therapeut. 233, 108014. https://doi.org/10.1016/j.pharmthera.2021.108014.
Pang, L., Liu, M., Chen, C., Huang, Z., Liu, S., Man, C., Jiang, Y., Zhang, W., Yang, X., 2024a. Effects of ultrasound pretreatment on the structure, IgE binding capacity, functional properties and bioactivity of whey protein hydrolysates via multispectroscopy and peptidomics revealed. Ultrason. Sonochem. 110, 107025. https://doi.org/10.1016/j.ultsonch.2024.107025.
Pang, L., Zhang, W., Chen, C., Li, R., Huang, Z., Liu, S., Zhao, K., Zhao, F., Yang, X., Jiang, Y., 2024b. Effect of enzymatic hydrolysis and ultrasound pretreatment on allergenicity, functional properties and bioactivity of whey protein isolates. Lwt 210, 116891. https://doi.org/10.1016/j.lwt.2024.116891.
Pang, L., Huang, Z., Li, R., Su, Y., Shi, J., Yang, X., Jiang, Y., 2025. Reducing the allergenicity of whey proteins while improving their functional properties and bioactivity using combined enzymes. Food Chem. 481, 144052. https://doi.org/10.1016/j.foodchem.2025.144052.
Patil, P., Mandal, S., Tomar, S.K., Anand, S., 2015. Food protein-derived bioactive peptides in management of type 2 diabetes. Eur. J. Nutr. 54 (6), 863–880. https://doi.org/10.1007/s00394-015-0974-2.
Patil, P.J., Usman, M., Zhang, C., Mehmood, A., Zhou, M., Teng, C., Li, X., 2022. An updated review on food-derived bioactive peptides: focus on the regulatory requirements, safety, and bioavailability. Compr. Rev. Food Sci. Food Saf. 21 (2), 1732–1776. https://doi.org/10.1111/1541-4337.12911.
Pellegrini, A., Thomas, U., Bramaz, N., Hunziker, P., Von Fellenberg, R., 2001. Isolation and identification of three bactericidal domains in the bovine alpha-lactalbumin molecule. Biochim. Biophys. Acta (BBA) - General Subjects 1526 (2), 131–140. https://doi.org/10.1016/S0304-4165(01)00137-8.
Peng, X., Li, Y., Yu, J., Gao, Y., Zhao, X., Jia, N., 2024. Assessment of the impact of whey protein hydrolysate on myofibrillar proteins in surimi during repeated freeze–thaw cycles: quality enhancement and antifreeze potential. Food Chem. 460, 140552. https://doi.org/10.1016/j.foodchem.2024.140552.
Pepe, G., Tenore, G.C., Mastrocinque, R., Stusio, P., Campiglia, P., 2013. Potential anticarcinogenic peptides from bovine milk. J. Amino Acids 2013, 939804. https://doi.org/10.1155/2013/939804.
Phelan, M., Aherne, A., Fitz Gerald, R.J., O'Brien, N.M., 2009. Bioactive peptides from milk proteins: a review. Int. Dairy J. 19 (11), 643–654. https://doi.org/10.1016/j.idairyj.2009.06.005.
Pihlanto-Leppälä, A., 1997. Bioactive peptide derived from in vitro proteolysis of bovine β-lactoglobulin and its effect on smooth muscle. J. Dairy Res. 64 (1), 149–155. https://doi.org/10.1017/S0022029996001926.
Pihlanto-Leppälä, A., 2001. Bioactive peptides derived from bovine whey proteins. Trends Food Sci. Technol. 11 (9–10), 347–356. https://doi.org/10.1016/s0924-2244(01)00003-6.
Piotto, S.P., Sessa, L., Concilio, S., Iannelli, P., 2012. Yadamp: yet another database of antimicrobial peptides. Int. J. Antimicrob. Agents 39, 346–351. https://doi.org/10.1016/j.ijantimicag.2011.12.003.

Pipaliya, R., Basaiawmoit, B., Sakure, A.A., Maurya, R., Bishnoi, M., Kondepudi, K.K., Hati, S., 2025. Peptidomics and molecular dynamics on bioactive peptides produced and characterized from the fermented whey of "Panchali" sheep of West India. Food Chem. 468, 142466. https://doi.org/10.1016/j.foodchem.2024.142466.

Pontonio, E., Montemurro, M., De Gennaro, G.V., Miceli, V., Rizzello, C.G., 2021. Antihypertensive peptides from ultrafiltration and fermentation of the ricotta cheese exhausted whey: design and characterization of a functional ricotta cheese. Foods 10 (11), 2573. https://doi.org/10.3390/foods10112573.

Power, O., Hallihan, A., Jakeman, P., 2009. Human insulinotropic response to oral ingestion of native and hydrolysed whey protein. Amino Acids 37 (2), 333–339. https://doi.org/10.1007/s00726-008-0156-0.

Pulendran, B., Davis, M.M., 2020. The science and medicine of human immunology. Science 369, eaay4014.

Punia, H., Tokas, J., Malik, A., Sangwan, S., Baloda, S., Singh, N., Singh, S., Bhuker, A., Singh, P., Yashveer, S., Agarwal, S., Mor, V.S., 2020. Identification and detection of bioactive peptides in milk and dairy products: remarks about agro-foods. Molecules 25, 3328. https://doi.org/10.3390/molecules25153328.

Purohit, K., Reddy, N., Sunna, A., 2024. Exploring the potential of bioactive peptides: from natural sources to therapeutics. IJMS 25, 1391. https://doi.org/10.3390/ijms25031391.

Qian, J., Chen, D., Zhang, Y., Gao, X., Xu, L., Guan, G., Wang, F., 2023. Ultrasound-assisted enzymatic protein hydrolysis in food processing: mechanism and parameters. Foods 12, 4027. https://doi.org/10.3390/FOODS12214027.

Qian, Z.Y., Jollès, P., Migliore-Samour, D., Schoentgen, F., Fiat, A.M., 1995. Sheep κ-casein peptides inhibit platelet aggregation. Biochim. Biophys. Acta (BBA) – General Subjects 1244 (2–3), 411–417. https://doi.org/10.1016/0304-4165(95)00047-F.

Quintieri, L., Luparelli, A., Caputo, L., Schirinzi, W., De Bellis, F., Smiriglia, L., Monaci, L., 2025. Unraveling the biological properties of whey peptides and their role as emerging therapeutics in immune tolerance. Nutrients 17 (6), 938. https://doi.org/10.3390/nu17060938.

Rackerby, B., Le, H.N.M., Haymowicz, A., Dallas, D.C., Park, S.H., 2024. Potential prebiotic properties of whey protein and glycomacropeptide in gut microbiome. Food Sci. Anim. Resour. 44, 299–308.

Rakvaag, E., Fuglsang-Nielsen, R., Bach Knudsen, K.E., Purup, S., Larsen, M.K., Hermansen, K., Gregersen, S., 2019. Whey protein combined with low dietary fiber improves lipid profile in subjects with abdominal obesity: a randomized, controlled trial. Nutrients 11 (9), 2091. https://doi.org/10.3390/nu11092091.

Ramani, A., Hazra, T., Mudgil, S., Mudgil, D., 2024. Emerging potential of whey proteins in prevention of cancer. Food Human. 2, 100199. https://doi.org/10.1016/j.foohum.2023.12.007.

Regueiro, U., López-López, M., Varela-Fernández, R., Sobrino, T., Diez-Feijoo, E., Lema, I., 2022. Immunomodulatory effect of human lactoferrin on toll-like receptors 2 expression as therapeutic approach for keratoconus. IJMS 23, 12350.

Ren, Q., Ma, X., Keijzer, P., Bastiaan-Net, S., Wichers, H.J., Hettinga, K.A., 2025. Comparative in vitro intestinal transport and bioactivity of peptides from wet- and dry-heated goat milk proteins after simulated infant digestion. Food Chem. 495, 146326. https://doi.org/10.1016/J.FOODCHEM.2025.146326.

Ren, X., Wu, C., Hong, H., Luo, Y., Tan, Y., 2025. Behavioral and proteomics studies on the regulation of response speed in mice by whey protein hydrolysate intervention. Nutrients 17 (15), 2500. https://doi.org/10.3390/nu17152500.

Renaud, V., Faucher, M., Dubois, M.J., Pilon, G., Varin, T., Marette, A., Bazinet, L., 2023. Impact of a whey protein hydrolysate treated by electrodialysis with ultrafiltration membrane on the development of metabolic syndrome and the modulation of gut microbiota in mice. Int. J. Mol. Sci. 24 (16), 12968. https://doi.org/10.3390/ijms241612968.

Rendon-Rosales, M.A., Torres-Llanez, M.J., González-Córdova, A.F., Hernández-Mendoza, A., Mazorra-Manzano, M.A., Vallejo-Cordoba, B., 2019. In vitro antithrombotic and hypocholesterolemic activities of milk fermented with specific strains of Lactococcus lactis. Nutrients 11 (9), 2150. https://doi.org/10.3390/nu11092150.

Riaz Rajoka, M.S., Thirumdas, R., Mehwish, H.M., Umair, M., Khurshid, M., Hayat, H.F., Phimolsiripol, Y., Pallarés, N., Martí-Quijal, F.J., Barba, F.J., 2021. Role of food antioxidants in modulating gut microbial communities: novel understandings in intestinal oxidative stress damage and their impact on host health. Antioxidants 10, 1563.

Rival, S.G., Boeriu, C.G., Wichers, H.J., 2001. Caseins and casein hydrolysates. 2. Antioxidative properties and relevance to lipoxygenase inhibition. J. Agric. Food Chem. 49 (1), 295–302. https://doi.org/10.1021/jf000700b.

Roobab, U., Abida, A., Chacha, J.S., Athar, A., Madni, G.M., Ranjha, M.M.A.N., Rusu, A.V., Zeng, X.A., Aadil, R.M., Trif, M., 2022. Applications of innovative non-thermal pulsed electric field technology in developing safer and healthier fruit juices. Molecules 27, 4031. https://doi.org/10.3390/MOLECULES27134031.

Rosa, L.S., Santos, M.L., Abreu, J.P., Rocha, R.S., Esmerino, E.A., Freitas, M.Q., Teodoro, A.J., 2023. Probiotic fermented whey-milk beverages: effect of different probiotic strains on the physicochemical characteristics, biological activity, and bioactive peptides. Food Res. Int. 164, 112396. https://doi.org/10.1016/j.foodres.2022.112396.

Roth-Walter, F., Afify, S.M., Pacios, L.F., Blokhuis, B.R., Redegeld, F., Regner, A., Petje, L.-M., Fiocchi, A., Untersmayr, E., Dvorak, Z., Hufnagl, K., Pali-Schöll, I., Jensen-Jarolim, E., 2021. Cow's milk protein β-lactoglobulin confers resilience against allergy by targeting complexed iron into immune cells. J. Allergy Clin. Immunol. 147, 321–334.e4.

Saadi, S., Makhlouf, C., Nacer, N.E., Halima, B., Faiza, A., Kahina, H., Wahiba, F., Afaf, K., Rabah, K., Saoudi, Z., 2024. Whey proteins as multifunctional food materials: recent advancements in hydrolysis, separation, and peptidomimetic approaches. Compr. Rev. Food Sci. Food Saf. 23, 1–40. https://doi.org/10.1111/1541-4337.13288;JOURNAL:JOURNAL:15414337;WGROUP:STRING:PUBLICATION.

Sahna, K.O., Cakir, B., Tunali-Akbay, T., 2022. Antiproliferative activity of whey and casein bioactive peptides on breast cancer: an in vitro and in silico study. Int. J. Pept. Res. Ther. 28, 128. https://doi.org/10.1007/s10989-022-10436-8.

Sakkas, L., Lekaki, E., Moatsou, G., 2022. Whey protein hydrolysates of sheep/goat origin produced by the action of trypsin without pH control: degree of hydrolysis, antihypertensive potential and antioxidant activities. Foods 11 (14), 2103. https://doi.org/10.3390/foods11142103.

Sánchez, A., Vázquez, A., 2017. Bioactive peptides: a review. Food Qual. Saf. 1, 29–46. https://doi.org/10.1093/fqs/fyx006.

Santos, J.D.dos, Martins, L.C., Silva, T.F., 2021. Production of multifunctional peptides from casein using extended fermentation and their effects on MMP inhibition and antibacterial activity. Int. J. Peptide Res. Ther. 27, 305–314. https://doi.org/10.1007/s10989-020-10103-1.

Sar, T., Bogovic Matijasic, B., Danilovic, B., Gamero, A., Gandía, M., Krausova, G., Martínez-Villaluenga, C., Peñas, E., Bagherzadehsurbagh, E., Cemali, Ö., Santa, D., Künili, I.E., Kesenkas, H., Chassard, C., Pracer, S., Vergères, G., Ergün, B.G., 2025. A systematic review of health promoting effects of consumption of whey-based fermented products on adults. Front. Nutr. 12, 1651365. https://doi.org/10.3389/fnut.2025.1651365.

Saubenova, M., Oleinikova, Y., Rapoport, A., Maksimovich, S., Yermekbay, Z., Khamedova, E., 2024. Bioactive peptides derived from whey proteins for health and functional beverages. Fermentation 10, 359. https://doi.org/10.3390/FERMENTATION10070359.

Segura Campos, M., Chel Guerrero, L., Betancur Ancona, D., Hernández Escalante, V.M., 2011. Bioavailability of bioactive peptides. Food Rev. Int. 27 (3), 213–226. https://doi.org/10.1080/87559129.2011.563395.

Shadrack, S.M., Wang, Y., Mi, S., Lu, R., Zhu, Y., Tang, Z., McClements, D.J., Cao, C., Xu, X., Li, W., Yuan, B., 2025. Enhancing bioavailability and functionality of plant peptides and proteins: a review of novel strategies for food and pharmaceutical applications. Food Chem. 485, 144440. https://doi.org/10.1016/j.foodchem.2025.144440.

Shahidi, F., Saeid, A., 2025. Bioactivity of marine-derived peptides and proteins: a review. Mar. Drugs 23, 157. https://doi.org/10.3390/md23040157.

Shahidi, F., Zhong, Y., 2008. Bioactive peptides. J. AOAC Int. 91 (4), 914–931. https://doi.org/10.1093/jaoac/91.4.914.

Shao, Y. hong, Zou, L., Xiong, Z. hao, Su, L., xia, Tu, Z. cai, Liu, J., 2024. Simulated in vitro gastrointestinal digestion of β-lactoglobulin treated by ultrasound: detection of peptides profile and the antioxidant activity. Food Res. Int. 175, 113763. https://doi.org/10.1016/J.FOODRES.2023.113763.

Shen, W., Matsui, T., 2017. Current knowledge of intestinal absorption of bioactive peptides. Food Funct. 8 (12), 4306–4314. https://doi.org/10.1039/C7FO01185G.

Shi, C., Liu, J., Deng, Y., Luo, S., Liu, C., Liu, Y., 2025. Enhanced calcium and thermal stability of whey protein hydrolysate stabilized emulsions by ultrasound-assisted glycosylation. Ultrason. Sonochem. 119, 107413. https://doi.org/10.1016/j.ultsonch.2025.107413.

Silva, S.O., Mafra, A.K.C., Pelissari, F.M., Rodrigues de Lemos, L., Molina, G., 2025. Biotechnology in agro-industry: valorization of agricultural wastes, by-products and sustainable practices. Microorganisms 13 (8), 1789. https://doi.org/10.3390/microorganisms13081789.

Singh, B.P., 2025. Harnessing probiotic fermentation for the generation of food-derived bioactive peptides: current status and prospects. Eur. Food Res. Technol. 2061–2076. https://doi.org/10.1007/S00217-025-04774-2.

Singh, N., Gaur, S., 2024. New insights into multifunctional aspects of milk derived bioactive peptides: a review. Food Chem. Adv. 4, 100628. https://doi.org/10.1016/j.focha.2024.100628.

Singh, B.P., Aluko, R.E., Hati, S., Solanki, D., 2022. Bioactive peptides in the management of lifestyle-related diseases: current trends and future perspectives. Crit. Rev. Food Sci. Nutr. 62 (17), 4593–4606. https://doi.org/10.1080/10408398.2021.1877109.

Solieri, L., Valentini, M., Cattivelli, A., Sola, L., Helal, A., Martini, S., Tagliazucchi, D., 2022. Fermentation of whey protein concentrate by streptococcus thermophilus strains releases peptides with biological activities. Process Biochem. 121, 590–600. https://doi.org/10.1016/J.PROCBIO.2022.08.003.

Soltani, M., Rabbani, M., Mofid, V., Mohammad Mortazavian, S.A., 2025. Whey permeate-based isotonic beverage formulation containing whey protein probiotic and peptides Co-encapsulated in alginate Matrix. J. Food Sci. Technol. 22 (162). https://doi.org/10.22034/FSCT.22.162.16.

Stožinić, M.V., Zarić, D.B., Rakin, M.B., Lončarević, I.S., Pajin, B.S., Bulatović, M.L., 2024. Impact of whey bioactive hydrolysates on the quality of fat fillings for confectionery products. Food Feed Res. 51 (2), 189–198.

Sumny, E.H., Kempka, A.P., 2025. Bioactive whey peptides: a bibliometric and functional overview. Discov. Food 2025 51 (5), 1–32. https://doi.org/10.1007/S44187-025-00430-6.

Sun, Y., Wang, X., Li, L., Zhong, C., Zhang, Y., Yang, X., Li, M., Yang, C., 2024. The role of gut microbiota in intestinal disease: from an oxidative stress perspective. Front. Microbiol. 15.

Sutantawong, S., Kim, B.J., Kuhfeld, R.F., Qu, Y., Dallas, D.C., 2024. Release of bioactive peptides from whey proteins across digestion in adult humans and a static in vitro digestion model. J. Funct.Foods 122, 106540. https://doi.org/10.1016/J.JFF.2024.106540.

Sutantawong, S., Jin Kim, B., Yunyao, Q., Dallas, D.C., 2025. Survival of intact bovine whey proteins across in vivo and simulated static in vitro models of the adult gastrointestinal tract. Food Chem. 465, 142013. https://doi.org/10.1016/J.FOODCHEM.2024.142013.

Sweitzer, N.K., 2003. What is an angiotensin converting enzyme inhibitor? Circulation 108. https://doi.org/10.1161/01.CIR.0000075957.16003.07.

Tadesse, S.A., Emire, S.A., 2020. Production and processing of antioxidant bioactive peptides: a driving force for the functional food market. Heliyon 6, e04765. https://doi.org/10.1016/j.heliyon.2020.e04765.

Taesuwan, S., Jirarattanarangsri, W., Wangtueai, S., Hussain, M.A., Ranadheera, S., Ajlouni, S., Zubairu, I.K., Naumovski, N., Phimolsiripol, Y., 2024. Unexplored opportunities of utilizing food waste in food product development for cardiovascular health. Cur. Nutr. Rep. 13 (4), 896–913. https://doi.org/10.1007/s13668-024-00571-7.

Taghipour, M.J., Ezzatpanah, H., Ghahderijani, M., 2023. In vitro and in silico studies for the identification of anti-cancer and antibacterial peptides from camel milk protein hydrolysates. PLoS One 18 (7), e0288260. https://doi.org/10.1371/journal.pone.0288260.

Tai, C.S., Chen, Y.Y., Chen, W.L., 2016. β -lactoglobulin influences human immunity and promotes cell proliferation. BioMed Res. Int. 2016, 1–12.

Tenenbaum, M., Dugardin, C., Moro, J., Auger, J., Baniel, A., Boulier, A., Cudennec, B., 2023. In vitro comparison of whey protein isolate and hydrolysate for their effect on glucose homeostasis markers. Food Funct. 14 (9), 4173–4182. https://doi.org/10.1039/d3fo00467h.

Teschemacher, H., Koch, G., Brantl, V., 1997. Milk protein-derived opioid receptor ligands. Biopolymers 43, 99–117.

Thampy, A., Kumar, M.K.P., Peddha, M.S., Reddy, M., 2024. The effectiveness of whey proteins in prevention and treatment of cancer: a review. Crit. Rev. Food Sci. Nutr. 64 (8), 2088–2104. https://doi.org/10.1080/10408398.2022.2121256.

Tovar-Jiménez, X., Téllez-Jurado, A., Gómez-Aldapa, C.A., Mercado-Flores, Y., Arana-Cuenca, A., 2021. Antioxidant and antihypertensive activity of bovine whey protein concentrate enzymatic hydrolysates. Biotecnia 23 (1). https://doi.org/10.18633/biotecnia.v23i1.1321.

Tsafack, P.B., Li, C., Tsopmo, A., 2022. Food peptides, gut microbiota modulation, and antihypertensive effects. Molecules 27 (24), 8806. https://doi.org/10.3390/molecules27248806.

Tulipano, G., Faggi, L., Nardone, A., Cocchi, D., Caroli, A.M., 2015. Characterisation of the potential of β-lactoglobulin and α-lactalbumin as sources of bioactive peptides affecting incretin function: in silico and in vitro comparative studies. Int. Dairy J. 48, 66–72. https://doi.org/10.1016/j.idairyj.2015.01.008.

Udenigwe, C.C., Aluko, R.E., 2012. Food protein-derived bioactive peptides: production, processing, and potential health benefits. J. Food Sci. 77 (1), 11–24. https://doi.org/10.1111/j.1750-3841.2011.02455.x.

Van der Sluis, M., De Koning, B.A.E., De Bruijn, A.C.J.M., Velcich, A., Meijerink, J.P.P., Van Goudoever, J.B., Büller, H.A., Dekker, J., Van Seuningen, I., Renes, I.B., Einerhand, A.W.C., 2006. Muc2-deficient mice spontaneously develop colitis, indicating that MUC2 is critical for colonic protection. Gastroenterology 131, 117–129.

Vashisht, P., Singh, L., Mahanta, S., Verma, D., Sharma, S., Saini, G.S., Sharma, A., Chowdhury, B., Awasti, N., Gaurav, M.,S., Chauhan, D.S., 2024. Pulsed electric field processing in the dairy sector: a review of applications, quality impact and implementation challenges. Int. J. Food Sci. Technol. 59, 2122–2135. https://doi.org/10.1111/IJFS.16979.

Vega-Bautista, A., de la Garza, M., Carrero, J.C., Campos-Rodríguez, R., Godínez-Victoria, M., Drago-Serrano, M.E., 2019. The impact of lactoferrin on the growth of intestinal inhabitant bacteria. Int. J. Mol. Sci. 20, 4707.

Wang, K., Fu, Z., Li, X., Zhang, L., Li, J., Li, Y., Mao, X., Yu, L., 2022. Whey protein hydrolysate alleviated atherosclerosis and hepatic steatosis by regulating lipid metabolism in apoE− mice fed a Western diet. Food Res. Int. 157, 111419. https://doi.org/10.1016/j.foodres.2022.111419.

Wijesekara, T., Abeyrathne, E.D.N.S., Ahn, D.U., 2024. Effect of bioactive peptides on gut microbiota and their relations to human health. Foods 13, 1853. https://doi.org/10.3390/foods13121853.

Wong, K.F., Middleton, N., Montgomery, M., Dey, M., Carr, R.I., 1998. Immunostimulation of murine spleen cells by materials associated with bovine milk protein fractions. J. Dairy Sci. 81, 1825–1832.

World Health Organization, 2025. Diabetes. World Health Organization. https://www.who.int/news-room/fact-sheets/detail/diabetes.

Xiao, R., Carter, J.A., Linz, A.L., Ferguson, M., Badger, T.M., Simmen, F.A., 2006. Dietary whey protein lowers serum C-peptide concentration and duodenal SREBP-1c mRNA abundance, and reduces occurrence of duodenal tumors and colon aberrant crypt foci in azoxymethane-treated male rats. J. Nutr. Biochem. 17, 626–634.

Xie, D., Shen, Y., Su, E., Du, L., Xie, J., Wei, D., 2022. The effects of angiotensin I-converting enzyme inhibitory peptide VGINYW and the hydrolysate of α-lactalbumin on blood pressure control, oxidative stress and gut microbiota of spontaneously hypertensive rats. Food Funct. 13 (5). https://doi.org/10.1039/D1FO03570C.

Xu, S., Fan, F., Liu, H., Cheng, S., Tu, M., Du, M., 2020. 'Novel anticoagulant peptide from lactoferrin binding thrombin at the active site and exosite-I. J. Agric. Food Chem. 68 (10), 3132–3139. https://doi.org/10.1021/acs.jafc.9b08094.

Xue, H., Han, J., He, B., Yi, M., Liu, X., Song, H., Li, J., 2021. Bioactive peptide release and the absorption tracking of casein in the gastrointestinal digestion of rats. Food Funct. 12, 5157–5170. https://doi.org/10.1039/D1FO00356A.

Yami, H.A., Tahmoorespur, M., Javadmanesh, A., Tazarghi, A., Sekhavati, M.H., 2023. The immunomodulatory effects of lactoferrin and its derived peptides on NF-κB signaling pathway: a systematic review and meta-analysis. Immun. Inflamm. Dis. 11, e972.

Ye, H., Tao, X., Zhang, W., Chen, Y., Yu, Q., Xie, J., 2022. Food-derived bioactive peptides: production, biological activities, opportunities and challenges. J. Fut. Foods 2, 294–306. https://doi.org/10.1016/j.jfutfo.2022.08.002.

Yu, Y.-J., Amorim, M., Marques, C., Calhau, C., Pintado, M., 2016. Effects of whey peptide extract on the growth of probiotics and gut microbiota. J. Funct.Foods 21, 507–516.

Yu, X.-C., Li, Z., Liu, X.-R., Hu, J.-N., Liu, R., Zhu, N., Li, Y., 2021. The antioxidant effects of whey protein peptide on learning and memory improvement in aging mice models. Nutrients 13 (6), 2100. https://doi.org/10.3390/nu13062100.

Yukalo, V.G., Datsyshyn, K.Y., Shkilna, M.B., 2022. Low allergenic fermented drink enriched with bioactive peptides of whey proteins. Scientific Messenger of LNU of Veterinary Medicine and Biotechnologies. Series. Food Technol. 24 (97), 20–26. https://doi.org/10.32718/nvlvet-f9704.

Zaky, A.A., Simal-Gandara, J., Eun, J.-B., Shim, J.-H., Abd El-Aty, A.M., 2022. Bioactivities, applications, safety, and health benefits of bioactive peptides from food and by-products: a review. Front. Nutr. 8, 815640. https://doi.org/10.3389/fnut.2021.815640.

Zhang, W., Xiao, D., Mao, Q., Xia, H., 2023. Role of neuroinflammation in neurodegeneration development. Signal Transduct. Target. Ther. 8 (1), 267. https://doi.org/10.1038/s41392-023-01486-5.

Zhao, C., Ashaolu, T.J., 2020. Bioactivity and safety of whey peptides. LWT 134, 109935. https://doi.org/10.1016/j.lwt.2020.109935.

Zhao, Q., Chen, Y.-Y., Xu, D.-Q., Yue, S.-J., Fu, R.-J., Yang, J., Xing, L.-M., Tang, Y.-P., 2021. Action mode of gut motility, fluid and electrolyte transport in chronic constipation. Front. Pharmacol. 12.

Zhao, C., Chen, N., Ashaolu, T.J., 2023. Prebiotic and modulatory evidence of lactoferrin on gut health and function. J. Funct.Foods 108, 105741.

Zhao, J., Guo, J., Xu, X., Song, X., Kong, W., Liu, M., Zhang, T., Zeng, X., Wu, Z., Tao, M., Luo, H., Su, Y., Ling, N., Pan, D., Guo, Y., 2025. Physiological and hypotensive effects of whey protein-derived angiotensin converting enzyme inhibitory peptides in spontaneously hypertensive rats. Food Biosci. 68, 106321. https://doi.org/10.1016/j.fbio.2025.106321.

Zioudrou, C., Streaty, R.A., Klee, W.A., 1979. Opioid peptides derived from food proteins. The exorphins. J. Biol. Chem. 254 (7), 2446–2449. https://doi.org/10.1016/s0021-9258(17)30243-0.

Złotkowska, D., Stachurska, E., Fuc, E., Wróblewska, B., Mikołajczyk, A., Wasilewska, E., 2021. Differences in regulatory mechanisms induced by β-lactoglobulin and κ-casein in cow's milk allergy mouse model–in vivo and ex vivo studies. Nutrients 13, 349.

Further reading

Aslanidis, A., Karlstetter, M., Scholz, R., Fauser, S., Neumann, H., Fried, C., Pietsch, M., Langmann, T., 2015. Activated microglia/macrophage whey acidic protein (AMWAP) inhibits NFκB signaling and induces a neuroprotective phenotype in microglia. J. Neuroinflammation 12, 77. https://doi.org/10.1186/s12974-015-0296-6.

Asledottir, T., Picariello, G., Mamone, G., Ferranti, P., Røseth, A., Devold, T.G., Vegarud, G.E., 2019. Degradation of β-casomorphin-7 through in vitro gastrointestinal and jejunal brush border membrane digestion. J. Dairy Sci. 102 (10), 8622–8629. https://doi.org/10.3168/jds.2019-16771.

Ayabe, T., Shinohara, M., Kita, M., Takahashi, C., Saito, J., Furuyashiki, T., Ano, Y., 2024. Supplementation with whey peptide rich in β-lactolin improves trait anxiety and subjective stress in healthy adults: a randomized, double-blind, placebo-controlled study. Sci. Rep. 14 (1), 23444. https://doi.org/10.1038/s41598-024-73780-3.

EFSA (European Food Safety Authority), 2009. Review of the potential health impact of β-casomorphins and related peptides. EFSA Sci. Rep. 231, 1–107. https://doi.org/10.2903/j.efsa.2009.231r.

Flis, Z., Molik, E., Ptak, A., Szatkowski, P., 2025. Natural bioactive compounds in sheep milk: potential biomedical applications. Curr. Issues Mol. Biol. 47 (6), 456. https://doi.org/10.3390/cimb47060456.

Geoffroy, T.R., Thibodeau, J., Faucher, M., Langevin, M.E., Lutin, F., Bazinet, L., 2022. Relationship between feed concentration and bioactive cationic peptide recovery: impact on ecoefficiency of EDUF at semi-industrial scale. Sep. Purif. Technol. 286, 120403. https://doi.org/10.1016/j.seppur.2021.120403.

Katayama, S., Nakamura, S., 2019. Emerging roles of bioactive peptides on brain health promotion. Int. J. Food Sci. Technol. 54 (6), 1949–1955. https://doi.org/10.1111/ijfs.14076.

Kita, M., Obara, K., Kondo, S., Umeda, S., Ano, Y., 2018. Effect of supplementation of a whey peptide rich in tryptophan-tyrosine-related peptides on cognitive performance in healthy adults: a randomized, double-blind, placebo-controlled study. Nutrients 10 (7), 899. https://doi.org/10.3390/nu10070899.

Kita, M., Kobayashi, K., Obara, K., Koikeda, T., Umeda, S., Ano, Y., 2019. Supplementation with whey peptide rich in β-lactolin improves cognitive performance in healthy older adults: a randomized, double-blind, placebo-controlled study. Front. Neurosci. 13, 448995.

Mansinhbhai, C.H., Sakure, A., Liu, Z., Maurya, R., Das, S., Basaiawmoit, B., 2023. Anti-inflammatory, ACE inhibitory, antioxidative activities and release of novel antihypertensive and antioxidative peptides from whey protein hydrolysate with molecular interactions. J. Am. Nutr. Assoc. 42 (4), 371–385. https://doi.org/10.1080/07315724.2022.2052201.

Park, Y.W., Nam, M.S., 2015. Bioactive peptides in milk and dairy products: a review. Korean journal for food science of animal resources 35 (6), 831–840. https://doi.org/10.5851/kosfa.2015.35.6.831.

Shi, C., Deng, Y., Wang, Z., Luo, S., Liu, C., Liu, Y., 2024. Investigating the thermal stability and calcium resistance of O/W emulsions prepared with glycosylated whey protein hydrolysates modified by different saccharides. Food Chem. 454, 139805. https://doi.org/10.1016/j.foodchem.2024.139805.

Sun, X., Acquah, C., Aluko, R.E., Udenigwe, C.C., 2020. Considering food matrix and gastrointestinal effects in enhancing bioactive peptide absorption and bioavailability. J. Funct.Foods 64, 103680. https://doi.org/10.1016/j.jff.2019.103680.

Suzuki, K., Okamatsu, Y., Uchida, R., Sasahara, I., Takeshita, M., Sato, W., Murakami, H., 2024. Effect of whey protein-derived decapeptide on mood status and blood variables in healthy adults: a randomized, double-blind, placebo-controlled cross-over trial. Eur. J. Nutr. 63 (7), 2789–2799. https://doi.org/10.1007/s00394-024-03464-1.

Zumbro, E.L., Rao, M., Balcom-Luker, S., Broughton, K.S., LeMieux, M.J., 2021. Whey protein supplementation improves the glycemic response and may reduce non-alcoholic fatty liver disease related biomarkers in women with polycystic ovary syndrome (PCOS). Nutrients 13 (7), 2451. https://doi.org/10.3390/nu13072451.

Chapter 20

Lactose Intolerance

Simona Dominici[1], Nico Donati[2], Mirko Tarocchi[2], Filippo Fassio[3] and Maria Sole Facioni[4, 5]
[1]ELLEFREE S.r.l., Polo Tecnologico Lucchese, Lucca, Italy; [2]Gastroenterology Unit, Department of Experimental and Clinical Biomedical Sciences, "Mario Serio", University of Florence, Florence, Italy; [3]SOSD Allergology and Clinical Immunology, San Giovanni di Dio Hospital, Azienda USL Toscana Centro, Florence, Italy; [4]AILI-Associazione Italiana Latto-Intolleranti Aps, Lucca, Italy; [5]Department of Clinical Medicine and Surgery, University of Naples "Federico II," Naples, Italy

1. Introduction

Lactose, a disaccharide composed of glucose and galactose and naturally occurring in mammalian milk, constitutes a fundamental carbohydrate in human nutrition and plays a pivotal role in the contemporary food and pharmaceutical industries. The wide-ranging functional versatility of this sugar is due to its unique physicochemical properties, which include the existence of two isomeric forms (α-lactose and β-lactose) and their distinct temperature-dependent solubility and crystallization behaviors. These characteristics enable lactose to serve in multiple technological capacities, such as a sweetener, Maillard reaction promoter, encapsulation matrix, cryoprotectant, and fermentation substrate. Consequently, lactose is a fundamental component in a wide range of products, encompassing baked goods, ice creams, processed meats, infant formulas, and various dairy derivatives.

Despite its functional and nutritional value, the utility of lactose is significantly constrained by the global prevalence of lactose intolerance, a condition arising from reduced or absent activity of the intestinal enzyme lactase (β-galactosidase). This enzymatic deficiency has been shown to impair lactose hydrolysis in the small intestine, resulting in gastrointestinal symptoms upon ingestion in affected individuals.

This chapter provides a comprehensive examination of lactose, encompassing both its industrial importance and the physiological and clinical dimensions associated with its consumption. The discussion commences with a comprehensive examination of the structural and physicochemical characteristics of lactose, subsequently progressing to an investigation of its multifarious industrial applications. In the ensuing discussion, the genetic bases of lactose intolerance are addressed, alongside the broader nutritional and public health implications of this condition.

A critical section is dedicated to differentiating lactose intolerance from cow's milk protein allergy—two frequently conflated conditions with fundamentally distinct immunopathological mechanisms, clinical manifestations, and diagnostic pathways. The chapter also presents an overview of both established and emerging diagnostic modalities for lactose intolerance, including hydrogen breath testing and genetic screening, while emphasizing the necessity of accurate differential diagnosis in the context of overlapping gastrointestinal disorders such as celiac disease and irritable bowel syndrome.

In the subsequent sections, evidence-based dietary management strategies are discussed, focusing on the role of naturally low-lactose foods, the development and regulation of lactose-free alternatives, and the essential role of standardized food labeling. The final part of the chapter examines current and prospective therapeutic interventions, detailing the mechanisms and formulations of lactase enzyme supplementation, and evaluating the potential of microbiota-targeted approaches—including probiotics, prebiotics, and postbiotics—to enhance lactose tolerance and improve patient outcomes.

The integration of biochemical, clinical and technological perspectives is intended to provide the advanced knowledge necessary to address the multifactorial challenges posed by lactose, thereby supporting informed decision-making in both product innovation and personalized dietary guidance.

Whey. https://doi.org/10.1016/B978-0-443-27648-4.00015-3

2. Lactose as a functional carbohydrate: Physicochemical properties and technological applications

Lactose plays a crucial role in human nutrition, serving as an indispensable ingredient or additive in a vast array of food and pharmaceutical applications. Lactose is derived from whey, a by-product of cheese and dairy production, in an industrial context. Ultrafiltration is employed in the processing of the whey to separate and remove the proteins, thereby leaving a permeate that is rich in lactose content. Subsequently, the permeate is subjected to concentration by evaporation or reverse osmosis, thereby increasing its lactose content. Once a supersaturated solution is attained, the process of crystallization is induced by the application of a cooling mechanism. The resulting lactose crystals are subsequently separated from one another through the processes of centrifugation or decantation, and they are subsequently subjected to drying techniques such as flash or fluidized bed drying, in a manner conducive to their subsequent utilization in an industrial setting (Dominici et al., 2022). Two isomeric forms of lactose, α-lactose and β-lactose, occur naturally. The distinction between the two forms is based on their stereochemical configuration at the C1 substituent group in the glucose moiety. The anomers display disparate chemical and physical characteristics, which affect their functionality within diverse formulations. In aqueous solutions, α-lactose and β-lactose interconvert through the process known as mutarotation, which involves the transition of the glucose component through an open-chain form. In an equilibrium state, approximately 37% of the dissolved lactose exists in the α-form, while 63% exists as β-lactose, with the remaining balance influenced by temperature and solution conditions. The rate of mutarotation is dependent on pH, with the slowest conversion occurring around pH 5 (Portnoy et al., 2021; Illanes et al., 2016).

The solubility and crystallization behavior of lactose is significantly influenced by temperature. As temperature increases, the solubility of lactose rises, and the equilibrium between α-lactose and β-lactose shifts. At temperatures below 93.5°C, the predominant form is α-lactose monohydrate. This form is distinguished by its stability, hardness, and reduced hygroscopicity, rendering it appropriate for a multitude of applications that necessitate minimal moisture absorption. Conversely, at elevated temperatures, β-lactose assumes the dominant form due to its enhanced solubility and sweetness. The formation of β-anhydrous lactose at temperatures exceeding 93.5°C results in a higher degree of hygroscopicity compared to α-lactose monohydrate, accompanied by a greater solubility in water. This renders it advantageous in applications that demand a higher solubility, but less suitable for products that require long-term stability under humid conditions (Dominici et al., 2022; Sánchez-García et al., 2021). Amorphous lactose, frequently designated as lactose glass, represents another variant produced by rapid drying processes that impede crystallization. This form contains both α- and β-anomers, yet it is inherently unstable. In moist conditions, amorphous lactose readily transitions to α-lactose monohydrate, which is more thermodynamically stable. These transitions and characteristics highlight the necessity of meticulous control over processing conditions to guarantee the optimal form and functionality of lactose in its intended application. The versatility of lactose, coupled with its sensitivity to environmental and processing factors, renders it a unique and valuable component in a diverse range of industries (Huppertz et al., 2016; Hellrup et al., 2015).

Lactose has a wide range of applications in the food industry and is used in the production of many different types of food, thanks to its physical and chemical characteristics.

In particular, lactose can be defined as: sweetener, browning agent, encapsulating agent, anti-freezing agent, shelf-life extender, fermentation substrate (Facioni et al., 2020).

Lactose is one of the least sweet of the main commercial sweeteners, with a sweetness of between 0.2 and 0.4 compared to sucrose.

Lactose is a significant sweetener in a range of baked products, including pastries, cakes, and other baked goods. Its use in such products contributes to their palatability, flavor enhancement, and textural properties. Moreover, lactose can be employed as a replacement for sucrose. The substitution of 15%–20% of sucrose with lactose does not typically affect the acceptability of the food product, improving its palatability, mouthfeel and viscosity (Stourman 2023; Carocho et al., 2017). In addition, lactose, as a reducing sugar, participates in the Maillard reaction, whereby amino acids react with reducing sugars to form a variety of chemical compounds, which are responsible for the characteristic browning and flavor development observed in baked goods (Nielsen et al., 2022).

Lactose is also involved in the encapsulation process, whereby ingredients are protected by a coating. Indeed, due to its intrinsic properties, lactose forms an amorphous layer that absorbs moisture, creating a crystalline capsule that improves dispersion in powdered foods (Li et al., 2017). Moreover, lactose functions as an anti-freezing agent, as evidenced by its role in the production of ice cream. During the freezing phase, lactose influences the consistency of the product. The addition of more lactose lowers the freezing temperature, resulting in softer ice cream at the same serving temperature (Portnoy et al., 2021). The incorporation of lactose into baked goods serves to prolong the shelf life of the product. This is achieved by the adsorption of volatile compounds and coloring agents, the enhancement of texture and softness, the reduction of humidity, and the reaction with proteins to maintain freshness (Melini et al., 2018; Taglieri et al., 2021).

Another important role of lactose is its involvement in the fermentation process in dairy and cured meats. In cured meats, it provides essential support to starter microorganisms, facilitating the production of lactate that enhances flavor, reduces sugar metabolism, acidifies the product, accelerates drying, and shortens processing time. In dairy, strains such as *S. thermophilus* and *L. delbrueckii* subsp. *bulgaricus* metabolize milk lactose, lowering the pH and causing the protein coagulation that is essential to the cheese-making process (Sørensen et al., 2016).

In conclusion, it has been demonstrated that lactose is utilized extensively in the food industry, as it exhibits a preponderance of favorable characteristics and a comparatively modest cost (Table 20.1).

3. Lactose intolerance: Molecular basis, nutritional implications, and genetic determinants

Lactose is a disaccharide found mainly in mammalian milk, where it constitutes approximately 2%–8% by weight. This sugar is unique to milk and has an osmotic function, attracting water to maintain the fluidity of milk. The lactose content of milk is inversely proportional to the protein and fat content, and varies between mammalian species without significant differences, except for sea lions and green monkeys (Table 20.2) (Facioni et al., 2020; Misselwitz et al., 2019).

It is produced exclusively during lactation at the level of the mammary gland. During this period, the specificity of the enzyme galactosyltransferase, which typically catalyzes the addition of galactose to glycoprotein saccharide portions, undergoes a modification. This modification occurs through the production of the protein α-lactalbumin, which is regulated by the lactation-promoting hormone prolactin.

TABLE 20.1 Key characteristics of lactose.

Positive	Negative
Sweetener	Solubility
Browning agent	Stickiness
Encapsulating agent	Caking
Anti-freezing agent	
Shelf-life extender	
Fermentation substrate	

Modified from Dominici et al. (2022).

TABLE 20.2 Content of lactose in different types of most common milk.

Milk	Lactose (g/100g)
Sea lion	0.3
Buffalo	4.1
Goat	4.1
Cow	4.8
Sheep	4.8
Human	7
Donkey	7.4
Green monkey	10.2

Modified from Facioni et al. (2020)

The synthesis of the protein α-lactalbumin results in the formation of the enzyme α-lactalbumin-galactosyltransferase, otherwise known as lactose synthase. This enzyme is responsible for the synthesis of the disaccharide lactose, which is formed by the joining of a molecule of UDP-D-galactose and D-glucose, accompanied by the loss of UDP and the formation of a β(1–4) glycosidic bond (Sadovnikova, 2021). Chemically, it is composed of: glucose and galactose, both important sugars for the organism.

In particular, glucose serves as the primary energy source for cellular metabolism. Glycolysis, which occurs in the cytoplasm, involves the breakdown of glucose to yield two ATP molecules and intermediates that are utilized in the Krebs cycle within the mitochondria. The complete oxidation of glucose has the potential to generate up to 36 ATP molecules.

Furthermore, glucose metabolism is linked to the pentose phosphate pathway, which produces NADPH. This molecule is essential for the synthesis of lipids and the antioxidant defense mechanism, as well as ribose-5-phosphate, which is a precursor for nucleotide synthesis. Moreover, in the event of surplus glucose, this is stored in the form of glycogen within the liver and muscles. This can be mobilized during periods of fasting or intense physical exertion in order to maintain a state of energy equilibrium (Chandel, 2021; Ugidos-Rodriguez et al., 2018).

The regulation of glucose levels is controlled by the hormones insulin and glucagon. The hormone insulin is responsible for facilitating the uptake of glucose into cells. Conversely, the hormone glucagon stimulates the release of glucose from glycogen stores and plays a role in supporting the production of new glucose through gluconeogenesis (Nakrani et al., 2023).

It is notable that the human brain, which accounts for approximately 20% of the body's total energy expenditure, relies almost exclusively on glucose as its primary energy source. Similarly, skeletal muscles utilize glucose as a rapid source of energy during physical activity, while red blood cells depend entirely on glycolysis for ATP production due to the absence of mitochondria in these cells (Dienel, 2019).

Galactose is a crucial component in the synthesis of glycolipids and glycoproteins, which are indispensable for maintaining cell membranes and facilitating intercellular communication. The conversion of galactose to glucose-1-phosphate via the Leloir pathway then permits entry into the pathway of glucose metabolism (Ugidos-Rodriguezet al., 2018).

It is evident that galactose is of great importance for growth, especially in the context of neonatal development. Its function in the synthesis of glycolipids and glycoproteins is indispensable for the construction of cell membranes and the maintenance of the nervous system. Glycolipids, such as cerebrosides and gangliosides, are fundamental for the formation of myelin, which expedites nerve transmission. Furthermore, glycoproteins provide vital support for cell communication and membrane stability (Yu et al., 2014).

Following the intake of lactose, the enzyme lactase, also known as lactase-phlorizin hydrolase, a glycoside hydrolase, catalyzes the breakdown of lactose into its constituent components.

The hydrolysis of lactose takes place in the small intestine, specifically at the brush border of enterocytes. Lactase (a β-galactosidase enzyme) interacts with lactose at its active site, where it cleaves the β-1,4 glycosidic bond between glucose and galactose, releasing the two resulting monosaccharides.

Glucose and galactose are then absorbed by enterocytes via specific transporters: SGLT1 (sodium-glucose co-transporter) is responsible for the transport of glucose, while GLUT2 is involved in galactose transport (Sadovnikova et al., 2021).

Lactase activity declines physiologically with age. In particular, it is observed that by the age of two or 3 years, it tends to decline to very low levels around the age of 10 years, if genetically programmed.

In instances of lactase non-persistence (LNP), wherein there is a reduction in lactase activity, undigested lactose reaches the colon, where it is fermented by the gut microbiota. The fermentation process leads to the production of typically gastrointestinal symptoms, including bloating, abdominal discomfort, diarrhea and/or constipation. However, some individuals also present with extra-intestinal manifestations, such as headaches, nausea, chronic fatigue and urogenital infections (Facioni et al., 2020; Di Stefano et al., 2022).

Lactose intolerance is the symptomatic condition that characterizes individuals who are unable to digest lactose due to a partial or total deficiency of the enzyme lactase.

Lactase deficiency is a physiological phenomenon that occurs in many populations after weaning. However, in other populations, genetic mutations associated with lactase persistence enable individuals to maintain lactase activity. The ability to digest lactose in adulthood (lactase persistence) is an evolutionary adaptation that is particularly prevalent in populations with a long history of dairy farming, as evidenced by its presence during the Neolithic era (Misselwitz et al., 2019; Fassio et al., 2018; Ségurel et al., 2017).

Consequently, the prevalence of this defect is estimated to be approximately 40%–60% in Southern Europe, 30% in Central Europe, and 5%–10% in Northern Europe (Evershed et al., 2022).

The highest prevalence of individuals with lactose malabsorption is observed in South America, with an estimated 60%–70% of the population affected, followed by Africa, where the prevalence ranges between 60% and 80%. Southeast Asia also exhibits a high prevalence, with an estimated 90% of the population affected. In North America, conversely, the prevalence is relatively low, with an estimated average of approximately 45% in the USA and Canada. Overall, the global prevalence of lactose intolerance is estimated to be approximately 70% (Ugidos-Rodriguezet al., 2018; EFSA, 2010; Liebert et al., 2017).

4. Differentiating Cow's milk protein allergy and lactose intolerance: Mechanisms, diagnosis, and clinical implications

Lactose Intolerance and cow's milk protein allergy (CMPA) are often confused due to their shared association with dairy consumption. However, they are fundamentally different conditions in terms of pathophysiology, immune system involvement, and diagnostic approach (Lomer et al., 2008). While lactose intolerance results from an enzymatic deficiency leading to digestive symptoms, CMPA is an immune-mediated hypersensitivity reaction that can have severe systemic effects (Venter et al., 2016). Understanding these differences is crucial, particularly in the context of allergy testing, where IgE-mediated responses play a central role (Nowak-Wegrzyn et al., 2017).

The fundamental distinction lies in the underlying mechanisms. Lactose intolerance is a non-immune-mediated condition caused by a deficiency in lactase, the enzyme responsible for breaking down lactose into its absorbable monosaccharide components, glucose and galactose. In contrast, CMPA is an immune reaction to proteins found in cow's milk, such as casein, beta-lactoglobulin, and alpha-lactalbumin. This immune response can be mediated by immunoglobulin E (IgE) or by non-IgE mechanisms, with differing clinical presentations and diagnostic pathways (Fiocchi et al., 2010).

IgE-mediated CMPA is an immediate hypersensitivity reaction in which the immune system recognizes milk proteins as foreign antigens and produces allergen-specific IgE antibodies. These antibodies bind to high-affinity FcεR receptors on mast cells and basophils. Upon subsequent exposure to the allergen, cross-linking of bound IgE triggers cellular degranulation, leading to the release of histamine, prostaglandins, and leukotrienes, which cause rapid allergic symptoms. These symptoms typically appear within minutes to 2 hours after ingestion and may include urticaria, angioedema, respiratory distress, anaphylaxis, or gastrointestinal manifestations such as nausea and vomiting (Muraro et al., 2014; Skripak et al., 2007).

In contrast, non-IgE-mediated CMPA involves a delayed hypersensitivity response, where T cells play a central role instead of IgE antibodies. This leads to chronic inflammation, primarily affecting the gastrointestinal tract. Symptoms often emerge several hours to days after ingestion and include diarrhea, vomiting, gastroesophageal reflux, colic, and, in severe cases, food protein-induced enterocolitis syndrome (FPIES), which can cause profound vomiting, lethargy, and dehydration (Nowak-Wegrzyn et al., 2017).

Distinguishing between CMPA and lactose intolerance requires different diagnostic approaches. IgE-mediated CMPA is typically diagnosed using specific IgE blood tests and skin prick tests (SPT). The former measures the presence of IgE antibodies against cow's milk proteins in the bloodstream, while the latter evaluates cutaneous reactivity by introducing small amounts of milk protein extract into the skin. A positive result, indicated by a wheal and flare reaction, suggests IgE-mediated allergy. However, a positive IgE test alone does not confirm clinical allergy and must be interpreted alongside the patient's history (Skripak et al., 2007).

For non-IgE-mediated CMPA, the absence of detectable IgE antibodies makes diagnosis more challenging, often relying on clinical observation. In many cases, an elimination diet followed by a controlled reintroduction of the suspected allergen is used. The oral food challenge (OFC) remains the gold standard for diagnosis, conducted in a hospital setting under strict medical supervision to monitor for adverse reactions. IgE-mediated reactions appear rapidly, while non-IgE symptoms develop more gradually, requiring extended monitoring (Nowak-Wegrzyn et al., 2017). On the other hand, lactose intolerance cannot be diagnosed using IgE testing since it does not involve the immune system.

Misinterpretation of symptoms can lead to unnecessary dietary restrictions, particularly in children. A patient with lactose intolerance may experience bloating, flatulence, and diarrhea after consuming dairy, but these symptoms stem from osmotic imbalances and bacterial fermentation rather than immune activation (Lomer et al., 2008). In contrast, a patient with CMPA, especially the IgE-mediated form, is at risk of severe allergic reactions, requiring complete avoidance of cow's milk and potential emergency interventions such as epinephrine administration in cases of anaphylaxis (Muraro et al., 2014).

A clear distinction between lactose intolerance and CMPA is essential to ensure appropriate clinical management and prevent complications due to misdiagnosis. The use of specific diagnostic tools, detailed clinical history assessment, and tailored approaches help avoid unnecessary dietary exclusions while safeguarding individuals at risk of severe allergic reactions.

5. Diagnostic approaches to lactose intolerance: Hydrogen and 13C breath tests, genetic testing, and historical methods

Due to the nonspecific nature of symptoms such as bloating, diarrhea, and abdominal pain, a detailed medical history is essential to support the hypothesis of lactose intolerance and distinguish it from other gastrointestinal and IBS-like disorders (Borghini et al., 2017). History should explore various aspects, including the timing of symptoms in relation to lactose intake, symptom patterns, and family comorbidity, making it essential to avoid unnecessary treatments or inappropriate dietary restrictions (Micic et al., 2019). When clinical history suggests lactose intolerance, various diagnostic tools are available to confirm the diagnosis. Lactase deficiency impairs lactose absorption in the intestine, leading to symptoms caused not only by an increased osmotic load and subsequent water retention in the gut, but also by fermentation of unabsorbed lactose (Catanzaro et al., 2021).

When the disaccharide is not digested, it undergoes fermentation by anaerobic bacteria in the colon. This process produces gases such as carbon dioxide (contributing to bloating), methane (CH_4), and hydrogen (H_2), as well as short-chain fatty acids like propionic and butyric acids, which can lead to osmotic diarrhea. Gasses such as H_2 and CH_4 accumulate in the intestine, but also pass through the intestinal wall into the bloodstream and here transported to the lungs, where they are excreted in the exhaled breath (Usai-Satta et al., 2022).

H_2-Lactose Breath Test is the gold standard and most frequently used method for diagnosing this condition. It offers several advantages, as it is non-invasive, cost-effective, and simple to administer and read, with a high sensitivity (70%–100%) and specificity (69%–100%) (Robles et al., 2020; Romagnuolo et al., 2002).

The hydrogen breath test involves measuring baseline exhaled hydrogen after fasting, followed by ingestion of a lactose solution containing 25–50 g of lactose in adults. The recommended test duration is 3–5 h, although it can be shortened if malabsorption and intolerance are confirmed early. Standard measurement intervals are every 30 min to assess malabsorption and intolerance. Longer intervals (up to 60 min) may be useful for evaluating malabsorption, while shorter intervals (10–15 min) may help establish a temporal correlation between hydrogen production and the onset of symptoms, thus supporting the diagnosis of intolerance. An increase in hydrogen concentration of $\geq$20 parts per million (ppm) above baseline at any time point during the test is considered indicative of lactose maldigestion or malabsorption (Hammer et al., 2022). However, 2.5%–15% of hydrogen breath test results may yield false negatives due to various factors. These include conditions that impair lactose absorption, such as active celiac disease, the presence of small intestinal bacterial overgrowth (SIBO), or cases involving individuals who do not produce hydrogen. Additionally, elevated methane (CH_4) colonic production can reduce detectable hydrogen levels, as CH_4-producing bacteria consume hydrogen during fermentation (Kerber et al. 2007; Eisenmann et al. 2008; Lee et al. 1996).

In such cases, the administration of a genetic test may be beneficial. A substantial body of research conducted in recent decades has identified a genetic predisposition to lactose intolerance. The most common cause is a single-nucleotide polymorphism (SNP) located 13.9 kb upstream of the lactase (LCT) gene. This SNP, known as C-13910T, lies within a regulatory region: the T allele is associated with continued expression of the LCT gene (lactase persistence), whereas individuals with the CC genotype typically exhibit lactase non-persistence, in caucasian population (Enattah et al. 2007).

The test consists of DNA isolation, obtained through a blood sample or by oral swab, and subsequent analysis of polymorphisms (Santonocito et al., 2015). The presence of a predisposing polymorphism (C/C or G/G) does not necessarily imply intolerance nor does it allow us to predict if and when this will develop in future, but it can be helpful in cases where H_2 breath test cannot be used or to confirm the hypothesis of lactose intolerance (Catanzaro et al., 2021).

The Quick Lactase Test is an invasive diagnostic procedure that evaluates lactase activity using duodenal biopsy samples incubated with lactose. Enzyme activity is indicated by a color change: dark blue represents normal activity, light blue indicates partial activity, and no color change signifies lactase deficiency. Studies, such as by *Ojetti* et al., have demonstrated strong concordance between this test and the hydrogen breath test, particularly in patients with positive breath test results (Ojetti et al., 2008). However, this test is especially valuable when symptoms continue despite a negative hydrogen breath test, as it can detect hypolactasia in such cases. Its main drawbacks include its invasive nature, higher cost, the requirement for immediate sample processing, and the potential for inaccurate results if the biopsy specimen is insufficient. Consequently, this test is generally reserved for select cases where non-invasive methods yield inconclusive results (Mattar et al., 2013).

The Lactose Tolerance Test measures blood glucose after ingesting 50 g of lactose after 30 min, 60 min and 120 min. A rise indicates lactose digestion; no rise suggests intolerance (Swagerty et al., 2002). Its use is limited due to low accuracy and a 20% rate of false results, which can be affected by factors such as gastric emptying and glucose metabolism (Domínguez-Jiménez et al., 2014). Ghoshal et al. advocate for its use when the hydrogen breath test is unavailable, particularly in hydrogen non-producers. However, the test is considered unreliable in diabetic patients. *Varjù* et al. suggest using it in combination with a hydrogen breath test due to a false positive rate of 9.8% (Ghoshal et al., 2009; Varjú et al., 2020).

Another diagnostic approach is the lactose breath test using the stable isotope C13, which offers an advanced method to evaluate lactase enzyme activity in the small intestine. This test is notable for its high specificity and sensitivity, making it especially valuable in cases where conventional tests like the hydrogen breath test (H_2) may produce false-negative results (Hiele et al., 1988). A 2024 study demonstrated that the C13-lactose breath test has a sensitivity of 93% and a specificity of 70%, outperforming the hydrogen (H_2) breath test in diagnostic accuracy. This makes it particularly useful for patients with lactase non-persistence who receive negative results on the H_2 breath test (Balsiger et al. 2024).

6. Lactose-intolerance diet management: The role of aged cheeses, fermented foods, and lactose-free alternatives

The primary treatment for lactose intolerance involves preventing gastrointestinal symptoms by reducing or eliminating lactose from the diet. To manage their symptoms, individuals with lactose intolerance should avoid high-lactose foods like fresh milk and dairy products, while maintaining adequate nutrient intake through alternative food sources (Gille et al., 2018). It should also be noted that lactose is a common additive in many processed foods; the so-called "hidden lactose" is used for its texture and flavor enhancing properties (Van der Merwe et al., 2020).

Lactose is widely used as an excipient in oral medications within the pharmaceutical industry. Although the lactose content in solid oral dosage forms is typically low, usually less than 2 g per day, and generally well tolerated, alternative medications may be required for individuals with severe lactose intolerance (Mill et al., 2018). The appropriate lactose restriction depends on the severity of the patient's symptoms. Most individuals with lactose intolerance can tolerate up to 5 g of lactose per single serving in their diet (Fassio et al., 2018).

To support the management of lactose intolerance, alternative foods that are naturally lactose-free, such as vegetables, fruit, meat, fish, legumes. Also, alternative foods, like plant-based drinks (e.g., soy, almond, and rice milk) and soy products (e.g., tofu, edamame) are naturally lactose free.

In addition, the food industry has developed many "low-lactose" and "lactose-free" products using different processes to remove lactose from lactose-rich dairy foods, using physically techniques like ultrafiltration or chromatographic separation, following by subsequent hydrolysis of the remaining molecules, resulting in a reduction of the lactose concentration by tens or hundreds of times, making it suitable for intolerant patients (Dekker et al., 2019). Also, most aged hard cheeses naturally contain very little, if any lactose (Facioni et al., 2021). Alternatively, lactose-free milk can be produced through enzymatic hydrolysis, in which microbial β-galactosidase breaks down lactose into its constituent monosaccharides, glucose and galactose (Harju et al., 2012). However, many countries do not have laws regulating commercialization of lactose-free products, which poses the problem of mislabeling and causation of symptoms (Fassio et al., 2018).

Exogenous lactase, available in supplement form, helps break down lactose into its monosaccharides, glucose and galactose, to improve absorption. Yet, these supplements may not fully eliminate symptoms, as lactose digestion can remain incomplete. Their effectiveness depends on several factors, including the amount of dairy consumed and the timing of enzyme intake. Therefore, lactase supplements should be considered a complement to dietary restriction, rather than a substitute for it (Di Pierro et al., 2015). Dairy products represent a major source of calcium, essential for both adults and, even more so, for children (Dror et al., 2014).

Although further research is needed to fully understand the long-term impact of dairy reduction on mineral balance and overall health, individuals with lactose intolerance should aim to eliminate lactose—preferably by choosing lactose-free dairy products—rather than excluding dairy entirely, in order to preserve the nutritional benefits of these foods.

Therefore, consideration should be given to a temporary lactose-free diet with the goal of reducing symptoms, followed by a gradual introduction of dairy products as tolerated.

(Catanzaro et al., 2021).

As detailed in Section 4, lactose is extensively utilized within the food industry owing to its favorable physicochemical properties, which contribute to its widespread inclusion as both an ingredient and functional additive. For individuals withlactose intolerance, rigorous scrutiny of food labeling remains the sole reliable means of avoiding inadvertent lactose consumption. In this context, the advancement and standardization of front-of-pack (FOP) labeling represent a promising approach to enhance transparency and support informed decision-making among lactose intolerance consumers (Facioni et al., 2020). Recent empirical evidence suggests that individuals with lactose intolerance not only rely on the ingredient list but also actively seek out certification symbols displayed on the FOP. The presence of a standardized certification mark, such as *Lfree*, has been shown to facilitate a more efficient and secure shopping experience by reducing the cognitive burden associated with label interpretation. This labeling strategy may therefore play a critical role in safeguarding dietary adherence and improving the quality of life for consumers managing lactose intolerance (Facioni et al., 2022).

7. Lactase supplements in lactose intolerance: Mechanisms, formulations, and innovations in enzyme delivery

Lactase supplements constitute an efficacious therapeutic option for the management of lactose intolerance. Contemporary formulations, incorporating extended-release and delayed-release products, are effecting a transformation in the management of this condition by offering customized solutions. Continued innovation in release technologies and enzyme formulations promises to further enhance the quality of life of lactose intolerance patients by enabling them to include lactose in their diet without experiencing gastrointestinal discomfort.

It is important to note that these products should not be regarded as a "cure" for intolerance. Rather, they are designed to be used as a supplementary measure for sporadic instances or when there is a desire to indulge in a treat.

Lactase supplements contain the lactase enzyme, which catalyzes the breakdown of lactose into glucose and galactose. These supplements are typically formulated as tablets or capsules, and are designed to be ingested either before or during the consumption of foods containing lactose. The primary function of lactase supplements is to provide an adequate dose of lactase, thereby facilitating the digestion of lactose in individuals with a deficiency or reduced level of enzyme activity. This can significantly reduce the gastrointestinal symptoms associated with lactose intolerance (Baijal et al., 2020).

Enzyme mixtures are preparations containing multiple enzyme types, including lactase, amylase, protease and lipase. These mixtures are frequently employed in the management of more complex digestive disorders that involve different classes of nutrients. Despite the potential benefits of enzyme mixtures, lactase-only supplements are often the preferred treatment option for individuals suffering from lactose intolerance (Ianiro et al., 2016). Lactase is typically included in supplements as β-galactosidase or as lactase derived from various sources, including lactic acid bacteria (e.g., *Streptococcus thermophilus and Lactobacillus bulgaricus*), fungi (e.g., *Aspergillus oryzae*), or yeasts (e.g., *Kluyveromyces lactis*) (Dutra Rosolen et al., 2015). The activity of exogenous lactase supplements is influenced by several factors. Understanding these influences can help optimize lactase activity for individuals with lactose intolerance:

1. Temperature: the optimal working temperature for most lactase supplements is around body temperature (37–38°C), but some commercial formulations are also effective at lower temperatures, such as 15°C. Most commercial lactase enzymes, including those derived from *Kluyveromyces lactis*, show 55%–61% of their maximum activity at 37°C and are highly effective at hydrolyzing lactose in milk when preincubated at 38°C for 15–20 min. This temperature closely matches the human digestive tract, making it ideal for supplement activity during digestion
2. pH: commercial lactase products generally show significant activity between pH 3.0 and 6.5, but are often acid-sensitive and may lose activity in the stomach unless specially formulated. Deviations from this pH range can lead to enzyme denaturation and loss of functionality.
3. Enzyme Inhibition: various substances, known as inhibitors, can interfere with enzyme activity by disrupting the catalytic processes at the enzyme's active site, either directly or indirectly (O'Connell et al. 2006).

Due to these influences, under such circumstances, lactose hydrolysis is impaired and can result in gastrointestinal symptoms commonly associated with lactose intolerance. In this regard, it is important to consider these factors during a supplement lactase formulation.

Lactase supplementation is available in various formulations, each characterized by distinct release mechanisms that influence enzymatic availability and functional duration. The most widely used include:

- Immediate-release (classic-acting) formulations, which are engineered to exert enzymatic activity over a sustained but relatively short period, aligning with the digestion of lactose during mealtime;
- Prolonged-release formulations, which facilitate the gradual release of lactase throughout the gastrointestinal tract, offering an extended duration of activity, typically around 5–6 h. This controlled release profile optimizes enzymatic efficacy over time and minimizes the need for frequent dosing across multiple meals.

Moreover, the timing of lactase administration plays a critical role in determining its effectiveness. Traditional protocols recommend consumption approximately 30 to 5 min prior to lactose ingestion; however, advances in formulation technology have introduced products capable of maintaining efficacy even when taken shortly after a lactose-containing meal.

The functional window of exogenous lactase activity is generally around 60 min, though this may vary depending on several physiological and dietary factors. These include the physical consistency of the ingested food, the overall composition of the meal, gastric pH, and the rate of gastric emptying, all of which can modulate the bioavailability and enzymatic performance of the supplement (Morsanuto et al., 2020; Ibba et al., 2014).

In summary, exogenous lactase supplementation presents a valuable therapeutic strategy for the management of lactose intolerance, offering several clinically relevant advantages:

- Symptomatic relief: pre-prandial administration of lactase has been shown to enhance the enzymatic hydrolysis of dietary lactose, thereby mitigating the onset of typical gastrointestinal symptoms such as bloating, flatulence, abdominal pain, and diarrhea.
- Increased dietary flexibility: the effective use of lactase supplements enables individuals with lactose intolerance to consume dairy and other lactose-containing products with reduced risk of adverse reactions. This expanded dietary tolerance contributes to improved nutritional adequacy and supports psychosocial well-being, particularly in social or cultural contexts where dairy consumption is customary.
- Practicality and market availability: lactase preparations are widely accessible and commercially available in diverse formulations (e.g., tablets, capsules, liquid drops), offering flexibility in use and integration into daily dietary practices.

8. The role of probiotics, prebiotics, and postbiotics in lactose intolerance

The management of lactose intolerance has traditionally focused on dietary restriction of lactose-containing foods and the use of exogenous lactase supplements. However, emerging evidence suggests that modulation of the gut microbiota through probiotics, prebiotics, and postbiotics may provide complementary strategies to improve symptoms and enhance lactose digestion (Fassio et al., 2018).

Probiotics, live microorganisms that confer health benefits when administered in adequate amounts, have been extensively studied for their potential to assist in lactose digestion. Certain strains, particularly of the genera *Lactobacillus* and *Bifidobacterium*, possess intrinsic β-galactosidase activity, which facilitates the breakdown of lactose into glucose and galactose directly within the gastrointestinal lumen (He et al., 2008). Clinical trials have demonstrated that fermented dairy products containing viable probiotic cultures, such as yogurt, are better tolerated than unfermented milk, likely due to this enzymatic activity and the slower gastrointestinal transit time induced by the semi-solid matrix (Savaiano et al. 2014). The daily consumption of selected probiotic strains may also contribute to a more stable and diverse gut microbiota, which is associated with improved colonic fermentation patterns and reduced gas production.

Prebiotics, in contrast, are non-digestible food components, primarily oligosaccharides such as fructooligosaccharides (FOS) and galactooligosaccharides (GOS), that selectively stimulate the growth and activity of beneficial bacteria in the colon. Although their direct effect on lactase activity is limited, prebiotics can enhance the survival and functionality of endogenous and supplemented probiotics, potentially leading to better lactose metabolism over time (Slavin 2013). In lactose-intolerant individuals, prebiotic supplementation may reduce gastrointestinal symptoms by modulating gut fermentation and improving the composition of the microbiota toward a more saccharolytic profile.

More recently, postbiotics (non-viable bacterial products or metabolic by-products derived from probiotic fermentation) have garnered attention for their therapeutic potential. Unlike probiotics, postbiotics are not reliant on bacterial survival in the gastrointestinal tract, offering greater stability and safety. Bioactive compounds such as short-chain fatty acids (SCFAs), peptides, and polysaccharides produced during fermentation may exert anti-inflammatory effects, improve gut barrier function, and modulate intestinal motility, all of which can impact symptom severity in lactose-intolerant patients (Aguilar-Toalá et al. 2021). Although research in this area is still emerging, postbiotics represent a promising adjunct to traditional lactose intolerance management.

A comprehensive approach that incorporates dietary management with microbiota modulation may provide synergistic benefits. In a 2018 review, it emphasized the role of intestinal microbiota adaptation in modulating lactose intolerance symptoms, suggesting that both probiotic supplementation and natural bacterial shifts through dietary habits can influence lactose tolerance thresholds over time. The review also noted that individual variability in microbiota composition may partly explain the heterogeneous clinical presentations of lactose intolerance and the varying response to microbiota-targeted interventions (Fassio et al., 2018).

Taken together, these insights suggest that a microbiota-centered strategy, including probiotics, prebiotics, and postbiotics, may enhance not only the digestive capacity for lactose but also the overall quality of life in affected individuals. Ongoing research will help refine strain-specific effects, optimal dosages, and long-term efficacy in this context.

9. Lactose intolerance and its clinical intersections: Celiac disease, irritable bowel syndrome, and micronutrient deficiencies

Lactose intolerance often presents with symptoms that overlap with other gastrointestinal disorders, making differential diagnosis challenging.

Celiac disease, a common autoimmune enteropathy in genetically predisposed individuals, is triggered by the ingestion of gluten—a protein found in wheat, rye, and barley. This immune response leads to villous atrophy in the small intestine, resulting in impaired nutrient absorption (Caio et al., 2019).

Lactase deficiency is frequently observed in individuals with celiac disease, and in some cases, lactose intolerance may represent the only initial manifestation of undiagnosed celiac disease. Common symptoms such as abdominal pain, bloating, and diarrhea can overlap in both conditions. Because symptom relief often follows dietary restriction, clinicians may be inclined to over diagnose lactose intolerance, potentially overlooking celiac disease, particularly in pediatric patients (Grimheden et al., 2012).

In untreated celiac disease, atrophy of the intestinal villi damages the apical brush border of enterocytes, where the enzyme lactase is normally expressed. This damage can lead to a partial lactase deficiency, resulting in lactose malabsorption and secondary lactose intolerance (transient form of lactose intolerance) (Ojetti et al., 2005). However, lactose intolerance is frequently reported in celiac disease patients who are considered unresponsive to a gluten-free diet, an issue affecting approximately 20% of individuals with celiac disease. Among these non-responsive patients, lactose intolerance is a common contributing factor, identified in up to 8% of cases (Leffler et al., 2007).

After starting a gluten-free diet for the treatment of celiac disease and following mucosal healing, most patients are able to tolerate lactose within one to 2 months. The hydrogen breath test has been proposed as a useful tool to monitor mucosal recovery in treated celiac patients (Lerner, 2018).

Lactose malabsorption can play a key role also in Irritable Bowel Syndrome (IBS). IBS is the most prevalent functional gastrointestinal disorder which exact pathophysiology is not well understood, although proposed mechanisms include alterations in visceral hypersensitivity, disordered gut-brain interaction, intestinal microbiome, enteric sensory and motor dysfunction, altered pain processing, and immune dysregulation (Drossman, 2016).

IBS diagnosis is based on clinical evaluation using the Rome IV Criteria, which require recurrent abdominal pain related to defecation or a change in bowel habits, with symptoms starting at least 6 months before diagnosis and persisting for the last 3 months (Palsson et al., 2020).

Diet plays a key role in IBS and has been reported to be a symptom trigger in approximately 80% of patients (Böhn et al., 2013). In particular, IBS patients often experience symptom improvement when excluding FODMAPs (Fermentable Oligosaccharides, Disaccharides, Monosaccharides and Polyols), short-chain carbohydrates that are poorly absorbed in the gastrointestinal tract. These compounds increase osmotic load and lead to excess gas production through bacterial fermentation, and lactose is one of the common FODMAPs involved (Halmos et al., 2014).

Given the overlap in clinical symptoms, lactose intolerance should be evaluated in patients with IBS. A meta-analysis conducted by Péter Varjú et al. found that lactose intolerance is more prevalent in IBS patients compared to healthy controls, whereas lactose maldigestion (i.e., lactase enzyme deficiency) occurs at similar rates in both groups. This suggests that IBS symptoms may heighten sensitivity to lactose, even without actual maldigestion (Varjú et al., 2018).

In a single-center study conducted in Italy, 79.9% of patients diagnosed with IBS tested positive for lactose intolerance using the hydrogen breath test, compared to only 25% in the control group. After 1 month of appropriate treatment, symptoms in lactose-intolerant patients improved significantly, suggesting that many individuals initially diagnosed with IBS may actually suffer from lactose intolerance (Catanzaro et al., 2021).

The characteristic symptoms of lactose intolerance can also be mistaken for other conditions such as gluten sensitivity, small intestinal bacterial overgrowth (SIBO), and other forms of dysbiosis. Therefore, accurate differential diagnosis and careful clinical evaluation by the physician are essential (Borghini et al., 2017).

In addition, it is important to consider eventual deficiency due to exclusion from the diet. In particular, dairy products are a convenient source of essential nutrients that support bone health, including proteins; vitamins such as A, B2, B5, B12, K2, and D; and minerals like calcium, potassium, magnesium, phosphorus, zinc, and iodine (Dekker et al., 2019).

There is evidence of significant advances in height, body weight, bone mineral content, and bone mineral density in children who consume dairy products (Cheng et al., 2005). In individuals with lactose intolerance, inadequate calcium intake has been linked to lower bone mineral density, raising concerns about an increased risk of osteoporosis and fractures later in life. According to the National Medical Association, calcium requirements are the same for males and females during the first 50 years of life: 700 mg/day for ages 1–3 years; 1000 mg/day for ages 4–8 years and 19–50 years. The highest recommended intake occurs during adolescence, when maximal bone growth takes place (1300 mg/day for ages 9–18 years). Calcium needs diverge with the onset of menopause, with recommended intake for women increasing to 1200 mg/day. This level remains consistent for both sexes from age 70 onwards, with a daily recommended allowance of 1200 mg to help prevent osteoporosis (Facioni et al., 2020).

Studies on nutrient intake in lactose-intolerant individuals show that, compared to lactose-tolerant people, they consume lower amounts of calcium. Their average daily intake ranges from 388 to 739 mg, which is below the Recommended Dietary Allowance (RDA) of 1000 mg per day (Barr, 2013; Nicklas et al., 2011).

Interestingly, from observational studies it has emerged that the avoidance of dairy products was associated with poor bone health, higher blood pressure, and an increased risk of diabetes mellitus (Huncharek et al., 2008; Miller et al., 2000; Pasin, 2015).

For individuals with lactose intolerance who avoid dairy products, including lactose-free options, consuming calcium-rich foods with high bioavailability (such as green vegetables like kale and broccoli) as well as calcium-fortified foods like plant-based beverages and breakfast cereals, or calcium supplementation (e.g., calcium formate), represent reasonable alternative sources of calcium (Melse-Boonstra, 2020).

10. Conclusion

Lactose and lactose intolerance represent two complementary facets of a central issue in the fields of food science, nutrition, and clinical care. On one hand, lactose is an extremely versatile disaccharide, widely utilized in both the food and pharmaceutical industries due to its unique physicochemical properties. It serves as a mild sweetener, a Maillard reaction facilitator, a cryoprotectant, an encapsulating agent, and a fermentative substrate. These characteristics make lactose indispensable in a variety of products, including baked goods, ice creams, cured meats, and dairy derivatives. On the other hand, lactose intolerance imposes significant dietary restrictions on a growing segment of the population, necessitating a personalized and multidimensional approach to management.

Effective lactose intolerance management cannot be reduced to the indiscriminate elimination of all lactose-containing foods, as this may lead to nutritional imbalances, particularly in terms of calcium, vitamin D, and high-quality proteins. The most effective strategies involve a combination of low-lactose diets, the informed use of naturally lactose-free or technically "lactose-free" products, and personalized lactase supplementation, available in immediate, prolonged, or delayed-release formulations. In parallel, meticulous label reading, supported by reliable certification schemes such as the *Lfree* label, is essential. Clear and harmonized regulatory frameworks for food labeling are critical to ensuring safety for lactose-intolerant consumers.

In addition to dietary and enzymatic strategies, growing interest is being directed toward the modulation of the gut microbiota through probiotics, prebiotics, and postbiotics. These interventions show promise in improving lactose tolerance through synergistic mechanisms, including endogenous β-galactosidase production, enhanced colonic fermentation profiles, reduced intestinal inflammation, and improved epithelial barrier function. An integrated approach that considers interindividual variability and microbiota composition differences may represent the future of nutritional management in lactose intolerance.

In conclusion, while lactose continues to play a vital role in the food industry, its widespread presence necessitates awareness and adaptive strategies for individuals with lactose intolerance. The combination of technological innovation, nutritional education, personalized dietary support, and microbiota-based interventions currently represents the most comprehensive paradigm for promoting health, safety, and quality of life among lactose-intolerant individuals.

References

Aguilar-Toalá, J.E., Garcia-Varela, R., Garcia, H.S., Mata-Haro, V., González-Córdova, A.F., Vallejo-Cordoba, B., Hernández-Mendoza, A., 2021. Postbiotics: an evolving term within the functional foods field. Trends Food Sci. Technol. 108, 105–114. https://doi.org/10.1016/j.tifs.2020.12.002.

Baijal, R., Tandon, R.K., 2020. Effect of lactase on symptoms and hydrogen breath levels in lactose intolerance: a crossover placebo-controlled study. JGH Open 5 (1), 143–148. https://doi.org/10.1002/jgh3.12463.

Balsiger, L.M., Houben, E., Vanuytsel, T., van Ranst, M., Tack, J., Verbeke, K., 2024. Added value of 13C analysis in breath tests in H2-negative subjects to diagnose lactose malabsorption: a proof of concept study. Dig. Dis. Sci. 69 (6), 2147–2153.

Barr, S.I., 2013. Perceived lactose intolerance in adult Canadians: a national survey. Appl. Physiol. Nutr. Metab. 38 (8), 830–835.

Böhn, L., Störsrud, S., Törnblom, H., Bengtsson, U., Simrén, M., 2013. Self-reported food-related gastrointestinal symptoms in IBS are common and associated with more severe symptoms and reduced quality of life. Am. J. Gastroenterol. 108, 634–641.

Borghini, R., Donato, G., Alvaro, D., Picarelli, A., 2017. New insights in IBS-like disorders: Pandora's box has been opened; a review. Gastroenterol. Hepatol. Bed Bench 10 (2), 79–89.

Caio, G., Volta, U., Sapone, A., Leffler, D.A., De Giorgio, R., Catassi, C., Fasano, A., 2019. Celiac disease: a comprehensive current review. BMC Med. 17 (1), 142.

Carocho, M., Morales, P., Ferreira, I.C.F.R., 2017. Sweeteners as food additives in the XXI century: a review of what is known, and what is to come. Food Chem. Toxicol. https://doi.org/10.1016/j.fct.2017.06.046.

Catanzaro, R., Sciuto, M., Marotta, F., 2021a. Lactose intolerance: an update on its pathogenesis, diagnosis, and treatment. Nutr. Res. 89, 23–34.

Catanzaro, R., Sciuto, M., Marotta, F., 2021b. Lactose intolerance—old and new knowledge on pathophysiological mechanisms, diagnosis, and treatment. SN Compr. Clin. Med. 3, 499–510.

Catanzaro, R., Sciuto, M., Singh, B., Pathak, S., Marotta, F., 2021c. Irritable bowel syndrome and lactose intolerance: the importance of differential diagnosis. A monocentric study. Minerva Gastroenterol. 67 (1), 72–78.

Chandel, N.S., 2021. Glycolysis. Cold Spring Harbor Perspect. Biol. 13 (5), a040535. https://doi.org/10.1101/cshperspect.a040535.

Cheng, S., Lyytikainen, A., Kroger, H., Lamberg-Allardt, C., Alen, M., Koistinen, A., Wang, Q.J., Suuriniemi, M., Suominen, H., Mahonen, A., et al., 2005. Effects of calcium, dairy product, and vitamin D supplementation on bone mass accrual and body composition in 10–12-year-old girls: a 2-year randomized trial. Am. J. Clin. Nutr. 82, 1115–1126.

Dekker, P.J.T., Koenders, D., Bruins, M.J., 2019. Lactose-free dairy products: market developments, production, nutrition and health benefits. Nutrients 11, 551.

Di Pierro, F., Bertuccioli, A., Marini, E., Ivaldi, L., 2015. A pilot trial on subjects with lactose and/or oligosaccharides intolerance treated with fixed mixture of pure and enteric-coated α- and β-galactosidase. Clin. Exp. Gastroenterol. 8, 95–100.

Di Stefano, M., Brondino, N., Bonaso, V., Miceli, E., Lapia, F., Grandi, G., Pagani, E., Corazza, G.R., Di Sabatino, A., 2022. The perception of lactose-related symptoms of patients with lactose malabsorption. Int. J. Environ. Res. Publ. Health 19 (16), 10234. https://doi.org/10.3390/ijerph191610234.

Dienel, G.A., 2019. Brain glucose metabolism: integration of energetics with function. Physiol. Rev. 99 (1), 949–1045. https://doi.org/10.1152/physrev.00062.2017.

Domínguez-Jiménez, J.L., Fernández-Suárez, A., Ruiz-Tajuelos, S., Puente-Gutiérrez, J.J., Cerezo-Ruiz, A., 2014. Lactose tolerance test shortened to 30 minutes: an exploratory study of its feasibility and impact. Rev. Esp. Enferm. Dig. 106, 381–385.

Dominici, S., Marescotti, F., Sanmartin, C., Macaluso, M., Taglieri, I., Venturi, F., Zinnai, A., Facioni, M.S., 2022. Lactose: characteristics, food and drug-related applications, and its possible substitutions in meeting the needs of people with lactose intolerance. Foods 11, 1486. https://doi.org/10.3390/foods11101486.

Dror, D.K., Allen, L.H., 2014. Dairy product intake in children and adolescents in developed countries: trends, nutritional contribution, and a review of association with health outcomes. Nutr. Rev. 72, 68–81.

Drossman, D.A., 2016. Functional gastrointestinal disorders: history, pathophysiology, clinical features, and Rome IV. Gastroenterology 150, 1262–1279.

Dutra Rosolen, M., Gennari, A., Volpato, G., Volken de Souza, C.F., 2015. Lactose hydrolysis in milk and dairy whey using microbial β-galactosidases. Enzym. Res. 806240. https://doi.org/10.1155/2015/806240.

EFSA Panel on Dietetic Products, Nutrition and Allergies (NDA), 2010. Scientific opinion on lactose thresholds in lactose intolerance and galactosaemia. EFSA J. 8, 1777. https://doi.org/10.2903/j.efsa.2010.1777.

Eisenmann, A., Amann, A., Said, M., Datta, B., Ledochowski, M., 2008. Implementation and interpretation of hydrogen breath tests. J. Breath Res. 2, 046002.

Enattah, N.S., Trudeau, A., Pimenoff, V., Maiuri, L., Auricchio, S., Greco, L., Rossi, M., Lentze, M., Seo, J., Rahgozar, S., et al., 2007. Evidence of still-ongoing convergent evolution of the lactase persistence T-13910 alleles in humans. Am. J. Hum. Genet. 81, 615–625.

Evershed, R.P., Davey Smith, G., Roffet-Salque, M., Timpson, A., Diekmann, Y., Lyon, M.S., Cramp, L.J.E., Casanova, E., Smyth, J., Whelton, H.L., Dunne, J., Brychova, V., Šoberl, L., Gerbault, P., Gillis, R.E., Heyd, V., Johnson, E., Kendall, I., Manning, K., Marciniak, A., Outram, A.K., Vigne, J.D., Shennan, S., Bevan, A., Colledge, S., Allason-Jones, L., Amkreutz, L., Anders, A., Arbogast, R.M., Bălăşescu, A., Bánffy, E., Barclay, A., Behrens, A., Bogucki, P., Carrancho Alonso, Á., Carretero, J.M., Cavanagh, N., Claßen, E., Collado Giraldo, H., Conrad, M., Csengeri, P., Czerniak, L., Dębiec, M., Denaire, A., Domboróczki, L., Donald, C., Ebert, J., Evans, C., Francés-Negro, M., Gronenborn, D., Haack, F., Halle, M., Hamon, C., Hülshoff, R., Ilett, M., Iriarte, E., Jakucs, J., Jeunesse, C., Johnson, M., Jones, A.M., Karul, N., Kiosak, D., Kotova, N., Krause, R., Kretschmer, S., Krüger, M., Lefranc, P., Lelong, O., Lenneis, E., Logvin, A., Lüth, F., Marton, T., Marley, J., Mortimer, R., Oosterbeek, L., Oross, K., Pavúk, J., Pechtl, J., Pétrequin, P., Pollard, J., Pollard, R., Powlesland, D., Pyzel, J., Raczky, P., Richardson, A., Rowe, P., Rowland, S., Rowlandson, I., Saile, T., Sebők, K., Schier, W., Schmalfuß, G., Sharapova, S., Sharp, H., Sheridan, A., Shevnina, I., Sobkowiak-Tabaka, I., Stadler, P., Stäuble, H., Stobbe, A., Stojanovski, D., Tasić, N., van Wijk, I., Vostrovská, I., Vuković, J., Wolfram, S., Zeeb-Lanz, A., Thomas, M.G., 2022. Dairying, diseases and the evolution of lactase persistence in Europe. Nature 608 (7922), 336–345. https://doi.org/10.1038/s41586-022-05010-7.

Facioni, M.S., Dominici, S., Farace, B., Apicella, A., Romano, M.F., Tarabella, A., 2022. A survey on consumers with lactose intolerance in Italy: their perception, habits and needs when shopping for groceries. EC Nutrition 17 (9).

Facioni, M.S., Dominici, S., Marescotti, F., Covucci, R., Taglieri, I., Venturi, F., Zinnai, A., 2021. Lactose residual content in PDO cheeses: novel inclusions for consumers with lactose intolerance. Foods 10 (9), 2236.

Facioni, M.S., Raspini, B., Pivari, F., Dogliotti, E., Cena, H., 2020. Nutritional management of lactose intolerance: the importance of diet and food labelling. J. Transl. Med. 18 (1), 260. https://doi.org/10.1186/s12967-020-02429-2.

Fassio, F., Facioni, M.S., Guagnini, F., 2018. Lactose maldigestion, malabsorption, and intolerance: a comprehensive review with a focus on current management and future perspectives. Nutrients 10, 1599. https://doi.org/10.3390/nu10111599.

Fiocchi, A., Brozek, J., Schünemann, H., Bahna, S.L., von Berg, A., Beyer, K., World Allergy Organization, 2010. World Allergy Organization (WAO) diagnosis and rationale for action against cow's milk allergy (DRACMA) guidelines. World Allergy Organ. J. 3 (4), 57–161.

Ghoshal, U.C., Kumar, S., Chourasia, D., Misra, A., 2009. Lactose hydrogen breath test versus lactose tolerance test in the tropics: does positive lactose tolerance test reflect more severe lactose malabsorption? Trop. Gastroenterol. (Apr–Jun) 30 (2), 86–90.

Gille, D., Walther, B., Badertscher, R., Bosshart, A., Brügger, C., Brühlhart, M., Gauch, R., Noth, P., Vergères, G., Egger, L., 2018. Detection of lactose in products with low lactose content. Int. Dairy J. 83, 17–19.

Grimheden, P., Anderlid, B.M., Gåfvels, M., Svahn, J., Grahnquist, L., 2012. Lactose intolerance in children is an overdiagnosed condition. Risk of missing intestinal diseases such as IBD and celiac disease. Lakartidningen 109 (5), 218–221.

Halmos, E.P., Power, V.A., Shepherd, S.J., Gibson, P.R., Muir, J.G., 2014. A diet low in FODMAPs reduces symptoms of irritable bowel syndrome. Gastroenterology 146 (1), 67–75.e5.

Hammer, H.F., Basilisco, G., Borrelli, O., Nakov, R., Fox, M.R., Hammer, J., Keller, J., Lopetuso, L., Salvatore, S., Benninga, M., Pohl, D., Thapar, N., Sonyi, M., Dumitrascu, D., Hauser, B., Herszenyi, L., 2022. European guideline on indications, performance, and clinical impact of hydrogen and methane breath tests in adult and pediatric patients: European Association for Gastroenterology, Endoscopy and Nutrition, European Society of Neurogastroenterology and Motility, and European Society for Paediatric Gastroenterology Hepatology and Nutrition consensus. United Eur. Gastroenterol. J. 10 (1), 15–40.

Harju, M., Kallioinen, H., Tossavainen, O., 2012. Lactose hydrolysis and other conversions in dairy products: technological aspects. Int. Dairy J. 22.

He, T., Priebe, M.G., Zhong, Y., Huang, C., Harmsen, H.J., Raangs, G.C., Vonk, R.J., 2008. Effects of yogurt and bifidobacteria supplementation on the colonic microbiota in lactose-intolerant subjects. J. Dairy Sci. 91 (2), 349–359.

Hellrup, J., Alderborn, G., Mahlin, D., 2015. Inhibition of recrystallization of amorphous lactose in nanocomposites formed by spray-drying. J. Pharmaceut. Sci. 104 (11), 3760–3769. https://doi.org/10.1002/jps.24583.

Hiele, M., Ghoos, Y., Rutgeerts, P., Vantrappen, G., Carchon, H., Eggermont, E., 1988. The 13C lactose breath test using naturally labelled lactose for the diagnosis of lactose malabsorption. J. Lab. Clin. 112 (2), 193–200.

Huncharek, M., Muscat, J., Kupelnick, B., 2008. Impact of dairy products and dietary calcium on bone-mineral content in children: results of a meta-analysis. Bone 43 (2), 312–321.

Huppertz, T., Gazi, I., 2016. Lactose in dairy ingredients: effect on processing and storage stability. J. Dairy Sci. 99, 6842–6851. https://doi.org/10.3168/jds.2015-10033.

Ianiro, G., Pecere, S., Giorgio, V., Gasbarrini, A., Cammarota, G., 2016. Digestive enzyme supplementation in gastrointestinal diseases. Curr. Drug Metabol. 17 (2), 187–193. https://doi.org/10.2174/138920021702160114150137.

Ibba, I., Gilli, A., Boi, M.F. e, Usai, P., 2014. Effects of exogenous lactase administration on hydrogen breath excretion and intestinal symptoms in patients presenting lactose malabsorption and intolerance. Biomed Res. Int. 2014, 680196. https://doi.org/10.1155/2014/680196.

Illanes, A., Guerrero, C., Wilson, L., Conejeros, R., Scott, F., 2016. Lactose: production and upgrading. In: Lactose-Derived Prebiotics: A Process Perspective. Elsevier Academic Press, Cambridge, MA, USA, pp. 1–33. https://doi.org/10.1016/C2014-0-03393-X.

Kerber, M., Oberkanins, C., Kriegshäuser, G., Kollerits, B., Dossenbach-Glaninger, A., Fuchs, D., Ledochowski, M., 2007. Hydrogen breath testing versus LCT genotyping for the diagnosis of lactose intolerance: a matter of age? Clin. Chim. Acta 383, 91–96.

Lee, M., Barrie, S.J., 1996. Breath testing in intestinal disaccharidase deficiency and bacterial overgrowth of the small intestine. J. Nutr. Environ. Med. 6, 43–54.

Leffler, D.A., Dennis, M., Hyett, B., Schuppan, D., Kelly, C.P., 2007. Etiologies and predictors of diagnosis in nonresponsive celiac disease. Clin. Gastroenterol. Hepatol. 5 (4), 445–450.

Lerner, A., Wusterhausen, P., Ramesh, A., Matthias, T., 2018. Celiac disease and lactose intolerance. Int. J. Celiac Dis. 6, 68–70.

Li, R., Roos, Y.H., Miao, S., 2017. Characterization of mechanical and encapsulation properties of lactose/maltodextrin/WPI matrix. Food Hydrocoll. 63, 149–159. https://doi.org/10.1016/j.foodhyd.2016.08.033.

Liebert, A., López, S., Jones, B.L., Montalva, N., Gerbault, P., Lau, W., Thomas, M.G., Bradman, N., Maniatis, N., Swallow, D.M., 2017. World-wide distributions of lactase persistence alleles and the complex effects of recombination and selection. Hum. Genet. 136 (11–12), 1445–1453. https://doi.org/10.1007/s00439-017-1847-y.

Lomer, M.C., Parkes, G.C., Sanderson, J.D., 2008. Review article: lactose intolerance in clinical practice–myths and realities. Aliment. Pharmacol. Ther. 27 (2), 93–103.

Mattar, R., Basile-Filho, A., Kemp, R., dos Santos, J.S., 2013. Comparison of quick lactose intolerance test in duodenal biopsies of dyspeptic patients with single nucleotide polymorphism LCT-13910C>T associated with primary hypolactasia/lactase-persistence. Acta Cir. Bras. 28 (Suppl. 1), 77–82.

Melini, V., Melini, F., 2018. Strategies to extend bread and GF bread shelf-life: from sourdough to antimicrobial active packaging and nanotechnology. Fermentation 4 (1), 9. https://doi.org/10.3390/fermentation4010009.

Melse-Boonstra, A., 2020. Bioavailability of micronutrients from nutrient-dense whole foods: zooming in on dairy, vegetables, and fruits. Front. Nutr. 7.

Micic, D., Rao, V.L., Rubin, D.T., 2019. Clinical approach to lactose intolerance. JAMA 322 (16), 1600–1601. https://doi.org/10.1001/jama.2019.14740.

Mill, D., Dawson, J., Johnson, J.L., 2018. Managing acute pain in patients who report lactose intolerance: the safety of an old excipient re-examined. Adv. Drug Saf. 9, 227–235.

Miller, G.D., DiRienzo, D.D., Reusser, M.E., McCarron, D.A., 2000. Benefits of dairy product consumption on blood pressure in humans: a summary of the biomedical literature. J. Am. Coll. Nutr. 19 (Suppl. 2), 147S–164S.

Misselwitz, B., Butter, M., Verbeke, K., Fox, M.R., 2019. Update on lactose malabsorption and intolerance: pathogenesis, diagnosis and clinical management. Gut 68 (11), 2080–2091. https://doi.org/10.1136/gutjnl-2019-318404.

Morsanuto, V., Galla, R., Molinari, C. e, Uberti, F., 2020. A new palmitoylethanolamide form combined with antioxidant molecules to improve its effectiveness on neuronal aging. Brain Sci. 10 (7), 457. https://doi.org/10.3390/brainsci10070457.

Muraro, A., Werfel, T., Hoffmann-Sommergruber, K., Roberts, G., Beyer, K., Bindslev-Jensen, C., Cardona, V., Dubois, A., du Toit, G., Eigenmann, P., Fernandez Rivas, M., Halken, S., Hickstein, L., Høst, A., Knol, E., Lack, G., Marchisotto, M.J., Niggemann, B., Nwaru, B.I., Papadopoulos, N.G., Poulsen, L.K., Santos, A.F., Skypala, I., Schoepfer, A., Van Ree, R., Venter, C., Worm, M., Vlieg-Boerstra, B., Panesar, S., de Silva, D., Soares-Weiser, K., Sheikh, A., Ballmer-Weber, B.K., Nilsson, C., de Jong, N.W., Akdis, C.A., 2014. EAACI Food Allergy and Anaphylaxis Guidelines Group. EAACI food allergy and anaphylaxis guidelines: diagnosis and management of food allergy. Allergy 69 (8), 1008–1025. https://doi.org/10.1111/all.12429.

Nakrani, M.N., Wineland, R.H., Anjum, F., 2023. Physiology, glucose metabolism. In: StatPearls [Internet]. StatPearls Publishing, Treasure Island (FL). PMID: 32809434.

Nicklas, T.A., Qu, H., Hughes, S.O., He, M., Wagner, S.E., Foushee, H.R., Shewchuk, R.M., 2011. Self-perceived lactose intolerance results in lower intakes of calcium and dairy foods and is associated with hypertension and diabetes in adults. Am. J. Clin. Nutr. 94 (1), 191–198.

Nielsen, S.D., Knudsen, L.J., Bækgaard, L.T., Rauh, V., Larsen, L.B., 2022. Influence of lactose on the Maillard reaction and dehydroalanine-mediated protein cross-linking in casein and whey. Foods 11 (7), 897. https://doi.org/10.3390/foods11070897.

Nowak-Wegrzyn, A., Chehade, M., Groetch, M.E., Spergel, J.M., Wood, R.A., Allen, K., 2017. International consensus guidelines for the diagnosis and management of food protein-induced enterocolitis syndrome. J. Allergy Clin. Immunol. 139 (4), 1111–1126.

O'Connell, S., Walsh, G., 2006. Physicochemical characteristics of commercial lactases relevant to their application in the alleviation of lactose intolerance. Appl. Biochem. Biotechnol. 134, 179–191. https://doi.org/10.1385/ABAB:134:2:179.

Ojetti, V., La Mura, R., Zocco, M.A., Cesaro, P., De Masi, E., Mazza, A.L., et al., 2008. Quick test: a new test for the diagnosis of duodenal hypolactasia. Dig. Dis. Sci. 53, 1589–1592.

Ojetti, V., Nucera, G., Migneco, A., Gabrielli, M., Lauritano, C., Danese, S., Zocco, M.A., Nista, E.C., Cammarota, G., De Lorenzo, A., Gasbarrini, G., 2005. High prevalence of celiac disease in patients with lactose intolerance. Digestion 71 (2), 106–110.

Palsson, O.S., Whitehead, W., Törnblom, H., Sperber, A.D., Simrén, M., 2020. Prevalence of Rome IV functional bowel disorders among adults in the United States, Canada, and the United Kingdom. Gastroenterology 158, 1262–1273.

Pasin, G., Comerford, K.B., 2015. Dairy foods and dairy proteins in the management of type 2 diabetes: a systematic review of the clinical evidence. Adv. Nutr. 6 (3), 245–259.

Portnoy, M., Barbano, D.M., 2021. Lactose: use, measurement, and expression of results. J. Dairy Sci. 104, 8314–8325. https://doi.org/10.3168/jds.2020-18706.

Robles, L., Priefer, R., 2020. Lactose intolerance: what your breath can tell you. Diagnostics 10, 412.

Romagnuolo, J., Schiller, D., Bailey, R.J., 2002. Using breath tests wisely in a gastroenterology practice: an evidence-based review of indications and pitfalls in interpretation. Am. J. Gastroenterol. 97, 1113–1126.

Sadovnikova, A., Garcia, S.C., Hovey, R.C., 2021. A comparative review of the cell biology, biochemistry, and genetics of lactose synthesis. J. Mammary Gland Biol. Neoplasia 26 (2), 181–196. https://doi.org/10.1007/s10911-021-09490-7.

Sánchez-García, Y.I., Gutiérrez-Méndez, N., Salmerón, I., Ramos-Sánchez, V.H., Leal-Ramos, M.Y., Sepúlveda, D.R., 2021. Mutarotation and solubility of lactose as affected by carrageenans. Food Res. Int. 142, 110204. https://doi.org/10.1016/j.foodres.2021.110204.

Santonocito, C., Scapaticci, M., Guarino, D., Annicchiarico, E.B., Lisci, R., Penitente, R., et al., 2015. Lactose intolerance genetic testing: is it useful as routine screening? Results on 1426 south-central Italy patients. Clin. Chim. Acta 439, 14–17.

Savaiano, D.A., 2014. Lactose digestion from yogurt: mechanism and relevance. Am. J. Clin. Nutr. 99 (5 Suppl. 1), 1251S–1255S.

Ségurel, L., Bon, C., 2017. On the evolution of lactase persistence in humans. Annu. Rev. Genom. Hum. Genet. 18, 297–319. https://doi.org/10.1146/annurev-genom-091416-035340.

Skripak, J.M., Matsui, E.C., Mudd, K., Wood, R.A., 2007. The natural history of IgE-mediated cow's milk allergy. J. Allergy Clin. Immunol. 120 (5), 1172–1177.

Slavin, J., 2013. Fiber and prebiotics: mechanisms and health benefits. Nutrients 5 (4), 1417–1435. https://doi.org/10.3390/nu5041417.

Sørensen, K.I., Curic-Bawden, M., Junge, M.P., Janzen, T., Johansen, E., 2016. Enhancing the sweetness of yoghurt through metabolic remodeling of carbohydrate metabolism in *Streptococcus thermophilus* and *Lactobacillus delbrueckii* subsp. *bulgaricus*. Appl. Environ. Microbiol. 82 (12), 3683–3692. https://doi.org/10.1128/AEM.00462-16.

Stourman, N., 2023. Lactose in food technology: innovations and applications. J. Food Technol. Preserv. 7 (4), 186. https://doi.org/10.35841/2591-796X-7.4.186.

Swagerty, D.L., Walling, A.D., Klein, R.M., 2002. Lactose intolerance. Am. Fam. Physician 65, 1845–1850.

Taglieri, I., Macaluso, M., Bianchi, A., Sanmartin, C., Quartacci, M.F., Zinnai, A., Venturi, F., 2021. Overcoming bread quality decay concerns: main issues for bread shelf life as a function of biological leavening agents and different extra ingredients used in formulation. A review. J. Sci. Food Agric. 101 (5), 1732–1743. https://doi.org/10.1002/jsfa.10816.

Ugidos-Rodriguez, S., Matallana-Gonzalez, M.C., Sanchez-Mata, M.C., 2018. Lactose malabsorption and intolerance: a review. Food Funct. 9, 4056–4068. https://doi.org/10.1039/c8fo00555a.

Usai-Satta, P., Lai, M., Oppia, F., 2022. Lactose malabsorption and presumed related disorders: a review of current evidence. Nutrients 14 (3), 584. https://doi.org/10.3390/nu14030584.

Van der Merwe, J., Steenekamp, J., Steyn, D., Hamman, J., 2020. The role of functional excipients in solid oral dosage forms to overcome poor drug dissolution and bioavailability. Pharmaceutics 12 (5), 393.

Varjú, P., Gede, N., Szakács, Z., Hegyi, P., Cazacu, I.M., Pécsi, D., Fábián, A., Szepes, Z., Vincze, Á., Tenk, J., Balaskó, M., Rumbus, Z., Garami, A., Csupor, D., Czimmer, J., 2018. Lactose intolerance but not lactose maldigestion is more frequent in patients with irritable bowel syndrome than in healthy controls: a meta-analysis. Neuro Gastroenterol. Motil. 31 (5), e13527.

Varjú, P., Ystad, B., Gede, N., Hegyi, P., Pécsi, D., Czimmer, J., 2020. The role of small intestinal bacterial overgrowth and false positive diagnosis of lactose intolerance in southwest Hungary-A retrospective observational study. PLoS One 15 (5), e0230784. https://doi.org/10.1371/journal.pone.0230784.

Venter, C., Patil, V., Grundy, J., Glasbey, G., Twiselton, R., Arshad, S.H., Dean, T., 2016. Prevalence and cumulative incidence of food hypersensitivity in the first 10 years of life. Pediatr. Allergy Immunol. 27 (5), 452–458. https://doi.org/10.1111/pai.12564.

Yu, R.K., Itokazu, Y., 2014. Glycolipid and glycoprotein expression during neural development. In: Yu, R., Schengrund, C.L. (Eds.), Glycobiology of the Nervous System, Advances in Neurobiology, vol. 9. Springer, New York, NY. https://doi.org/10.1007/978-1-4939-1154-7_9.

Part 4

Utilization of whey as a dairy waste

Chapter 21

Whey as a dairy waste: Treatments in effluent plant and disposal methods in environment

Mohan Li[1, 2, 3], Yumeng Zhang[1], Xiqing Yue[1] and Junhua Shao[1]

[1]*College of Food Science, Shenyang Agricultural University, Shenyang, China;* [2]*State Key Laboratory of Dairy Biotechnology, Shanghai Engineering Research Center of Dairy Biotechnology, Dairy Research Institute, Bright Dairy & Food Co. Ltd., Shanghai, China;* [3]*School of Life Sciences & Biotechnology, Shanghai Jiao Tong University, Shanghai, China*

1. Introduction

Faced with the growing demand for milk and its processed livestock products, dairy products are expanding on a global scale. It is mainly attributed to the fact that the utilization of dairy products assists in ameliorating the production of metabolic diseases in the human body (Saubenova et al., 2024). However, dairy processing is one of the main ways of being a serious consumer of water resources and producing large quantities of wastewater (Nazos et al., 2023; Bhuvanendran et al., 2024a). The 80%–95% of the volume of milk required in the cheese production process generates whey wastewater and retains more than half of the original milk nutrients (Brandelli et al., 2015; Uribarrena et al., 2024). Whey is a yellow-green liquid component of the organic matter remaining after the removal of casein and most of the fat from the curd in cheese making (Rocha-Mendoza et al., 2021). Depending on the processing method and raw material, the main composition and color of whey can vary. The whey residue contains mainly lactose, protein, and small amounts of nutrients such as minerals (sodium, potassium, and calcium salts), vitamins, and fats, in addition to a high content of water (Boscaini et al., 2023).

Whey has been evaluated as a valuable industrial waste based on the presence of highly nutritious compounds in it and its promising applications (El-Aidie and Khalifa, 2024; Castro-Muñoz et al., 2022). Lactose, as the most significant component of whey solids, is commonly used to regulate the balance of intestinal flora. Meanwhile, the existence of active sites such as glycosidic and carbon bonds in lactose has enabled more lactose to be converted into derivatives such as oligogalactose for a wider range of applications (Rocha and Guerra, 2020; Liu et al., 2023). The whey proteins (WPs) are obtained after ultrafiltration of whey and are isolated to obtain β-lactoglobulin (β-lg), α-lactalbumin (α-la), etc. (Zeng et al., 2024). All of them are vital sources of essential branched chain amino acids and used for health enhancement at all ages (Yigit et al., 2023). Because of the very existence of whey-loaded nutrients, there are many challenges that may be faced in processing whey.

As a matter of fact, whey, the most serious pollutant produced during processing, transport, and distribution, may be released directly into the environment. The direct disposal of whey is a major loss of potential nutrients and energy (Zandona et al., 2021). Firstly, the richness of organic nutrients in whey leads to high levels of biological oxygen demand (BOD) and chemical oxygen demand (COD). Untreated discharges to rivers lead to excessive oxygen depletion in the water and eventually to eutrophication (Macedo et al., 2021). Direct dumping on land is another direct whey disposal route that is not only harmful to crop growth but also produces volatile compounds that pollute the air (Ayed et al., 2023). Similarly, dairy production is a substantial source of greenhouse gas emissions (Hassoun et al., 2024). Consequently, management strategies for whey wastewater have received much attention in order to reduce pollution to the environment and to minimize the cost of wastewater treatment for dairies.

Whey. https://doi.org/10.1016/B978-0-443-27648-4.00002-5

Typically, the removal of soluble organic matter as well as suspended matter from whey can be achieved using conventional physicochemical methods (precipitation, coagulation, adsorption, etc.) (Adesra et al., 2021). This treatment process is gradually being replaced by the emerging electrochemical technology due to its higher cost and the requirement of additional chemical reagents (Shah et al., 2024). Nevertheless, aerobic and anaerobic biological treatments are the main forms of whey treatment at present, which are mainly founded on the growth of certain microorganisms under aerobic or anaerobic conditions of organic materials in wastewater (Bazrafshan et al., 2021). More importantly, anaerobic digestion and codigestion processes with other substrates are sustainable technological means to minimize wastage and generate bioenergy such as biogas (Charalambous et al., 2023; Stanchev et al., 2020). This paper provides an overview of the whey production process and aspects of its composition and physicochemical properties. Also, the focus is on outlining the common treatment of whey as a waste in effluent plants and in its environment. With these considerations, this helps the dairy industry to enhance the environmental impacts of whey and provide effective solutions for sustainable management.

2. The main components of whey wastewater composition and properties

2.1 Whey wastewater generation

The liquid stream obtained after cheese and casein production in the dairy industry is the main source of whey. Cheese, whey and their by-products from milk processing are shown in Fig. 21.1. The yield of whey can reach up to 200 million tonnes per year, while the total yield is growing at a rate of 1%–2% per year as the demand for cheese expands. In fact, for each kilogram of cheese produced there is an addition of 9 L of whey, of which 50% is wasted as waste (Aranda-Jaramillo et al., 2023). Depending on the cheese production process, sweet whey and sour whey are obtained in yellow, yellow-green or even light blue colors. Sweet whey is usually produced in the presence of rennet and maintains a pH around 6.5. Acid whey, the residue derived from the addition of organic or inorganic acids to milk, can have a pH as low as about 3.8. Apart from the difference in acidity between the two types of whey, sweet whey exhibits a lower mineral content (2.5–4.7 g/L) while acid whey has a lower lactose content (44–46 g/L) (Soumati et al., 2023). Beyond that, the whey produced is enabled to reuse the primary whey as an ingredient for white milk cheese, whey cheese processing (Fancello et al., 2024). At this point, the secondary whey produced contains about 60% of the dry matter content of the primary whey, with a much lower protein concentration (Asunis et al., 2020; Hassoun et al., 2024).

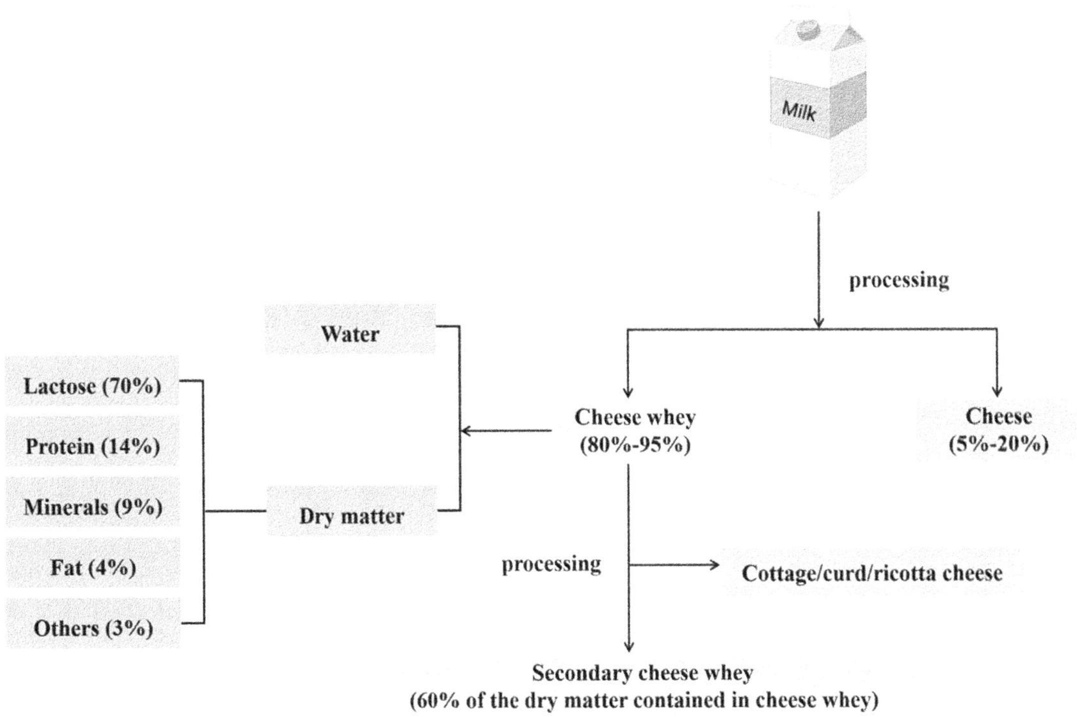

FIGURE 21.1 Cheese, cheese whey, and their by-products from milk processing.

2.2 Properties and composition of whey wastewater

Whey is considered a good source of lactose, WPs, essential minerals, soluble vitamins, and lipids and has a lactose content of 45 g/L (Smithers, 2008; Zotta et al., 2020). The high content of lactose represents a significant impediment to the direct discharge of whey. If lactose can be efficiently recycled and incorporated into foods after fermentation and conversion in bacterial fibers and fatty acid esters, food texture and flavor can be added. However, to enhance digestibility, it usually needs to be hydrolyzed to galactose before being applied (Walzem et al., 2002; Bhuvaneshwari et al., 2022). Another crucial factor affecting whey emissions relates to the production of proteins that are poorly susceptible to direct degradation (Lappa et al., 2019). WPs consists mainly of β-lg (55%), α-la (20%), bovine serum albumin, and immunoglobulins. Additionally, lactoferrin, casein peptides, and milk fat globule membrane proteins may also be contained (Tsermoula et al., 2021; Mehra et al., 2021). The high concentrations of phosphorus and nitrogen in whey constitute the major minerals, with possibly small amounts of sodium, potassium and calcium ions, broadly used in the production of fertilizers. The presence of oil and grease in wastewater discharged directly may form a film on the surface of the water, thereby preventing the passage of oxygen, requiring pretreatment of the effluent. Nevertheless, the main composition of whey can be varied in the source of milk, type of cheese, and adjustments in the processing (Ganju and Gogate, 2017). As a result of the use of different acids and bases during cheese preparation, the various whey obtained fluctuates accordingly in terms of pH.

During cheese production, 55% of the organic and inorganic nutrients in milk can be retained in the whey (Addai et al., 2020). Owing to the high content of organic compounds in whey, the direct discharge of 50% of wastewater globally not only leads to a significant loss of nutrients, but also creates a serious environmental hazard to the industry (Caltzontzin-Rabell et al., 2024; Casallas-Ojeda et al., 2021). As a matter of fact, the rapid depletion of dissolved oxygen in untreated whey after its direct discharge into rivers or oceans is detrimental to the survival of organisms. Typically for dairy wastewater, BOD and COD, which are the oxygen requirements of microorganisms and chemically decomposed organic compounds, act as important parameters for assessing wastewater treatment (Uald Lamkaddam et al., 2023). The limitation of whey in discharge and subsequent application lies precisely in the high contamination caused by the presence of BOD (40–60 g/L) and COD (50–80 g/L) in it. Taking on the primary burden of high organic loads in whey, 1 kg of lactose content in the discharge can generate 1.3 kg of COD (Kumar Awasthi et al., 2022). Lately, a study was conducted to assess the environmental impacts of wastewater indicators as well as discharges in the dairy industry (Ekka et al., 2021). In dairies without pretreatment, dairy wastewater can carry COD and BOD up to 13,995 and 8501 mg/L respectively, which seriously exceeds the standard limits for permissible discharges. The nitrogen and phosphorus levels reached 319 and 30.8 mg/L, respectively. The presence of these compounds in the wastewater proved to be associated with severe toxic accumulation. Consequently, effective treatment of whey generated from the dairy plant needs to be implemented subsequently to mitigate the impact on the environment.

3. Treatment of whey in effluent plant and the environment

Whey is recognized as a major load to effluent plant and the friendly environment because of its high concentration of BOD and COD as well as a variety of residual nutrients comprising of organic matter (Bhuvanendran et al., 2024b). Currently, many effluent plants lack a core knowledge of energy requirements, where the energy produced by whey in the digestion pathway is directly burned or discarded and cannot be fully utilized. Meanwhile, whey being an organic waste, direct discharge on soil is unfavorable for crop yield with serious environmental hazards (Yadav et al., 2015). Whey disposal is gradually coming into the limelight under strict laws in various countries. Overall, effective treatment processes for whey waste streams are the current target for improvement for promoting sustainable energy generation along with reducing unnecessary energy consumption. The major disposal of whey in effluent plants and in the environment are shown in Fig. 21.2.

3.1 Physicochemical techniques

The high content of proteins, lactose, and lipids in whey wastewater can be removed by physicochemical methods. Traditional physicochemical processes are achieved by precipitation, dissolved air flotation, aeration and degassing. The use of membrane filtration (nanofiltration, reverse osmosis and dialysis) further enhances the filtration of whey wastewater. However, membrane filtration as well as coagulation and adsorption are hardly comprehensive pretreatment measures. Indeed, the formation of sludge over a long period of time results in clogging of the membrane pores, thereby impacting the wastewater passage rate. Furthermore, the addition of chemicals such as flocculants, coagulants affects the

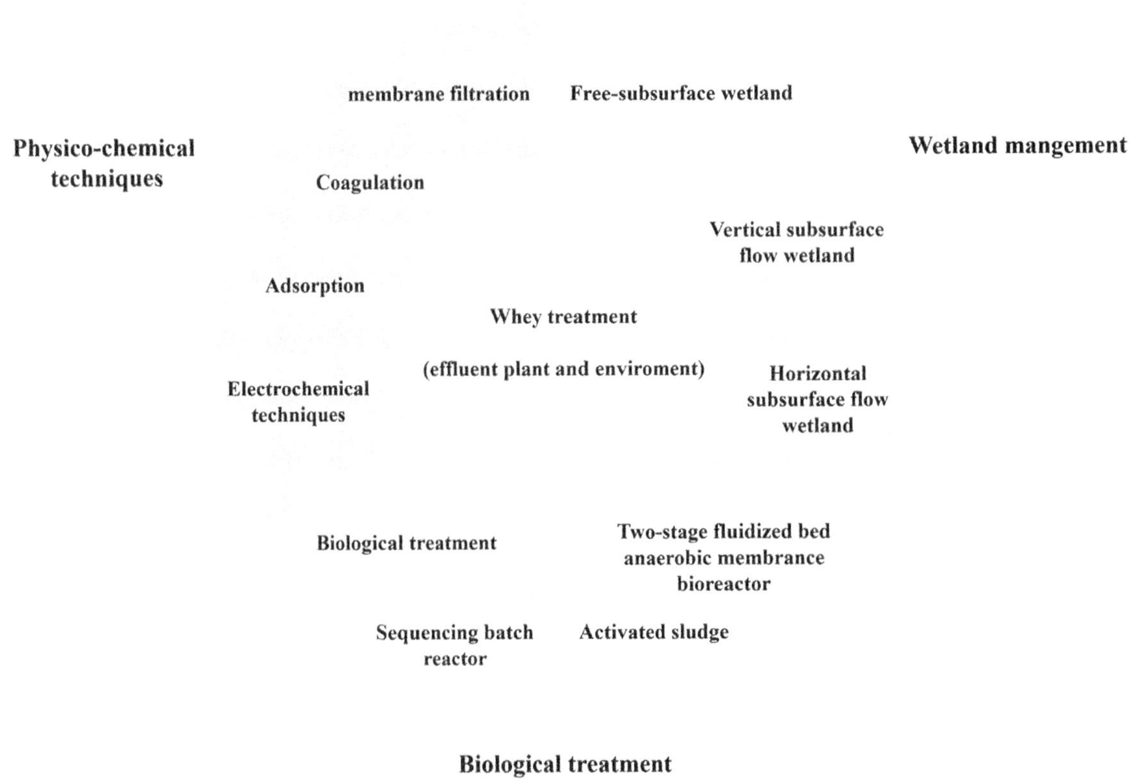

FIGURE 21.2 The major disposal of whey in effluents plants and in the environment.

stability of the particles in the wastewater, making it harder to achieve the desired treatment results and even possibly generating secondary pollution (Ahmad et al., 2019). In another advanced oxidation process widely used for the removal of organic compounds, it is achieved based on the simultaneous combination of coagulation and oxidation functions. The removal of recalcitrant organic compounds in whey is targeted more effectively under conditions such as ultraviolet light, ozone, and further reduces the surrounding toxicity (Mousazadeh et al., 2021).

Electrochemical techniques such as electrodeposition, electroflocculation (EC), and electrooxidation (EO) have gradually been recognized as promising alternatives for the removal of organic charges in whey treatment (Chandra et al., 2018). EC works mainly by applying a potential between electrodes immersed in a solution, with hydroxyl ions and hydrogen are formed at the cathode. Optimization of the conditions of the EC reactor (time, pH, current intensity) and other variables dramatically reduced the COD as well as turbidity indicators in the wastewater to 97% and 99% (Tezcan Un et al., 2014; Melchiors et al., 2016). Compared with other technologies, EC has more challenges in treating the more difficult to degrade organic compounds in whey. Whereas EO technology proved to be more effective in the oxidative decomposition of organic compounds, but simultaneously has the limitation of period of a long period of duration. Evidently, the EC process in combination with other technologies is an excellent solution strategy for the sustainable treatment of whey (Elia et al., 2023). explored the COD removal from goat cheese whey under optimal reaction conditions for EC, EO alone and combined treatment. Comparing the EC/EO treatment alone for 2 h and the combined treatment technique for 90 min, the COD removal rate sharply increased from 11.7% to 82.4%. It was found that the majority of COD degradation was achieved in 90 min compared with EO, which requires up to 21 h (Linares-Hernández et al., 2010). The combined EC and EO technology was capable of removing up to 80% of suspended solids and degradation of organic compounds by more than 75% (Saeed et al., 2023). It is of concern that most of these electrochemical techniques are currently done on a laboratory scale, setting them apart from the results of real whey treatments. More importantly, most of the coreactors focus on the removal of specific pollutants, and overall wastewater treatment applications should receive more attention.

3.2 Aerobic digestion

Biological treatment methods for whey are mainly focused on the degradation processes of lipids and other organic substrates by living microorganisms under aerobic or anaerobic conditions. The production of odorous substances such as volatile fatty acids decreases at the same time through the fermentation process of the corresponding microorganisms (Porwal et al., 2015). Practically, the production of higher economic value products by microorganisms through metabolic pathways stands out as a distinctive feature that is more preferable to physicochemical treatment techniques (Abdel-Fatah, 2023). Activated sludge, sequencing batch reactor (SBR), membrane bioreactor, aeration lagoon, and rotating biological contactor are commonly employed technologies for aerobic processes (Sar et al., 2022; Sakkas et al., 2014).

One common means of whey treatment in factories is the aerobic activated sludge process, an excellent way to reduce organic matter (Murshid et al., 2023). The microorganisms were introduced into the whey wastewater environment and waited for them to grow and then isolated. However, the conventional process produces a sludge content vulnerable to secondary contamination. Moreover, settling of excessive sludge contributes to reduced biomass recovery. Initially, the favorable stability of SBR was utilized as an effective treatment unit for whey. It is closed system and better to avoid the occurrence of contamination from external microorganisms (Nath et al., 2016; Jena et al., 2020) introduced dairy industrial wastes into SBR and microorganisms in activated sludge acted as an integral component in degrading pollutants in SBR. COD degradation rates of 90% and 80% were achieved under 8 h anoxic and 4 h aerobic continuous reaction, respectively. At the same time, suitable pelletization of activated sludge in the reactor was found to be conducive to less generation of phosphate and nitrate in the aerobic reaction. Further optimization was carried out by adding intermittent aeration periods to obtain an Intermittently aerated SBR (Rodriguez-Sanchez et al., 2024).

The presence of high nutrients in whey wastewater for biological nitrogen removal is an essential factor in the process. In the conventional whey digestion process ammonia was first converted to nitrite by bacteria, with subsequent dependence on oxygen and oxidizing bacteria for conversion to nitrite (Xi et al., 2022). The lower the COD/N ratio in an SBR, the higher the digestive activity of the microorganisms. Bucci et al. (2020) studied that by adding nitrogen nutrients for use in a bioaerobic process, more appropriate COD/N ratios confirmed higher nitrification activity. It was also found that the reduction of organic loads in organic wastewater was more beneficial for microorganisms to carry out digestion. Actually, most aerobic treatments are not suitable for high concentrations of whey effluents, owing to the fact that the transfer of oxygen does not take place smoothly. Hence, pretreatment before aerobic digestion is considered to be a very necessary process. Normally, adequate dilution of whey before microbial fermentation becomes feasible (Liu et al., 2018). The removal of some organic compounds, suspended solids, etc., using physicochemical means such as $Ca(OH)_2$, NaOH, and iron salts and aluminum salts is equally effective, reducing the COD content of the raw whey by approximately 50% (Prazeres et al., 2013; Rivas et al., 2011). Recently Altiparmaki et al. (2022), explored and found that hydrothermal treatment at high temperatures and pressures of 120–200°C for 30–120 min pretreatment technique assisted in the reduction of COD from 75 to 62.2 g/L. Nevertheless, the degradation of organic compounds was accompanied by a significant rise in nitrate and phosphate, which warrants further improvement.

Aerobic processes tend to possess low efficiency as a consequence of high lactose content and low buffering capacity. Meanwhile, aerobic digestion alone is insufficient to completely degrade all organic compounds in most of the digestion and relies on a combination of aerobic and anaerobic technologies for treatment. Frigon et al. (2009) studied the treatment of whey in SBR digesters under psychrophilic (20°C) anaerobic conditions followed by low oxygen concentrations, with an overall COD removal of above 97% after 3 days.

3.3 Anaerobic digestion

Anaerobic treatment offers an effective strategy to utilize the high organic content of whey to produce biofuels. Actually, the lack of nitrogen content in whey prevents its aerobic digestion. Anaerobic digestion of whey has been shown to be the most dominant way to reduce the high COD (Antonopoulou et al., 2021). Anaerobic digestion characterizes the hydrolysis of complex organic matter in the presence of extracellular enzymes, using anaerobic microorganisms to progressively produce hydrogen, acetic acid and methane (Carvalho et al., 2013). Previously, anaerobic digestion of whey for the preparation of hydrogen has been extensively studied for its significance in terms of energy saving and pollution reduction as an economically viable option (Rasul et al., 2022). Hydrogen was demonstrated to be acquired efficiently through whey digestion as a clean alternative energy source for power generation, fuel, etc. (Gomez-Romero et al., 2014; Policastro et al., 2023). Apart from that, the generation of biogas containing methane from whey anaerobic digestion represents the current mainstream of interest. Biogas generated from whey anaerobic treatment is normally realized using Up-flow anaerobic sludge blanket and Up-flow anaerobic filter devices. More recently, the two-stage process has gained more attention in order to meet higher biogas volumes as well as more efficient elimination rates of organic matter (Batista et al., 2023). It was found that the process technology of two-stage fluidized bed anaerobic membrane bioreactor is feasible, while it was also demonstrated that the COD removal efficiency was as high as 89%, and that the fluidized state of sulphide granular activated carbon improved the reactor's performance (Parihar et al., 2024).

Notably, the outcome of such conventional utilization of whey for biogas production in effluent plants can be constrained by multiple factors. The digestive performance of a single digestive substrate suffers from an imbalance in its carbon to nitrogen ratio, while its biogas production is less efficient and productive (Miranzadeh et al., 2024). In response to this drawback, the current advocacy of utilizing whey with other organic wastes as a substrate to participate in the codigestion process. Iglesias-Iglesias et al. (2021) implemented codigestion of sewage sludge and whey in a continuously stirred reactor to create medium-chain carboxylic acids, which are predominantly caproic acids, as well as volatile fatty acids. Both served as an essential source for the production of polyhydroxyalkanoates, effectively confirming the feasibility of codigestion to produce organic acid streams. In another study Hallaji et al. (2019), compared the activity of activated sludge, fruit wastes, and whey anaerobic codigestion in several ways. An improvement in methane production and efficiency was found for whey mixed with either or both, with the best results in the presence of all three (85% sludge, 15% fruit waste and whey) and a 9% increase in digested COD. A major reason for this was that the carbon to nitrogen ratio at this time was more appropriate for the growth of methanogenic bacteria and hence the production of methane. Overall, most of the livestock manure, agricultural residues, and wasted food were employed as external waste participation (Bella and Venkateswara Rao, 2022; Bella et al., 2023). More importantly, the microorganisms involved in the anaerobic digestion of whey actually represent a very extremely small fraction. Consequently, in order to explore the close link between the generating capacity of a biogas system and the composition of the microbial community, research has been carried out to isolate cocultures of *Tissierella praeacuta*, which is suitable for anaerobic digestion, as well as the ammonia-resistant bacterium *M. bourgensis MS2*, from the flora, using a combination of multilogistics (Agostini et al., 2024). Coculture with isolated bacteria showed bioaugmentation results, with a noticeable enhancement of methane production by 26%–34%.

Of greater significance is the fact that the nutrients required for the growth of photosynthetic microorganisms (microalgae), consisting mainly of carbon, nitrogen and phosphorus, are exactly what whey possesses. Microalgae have been proved to be directly applicable for bioenergy production such as biogas (Bhatia et al., 2021). Consequently, the cultivation of whey as a substrate for microalgal systems not only improves biogas production, but also efficiently processes whey and generates sustainable energy. *Chlorella vulgaris*, *Spirulina*, and *Scenedesmus* sp. are microalgal species commonly exploited for whey bioremediation. Whey serves as an alternative source for acquiring microalgal biomass and improving the metabolic processes of microalgae (Braun et al., 2024). From an economic point of view, the cost of the culture medium is substantially reduced, making it undoubtedly a favorable option for microalgae acquisition. Nevertheless, the higher content of organic matter included in the whey wastewater presents difficulties to be metabolized efficiently by the microalgae. In this case, the dilution of whey using the addition of 50% water followed by the cultivation of microalgae is a usual method (Pereira et al., 2019; Patel et al., 2020) have sought to avoid culturing microalgae at the expense of consuming large quantities of water, alleviated by multiple chemical pretreatment steps of whey. The study showed that the organic and inorganic degradation of the whey matrix reached more than 90% after 9 days of microalgae cultivation. In another study, a coculture of microalgae and bacteria was exploited to further the goal of sustainable production (Makut et al., 2019). By screening for strains that were whey resistant and adapted to the culture conditions at the refinery, it was shown that the experimental group in the coculture had higher biomass titers than the microalgae culture alone. Additionally Salati et al. (2017) cultured microalgae progressively in batches at various carbon concentrations as well as under environmental conditions, which reduces the greater impact of high concentrations of organic matter. Except for gaining high energy yield, culture of *Arthrospira platensis* SAG 21.99 with whey as substrate gave rise to production of proteins with antiinflammatory, antioxidant and other activities. Studies of microcultures of *Spirulina platensis* under 0.8% $v\ v^{-1}$whey substrate managed to reach a yield of 1.57 g/L while producing 4.7 mg/mL of phycobiliprotein, which is clearly superior to findings under nutrition alone (Cavallini et al., 2024). Overall, mixed-nutrient cultures involving whey have been widely used by microalgae as an alternative strategy to improve biomass yield and rate and enhance the ability to prepare valuable products. Indeed, if the biogas could be further purified to methane, it would help its application in a wider range of fields. Rincón-Pérez et al. (2021) have demonstrated a 21% enhancement of methane production rates for anaerobic digestion of microalgae in an anaerobic sequential batch reactor. Moreover, the codigestion reaction rate is faster with relatively weakly alkaline microalgae (pH 7.5).

Of note, the use of pretreatment before anaerobic digestion can be effective in enhancing biogas production for whey with high organic matter content and complex composition. The common pretreatments involve physical, chemical, biological treatments or a combination of these options to diminish the production of inhibitory substances during the digestion process, favoring the further evolution of the field of anaerobic digestion. Bella et al. (2023) provided the first comprehensive evaluation of three pretreatment options, physical (ultrasonication), chemical (ozone), and biological treatment (enzymes), for the upgrading of biogas productivity and effective separation of organic matter. Compared with unpretreated whey, the three options enhanced methane production rate by 66.3%, 64.18%, and 70.7%, respectively. It was a consequence of enzyme hydrolysis more conducive to rapid microbial metabolism. Furthermore, the high intensity and continuous ultrasound contributed more to the rapid decomposition of organic matter, as evidenced by even a better COD degradation rate.

More recently, the supplementation of anaerobic digestion of whey mono-substrate has been progressively adopted in the form of additives to enhance biogas production. Madrigal et al. (2022) extracted biochar (0.5–2.0 g) from farm waste (cow dung) to add to pure whey wastewater that showed a better degradation of organic matter and sustainable energy under favorable conditions at pH 8.5.

Other than energy recovery in anaerobic digestion of whey, volatile fatty acids are formed in large quantities as intermediate products. It acts as a substrate in the anaerobic reaction for methanogens to grow and thus produce methane. Nevertheless, as the rate of whey feed is higher than its specific residence time in the digester, the volatile fatty acids gradually accumulate. At this point, the accumulation of volatile fatty acids is above the consumption of methanogens and the microenvironment shifts to slightly acidic (Charalambous et al., 2020; Domingos et al., 2018). The biological activity of methanogens can be greatly reduced, destabilizing the anaerobic digestion of whey. For this reason Casallas-Ojeda et al. (2024) have revealed that the rate of anaerobic digestion of whey is considerably higher in thermophilic conditions at 35°C than in chilled conditions at 20°C. Meanwhile, organic loading was a more prominent influence on the accumulation of volatile fatty acids. The digestion process has been optimized in terms of temperature, degree of organic loading, the composition of the raw material and other aspects of the anaerobic digestion of whey.

3.4 Wetland management

Practically, it has been found that approximately 4–11 million tonnes of whey wastewater are directly discharged and deposited in the environment every year in the cheese industry, damaging the ecosystem severely (Mahmoudi et al., 2024). Also, direct treatment of whey wastewater releasing unpleasant odors can be very cumbersome. An efficient and cost-effective artificial wetland system interweaving vegetation, soil and microorganisms has been applied as a treatment option targeting the reduction of environmental hazards from whey (Comino et al., 2011; Sultana et al., 2016). Phytoremediation is the transfer of ambient oxygen into the substrate with the help of plants, causing the growth of corresponding microorganisms in the root environment, which ultimately benefits the degradation of harmful substances in whey (Kataki et al., 2021; Yazdani and Golestani, 2019).

In fact, the presence of substrate enables better survival of microorganisms while acting as a catalyst for the degradation of organic components in whey wastewater. Traditionally, sand, gravel, limestone, etc. are utilized as filler substrates for artificial wetland systems (Parde et al., 2021). Apart from that Kotsia et al. (2024), have investigated the replacement of conventional gravels by potential resources (recycled construction and demolition wastes, rock-processing residues), achieving more than 79% removal of BOD and COD. The turbidity and suspended solids content were markedly reduced and showed no toxic effects on the environment. Owing to the ability to receive most of the organic matter during the first stage of the process in a vertical flow wetland, a higher removal rate of organic compounds was observed. Nevertheless, more and more research is currently being conducted on the utilization of organic waste substrates (biochar, seaweeds, rice husks, etc.) as a better application approach to putting the waste to valuable use (Wang et al., 2023). More importantly, macrophyte species are also a major influence on a wetland system's capacity to serve as a suitable growing environment for microorganisms as well as increase solids deposition. Seasonal and climatic variations similarly possess potential impact on microorganisms in whey and artificial wetlands (Allen et al., 2023). To further ameliorate the deficiencies of conventional wetland systems, hybrid wetland systems that combine aerobic and anaerobic treatment are being studied more and more extensively. Nocetti et al. (2024b) demonstrated the highest surface COD removal of 45.04/20.15 $g\,m^{-2}\,d^{-1}$ using a hybrid system of vertical subsurface flow wetland (VF) and a horizontal subsurface flow wetland (HSSF). Denitrification was shown to be beneficial for total nitrogen removal from whey at the HSSF stage by a pilot-scale hardware system. Indeed, the pilot-scale hardware system was applied to identify the role position of each phase in the mixed wetland treatment. Nocetti et al. (2024a) discovered over 90% removal of organic compounds in whey treated for 9 months with a tandem system of VF, HSFF, and free-surface wetlands. With the assistance of the HSFF system, it was clear that difficult to be degraded compounds were degraded at a higher rate in the VF stage while total nitrogen removal was higher in the free-surface stage.

4. Conclusion

In the face of growing global demand for livestock products, whey continues to be produced as a major outflow from dairy production. By analyzing the main components of whey, it gradually serves as a valuable by-product on account of the presence and wide application of proteins, carbohydrates and various nutrients. Apart from the utilizable nutrients in whey, the contamination created by the loaded high content of organic compounds needs to be dealt with in a timely manner. As a matter of fact, effective waste management techniques are essential for the conservation of the environment and for the disposal costs of food companies. To date, physicochemical and biological methods are the common treatment processes applied in effluent plants after whey entry. Nevertheless, biological treatment in aerobic or anaerobic is of more interest when viewed from the perspective of generating valuable bioenergy such as biogas from whey and reducing the carbon footprint. This chapter takes anaerobic codigestion, pretreatment and as substrate for microalgae in terms of improving the quality and yield of sustainable energy recovery from whey. Wetland treatment is also presented as another promising alternative to improve organic removal efficiency. Improvements in current methods, however, remain in the laboratory on a small scale and require further consideration of the feasibility of the technology and the economic conditions for implementation. Overall, the treatment of whey in effluent plants and in the environment is a major research area facing the dairy industry. This is not only an opportunity for the sustainability of valuable resources, but also a vital goal to reduce environmental damage to the survival of various organisms.

References

Abdel-Fatah, M.A., 2023. Integrated management of industrial wastewater in the food sector. Sustainability 15, 16193.

Addai, F.P., Lin, F., Wang, T., Kosiba, A.A., Sheng, P., Yu, F., Gu, J., Zhou, Y., Shi, H., 2020. Technical integrative approaches to cheese whey valorization towards sustainable environment. Food Funct. 11, 8407–8423.

Adesra, A., Srivastava, V.K., Varjani, S., 2021. Valorization of dairy wastes: integrative approaches for value added products. Indian J. Microbiol. 61, 270–278.

Agostini, S., Bucci, L., Doni, D., Costantini, P., Gupte, A., Müller, B., Sibilla, F., Basaglia, M., Casella, S., Kougias, P.G., Campanaro, S., Favaro, L., Treu, L., 2024. Bioaugmentation strategies based on bacterial and methanogenic cultures to relieve stress in anaerobic digestion of protein-rich substrates. Renew. Energy 225, 120270.

Ahmad, T., Aadil, R.M., Ahmed, H., Rahman, U.U., Soares, B.C.V., Souza, S.L.Q., Pimentel, T.C., Scudino, H., GuimarãES, J.T., Esmerino, E.A., Freitas, M.Q., Almada, R.B., Vendramel, S.M.R., Silva, M.C., Cruz, A.G., 2019. Treatment and utilization of dairy industrial waste: a review. Trends Food Sci. Technol. 88, 361–372.

Allen, C.R., Burr, M.D., Camper, A.K., Moss, J.J., Stein, O.R., 2023. Seasonality, C:N ratio and plant species influence on denitrification and plant nitrogen uptake in treatment wetlands. Ecol. Eng. 191, 106946.

Altiparmaki, G., Kourletakis, P., Moustakas, K., Vakalis, S., 2022. Assessing the effect of hydrothermal treatment (HT) severity on the fate of nitrates and phosphates in dairy wastewater. Fuel 312, 122866.

Antonopoulou, G., Ntaikou, I., Bebelis, S., Lyberatos, G., 2021. On the evaluation of filtered and pretreated cheese whey as an electron donor in a single chamber microbial fuel cell. Biomass Convers. Biorefinery 11, 633–643.

Aranda-Jaramillo, B., Leon-Becerril, E., Aguilar-Juarez, O., Castro-Munoz, R., Garcia-Depraect, O., 2023. Feasibility study of biohydrogen production from acid cheese whey via lactate-driven dark fermentation. Fermentation-Basel 9, 644.

Asunis, F., De Gioannis, G., Dessì, P., Isipato, M., Lens, P.N.L., Muntoni, A., Polettini, A., Pomi, R., Rossi, A., Spiga, D., 2020. The dairy biorefinery: Integrating treatment processes for cheese whey valorisation. J. Environ. Manag. 276, 111240.

Ayed, L., M'Hir, S., Asses, N., 2023. Sustainable whey processing techniques: innovations in derivative and beverage production. Food Biosci. 53, 102642.

Batista, L.P.P., Paulinetti, A.P., Júnior, A.D.N.F., Albanez, R., Ratusznei, S.M., Etchebehere, C., Lovato, G., Rodrigues, J.A.D., 2023. Two-stage thermophilic anaerobic digestion of cheese whey: process optimization, comparison with single-stage, and full-scale estimation. Chem. Eng. Process. Process Intens. 183, 109260.

Bazrafshan, N., Dadashi Firouzjaei, M., Elliott, M., Moradkhani, A., Rahimpour, A., 2021. Preparation and modification of low-fouling ultrafiltration membranes for cheese whey treatment by membrane bioreactor. Case Stud. Chem. Environ. Eng. 4, 100137.

Bella, K., Pilli, S., Venkateswara Rao, P., 2023. A comparison of ultrasonic, ozone, and enzyme pre-treatments on cheese whey degradation for enhancement of anaerobic digestion. J. Environ. Manag. 340, 117960.

Bella, K., Venkateswara Rao, P., 2022. Anaerobic co-digestion of cheese whey and septage: effect of substrate and inoculum on biogas production. J. Environ. Manag. 308, 114581.

Bhatia, S.K., Mehariya, S., Bhatia, R.K., Kumar, M., Pugazhendhi, A., Awasthi, M.K., Atabani, A.E., Kumar, G., Kim, W., Seo, S.-O., Yang, Y.-H., 2021. Wastewater based microalgal biorefinery for bioenergy production: progress and challenges. Sci. Total Environ. 751, 141599.

Bhuvaneshwari, S., Majeed, F., Jose, E., Mohan, A., 2022. Different treatment methodologies and reactors employed for dairy effluent treatment - a review. J. Water Process Eng. 46, 102622.

Bhuvanendran, R.K., Jagadeesan, B.K., Karthigeyan, J., Bhuvaneshwari, S., Vallinayagam, S., Prasannakumari, A.S.N., 2024a. Remediation and management techniques for industrial dairy wastewater and sludge: a review. J. Mater. Cycles Waste Manag. 26 (5), 2634–2655.

Bhuvanendran, R.K., Jagadeesan, B.K., Karthigeyan, J., Bhuvaneshwari, S., Vallinayagam, S., Prasannakumari, A.S.N., 2024b. Remediation and management techniques for industrial dairy wastewater and sludge: a review. J. Mater. Cycles Waste Manag. 26, 2634–2655.

Boscaini, S., Skuse, P., Nilaweera, K.N., Cryan, J.F., Cotter, P.D., 2023. The Whey to good health: whey protein and its beneficial effect on metabolism, gut microbiota and mental health. Trends Food Sci. Technol. 133, 1–14.

Brandelli, A., Daroit, D.J., CorrêA, A.P.F., 2015. Whey as a source of peptides with remarkable biological activities. Food Res. Int. 73, 149–161.

Braun, J.C.A., Balbinot, L., Beuter, M.A., Rempel, A., Colla, L.M., 2024. Mixotrophic cultivation of microalgae using agro-industrial waste: tolerance level, scale up, perspectives and future use of biomass. In: Algal Research-Biomass Biofuels and Bioproducts, 80, 103554.

Bucci, P., Coppotelli, B., Morelli, I., Zaritzky, N., Caravelli, A., 2020. Simultaneous heterotrophic nitrification and aerobic denitrification of wastewater in granular reactor: microbial composition by next generation sequencing analysis. J. Water Process Eng. 36, 101254.

Caltzontzin-Rabell, V., Feregrino-PéREZ, A.A., GutiéRREZ-Antonio, C., 2024. Bio-upcycling of cheese whey: transforming waste into raw materials for biofuels and animal feed. Heliyon 10, e32700.

Carvalho, F., Prazeres, A.R., Rivas, J., 2013. Cheese whey wastewater: characterization and treatment. Sci. Total Environ. 445–446, 385–396.

Casallas-Ojeda, M., Perez-Esteban, N., Cabeza, I., Cobo, M., Olaya-Rincon, M., Caicedo-Concha, D.M., Astals, S., 2024. Understanding the acidification risk of cheese whey anaerobic digestion under psychrophilic and mesophilic conditions. Heliyon 10, e26476.

Casallas-Ojeda, M., Torres-Guevara, L.E., Caicedo-Concha, D.M., Gomez, M.F., 2021. Opportunities for waste to energy in the milk production industry: perspectives for the circular economy. Sustainability 13, 12892.

Castro-Muñoz, R., DíAZ-Montes, E., Gontarek-Castro, E., Boczkaj, G., Galanakis, C.M., 2022. A comprehensive review on current and emerging technologies toward the valorization of bio-based wastes and by products from foods. Compr. Rev. Food Sci. Food Saf. 21, 46–105.

Cavallini, A., Torre, S., Usai, L., Casula, M., Fais, G., Nieri, P., Concas, A., Lutzu, G.A., 2024. Effect of cheese whey on phycobiliproteins production and FAME profile by Arthrospira platensis (Spirulina): promoting the concept of a circular bio-economy. Sustain. Chem. Pharm. 40, 101625.

Chandra, R., Castillo-Zacarias, C., Delgado, P., Parra-SaldíVAR, R., 2018. A biorefinery approach for dairy wastewater treatment and product recovery towards establishing a biorefinery complexity index. J. Clean. Prod. 183, 1184–1196.

Charalambous, P., Constantinou, D., Samanides, C.G., Vyrides, I., 2023. Enhancing biogas production from cheese whey using Zero-Valent Iron: a comparative analysis of batch and semi-continuous operation modes. J. Environ. Chem. Eng. 11, 111278.

Charalambous, P., Shin, J., Shin, S.G., Vyrides, I., 2020. Anaerobic digestion of industrial dairy wastewater and cheese whey: performance of internal circulation bioreactor and laboratory batch test at pH 5-6. Renew. Energy 147, 1–10.

Comino, E., Riggio, V., Rosso, M., 2011. Mountain cheese factory wastewater treatment with the use of a hybrid constructed wetland. Ecol. Eng. 37, 1673–1680.

Domingos, J.M.B., Puccio, S., Martinez, G.A., Amaral, N., Reis, M.A.M., Bandini, S., Fava, F., Bertin, L., 2018. Cheese whey integrated valorisation: production, concentration and exploitation of carboxylic acids for the production of polyhydroxyalkanoates by a fed-batch culture. Chem. Eng. J. 336, 47–53.

Ekka, B., Dejus, S., Juhna, T., 2021. Case study on the dairy processing industries and their wastewater generation in Latvia. J. Dairy Res. 88, 425–428.

El-Aidie, S.A.M., Khalifa, G.S.A., 2024. Innovative applications of whey protein for sustainable dairy industry: environmental and technological perspectives-A comprehensive review. Compr. Rev. Food Sci. Food Saf. 23, e13319.

Elia, S., Stylianou, M., Agapiou, A., 2023. Combined EC/EO processes for treating goat cheese whey wastewater. Sustain. Chem. Pharm. 32, 100963.

Fancello, F., Zara, G., Hatami, F., Scano, E.A., Mannazzu, I., 2024. Unlocking the potential of second cheese whey: a comprehensive review on valorisation strategies. Rev. Environ. Sci. Biotechnol. 23, 411–441.

Frigon, J.C., Breton, J., Bruneau, T., Moletta, R., Guiot, S.R., 2009. The treatment of cheese whey wastewater by sequential anaerobic and aerobic steps in a single digester at pilot scale. Bioresour. Technol. 100, 4156–4163.

Ganju, S., Gogate, P.R., 2017. A review on approaches for efficient recovery of whey proteins from dairy industry effluents. J. Food Eng. 215, 84–96.

Gomez-Romero, J., Gonzalez-Garcia, A., Chairez, I., Torres, L., GarcíA-PeñA, E.I., 2014. Selective adaptation of an anaerobic microbial community: biohydrogen production by co-digestion of cheese whey and vegetables fruit waste. Int. J. Hydrogen Energy 39, 12541–12550.

Hallaji, S.M., Kuroshkarim, M., Moussavi, S.P., 2019. Enhancing methane production using anaerobic co-digestion of waste activated sludge with combined fruit waste and cheese whey. BMC Biotechnol. 19, 19.

Hassoun, A., Tarchi, I., AïT-Kaddour, A., 2024. Leveraging the potential of fourth industrial revolution technologies to reduce and valorize waste and by-products in the dairy sector. Curr. Opin. Green Sustain. Chem. 47, 100927.

Iglesias-Iglesias, R., Portela-Grandio, A., Treu, L., Campanaro, S., Kennes, C., Veiga, M.C., 2021. Co-digestion of cheese whey with sewage sludge for caproic acid production: role of microbiome and polyhydroxyalkanoates potential production. Bioresour. Technol. 337, 125388.

Jena, J., Narwade, N., Das, T., Dhotre, D., Sarkar, U., Souche, Y., 2020. Treatment of industrial effluents and assessment of their impact on the structure and function of microbial diversity in a unique Anoxic-Aerobic sequential batch reactor (AnASBR). J. Environ. Manag. 261, 110241.

Kataki, S., Chatterjee, S., Vairale, M.G., Dwivedi, S.K., Gupta, D.K., 2021. Constructed wetland, an eco-technology for wastewater treatment: a review on types of wastewater treated and components of the technology (macrophyte, biolfilm and substrate). J. Environ. Manag. 283, 111986.

Kotsia, D., Sympikou, T., Topi, E., Pappa, F., Matsoukas, C., Fountoulakis, M.S., 2024. Use of recycled construction and demolition waste as substrate in constructed wetlands for the wastewater treatment of cheese production. J. Environ. Manag. 362, 121324.

Kumar Awasthi, M., Paul, A., Kumar, V., Sar, T., Kumar, D., Sarsaiya, S., Liu, H., Zhang, Z., Binod, P., Sindhu, R., Kumar, V., Taherzadeh, M.J., 2022. Recent trends and developments on integrated biochemical conversion process for valorization of dairy waste to value added bioproducts: a review. Bioresour. Technol. 344, 126193.

Lappa, I.K., Papadaki, A., Kachrimanidou, V., Terpou, A., Koulougliotis, D., Eriotou, E., Kopsahelis, N., 2019. Cheese whey processing: integrated biorefinery concepts and emerging food applications. Foods 8.

Linares-Hernández, I., Barrera-DíAZ, C., Bilyeu, B., JuáREZ-GarcíAROJAS, P., Campos-Medina, E., 2010. A combined electro-coagulation–electrooxidation treatment for industrial wastewater. J. Hazard Mater. 175, 688–694.

Liu, P., Chen, Y.H., Ma, C.Q., Ouyang, J., Zheng, Z.J., 2023. β-Galactosidase: a traditional enzyme given multiple roles through protein engineering. Crit. Rev. Food Sci. Nutr. 65 (7), 1306–1325.

Liu, P., Zheng, Z., Xu, Q., Qian, Z., Liu, J., Ouyang, J., 2018. Valorization of dairy waste for enhanced D-lactic acid production at low cost. Process Biochem. 71, 18–22.

Macedo, A., Azedo, D., Duarte, E., Pereira, C., 2021. Valorization of goat cheese whey through an integrated process of ultrafiltration and nanofiltration. Membranes 11, 477.

Madrigal, G., Huaraya, M., Sancho, T., Mendieta, O., Jaimes-EstéVEZ, J., 2022. Biochar from bovine manure as a sustainable additive to improve the anaerobic digestion of cheese whey. Bioresour. Technol. Rep. 20, 101258.

Mahmoudi, A., Hannachi, C., Mhiri, F., Hamrouni, B., 2024. Performances of constructed wetland system to treat whey and dairy wastewater during a macrophytes life cycle. Desalin. Water Treat. 318, 100364.

Makut, B.B., Das, D., Goswami, G., 2019. Production of microbial biomass feedstock via co-cultivation of microalgae-bacteria consortium coupled with effective wastewater treatment: a sustainable approach. Algal Res. 37, 228–239.

Mehra, R., Kumar, H., Kumar, N., Ranvir, S., Jana, A., Buttar, H.S., Telessy, I.G., Awuchi, C.G., Okpala, C.O.R., Korzeniowska, M., Guiné, R.P.F., 2021. Whey proteins processing and emergent derivatives: an insight perspective from constituents, bioactivities, functionalities to therapeutic applications. J. Funct.Foods 87, 104760.

Melchiors, M.S., Piovesan, M., Becegato, V.R., Becegato, V.A., Tambourgi, E.B., Paulino, A.T., 2016. Treatment of wastewater-from the dairy industry using electroflocculation and solid whey recovery. J. Environ. Manag. 182, 574–580.

Miranzadeh, M.B., Jafarsalehi, M., Akram, J., Ebrahimi, M., Mazaheri, A., Mashayekh, M., 2024. Boosting biogas production in the wastewater treatment plants: a narrative review on co-digestion of sewage sludge with internal and external organic waste. Bioresour. Technol. Rep. 26, 101856.

Mousazadeh, M., Niaragh, E.K., Usman, M., Khan, S.U., Sandoval, M.A., Al-Qodah, Z., Khalid, Z.B., Gilhotra, V., Emamjomeh, M.M., 2021. A critical review of state-of-the-art electrocoagulation technique applied to COD-rich industrial wastewaters. Environ. Sci. Pollut. Control Ser. 28, 43143–43172.

Murshid, S., Antonysamy, A., Dhakshinamoorthy, G., Jayaseelan, A., Pugazhendhi, A., 2023. A review on biofilm-based reactors for wastewater treatment: recent advancements in biofilm carriers, kinetics, reactors, economics, and future perspectives. Sci. Total Environ. 892, 164796.

Nath, A., Veraszto, B., Basak, S., Koris, A., Kovacs, Z., Vatai, G., 2016. Synthesis of lactose-derived Nutraceuticals from dairy waste whey-a review. Food Bioprocess Technol. 9, 16–48.

Nazos, T.T., Stratigakis, N.C., Spantidaki, M., Lagouvardou Spantidaki, A., Ghanotakis, D.F., 2023. Characterization of cheese whey effluents and investigation of their potential to be used as a nutrient substrate for Chlorella biomass production. Waste Biomass Valoriz. 14, 3643–3655.

Nocetti, E., Hadad, H.R., Di Luca, G.A., Mufarrege, M.D.L.M., Maine, M.A., 2024a. Performance of a multi-stage hybrid wetland system for the treatment of a dairy effluent. J. Water Process Eng. 58, 104797.

Nocetti, E., Hadad, H.R., Maine, M.A., Di Luca, G.A., Mufarrege, M.D.L.M., 2024b. Effect of pollutant loading rate and macrophyte uptake on the performance of a pilot-scale hybrid wetland system for the final treatment of dairy wastewater. Ecol. Eng. 204, 107290.

Parde, D., Patwa, A., Shukla, A., Vijay, R., Killedar, D.J., Kumar, R., 2021. A review of constructed wetland on type, treatment and technology of wastewater. Environ. Technol. Innov. 21, 101261.

Parihar, R.K., Bhandari, K., Burnwal, P.K., Ghosh, S., Chaurasia, S.P., Midda, M.O., 2024. Advancing dairy wastewater treatment: exploring two-stage fluidized bed anaerobic membrane bioreactor for enhanced performance, fouling, and microbial community analysis. J. Water Process Eng. 58, 104917.

Patel, A.K., Joun, J., Sim, S.J., 2020. A sustainable mixotrophic microalgae cultivation from dairy wastes for carbon credit, bioremediation and lucrative biofuels. Bioresour. Technol. 313, 123681.

Pereira, M.I.B., Chagas, B.M.E., Sassi, R., Medeiros, G.F., Aguiar, E.M., Borba, L.H.F., Silva, E.P.E., Andrade Neto, J.C., Rangel, A.H.N., 2019. Mixotrophic cultivation of Spirulina platensis in dairy wastewater: effects on the production of biomass, biochemical composition and antioxidant capacity. PLoS One 14, e0224294.

Policastro, G., Cesaro, A., Fabbricino, M., 2023. Photo-fermentative hydrogen production from cheese whey: engineering of a mixed culture process in a semi-continuous, tubular photo- bioreactor. Int. J. Hydrogen Energy 48, 21038–21054.

Porwal, H.J., Mane, A.V., Velhal, S.G., 2015. Biodegradation of dairy effluent by using microbial isolates obtained from activated sludge. Water Resour. Ind. 9, 1–15.

Prazeres, A.R., Carvalho, F., Rivas, J., Patanita, M., Dôres, J., 2013. Pretreated cheese whey wastewater management by agricultural reuse: chemical characterization and response of tomato plants Lycopersicon esculentum Mill. under salinity conditions. Sci. Total Environ. 463–464, 943–951.

Rasul, M.G., Hazrat, M.A., Sattar, M.A., Jahirul, M.I., Shearer, M.J., 2022. The future of hydrogen: challenges on production, storage and applications. Energy Convers. Manag. 272, 116326.

Rincón-Pérez, J., Celis, L.B., Morales, M., Alatriste-MondragóN, F., Tapia-RodríGUEZ, A., Razo-Flores, E., 2021. Improvement of methane production at alkaline and neutral pH from anaerobic co-digestion of microalgal biomass and cheese whey. Biochem. Eng. J. 169, 107972.

Rivas, J., Prazeres, A.R., Carvalho, F., 2011. Aerobic biodegradation of precoagulated cheese whey wastewater. J. Agric. Food Chem. 59, 2511–2517.

Rocha-Mendoza, D., Kosmerl, E., Krentz, A., Zhang, L., Badiger, S., Miyagusuku-Cruzado, G., Mayta-Apaza, A., Giusti, M., Jimenez-Flores, R., Garcia-Cano, I., 2021. Acid whey trends and health benefits. J. Dairy Sci. 104, 1262–1275.

Rocha, J.M., Guerra, A., 2020. On the valorization of lactose and its derivatives from cheese whey as a dairy industry by-product: an overview. Eur. Food Res. Technol. 246, 2161–2174.

Rodriguez-Sanchez, A., Gil-Pulido, B., Leonard, P., Finnegan, W., Zhan, X., Dobson, A.D.W., O'Leary, N., 2024. Treatment of dairy wastewater using intermittent-aeration sequencing batch reactor at pilot-scale. Bioresour. Technol. Rep. 26, 101864.

Saeed, O.F., Hameed, K.W., Abbar, A.H., 2023. Treatment of vegetable oil refinery wastewater by sequential electrocoagulation-electrooxidation process. J. Environ. Manag. 342, 118362.

Sakkas, L., Moutafi, A., Moschopoulou, E., Moatsou, G., 2014. Assessment of heat treatment of various types of milk. Food Chem. 159, 293–301.

Salati, S., D'Imporzano, G., Menin, B., Veronesi, D., Scaglia, B., Abbruscato, P., Mariani, P., Adani, F., 2017. Mixotrophic cultivation of Chlorella for local protein production using agro-food by-products. Bioresour. Technol. 230, 82–89.

Sar, T., Harirchi, S., Ramezani, M., Bulkan, G., Akbas, M.Y., Pandey, A., Taherzadeh, M.J., 2022. Potential utilization of dairy industries by-products and wastes through microbial processes: a critical review. Sci. Total Environ. 810, 152253.

Saubenova, M., Oleinikova, Y., Rapoport, A., Maksimovich, S., Yermekbay, Z., Khamedova, E., 2024. Bioactive peptides derived from whey proteins for health and functional beverages. Fermentation-Basel 10, 359.

Shah, A.A., Walia, S., Kazemian, H., 2024. Advancements in combined electrocoagulation processes for sustainable wastewater treatment: a comprehensive review of mechanisms, performance, and emerging applications. Water Res. 252, 121248.

Smithers, G.W., 2008. Whey and whey proteins—from 'gutter-to-gold. Int. Dairy J. 18, 695–704.

Soumati, B., Atmani, M., Benabderrahmane, A., Benjelloun, M., 2023. Whey Valorization Innovative strategies for sustainable development and value-added product creation. J. Ecol. Eng. 24, 86–104.

Stanchev, P., Vasilaki, V., Egas, D., Colon, J., Ponsá, S., Katsou, E., 2020. Multilevel environmental assessment of the anaerobic treatment of dairy processing effluents in the context of circular economy. J. Clean. Prod. 261, 121139.

Sultana, M.-Y., Mourti, C., Tatoulis, T., Akratos, C.S., Tekerlekopoulou, A.G., Vayenas, D.V., 2016. Effect of hydraulic retention time, temperature, and organic load on a horizontal subsurface flow constructed wetland treating cheese whey wastewater. J. Chem. Technol. Biotechnol. 91, 726–732.

Tezcan Un, U., Kandemir, A., Erginel, N., Ocal, S.E., 2014. Continuous electrocoagulation of cheese whey wastewater: an application of Response Surface Methodology. J. Environ. Manag. 146, 245–250.

Tsermoula, P., Khakimov, B., Nielsen, J.H., Engelsen, S.B., 2021. WHEY - the waste-stream that became more valuable than the food product. Trends Food Sci. Technol. 118, 230–241.

Uald Lamkaddam, I., Vega, E., ColóN, J., Ponsá, S., Llenas, L., Mora, M., 2023. Progressive freeze concentration of cheese whey for protein and lactose recovery. Int. Dairy J. 139, 105572.

Uribarrena, M., Rovira-Cal, E., Urbina, L., SuáREZ, M.J., Aymerich, E., Guerrero, P., De La Caba, K., Etxabide, A., 2024. Valorization of cheese whey: closing the loop from protein extraction to whey protein film composting. Green Chem. 26, 4103–4111.

Walzem, R.L., Dillard, C.J., German, J.B., 2002. Whey components: millennia of evolution create functionalities for mammalian nutrition: what we know and what we may be overlooking. Crit. Rev. Food Sci. Nutr. 42, 353–375.

Wang, J.-F., Cai, Z.-X., Li, Y.-H., Sun, Y.-Y., Wu, H.-M., Song, X.-S., Zhu, H., 2023. Microbiota and genetic potential for reducing nitrous oxide emissions by biochar in constructed wetlands. Sci. Total Environ. 903, 166489.

Xi, H., Zhou, X., Arslan, M., Luo, Z., Wei, J., Wu, Z., Gamal El-Din, M., 2022. Heterotrophic nitrification and aerobic denitrification process: promising but a long way to go in the wastewater treatment. Sci. Total Environ. 805, 150212.

Yadav, J.S.S., Yan, S., Pilli, S., Kumar, L., Tyagi, R.D., Surampalli, R.Y., 2015. Cheese whey: a potential resource to transform into bioprotein, functional/nutritional proteins and bioactive peptides. Biotechnol. Adv. 33, 756–774.

Yazdani, V., Golestani, H.A., 2019. Advanced treatment of dairy industrial wastewater using vertical flow constructed wetlands. Desalination Water Treat. 162, 149–155.

Yigit, A., Bielska, P., Cais-Sokolinska, D., Samur, G., 2023. Whey proteins as a functional food: health effects, functional properties, and applications in food. J. Am. Nutr. Assoc. 42, 758–768.

Zandona, E., Blazic, M., Jambrak, A.R., 2021. Whey utilisation: sustainable uses and environmental approach. Food Technol. Biotechnol. 59, 147–161.

Zeng, X., Wang, Y., Yang, S., Liu, Y., Li, X., Liu, D., 2024. The functionalities and applications of whey/whey protein in fermented foods: a review. Food Sci. Biotechnol. 33, 769–790.

Zotta, T., Solieri, L., Iacumin, L., Picozzi, C., Gullo, M., 2020. Valorization of cheese whey using microbial fermentations. Appl. Microbiol. Biotechnol. 104, 2749–2764.

Chapter 22

Whey: A potential source of microorganisms and production of value-added products

Valeria Caltzontzin Rabell, David Gael Zamorano-Maldonado, Carlos Eduardo Guzmán-Martínez, Christopher Alexis Cedillo-Jiménez and Ana Angélica Feregrino-Pérez
Engineering Faculty, Autonomous University of Queretaro, Querétaro, Mexico

1. Introduction

Cheese whey (CW) is a by-product generated during the production of cheese from milk of various bovines, including cows, goats, sheep, buffaloes, dromedary camels and even donkeys (Bittante et al., 2022). This by-product is important due to the large quantities produced annually -approximately 200 million tonnes- and its relatively low rate of upcycling, with only 50% used for food or pharmaceutical applications (Barba, 2021). Depending on the cheese production method, CW can be classified into two types: sweet (pH $>$ 6.0) or acidic (pH $<$ 5.8) (Bintsis and Papademas, 2023).

Due to its rich nutrient composition, which includes lactose, proteins, lipids and mineral salts, CW has the potential to have an environmental impact if not valorized properly. Some of the current upcycling strategies that have been proposed for CW are producing other types of cheese like ricotta or whey butter, animal feed, as food ingredient or as protein powder. Extracted proteins have also shown activity against oxidative stress, antifungal activity and antioxidant capacity. CW can also be used as a growing medium for microorganisms (like bacteria), mushrooms and microalgae, which can produce polysaccharides, prebiotics, carotenoids or chlorophyll (Barba, 2021). It is important to mention that CW has its own microbiota that can be inactivated when undergoing a high temperature treatment (Bintsis and Papademas, 2023).

This microbiota could also produce molecules of interest, therefore it is important to research and know more about them to obtain the beneficial and desired products. This chapter includes a review of diverse papers which describe whey as a potential source of microorganisms and value-added products. In the first section of this chapter, the microbial composition of whey is explored as well as identification and quantification methods. In the second section, the potential of CW microorganisms is described, including probiotic production and its application considering fermentation as an important method. The third section includes value-added products from cheese whey, including microorganisms used in agriculture and the production of enzymes, secondary metabolites and antioxidant and/or antimicrobial compounds.

2. Whey microbial composition

CW can contain a diverse range of microorganisms, some beneficial and others potentially harmful. These microorganisms play a crucial role in fermentation and flavor development and can also influence whey's quality and functional properties. However, the presence of potentially harmful microorganisms poses a risk to food safety. Therefore, understanding whey microbiota is not only essential to explore new opportunities in the food and biotechnology industries, but also for ensuring food safety.

Many of the microorganisms found in whey are derived from the milk used in cheese production. If this milk does not undergo proper cleaning, the resulting cheese and whey could potentially harbor harmful pathogens (Varnam and Sutherland, 2001). This underscores the critical importance of stringent hygiene practices in the food production process. Table 22.1 shows microorganisms present in different cheeses and whey (Bintsis, 2021; Marth and Steele, 2001).

Whey. https://doi.org/10.1016/B978-0-443-27648-4.00012-8

TABLE 22.1 **Microorganisms found in cheeses and whey (Bintsis, 2021; Marth and Steele, 2001).**

Bacteria	Molds	Yeast
• *Lactococcus lactis*	• *Penicillium camemberti*	• *Candida catenulate*
• *Lactococcus cremoris*	• *Penicillium roqueforti*	• *Candida etchellsii*
• *Streptococcus thermophilus*	• *Penicillium commune*	• *Candida glaebosa*
• *Streptococcus salivarius*	• *Penicillium carneum*	• *Candida lambica*
• *Leuconostoc lactis*	• *Penicillium paneum*	• *Candida parapsilosis*
• *Leuconostoc mesenteroides*	• *Geotrichum cadidum*	• *Candida zeylanoides*
• *Lactobacillus delbrueckii*	• *Cladosporium*	• *Pichia kudriavsevii*
• *Lactobacillus helveticus*	• *Chaetomium*	• *Rhodotorula*
• *Lactobacillus acidophilus*	• *Aspergillus*	• *Saccharomyces cerevisiae*
• *Lactobacillus casei*		• *Candida intermedia*
• *Lactobacillus kefir*		• *Candida sake*
• *Propionibacterium freudenreichii*		• *Kazachstania unispora*
• *Propionibacterium jensenii*		• *Pichia fermentans*
• *Propionibacterium thoenii*		• *Pichia guillermondii*
• *Propionibacterium acidipropionici*		• *Saturnispora mendoncae*
• *Brevibacterium linens*		• *Torulaspora delbrueckii*
• *Streptococcus faecalis*		• *Candida mogii*
• *Streptococcus faecium*		• *Candida incospicua*
• *Enterococcus faecalis*		• *Candida xylopsoci*
• *Enterococcus faecium*		• *Galactomyces candidus*
• *Bifidobacterium longum*		• *Pichia farinose*
• *Bifidobacterium bifidum*		• *Pichia kudriavzevii*
• *Bifidobacterium animalis*		• *Pichia membranefasciens*
		• *Zygosaccharomyces rouxii*
		• *Yarrowia lipolytica*
		• *Trichosporon cutaneum*

The main microorganisms found in CW can perform the homofermentative and heterofermentative processes. Examples of those employed in hard cheese production are *Propionibacterium shermanii, Lactobacillus casei, Lactococcus lactis,* and *L. helveticus*. On the other hand, there are *Streptococcus lactis, Leuconostoc, Brevibacterium linens, Penicillium candidum, Penicillium roqueforti, Penicillium Camemberti,* and *Geotrichum cadidum* are employed in soft cheese manufacture (Besediuk et al., 2024). Fungal species that could be present in whey are: *Yarrowia lipolytica, Debaryomyces hansenii, Kodamaea ohmeri, Kluyveromyces marxianus, Penicillium kudravzevi, Kluyveromyces marxianus, Kluyveromyces lactis, Alternaria, Cladosporium, Debaryomyces, Chaetomium, Aspergillus,* and *Trichosporon cutaneum* (De Souza et al., 2021).

Next, a brief description of some microorganisms in whey:

- *Propionibacterium shermanii:* Gram-positive, facultative anaerobic bacteria. This microorganism produces lactic acid and propionic acid. Besides, it synthesizes large amounts of vitamin B12 (Lee et al., 1974).
- *Lactobacillus casei:* This group is constituted by *L. paracasei, L. casei*, and *L. rhamnosus*. They can produce bioactive metabolites, which can confer host benefits when they are consumed (Hill et al., 2018).
- *Lactococcus lactis:* is a strictly homolactic fermentative bacteria and produces lactic acid. Besides, it has shown the potential of bacteriocin production as a bio-preservative against Listeria monocytogenes (Song et al., 2017).
- *Lactobacillus helveticus* is a lactic acid bacteria (LAB). It is gaining importance as a health-promoting culture in probiotic and nutraceutical food products. It has the potential to produce bioactive peptides or bacteriocins (Giraffa, 2014).
- *Streptococcus lactis* is a Gram-positive bacteria without spores and facultative anaerobic. It is a lactic acid bacteria. It is essential not only for cheese production but also for yogurt manufacturing (Wood, 2012).
- *Leuconostoc:* This bacteria produces gas (CO_2) and aroma compounds in dairy products (Hemme and Foucaud-Scheunemann, 2004).
- *Brevibacterium linens* is a Gram-positive, aerobic, and non-sporulating bacteria that is used to produce Munster, Limburger, and Epoisses cheeses (Fox et al., 2017).
- *Penicillium candidum* is a filamentous fungus that produces a white film on cheeses such as Brie and Camembert. It produces lipases and proteases (Fox et al., 2017).
- *Penicillium roqueforti* is a filamentous fungus that makes blue cheeses, such as Roquefort, Stilton, and Gorgonzola. It can produce penicillin and release lipases to hydrolyze fats into free fatty acids, giving blue cheese its aroma (Gillot et al., 2017).

- *Penicillium camemberti:* This fungus produces the white coating over Camembert-type cheeses. It also produces ammonia, associated with mature cheeses (Fox et al., 2017).
- *Geotrichum cadidum* is used to ripen some goat and soft cheeses. However, it could also function as an opportunistic pathogen (Boutrou and Guéguen, 2005).
- *Yarrowia lipolytica:* This yeast is recognized for its enzymatic activities. Some patents have proposed using this species to accelerate ripening. It shows proteolytic activity (Suzzi et al., 2001).
- *Debaryomyces hansenii:* It is a yeast with high osmotic tolerance. It produces lipases and proteases, which contribute to cheese ripening. Besides, it supports cheese conservation against fungal invasions (Deak, 2007).

Beneficial microorganisms, known as probiotics, play a crucial role in maintaining health. Lactic bacteria, a type of probiotic, are particularly noteworthy for their ability to improve intestinal health and modulate the immune system. Importantly, these bacteria also have the unique ability to inhibit pathogen microorganisms, thereby preventing potential illnesses (Hill et al., 2014).

Regarding pathogenic microorganisms, *Escherichia coli*, *Salmonella*, *Staphylococcus aureus*, and *Listeria monocytogenes* can pose a health risk if whey is consumed without proper heat treatment or if used as an ingredient in products that will not undergo cooking.

For example, *Escherichia coli* pathogen can cause serious illnesses such as hemorrhagic diarrhea and hemolytic uremic syndrome, especially in children and older people (Jay et al., 2008). *Salmonella* can cause salmonellosis, an infection that manifests with severe gastrointestinal symptoms. Salmonella contamination can occur during whey production and processing (Doyle et al., 2020). *Listeria monocytogenes* is a pathogen that can grow at refrigeration temperatures and is known to cause listeriosis, a severe disease that can be fatal, especially in immunocompromised and elder people, and pregnant women. This pathogen can contaminate whey during processing or storage (Ryser and Marth, 2007). *Staphylococcus aureus* can produce toxins that cause food poisoning. This pathogen can contact whey through contaminated food handlers or equipment (Hennekinne et al., 2012). In the same way, certain molds, such as *Penicillium* and *Aspergillus*, can cause spoilage and pose a risk if not adequately controlled (Jay et al., 2008). Therefore, strict microbiological control and pasteurization techniques are essential to ensure food safety.

Furthermore, the detailed study of the microbiota in whey is crucial to ensure the quality and safety of derived products and to open the door to innovations in the production of new functional foods and dietary supplements. Identifying and quantifying strains with probiotic properties or the ability to produce specific enzymes can lead to developing products with health benefits or specialized industrial applications.

2.1 Methods for identification and quantification of microorganisms

Qualitative and quantitative methods can give more information regarding the type and quantity of microorganisms found in CW. Qualitative methods for microorganism identification can be divided into phenotypic, genotypic, and immunological methods (Table 22.2). The adequate technique depends on the microorganism's type and the information the researcher needs.

Quantitative methods include plate count, direct, indirect, and molecular techniques (Table 22.3).

The identification and quantification of microorganisms in cheese whey will allow to know its composition and thus propose strategies for its use. In this sense, and knowing the microorganisms that exist in this by-product, there are several potential uses, for example, its use to produce probiotics. The next section describes how CW can be used, as well as the benefits it has and its application in the food industry.

3. Potential of microorganisms present in cheese whey

The microbial diversity present in whey is related to the complexity of its composition derived from its source product, milk. It also contains numerous antimicrobial molecules derived from casein peptides, as well as enzymes, hormones, and antibodies (Singh et al., 2021). CW has microorganisms that can be used for producing new and alternative foods. This section describes the production of probiotics from CW, their benefits and its application as functional foods.

3.1 Production of probiotics

CW is used as a medium for obtaining LAB that are naturally present in it and, due to this distinctive characteristic of its microbiological composition, whey is an excellent growth medium for the proliferation of probiotic microorganisms. It serves as a substrate to produce various products through fermentation processes. Although not all LAB are considered

TABLE 22.2 Qualitative methods described for identifying cheese whey microorganisms.

Phenotypic
Microscopy. Allows seeing cellular morphology to make a phenotypic identification. It is divided into Optical microscopy (Welsh, 2019), Fluorescence microscopy (Kubitscheck, 2017), Transmission electron microscopy (Bozzola and Russell, 1999) and Atomic force microscopy (Dufrêne, 2002).
Cultivation in selective and differential media. Essential for isolating and identifying microorganisms. Employs culture media with inhibitory agents that inhibit the growth of some microorganisms but allow the development of others. These agents include antibiotics, dyes, and salts. These culture media allow the identification of microorganisms based on their metabolic characteristics. Some examples of these selective and differential media are: MacConkey Agar, Sabouraud Agar and mannitol-Salt Agar (Leber, 2020).
Biochemical tests. Based on microorganisms' metabolic and enzymatic capacities, mainly bacteria. Often combined with other phenotypic and molecular tests for complete identification. Some examples are Catalase Test, Oxidase Test, indole Test, Citrate Test, the Voges-Proskauer test and Urease Test (Al-Joda and Jasim, 2021).
Genotypic
Polymerase chain reaction (PCR). Molecular biology technique widely used to amplify specific DNA sequences. It is based on DNA replication and consists of repeated cycles of three fundamental steps (denaturation, hybridization (alignment) and extension (elongation). The cycles are repeated, resulting in exponential amplification of the specific region of DNA. The goal is to determine the presence or absence of a specific genetic sequence characteristic of a microorganism. The basic process involves selection of specific Primers, DNA amplification and detection (Kadri, 2019).
DNA sequencing: Determines the order of nucleotides (adenine, thymine, cytosine, and guanine) in a DNA molecule. It allows the determination of the nucleotide sequence in specific segments of DNA (Sanger et al., 1977).
Immunilogical
Immunofluorescence: Uses antibodies labeled with fluorochromes, which emit visible light when excited with light of a specific wavelength. It is used to identify and locate specific microorganisms in clinical and environmental samples (Bohlool and Schmidt, 1980).

TABLE 22.3 Quantitative methods described for identifying cheese whey microorganisms.

Plate count
Colony forming unit count (CFU). Based on the premise that each visible colony on an agar plate comes from a single viable microorganism (or colony-forming unit) that has multiplied to form a visible colony. Counting these colonies provides an estimate of the number of microorganisms in the original sample (Postgate, 1969).
Direct methods
Direct microscopy. Used to quantify microorganisms in liquid or solid samples by counting visible cells in microscopic fields (Madigan et al., 2008).
Flow cytometry. This technique relies on a flow cytometer to detect and measure properties such as size, internal complexity (granularity), and fluorescence of individual cells as they pass through a laser or other light source (Tung et al., 2007).
Indirect methods
Turbidity. Used to evaluate the concentration of particles suspended in a liquid, such as microorganisms, cells, or inorganic particles. Turbidity is an optical property of liquids that causes light to be scattered and absorbed rather than transmitted in a straight line. This method is widely used in microbiology, biotechnology, water treatment, and the food industry (Madigan et al., 2008).
Viability test. Used to evaluate the ability of cells or microorganisms to sustain life, to grow, and reproduce in each environment. These tests are essential in microbiology, cell biology, and biochemistry to determine the effectiveness of antimicrobial treatments, culture conditions, and the quality of biological products (Riss et al., 2013).
Molecular techniques
Quantitative PCR (qPCR). Also known as real-time PCR; unlike conventional PCR, qPCR measures the amount of amplified DNA in real-time during each reaction cycle, providing both qualitative and quantitative data (Heid et al., 1996).
Next-generation sequencing. In this technique, the relative abundance of each taxon present in the sample is quantified, generally expressed as a percentage of the total reads (Kallastu et al., 2023).

probiotic, the most known are bifidobacteria, *Lactobacilli*, and non-pathogenic yeasts. It is important to highlight that for a microorganism to be considered probiotic, it is necessary to characterize its genus, species, and strain to ensure its safety and innocuity (Parra Huertas, 2010).

In recent years, whey has been considered a potential vector to produce probiotic biomass, acting as a natural and induced growth medium for yeast strains. These strains not only give distinctive aroma and flavor characteristics to fermented foods but also enhance their value as functional foods (Suárez et al., 2021). Although most probiotic strains fall within the domain of LAB, there are also some yeasts and molds that, due to their characteristics, can be used as probiotics. Table 22.4 details the most commonly used genera in functional foods as probiotic microorganisms.

Probiotics are generally defined by the FAO/WHO (2001) as "live microorganisms which, when administered in adequate amounts, confer a health benefit." Consequently, since whey is composed of these types of microorganisms, it becomes a potential ingredient to produce functional foods that offer multiple health benefits to consumers. However, for a food product to be considered probiotic, it must meet a minimum requirement of colony-forming units (CFU). Although this requirement varies by country according to its legislation, it is stipulated that probiotic foods should contain between $10^6 \mathrm{CFU}/g$ y $10^{10} \mathrm{CFU}/g$ in order to be recognized as "GRAS" (Generally Recognized as Safe) (Champagne et al., 2011). In recent years, assorted studies have focused on the correlation between microbiota and how hosts require these beneficial microorganisms to improve their health status. Fig. 22.1 outlines the main pathologies where different authors have highlighted the beneficial effects exerted by probiotic microorganisms (Castañeda Guillot, 2018).

TABLE 22.4 Genera most used as probiotics in functional foods.

	Genera
Bacteria	*Lactobacillus, Bifidobacterium, Lactococcus, Carnobacterium, Enterococcus, Streptococcus, Pediococcus, Propionibacterium, Leuconostoc and Bacillus.*
Fungi	*Aspergillus*
Yeasts	*Saccharomyces*

Recovered from Amara and Shibl (2015).

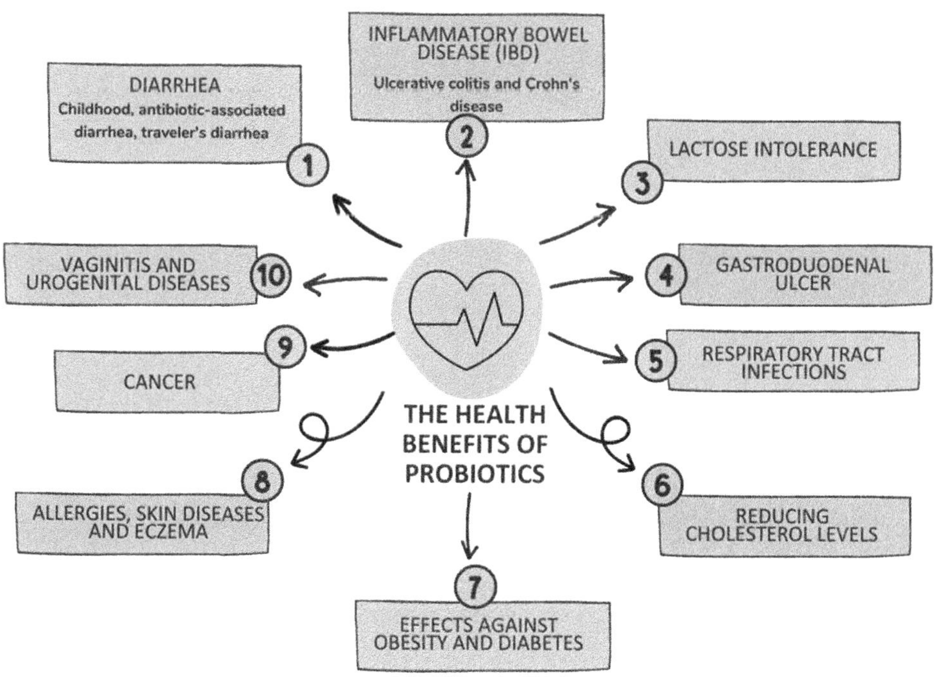

FIGURE 22.1 Main pathologies targeted through the implementation of probiotics in food. *Taken from a compilation made by Castañeda Guillot (2018).*

3.2 Applications in functional foods

The primary way for probiotics to reach the intestinal tract and exert positive effects on the host is through the consumption of functional foods. These are foods that, by providing essential nutrients, offer health benefits due to their specific components that enhance the physiological functions of any organism that consumes them (Barazarte Barazarte et al., 2015). Functional foods provide protection to probiotic microorganisms against adverse conditions they face to survive when administered orally, such as tolerating bile, acid, and gastric enzymes, and subsequently adhering to the intestinal epithelium (Huang and Adams, 2004). These foods must ensure the viability of microorganisms within a range of $> 10^6 CFU/g$ during storage and throughout their shelf life (Sánchez et al., 2015). Table 22.5 lists some examples of the main applications of whey in functional foods, highlighting the use of probiotic microorganisms.

Due to its nutritional richness and its significance as an economical source of protein, whey is used in dairy products and derivatives through whey processing. This results in the production of fermented beverages, lactic acid, cottage cheese, soft cheese, baby formulas, ethanol production, as well as applications as an additive in bakery products, sauces, jams, meat preservatives, nutraceuticals, and dietary products (Rocha and Guerra, 2020). Finally, there are applications in the use of gums, biodegradable packaging, products that inhibit microbial growth, and highly proteic products (Guel García et al., 2018).

Regarding the potential that whey has shown to develop functional food, fermentation is an interesting process that promises health benefits and the development of flavor. Fermentation is an anaerobic glycolysis process through which pyruvic acid is produced without oxygen. There are distinct types of fermentation (alcoholic, acetic, lactic, butyric, butanedioic, propionic, etc.), and they are carried out by various microorganisms such as bacteria, yeasts, and filamentous fungi, either naturally present in whey or intentionally added. Fermentation is used as a food preservation technique, extending the shelf life of foods. This process enhances the flavor, aroma, texture, and nutritional value of the food (Robledo-Márquez et al., 2021). Specifically, whey primarily contains microorganisms such as *Lactobacillus* and *Bifidobacterium* (Ntougias et al., 2015). These types of microorganisms engage in lactic fermentation by metabolizing the carbohydrates present in the substrate (such as lactose in whey) to obtain energy, producing lactic acid as the final product (Emkani et al., 2022). Lactic acid bacteria, particularly the genus *Lactobacillus*, are the most used probiotics due to their "generally recognized as safe" status and their potential health benefits. In this regard, it has been demonstrated that fermented foods are rich sources of new strains of these microorganisms with probiotic characteristics (Liu et al., 2022).

There are multiple benefits to consuming fermented beverages for including probiotics and prebiotics in the diet. Probiotics offer health benefits when consumed in adequate amounts, as they stimulate the protective functions of the digestive system. Prebiotics, on the other hand, are indigestible ingredients that promote the growth of beneficial bacteria in the colon, enhancing the host's health potential. Additionally, they have demonstrated various functions, including

TABLE 22.5 Main applications of whey in functional probiotic foods.

Functional food	Main genres present	Characteristics	References
Whey cheese	*Lactobacillus and enterococos*	Produced thanks to the acidification of whey, which causes proteins to precipitate	Mangione et al. (2023)
Whey-based drink and energy drink	*Lactococcus lactis, Streptococcus*	Fruit pulp is usually added to give it a better flavor.	Rodríguez-Basantes et al. (2020)
Fermented whey drinks	*L. rhamnosus, B. bifidum, Propionibacterium, L. casei, L. bulgaricus, S. thermophilus*	In addition to using fruit pulp, they add other types of ingredients to give the drink more body, such as starter cultures.	Rodríguez González et al. (2020); Colominas-Aspuro et al. (2021)
Whey powder	*Lactobacillus, Lactococcus, Leuconostoc, Streptococcus*	A whey concentrate is obtained as a food supplement and protein	Zeng et al. (2023)
Ice cream	*Lactobacillus, Lactococcus, Leuconostoc, Streptococcus*	A whey concentrate was added to the ice cream as a substitute for raw materials of higher economic value.	Cordoví-Carmenates et al. (2021)

competition against pathogenic bacteria, immunomodulation, and clinical treatment of conditions such as diarrhea, inflammatory diseases, and certain types of cancer (Pérez-Armendáriz and Cardoso-Ugarte, 2020).

3.3 Other food products from cheese whey

Due to its high protein content, bioactive peptides and functional properties, CW has been used to create diverse products across different sectors of the food industry. One of the main uses for CW is whey protein concentrates and whey protein isolates, which are mainly used as nutritional supplements, sports drinks, and functional foods thanks to their digestibility and high nutritional value. These products have essential amino acids and bioactive compounds beneficial for general health (Barone et al., 2020).

Furthermore, CW has been added to high-protein yogurts and fortified dairy beverages to enhance nutritional profile while improving textural and organoleptic properties. These products can contain fruits or flavors to increase their appeal (Shukla et al., 2024). Energy bars have also been included in products with isolated forms of CW and are targeted at active consumers (Barakat and Alfheeaid, 2023). Regarding bakery and sauce industries, CW derivatives can be used as gelling agents or emulsifiers, which improve sensory attributes and shelf of life of the final product (Li et al., 2020). Finally, research has enabled the development of fat replacers from CW proteins to be used in low-fat cheeses, dressings and ice creams (Li et al., 2022). It is clear the need to explore CW in new food applications since it not only supports the development of functional and health-promoting products but also reinforces the principles of circular economy by maximizing the value extracted from whey.

4. Value-added products derived from cheese whey

As mentioned in the section above, whey and its microorganisms can be used to produce different food products. However, it can also be used for producing other non-related food products.

4.1 Agronomic applications of whey: Biofertilizers and biocontrol

Fresh and fermented whey, obtained after cheese production in both whole and fractionated forms, has been studied for its potential application as a plant biostimulant to enhance crop production. Benítez de la Torre et al. (2023) found that raw whey concentrations above 50% strongly inhibit alfalfa seed germination, likely due to the sodium chloride content causing adverse physiological effects. Additionally, no significant differences in early vegetative development were observed with whey concentrations between 0% and 20%, and a similar trend was seen in maize plants. Consequently, most studies have focused on whey as a fermentation substrate, with its resulting biomass and metabolites demonstrating substantial agronomic potential. This section presents case studies highlighting whey's applications in enhancing crop protection, boosting production, and improving soil quality, primarily through the generation of plant growth-promoting bacteria (PGPB) as biofertilizers and as biocontrol agents.

4.1.1 Integrated bioprocesses for agricultural bioproducts

An integrated bioprocess for producing high value-added whey-based products for crop systems was reported by Caballero et al. (2020). In this study, whey fermentation was carried out using *Lactobacillus rhamnosus (L. rhamnosus)*, resulting in the production of bacterial biomass, lactic acid, peptides, and free amino acids, which were subsequently evaluated for their biostimulant and biocontrol abilities for soil and plants. The fermentation process yielded 42 g/L of lactic acid, 13.54% (dry weight) of low molecular weight peptides (300–1000 Da), and the concentration of *L. rhamnosus* biomass increased from 0.07 g/L to 3.2 g/L (dry weight). *L. rhamnosus* was proven to be effective as a biocontrol agent against *Phytophthora cactorum, Phytophthora cinnamomic* and *Pseudomonas syringae*. Additionally, 7 days after the application of lactic acid, a stimulatory effect on soil microorganisms was observed, influencing the bacterial community composition: Bacillaceae and Veillonellaceae increased to 13.8% and 15.2%, respectively, while these families were absent in the control samples. An enrichment of Pseudomonadaceae, Rhizobiaceae and Microbacteriaceae was also observed. The dehydrogenase activity, a key indicator of soil microbial oxidative metabolism, was stimulated by the protein hydrolyzate treatment by 60%–80% when applied at concentrations of 0.5% and 1% v/w. An integrative approach to obtain various products for agricultural applications was performed by Caballero et al. (2020), and it represents one of the most comprehensive analyses. In the following cases, specific products or processes that might also be used for agricultural purposes are presented.

4.1.2 Plant growth-promoting bacteria (PGPB) and biofertilizers

PGPBs are utilized as biofertilizers because their increase of soil fertility increasing the availability of nutrients such as solubilizing phosphates, fixing atmospheric nitrogen, producing metabolites that stimulate plants resulting in the enhancement of crop performance (Daniel et al., 2022).

The production of *Rhizobium meliloti*, a PGPB for nitrogen fixation and the solubilization of phosphorus and zinc, using whey was reported by Bissonnette et al. (1986). Cell accumulation during the fermentation process was found to be comparable to the yield achieved with yeast-mannitol broth. Although this early research explored whey for PGPB production, this approach has not been widely pursued in subsequent studies. A co-substrate of composting leachate enriched with 23% whey was used by Santiago-Badillo et al. (2021) in a fermentation process to stimulate the growth of *Bacillus species*, recognized for phosphate solubilization. Maximum cell densities of *Bacillus megaterium* and *Bacillus cereus* reached 9 log CFU/mL after 144 h. Sporulation rates also increased significantly, with *Bacillus megaterium* reaching 89% (compared to 23% in the control) and *Bacillus cereus* reaching 90% (compared to 47%), indicating improved efficacy of these strains as biofertilizers.

An experiment to cultivate the cyanobacterium *Spirulina platensis* in both synthetic media and whey supplemented with 0.5 g/L of K_2HPO_4 and 1.5 g/L of urea was conducted by Zapata et al. (2021). Biomass production between whey and synthetic media was found to be statistically insignificant. However, statistically significant differences in phytohormone production were observed, with notable increases in the whey medium. Compared to synthetic media, increases of up to 1200-fold for indole acetic acid, 1000-fold for phenylacetic acid, 132-fold for salicylic acid, 7-fold for 1-aminocyclopropane-1-carboxylic acid, and 20-fold for abscisic acid were recorded. These findings provide strong evidence for the potential use of whey in *S. platensis* cultivation, not only as a means of waste utilization but also for its advantage in producing phytohormones and enhancing its value as a biostimulant.

4.1.3 Plant growth-promoting bacteria (PGPB) and biocontrol

PGPBs are utilized as biocontrol agents to antagonize plant pathogens through nutrient competition, toxin inhibition, and the induction of host plant defense mechanisms. These agents are primarily applied to soil in agricultural practices and have also been explored for use in post-harvest treatments (Morales-Cedeño et al., 2021). Palese et al. (2021) investigated the effects of aerated compost tea made with whey from ricotta cheese, using a forced air-blowing system. This compost tea demonstrated significant in vitro inhibition of *Rhizoctonia solani, Fusarium oxysporum,* and *Botrytis cinerea.* Additionally, when kohlrabi plants infected with *Rhizoctonia solani* were treated with a compost tea formulation containing 3:35:7 L of whey, water, and compost, a notable suppression of damping-off symptoms was seen. Earlier, Pane et al. (2012) prepared compost teas from various waste materials, using water and diluted whey derived from buffalo mozzarella cheese (1:5 v/v) as extractants and fermenting them with forced air. The whey-based compost tea significantly reduced the severity of corky root disease caused by *Pyrenochaeta lycopersici*, with disease severity lowered to 40.9% compared to 77.3% in the control. Additional reductions in disease severity were also noted for *Botrytis cinerea* and *Alternaria alternata* in both whey-based and control compost teas, suggesting that whey may enhance microbial and metabolite composition, thereby improving its efficacy against specific pathogens compared to the control.

4.1.4 Whey as stabilization excipient for biofertilizers and biocontrol products

A critical issue with using PGPBs obtained in the laboratory for application in productive soils was reported by O'Callaghan et al. (2022). Desiccation, pH fluctuations, and temperature variations are some of the environmental factors present in crop settings that may reduce the efficacy of PGPBs, particularly gram-negative bacteria.

The effect of whey-based excipients on protecting the PGPBs, *Herbaspirillum frisingense, Azospirillum brasilense,* and *Rhizobium leguminosarum* during drying and storage, to enhance their viability in soil applications and improve their efficacy as biofertilizers, was investigated by Chaparro-;Rodríguez et al. (2023). It was found that the whey protein excipient significantly improved bacterial survival, *H. frisingense* showed a reduction in viability loss to 2.26 log CFUs compared to over six log CFU in the control. An improvement in storage stability over 90 days at 18°C was also observed, with a viability reduction of only 1.22 to 1.51 log CFU for *A. brasilense* and *R. leguminosarum*, maintaining levels between 5 and 7 log CFU/g of dry capsule. However, at storage temperatures of 28°C, a considerable viability loss higher than 5 log CFU was observed, indicating that elevated storage temperatures negatively impact stability. This finding underscores the importance of producing a high microbial load while carefully managing storage conditions, as cooler temperatures are essential for maximizing viability.

Several mechanisms by which PGPBs function as biofertilizers and provide biocontrol have been documented and proposed as competitive alternatives to conventional products, potentially mitigating negative environmental impacts (Glick, 2012). However, the use of whey may result in negative effects when applied undiluted to plants. Increased electrical conductivity can raise salinity, potentially damaging roots. Acidic pH may elevate the availability of heavy metals in the soil to toxic levels over the long term and accelerate corrosion of metallic components in irrigation systems. Sodium accumulation from whey, a by-product of the cheese industry, can also negatively affect soil structure. Therefore, ongoing studies are essential to optimize compost proportions to manage electrical conductivity and pH effects, assess long-term impacts on soil structure and microbiota, and develop strategies to reduce corrosion in irrigation systems.

4.2 Production of enzymes, secondary metabolites, and bioactive compounds

CW has protein particles that include alpha lactalbumin, beta-lactoglobulins, lactoferrins, lactoperidases, immunoglobulins, casein, macropeptides, a wide variety of growth factors (Li et al., 2022) and in hydrolyzed form various peptides with biological activity, in addition to lipids, carbohydrates, and vitamins (Illanes, 2011; Poveda, 2013). Additionally, whey is considered a source of antioxidants (Zhao and Ashaolu., 2020). Table 22.6 shows derivates and function of compounds found in cheese whey.

Regarding enzyme production and exopolysaccharides from cheese whey, it represents a biotechnological strategical way to generate value added products. Although these compounds are obtained in little quantities, they have great impact due to their functionality and high price in the market (Li et al., 2020). Particularly, enzymes from CW like lactase, are essential in food-industry and pharmaceuticals. Lactase is used to elaborate lactose-free products and as a tool to obtain prebiotics like galactooligosaccharides (Rueda Acosta, 2019). On the other hand, exopolysaccharides (EPS) are extracellular polymers synthesized by lactic bacteria during fermentation. EPS can be used as gelling agents, stabilizers and biosorbents. Due to its functionality and relatively scarceness, EPS are considered as high value-added products, not only improving sensorial characteristics of foods, but giving bioactive properties like antioxidant activity (Li et al., 2020).

These developments upcycle CW not only as a source of nutrients but as a platform to bio produce premium compounds with great market potential, favoring a biorefinery focus and circular economy in the dairy industry.

4.2.1 Polysaccharides and lactic acid

As can be seen in Table 22.6, lactose (a disaccharide composed of glucose and galactose) represents the majority component in whey, which allows the implementation of biotechnological processes in various industries, including the food industry to obtain products such as polyesters, polysaccharides, prebiotics, enzymes, and lactic acid among others. For example, lactose can be bio-converted into heteroexopolysaccharides (EPS) through a single-step fermentation using an EPS-producing strain, *Lactobacillus plantarum*, which has a high Beta-Galactosidase activity, resulting in an EPS consisting of galactose, glucose, mannose and arabinose, as well as functional groups that give it thermostable, biosorption, gelling, thickening and highly negative charge properties, generating a compound with high use value in various industries (Li et al., 2020).

To obtain prebiotics such as galactooligosaccharides (GOS), enzymatic hydrolysis techniques and fermentation from the lactose present in whey have been used. Prebiotics stimulate the growth of the microbial gastric flora, generating beneficial effects for the consumer (Kaur et al., 2020, Iqbal et al., 2022; Ambrogi et al., 2023). The most efficient, economical, and environmentally friendly process for obtaining GOS is via a transgalactosylation reaction catalyzed by Beta-galactosidase (Iqbal et al., 2022; Ambrogi et al., 2023). Beta-galactosidase, commonly known as lactase, is a hydrolytic enzyme found in microorganisms such as bacteria, yeasts, and fungi, the latter two being widely used in industry for their high-efficiency characteristics and resistance to wide ranges of temperature and pH (Movahedpour et al., 2022). In 2019, Rueda-Acosta used CW as a carbon source for lactase production, obtaining concentrations higher than 4% using microorganisms (*Lactobacillus case*). This is supported by the work of Yadav et al., which indicates that the biotransformation of whey into high-quality proteins is one of the best options for lactase-producing microorganisms that use lactose for their cell growth, given the high concentrations of organic carbon, high-quality amino acids, vitamins, and minerals (Yadav et al., 2015). Lactase can prevent lactose crystallization in agro-industrial processes such as condensation, sweetening, and production of frozen dairy products. Additionally, it is used to treat lactose intolerance problems in addition to the output of GOS (Fernandes and Carvalho, 2016).

Organic acids, volatile aromatic substances, fatty acids, and biopeptides are also valuable metabolites that can be obtained by using low-cost byproducts that are sources of carbon and nitrogen, such as CW, allowing its valorization (Yadav et al., 2015; Karim et al., 2020). Zhou et al. (2019) obtained ethane and galactonic acid through a two-step

TABLE 22.6 Components of whey generally identified, as well as some nutrients and compounds derived from it.

Derivates or byproducts	Function	Method	References
		Total solids (6% and 7%): Lactose (70%–72%)	
Polyhydroxyalkanoates (PHAs)	Polyolefins	Fermentation with microbial mixture.	Colombo et al. (2016)
Hetero-exopolysaccharide (EPS)	Gelling and thickener agent	Fermentation with microorganisms (*Lactobacillus plantarum*) with high beta-galactosidase activity	Li et al. (2020)
Galactooligosaccharides (GOS)	Prebiotic	Enzymatic hydrolysis. Fermentation. Synthesis of GOS from lactose using microbial sources (bacterial, yeast, or fungi)	Singh et al. (2021); Bolognesi et al. (2022)
Lactic acid	Organic acid with antimicrobial peptides (pediocin PA-1)	Extractive fermentation with microorganisms *Lactobacillus* spp, *Pediococcus* spp	Bernardo et al. (2016); Catone et al. (2021)
Lactase	Hydrolytic enzyme of lactose in milk to obtain monosaccharides	Microbial enzyme production. Inoculation with GRAS microorganisms (facultative heterofermentative bacteria) (*Lactobacillus casei*)	Rueda Acosta (2019)
Alcohols	Various functions	Fermentation with microorganisms	Zhou et al. (2019); Karim and Aider (2022)

bioprocess. Using a hydrolysis pretreatment, glucose and galactose were obtained from lactose, followed by a fermentation using *Saccharomyces cerevisiae* to obtain ethanol from glucose selectively and by oxidation of galactose mediated by *Gluconobacter oxydans*, galactonic acid. Whereas ethanol, 2-phenyl ethanol, and isoamyl alcohol, in addition to organic acids (lactic, acetic, citric, and propionic), have been obtained using electro-activation techniques of fermentation induced by *Kluyveromyces marxianus*, and whey as a substrate (Karim and Aider, 2022).

CW has approximately 0.1%–0.8% lactic acid. However, directed fermentation technologies can increase this value (Chandrapala et al., 2016; Turner et al., 2017). Using a mixed culture of microorganisms produced a concentration of 20.1 g/L of LA (Luongo et al., 2019). Catone et al. obtained a concentration of 27.58 g/L of lactic acid using a lactic bacteria (*Lactobacillus casei*) (Catone et al., 2021). *Pedioccocus* strains have proven to be more efficient in the production of lactic acid. Juodeikiene et al. (2016) indicate that the highest lactic acid production was received with an acid-tolerant *Pedioccocus strain (P. acidilactici)* with a production of 56.22 g/L of lactic acid. Verma et al. (2023) used a beneficial bacteria (*Pediococcus pentosaceus*) that was inoculated in panner whey, showing a high conversion rate of lactose to lactic acid (42.12%) and a production of 15.5 g/L of lactic acid after 48 h of incubation, in addition to the presence of pediocin PA-1, a peptide with antimicrobial activity, which ensured safety in the production of lactic acid. According to these results, implementing advances in fermentation techniques and a rich source of carbohydrates, minerals, and proteins such as whey allows for a sustainable, cheap, and environmentally friendly production of lactic acid (Huang et al., 2023).

5. Concluding remarks

This chapter reviewed the microorganisms that can be found in cheese whey, highlighting those with probiotic activity. Identifying these microorganisms in a precise way while being cost-effective remains a challenge. Therefore, it is vital to know the varieties of the techniques and their appropriate applications, ensuring the safety of the developed products. The diverse microbial population in CW, offers opportunities for upcycling, transforming it not only as a source of probiotics, but also into prebiotics and enzymes. These value-added products restore CW value, lost when discarded and positioning it as a valuable resource rather than just waste. Additionally, using whey to culture beneficial agricultural microorganisms underscores its potential in sustainable agricultural practices. To fully harness CW potential, further research is needed to develop more efficient methods for identification and quantification of microorganisms. Understanding the complete composition of this by-product is vital for ensuring food safety and product reliability. These insights can promote innovation in food, pharmaceutical and agricultural industries, enabling the creation of functional products with health benefits. Upcycling CW promises substantial environmental and economic impacts. By reducing waste and using resources more efficiently, this approach supports the development of sustainable solutions. It is an example of how to use natural resources to promote sustainability and innovation across different types of industries.

References

AL-Joda, B.M.S., Jasim, A.H., 2021. Biochemical testing revision is used to identify several kinds of bacteria. J. Univ. Babylon Pure Appl. Sci. 29 (2), 168–176.

Amara, A.A., Shibl, A., 2015. Role of Probiotics in health improvement, infection control and disease treatment and management. Saudi Pharm. J. 23 (2), 107–114. https://doi.org/10.1016/j.jsps.2013.07.001.

Ambrogi, V., Bottacini, F., Cao, L., Kuipers, B., Schoterman, M., van Sinderen, D., 2023. Galacto-oligosaccharides as infant prebiotics: production, application, bioactive activities and future perspectives. Crit. Rev. Food Sci. Nutr. 63 (6), 753–766. https://doi.org/10.1080/10408398.2021.1953437.

Barba, F.J., 2021. An integrated approach for the valorization of cheese whey. Foods 10, 564. https://doi.org/10.3390/foods10030564.

Barakat, H., Alfheeaid, H.A., 2023. Date palm fruit (Phoenix dactylifera) and its promising potential in developing functional energy bars: review of chemical, nutritional, functional, and sensory attributes. Nutrients 15 (9), 2134. https://doi.org/10.3390/nu15092134.

Barazarte Barazarte, H.E., Sangronis, E., Moreno Pé-rez, I.M., Garmendia Guedez, C.A., Mujica, Y.J., 2015. Laminados de guayaba (*Psidium guajava* L.) enriquecidos con inulina y calcio. Arch. Latinoam. Nutr. 65 (4), 225–233.

Barone, G., Moloney, C., O'Regan, J., Kelly, A.L., O'Mahony, J.A., 2020. Chemical composition, protein profile and physicochemical properties of whey protein concentrate ingredients enriched in α-lactalbumin. J. Food Compos. Anal. 92, 103546. https://doi.org/10.1016/j.jfca.2020.103546.

Benítez de la Torre, A., Pérez-Ramírez, E., Morales-García, Y.E., Muñoz-Rojas, J., Díaz-Ruíz, R., Morales-Almora, P., 2023. Effect of using sweet whey in irrigation of alfalfa and corn. Revista mexicana de ciencias agrícolas 14 (29), e3532. https://doi.org/10.29312/remexca.v14i29.3532.

Bernardo, M.P., Coelho, L.F., Sass, D.C., Contiero, J., 2016. L-(+)-Lactic acid production by Lactobacillus rhamnosus B103 from dairy industry waste. Braz. J. Microbiol. 47, 640–646. https://doi.org/10.1016/j.bjm.2015.12.001.

Besediuk, V., Yatskov, M., Korchyk, N., Kucherova, A., Maletskyi, Z., 2024. Whey-from waste to a valuable resource. J. Agric. Food Res. 101280. https://doi.org/10.1016/j.jafr.2024.101280.

Bintsis, T., 2021. Yeasts in different types of cheese. AIMS Microbiol. 7 (4), 447. https://doi.org/10.3934/microbiol.2021027.

Bintsis, T., Papademas, P., 2023. Sustainable approaches in whey cheese production : a review. Dairy 4, 249–270. https://doi.org/10.3390/dairy4020018.

Bissonnette, N., Lalande, R., Bordeleau, L.M., 1986. Large-scale production of Rhizobium meliloti on whey. Appl. Environ. Microbiol. 52 (4), 838–841. https://doi.org/10.1128/aem.52.4.838-841.1986.

Bittante, G., Amalfitano, N., Bergamaschi, M., Patel, N., Haddi, M., Benabid, H., Pazzola, M., Vacca, G.M., Tagliapietra, F., Schiavon, S., 2022. Composition and aptitude for cheese-making of milk from cows, buffaloes, goats, sheep, dromedary camels, and donkeys. J. Dairy Sci. 105, 2132–2152. https://doi.org/10.3168/jds.2021-20961.

Bohlool, B.B., Schmidt, E.L., 1980. The immunofluorescence approach in microbial ecology. In: Advances in Microbial Ecology. Springer US, Boston, MA, pp. 203–241. https://doi.org/10.1007/978-1-4615-8291-5_6.

Bolognesi, L.S., Gabardo, S., Dall Cortivo, P.R., Ayub, M.A.Z., 2022. Biotechnological production of galactooligosaccharides (GOS) using porungo cheese whey. Food Sci. Technol. 42. https://doi.org/10.1590/fst.64520.

Boutrou, R., Guéguen, M., 2005. Interests in Geotrichum candidum for cheese technology. Int. J. Food Microbiol. 102 (1), 1–20. https://doi.org/10.1016/j.ijfoodmicro.2004.12.028.

Bozzola, J.J., Russell, L.D., 1999. Electron Microscopy: Principles and Techniques for Biologists. Jones & Bartlett Learning.

Caballero, P., Rodríguez-Morgado, B., Macías, S., Tejada, M., Parrado, J., 2020. Obtaining plant and soil biostimulants by waste whey fermentation. Waste Biomass Valoriz. 11 (6), 3281–3292. https://doi.org/10.1007/s12649-019-00660-7.

Castañeda Guillot, C., 2018. Probiotics, update: an update. Cuban J. Pediatr. 90 (2), 286–298. https://doi.org/10.1016/j.jped.2014.08.005.

Catone, M.V., Palomino, M.M., Legisa, D.M., Martin, J.F., García, V.M., Ruzal, S.M., Allievi, M.C., 2021. Lactic acid production using cheese whey based medium in a stirred tank reactor by a ccpA mutant of Lacticaseibacillus casei. World J. Microbiol. Biotechnol. 37, 61–62. https://doi.org/10.1007/s11274-021-03028-z.

Champagne, C.P., Ross, R.P., Saarela, M., Hansen, K.F., Charalampopoulos, D., 2011. Recommendations for the viability assessment of probiotics as concentrated cultures and in food matrices. Int. J. Food Microbiol. 149 (3), 185–193. https://doi.org/10.1016/j.ijfoodmicro.2011.07.005.

Chandrapala, J., Chen, G.Q., Kezia, K., Bowman, E.G., Vasiljevic, T., Kentish, S.E., 2016. Removal of lactate from acid whey using nanofiltration. J. Food Eng. 177, 59–64. https://doi.org/10.1016/j.jfoodeng.2015.12.019.

Chaparro-Rodríguez, M., Estrada-Bonilla, G., Rosas-Pérez, J., Gómez-Álvarez, M., Cruz-Barrera, M., 2023. Hydrogel capsules as new approach for increasing drying survival of plant biostimulant gram-negative consortium. Appl. Microbiol. Biotechnol. 107, 6671–6682. https://doi.org/10.1007/s00253-023-12699-7.

Colombo, B., Sciarria, T.P., Reis, M., Scaglia, B., Adani, F., 2016. Polyhydroxyalkanoates (PHAs) production from fermented cheese whey by using a mixed microbial culture. Bioresour. Technol. 218, 692–699. https://doi.org/10.1016/j.biortech.2016.07.024.

Colominas-Aspuro, A.M., Rodríguez, D., Nodarse, D., 2021. Whey beverage fermented with probiotic bacteria, with added rice flour and orange flavor. Food Sci. Technol. 31 (1), 1–7.

Cordoví-Carmenates, M., Wilson-González, F.E., Galindo-Llanes, P.Á., M'Boumba-Rodríguez, A., 2021. Utilization of concentrated cheese whey in the mix for the production of cream ice cream. Yachasun - Multidiscipl. Peer-Rev. Sci. J. 5 (9 Ed. esp.). https://doi.org/10.46296/yc.v5i9ucedespsoct.0121. Article 9 Ed. esp.

Daniel, A.I., Fadaka, A.O., Gokul, A., Bakare, O.O., Aina, O., Fisher, S., Burt, A.F., Mavumengwana, V., Keyster, M., Klein, A., 2022. Biofertilizer: the future of food security and food safety. Microorganisms 10 (6), 1220. https://doi.org/10.3390/microorganisms10061220.

De Souza, T.P., Evangelista, S.R., Passamani, F.R.F., Bertechini, R., de Abreu, L.R., Batista, L.R., 2021. Mycobiota of Minas artisanal cheese: safety and quality. Int. Dairy J. 120, 105085. https://doi.org/10.1016/j.idairyj.2021.105085.

Deak, T., 2007. Handbook of Food Spoilage Yeasts. CRC press.

Doyle, M.P., Diez-Gonzalez, F., Hill, C. (Eds.), 2020. Food Microbiology: Fundamentals and Frontiers. John Wiley & Sons. https://doi.org/10.1128/9781555819972. Print ISBN:9781555819965. Online ISBN:9781683670476.

Dufrêne, Y.F., 2002. Atomic force microscopy is a powerful tool in microbiology. J. Bacteriol. 184 (19), 5205–5213. https://doi.org/10.1128/JB.184.19.5205-5213.2002.

Emkani, M., Oliete, B., Saurel, R., 2022. Effect of lactic acid fermentation on legume protein properties, a review. Fermentation 8 (6). https://doi.org/10.3390/fermentation8060244. Article 6.

Fernandes, P., Carvalho, F., 2016. Microbial Enzimes for the Food Industry. Biotechnology of Microbial Enzymes: Production, Biocatalysis and Industrial Applications. Elsevier Inc. https://doi.org/10.1016/B978-0-12-803725-6.00019-4.

Fox, P.F., Guinee, T.P., Cogan, T.M., McSweeney, P.L., 2017. Fundamentals of Cheese Science.

Gillot, G., Jany, J.L., Poirier, E., Maillard, M.B., Debaets, S., Thierry, A., Coton, M., 2017. Functional diversity within the Penicillium roqueforti species. Int. J. Food Microbiol. 241, 141–150. https://doi.org/10.1016/j.ijfoodmicro.2016.10.001.

Giraffa, G., 2014. *Lactobacillus helveticus*: importance in food and health. Front. Microbiol. 5, 338.

Glick, B.R., 2012. Plant growth-promoting bacteria: mechanisms and applications. Scientifica 2012, 963401. Available: https://pubmed.ncbi.nlm.nih.gov/24278762/.

Guel García, G.P., Hernández Mendoza, J.L., Rodríguez Castillejos, G., 2018. Use of bacteria obtained from whey and its potential use as probiotics in the food industry. A short review. Revista Boliviana de Química 35 (1), 40–45. Online ISSN 0250-5460.

Heid, C.A., Stevens, J., Livak, K.J., Williams, P.M., 1996. Real-time quantitative PCR. Genome Res. 6 (10), 986–994.

Hemme, D., Foucaud-Scheunemann, C., 2004. Leuconostoc, characteristics, use in dairy technology, and prospects in functional foods. Int. Dairy J. 14 (6), 467–494. https://doi.org/10.1016/j.idairyj.2003.10.005.

Hennekinne, J.A., De Buyser, M.L., Dragacci, S., 2012. Staphylococcus aureus and its food poisoning toxins: characterization and outbreak investigation. FEMS Microbiol. Rev. 36 (4), 815–836. https://doi.org/10.1111/j.1574-6976.2011.00311.x.

Hill, C., Guarner, F., Reid, G., Gibson, G.R., Merenstein, D.J., Pot, B., et al., 2014. Expert consensus document: the International Scientific Association for Probiotics and Prebiotics consensus statement on the scope and appropriate use of the term probiotic. Nat. Rev. Gastroenterol. Hepatol. 11, 506–514. https://doi.org/10.1038/nrgastro.2014.66.

Hill, D., Sugrue, I., Tobin, C., Hill, C., Stanton, C., Ross, R.P., 2018. The Lactobacillus casei group: history and health-related applications. Front. Microbiol. 9, 2107. PMCID: PMC6160870 PMID: 30298055.

Huang, J., Wang, J., Liu, S., 2023. Advanced fermentation techniques for lactic acid production from agricultural waste. Fermentation 9, 765. https://doi.org/10.3390/fermentation9080765.

Huang, Y., Adams, M.C., 2004. In vitro assessment of the upper gastrointestinal tolerance of potential probiotic dairy propionibacteria. Int. J. Food Microbiol. 91 (3), 253–260. https://doi.org/10.1016/j.ijfoodmicro.2003.07.001.

Illanes, A., 2011. Whey upgrading by enzyme biocatalysis. Electron. J. Biotechnol. 14 (6), 9. http://www.scielo.cl/scielo.php?script=sci_arttext&pid=S0717-34582011000600009&lng=es&tlng=en.

Iqbal, M.W., Riaz, T., Mahmood, S., Liaqat, H., Mushtaq, A., Khan, S., Amin, S., Qi, X., 2022. Recent Advances in the Production, Analysis, and Application of Galacto-Oligosaccharides. Food Rev. Int. 39 (8), 5814–5843. https://doi.org/10.1080/87559129.2022.2097255.

Jay, J.M., Loessner, M.J., Golden, D.A., 2008. Modern Food Microbiology. Springer Science & Business Media.

Juodeikiene, G., Zadeike, D., Bartkiene, E., Klupsaite, D., 2016. Application of acid tolerant Pedioccocus strains for increasing the sustainability of lactic acid production from cheese whey. LWT-Food Sci. Technol. 72, 399–406. https://doi.org/10.1016/j.lwt.2016.05.023.

Kadri, K., 2019. Polymerase chain reaction (PCR): principle and applications. Synth. Biol.-New Interdiscipl. Sci. 1–17. https://doi.org/10.5772/intechopen.86491.

Kallastu, A., Malv, E., Aro, V., Meikas, A., Vendelin, M., Kattel, A., Kazantseva, J., 2023. Absolute quantification of viable bacteria abundances in food by next-generation sequencing: quantitative NGS of viable microbes. Curr. Res. Food Sci. 6, 100443. https://doi.org/10.1016/j.crfs.2023.100443.

Karim, A., Aïder, M., 2022. Bioconversion of electro-activated lactose, whey and whey permeate to produce single cell protein, ethanol, aroma volatiles, organic acids and fat by Kluyveromyces marxianus. Int. Dairy J. 129, 105334. https://doi.org/10.1016/j.idairyj.2022.105334.

Karim, A., Gerliani, N., Aïder, M., 2020. Kluyveromyces marxianus: an emerging yeast cell factory for applications in food and biotechnology. Int. J. Food Microbiol. 333, 108818. https://doi.org/10.1016/j.ijfoodmicro.2020.108818.

Kaur, R., Panwar, D., Panesar, P.S., 2020. Biotechnological approach for valorization of whey for value-added products. In: Food Industry Wastes: Assessment and Recuperation of Commodities. Elsevier, pp. 275–302. https://doi.org/10.1016/B978-0-12-817121- 9.00013-9.

Kubitscheck, U. (Ed.), 2017. Fluorescence Microscopy: From Principles to Biological Applications. John Wiley & Sons.

Leber, A.L., 2020. Clinical Microbiology Procedures Handbook. John Wiley & Sons. https://doi.org/10.1128/9781555818814. Print ISBN:9781555818807, Online ISBN:9781683670766.

Lee, I.H., Fredrickson, A.G., Tsuchiya, H.M., 1974. Diauxic growth of propionibacterium shermanii. Appl. Microbiol. 28 (5), 831–835. https://doi.org/10.1128/am.28.5.831-835.1974.

Li, C., Ding, J., Chen, D., Shi, Z., Wang, L., 2020. Bioconversion of cheese whey into a hetero-exopolysaccharide via a one-step bioprocess and its applications. Biochem. Eng. J. 161, 107701. https://doi.org/10.1016/j.bej.2020.107701.

Li, H., Zhang, L., Jia, Y., Yuan, Y., Li, H., Cui, W., Yu, J., 2022. Application of whey protein emulsion gel microparticles as fat replacers in low-fat yogurt: applicability of vegetable oil as the oil phase. J. Dairy Sci. 105 (12), 9404–9416. https://doi.org/10.3168/jds.2022-22314.

Li, M., Dong, Y., Li, W., Shen, X., Abdlla, R., Chen, J., Yue, X., 2022. Characterization and comparison of whey proteomes from bovine and donkey colostrum and mature milk. Lwt 158, 113113. https://doi.org/10.1016/j.lwt.2022.113113.

Liu, C., Xue, W., Ding, H., An, C., Ma, S., Liu, Y., 2022. Probiotic potential of Lactobacillus strains isolated from fermented vegetables in Shaanxi, China. Front. Microbiol. 12. https://doi.org/10.3389/fmicb.2021.774903.

Luongo, V., Policastro, G., Ghimire, A., Pirozzi, F., Fabbricino, M., 2019. Repeated-batch fermentation of cheese whey for semi-continuous lactic acid production using mixed cultures at uncontrolled PH. Sustainability 11, 3330. https://doi.org/10.3390/su11123330.

Madigan, M.T., Martinko, J.M., Dunlap, P.V., Clark, D.P., 2008. Brock biology of microorganisms 12th ed. Int. Microbiol. 11 (1).

Mangione, G., Caccamo, M., Natalello, A., Licitra, G., 2023. Literature review by graduate students: history, production technologies, and characteristics of ricotta cheese. J. Dairy Sci. 106 (6), 3807–3826. https://doi.org/10.3168/jds.2022-22460.

Marth, E.H., Steele, J. (Eds.), 2001. Applied Dairy Microbiology. CRC Press.

Morales-Cedeño, L.R., Orozco-Mosqueda, M. del C., Loeza-Lara, P.D., Parra-Cota, F.I., de los Santos-Villalobos, S., Santoyo, G., 2021. Plant growth-promoting bacterial endophytes as biocontrol agents of pre- and post-harvest diseases: fundamentals, methods of application and future perspectives. Microbiol. Res. 242, 126612. https://doi.org/10.1016/j.micres.2020.126612.

Movahedpour, A., Ahmadi, N., Ghalamfarsa, F., Ghesmati, Z., Khalifeh, M., Maleksabet, A., Shabaninejad, Z., Taheri-Anganeh, M., Savardashtaki, A., 2022. β-Galactosidase: from its source and applications to its recombinant form. In: Biotechnology and Applied Biochemistry, 69. John Wiley and Sons Inc, pp. 612–628. https://doi.org/10.1002/bab.2137.

Ntougias, S., Tsiamis, G., Soultani, D., Melidis, P., 2015. Dominance of rumen microorganisms during cheese whey acidification: acidogenesis can be governed by a rare Selenomonas lacticifex-type fermentation. Appl. Microbiol. Biotechnol. 99 (21), 9309–9318. https://doi.org/10.1007/s00253-015-6827-3.

O'Callaghan, M., Ballard, R.A., Wright, D., 2022. Soil microbial inoculants for sustainable agriculture: limitations and opportunities. Soil Use Manage. 38 (3), 1340–1369. https://doi.org/10.1111/sum.12811.

Palese, A.M., Pane, C., Villecco, D., Zaccardelli, M., Altieri, G., Celano, G., 2021. Effects of organic additives on chemical, microbiological and plant pathogen suppressive properties of aerated municipal waste compost teas. Appl. Sci. 11 (16), 7402. https://doi.org/10.3390/app11167402.

Pane, C., Celano, G., Villecco, D., Zaccardelli, M., 2012. Control of Botrytis cinerea, Alternaria alternata and Pyrenochaeta lycopersici on tomato with whey compost-tea applications. Crop Prot. 38, 80–86. https://doi.org/10.1016/j.cropro.2012.03.012.

Parra Huertas, R.A., 2010. Review. Lactic acid bacteria: functional role in foods. Biotechnol. Agric. Agroind. Sector 8 (1), 93–105.

Pérez-Armendáriz, B., Cardoso-Ugarte, G.A., 2020. Traditional fermented beverages in Mexico: biotechnological, nutritional, and functional approaches. Food Res. Int. 136, 109307. https://doi.org/10.1016/j.foodres.2020.109307.

Postgate, J.R., 1969. Chapter XVIII viable counts and viability. In: Methods in Microbiology, 1. Academic Press, pp. 611–628. https://doi.org/10.1016/S0580-9517(08)70149-1.

Poveda, E., 2013. Suero lácteo, generalidades y potencial uso como fuente de calcio de alta biodisponibilidad. Rev. Chil. Nutr. 40 (4), 397–403. https://doi.org/10.4067/S0717-75182013000400011.

Riss, T.L., Moravec, R.A., Niles, A.L., et al., 2013. Cell viability assays [Updated 2016 Jul 1]. In: Markossian, S., Grossman, A., Arkin, M., et al. (Eds.), Assay Guidance Manual [Internet]. Eli Lilly & Company and the National Center for Advancing Translational Sciences, Bethesda (MD), 2004-. Available from: https://www.ncbi.nlm.nih.gov/books/NBK144065/.

Robledo-Márquez, K., Ramírez, V., González-Córdova, A.F., Ramírez-Rodríguez, Y., García-Ortega, L., Trujillo, J., 2021. Research opportunities: traditional fermented beverages in Mexico. Cultural, microbiological, chemical, and functional aspects. Food Res. Int. 147, 110482. https://doi.org/10.1016/j.foodres.2021.110482.

Rocha, J.M., Guerra, A., 2020. On the valorization of lactose and its derivatives from cheese whey as a dairy industry by-product: an overview. Eur. Food Res. Technol. 246 (11), 2161–2174. https://doi.org/10.1007/s00217-020-03580-2.

Rodríguez González, D., Colominas Aspuro, A.M., Rodríguez Fuertes, W.S., Hernández Monzón, A., 2020. Fermented whey beverage with the addition of wheat bran and guava pulp (Psidium guajava L.). Chem. Technol. 40 (2), 428–441.

Rodríguez-Basantes, A.I., Abad-Basantes, C.A., Pérez-Martínez, A., Diéguez-Santana, K., 2020. Development of a beverage based on whey and Theobroma grandiflorum pulp. Biotechnol. Agric. Agroind. Sector 18 (2), 166–175. https://doi.org/10.18684/bsaa(18)166-175.

Rueda Acosta, M.E., 2019. Optimización de las condiciones de cultivo de lactobacillus casei ATCC 393 empleando suero de leche como fuente de carbono para la producción de lactasa (Bachelor's thesis, Quito: Universidad de las Américas, 2019). http://dspace.udla.edu.ec/handle/33000/10726.

Ryser, E.T., Marth, E.H. (Eds.), 2007. Listeria, Listeriosis, and Food Safety. CRC press.

Sánchez, M.T., Ruiz, M.A., Morales, M.E., 2015. Probiotic microorganisms and health. Ars Pharm. 56 (1), 45–59. https://doi.org/10.4321/S2340-98942015000100007.

Sanger, F., Nicklen, S., Coulson, A.R., 1977. DNA sequencing with chain-terminating inhibitors. Proc. Natl. Acad. Sci. 74 (12), 5463–5467. https://doi.org/10.1073/pnas.74.12.5463. PMID: 271968 PMCID: PMC431765.

Santiago-Badillo, T.P., Pham, T.T.H., Nadeau, M., et al., 2021. Production of plant growth–promoting bacteria inoculants from composting leachate to develop durable agricultural ecosystems. Environ. Sci. Pollut. Control Ser. 28, 29037–29045. https://doi.org/10.1007/s11356-019-06135-5.

Shukla, V., Villarreal, M., Padilla-Zakour, O.I., 2024. Consumer acceptance and physicochemical properties of a yogurt beverage formulated with upcycled yogurt acid whey. Beverages 10 (1), 18. https://doi.org/10.3390/beverages10010018.

Singh, P., Rao, P.S., Sharma, V., Arora, S., 2021. Physico-chemical aspects of lactose hydrolysed milk system along with detection and mitigation of maillard reaction products. Trends Food Sci. Technol. 107, 57–67. https://doi.org/10.1016/j.tifs.2020.11.030.

Song, A.A.L., In, L.L., Lim, S.H.E., Rahim, R.A., 2017. A review on Lactococcus lactis: from food to factory. Microb. Cell Fact. 16, 1–15. https://doi.org/10.1186/s12934-017-0669-x.

Suárez, W.A.B., Díaz, R.E.F., Sánchez, A.A.M., 2021. Production of probiotic biomass and lactic acid from sweet whey. Environ. J. Water Air Soil 11 (2). https://doi.org/10.24054/19009178.v2.n2.2020.4673. Article 2.

Suzzi, G., Lanorte, M.T., Galgano, F., Andrighetto, C., Lombardi, A., Lanciotti, R., Guerzoni, M.E., 2001. Proteolytic, lipolytic, and molecular characterization of Yarrowia lipolytica isolated from cheese. Int. J. Food Microbiol. 69 (1–2), 69–77. https://doi.org/10.1016/S0168-1605(01)00574-8.

Tung, J.W., Heydari, K., Tirouvanziam, R., Sahaf, B., Parks, D.R., Herzenberg, L.A., Herzenberg, L.A., 2007. Modern flow cytometry: a practical approach. Clin. Lab. Med. 27 (3), 453–468.

Turner, T.L., Kim, E., Hwang, C., Zhang, G.C., Liu, J.J., Jin, Y.S., 2017. Conversion of lactose and whey into lactic acid by engineered yeast. J. Dairy Sci. 100 (1), 124–128. https://doi.org/10.3168/jds.2016-11784.

Varnam, A., Sutherland, J.P., 2001. Milk and Milk Products: Technology, Chemistry and Microbiology, vol. 1. Springer Science & Business Media.

Verma, S.K., Iram, D., Sansi, M.S., Pandey, K.K., Vij, S., Sood, S.K., 2023. Sustainable utilization of dairy waste paneer whey by pediococcus pentosaceus NCDC 273 for lactic acid production. Biocatal. Agric. Biotechnol. 47, 102588. https://doi.org/10.1016/j.bcab.2022.102588.

Welsh, C.T., 2019. Microbiology: A Laboratory Manual. Pearson.

Wood, B.J., 2012. Microbiology of Fermented Foods. Springer Science & Business Media.

Yadav, J.S.S., Yan, S., Pilli, S., Kumar, L., Tyagi, R.D., Surampalli, R.Y., 2015. Cheese whey: a potential resource to transform into bioprotein, functional/nutritional proteins and bioactive peptides. Biotechnol. Adv. 33 (6), 756–774. https://doi.org/10.1016/j.biotechadv.2015.07.002.

Zapata, D., Arroyave, C., Cardona, L., Aristizábal, A., Poschenrieder, C., Llugany, M., 2021. Phytohormone production and morphology of Spirulina platensis grown in dairy wastewaters. Algal Res. 59, 102469. https://doi.org/10.1016/j.algal.2021.102469.

Zeng, X., Wang, Y., Yang, S., Liu, Y., Li, X., Liu, D., 2023. The functionalities and applications of whey/whey protein in fermented foods: a review. Food Sci. Biotechnol. 33 (4), 769–790. https://doi.org/10.1007/s10068-023-01460-5. PMID: 38371680; PMCID: PMC10866834.

Zhao, C., Ashaolu, T.J., 2020. Bioactivity and safety of whey peptides. LWT Food Sci. Technol. 134, 109935. https://doi.org/10.1016/j.lwt.2020.109935.

Zhou, X., Hua, X., Huang, L., Xu, Y., 2019. Bio-utilization of cheese manufacturing wastes (cheese whey powder) for bioethanol and specific product (galactonic acid) production via a two-step bioprocess. Bioresour. Technol. 272, 70–76. https://doi.org/10.1016/j.biortech.2018.10.001.

Chapter 23

Whey: Production of bioplastics and bioenergy

Valeria Caltzontzin Rabell, Juan Manuel Vera-Morales, Sergio Iván Martínez-Guido, Araceli Guadalupe Romero-Izquierdo and Claudia Gutiérrez-Antonio
Engineering Faculty, Amealco Campus, Autonomous University of Queretaro, Querétaro, Mexico

1. Introduction

Due to population growth, the need for optimizing resources use has become vital. Therefore, the upcycling of waste, particularly food waste, has gained interest. The cheese industry annually generates up to 145 million tons of whey, reusing almost 54% for mainly food purposes (Lavelli and Beccalli, 2022). Cheese whey (CW) is mainly classified as sweet and acid depending on its pH. Sweet CW has a pH of 6–7, while acid whey has a pH of 4.5–5.8. This is due to the coagulation process: a coagulation with rennet produces sweet CW, whereas a coagulation with mild acids. Another difference is that lactose is usually lower in acid CW (Figueroa Pires et al., 2021). Regarding its composition, CW is mainly comprised by saccharides (4%–5%) such as lactose, proteins (0.6%–0.8%), and lipids (0.4%–0.5%). This rich composition is the main reason of its pollutant activity. However, it also brings an opportunity to be upcycled as diverse bioproducts (Ponist et al., 2021) such as foods and biomolecules; however, it has also reported its use to produce bioplastics and bioenergy.

Transforming CW into bioplastics and bioenergy promises an opportunity to address environmental and economic challenges. Whey has polysaccharides and proteins, which make it an ideal substrate to produce both products. Bioplastics are an alternative to fossil fuel–based plastics and are easier to biodegrade; also they can be used in packaging, or even medical applications (Figueroa-Pires et al., 2024). This chapter describes in Section 2 bioplastics, what are they, how they can be generated from whey, and some of the reported applications. In the next section, a description of different types of bioenergy obtained from this by-product, as well as an economic and environmental impact of this alternative energy, is described.

2. Bioplastics from whey

According to Thakur et al. (2018), 34 million tons of plastic waste are generated each year, with 93% of this waste being dumped into oceans and landfills, directly impacting human health. This issue is exacerbated by the 2.8 kg of CO_2 and other toxic compounds emitted when burning each kilogram of plastic waste. In food packaging, the use of conventional plastics presents the risk of hazardous chemicals seeping into food and the environment, which is a significant challenge due to their nonbiodegradability. Therefore, bioplastics offer a sustainable alternative to the generation of nonbiodegradable plastic waste.

Bioplastics are primarily composed of organic materials, including cellulose, starch, and other carbohydrates, which contribute to their biodegradability. This biodegradation occurs through the activity of microorganisms such as bacteria, algae, and fungi (Kale et al., 2017), with processes that depend on environmental conditions such as temperature, water, oxygen, and even the chemical composition of each polymer. However, these biodegradable processes result in very low CO_2 emissions compared to conventional plastic disposal. According to Kumar et al. (2024), different compounds such as starch, polyhydroxyalkanoates, polylactic acid, and polybutylene succinate, as well as their blends with active agents, have been studied by researchers around the world.

Whey. https://doi.org/10.1016/B978-0-443-27648-4.00007-4

Polysaccharide-based bioplastics present advantages due to their availability, biodegradability, and affordability. However, these polymers are usually sensitive to microfractures, moisture, and a phenomenon called retrogradation, the spontaneous gelatinization of starches. On the other hand, protein-based biopolymers are more resistant to fractures and moisture and present better flexibility. Nevertheless, these are not resistant to water diffusion. Currently, the better options are formulations based on a mixture of polysaccharide and protein-based bioplastics, even adding lipid compounds for better water contention and resistance (Bhatia et al., 2021).

Bioplastics or their precursors can be produced by a variety of microorganisms, such as bacteria, microalgae, or fungi. These microorganisms can also be fed from diverse substrates, from photosynthesis and CO_2 to sugars, proteins, and organic acids. The metabolic routes and downstream chemical reactions to obtain bioplastics can also vary significantly depending on the selected microorganism, the available substrate, and the desired final properties of the material.

Although there are a wide array of possibilities for bioplastic synthesis processes, a major limitation for large-scale production is the cost and availability of carbon and energy sources for microorganisms. One of the most promising strategies to mitigate these costs is the utilization of nutrient-rich effluents from other industries and, ideally, to also provide other advantages such as reducing culture contamination due to specific physiochemical conditions or providing bioplastics certain properties due to the substances present in the effluent.

For example, within the dairy industry, whey is the main effluent. This casein-rich stream also contains lactose, sugar, and other residual components and additives, such as amino acids, lipids, and salts. Whey is commonly transformed into other products, such as whey powder, by spray-drying or ethanol via alcoholic fermentation. However, the high amount of whey currently produced worldwide, and the salts added to obtain certain types of dairy products limit its revalorization. Although many microorganisms have the potential to create biopolymers, only a few of them have proven commercial potential using whey as culture media. Also, among the multiple biopolymer molecules, polyhydroxyalkanoates (PHAs) and polylactic acid (PLA) seem to be the best suited to obtain from whey.

2.1 PHAs

PHAs are linear polyesters produced by different microorganisms as intracellular energy reservoirs. The production of PHAs is a complex process depending on many factors, mainly the carbon sources available to microorganisms and the presence of other nutrients, such as nitrogen and phosphate.

Notable microorganisms suited for high production of PHAs on whey are presented in Table 23.1. *Haloferax mediterranei* is a halophyte microorganism commonly used to produce PHAs. However, although capable of consuming lactose, *H. mediterranei* does not accumulate high amounts of PHAs when cultured on whey. *Alcaligenes latus* is another common PHA producer that, unlike *H. mediterranei*, has shown good production levels on whey. A two-step fermentation has been proposed to first convert the lactose into a more suited carbon source, either via acid fermentation (e.g., with *Acetobacter pasteurianus*) or using an enzymatic pretreatment. This can also be achieved using a microorganism polyculture or even genetically engineered microorganisms such as *E. coli*. These strategies, although functional, can increase the cost of the process, even losing the economic advantages of using a whey effluent.

Once the microbial culture reaches an adequate concentration and PHAs are accumulated, the downstream process consists of the separation of the biomass from the media, usually by centrifugation but sometimes via filtration or air flotation, depending on the specific characteristics of the culture. When the biomass is separated, a drying step is required. This step can also affect the resulting biopolymer as well as the overall cost of the full process, so it must be chosen according to the specific requirements. Various mechanical (milling, homogenization, ultrasonication, etc.) and chemical (enzymatic digestion, alkaline treatments, etc.) subsequent treatments can be used to break the cell walls and release the desired PHAs. The final step requires the separation of the PHAs from the rest of the biological material. The most used separation process involves the use of halogenated solvents such as chloroform and dichloromethane to carry on the PHAs; then the addition of organic solvents such as ethanol or acetone precipitates the biopolymers. The resulting mixture of solvents represents a high hazard, and its treatment for recovery is often expensive. Different strategies have been proposed to replace the use of chlorinated solvents, such as the use of nonchlorinated azeotropes, supercritical fluids, ionic liquids, or chemical and biological dissolution of non-PHAs organic matter, but currently neither of these has shown scalable results (Zikmanis et al., 2020).

2.2 PLA

PLA is a polymer obtained from a dimer called lactide, produced from a chemical conversion of lactic acid. Due to the existence of D and L isomers of lactic acid, a combination of three different forms of the dimer is possible. The ratio of

TABLE 23.1 Examples of microorganisms and proposed pretreatments to obtain PHAs from whey.

Substrate	Microorganism	Pretreatment	Yield (g/L)	References
Cheese whey mother liquor	*Paracoccus homiensis*	Centrifugation and filtration. pH stabilization at 7	1.1	Mozejko-Ciesielska et al. (2022)
Cheese whey	Mixed culture selected from sludge from municipal waste	Acidogenic fermentation	7.8	Carvalheira et al. (2022)
Scotta cheese whey	*Citrobacter freundii* *Leuconostoc* spp.	pH stabilization at 7	1.065	Bosco et al. (2021)
Cheese whey	*Bacillus* sp.	Acidification with *Acetobacter pasteurianus*	0.411	Chang et al. (2021)
Cheese whey	Native mixed culture selected from the cheese whey	Filtering and dissolution at 10% whey	0.389	Ponmanian and Sivashanmugam (2024)
Cheese whey	Mixed culture selected from a saline river	Acidogenic fermentation		Grana et al. (2024)
Deproteinized cheese whey	*Ralstonia eutropha*	Acidification for protein precipitation and subsequent centrifugation	4.71	Patil et al. (2024)

these three forms to the polymerized material affects its properties and degradation processes, although these proportions can be controlled depending on the specific production processes of this bioplastic. PLA can also be polymerized along other biopolymers such as polycaprolactone and polyethylene glycol to adjust its properties for different purposes. PLA has been compared to polystyrene due to its potential properties as a food packaging material, since PLA presents excellent flavor and aromas barriers, and high grease and oil resistance. Since PLA can be degraded by animal metabolism, it has been also highly praised as the best option for biomedical applications (Ebrahimi and Ramezani Dana, 2022).

PLA is more sensitive to impurities in the medium than PHAs; hence, the use of effluents for its production is often limited by required pretreatments. PLA production consists of two steps. The first step is the production of lactic acid from sugar via microorganism fermentation.

Various microorganisms can synthesize lactic acid: fungi, *Lactobacillus* sp., *Bacillus coagulans*, etc. A lot of these species generate lactic acid in a heterolactic fermentation, producing also ethanol, acetic acid, and CO_2. Due to the purity requirements in downstream processes, a homolactic fermentation is desirable, and due to the eventual reduction of pH in the media, only selected species with good adaptability have real potential to be used in biopolymer synthesis, even few if an effluent such as whey is intended to be used. Most microorganisms used in lactic fermentation are, therefore, from the *Lactobacillus* family. Table 23.2 presents some examples on the production of lactic acid from whey with the purpose of using it for PLA synthesis.

Lactic acid purification for its subsequent transformation in PLA remains a costly process, both economically and environmentally, as it typically requires an alkali addition that generates a nonecofriendly effluent. There are potential options, such as the replacement of alkali salts with more ecological substances, or the use of membrane-based bioreactors. Once the lactic acid is obtained and purified, the dimer lactide is produced and transformed by either condensation and subsequent polymerization, or a direct ring opener polymerization, to obtain PLA. Although condensation is a cheap process, it often produces polymers with low molecular weight. A direct biosynthesis process has been obtained in the laboratory using metabolically engineered microorganisms (Tripathi et al., 2021).

When aiming to integrate bioplastics into everyday life, the goal is to gain a comprehensive understanding of these raw materials and their potential for large-scale use in bioplastic production for food packaging. In this regard, Acquavia et al. (2024) note that the use of bioplastics for food packaging has been steadily increasing in recent years. Examples of bioplastic implementation include the use of orange peels in biofilm packaging, as reported by Yaraddodi et al. (2022); biochar plastics derived from walnuts, as reported by Alfattani et al. (2022); and dye adsorbents based on fenugreek seeds, as noted by Taqui et al. (2021).

Similarly, Mavai et al. (2024) reviewed the use of polylactic acid blended with polysaccharides such as gum and cellulose to improve the water barrier, thermal, crystallization, degradability, and mechanical properties of bioplastics for

TABLE 23.2 Recent studies on lactic acid production from whey to produce PLA.

Microorganism	Pretreatment	Lactic acid yield (g/L)	References
Lactobacillus paracasei *Lactobacillus casei* *Lactobacillus delbrueckii*	Powdered and rehydrated whey	87 48 60	Garavand et al. (2022)
Kluyveromyces lactis *Enterococcus faecalis*	Deproteinized cheese whey	35.7 50.6	Gordillo-Andia et al. (2024)
Lactobacillus sp.	Centrifugation	29.5	Carlozzi et al. (2021)
Lactobacillus delbrueckii Engineered *Lactococcus lactis*	pH maintained at 6.5 during the fermentation		Sahoo and Jayaraman (2019)
Lactobacillus acidophilus	ZnO nanoparticles added		Kochubei-lytvynenko et al. (2024)

smart food packaging. The review concludes that the addition of these polysaccharides reduces production costs and helps in manufacturing packaging materials with superior quality. The incorporation of gum and cellulose enhances both mechanical and barrier properties, compared to using polylactic acid alone, making it an ideal sustainable packaging material for food applications. An edible film made from glycerol, $CaCO_3$, corn oil, and 15% CW has been described by Moentamaria et al. (2023); this film has a strength that meets the tensile strength standards of edible films according to Japanese Industrial Standards. Another study done by Chalermthai et al. (2021), where the life cycle assessment of bioplastic production from whey protein was done, showed that producing this film had the lowest impact regarding global warming and water compared to other plastics.

Films made with proteins can be made using whey protein. These proteins need denaturation or pH adjustments and are suitable for meat due to an affinity for hydrophilic surfaces. Figueroa-Pires et al. (2024) review different works that use CW to produce films. They conclude that further research must be conducted to improve the films, mostly in improving hydrophobicity as well as analyzing the potential carrying of antimicrobials, antioxidants, or other biomolecules. Some other opportunities for this films to address are as follows: (1) the industrial making of the films, since they have been only made in laboratory conditions; (2) the expensive equipment that is needed to pretreat the proteins and the further sterilization of the product; and (3) the acceptation from the consumer since they might be skeptical about new products (Figueroa-Pires et al., 2024).

Therefore, while conventional plastic packaging poses significant negative effects on human health and the environment, biodegradable bioplastics offer a sustainable alternative. However, the widespread adoption of bioplastics faces challenges related to their properties, costs, production processes, and market alignment. Overcoming these challenges through research, innovation, and policy support is crucial for improving their competitiveness and ensuring sustainable economic growth. Additionally, advancing bioplastic production through novel approaches, such as metabolic engineering, is essential for a cleaner environment and safer food packaging.

Mostafa et al. (2018) reported that nearly 1.7 million tons of biodegradable plastics were produced worldwide in 2014, with a 72% increase by 2018. However, the production of biodegradable plastics remains limited due to high manufacturing costs and the low level of mechanical and technological development. According to Islam et al. (2024), bioplastics are expected to contribute to the "Sustainable Development Goals," particularly by using less-toxic reagents during production, developing new recycling and degradation pathways, and shifting away from fossil fuels, especially if integrated into circular economies. Nonetheless, their environmental impacts are still a contentious issue. These impacts include concerns related to land use, pesticide and fertilizer usage, eutrophication, and the formation of microplastics and nanoparticles. The lack of standardized labeling and inadequate waste collection and processing further exacerbate these risks. Therefore, a comprehensive assessment of the environmental impacts of bioplastics, particularly during production and end-of-life management, is essential. Including life cycle analysis in the development of new technologies or improvements to existing ones is crucial in selecting alternatives that not only have social benefits but also align with all three pillars of sustainability.

3. Bioenergy from whey

Due to high content of biological and chemical oxygen demand, until 30–50 mg/L and 60–80 mg/L, respectively, cheese whey produced as waste stream from dairy industries worldwide is a feasible alternative to produce bioenergy (Osorio-González et al., 2022); moreover, its production has increased significantly, estimating that is generated up 9–10 L of CW of per kg of cheese produced (Akbi et al., 2017), generating more than 183 million tons per year of this residue at global level (Osorio-González et al., 2022). These data describe the importance of implementing suitable environmental regulations for this residue, boosting its use as raw material to avoid water-bodies eutrophication and the increase of soil acidity (Akbi et al., 2017; Asunis et al., 2020; Osorio-González et al., 2022), proposing new schemes of circular economy contributing to sustainability. In this sense, novel processes for CW conversion into bioenergy have been reported. In this section are highlighted papers about processes to obtain biogas, biohydrogen, and bioethanol from CW.

3.1 Biogas production

Biogas is a renewable mixture of methane, carbon dioxide, and other gaseous traces, such as ammonia and hydrogen sulfide, which is produced from the anaerobic digestion process, using a wide range of biomasses (Sher et al., 2024); likewise, it is worth to mention that the biogas composition varies according to the biomass compounds and the operational parameters of the process (Asunis et al., 2020; Sher et al., 2024). Specifically, in the case of CW, its high organic load allows its use as feedstock in the biodigestion process. However, in accordance with Asunis et al. (2020), because its acidity levels promote the accumulation of volatile fatty acids, the anaerobic medium is acidified, inhibiting the methanogenic activity; thus the methane yield and the process stability decrease, while the hydraulic retention time (HRT) increases in more than 5 days for ensuring the continuous operation. These drawbacks are solved by adding an alkaline solution, diluting the CW, or codigesting it with mixtures of substrates characterized for high buffering capacity.

Carlini et al. (2015) proposed the biogas production from a mixture of CW and poultry manure in a relation of 3:1, 1:1 and 1:3, into a batch process at 37°C, pH 6.5–7.5 and HRT of 35 days. According to the experimental results, at 1:1 relation was obtained until 223 mL of biogas per g of volatile solids, with 65% of methane in the biogas, which means a plant of anaerobic digestion to produce 65 kW from organic waste. Then, Asunis et al. (2020) presented some works about the codigestion of CW with biomasses, such as fish ensilage, poultry manure, sewage sludge, dairy cattle manure, and cattle slurry, comparing the reactor type, the operational parameters (temperature, pH, and HRT), and the methane production. According to the results, 392 L methane per kg of volatile solids were obtained into a continuous stirred reactor at 35°C, pH of 6.4–7.7, and HRT of 15.6, using a mixture of 85% CW and 15% dairy manure (v/v). Kovalev et al. (2023) conducted a study evaluating the energy recovery, using the heating value as a parameter, from two-stage mesophilic-thermophilic anaerobic digestion of CW. First, the CW was diluted at three chemical oxygen demand (COD) levels: 6.8, 9.2, and 13.8 g oxygen per L; then, diluted CW was inserted to mesophilic acidogenic reactor with 10 h as HRT. The effluent of the last reactor was entered into three thermophilic methanogenic reactors, operated at 72, 48, and 24 h as HRT; using a polyurethane foam to immobilize the anaerobic acidogenic and methanogenic sludge. According to the results, the maximum energy production rate was 53.2 kJ per L per day at COD of 13.8 g oxygen per L, 10 and 48 h, as HRT in first and second reaction stage, respectively.

In the same year, Blanco et al. (2023) presented a sequential fermentative-methanogenic process in fixed-film reactors to achieve a low-cost stable methanogenic process from CW, boosting hydrogen and methane production. In the system, the processing fermented CW without applying alkalization methods in methanogenesis phase was technically feasible, achieving 150–170 NmL methane per g COD. Also, hydrogen production was favored by maintaining the fermentation pH values between 5 and 5.5. Then, other two-stage system for anaerobic digestion using CW was studied by Batista et al. (2023), wherein was used an anaerobic sequencing batch and fed-batch biofilm reactor at 55°C, achieving a methane productivity of 142.9 mol methane/m3 day. Also, to know the system's effectiveness, the same equipment was proved with a single stage. According to the results, both systems allow hydrogen and methane production. However, the single-stage system is 42% cheaper than the two-stages, generating 9.89 MJ/kg COD and 5.14 MJ/kg COD, respectively. Then, in 2024, an extensive study of 24 samples of CW collected from 17 farms was carried out in the Canary Islands (Ramos-Suárez et al., 2024). The collected samples involved different industries (industrial and artisanal factories) and different kinds of milk (cow, goat, sheep, and their mixtures), along with different posttreatments to assess the biogas potential in this region. According to the results, the artisan CW obtained the higher methane potential, in comparison with the industrial CW, without any difference between the milk type. The biogas production was carried out using a commercial biogas batch fermentation system.

Recently, another study by Casallas-Ojeda et al. (2024) reported biogas production using mixtures of CW and dairy manure at 20°C (psychrophilic conditions). The primary purpose of the study was analyzing the influence of different mixtures CW: dairy manure and different inoculum-to-substrate ratios on biomethane production. According to the results, the best mixture CW: dairy manure was 65:35 (weight basis) with an organic load of 0.6 g volatile solids/L. A life cycle analysis was conducted to assess the technical feasibility of incorporating the generated biogas to replace propane gas and electricity needs in cheese production. The analysis indicates that biogas substitution could reduce total emissions from the cheese factory by 43.6%.

A life cycle analysis was conducted to assess the technical feasibility of incorporating the generated biogas to replace propane gas and electricity in cheese production. The analysis indicates that substituting biogas could reduce total emissions from the cheese factory by 43.6%.

3.2 Biohydrogen production

Hydrogen produced from waste biomass is considered a feasible alternative to produce energy and generate a positive environmental impact, facing its conventional production routes, frequently derived from fossil sources (Asunis et al., 2020; Osorio-González et al., 2022). In the case of hydrogen production from CW, some processes have been reported, such as dark fermentation, biophotolysis (direct and indirect), and bioelectrochemical (microbial electrocells). A broad revision about the these processes can be found in Asunis et al. (2020) and Osorio-González et al. (2022). Regarding these processes, the dark fermentation has been extensively studied; it involves three main stages (Asunis et al., 2020): (1) lactose hydrolysis into glucose and galactose, (2) conversion of monomeric sugars into lactate by homolactic microorganisms, and (3) lactate conversion into H_2 and VFAs by fermentative fermentation. Next, some works for hydrogen production using mainly dark fermentation are presented.

Ferreira-Rosa et al. (2014) analyzed the effects of various HRT's and different sources of inoculum into two anaerobic fluidized bed reactors, using as substrate CW. According to the results, the maximum hydrogen yield was 1.33 mol H_2/mol lactose obtained at HRT of 4 h using swine sludge as inoculum; also, under the same conditions, the reactor produced 1.22 mol ethanol/L lactose. In this context, Ramos and Silva (2017) published a report about the hydrogen production from CW using an expanded granular sludge reactor (EGSB), such is a biofilm reactor, operated with a carrier material for improving the bacterial adhesion, increasing the biomass concentration into the reactor. The purpose of the work was to evaluate the influence of the HRT and CW concentration on the reactor performance. Based on the experimental data, reducing the HRT enhances H_2 yield, reaching up to 0.64 mmol H_2/COD at an HRT of 4 h. Additionally, the efficient immobilization of the microbial community was shown to further increase hydrogen production.

Blanco et al. (2019) reported the potential of dark fermentation to hydrogen production from CW, using an acidogenic anaerobic structured-bed reactor operated at 25°C, 24 h as HRT and 24 kg COD/m3*day as organic loading rate. According with results, the hydrogen yield was 1.4 mol H_2/mol—consumed lactose; also, a kinetic model to describe the hydrogen performance was determined, highlighting that roughly 74.5% of total hydrogen volume into the reactor is produced from lactate and acetate. Ramos and Silva (2018) studied the cofermentation of vinasse and CW using two thermophilic (55°C) anaerobic fluidized bed reactors, varying the relation between vinasse to CW, and decreasing the HRT. According to experimental results, the maximum hydrogen yield was 1.01 mmol H_2/g COD at 6 h of HRT; also, it is worth mentioning that increasing the CW concentration, regarding the vinasse values, the hydrogen yield decreases.

Asunis et al. (2019) carried out tests for batch dark fermentation using sheep CW without inoculum, varying pH's, and HRT's. According to experimental details, dark fermentation involves two main stages; in the first one, the carbohydrates are converted to lactic acid, while, in the second one, the formed lactic acid is degraded to short-chain fatty acids, such as acetic, butyric, and propionic acids, along with hydrogen. The maximum hydrogen produced was 162 L/kg-total organic compounds at pH 6, after 45 h. Also, with these ideas, the operational parameters into the reactor could be oriented to form any product. Other report about reactor test for hydrogen production was presented by Dessì et al. (2020), wherein was proved the hydrogen generation from CW with simultaneous production and extraction of volatile fatty acids (VFAs), using an upflow anaerobic sludge blanket (UASB) reactor, which has good performance for high organic load streams. The reactor was operated at 20 and 35°C, and pH values of 5.0 and 4.5; adding a recirculation loop by a silicone tube coil submerged in water, to extract the VFAs. According to the results, 0.9 mol H_2/g glucose was obtained at 35°C and pH 5.0 without the VFAs extraction, while with VFAs extraction, low hydrogen yield was obtained (0.3 mol H_2/mol glucose), preferencing the butyric acid extraction.

Rao and Basak (2021) reported the dark fermentation for hydrogen production from CW using *Enterobacter aerogenes* cells into a bioreactor, to experimental and theoretical levels. The maximum hydrogen production was 24.7 mLH_2/h at 32.5 g/L CW at 31°C and pH 6.5; also, with experimental results, a mathematical model was constructed using the

Monod equation to describe the kinetics for substrate consumption and the cell growing under optimum conditions. Another work was reported by Camacho et al. (2022), wherein was analyzed the environmental performance of a two-stage wastewater treatment process: dark fermentation and anaerobic digestion; using mixtures of sugar beet molasses, CW, wine vinasses, and wastewater treatment plant sludge, for producing hydrogen. The analysis methodology was life cycle analysis. According to the results, the process from CW has zero emissions, compared with the other raw materials, being energy self-sufficient.

Aranda-Jaramillo et al. (2023) investigated the feasibility of hydrogen production from CW via lactate-driven dark fermentation, at pH 5.8. In accordance with experimental results, it is technically feasible to use CW without dilution by this process, achieving until 44.5 NmL/g COD fed. More recently, Sarkar et al. (2024) reported the continuous hydrogen production using a tubular biofilm reactor, into an acidogenic fermentation medium from CW. The experiment was carried out for 75 days, analyzing the formation of VFAs, the hydrogen yield and the continuous replacement of CW. According to the results, continuous hydrogen production is feasible, reaching up to 3.88 mL H_2/h, feeding at 8 mL/min of CW; also, it is worth mentioning that VFAs formation was influenced by the substrate flow rate.

3.3 Bioethanol production

Bioethanol production via fermentation from CW has been considered as a feasible alternative in comparison with other biomasses, such as corn, sugarcane, starches, food wastes, or lignocellulosic materials, due to its high organic load and its improper disposal, being the used strain, a key factor for achieving its good performance. Bioethanol production by fermentation pathway requires yeast to convert the lactose to ethanol, such as *Saccharomyces cerevisiae* and *Candida pseudrotropicalis*, which have four times more tolerance to ethanol concentrations, increasing the CW conversion; other yeast capable to form biofilms and tolerate the ethanol concentration is *Kluyveromyces marxianus*, which hydrolyze lactose, allowing high ethanol yield. An extensive revision about yeasts to produce bioethanol from CW is presented in Asunis et al. (2020) and Osorio-González et al. (2022). Reports about the ethanol production from CW have proved that the continuous fermentation is superior to batch processes, resulting in high ethanol yield, but it is required a viable strategy to retain the microorganisms in the medium, for instance, cell immobilization, cell recycle, and membrane retention, achieving until 4.5 g ethanol/L-CW*h (Asunis et al., 2020); also, regarding the key operational parameters, the CW conversion depends on substrate concentration, pH, yeast efficacy, and temperature. Some reports about ethanol production from CW via fermentation are presented next.

Das et al. (2016) presented the direct fermentation of CW powder, using an initial concentration of 150 g/L at 35°C, pH 4.5, and *K. marxianus*. According to the results, the maximum ethanol yield was 43.71 g/L in 72 h; also, it was determined that ethanol conversion into the system follows the Luedeking-Piret equation.

Beniwal et al. (2018) studied the immobilization of beta-galactosidase from dairy yeast on silicon dioxide–based nanoparticles; recycled it 15 times for the hydrolysis of lactose at 37°C and pH 7.0. Then, fermentation was carried out, obtaining up to 63.9 g/L of ethanol using concentrated CW. Another fermentation process from CW was reported by Kallarakkal et al. (2021), wherein a strain named *Oscillatoria* sp. grew up into a nutritive medium with CW water; then, the strain was fermented by ABE (acetone, butanol, and ethanol) fermentation route. According to the results, the butanol yield was increased until 1.6 times, regarding other experiments without CW. Recently, Zhubanova et al. (2024) published a revision paper about the importance of using CW for bioethanol production by novel fermentation technologies using multiple approaches, highlighting the potential reduction of environmental impacts for the use of CW as raw material into the fermentation process.

As mentioned before, there are several conversion processes to produce both bioenergy as well as value-added products from CW. In particular, the production of bioenergy has captured the attention of the scientific community, due to the possibility of being used in the same farms where the waste is generated. In the literature, it has been reported that the conversion of whey to bioethanol, hydrogen, biogas, methane, hydrogen and methane, biodiesel, electricity, and biocrude oil (Table 23.3).

3.4 Economic and environmental impact of the bioenergy used derived from whey

From Table 23.3, it can be observed that the biofuels production yield is lower than 30% in all cases when whey is used alone as raw material. The yields increasing when mixtures with other biomasses (such as manure, grape marc, food waste, among others) are used, or pretreated whey (permeate, deproteinized, or powder) are applied. Among these reported works, the use of hydrothermal liquefaction highlights, since no pretreatment is used, and the mass yield to

TABLE 23.3 Reported studies on bioenergy production derived from whey.

Raw material	Product	Yield	Conversion process	Reference
Cheese-whey powder	Ethanol	0.54 g ethanol/g sugar	Batch ethanol fermentation using *Kluyveromyces marxianus*	Ozmihci and Kargi (2007)
Cheese industrial waste		0.12 g ethanol/g substrate	Fermentation with *Kluyveromyces marxianus*	Hadiyanto et al. (2013)
Cheese whey powder		3.9% (vol/vol)	Free as well as alginate immobilized *Escherichia coli* and *E. coli* expressing *Vitreoscilla hemoglobin* (VHb) in both shake flask and fermenter culture	Sar et al. (2016)
Whey permeates		14.9%–15.9% v/v	Fermentation with *Saccharomyces cerevisiae*	Parashar et al. (2016)
Deproteinized cheese whey		63.9 g/L on concentrated cheese whey (150 g/L)	Bioconversion of hydrolyzed whey into ethanol, coimmobilization strategy was used using *Saccharomyces cerevisiae* and *Kluyveromyces marxianus* for fermentation	Beniwal et al. (2018)
Cheese whey powder	Hydrogen	3.1 ± 0.04 (mol H_2/mol lactose), pH: 6.0, $[S_0]$: 15 g/L	Anaerobic granular sludge from a full-scale upflow anaerobic sludge blanket (UASB) reactor varying substrate initial concentration and pH	Davila-Vazquez et al. (2008)
Cheese whey wastewater		3.5 mol H_2/mol $lactose_{consumed}$	Dark fermentation	Azbar et al. (2009)
Cheese whey wastewater		2.04 mol H_2/mol lactose	Dark fermentation by free and immobilized *Enterobacter aerogenes* MTCC 2822 followed by photofermentation of VFAs	Rai et al. (2012)
Cheese whey powder		20 g/L total sugar concentration	Thermophilic dark fermentation	Kargi et al. (2012)
Cheese whey		94.2 L H_2 kg_{VS}^{-1}	Integration of dark fermentation and biocatalyzed electrolysis	Moreno et al. (2015)
Cheese whey using buffalo manure as buffering agent		152.2 (±43.9) mL H_2/g VS	Thermophilic dark fermentative	Ghimire et al. (2017)
Cheese whey		2.0 L H_2 L^{-1} d^{-1}	Fermentative hydrogen production	Dessi et al. (2020)
Cattle slurry and cheese whey	Biogas	621 L/kg volatile solid	Anaerobic codigestion	Comino et al. (2012)
Sewage sludge with a thermal dried mixture of food waste, cheese whey and olive mill wastewater		Max composition 69.5% of methane in the biogas	Anaerobic codigestion	Maragkaki et al. (2018)
Industrial dairy wastewater and cheese whey		69.8 m^3 per influent COD/L per day	Anaerobic digestion	Charalambous et al. (2020)

Agro-industrial waste with cheese whey		636 ± 1 Ncm^3 gVS^{-1} with 62 ± 1% CH_4	Anaerobic codigestion	Ivanchenko et al. (2021)
Cheese whey and dairy manure		562 mL CH_4/g VS	Anaerobic codigestion at psychrophilic conditions	Casallas-Ojeda et al. (2024)
Grape marc and cheese whey	Methane	0.363 m^3 CH_4 kg^{-1} volatile solids	Solid-state anaerobic digester	Kassongo et al. (2020)
Microalgal biomass and cheese whey		1000 N mL/L-d	Anaerobic codigestion	Rincón-Pérez et al. (2021)
Cheese whey	Hydrogen and methane	Hydrogen: 0.78 ± 0.05 mol H_2/mol glucose consumed Methane: 6.7 L CH_4/L of influent	Mesophilic fermentative hydrogen production Continuous mesophilic anaerobic bioreactor.	Venetsaneas et al. (2009)
Cheese whey powder		28 L H_2/L/d	Two stage fermentation processes	Cota-Navarro et al. (2011)
Cheese whey permeate	Microalgae (raw material for biofuels)	Productivities of roughly 40 mg $L^{-1}d^{-1}$ with the alga *S. obliquus*	40% v/v, obtaining final biomass concentrations of 4.9 g L^{-1} and lipid	Barbera et al. (2018)
Cheese whey	Electricity	Maximum power density 1800 ± 120 W/m^2	Anaerobic bioremediation with the added benefit of generating electricity in microbial fuel cells (MFCs)	Kassongo and Togo (2011)
Dairy waste		850 mV and 28 μA	Microbial fuel cells (MFC) in the treatment of dairy wastewater	Afaf et al. (2021)
Acid whey	Biocrude oil	26.53% mass	Hydrothermal liquefaction	Sudibyo et al. (2021)
Ricotta cheese whey	Biodiesel	Lipids: 14.4 and 9.9 g L^{-1} biodiesel yield: 32.6%	The microorganisms were selected in terms of biomass and lipid production Transesterification	Carota et al. (2017)

biocrude oil is 26.53% mass. Moreover, the use of microbial to produce lipids from whey was explored, being the final biofuel biodiesel; in this case, the yield is significant.

The reported studies show that the conversion of whey is feasible from technical point of view, but they suggested that low yields indicate that economic feasibility remains as a challenge. Nevertheless, the environmental impact of valorization whey is not considered in previous studies. Due to this, it is necessary to realize the analysis considering economic and environmental aspects, to propose a sustainable valorization option. In this regard, scarce works have been found, which are presented in Table 23.4.

From Table 23.4, it can be observed that in all studies, it is reported that environmental benefits are generated due to the valorization of whey. The specific values depend on the type of bioenergy produced, and the final use of it. In most cases, the valorization is performed to produce biogas, which is also used to produce heat and electricity that is employed in the same production process. In one study, hydrogen production followed by methane production is addressed with important reductions on the environmental impact. Moreover, in these studies, a significant reduction in carbon footprint

TABLE 23.4 Reported studies on technical, economic, and environmental analysis regard the whey valorization.

Raw material	Product	Yield	Conversion process	Economic and environmental data	References
Waste water of milk processing	Biogas and biohydrogen	6.28 g of biogas per 1 kg milk	Anaerobic digestion and reforming of biogas	Overall energy efficiency of hydrogen and biogas production from biogas is between 19% and 70% Electric consumption of the system is 0.6 kWh/m^3 biogas.	Coskun et al. (2012)
Manure, cheese whey, maize silage, and fodder beet	Biogas, electricity, and heat	Biogas: 289 Nm^3/MWh Electricity: 422 kWh/MWh Heat: 578 kWh/MWh	Anaerobic digestion and combined heat and power plant	Global warming potential which can be reduced by up to 50% Economic analysis is not assessed	Whiting and Azapagic (2014)
Agricultural waste, delactosed whey permeate	Renewable diesel	75.4%	Yeast fermentation with hydrothermal liquefaction	Minimum fuel selling price of $4.78 per gallon of renewable diesel, and greenhouse gas emissions of 30.0 g CO_2-eq MJ^{-1}	Summers et al. (2015)
Dairy industry wastewater	Hydrogen and methane	H_2: 105 mLg^{-1} COD CH4: 590 mLg^{-1} COD	Two-stage serial CSTR system	0.098 $L/substrate/day Reduction in chloride, nitrate, ammoniacal nitrogen, and sulfate	Kothari et al. (2017)
Milk-whey and potato stem	Biogas and energy generation	0.16–0.18 m^3/kg feedstock 13,277 kW	Anaerobic codigestion	Economic feasibility (33.93 M.USD) if raw materials prices are 0.01 USD/L (milk whey) and 0.04 USD/kg (potato stem), and a fertilizer selling market price is 30 USD/t Environmental impact is not assessed	Martínez-Ruano et al. (2019)
Cheese whey and dairy manure	Biogas	562 mL CH_4/g VS	Anaerobic codigestion at psychrophilic conditions	Reintroduction of waste into the dairy chain contributes to satisfying some of the energy needs of the cheese production process. This reintroduction would reduce the carbon footprint by at least 4%–3.6% Economic analysis is not assessed	Casallas-Ojeda et al. (2024)
Cheese whey	Electricity and heat	Electricity: 38 kW Heat: 350 kW	Anaerobic digestion plus combined heat and power plant	64% reduction in levelized cost of energy and 55% decrease in carbon footprint	Sifakis et al. (2024)

is observed. It highlights the production of renewable diesel through yeast fermentation followed by hydrothermal liquefaction. Even in this kind of process, there are environmental benefits. Therefore, it is required to implement valorization processes of whey to guarantee the sustainable development of the sector.

On the other hand, economic feasibility is not reachable in all the cases. In some cases, the operating costs are reduced because of the use of bioenergy into the same facilities. To achieve profitability, it is important to consider the use of biorefinery schemes for valorization of whey. A biorefinery is a processing scheme that allows the complete conversion of one biomass to produce at least one bioenergy type, one biofuel, and one value-added product. In this manner, the economic feasibility of biomass conversion depends on several products. In this line, some efforts have been reported. One study is dedicated to the proposal of a biorefinery to produce milk, meat, corn ethanol, soybean biodiesel, and biogas (Aguirre-Villegas et al., 2015). The other research presents a review focused on the bioenergy production from dairy manure (Zhu et al., 2021). Nevertheless, in both studies, the whey is not considered as internal raw material for the biorefinery. On the other hand, the work of Asunis et al. (2020) reported a dairy biorefinery, where the treatment of CW valorization processes is included. Authors indicated that bioenergy can be obtained as well as biochemicals. The authors highlighted the use of dark fermentation as a core process for this kind of facility.

4. Concluding remarks

Cheese whey is an abundant by-product of the dairy industry, which is considered a highly polluting waste due to its high organic content. As mentioned in this chapter, CW has the potential to produce bioplastics and bioenergy due to the molecules it has. An upcycling of this waste could potentially reduce the carbon footprint dairy industry has as well as generating new raw materials that can be used to generate innovative products. The new products and energy can also support sustainable agriculture, while promoting bio-based industries. It is important that further research is carried out to assess the cost and scalability of the proposed processes. Furthermore, the integration of these strategies into biorefinery models is needed to maximize economic feasibility. It is also suggested to explore technologies such as metabolic engineering of microorganisms to obtain better yields. Life cycle assessments are also encouraged to truly obtain sustainable products from waste. Finally, it is important to change the paradigm and see this highly available waste as a new raw material that can be used to generate new products.

References

Acquavia, M.A., Benítez, J.J., Guzmán-Puyol, S., Porras-Vázquez, J.M., Hierrezuelo, J., Grifé-Ruiz, M., Romero, D., Di-Capua, A., Bochicchio, R., Laurenza, S., Bianco, G., Heredia-Guerrero, J.A., 2024. Enhanced extraction of bioactive compounds from tea waste for sustainable polylactide-based bioplastic applications in active food packaging. Food Packag. Shelf Life 46, 101410. https://doi.org/10.1016/j.fpsl.2024.101410.

Afaf, J., Al-saned, O., Kitafa, B.A., Badday, A.S., 2021. Microbial fuel cells (MFC) in the treatment of dairy wastewater. In: IOP Conf. Series: Materials Science and Engineering, p. 1067. https://doi.org/10.1088/1757-899X/1067/1/012073.

Aguirre-Villegas, H.A., Passos-Fonseca, T.H., Reinemann, D.J., Armentano, L.E., Wattiaux, M.A., Cabrera, V.E., Norman, J.M., Larson, R., 2015. Green cheese: partial life cycle assessment of greenhouse gas emissions and energy intensity of integrated dairy production and bioenergy systems. J. Dairy Sci. 98 (3), 1571–1592. https://doi.org/10.3168/jds.2014-8850.

Akbi, A., Saber, M., Aziza, M., Yassaa, N., 2017. An overview of sustainable bioenergy potential in Algeria. Renew. Sustain. Energy Rev. 72, 240–245. https://doi.org/10.1016/j.rser.2017.01.072.

Alfattani, R., Shah, M.A., Siddqui, M.I.H., Ali, M.A., Alnaser, I.A., 2022. Bio-char characterization produced from walnut shell biomass through slow pyrolysis: sustainable for soil amendment and an alternate bio-fuel. Energies 15 (1), 1. https://doi.org/10.3390/en15010001.

Aranda-Jaramillo, B., León-Becerril, E., Aguilar-Juárez, O., Castro-Muñoz, R., García-Depraect, O., 2023. Feasibility study of biohydrogen production from acid cheese whey via lactate-driven dark fermentation. Fermentation 9. https://doi.org/10.3390/fermentation9070644.

Asunis, F., De Gioannis, G., Dessì, P., Isipato, M., Lens, P.N.L., Muntoni, A., Polettini, A., Pomi, R., Rossi, A., Spiga, D., 2020. The dairy biorefinery: Integrating treatment processes for cheese whey valorisation. J. Environ. Manag. 276, 111240. https://doi.org/10.1016/j.jenvman.2020.111240.

Asunis, F., De Gioannis, G., Isipato, M., Muntoni, A., Polettini, A., Pomi, R., Rossi, A., Spiga, D., 2019. Control of fermentation duration and pH to orient biochemicals and biofuels production from cheese whey. Bioresour. Technol. 289, 121722. https://doi.org/10.1016/j.biortech.2019.121722.

Azbar, N., Dokgöz, F.T.Ç., Peker, Z., 2009. Optimization of basal medium for fermentative hydrogen production from cheese whey wastewater. Int. J. Green Energy 6 (4), 371–380. https://doi.org/10.1080/15435070903107049.

Barbera, E., Bertucco, A., Kumar, S., 2018. Nutrients recovery and recycling in algae processing for biofuels production. Renew. Sustain. Energy Rev. 90, 28–42. https://doi.org/10.1016/j.rser.2018.03.004.

Batista, L.P.P., Paulinetti, A.P., Júnior, A.D.N.F., Albanez, R., Ratusznei, S.M., Etchebehere, C., Lovato, G., Rodrigues, J.A.D., 2023. Two-stage thermophilic anaerobic digestion of cheese whey: process optimization, comparison with single-stage, and full-scale estimation. Chem. Eng. Process. Process Intensif. 183, 109260. https://doi.org/10.1016/j.cep.2022.109260.

Beniwal, A., Saini, P., Kokkiligadda, A., Vij, S., 2018. Use of silicon dioxide nanoparticles for β-galactosidase immobilization and modulated ethanol production by co-immobilized K. marxianus and S. cerevisiae in deproteinized cheese whey. LWT 87, 553–561. https://doi.org/10.1016/J.LWT.2017.09.028.

Bhatia, S.K., Otari, S.V., Jeon, J.M., Gurav, R., Choi, Y.K., Bhatia, R.K., Pugazhendhi, A., Kumar, V., Rajesh Banu, J., Yoon, J.J., Choi, K.Y., Yang, Y.H., 2021. Biowaste-to-bioplastic (polyhydroxyalkanoates): conversion technologies, strategies, challenges, and perspective. Bioresour. Technol. 326 (January), 124733. https://doi.org/10.1016/j.biortech.2021.124733.

Blanco, V.M.C., Fuess, L.T., Araujo, M.N., Zaiat, M., 2023. Biodigestion of easily-acidifying cheese whey in a sequential fermentative-methanogenic process: strategies for boosting energy production and minimizing alkalinization costs. J. Water Process Eng. 55, 104245. https://doi.org/10.1016/j.jwpe.2023.104245.

Blanco, V.M.C., Oliveira, G.H.D., Zaiat, M., 2019. Dark fermentative biohydrogen production from synthetic cheese whey in an anaerobic structured-bed reactor: performance evaluation and kinetic modeling. Renew. Energy 139, 1310–1319. https://doi.org/10.1016/j.renene.2019.03.029.

Bosco, F., Cirrincione, S., Carletto, R., Marmo, L., Chiesa, F., Mazzoli, R., Pessione, E., 2021. Pha production from cheese whey and "scotta": comparison between a consortium and a pure culture of leuconostoc mesenteroides. Microorganisms 9 (12). https://doi.org/10.3390/microorganisms9122426.

Camacho, C.I., Estévez, S., Conde, J.J., Feijoo, G., Moreira, M.T., 2022. Dark fermentation as an environmentally sustainable WIN-WIN solution for bioenergy production. J. Clean. Prod. 374, 134026. https://doi.org/10.1016/j.jclepro.2022.134026.

Carlini, M., Castellucci, S., Moneti, M., 2015. Biogas production from poultry manure and cheese whey wastewater under mesophilic conditions in batch reactor. Energy Proc. 82, 811–818. https://doi.org/10.1016/j.egypro.2015.11.817.

Carlozzi, P., Giovannelli, A., Traversi, M.L., Touloupakis, E., 2021. Poly(3-hydroxybutyrate) bioproduction in a two-step sequential process using wastewater. J. Water Proc. Eng. 39 (June), 101700. https://doi.org/10.1016/j.jwpe.2020.101700.

Carota, E., Crognale, S., D'Annibale, A., Gallo, A.M., Stazi, S.R., Petruccioli, M., 2017. A sustainable use of Ricotta Cheese Whey for microbial biodiesel production. Sci. Total Environ. 584–585, 554–560. https://doi.org/10.1016/j.scitotenv.2017.01.068.

Carvalheira, M., Hilliou, L., Oliveira, C.S.S., Guarda, E.C., Reis, M.A.M., 2022. Polyhydroxyalkanoates from industrial cheese whey: production and characterization of polymers with differing hydroxyvalerate content. Curr. Res. Biotechnol. 4 (November 2021), 211–220. https://doi.org/10.1016/j.crbiot.2022.03.004.

Casallas-Ojeda, M., Cabeza, I., Sanchez, N., Caicedo-Concha, D.M., Astals, S., 2024. Cheese whey and dairy manure anaerobic co-digestion at psychrophilic conditions: technical and environmental evaluation. Environ. Res. 251 (Part 1), 118525. https://doi.org/10.1016/j.envres.2024.118525.

Chalermthai, B., Giwa, A., Schmidt, J.E., Taher, H., 2021. Life cycle assessment of bioplastic production from whey protein obtained from dairy residues. Bioresour. Technol. Rep. 15, 100695. https://doi.org/10.1016/j.biteb.2021.100695.

Chang, Y.C., Reddy, M.V., Imura, K., Onodera, R., Kamada, N., Sano, Y., 2021. Two-stage polyhydroxyalkanoates (Pha) production from cheese whey using acetobacter pasteurianus c1 and bacillus sp. cyr1. Bioengineering 8 (11). https://doi.org/10.3390/bioengineering8110157.

Charalambous, P., Shin, J., Shin, S.G., Vyrides, I., 2020. Anaerobic digestion of industrial dairy wastewater and cheese whey: performance of internal circulation bioreactor and laboratory batch test at pH 5-6. Renew. Energy 147 (1), 1–10. https://doi.org/10.1016/j.renene.2019.08.091.

Comino, E., Riggio, V.A., Rosso, M., 2012. Biogas production by anaerobic co-digestion of cattle slurry and cheese whey. Bioresour. Technol. 114, 46–53. https://doi.org/10.1016/j.biortech.2012.02.090.

Coskun, C., Bayraktar, M., Oktay, Z., Dincer, I., 2012. Investigation of biogas and hydrogen production from waste water of milk-processing industry in Turkey. Int. J. Hydrogen Energy 37 (21), 16498–16504. https://doi.org/10.1016/j.ijhydene.2012.02.174.

Cota-Navarro, C.B., Carrillo-Reyes, J., Davila-Vazquez, G., Alatriste-Mondragón, F., Razo-Flores, E., 2011. Continuous hydrogen and methane production in a two-stage cheese whey fermentation system. Water Sci. Technol. 64 (2), 367–374. https://doi.org/10.2166/wst.2011.631.

Das, B., Sarkar, S., Maiti, S., Bhattacharjee, S., 2016. Studies on production of ethanol from cheese whey using Kluyveromyces marxianus. Mater. Today Proc. 3, 3253–3257. https://doi.org/10.1016/j.matpr.2016.10.006.

Davila-Vazquez, G., Alatriste-Mondragón, F., de León-Rodríguez, A., Razo-Flores, E., 2008. Fermentative hydrogen production in batch experiments using lactose, cheese whey and glucose: influence of initial substrate concentration and pH. Int. J. Hydrogen Energy 33 (19), 4989–4997. https://doi.org/10.1016/j.ijhydene.2008.06.065, 2008.

Dessì, P., Asunis, F., Ravishankar, H., Cocco, F.G., De Gioannis, G., Muntoni, A., Lens, P.N.L., 2020. Fermentative hydrogen production from cheese whey with in-line, concentration gradient-driven butyric acid extraction. Int. J. Hydrogen Energy 45 (46), 24453–24466. https://doi.org/10.1016/j.ijhydene.2020.06.081.

Ebrahimi, F., Ramezani Dana, H., 2022. Poly lactic acid (PLA) polymers: from properties to biomedical applications. Int. J. Polym. Mater. Polym. Biomater. 71 (15), 1117–1130. https://doi.org/10.1080/00914037.2021.1944140.

Figueroa-Pires, A., Garcia Marnotes, N., Díaz Rubio, O., Cobos Garcia, A., Dias Pereira, C., 2021. Dairy by-products: a review on the valorization of whey and second cheese whey. Foods 10, 1067. https://doi.org/10.3390/foods10051067.

Figueroa-Pires, A., Díaz, O., Cobos, A., Dias-Pereira, C., 2024. A review of recent developments in edible films and Coatings-Focus on whey-based materials. Foods 13, 2638. https://doi.org/10.3390/foods1316.

Garavand, F., Rouhi, M., Jafarzadeh, S., Khodaei, D., Cacciotti, I., Zargar, M., Razavi, S.H., 2022. Tuning the Physicochemical, structural, and antimicrobial attributes of whey-based poly (L-Lactic acid) (PLLA) films by Chitosan nanoparticles. Front. Nutr. 9, 880520. https://doi.org/10.3389/fnut.2022.880520.

Ghimire, A., Luongo, V., Frunzo, L., Pirozzi, F., Lens, P.N.L., Esposito, G., 2017. Continuous biohydrogen production by thermophilic dark fermentation of cheese whey: use of buffalo manure as buffering agent. Int. J. Hydrogen Energy 42 (8), 4861–4869. https://doi.org/10.1016/j.ijhydene.2016.11.185.

Gordillo-Andia, C., Almirón, J., Barreda-Del-Carpio, J.E., Roudet, F., Tupayachy-Quispe, D., Vargas, M., 2024. Influence of Kluyveromyces lactis and Enterococcus faecalis on obtaining lactic acid by cheese whey fermentation. Appl. Sci. (Switzerland) 14 (11). https://doi.org/10.3390/app14114649.

Grana, M., Marreiros, B.C., Carvalheira, M., Ficara, E., Reis, M.A.M., 2024. Polyhydroxyalkanoates production from cheese whey under near-seawater salinity conditions. New Biotechnol. 84 (September), 53–63. https://doi.org/10.1016/j.nbt.2024.09.005.

Hadiyanto, H., Ariyanti, D., Aini, A., Pinundi, D., 2013. Batch and fed-batch fermentation system on ethanol production from whey using Kluyveromyces marxianus. Int. J. Renew. Energy Dev. 2 (3), 127–131. https://doi.org/10.14710/ijred.2.3.127-131.

Islam, M., Xayachak, T., Haque, N., Lau, D., Bhuiyan, M., Pramanik, B.K., 2024. Impact of bioplastics on environment from its production to end-of-life. Process Saf. Environ. Prot. 188, 151–166. https://doi.org/10.1016/j.psep.2024.05.113.

Ivanchenko, A., Yelatontsev, D., Savenkov, A., 2021. Anaerobic co-digestion of agro-industrial waste with cheese whey: impact of centrifuge comminution on biogas release and digestate agrochemical properties. Biomass Bioenergy 147, 106010. https://doi.org/10.1016/j.biombioe.2021.106010.

Kale, G., Kijchavengkul, T., Auras, R., Rubino, M., Selke, S.E., Singh, S.P., 2017. Compostability of bioplastic packaging materials: an overview. Macromol. Biosci. 7 (3), 255–277. https://doi.org/10.1002/mabi.200600168.

Kallarakkal, K.P., Muthukumar, K., Alagarsamy, A., Pugazhendhi, A., Naina Mohamed, S., 2021. Enhancement of biobutanol production using mixotrophic culture of Oscillatoria sp. in cheese whey water. Fuel 284, 119008. https://doi.org/10.1016/j.fuel.2020.119008.

Kargi, F., Eren, N.S., Ozmihci, S., 2012. Hydrogen gas production from cheese whey powder (CWP) solution by thermophilic dark fermentation. Int. J. Hydrogen Energy 37 (3), 2260–2266. https://doi.org/10.1016/j.ijhydene.2011.11.018.

Kassongo, J., Shahsavari, E., Ball, A.S., 2020. Renewable energy from the solid-state anaerobic digestion of grape marc and cheese whey at high treatment capacity. Biomass Bioenergy 143, 105880. https://doi.org/10.1016/j.biombioe.2020.105880.

Kassongo, J., Togo, C.A., 2011. Performance improvement of whey-driven microbial fuel cells by acclimation of indigenous anodophilic microbes. Afr. J. Biotechnol. 10 (40), 7846–7852.

Kochubei-lytvynenko, O., Bilyk, O., Vysotskyi, O., Zabroda, A., 2024. Consideration of technological and safety aspects of using zinc oxide nanoparticles for intensifying whey. Food Prod. Technol. 3 (78), 39–45. https://doi.org/10.15587/2706-5448.2024.310354.

Kothari, R., Kumar, V., Pathak, V.V., Tyagi, V.V., 2017. Sequential hydrogen and methane production with simultaneous treatment of dairy industry wastewater: bioenergy profit approach. Int. J. Hydrogen Energy 42 (8), 4870–4879. https://doi.org/10.1016/j.ijhydene.2016.11.163.

Kovalev, A.A., Mikheeva, E.R., Katraeva, I.V., Kovalev, D.A., Kozlov, A.M., Yu, V.,L., 2023. Bioenergy recovery from two-stage mesophilic-thermophilic anaerobic digestion of cheese whey. Int. J. Hydrogen Energy 48, 4676–4685. https://doi.org/10.1016/j.ijhydene.2022.11.003.

Kumar, S., Dubey, N., Kumar, V., Choi, I., Jeon, J., Kim, M., 2024. Combating micro/nano plastic pollution with bioplastic: sustainable food packaging, challenges, and future perspectives. Environ. Pollut. 363, 125077. https://doi.org/10.1016/j.envpol.2024.125077.

Lavelli, V., Beccalli, M.P., 2022. Cheese whey recycling in the perspective of the circular economy: modeling processes and the supply chain to design the involvement of the small and medium enterprises. Trends Food Sci. Technol. 126, 86–98. https://doi.org/10.1016/j.tifs.2022.06.013.

Maragkaki, A.E., Vasileiadis, I., Fountoulakis, M., Kyriakou, A., Lasaridi, K., Manios, T., 2018. Improving biogas production from anaerobic co-digestion of sewage sludge with a thermal dried mixture of food waste, cheese whey and olive mill wastewater. Waste Manag. 71, 644–651. https://doi.org/10.1016/j.wasman.2017.08.016.

Martínez-Ruano, J.A., Restrepo-Serna, D.L., Carmona-Garcia, E., Poveda Giraldo, J.A., Aroca, G., Cardona, C.A., 2019. Effect of co-digestion of milk-whey and potato stem on heat and power generation using biogas as an energy vector: techno-economic assessment. Appl. Energy 241, 504–518. https://doi.org/10.1016/j.apenergy.2019.03.005.

Mavai, S., Bains, A., Sridhar, K., Rashid, S., Elossaily, G.M., Ali, D., Chawla, P., Sharma, M., 2024. Formulation and application of poly lactic acid, gum, and cellulose-based ternary bioplastic for smart food packaging: a review. Int. J. Biol. Macromol. 268, 131687. https://doi.org/10.1016/j.ijbiomac.2024.131687.

Moentamaria, D., Irfin, Z., Maya, F., Kusuma, H.S., 2023. Optimization of making edible films based on glycerol, corn oil and whey cheese using rotatable central composite design. Egypt. J. Chem. 66, 199–205. https://doi.org/10.21608/ejchem.2023.180016.7307.

Moreno, R., Escapa, A., Cara, J., Carracedo, B., Gómez, X., 2015. A two-stage process for hydrogen production from cheese whey: integration of dark fermentation and biocatalyzed electrolysis. Int. J. Hydrogen Energy 40 (1), 168–175. https://doi.org/10.1016/j.ijhydene.2014.10.120.

Mozejko-Ciesielska, J., Marciniak, P., Moraczewski, K., Rytlewski, P., Czaplicki, S., Zadernowska, A., 2022. Cheese whey mother liquor as dairy waste with potential value for polyhydroxyalkanoate production by extremophilic Paracoccus homiensis. Sustain. Mater. Technol. 33 (June), e00449. https://doi.org/10.1016/j.susmat.2022.e00449.

Osorio-González, C.S., Gómez-Falcon, N., Brar, S.K., Ramírez, A.A., 2022. Cheese whey as a potential feedstock for producing renewable biofuels: a review. Energies 15. https://doi.org/10.3390/en15186828.

Ozmihci, S., Kargi, F., 2007. Kinetics of batch ethanol fermentation of cheese-whey powder (CWP) solution as function of substrate and yeast concentrations. Bioresour. Technol. 98 (16), 2978–2984. https://doi.org/10.1016/j.biortech.2006.10.005.

Parashar, A., Jin, Y., Mason, B., Chae, M., Bressler, D.C., 2016. Incorporation of whey permeate, a dairy effluent, in ethanol fermentation to provide a zero waste solution for the dairy industry. J. Dairy Sci. 99 (3), 1859–1867. https://doi.org/10.3168/jds.2015-10059.

Patil, T.D., Ghosh, S., Agarwal, A., Patel, S.K.S., Tripathi, A.D., Mahato, D.K., Kumar, P., Slama, P., Pavlik, A., Haque, S., 2024. Production, optimization, scale up and characterization of polyhydoxyalkanoates copolymers utilizing dairy processing waste. Sci. Rep. 14 (1), 1–11. https://doi.org/10.1038/s41598-024-52098-0.

Ponist, J., Samesova, D., Schwarz, M., 2021. Methods of processing whey waste from dairies. A review. Environ. Prot. Eng. 47, 67–84. https://doi.org/10.37190/epe210405.

Ponmanian, M., Sivashanmugam, P., 2024. Biotransformation of Cheese Whey Effluent: A Dual Approach for Wastewater Treatment and Polyhydroxyalkanoates Production Using Mixed Microbial Cultures. https://doi.org/10.2139/ssrn.4987377. https://ssrn.com/abstract=4987377.

Rai, P.K., Singh, S.P., Asthana, R.K., 2012. Biohydrogen production from cheese whey wastewater in a two-step anaerobic process. Appl. Biochem. Biotechnol. 167, 1540–1549. https://doi.org/10.1007/s12010-011-9488-4.

Ramos, L.R., Silva, E.L., 2017. Improving EGSB reactor performance for simultaneous bioenergy and organic acid production from cheese whey via continuous biological H2 production. Biotechnol. Lett. 39, 983–991. https://doi.org/10.1007/s10529-017-2323-4.

Ramos, L.R., Silva, E.L., 2018. Continuous hydrogen production from cofermentation of sugarcane vinasse and cheese whey in a thermophilic anaerobic fluidized bed reactor. Int. J. Hydrogen Energy 43, 13081–13089. https://doi.org/10.1016/j.ijhydene.2018.05.070.

Ramos-Suárez, J.L., Álvarez-Méndez, S.J., Ritter, A., González, J.M., Pérez, A.C., 2024. A comprehensive evaluation of cheese whey to produce biogas in the Canary Islands. Biomass Bioenergy 180, 107008. https://doi.org/10.1016/j.biombioe.2023.107008.

Rao, R., Basak, N., 2021. Optimization and modelling of dark fermentative hydrogen production from cheese whey by Enterobacter aerogenes 2822. Int. J. Hydrogen Energy 46, 1777–1800. https://doi.org/10.1016/j.ijhydene.2020.10.142.

Rincón-Pérez, J., Celis, L.B., Morales, M., Alatriste-Mondragón, F., Tapia-Rodríguez, A., Razo-Flores, E., 2021. Improvement of methane production at alkaline and neutral pH from anaerobic co-digestion of microalgal biomass and cheese whey. Biochem. Eng. J. 169, 107972. https://doi.org/10.1016/j.bej.2021.107972.

Rosa, P.R.F., Santos, S.C., Sakamoto, I.K., Varesche, M.B.A., Silva, E.L., 2014. Hydrogen production from cheese whey with ethanol-type fermentation: effect of hydraulic retention time on the microbial community composition. Bioresour. Technol. 161, 10–19. https://doi.org/10.1016/j.biortech.2014.03.020.

Sahoo, T.K., Jayaraman, G., 2019. Co-culture of Lactobacillus delbrueckii and engineered Lactococcus lactis enhances stoichiometric yield of d-lactic acid from whey permeate. Appl. Microbiol. Biotechnol. 103 (14), 5653–5662. https://doi.org/10.1007/s00253-019-09819-7.

Sar, T., Stark, B.C., Yesilcimen Akbas, M., 2016. Effective ethanol production from whey powder through immobilized E. coli expressing Vitreoscilla hemoglobin. Bioengineered 8 (2), 171–181. https://doi.org/10.1080/21655979.2016.1218581.

Sarkar, O., Rova, U., Christakopoulos, P., Matsakas, L., 2024. Continuous biohydrogen and volatile fatty acids production from cheese whey in a tubular biofilm reactor: substrate flow rate variations and microbial dynamics. Int. J. Hydrogen Energy 59, 1305–1316. https://doi.org/10.1016/j.ijhydene.2024.02.041.

Sher, F., Smječanin, N., Hrnjić, H., Karadža, A., Omanović, R., Šehović, E., Sulejmanović, J., 2024. Emerging technologies for biogas production: a critical review on recent progress, challenges and future perspectives. Process Saf. Environ. Prot. 188, 834–859. https://doi.org/10.1016/j.psep.2024.05.138.

Sifakis, N., Savvakis, N., Petropoulou, M., Arampatzis, G., 2024. Techno-economic optimization of a novel industrial hybrid renewable energy system based on the waste-to-X principle. Energy Convers. Manag. 313, 118613. https://doi.org/10.1016/j.enconman.2024.118613.

Sudibyo, H., Wang, K., Tester, J.W., 2021. Hydrothermal liquefaction of acid whey: Effect of feedstock properties and process conditions on energy and nutrient recovery. ACS Sustain. Chem. Eng. 9 (34), 11403–11415. https://doi.org/10.1021/acssuschemeng.1c03358.

Summers, H.M., Ledbetter, R.N., McCurdy, A.T., Morgan, M.R., Seefeldt, L.C., Jena, U., Hoekman, S.K., Quinn, J.C., 2015. Techno-economic feasibility and life cycle assessment of dairy effluent to renewable diesel via hydrothermal liquefaction. Bioresour. Technol. 196, 431–440. https://doi.org/10.1016/j.biortech.2015.07.077.

Taqui, S.N., Mohan, C.S., Goodarzi, M.S., Elkotb, M.A., Khatoon, B.A., Soudagar, M.E.M., Koki, I.B., Elfasakhany, A., Khalifa, A.S., Ali, M.A., Saifullah, Z., Siddiqui, M.I.H., Safaei, M.R., Saleel, C.A., 2021. Sustainable adsorption method for the remediation of crystal violet dye using nutraceutical industrial fenugreek seed spent. Appl. Sci. 11 (16). https://doi.org/10.3390/app11167635.

Thakur, S., Chaudhary, J., Sharma, B., Verma, A., Tamulevicius, S., Thakur, V.K., 2018. Sustainability of bioplastics: opportunities and challenges. Curr. Opin. Green Sustainable Chem. 13, 68–75. https://doi.org/10.1016/j.cogsc.2018.04.013.

Tripathi, N., Misra, M., Mohanty, A.K., 2021. Durable polylactic acid (PLA)-Based sustainable engineered blends and biocomposites: recent developments, challenges, and opportunities. ACS Eng. Au 1 (1), 7–38. https://doi.org/10.1021/acsengineeringau.1c00011.

Venetsaneas, N., Antonopoulou, G., Stamatelatou, K., Kornaros, M., Lyberatos, G., 2009. Using cheese whey for hydrogen and methane generation in a two-stage continuous process with alternative pH controlling approaches. Bioresour. Technol. 100 (15), 3713–3717. https://doi.org/10.1016/j.biortech.2009.01.025.

Whiting, A., Azapagic, A., 2014. Life cycle environmental impacts of generating electricity and heat from biogas produced by anaerobic digestion. Energy 70, 181–193. https://doi.org/10.1016/j.energy.2014.03.103.

Yaradoddi, J.S., Banapurmath, N.R., Ganachari, S.V., Soudagar, M.E.M., Sajjan, A.M., Kamat, S., Mujtaba, M.A., Shettar, A.S., Anqi, A.E., Safaei, M. R., Elfasakhany, A., Siddiqui, M.I.H., Ali, M.A., 2022. Bio-based material from fruit waste of orange peel for industrial applications. J. Mater. Res. Technol. 17, 3186–3197. https://doi.org/10.1016/j.jmrt.2021.09.016.

Zhu, Q.-L., Wu, B., Pisutpaisal, N., Wang, Y.-W., Ma, K.-d., Dai, L.C., Qin, H., Tan, F.-R.-., Maeda, T., Xu, Y.-s., Hu, G.-Q., He, M.X., 2021. Bioenergy from dairy manure: technologies, challenges and opportunities. Sci. Total Environ. 790, 148199. https://doi.org/10.1016/j.scitotenv.2021.148199.

Zhubanova, A., Akimbekov, N., Tastambek, K., Abdieva, G., Ualieva, P., Malik, A., Altynbay, N., Xiaohui, Q., 2024. Whey to ethanol: unlocking the potential of cheese whey for sustainable bioethanol production. Eng. Sci. 27, 998. https://doi.org/10.30919/es998.

Zikmanis, P., Kolesovs, S., Semjonovs, P., 2020. Production of biodegradable microbial polymers from whey. Bioresour. Bioprocess. 7 (1). https://doi.org/10.1186/s40643-020-00326-6.

Part 5

Future/Recent trends

Chapter 24

Future developments of cheese whey management

Alejandra Ramírez-Hernández, Claudia Gutiérrez-Antonio, Ana Angélica Feregrino-Pérez and Juan Fernando García-Trejo
Engineering Faculty, Autonomous University of Queretaro, Campus Juriquilla, Querétaro, Mexico

1. Introduction

In the year 2050, the world population will reach 9.1 billion people; this increment will lead to an augment of the energy and material resources required by society (OECD/FAO, 2025). In particular, food production needs to be significantly augmented; the forecast for food production indicates that an increase of 70% needs to be achieved between 2005 and 2050 (OECD/FAO, 2025). An important aspect to mention is that this food production increase is moving not just to satisfy the amounts, but also the nutritional requirements. This new direction is motivated by society's awareness of self-care and quality of life, for which food must be nutritious, in sufficient quantity, and healthy. In this context, cheese is an important food derived from milk with high nutritional value, delicious flavor, and is produced worldwide in different kinds.

According to the Food and Agricultural Organization (FAO), cheese is one of the main agricultural products worldwide (Carvalho et al., 2013). Cheese is the name for a group of fermented milk-based food products, which are fabricated in a wide range of flavors and forms throughout the world (Fox and McSweeney, 2025). Indeed, a great diversity of cheeses is produced worldwide (McSweeney et al., 2025); nevertheless, no one really knows how many varieties of cheese there is in the world because many very similar cheeses have different names (Fox and McSweeney, 2025). According to the Organization for Economic Cooperation and Development, cheese is the most important processed dairy product consumed in terms of milk solids (OECD, 2023). Moreover, the per capita consumption of cheese is expected to grow up to 17.6% in 2032 (OECD/FAO, 2025), considering the human eating habits and the good nutritional qualities of cheese (Soares et al., 2021).

The production of cheese dates back more than 4000 years, until the 1800s when the cheese industry was born (IDFA, 2025). Due to this, the cheese production has experienced significant progress over time (Soares et al., 2021). In general, cheese is produced from milk, which is processed through different unit operations shown in Fig. 24.1. Based on Fig. 24.1, it can be observed that the first step is the standardization of milk; in this operation, the content of fat of milk is adjusted in order to fulfill the final specifications of the different products. Then, the standardized milk is pasteurized, and then cooled down to the required temperature to start the inoculation culture. After that, the cutting of cheese is performed; in this stage, the amount and quality of rennet, the time, and the temperature are the main variables involved. Another key stage is the texturing curve, where the main variables are the size of the cut as well as the time. The cutting and texturing curd are the most important stages to produce a cheese with adequate quality. Later, the cheese whey is separated from the solid cheese. Final stages include the addition of coagulants, storage, aging, and packaging.

According to Soares et al. (2021), the transformations that occurred during milk processing, such as the generation of effluent from sanitation and cleaning, have great potential to cause environmental impacts. These effluents have a high organic content in large water volumes. In particular, as can be observed in Fig. 24.1, whey separation is an important step in the cheese production; here, the solid fraction is separated to continue its processing. The high organic content of the cheese whey makes it an effluent with high environmental impact; it is estimated that per 1 kg of cheese, approximately 9 kg of whey are generated (Hausjell et al., 2019).

Whey. https://doi.org/10.1016/B978-0-443-27648-4.00025-6

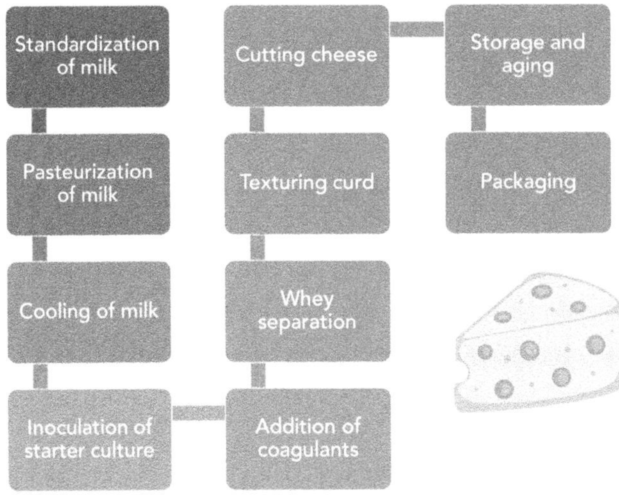

FIGURE 24.1 General process for cheese production. *Own elaboration based on information from Kumar Sachan and Karnwal (2022), with the cheese illustration taken from Pixabay (2025).*

Cheese whey is the liquid remaining following the precipitation and removal of milk casein during cheese-making (González Siso, 1996). Cheese whey is composed of lactose, soluble proteins, lipids, mineral salts, B group vitamins, as well as lactic and citric acids (González Siso, 1996). Moreover, the values of biochemical oxygen demand (BOD) range from 30,000 to 60,000 mg/L and chemical oxygen demand (COD) from 50,000 to 100,000 mg/L; as a reference, these values are equivalent to 100–175 times the organic matter contained in the same volume of domestic wastewater (Paiva Anciens Ramos et al., 2021). Due to its composition, cheese whey has a strong impact on the environment when it is not properly managed.

In the past, cheese whey was mainly disposed of in land or directly discharged to rivers, lakes, or the ocean (Prazeres et al., 2012). González Siso (1996) reported that 50% of the total world cheese whey produced was treated and transformed. Nevertheless, nowadays it is reported that the same percentage (50%) of the total cheese whey is discharged to the environment without any treatment (Asunis et al., 2020). Regarding this matter, public awareness of sustainability can change the growing trend of cheese production, since the production of milk and dairy products contributes significantly to greenhouse gas emissions. Due to this, new strategies must be implemented to manage this waste.

In brief, whey has generally been considered an unwanted by-product by cheesemakers (de Paiva Anciens Ramos et al., 2021). The high volumes and composition of cheese whey have driven the search of treatments options. In those processes, the main objective is to neutralize the threat that this effluent represents. The treatment processes include mainly biological treatments, physicochemical treatments, and direct land application (Prazeres et al., 2012). Nevertheless, these technologies are not generally adopted due to high costs, operational difficulties, and limited effectiveness (de Paiva Anciens Ramos et al., 2021); this complicates the establishment of cheese whey treatment systems. Therefore, the paradigm needs to be changed. Cheese whey should be considered a valuable resource that can be converted into new marketable products. Indeed, the compounds present in cheese whey, which make it harmful to the environment, are valuable resources that can be recovered or used to generate different kinds of products. This new approach can solve the pollution problem, generated due to its improperly disposal, and use this effluent as a new raw material to generate value-added products and bioenergy. This approach is also aligned with the circular economy principles, leading the cheese industry to its sustainable development. Therefore, this chapter presents and discusses the treatments and valorization strategies for cheese whey valorization, with the aim of incorporating circular economy principles into the cheese industry that guarantee its sustainable development.

2. Treatments technologies for cheese whey

Several processes have been reported for the treatment of cheese whey. It is important to mention that due to the composition of cheese whey, its treatment is required before being disposed of in the drainage system. The treatment processes include mainly biological treatments, physicochemical treatments, and direct land application (Prazeres et al., 2012). A general overview of these treatment technologies is shown in Fig. 24.2. Next, dilution, coagulation, and flocculation, as well as constructed wetlands strategies, are discussed as treatment technologies for cheese whey.

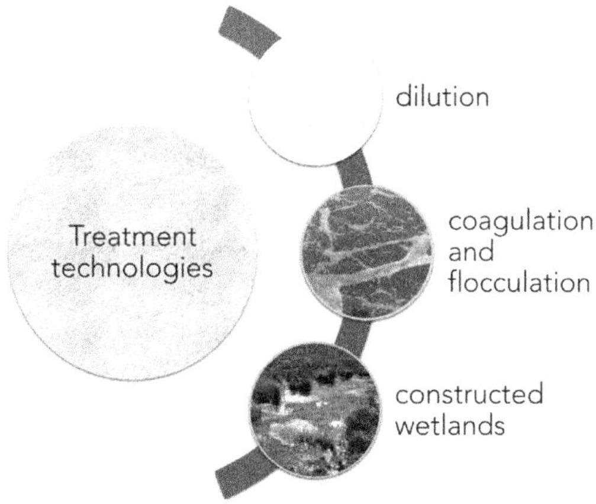

FIGURE 24.2 Treatment technologies for cheese whey. *Own elaboration based on information from Prazeres et al. (2012) with illustrations taken from Pixabay (2025).*

2.1 Dilution

Cheese whey dilution is considered as an alternative for its disposal; in this case, the blending of cheese whey is performed with less contaminant's effluents, such as domestic wastewater (Minhalma et al., 2007). This strategy allows for the decrease in the organic matter content by looking for its treatment in municipal wastewater treatment plants. Nevertheless, cheese whey can affect the stability of microorganisms present in treatment processes performed in those municipal wastewater facilities. Therefore, this alternative seems logical, but it is not the best option.

On the other hand, in small factories, a common strategy is the disposal of cheese whey in water reservoirs, such as lakes and rivers. This is another type of dilution. However, consequences include a strong stench and eutrophication caused by the high nitrogen and phosphorus content in wastewater (Patsialou et al., 2024). In 2025, Ramirez-Flores et al. (2025) reported that the dilution of treated cheese whey with water (1:20 ratio) was effective as a nutrient solution for the cultivation of lettuce, cabbage, and tomato. According to the authors, the crops have similar growth rates to those obtained with commercial nutritive solutions. In this line, more research on the effect of this mixture on other edible crops could help to increase the use of this valuable resource.

2.2 Precipitation processes

Coagulation and flocculation processes are physicochemical treatments meant to the reduction of contaminant indicators, such as organic matter, turbidity and suspended solids (Prazeres et al., 2012). Therefore, technologies such as coagulation and flocculation processes have been implemented as cheese whey treatment technology.

In coagulation-flocculation treatments, the use of $FeSO_4$, $FeCl_3$, and $Al_2(SO_4)_3$ has been reported, while the use of precipitation has been analyzed with lime and NaOH (Carvalho et al., 2013). For both types of treatments, the residual chemical demand of oxygen lies between 7 and 9.5 g/L (Carvalho et al., 2013). Results indicated that the chemical oxygen demand is significantly reduced. However, it is required to analyze if the residual stream can or cannot be disposed of in the drainage system to be treated in municipal wastewater plants.

2.3 Constructed wetlands

Wetlands are an alternative treatment system to conventional treatment systems, such as the activated sludge process; this technology exhibits high treatment efficiency for organics and suspended solids (Vymazal, 2019). The constructed wetlands are engineered systems designed and constructed to utilize the natural processes involving wetland vegetation, soils, and their associated microbial assemblages to assist in treating wastewater (Vymazal, 2019). In other words, we are taking advantage of the ecosystem services for the treatment of cheese whey.

According to Carvalho et al. (2013), the use of constructed wetlands allows to significantly reduced the chemical demand for oxygen until 0.16 g/L. After the treatment, the water can be used in agriculture, but it is not feasible to be directly discharged. This is an important result, which remarks the importance of the proper management of cheese whey.

3. Valorization technologies for cheese whey

On the other hand, several processes have been reported for the valorization of cheese whey. In this type of technology, the main objective is not to remove the hazardous components present in cheese whey, but its conversion to obtain marketable products. These products can include biofuels, bioenergy, chemicals, and value-added products. These alternatives are mainly biochemical and biological processes, which will be described next, Fig. 24.3.

3.1 Anaerobic digestion

Anaerobic digestion is a biological process where microorganisms transform the molecules present in cheese whey in the absence of oxygen (Prakash Dalbanjan et al., 2025). As a result, the cheese whey is treated, and valuable products are generated, such as bioethanol and biogas (Tugume et al., 2025). Nevertheless, methane yields are lows and the process is not stable due to the acidification caused by the accumulation of volatile fatty acids (Asunis et al., 2020). Indeed, the high organic load in raw cheese whey makes the aerobic digestion inappropriate (Prazeres et al., 2012) as a valorization strategy.

Due to this, the efforts have been led in two main directions. The first one is the co-digestion of cheese whey with other materials, such as biochar from bovine manure (Madrigal et al., 2022), grease interceptor waste with pulped food waste (Bolen et al., 2022), septage (Bella and Venkateswara Rao, 2022), glycerol (de Souza Almeida et al., 2023), hydrochar (Liakos et al., 2023), and dairy manure (Casallas-Ojeda et al., 2024a), among others. In these studies, the maximum production rates are reported for dairy manure (562 $mLCH_4/gVS$), biochar (358.4 $mLCH_4/gVS$), and septage (342.22 $mLCH_4/gVS$).

The second line to improve the anaerobic digestion of cheese whey relies on the use of pretreatments, which help to improve the initial conditions of cheese whey. In this line, it has been reported the use of scrap or powder zero-valent iron (ZVI) (Charalamboous and Vyrides, 2021), chain elongation technologies (Venkateswa Reddy et al., 2022), ultrasonic, ozone, and enzymes (Bella et al., 2023). The application of these strategies allows to increase between 10% and 50% the methane production.

Other studies reported the use of bacterial and archeal populations (Pagliano et al., 2018), partitioned up-flow anaerobic digestion tanks (Fagbohungbe et al., 2019), and two-stage thermophilic anaerobic digestion (Poloni Batista et al., 2023). Recently, it was reported that the degradation rate of cheese whey under mesophilic conditions was faster than under psychrophilic conditions (Casallas-Ojeda et al., 2024b). In this context, the analysis of more substrates for co-digestion, as well as the use of pretreatments and new conversion technologies, could help to increase the use of this valuable resource.

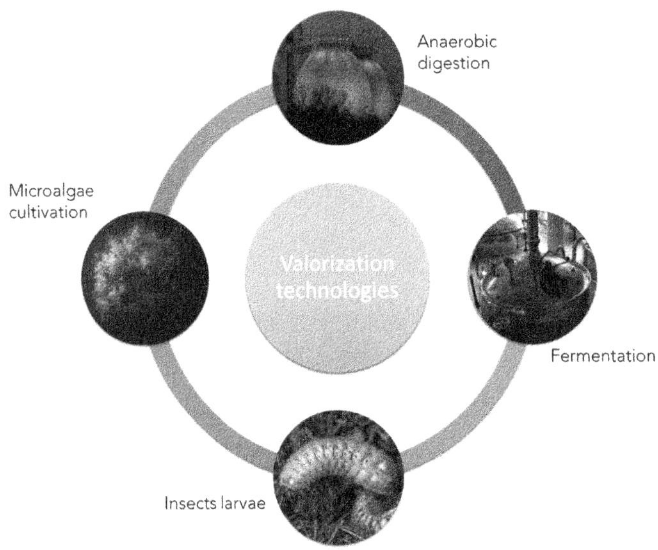

FIGURE 24.3 Valorization technologies for cheese whey. *Own elaboration with illustrations taken from Pixabay (2025).*

3.2 Fermentation

Fermentation is a process where microorganisms (fungi, yeasts, and/or bacteria) are cultivated on solid and moist substrates to act as nutrient sources for the conversion of organic waste, used as substrate, into high-value-added products (Handa and Rajamani, 2023). Here it is important to highlight that the direct fermentation of cheese whey is not economically feasible because the low lactose content results in low ethanol titer (2%–3% v/v), making the distillation process too expensive (Guimarães et al., 2010). Thus, fermentation must be performed with a solution with high lactose concentration; this can be achieved through an ultrafiltration or reverse osmosis process, or through the blending with other effluents with high content of sugars (Guimarães et al., 2010). Due to the aforementioned, research efforts have been focused on the production of hydrogen, bioethanol, and value-added products.

Regarding bioethanol production, the research on this topic dates back to 1987. In that year, it was reported that the fermentation of cheese whey employing *Kluyveromyces marxianus* CBS 397 for ethanol production (Wang et al., 1987). From that date, many studies have been reported using cultures of *Kluyveromyces marxianus* with different substrates such as cheese whey podwer (Kargi and Ozmihci, 2006; Ozmihci and Kargi, 2007a, 2007b; Dragone et al., 2011; Das et al., 2016, 2017), cheese whey (Hadiyanto et al., 2014), cheese whey permeate (Diniz et al., 2014), cheese whey lactose concentrate (Lukjanenko et al., 2014), lactose with different salts (Lukjanenko et al., 2014), ricotta cheese whey (Sansonetti et al., 2009), sugar beet thick juice diluted with crude cheese whey (Oda et al., 2010). In these studies, bioethanol is obtained between 77% and 97% of the theoretical ethanol value, with maximum concentrations reported from 8 to 43.7 g/L. Another approach reported is the fermentation of cheese whey using *Candida pseudotropicalis* (Ghaly and El-Taweel, 1995).

Regarding the hydrogen production, several works have been reported, with the use of dark fermentation being the main technology. Dark fermentation is a fermentative process in which organic and inorganic substances are converted to biohydrogen by anaerobic bacteria in the absence of light (Saravanan et al., 2021). These studies were focused on the use of thermophilic dark fermentation (Kargi et al., 2012a), comparison of thermophilic and mesophilic dark fermentations (Kargi et al., 2012b), the influence of pH on dark fermentation (De Gioannis et al., 2014), integration of dark fermentation with biocatalyzed electrolysis (Moreno et al., 2015), the use of parametric response surface to optimize hydrogen yield (Akhlaghi et al., 2017), use of *Enterobacter asburiae* and variation of operating conditions (Alvarez-Guzmán et al., 2020), start-up strategies for thermophilic dark fermentation (Lovato et al., 2021), the use of *Enterobacter aerogenes* for dark fermentation of cheese whey (Rao and Basak, 2021), improvement of operating conditions for hydrogen production (Polettini et al., 2022), continuous production from cheese why in a single stage (Aranda-Jaramillo et al., 2025), and the use of carbon felt to increase hydrogen yield (Palanivel and Samsudeen, 2025). In all cases, wide microbial species are reported in those studies. It is important to mention that the hydrogen yields were gradually increased, reaching a maximum value of 371 $LH_2/kg\ TOC_{whey}$. Also, in a study where vinasses are cofermented, the hydrogen yield is increased until 82%. Moreover, conventional fermentation has also been reported for hydrogen production. In this case, it has been reported the use of psycrophilic bacteria (*Rahnella aquatilis*) (Dębowski et al., 2014), cofermentation of cheese whey and glucose for hydrogen and ethanol production in a continuous reactor (Ferreria Rosa et al., 2014), cofermentation of vinasse and cheese whey under thermophilic conditions (Rodrigues-Ramos and Luiz Silva, 2018), the fermentation of cheese whey and glycerol in an anaerobic fluidized bed reactor (de Souza Almeida et al., 2022), as well as the fermented acidic cheese whey using lactate (Ordoñez-Frías et al., 2024).

In addition, the fermentation of cheese whey has also been directed to its conversion to value-added products. In this line, it can be listed the production of lactobionic and gluconic acids (Alonso et al., 2015), antilisterial bacteriocin-like substances (Chaves de Lima et al., 2017), polyhydroxyalkanoates (Gouveia et al., 2017), bio-plastic precursors (Girotto et al., 2017), whey spirits and white whiskey (Risner et al., 2018), succinid acid (Uysal and Hamamci, 2021), antifungal agent (Dopazo et al., 2023), volatile fatty acids (Petitta et al., 2025; Rovira-Cal et al., 2025), single cell protein (Amini et al., 2025; Moscariello et al., 2025), innovative fermented products (Mangiapelo et al., 2025). This line of research is interesting since it allows new value-added products that can increase the profitability of the dairy industry.

3.3 Microalgae cultivation

Microalgae are photosynthetic organisms characterized as unicellular or simple multicellular, whose size ranges between 2 and 200 mm (Barroca et al., 2023). Recently, a diverse range of microalgae applications have been achieved, which include the production of biofuel, the production of biomass for human and animal nutrition, soil conditioning, bioremediation, the formulation of cosmetic products and pharmaceuticals, as well as high-technology applications (Silva et al., 2019). Microalgae are characterized for its high growth rate, in spite of their small size. Moreover, they can be cultivated in waters of different qualities, including those residuals. In this context, the use of cheese whey has been reported for microalgae cultivation. Moreover, the high compatibility of microalgae cultivation with conventional wastewater treatment processes, together with the development of supporting technologies, facilitates the establishment of industrial-scale microalgae-based whey wastewater treatment systems (Chen et al., 2025).

Based on the literature review, the cultivation of *Chlorella* in solutions of cheese whey is the most studied. The microalgae species reported included *Scenedesmus obliquus* (Girard et al., 2014), *Scenedesmus* sp. (Borges et al., 2016), *Chlorella sorokiniana* (Riaño et al., 2016), *Scenedesmus obliquus* and *Chlorella protothecoides* (Girard et al., 2017), *Tetradesmus obliquus* (Bentahar et al., 2019; Suwal et al., 2019), *Desmodesmus* sp. (Abril Bonett et al., 2020), *Arthrospira* (Spirulina) *platensis* (Vieira Salla et al., 2016; Athanasiadou et al., 2023; Cavallini et al., 2024), *Spirulina platensis*, *Chlorella homosphaera* and *Scenedesmus obliquus* (Arenhart Braun et al., 2024), groups of *Scenedesmus*, *Monoraphidium*, diatoms and filamentous green algae (Nham et al., 2024), *Chlorella* sp. (Stratigakis et al., 2024). In all cases, cheese whey (raw, permeate, or powder) is used in diluted solutions. The main applications comprise the extraction of value-added products as well as the production of biofuels, mainly biodiesel. Nevertheless, other biofuels can be produced and value-added products can be extracted from microalgae.

3.4 Black soldier fly cultivation

Black soldier fly is an insect that has attracted the attention of the scientific community. BSFL, *Hermetia illucens*, have proven to convert organic waste into high-quality nutrients for pet foods, fish and poultry feeds, as well as residue fertilizer for soil amendment (Siddiqui et al., 2022). For the valorization of cheese whey, just two works have been identified. Caltzontzin-Rabell et al. (2024a) reported the feasibility of using cheese whey in the cultivation of BSFL; indeed, the incorporation of cheese whey allows for an increase in the weight, where the main fractions were fats (42.48%) and protein (34.98%). Later, the cheese whey management was evaluated in the cultivation of BSFL (Caltzontzin-Rabell et al., 2024b). The results indicated that the management of cheese whey affects the weight and fat accumulation in BSFL. It is important to mention that this valorization strategy is feasible, allowing the conversion of cheese whey into biomass for larvae. This biomass can be further processed to generate biofuels as well as value-added products; indeed, the BSFL can be directly used as feed for chickens. Moreover, these insects tolerate the waste without dilution, which makes a promising strategy forward its complete valorization.

4. Future trends

Worldwide, cheese is one of the main food products. There are many varieties of cheese produced from milk of different species. Due to its properties and benefits to human health, the consumption of cheese represents an important economic contribution to the dairy milk sector. In the production of cheese, there are many residues generated, with the cheese whey being one of the most abundant.

Cheese whey is composed of lactose, soluble proteins, lipids, mineral salts, B group vitamins, as well as lactic and citric acids (González Siso, 1996). Moreover, cheese whey has values of biochemical oxygen demand (BOD) ranging from 30,000 to 60,000 mg/L and chemical oxygen demand (COD) from 50,000 to 100,000 mg/L (de Paiva Anciens Ramos et al., 2021). The high content of organic matter present in cheese whey, along with its elevated volumes, complicates the direct discharge of this residue, making its treatment mandatory. Cheese has been produced for many years, and nowadays its environmental impact has captured the attention of society as well as the scientific community. This is due to the fact that from 1996 to the present day, only 50% of the total world cheese whey produced is discharged to the environment without any treatment (González Siso, 1996; Asunis et al., 2020).

Thus, the first technologies were focused on the treatment of cheese whey. In treatment technologies, the main objective is the reduction of the organic matter content, so that the resulting effluent can be discharged with reduced environmental impact. In this type of technology, it can be listed the dilution, constructed wetlands, and coagulation with flocculation. Dilution is the less effective technology; the organic content of the resulting effluent is decreased, but the compounds present affect the performance of wastewater treatment systems. On the other hand, constructed wetlands and coagulation-flocculation technologies are successful in the reduction of the chemical oxygen demand of cheese whey. The main drawbacks of these technologies are their high cost, which makes it difficult its adoption specially in small farms where cheese is produced. This fact is of great relevance considering that artisan cheese is experiencing a resurgence, as part of society's search for healthy, unique, and local products (The Guardian, 2023). In this context, the paradigm needs to change. Cheese whey must be considered as a valuable resource rich in lactose, soluble proteins, lipids, mineral salts, B group vitamins, as well as lactic and citric acids (González Siso, 1996). Therefore, cheese whey is an important raw material from which several products can be obtained. Considering the valorization of cheese whey, several studies have reported its anaerobic digestion, fermentation, as well as its use as substrate for microalgae and BSFL.

Regarding anaerobic digestion technology, the direct conversion of cheese whey is not appropriate. Due to this, the efforts have been focused on the co-processing of cheese whey with other materials, such as biochar, hydrochar, dairy manure, septage, and grease; in this case, costs associated with the collection and/or production of a second raw material for the process must be considered. Other studies are focused on the use of pretreatments, which certainly allows for increasing the methane production between 10% and 50%; here, the cost-benefit must be analyzed since any additional unit operation increases both capital and operating costs. The use of anaerobic digestion for cheese whey processing allows to generate biogas and digestate. Nevertheless, the composition of cheese whey makes it difficult to process, leading to additional operating costs due to its mix with other materials or the addition of pretreatment operations. Both strategies increase capital costs, operating costs, and also the environmental impact; these aspects complicate the potential adoption is small factories.

Fermentation is another technology reported for the valorization of cheese whey. In this technology, the studies are driven for the production of bioethanol, hydrogen as well as value-added products. For the production of bioethanol, *Kluyveromyces marxianus* is the most used yeast in conventional fermentation, reaching concentrations between 8 and 43.7 g/L. On the other hand, dark fermentation is the most reported technology for the production of biohydrogen from cheese whey. In this line, the maximum yield reported is 371 LH_2/kg TOC_{whey}. Moreover, the production of hydrogen has also been addressed considering conventional co-fermentation with vinasse, glucose and glycerol. Finally, the production of value-added products from cheese whey fermentation has led the research. Several works, recently published, reported the production of proteins, alpha-hydroxy acids, antifungal agents, as well as fermented products. The use of fermentation to process cheese whey allows to generate bioethanol, biohydrogen, or value-added products. Similar to the case of anaerobic digestion, pretreatments are required for cheese whey or the use of secondary raw materials in order to increase the yields of each specific product. Also, in all cases just one main product is obtained: a biofuel or a value-added product. An important aspect, usually not reported, is the composition and quantity of the waste derived in the fermentation process. This aspect is relevant in the pursuit of decreasing the environmental impact associated with cheese production.

Regarding the use of cheese whey as substrate for microalgae cultivation, the studies are focused on the feasibility of adaption of these species. Due to this, diluted solutions containing cheese whey, as raw, permeate or powder, are used. The microalgae species studied include *Chlorella*, *Scenedesmus*, *Tetradesmus* and *Spirulina* platensis. The main applications comprise the extraction of value-added products or the production of biofuels, mainly biodiesel. This biological treatment has two main aspects that need to be considered. The first one is the fact that diluted solutions are required in order to be technically feasible the microalgae cultivation; this aspect increase the water footprint associated to its valorization. The second one relies on the processing cost of microalgae once is ready to be harvested; in other words, the process associated to the extraction and separation of the main components present in the microalgae (Vasistha et al., 2021). Due to this, Kumar et al. (2025) indicated the necessity of developing novel cell disruption technologies, economically viable downstream processing, as well as the integration with waste management and the circular economy.

Recently, the use of BSFL cultivation has captured the attention of the scientific community due to its affinity for organic waste processing (Siddiqui et al., 2022). The BSFL decomposes many types of food waste and converts it into valuable biomass, which is composed of proteins, lipids, chitin, and vitamins (Ramu Ganesan et al., 2024). This insect has been reported as a valorization strategy for poultry manure, swine manure, cow manure, food manufacturing byproducts, biogas digestate, vegetable waste, fresh fruit waste, human and animal waste (Singh and Kumari, 2019). Indeed, recently, it has been reported that the technical feasibility of using cheese whey as a substrate for BSFL cultivation. Moreover, the study also reveals that the management of the cheese whey before it is fed to BSFL affects its weight and fat accumulation. In these studies, cheese whey is employed as a substitute for water in the Gainesville Diet. Thus, it is an important advance to decrease the water footprint, but it also requires the substitution of the edible grains that are part of the Gainesville diet.

Considering the advances in treatment and valorization technologies, it is clear that new approaches must be developed. The proposed technologies allow the obtaining of just one product, which can be a biofuel or a value-added product. Except for the BSFL cultivation, in all cases, cheese whey needs to be diluted or pretreated in order to be effectively converted. This issue can potentially limit the adoption in small farms, due to the high costs of technologies and small processed quantities. The analyses show that valorization technologies for cheese whey processing must be feasible from technical, environmental, and economic points of view. In this context, biorefineries for cheese whey valorization rise as a promising technology.

A biorefinery is defined as a set of conversion processes where at least one biofuel, one type of bioenergy, and one value-added product are obtained. This processing scheme allows to generate products with different commercial values.

Biofuels are usually generated in elevated amounts, but their price is low; on the other hand, value-added products are generated in relatively small amounts, but they have elevated prices in the market. In addition, the bioenergy produced in the process (heat and/or electricity) can be used to satisfy, at least partially, the energy requirements. Indeed, according to Kumar and Verma (2021), the development of biorefineries will have a positive impact on the strengthening economy, environment, as well as providing job opportunities at the local, national, and international levels. Moreover, the use of waste as raw materials for a biorefinery processing scheme has gained attention due to its alignment with the circular economy principles. In this context, Daimary et al. (2025) indicate that the valorization and conversion of biowaste through a biorefinery approach is an eco-friendly method of transforming biowaste into usable products along with commercially viable value-added products that address energy, chemicals, and environmental concerns. To our knowledge, just one biorefinery has been reported for the valorization of cheese whey. Caltzontzin-Rabell and Gutiérrez-Antonio (2024) proposed a biorefinery for the conversion of cheese whey into biodiesel and an antimicrobial extract. For this, cultivation of BSFL has been used for the processing of non-diluted cheese whey. The cultivation of other insects must also be explored in order to have a wide portfolio of valorization technologies.

In addition to the use of biorefineries, the application of process intensification strategies, as well as energy integration, is desirable. Process intensification allows for compact production processes with minor space and energy requirements, and improved yields. Process intensification equipment can be derived from enhanced unit operations, which significantly decrease the size of the equipment, or from the combination of at least two unit operations, usually called hybrid technologies. On the other hand, energy integration enables the use of the circulating energy in the process to decrease the use of fuels and cooling agents; moreover, energy integration contributes to a significant decrease in carbon dioxide emissions derived from the utilities avoided. Some studies of biorefineries for other raw materials reported decreasing in operating costs and carbon dioxide emissions (Carrasco-Suárez et al., 2022; Romero-Izquierdo et al., 2022, 2023). Moreover, the incorporation of renewable energies needs to be explored. For instance, solar radiation can be used to generate thermal and electrical energy, while a micro turbine can take advantage of the kinetic energy contained in emissions to the atmosphere. In this line, the waste derived from the valorization technologies can also be processed to generate, at least, energy that can be used to satisfy the energy requirements of the process.

Finally, the application of optimization strategies is mandatory both for the production process and the supply chain. In this context, statistical, deterministic, and stochastic optimization strategies are powerful tools, depending on the specific problem to be solved. If experimental determinations are carried out in order to determine the optimal operating conditions of a technology, then statistical optimization is the most appropriate strategy. Moreover, if the modeling of the process is well known and it is possible to develop an objective function that is continuous and whose derivative exists, deterministic optimization is the tool to be used. Finally, if the modeling of the process is well known but its objective function is not defined, or even if it has no closed form, then stochastic or evolutionary optimization is the best alternative. In some cases, hybrid strategies can be applied, such as the use of a stochastic procedure to reach the neighborhood of the optimal value, and then switch to a deterministic optimization to refine the optimal value.

In brief, the valorization technologies to process cheese whey must be sustainable. This will help to take advantage of the interesting composition of this waste to develop new products that can be incorporated into the market. This will lead to the development of a sustainable production process for cheese production along with the use of its most abundant waste. In this way, the dairy sector will move forward to its sustainable growth.

5. Conclusions

Cheese whey is the most abundant byproduct generated in the production of cheese. Considering the resurgence that artisan cheese is experiencing, as part of the search of society for healthy, unique, and local products, the adequate management of cheese whey is mandatory. Treatments technologies as constructed wetlands and coagulation-flocculation, are successful in decreasing the organic matter content. Nevertheless, their operating and investment costs are too elevated for small farms. Regarding the valorization technologies, most of them allow the use of cheese whey diluted or pretreated. In these technologies, biofuels or value-added products can be obtained. This combination of issues makes it difficult to implement these technologies, especially in small farms. Due to this, the use of biorefineries, along with energy integration and process intensification strategies, is a promising approach to reach technical, economic, and environmental feasibility in the valorization of cheese whey. Indeed, biorefineries are considered the most viable strategy to take advantage of the nutrient-rich composition of cheese whey. This will allow the implementation of circular economy principles in the cheese production, through the implementation of optimization, process intensification, energy integration, and renewable energies. In this way, we can continue enjoying the benefits of cheese all over the world, while moving forward with sustainable development.

Acknowledgments

The authors appreciate the financial support provided by Universidad Autónoma de Querétaro for the development of this work. Also, A. Ramírez-Hernández benefited from tuition fees for her postgraduate studies from the Facultad de Ingeniería at Universidad Autónoma de Querétaro.

References

Akhlaghi, M., Boni, M.R., De Gioannis, G., Muntoni, A., Polettini, A., Pomi, R., Rossi, A., Spiga, D., 2017. A parametric response surface study of fermentative hydrogen production from cheese whey. Bioresour. Technol. 244 (Part 1), 473–483. https://doi.org/10.1016/j.biortech.2017.07.158. ISSN 0960-8524.

Almeida, P.de S., Menezes, C.A.de, Pereira Camargo, F., Sakamoto, I.K., Varesche, M.B.A., Silva, E.L., 2022. Producing hydrogen from the fermentation of cheese whey and glycerol as cosubstrates in an anaerobic fluidized bed reactor. Int. J. Hydrogen Energy 47 (31), 14243–14256. https://doi.org/10.1016/j.ijhydene.2022.02.176. ISSN 0360-3199.

Almeida, P.de S., de Menezes, C.A., Pereira Camargo, F., Sakamoto, I.K., Lovato, G., Rodrigues, J.A.D., Varesche, M.B.A., Silva, E.L., 2023. Biomethane recovery through co-digestion of cheese whey and glycerol in a two-stage anaerobic fluidized bed reactor: effect of temperature and organic loading rate on methanogenesis. J. Environ. Manag. 330, 117117. https://doi.org/10.1016/j.jenvman.2022.117117. ISSN 0301-4797.

Alonso, S., Rendueles, M., Díaz, M., 2015. Simultaneous production of lactobionic and gluconic acid in cheese whey/glucose co-fermentation by Pseudomonas taetrolens. Bioresour. Technol. 196, 314–323. https://doi.org/10.1016/j.biortech.2015.07.092. ISSN 0960-8524.

Alvarez-Guzmán, C.L., Cisneros-de la Cueva, S., Balderas-Hernández, V.E., Smoliński, A., De León-Rodríguez, A., 2020. Biohydrogen production from cheese whey powder by Enterobacter asburiae: effect of operating conditions on hydrogen yield and chemometric study of the fermentative metabolites. Energy Rep. 6, 1170–1180. https://doi.org/10.1016/j.egyr.2020.04.038. ISSN 2352-4847.

Amini, A., Remelli, A., Mojarad, V.A., El Achkar, M., Dell'Anno, M., Rossi, L., Mirra, G., Di Giancamillo, A., Scalvini, F.G., Tedeschi, G., Parati, K., Turolla, A., 2025. From cheese whey to single-cell protein production: by-product valorization through acidogenic fermentation and purple phototrophic bacteria. Bioresour. Technol. 435, 132894. https://doi.org/10.1016/j.biortech.2025.132894. ISSN 0960-8524.

Aranda-Jaramillo, B., García-Depraect, O., Aguilar-Juárez, O., Archundia-Amador, M., León-Becerril, E., 2025. Continuous hydrogen production from cheese whey in a single-stage lactate-driven dark fermentation reactor. J. Taiwan Inst. Chem. Eng. 177, 106307. https://doi.org/10.1016/j.jtice.2025.106307.

Asunis, F., De Gioannis, G., Dessì, P., Isipato, M., Lens, P.N.L., Muntoni, A., Polettini, A., Pomi, R., Rossi, A., Spiga, D., 2020. The dairy biorefinery: integrating treatment processes for cheese whey valorisation. J. Environ. Manag. 276, 111240. https://doi.org/10.1016/j.jenvman.2020.111240. ISSN 0301-4797.

Athanasiadou, V., Klontza, E.E., Dimitriou-Christidis, P., Fountoulakis, M., Lekkas, D.F., 2023. Evaluation of Arthrospira (Spirulina) platensis growth on cheese whey in the context of circular economy. Sustain. Chem. Pharm. 34, 101173. https://doi.org/10.1016/j.scp.2023.101173. ISSN 2352-5541.

Barroca, M.J., Florença, S.G., Martins, C.B., Guiné, R.P.F., 2023. Chapter 48 - strategy and design of innovation policy road mapping for microalgae-based foods. In: Jacob-Lopes, E., Isabel Queiroz, M., Manzoni Maroneze, M., Queiroz Zepka, L. (Eds.), Handbook of Food and Feed from Microalgae. Academic Press, pp. 625–642. https://doi.org/10.1016/B978-0-323-99196-4.00045-0.

Batista, L.P.P., Paula Paulinetti, A., Júnior, A.D.N.F., Albanez, R., Ratusznei, S.M., Etchebehere, C., Lovato, G., Rodrigues, J.A.D., 2023. Two-stage thermophilic anaerobic digestion of cheese whey: process optimization, comparison with single-stage, and full-scale estimation. Chem. Eng. Process. Process Intensif. 183, 109260. https://doi.org/10.1016/j.cep.2022.109260. ISSN 0255-2701.

Bella, K., Venkateswara Rao, P., 2022. Anaerobic co-digestion of cheese whey and septage: effect of substrate and inoculum on biogas production. J. Environ. Manag. 308, 114581. https://doi.org/10.1016/j.jenvman.2022.114581. ISSN 0301-4797.

Bella, K., Pilli, S., Venkateswara Rao, P., 2023. A comparison of ultrasonic, ozone, and enzyme pre-treatments on cheese whey degradation for enhancement of anaerobic digestion. J. Environ. Manag. 340, 117960. https://doi.org/10.1016/j.jenvman.2023.117960. ISSN 0301-4797.

Bentahar, J., Doyen, A., Beaulieu, L., Deschênes, J.-S., 2019. Acid whey permeate: an alternative growth medium for microalgae Tetradesmus obliquus and production of β-galactosidase. Algal Res. 41, 101559. https://doi.org/10.1016/j.algal.2019.101559. ISSN 2211-9264.

Bolen, T.J., Hasan, M., Conway, T., Stéphane Yaméogo, D.D., Sanchez, P., Rahman, A., Azam, H., 2022. Feasibility assessment of biogas production from the anaerobic co-digestion of cheese whey, grease interceptor waste and pulped food waste for WRRF. Energy 254 (Part A), 124144. https://doi.org/10.1016/j.energy.2022.124144. ISSN 0360-5442.

Bonett, J.E.A., Geraldino, P.de S., Gomes Cardoso, P., Coelho, F.de F., Duarte, W.F., 2020. Isolation of freshwater microalgae and outdoor cultivation using cheese whey as substrate. Biocatal. Agric. Biotechnol. 29, 101799. https://doi.org/10.1016/j.bcab.2020.101799. ISSN 1878-8181.

Borges, W.da S., Araújo, B.S.A., Moura, L.G., Filho, U.C., de Resende, M.M., Cardoso, V.L., 2016. Bio-oil production and removal of organic load by microalga Scenedesmus sp. using culture medium contaminated with different sugars, cheese whey and whey permeate. J. Environ. Manag. 173, 134–140. https://doi.org/10.1016/j.jenvman.2015.11.015. ISSN 0301-4797.

Braun, J.C.A., Balbinot, L., Beuter, M.A., Rempel, A., Colla, L.M., 2024. Mixotrophic cultivation of microalgae using agro-industrial waste: tolerance level, scale up, perspectives and future use of biomass. Algal Res. 80, 103554. https://doi.org/10.1016/j.algal.2024.103554. ISSN 2211-9264.

Caltzontzin-Rabell, V., Gutiérrez-Antonio, C., 2024. Revalorization of cheese whey through a biorefinery scheme. Chem. Eng. Trans. 109, 241–246. https://doi.org/10.3303/CET24109041. ISSN 2283-9216.

Caltzontzin-Rabell, V., Escobar-Ortiz, A., Gutiérrez-Antonio, C., Feregrino-Pérez, A.A., García-Trejo, J.F., 2024a. Revaluation process of cheese whey through the cultivation of black soldier fly larvae (Hermetia illucens). Eng. Rep. 6 (7), e12853. https://doi.org/10.1002/eng2.12853. ISSN 2577-8196.

Caltzontzin-Rabell, V., Feregrino-Pérez, A.A., Gutiérrez-Antonio, C., 2024b. Bio-upcycling of cheese whey: transforming waste into raw materials for biofuels and animal feed. Heliyon 10 (12), e32700. https://doi.org/10.1016/j.heliyon.2024.e32700. ISSN 2405-8440.

Carrasco-Suárez, M.T., Romero-Izquierdo, A.G., Gutiérrez-Antonio, C., Gómez-Castro, F.I., Hernández, S., 2022. Production of renewable aviation fuel by waste cooking oil processing in a biorefinery scheme: intensification of the purification zone. Chem. Eng. Process. Process Intensif. 181, 109103. https://doi.org/10.1016/j.cep.2022.109103. ISSN 0255-2701.

Carvalho, F., Prazeres, A.R., Rivas, J., 2013. Cheese whey wastewater: characterization and treatment. Sci. Total Environ. 445–446, 385–396. https://doi.org/10.1016/j.scitotenv.2012.12.038. ISSN 0048-9697.

Casallas-Ojeda, M., Cabeza, I., Sanchez, N., Caicedo-Concha, D.M., Astals, S., 2024a. Cheese whey and dairy manure anaerobic co-digestion at psychrophilic conditions: technical and environmental evaluation. Environ. Res. 251 (Part 1), 118525. https://doi.org/10.1016/j.envres.2024.118525. ISSN 0013-9351.

Casallas-Ojeda, M., Perez-Esteban, N., Cabeza, I., Cobo, M., Olaya-Rincon, M., Caicedo-Concha, D.M., Astals, S., 2024b. Understanding the acidification risk of cheese whey anaerobic digestion under psychrophilic and mesophilic conditions. Heliyon 10 (5), e26476. https://doi.org/10.1016/j.heliyon.2024.e26476. ISSN 2405-8440.

Cavallini, A., Torre, S., Usai, L., Casula, M., Fais, G., Nieri, P., Concas, A., Lutzu, G.A., 2024. Effect of cheese whey on phycobiliproteins production and FAME profile by Arthrospira platensis (Spirulina): promoting the concept of a circular bio-economy. Sustain. Chem. Pharm. 40, 101625. https://doi.org/10.1016/j.scp.2024.101625. ISSN 2352-5541.

Charalambous, P., Vyrides, I., 2021. In situ biogas upgrading and enhancement of anaerobic digestion of cheese whey by addition of scrap or powder zero-valent iron (ZVI). J. Environ. Manag. 280, 111651. https://doi.org/10.1016/j.jenvman.2020.111651. ISSN 0301-4797.

Chaves de Lima, E.de L., Fernandes, J.de M., Cardarelli, H.R., 2017. Optimized fermentation of goat cheese whey with Lactococcus lactis for production of antilisterial bacteriocin-like substances. LWT 84, 710–716. https://doi.org/10.1016/j.lwt.2017.06.040. ISSN 0023-6438.

Chen, S., Radican, E., Zhu, H., Xiao, Z., Qiao, M., Luo, Y., 2025. Toward a circular economy: integrative approaches to microalgal bioremediation of whey wastewater and challenges to industrial scale-up. Bioresour. Technol. 436, 133032. https://doi.org/10.1016/j.biortech.2025.133032. ISSN 0960-8524.

Christoforos Stratigakis, N., Nazos, T.T., Chatzopoulou, M., Mparka, N., Spantidaki, M., Lagouvardou-Spantidaki, A., Ghanotakis, D.F., 2024. Cultivation of a naturally resilient Chlorella sp.: a bioenergetic strategy for valorization of cheese whey for high nutritional biomass production. Algal Res. 82, 103616. https://doi.org/10.1016/j.algal.2024.103616. ISSN 2211-9264.

Daimary, N., Deb, B., Roy, B., Kumar Ranjan, R., Mukherjee, A., 2025. Sustainable biorefinery approach for the transformation of biowaste into biofuels and chemicals for the circular economy: a review. Sustain. Energy Technol. Assessments 82, 104457. https://doi.org/10.1016/j.seta.2025.104457. ISSN 2213-1388.

Das, B., Sarkar, S., Maiti, S., Bhattacharjee, S., 2016. Studies on production of ethanol from cheese whey using Kluyveromyces marxianus. Mater. Today Proc. 3 (10), 3253–3257. https://doi.org/10.1016/j.matpr.2016.10.006. ISSN 2214-7853.

Das, B., Das, M., Bhattacharjee, S., Bhattacharjee, C., 2017. Ethanol production from deproteinized cheese whey powder in a batch fermentation process: optimization of process and kinetic modelling. Desalination Water Treat. 64, 198–206. https://doi.org/10.5004/dwt.2017.20238. ISSN 1944-3986.

De Gioannis, G., Friargiu, M., Massi, E., Muntoni, A., Polettini, A., Pomi, R., Spiga, D., 2014. Biohydrogen production from dark fermentation of cheese whey: influence of pH. Int. J. Hydrogen Energy 39 (36), 20930–20941. https://doi.org/10.1016/j.ijhydene.2014.10.046. ISSN 0360-3199.

Dębowski, M., Korzeniewska, E., Filipkowska, Z., Zieliński, M., Kwiatkowski, R., 2014. Possibility of hydrogen production during cheese whey fermentation process by different strains of psychrophilic bacteria. Int. J. Hydrogen Energy 39 (5), 1972–1978. https://doi.org/10.1016/j.ijhydene.2013.11.082. ISSN 0360-3199.

Diniz, R.H.S., Rodrigues, M.Q.R.B., Fietto, L.G., Passos, F.M.L., Silveira, W.B., 2014. Optimizing and validating the production of ethanol from cheese whey permeate by Kluyveromyces marxianus UFV-3. Biocatal. Agric. Biotechnol. 3 (2), 111–117. https://doi.org/10.1016/j.bcab.2013.09.002. ISSN 1878-8181.

Dopazo, V., Illueca, F., Luz, C., Musto, L., Moreno, A., Calpe, J., Meca, G., 2023. Revalorization by lactic acid bacterial fermentation of goat whey from cheese industry as a potential antifungal agent. Food Biosci. 53, 102586. https://doi.org/10.1016/j.fbio.2023.102586. ISSN 2212-4292.

Dragone, G., Mussatto, S.I., Almeida e Silva, J.B., Teixeira, J.A., 2011. Optimal fermentation conditions for maximizing the ethanol production by Kluyveromyces fragilis from cheese whey powder. Biomass Bioenergy 35 (5), 1977–1982. https://doi.org/10.1016/j.biombioe.2011.01.045. ISSN 0961-9534.

Fagbohungbe, M.O., Onyeri, C.A., Semple, K.T., 2019. Co-fermentation of whey permeates and cattle slurry using a partitioned up-flow anaerobic digestion tank. Energy 185, 567–572. https://doi.org/10.1016/j.energy.2019.07.051. ISSN 0360-5442.

Fox, P.F., McSweeney, P.L.H., 2025. Chapter 1 - cheese: an overview. In: McSweeney, P.L.H., Cotter, P.D., Everett, D.W., Govindasamy-Lucey, R. (Eds.), Cheese, fifth ed. Academic Press, pp. 5–20. https://doi.org/10.1016/B978-0-443-15956-5.00035-X.

Ganesan, A.R., Mohan, K., Kandasamy, S., Surendran, R.P., Kumar, R., Rajan, D.K., Rajarajeswaran, J., 2024. Food waste-derived black soldier fly (Hermetia illucens) larval resource recovery: a circular bioeconomy approach. Process Saf. Environ. Prot. 184, 170–189. https://doi.org/10.1016/j.psep.2024.01.084. ISSN 0957-5820.

Ghaly, A.E., El-Taweel, A.A., 1995. Effect of micro-aeration on the growth of Candida pseudotropicalis and production of ethanol during batch fermentation of cheese whey. Bioresour. Technol. 52 (3), 203–217. https://doi.org/10.1016/0960-8524(95)00026-B. ISSN 0960-8524.

Girard, J.-M., Roy, M.-L., Ben Hafsa, M., Gagnon, J., Faucheux, N., Heitz, M., Tremblay, R., Deschênes, J.-S., 2014. Mixotrophic cultivation of green microalgae Scenedesmus obliquus on cheese whey permeate for biodiesel production. Algal Res. 5, 241–248. https://doi.org/10.1016/j.algal.2014.03.002. ISSN 2211-9264.

Girard, J.-M., Tremblay, R., Faucheux, N., Heitz, M., Deschênes, J.-S., 2017. Phycoremediation of cheese whey permeate using directed commensalism between Scenedesmus obliquus and Chlorella protothecoides. Algal Res. 22, 122–126. https://doi.org/10.1016/j.algal.2016.12.013. ISSN 2211-9264.

Girotto, F., Cristina Lavagnolo, M., Pivato, A., Cossu, R., 2017. Acidogenic fermentation of the organic fraction of municipal solid waste and cheese whey for bio-plastic precursors recovery – effects of process conditions during batch tests. Waste Manag. 70, 71–80. https://doi.org/10.1016/j.wasman.2017.09.015. ISSN 0956-053X.

González Siso, M.I., 1996. The biotechnological utilization of cheese whey: a review. Bioresour. Technol. 57 (1), 1–11. https://doi.org/10.1016/0960-8524(96)00036-3. ISSN 0960-8524.

Gouveia, A.R., Freitas, E.B., Galinha, C.F., Carvalho, G., Duque, A.F., Reis, M.A.M., 2017. Dynamic change of pH in acidogenic fermentation of cheese whey towards polyhydroxyalkanoates production: impact on performance and microbial population. New Biotechnol. 37 (Part A), 108–116. https://doi.org/10.1016/j.nbt.2016.07.001. ISSN 1871-6784.

Guimarães, P.M.R., Teixeira, J.A., Domingues, L., 2010. Fermentation of lactose to bio-ethanol by yeasts as part of integrated solutions for the valorisation of cheese whey. Biotechnol. Adv. 28 (3), 375–384. https://doi.org/10.1016/j.biotechadv.2010.02.002. ISSN 0734-9750.

Hadiyanto, D.A., Aini, A.P., Siti Pinundi, D., 2014. Optimization of ethanol production from whey through fed-batch fermentation using Kluyveromyces marxianus. Energy Proc. 47, 108–112. https://doi.org/10.1016/j.egypro.2014.01.203. ISSN 1876-6102.

Handa, A., Rajamani, P., 2023. Chapter 27 - waste management and environment. In: Bhatt, A.K., Bhatia, R.K., Bhalla, T.C. (Eds.), Basic Biotechniques for Bioprocess and Bioentrepreneurship. Academic Press. https://doi.org/10.1016/B978-0-12-816109-8.00027-1. Pages 391-413.

Hausjell, J., Miltner, M., Herzig, C., Limbeck, A., Saracevic, Z., Saracevic, E., Weissensteiner, J., Molitor, C., Halbwirth, H., Spadiut, O., 2019. Valorisation of cheese whey as substrate and inducer for recombinant protein production in *E. coli* HMS174(DE3). Bioresour. Technol. Rep. 8, 100340. https://doi.org/10.1016/j.biteb.2019.100340. ISSN 2589-014X.

IDFA, 2025. International Dairy Foods Association, History of Cheese. Available from: https://www.idfa.org/history-of-cheese.

Kargi, F., Ozmıhcı, S., 2006. Utilization of cheese whey powder (CWP) for ethanol fermentations: effects of operating parameters. Enzym. Microb. Technol. 38 (5), 711–718. https://doi.org/10.1016/j.enzmictec.2005.11.006. ISSN 0141-0229.

Kargi, F., Seza Eren, N., Ozmihci, S., 2012a. Bio-hydrogen production from cheese whey powder (CWP) solution: comparison of thermophilic and mesophilic dark fermentations. Int. J. Hydrogen Energy 37 (10), 8338–8342. https://doi.org/10.1016/j.ijhydene.2012.02.162. ISSN 0360-3199.

Kargi, F., Seza Eren, N., Ozmihci, S., 2012b. Hydrogen gas production from cheese whey powder (CWP) solution by thermophilic dark fermentation. Int. J. Hydrogen Energy 37 (3), 2260–2266. https://doi.org/10.1016/j.ijhydene.2011.11.018. ISSN 0360-3199.

Kumar, B., Verma, P., 2021. Biomass-based biorefineries: an important architype towards a circular economy. Fuel 288, 119622. https://doi.org/10.1016/j.fuel.2020.119622. ISSN 0016-2361.

Kumar, S., Raj, R., Tripathi, A., Das, S., Bhowmick, G.D., Ghangrekar, M.M., 2025. Exploring innovative techniques in algal lipid extraction: pioneering paths to sustainable biofuel production. Biocatal. Agric. Biotechnol. 66, 103583. https://doi.org/10.1016/j.bcab.2025.103583. ISSN 1878-8181.

Kumar Sachan, R.S., Karnwal, A., 2022. Chapter 13 - advancement in cheese production technology. In: Singh, J., Vyas, A. (Eds.), Advances in Dairy Microbial Products. Woodhead Publishing, pp. 191–208. https://doi.org/10.1016/B978-0-323-85793-2.00023-0.

Liakos, D., Altiparmaki, G., Kalampokidis, A., Lekkas, D.F., Vakalis, S., 2023. The role of hydrochar on the production of biogas and volatile fatty acids during anaerobic digestion of cheese whey wastewater. Sustain. Chem. Pharm. 35, 101153. https://doi.org/10.1016/j.scp.2023.101153. ISSN 2352-5541.

Lovato, G., Augusto, I.M.G., Júnior, A.D.N.F., Albanez, R., Ratusznei, S.M., Etchebehere, C., Zaiat, M., Rodrigues, J.A.D., 2021. Reactor start-up strategy as key for high and stable hydrogen production from cheese whey thermophilic dark fermentation. Int. J. Hydrogen Energy 46 (54), 27364–27379. https://doi.org/10.1016/j.ijhydene.2021.06.010. ISSN 0360-3199.

Lukjanenko, J., Kovtuna, K., Scherbaka, R., Vigants, A., 2014. Bioethanol and biomass production by Kluyveromyces marxianus during lactose fermentation at different salts and substrate concentrations. J. Biotechnol. 185 (Supplement), S122. https://doi.org/10.1016/j.jbiotec.2014.07.417.

Madrigal, G., Huaraya, M., Sancho, T., Mendieta, O., Jaimes-Estévez, J., 2022. Biochar from bovine manure as a sustainable additive to improve the anaerobic digestion of cheese whey. Bioresour. Technol. Rep. 20, 101258. https://doi.org/10.1016/j.biteb.2022.101258. ISSN 2589-014X.

Mangiapelo, L., Costanzi, E., Blasi, F., Trovarelli, V., Terzo Cenci-Goga, B., Cossignani, L., Ianni, F., 2025. Valorization of second cheese whey enriched with apple pomace for the production of innovative fermented products. Innov. Food Sci. Emerg. Technol. 104. https://doi.org/10.1016/j.ifset.2025.104086. ISSN 1466-8564.

McSweeney, P.L.H., Fox, P.F., 2025. Chapter 30 - diversity and classification of cheese varieties: an overview. In: McSweeney, P.L.H., Cotter, P.D., Everett, D.W., Govindasamy-Lucey, R. (Eds.), Cheese, fifth ed. Academic Press, pp. 771–797. https://doi.org/10.1016/B978-0-443-15956-5.00046-4.

Minhalma, M., Magueijo, V., Queiroz, D.P., Norberta de Pinho, M., 2007. Optimization of "Serpa" cheese whey nanofiltration for effluent minimization and by-products recovery. J. Environ. Manag. 82 (2), 200–206. https://doi.org/10.1016/j.jenvman.2005.12.011. ISSN 0301-4797.

Moreno, R., Escapa, A., Cara, J., Carracedo, B., Gómez, X., 2015. A two-stage process for hydrogen production from cheese whey: integration of dark fermentation and biocatalyzed electrolysis. Int. J. Hydrogen Energy 40 (1), 168–175. https://doi.org/10.1016/j.ijhydene.2014.10.120. ISSN 0360-3199.

Moscariello, C., Matassa, S., Oliva, A., Lens, P.N.L., Esposito, G., Pirozzi, F., Papirio, S., 2025. One-stage single cell protein production from hemp (Cannabis sativa L.) biomass residues and cheese whey through sequential anaerobic-aerobic fermentation. Waste Manag. 193, 462–471. https://doi.org/10.1016/j.wasman.2024.12.029. ISSN 0956-053X.

Nham, Q., Legrand, C., Lindehoff, E., 2024. Microalgal production and nutrient recovery under mixotrophic mode using cheese whey permeate. Bioresour. Technol. 410, 131250. https://doi.org/10.1016/j.biortech.2024.131250. ISSN 0960-8524.

Oda, Y., Nakamura, K., Shinomiya, N., Ohba, K., 2010. Ethanol fermentation of sugar beet thick juice diluted with crude cheese whey by the flex yeast Kluyveromyces marxianus KD-15. Biomass Bioenergy 34 (8), 1263–1266. https://doi.org/10.1016/j.biombioe.2010.03.014. ISSN 0961-9534.

OECD, 2023. OECD-FAO Agricultural Outlook 2023-2032 Report. Available from: https://www.oecd.org/en/publications/oecd-fao-agricultural-outlook-2023-2032_08801ab7-en/full-report.html.

OECD/FAO, 2025. OCDE-FAO Agricultural Outlook 2024-2033. OECD Publishing, Paris/FAO, Rome. https://doi.org/10.1787/2b0c9d81-es.

Ordoñez-Frías, E.J., Muñoz-Páez, K.M., Buitrón, G., 2024. Biohydrogen production from fermented acidic cheese whey using lactate: reactor performance and microbial ecology analysis. Int. J. Hydrogen Energy 52 (Part D), 389–403. https://doi.org/10.1016/j.ijhydene.2023.06.307. ISSN 0360-3199.

Ozmihci, S., Kargi, F., 2007a. Effects of feed sugar concentration on continuous ethanol fermentation of cheese whey powder solution (CWP). Enzym. Microb. Technol. 41 (6–7), 876–880. https://doi.org/10.1016/j.enzmictec.2007.07.015. ISSN 0141-0229.

Ozmihci, S., Kargi, F., 2007b. Ethanol fermentation of cheese whey powder solution by repeated fed-batch operation. Enzym. Microb. Technol. 41 (Issues 1–2), 169–174. https://doi.org/10.1016/j.enzmictec.2006.12.016. ISSN 0141-0229.

Pagliano, G., Ventorino, V., Panico, A., Romano, I., Robertiello, A., Pirozzi, F., Pepe, O., 2018. The effect of bacterial and archaeal populations on anaerobic process fed with mozzarella cheese whey and buttermilk. J. Environ. Manag. 217, 110–122. https://doi.org/10.1016/j.jenvman.2018.03.085. ISSN 0301-4797.

Palanivel, P.D., Samsudeen, N.M., 2025. Augmenting dark fermentation with carbonaceous material for enhanced biohydrogen yield and substrate conversion in whey wastewater treatment. Int. J. Hydrogen Energy 145, 45–56. https://doi.org/10.1016/j.ijhydene.2025.06.007. ISSN 0360-3199.

Patsialou, S., Tsakona, I.A., Vayenas, D.V., Tekerlekopoulou, A.G., 2024. Biological treatment of second cheese whey using marine microalgae/cyanobacteria-based systems. Eng. Proc. 81 (1), 4. https://doi.org/10.3390/engproc2024081004.

Petitta, C., Tucci, M., Daghio, M., Capelli, C., Viti, C., Adessi, A., di Palma, L., Cruz Viggi, C., Aulenta, F., 2025. Supplementation with electrically conductive materials unlocks lactic acid conversion into volatile fatty acids during cheese whey fermentation. J. Environ. Chem. Eng. 13 (4), 117197. https://doi.org/10.1016/j.jece.2025.117197. ISSN 2213-3437.

Polettini, A., Pomi, R., Rossi, A., Zonfa, T., De Gioannis, G., Muntoni, A., 2022. Continuous fermentative hydrogen production from cheese whey – new insights into process stability. Int. J. Hydrogen Energy 47 (49), 21044–21059. https://doi.org/10.1016/j.ijhydene.2022.04.229. ISSN 0360-3199.

Prakash Dalbanjan, N., Korgaonkar, K., Jayawant Kadapure, A., Basavaraj Halladamani, S., Ramangouda, G., Praveen Kumar, S.K., 2025. Green Energy from Waste: Evaluating the Sustainability of Anaerobic Biofuel Technologies, the Microbe, 7. https://doi.org/10.1016/j.microb.2025.100410. ISSN 2950-1946.

Prazeres, A.R., Carvalho, F., Rivas, J., 2012. Cheese whey management: a review. J. Environ. Manag. 110, 48–68. https://doi.org/10.1016/j.jenvman.2012.05.018. ISSN 0301-4797.

Ramirez-Flores, K.A., Colina-Andrade, G.J., Teran-Hilares, R., Tejada-Meza, K., 2025. Treated cheese whey as a promising nutrient solution for hydroponic cultivation of lettuce, cabbage and tomato. Waste Manag. Bull. 3 (3), 100229. https://doi.org/10.1016/j.wmb.2025.100229. ISSN 2949-7507.

Ramos, G.L.D.P.A., Guimarães, J.T., Pimentel, T.C., Cruz, A.G.da, Quitério de Souza, S.L., Vendramel, S.M.R., 2021. Chapter 19 - whey: generation, recovery, and use of a relevant by-product. In: Bhat, R. (Ed.), Valorization of Agri-Food Wastes and By-Products. Academic Press, pp. 391–414. https://doi.org/10.1016/B978-0-12-824044-1.00030-1.

Rao, R., Basak, N., 2021. Optimization and modelling of dark fermentative hydrogen production from cheese whey by Enterobacter aerogenes 2822. Int. J. Hydrogen Energy 46 (2), 1777–1800. https://doi.org/10.1016/j.ijhydene.2020.10.142. ISSN 0360-3199.

Riaño, B., Blanco, S., Becares, E., García-González, M.C., 2016. Bioremediation and biomass harvesting of anaerobic digested cheese whey in microalgal-based systems for lipid production. Ecol. Eng. 97, 40–45. https://doi.org/10.1016/j.ecoleng.2016.08.002. ISSN 0925-8574.

Risner, D., Shayevitz, A., Haapala, K., Meunier-Goddik, L., Hughes, P., 2018. Fermentation and distillation of cheese whey: carbon dioxide-equivalent emissions and water use in the production of whey spirits and white whiskey. J. Dairy Sci. 101 (Issue 4), 2963–2973. https://doi.org/10.3168/jds.2017-13774. ISSN 0022-0302.

Rodrigues Ramos, L., Luiz Silva, E., 2018. Continuous hydrogen production from cofermentation of sugarcane vinasse and cheese whey in a thermophilic anaerobic fluidized bed reactor. Int. J. Hydrogen Energy 43 (29), 13081–13089. https://doi.org/10.1016/j.ijhydene.2018.05.070. ISSN 0360-3199.

Romero-Izquierdo, A.G., Gutiérrez-Antonio, C., Gómez-Castro, F.I., Hernández, S., 2022. Synthesis and intensification of a biorefinery to produce renewable aviation fuel, biofuels, bioenergy and chemical products from Jatropha Curcas fruit. IET Renew. Power Gen. 16 (14), 2988–3008. https://doi.org/10.1049/rpg2.12388. ISSN 1752-1424.

Romero-Izquierdo, A.G., Gutiérrez-Antonio, C., Gómez-Castro, F.I., Hernández, S., García-Trejo, J.F., 2023. Computer aided-design of energy efficiency tools on a microalgae biorefinery scheme. In: Kokossis, A.C., Georgiadis, M.C. (Eds.), Efstratios Pistikopoulos, Computer Aided Chemical Engineering, 52. Elsevier, pp. 337–342. https://doi.org/10.1016/B978-0-443-15274-0.50054-8. ISSN 1570-7946.

Rosa, P.R.F., Christine Santos, S., Luiz Silva, E., 2014. Different ratios of carbon sources in the fermentation of cheese whey and glucose as substrates for hydrogen and ethanol production in continuous reactors. Int. J. Hydrogen Energy 39 (3), 1288–1296. https://doi.org/10.1016/j.ijhydene.2013.11.011. ISSN 0360-3199.

Rovira-Cal, E., Besga, L., Pérez-Lorenzo, E., Fernández-Arévalo, T., Aymerich, E., 2025. Cheese whey fermentation for selective VFA production: operational parameters assessment and scale-up. J. Environ. Chem. Eng. 13 (3). https://doi.org/10.1016/j.jece.2025.116839, 116839, ISSN 2213-3437.

Salla, A.C.V., Cláudia Margarites, A., Seibel, F.I., Carlos Holz, L., Barbosa Brião, V., Bertolin, T.E., Colla, L.M., Costa, J.A.V., 2016. Increase in the carbohydrate content of the microalgae Spirulina in culture by nutrient starvation and the addition of residues of whey protein concentrate. Bioresour. Technol. 209, 133–141. https://doi.org/10.1016/j.biortech.2016.02.069. ISSN 0960-8524.

Sansonetti, S., Curcio, S., Calabrò, V., Iorio, G., 2009. Bio-ethanol production by fermentation of ricotta cheese whey as an effective alternative non-vegetable source. Biomass Bioenergy 33 (12), 1687–1692. https://doi.org/10.1016/j.biombioe.2009.09.002. ISSN 0961-9534.

Saravanan, A., Senthil Kumar, P., Shiong Khoo, K., Show, P.-L., Femina Carolin, C., Fetcia Jackulin, C., Jeevanantham, S., Karishma, S., Show, K.-Y., Lee, D.-J., Chang, J.-S., 2021. Biohydrogen from organic wastes as a clean and environment-friendly energy source: production pathways, feedstock types, and future prospects. Bioresour. Technol. 342, 126021. https://doi.org/10.1016/j.biortech.2021.126021. ISSN 0960-8524.

Siddiqui, S.A., Ristow, B., Rahayu, T., Putra, N.S., Yuwono, N.W., Nisa, K., Mategeko, B., Smetana, S., Saki, M., Nawaz, A., Nagdalian, A., 2022. Black soldier fly larvae (BSFL) and their affinity for organic waste processing. Waste Manag. 140, 1–13. https://doi.org/10.1016/j.wasman.2021.12.044. ISSN 0956-053X.

Silva, J., Alves, C., Pinteus, S., Reboleira, J., Pedrosa, R., Bernardino, S., 2019. Chapter 3.10 - Chlorella. In: Nabavi, S.M., Silva, A.S. (Eds.), Non-vitamin and Nonmineral Nutritional Supplements. Academic Press, pp. 187–193. https://doi.org/10.1016/B978-0-12-812491-8.00026-6.

Singh, A., Kumari, K., 2019. An inclusive approach for organic waste treatment and valorisation using Black Soldier Fly larvae: a review. J. Environ. Manag. 251, 109569. https://doi.org/10.1016/j.jenvman.2019.109569. ISSN 0301-4797.

Soares, B.B., Costa Alves, E., Almeida Neto, J.A.de, Rodrigues, L.B., 2021. Chapter 8 - environmental impact of cheese production. In: Galanakis, C.M. (Ed.), Environmental Impact of Agro-Food Industry and Food Consumption. Academic Press, pp. 169–187. https://doi.org/10.1016/B978-0-12-821363-6.00009-6.

Suwal, S., Bentahar, J., Marciniak, A., Beaulieu, L., Deschênes, J.-S., Doyen, A., 2019. Evidence of the production of galactooligosaccharide from whey permeate by the microalgae Tetradesmus obliquus. Algal Res. 39, 101470. https://doi.org/10.1016/j.algal.2019.101470. ISSN 2211-9264.

The Guardian, 2023. Not Orange and Plastic-Wrapped: US Artisanal Cheese Boosts Dairy and Goes Green. Available from: https://www.theguardian.com/food/2023/jul/27/us-artisinal-cheese-farming.

Tugume, M., Ibrahim, M.G., Nasr, M., 2025. Valorization of cheese whey wastewater to achieve sustainable development goals. Renew. Sustain. Energy Rev. 211, 115273. https://doi.org/10.1016/j.rser.2024.115273. ISSN 1364-0321.

Uysal, U., Hamamcı, H., 2021. Succinic acid production from cheese whey via fermentation by using alginate immobilized Actinobacillus succinogenes. Bioresour. Technol. Rep. 16, 100829. https://doi.org/10.1016/j.biteb.2021.100829. ISSN 2589-014X.

Vasistha, S., Khanra, A., Clifford, M., Rai, M.P., 2021. Current advances in microalgae harvesting and lipid extraction processes for improved biodiesel production: a review. Renew. Sustain. Energy Rev. 137, 110498. https://doi.org/10.1016/j.rser.2020.110498. ISSN 1364-0321.

Venkateswar Reddy, M., Uttej Nandan Reddy, V., Young-Cheol Chang, 2022. Integration of anaerobic digestion and chain elongation technologies for biogas and carboxylic acids production from cheese whey. J. Clean. Prod. 364, 132670. https://doi.org/10.1016/j.jclepro.2022.132670. ISSN 0959-6526.

Vymazal, J., 2019. Constructed wetlands for wastewater treatment. In: second ed. Brian Fath, Encyclopedia of Ecology. Elsevier. https://doi.org/10.1016/B978-0-12-409548-9.11238-2. Pages 14-21.

Wang, C.-J., Jayanata, Y., Bajpai, R.K., 1987. Effect of multiple substrates in ethanol fermentations from cheese whey. J. Ferment. Technol. 65 (3), 249–253. https://doi.org/10.1016/0385-6380(87)90084-7. ISSN 0385-6380.

Chapter 25

Membrane filtration technology and cheese whey: Applications, challenges and opportunities

Cláudia Andréia Gräff[1, 2], Marcy Heli Paiva Rodrigues[1], Claucia Fernanda Volken de Souza[1, 2] and Daniel Neutzling Lehn[1, 2]

[1]*Food Biotechnology Laboratory, University of Vale do Taquari - Univates, Lajeado, Brazil;* [2]*Graduate Program in Biotechnology, University of Vale do Taquari - Univates, Lajeado, Brazil*

1. Introduction

Membrane filtration technology is employed for the selective separation and concentration of commercially valuable components in cheese whey, primarily proteins, based on their molecular sizes (Chen et al., 2023; Wen-qiong et al., 2019). These separation process can be classified into microfiltration, ultrafiltration, nanofiltration, and reverse osmosis. This separation is achieved using a physical barrier and an applied force generated by hydraulic pressure (Castro-Muñoz and Fíla, 2018). It represents a technological advancement for the dairy industry, enabling the recovery of whey proteins in their native form (Chen et al., 2021). To maximize the use of cheese whey components, membrane characteristics (composition and geometric configuration) are crucial factors that must be considered. The choice of membrane material also influences the durability and cost of the system (Cunha et al., 2022). In cheese whey processing, polymeric membranes are used for microfiltration, ultrafiltration, nanofiltration, and reverse osmosis, while inorganic membranes are mainly applied in microfiltration, ultrafiltration, and nanofiltration (Reig et al., 2021).

Filtration modules, regardless of membrane type, operate in two modes: dead-end and crossflow. In dead-end filtration, the feed is directed straight onto the membrane, producing only permeate. In crossflow filtration, the feed flows parallel to the membrane surface, yielding both permeate and retentate. Crossflow is preferable for feeds with high solids or fouling potential, as it reduces fouling and concentration polarization (Yogarathinam et al., 2022). Integrating multi-stage processes—where the feed passes sequentially through microfiltration, ultrafiltration, and nanofiltration membranes—enhances whey protein recovery and purity while minimizing nutrient loss. These stages also can reduce fouling, improve flow, and decrease energy consumption, conducting the process in a more sustainable way (Blais et al., 2022; Macedo et al., 2021a; Reig et al., 2021).

Similarly, diafiltration can be integrated into ultrafiltration and nanofiltration to further purify whey proteins by removing residual lactose and mineral salts, resulting in WPI with up to 95% (w/w) protein. In diafiltration, the retentate is diluted with a solvent—typically water or buffer—followed by repeated filtration of the diluted retentate. This process can be performed repeatedly, continuously or in batch mode (Macedo et al., 2021b; Reig et al., 2021). According to Cabral et al. (2019), achieving an 80% (w/w) protein in WPC requires a combination of ultrafiltration and diafiltration. Diafiltration can facilitates the removal of mineral salts and lactose, preventing excessive solute accumulation during operation.

In this context, the objective of this chapter is to describe the application of membrane filtration technology for the comprehensive use of cheese whey, a by-product of the dairy industry, in the production of high-value-products. The chapter focuses on the challenges and future perspectives of these technologies, emphasizing its role in industry to promote circular economy practices through economic and environmental sustainability.

Whey. https://doi.org/10.1016/B978-0-443-27648-4.00027-X

2. Characteristics of membranes for cheese whey filtration

The membranes used in cheese whey processing differ primarily in pore size and molecular weight cut-off (MWCO, typically expressed in Da), which determine the molecular weight of whey constituents that are retained or permeate through the membrane, as illustrated in Fig. 25.1. These membranes allow for the removal of water, fat, lactose, sodium chloride, and other mineral salts from cheese whey while concentrating proteins. As a result, products such as concentrated, defatted, delactosed, desalted, and demineralized whey, as well as whey protein concentrate (WPC) (containing at least 35% w/w protein) and whey protein isolate (WPI) (containing at least 90% w/w protein) can be obtained (Kelly, 2019; Reig et al., 2021).

The characteristics of membranes, such as material composition, surface and pore areas, shape and geometry, as well as pore morphology and size, directly influence the efficiency of cheese whey filtration. These structural differences affect factors such as liquid flow dynamics, selective rejection of solutes (e.g., proteins, sugars, and minerals), permeability, clogging resistance, and ease of cleaning (Al-Mutwalli et al., 2020; Reig et al., 2021; Schopf et al., 2021).

Organic membranes are widely used in cheese whey filtration due to their flexibility, relatively low cost, and good performance in terms of selectivity and permeability. Their manufacturing involves polymers such as polyethersulfone, polysulfone, polyamides, polypropylene, regenerated cellulose acetate, and fluorinated polymers. Highly hydrophobic and rough-surfaced membranes, such as fluorinated polymeric membranes, enhance solute rejection by promoting gel layer formation during filtration. In contrast, hydrophilic membranes, such as regenerated cellulose acetate, minimize clogging. Polyethersulfone membranes offer a balanced combination of moderate hydrophilicity and uniform porosity, optimizing both permeability and solute rejection (Damar et al., 2020).

Polyethersulfone membranes are extensively employed in cheese whey ultrafiltration. Bejarano-Toro et al. (2022) used a polyethersulfone ultrafiltration membrane (10 kDa, spiral module) to concentrate proteins from sweet whey, achieving

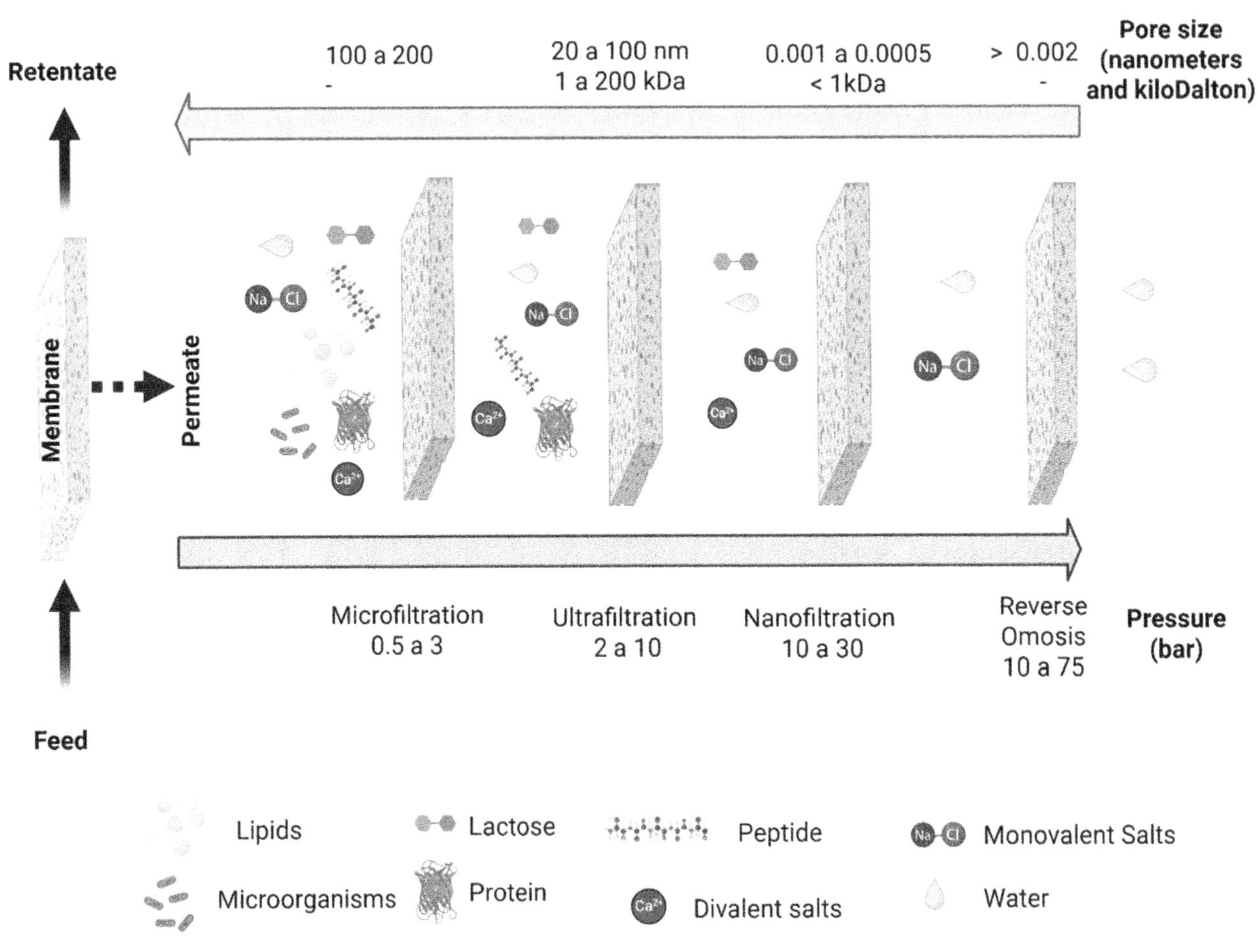

FIGURE 25.1 Key characteristics of membrane separation processes for crossflow filtration.

retention coefficients (Rp) ranging from 0.60 to 0.73, depending on operational conditions. The permeate flow (Jp) was influenced by transmembrane pressure, reaching values above 41 $L/m^2/h$. The resulting WPC contained 18.2% (w/w) total solids, of which 45% were proteins. Similarly, Rama et al. (2021) also used a polyethersulfone ultrafiltration membrane (10 kDa, spiral module) with crossflow pumping to separate lactose and concentrate sweet whey proteins. However, lactose separation was inefficient, with rejection rates around 28%, indicating that part of the sugar remained trapped on the membrane surface due to concentration polarization caused by protein aggregates. Consequently, the retentate's protein content reached only 22% (w/w), failing to meet WPC specifications.

While processing sheep's milk whey proteins, Sánchez-Moya et al. (2020) evaluated three distinct ultrafiltration membranes (polysulfone (25 kDa) and polyethersulfone (10 and 5 kDa)) for their ability to retain proteins while allowing lactose passage, aiming to obtain a protein concentrate and a lactose-rich permeate. The ultrafiltration system featured a flat-sheet membrane module with an effective area of 8.48 cm^2 and a 0.8 L feed tank. Operating pressures ranged from 4 to 10 bar, and permeated fluxes along with lactose and protein rejection coefficients were determined. The 25 kDa membrane exhibited the highest permeate flux due to its larger molecular weight cut-off. Regarding lactose separation, this membrane achieved an 85% rate, whereas the 10 and 5 kDa membranes reached 80%. Protein retention was 100% across all membranes, ensuring a pure protein concentrate in the retentate.

Ceramic membranes, primarily composed of zirconium, aluminum, or titanium oxides, stand out for their thermal, chemical, and clogging resistance, making them suitable for long-duration filtration processes. Although more expensive, ceramic membranes are highly durable and easier to clean, making them an attractive option for cheese whey processing (Bortot Coelho et al., 2023). Hinkova et al. (2016) employed tangential filtration of sweet and salty whey using tubular ceramic membranes with various pore sizes and compositions—aluminum oxide, zirconium oxide, and titanium oxide. Their findings demonstrated the superior performance of inorganic membranes in whey protein retention and partial lactose rejection. The 5 kDa ultrafiltration membrane achieved protein rejection rates of 95%–98%, making it ideal for WPC production. Meanwhile, the 500 nm microfiltration membrane exhibited high permeability, reaching fluxes of up to 59 $L/m^2/h$, while minimizing lactose losses. However, polarization layer accumulation on the membrane surface impacted its efficiency, particularly for lower porosity membranes.

According to Al-Mutwalli et al. (2020), ultrafiltration of whey using ceramic membranes required less water during diafiltration. This aspect is relevant for the economic and environmental sustainability of membrane-based whey processing, aligning with industry efforts to reduce water and energy consumption during filtration and develop chemical regeneration techniques to extend membrane lifespan (Aydiner et al., 2014; Arsic and Vukovic, 2021; Macedo et al., 2021b).

To optimize dairy industry by-product utilization through integrated filtration stages, Sluková et al. (2016) processed sweet whey combining ultrafiltration (50 kDa, tubular, ceramic coated with zirconium oxide) and nanofiltration (spiral, polypiperazine thin-film composite) membranes. Ultrafiltration recovered 81% of the proteins in the retentate, while nanofiltration retained 93% of the lactose, demonstrating the efficiency of membranes in whey fractionation and industrial resource management. Similarly, Brião et al. (2024) integrated membrane filtration technologies, employing spiral ultrafiltration (10 kDa, polyethersulfone), nanofiltration (200 Da, polyamide), and reverse osmosis (99.3% salt rejection) to recover proteins and lactose from sweet whey while reusing the water from diafiltration. Ultrafiltration concentrated proteins in the retentate, with a rejection rate exceeding 93%, yielding WPC with 88% protein content on a dry basis. Nanofiltration purified lactose to 90% purity, reducing monovalent mineral salts by up to 80%. Furthermore, the water from the diafiltration—recovered through reverse osmosis—showed similar quality to potable water, enabling its reuse without compromising the quality of the obtained WPC.

Another promising aspect of membrane filtration technology for whey component fractionation is the development of hybrid membranes that combine organic and inorganic materials, enhancing resistance and selectivity. Moreover, membrane surface modifications, such as the introduction of hydrophilic layers, can reduce biofilm formation and improve cleaning efficiency (Samree et al., 2020).

3. Membrane performance and fouling factors

When pressure is applied to the filtration system, the separation of target molecules is driven by permeate flow, which is limited by membrane resistance. As solids rejected due to size exclusion accumulate at the feed-membrane interface, the resistance to solvent passage increases (Haribabu et al., 2020; Levitsky et al., 2021). This phenomenon, known as concentration polarization, is reversible, especially in tangential flow filtration (Rama et al., 2021; Krantz et al., 2022). Over time, concentration polarization contributes to membrane surface fouling, occurring already within the first minutes of filtration (Aghajani et al., 2018; Corbatón-Báguena et al., 2017). This process consists of the formation of a gel or cake layer on the outer membrane surface, which acts as a secondary membrane that alters the flux of mass transfer and consequently increases the rejection of components (Brião et al., 2017). Adjustments to fluid dynamics can help mitigate this effect (Aslam et al., 2022; Villecco et al., 2020). Operating pressure directly influences permeate flow behavior, while tangential velocity minimizes concentration polarization in the retentate stream, potentially affecting both permeate flux and separation efficiency (Brião et al., 2017).

Damar et al. (2020) state that ultrafiltration membrane surface characteristics and morphology primarily govern solute-membrane interactions, alongside other factors such as whey composition, pH, temperature, and operational filtration parameters. The same authors suggest that the zeta potential may be a distinctive surface property for hydrophobic membranes interacting with whey. Wang et al. (2024) observed that higher sodium chloride concentrations enhance protein structuring and reduce zeta potential, thereby increasing membrane fouling. Tanudjaja et al. (2023) analyzed 88 articles published between 1993 and 2023 on ultrafiltration and microfiltration membrane fouling, identifying gaps in research regarding the zeta potential, despite its relevance to mechanisms governing membrane fouling. However, using machine learning models, the authors found that salt concentration and membrane configuration had the least impact on fouling indicators.

Operational conditions, such as system pressure and solid concentration in the feed solution, affect permeate flux. Permeate flow rate decreases as system pressure increases (Aslam et al., 2022). In ultrafiltration, for example, lower pressures (e.g., around 1 bar) can yield higher permeate fluxes than those at 3 bar (Prabhuzantye et al., 2019; Aslam et al., 2022; Rastogi et al., 2022). Excessively high transmembrane pressure can compress the gel layer, leading to maximal resistance due to particle deposition (Rastogi et al., 2022). Fouling rate correlates with variations in membrane permeability due to increased mass transfer resistance (Villecco et al., 2020).

Similarly, prolonged filtration time increases the concentration of fouling agents, leading to greater thickness and compaction of the fouling layer (Krantz et al., 2022; Damar et al., 2020). This intensifies interactions between proteins and the filtration medium, leading to significant protein deposition on the membrane surface (Wang et al., 2024). Extended filtration time also promotes pore fouling due to increased transmembrane pressure, as reported by Corbatón-Báguena et al. (2017). The authors modeled permeate flux by combining the two primary fouling mechanisms, namely complete pore blockage and cake layer formation, with a time-dependent pore-blocking parameter, applying Hermia's equation and its adaptations to cross-flow ultrafiltration data.

Whey composition is a critical variable in fouling mechanisms due to protein concentration. The main whey proteins (beta-lactoglobulin, alpha-lactalbumin, and bovine serum albumin) act as adhesive foulants, physically adsorbing onto the membrane surface and potentially penetrating the pores (Aslam et al., 2022). Hydrophobic and ionic interactions between proteins and the active surface groups of a membrane are significant contributors to fouling (Damar et al., 2020). Beta-lactoglobulin (BLG), the major whey protein, is cited as the primary fouling agent (Aslam et al., 2022). The BLG content varies across whey from different animal species, being highest in sheep whey (61.1%) and absent in camel whey (Khan et al., 2019). However, most research on whey protein concentration utilizes bovine whey (Rodrigues et al., 2024), reinforcing the idea that most anti-fouling approaches are developed based on models that also utilize bovine whey. The type of cheese and its production method can also influence separation characteristics. The diversity of filtration methods for fouling assessment and the lack of standardized reporting in scientific literature often hinder meaningful comparisons between findings (Tanudjaja et al., 2023).

In whey ultrafiltration, dominant factors hindering fractionation include pore blocking and cake formation, which exceed the resistance caused by internal fouling, as most whey particles are larger than membrane pore diameters (Corbatón-Báguena et al., 2017; Damar et al., 2020). Wang et al. (2024) reported that protein-protein interactions play a greater role in cake formation than protein-membrane interactions. The authors attributed this to conformational changes altering the effective hydrodynamic radius of the protein and its interactions with the membrane and medium, a phenomenon promoted by the presence of magnesium salts.

3.1 Techniques for preventing and reversing fouling

The industrial cleaning process generally follows the membrane manufacturer's guidelines and is similar to the Clean-in-Place (CIP) procedures used in the dairy industry. This process is carried out at temperatures close to 40°C, employing a 1% sodium hydroxide alkaline solution, followed by rinsing with water until a neutral pH is reached. Subsequently, an acidic cleaning step with 0.25% nitric acid is performed, followed by another water rinse until neutral pH is achieved, and a final alkaline cleaning step to restore flow rate (Villecco et al., 2020; Rama et al., 2021).

Studies on ultrafiltration membrane cleaning with satisfactory results highlight the use of anionic biosurfactants for protein removal by degrading protein structures. The optimal pH range for surfactant application is 9.8, which enhances the interaction between proteins and the cleaning agent, effectively removing deposits. However, the high cost of biosurfactants is a drawback, despite their alignment with sustainability practices and the reduction of harmful chemical use (Aghajani et al., 2018). Citric acid and phosphoric acid have also been reported as effective in removing inorganic fouling, which typically occurs in the absence of proteins due to the precipitation of insoluble phosphates. Enzymatic cleaning at pH 7.0 using proteases and lipases for organic deposit removal causes less damage to membranes (Chen et al., 2019). For biofouling removal, characterized by biofilm formation on the membrane surface, sodium azide can be used. Villeneuve et al. (2021) observed an inhibitory effect on *Streptococcus* and *Bacillus* biofilms when sodium azide was added to the feed solution in a crossflow filtration system with polyethersulfone (PES) membranes, maintaining permeate flow.

Membrane surface roughness facilitates the accumulation of fouling agents, and was identified as the primary factor contributing to irreversible fouling in the study by Damar et al. (2020). Internal fouling results from physicochemical solute-membrane interactions and occurs with solutes smaller than or close to the membrane pore size, reducing their effective radius or causing complete blockage (Aghajani et al., 2018; Corbatón-Báguena et al., 2017). This type of fouling is often irreversible, leading to partial or total pore blockage.

A promising alternative to slow the progression of concentration polarization and both external and internal membrane fouling is the water hammer technique, which involves the induction of periodic pressure spikes. This natural phenomenon is characterized by a sudden knocking sound when abrupt changes in fluid velocity generate shock waves. When applied under controlled conditions, it does not affect the pores of spiral-wound ultrafiltration membranes. The counterpressure generated by the liquid velocity surge (water hammer effect) promotes viscous shear that disrupts the deposition of foulants on the membrane surface (Aslam et al., 2022). According to Krantz et al. (2022), despite the benefits of this technique, further research is needed, particularly on its application to hollow fiber and tubular ultrafiltration membranes.

Another option for mitigating fouling is the application of pulsed flow, which has shown promising results when combined with alkaline cleaning in hollow fiber and spiral-wound PES membranes. In this techique, spacer filaments are subjected to flow disturbances along the length of a spiral-wound membrane, dislodging fouling deposits that are typically resistant to convencional cleaning processes and other methods described in the literature (Kürzl and Kulozik, 2023). The use of encapsulated transglutaminase applied to cheese whey prior to ultrafiltration has demonstrated a 9% reduction in fouling and an increase in protein recovery from 69% to 85%. The resistance of the gel layer decreased after enzymatic treatment, likely due to the enzyme catalyzing protein cross-linking, forming high-molecular-weight polymers (Rastogi et al., 2022). A recent approach for understanding fouling mechanisms and improving membrane filtration technology design involves artificial intelligence-based modeling, with particular emphasis on preventing and reducing membrane fouling (Tuo et al., 2025).

3.2 Artificial intelligence as an ally in preventing fouling

The main challenges in membrane technology include their susceptibility to fouling, low permeability rates, and reduced flow rates. Initially, mathematical modeling aimed to describe membrane fouling phenomena in cheese whey ultrafiltration based on experimental data. Today, artificial intelligence (AI) seeks to expedite the development of new membranes (Ignacz et al., 2024).

Machine learning, a subset of AI, integrates computer science, mathematics, and statistics. This approach is based on training an algorithm with normalized datasets containing input and output variables. A typical process begins with data collection from published sources or laboratory experimental databases. Data undergo pre-processing such as normalization and cleaning, removing duplicates or missing values, and are then split into training and testing datasets. Through autonomous learning, the technique aims to identify patterns, predict scenarios, explain phenomena, and propose adjustments to parameters. The trained model is validated by comparing it against the test dataset, with accuracy, precision, and sensitivity serving as performance metrics (Fig. 25.2) (Tuo et al., 2025; Yogarathinam et al., 2024; Gaudio et al., 2023b; Jawad et al., 2021; Dansawad et al., 2023).

Regression-based learning algorithms can predict relationships between input and output variables, making them useful for predictive modeling. Villeco et al. (2020) developed a control system that enhanced the efficiency of ultrafiltration by inducing pressure oscillations based on fuzzy logic. To optimize ultrafiltration, the system modulates the transmembrane pressure (TMP) using an n-pulse function, meaning oscillations at relatively low levels of operating pressure. The system may consist of a dedicated algorithm combined with fuzzy logic or an artificial neural network that monitors fouling and adjusts controlled variables accordingly. The findings (Fig. 25.3) indicated an increase in ultrafiltration time based on permeate flow rates with the application of a pulse function to the transmembrane pressure inducing a cleaning effect, with a maximum of four pulsations, which significantly extended operating time, before traditional cleaning was required to conclude the process.

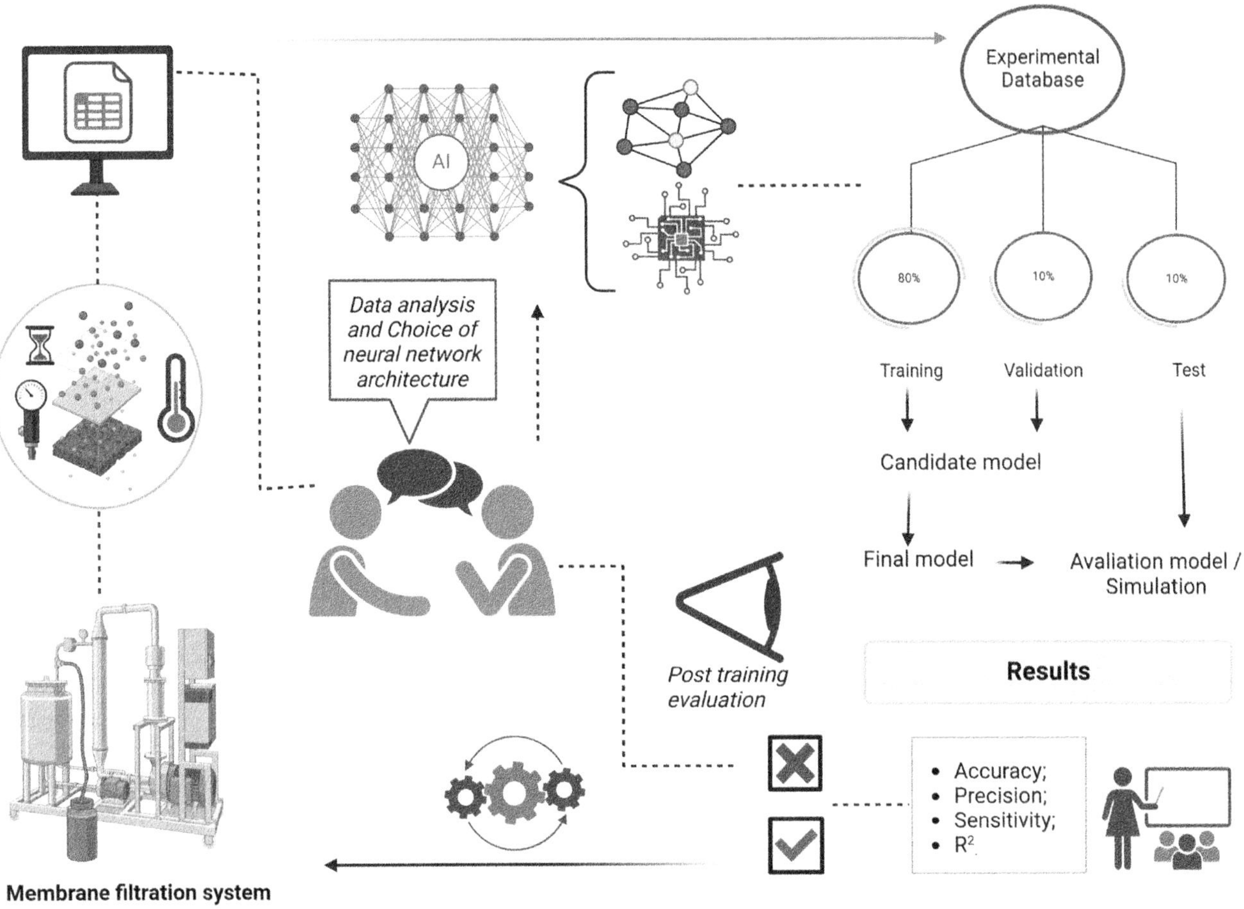

FIGURE 25.2 Schematic diagram of a neural network algorithm applied to membrane separation processes.

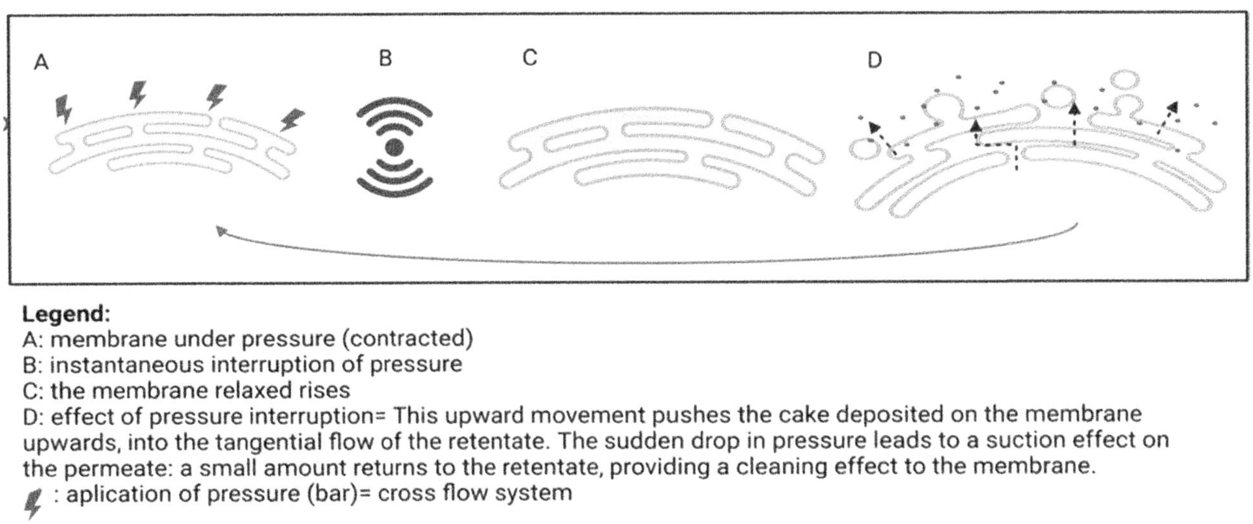

FIGURE 25.3 Effect of pressure interruption on the cake layer on the membrane surface.

Haribabu et al. (2020) applied computational fluid dynamics using a multiphase mixture model that disregarded fouling and focused instead on concentration polarization. Their findings highlighted the importance of accurately mapping the geometry of the system when modeling concentration polarization effects, as simplified 2D geometry simulations qualitatively aligned with experimental data. However, 3D geometry simulations provided highly accurate predictions of experimental fluxes, whereas 2D geometry failed to predict permeate flow. Particle size was incorporated into the model to assess osmotic pressure and drag coefficient, with the relationship between these factors helping determine the permeate flux.

According to Jawad et al. (2021), artificial neural networks are the most frequently employed technique for precise membrane process modeling using AI. This method establishes relationships between input (independent) and output (dependent) variables, demonstrating high performance in predicting membrane fouling, particularly in water and wastewater treatment systems. Well-known artificial neural network models, such as multilayer perceptron and radial basis function networks, can successfully predict membrane fouling with an R^2 of 0.99 and near-zero error.

Machine learning analyses using random forest and neural network models have indicated that pore size and transmembrane pressure are dominant factors in fouling during ultrafiltration and microfiltration (Tanudjaja et al., 2023). For permeate flux, pore size and protein concentration in the solution are the most influential variables, whereas transmembrane pressure is the primary factor for protein rejection. Meanwhile, membrane composition was identified as the most significant parameter in studies involving non-BSA proteins, corroborating findings from Damar et al. (2020). Regardless of the protein used, pore size remains the most critical parameter in protein rejection.

Progressive learning stages tend to improve predictive models. In Tuo et al. (2025), incorporating the Synthetic Minority Over-sampling Technique (SMOTE) algorithm improved predictive performance compared to processing the same dataset without expansion. The study reported determination coefficients of 0.833 for flux and 0.723 for protein rejection, demonstrating the robustness of the model using SMOTE combined with a one-dimensional convoluted neural network (ODCNN).

With regards to the costs associated with implementing AI-based solutions in real industrial processes, there is still limited data to justify the necessary investments (Jawad et al., 2021). A comparative analysis of computational costs and resources (experimental setup, dataset size for training, and generalization capabilities) could highlight the advantages of AI applications. Further studies are needed to compare the most widely used techniques in terms of accuracy, processing time, and artificial neural network modeling.

Thus, just as advances in instrumental techniques in physics and analytical chemistry have progressively elucidated the mechanisms involved in membrane fouling, the development of novel artificial neural networks holds significant potential for improving membrane process simulations, optimization, and control.

4. Trends in applications involving the purification/concentration of whey-derived substances: Modeling new applications

The increasing market value of whey protein has led to a reassessment of milk heat treatments. However, even under moderate heating conditions, the concentration of native protein in whey remains low (between 0.55% and 0.75%), highlighting the appeal of membrane filtration technology to concentrate these compounds of interest. In addition to the major whey proteins, beta-lactoglobulin and alpha-lactalbumin, whey contains several smaller biologically active proteins, such as immunoglobulins, lactoferrin, lactoperoxidase, and other enzymes (Ostertag et al., 2023; Poonia and Petkoska, 2023; Saadi et al., 2024; Krolitzki et al., 2024).

Given the heterogeneity and characteristics of whey and its derivatives, strategies for purifying whey-derived substances require knowledge derived from experimental data and proper planning. Membrane separation is proposed as an environmentally friendly alternative to other processes, as it does not involve the use of harmful chemicals and is easily scalable, making it a promising technology for industrial applications in substance purification. Membrane separation operations are frequently used for the purification or concentration of whey components of interest, differing in terms of purity level, membrane type, and target compound. In these processes, several operational parameters are relevant, including transmembrane pressure, circulation velocity within the system, and operating temperature (Kravtsov et al., 2021; Alfano et al., 2022; Mamay et al., 2023).

Both ceramic and polymeric membranes can be employed. Ceramic membranes offer high efficiency and durability but come with a higher cost, whereas polymeric membranes, despite their lower acquisition cost, are more susceptible to contamination, which reduces their lifespan. Typically, whey microfiltration is used as an initial step to remove coarse impurities (such as fat) through ceramic membranes. Subsequent ultrafiltration enables the concentration and purification of compounds such as lactoferrin; however, this process requires diafiltration, which enhances protein removal from whey but demands higher water consumption. A drawback of this technique is concentration polarization and the formation of deposits, which affect membrane permeability and selectivity, as well as the dependence on specific conditions to maximize operational yield. As with diafiltration, integration with other methods, such as magnetic separation for specific compounds, can enhance purity and add value to the separated components (Krolitzki et al., 2024; Xu et al., 2025).

An alternative to improving the purification of whey-derived substances is the combination/integration of processes such as microfiltration, ultrafiltration, nanofiltration, reverse osmosis, and diafiltration. The combination of ultrafiltration with nanofiltration and diafiltration allows for the separation of whey proteins and the concentration of lactose. This approach also facilitates water recovery for reuse in the process, achieving protein purification efficiencies of 85% and lactose concentration levels exceeding 90%. Techniques such as reverse osmosis after diafiltration have demonstrated potential for reusing up to 66% of water as potable, promoting sustainability in the purification process. Focusing on ultrafiltration, the physicochemical properties of membranes (such as hydrophobicity, roughness, and surface charge) directly influence retention efficiency and the risk of fouling. Hydrophilic membranes generally perform better in reducing irreversible fouling. Ultrafiltration membranes made from different active materials (such as polysulfone and regenerated cellulose) vary in performance. For instance, regenerated cellulose membranes exhibit higher flux recovery after cleaning due to their lower tendency for fouling (Damar et al., 2020; Brião et al., 2024).

By utilizing ultrafiltration with 10 kDa membranes, Damar et al. (2020) obtained protein concentrations of 2.88 mg/mL beta-lactoglobulin and 0.78 mg/mL alpha-lactalbumin in the retentate. In contrast, the study by Brião et al. (2024), which employed an integrated approach combining nanofiltration and diafiltration, achieved protein concentrates with 88% purity, representing a significant improvement over ultrafiltration alone. Regarding lactose concentration, its purity can be enhanced through the combination of nanofiltration and diafiltration, reaching up to 90%, and surpassing methods mashed solely on ultrafiltration. Additionally, water recovery during diafiltration contributes to a considerable reduction in water consumption while maintaining the quality of the resulting concentrates. The application of ultrafiltration to concentrate bovine whey has proven to be an effective technique for reducing the volume of whey requiring thermal treatment for protein precipitation, with an estimated energy requirements reduction of over 70% (Pires et al., 2024).

Regarding experimental studies on substance purification, Krolitzki et al. (2024) investigated lactoferrin purification from acid whey using ultrafiltration and magnetic separation, emphasizing the importance of membrane selectivity in minimizing the retention of non-target proteins. In this study, ultrafiltration significantly increased protein concentration before the magnetic separation step, which facilitated the selective removal of the target protein. High selectivity for lactoferrin was observed, with a yield of 60% and a purity of 78%.

Mamay et al. (2023) studied the extraction of micellar casein and whey proteins from skim milk, highlighting the influence of operational parameters on microfiltration efficiency. The choice between ceramic and polymeric membranes affected both the permeation rate and the lifespan of the system. In their study, for the microfiltration stage applied to skim milk, membrane selection directly influenced separation efficiency, with ceramic membranes (despite their higher cost) proving more resistant to fouling and achieving higher permeate flow rates.

Meanwhile, Gaudio et al. (2023a) proposed an integrated multiple membrane system to valorize by-products from the dairy industry, combining ultrafiltration and vacuum evaporation to concentrate proteins from ricotta whey (Scotta, or second whey), with emphasis on reducing energy consumption and mitigating environmental impact. In this integrated system, the overall water removal reached 84.9%, with a progressive increase in protein concentration throughout the process. Considering the studies mentioned, the main limitations include the need for strict control of operational parameters to prevent membrane clogging, variations in separation efficiency depending on the feed material's properties, and the complexity of system maintenance. Specifically for whey processing, its compositional complexity may affect membrane performance under real operating conditions. Further studies are needed to assess the impact of these components on fouling and membrane selectivity in different scenarios.

Bearing in mind studies aimed at recovering lactose from cheese whey, the combination of ultrafiltration, nanofiltration, and diafiltration processes emerge as a viable alternative to mitigate separation issues commonly associated with the presence of lactic acid and calcium in acid whey, which hinder lactose purification and crystallization. Research indicates that applying these processes to sweet whey is more effective for the purpose of lactose purification, achieving purity levels above 90% and concentrations of up to 200 g/L. However, diafiltration may lead to lactose loss during processing. Ultrafiltration, then, has been found to be the most suitable technique for separating lactose from sweet whey, whereas nanofiltration combined with diafiltration is usually recommended for lactose separation from acid whey (Roselli et al., 2024; Hofmann and Hamel, 2023; Kravtsov et al., 2021).

Other trends aimed at improving the purification of whey-derived compounds involve specific whey treatment steps, such as enzymatic hydrolysis, to enhance membrane separation performance. These pre-treatments enable the production of valuable compounds, such as bioactive peptides, while also improving the efficiency of separation processes by reducing membrane fouling. Rosseto et al. (2025) employed transglutaminase to enhance ultrafiltration performance. The study reported that transglutaminase is an effective tool for optimizing cheese whey ultrafiltration by reducing fouling, increasing permeate flux in polyethersulfone membranes, improving protein recovery, and enhancing process efficiency, with potential large-scale applications in the food industry. Krunic and Rakin (2025) investigated the generation of bioactive peptides through enzymatic reactions in whey prior to ultrafiltration. Their study found that hydrolysis with proteinase K produced the highest proportion (62%) of peptides with molecular weights below 3 kDa, along with the highest antioxidant and acetylcholinesterase inhibitory activities. Ultrafiltration following hydrolysis further enhanced the bioactivity of these peptides, particularly thermolysin, which required an additional ultrafiltration step.

Alternatives to improve whey treatment via membrane separation include membrane surface modifications and ionic regulation of whey proteins. Li et al. (2025) tested ultrafiltration membranes composed of sulfonated block copolymers (mPEG-b-SPAN) for whey protein recovery in high-salinity environments. Their findings indicated that the incorporation of sulfonic groups and mPEG segments significantly improved selectivity and resistance to membrane fouling. Wenqiong et al. (2025) examined the effects of ionic regulation on whey protein structure during ultrafiltration to enhance process efficiency. Ionic regulation (addition of Na^+, Zn^{2+}, and K^+ ions at specific intervals) increased protein-membrane repulsion, reducing the formation of the fouling cake layer, and facilitating protein removal during washing, making membrane cleaning more efficient and less reliant on harsh chemicals. This technique allows for the recovery of high-value proteins, prevents losses due to membrane clogging, and enables the production of purer and more concentrated proteins for premium applications, ultimately reducing operational costs by minimizing frequent replacement of filtration membranes.

With regards to the utilization of computational tools for simulating and predicting membrane separation behavior, various approaches have been reported, emphasizing that ultrafiltration system modeling is essential for optimizing compound separation from complex matrices such as cheese whey. Modeling is particularly useful for predicting and mitigating the effects of membrane fouling and concentration polarization, thereby reducing product losses. The selection of an optimal model depends on the specific separation objectives and practical constraints of the ultrafiltration system under analysis. For industrial applications, the combination of various modeling techniques can yield more accurate predictions and improve process optimization (Tuo et al., 2025; Casado-Coterillo et al., 2023; Gaudio et al., 2023b; Haribabu et al., 2020; Corbatón-Báguena et al., 2017).

Casado-Coterillo et al. (2023) evaluated the simultaneous separation of lactose and lactic acid using commercial nanofiltration membranes, applying the Kedem-Katchalsky and Spiegler-Kedem models. The best fit for experimental data was obtained with the Spiegler-Kedem model for rejection and flux parameters, providing a more realistic estimation of process efficiency. The commercial AFC30 membrane (PCI Membranes Filtration Group) demonstrated the highest efficiency, achieving lactose rejection above 98% while maintaining lactic acid rejection below 37%, thereby facilitating separation and enhancing the industrial valorization of cheese whey. Haribabu et al. (2020) applied computational fluid dynamics (CFD) using a multiphase mixture model to predict solid concentration and velocity distributions near the membrane. Their model closely matched experimental data for the studied flow modes without incorporating additional resistance models, such as pore blocking, surface adsorption, or gel/cake formation. Due to the highly compact globular structure of whey proteins, the study employed the hard sphere theory to model suspension permeability and osmotic pressure. The validated model accurately predicted permeate flux without empirical parameter adjustments. Results indicated that concentration polarization was the primary factor in flux decline, suggesting the need for additional fouling models in these conditions. While phenomenological models provide fundamental insights into solute transport, CFD-based approaches offer a detailed spatial analysis of flow dynamics.

The use of artificial neural networks to predict the efficiency of ultrafiltration in a crossflow filtration system, with optimization based on genetic algorithms, to control permeate flux decline was studied by Gaudio et al. (2023b). The neural networks were trained on an experimentally derived dataset from laboratory-scale hollow fiber membrane modules used for ricotta whey (scotta) valorization. The neural network with three input variables demonstrated the best performance, preventing permeate flux decline with a mean squared error of 2.89×10^{-13}. Meanwhile, a hybrid model combining neural networks and theoretical models yielded superior predictions of dynamic process behavior and operational variable control. Although artificial neural networks provide a robust alternative for predicting system performance under varying operational conditions, their effectiveness depends on access to extensive experimental data for proper training.

Rezapour et al. (2025) studied the use of graphene oxide-modified nanofiltration membranes for sweet whey demineralization, proposing a convolutional neural network model to predict dynamic permeate flux behavior and optimize operational conditions. The proposed model demonstrated high accuracy, with an R^2 value of 1.00 for both training and test datasets and a mean squared error of 0.44, indicating excellent correlation between predicted and experimental values. The proposed model successfully identified optimal conditions for whey demineralization, comprising a promising tool for industrial-scale performance prediction. However, while laboratory-scale results have been promising, industrial scalability must consider the long-term stability of modified membranes as well as production and maintenance costs.

Saltik et al. (2017) applied dynamic modeling to ultrafiltration membranes, incorporating the concept of membrane resistance, divided into static and dynamic components, to represent permeate flux behavior over time. Static resistance reflects intrinsic membrane properties, while dynamic resistance models fouling accumulation, influenced by transmembrane pressure, permeate flux, and protein concentration in the retentate. The model was validated with industrial data; however, its complexity may limit applications at larger scales or in multi-unit membrane systems. The model effectively described membrane performance, including gel layer formation and concentration polarization. By integrating physical process aspects with data-driven parameter estimation techniques, this approach can enable both real-time control and optimization applications.

Thus, considering whey purification processes and computational modeling tools, the selection of appropriate techniques and models depends on the specific objectives and practical constraints of membrane separation systems. For industrial applications, combining different modeling strategies and whey pre-treatment techniques can enhance prediction accuracy and optimize the overall process.

5. Future perspectives on membrane filtration technology for cheese whey

New biomaterials are being developed to refine the utilization of cheese whey in the dairy industry by employing membrane filtration technology with specific properties. However, the control of operational parameters in these bioprocesses still requires honing to ensure scalability and to make industrial implementation viable. In this context, Al-Mutwalli et al. (2024) immobilized the enzyme β-galactosidase on ceramic ultrafiltration membranes (15 kDa, tubular and disc modules) made of zirconium oxide and titanium oxide, to integrate protein separation and lactose hydrolysis in a single process. The enzymatic immobilization methodology—surface activation with gelatin and cross-linking with glutaraldehyde—exemplifies how advancements in surface chemistry can be applied to process engineering in the dairy industry. Similarly, Rosseto et al. (2024) described the application of transglutaminase as a pre-treatment in the membrane filtration process of cheese whey, increasing the molecular weight of proteins, and thereby facilitating their retention during ultrafiltration. This enhances filtration efficiency, reducing membrane fouling and increasing permeate flux. Additionally, enzymatic pre-treatment with transglutaminase can improve the functional properties of whey proteins, such as solubility, thermal stability, and gel-forming capacity, expanding their industrial applications.

Another emerging technology has been the combination of flat-sheet microfiltration (0.2 μm) and ultrafiltration (200 kDa) membranes—both polymeric membranes made of polyvinylidene fluoride—with progressive freeze concentration for phosphorus recovery from cheese whey (Hidayat et al., 2023). The membranes perform the initial separation of proteins, fat, and other solids from the whey, while freeze concentration selectively removes water and concentrates phosphorus. This method is efficient, preserves the quality of recovered components, and operates with lower energy consumption compared to traditional processes such as thermal evaporation. Phosphorus is an essential nutrient for plant growth and food production; however, its availability is challenged by dependence on finite reserves of phosphate rocks concentrated in few territories. The hybrid process proposed by Hidayat et al. (2023) offers a viable alternative, utilizing industrial by-products, such as cheese whey to extract and recycle phosphorus, reducing reliance on non-renewable sources and promoting a circular economy.

The utilization of near-infrared (NIR) spectroscopy as an analytical tool for continuous monitoring of protein concentrations in industrial processes, as proposed by Tonolini et al. (2023), is another interesting innovation. Implementing this technique allows for real-time adjustments of operational parameters, reducing losses, optimizing resource utilization, and improving product quality. This increases the efficiency and sustainability of whey protein fractionation, including beta-lactoglobulin and alpha-lactalbumin. Moreover, the use of this analytical tool eliminates the need for time-consuming conventional chemical analyses, such as high-performance liquid chromatography. Tonolini et al. (2023) integrated NIR spectroscopy with regression models and data filtering techniques, enabling the capture of dynamic variations in process performance and providing a detailed and immediate overview of system conditions.

These new materials and techniques not only improve the efficiency of selective separation and concentration of different whey components but also reduce operational costs and improve the economic and environmental sustainability of the process by minimizing the need for frequent membrane cleaning and replacement. Consequently, these emerging processes expand the potential for transforming cheese whey into high-value products while offering practical solutions for sustainable resource management.

6. Conclusions

Membrane filtration technology has become an essential tool for the effective utilization of cheese whey, transforming this industrial by-product into high-value products such as whey protein concentrates and isolates (WPC and WPI), as well as purified lactose. The versatility of combined separation processes enables the selective recovery of components. However, challenges such as membrane fouling and concentration polarization continue to require innovative solutions to ensure efficiency and sustainability.

Current trends in this field focus on the research and development of hybrid and surface-modified membranes to enhance resistance, selectivity, and reduce propensity to fouling. Additionally, the integration of enzymatic pre-treatments has shown potential to improve filtration efficiency and the quality of the final product. The application of artificial intelligence and computational modeling, including neural networks and fluid dynamics simulations, represents a promising approach to optimize operational parameters, predict fouling formation, and reduce costs, although its industrial scalability still requires validation.

Future research directions in membrane-based separation will likely focus on developing more sustainable processes, such as water recovery and decreasing energy consumption, to further drive innovation. The combination of durable ceramic membranes with advanced real-time monitoring techniques presents a promising strategy for improving operational efficiency. In this context, membrane filtration not only significantly contributes to enhancing the value of cheese whey but also establishes itself as a sustainable and economically viable production strategy, aligning with global sustainability demands.

References

Aghajani, M., Rahimpour, A., Amani, H., Taherzadeh, M.J., 2018. Rhamnolipid as new bio-agent for cleaning of ultrafiltration membrane fouled by whey. Eng. Life Sci. 18 (5), 272–280. https://doi.org/10.1002/elsc.201700070.

Alfano, A., D'ambrosio, S., Cimini, D., Falco, L., D'Agostino, M., Finamore, R., Schiraldi, C., 2022. No waste from waste: membrane-based fractionation of second cheese whey for potential nutraceutical and cosmeceutical applications, and as renewable substrate for fermentation processes development. Fermentation 8 (10), 514. https://doi.org/10.3390/fermentation8100514.

Al-Mutwalli, S.A., Mehmet, D., Koseoglu-Imer, D.Y., 2020. Performance evaluation of ceramic membrane on ultrafiltration and diafiltration modes for efficient recovery of whey protein. J. Membr. Sci. Res. 6 (2), 138–146. https://doi.org/10.22079/jmsr.2019.115152.1295.

Al-Mutwalli, S.A., Dilaver, M., Uru, S.K., Koseoglu-Imer, D.Y., Lipnizki, F., 2024. β-galactosidase immobilization on ceramic ultrafiltration membrane for simultaneous lactose hydrolysis and protein separation. J. Water Process Eng. 64, 105619. https://doi.org/10.1016/j.jwpe.2024.105619.

Arsic, S., Vukovic, P., 2021. Analysis of profitability of integrated production protein concentrate, lactose powder whey. Manag. Econom. Eng. Agric. Rural Dev. 21 (1), 65–69. Available from: http://managementjournal.usamv.ro/pdf.

Aslam, M., Wicaksana, F., Farid, M., Wong, A., Krantz, W.B., 2022. Mitigation of membrane fouling by whey protein via water hammer. J. Membr. Sci. 642, 119967. https://doi.org/10.1016/j.memsci.2021.119967.

Aydiner, C., Sen, U., Topcu, S., Ekinci, D., Altinay, A.D., Koseoglu-Imer, D.Y., Keskinler, B., 2014. Techno-economic viability of innovative membrane systems in water and mass recovery from dairy wastewater. J. Membr. Sci. 458, 66–75. https://doi.org/10.1016/j.memsci.2014.01.058.

Bejarano-Toro, E., Sepúlveda-Valencia, J.U., Sandoval, E.R., 2022. Use of ultrafiltration technology to concentrate whey proteins after white cheese manufacturing. Rev. Fac. Nac. Agron. 75 (2), 9961–9969. https://doi.org/10.15446/rfnam.v75n2.98600.

Blais, H.N., Schroën, K., Tobin, J.T., 2022. A review of multistage membrane filtration approaches for enhanced efficiency during concentration and fractionation of milk and whey. Int. J. Dairy Technol. 75 (4), 749–760. https://doi.org/10.1111/1471-0307.12884.

Bortot Coelho, F.E., Magnacca, G., Boffa, V., Candelario, V.M., Luiten-Olieman, M., Zhang, W., 2023. From ultra to nanofiltration: a review on the fabrication of ZrO_2 membranes. Ceram. Int. 49 (6), 8683–8708. https://doi.org/10.1016/j.ceramint.2023.01.076.

Brião, V.B., Mossmann, J., Seguenka, B., Graciola, S., Piccin, J.S., 2024. Integrating whey processing: ultrafiltration, nanofiltration, and water reuse from diafiltration. Membranes 14 (9), 191. https://doi.org/10.3390/membranes14090191.

Brião, V.B., Seguenka, B., Zanon, C.D., Milani, A., 2017. Cake formation and the decreased performance of whey ultrafiltration. Acta Sci. Technol. 39 (5), 517. https://doi.org/10.4025/actascitechnol.v39i5.27585.

Cabral, S.R., Brito, B., Pereira, M., Figueiredo, A.S., Martins, A.L., Pinho, M., 2019. Optimization of cheese whey ultrafltration/diafltration for the production of beverage liquid protein concentrates with lactose partially removed. J. Membr. Sci. Res. 5 (2), 172–177. https://doi.org/10.22079/jmsr.2018.92367.1208.

Casado-Coterillo, C., Díaz-Guridi, P., Otero, J.A., Ibáñez, R., 2023. Modeling of lactic acid rejection from lactose in acidified cheese whey by nanofiltration. J. Dairy Sci. 106 (7), 4533–4544. https://doi.org/10.3168/jds.2022-22502.

Castro-Muñoz, R., Fíla, V., 2018. Membrane-based technologies as an emerging tool for separating high-added-value compounds from natural products. Trends Food Sci. Technol. 82, 8–20. https://doi.org/10.1016/j.tifs.2018.09.017.

Chen, G.Q., Qu, Y., Gras, S.L., Kentish, S.E., 2023. Separation technologies for whey protein fractionation. Food Eng. Rev. 15 (3), 438–465. https://doi.org/10.1007/s12393-022-09330-2.

Chen, G.Q., Artemi, A., Lee, J., Gras, S.L., Kentish, S.E., 2019. A pilot scale study on the concentration of milk and whey by forward osmosis. Separ. Purif. Technol. 215, 652–659. https://doi.org/10.1016/j.seppur.2019.01.050.

Chen, Y., Callanan, M., Shanahan, C., Tobin, J., Gamon, L.F., Davies, M.J., Giblin, L., Brodkorb, A., 2021. The use of membrane filtration to increase native whey proteins in infant formula. Dairy 2 (4), 515–529. https://doi.org/10.3390/dairy204004.

Corbatón-Báguena, M.-J., Álvarez-Blanco, S., Vincent-Vela, M.-C., 2017. Ultrafiltration of whey: membrane performance and modelling using a combined pore blocking-cake formation model. J. Chem. Technol. Biotechnol. 93 (7), 1891–1900. https://doi.org/10.1002/jctb.5446.

Cunha, T.M.P., Canella, M.H.M., Haas, I.C.S., Amboni, R.D.M.C., Prudencio, E.S., 2022. A theoretical approach to dairy products from membrane processes. Food Sci. Technol. 42, e12522. https://doi.org/10.1590/fst.12522.

Damar, I., Cinar, K., Gulec, H.A., 2020. Concentration of whey proteins by ultrafiltration: comparative evaluation of process effectiveness based on physicochemical properties of membranes. Int. Dairy J. 111, 104823. https://doi.org/10.1016/j.idairyj.2020.104823.

Dansawad, P., Li, Y., Li, Y., Zhang, J., You, S., Li, W., Yi, S., 2023. Machine learning toward improving the performance of membrane-based wastewater treatment: a review. Adv. Membr. 3, 100072. https://doi.org/10.1016/j.advmem.2023.100072.

Gaudio, M.T., Curcio, S., Chakraborty, S., 2023a. Design of an integrated membrane system to produce dairy by-product from waste processing. Int. J. Food Sci. Technol. 58 (4), 2104–2114. https://doi.org/10.1111/ijfs.15986.

Gaudio, M.T., Curcio, S., Chakraborty, S., Calabrò, V., 2023b. Application of artificial neural networks for modelling and control of flux decline in cross-flow whey ultrafiltration. Processes 11 (4), 1287. https://doi.org/10.3390/pr11041287.

Haribabu, M., Dunstan, D.E., Martin, G.J., Davidson, M.R., Harvie, D.J., 2020. Simulating the ultrafiltration of whey proteins isolate using a mixture model. J. Membr. Sci. 613, 118388. https://doi.org/10.1016/j.memsci.2020.118388.

Hidayat, I., Paredes, L., Binder, P.M., Guerra-Gorostegi, N., Mora, M., Ponsá, S., Oatley-Radcliffe, D.L., Llenas, L., 2023. A novel hybrid membrane process coupled with freeze concentration for phosphorus recovery from cheese whey. Membranes 13 (4), 450. https://doi.org/10.3390/membranes13040450.

Hinkova, A., Bubnik, Z., Henke, S., Pour, V., Zidova, P., Sarka, E., Hassan, N., Kadlec, P., 2016. Cheese whey tangential filtration using tubular mineral membranes. Chem. Pap. 70 (3), 325–332. https://doi.org/10.1515/chempap-2015-0191.

Hofmann, K., Hamel, C., 2023. Screening and scale-up of nanofiltration membranes for concentration of lactose and real whey permeate. Membranes 13 (2), 173. https://doi.org/10.3390/membranes13020173.

Ignacz, G., Bader, L., Beke, A.K., Ghunaim, Y., Shastry, T., Vovusha, H., Carbone, M.R., Ghanem, B., Szekely, G., 2024. Machine learning for the advancement of membrane science and technology: a critical review. J. Membr. Sci. 713, 123256. https://doi.org/10.1016/j.memsci.2024.123256.

Jawad, J., Hawari, A.H., Zaidi, S.J., 2021. Artificial neural network modeling of wastewater treatment and desalination using membrane processes: a review. Chem. Eng. J. 419, 129540. https://doi.org/10.1016/j.cej.2021.129540.

Kelly, P., 2019. Manufacture of whey protein products: concentrates, isolate, whey protein fractions and microparticulated. In: Whey Proteins: From Milk to Medicine. Elsevier Inc., pp. 97–122. https://doi.org/10.1016/b978-0-12-812124-5.00003-5

Khan, I.T., Nadeem, M., Imran, M., Ullah, R., Ajmal, M., Jaspal, M.H., 2019. Antioxidant properties of milk and dairy products: a comprehensive review of the current knowledge. Lipids Health Dis. 18 (1). https://doi.org/10.1186/s12944-019-0969-8.

Krantz, W.B., Filicia, W., Aslam, M., Wong, A., Farid, M., 2022. Development and validation of a model for mitigating particulate fouling in ultrafiltration using water-hammer. J. Membr. Sci. 666, 121098. https://doi.org/10.1016/j.memsci.2022.121098.

Kravtsov, V.A., Kulikova, I., Anisimov, G.S., ЕвдокиМов, И.А., ХраМцов, А.Г., 2021. Variety of dairy ultrafiltration permeates and their purification in lactose production. IOP Conf. Ser. 677 (3), 032001. https://doi.org/10.1088/1755-1315/677/3/032001.

Krolitzki, E., Ostertag, F., Schwaminger, S.P., Hinrichs, J., Berensmeier, S., 2024. Purification of lactoferrin through a sequential filtration and magnetic separation process: a processing example for acid whey valorization. Future Foods 9, 100342. https://doi.org/10.1016/j.fufo.2024.100342 pp.

Krunic, T., Rakin, M., 2025. Combination of controlled enzymatic hydrolysis and ultrafiltration as the technique for bioactive peptide production intended for the food industry. Innov. Food Sci. Emerg. Technol., 103979 https://doi.org/10.1016/j.ifset.2025.103979.

Kürzl, C., Kulozik, U., 2023. Comparison of the efficiency of pulsed flow membrane cleaning in hollow fibre (HFM) and spiral-wound microfiltration membranes (SWM). Food Bioprod. Process. 139, 166–177. https://doi.org/10.1016/j.fbp.2023.03.012.

Levitsky, I., Tavor, D., Gitis, V., 2021. Microbubbles and organic fouling in flat sheet ultrafiltration membranes. Separ. Purif. Technol. 268, 118710. https://doi.org/10.1016/j.seppur.2021.118710.

Li, J., Wang, K., Liu, F., Ji, M., Ge, Z., Xie, M., 2025. Charge-enhanced sulfonated block copolymer composite membrane for whey recovery. Separ. Purif. Technol., 132081 https://doi.org/10.1016/j.seppur.2025.132081.

Macedo, A., Azedo, D., Duarte, E., Pereira, C., 2021a. Valorization of goat cheese whey through an integrated process of ultrafiltration and nanofiltration. Membranes 11 (7), 477. https://doi.org/10.3390/membranes11070477.

Macedo, A., Bilau, J., Cambóias, E., Duarte, E., 2021b. Integration of membrane processes for by-product valorization to improve the eco-efficiency of small/medium size cheese dairy plants. Foods 10 (8), 1740. https://doi.org/10.3390/foods10081740.

Mamay, D., Babenyshev, S., Mamay, A., 2023. Membrane process for the extraction of casein and whey proteins from skim milk. E3S Web Conf. 462, 01020. https://doi.org/10.1051/e3sconf/202346201020.

Ostertag, F., Krolitzki, E., Berensmeier, S., Hinrichs, J., 2023. Protein valorisation from acid whey—screening of various micro- and ultrafiltration membranes concerning the filtration performance. Int. Dairy J. 146, 105745. https://doi.org/10.1016/j.idairyj.2023.105745.

Pires, A., Bozek, A., Pietruszka, H., Szkolnicka, K., Gomes, D., Díaz, O., Cobos, A., Pereira, C., 2024. Whey cheeses containing probiotic and bioprotective cultures produced with ultrafiltrated cow's whey. Foods 13 (8), 1214. https://doi.org/10.3390/foods13081214.

Poonia, A., Petkoska, A.T. (Eds.), 2023. Whey Valorization: Innovations, Technological Advancements and Sustainable Exploitation. Springer.

Prabhuzantye, T., Khaire, R.A., Gogate, P.R., 2019. Enhancing the recovery of whey proteins based on application of ultrasound in ultrafiltration and spray drying. Ultrason. Sonochem. 55, 125–134. https://doi.org/10.1016/j.ultsonch.2019.03.008.

Rama, G.R., Timmers, L.F.S.M., Volken de Souza, C.F., 2021. Ultrafiltration of cheese whey: achieving high protein rejection and sustaining membrane efficiency. J. Food Process. Preserv. 45 (11), e15908. https://doi.org/10.1111/jfpp.15908.

Rastogi, Y., Priya, Gogate, P.R., 2022. Intensified recovery of whey proteins using combination of enzyme in free or immobilized form with ultrafiltration. Chem. Eng. Process. Process Intensif. 179, 109076. https://doi.org/10.1016/j.cep.2022.109076.

Reig, M., Vecino, X., Cortina, J.L., 2021. Use of membrane technologies in dairy industry: an overview. Foods 10 (11), 2768. https://doi.org/10.3390/foods10112768.

Rezapour, M., Esmaiili, M., Mahmoudian, M., Sarand, A.B., 2025. Design and modeling of a nanocomposite system for demineralization of sweet whey. Sci. Rep. 15 (1), 4824. https://doi.org/10.1038/s41598-025-89009-w.

Rodrigues, M.H.P., Gräff, C.A., Tupuna-Yerovi, D.S., Schmitz, C., Camargo, J., Timmers, L.F.S., Lehn, D.N., Volken de Souza, C.F., 2024. The bioactive potential of cheese whey peptides from different animal origins (bovine, goat, sheep, buffalo, and camel): a systematic review and meta-analysis. Food Res. Int. 115053. https://doi.org/10.1016/j.foodres.2024.115053.

Roselli, M., Onesti, R., Boi, C., Bandini, S., 2024. Recovery of lactose from acid whey by nanofiltration: an experimental study. Separ. Purif. Technol. 353, 128303. https://doi.org/10.1016/j.seppur.2024.128303.

Rosseto, M., Rigueto, C.V.T., Gomes, K.S., Krein, D.D.C., Loss, R.A., Dettmer, A., Richards, N.S.P.S., 2024. Whey filtration: a review of products, application, and pretreatment with transglutaminase enzyme. J. Sci. Food Agric. 104 (6), 3185–3196. https://doi.org/10.1002/jsfa.13248.

Rosseto, M., Rigueto, C.V.T., Gomes, K.S., Menezes, J.F., Bucior, D., Dettmer, A., dos Santos Richards, N.S.P., 2025. Process optimization and characteristics of enzymatically cross-linked and ultrafiltrated whey. Int. Dairy J. 160, 106076. https://doi.org/10.1016/j.idairyj.2024.106076.

Saadi, S., Makhlouf, C., Nacer, N.E., Halima, B., Faiza, A., Kahina, H., Wahiba, F., Afaf, K., Rabah, K., Saoudi, Z., 2024. Whey proteins as multifunctional food materials: recent advancements in hydrolysis, separation, and peptidomimetic approaches. Compr. Rev. Food Sci. Food Saf. 23 (1). https://doi.org/10.1111/1541-4337.13288.

Saltık, M.B., Özkan, L., Jacobs, M., van der Padt, A., 2017. Dynamic modeling of ultrafiltration membranes for whey separation processes. Comput. Chem. Eng. 99, 280–295. https://doi.org/10.1016/j.compchemeng.2017.01.035.

Samree, K., Srithai, P., Kotchaplai, P., Thuptimdang, P., Painmanakul, P., Hunsom, M., Sairiam, S., 2020. Enhancing the antibacterial properties of PVDF membrane by hydrophilic surface modification using titanium dioxide and silver nanoparticles. Membranes 10 (10), 289. https://doi.org/10.3390/membranes10100289.

Sánchez-Moya, T., Hidalgo, A.M., Ros-Berruezo, G., López-Nicolás, R., 2020. Screening ultrafiltration membranes to separate lactose and protein from sheep whey: application of simplified model. J. Food Sci. Technol. 57 (9), 3193–3200. https://doi.org/10.1007/s13197-020-04350-4.

Schopf, R., Schmidt, F., Linner, J., Kulozik, U., 2021. Comparative assessment of tubular ceramic, Spiral wound, and hollow fiber membrane microfiltration module systems for milk protein fractionation. Foods 10, 692. https://doi.org/10.3390/foods10040692.

Sluková, M., Hinková, A., Henke, S., Smrž, F., Lukačíková, M., Pour, V., Bubník, Z., 2016. Cheese whey treated by membrane separation as a valuable ingredient for barley sourdough preparation. J. Food Eng. 172, 38–47. https://doi.org/10.1016/j.jfoodeng.2015.03.023.

Tanudjaja, H.J., Ng, A.Q.Q., Chew, J.W., 2023. Understanding single-protein fouling in micro- and ultrafiltration systems via machine-learning-based models. Ind. Eng. Chem. Res. 62 (19), 7610–7621. https://doi.org/10.1021/acs.iecr.3c00275.

Tonolini, M., Van den Berg, F.W.J., Skou, P.B., Sørensen, K.M., Engelsen, S.B., 2023. Near-infrared spectroscopy as a process analytical technology tool for monitoring performance of membrane filtration in a whey protein fractionation process. J. Food Eng. 350, 111487. https://doi.org/10.1016/j.jfoodeng.2023.111487.

Tuo, J., Zha, M., Li, H., Xie, D., Wang, Y., Sheng, G.P., Wang, Y., 2025. Predictive modeling and insight into protein fouling in microfiltration and ultrafiltration through one-dimensional convolutional models. Separ. Purif. Technol. 352, 128237. https://doi.org/10.1016/j.seppur.2024.128237.

Villecco, F., Aquino, R.P., Calabrò, V., Corrente, M.I., d'Amore, M., Grasso, A., Naddeo, V., 2020. Fuzzy-assisted ultrafiltration of whey by-products recovery. Euro-Mediterr. J. Environ. Integr. 5 (1). https://doi.org/10.1007/s41207-019-0138-5.

Villeneuve, W., Bérubé, A., Chamberland, J., Pouliot, Y., Labrie, S., Doyen, A., 2021. Contribution of biofouling to permeation flux decline and membrane resistance changes during whey ultrafiltration. Int. Dairy J. 117, 105010. https://doi.org/10.1016/j.idairyj.2021.105010.

Wang, W., Zhou, J., Li, J., Cong-Cong, T., 2024. The mechanism of whey protein on membrane surface fouling during ultrafiltration process. Food Biophys. https://doi.org/10.1007/s11483-023-09811-9.

Wenqiong, W., Jianju, L., Jiyang, Z., Zhangwei, H., Congcong, T., 2025. The effect of ionic regulation on the structure of whey protein during ultrafiltration process and hydraulic reverse cleaning efficiency. Int. J. Biol. Macromol. 286, 138361. https://doi.org/10.1016/j.ijbiomac.2024.138361.

Wen-qiong, W., Yun-chao, W., Xiao-feng, Z., Rui-xia, G., Mao-lin, L., 2019. Whey protein membrane processing methods and membrane fouling mechanism analysis. Food Chem. 289, 468–481. https://doi.org/10.1016/j.foodchem.2019.03.086.

Xu, Z., Ma, G., Rana, D., Matsuura, T., Lan, C.Q., 2025. Engineering polyvinylidene fluoride membranes using charge tunable dendrimer polyamidoamine to enable microporous membranes for protein separation. Separ. Purif. Technol. 354, 128887. https://doi.org/10.1016/j.seppur.2024.128887.

Yogarathinam, L.T., Velswamy, K., Gangasalam, A., Ismail, A.F., Goh, P.S., Narayanan, A., Abdullah, M.S., 2022. Performance evaluation of whey flux in dead-end and cross-flow modes via convolutional neural networks. J. Environ. Manag. 301, 113872. https://doi.org/10.1016/j.jenvman.2021.113872.

Yogarathinam, L.T., Usman, J., Abba, S.I., Lawal, D., Baig, N., Aljundi, I.H., 2024. Evaluating nano-metal oxide mixed matrix membranes for whey protein separation using hybrid intelligent optimization learning. Chem. Eng. Res. Des. 205, 388–400. https://doi.org/10.1016/j.cherd.2024.04.009.

Chapter 26

Novel whey-based products

Lara Barnum, Siyu Liu and Alice Marciniak
Department of Food Science, University of Guelph, Guelph, ON, Canada

1. Introduction

The dairy industry produces significant quantities of by-products, which are often disposed of as waste, including whey from cheese making (Alane et al., 2017; Baba et al., 2016; Pires et al., 2021). Whey, the remaining components of milk after coagulation, is generated quickly, 9–10 L resulting from 1 kg of cheese being produced (Bandara et al., 2023; Rincón-Catalán et al., 2022). The disposal of whey if untreated poses a significant hazard to the environment due to its high biochemical and chemical oxygen demand, and when disposed of in aquatic environments, it will reduce the water oxygen content, potentially harming aquatic life and the ecosystem (Bandara et al., 2023; Hughes et al., 2019; Jan et al., 2021; Pires et al., 2021).

Now, the functional and bioactive components of whey are being recognized, especially the proteins and sugar present for their potential uses. After cheese making, whey still retains approximately 55% of the nutrients as well as roughly 20% of the protein content of the milk, however the exact composition of the whey will depend on the type of milk, type of cheese produced, and any other factors that normally affect the composition of milk will also impact the composition of the whey, such as seasonal cycles (Baba et al., 2016; Bandara et al., 2023; Pires et al., 2021). Whey can be categorized further into acid whey from acid coagulation, and sweet whey, from rennet coagulated cheese production. While their pH's vary, acid whey being more acidic, they both have similar amounts of fat, while sweet whey has a higher lactose, mineral, and protein content. Nonetheless the nutrients and organic compound composition of both provide opportunities for utilization and potential valorization of this waste stream. One of the most common uses for whey outside of disposal include dehydration in production of whey powders, whey protein concentrates, and whey protein isolates, which have applications in the food industry (Pires et al., 2021), as well as being desirable as nutritional supplement for supporting muscle growth and recovery (Cintineo et al., 2018).

Overall, the composition of whey provides additional opportunities for its use and valorization in novel ways in recent years. One such application is taking advantage of its nutritional value and protein content in the production foods for the treatment of conditions such as Moderate Acute Malnutrition (MAM) (Stobaugh, 2018), and diabetes (Zhou et al., 2021). Additionally, the presence of lactose, a fermentable sugar, invites the possibility of producing fermented products. Fermented foods and beverages have increased in popularity in recent years due to their health benefiting properties, which provides an excellent opportunity for the use of whey in fermented beverages (Sabokbar and Khodaiyan, 2015). Under the right conditions, the lactose in whey can also be fermented into ethanol, and this has been used as of late to produce alcoholic beverages and spirits, of which there are several commercially available (Hughes et al., 2019). There is also the potential for whey providing interesting alternatives to existing processes for chemical synthesis and even energy production, providing a more environmentally sustainable avenue for production.

2. Whey based functional foods

As discussed in previous chapters, whey has a high protein content and bioactivity, which make it a valuable ingredient in supplements and other functional foods (Cintineo et al., 2018). The protein is desirable as nutritional supplements for supporting muscle growth and recovery, which can lead to improved overall health. Evidence showing that consumption of protein before, immediately after, and post workouts can significantly increase muscle protein synthesis (Cintineo et al., 2018), and this may have contributed to the rise in popularity of whey protein-based workout powders such as those

Whey. https://doi.org/10.1016/B978-0-443-27648-4.00026-8

boasting whey protein concentrates and whey protein isolates. These however are not the only applications of whey in functional foods.

2.1 Treatment of moderate acute malnutrition

Whey-containing-supplements can be made to aide children in recovery from moderate acute malnutrition and enhance their growth. MAM presents through rapid weight loss, often caused by a lack of important nutrients in the diet, such as protein (Manary and Sandige, 2008). Affecting roughly 33 million children, MAM increases their risk of developing illnesses, disabilities and death compared with healthy children (Manary and Sandige, 2008; Stobaugh, 2018). The treatment for MAM involves nutritional supplementation with foods containing the necessary nutrients and are often flour blends comprising cereal and legumes or peanut-based Ready-to-Use Supplementary Foods (RUSF's) (Manary and Sandige, 2008; Stobaugh, 2018) (Fig. 26.1). However, recent studies have shown that whey proteins can be used in supplements to treat MAM in children with success (Stobaugh et al., 2016).

The structure of whey proteins, consisting of branched-chain amino acids, makes it an excellent candidate for treatment of MAM. Its metabolism conserved the need to be broken down, which allows for easier recovery of lean tissue mass—a critical part of recovery from malnutrition (Stobaugh, 2018). Stobaugh et al. (2016) conducted a double blind, randomized controlled clinical effectiveness trial with children suffering from MAM from southern Malawi. They tested the effectiveness of soy versus their novel whey RUSF's on improving mid-upper arm circumference, a metric for identifying malnourished children. Both versions of the RUSF's had similar features including packaging, smell, taste, and color as they used the same ingredients in similar quantities, aside from extruded soy flour being substituted for whey permeate and whey protein isolate in the novel whey version. The results showed that children who received the whey-based RUSF had a significantly higher recovery rate (83.9%) in comparison to the soy-based alternative (80.5%). Beyond recovery rate, better growth markers such as a higher mid-upper arm circumference was observed in the children receiving the whey RUSF (Stobaugh et al., 2016).

Incorporating animal-based proteins including both whey powder and egg white into RUSF's may improve the overall amino acid profile and value of the protein. Azimi et al. (2020) found that with the use of egg white and whey RUSF there was a significant increase in weight and height observed. 92% of children in the egg white and RUSF group recovered from MAM, compared with only 12% in the untreated control group. In addition to improved growth indicators, they also found improved clinical outcomes for diarrhea and fever in their RUSF treated group (Azimi et al., 2020).

Overall, these studies show the importance of the use of whey protein in malnutrition treatment as it provides opportunities for more biologically valuable protein supplementation, especially when compared with plant protein sources for the same purpose.

2.2 Antidiabetic ice pops

Studies have shown that whey proteins and their derived protein hydrolysates and peptides demonstrate antidiabetic effects, in tests on mice and rat models as well as clinical trials with human subjects (Zhou et al., 2021). Whey proteins

FIGURE 26.1 Delivery of peanut-based RUSFs to village in East Africa. *From "Unloading food aid from a WFP helicopter in Barada, Mozambique, March 26, 2019" by DFID—UK Department for International Development is licensed under CC BY 2.0.*

FIGURE 26.2 Jamun fruit. *From* Ripe jamun fruits *by Rajesh Dangi, CC BY 2.5 via Wikimedia Commons (Dangi, 2007).*

have a positive effect on the treatment of type 2 diabetes mellitus, which is the most common form of diabetes, accounting for over 90% of diabetes cases. This form of diabetes is characterized by insulin resistance or not producing enough insulin, which can lead to additional and more severe conditions. Whey protein consumption has been reported to reduce fasting plasma glucose and enhance insulin secretion among other beneficial effects (Zhou et al., 2021). Due to these characteristics, the incorporation of whey is an interesting method for producing functional foods to aid in diabetes management, partially due to being rich in whey proteins.

Jan et al. (2021) made use of whey as well as another antidiabetic active ingredient, juice of jamun fruit to create antidiabetic ice pops. Jamun is a seasonal, native to India, fruit with a variety of beneficial qualities, including antioxidant, antifungal, antibacterial, antiinflammatory, and most importantly for this research, antidiabetic (Fig. 26.2). Whey also provides additional desirable qualities to the ice pops as it is rich in essential nutrients. One challenge to the use of these ingredients is that both whey and jamun juice are highly perishable, being a dairy product and fresh fruit juice, respectively. Blending them together and forming an ice pop allows for extended shelf life and stability while retaining the essential properties of the ingredients. Jan et al. (2021) tested the antidiabetic effect of the ice pops on a rat model and found that the ice pops were able to block alpha-amylase activity, which reduces the speed sugars will be absorbed, and can aid in reducing glucose spikes, which is important for diabetics. Overall, they created a healthy ice pop with jamun juice and whey, which had the potential to aid those with diabetes.

3. Whey based beverages

The popularity of research in whey-based beverages may stem from the functionality of the whey in this form. Products can be made that are desirable in flavor as well as nutritional content by determining the appropriate combination of whey, flavors such as fruit juice and pulp, and in some cases through fermentation (Panghal et al., 2018). Whey- based beverages have been reported to be refreshing, thirst quenching, while being less acidic and healthier than other drinks (Panghal et al., 2018). Whey beverages come in many forms, fruit flavored, fermented, alcoholic and more.

3.1 Fermented whey beverages

Recently, public interest in fermented foods and beverages has significantly increased due to their potential health benefits (Sabokbar and Khodaiyan, 2015). One desirable characteristic of these foods is that they are probiotics, which contain microorganisms that provide a health benefit when consumed (Sabokbar and Khodaiyan, 2015). Probiotic beverages, and those specifically containing whey have increased attention due to their potential health benefits (M'Hir et al., 2019).

Whey makes an excellent base for functional beverages due to its high protein content and nutritional value, as well as providing lactose, an important sugar for fermentation by kefir grains or other cultures in production of fermented beverages (Sabokbar and Khodaiyan, 2015). However, whey alone has an unappealing taste, poor texture and mouthfeel, excessive acidity especially in acid whey, and low sweetness (Arsić et al., 2018). The quality, nutritional value, and sensory acceptability of whey-based beverages can be improved by the addition of probiotic microorganisms through kefir grains or lactic acid bacteria cultures, as well as other additives (Arsić et al., 2018; M'Hir et al., 2019). Additives can include fruits, vegetables, other dairy products such as milk, and even cereal grains such as oats (Bandara et al., 2023).

Before inoculation, the whey and other ingredients can be individually pasteurized then mixed, or mixed then pasteurized together to inactivate any naturally occurring microorganisms present (Li et al., 2018; Sabokbar et al., 2015). Ingredients can also be homogenized at this step if necessary (Li et al., 2018). Next the beverage can be inoculated with culture or kefir grains for fermentation.

Kefir is a dairy beverage that is naturally carbonated through fermentation by kefir grains. Kefir grains, small cauliflower-shaped granules in kefir beverages, are composed of sugars, proteins, and various microorganisms, such as lactic acid producing bacteria and yeast (Fig. 26.3) (M'hir et al., 2023; Sabokbar and Khodaiyan, 2015). These grains can be incorporated in various types of milk including cow, sheep, and goat milk, and the bacteria and yeast are responsible for the fermentation of the kefir beverage, producing acetic acid, carbon dioxide, and lactic acid (Li et al., 2018), and researchers often use the kefir grains to ferment whey-based beverages (Sabokbar and Khodaiyan, 2015). Some of the microorganisms in kefir are probiotic microorganisms that also aide in inhibition of both food spoilage and pathogenic microorganisms, which help in extending the products shelf life (M'hir et al., 2023).

Once inoculated, the beverage is incubated at a specific temperature for a specified time and allowed to ferment. During fermentation the microorganisms present from the culture or kefir grains release enzymes that break down the lactose present in the whey, as well as proteins and fat, which will produce alcohols, lactic acid, peptides, and volatile fatty acids, which are responsible for the aromas, flavors, shelf-life extension, and nutritional qualities of fermented whey (Bandara et al., 2023).

Fermentation is essential in producing fermented whey beverages as in this stage biochemical changes occur and will vary depending on the culture, especially between the choice of lactic acid bacteria and yeast (Bandara et al., 2023). The most common cultures used in production of fermented whey beverages include lactic acid bacteria, which have lactase enzyme (β-galactosidase) that break down lactose into glucose and galactose, its two monosaccharide components. The glucose is then fermented by homo-fermentative lactic acid bacteria to convert it first into pyruvate by the glycolytic pathway through a series of steps and associated enzymes, then finally into lactic acid by lactate dehydrogenase. Lactic acid can also be produced by hetero-fermentative microorganisms, and in this process, there are coproducts produced such as carbon dioxide, acetic acid, and/or ethanol (Bandara et al., 2023). These processes are shown below in Fig. 26.4.

During fermentation, proteolysis also occurs in which whey proteins can be partially hydrolyzed by protease and peptidase enzymes from the lactic acid bacteria. This also contributes to the aroma and flavor compounds that are produced during fermentation (Bandara et al., 2023).

Finally, the beverages are either frozen or simply kept at an appropriate temperature before analysis and sensory testing (Li et al., 2018; Sabokbar et al., 2015). The steps of the production process are shown below in Fig. 26.5.

The most common method of improving the sensory acceptability of fermented whey beverages is the addition of fruits, including pomegranate, apple, date, mulberry, mango, as well as vegetables such as carrot (Bandara et al., 2023). Saps such as palm sap and even cereal grains such as oats can be included in fermented whey beverages as well, for improved nutritional value and health benefits. Additionally, other dairy ingredients such as milk and milk ingredients can be included to improve texture and mimic existing fermented dairy beverages. In addition to the improved sensory qualities, these additions also add nutritional value, such as increased proteins and fats (Bandara et al., 2023).

FIGURE 26.3 Kefir grains with jar of kefir. *From* Kefir-jadro-upravene *by Svorad, CC BY-SA 4.0, via Wikimedia Commons (Svorad, 2018).*

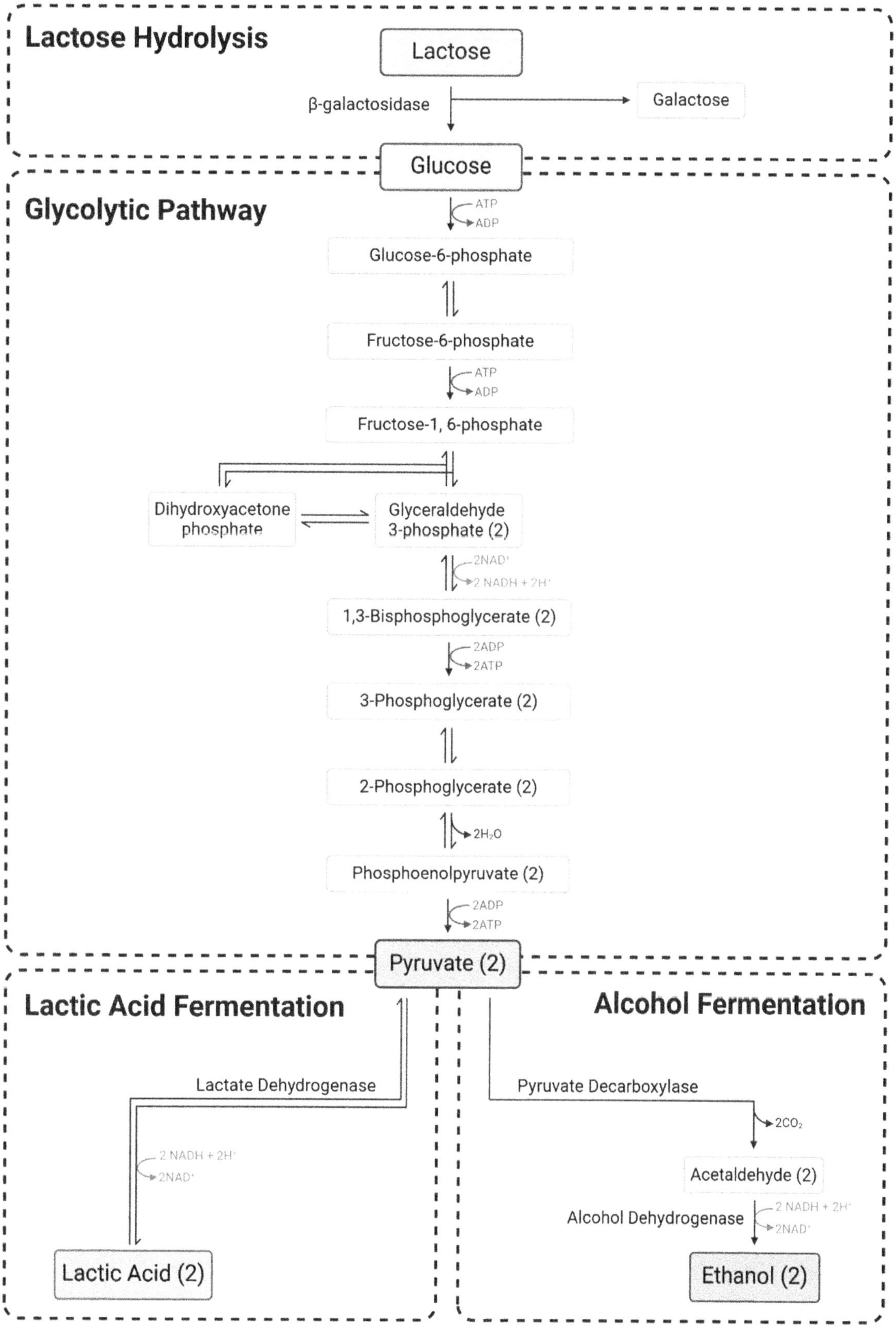

FIGURE 26.4 Pathways for bioconversion of lactose to lactic acid and ethanol. *Adapted from Bandara et al. (2023).*

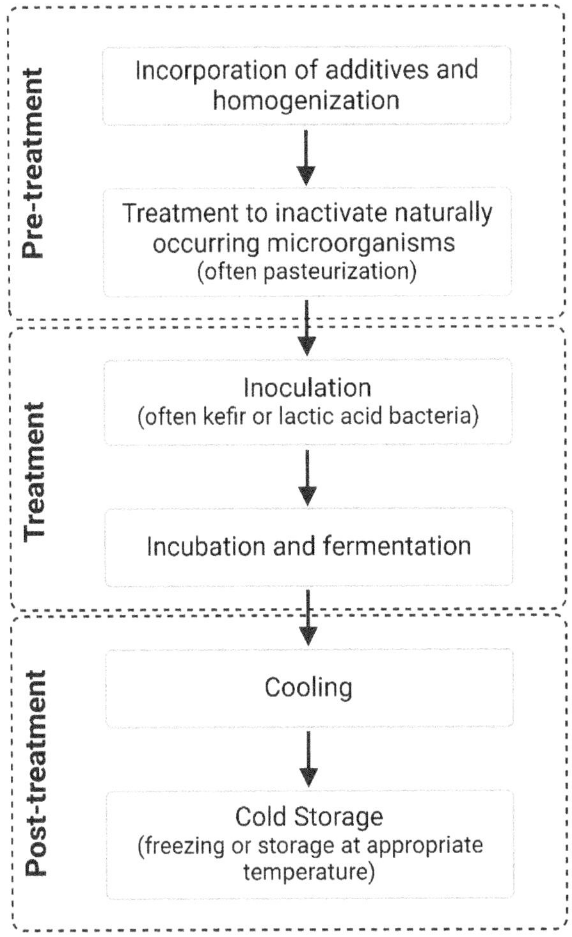

FIGURE 26.5 Process for production of fermented whey-based beverage. *Adapted from Bandara et al. (2023).*

In one study apple juice was combined with whey and inoculated with kefir grains at amounts between 2% and 8% (w/v) (Sabokbar et al., 2015). For the fruit additive, apples were selected due to their desirable flavor, aiding in improving the sensory qualities of the beverage, as well as their containing polyphenols, which are antioxidants (Sabokbar et al., 2015). Antioxidants can improve not only the shelf life of the fermented product, but also prevent off-flavors from developing, as well as having health benefits as they are responsible for a reduction of the oxidative damage on the human body (M'Hir et al., 2019). Fruit juices also improve the nutritional value of the beverages as they are rich in calcium and vitamins and are able to act as a medium for the growth of beneficial probiotic bacteria (Sabokbar et al., 2015).

Similar to apples, mulberries have prominent levels of phenolics, flavonoids, and also contain antioxidants, however, they have thin skin, which causes challenges for transportation and storage (Li et al., 2018). Due to this they are processed into other products with longer shelf lives, which includes fermented beverages. A study added mulberry juice to whey, also making use of kefir to produce a probiotic fermented beverage. They reported significant changes in the volatile flavor compounds present in the beverage after fermentation and determined the dominant volatile flavor compounds detected to be alcohols and ethyl esters, which are responsible for alcoholic and fruity aromas, respectively. Most compounds detected were similar in other fermented whey fruit juice-based beverages (Li et al., 2018). Other fruit juice additions to whey-based beverages fermented by kefir grains have been studied such as pomegranate juice, syrup made from dates, as well as mango juice and flavor, and a combination of palm sap and cow's milk (M'hir et al., 2019, 2023; Pereira et al., 2015; Sabokbar and Khodaiyan, 2015). These beverages were reported to have acceptable sensory properties, added nutritional value, and desirable antioxidant activities among other health benefiting properties.

While beverages can be fermented by addition of kefir, others make use of specific lactic acid bacteria strains or cultures containing multiple of these (AbdulAlim et al., 2018; Arsić et al., 2018; Sharma et al., 2017). As described before, lactic acid bacteria exist in kefir but can also be added directly to beverages for fermentation, or as parts of other cultures. Lactic acid bacteria are commonly added to dairy to produce various fermented products, and several lactic acid bacteria and yeast can be used to produce lactic beverages, which are an efficient mode for providing beneficial probiotic microorganisms to the human body (AbdulAlim et al., 2018). One study also made use of mulberry juice for a fermented whey-based beverage but instead of kefir, a starter culture was used, which contains strains of lactic acid bacteria. Increases in phenolic content, anthocyanin content, acidity and antioxidant activity were reported with increasing mulberry juice in the fermented beverage (AbdulAlim et al., 2018).

Other beverages have been found to also provide improved sensory qualities, nutritional value, and health benefits through fermentation by lactic acid bacteria in whey-based beverages containing carrot (Arsić et al., 2018). It has also been reported that the incorporation of high concentrations of juice additives can lead to a less viscous product, which may be due to a dilution of the proteins present in the beverage (Arsić et al., 2018). However, the addition of plant derived juices such as those from fruits and vegetables are still a promising avenue for making sensorily acceptable and nutritionally improved fermented whey beverages (Bandara et al., 2023).

Aside from fruits and vegetables, lactic acid bacteria have also been used to ferment whey-based beverages including cereal grains such as oats. Oats have a variety of health benefits such as being a source of beta-glucan, which is a prebiotic and can be related to reducing risk of cancer and cardiovascular disease (Sharma et al., 2017). They reported the main descriptions of the resulting beverage from sensory analysis were sweetness, acidity, and an unpleasant pungent flavor, which was due to the oats fermenting. Sharma et al. (2017) found that the pungent odor was responsible for a decrease in acceptability and determined mixing 1:1 with mango pulp improved acceptability.

Unlike the addition of fruit and vegetable juices, incorporating other dairy products such as milk can enrich the casein content of the fermented whey beverages, making their characteristics more like existing fermented milk drinks (Skryplonek, 2018). This addition can also increase the protein and fat content of the beverage, which creates a thicker consistency, and richer, creamier mouthfeel respectively (Bandara et al., 2023).

In a study, Janiaski et al. (2016) tested a variety of yogurt and whey-based beverages. The best performing whey beverages were fermented and contained additional milk products, using low-fat or full fat milk, as well as a fruit preparation of strawberries, among other ingredients, and lactic acid bacteria culture (Janiaski et al., 2016). In a sensory trial, it was reported that the fermented whey beverages were in the top four most liked beverages tested, as likable as two of the low-fat yogurt beverages tested in comparison and were preferred significantly over the nonfermented whey beverages. This shows potential for whey-based fermented beverages, as yogurt is considered to be higher in quality over whey, and is therefore more expensive, so a whey-based replacement could reduce cost of producing comparable products (Janiaski et al., 2016). Other studies have found similar improvements in fermented whey-based beverages through the addition of dairy products such as UHT milk, condensed milk, or skimmed milk powder, and buttermilk (Skryplonek, 2018; Skryplonek and Jasińska, 2015).

One example of a commercially available fermented whey beverage is produced by A. Vogel, a health food producer founded in Switzerland (https://shop.avogel.ca/). "Molkosan BERRY" is produced from whey from organic Swiss milk, which is fermented with lactic acid bacteria and flavored with pomegranate.

3.2 Whey based alcoholic beverages

Another type of fermentation that can be achieved in whey is alcohol fermentation. As described earlier due to the composition of whey it is a good material for fermentation by lactic acid producing microorganisms, but this is also good for alcohol fermenters, which allows for the production of whey-based alcoholic beverages. The bioconversion would be following the hetero-fermentative pathway as shown earlier in Fig. 26.4.

Ethanol can be produced from whey and whey permeate by the industrial method known as the Carbery method (Hughes et al., 2019). The first step involves receiving the whey or whey permeate, which should be cooled before inoculation. Unlike other industrial spirit production methods, in the Carbery method, sugar conversion steps such as mashing or cooking are not necessary since the whey is readily fermentable (Hughes et al., 2019). Similar to other industrial methods to produce ethanol, it makes use of microorganisms to ferment the sugars into ethanol. *Saccharomyces cerevisiae*, a yeast often used for ethanol production, is unable to directly ferment lactose from the whey. Other species such as *Kluyveromyces marxianus* and *Kluyveromyces lactis* can directly break down and ferment lactose (Hughes et al., 2019; Macwan et al., 2016).

The level of lactose present in whey can only produce a beverage containing 2%–3% v/v ethanol, an appropriate alcohol content for beer (Hughes et al., 2019). Higher amounts of lactose could be used through whey permeate, which is concentrated, however the concentration of ethanol achievable is limited by the sensitivity of the non-*Saccharomyces cerevisiae* yeast species to ethanol and solute concentrations (Hughes et al., 2019). Higher ethanol concentrations from fermentation, such as in wine, are achievable through the use of *Saccharomyces cerevisiae* in combination with a lactic acid bacteria to hydrolyze the lactose into glucose and galactose (Yang et al., 2024). In producing higher alcohol content distilled spirits, fermentation would be followed by extractive distillation to isolate and concentrate the alcohol, while separating out other volatile compounds present (Yang et al., 2024). The resulting material would then be treated like other distilled spirits and is processed further. For all resulting products filtration and/or flavor additions can be made followed by packaging and distribution (Hughes et al., 2019; Macwan et al., 2016).

Some methods to aid in production of whey beverages with a low alcohol content can include concentration, deproteinization, fermentation of the lactose present by microorganisms, and/or addition of sucrose (Macwan et al., 2016). Whey concentration would provide a higher concentration of lactose to be fermented, which would result in a higher alcohol concentration after fermentation (Bandara et al., 2023). Addition of sucrose would have a similar effect in providing more fermentable sugars for the yeast. Fermenting the lactose present would allow for the use of microorganisms that cannot usually ferment lactose itself such as *Saccharomyces cerevisiae*, which may increase the achievable alcohol content (Bandara et al., 2023; Macwan et al., 2016; Yang et al., 2024). The alcoholic beverages that can be produced from whey can be divided into low-alcohol, whey beer, whey wine, and whey spirits.

3.2.1 Low-alcohol

For low-alcohol whey beverages, containing less than 1.5% w/w alcohol, the production process includes whey deproteinization, concentration, fermentation by yeast or addition of sucrose to reach necessary alcohol content, addition of flavorings and sweeteners, and finally bottling (Macwan et al., 2016). Not all of the lactose present is fermented to alcohol, some is fermented to lactic acid as well, giving these beverages a refreshing sour flavor (Macwan et al., 2016). One example of this type of beverage is Milone, which contains less that 1% alcohol and is produced following standard brewing techniques (Holsinger et al., 1974). First the whey is inoculated with kefir culture and fermented to achieve specific levels of lactic acid and lactose before extracts were added and whey proteins precipitated out. Following filtration, the whey is then inoculated with lactose fermenting yeast, fermented again, and sweetened before bottling (Holsinger et al., 1974).

3.2.2 Beer

Pasta et al. (2024) made use of ricotta whey to replace some water and sugar to produce two types of beer, which underwent sensory analysis. The whey from ricotta is essentially deproteinized whey, which contains residual lactose and salt. Making use if the whey they produced a sweet milk stout making use of pilsner malt, brown malt, chocolate malt, water, ricotta whey, and cascade hops as well as *Saccharomyces cerevisiae* for fermentation. The resulting milk stout had a 4.0% v/v alcohol content. The second beer produced was a Gose-style beer made from pilsner malt, water, ricotta whey, lactic acid, Saphir hops and *Saccharomyces cerevisiae* for fermentation as well as *Lachancea* spp. strain WildBrew Philly Sour™. The final Gose-style beer had an alcohol content of 4.64% v/v. Both beers produced were carbonated and bottled. Sensory analysis by a trained panel determined that the beers exhibited qualities expected of the commercial beers of their respective styles (Pasta et al., 2024).

Beer made from whey can be produced with or without malt (Macwan et al., 2016). Malt in typical beer production is usually made by drying the germinated barley then heating to 80–85°C (Lawton and Alcaine, 2019). The malt is the source of starches broken down by α-amylase to get glucose for fermentation by yeast, however whey already contains fermentable sugar from lactose, so malt is not essential (Lawton and Alcaine, 2019).

Lawton and Alcaine (2019) made use of enzymes to hydrolyze the lactose to simpler sugars available to *Saccharomyces cerevisiae* for alcohol fermentation. They were made use of acid whey and barley meal to produce their beverages, one that was mashed for 8 h and produced an alcohol concentration of 3.23% v/v and another having a concentration of 3.37% v/v without mashing. This also shows that a mashing step is not required in producing beer with acid whey, which simplifies the production process when it can be made directly from acid whey without sugar being added for fermentation (Lawton and Alcaine, 2019).

The research by Pasta et al. (2024) and Lawton and Alcaine (2019) show the opportunities for making use of ricotta whey and acid whey from strained yogurt production, which may otherwise be waste, as the base for fermentation in the production of novel beers. However, there are some challenges in the production of beers fermenting whey, such as a loss of beer foam, as well as undesirable flavors and odors due to whey proteins, which indicate the need for further research in this area (Macwan et al., 2016).

One commercially available whey-based beer is a whey stout, produced by Bruny Island Cheese Co in Australia (https://brunyislandcheese.com.au/). Cow's milk whey from cheese making is used to enrich the beer, however a step to hydrolyze the lactose is not included and standard brewers' yeast is used, so the lactose remains in the beer giving it an enhanced sweetness and texture.

3.2.3 Wine

Wine produced from whey often makes use of fruit aromas for flavoring and has a relatively low alcohol content of generally 10–11% v/v (Macwan et al., 2016). The steps involved in producing wine from whey are similar to those for other whey alcohol production and incorporates steps from traditional wine making. The steps for whey-wine making include clearing, deproteinization, lactose hydrolysis, cooling, fermentation, decanting, aging, filtering, and bottling (Macwan et al., 2016).

Yang et al. (2024) produced a whey wine using jujube, mulberries, raisins, wolfberries, sugar, and whey. For their yeast they used *Saccharomyces cerevisiae* with edible lactic acid bacteria to hydrolyze the lactose from the whey for fermentation by the yeast. They found an increase in total acid, alcohol content, polyphenol, flavonoids, and antioxidant activity during fermentation of the whey wine. Their research also observed the changes in flavor compounds and microbial metabolic network due to fermentation, which provides new insights into the use of whey in wine (Yang et al., 2024).

3.2.4 Spirits

The first step of producing whey spirit is acquiring whey with consistent and appropriate composition to provide desirable aroma compounds (Hughes et al., 2019). Next the whey is pretreated to remove residual casein, whey cream, whey proteins, and other compounds that may interfere with downstream processing. As previously mentioned, the Carbery method is used to convert the lactose present in whey into ethanol. Like standard industrial ethanol production methods, microbial fermentation is used to produce ethanol from the sugars present. As previously stated, lactose is not readily fermentable by regular brewer's yeast, *Saccharomyces cerevisiae,* so it must be hydrolyzed before fermentation. Alternative yeast can be used such as *Kluyveromyces marxianus*, which produces β-galactosidase enzyme, which hydrolyzes lactose into glucose and galactose, which the yeast will subsequently convert to ethanol. Alternatively, lactase, a β-galactosidase enzyme, can be added directly to the whey without microorganisms to hydrolyze the lactose. This would then leave the glucose and galactose free to be fermented by *Saccharomyces cerevisiae*. Fermentation is followed by extractive distillation, which concentrates and separates the ethanol and, unlike other spirit production, a demethylizer is not required as very little methanol is produced during fermentation. After distillation, the whey spirit is processed downstream such as other distilled spirits in filtration, dilution, flavoring, packaging, and distributing the spirit (Hughes et al., 2019) (Fig. 26.6).

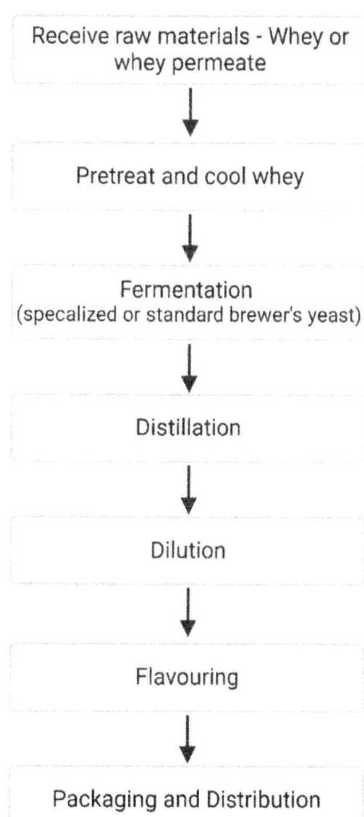

FIGURE 26.6 Carbery method for production of whey-based spirits. *Adapted from Hughes et al. (2019).*

Risner et al. (2018) produced a distilled whey spirit and compared the environmental impact with white whiskey. Their goal was to do a process-based life cycle analysis comparing water usage and carbon dioxide emissions in the production processes. White whiskey is an appropriate spirit for comparison because both are clear, unaged spirits. In the white whiskey the malted barley is the fermentable material, compared with whey in the distilled whey spirit. Distilled whey spirits and white whiskey were selected in this study as distilled whey spirits can act as a clear spirit substitute with a reduced environmental impact. The environmental impact can be reduced both through the removal of whey from waste streams and preventing landfill disposal of whey. White whiskey was fermented with *Saccharomyces cerevisiae* while the distilled whey spirit used *Kluyveromyces marxianus*. Overall, they found that there was a decrease in CO_2 emissions and less water required to produce the distilled whey spirit as compared with the white whiskey. Since the production methods for distilled whey spirits and whey whiskey are similar, it would be feasible for a white whiskey producing craft distillery to easily change to produce distilled whey spirits without extra costs (Risner et al., 2018).

Risner et al. (2019) compared the composition of volatile aroma compounds of the distillates of whey spirits produced from acid whey and sweet whey. As discussed in previous chapters, whey is generally divided into sweet whey, generated alongside production of rennet cheeses such as cheddar, and acid whey made during production of acid-precipitated cheeses such as cottage cheese. With the aim of determining if both types of whey can be fermented and distilled similarly, Risner et al. (2019) determined the volatile aroma compounds of each distillate and any differences therein. For fermentation of both, Kluyveromyces *marxianus* was used and the process of production is similar to the previously described process. Ultimately the volatile aromatic compounds of sweet whey and acid whey distillates were found to be unique with significant differences that may impact the sensory properties of the spirits (Risner et al., 2019). This work shows the importance of the whey source on producing distilled whey spirits.

There are a number of commercially available distilled whey spirits on the market:

- Sheep Whey Vodka and Sheep Whey Gin are produced in Australia by Hartshorn distillery (https://hartshorndistillery.com.au/) using whey from cheese produced by Grandvewe Sheep Cheese, their sheep farm. Their Sheep Whey Vodka won World's Best Vodka from World Vodka Awards (2018).
- Bertha's Revenge is a Gin produced by Ballyvolane House Spirits Company in Ireland (https://ballyvolanespirits.ie/). They make use of cow's milk whey from Irish dairy farmers and yeast capable of fermenting lactose to alcohol. They have also produced a variety of flavors of spirits with Bertha's Revenge Gin as the base.
- Black Cow vodka is made in the United Kingdom from cheese whey from a family farm (https://www.blackcow.co.uk/). The milk from the farm is used to produce cheese, the whey, which otherwise would have gone to waste, is not being used by their secret distilling process to make a smooth pure milk vodka.
- Wheyward Spirit is an 80 proof/40% alcohol whey spirit produce in the United States (https://www.wheywardspirit.com/). They produce both a clear distilled whey spirit as well as "Wheyskey," which is barrel aged in recoopered oak barrels. This allows the spirit to take on new flavors (Fig. 26.7).

FIGURE 26.7 Some whey-based spirits on the market. From left to right: Bertha's revenge gin, black cow vodka, sheep whey vodka. *Adapted from "Commercially available whey-based spirits: Bertha's revenge gin, black cow vodka, Vermont white vodka, and sheep whey vodka" (Hughes et al., 2019).*

There are also whey spirits produced in Canada:

- Yongehurst distillery in Canada produces a milk whey vodka (https://yongehurst.com/). Using a specific strain of yeast and whey from a dairy they ferment and distill it to produce a unique vodka.
- In Quebec, Canada Famille Migneron de Charlevoix produces Charlevoyou Petit-Lait whey spirits and whey gin (https://famillemigneron.com/). Their whey spirits are produced to use the whey from their cheese making.

3.3 Nonfermented whey beverages

Whey is also a valuable ingredient in nonfermented beverages. Due to its components, whey has been shown to aide in treatment of varying chronic conditions, such as diarrhea and bile illness making it an advantageous addition to a beverage (Baba et al., 2016). To understand how whey contributes to nonfermented beverages, one study tested different concentrations of whey from cheese and paneer to determine the effect they have on the nutritional profile, the antioxidant properties as well as its shelf life including the microbial stability. Similarly to the fermented beverages, the addition of fruit juices can be made for several reasons. In this case pineapple juice was used as it is rich in antioxidants, which have several health benefits. Pineapple juice also has a pH low enough to aide in prevention of aggregation of whey proteins, which if occurs leads to instability of the beverage. They pineapple juice was blended with the whey at different concentrations, and ultimately, they found that the cheese whey beverage has higher protein and microbial count, but the paneer whey beverage has a higher mineral content, better shelf life and stability (Baba et al., 2016). Similar beverages have also been made making use of mango and ginger to significantly improve the nutritional value as well as antibacterial, therapeutic, and prophylactic qualities of the beverage (Alane et al., 2017).

In one study, Singh et al. (2014) produced a lemon-based beverage from lactose hydrolyzed cheese whey. To do this they used a lactase enzyme that breaks down the disaccharide lactose into its monosaccharide subunits glucose and galactose (Moyer, 2007). Hydrolyzing the lactose in the whey has been found to enhance whey desirability in a number of ways, including enhancing the flavor, improving digestibility, and increasing sweetness (Singh et al., 2014). They found the optimal conditions for this enzymatic reaction and were able to produce an acceptable lemon-based beverage from the hydrolyzed whey.

4. Environmentally sustainable alternatives from whey

Beyond the applications in the food industry, whey has potential uses in providing more environmentally sustainable alternatives in power generation, chemical synthesis, and more.

4.1 Power generation: fuel and electricity

We use some form of energy in most parts of our daily lives, however the sources of this energy release CO_2, which pollutes the environment. Due to this, the search for more sustainable energy sources, which do not rely on fossil fuels has been an important focus of many nations (Mahato et al., 2022). Scientists are developing new renewable green energy technologies, to reduce environmental impact, which either release less CO_2 than their existing alternatives, or none at all. Much of the research has been on solar technologies, biogas, wind power and hydropower, to name a few (Mahato et al., 2022). Bioenergy is another interesting source of renewable energy. It is promising due to its production being from biodegradable biomass such as a variety of wastes from livestock, agriculture, forestry, industry and more (Rincón-Catalán et al., 2022). Biological or chemical processes are used to process the waste to produce energy, either in the form of electricity or heat, or biofuels. Biofuels can be separated into gaseous such as biomethane and biohydrogen, liquids such as bioethanol and biodiesel, and solids (Rincón-Catalán et al., 2022).

Alcohol that is produced through biological sources is referred to as bioethanol (Zou and Chang, 2022). It can be made through the fermentation of sugars, starches, cellulose, and waste products such as whey from cheese making. Ethanol combustion is more efficient and releases less greenhouse gas than gasoline, which makes it a desirable fuel (Zou and Chang, 2022). As previously described, due to the composition of whey it is an ideal material for fermentation, and this applies to fermentation into biofuel as well. The lactose present in whey can be used to form ethanol by alcohol fermentation. Tesfaw et al. (2021) used *Kluyveromyces marxianus* due to its ability to ferment lactose in whey into ethanol. They tested crude whey, which is whey that has not been pretreated through sterilization, deproteinization, or defatting, and this allows for the comparison of these processing steps. It was shown that both pH-adjusted and non-adjusted processing have the same efficacy in ethanol production from the whey, and additions such as molasses, yeast

FIGURE 26.8 Europe's largest bioethanol plant in Zeitz, Germany. *Adapted from* Bioethanol plant in Zeitz *by High Contrast, CC BY 3.0, via Wikimedia Commons (High Contrast, 2009).*

extract, peptone, and ammonium sulfate were found to increase the production of ethanol too (Tesfaw et al., 2021). Overall, this work shows the potential of using less processed whey in bioethanol production, which could reduce costs and time especially at a larger scale such as in a bioethanol plant. Tomaszewska and Białończyk (2016) on the other hand used β-galactosidase to hydrolyze the lactose in the whey so they could use *Saccharomyces cerevisiae* yeasts for the fermentation. Additionally, a bioreactor with direct contact membrane distillation was utilized to separate the ethanol from the fermentation material as it is produced with the aim of increasing the ethanol productivity. Direct contact membrane distillation is a process where volatile compounds evaporate through the membrane that is in contact with the feed and are then condensed. They ultimately determined that with deproteinized whey enriched with sucrose in a bioreactor with membrane distillation the theoretical ethanol production efficiency would be 1.9 times higher than without the membrane distillation (Tomaszewska and Białończyk, 2016) (Fig. 26.8).

While much research on the use of whey is directed to cheese whey, these methods differ significantly from ricotta whey, as ricotta whey composition differs. Ricotta whey contains a lower whey protein content due to the whey protein precipitation method, which also involves the addition of organic acids and salts raising their concentrations in the resulting whey (Carota et al., 2017). In one study, ricotta cheese whey was used to produce biodiesel, which is commonly made from renewable sources such as plant oils and animal fats (though less common) and comprises fatty acids alkyl monoesters. Where standard diesel has significant environmental and health hazards, biodiesel is a renewable, biodegradable, and nontoxic fuel. The cost of plant oils, and the issues of "food vs. fuel" have created the need for alternative methods of production, such as the use of microbial oils. Oleaginous yeasts are one of several microorganisms that can accumulate significant amounts of intracellular lipids, and oleaginous yeasts specifically have properties that make them suitable for industrial biodiesel production, including the ability to be scaled up. This research found that a strain of oleaginous yeasts was able to grow adequately in ricotta cheese whey to produce significant lipid quantities of a composition that are able similar to well established feedstocks for production of biodiesel, laying the groundwork for its production from ricotta cheese whey (Carota et al., 2017).

Another way to produce energy from whey is through a biofuel cell, which is a device that can allow for the conversion of organic matter into electrical energy (Choi et al., 2020). One such biofuel cell is a microbial fuel cell, in which microorganisms are used to generate electricity from wastewater by the degradation of the organic matter present (Rincón-Catalán et al., 2022). The microbial fuel cell has two chambers separated by a permeable membrane with an electrode in each that are attached by a cable to create a complete circuit. Under anaerobic conditions, microorganisms grow and oxidize the organic matter in the anode chamber and release electrons, protons, and CO_2. The protons move through the membrane to the cathode chamber, which is under aerobic conditions, while electrons move to the anode. The electrons can then travel through the cable to the cathode and with the protons and oxygen present form water and generate an electrical current (Rincón-Catalán et al., 2022).

Whey from cheese making is a cost-effective source of organic matter containing nutrients for the growth of microorganisms in the fuel cells (Mahato et al., 2022). Since it is generally disposed of in substantial amounts daily, it is easily available and its use for generating electricity would reduce the environmental impact of its disposal. Mahato et al. (2022) made use of wastewater containing cheese whey as the only source of nutrients for the microbes to generate electricity in a small microbial fuel cell. Seven fuel cells connected in series were able to produce an open circuit potential of 7V, enough to run a 4V LED strip lights for 3 h (Mahato et al., 2022). Rincón-Catalán et al. (2022) coupled anaerobic digestion carried out in a bioreactor with microbial fuel cell as a final polishing step to improve the energy production. In the anaerobic digestion, microbes breakdown the organic matter, in this case cheese whey, and can produce methane, a gaseous biofuel, from the whey without the presence of oxygen. This process is like that previously discussed to produce bioethanol. They then took the material left over after the digestion process with the microbial fuel cell to generate electricity. They found the maximum voltage generated by the microbial fuel cell was right at the beginning and was 344 mV. This shows the potential to use microbial fuel cells to produce electricity from the waste of anaerobic digestion of cheese whey for biofuel production, providing an even more sustainable process (Rincón-Catalán et al., 2022).

This can also be achieved in a biofuel cell without the use of microorganisms, such as in an enzymatic fuel cell. Where in a microbial fuel cell the living cells must be maintained, enzymatic fuel cells make use of redox-active enzyme reactions to generate power without the same maintenance (Choi et al., 2020). In one study, the potential for this system with cheese whey as the fuel source was successfully demonstrated using a cellobiose dehydrogenase enzyme (Choi et al., 2020).

4.2 Aroma compounds

Fermentation of whey can be used to produce volatile compounds aside from ethanol and methanol, and these include compounds responsible for aroma (Izawa et al., 2015). Aroma compounds have applications in various industries, including chemical, cosmetics, food, and pharmaceuticals. While aroma compounds are generally produced by extraction from plants and animals, or by chemical synthesis, these are expensive and environmentally harmful. One study used microbial fermentation, testing seven strains of *Wickerhamomyces pijperi*, a lesser studied yeast, for fermentation in glucose-added whey to produce various aroma compounds. They identified 12 aroma compounds in the fermentation broth, with the major components being acetaldehyde, ethyl acetate, and isoamyl alcohol, however the amounts of compounds produced varied based on the media, whether glucose added whey was used, and culture temperatures (Izawa et al., 2015). Overall, this work shows that whey can be used in an alternate method to generate aroma compounds by microbial fermentation.

4.3 Disinfectant

In the fresh cut produce industry, the most common disinfectant is chlorine, which can pose environmental and health related ramifications, however, alternative sanitizers to chlorine are costly and may cause undesirable changes to foods when applied (Santos et al., 2019). So, the development of new, lower-cost disinfectants has been of interest toresearchers. One study made use of fermented mixed ewe, goat, and cow milk whey with mesophilic-lactic acid bacteria to produce a disinfectant fermented whey solution with antibacterial activity against common bacteria, *Escherichia coli*, *Salmonella enterica*, and *Listeria monocytogenes* species, among others. The produced disinfectant was tested on shredded lettuce and was compared with chlorine and water. The fermented whey disinfectant was shown to improve the microbial quality of the lettuce as well as having similar quality indicators as in chlorine. This shows the potential for whey in the production of a lower cost, environmentally safe, chlorine alternative disinfectant for minimally processed foods such as shredded lettuce (Santos et al., 2019).

4.4 Biodegradable plastic

Traditional plastics are nondegradable and pose a significant environmental threat, so the development of biodegradable polymers such as polyhydroxyalkanoates (PHAs), which are plastics made up of polyesters and *R*-hydroxyalkanoic acids, have become a research area of interest (Basnett et al., 2017). A range of bacterial species are able to produce PHAs through a group of enzymes that catalyze the polymerization reaction required, called PHA synthases (Basnett et al., 2017). In one study, cheese whey powder, reconstituted in distilled water was used as the feed for fermentation by an acid producing microbial culture in a bioreactor to produce a volatile fatty acid-rich effluent (Domingos et al., 2018). The volatile fatty acids were concentrated then this solution was used as the feed to produce PHAs in a bioreactor in a pure culture fed-batch system. Ultimately continuous production of volatile fatty acids was achieved making use of cheese whey, and were able to achieve PHA production comparable with other production methods, which shows there were no inhibitory effects from the use of cheese whey as the feed, and showed that their method for production was feasible and effective (Domingos et al., 2018). This study shows the potential to produce biodegradable plastics by fermentation of cheese whey, producing a more environmentally friendly product from environmentally harmful waste.

4.5 Soil amendment

Acid whey has potential for use as fertilizers due to the high organic compound content, which can improve crop quality, yield, and growth (Fabian et al., 2024). In farming, the majority of the fertilizers used do not reach the desired plants as they are often washed away with rain and watering and can pollute the surrounding areas. This, however, can be mitigated by use in a hydrogel with good swelling properties, which, with absorbed water, would be released gradually in the soil. This method is limited by whey-based hydrogels being degraded quickly by microbes in soil, so one study investigated the addition of a biodegradable polymer to extend the microbial degradation period. Making use of acid whey from cream cheese production, as well as cellulose derivatives, hydroxyethylcellulose, and polyvinyl alcohol to create a biopolymeric hydrogel, using the whey as the polymer dissolution medium, a biopolymer hydrogel was produced. A significant increase in swelling properties with addition of the hydrogels depending on polyvinyl alcohol content was found, with an increase in water retention capacity of 19%. Overall, the incorporation of poly vinyl alcohol resulted in increased stability and swelling properties of the acid whey-based hydrogel in soil, increasing the effectiveness of acid whey in soil supplementation (Fabian et al., 2024).

5. Conclusion

Overall, novel uses of whey show its growing role in providing nutritional value in supplement foods and beverages, the presence of milk sugars for fermentation in ethanol production, and aiding potential for developing environmentally sustainable practices in various industries such as power generation, disinfectants, plastics, and soil amendments. The composition of whey lends itself to nutritional supplementation as well as providing a readily fermentable sugar source for the production of fermented beverages and alcoholic beverages and spirits. The use of whey in these ways may help to alleviate the impact of dairy waste on our environment. Additionally, sustainable alternatives for other products produced from whey are not only often release lower CO_2 than the standard versions but also attain a lower carbon footprint due to the use of the whey, which otherwise may be dumped and pollute the environment. Sustainable, environmentally friendly whey-based alternatives show the potential uses of whey outside of just its nutritional value and fermentability in the production of foods and beverages. Due to its composition, whey is not only suitable as a feed, having a high organic content, in production of biofuels, electricity, and plastic production, but also as a fertilizer as a component of a hydrogel soil amendment, a disinfectant chlorine alternative, and also has potential in aroma compound production.

References

AbdulAlim, T.S., Zayan, A.F., Campelo, P.H., Bakry, A.M., 2018. Development of new functional fermented product: mulberry-whey beverage. J. Nutr. Food Res. Technol. 1. https://doi.org/10.30881/jnfrt.00013.

Alane, D., Raut, N., Kamble, D.B., Bhotmange, M., Dinkar, C., Kamble, B., 2017. Studies on preparation and storage stability of whey based mango herbal beverage. Int. J. Chem. Stud. 5.

Arsić, S., Bulatović, M., Zarić, D., Kokeza, G., Subić, J., Rakin, M., 2018. Functional fermented whey carrot beverage-qualitative, nutritive and techno-economic analysis. Rom. Biotechnol. Lett. 23 (2), 13496–13504.

Azimi, F., Esmaillzadeh, A., Alipoor, E., Moslemi, M., Yaseri, M., Hosseinzadeh-Attar, M.J., 2020. Effect of a newly developed ready-to-use supplementary food on growth indicators in children with mild to moderate malnutrition. Public Health 185. https://doi.org/10.1016/j.puhe.2020.06.025.

Baba, W.N., Din, S., Punoo, H.A., Wani, T.A., Ahmad, M., Masoodi, F.A., 2016. Comparison of cheese and paneer whey for production of a functional pineapple beverage: nutraceutical properties and Shelf life. J. Food Sci. Technol. 53. https://doi.org/10.1007/s13197-016-2218-8.

Bandara, T.A., Munasinghe-Arachchige, S.P., Gamlath, C.J., 2023. Fermented whey beverages: a review of process fundamentals, recent developments and nutritional potential. Int. J. Dairy Technol. 76 (4), 737–757. https://doi.org/10.1111/1471-0307.12993.

Basnett, P., Ravi, S., Roy, I., 2017. Natural bacterial biodegradable medical polymers: polyhydroxyalkanoates. In: Science and Principles of Biodegradable and Bioresorbable Medical Polymers: Materials and Properties. https://doi.org/10.1016/B978-0-08-100372-5.00008-8.

Carota, E., Crognale, S., D'Annibale, A., Gallo, A.M., Stazi, S.R., Petruccioli, M., 2017. A sustainable use of Ricotta Cheese Whey for microbial biodiesel production. Sci. Total Environ. 584–585. https://doi.org/10.1016/j.scitotenv.2017.01.068.

Choi, H.S., Yang, X., Kim, D.S., Yang, J.H., Han, S.O., Park, C., Kim, S.W., 2020. Power generation from cheese whey using enzymatic fuel cell. J. Clean. Prod. 254. https://doi.org/10.1016/j.jclepro.2020.120181.

Cintineo, H.P., Arent, M.A., Antonio, J., Arent, S.M., 2018. Effects of protein supplementation on performance and recovery in resistance and endurance training. Front. Nutr. 5, 400140. https://doi.org/10.3389/fnut.2018.00083.

Dangi, R., 2007. Ripe jamun fruits (Photograph). Wikimedia Commons. https://commons.wikimedia.org/wiki/File:Ripe_jamun_fruits.jpg (CC BY 2.5).

Domingos, J.M.B., Puccio, S., Martinez, G.A., Amaral, N., Reis, M.A.M., Bandini, S., Fava, F., Bertin, L., 2018. Cheese whey integrated valorisation: production, concentration and exploitation of carboxylic acids for the production of polyhydroxyalkanoates by a fed-batch culture. Chem. Eng. J. 336. https://doi.org/10.1016/j.cej.2017.11.024.

Fabian, D.R.C., Durpekova, S., Dusankova, M., Hanusova, D., Bergerova, E.D., Sedlacik, M., Skoda, D., Sedlarik, V., 2024. Renewable whey-based hydrogel with polysaccharides and polyvinyl alcohol as a soil amendment for sustainable agricultural application. Int. J. Biol. Macromol. 259. https://doi.org/10.1016/j.ijbiomac.2023.129056.

High Contrast, 2009. Bioethanol Plant in Zeitz (Photograph). Wikimedia Commons. Available from: https://commons.wikimedia.org/wiki/File:Bioethanol_plant_in_Zeitz.jpg (CC BY 3.0).

Holsinger, V.H., Posati, L.P., DeVilbiss, E.D., 1974. Whey beverages: a review. J. Dairy Sci. 57, 849–859. https://doi.org/10.3168/jds.S0022-0302(74)84976-3.

Hughes, P., Risner, D., Meunier Goddik, L., 2019. Whey to vodka. In: Whey—Biological Properties and Alternative Uses. https://doi.org/10.5772/intechopen.81679.

Izawa, N., Kudo, M., Nakamura, Y., Mizukoshi, H., Kitada, T., Sone, T., 2015. Production of aroma compounds from whey using Wickerhamomyces pijperi. AMB Express 5. https://doi.org/10.1186/s13568-015-0108-5.

Jan, Y., Malik, M., Sharma, V., Yaseen, M., Bora, J., Haq, A., 2021. Utilization of whey for formulation of whey jamun juice ice pops with antidiabetic potential. Food Front 2. https://doi.org/10.1002/fft2.90.

Janiaski, D.R., Pimentel, T.C., Cruz, A.G., Prudencio, S.H., 2016. Strawberry-flavored yogurts and whey beverages: what is the sensory profile of the ideal product? J. Dairy Sci. 99. https://doi.org/10.3168/jds.2015-10097.

Lawton, M.R., Alcaine, S.D., 2019. Leveraging endogenous barley enzymes to turn lactose-containing dairy by-products into fermentable adjuncts for Saccharomyces cerevisiae-based ethanol fermentations. J. Dairy Sci. 102. https://doi.org/10.3168/jds.2018-15586.

Li, B., Gao, X., Li, N., Mei, J., 2018. Fermentation process of mulberry juice-whey based Tibetan kefir beverage production. Czech J. Food Sci. 36. https://doi.org/10.17221/468/2017-CJFS.

Macwan, S.R., Dabhi, B.K., Parmar, S.C., Aparnathi, K.D., 2016. Whey and its utilization. Int. J. Curr. Microbiol. Appl. Sci. 5. https://doi.org/10.20546/ijcmas.2016.508.016.

Mahato, J., Miah, M., Shovon, M.S., Roy, N., Easmin, M.S., Sharma, S.C.D., 2022. Electricity production by microbial fuel cell using cheese whey wastewater of the dairy industry in Rajshahi, Bangladesh. Chem. Biochem. Eng. Q. 35. https://doi.org/10.15255/CABEQ.2021.1922.

Manary, M.J., Sandige, H.L., 2008. Management of acute moderate and severe childhood malnutrition. BMJ 337. https://doi.org/10.1136/bmj.a2180.

Moyer, S.M., 2007. Lactose intolerance. In: Pediatric Clinical Advisor. https://doi.org/10.1016/B978-032303506-4.10186-5.

M'Hir, S., Rtibi, K., Mejri, A., Ziadi, M., Aloui, H., Hamdi, M., Ayed, L., 2019. Development of a novel whey date beverage fermented with kefir grains using response surface methodology. J. Chem. 2019. https://doi.org/10.1155/2019/1218058.

M'hir, S., Ziadi, M., Mejri, A., Ayed, L., 2023. Mixture of whey-milk and palm sap for novel kefir beverage using simplex-centroid mixture design. Kuwait J. Sci. 50. https://doi.org/10.1016/j.kjs.2023.04.008.

Panghal, A., Patidar, R., Jaglan, S., Chhikara, N., Khatkar, S.K., Gat, Y., Sindhu, N., 2018. Whey valorization: current options and future scenario—a critical review. Nutr. Food Sci. 48 (3), 520–535. https://doi.org/10.1108/NFS-01-2018-0017.

Pasta, C., Caccamo, M., Petriglieri, R., Difalco, A., Farina, G., Belvedere, G., Marino, G., Marino, V.M., Garavaldi, A., Musi, V., Alcaine, S.D., 2024. Sicilian whey: utilization of ricotta whey in the production of value-added artisanal beers. Fermentation 10. https://doi.org/10.3390/fermentation10010019.

Pereira, C., Henriques, M., Gomes, D., Gomez-Zavaglia, A., de Antoni, G., 2015. Novel functional whey-based drinks with great potential in the dairy industry. Food Technol. Biotechnol. 53. https://doi.org/10.17113/ftb.53.03.15.4043.

Pires, A.F., Marnotes, N.G., Rubio, O.D., Garcia, A.C., Pereira, C.D., 2021. Dairy by-products: a review on the valorization of whey and second cheese whey. Foods 10 (5), 1067. https://doi.org/10.3390/foods10051067.

Rincón-Catalán, N.I., Pérez-Fabiel, S., Mejía-González, G., Herrera-López, D., Castro-Chan, R., Cruz-Salomón, A., Sebastian, P.J., 2022. Power generation from cheese whey treatment by anaerobic digestion and microbial fuel cell. Waste Biomass Valori. 13. https://doi.org/10.1007/s12649-022-01720-1.

Risner, D., Shayevitz, A., Haapala, K., Meunier-Goddik, L., Hughes, P., 2018. Fermentation and distillation of cheese whey: carbon dioxide-equivalent emissions and water use in the production of whey spirits and white whiskey. J. Dairy Sci. 101. https://doi.org/10.3168/jds.2017-13774.

Risner, D., Tomasino, E., Hughes, P., Meunier-Goddik, L., 2019. Volatile aroma composition of distillates produced from fermented sweet and acid whey. J. Dairy Sci. 102. https://doi.org/10.3168/jds.2018-14737.

Sabokbar, N., Khodaiyan, F., 2015. Characterization of pomegranate juice and whey based novel beverage fermented by kefir grains. J. Food Sci. Technol. 52. https://doi.org/10.1007/s13197-014-1412-9.

Sabokbar, N., Moosavi-Nasab, M., Khodaiyan, F., 2015. Preparation and characterization of an apple juice and whey based novel beverage fermented using kefir grains. Food Sci. Biotechnol. 24. https://doi.org/10.1007/s10068-015-0278-6.

Santos, M.I.S., Fradinho, P., Martins, S., Lima, A.I.G., Ferreira, R.M.S.B., Pedroso, L., Ferreira, M.A.S.S., Sousa, I., 2019. A novel way for whey: cheese whey fermentation produces an effective and environmentally-safe alternative to chlorine. Appl. Sci. 9. https://doi.org/10.3390/app9142800.

Sharma, P., Trivedi, N., Gat, Y., 2017. Development of functional fermented whey–oat-based product using probiotic bacteria. 3 Biotech 7. https://doi.org/10.1007/s13205-017-0906-3.

Singh, S., Khemariya, P., Rai, A., 2014. Process optimization for the manufacture of lemon based beverage from hydrolyzed whey. J. Food Sci. Technol. 51. https://doi.org/10.1007/s13197-011-0563-1.

Skryplonek, K., 2018. The use of acid whey for the production of yogurt-type fermented beverages. Mljekarstvo 68, 139–149. https://doi.org/10.15567/mljekarstvo.2018.0207.

Skryplonek, K., Jasińska, M., 2015. Fermented probiotic beverages based on acid whey. Acta Sci. Pol. Technol. Aliment. 14. https://doi.org/10.17306/J.AFS.2015.4.39.

Stobaugh, H., 2018. Maximizing recovery and growth when treating moderate acute malnutrition with whey-containing supplements. Food Nutr. Bull. 39. https://doi.org/10.1177/0379572118774492.

Stobaugh, H., Ryan, K.N., Kennedy, J.A., Grise, J.B., Crocker, A.H., Thakwalakwa, C., Litkowski, P.E., Maleta, K.M., Manary, M.J., Trehan, I., 2016. Including whey protein and whey permeate in ready-to-use supplementary food improves recovery rates in children with moderate acute malnutrition: a randomized, double-blind clinical trial. Am. J. Clin. Nutr. 103. https://doi.org/10.3945/ajcn.115.124636.

Svorad, 2018. Kefir-jadro-upravene (Photograph). Wikimedia Commons. Available from: https://commons.wikimedia.org/wiki/File:Kefir-jadro-upravene.jpg (CC BY-SA 4.0).

Tesfaw, A., Oner, E.T., Assefa, F., 2021. Evaluating crude whey for bioethanol production using non-Saccharomyces yeast, Kluyveromyces marxianus. SN Appl. Sci. 3. https://doi.org/10.1007/s42452-020-03996-1.

Tomaszewska, M., Białończyk, L., 2016. Ethanol production from whey in a bioreactor coupled with direct contact membrane distillation. In: Catalysis Today. https://doi.org/10.1016/j.cattod.2016.01.059.

Yang, Y., Wang, H., Shuang, Q., Xia, Y., 2024. Novel insights into flavor formation in whey fermented wine: a study of microbial metabolic networks. LWT 197. https://doi.org/10.1016/j.lwt.2024.115911.

Zhou, X., Chai, L., Wu, Q., Wang, Y., Li, S., Chen, J., 2021. Anti-diabetic properties of bioactive components from fish and milk. J. Funct. Food. 85, 104669. https://doi.org/10.1016/j.jff.2021.104669.

Zou, J., Chang, X., 2022. Past, present, and future perspectives on whey as a promising feedstock for bioethanol production by yeast. J. Fungi. 8 (4), 395. https://doi.org/10.3390/jof8040395.

Chapter 27

Sensory aspects of whey and whey products

Michele Rosset[1], Mariana Buranelo Egea[2] and Tatiana Colombo Pimentel[3]
[1]*Paraná Federal Institute, Colombo, Brazil;* [2]*Goiano Federal Institute, Rio Verde, Brazil;* [3]*Paraná Federal Institute, Paranavaí, Brazil*

1. Introduction

Whey proteins are considered a high-quality source of nutrients. They are notable for containing various bioactive compounds produced from whey by enzymatic hydrolysis during fermentation and gastrointestinal digestion. They are abundant in essential amino acids, including cysteine, branched-chain amino acids like leucine, valine, and isoleucine, and bioactive peptides. Due to their antioxidant, anti-inflammatory, antihypertensive, anti-obesity, and appetite-suppressing properties, whey proteins show strong potential as functional food ingredients (Patel, 2015; Yiğit et al., 2023). Whey proteins also show functional properties that play an essential role in food processing, acting as emulsifiers, fat replacers, gelling agents, and encapsulating agents while also enhancing the sensory and textural qualities of food products (Whitson et al., 2010; Yiğit et al., 2023).

Sensory quality is crucial to food product development, providing valuable insights into consumer preferences and overall product quality. The sensory properties of whey solids, particularly their flavor, are recognized as a major limiting factor to their utilization in food formulations. Among the main concerns is the development of *off-flavors* in dried whey products, predominantly resulting from lipid oxidation and Maillard reaction pathways (Mortenson et al., 2008). However, recent studies have been published demonstrating that the addition of different forms of whey protein (concentrates and isolates) did not alter the sensory characteristics of the food products or even improve them (Elssadig et al., 2020; Komeroski et al., 2021; Costa et al., 2022; Borooah et al., 2024; Hércules et al., 2024). Whey protein can be added to an infinite number of products, including dairy products, meat products, emulsified products, protein bars, and even film coatings, and serve as a material for encapsulating molecules (Fig. 27.1).

1.1 Dairy products

Considering that yogurt is one of the most consumed dairy products in the world, the addition of ingredients with bioactive properties is an alternative to increasing the consumption of functional foods. However, the incorporation of extracts can influence the sensory characteristics of the product (Table 27.1). Thus, nanoencapsulation can be a viable alternative to reduce the negative impacts on the sensory quality of the product. The use of whey as an encapsulating agent was reported by Al-Subhi (2025). The author investigated the impact of yellow *Capsicum Annuum* extract (YCAE) encapsulated with whey protein on the sensory characteristics of yogurts. A control and four formulations with YCAE were investigated. The sensory assessment was conducted by 15 panelists. The assessment emphasized flavor (valued at 45 points), body and texture (40 points), and color and appearance (15 points). The control sample showed the highest taste scores, followed by samples with 250 and 500 mg YCAE. In contrast, incorporating free YCAE at the same proportions resulted in decreased taste ratings, indicating that its direct addition may compromise flavor when compared to the control or the encapsulated forms. After 21 days, all samples exhibited a decline in taste ratings. However, the encapsulated samples showed better scores in flavor than samples with YCAE-free incorporation. In addition, the results showed that nanoencapsulation of YCAE with whey protein isolate contributed to maintaining the quality of the body and

Whey. https://doi.org/10.1016/B978-0-443-27648-4.00028-1

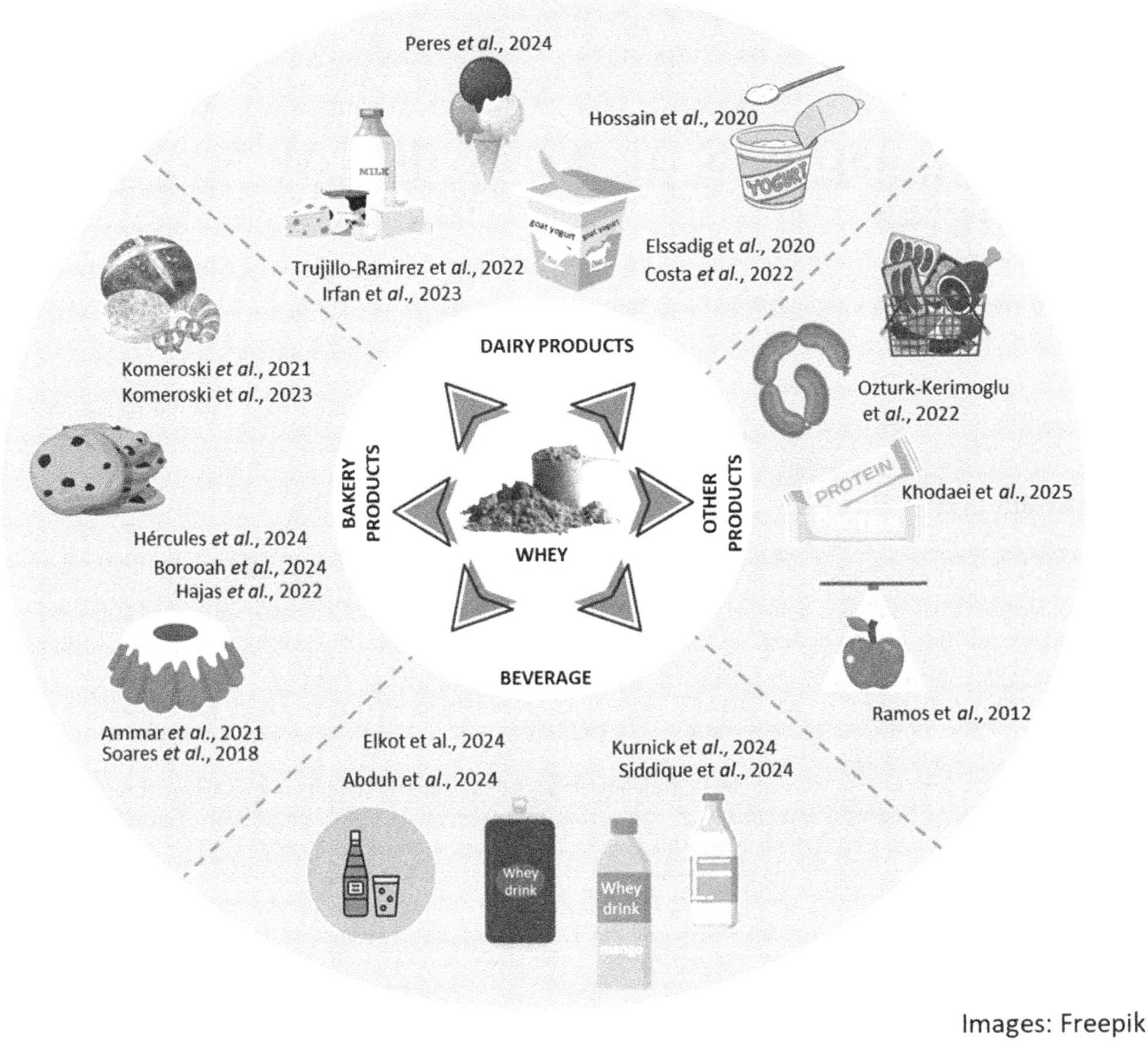

FIGURE 27.1 Diagram demonstrating possible applications of whey protein as an ingredient in food products.

texture of the yogurts, suggesting that encapsulating YCAE improves the taste, body, texture, color, and overall acceptability of yogurt.

The addition of whey in dairy products can improve their sensory attributes (Elssadig et al., 2020; Costa et al., 2022; Kant et al., 2024). Kant et al. (2024) investigated the effects of inulin and whey protein concentrate (WPC) as fat replacers in low-fat paneer, which is an indigenous dairy product used in many Indian recipes. They developed four formulations: Control sample, Inulin (0.25 g/g milk fat replaced), WPC (0.30 g/g milk fat replaced), and Inulin + WPC (60:40) (0.27 g/g milk fat replaced). The authors observed that the incorporation of 0.25 g/g was the most acceptable sample. Therefore, the addition of whey improved the acceptance of the products.

Goat milk offers superior digestibility, an ideal composition of fatty acids and proteins, and a rich content of bioactive compounds, which together contribute to its potential in treating or preventing certain medical conditions (Csapóné Riskó and Csapó, 2019; Jesus et al., 2019; Elssadig et al., 2020). It is also a valuable alternative for individuals who are allergic to cow's milk, as its proteins are more easily digested and produce fewer allergenic peptides. Furthermore, the fat in goat milk is more digestible and serves as an excellent source of energy, supporting various metabolic processes and helping to combat metabolic diseases. Goat milk and its derived products can act as functional foods, promoting nutrition and health for both young and elderly individuals, particularly those with cow milk allergies (Elssadig et al., 2020). Although goat

TABLE 27.1 Main studies assessing the sensory aspects of whey products.

Food product	Formulations	Sensory assays	Main results	References
Sausages	SF: Standard-fat control (20% fat +20% water) RF: Reduced fat control (10% fat +30% water) MWP1: *Samples containing 10% fat* +5% MWP + 25% water MWP2: Samples containing 10% fat +10% MWP + 20% water	10 trained panelists Triplicate at room temperature Sausage color (1 = extremely undesirable, 7 = extremely desirable), hardness *(1 = extremely soft, 7 = extremely* hard), juiciness (1 = extremely dry, 7 = extremely juicy), flavor intensity (1 = extremely low intensity, 7 = extremely high intensity) and overall impression (1 = extremely undesirable, 7 = extremely desirable) were evaluated using a seven-point unipolar scale. Presence of *off-flavor*	There are no significant differences between color, flavor intensity, and overall impression scores. Absence of *off-flavor* for any sausage samples 10% MWP: ↓hardness and ↑juiciness	Ozturk-Kerimoglu et al. (2022)
Low-fat paneer	S1: 0.5% fat-containing milk S2: Inulin 0.25 g/g S3: 0.30 g/g WPC S3: inulin (0.25 g/g) + WPC (0.30 g/g) S4: inulin (0.27 g/g) + WPC (0.27 g/g)	10 trained sensory panelists Scale: 100 points Flavor Body and Texture Color Appearance	Acceptability: S2>S1>S3>S4 Flavor, body, and texture quality: ↓during storage (7 days) Highest score in color and appearance: S2	Kant et al. (2024)
Yogurt with yellow *Capsicum Annuum* extract (YCAE)	Control T1 = 250 mg YCAE T2 = 500 mg YCAE T3 = 250 mg YCAE encapsulated in whey protein T4 = 500 mg, YCAE encapsulated in whey protein nanoparticles	15 panelists Flavor (worth 45 points) Body and texture (40 points) Color and appearance (15 points)	Taste: Control > T3>T4 After 21 days: ↓ a taste for all samples Overall acceptability: T3 and T4	Al-subhi (2025)
Goat milk yogurt	Whole yogurt (WY) Skim yogurt (SY) Skim yogurt inulin (SIY) Skim yogurt maltodextrin (SMY) Skim yogurt whey protein (SWY)	119 untrained panelists Sensory attributes: appearance, color, aroma, flavor, texture, viscosity, and overall acceptability 9-point category scale (1 = extremely dislike to 9 = extremely like) Purchase intention: 5-point scale (1 = definitely would not buy to 5 = definitely would buy)	SIY, SMY, and SWY: positive sensory scores for appearance, color, aroma, texture, and viscosity No difference in overall acceptability No difference in purchase intention	Costa et al. (2022)
Goat milk yogurt	0%, 5%, 10% of whey protein	30 semi-trained panelists Sensory attributes: color, texture, flavor, taste, and overall acceptability 5-point hedonic scale (5 = like extremely to 1 = dislike extremely	↑ sensory quality of all attributes	Elssadig et al. (2020)

Continued

TABLE 27.1 Main studies assessing the sensory aspects of whey products.—cont'd

Food product	Formulations	Sensory assays	Main results	References
Boru-type Künefe cheeses	Control: 2% fat milk T1: 0.5% whey protein/1.5% fat milk T2: 1% whey protein/1% fat milk T3: 1.5% whey protein/0.5% fat milk	11 trained panelists Sensory attributes: appearance (scale 0–5), texture (scale 0–5), odor (scale 0–5), taste (scale 0–5), overall acceptability (scale 0–20)	The best overall sensory quality: 0.5% whey protein/1.5% fat milk	Esen and Güzeler (2023)
Yogurt	Control: full-fat (3.5 g/100 g) Reduced-fat (<0.5 g/100 g)	36 untrained panelists Difference-from-Control test Sensory attributes: creaminess, smoothness, and thickness		Hossain et al. (2020)
Mango ice cream	82% of pasteurized mango pulp, 7% of pasteurized passion fruit, 6% of *whey protein*, and 5% of coconut oil	First group: 157 panelists not affected by cancer 9-point hedonic scale: 1 "disliked extremely" to 9 "liked extremely" Sensory attributes: aroma, appearance, flavor, texture, and overall impression Degree of sweetness: JAR - Just about right Five points for purchase intention: 1 = definitely would not buy it, 5 = definitely would buy it Second group: 75 patients undergoing chemotherapy CATA method (9-point structured hedonic scale, ranging from "dislike it extremely" to "like it extremely")	First group: ↑ acceptability Appearance: 7.87 Aroma: 8.17 Flavor: 8.50 Texture: 8.30 Global impression = 8.40 87.3% of the panelists considered the sweetness as JAR Certainly Purchase: 68,2% Probably would buy it: 27.4% Second group: Frequent terms: Yellow color (79%); creamy (74%); sweet (69%); mango aroma/flavor (61%); passion fruit aroma/flavor (51%); firm (39%)	Peres et al. (2024)
Buffalo milk ice cream	Control: 4% protein/10% milk fat High protein ice cream: 6%, 8%, 10%	6 trained panelists Descriptive sensory Sensory attributes: color, appearance, flavor, body, texture	↓ body and texture score within the protein concentration	Roy et al. (2022)
Biscuits	WPC as a wheat flour replacer 10%, 20%, 30%, 40%, 50%, 60% and 70%.	20 semi-trained panelists Sensory attributes: color, appearance, texture, taste, flavor, and overall acceptability 9-point hedonic scale: 1 "disliked extremely" to 9 "liked extremely" SWOT analysis was performed by a group of experts (n = 10)	Best appearance: 30% and 40% of WPC Best color, taste, and flavor: 40% of WPC Best overall acceptability: 30% of WPC	Borooah et al. (2024)
Cookies	11.9% of WPI	13 panelists for green lentil cookies 16 panelists for red lentil cookies Simplified profile analysis Sensory descriptors according to ISO 11035	WPI ↓ "lentil" and "baked" aromas	Hajas et al. (2022)

Cookies	S- standard NC- casein and reduce sugar NW- whey protein/reduced sugar CC- casein/cocoa/reduce sugar CW- whey protein/cocoa/reduced sugar	52 untrained panelists Appearance, texture, color, taste, global acceptance, purchase intention, and acceptability index Hedonic scale of nine points: 1 = disliked extremely, 9 = like extremely Five points for purchase intention: 1 = certainly not buy it, 5 = would certainly buy it	Appearance: S sample: 5.75 CW sample: 6.04 NW sample: 7.69 Color: S sample: 6.23 CW sample: 6.42 NW sample: 7.82 Flavor: S sample: 7.69 CW sample: 6.63 NW sample: 5.84 Global acceptance: S sample: 7.35 CW sample: 6.38 NW sample: 6.17 Purchase intention: S: 3.96 CW: 3.31 NW 3.11	Hércules et al. (2024)
Cookies	Whey protein: 9.15%–16.88%	Sorting task: 102 consumers Acceptability test: 102 consumers Smell, color, flavor, and general acceptance Nine-point hedonic scale	Greater acceptability: F2 (13.18%), F7 (11.87%), F12 (10.91%)	Luque-Vilca et al. (2024)
Bread	WF: 100 g of wheat flour CCPF: 50 g of cassava starch and 50 g of chickpea flour WP10: 10 g of whey protein WP20: 20 g whey protein WP30: 30 g whey protein	Non-celiac and untrained adult panelists 11 untrained Celiac panelists Acceptability of attributes (appearance, texture, taste, odor, color) Overall acceptance Hedonic evaluation of nine points scale (1 = Dislike extremely, 9 = Like extremely) Purchase intention: five points scale, (1 = I certainly would not buy it, 5 = I would certainly buy it) Acceptability index	Greater acceptability: WP10: 10 g of whey protein	Komeroski et al. (2021)

milk derivatives have high nutritional value, sensory acceptability is still considered low compared to cow milk derivatives, which may be related to the presence of short-chain fatty acids, such as caproic, caprylic, and capric acids, which give an unpleasant "goaty" taste to the product. The addition of fat substitutes may be an alternative to improve the quality and sensory acceptability of these products.

Costa et al. (2022) evaluated the effects of fat replacers (inulin, maltodextrin, and whey protein) on physicochemical, color, textural, rheological properties, and sensory acceptance of low-fat cupuassu goat milk yogurts. The analysis was performed for 28 days at 4°C. Inulin (5% w/vol), maltodextrin (5% w/vol), and whey protein isolate (5% w/vol) were added depending on the treatment. Five samples of cupuassu goat milk yogurts were performed: whole yogurt (WY), skimmed yogurt (SY), skimmed yogurt with inulin (SIY), skimmed yogurt with maltodextrin (SMY), and skimmed yogurt with whey protein (SWY). The sensory acceptance was evaluated by 119 untrained participants (72 women, 47 men) ranging from 19–63 years old (mean = 24, SD = 8). The panelists evaluated the appearance, color, aroma, flavor, texture, viscosity, and overall acceptability of each sample based on a 9-point category scale (1 = extremely dislike to 9 = extremely like). Additionally, purchase intention was evaluated using a 5-point scale ranging from 1 = definitely would not buy to 5 = definitely would buy. It is known that the acceptability of food is related to its flavor, which also comes from the type and quantity of fat. SY showed the lowest score for all tested sensory parameters (appearance = 4.97, color = 5.31, Aroma = 4.24, flavor = 3.04, texture = 4.18, viscosity = 4.24, overall acceptability = 3.50, and purchase = 1.62). The other formulations exhibited no statistical difference among them for flavor and overall acceptability parameters, indicating that the addition of fat substitutes compensated for the sensory attributes of the low-fat yogurt formulations. The samples added with 5% of whey protein isolate exhibited the highest scores for aroma (6.28) and flavor (4.93). Furthermore, the yogurt with whey protein showed the highest overall acceptability (5.21) compared to the samples added with maltodextrin (4.81) and inulin (5.03). Regarding purchase intention, no statistical difference was observed among the formulations, demonstrating that the addition of fat substitutes such as whey protein can be used as a technological strategy to produce low-fat goat milk derivatives, maintaining the sensory quality.

Elssadig et al. (2020) also evaluated the sensory quality of goat's yogurt formulated with different concentrations of whey protein powder (0%, 5%, and 10%). The sensory assessment was performed with 30 semi-trained panelists who evaluated the sensory attributes (color, texture, flavor, taste, and overall acceptability) using a 5-point hedonic scale. Scores were measured from "like extremely" to "dislike extremely," corresponding to ratings from 5 to 1. The incorporation of whey protein powder enhanced all the sensory attributes assessed. On day 0, the overall acceptability scores were 4.03 for the control sample, 4.40 for the sample with 5% whey protein powder, and 4.60 for the sample with 10% whey protein powder.

According to Esen and Güzeler (2023), reducing fat content and incorporating whey protein had little impact on the appearance, texture, odor, taste, and overall acceptability of Boru-type Künefe cheeses throughout 90 days of storage. The similarity in sensory attributes between the treated cheeses and the control sample supports the viability of fat reduction combined with WPC addition. All samples were judged as acceptable products by the panelists during storage.

Roy et al. (2022) evaluated the effects of whey protein isolate (6%, 8%, and 10%) on the sensory attributes of buffalo milk ice cream. No changes were observed between buffalo milk ice cream samples added with whey (8% and 10%) for color and appearance scores. The flavor score of 8% and 10% whey protein ice cream samples was significantly lower than control (4%) and 6% whey protein samples, which may be attributed to the characteristic "whey" flavor of whey protein isolate, which resulted in a reduction of flavor scores when used at high concentrations. Therefore, the authors observed that the sensory quality of buffalo milk ice cream decreased with an increase in whey protein concentration (8% and 10%). Optimizing the whey concentration is of paramount importance to avoid negative impacts on the sensory profile of the products.

In another study, Peres et al. (2024) developed a mango ice cream with whey protein and evaluated its sensory acceptance by not affected by cancer (n = 157) and 75 patients undergoing chemotherapy. The formulation was composed of mango, passion fruit, whey protein (6%), and unflavored coconut oil. The sensory analysis used the 9-point hedonic scale for aroma, appearance, flavor, texture, and overall acceptability. In addition, the "Check All That Apply" (CATA) method was performed on patients undergoing chemotherapy. Among healthy consumers, 96.8% approved the flavor, 87.26% considered the sweetness level ideal, and more than 95% indicated purchase intention. Among patients, 90% demonstrated positive acceptance, and 32% reported relief from nausea after consumption. The results revealed that the addition of whey protein contributed to the acceptance of the product, evidencing its potential as a functional food for cancer patients and the general population.

Another way to replace fat is to use microparticulate whey proteins (MWPs), which are formed by native whey proteins throughout the thermal denaturation and aggregation process. Hossain et al. (2020) investigated the influence of different forms and sizes of MWPs in powder and liquid-extruded microparticulate whey proteins (eMWPs) on the physicochemical and sensory properties of reduced-fat (<0.5 g/100 g), plain-type stirred yogurt. The results showed that the reduced-fat yogurts with eMWPs were highly ranked compared to the reference yogurt (full-fat, with 3.5 g/100 g).

1.2 Bakery products

Bakery products are among the most consumed foods worldwide, enjoyed by people of different age groups due to their convenience, accessibility, and sensory characteristics. The increased demand for foods with functional properties has driven the development of foods enriched with proteins (Komeroski and de Oliveira, 2023). The proteins of whey may be interesting in bakery products because they have favorable technological properties such as easy solubility, emulsification, and anti-allergic action, in addition to improving sensory characteristics (Table 27.1) (Soares et al., 2018; Ammar et al., 2021; Hajas et al., 2022; Komeroski and de Oliveira, 2023; Borooah et al., 2024).

Borooah et al. (2024) replaced whole wheat flour with WPC in biscuit formulation and observed the effects on the sensory characteristics. The sensory attributes, such as color, appearance, texture, taste, flavor, and overall acceptability, were evaluated using a 9-point hedonic scale. The samples were coded randomly and were presented to 20 semi-trained panelists. Furthermore, the SWOT analysis was performed by a group of experts (n = 10) proficient in baking and product reviewing. All the samples showed scores either above or at par with those of the control sample and were well above the limit of acceptance (≥5, denoting neither-like nor-dislike) in a 9-point hedonic scale. The addition of WPC at 30% and 40% showed a better appearance score (8.10), while 40% of WPC incorporation resulted in better color (8.10), taste (8.00), and flavor (8.08) aspects than the other samples. Regarding overall acceptability, the incorporation of 30% of WPC showed the highest score (8.18). It was not significantly different from the samples with 10% (7.62), 20% (7.95), 40% (7.91), and 50% (7.83) of WPC addictions, indicating that the replacement of whole wheat flour by WPC still 50% (w/w) can improve the sensorial properties of biscuit.

Hajas et al. (2022) investigate the impact of different combinations of WPI (11.9%), Inulin (6.0%) as dietary fiber, and xylitol (27.9%) as a sweetener in the enrichment of green- and red-lentil-based gluten-free cookies. A simplified profile analysis was used during the sessions, and the sensory descriptors were established according to ISO:11035. The values of sensory attributes were evaluated on a 1–100 structured linear scale. The sensory descriptors used on the test were as follows: surface homogeneity, surface color, baked aroma, lentil aroma, lentil taste, sweet taste, hardness, crunchiness, and crumbliness. The panelists were asked to rank samples according to the decrease in overall liking. The results showed that the incorporation of WPI resulted in less intensive "lentil" and "baked" aromas (mostly for red lentil cookies).

Hércules et al. (2024) used milk proteins as an alternative for reducing sugars in cookies, proposing a healthier product with maintenance of sensory quality. The acceptability, purchase intention, and acceptability index were evaluated. The test was conducted with 52 untrained panelists. The appearance, texture, color, taste, and global acceptance were assessed according to a hedonic scale of nine points. The purchase intention was evaluated using a scale of five points (1 = Certainly not buy it, 5 = Would certainly buy it). The acceptability index of samples was calculated by formula AI (%) = A × 100/B, where A is the average grade obtained for each treatment, and B is the maximum grade given for each treatment.

Regarding appearance, the best score was obtained with whey protein (7.69). Color is one of the most important quality properties for acceptability that influences consumer perception. The addition of proteins can improve the Maillard reaction, resulting in dark color cookies. However, in this study, the score of color cookies was considered satisfactory (7.82). The addition of whey protein did not influence the texture. It showed a similar score to the control sample (7.01) and whey samples (6.27 and 6.55 for whey protein and reduced sugar and whey protein, cocoa, and reduced sugar, respectively). Regarding the taste, the control sample showed the highest taste mean (7.69), with a statistically significant difference among the sample whey protein, cocoa, and reduced sugar (6.63), but did not differ statistically when compared to whey protein and reduced sugar sample (5.84).

Luque-Vilca et al. (2024) verified the effects of sustainable ingredients, such as cañihua flour, whey, and potato starch, in gluten-free cookie development. The combination of cañihua flour, whey, and potato starch resulted in a gluten-free alternative that is both nutritious and appealing in terms of sensory qualities, making it suitable for individuals with dietary restrictions. The use of 38.51% cañihua flour, 10.91% whey, and 25.69% potato starch exhibited superior sensory performance. It presented excellent texture, aroma, flavor, color, and overall appearance, reflecting its high sensory and physicochemical quality, such as protein 4.83%.

The incorporation of whey protein improved bakery product nutritional quality, besides showing sensory quality, providing a healthy and promising alternative for those who follow a gluten-free diet. Komeroski et al. (2021) evaluated the effect of whey protein, cassava, and chickpea flour on the physical, nutritional, and sensory characteristics of breads. Five formulations were developed with different proportions of cassava, chickpea, and whey protein (standard, 0%, 10%,

20%, and 30%). The samples were submitted to sensory analysis, purchase intention, and acceptability index by Celiac and non-Celiac groups on different days. The attributes evaluated were appearance, taste, odor, color, and texture. A hedonic scale with nine points was used, and a five-point scale, anchored by the extremes (1 = I certainly would not buy it, 5 = I would certainly buy it), was used to measure purchase intention respectively. The products added with whey protein showed technological quality similar to wheat bread. All evaluated samples showed satisfactory results in sensory analysis performed by Celiac and non-Celiac judges.

1.3 Meat products

The properties of milk proteins that are of most importance for the formulation of meat products are related to the immobilization of water, texture and consistency control, color improvement, and enhancement of sensory properties. Their gelation characteristics and their high water and fat-binding abilities are especially important. WPC has the advantage over non-fat dry milk in that it does not impart a scalded milk flavor to processed meat products (El-Magoli et al., 1996).

Ozturk-Kerimoglu et al. (2022) investigated the impacts of the MWP) on emulsified beef sausages on the proximate composition, techno-functional, instrumental, microstructural, sensory, and oxidative characteristics as partial fat replacers in emulsified beef sausage formulations. Four formulations were used in this study: one standard treatment (ST) comprised 20% beef fat and 20% water of the meat weight, one reduced-fat control treatment (RF) comprised 10% beef fat and 30% water of the meat weight, and two treatments with addiction of MWP addition: MWP1 and MWP2, were prepared by 10% beef fat +5% MWP + 25% water and 10% beef fat +10% MWP + 20% water, respectively.

The sensory evaluation was undertaken by 10 trained panelists (50% male/female, age range 22–40) who all regularly consume different types of emulsified sausages. The panelists were trained to identify and scale visual and textural attributes, as well as flavor and rancid odor (off-flavor), during three training sessions. The analysis was carried out in a test room equipped with cabinets and individual overhead lighting. The evaluation was performed in triplicate at room temperature. The authors evaluated the sausage color (1 = extremely undesirable, 7 = extremely desirable), hardness (1 = extremely soft, 7 = extremely hard), juiciness (1 = extremely dry, 7 = extremely juicy), flavor intensity (1 = extremely low intensity, 7 = extremely high intensity) and overall impression (1 = extremely undesirable, 7 = extremely desirable) using a seven-point unipolar scale. In addition, the presence of off-flavor was also evaluated. Water at room temperature and saltless bread were served to clean the palate between individual evaluations.

No significant differences were observed for color (6.3–6.6), flavor intensity (6.0–6.4), and overall acceptability (5.8–6.2) among the evaluated sausage samples. In addition, no *off-flavors* were detected by the panelists in any of the samples. It is important to consider that lipid oxidation represents a major limiting factor affecting the quality and consumer acceptability of meat and meat products. In this context, the reduction of fat content combined with the incorporation of modified whey protein (MWP) into the sausage formulation contributed to improved oxidative stability, highlighting the promising potential of MWP as a natural antioxidant agent.

The antioxidant effect of whey protein could be attributed to forming antioxidative Maillard components that arise from lactose. WPCs usually contain higher concentrations of other components, such as lactose, which, upon cooking, could lead to the formation of antioxidative Maillard components. These Maillard reaction products would conceivably enhance the antioxidative potential of WPCs (Ulu, 2004). In addition, free amino acids such as tryptophan, phenylalanine, and cysteine, present in the peptide fractions of whey proteins, may play a significant role in enhancing oxidative stability, given the documented antioxidant properties of these protein constituents (Corrochano et al., 2018).

About hardness and juiciness, the sensory scores ranged between 4.1–5.3 and 4.8–5.4, respectively. The formulation with 10% beef fat and 30% water showed higher hardness and lower juiciness scores than other treatments ($P < .05$). According to the authors, this may have been due to the high protein content and strong interactions between the protein chains. The incorporation of MWP with a concentration of 10% reduced hardness and increased juiciness scores compared to the reduced-fat control treatment, comprised of 10% beef fat and 30% water of the meat weight. Thus, the authors conclude that the incorporation of MWP into the reduced-fat formulation could structurally weaken the protein matrix and could play an important role as a lubricant that provides a softer texture and more juiciness.

1.4 Whey protein emulsions

The search for healthier foods has driven the food industry to develop products with lower fat content without compromising sensory quality (Trujillo-Ramirez et al., 2022). In this context, whey protein (WP) has stood out as a promising functional ingredient in the formulation of food emulsions (Table 27.2).

TABLE 27.2 Main studies assessing the sensory aspects of whey protein (WP) products with emphasis on emulsified foods.

Food product	Objectives/Formulations	Assays	Main results	References
Coarse canola oil-in-water (O/W) emulsion applied in yogurt (substituted 50% of the milk fat)	WP hydrolyzate and citrus pectin (1% of the total weight of the biopolymer; WP hydrolyzate/CP mass ratio 6:1; pH 4.25) subsequently homogenized to obtain emulsions with different average droplet sizes (4000, 3000, 120 and 60 nm).	80 habitual yogurt consumers Just-about-right scales with five descriptive anchors (-2 = much unacceptable, -1 unacceptable, 0 neutral, +1 acceptable and +2 much acceptable) for six attributes ("aroma", "taste", "after taste", "viscosity", "creaminess" and "overall acceptability")	↑ acceptance of creaminess and overall acceptance when the emulsion droplet size was smaller. Aroma was clearly centered on the "ideal" perception and acceptance was higher when the product used emulsion. Not affect the aroma, aftertaste, or flavor of the yogurt prototype.	Trujillo-Ramirez et al. (2022)
Avocado salad dressing enriched with omega-6	Avocado (40%, 60%, or 80%), WP concentrate (3%, 6.5%, or 10%), and mint extract (1%, 1.5%, or 2%)	20 trained panelists Hedonic scales with descriptive anchors from 1 to 5 for attributes: taste (sour flavor of vinegar and avocado flavor), viscosity (extremely thin to extremely thick), color (white to green), texture (fat texture in mouth), and overall quality (horrible to outstanding).	↑ sensory note with increasing WP. Values from 3.11 to 5 were found. No negative effect was observed on taste and flavor attributes with applying WP concentrate as an emulsifier at concentrations less than 6.5%.	Mooliani and Nouri (2021)
Olive oil emulsion and WP isolate applied in cheddar cheese (0%, 5%, 10%, 15%, and 20%)	Extra virgin olive oil and WP isolate (oil-in-water) emulsion with 5%:0.5% (v/v) ratio, respectively	15 semi-trained panelists 9-point hedonic scale for color, texture, taste, flavor, appearance, and overall acceptability of processed Cheddar cheese	↑ score for taste, aroma, texture/appearance, and overall acceptability of cheese Processed Cheddar cheese with 5% WP emulsion showed higher values of flavor (~7.6), aroma (~7.1), and texture/appearance (~7.5), 5% and 10% WP emulsion resulted in "slightly olive", "buttery" and "salty" flavors.	Irfan et al. (2023)
Carrot soup microemulsified with inulin and milk WP concentrate	Soup base with 351 g of carrots in 250 mL of evaporated milk. Product prepared with 20 mL of the base, 1 g of olive oil, 0.1 g of onion and 0.01 g of garlic enriched with 100% inulin or WPC or 50% of each.	18 panelists A request was made for the most desired formulation, which was later evaluated on the hedonic scale regarding flavor, color, texture, and acceptability attributes.	Highest desirability score for soup containing equal amounts of Inulin and WPC (70%). 70% and 35% of panelists liked the color and texture. 30% and 22% liked the taste of the soup very much or very much. 17% and 40% liked the soup very much or very much (sensory acceptability).	Fabela Morón et al. (2020)

Continued

TABLE 27.2 Main studies assessing the sensory aspects of whey protein (WP) products with emphasis on emulsified foods.—cont'd

Food product	Objectives/Formulations	Assays	Main results	References
Food-grade emulsion powder enriched with soy milk (SM) and whey protein (WP) permeates	Calorie distribution from carbohydrate: protein: lipid = 55:15:30), where A was 100% soy milk, B was 100% whey protein permeates, C was 70% soy milk and 30% whey protein permeates, and D was 50% soy milk and whey protein permeates each. Maltodextrin and residual lactose from WP permeate were calculated as carbohydrate sources. Soy milk and sunflower oil have been estimated as sources of lipids.	15 semi-trained panelists A 9-point hedonic scale (where 1 = extremely dislike, 2 = very dislike, 3 = moderately dislike, 4 = slightly dislike, 5 = neither like nor dislike, 6 = slightly like, 7 = moderately like 8 = very like, and 9 = extremely like) was used to evaluate color, flavor, texture, and overall acceptability	Formulations with different colors depending on the ingredients were especially low when a higher amount of SM was used, as in Sample A. ↑ scores for taste (6.9 and 7.0) and overall acceptability (6.8 each), for samples C and D, respectively.	Nowshin et al. (2023)
Lemon mousse	Lemon mousse formulations contain gelatine (1%–2% w/w), citric acid (0.9%–1.5% w/w), and whey protein isolate (WPI) (8%–18% w/w) concentration.	Descriptive analysis Ten trained panelists 23 terms describing the attributes within taste, flavor, texture, mouthfeel, and aftertaste of the product were evaluated according to the intensity of a 15 cm line with anchors Consumer testing 30 panellists Attributes of appearance, taste, texture, desire to eat, degree of salivation, and overall liking were assessed using a 15 cm scale with anchors at both ends of the liking scale being "very disliked"/"very weak"/"not salivating at all" and "very liked"/"very strong"/"extremely salivating", with a description of "neither liked nor disliked" in between.	↑ glossy appearance, smooth and sticky texture, melt-in-mouth sensation, sweet and creamy taste Consumer testing: 3D-printed lemon mousses were well received by consumers, regardless of the multisensory layer arrangement	Chow et al. (2021)
WP spread as a substrate for β-glucan delivery.	WP (30 g/100 mL) and 250 g/100 g de β-glucan.	100 untrained panelists 9-point hedonic scale ranging from 1 (disliked extremely) to 9 (liked very much) with a calculation of the acceptability index was used to evaluate the overall impression and the acceptance of the aroma and color attributes. Salt and sweetness taste intensity was evaluated using a nine-point verbal structured scale (1 = low intensity, 9 = high intensity).	Score of 6–7 in all attributes evaluated (overall impression, color, aroma), indicating that consumers liked the products slightly to moderately. The addition of NaCl did not result in differences in the acceptance of the products in terms of overall impression and color, nor in the intensity of the salty flavor.	Zanon et al. (2020)

WP, whey protein; ↑ increase.

Fat is essential in the texture, flavor, and stability of many foods, such as yogurts, cheeses, spreads, and sauces. However, its partial or total replacement by protein emulsions represents a viable alternative to meet the demand for lighter products (Mooliani and Nouri, 2021).

Whey protein-based emulsified ingredients have been used as fat replacers, maintaining or even improving the sensory acceptability of foods when droplet size is controlled (Trujillo-Ramirez et al., 2022). Trujillo-Ramirez et al. (2022) demonstrate that smaller droplets resulted in better sensory acceptance in the product where it was added (yogurt), attributed to the softer texture and final microstructure. In addition, WP appears to help decrease sedimentation in samples with lower stability, which is undesirable and not recommended for salad dressings (Mooliani and Nouri, 2021).

Furthermore, isolated WP in emulsified foods, such as mousses or emulsions subsequently added to cheeses, can increase the pH due to the buffering capacity of isolated WP in processed foods to improve their conservation. This decrease impacts the perception of acidity, improving acceptability (Chow et al., 2021; Irfan et al., 2023). On the other hand, when olive oil and WP emulsions were added to cheeses in larger amounts (15%–20% of cheese mass), they can induce bitter and umami flavors, possibly due to the accumulation of hydrophobic peptides (Irfan et al., 2023).

Compared to other protein sources, such as soy protein, whey protein tends to be better accepted in terms of flavor, color, and overall acceptability (Nowshin et al., 2023), which reinforces its applicability to improve the acceptability score in food products (Fabela Morón et al., 2020; Zanon et al., 2020). Thus, using whey protein in food emulsions is an effective strategy for developing products with an improved nutritional profile while maintaining the sensory appeal essential for acceptance by the modern consumer.

1.5 Whey protein (WP) beverages

Table 27.3 presents the studies assessing the sensory aspects of whey protein (WP) products, emphasizing beverages. Flavoring can be done with fresh or frozen fruits or pulp or even with their powdered ingredients. In general, flavoring interferes with all attributes of the whey beverage, including color, odor, flavor, texture, and general appearance, improving its acceptability (Kurnick et al., 2024). Flavoring initially influences the color, causing the whey beverage to change from a whitish color to a color due to the natural pigments of the ingredients used in this flavoring, making the beverage more attractive (Saleem et al., 2025).

Beverages containing whey protein are associated with negative sensory attributes, including undesirable qualities in the mouth, which can be very bothersome to consumers, limiting consumption and acceptance (Giles et al., 2025). These undesirable flavors may be due to (i) unbalanced glucose: lactose ratio, (ii) the high mineral content in whey, or (iii) the positive charge of whey proteins (Kurnick et al., 2024), which in turn are characterized by notes of soap and potato or broth (Abduh et al., 2024). In addition, these flavors may increase with storage time, mainly due to the increased oxidation of free fat during this period. Therefore, it is necessary to implement strategies such as flavoring or fermenting beverages to achieve better acceptability. Whey beverages fermented with probiotic cultures showed better sensory properties (Skryplonek et al., 2019). At the same time, the thermal processing of whey can contribute to the darkening of the beverage due to processes such as enzymatic browning or caramelization (Bansal et al., 2019; Siddique et al., 2024). As for texture, depending on the amount of WP added, a positive aspect identified in the beverages is that they have more body, resulting from greater viscosity, which consumers can perceive favorably (Giles et al., 2025).

1.6 Other whey protein products

WP can be added to bars to form high-protein or rich-protein products (Khodaei et al., 2024). In addition to increasing the nutritional value of the products, when WP is added to protein bars a positive sensory impact is observed when used in balanced proportions (6%–15%) (6%–15%) (Mathur et al., 2020; Jabeen et al., 2021; Małecki et al., 2022; Khodaei et al., 2024), while higher values appear to result in a bitter aftertaste (Jabeen et al., 2021). Furthermore, whey protein seems to be a protein ingredient that improves the flavor and texture of bars, especially when compared to soy and insect proteins (Małecki et al., 2022).

As for texture, the studies report a softer or firmer texture (Khodaei et al., 2024). This may occur due to the formation of gelled protein networks of globular proteins such as β-lactoglobulin and α-lactalbumin (Wang and Guo, 2019). These protein molecules undergo denaturation in heat or agitation, exposing hydrophobic groups forming a three-dimensional network that (i) retains water, increasing water retention capacity and increasing perceived moisture, making the texture softer or less dry; (ii) emulsifies lipids, resulting in a more uniform and creamy texture; (iii) interacts with carbohydrates, forming viscoelastic structures, resulting in a pleasant chewiness; or (iv) retains other molecules, contributing to a firmer or more cohesive texture (Savadkoohi and Kasapis, 2016; Wang and Guo, 2019; Vilotte, 2020).

TABLE 27.3 Main studies assessing the sensory aspects of whey protein (WP) products with emphasis on beverages.

Food product	Objectives/Formulations	Assays	Main results	References
Fermented whey drinks with either natural or artificial flavoring agents	whey added 4% (v/v) High Fructose Syrup (HFS) and natural orange juice and artificial orange syrup at concentrations of 15% and 12%, respectively, was fermented using 2.5% (v/v) *L. casei* and 2.5% (v/v) *L. plantarum* starters.	40 untrained panelists Sensory questionnaire covering taste, color, off-flavor, and aroma The results were then interpreted across three scales: less than normal, normal (best result), and more than normal.	Color: the addition of natural or artificial flavoring contributed to acceptability. Aroma: fermented whey had a more acidic flavor. Off flavor: unpleasant flavor categorized as excessive, with 77.50% of tasters selecting the "too much" scale.	Abduh et al. (2024)
Dairy drink fermented with chia seed and acerola syrup	Dairy base: sweet whey from curd cheese manufactured was fermented with lyophilized mixed culture DVS ABT-4 (Chr Hansen®) Dairy base varied from 55%–70%(70% pasteurized whole milk + 30% whey + 3% whole milk powder), syrup varied from 28%–44% (80% acerola pulp + 20% crystal sugar), and chia varied from 1%–2% (100% of the seeds.)	120 untrained panelists Hedonic scale (ranging from 1- "I disliked it very much" to 9- "I liked it very much"), in which the parameters of texture, flavor, aroma, and global impression. 5-point product purchase intention (ranging from 1–"I would certainly not buy it" to 5–"I would certainly buy it")	All formulations presented a median of 7 in color, flavor, texture, and overall acceptance. In the purchase intention, all formulations presented a median of 4. Acceptance index of 80%	(de Costa et al., 2022)
Whey-based sports beverage fortified with *Spirulina platensis*	Fermented ricotta cheese whey-based beverage supplemented with various *S. platensis* powder concentrations (0.25%, 0.5%, and 0.75% w/w) cooperated with a mixture of lemon and peppermint juice 10% and fermented by probiotic culture.	25 panelists The flavor score was 50, the body and fluidity score was 35, and the color and appearance score were 15, for a total of 100 points.	Soft texture and slightly whitish color. Flavorings are important to mask the residual flavor.	Elkot et al. (2024)
Beverage based on soursop (*Annona muricata* L.), inulin, and whey powders	Beverages made from a mixture of soursop (*Annona muricata* L.) (0% and 2%), whey (0% and 2%) and/or Inulin (0% and 1%) using a selected strain of Lactobacillus	100 untrained panelists 7-point hedonic scale, where 1- indicates dislike very much and 7 indicates like very much in which the parameters of texture, flavor, aroma, and global impression.	None difference between the attributes of color, odor, flavor and general acceptance. Beverages are neither liked nor disliked and are liked slightly.	Fernández-Huazo et al. (2024)
Beverages using sheep and goat cheese whey and secondary whey	Sheep cheese whey (SCW), goat cheese whey (GCW), sheep secondary whey (SSW), and goat secondary whey (GSW), natural or sweetened with 6% sugar, were added with stabilizers (3% inulin, 0.3% high-methoxylated pectin, 0.3% low-methoxylated pectin, 0.2% sodium citrate, or 0.3% of a commercial hydrocolloid blend). Non-fermented flavors: dry instant coffee, vanillin aroma, and sugar. Fermented flavors: lemon juice, mango, orange aroma, and sugar	120 consumers structured 9-point hedonic score to indicate their acceptance regarding aroma, flavor, acidity, and overall liking 5 semi-trained panelists Check-all-that-apply (CATA) was used with 14 terms: milk/whey products and flavors used in the samples (sweet, salty, acid, bitter, astringent, viscous, lumps, homogeneous, milk, liquid, rancid, yogurt/fermented milk, vanillin, and coffee)	Good acceptability regardless of whether the whey was from goat or sheep. The sensory evaluation showed that the ingredients/additives used masked the unpleasant taste of the WP.	Kurnick et al. (2024)

WP beverages supplemented with pomegranate, beetroot, and carrot concentrates functional	WP beverage was added with 3% concentrates of pomegranate (PCWb), beetroot (BCWb), and carrot (CCWb)	10 sensory experts 9-point hedonic scale evaluating color, flavor, taste, appearance, and overall acceptability	Color varied between 8.0 (BCWb) and 8.8 (PCWb) Flavor varied between 7.0 (blend) and 8.7 (PCWb)	Saleem et al. (2025)
Beverages produced by Co-spray drying WP isolate with polysaccharides	Maltodextrin (M) (4.76%), xanthan gum (XG) (0.15%–0.3%) and/or guar gum (GG) (0.15%–0.5%) co-spray dried with WP isolate (WPI).	10 screened and trained sensory panel Unstructured line scales (scaled 0–100) with appropriate anchors were used to evaluate 26 attributes; 2 appearance, 4 aroma, 8 taste and flavor, 5 mouthfeel, 7 after effects	Aroma and flavor: milky or slightly metallic descriptor Texture and body: higher viscosity Overall acceptance was good	Giles et al. (2025)
Probiotic beverage using whey and pineapple	WP and pineapple juice and coded as A (0:100), B (15:85), C (25:75) and D (35:65), with 8% sugar and *Lactobacillus acidophilus* LA-5 (ATCC 4356)	15 panelists 5-point hedonic scale (5 = Like a lot, 4 = Like a little, 3 = Neither like or dislike, 2 = Dislike a little, and 1 = Dislike a lot.) to evaluate color & appearance, flavor, mouth feel, sweetness and overall acceptability	Less satisfactory color and appearance score (>4) was recorded on day 42 Flavor score of selected beverages increased up to day 28 and then decreased slightly with the advancing storage period. Sweetness decreased with storage.	Islam et al. (2021)
Probiotic fermented beverages based on acid whey	Pasteurized acid whey (50%) with UHT milk (0%–50%), unsweetened condensed milk (0%–50%) or fermented skimmed milk powder (5%) with probiotic cultures of *Lactobacillus acidophilus* LA-5 or *Bifidobacterium animalis* ssp. lactis BB-12	6 trained painelists 5-point hedonic scale and included assessment of appearance, consistency, taste, and flavor. The scale ranged from 1 (very poor quality) to 5 (very good quality). We used the mean scores for each attribute to calculate overall sensory quality—a weighted average of the scores, where the weights reflected the influence of each factor on a determination of overall product quality. Taste weighted 0.30, flavor of 0.30, consistency of 0.25, and appearance of 0.15.	The flavor was described as clear and refreshing, more pronounced in beverages with *L. acidophilus.* Whey combined with milk and powdered or condensed milk significantly improved the texture, giving a thicker and more uniform consistency, similar to yogurt.	Skryplonek et al. (2019)
Tejate, a Mexican traditional maize–cocoa beverage	Formulation with 20 g of cacao flowers, 30 g of mamey sapota fruit seeds, and 100 g of cocoa beans per kg of maize nixcamalized with 6% of ash was added to Whey protein concentrate (80% of protein) or soy protein isolate (88% of protein) were added to the at 1%, 2%, and 2.5%	18 consumers of maize and coconut traditional beverages triangular sensory test was carried out to identify the protein-rich tea beverage with similar sensory attributes than the traditional tejate formulation	Similar formulations	Pérez-Ramírez et al. (2021)
Beverage based on whey and β-glucans	Formulation with 75%, 85%, and 90% whey and 25%, 15%, and 10% blackberry concentrate, with a fixed amount of 0.243 g β-glucans in 100 mL	100 panelists 7-point hedonic scale to rate the level of acceptability for evaluating attributes of appearance, color, odor, flavor, viscosity, acidity, sweetness, and general acceptance	WP changed the color of the drink Odor and flavor were frowned upon and are improved when fruit pulp is added.	Santana et al. (2024)

Continued

TABLE 27.3 Main studies assessing the sensory aspects of whey protein (WP) products with emphasis on beverages.—cont'd

Food product	Objectives/Formulations	Assays	Main results	References
Chia seeds added to mango-whey beverage	Untreated control beverage (85 mL of WP, 2 g of chia and sugar each, 5 g mango pulp, 0.1 g guar gum) (T0), application of chemicals (Sodium benzoate 0.25 g and potassium metabisulfite 0.75 g) (T1), pasteurization (T2), and microwave heating (at 4 different time lengths: T3, T4, T5, and T6)	80 panelists 9-point hedonic scale to evaluate sensory qualities such as color, flavor, texture, taste, and overall acceptability.	Whey interfered with color, especially when the duration of heat treatment was increased.	Siddique et al. (2024)
Functional drink based on whey with sweet lime juice (*Citrus limetta*), emblica (*Emblica officinalis*) and ginger	Thirty different formulations were used where the range of whey was 70%–80% v/v, sweet lime juice was 15%–25%, juices of emblica and ginger were 1%–2%, and the rest part was made up by sugar and salt (0.2%–2.5%).	15 Semi-trained panel 9-point hedonic scale to evaluate Taste, Color, Aroma, Aftertaste, Appearance, Mouthfeel, and Overall acceptability	Fresher, more natural aroma and flavor compared to the heat-treated version. Less browning and "cooked" flavor formation compared to heat treatment.	Bansal et al. (2019)
Whey drink with acerola	9 acerola-flavored milk drinks containing whey and processed by ohmic (OH) or conventional (CONV) heating. Milk + whey mixture (70:30), acerola (30%), sugar (9.9%), xanthan gum (0.1%). Varies according to voltages (6–15 V/cm) and frequencies (500–2000 Hz).	100 Consumers 9-points scale evaluation, associative words and Q methodology (innovation, feelings, attributes).	OH products with higher voltage/frequency were better accepted and associated with attributes such as "sweet flavor", "acerola aroma", "refreshing" and "bright color"—while those with lower intensity or CONV presented negative descriptors such as "cheese flavor", "watery" and "viscous"	Coimbra et al. (2020)

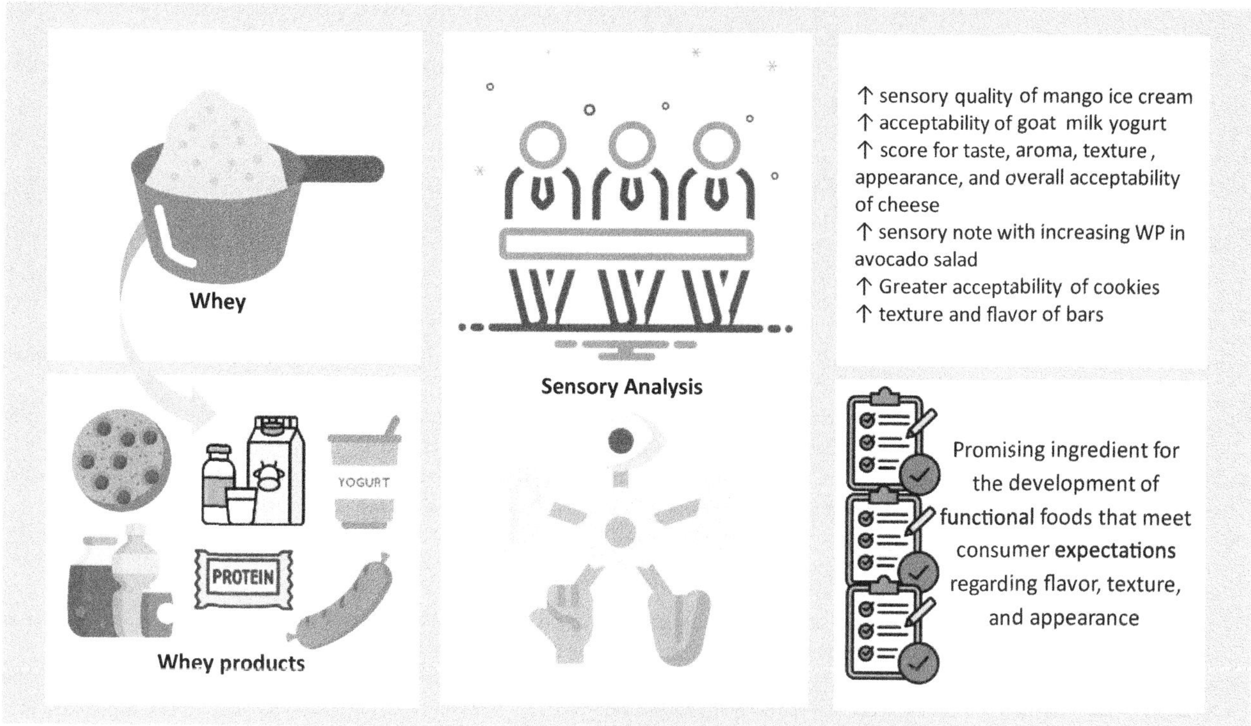

FIGURE 27.2 Summarization of the utilization of whey in food products.

The same behavior was found for other products, such as coatings and films developed using WP. WP can contribute to the texture when added in low quantities and does not negatively impact the flavor (Ramos et al., 2012). On the other hand, WP in high concentrations can result in a perceived milky, sweet, or metallic aftertaste that is worse when the whey is concentrated or hydrolyzed (Kandasamy et al., 2021).

A summarization of the use of whey proteins in food products with provided in Fig. 27.2.

2. Conclusion

The incorporation of whey proteins into food products offers significant functional advantages, particularly in sensory attributes. Despite minor changes in appearance, such as color change, whey proteins generally improve or maintain desirable characteristics, such as texture, flavor, and overall acceptability. Their ability to act as emulsifiers, fat replacers, gelling agents, and encapsulating agents contributes not only to the stability and structure of the food matrix but also to sensory quality. However, formulation optimization is necessary to reduce the deleterious effects associated with lipid oxidation and Maillard reactions during storage. Overall, whey proteins represent a promising ingredient for the development of functional foods that meet consumer expectations regarding flavor, texture, and appearance.

References

Abduh, S.B.M., et al., 2024. Sensorial attributes of fermented whey drinks with either natural or artificial flavoring agents. IOP Conf. Ser. Earth Environ. Sci. 1364 (1). https://doi.org/10.1088/1755-1315/1364/1/012075.

Al-subhi, F.M.M., 2025. Development, characterization, and prolonged shelf life of functional yogurt enriched with nanoencapsulated yellow capsicum annuum extract using whey protein. Appl. Food Res. 5 (1), 100758. https://doi.org/10.1016/j.afres.2025.100758.

Ammar, I., et al., 2021. Optimization of gluten-free sponge cake fortified with whey protein concentrate using mixture design methodology. Food Chem. 343 (February 2020). https://doi.org/10.1016/j.foodchem.2020.128457.

Bansal, V., et al., 2019. Effect of high pressure processing (HPP) on microbial safety, physicochemical properties, and bioactive compounds of whey-based sweet lime (whey-lime) beverage. J. Food Meas. Char. 13 (1), 454–465. https://doi.org/10.1007/s11694-018-9959-1.

Borooah, R., et al., 2024. Novel biscuits using whey protein and Ocimum gratissimum to mitigate protein burden: quality parameters, in vitro protein digestibility, SWOT analysis and implication on public health. Food Human. 3 (April), 100318. https://doi.org/10.1016/j.foohum.2024.100318.

Chow, C.Y., et al., 2021. Printability, stability and sensory properties of protein-enriched 3D-printed lemon mousse for personalised in-between meals. Food Hydrocoll. 120 (January), 1–12. https://doi.org/10.1016/j.foodhyd.2021.106943.

Coimbra, L.O., et al., 2020. Are ohmic heating-treated whey dairy beverages an innovation? Insights of the Q methodology. Lwt 134 (August). https://doi.org/10.1016/j.lwt.2020.110052.

Corrochano, A.R., et al., 2018. Invited review: whey proteins as antioxidants and promoters of cellular antioxidant pathways. J. Dairy Sci. 101 (6), 4747–4761. https://doi.org/10.3168/jds.2017-13618.

de Costa, J.A., et al., 2022. Dairy drink fermented with chia seed and acerola syrup: physicochemical, microbiological, and sensory characterization. Ciencia Anim. Bras. 23. https://doi.org/10.1590/1809-6891v23e-72395e.

Costa, M.P., et al., 2022. Rheological, physical and sensory evaluation of low-fat cupuassu goat milk yogurts supplemented with fat replacer. Food Sci. Anim. Res. 42 (2), 210–224. https://doi.org/10.5851/kosfa.2021.e64.

Csapóné Riskó, T., Csapó, Z., 2019. Goat keeping and goat milk products in human nutrition—review. Appl. Stud. Agribus. Commer. 13 (1–2), 24–36. https://doi.org/10.19041/apstract/2019/1-2/3.

El-Magoli, S.B., Laroia, S., Hansen, P.M.T., 1996. Flavor and texture characteristics of low fat ground beef patties formulated with whey protein concentrate. Meat Sci. 42 (2), 179–193. https://doi.org/10.1016/0309-1740(95)00032-1.

Elkot, W.F., et al., 2024. Development and characterization of a novel flavored functional fermented whey-based sports beverage fortified with Spirulina platensis. Int. J. Biol. Macromol. 258 (P2), 128999. https://doi.org/10.1016/j.ijbiomac.2023.128999.

Elssadig, L.S., Elgasim, E.A., Babiker, E.E., 2020. Quality attributes of concentrated yogurt (Labneh) of goats' milk formulated with whey protein powder during storage. Int. J. Res. -GRANTHAALAYAH 8 (7), 185–191. https://doi.org/10.29121/granthaalayah.v8.i7.2020.679.

Esen, M.K., Güzeler, N., 2023. The effects of the use of whey protein as a fat replacer on the composition, proteolysis, textural, meltability, microstructural, and sensory properties of reduced-fat Boru-type Künefe cheese during storage. Int. Dairy J. 137, 105519. https://doi.org/10.1016/j.idairyj.2022.105519.

Fabela Morón, M.F., et al., 2020. Evaluación de la calidad sensorial y nivel de aceptación de formulaciones de sopa de zanahoria microemulsionada con Inulina y Concentrado de Proteína de Suero de Leche (WPC). Agro Productividad 13 (7), 101–106. https://doi.org/10.32854/agrop.vi.1688.

Fernández-Huazo, S., et al., 2024. Development and characterization of functional beverage based on soursop (*Annona muricata L.*), inulin, and whey powders. Appl. Food Res. 4 (2). https://doi.org/10.1016/j.afres.2024.100472.

Giles, H., et al., 2025. Co-spray drying whey protein isolate with polysaccharides provides additional lubrication impacting the sensory profile of model beverages. Food Hydrocoll. 160 (P1), 110778. https://doi.org/10.1016/j.foodhyd.2024.110778.

Hajas, L., et al., 2022. Development of protein- and fiber-enriched, sugar-free lentil cookies: impact of whey protein, inulin, and xylitol on physical, textural, and sensory characteristics. Foods 11 (23). https://doi.org/10.3390/foods11233819.

Hércules, S.T.T., et al., 2024. Milk proteins as alternative ingredients replacing sugar and their effects on chemical, technological and sensory quality of cookies. J. Culin. Sci. Technol. 1–18. https://doi.org/10.1080/15428052.2024.2361711.

Hossain, M.K., et al., 2020. The impact of extruded microparticulated whey proteins in reduced-fat, plain-type stirred yogurt: characterization of physicochemical and sensory properties. Lwt 134 (January). https://doi.org/10.1016/j.lwt.2020.109976.

Irfan, S., et al., 2023. Physicochemical, microbial, and functional attributes of processed Cheddar cheese fortified with olive oil–whey protein isolate emulsion. Food Sci. Nutr. 11 (3), 1247–1256. https://doi.org/10.1002/fsn3.3159.

Islam, M.Z., et al., 2021. Development of probiotic beverage using whey and pineapple (Ananas comosus) juice: sensory and physico-chemical properties and probiotic survivability during in-vitro gastrointestinal digestion. J. Agric. Food Res. 4 (November 2020), 100144. https://doi.org/10.1016/j.jafr.2021.100144.

Jabeen, S., et al., 2021. Formulation and characterization of protein-energy bars prepared by using dates, apricots, cheese and whey protein isolate. Food Sci. Technol. 41 (June), 197–207. https://doi.org/10.1590/fst.12220.

Jesus, M.S., et al., 2019. Bioactive compounds recovery optimization from vine pruning residues using conventional heating and microwave-assisted extraction methods. Ind. Crop. Prod. 132 (January), 99–110. https://doi.org/10.1016/j.indcrop.2019.01.070.

Kandasamy, S., et al., 2021. Application of whey protein-based edible films and coatings in food industries: an updated overview. Coatings 11 (9). https://doi.org/10.3390/coatings11091056.

Kant, R., et al., 2024. A comprehensive evaluation of inulin and whey protein concentrate as fat replacers in low-fat paneer: physicochemical, textural and sensory studies. Food Human. 2 (December 2023), 100240. https://doi.org/10.1016/j.foohum.2024.100240.

Khodaei, D., Noci, F., Ryan, L., 2024. Optimizing protein bars with whey protein isolate, pea protein isolate, and blue whiting (*micromesistius poutassou*) fish protein hydrolysate: a simplex-centroid mixture design study. Food Sci. Nutr. https://doi.org/10.1002/fsn3.4701.

Komeroski, M.R., et al., 2021. Effect of whey protein and mixed flours on the quality parameters of gluten-free breads. Int. J. Gastron. Food Sci. 24 (May). https://doi.org/10.1016/j.ijgfs.2021.100361.

Komeroski, M.R., de Oliveira, V.R., 2023. Influence of the amount and type of whey protein on the chemical, technological, and sensory quality of pasta and bakery products. Foods 12 (2801), 1–16. https://doi.org/10.3390/foods12142801.

Kurnick, J.V., et al., 2024. Development of value-added beverages using sheep and goat cheese whey and secondary whey. Int. Dairy J. 152. https://doi.org/10.1016/j.idairyj.2024.105886.

Luque-Vilca, O.M., et al., 2024. Utilization of sustainable ingredients (cañihua flour, whey, and potato starch) in gluten-free cookie development: analysis of technological and sensorial attributes. Foods 13 (10). https://doi.org/10.3390/foods13101491.

Małecki, J., Tomasevic, I., Sołowiej, B.G., 2022. The influence of the syrup type on rheology, color differences, water activity, and nutritional and sensory aspects of high-protein bars for sportsmen. J. Food Qual. 2022. https://doi.org/10.1155/2022/2317676.

Mathur, M., et al., 2020. Development and optimization of ingredients for multigrain fibre and protein enriched composite bars using sensory evaluation. Pharm. Innov. 9 (1), 143–149.

Mooliani, H., Nouri, M., 2021. Optimization of oxidative, physical stability and microbial characteristics of salad dressing emulsions based on avocado and whey protein combined with mint (*Mentha spicata L.*) extract. J. Food Meas. Char. 15 (6), 5713–5724. https://doi.org/10.1007/s11694-021-01131-1.

Mortenson, M.A., Vickers, Z.M., Reineccius, G.A., 2008. Flavor of whey protein concentrates and isolates. Int. Dairy J. 18 (6), 649–657. https://doi.org/10.1016/j.idairyj.2007.12.003.

Nowshin, H., et al., 2023. Nutritional, physical and sensorial properties of food-grade emulsion powder enriched with soy milk and whey protein permeates. Food Human. 1 (March), 769–776. https://doi.org/10.1016/j.foohum.2023.07.025.

Ozturk-Kerimoglu, B., et al., 2022. Chemical, technological, instrumental, microstructural, oxidative and sensory properties of emulsified sausages formulated with microparticulated whey protein to substitute animal fat. Meat Sci. 184 (September 2021), 108672. https://doi.org/10.1016/j.meatsci.2021.108672.

Patel, S., 2015. Emerging trends in nutraceutical applications of whey protein and its derivatives. J. Food Sci. Technol. 52 (11), 6847–6858. https://doi.org/10.1007/s13197-015-1894-0.

Peres, A.P., et al., 2024. Sensory acceptance of mango ice cream added with whey protein by consumers affected and not affected by cancer. Braz. J. Food Technol. 27, 1–12.

Pérez-Ramírez, I.F., et al., 2021. Chemical and sensorial characterization of Tejate, a Mexican traditional maize-cocoa beverage, and improvement of its nutritional value by protein addition. J. Food Sci. Technol. 58 (9), 3548–3560. https://doi.org/10.1007/s13197-021-05073-w.

Ramos, Ó.L., et al., 2012. Edible films and coatings from whey proteins: a review on formulation, and on mechanical and bioactive properties. Crit. Rev. Food Sci. Nutr. 52 (6), 533–552. https://doi.org/10.1080/10408398.2010.500528.

Roy, S., et al., 2022. Quality attributes of high protein ice cream prepared by incorporation of whey protein isolate. Appl. Food Res. 2 (1), 100029. https://doi.org/10.1016/j.afres.2021.100029.

Saleem, M., et al., 2025. Nutritional, physicochemical, and antioxidant characterization of pomegranate, beetroot, and carrot concentrates supplemented functional whey beverages. Food Chem. X 25 (September 2024), 102206. https://doi.org/10.1016/j.fochx.2025.102206.

Santana, A.H., et al., 2024. 'Sensorial evaluation and physical chemical characterization of a beverage based on whey and β-glucans as a potential prevention of nonalcoholic steatohepatitis'. Food Nutr. Sci. 15 (08), 770–789. https://doi.org/10.4236/fns.2024.158050.

Savadkoohi, S., Kasapis, S., 2016. High pressure effects on the structural functionality of condensed globular-protein matrices. Int. J. Biol. Macromol. 88, 433–442. https://doi.org/10.1016/j.ijbiomac.2016.04.012.

Siddique, F., et al., 2024. Comparison of chemically treated, pasteurized, and microwave-treated (at different time durations) chia seeds added to mango-whey beverage, during different storage periods, for physicochemical and sensory parameters. J. Food Qual. 2024. https://doi.org/10.1155/2024/6688945.

Skryplonek, K., Dmytrów, I., Mituniewicz-Małek, A., 2019. Probiotic fermented beverages based on acid whey. J. Dairy Sci. 102 (9), 7773–7780. https://doi.org/10.3168/jds.2019-16385.

Soares, J.P., et al., 2018. Efeito da adição de proteína do soro do leite como substituto do trigo na formulação de bolos sem adição de açúcar. The effect of the addition of whey protein as a substitute for wheat in the development.

Trujillo-Ramirez, D., et al., 2022. Impact of the droplet size of canola oil-in-water emulsions on the rheology and sensory acceptability of reduced-milk fat stirred yogurt. J. Food Sci. Technol. 59 (12), 4853–4862. https://doi.org/10.1007/s13197-022-05573-3.

Ulu, H., 2004. Effect of wheat flour, whey protein concentrate and soya protein isolate on oxidative processes and textural properties of cooked meatballs. Food Chem. 87, 523–529. https://doi.org/10.1016/j.foodchem.2004.01.002.

Vilotte, A., 2020. Aggregation, Gelation and Hydrodynamic Spinning of Whey Proteins. Université Grenoble Alpes.

Wang, C., Guo, M., 2019. Whey protein structure and denaturation and interactions with other food components. In: Guo, M. (Ed.), Whey Protein Production, Chemistry, Functionality, and Applications. Wiley, pp. 67–101. https://doi.org/10.1002/9781119256052.ch4.

Whitson, M.E., Miracle, R.E., Drake, M.A., 2010. Sensory characterization of chemical components responsible for cardboard flavor in whey protein. J. Sensory Stud. 25 (4), 616–636. https://doi.org/10.1111/j.1745-459X.2010.00289.x.

Yiğit, A., et al., 2023. 'Whey proteins as a functional food: health effects, functional properties, and applications in food. J. Am. Nutr. Assoc. 42 (8), 758–768. https://doi.org/10.1080/27697061.2023.2169208.

Zanon, E.O., et al., 2020. Development of a whey protein spread enriched with β-glucan: an alternative for whey valorization. J. Sci. Food Agric. 100 (4), 1711–1717. https://doi.org/10.1002/jsfa.10186.

Index

'*Note:* Page numbers followed by "f" indicate figures and "t" indicate tables.'

C

D

E

M

N

O

P

Q

R

S

Y

Z

Printed and bound by CPI Group (UK) Ltd, Croydon, CR0 4YY
14/09/2026
02216413-0002